Geo-Resources

Geo-Resources

A Festschrift for Octagenerian Scholar Professor Virendra Krishna Verma

Editors

Prof. K.L. Shrivastava
Department of Geology
Jai Narain Vyas University
Jodhpur, India

Prof. Arun Kumar
Department of Earth Science
Manipur University
Imphal, India

Published by:

Scientific Publishers (India)
5 A, New Pali Road, P.O. Box 91
Jodhpur 342 001 (India)

E-mail: info@scientificpub.com
Website: www.scientificpub.com

Branch Office
Scientific Publishers (India)
4806/24, Ansari Road, Daryaganj
New Delhi - 110 002 (India)

ISBN: 978-81-7233-895-4

Laser Typeset : Rajesh Ojha
Printed in India

Hon'ble Reviewers

Prof. U. Aswathanarayan
Director, MICWM
B-16, Santishikha
Somajigauda
HYDERABAD - 500 082

Prof. K.S. Valdiya
Jawaharlal Nehru Centre
Ad. Scientific Research
Jakkur
BANGALORE - 560 004

Dr. S. Viswanathan
Former Director, AMD
B-203, United Avenue Apt.
7-1-29, Ameerpet
HYDERABAD- 500 001

Dr. A. Arunachalam
Director General
ICAR
Krishi Bhawan
NEW DELHI 110 001

Prof. O.P. Verma
Founder President
Indian Geological Congress
ROORKEE - 247 667

Prof. S.K. Tandon
Pro Vice Chacellor
Delhi University
DELHI - 110 007

Prof. K.L. Rai
45 Chatrapati Shivajee Nagar
Kotar Road
BHOPAL - 462 016

Dr. Harsh Gupta
President Geol. Soc. India
NDMA Bhawan
NEW DELHI - 110 021

Prof. M.K. Pandya
Emeritus Professor
Department of Geology
MLS University
UDAIPUR - 313 001

Prof. S.C. Mukhopadhyay
Emeritus Professor
Department of Geography
Kolkata University
KOLKATA - 700 019

Prof. R.S. Sharma
Emeritus Professor
Department of Geology
Rajasthan University
JAIPUR - 302 001

Prof. Saviner Singh
Emeritus Professor
Department of Geography
Allahabad University
ALLAHABAD - 211 002

Prof. B. Satyanarayana
1-8-702/62/17
Behind Sankar Mutt
HYDERABAD - 500 044

Dr. N.K. Dutta
Director General (Retd.)
Geological Survey of India
9, Utkarsh Nagar
NAGPUR - 440 001

Dr. V.K. Gaur
Distinguished Scientist
Nat. Geophy. Res. Inst.
Uppal Road
HYDERABAD - 500 007

Prof. S. Acharya
Former Vice Chancellor U.U
155, Vip Coloney Nayapalli
BHUBANESHWAR – 751 015

Prof. P.S. Saklani
Former Vice Chancellor KU
Emeritus Professor Delhi University
Delhi – 110 007

Prof. G.R. Jakhar
Vice Chancellor
MGS, University
Bikaner – 334 001

Dr. S.M. Ramasamy
Vice Chancellor
Gandhigram Rural University
Gandhigram – 624 302

Prof. A.S. Faroda
Former Vice Chancellor
M.P.T. University
Udaipur – 313 001

Prof. R.N. Sharma
Vice Chancellor
IASE University
Sardarsahr – 331 403

Prof. M.P. Singh
Pro Vice Chancellor,
Lucknow University
Lucknow – 226 007

Prof. Damodar Sharma
Former Vice Chancellor
Rajasthan Technical University
Kota – 324 010

Prof. D.C. Gupta
Pro Vice Chancellor,
Barkatullah University
Bhopal – 462 026

Dr. S.B. Chauhan
Former Jt. Advisor
Planning Commission
D6, 6021/6 Vasant Kunj
New Delhi – 110 070

Dr. A.K. Sarangi
General Manager,
Uranium Corporation India Ltd
Jaduguda

Dr. C.P Sisodia
Dy. Director Gen. (Retd.)
Geological Survey of India
94/38 Agarwal Form
Mansarovar
Jaipur – 302 001

Dr. R.H. Swakar
Geol. Soc. India
5th Block Jai Nagar
Bangalore – 560 040

Prof. S. Ibotombi
Chairman
Department of Earth Science
Manipur university
Imphal – 795 003

Prof. S.C. Mathur
Chairman,
Department of Geology
JNV University
Jodhpur – 342 005

Prof. S. Kumar
Emeritus Professor
Centre of Ad. Study in Geology
Lucknow University
Lucknow – 226 007

Prof. Anand Mohan
Registrar,
Dyalbagh University
Dayalbagh, Agra – 282 005

Prof. G.M. Bhat
Chairman,
Department of Geology
Jammu University
Jammu – 180 006

Prof. P.K. Kathal
Dean,
Centre of Ad. Study in Geology
Dr. H.S. Gaur University
Sagar – 470 003

Dr. D.K. Sinha
Scientist H
Atomic Mineral Directorate
Begumpet
Hyderabad – 500 001

Dr. Satendra
IFS
Director, NIDM
IIPA Campus, IP State
New Delhi – 110 002

Prof. Pankaj K. Srivastava
Dean,
Department of Geology
Jammu University
Jammu – 180 006

Dr. Arun Kumar Shandilya
Department of Geology
Dungar College
MGS University
Bikaner – 334 003

Prof. Vinod Sharma
Dean,
Indian Gandhi Dev. Res. Institute
Goregaon (East)
Mumbai – 400 065

Prof. Vibhuti Rai
Department of Geology
Lucknow University
Lucknow – 226 001

Prof. Anita Pande
Department of Geography
DBS Campus, Kumaon University
Nainital – 263 001

Dr. Beena Tripathi
Department of Geology
JNV University
Jodhpur – 342 005

Prof. P.K.Verma
Director General MPCST
Vigyan Bhavan
Bhopal – 462 003

Prof. Santosh Kumar
Department of Geology
Kumaun University
Nainital – 263 002

Dr. Ramesh Kumar
Scientist H,
Atomic Mineral Directorate
Hyderabad – 500 001

Prof. Sunil Parihar
Head,
Department of Mining Engg.
JNV University
Jodhpur – 342 001

Dr. Ashutosh Shrivastava
Hydrogeo Scientist
Rajasthan State Ground Water
Jaipur – 302 001

Prof. Kamal K. Agarwal
Chairman,
Centre of Ad. Study in Geology
Lucknow University
Lucknow – 226 007

Prof. M.K. Pandit
Department of Geology
Rajasthan University
Jaipur – 302 001

Dr. D.R. Patel
Director,
Department of Geology and Mines
Raipur – 492 001

Prof. R.B. Singh
Department of Geography
Delhi School of Economics
Delhi University
Delhi – 110 007

Prof. V.V. Nageswara
Department of Geology
Andhra University
Visakhapatnam

Dr. Virendra Gaur
Department of Geology
JNV University
Jodhpur – 342 005

Dr. P.K. Jain
School of Earth Science
Jiwaji University
Gwalior – 474 001

Prof. H.S. Sharma
Prof of Geography (Retd.)
Rajasthan University
A-3, Santiniketan, Tonk Road
Jaipur – 302 015

Prof. R.N. Sahu
Head,
Department of Geoinfomatics
Barkatullah University
Bhopal – 462 026

Prof. N.C. Pant
Department of Geology
Centre of Ad. Study
Delhi University
Delhi – 110 007

Dr. S.K Trivedi
Department of Geology
JNV University
Jodhpur – 342 005

Prof. Rana Pratap
Department of Geography
Magadh University
Bodhigaya – 824 234

Dr. Sishir Sharma
Department of Geology
Dungar College
MGS University
Bikaner – 334 003

Dr. Ashok Kumar
IFS
APCCF
Tripura (Assam)

Prof. K.K. Sharma
Department of Geology
Government College
Sirohi – 307 011

Dr. S.R. Jakhar
Department of Geology
JNU University
Jodhpur – 342 005

Dr. S.S. Deb
Chief Geoscientist
Oil India India Limited
Basni
Jodhpur – 342 005

Prof. Sreeup Goswami
Chairman,
Department of Geology
Ravenshaw University
Cuttack – 753 003

Dr. Alok Tripathi
Chief Geologist,
Oil Natural Gas Commission Ltd.
Cinnamara
Jorhat – 785 704

Prof. Ashok Srivastava
Department of Geology
Amravati University
Amravati – 444 602

Dr. Devesh Khandelwal
Department of Geology
Dungar College
MGS University
Bikaner – 334 003

Prof. Rameshwar Bali
Department of Geology
Centre of Ad. Study
University of Lucknow
Lucknow – 226 007

Dr. Meenal Mishra
Department of Geology
India Gandhi Open University
New Delhi – 110 001

Dr. Mamta Chauhan
Scientist
Indian Space Res. Organization
Ahmadabad – 380 015

Dr. Anju Mathur
HOD,
Vyas Inst. of Technology
Pal Road
Jodhpur – 342 005

Dr. P.S. Dhinwa
Sr. Scientist
Space Appl. Centre, ISRO
Ahmadabad – 380 00

Prof. G.K. Panda
Department of Geography
Utkal University
Bhubaneswar – 751 004

Prof. S.N. Lal
Department of Geography
Kumaon University
Nanital – 263 001

Dr. Rajneesh V. Khilnani
Director
Hydrogeo Survey Pvt. Ltd.
C- 103, Sastri Nagar
Jodhpur – 342 003

Dr. Mahesh P.Srivastava
Vice President
Lafarge (Cement)
Gurgaon – 112 001

Prof. Sakir Ali
Centre for Ad. Study in Geology
Dr. Hari Singh Gaur University
Sagar – 470 003

Prof. L. Elango
Director
Department of Geology
Anna University
Chennai – 600 025

Prof. S.N. Mohapatra
Head
School of Studies in Earth Science
Jiwaji University
Gwalior – 474 001

Dr. Prakhar Kumar
Scientist H
Atomic Mineral Directorate
Nagpur – 440 001

Dr. T.S. Sharma
Director,
Remote Sensing Centre
Department of Sci. & Tech.
Pal Road
Jodhpur – 342 001

Dr. P.K. Jain
Dy. Mineral Economist
Indian Bureau of Mines
Indira Bhavan
Nagpur – 440 001

Prof. Harsh Bhu
Department of Geology
MLS University
Udaipur – 313 001

Dr. Vikas Mohan
Dy. Gen Manager
Reliance Cement Pvt. Ltd.
C-Scheme
Jaipur – 302 001

Prof. Arun Kumar
Department of Earth Science
Manipur University
Imphal – 795 003

Prof. K.L. Shrivastava
Department of Geology
J.N.V. University
Jodhpur – 342 005

Preface

Geo-Resources have supplied useful or valuable material for human consumption long before they become objects of scientific curiosity or commercial exploitation. Even in prehistoric times, the development of material culture was inconceivable without the use of geo-resources. In modern time the study of geo-resources has evolved into an applied science employing detailed field observations, sophisticated laboratory techniques for additional information and computer modeling to build complex hypothesis.

The science of geo-resources has greatly reflected the information explosion. The most common text books suffered from either faster obsolescence or considerable revision. Many new journals exclusively devoted to geo-resources, or even to sub-disciplines of geo-resources, started. Accent on publication of thematic mostly multi-authored volumes appeared. The present edited volume is also one of the result and upshots of the same race of geo-resources against time.

The volume set out to present various aspects of a very broad field of geo-resources, at a time when the competitively growing economy of the nation, in other words, the pressing needs of the society, are compelling the discipline to grow, and pile of data is accumulating and opinions changing very rapidly.

The purpose of this edited volume, at first is to present new researches and to emphasize the many areas in which problem still persists with a view to contributing and encouraging further research. The second purpose is to present review articles summarizing our present understanding authoritatively, and in so doing, demonstrate almost all aspects of the subject selected.

The volume has been divided broadly into four sections and contain sixty five papers, *intoto*. The first section is on metallic and non-metallic mineral resources. It includes twenty four papers; eight on metallic and sixteen on non-metallic mineral resources. The metallic part covers ferrous metal, base metal and nobel metal resources and non-metallic part covers zeolites, fluorite, wollastonite, talc, gemstone, pyrite, phosphorite, clay, lime stone and sandstone deposits. The second section is on geo-energy resources. It includes ten papers on atomic minerals, hydrocarbons and coals. The third section includes thirteen papers on water resources. It includes exploration, quality and various aspects of water resource management and case studies. The last section is having eighteen papers on land resources, geohazards and environment.

All the papers in the volume were reviewed by the editors and invariably scrutinized and approved by at least two external referees, before being finally accepted by the editorial board. The editors have regretfully been forced not to include as many as 19 papers submitted for the consideration in this volume, because more than one referee advised against it. Still, we unhesitatingly own the responsibility for errors of omission and commission, although every care has been taken. We would not find it truly rewarding unless the volume delivered the desired good to the specialized scientist, teachers, scholars, students and those associated with industry.

Amongst the authors and reviewers, a large number of friends, colleagues and students of professor Vinay Krishan Verma are missing in the volume, who were not in position to contribute a paper because of temporary hindrance or engagement in research. Many of them expressed their regret at not being able to participate in this volume, on behalf of all of them we edit this book as dedication to Professor Vinay Krishan Verma, on occasion of his 80th birthday.

K. L. Shrivastava
Arun Kumar

Acknowledgement

Our contribution in present endeavour was channelized at various stages and in multifold modes so that our efforts culminated in the form of the present volume. In the present case where the much derided inertia was shaken off and momentum imparted by many, from near ones including our colleagues and students in the university, to peer, professional colleagues and contemporaries positioned far away, we thank them from the innermost recesses of our heart.

We place on record our sincere thanks to all the authors for responding to our request and for contributing the papers to this special volume. The following authors and geoscientists of distinction have also extended their great support without which this initiative have been difficult to sustain: Prof. M. K. Pandya, Prof. K. Surya Prakash Rao, Prof. J. P. Shrivastava, Prof. D. C. Gupta, Prof. S. C. Mukhopadhyay, Prof. Hema Achyuthan, Prof. U. C. Singh, Dr. Maurizio Gorla, Prof. Anil Pophare, Prof. Madhumita Das, Dr. Arun Kumar Shandilya (Bikaner), Prof. S. C. Mathur, Prof. S. N. Mohapatra, Prof. S. N. Lal, Prof. T. K. Pandya, Prof. R. S. N. Sastri, Dr. L. K. Sinha, Dr. D. K. Chadha, Dr. S. D. Dogra, Dr A. K. Sarangi, Dr. P. K. Jain (Nagpur), Prof. P. K. Verma, Prof. H. B. Srivastava, Prof. P. K. Jain (Gwalior), Prof. Saibam Ibotombi, Prof. Vibhuti Rai, Dr. K. K. Sharma, Prof. Sreeup Goswami and Prof. G. K. Panda. The authors are also thanked for their tolerant attitude to the editorial revisions.

We express our gratitude to all the Hon'ble reviewers and board of editors, whose names are listed separately, for giving their valuable time and expertise to review the papers critically and giving constructive suggestions which have helped in greatly improving the value of the contributions. We are specially grateful to Dr. Harsh Gupta, Prof. K. S. Valdiya, Prof. S. K. Tandon, Dr. A. Arunachalam, Prof. U. Aswathnarayana, Dr. S. Viswanathan, Prof. Savinder Singh, Prof. K. L. Rai, Prof. Ernst Hegner, Dr. V. P. Dimri, Prof. Petricia Vickers-Rich, Dr. P. S. Parihar, Dr. N. K. Dutta, Dr. Vaidyanadhan, Dr. B. K. Bansal, Dr. Fareeduddin, Dr. S. K. Bhushan, Prof. P. S. Saklani, Prof. K. K. Singh, Prof. Baleshwar Thakur, Prof. B. Satyanarayan, Prof. S. S. Thigale, Prof. V. C. Tiwari, and Prof. G. M. Bhat.

We are greatly indebted to the following geoscientists who gave their friendly and valuable suggestions and help in preparation of this volume: Prof. M. K. Pandit, Prof.Vinod Agarwal, Prof. A. S. Sheoran, Dr. Ajay (ISRO), Prof. M. S.

Sisodia, Prof. T. K. Biswal, Prof. A. K. Sen, Prof. P. K. Kathal, Prof. R. K. Trivedi, Prof. R. P. Tiwari, Prof. Devesh Walia, Dr. D. K. Sinha, Dr. Virendra Gaur, Prof. S. P. Singh, Prof. Viqar Hussain, Dr. L. K. Nanda, Dr. S. K. Trivedi, Prof. Anil Maheswari, Prof. Harsh Bhu, Prof. S. K. Parihar, Dr. Trilochan Singh, Dr. Beena Tripathi, Prof. Sakir Ali Adil, Prof. K. K. Agarwal, Dr. S. R. Jakhar, Dr. K. Ramesh Kumar, Prof. R. N. Sahu, Prof. L. P. Chaurasia, Prof. Arun Shandilya (Sagar), Dr. Prakhar Kumar, Dr. Sishir Sharma, Dr. Rameshwar Bali, Dr. Arun Vyas, Dr. Rajneesh V. Khilnani, Dr. Devesh Khandelwal, Dr. T. S. Sharma and Dr. R. G. Goswami.

We are thankful to all of our research students, present and former, for providing us assistance. The assistance is gratefully acknowledged.

We both are indebted to our families for their endless patience, continuous support and undying encouragement that has always been a continuous source of motivation. KLS is thankful to his wife Dr. Raka Srivastava and daughters Krati and Srasti and AK is thankful to his wife Mrs. Manisha and both the children. Under most stressful situations they all had been kind and generous to inspire us to raise our potential.

The completion of this book would not have been possible without trust and confidence best owned by us in Scientific Publishers (India), Jodhpur specially Mr. Pawan Kumar, Mr. Tanay Sharma and Mr. Rajesh Ojha. Our cordial thanks are due to them for their patient assistance in rising the technical quality of the book.

K. L. SHRIVASTAVA
ARUN KUMAR

Professor Virendra Krishan Verma — An Appreciation

Professor Virendra Krishan Verma is an eminent great academisian and administrator of distinction, geoscientist in the country gifted with modesty and gentleness. Son of Late Shrimati Champa Devi and Professor R.S. Varma, a renowned mathematician and operational research scientist, Virendra Krishna was born on June 30th, 1934 in village, Nai Bazar, in the district of Azamgarh in eastern Uttar Pradesh. He had his entire education at Lucknow and had a brilliant scholastic record standing First class First both at Graduation and Post graduation levels from University of Lucknow. Soon after, in December 1956 he moved to the rugged terrains of Geology as an Assistant Geologist (re-designated later as Scientific Officer, SB) in the Atomic Mineral Division, Department of Atomic Energy, Government of India and continued till 1963. During this period he was awarded two advance merit increments. Belonging to education driven family, he joined the University of Jodhpur (1963) where he also earned his PhD degree in 1964. He moved to the University of Delhi in 1966 as Reader in Geology where he was set to gain an uncommon experience to establish a newly created Department of Geology. He took over as Professor of Geology in 1976 and formally retired after two extensions in 1999 from the University service. He steered the activities as Head of the Department for two consecutive terms from 1973 to 1976 and 1976 to 1979 and then again for the third term from 1993 to 1994. It is his vision and team leadership that the department is now elevated to the level of Centre of Advance Study.

Likewise at a young age he was called upon to act as Dean to lead the Faculty of Science twice in 1978-1979 and 1993-1994. Keeping abreast with the prevailing need for (i) developing interdisciplinary approach and (ii) providing sufficient space for vocational studies as Dean, he reconstructed and reoriented several courses of studies like Physical Sciences (combining relevant portions of Physics and Chemistry), Life Sciences (including Anthropology, Botany and Zoology), Earth Sciences (relevant Geology and Geography and other earth related fields), Mathematical Sciences (clubbing Mathematics and Statistics), and Behavioral Sciences (Psychology and Management) to match with appropriate requirement for various streams in a manner not wholly or partly covered in the other two disciplines.

Beside academic activities he was assigned some administrative responsibilities too such as Warden, Jubilee Hall and Provost, International Student House and

often to represent university at the governing bodies of its colleges reflecting academic, professional and administrative acumen.

Professor Verma has worked on several research projects sponsored by University Grants Commission (UGC), New Delhi, Indian Council of Medical Research (ICMR), New Delhi, Defence Terrain Research Laboratory of Defence Research and Development Organisation (DRDO), New Delhi and Space Application Centre, Ahmedabad of the Indian Space Research Organisation (ISRO).

Professor Verma's research pursuit had significant contribution in the field of Earth Sciences: (a) *quantification of sediment attributes* and statistical treatment thereof in as early as in late 1950s, an uncommon research area at that time (b) perceptive insight on Central India Gondwana Geology with the *discovery of Voidal Concretions* in the late Triassic Pachmarhi sandstone a sequence much older than reported before anywhere in the world and on a digenetically driven chromatic differentiation and segregation imprints thereon, (c) *in situ* development of *inorganic calcareous nodules* in the Lower and Middle Siwalik rocks formed *around centers of nucleations,* (d) *itacolumites flexibility* due to *intergranular voids developed on digenesis,* (e) clues to reveal that Jabalpur *Lamenta Green sands are intra-trappean piedmont terrestrial deposits* and *not of marine origin inter-trappeans.* (f) upgrading bentonites for use in suspension of magnesium trisilicate as disintegrants and binding agents in sulpha-gunidinc IP and ido hydrooxyquinolin, *opens a new vista of research activity in medicinal geology,* (g) the South Delhi Kaolin on beneficiation possesses lower abrasive and higher cleansing property for dentifrice formulation.

Professor Verma was greatly influenced by Professor W. D. West during summer institute held at the University of Saugor in mid 60's and developed interest in geomorphology. Coupled with ongoing sedimentological work he was doing, he embarked on (i) *river and basin morphometry* studies; (ii) subsequently *designed two new quasi—quantitative parameters: platform cliff-ratio* and *dissection percentage* for Western California Pacific Marine Terraces correlation; (iii) formulated a *visual method for stream tortuousity* appraisal; and (iv) made *significant* contributions to the *Satpura geomorphology*—reconstructing *its denudational chronology,* discerning its regional neotectonics as an influence to explain the *terrain youthfulness,* now well attested by Jabalpur (1997) seismicity in hitherto regarded stable landmass.

His research interest in environmental studies inspired by Professor A S Paintal are primarily a spin off of applied geomorphology and comprise (a) *computing-landscape aesthetism terrain attributes* for *tourism development,* (b) introducing *visual technique* for digitizing Landsat scenes and also correlate composite spectoral reflectance (CSPR) with morphometric attributes, (c) *investigating landslide occurrences*—causative factors, and (d) identified the *role of geoscientists in alternate energy resources development:* (i) discerning ocean's cold air and hinterland valleys bottlenecks formed by mountain passes as appropriate sites for wind energy; (ii) siting *wave energy concentration headlands* in coastal areas; and (iii) *tapping frictional energy locked along dynamically active lineaments in* seismic prone areas that may also dampen earth's tremor as and when occurs. Currently he is interested in climate studies. He distinguishes global warming due to excessive CO_2 emission in the atmosphere and ozone layer puncturing as a process and climate change a product, not necessarily always related. He strongly advocates for geogenic-centered climate change reasoning, based on pulsation in sun spot activity, increased volcanicity and intensity and frequency of seismicity at times concomitant release of

nutrient a higher potent GS gas, deviation in earth's spin axis, ongoing Himalaya uplift (bringing in shorter heavy downpour period and longer moderate summer and winter spreads with fewer days of harsh and extreme temperatures), sea level and temperature rise attributable to the sea floor spreading on basaltic ejection, glacial melt untenable as a long journey from the Himalaya which would facilitate underground water percolating and/or evaporation in warm conditions at much faster rate than surface runoff, no drought like or alarming floods in recent past despite several glacier-recessions melt, and hydrologic cycle practically stable.

Professor Virendra Krishan Verma

Professor Verma has to his credit 130 research publications. He has authored five original text books:

1981. Text Book of Sedimentary Petrology, Dehradun: International Book Distributors, p. 134; 2001. Revised and rewritten as Sedimentology, New Delhi: Harman Publishing House, p. 190.

1981. Annotated Bibliography on Geomorphology of the Himalaya, Publication No. 2, Department of Geology, University of Delhi, Delhi, p. 26.

1989. Avsaadi Shail-Vigyan, Chandigarh: Haryana Sahitya Academy, p. 168.

1999. Lectures on Geomorphology for M.A./ M.Sc. students. Delhi: Pilgrim Pvt. Ltd., for PG students p. 485

2013. Environmental Geosciences (under publication)

and edited one book and 7 monographs/compilations:

1973. Recent Researches in Geology, Vol. I, (Professor A. G. Jhingran, Commemorative Volume), Delhi: Hindustan Publishing Corporation.

1975. Recent Researches in Geology, Vol. II, (Professor A. G. Jhingran, Commemorative Volume), Delhi: Hindustan Publishing Corporation.

1975. Proceedings, Symposium on Sediment, Sedimentation and Sedimentary Environment, Delhi: Delhi University Press, p. 390.

1978. Recent Researches in Geology, Vol. Ill, (Professor I. C. Pande Commemorative Volume), Delhi: Hindustan Publishing Corporation.

1978. Proceedings, Symposium on Morphology and Evolution of Landforms, Delhi: Delhi University Press, p. 267.

1978. Recent Researches in Geology, Vol. VII, (Professor A. G. Jhingran Memorial Volume), Delhi: Hindustan Publishing Corporation.

1982. Himalaya: Landforms and Processes, New Delhi, Today and Tomorrow, p166.

Besides by University of Delhi, Professor Verma has also been a recognized Research Supervisor for guiding doctoral work by other institutions in India and abroad, viz (a) Meerut University (now CCS University), (b) H N B University of Garhwal, Shrinagar (c) North Eastern Hill University, Kohima Campus (d) Patrice Lumumba Peoples Friendship University, Moscow (Russia) (e) Tribhuvan University, Kathmandu, Nepal (f) Jawaharlal Nehru Technological University, Hyderabad and (g) Dr B. R. Ambedkar University, Agra.

He successfully supervised 24 Ph.Ds. and 12 M.Phils. dissertations and was also a co- guide to M.S. (Orthopedic Surgery) Thesis from Aligarh Muslim University.

Professor Verma is widely travelled Geoscientist. He held several Visiting assignments that include: (i) Associate Professor at the California Institute of Technology, Pasadena, Los Angeles, USA in 1971-72; (ii) Scholar Northwestern University, Evanston, Illinois (USA) in 1972 both (i &ii above) under Ford Foundation Faculty Exchange Programme; (iii) Scientist at the invitation of Centre de la National Recherches de Scientifique (CNRS), Paris (France) in 1975; (iv) Professor, Patrice Lumumba People's Friendship University, Moscow, Russia under Indo-Soviet Cultural Exchange Programme in 1976; (v) Professor, Tribhuvan University, Kathmandu (Nepal) as Colombo Plan Expert on deputation by the

Ministry of External Affairs, Government of India under Technical Cooperation Mission during 1986-88; (vi) on invitation to Macquairie University, Australia for technical discussions in 1973 (vii) Visiting Professor at the University of Madras, Chennai in 1977 and University of Rajasthan, Jaipur in 1993. He also gave lecture at Ohio Weslyan University. Delaware (USA), University of Saugor, Sagar in 1976 in UGC sponsored programme, Space Applications Centre, Ahmedabad in 1982, and at Andhra University, Visakhapatnam in 1999 under UGC's SAP programme.

Professor Verma for his distinguished contribution in the field of Earth Science and Environmental Sciences has been conferred with several honours and distinctions:

A. Life-Time Achievement Award - 2012, Society of Earth Scientists.

Professor KP Rode Memorial Lecture, 93rd Indian Science Congress Association (ISCA), Hyderabad session 2006

Professor M.S. Sadasivaih Endowment Lectures, 1996, Karnataka University, Dharwad

Platinum Jubilee Lecture (Earth System Sciences), 83rd ISCA session, Patiala, 1996

B. Fellow: National Academy of Science (India) since 1975

C. Member of the Punjab State Council of Science and technology, Chandigarh (2001-2003).

D. Served the governing body of Wadia Institute of Himalayan Geology now at Dehradun for several years. In association with the then stalwarts Professor D.N. Wadia and Professor A.G. Jhingran he assisted in the establishment of the Institute.

E. Worked on Advisory Panel (Geology) of the Commission for Scientific and Technical Terminology, Government of India for several years.

He also remained associated with academic bodies of several universities in India. Even now he is academically associated with the Indira Gandhi National Open University (IGNOU) and Consortium for Educational Communication in distance learning.

F. He has been the President of

(i) Indian Association of Sedimentologists 1986-1990 and gave Presidential Address on 'Sedimentology in India : Development, Growth and Prospects'; one of its founders he earlier served as its General Secretary and Foreign Secretary.

(ii) Indian Institute of Geomorphologists 1999-2000 and addressed on 'Geomorphology of India, New Millenium : Some Problems and Prognoses.'

He has unprecedented distinction of being the ***President of two sections*** of Indian Science Congress Association viz., (i) Geology and Geography (now Earth System Sciences) Section of the 72nd ISCA session 1985, discussed on "High Altitude Geo-environment and Resources : Problems and Prognoses". (ii) Environmental Science Section 95th, ISCA Session, 2007-08 and delivered Presidential Address on 'Indian Geo-Enviro-Informatics Convergence.' (iii) He was also Convener, Forum of Environmental Sciences ' 1997-98 and 1998-99 where he strived hard for raising the

forum to a full-fledged section as now it is. He delivered Conveners address on Geo-informatics Syndrome (1998) and Interactive areas in Environmental Sciences: Challenges and Opportunities (1999)

His association with Indian Science Congress began, over six decades back, in 1953 as a student volunteer and subsequently held several positions of responsibility. Besides being President of its two section as above, he was Recorder of the Geology and Geography section (1978-80) and before that as an elected sectional committee member (1977-78). He was Local Sectional Secretary, Geology and Geography section 1974-75 and of Publication, 1985-86, Local Secretary 84th ISCA session 1996-97. Professor Verma served the ISCA Executive Committee (2002-03) and Council (2005-06 and 2006-07). The General President ISCA nominated him as External member of the Endowment Committee for 2001-02 and as member of Publication Committee for 2005-06 and Chairperson for a Plenary session on Innovative Water Technology for Poor' during Ahmedabad 92nd session 2005. Professor Verma is presently Director, Environment Division, Vogue Construction and Consultancy Services Pvt. Limited, New Delhi.

It remained a matter of great honour for both of us to be an associate of Prof. V. K. Verma in the field of Earth System Science and in every part of the personal life. We feel proud to dedicate this book on 'Georesources' to Prof. V. K. Verma. Professor Verma has led a rich and successful academic life and has driven to forge the growth of Earth and Environmental Sciences through common linkages cutting edges. His firm belief in truth makes him strong and courageous to face the challenges ahead. His passion, devotion and constructive suggestions in the academic field remain unparallel. He is endowed with a combination of successful professional (field geologist), an adorable teacher, researcher and an admirable tireless innovative thinker with humility and concern for others leading to innumerable friends as they are who prize their association with him. He is a valued trusted friend, philosopher and guide to his associates. By joining hands with all the contributors, reviewers and members of editorial board we wish him all the best for future.

K.L. Shrivastava
Arun Kumar

Contents

Part - I:

METALLIC AND NON-METALLIC MINERAL RESOURCES

I-1

Geology of Khursipar- Manegaon Deposit of Titanoferous Magnetite, Amgaon Group, Bastar Craton, India

H.S. Kale[1*], G.G. Wadpalliwar[1] and S.N. Mohapatra[2]

[1]Department of Geology, RTM, Nagpur university Nagpur
[2]School of Studies in Earth Science, Jiwaji University, Gwalior
E-mail: drhskale_15geo@rediffmail.com

Abstract

This paper dealt with the evaluation of the titanoferous magnetite on the basis of petrography, ore mineralogy and geochemistry. This deposit forms part of Amgaon Group which overlies Dongargarh Super Group. The deposit of titano-magnetite is associated with amphibolites metagabbros and its igneous counterparts. The Ti-magnetite ore zone forms top portion of the elongated mounds trending north-south direction. These are emplaced along a narrow rectilinear zone, marking contact between the granite and granitic gneiss. The EPMA studies of titanoferous magnetite revealed that, the principal ore mineral is hematite and magnetite while ilmenite forms as main titanium bearing mineral.

Major Oxides assayed varies from, FeO % (49.54-57.33), Al_2O_3 % (2.13-3.95), SiO_2% (1.17-3.89), TiO_2% (9.19-19.01), V_2O_5 % (0.30-1.51), MgO% (0.33-1.98), MnO % (0.20-0.26), P_2O_5 % (0.02-0.03), Na_2O% (0.06-0.17) and average CaO is 0.01%. The trace elements present are Ni 12-200ppm, Co 250-850ppm, Mo ‹2ppm, W ‹50ppm, Cr 5-125 ppm, Ag 0.5-0.8ppm, Ga 0.52-0.92 ppm, Nb 21.11-70 ppm, Ta ‹1000.01-1587.09 ppm and Be ‹51.23-90.89ppm.

Key Words : Titanoferous Magnetite, Amgaon Group, Baster Craton, Khursipur - Mangegaon Deposit.

Introduction

The area under investigation covers a part of early Precambrian of Amgaon Group which overlies Dongargarh Super Group (Mohabey and Bhake, 1994), includes in Survey of India Toposheet No. 64 C/3. Physiographically the area is gently undulating developed on granitic gneiss and granite and is covered by alluvium. Then it is traversed by metabasics dykes. The titanomagnetite rich zone associated with gabbroic rock. Magmatic Fe-Ti oxide ore are commonly associated with or hosted in mafic intrusion or Proterozoic anorthosite complex (Lister 1966, Force 1991 and Bateman 1951). These ore are important sources of Fe, Ti, V (Duchesen, 1999) and are generally characterized by complex field and textural relation (Kwan-Nang-Pang, et al, 2008).

The field evidence indicate that they occur as lenses or layers of massive Fe-Ti oxide and that are in sharp contact with their host rocks (Willemse, 1969 and Duchesen, 1999). These are emplaced along a narrow rectilinear zone, marking contact between the granite and granitic gneiss. The ore zone varies in thickness from about 5 to 40 mt. A unique feature of Fe-Ti oxide ore and associated intrusive rocks is their magmatic origin (Lindsley, 1992 and Frost, 1991). The occurrences of V-Ti magnetite deposit of India have been reported from Maulabhanj and Dhekanal district (Chakraborty and Mallick,1968), Nuasahi of the Keonjhar district (Mohanti and Paul 2008), Hatticher, Joshipur, Kumhardubi and Udala areas of the Mayurbhanj District of Odisha State (Roy, 1956: Nanda, 1973: Das and Mukharjee, 1984)

and Dublabera of Sinhgbhum district Jharkhand (Shrivastava, 1963: Dune and Day 1937: Dasgupta, 1969: Vasudev and Shriwasan 1979). Similar occurrences of such deposit include those in the Ushushwana and Rooiwater Complex in the South Africa (Winter 1965 & Reynolds 1978) and Panzhihua gabbroic layered intrusion and associated Fe-Ti-V oxide deposit at Sichuan province in South-West China.

Geological setup

The rocks of the area belong to the Early Precambrian, Amgaon Group and overlying Dongargarh Super-Group (Broderick, 1917: Sarkar, 1957: Sarkar, 1958: Sarkar 1967). They consist of mainly granitic gneisses and granites which are traversed by metabasics as a dyke, which comprises metagabbros and its igneous facies (Bhake, 1993 and Bhattachrji, 1932.) The titanomagetite rich bodies' shows massive ore zone forms top portion of the elongated mounds extending in general north-south orientation. However, it shows pinch and swell character with possible linear mineralization, the in low lying area between the mounds. On a gross scale the area is characterized by N-S structural trend which is defined by the attitude of foliation planes of gneisses and above all by the persistent linearity of the metabasic dykes along this direction (Fig. 1. C and D). This structural trend thus is a part of the lineament in this region of the Indian Precambrian Shield. The three sets of joint are quite prominent (Fig. 2.F) which are found in almost all the rocks but are more pronounced in the granites. The other feature of the dykes is their offsetting along faults which trend in almost E-W direction. In their origin these joints are clearly related to the faults, the joints of first and second set have trends similar to the faulting whereas those of the third set have conjugate attitude with the first set. The characteristics of the rocks and geochronology in this part of the shield are indicated by the following geological column worked out by Sarkar et al. (1980) (Table 1).

Table 1. Regional Stratigraphic Succession of the area (Sarkar et al., 1980)

Chhattisgarh	Raipur Group
Supergroup	Chandrapur Sandstone
	------Unconformity------
	Khairagarh orogenic phase (C.900 Ma?)
Khairagarh Group	Mangikhuta Volcanics
	Karutola Formation
	Sitagota Volcanics (1367 Ma)
	(Intertrappean Shale (1686 Ma)
	Bortalao Formation
	Basal Shale (1534 Ma)
	------Unconformity------
	Dongargarh granite (< C.2200 Ma)
Nandgaon Group	Pitepani volcanic
	Bijli rhyolites (C.2200 Ma)
	Amgaon Orogeny, metamorphism
	and granitisation (>C.2300 Ma ?)
Amgaon Group	Quartz-sericite schist, felspathic quartzite, garnet epidote quartzite, hornblend biotite quartzite, quartz- feldspar biotite gneiss, hornblend schist and amphibolite.

Granite and granitic gneisses

The granite gneisses which belong to the Amgaon Orogenic phase and regional metamorphism occur in the western part of the area. These gneisses have been formed by granitization of the underlying metasediments. The gneisses have pink K-feldspars as augens, oval in shape which lies along the foliation planes (Fig. 2.E). Generally the foliation is seen to wrap around the feldspar augens. Also coarse quartz grains are stretched and fragmented along the foliation planes. The gneisses ranging from white to grey in color are medium to coarse grained and consist of quartz and feldspars with minor amount of chlorite and sphene. Lineations as defined by streaks of biotite and stretched quartz are developed to varying extent which lies in the schistosity plane of the gneisses (Fig.2.D and 3.B).

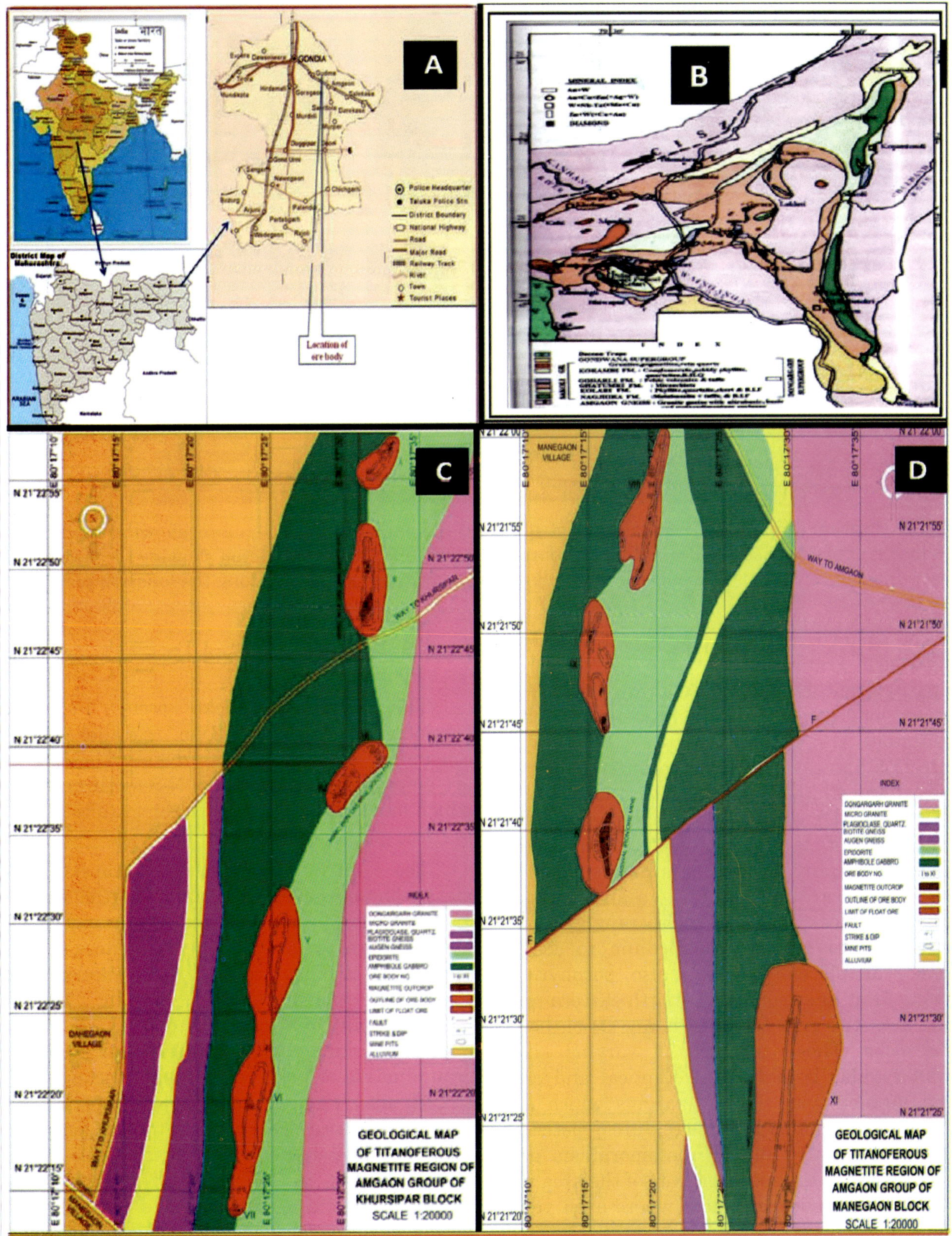

Fig. 1. (A) Map showing of global location of the study area. (1B) Geological map of Sakoli fold Belt. (1C) Geological map of Khursipar. (1D) Geological map of Manegaon.

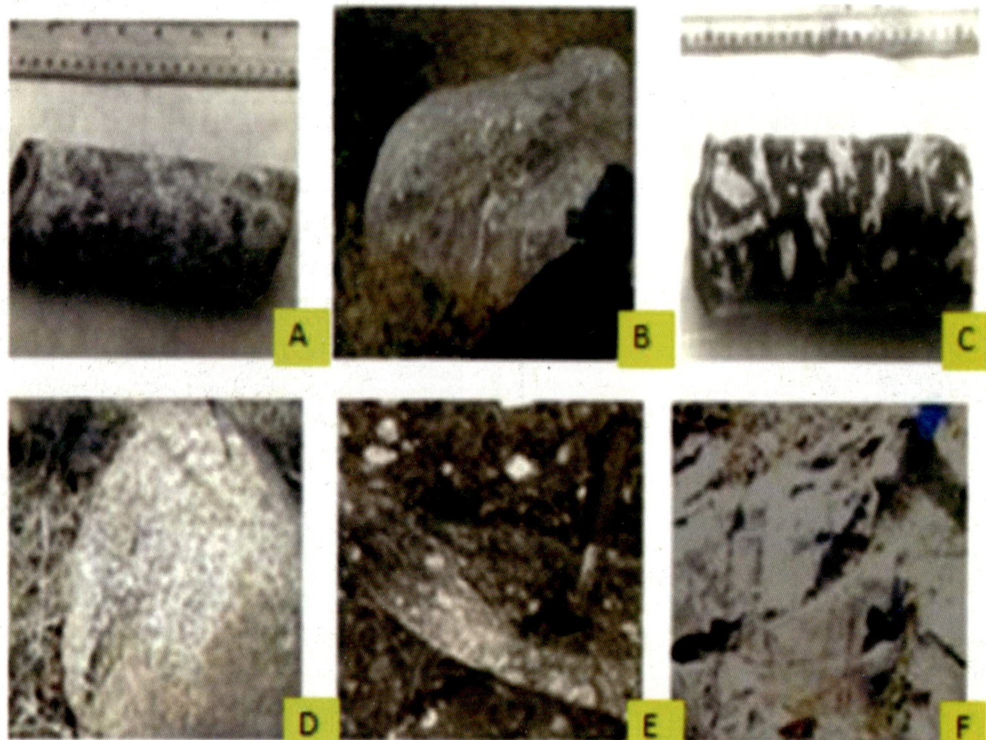

Fig. 2. Photograph of the lithounits of study area (A) and **(B)** Mottled gabbro: showing cumulates with in rim of plagioclase **(C)** Pegmatoidal gabbro core sample **(D)** Gabbro outcrop showing megascopically observing ophitic texture **(E)** Augen gneiss: plagioclase porphyroblast lying along the foliation plane **(F)** Titanomagetite Outcrop: showing joint pattern

The eastern part of the area is occupied by Dongargarh granite. It is large batholiths of composite nature with separate forcefully intruded magmatic bodies and numerous patches of granitoid. This batholithic mass extends in N-S direction and covers between the Sakoli belt in the west and Chhattisgarh basin in the east (Fig 1.B, C, and D). It was emplaced during the culmination of tectonics associated with the Nandgaon orogeny. In this area outcrops of the granitic rocks are marked by tors and rounded hummocky blocks. These rocks ranging from granodiorite to granite are medium to coarse grained, at places they are also porphyritic, showing faint lineation. In the north eastern part of the area the granites have many xenoliths of basic rocks. These xenoliths of varying sizes exhibit regular borders with elliptical and oval shape marked with chilled margins which are silicified. The primary igneous texture of both the rocks is obliterated due to metamorphism and fluid activity. In the southern part of the area the granite is epidotised at places where even veins of epidote are seen.

Host Rock of Fe-Ti-V Oxide Ore

The metabasic rocks occur in dykes emplaced in a weak zone, parallel to the lineament which occurs as dykes. These dykes are emplaced along a narrow N-S rectilinear zone which marks the contact between granite and the granitic gneisses.

The dyke rocks are comprised of metagabbros and its igneous facies the western part the amphibole gabbro. These metabasics have constant increase in the coarseness of grain towards west. They are mainly melanocratic and some of them show ophitic texture megascopically where thick clots of ferromagnesian prophyroblasts are set in a groundmasss of light coloured plagioclases (Fig. 2.A, B and C). Further, the concentration of plagioclase decreases from east to west across the ore body. Again, epidote and calcite veins become more common in the west. The rocks show different varieties of textures like Blasto ophitic, Spinifex, Intergranular–Interlocking, Poikiolitic, Porphyroblastic, Granoblastic and Schistose (Foye, 1916 and Warren, 1908).

Fig. 3. Field Photograph of the study area

Titanomagnetite

The ore bodies occur in the form of elongated lenses in gabbroic rock which constitute narrow hillock in N-S direction (Fig. 3 G). The ore bodies have varying thickness which ranges from 2-40 mt. The ore contains chiefly oxides of iron, titanium and vanadium with minor quantity of sulphides of copper and other metal like Ni, Co, Zn and Ag also. Titanomagnetite occurs in weathered, massive, banded, and disseminated forms. Upper part is fairly soft due to weathering and it show brownish in color it is due to martitization of magnetite. Massive ore bodies are coarsely crystalline but along with fine grained layers are also present at some places. Sulphide mineralizations are found along slip planes. This is also weathered, fractured and bluish black to dark black in color. Banded ore are fine grained, grayish in color and fracture has probably developed due to alternate layering of magnetite and basics. Disseminated ore material is fairly common in host rock, some places crude foliation in magnetite reach amphibolites is observed. Distribution of ore bodies in basic host show a preferential distribution along narrow zone in a dyke. Float ore is fairly wide on the hill slope side as well as on top of the hills. The geological map (Fig 1.C and D) shows the distribution of rock types. The ore occurs in the following three situations. Weathered ore, Massive ore and Banded ore.

Table 2. Attitude of the ore bodies

Ore body No	length of Ore body (Mt)	Width of Ore body (Mt)	Dip of ore body	Characteristics features of the ore body
I	80	15	Dip towards west at 60^0 to 63^0	Massive magnetite, ore body flanked by foliated Gabbro
II	125	10-20	Dip towards west side at 65^0 to 68^0	Insitu magnetite and hematite to limonite ore mineral
III	20	5	Dip towards west at 55^0 to 60^0	Disseminated, Lensoid Massive magnetite is exposed in small patch,
IV	35	12	Dip not measured because no outcrop is found.	Massive ore in 3 thin Layers, band thickness not more than half meter, ore body enclosed by leucogabbro,
V	185	23	Dip not measured because no outcrop is found.	Fairly large weathered ore body,
VI	150	35	Dipping towards west 70^0	Massive ore become gradually banded, where it is contact with amphibolites. Magnetite body has lensoidal shape which is specially in massive in character
VII	35	20	Dip towards West at 70^0	Small exposures of massive magnetite
VIII	250	10	Dip towards West 70^0	Small isolated magnetite lenses, disseminated ore in gabbroic rocks, Massive ore band passes into banded ores. 5 thin layers of massive magnetite have been encountered. Some auxiliary thin layers of magnetite in anorthositic gabbro.
IX	141	20	The ore body seems to dip towards west with 85^0.	Exposures of massive magnetite are more prominent. Patchy/discontinuous nature of magnetite lenses
X	130	15	It dips westward at 65^0	Massive magnetites are confined to the crest of the hillocks. Massive magnetite band enclosed in a gabbroic mass. Fine grain gabbro with magnetite grains, then thin magnetite layers in medium grain gabbro, which passes into coarse gabbro with spinifex texture. To the west the main massive magnetite passes into disseminated magnetite ore. Shape of ore body is lensoidal.
XI	430	10-40	Nearly vertical towards West	Chiefly massive ore.

Massive ore: There are not many outcrops of massive magnetite which are seen only in some hillocks. The outcropping magnetite are mainly weathered which are fractured and loose mass. Massive magnetites are bluish black to dark black in colour and they are coarsely crystalline. In drill cores of massive magnetite interstitial mafic grains are visible. At places distinct amphibolitic layers are also seen. Slip planes and slicknessides have been favorable for sulphide mineralization in these massive ores.

Banded ore: Banded ores could be recognized only in fresh drill cores. These are fine grained and mostly grayish to brownish in colour. This banded feature has probably developed due to alternate layering of magnetites and basics. Disseminations of ore material are fairly common in the host rocks which attain the category of ore grade mineralization in many instances. At places crude foliation in magnetite rich amphibolites has been observed.

Weathered ore: In the upper part the ore body is fairly soft at places due to weathering. This ore during core drilling gets crushed and is recovered as cherry red, brown to black powder ore. The brownish color is due to martitization of magnetite.

Petrography

(A) Host Rocks: Mineralogical analysis has been carried out through optical microscope and Electron Probe Micro Analysis (EPMA). The ore bearing host rock is essentially gabbroic in composition broadly consist of plagioclase feldspar, hornblende, penninite, biotite and minor constituents are quartz, apatite, actinolite, chlorite and oxide mineral like dominantly subhedral magnetite coexisting with discreet grains ilmenite. The plagioclase feldspars vary from andesine to bytonite in composition. These plagioclases are thick clouded appearance with faint twin lamellae. Twinning is based on albite laws (Fig. 4.C). These twin lamellae very frequently show bending displacements and kink bands. (Fig. 4.A). At places plagioclases are traversed by microcalcite veins which are generally in the western parts of the dykes Another important feature which is fairly common is the occurrence of coarse euhedral crystals of plagioclase within the hornblende porphyroblasts (Fig. 4.D). Hornblende is another essential component of the rock and is strongly pleochroic. It is found in different modes. Primary hornblende is essentially present as porphyroblasts (Fig. 4E). Some smaller hornblende grains occur along the cleavage traces of the plagioclase. Further, inclusions of opaques – (Fe-Ti oxides) stripped pyroxenes and Mg rich biotite appear quite frequently in hornblende. Also, at places actinolitte and chlorites are found replacing the hornblende grains. Penninite is the variety of chlorite which is quite common and usually forms pseudomorphs of amphiboles. Moreover, some of the chlorite occupies the micro fractures which cut across the twin lamellae in plagioclases. The primary variety of biotite has inclusions of opaque's which are along the cleavage traces. The secondary biotite is seen replacing the hornblende grains. Quartz is essentially anhedral and most commonly occur in the inters-titial spaces between plagioclases and hornble-nde. The other frequently occurring variety is stretched quartz which has a wavy extinction. A recrystallized kind of quartz is aligned along the micro fractures and also present in the interstices. The other important accessory mineral is actinolite which surrounds the hornblende crystals. Apatites contain opaque inclusions and have subhedral outlines. Epidote is commonly associated with the saussuritized portions of plagioclases in small proportions except for amphibole schist, amphibole-calc-silicate rock where it is a dominant mineral and alters into allanite. Calcite is found at some places as thin veinlets within plagioclase, but occurs in an appreciable quantity in the amphibole-calc silicate rocks. Garnets as euhedral (Fig. 4.D) grains are traversed by cracks and are very abundant in the amphibole predominant rocks.

(B) Ore Petrography: Polished section of the Fe-Ti ore, that were earlier studied under reflected light microscopy were subjected to EPMA study. The quantities analysis were carried out for illmenite and Ti- Hematite/ Magnetite grains which were detected during the qualitative spectral scan. The standard used for the elemental peak calibration along the X-ray emission line for the quantitative analysis and PET and LLIF crystals were used (Fig. 6.A and 6.B). Black Scattered Electron (BSE) imaging at high magnification was employed for searching the illmenite grain. Quantitative and qualitative analytical technique were employed for analyzing their chemistry using wave-length dispersive (WDS) technique. PEAK SIGHT on line software program developed by CAMECA was use to acquire and process the data.

Minerals distribution of the iron ore from this area are; Hematite 45-47%, Magnetite, ilmenite 20-21%, Chlorite/chamosite 18-20%, Goethite/limonite 3-4%, Mica (muscovite, biotite) 3-4%, Quartz 2-3%, and Amphibole (tremolite), diopside ~1%. Black Scattered Electron (BSE) Image of the iron ore is revealed that, there is an intimate association between ilmenite and Ti-Hematite/magnetite ranging from coarse size (500 microns) to less than 5 microns. Hematite/magnetite is Ti bearing. Apart from Titanium, hematite/magnetite contains Vanadium. Almost every hematite/magnetite grain contains ilmenite in the form of fine to extremely fine elongated lamellae. Some of the hematite/magnetite grains carry blebs of Ilmenite (Fig.5.A and D). Most of the magnetic grain contain fine to extremely fine inclusion of Sillicate. The principal iron mineral is Hematite while ilmenite is main titanium bearing mineral and both are seen in major proportion. Goethite/limonite, magnetite are noticed in minor amounts. Chlorite/chamosite is the silicate mineral seen in major amounts while mica, quartz and amphibole are noticed in minor to traces. Almost Ti-free magnetite's show grey white color where as the Ti-rich magnetites are pinkish in color it display anisotrpism in reflected light. The magnetites are mostly free of exsolution textures. The fractures within the magnetite grains are filled by sulphides and silicates. The magnetite cleavages are invaded by secondary silicates and the interstitial spaces in the coarse magnetitic mesh are occupied by silicates and within them are the oxides along with globules of sulphides. Textural relations as revealed in the oxide-silicate assemblages are, the co-existing coarse grained magnetite and ilmenite form a net work with interstitial space filled by silicates. Where the magnetites are free from exsolved ilmenite or ulvospinel, this condition is attributable to early co-crystalization of magnetite and ilmenite (Wadpalliwar and Kale, 2012: Gonindaiah et al 2007: Lakshmipathaiah et al 2006: Vinayaka et al 2007). The magnetite's with exsolution textures exhibit the following features.

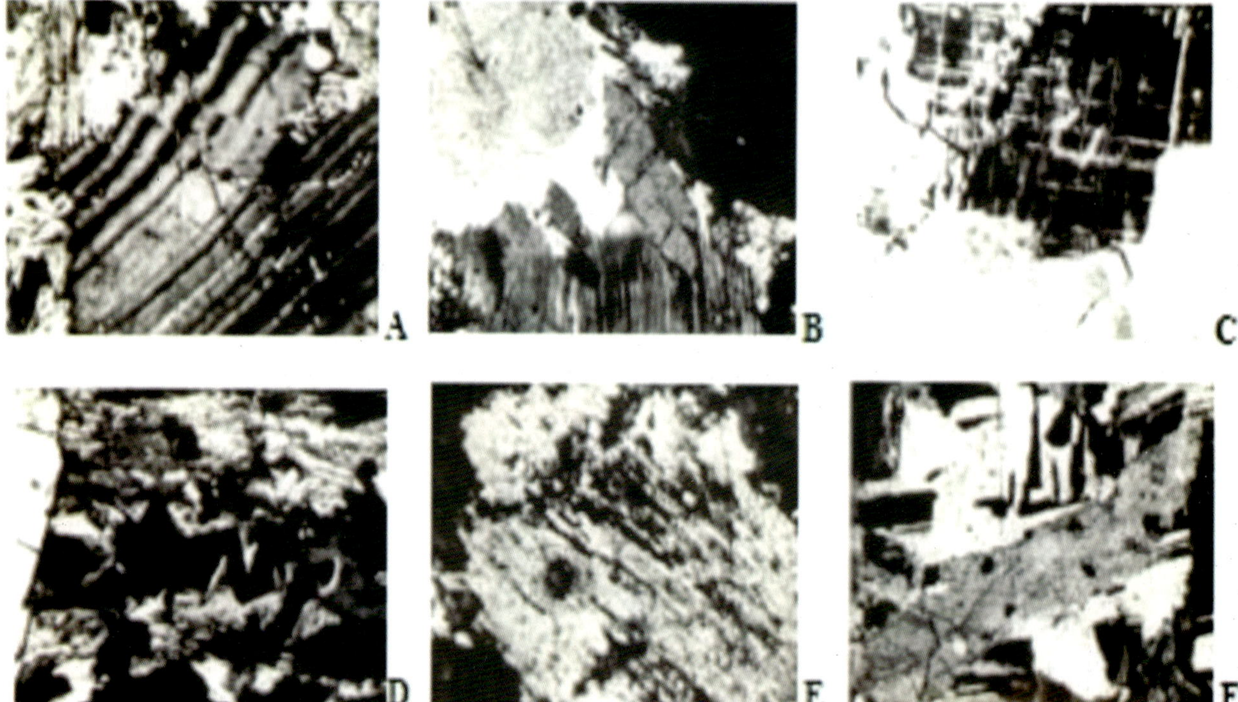

Fig. 4. (A) Thin section show banding and displacement of twin lamellae in plagioclase. Post deformation annealing crystallization of hornblende (100X) **(B)** Uralite gabbro with euhedral apatite at central portion (100X) **(C)** Plagioclase show both the twins Carlsbad and albite law (100X) **(D)** Abrupt termination of schistocity against garnet porphyroblast (100X) **(E)** Hornblende (dark color) needles within pyroxene (light color) are optically iso-oriented with the hornblende porphyroblast (100X) **(F)** Conjugated deformation and displacement of twins in plagioclase due to translation grinding. Secondary calcite vein cut through plagioclase mesh (100X)

Exsolved ilmenite is seen as coarse straight laths along two octahedral directions. Its along fine lamellae are arranged along three octahedral directions (Fig. 5.B and C). As vermicular elongated lamellae they occupy one (111) direction only and are closely associated with unvospinel along (100) (Fig. 5.F) "Trellis texture " formed by very fine, minute worm like ilmenites, concordant to and arranged within the framework of bigger ilmenite lamellae is also seen (Fig. 5.A). "Cloth texture" appears due to coarse, straight, relatively small laths of ulvospinel along (100) plane containing coherent wormlike ulvospinels within the bigger frame work. In the exsolved ilmenites bordering the magnetite grains. These may be diffusion products under slow cooling condition. Such ilmenites are visible also along the cracks within the silicates which confirm their post deformational diffusion into pressure release zones (Gjlsvik, 1957). Very small and irregular exsolved ilmenites were produced possibly in rapid cooling condition which again is confirmed by localized spinifex texture. Thinning out of the ilmenite lamellae occurring within magnetite, at the junction of two such lamellae, regularity of their occurrences and the sharpness of their outline, clearly indicate that these ilmenite lamellae are formed due to exsolution with falling temperature from solid solution of magnetite and ilemnite, the titanium rich crystallizing first (Shrivastava, 1963). As already described by Dunn and Dey (1937) and Roy (1955) at some places due to extensive fracturing ilmenite lamellae were separated and placed along microfractures. Fractures in magnetite filled by ilmenite due to diffusion. Thus it is clear that the oxides and the silicates were crystallized before deformation. Perhaps the metamorphism accompanying this deformation was responsible for remobilization of ilmenite and also some silicates which in filled the microfractures. In few magnetite grains spinels surrounded by black borders are seen. At places minute ulvospinel lamellae were seen partially homogenized with magnetite host. In banded ores the magnetite grains are elongated conformable with the silicate bands which occur alternately. These bandings seem to be incipient, but the preferred orientation of magnetite grains within the bands might have been due to rolling out during orogeny or due to tectonic stretching. Goethite/limonite present in minor amount and occurs as irregular patches. At few places, goethite shows colloform banding with hematite. Mica (Muscovite, biotite), quartz and amphibole (tremolite) are the other gangue minerals present in very minor to traces.

Sulphides occur only as minor minerals along fractures. They also occur as specks and stringers in the magnetites. Few sulphide minerals are identified in the oxide-sulphide-silicate assemblage, like phyrrhotite and chalcopyrite. (minerals of intermediate solid solution group and pyrite).

Geochemistry

Samples were collected systematically from the section of mine pit from bottom to top and selected outcrops, crush in steel jaw crusher that was cleaned extensively with de-ionized water and finally pulverized in agate mortars in order to minimize potential contamination of transitional metal in trace element analysis. The major oxides were determined by wave-length dispersive X-ray fluorescence spectrometry on fuse glass disks. The selected trace element Sc, Cr, Nb, Ta, Be, Co, Cu, Zn were determined by WD-XRFS on pressed powder pellets. The major oxide and selected trace element data are given in Table No. 3.

The geo-chemical data of titanoferous magnetite exihibits that SiO_2 content ranges from 1.17 to 3.89% whereas large compositional variation of TiO2 (9.19 to 19.01%) but Al_2O_3 has narrow range (2.13 to 3.95%). In the ore, rich concentration of FeO (total) which ranges from 49.54 to 57.33% and increases with increasing TiO_2. The Bivarient plot (Fig. 7 H) show strong co-relation between FeO and TiO_2. The average content of MgO is 1.09% which varies from 0.33 to 1.98% but shows positive co-relation with FeO (Fig. 7 A) and TiO_2 (Fig. 7 C) whereas negative trend between Al_2O_3 and SiO_2 (Fig. 7.B). The trace element V and Zn found in the titanoferous magnetite as a major element V_2O_5, ranges from 0.31 to 1.51% and ZnO from 0.52 to 2.07%, but there is no standard co-relation between MgO (Fig. 7.F and G). The MnO content varies from 0.20 to 0.26, Na2O = 0.06 to 0.17%, K2O = 0.01 to 0.08%, P2O5 = 0.02 to 0.03% and the average concentration of CaO is 0.01.

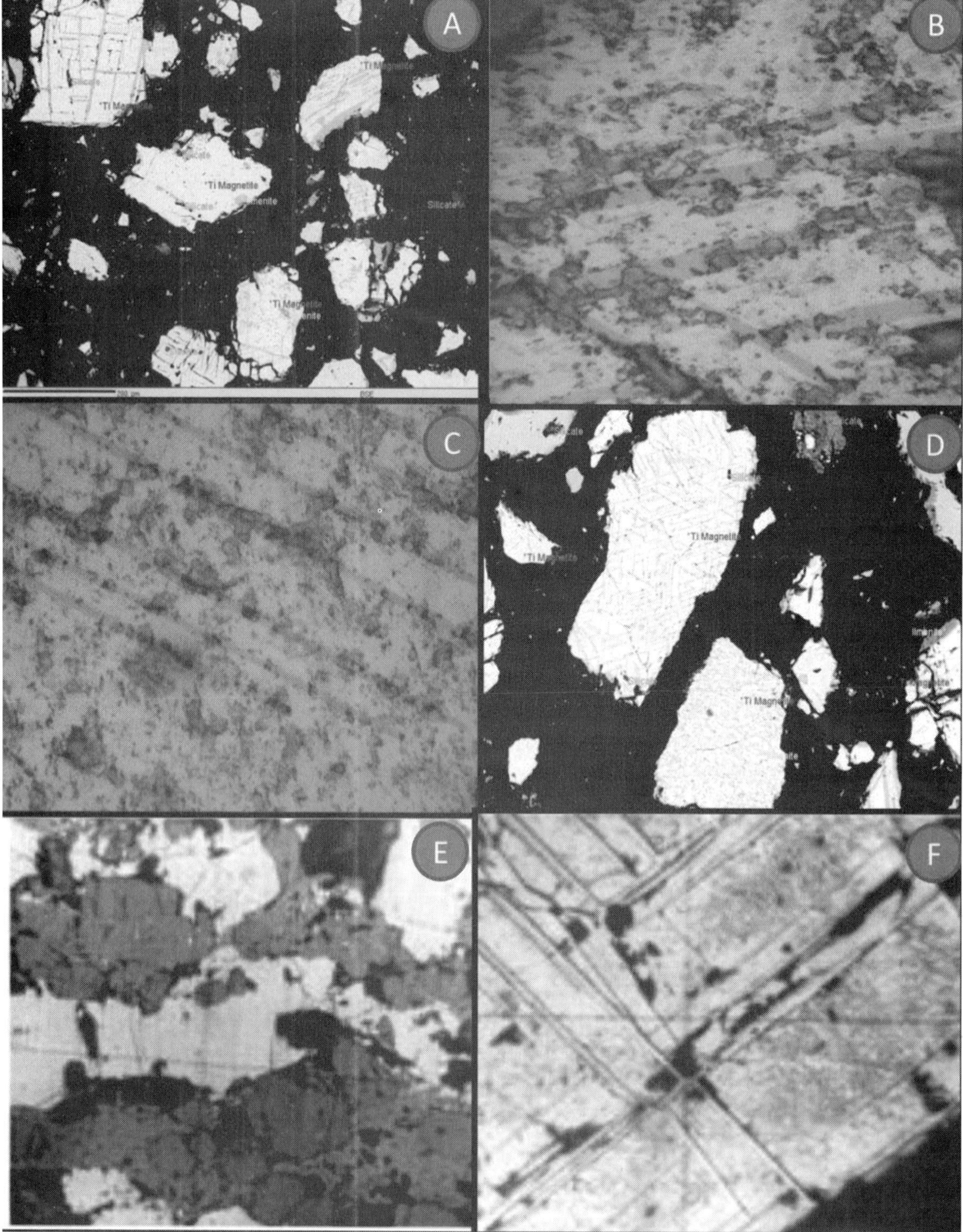

Fig. 5. Photomicrograph. A) Back Scattered Electron Image showing thin to thick elongated lamellae of Ilmenite within Ti-magnetite, most of the Ti-magnetite grains show this type of intergrowth texture. Fine inclusions of silicates are also noticed within Ti-magnetite grains. **D)** Back Scattered Electron Image showing extremely fine lamella (finer than 5 microns) of Ilmenite within Ti-magnetite grains. **E)** One set of small fine ilmenite lamellae along (111) coexists with smaller set of ulvospinel lamellae within magnetite grain. Inclusion of sulphides and silicates (150X) **C)** Three sets exsolution intergrowth of ilemenite along (111) titanomagnetite (150X) **F)** Two sets of ilmenite lamellae in magnetite the minute coherent worm like lamellae being contain within the framework of bigger lamellae along octahedral planes of magnetite (150X) **B)** Banded ore with alternate layers of oxides and sillicates grains perceptibly elongated parallel to the foliations., presence of sulphides (100X)

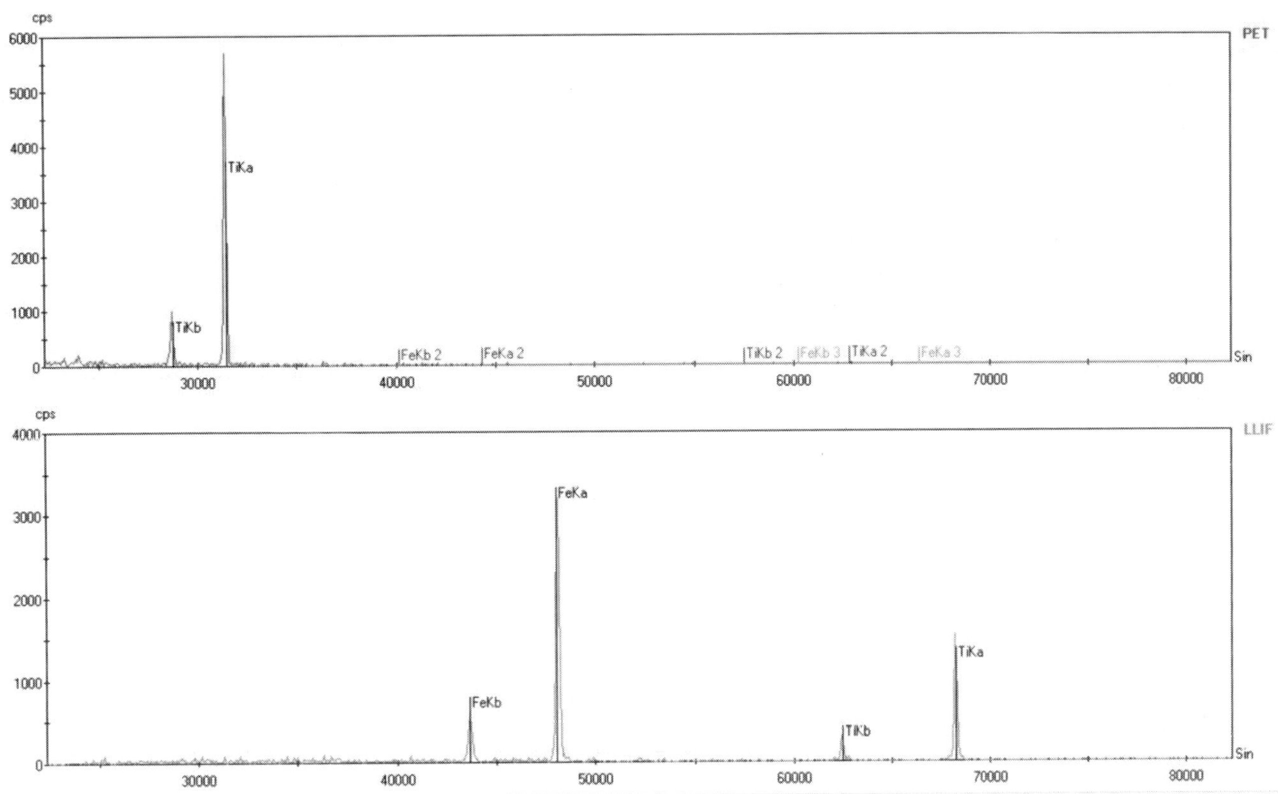

Fig. 6A. Qualitative analysis of Ilmenite grain.

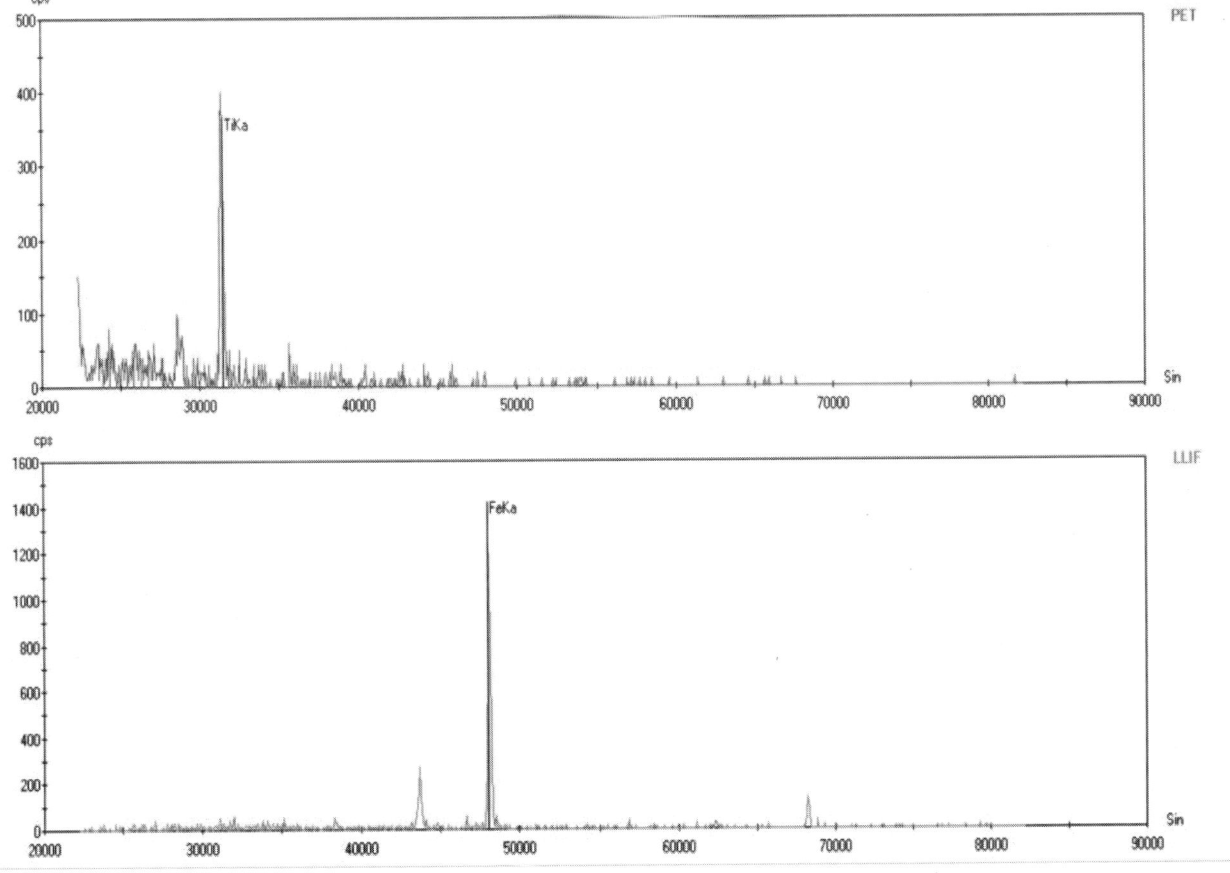

Fig.-6B. Qualitative analysis of Ti-Hematite/Magnetite grain.

Table 3. Major oxide and selected trace elements concentrations of the Khursipar and manegaon areas.

Major Oxide %	Khursipar Iron Ore Mine						Nicco Jaiswal Iron Ore Mine (Near Manegaon)						Manegaon Iron Ore Mine					
	Massive Ore		Babded/Foll iatedOre		Weathred Ore		Massive Ore		Babded/Foll iatedOre		Weathred Ore		Massive Ore		Babded/Foll iatedOre		Weathred Ore	
	KIOM -1	KIOM -2	KIOM -3	KIOM -4	KIOM -5	KIOM -6	NJIO M-1	NJIO M-2	NJIO M-3	NJI OM-4	NJI OM-5	NJIO M-6	MIOM -1	MIO M-2	MIO M-3	MIO M-4	MIO M-5	MIO M-6
SiO_2	3.11	1.11	1.12	2.39	3.47	2.21	1.87	2.90	2.60	2.90	3.89	2.58	1.17	2.41	2.66	1.89	1.88	1.51
TiO_2	15.05	19.01	16.98	14.99	8.01	18.61	11.85	14.10	17.11	13.11	10.08	17.11	14.56	11.98	17.89	14.11	9.19	18.01
Al_2O_3	1.50	1.79	1.29	3.08	3.95	3.59	2.78	2.73	2.73	2.73	2.60	3.84	3.18	3.55	2.13	2.13	2.66	3.14
FeO(T)	54.30	57.33	53.05	53.05	48.93	54.75	52.09	51.91	55.09	53.00	49.54	53.55	53.76	49.99	55.99	54.00	49.59	53.53
V_2O_5	0.39	1.51	0.43	0.55	0.29	1.134	0.40	1.41	0.30	0.55	1.21	1.11	0.41	1.48	0.43	0.50	0.39	1.11
ZnO	0.57	2.07	0.57	2.03	1.44	1.987	0.58	1.10	0.57	2.07	0.57	2.03	0.52	2.01	0.57	1.09	0.57	2.00
MnO	0.23	0.25	0.21	0.20	0.23	0.24	0.26	0.26	0.26	0.26	0.23	0.20	0.20	0.24	0.24	0.26	0.23	0.20
MgO	0.89	1.98	1.42	1.44	0.37	0.99	1.11	1.07	1.47	0.97	0.34	1.15	1.45	0.38	1.44	1.69	0.33	1.14
CaO	0.01	0.01	0.01	0.01	0.01	0.01	0.01	0.01	0.01	0.01	0.01	0.01	0.01	0.01	0.01	0.01	0.01	0.01
Na_2O	0.08	0.09	0.08	0.07	0.08	0.09	0.14	0.11	0.14	0.17	0.09	0.06	0.06	0.07	0.13	0.14	0.10	0.07
K_2O	0.03	0.04	0.03	0.01	0.02	0.03	0.07	0.07	0.08	0.08	0.03	0.01	0.01	0.04	0.08	0.08	0.03	0.02
P_2O_5	0.03	0.02	0.02	0.02	0.03	0.02	0.02	0.02	0.02	0.02	0.03	0.02	0.02	0.03	0.02	0.02	0.03	0.02
Trace Elements (PPM)																		
Sc	9.28	35.00	10.12	35.00	13.32	22.13	19.12	34.43	35.01	15.09	13.09	19.13	34.32	17.13	29.92	26.55	5.10	34.99
Cr	55.11	125.09	66.98	67.90	5.76	125.00	24.76	35.98	85.00	76.00	5.00	87.66	125.00	10.12	115.00	125.00	6.99	115.98
Co	375.12	850.00	510.23	650.11	250.00	566.98	440.44	278.01	750.09	690.98	256.98	750.94	809.98	250.00	845.76	850.65	250.00	849.00
Ni	25.98	199.99	70.76	90.98	15.74	31.10	44.45	41.23	175.98	28.98	12.76	12.00	101.11	20.65	80.98	200.00	12.00	45.55
Cu	5001.00	15000.0	5755.89	9876.90	4890.87	4976.12	5001.11	4567.09	15100.07	5098.0	4213.4	4613.49	12080.67	4098.08	6980.09	15000.0	4098.09	15000.0
Ga	0.50	0.90	0.65	0.89	0.51	0.92	0.55	0.76	0.89	0.75	0.56	0.60	0.91	0.57	0.60	0.89	0.52	0.68
Be	50.25	90.00	55.10	90.01	58.98	89.09	51.78	85.01	56.01	54.90	85.67	51.23	90.89	67.00	58.50	89.04	78.23	89.98
Ta	1001.11	1500.00	1112.98	1330.90	1112.12	1453.98	1000.01	1379.03	1450.7	1011.1	1340.9	1578.09	1456.09	1000.11	1122.20	1400.09	1234.43	1500.90
Nb	19.99	70.00	20.00	67.65	49.90	50.57	23.40	67.03	50.50	62.90	54.60	26.67	37.90	42.43	52.53	47.43	21.11	69.69

The concentration of Sc, Nb and Be very little variation, ranging from 5.10 to 35.01 ppm, 21.11 to 70.00 ppm, 51.23 to 90.89 ppm. Scandium shows positive co-relation versus MgO (Fig. 8.J) but Nb and Be do not show any standard co-relation (Fig 8N and P). Ta has rich concentration (1000.1 to 1578.09 but show same co-relation likes as Nb and Be (Fig. 8.O). The abundance of Ni and Co vary widely, which ranging from 12.00 to 200.00 ppm and 250.00 to 850.00 ppm respectively; both the trace elements show positive co-relation versus MgO wt% (Fig. 8.K and M). The average content of chromium is 69.90 ppm and it ranges from 5.00 to 125.09 ppm and show strong co-relation versus MgO wt% (Fig. 8.I). The trace element Cu has richest concentration (4098.08 to 15100.07 ppm) and show positive co-relation versus MgO (Fig. 8L).

Ore Genesis

The field geology, petrography and geochemistry suggests that the ore is formed as a result of fractional crystallization of Fe-Ti-V rich mafic magma as late magmatic segregation. The ore is essentially consisting of Ti magnetite with accessory co-precipitate granular illmenite. Sulphides are formed later than the Ti Magnetite. Hematite and goethite are formed during weathering due to dehydration and oxidation. Micro intergrowths of ulvospinel and illmenite are formed due to exsolution during sub-solidus cooling. The ore occurs as a layer typical of magmatic type of origin (Gordon, 1966) (Fig. 3G). In addition, the magnetite ore are associated with silicates. As a result of very extensive fractionation, magma evolved into Fe rich melts, which eventually become FeO saturated and as a result formed the Fe-Ti-V oxide rich ore of the triple group (Wager and Brown 1968). The oxide rich ore may have been derived from Fe rich tholeitic Magma (Fig. 9A). The fractional crystallization of magma eventually leads to saturation of Fe_2O_3 bearing phases (Irvine 1983, Wilson 1989. The titonoferous magnetite fills the space between or completely encloses the silicate minerals (Fig 4 & 5). These textural relationships support the case for invasion of dense. Fe- rich melt into a silicate crystal mush (Reynolds, 1985; Park 1961; Philpotts 1967; Kolkar 1982; Nystrom and Henrizuer 1994; Lyons 1988; Williams 1969; Von Gruenwaldt 1994)

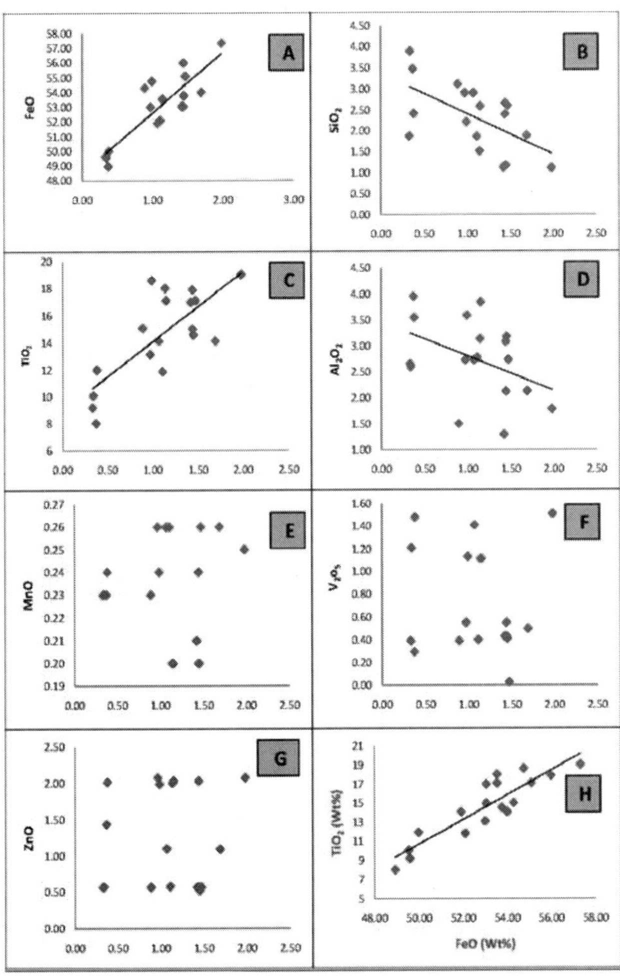

Fig. 7. Bivarient plots of MgO wt % versus Major Oxide (A, B, C, D, E, F, G) and is Bivarient plots of FeO wt % versus TiO_2.

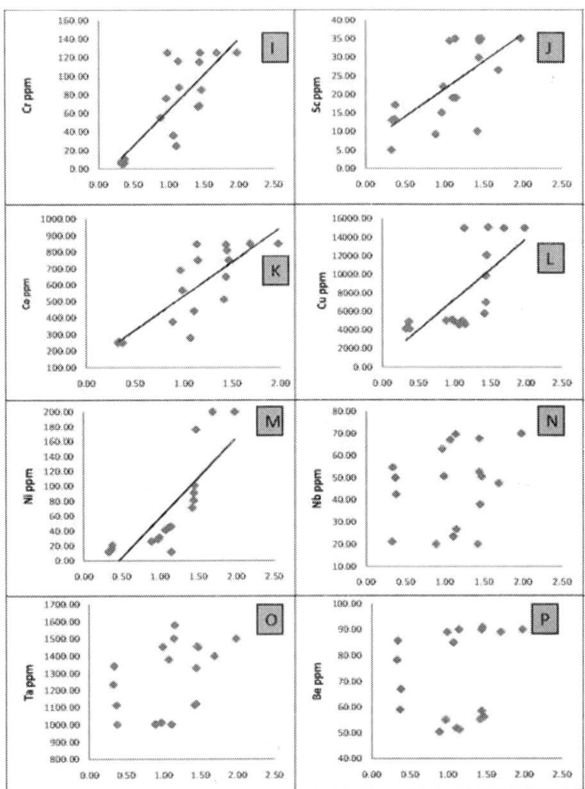

Fig. 8. Bivarient plots of MgO wt % versus Trace elements (I,J,K,L,M,N,O and P).

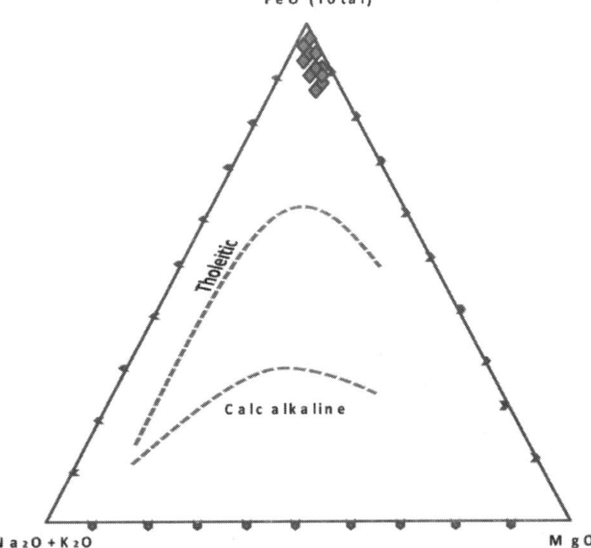

Fig. 9. AFM diagram showing the geochemical variation of the Ti-Fe oxide ore around Khursipar and Manegaon villeges of Amgaon group of Sakoli fold belt. The tholeiitic and calc alkaline trends after Wilson (1989)

Conclusion

The results of the study conclude that the ore is associated with gabbroid rocks. The Fe-Ti-

V oxide ore consist of Ti-magnetite, hematite, illmenite and Ti magnetite is altered into hematite and goethite. Titanoferous Magnetite formed as a result of fractional crystallization of Fe-Ti-V rich mafic magma as late magmatic segregation. The magma is rich in Fe and Ti but poor in SiO_2. The textural relationship between Fe-Ti-V rich oxide ore support crystallization of illmenite and magnetite from oxide ore melt. Hematite and goethite are formed during weathering due to oxidation and hydration. This mineral produced magnetic mineral concentrate around 90% with specific gravity over four and it is utilized as heavy media in a heavy media separation at minus 300 mesh size. It is used in ferrotitanium alloy industries after suitably blending with ilimenite or rutile and this ore is widely used in lining of Furness.

Acknowledgements

The authors gratefully acknowledge the Managing Director, Maharashtra State Mining Corporation limited, Government of Maharashtra undertaking and to the, Controller General, Indian Bureau of Mines, Nagpur for the facility and encouragement. Authors are also acknowledge the Head, Professor P.P. Kundal, Department of Geology, RTM, Nagpur University, Nagpur for his constructive suggestions.

References

Bateman, A. M. (1951). "The formation of late magmatic oxide ores". Economic Geology. Vol. 46. Pp. 404-426.

Bhake, S. S. (1993), "Petrography, Structure and Metamorphic history of the area NE of Ngzira, District Bhandara, Maharashtra." Unpublished Ph. D. Thesis. Nagpur University. Pp 203.

Bhattachrji, D. S. (1932), Genral report for 1928, Rec. Geol. Surv. Ind., Vol. 62. 63,66 Pp. 132-155,114-117 and 108-109 respectively.

Broderick T. M. (1917). "The relation of the titanoferous Magnetite of North Estern Minnesota to the Duluth Gabbro". Eco. Geo. Vol. 12. Pp 663-696.

Chakraborty, K.L. and Mallick (1968), "Mineralogy and temperature of formation of titaniferous magnetite ore associated with gabbro in Maulabhanj prabatn area Dhenkanal district, Orissa." Bul. Geo. Soc. of india Vol. 5. Pp 134-137.

D. D. Klemm, J. Henckel, R. M. Dehm, and Gerhard Von Gruenewaldt (1985) "The geochemistry of titanomagnetite in magnetite layers and their host rocks of the eastern Bushveld Complex" Economic Geology, July, Vol. 80, Pp. 1075-1088,

Das, S. K. and Mukharji, S. (1984), "Mineragraphic study of vanadium bearing titaniferous magnetite deposit of Patpur-Bisoj area, Mayurbhanj district, Orissa."Proc. Indian Geol. Cong. IIT Bombay. Pp 163-171.

Dasgupta, H. C. (1969), "Fe-Ti-V oxide ore associated with gabbro rock around Dublabera, Bihar." Qut. Jou. Geo. Min. Met. Soc. India. Vol. 41 (2). PP. 51-64.

Duchesen, J. C. (1999). Fe-Ti deposits in Rogaland anorthsites (South norway): geochemical characteristics and problems of interpretation. Mineralium Deposit, Vol. 34. Pp. 182-198.

Dunn J. A. and Day, A. K. (1937) "Vanadium bearing titaniferous iron ores in a Singhbhum and Mayurbhanj, India." Trans. Min. Gol. Inst. India. Vol. 11 (3). Pp. 117-194.

Forst, B. R. (1991). Magnetic petrology; factors that control the occurrence of magnetite in crustal rock. In: Lindsley, D.H. (ed.) Oxide Minerals: Petrologic and Magnetic Signification. Mineralogical society of America. Reviews in Mineralogy. Vol. 25, Pp. 489-509.

Foye W. G. (1916) "The relation of the titaniferous magnetite ores of Glamornan Township, Haliberton County Ontario to the associated scopolitic gabbros". Eco. Geo. Vol. 11. Pp 662-694.

Gjlsvik T. (1957) "Geochemical and mineralogical investigation of titaniferous iron ores, west coast of Norway". Eco. Geo. Vol. 52. Pp 482-498.

Gordon F. Lister, (1966) "The composition and origin of selected iron-titanium deposits". Eco. Geo. Vol. 61. Pp. 275-310.

Govindaiah, S. Jagadeesh, C and Suresh, B. (2007). Microstructural evolution of V-Ti magnetite layers of Devaranarasipura ultramafic-mafic complex in Shimoga schist belt of Dharwar cration, Indian, Ind. Mineral v. 41, no. 3., Pp. 144-152.

Irvine, T. N., Keith. D.W. and Todd, S. G. (1983). The J-M platinum-palladiun reef of the Stillwater complex, Montana 2. Origin by double-diffusive convective magma mixing and implifications for the Bushveld complex. Economic Geology, Vol. 78, Pp. 1287-1334.

Kolkar, A. (1982). Mineralogy and geochemistry of Fe-Ti and apatite (nelsonite) deposite and evolution of the liquid immiscibilty hypothesis. Economic Geology Vol. 77, Pp. 1146-1158.

Kwan-Nang Pang, Mei-Fu Zhao, Donald Lindsley, Donggao Zhao and jhon Malpas (2008). "Origin of Fe-Ti oxide ores in mafic Intrusion: Evidence from Panzhihua Intrusion SW- China" Jour of Petrology, vol. 49(2), Pp. 295-313.

Lakshmipathaiah, H.K. Govindaiah, S. and Srikantappa, C., (2006). Chemistry and subsolidus re-equilibrium of V-Ti magnetite ores in the southeastern part of Shimoga schist belt, Dharwar cration, Indian, Jour. Applied Geochemistry. Vol.8, No. 1 Pp. 11-24.

Lindsley, D. H. (1992). Experimental studies of oxide minerals. In Lindsley, D. H. (ed) Oxide Minerals: Petrologic and Magnetic Signification. Mineralogical society of America. Reviews in Mineralogy vol 25. Pp. 69-106.

Lister, G.F. (1966). The composition and origin of selected iron-titanium deposits Economic Geology., Vol. 61. Pp. 275-310.

Lyons, J.I. (1988). Volcanogenetic iron oxide deposites. Cerro de Mercado and vicinity, Durango. Mexico. Economic Geology Vol, 83. Pp 1886-1906.

Mei-fu Zhou, Poul T Robionsons, C. Michael Lesher, Reid R. Keays, Cheng-Jiang Zhang and Jhon Malpas (2005). "Geochemistry, Petrogenesis and Metalogenesis of Panzhihua Gabbroic layerd intrusion and associated Fe-Ti-V oxide deposits Sichuan Province SW China." Journal of petrology, vol. 46. Pp. 2253-2280.

Mohabey, N. K. and Bhake, S. S (1994), "Bimodal Magmatism in Amgaon group of rocks". Gondwana Geological Magzene, Vol. 8. Pp. 40-58.

Mohabey, N. K. and Bhake, S. S (1996), "Sakoli group –A part of a Greenstone belt of eastern Maharashtra, India". Gondwana Geological Magzene, Vo. 11. Pp. 45-60.

Mohanty, J. K. and Paul, A. K. (2008). "Fe-Ti oxide ore of the Mesoarchaean Nuhasahi Ultramafic complex, Orissa and its utilization potential." Jou. Geo. Soc. Of India V.72. Pp. 623-233.

Nanda, A. K. (1973), "Our resources of vanadium ore." Jou. Min. Met. And Fuels, Pp. 373-376.

Nystrom, J.O. & Henriquez., F. (1994) Magmatic features of iron ore of the Kiruna type chile and Swedem ore texture and magmatic geochemistry. Economic Geology Vol. 89, Pp. 820-839.

Park, C. F. (1961). A magmatic lava flow in Northern Chile. Economic Geology 56, 431-441.

Philphotts, A. R. (1967). Origin of certain iron-titanium oxide and apatite rock. Economic Geology Vol. 62. Pp. 303-315.

Ramiengar A. S. and Chayapathi N. (1977) "Vanidiferous Magnetite deposit near Masanikere, Channagiri Taluka, Shimoga District, Karan-ataka". Indian Mineral Vol. 31 No. 1 Jan- March Pp 1-5.

Reynolds, I. M. (1978). Mineralogical studies of South African titaniferous iron ores: their applications to extractive metallurgy. Transaction of the Geological society of Africa Vol. 81, Pp. 233-240.

Reynolds, I. M.(1985). The nature and origin titaniferious magnatite-rich layers in the upper zone of the Bushveld complex: arevive and synthesis. Economic Geology Vol. 80, Pp. 1089-1108.

Roy B. C. (1955) Records G. S. I. Vol. IXXBI, BULL. NO 14.

Roy S. (1956), "Origin of vanadium bearing titanoferous magnetite ore of Mayurbhanj India." Proc. Nat. inst. Sci. India Vol. 22. Pp. 285-292.

Roy, B. C. (1955) Records G. S. I. Vol. IXXBI, BULL. NO 14.

Sarkar S. N. (1957), " Stratigraphy and tectonics of the Dongargarh system : A new system in the Precambrians of Bhandara-durg-Balaghat area, Bombay and Madhya Pradesh". Jou. Sci. Engg. Research, IIT Kharagpur. Vol. 1. Pp. 237-268.

Sarkar, S. N. (1958), "Stratigraphy and tectonics of the Dongargarh system : A new system in the Precambrians of Bhandara-durg-Balaghat area, Bombay and Madhya Pradesh." Jou. Sci. Engg. Research, IIT Kharagpur. Vol. 2. Pp. 145-160.

Sarkar, S. N. (1967), "Precambrians Geochronology of Bhandara-Durg, India" Indian Gondwana Geological Magzene, Vo. 106 (6). Pp. 525-549.

Sarkar, S. N. (1980), "New data on geochronology of Precambrians of Bhandara-Durg, Central India". Indian Journal of Earth Sciences, Vol. 8. Pp. 131-151.

Shrivastava S. P. N. (1963) "Vanadiferous magnetite ore in the Dhalbhumgarh Subdivision, Singhbhum District, Bihar". Jour. of Mine Metal and Fuels. Pp 11-16.

Vasudeo, V. N. and Srinivasan, R. (1979) "Vanadium bearing titaniferous magnetite deposit of Karnataka, India." Jou. Geo. Soc. Indai Vol. 20. Pp. 170-178.

Vinayaka, S., Govindaiah, S. and Pathan, A. M. (2007). Textural featural of geochemistry and V-Ti magnetite layers in the ultramafic-mafic complex of Nuggihalli schist belt of Dharwar craton, India, Ind. Mineral. Vol. 33, no. 1, Pp133-143.

Von Gruenwaldt, G. (1994). Illmenite apatite enrichment in the upper zone of the Bhushveld complex: A major titanium rock phosphate resource. International geology revives, 35, Pp. 987-1000.

Wadpalliwar G. G. and Kale. H. S. (2012)."Ore mineralogy and chemistry of V-Ti Magnetite ore of the Amgaon group, Gondia distric, Maharastra, India. Indan Meneralogist." Vol. 46(1). Pp. 86-98.

Wager, L. R. & Brown, G. M. (1968). Layered Igneous Rock. Edinburgh Oliver & Boyed. 588 pp.

Warren C. H. (1908) Am. Jour. Sci. 4th series, Vol. 26. Pp 474-475.

Willemese, J. (1969). The vanadiferous magmatic iron ore of the Bushveld Igneous Complex . Economic Geology Monographs., 4. 187-207.

Williams, D. (1969). Ore deposits of volcanic affiliation. In: James. C. H. (ed). Sedimentary Ores Ancient and Modern (Revised). Geological Department. Leicester, Special Publication 1. 197-206.

Wilson, M. (1989). Igneous petrogenesis. London Unwin hyman. 494pp.

Winter, P.E. (1965). The. Ushushwana Igneous Complex Swazrland Geological Survey Bulletin 5. 29pp

Hon'ble Reviewer

1. **Prof. O.P. Verma**
   Founder President
   Indian Geological Congress
   ROORKEE - 247 667
2. **Prof. J.P. Srivastava**
   Centre of Ad. Study in Geology
   University of Delhi
   DELHI - 110 007

I-2

Major Oxide Geochemistry of Diabasic Rocks of Mineralized Ophiolite of Manipur and its Implication on Alkalic Basalt Lineage and Transition to Tholeiite Character

Ch. Mangi Khuman[1], Soibam Ibotombi[1], L. Romendro[2] and S. Subhamenon[1]

[1]Department of Earth Sciences, Manipur University, Imphal, Manipur
[2]Department of Geology, Thoubal College, Thoubal, Manipur
E-mail: ibotombi2002@yahoo.co.uk

Abstract

Diabasic dyke and sill rocks do occur in the Manipur Ophiolite Melange Zone (MOMZ) as isolated bodies within the peridotite serpentinites and within pelagic sediments mainly the shales and cherts. Petrographic studies indicate that these basic rocks are principally composed of olivine, orthopyroxene, clinopyroxene; and plagioclase laths and intergranular/intersertal clinopyroxene grains. Textural features of these diabasic rocks indicate no reaction rim enclosing olivine by pyroxene indicating cotectic relationship between the olivine and calcium poor pyroxene implying a possible alkali-basalt lineage. CIPW norms of these rocks derived from the major and minor elements data reveal undersaturated nature having normative nepheline probably for the least contaminated samples. However, many of the samples indicate normative quartz implying possible contamination with chert during their extrusion giving rise to tholleiite character. Probably, the diabasic rocks of Manipur ophiolite are of alkali basalt lineage which appears to have transitional tholeiite character due to assimilation of chert.

Keywords: Ophiolite, Geochemistry, Manipur, Alkali Basalt, Tholeiite Basalt.

Introduction

The Manipur Ophiolite Mélange Zone (MOMZ) is found to have a number of diabasic dykes and sills, the widths of which range generally between 3-8 meters. According to Penguin dictionary of Geology *diabase* is defined as slightly metamorphosed medium grained basic igneous rock. The dykes are generally found occurring within the host of either peridotitic serpentinites or within pelagic sediments, which are mostly shales and cherts. Some basic rocks are also found as exotic blocks floating on shales. The blocks are generally small having diameters of few meters only and commonly have flat surfaces (but the pillow lavas are generally rounded/elliptical). From the consideration of their shape, size, and field setting and from the similarity in grain size and mineralogy, these exotic blocks could be inferred to be the broken pieces of diabasic dykes, which have been displaced and injected as broken masses during the tectonic emplacement of ophiolite belt.

The unique feature of the dykes in the common ophiolite sequences is that they consist of 100% dyke rocks without intervening older country rock screens (Coleman, 1977). The occurrence of the remarkably continuous dyke swarms has led to the idea that they have been formed by continuous injection of basaltic liquid from a large magma chamber along a zone of extension. But such features are absent in the dykes of Manipur Ophiolite. The sporadically occurring dykes are only few meters thick and are very widely separated (in terms of hundreds or thousands of meters).

Good exposures of pillow lavas are not very frequent. But, there are some sporadic and camouflaged (with host shale) occurrences of very few isolated pillows as sub-spherical exotic blocks within pelagic shale. Those found are usually concentrated in the eastern as well as western margins of the MOMZ.

Each and every rock may be regarded as a chemical system, in which chemical changes can be brought about by various agencies. The chemical system and also the changes in the system are reflected in the mineralogy and texture of the resulting rocks. Hence, the various trends of the different parameters of the rock chemistry provide clues for the extraction of valuable information pertaining to the mineralogy and texture, and the consequent changes. Thus, the authours in this paper use the major oxides data as an important tool in trying to reveal the possible petrochemistry responsible for the generation and evolution of the diabasic rocks emphasizing on their alkalic basalt lineage and the consequent transition to tholeiitic character.

Geological and Tectonic Setting

The geological and tectonic setting of the westerly arcuate Indo-Myanmar Ranges (IMR) in general and its constituent Manipur hills in particular is attributed to the geological conditions of the extensional basin and the subsequent tectonic inversion leading to the subduction of the Indian plate beneath the Myanmar (Burma) plate thereby causing the consequent orogenesis of the IMR. The rock formations in Manipur are principally of flysch sediments of mainly of Tertiary age which constitutes about 65% of the total area. The flysch sediments are characterized as the Disangs (Mallet, 1876) and the Barails (Evans, 1932). The age of the Disangs is assigned to be Eocene to Upper Cretaceous while that of the Barails to be Oligocene to Upper Eocene. Disangs are a group of monotonous sequence of dark grey to black splintery shales which have sometimes intercalations of siltstone and fine to medium grained sandstones of light to brownish grey especially in the upper horizon. Barails are predominantly arenaceous sediments characterized by light grey to brown, fine to medium grained sandstones with minor to considerably thick interbands of shale. The Barails are uncon-formably overlain by the molasse sediments characterized as the Surmas (Evans, 1932) and Tipams (Mallet, 1876), whose age is estimated to be Miocene (Anon, 1974). The Manipur Ophiolite Mélange Zone (MOMZ), in which the present study is carried out, over thrusts the Disang-Barail flysch belt (Soibam, 2000, Khuman and Soibam 2010). The MOMZ is again overthrusted by an exposure of metamorphic rocks having a very limited extent on the extreme eastern part of the state. The metamor-phic complex known as the Naga Metamorphics is principally composed of low to medium grade phyllitic schists, quartzite, marble, granite-gneiss, etc. Brunnsch-weiler (1966) assigned them Pre-Mesozoic or older age. The different litho-units in this region are intensely deformed and juxtaposed against each other as an imbricate thrust system having NNE-SSW trend in this part of the IMR, in such a way that the older groups of rocks are on the eastern side while the younger ones are on the western side (Soibam, 1998, 2000; Soibam and Pradipchandra, 2006; Khuman and Soibam, 2010). The general geological, structural and tectonic features of Manipur are shown in Fig. 1.

The Ophiolite Belt in Manipur consists predominantly of the main ophiolitic body (ultramafic suite), pelagic sediments and various blocks of exotic rocks. The ultramafics having sporadic diabasic dykes cutting across them are found to have been sandwiched with pelagic shale and generally occupy the central portion of the belt and very few of the exotic rocks are found in this horizon. The ultramafic suite is flanked at the western and eastern margins by exotic rocks horizons. The host pelagic sediments particularly shale and cherty shale exposed along the eastern and western outer margins generally have small igneous intrusions in the form of dykes and sills. Out of the pelagic sediments, shale is exceedingly predominant and sometimes mixed with flysch like sediments. The associated exotic rocks in the host of mostly of pelagic shale exposed along the eastern and western margins of the Ophiolite Belt include a number of blocks of variable dimensions of diabasic dyke, pillow lava, conglomerate, gritty sandstone and rocks kindred with lava extrusion. The assembly of the different tectonic blocks of the whole range of the litho-units comprising of the ultramafic suite, pelagic sediments and exotic rocks of various kinds constitute the Manipur Ophiolite Mélange Zone. The ophiolite mélange zone in Manipur sector is wider on the northern side, which gradually tapers towards the south. The ophiolite suite extends more than 100 km as an almost continuous body in the NNE-SSW trend from the border of Nagaland state on the north upto Moreh on the south eastern part of Manipur. On the southern sector beyond Moreh area this ultramafic suite is exposed as widely isolated blocks upto the south eastern extremity of the state and beyond (in Myanmar). The peridotitic ultramafics are mainly of metamorphic harzburgite and lherzolite (Khuman and Soibam, 2010) with sporadic podiform chromite occurrences. Exposures of exotic blocks of pillow lavas floating on pelagic shale are more predominant on the western margin than on the eastern margin, and are better preserved (sometimes in the scale of a small hillock) and more frequent on the northern side where the MOMZ is wider. Although most of the exotic rocks float over pelagic shale, some smaller blocks of gritty sand are observed to float over the ophiolitic ultramafics as well, particularly on the eastern margin of the MOMZ. The radiolarian chert exotic blocks found on the western margin is sometimes intensely folded. On the eastern margin of the MOMZ the radiolarian chert exposures are found to be less deformed. They are either found juxtaposed along reverse faults with ultramafics or inter-bedded with pelagic cherty shale.

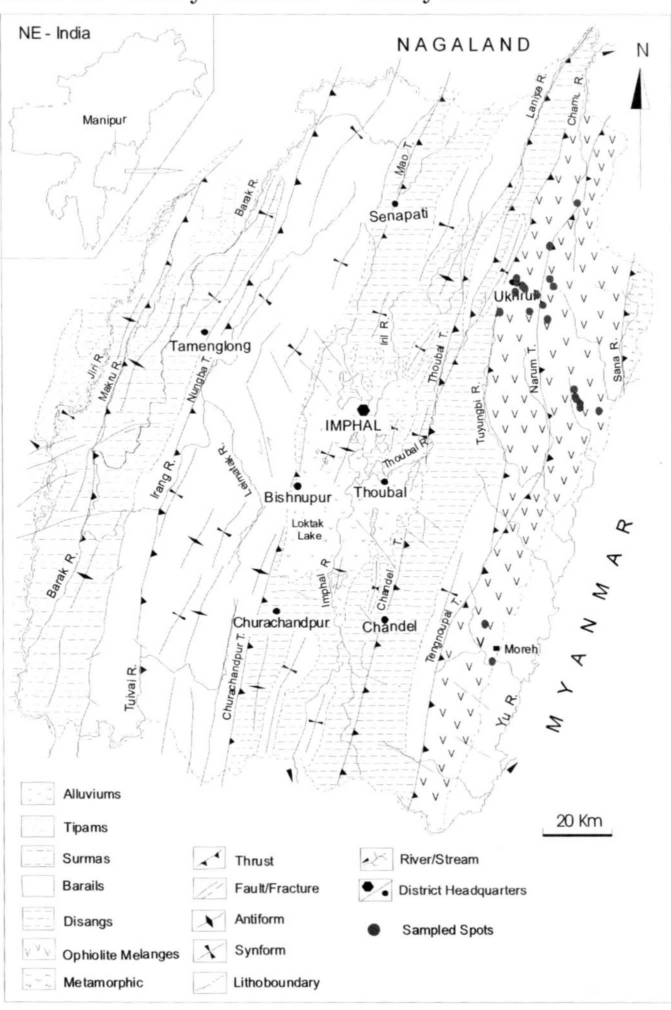

Fig. 1. Geological and structural map of Manipur (modified after Soibam, 1998)

Most of the earlier workers feel that the Nagaland-Manipur ophiolite belt is a dismembered one (Acharyya *et al.*, 1986, Vidyadharan *et al.*, 1986), thereby making the different units of the ophiolite sequence not exposed together at the same place. It may be argued, however, in this regard that the cumulate basics (gabbros) could have not been scraped off in the process of obduction and remained buried; but the heavier ultramafics are brought up on the surface, but not the lighter gabbros, is practically questionable. Therefore, considering the field setting, the authours are of the opinion that representative proportions of all the units of the ophiolite sequence are obducted as different blocks but in a jumbled manner so as to give rise to the MOMZ, where the not so well developed basic rocks have been represented poorly.

Petrography and Mineralogy

The dyke and sill rocks in Manipur Ophiolite Mélange Zone is fine (< 1mm) to medium (1-5 mm) grained. In some samples the colour is dark green and appears almost black; but most of the samples are light green with macroscopically visible leucocratic grains of plagioclase laths embedded within greenish masses. The sill rocks are identical with the dyke rock samples. The pillow lavas are generally greenish in colour and fine grained.

The basaltic dyke and sill rocks are generally composed of olivine, orthopyroxene, clinopyroxene; and plagioclase laths and intergranular/intersertal clinopyroxene grains (Fig. 2a). The olivine and pyroxene phenocrysts are mostly euhedral to sub-hedral and are least altered; and their sizes range from 0.2-0.5mm diameter in various samples. These grains constitute about 15-20% modal proportions. The subhedral plagioclase grains are cloudy or dirty albite and consist of minute crystals of epidote (Fig. 2b); and constitute about 60-65% modal proportions. The lengths of the laths generally range from 0.5-5mm. The interganular clinopyroxene grains are subhedral to anhedral and are found altered to actinolite (green amphibole), chlorite and epidote, the mixture of which is known as uralite (Deer *et al.*, 1978); and constitute about 15-20% modal proportions. In some samples there are limited magnetite grains as pseudomorphs most probably after olivine or pyroxene. Some samples have olivine and clinopyroxene grains over which plagioclases are epitaxially grown. In some samples crystals of pyroxene and olivine are almost absent but found to consist of equidimensional plagioclase phenocrysts as large as 5mm diameter; these samples have considerable proportion of magnetite, which appear to be pseudomorphs after mafic minerals. There are also some samples in which ilmenite grains are present. It is also observed that in a few of the samples there are phenocrysts of pyroxene found enclosing sub-ophitic plagioclase microlites; these pyroxene grains are again enclosed within larger plagioclase laths giving rise to the opposite diabasic texture. Vesicles present in the dyke rocks are generally of 0.02-0.04mm diameters, but there are few as large as 0.1mm. The vesicles are filled by secondary calcite (Fig. 2c). It is commonly observed in most of the samples that the plagioclase grains have been altered to calcite and/or have been replaced by secondary quartz. Most of the samples contain inclusions of small pieces of chert, the outer margin of which have been melted to quartz. The chert (quartz) grains are either corroded or surrounded by reaction zones (Fig. 2d).

The pillow lavas are found to consist mainly of plagioclase crystallites and intersertal minute grains of pyroxene, which are partially or wholly uralitised. The slender plagioclase crystallites have random to radial orientation giving rise to typical variolitic textures (Fig. 2e). There are some larger phenocrysts of pyroxene, major portions of which are uralitised. EPMA data and petrographic examination indicate that plagioclase crystallites are found to be of albite composition and associated with myriad minute crystals of epidote, and also that there are intersertal felty textured greenish and fibrous grains of uralite, which are mixtures of actinolite, chlorite and epidote.

The EPMA analysis of representative samples of dyke/sill and pillow lava of Manipur Ophiolite has been carried out at the Institute Instrumentation Centre (IIC), Indian Institute of Technology (IIT), Roorkee using JEOL-JXA-8600. The probes were run under the operating conditions of a probe current of 20nA, an acceleration voltage of 15kV and beam diameter of 0.5μm. From the data (Table 1), the plagioclase of dyke and pillow lava are found to be albite with the chemical composition of (Na, Fe, Ca, K) Al $(Si, Al)_3 O_8$. Whereas the chemical composition of actinolite present in uralite is found to be Ca_2

(Al, Na, Ca, Fe, Mg, K, Cr)$_5$ Si$_8$ O$_{22}$ (OH)$_2$ and that of chlorite is calculated as (Mg, Fe, Al, Mn)$_6$ (Si, Al)$_9$ O$_{10}$ (OH)$_8$. And epidote of the uralite is found to have two end members, one with composition of (Mg, Ca) $_2$ (Fe, Mg) (Al, Fe, Mn)$_2$ O (Si, Al, O$_4$) (Si$_2$O$_7$) OH, while that of the other member being (Ca,Mg) (Fe, Al) (Al, Mg, Mn)$_2$ O (SiO$_4$) (Si$_2$O$_7$) OH.

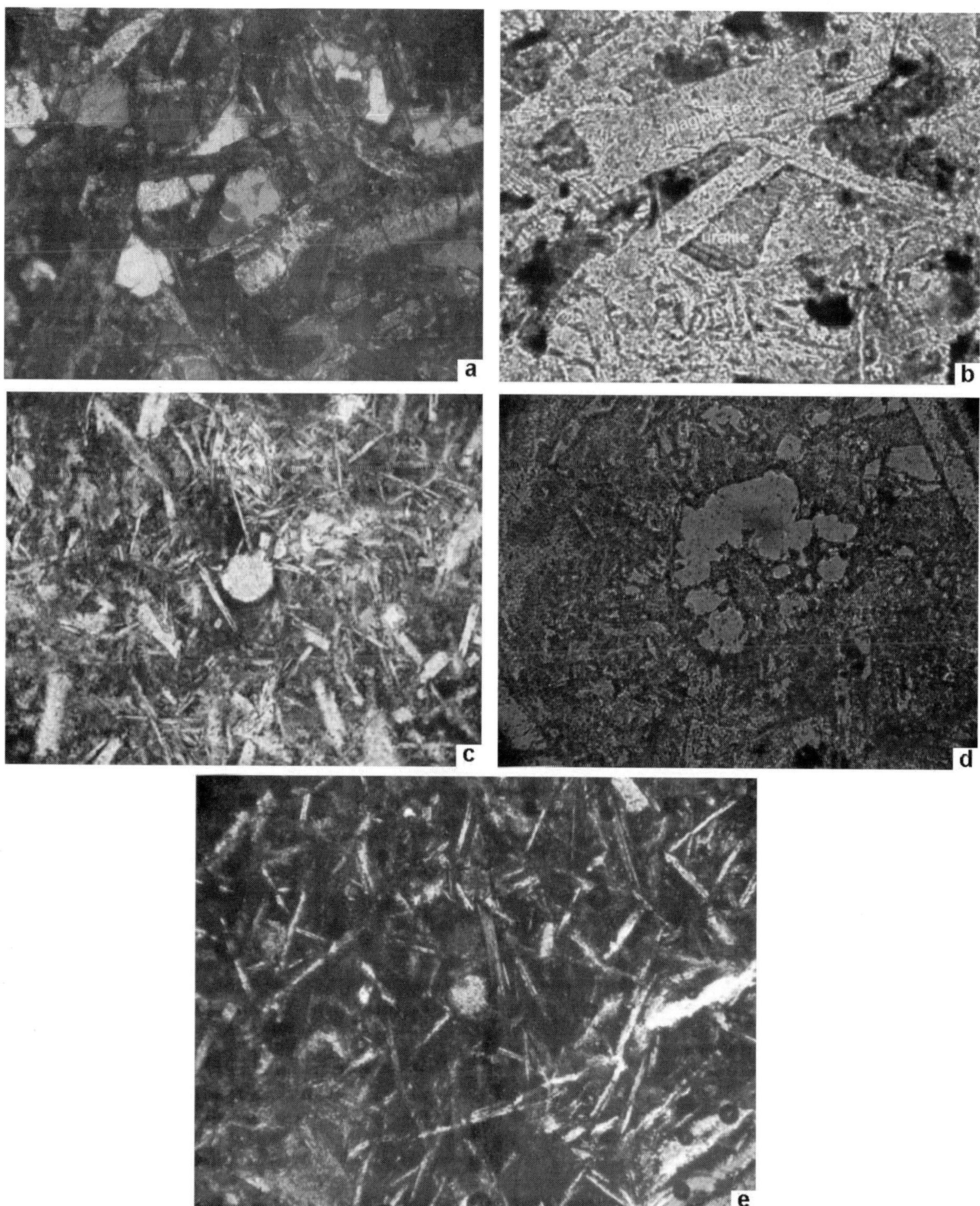

Fig. 2: Photomicrographs showing different textural features of dyke, sill and volcanic rocks. **a:** Earlier formed pyroxene and olivine grains (cross nicol). Width of the photo is 1.6mm; **b.** Sub-hedral plagioclase grains which are cloudy or dirty albite and consisting of minute crystal of epidote (polarized light). Width of the photo is 0.8mm; **c.** Typical vesicle of the dyke rock (cross nicol); **d.** Small secondary quartz the outer margin of which is corroded (cross nicol). Width of the photo is 0.4 mm, **e.** Variolitic texture with vesicle (polarized light). Width of the photo is 0.4 mm

Major Oxide Geochemistry

Six representative samples of the basic rocks found in different locations in Manipur Ophiolite Mélange Zone, were analyzed using X-Ray Fluorescence (XRF) spectrometer with recommended matrix corrections at the Department of Instrumentation and USIC, Guwahati University, Guwahati, for 10 major and minor element oxides. The major and minor element oxides data along with the derived CIPW norms of the diabasic rocks are shown in Table 1.

Table 1: Major and minor elements oxides in percent (XRF data) and CIPW norms of basic rocks of Manipur Ophiolite.

Samples / Major oxides	Sc10 (KW)	Su29 (X_2)	Su-43(B)	Su-66	Su-77(A)	Su-81(A)
SiO_2	50.41	47.22	44.73	45.19	51.19	48.00
Al_2O_3	13.17	13.22	13.32	13.36	13.38	15.67
Fe_2O_3(T)	7.77	9.81	9.18	10.92	13.11	11.05
MnO	0.11	0.12	0.11	0.15	0.11	0.28
CaO	6.35	4.82	8.53	5.96	5.83	3.29
MgO	7.14	6.22	5.37	4.59	3.99	6.42
Na_2O	6.22	6.12	6.48	3.51	4.76	4.57
K_2O	0.59	0.16	0.09	0.06	0.10	0.14
TiO_2	0.99	1.22	1.38	1.15	1.19	1.30
P_2O_5	0.17	0.19	0.42	0.01	0.02	0.11
LOI	7.30	11.01	11.06	15.37	6.20	9.58
Total	100.22	100.11	100.67	100.27	100.68	100.41
*NORMS						
Quartz	-	-	-	4.55	5.46	4.02
Orthoclse	3.34	1.11	0.55	0.55	0.55	0.55
Albite	41.37	47.16	28.56	29.87	40.40	38.77
Anorthite	6.39	8.06	6.95	20.29	14.73	8.89
Nepheline	5.96	2.25	14.34	-	-	-
Corundum	-	-	-	-	-	4.79
Diopside	19.33	53.21	26.29	4.29	8.37	-
Hypersthne	-	-	-	15.87	12.65	24.28
Olivine	10.38	23.95	7.98	-	-	-
Ilmenite	0.30	2.28	2.58	2.13	3.80	0.30
Apatite	0.34	0.33	1.01	-	-	3.02

*Norms calculated on conversion of all Fe_2O_3 (T) to FeO (T) and deletion of volatiles

From the CIPW norms of the basic rocks derived from the major and minor elements data, it is perceived that some samples are undersaturated (normative nepheline) and these are considered to be the samples which are least contaminated with chert, as observed from petrographic examination. These samples are found to have a range of 44.72-50.4 weight percent of SiO_2. Moreover, the sample S10 (KW) having the highest SiO_2 content out of the undersaturated samples, which happens to be the least contaminated one with considerable amount of cumulus olivine crystals. It is, therefore, reasonable to assume that the silica content of some of the original undersaturated melts could be as high as 50% and sample Su 43(B) having as high as 14.3% normative nepheline contains 44.73% SiO_2. On the other hand sample Su-66 having 4.5% normative quartz contains 45.19% SiO_2. As Su 66 contains lower Na_2O content (also for all other samples having normative quartz), the sample has been contaminated by chert and therefore, extra SiO_2 have been added to that of the original melt. Considerable assimilation of chert is evident from petrographic examination in this sample. But the extra SiO_2 cannot be very large because sample Su 43(B) having 44.73% SiO_2 is still undersaturated. Thus, only few percent which may be about 2-3% of SiO_2 must have been added to the undersaturated melts to become saturated. Therefore, the lower limit of SiO_2 content in some of the melts cannot be much lower than 4%. Hence, it can be assumed that SiO_2 contents of the melt of the basic rocks of Manipur Ophiolite cannot be beyond 43-50 weight percent.

Considering the possible variation of SiO_2 contents within a range of 43-50% and since, the weight percents of SiO_2 of all the samples of the basic rocks are also well within this range even after considerable assimilation with chert in some of the samples, SiO_2 variation could be used, despite some possible discrepancies, for interpretation of differentiation trend as well as classification of the type of magmatic lineage. Therefore, when total alkalis (% Na_2O+K_2O) versus silica (% SiO_2) are plotted over the fields of non-potassic volcanic rocks nomenclature of which are used as given by Cox *et al.* (1979), the basic rocks of Manipur Ophiolite are found to have been pigeonholed in the fields of alkali basalt and hawaiites (Fig. 3). The boundary between the alkalic and tholeiitic fields in the figure is extrapolated from MacDonald and Katsura (1964).

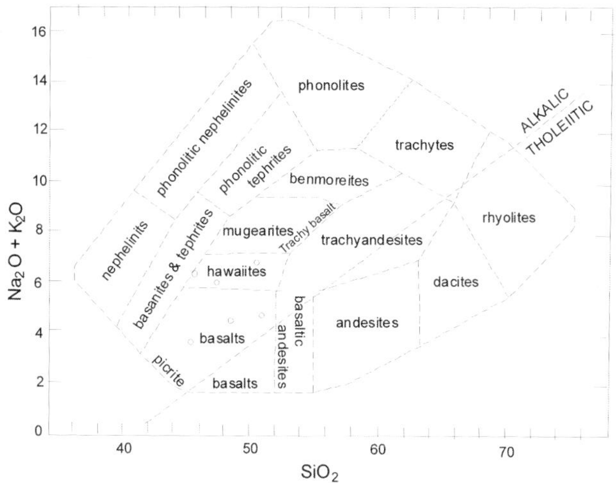

Fig. 3: Total alkali (wt % Na_2O+K_2O) vs wt % SiO_2 diagram. The basic rocks of Manipur Ophiolite are of alkali basalt lineage (adopted from Cox et al., 1979 and McDonald and Katsura, 1964).

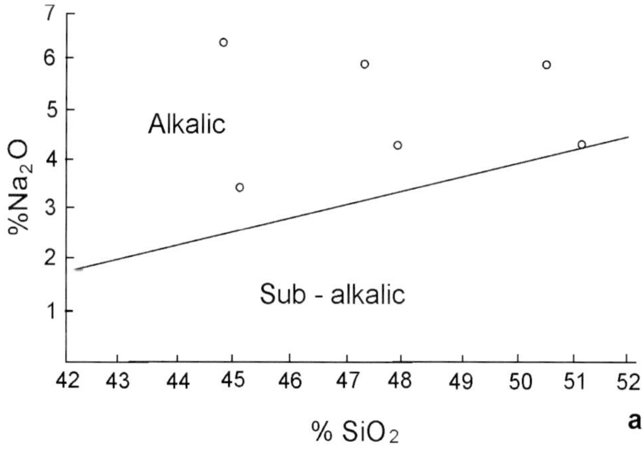

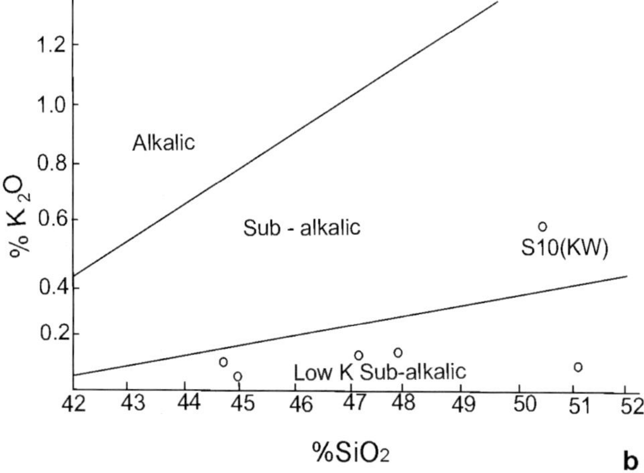

Fig. 4: Diagram of wt % K_2O and Na_2O separately vs wt % SiO_2. The basic rocks of Manipur Ophiolite plot in the alkali basalt field in 'a' (wt %Na_2O) and in the sub-alkali basalt field in 'b' (wt % K_2O)

From the weight percent data of K_2O and Na_2O separately versus weight percent SiO_2 diagrams shown in Figure 4 is found out. From the figure it is evident that the basic rocks of the study area plot in the alkali field in one diagram (% Na_2O) and in the sub-alkali field on the other (K_2O). Such rocks are termed as transitional basalts (Middlemost, 1975). The basic rocks of Manipur Ophiolite, therefore, could be inferred to be alkali basalts having a transitional character.

The Na_2O values are found to have been clustered at around two points, one at about 6% and the other at about 4%. From the CIPW norms calculation (Table 1), samples having higher values of Na_2O are found to have normative nepheline ranging from 2.5-14.3% indicative of undersaturated nature of the melt. On the other hand, samples having lower values of Na_2O are found to have normative quartz of about 5%.

Discussion

An essential and critical petrographic feature of magmas of tholeiitic lineage is the reaction relation between olivine and calcium poor pyroxene, often indicated by pyroxene riming olivine; and in strong contrast olivine and calcium poor pyroxene belonging to the alkali basalt lineage show a cotectic relationship and hence there is no reaction rim enclosing olivine by pyroxene (Wilkinson, 1967). The olivine crystals of the dyke/sill rocks of Manipur Ophiolite appear not to have reaction rim of pyroxene, and since found to co-exist with orthopyroxene, indicate cotectic relationship of olivine with pyroxenes. It is also known that absence of the reaction rim over olivine grains is primarily a consequence of the undersaturated nature of the residual liquid, compared with the normative quartz bearing residual liquid in tholeiites (Wilkinson, 1967). With appropriate cooling, tholeiitic pyroxenes may exolve various pyroxene lamellae (Wilkinson, 1967) which are not observed in the dyke/sill rocks of Manipur. At the same time even if contaminated by chert and secondary quartz is formed pseudomorphing after plagioclase laths, most of the samples have normative nepheline (Table 1). The corroded or surrounded by reaction zones over chert/quartz grains (Fig. 2d) indicate undersaturated nature of the melt. These observations from the petrographic examination, therefore, imply that the dyke/sill rocks of the Ophiolite Mélange Zone of Manipur appear to belong to the alkali basalt lineage.

But, the normative quartz of some samples of basic rocks of the ophiolite belt in Manipur appears to indicate saturated nature of the samples. This could have been caused by considerable assimilation of the melts with chert. Thus it can be assumed that the melts originally were deficient in silica (undersaturated), but because of the incorporation of assimilated chert, the melt could have become silica saturated. Chert assimilation into the mush of the crystals of the residual liquid could have caused a reaction in such a way that some of the few nepheline grains reacted with SiO_2, which resulted to the elimination of Na from the melt/rock. A possible mechanism by virtue of which Na may be eliminated from the melt/rock into the hydrothermal solution is postulated in the form of the following reaction:

$$\underset{\text{nepheline}}{4\,NaA1SiO_4} + 2\,SiO_2 + H_2O \rightarrow \underset{\text{albite}}{2\,NaA1Si_3O_8} + \underset{\text{removed in soln.}}{2\,NaOH + Al_2O_3}$$

The above reaction may attribute to the lost of Na from the melt/rock and hence could be the reason for low Na_2O contents for chert assimilated samples. The Al_2O_3 produced in the above reaction could have been compensated in the later crystallized phases like feldspar and uralitised pyroxene, as the Al_2O_3 contents in almost all the samples were consistent indicating that there has not been loss of alumina.

Al_2O_3 values in all the samples except in one (Su 81, which is a pillow) is found to cluster at around 13% (Table 1). The clustering of Al_2O_3 at certain value indicates that there has been very little effect of contamination and magmatic differentiation on the alumina content. The consistency of the value of alumina despite variable range of about 3-20% partial melting as suffered by the source rocks of spinel lherzolite (Khuman, 2009; Khuman and Soibam, 2010; Soibam and Khuman, 2011), compels us to assume that most of the Al, locked in the structures of primordial clinopyroxene and orthopyroxene are extracted in the melt even at the very low level of partial melting; and progressive melting leads to the extraction of other oxides other than alumina. Very little fractionation of Al-bearing phase (plagioclase) which attributed to make alumina content of not much variation in all the samples is evident from the fact of having almost similar modal proportion of plagioclase in all the samples of the basic rocks.

The samples with lower Na_2O value are found to have normative quartz and normative hypersthene and therefore, no normative olivine, contrary to samples with higher Na_2O value. This supports the validity of the above reaction by virtue of which the originally undersaturated melt becomes saturated. Therefore, the parent basaltic melts of Manipur ophiolite must have been of alkalic lineage.

If the assumption is made that most ophiolites formed at spreading centers under the cover of seawater and that high heat flow is maintained after their formation, it is possible to show that sub-sea geothermal systems operating at spreading centers are capable of producing hydrothermal metamorphism (Spooner and Fyfe, 1973) as well the above postulated nepheline-quartz reaction. High heat flow is symmetrically disposed on either side of present day spreading centers (Lee and Uyeda, 1965), and within the ridge axis geothermal gradients from 500 to 1400°C/km have been postulated (Spooner and Fyfe, 1973). Even though, considerable fluctuations in the heat flow is evident away from the spreading center, it is also commonly observed that the gradients gradually reduce away from the center and suggested that thermal gradients of 150°C/km are maintained at least 100km away from the axis. Therefore, it seems inevitable that sub-sea floor convection have been derived by virtue of which circulating hot waters at shallow depths (< 3 km) could interact with the newly formed dyke and sill rocks. As a result, the new oceanic crust near the spreading center have undergone hydrothermal metamorphism of greenschist facies (Soibam and Khuman, 2011). Therefore, within the temperature range of greenscist facies, which is about 250-450°C, the above postulated reaction could also have taken place.

Experimental work by Bischoff and Dickson (1975) on seawater-basalt at 200°C and 500 bars provided information on the possible ions involved in sub ocean floor hydrothermal alteration. It was found that Ca^{+2} continued to increase in the seawater. Heavy metals, such as Fe, Mn, Ni and Cu are leached under the conditions of the experiment and could concentrate in the fluid with high concentration values. The results of this experiment point to at least a partial explanation of the processes of alteration within ophiolites. The spilitic assemblage must represent the residual part of the basalt that has undergone hot seawater reaction. Na enrichment may be related more to the loss of Ca and other elements rather than addition of Na. In the experiment it was found that K was strongly leached from the basalt. Although pillow lava of Sakhalin and Kurile Island were erupted under submarine conditions, only a small part of these lavas are spilitic, suggesting that sea water reaction (i.e. Na from sea water) may not play an important role in the generation of spilitic rocks (Shilov, 1974). Thus, the postulated reaction along with spilitic metamorphism is quite possible.

The variation of SiO_2 content could be attributed to the different degrees of partial melting of source rock and gradual processes of fractional crystallization. As Na_2O contents of the samples having normative quartz of about 5% are found to cluster at a comparatively lower value of about 4%, which could be attributed to the loss of Na because of nepheline-quartz reaction on account of assimilation of chert in the melt, it is inferred that the original melts must have had lower SiO_2 and higher Na_2O contents. Hence, the samples before assimilation with chert could have been undersaturated with quartz (i.e. no normative quartz) but their SiO_2 contents are well within the postulated range of about 43-50%.

The dykes within the peridotitic serpentinites are found to have been assimilated considerably by chert. This shows that the ultramafic rocks have been raised to the sub-ocean floor indicating that diapirism of the upper mantle rocks still continued even after partial melting of the rocks have died out. From smoothness and other outward physical characteristics of the pillows as well as the sizes of the vesicles of the pillows Khuman and Soibam (2010) has established that the depth of ocean water must have ranged approximately 2000-3000m.

Conclusion

From the observations on petrography, mineralogy and major oxide geochemistry of the diabasic rocks of the ophiolite belt in Manipur and the consequent discussions enumerated above it is arrived at the conclusion that the melt of the basic rocks derived from partial melting of upper mantle materials was originally of alkali basalt lineage. The depth of the ophiolite basin ranged between 2000-3000m where the oceanic crust was thin. The melt either intruding or extruding at shallow depths of 0-3km, from the oceanic floor, interacted with circulating oceanic hot brine so that not only spilitic metamorphism occurred but also nepheline-quartz reaction took place on account of assimilation of chert in the melt. This interaction of the melt/rock at the spilitic metamorphism temperature of about 250-450°C with assimilated chert gave rise to the increase of SiO_2 percent in such a way that the rocks appear to range and transition from alkali basalt lineage to tholeiite character. Thus, the diabasic rocks of Manipur ophiolite are of alkali basalt lineage which appears to have transitional tholeiite character due to assimilation of chert.

References

Acharyya, S.K., Roy, D.K. and Mitra, N.D., 1986. Stratigraphy and palaeontology of the Naga Hills ophiolite belt. *Geol. Sur. India, Memoir*, 119, 6-12.

Anonymous, 1974. Geology and mineral resources of the states of India. *Geol. Sur. India, Misc. Publ.*, 30 (4), 124.

Bischoff, J.L. and Dickson, F.W., 1975. Sea-water basalt interaction at 200°C and 500 bars: implications for origin of seafloor heavy metal deposits and regulation of seawater chemistry. *Earth Planetary Science Letters*, 25, 385-397.

Brunnschweiler, R.O., 1966. On the geology of Indo-Burman Ranges. *J. Geol. Soc. Australia*, 13, 137-195.

Coleman, R.G., 1977. Ophiolites – Ancient Oceanic Lithosphere? Springer-Verlag, Berlin–Heidelberg, 229p.

Cox, K.G., Bell, J.D. and Pankhrust, R.J., 1979. The Interpretation of Igneous Rocks. Allen and Unwin, London 450p.

Deer, W.A., Howie, R.A. and Zussman, J., 1978. An Introduction to the Rock-Forming Minerals. ELBS Longman Group (FE), Hongkong, 528p.

Evans, P., 1932. Explanory notes to accompany a table showing the Tertiary succession in Assam. *Trans. Mining and Geol. Int. India*, 27, 155-260.

Khuman, M. Ch., 2009. Petrological and Geochemical Studies of the Ophiolite Belt in Parts of Chandel and Ukhrul Districts Manipur and Their Tectonic Significance. Unpublished PhD Thesis of Manipur University, Imphal, 238p.

Khuman, M. Ch., and Soibam, I., 2010. Ophiolite of Manipur - its field setting and petrotectonic significance. *Geological Society of India, Memoirs*, 75, 255-290.

Lee, W.H.K. and Uyeda, S., 1965. Review of heat flow data. In: Lee, W.H.K (ed.), Terrestrial Heat Flow. *Am. Geophysical Union, Monographs*, 8, 87-190.

Macdonald, G.A. and Katsura, T., 1964. Chemical composition of Hawaiian lavas. *J. Petrology*, 5, 82-133.

Mallet, F.R., 1876. On the coal-fields of Naga Hills bordering the Lakhimpur and Sibsagar districts of Assam. *Geol. Sur. India, Memoirs*, 12(2), 266-363.

Middlemost, E.A.K., 1975. The basalt clan. *Earth Sc. Review*, 11, 337-357.

Shilov, V.N., 1974. The pillow lavas of Sakhalin and Kurile Islands and their significance for the solution of the spilitic problems. In: Amstulz, G.C., (ed.), Spilites and Spilitic Rocks, Springer-Verlag, Berlin, 387-416.

Soibam, I. 1998. On the geology of Manipur. In Souvenir: IX Manipur Science Congress, 12-19.

Soibam, I., 2000. Structural and tectonic framework of Manipur. In Souvenir: X Manipur Science Congress, 26-37.

Soibam, I. and Khuman M, Ch., 2011. Basic rocks of Manipur Ophiolite in the Indo-Myanmar hill ranges, NE India and the petrotectonic significance. *J. Nepal Geological Society*, 43, 13-28.

Soibam, I. and Pradipchandra Singh, M., 2006. Analysis of drainage systems of Manipur and implications on the tectonics of the Indo-Myanmar Ranges. In: Saklani, P.S. (ed.), *Himalaya (Geological Aspects)*, 4, 281-302.

Spooner, E.T.C. and Fyfe, W.S., 1973. Sub-sea-floor metamorphism, heat and mass transfer. *Cont. Mineralogy and Petrology*, 42, 287-304.

Vidyadharan, K.T., Srivastava, R.K., Bhattacharyya, S., Joshi, A. and Jena, S. K., 1986, Distribution and description of major rock types. *Geol. Sur. India, Memoirs*, 119, 18-27.

Wilkinson, J.F.G., 1967. The petrography of basaltic rocks, In: Hess, H.H. and Poldervaart, A., (eds.), Basalts – the Poldervaart Treatise on Rocks of Basaltic Composition, Vol. 1, Interscience Publishers, New York, 163-214.

Hon'ble Reviewers

1. **R. Vaidyanadhan**
   Geological Society of India
   Gavipuram
   BANGALORE - 560 019
2. **Prof. B. Satyanarayana**
   1-8-702/62/17
   Behind Sarkar Mutt
   HYDERABAD - 500 044

I-3

Modern Analogue to Ore Forming Processes

K. L. Shrivastava and Virendra Gaur

Department of Geology, J. N. V. University, Jodhpur-342 005, India,
Email- klsgeology@yahoo.co.in

Abstract

As late as in the last quarter of the nineteenth century, the formation of economic mineral deposits have been observed in the modern environments. Later, the evolution of this very fast growing sub-discipline has now revolutionalized understanding of ore genesis. Carbonatites are, perhaps, most potentially single rock having diverse and rare mineral deposits. Existence of carbonate lava, one of the most discussed controversy of Earth Science, came to an end when outflow of carbonatite lava have been observed from a 'live' volcano at Oldoinyo Lengai, Tanjania. Later, some other carbonatite volcanoes have also been reported. Modern environment of formation of phosphorite at coast of Namibia and Chile-Peru have confirmed the formation of phosphorite by both ways; precipitation by upwelling in lagoon like and replacement of limestones during diagenesis on continental shelf. Volcanogenic Massive Sulphide (VMS) represent a large family of sulphide deposits occur in variety of tectonic settings but are typically related to precipitation of metals from hydrothermal solutions circulating in volcanically active submarine environment. 'Black smokers' at 21^0 N on East Pacific Rise are hot, metal charged, reduced and slightly acidic hydrothermal fluids that vent onto the sea floor along Mid-Oceanic Ridges representing modern analogue of the VMS. Salton Sea is typical example of Modern analogue to sedimentary Exhalative (SEDEX) sulphide deposits. Salton Sea hydrothermal activity shows a good combination of both, Syn-sedimentary exhalative and replacement process. Red Sea representing an intermediate stage between VMS and SEDEX deposits. With slow spreading rate at Mid-oceanic ridges, polymetallic node (or Manganese nodule) have been formed in recent geological past and are being formed even today either by hydrogeneous processes or diagenetic process. The mechanism involved may be hydrothermal, halmyrolitic or biogenic process or combination (s) of these mechanisms. The Central Indian Ocean Basin (CIOB) hosts the world's second largest deposits of high grade nodule, after the Central Belt of Pacific Ocean.

Keywords: Modern geoenvironments, Oregenesis, VMS, Black Smoker, Phosphogenesis, Polymetallic nodule.

Introduction

In the history of inception of our understanding of the genesis of economic mineral deposits, the neptunistic views, that water is the source of every thing living or dead, and the plutonic view that the basis of everything earthly was fire, have been advocated by two Greek thinkers, Thales (640BC) and Xenon (340 to 270 BC), respectively.

Perhaps the first truly scientific approach was expressed by Georgre Bauer (1494-1555), better known as Georgius Agricola, who considered that ore veins were formed by solutions circulating in the bowels of the Earth. Since ore was usually only in veins, the necessary richness and coarseness were achieved by the term, "Ore deposit", becoming virtually synonymous with vein fillings.

James Hutton (1726-1997), a leader of the plutonists, linked the formation of ore veins with internal (magmatic and Igneous molten) processes within the Earth. A. G. Werner 91749-1817), a leader of the Neptunists, favoured that veins of ore were formed from aqueous solutions flowing downward through cracks in rocks, eroding them and precipating ores in them. Interestingly, the two schools of contradictory views gradually subsided, but as late as in the first half of the nineteenth century, when the fact was better accepted that the ores can also be formed through the action of external geological processes, chiefly by sedimentation.

The concept of uniformitarianism, with James Hutton (1726-1797) as its postulator means that the physical world, as far back as the geological record can be read, did not differ importantly from the world as it exists today. Modern geological processes like erosion, deposition and volcanic action may have varied in intensity from time to time, but the interpretation of earth history does not require the intervention of conditions and processes different from those that now can be observed. Adverse to it, there are many cases in Geological interpretations where law of the "Present is key to the Past" can be very well accepted. For example, modern palaeontologists prefer to interpret many "functional morphology" on the basis of observations made on living forms of fossils. Even this branch of Palaeontology is now growing as 'Actuo-Palaeontology."

Modern ore forming processes could recognized as late as in the last quarter of the twentieth century. The sulphide concentration formation from modern volcanic cones in Japan, deposition of sulphide in pipes discharging geothermal drilling water in Salton Sea and the most popular among 'live' formation of ores, the deposition of sulphide ores from springs on the floor of Red Sea, which gained a milestone status in the history of the evolution of the genesis of ore deposits. The present paper is an exploration of the present status of knowledge on modern analogue of ore forming processes, a subdiscipline, which has revolutionalized the understanding of ore genesis.

Oldoinyo Lengai: Modern Analogue to Carbonatite

Carbonatites constitutes a distinctive group of Igneous rocks representing a distinctive ore type and environment of its formation. Initially considered to be petrological oddities the carbonatites are now becoming important economic target. They are granular Igneous rocks consisting primarily of calcite, dolomite, ankerite and other rock forming carbonate as principal constituents. Sub-ordinates are apatite, magnetite, silicates and other accessory phases. Apparently, they crystallize from a carbonate rich melt and intrude into the rocks in which they are found in close association with Igneous silicate rocks having low silica and high alkali and alkaline metal contents. They appear to be confined to zones of rifting in old stable shields of the continents. Carbonatites typically comprise of 80–99 wt% of carbonate, chiefly calcite soviet. Dolomitic, ankeritic and siderite manganiferous carbonatites show an increasing order of rareness. The combination of brecciation, multiple intrusions, assimilation and fenitization produces a very complicated overprinted fabric, many times.

Carbonatites occur as small plugs, stocks, ring dykes, cone sheets and carbonate breccia in association with alkali rocks. At places they also occur in the form of large irregular or sheet like bodies, sills or dyke swarms, adjacent to the alkali rock complexes or as an independent composite complex.

Carbonatites occur mainly in a continental environment and rarely in oceanic environments as on the Canary Islands. They are generally confined to large scale, intra-plate fractures, graben or rifts that correlate with the periods of extension, and may be associated with a broad zone of epiorogenic uplift. Ranging in age from the Proterozoic to Recent. Carbonatites rarely occur alone. A typical association includes carbonatite nephelinite (olivine poor), alkali peridotites-pyroxinite, ijolite-syenite and carbonatite-syenite complxes. The varities include calcium carbonate (soviet), dolomitic carbonatite (beforsite), ferroan or ankeritic calcite-rich carbonate (ferro-carbonatite), magnesite-olivin-apatite±phlogopite rocks, nephelinite, Syenite pyroxenite, peridotites and phonolite. Country rocks may be various lithologies and of different metamorphic grade. Single or varied types of carbonates usually differentiated with decreasing age, are commonly present.

Carbonatite began to be exploited for minerals since the second quarter of the last century when it was realized that they are the sole commercial source of Nb, the principal source of REEs, elevated concentration of zirconium, and tantalium, the largest source of Igneous phosphates and a rare source of strontianite. Large deposits of magnetite are also known (e.g. Palabora). In places magnetite, phlogopite and vermiculite are of importance. Due to the fact that the magnetites are titaniferous, too rich in Ti to constitute a good iron ore and too rich in Fe to constitute a valuable Ti ore. Additionally, a myrid of other metals and industrial minerals, such as vermiculite, barite, fluorite, ZrO_2, U and Ni-sulphates, U-Ta and Th-Nb rich pyrochlore can be extracted as economically valuable by-products. Thus carbonatites are potentially one of the valuable single rock type with such a diverse spectrum of mineralization.

There is now a general consensus among geoscientists that the parental magmas of carbonatites are mantle derived, and must be enriched in CO_2 (carbonated). Most carbonatite magmas are derived from carbonated silicate parental magmas by one of the three crystallization paths; (a) cooling of a magma to a liquidus-solidus terminal point where it precipitates silicates and CO_2 vapours, (b) cooling of a magma to a silicate-carbonate field boundary where silicates and carbonates are co-precipitated, or (c) cooling of calcite, except along the limiting field boundary where acarbonate–rich magma is exsolved without the co-precipitation of calcite, except along the limiting field boundary where the two major liquidus surfaces intersects. The Petrogenesis of economic carbonatites, is one of the most discussed controversy of economic geology. The largest carbonatite occurrence located on the Kola Peninsula (Russia), the carbonatite of Mount Pass (California) showing the highest REE abundences, the Palabora carbonatite (Transvaal) showing the widest spectrum of economic deposits.

The magma that erupted at Oldoinyo Lengai (Africa) is most unusual because it represented carbonatite composition. Present day observation of this volcano has put to an end one the most important controversy of earth science regarding existence of carbonatite lava (Fig. 1). Besides, there are other sites, later discovered, where extrusive carbonatites occur, i.e., at kerimasi, Mosonik and Shombole near by eastern rift valley and Kaiserstuhl in the Rhine graben, Germany (Shrivastava, et al., 2009).

Fig. 1. "Live volcano of carbonatite at Oldoinyo Lengai, Tanzania: An end to the controversy regarding existence of carbonatite lava (Photo: Martin Rietze/Barcroft Media)

Bell and Keller (1995) suggested that lava from Oldoinyo Lengai volcano is an extreme composition in a protracted range of melts associated with nephelinite carbonatite centers. Natrocarbonatite in fact, contains high proportion of alkalies dominated by sodium carbonate, made up of two minerals nyerecite and gregoryite. Both minerals are carbonates with significant amount of Na and K. they, being anhydrous quickly reacts with fluids. As a result the colour of the rock changes from black to white and texture changes from hard to soft and crumbly (Fig.1). Modern analogue of carbonates have greatly enhanced our understanding about petrogenesis of Carbonatites formed during geological past and also economic minerals originated along with this host.

Namibia and Chile-Peru Coast: Phosphogenesis in Modern Environment

Phosphorite sediments containing more than 15-20% P_2O_5 generally form along continental shelves and in shallow marginal marine settings such as lagoons and deltas. The process of formation are, therefore, essentially involve the upwelling of cold ocean water containing above-normal concentrations of metals, and their subsequent precipitation onto the continental shelves. In fact one of the earliest published accounts of the upwelling hypothesis, attributed to the Russian scientist A.V. Kazakov in the 1930s, applied to the formation of Phosphorite-rich sediments in a continental shelf setting. Deep sea drilling and ocean floor exploration have shown that phosphate concentrations are presently forming off-shore along the coastlines of Namibia and Chile-Peru, and are related to upwelling of deep, cold currents carrying small amounts of phosphate in solution (Fig. 2).

The formation of apatite/cellophane in phosphatic sediments has been considerable debate as to whether it has originated as a primary precipitate at the sea water-sediment interface (where the organic matter is decaying) or as a product of diagenesis during later compaction and dewatering of the sediments. The evidences existed for both the processes, satisfactorily.

High concentration of phosphate in Precambrian sea (Sheldon, 1980) is ofcourse, likely because of low-O_2, high CO_2 atmosphere, more acidic seawater with high solubility of phosphate. Unfortunately, only few, that too directly precipitated phosphate deposits occur during Precambrian times. It was because of lack of biologically mediated deposition. Cambrian explosion of life coincide with removing CO_2 (from ocean and atmosphere), increasing pH in shallow water environment which has greatly resulted in direct precipitation of apatite from a saturated water column. A more efficient form of concentration would have been through the decay of phosphate-dependent life forms and the formation of Phosphorite in association with organic-rich sediments. This is the reason why all large Phosphorite deposits are Phanerozoic in age. It is pertinent to note that the Precambrian-Cambrian boundary, for example, coincides with a period in Earth history when substantial deposition of phosphatic sediments took place (Cook and Shergold, 1984), good examples of which include the deposits of Mongolia, Southeast Asia and Australia.

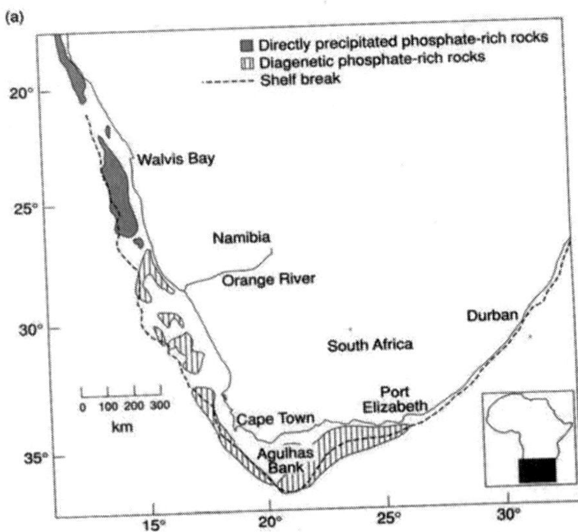

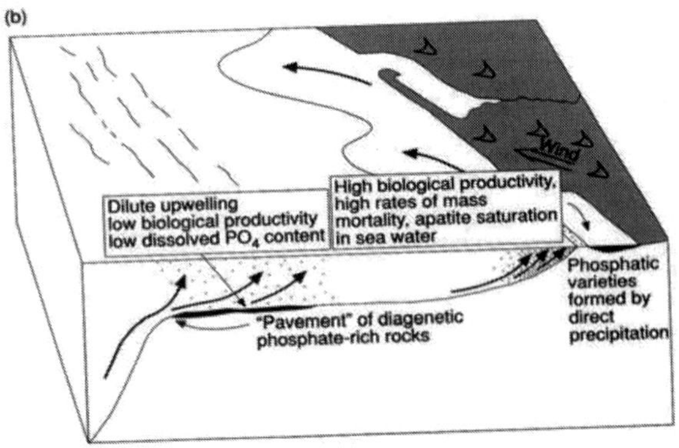

Fig. 2. Figure showing distribution of phosphorite deposits (upper) and phosphogenesis model (lower) of contrasting modes (after Birch, 1980) in the modern environment

Further, Mesozoic and Cenozoic represents an ideal environment for the formation of diagenetic apatite as the sediments are compacted and lithified. Morocco deposits are ideal example of cretaceous-palaeogene boundary for such deposition, however, Florida oolitic phosphate is also have been originated following the same way but subsequent reworked and allochthonous (Riggs, 1979). It is obvious that enhanced circulation and upwelling have played a role in phosphogenesis (Cook and McEthinny, 1979; Sheldon, 1980) with positive correlation between glacial events and phosphogenesis.

Permian Phosphoria formation possibly explains this mechanism (Robb, 2011).

At last, the study of present day formation of Phosphorite has helped to conveniently settle down the genetic model of phosphogenesis (Fig. 2). The nature and origin of Phosphorite forming off the western and southern continental margins of south Africa and Namibia have been studied by Birch (1980). Exploration of continental shelf in this region reveals two types of Phosphorite. One variety developed mainly along the Namibian coastline and comprises oolitic apatite/collphane ores derived by accretionary growth arising from the direct precipitation of phosphate from seawater. The other type, formed mainly along the south African coastline, consists of phosphatic replacement of fossiliferous limestone and is diagenetic in origin.

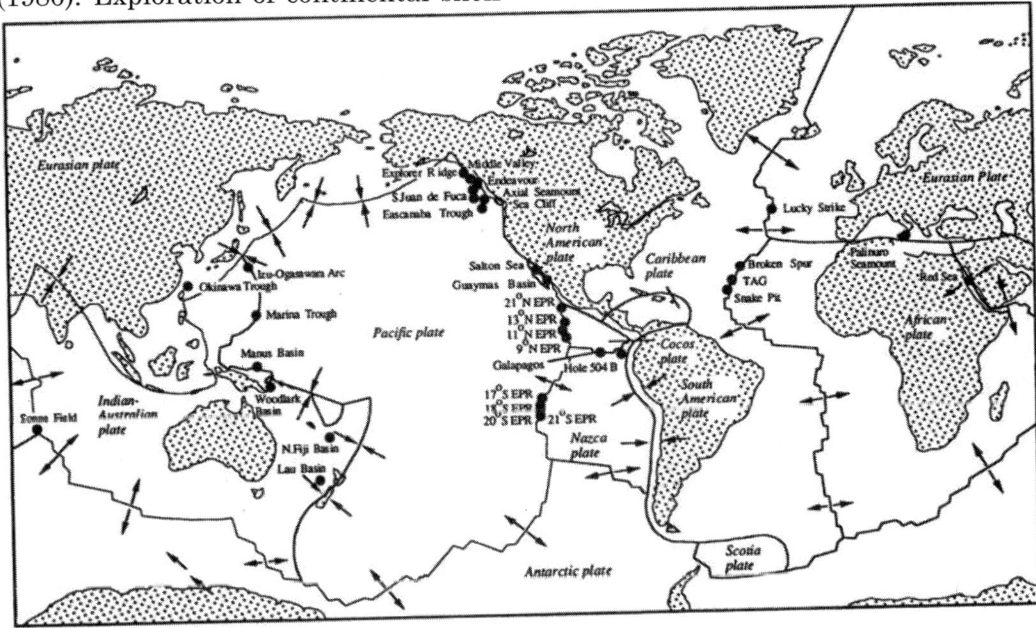

Fig. 3. Active Sulphide-generating Hydrothermal systems at the modern seafloor (after Mishra, 2000) on the basis of data after Herzig and Hannington, 1995)

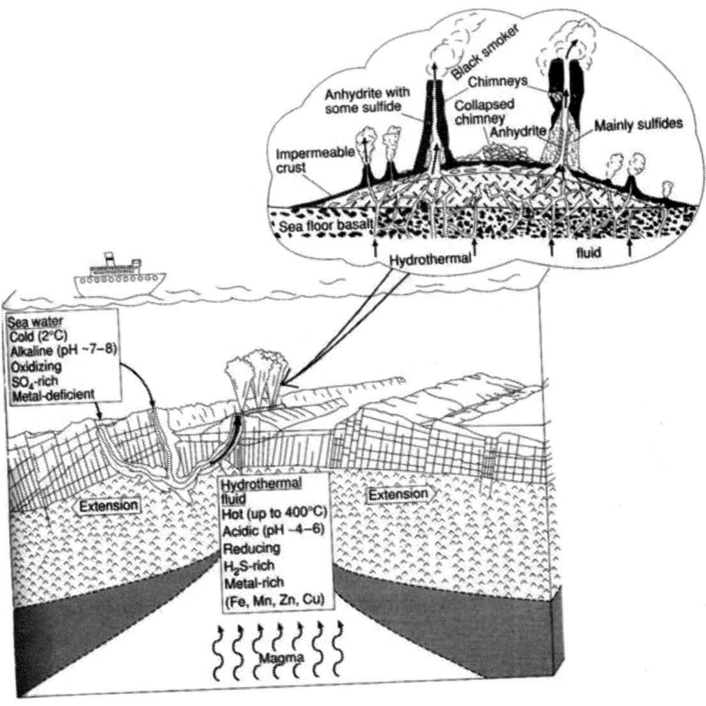

Fig. 4. Black smokers, white smokers chimneys in conceptual diagram of massive sulphide mineralization in mid-oceanic ridge environment (after Lydon, 1988; Scott, 1997, Robb, 2011)

A joint model for both types have been proposed by Robb (2011). Accordingly, diffuse upwelling along the outer reaches of the shelf results in reduced nutrient supply and biological productivity, and consequently the rate of phosphate release during decay of organism on the sea floor is also low. Apatite is under-saturated and cannot precipitated directly from the sea water, but subsequent diagenesis can lead to replacement of limestone by calcium phosphate. By contrast, in the near shore environment very high biological productivity in restricted shallow embayments and lagoons is maintained by wind-induced nutrient upwelling. Mass mortality of life forms in these restricted environments (especially during summer low tides) leads to the accumulation of abundant organic matter on the substrate with local increase in dissolved phosphate content to saturation levels. Robb (2011) has further added that the Varve-like layers of pure apatite as well as accretionary oolitic growth in the modern environment, provide the textural support for a mode of formation by direct precipitation of phosphate at the sea water sediment interface.

Black Smokers: Modern Analogue to VMS Deposits

The term "volcanogenic massive sulfide," or VMS, refers to a large family of mainly Cu-Zn (occasionally with minor Pb and Au) deposit types that formed during episodes of major orogenesis throughout Earth history. They are also referred as VHMS deposits, where the acryonym stands for "volcanic hosted massive sulfide." VMS deposits occur in a variety of tectonic settings, but are typically related to precipitation of metals from hydrothermal solutions circulating in volcanically active submarine environments.

Black smokers are described as hot (up to 400^{0}C), metal charged, reduced, and slightly acidic hydrothermal fluids that vent onto the sea floor, usually in zones of extension and active volcanism along mid-ocean ridges. The discovery in 1979 of 'black smokers" at 21^0 N on the East Pacific Rise (Francheteau et al., 1979) provided spectular insights into the nature of submarine hydrothermal activity and its potential impact on the understanding of ore deposition in such environments. Black smoker fluids usually vent through tube-like structures, called chimneys, that are built out of a mixture of anhydrite, barite, and sulfide minerals such as pyrite, pyrrhotite, chalcopyrite, and sphalerite, as well as gangue opaline silica.

The fluids originate essentially from cold (2^{0}C), alkaline, oxidizing, and metal-deficient sea water, circulate through the basaltic ocean crust and scavenge metals to form the hydrothermal fluids now observed at more than 100 black smoker sites in the Pacific Atlantic, and Indian Oceans as well as the Mediterranean Sea (Scott, 1977). Black smokers venting along mid ocean ridges represent a direct analogy for one specific type of the VMS family of deposits, namely the ophilite-hosted Cu-Zn deposits such as those of the Troodos Massif, Cyprus. An ophiolite represents a section of a mid-ocean ridge spreading centre that is preserved by obduction onto continental material, and is characterized by a sheeted dyke complex, pillow basalt, and pelagic sediment. A lens of massive sulfide ore, comprising mainly pyrite, chalcopyrite, and sphalerite is located at the ocean floor interface and this is underlain by a pipe-like zone of dissedimented sulfides and intense chloritic alteration representing the conduit along which fluids passed on their way to the ocean floor. The geometry of this ore deposit type is fairly typical of most other VMS deposits even though they may form in other tectonic settings such as island arcs, back-arc basins, and fore arc throughs, the other VMS deposits formed in other tectonic settings include Lokken in Norway and Outokumpu in Finland. Famous examples of VMS deposits formed in other tectonic settings include the greenstone belt hosted Kidd Creek and Mattabi deposits in the Archean Superior Provence of Canada, Besshi and numerous Kuroko-type deposits of the Japanese island arc, Rio Tinto and Neves Corvo in the Iberian Pyrite Belt of Spain and Portugal, the Buchans and Bathurst deposits of Newfoundland, Jerome in Arizona, and the Mount Lyell and Hellyer deposits in Tasmania (Robb, 2011).

Truly, in VMS deposits of Geological age there is clear zonation pattern of metals explained by the occurrence of sequence from Fe-Cu to Cu-Pb-Zn to Pb-Zn-Ba in both vertical and lateral directions. This sequence reflects, at the first glance, as variable solubility of these metals at progressively low temperatures. In fact White Smokers are seen in the modern environment when base metals could not be dissolved in fluids because of low temperature although they could transport sulphate complexes to precipitate barite and anhydrite. The dissolution and precipitation mechanism becomes very complex in some cases. For example biologically mediated precipitations by sulphate-reducing bacteria in exhalative vent are likely to be an important factor. Further, a different product would result from the same black smoker fluids if it is much less dense than the sea water. It will produce buoyant metal rich plume that will disperse metal to more distal ocean floor sediments and will finally result into low grade deposit. It is true that there are many classes of the VMS deposits but the modern analogue study has greatly enhanced our understanding to their genesis.

Red Sea: Intermediate stage to VMS and SEDEX Deposits

There is as such, no special or temporal link between SEDEX and VMS deposits, it is widely held they represent a continuous and are conceptually linked by the fact they formed by the same basic processes (Gilmour, 1976; Plimer, 1978; Guilbert and Park, 1986; Kirkham and Roscoe, 1993; Misra, 2000). These processes are active and can be studied in modern day environments.

The notion of a continuum between VMS and SEDEX deposit types is contentious. The principal VMS and SEDEX deposits apparently formed at different times of Earth history and indifferent tectonic settings. Also, studies have shown that some of the stratiform base metal deposits, previously thought to be syngenetic and related to exhalative venting of hydrothermal solutions, onto the sea floor, are more likely to be the product of replacement processes and, hence, are epigenetic in origin. A interpretation of this type has, for example, recently been made with respect to the Mount Isa deposits in Australia (Perkins, 1997). With this cautionary provison in mind, it is nevertheless convenient to discuss VMS and SEDEX deposits as a conceptual continuum, insofar as the principal ore-forming processes are concerned.

The Red Sea is known as an ocean basin in the early stage of its development having active geothermal venting and deposition of base metals in the 'lows' of the seafloor (Bischoff, 1969). The well discussed site of mineralization is "Atlantic II Deep," where rich Zn-Cu-Pb-Au mineralization is almost 50 million tons in the muddy sediments. On the one hand, the tectonic and geological setting of the Red Sea mineralization is analogous to mid-oceanic ridge VMS setting and on other hand, the absence of black smokers and the presence of both clastic and carbonate sediment hosts to the mineralization are features more similar to SEDEX type (Pirajno, 1992). Red Sea, hence, represent intermediate stage between two extremes in the range of VMS and SEDEX ore forming systems represented by black smokers (at 21^0 N on East Pacific Rise) and the active, metal charged geothermal system of the Salton sea.

Salton Sea: Modern Analogue to SEDEX Deposits

Sedimentary exhalative (SEDEX) represent largest and richest category of Lead-Zinc deposits of the world, to include HYC (Mc Arthur River), Mount Isa and Broken Hill (Eastern Australia), Sullivan (British Columbia), Aggeneys and Gamsberg (South Africa), Dariba-Rajpura-Sindesarkalan (India). These deposits are Palaeo to Mesoproterozoic in age. Palaeozioc deposits include Rammelsberg and Meggan (Germany), Navan (Ireland), Red dog (Alaska) and Howard's Pass (Canada). They, ofcourse, are dominated by lead and zinc with lesser copper but commonly Ba and Ag metal association and are related to hyderothermal fluids venting onto the sea floor but without any link to volcanism. These deposits are considered to formed within intercratonic rift basins, hosted by marine clastic and chemical sediments without association of volcanic rocks.

In Gulf of California, the Salton Sea is typical example of modern analogue to SEDEX deposits. The hyderothermal activity shows a good combination of both, syn-sedimentary exhalative and replacement process. It has resolved to enough degree, often controversial views, that prevail regarding the origin of sediment hosted stratiform deposits.

The Salton Sea has been percolating water along fault related pathways and through earlier formed lacustrine sediments to form a vigorous hydrothermal system. The hot metalliferous brines, also derived from Salton Sea waters and heated by a combination of the high prevailing geothermal gradient and active volcanism in the region, vent from numerous sites, and it is these that are of particular interest to the study of ore-forming processes (White, 1963; Skinner et al., 1969) (Fig. 5). The combination of well understood fluid compositional and circulation pathways, set in an active seismic and volcanic, intracontinental rift environment, has made the Salton Sea geothermal system a classic example of an ore-forming environment whose evolution can be observed, from which many inferences about ancient sediment-hosted ore forming processes can be made (Fig. 5). In addition to its obvious applicability to the understanding of SEDEX ore-forming processs, the processes observed in this environment might also be relevant to the formation of stratiform red-bed hosted Cu deposits such as the Kupferschiefeer ores of central Europe and the Central African Copperbelt (McKibben et. al., 1988; McKibben and Hardie, 1997).

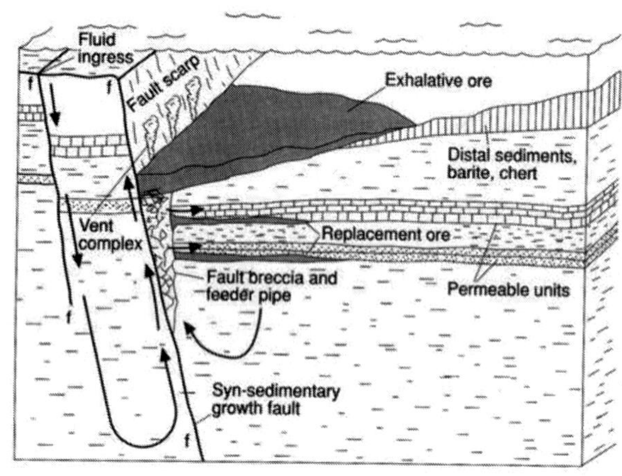

Fig. 5. Conceptual diagram for the formation of SEDEX sulphide ores (after, Goodfellow et. al., 1993; Mishra, 2000; Robb, 2011)

Central Indian Ocean Basin: Polymetallic Nodules

Polymetallic nodules, also called manganese nodules are rock concentrations formed of concentric layers of iron and manganese hydroxides around a core. The core may be microscopically small and is sometimes completely transformed into manganese minerals by crystallization (Roonwal, 1986). Nodules vary in size from tiny particles visible only under a microscope called micro-nodules to large pellets more than 20 centimeters across. However, most nodules are between 3 and 5 cm in diameter, about the size of Potatos (Fig. 6A and B). A new morphology named polynodule have been recently reported by Shrivastava et al. (2009). Roonwal and Friedrich (1981) have mentioned that the surface of nodules is generally smoth, sometimes rough, mammilated (knobby) or otherwise irregular. The nodules lie on the sea bottom sediment, generally, half buried but some are completely covered by sediments (Pattan and Parthiban, 2006).

Ore microscopy is better demonstrating that the concentric growth lines developed in the nodules (Fig.6 A and B). Occasionally spheroidal spore like structures can be seen in between the spaces of the filaments. No colonial type structures or distinct morphologies leading to identification of microbes were observed. Presence of colloform grey or white grey goethite (FeO.OH) is distinct being relatively brighter than dark grey manganese oxides of complex mineralogy (Fig. 6C and D).

MÖssbauer spectroscopic study of polymetallic nodule from IONB have been performed at 300K using constant acceleration spectrometer and ^{57}Co source in matrix (Shrivastava et al., 2009 b). The spectra (Fig. 6 E and F) were analysed by fitting appropriate number of discrete doublets and sexlets with Lorentzian peaks using a least square filling programme. It has confirmed the presence of hyderated iron oxide mineral possibly, goethite in the nodule.

There are many views which have been proposed for the formation of polymetallic nodules (Shrivastava et al., 2009b). First theory out of two most accepted ones is a hydrogeneous process in which concretions are formed by slow precipitation of the metallic components from sea water (Barnes and Dymond, 1967). A diagenetic process is the second theory, in which the manganese is remobilized in the sediment column and precipitates at the sediment–water interface. It is thought, in the first theory, to produce nodules with similar iron and manganese content and a relatively high grade of nickel, copper and cobalt. The nodules formed by the second theory are characteristically rich in manganese but poor in iron, nickel, copper and cobalt.

Apart from these two theories there are atleast three more mechanisms suggested in the literature for the formation of nodules (Shrivastava et. al., 2009b). First one is a hydrothermal process in which the metals derived from the hot springs which are normally associated with the volcanic actively. Halmyrolitic process, in which the metallic components come from the decomposition of basaltic debris by sea water, is the second mechanism. The third mechanism is a biogenic process in which the activity of micro-organism catalyzes the precipitation of metal hydroxides.

The Central Indian Ocean Basin (CIOB) hosts the world's second largest and high grade manganese nodule deposit, after the central belt of the Pacific Ocean (Roonwal, 1985). The crust below the manganese nodule field was formed between 60 and 49 Ma under fast, medium and slow spreading rates (Naquvi, 2005). More than 0.4 million km^2 of CIOB have been surveyed for the Manganese nodules. The survey also yielded information about the morphotectonic features of the Central Indian Ocean Basin, such as sea mounts, hills, ridges, normal and parallel lineaments. These features have distributed the topographic gradient (1:7000) of the sea floor. Spreading rate variation and formation of new oceanic crust during more than one episodes of India–Eurasia collision have influenced the Indian Ocean Nodule Field (IONF) which is divided into four zones from north to south on the basis of nature of the ridge parallel lineations. These zones are related to crustal accretion, rate of crust formation and crustal flexure. The IONF is characterized by both linear and isolated sea mounts (Mukhopadhyay et al., 2002; Naqvi, 2005).

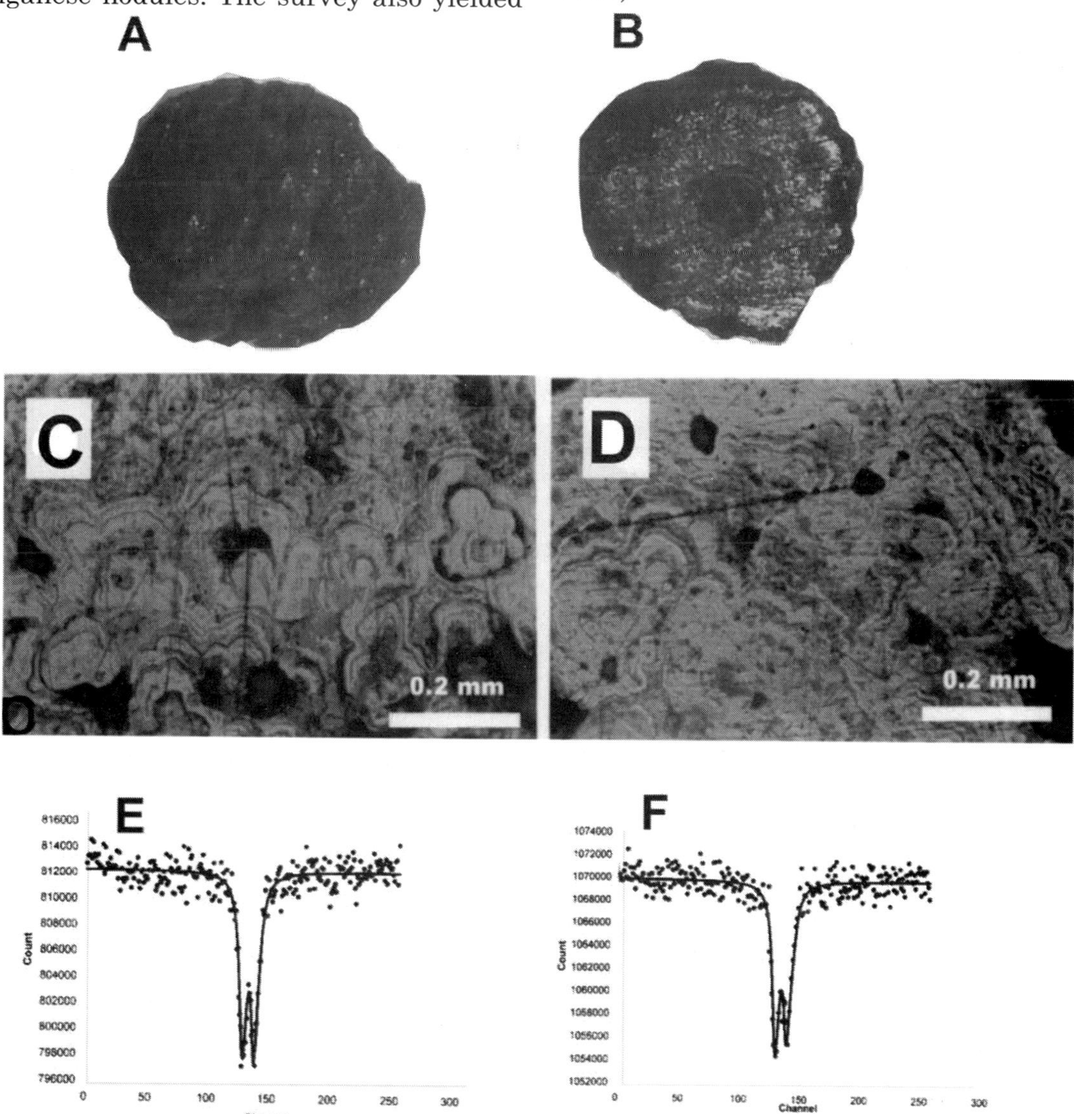

Fig.6 A: Top surface of polymetallic polynodule (natural size) B: Cross section of the same (natural size) C-D: Polished section under ore-microscope, displaying concentric growth lines, bright grey goethite is distinguished from dark grey manganese minerals. E-F: Typical ^{57}Fe Mossbauer spectra of polymetallic nodule at 300K. (All, after Shrivastava, 2009b)

Conclusion

Carbonatites are, perhaps, most potentially single rock having diverse and rare mineral deposits. Existence of carbonate lava, one of the most discussed controversy of Earth Science, came to an end when outflow of carbonatite lava have been observed from a 'live' volcano at Oldoinyo Lengai, Tanjania. It has given answers to many questions regarding petrogenesis of carbonatites in the Geological Past and also economic minerals originated along with this host. Later, some other carbonatite volcanoes have also been reported.

Phosphogenesis have been observed in the modern environment at coast of Namibia and Chile-Peru. Diffuse upwelling in continental shelf reduces nutrients, hence biological productivity, consequently the rate of phosphate released during decay of organism on the sea floor is low and phosphate can not be precipitated directly from Sea water but subsequent diagenesis lead to replace limestone by calcium phosphate. Adverse to it, lagoons are being maintained by wind-induced nutrient upwelling, where mass mortality, abundance of organic matter with dissolved phosphate reaches to the level of saturation. Hence, resulted into direct precipitation. Modern environment has confirmed both precipitation and replacement hypothesis of phosphogenesis.

Volcanogenic Massive Sulphide (VMS) represent a large family of sulphide deposits occur in variety of tectonic settings but are typically related to precipitation of metals from hydrothermal solutions circulating in volcanically active submarine environment. 'Black smokers' at 21°N on East Pacific Rise are hot, metal charged, reduced and slightly acidic hydrothermal fluids that vent onto the sea floor along Mid-Oceanic Ridges representing modern analogue of the VMS. Salton Sea is typical example of Modern analogue to sedimentary Exhalative (SEDEX) sulphide deposits. This category represent largest and richest category of Lead-Zinc deposits. Salton Sea hydrothermal activity shows a good combination of both, Syn-sedimentary exhalative and replacement process. Red Sea representing an intermediate stage between VMS and SEDEX deposits. On the one hand, the tectonic and geological setting of the Red Sea sulphide mineralization is analogous to mid-oceanic ridge, VMS setting and on the other hand, the absence of black smokers and presence of both clastic and carbonate hosts to the mineralization are features similar to SEDEX type.

With slow spreading rate at Mid-oceanic ridges, polymetallic nodule (or Manganese nodule) have been formed in recent geological past and are being formed even today either by hydrogeneous processes (slow precipitation of metallic components from the sea water) or diagenetic process (metal is mobilized in the sediment column and precipitates at the sediment-water interface). The mechanism involved may be hydrothermal, halmyrolitic or biogenic process or combination (s) of these mechanisms. The Central Indian Ocean Basin (CIOB) hosts the world's second largest deposits of high grade nodule, after the Central Belt of Pacific Ocean.

Modern analogue of ore forming processes has greatly enhanced our understanding regarding ore genesis. The ore genesis has now grown exponentially, hence, diserve tobe treated as a sub-discipline of Economic Geology. Authors will like to propose by coining a term for this subdiscipline as 'Actuo-Ore genesis" following the pattern of palaeontologists studying modern form of life to interpret fossils.

References

Barnes, S.S., Dymond, J.R.; 1967: Rate of accumulation of ferromanganese nodules. Nature, Vol. 213, pp. 1218- 1219.

Bell, K. and Keller, J.; 1995: "Carbonatite volcanism Oldoinyo Lengai and the petrogenesis of Natrocarbonatite" in IAVCEI Proc. In volcanogeny 4, Berlin. Editor K. Bell & J. Keller, Springer-verlag.

Birch, G. F.; 1980: "A model of penecontemporaneous phosphatization by diagenetic and authigenic mechanism from the western margin of Southern Africa." In Y.K. Bentor (ed.), Marine Phosphorites. Special Publication of the Society of Economic Palaeontologists and Mineralogists, 29, pp. 79-100.

Bischoff, J.L.; 1969: "Red Sea geothermal brine deposits: their mineralogy, geochemistry and genesis. In E.T. Degens and D.A. Ross (eds), Hot Brines and Recent heavy Metal Deposits in the Red Sea. Springer-Verlag, pp. 368-401.

Cook, P.J. and McElhinny, W.; 1997: "A reevaluation of the spatial and temporal distribution of sedimentary phosphate deposits in the light plate tectonics." Economic Geology, Vol. 13, pp. 315-330.

Cook, P.J. and Shergold, J. H.; 1984: "Phosphorous, phosphorites and skeletal evolution at the

Precambrian-Cambrian boundary. Nature, vol. 308, pp. 2331-2337.

Francheteau, J. and 14 others; 1979: "Massive deep-sea sulphide ore deposits discovered on the East Pacific Rise." Nature, vol. 277, pp. 523-528.

Gilmour, P.; 1976: "Some transitional types of mineral deposits in volcanic and sedimentary rocks." In K. H. Wolf (ed.), Handbook of Strata-bound and Stratiform Ore Deposits, Vol. 1. Elsevier, pp. 111-160.

Goodfellow, W. D., Lydon, J. W. and Turner, R. J. W.; 1993: "Geology and genesis of stratiform sediment-hosted (SedEx) zinc-lead-silver sulfide deposits." Geological Association Canada, Special Paper, 40, pp. 201-252.

Guilbert, J.M. and Park, C.F.; 1986: "The Geology of Ore Deposit." W. H. Freeman and Co., pp. 985.

Herzig, P.M. Hannington, M.D.; 1995: "Polymetallic massive sulfides at the modern seafloor: a review. Ore Geol. Rev., vol. 10, pp. 95-115.

Kirkham, R.V. and Roscoe, S. M.; 1993: "Atmospheric evolution and ore deposit formation" in S. Ishihara and 9 others (eds), Resource Geology, special Issue 15, pp. 1-17.

Lydon, J. W.; 1988: "Volcanogenic massive sulphide deposits, Part 2: Genetic models." Geoscience Canada, vol. 15, pp. 43-65.

Mishra, K.C.; 2000; "Understanding Mineral Deposits." Kluwer Academic Publishers, 845p.

Mukhopadhyay, R., Iyer, S.D. and Ghosh, A.K.; 2002 : The Indian ocean nodule field. Petrotectonic and ferromanganese deposits. Earth Sci. Rev. Vol. 60 pp. 67–130.

Naqvi, S.M.; 2005 : Geology and evolution of the Indian Plate (from Hadean to Holocean – 4 Ga to 4 Ka). Capital Pub. Comp., New Delhi. 450 p.

Pattan, J.N. and Parthiban, G.; 2006 : Composition and origin of buried ferromanganese nodules from Central Indian Ocean basin. Jour. Geol. Soc. Ind. Vol. 68. pp. 605–609.

Perkins, W.G.; 1997: "Mount Isa lead-zinc ore bodies: replacement lodes in a zoned syndeformational copper-lead-zinc system? Ore Geology Reviews, vol. 12 pp. 61-110.

Pirajno, F.; 1992: "Hydrothermal Mineral Deposits." Spriger-Verlag, 709p.

Plimer, I.R.; 1978: "Proximal and distal stratabound ore deposits." Mineralium Deposita, vol. 13, pp. 345-353.

Robb, L.; 2011: "Introduction to Ore Forming Processes" Blackwell Pub. 373 p.

Roonwal, G.S.; 1985: Mangenese nodules in the the Indian Ocean : Prospects and challenge; in 'The Ocean- realities and prospects.' (Ed. R.C. Sharma). Rajesh . New Delhi pp. 83-90.

Roonwal, G.S.; 1986: The Indian Ocean: Exploitable mineral and petroleum resources. Springer-Verlag. Berlin Heidelberg, Germany. 198p.

Roonwal, G.S., Friedrich, G.H.; 1981: Surface characters and growth patternin some manganese nodules from central Pacific. Ind. Jour. Mar. Sci., Vol. 10, pp. 57-60.

Scott, S. D.; 1997: "Submarine hydrothermal systems and deposits." In H.L. Barnes (ed.), Geochemistry of Hydrothermal Ore Deposits, 3rd edn John Wiley, pp. 797-875.

Sheldon, R. P.; 1980: "Episodicity of Phosphate deposition and deep ocean circulation: a hypothesis.: in Y.K Bentor (ed.), Marine Phosphorites. Special Publication, Society of Economic Palaeontlogists and Mineralogists, vol. 29, pp. 239-247.

Shrivastava, K. L., Guar, V., Dadhich, C. P., Gupta, R. P. and Tripathi, A.; 2009a: "Polymetallic polynodule from Indian ocean Nodule field" in 'Economic Mineralization' Editor K.L. Shrivastava, (ISBN: 978-81-7233-572-4) Scientific Publishers (India), pp. 254-263. Referees: Prof. G. S. Roonwal and Dr. V. K. Vaya.

Shrivastava, K. L., Sayeed, Y., Chauhan, M., and Hegner, E., 2009b: "Experimental, Petrological and field constraints on the petrogenesis of Economic Carbonatites with special reference to the Sarnu-Dandali Area, Northwestern India" in Economic Mineralization (Editor, K.L. Shrivastava) pp. 180-204. Sci. Pub. (India).

www.telegraph.co.uk: "Live volcano of carbonatite at Oldoinyo Lengai, Tanzania (Photo: Martin Rietze/ Barcroft Media).

Hon'ble Reviewers

1. **Prof. K.L. Rai**
   45, Chatrapati Shivajee Nagar
   Kolar Road
   BHOPAL - 462 016
2. **Prof. Rajesh K. Srivastava**
   Centre of Ad. Study in Geology
   Banaras Hindu University
   VARANASI - 221 005

I-4

ORE Mineral Deposits related to Ultramafic Rocks and their Genesis

T.K. Pandya and M.K. Pandya

Department of Geology, Mohanlal Sukhadia University, Udaipur, 313001
E mail: trilok.pandya.56@gmail.com

Abstract

Genetically the ore deposits of ultramafic category belongs to (1) Ore deposits of ultramafic-mafic association, (2) The mafic-ultramafic Iron-Nikel Sulphide-Platinoid association, (3) Copper-Nikel deposits of ultramafic volcanic rock association and (4) Ore deposits of Ophiolites association. After describing briefly the salient characteristic features such as mode of occurrence, host rocks, geological and tectonic setting, mineralogical features and global geological and geographical distribution along with Indian ore deposits; genetic models of the ore formation are presented.

The Chromite-Nikel-Platinoid ore deposits formed in (a) Layered intrusions, (b) Alpine type intrusions and (c) Individual intrusions respectively appear to have been injected into comparatively-stable tectonic environment, highly unstable environment and in zones of regional thurst faulting.

The mafic-ultramafic Iron-Nikel-Copper Sulphide-Platinoid associations are dominantly Sulphide-Oxide and minor Arsenoid and Sulfarsenides are neither bands of layered complexes nor intrusions of Alpine type form in Norite-Gabbro intrusions, Peridotites or Serpentinized ultramafic rocks. Copper-Nikel deposits of ultramafic-volcanic rock associations hosted by ultramafic high MgO volcanic flows called Komatiites the extrusive equivalents of Peridotites, Harzburgite and even Dunites.

Both Ophiolites and alpine Peridotites involve the tectonic emplacement of slices of upper mantle and oceanic crust, mafic and ultramafic rocks formed at ocean spreading centers or onto continental crust from mainly Chromite and Nikel deposits and rarely small Ni±Cu±Pt group metal deposits along with Fe-Ni-Co Arsenide, Magnetite, Chromite, Ilmenite and Rutile occurrences.

Genetic models for worlds well known ore deposits of Sudbury, Canada belonging to Fe-Ni-Cu Sulphide-Platinoids association include (1) Sulphide: Sillicate immiscibility and consequent magmatic differentiation in situ (2) Magmetic differentiation and subsequent injection, (3) Hydrothermal activity, (4) Sulphurization to a cooling melt with consequent abstraction of Fe-Ni-Cu etc. as the sulphides and (5) Meteoritic impact or impact of meteorite triggering terrestrial magmatism yielding the sulphide concentrations are critically discussed. Evidences on the most possible process of formation of deposits of Sudbury category are the magmatic segregation in situ at depth and injections of sulphide bearing silicate liquid and discrete sulphide liquids.

One notable feature of the occurrence of these ores is their remarkably frequent association with volcanic rocks particularly pillow basalts. Jensen and Bateman's (1979) concept of early and late magmatic processes of gravitative crystallization, differentiation and injection, residual liquid segregation and immisible liquid segregation apply with sound basis. There is no genetic controversy of ore deposits associated with ultramafic volcanic Komatiites as the ore deposits are stratabound within section of Komatiite flows. The Ophiolite associated ore deposits found in rafts or slices of three layers of oceanic crust, the "Steinmanns-Trinity" are assemblages of mafic to ultramafic rocks with mantle Peridotites at the base, successively overlain by plutonic rocks.

Keywords: Komatite, Ophiolite, Chromite-Nickel deposit, Ni-Cu-Pt deposit, Fe-Ni-Cu deposit

Introduction

As per Wyllie P.J. (1967) the ultramafic and ultrabasic rocks relate respectively to mineralogical and chemical classification of rocks. Rocks of ultramafic clan are those with color index more than 70; that is these rocks contain more than 70% of mafic minerals. Ultrabasic rocks are those containing less than 45% SiO_2. Most ultramafic rocks are also ultrabasic and most ultrabasic rocks are also ultramafic but there are exceptions.

In most of the texts of Economic Mineral Deposits the ultramafic and mafic rocks are invariably dealt together and categorised under the caption of mineral deposits of "Mafc-Ultramafic Association" due to close association of these two rock groups and because of their inter related processes of formation.

The principal ores of Mafic-Ultramafic rock association are, (1) Chromite ores, (2) Fe $\pm$ Ti $\pm$ V oxide ores, (3) Cu-Ni ($\pm$ Co) sulfide ores and (4) ores of platinum group elements termed platinoids which include (platinum, palladium, rhodium, ruhenium, osmium and iridium) either as native elements, alloys or as sulfides. The category of ultramafic rocks also include Carbonatite hosted REE ($\pm$ Cu) deposits and Kimberlite-Lamproites bearing diamond deposits amongst special lithology types.

Indirectly, the Mafic-Ultramafic assemblages serve as "Source Provinance" through endogeneous recycling (Mukherjee, A; 1999) for (a) Archean Au deposits notionally-recycled from the mafics in greenstone belts and Au and Sb mineralization from – hydrothermally altered Komatiites, (b_1) sediment hosted Cu deposits derived from Cu rich basalts by hydrothermal fluids in rift settings, (c) volcanogenic massive Cu+Zn ($\pm$ Au) sulfide deposits formed through scavanging of ocean/bacarc crusts by sea water turned hydrothermal fluids. Additionally through exogenous recycling (d) some mafic-ultramafic rocks constitute potential, protore source for 'Silicate Ni and' 'Laterite Au, Ni, Co and rarely platinum group elements in weathered crust.

The present paper does not deal with ore deposits which are associated with Kimbelites, Carbonatites, and Anorthosites along with those, mentioned under the indirect category and formed by metamorphic processes.

II. ORE DEPOSITS OF ULTRAMAFIC-MAFIC ASSOCIATION

A. CHROMIUM-NICKEL-PLATINOID ASSOCIATION

The ultramafic rocks yielding above metals occur in two main environments, (i) Layered Intrusions and, (ii) Alpine-type Intrusions (Fig. 1).

(i) Layered Intrusions

Most (through not all) of intrusions of this plutonic type belong to Precambrian and or lower Paleozoic age are large sized (Bushveld complex measuring 375 km east-west by 300 km north-south, extending over 67,000 km^2) and exhibit the forms ranging from saucer shaped to steep sided cone (Fig. 1). They are characteristically layered compositions of layers ranging from peridotites at the base to granites or granophyres at the top. Where a chilled border phase is present, this usually has the composition of

Gabbro. The layers which show high degree of perfection and continuity result chiefly from differentiation in situ, although some of them have formed due to concordant intrusions of later age.

(ii) Alpine Type Intrusions

These intrusions are much more prominent in geologically and physiographically young terrains and are generally conspicuously associated with basaltic and andesitic volcanic rocks and their derivatives. This suggests the intrusion in this case may be shallow subvolcanic rather than the result of deep plutonic activity. Most of the worlds alpine-type occurrences are Paleozoic or younger, (many of them Mesozoic or Tertiary).

(Iii) Individual Intrusions

Individual intrusions are small and most of these are conspicuously elongated and lens shaped. Large number of these small bodies occurs as components of narrow belts-serpentine belts. Such belts are associated with zones of regional thrust faulting related in turn to the trend of earliest geosynclines. These rocks are usually layered, but the layers rarely show the perfection and continuity. These short size lensoid bodies range in composition from dunite to gabbro. The layering may result probably due to flow.

(iv) Composition of Ultramafics

The mean composition of any large layered intrusion approximates to that of a gabbro, whereas, that of an alpine-type intrusion is distinctly more basic-that type of peridotite or picrite. Peridotites and related rocks are although present in both types of intrusions, bulk much more in the alpine type than the layered types.

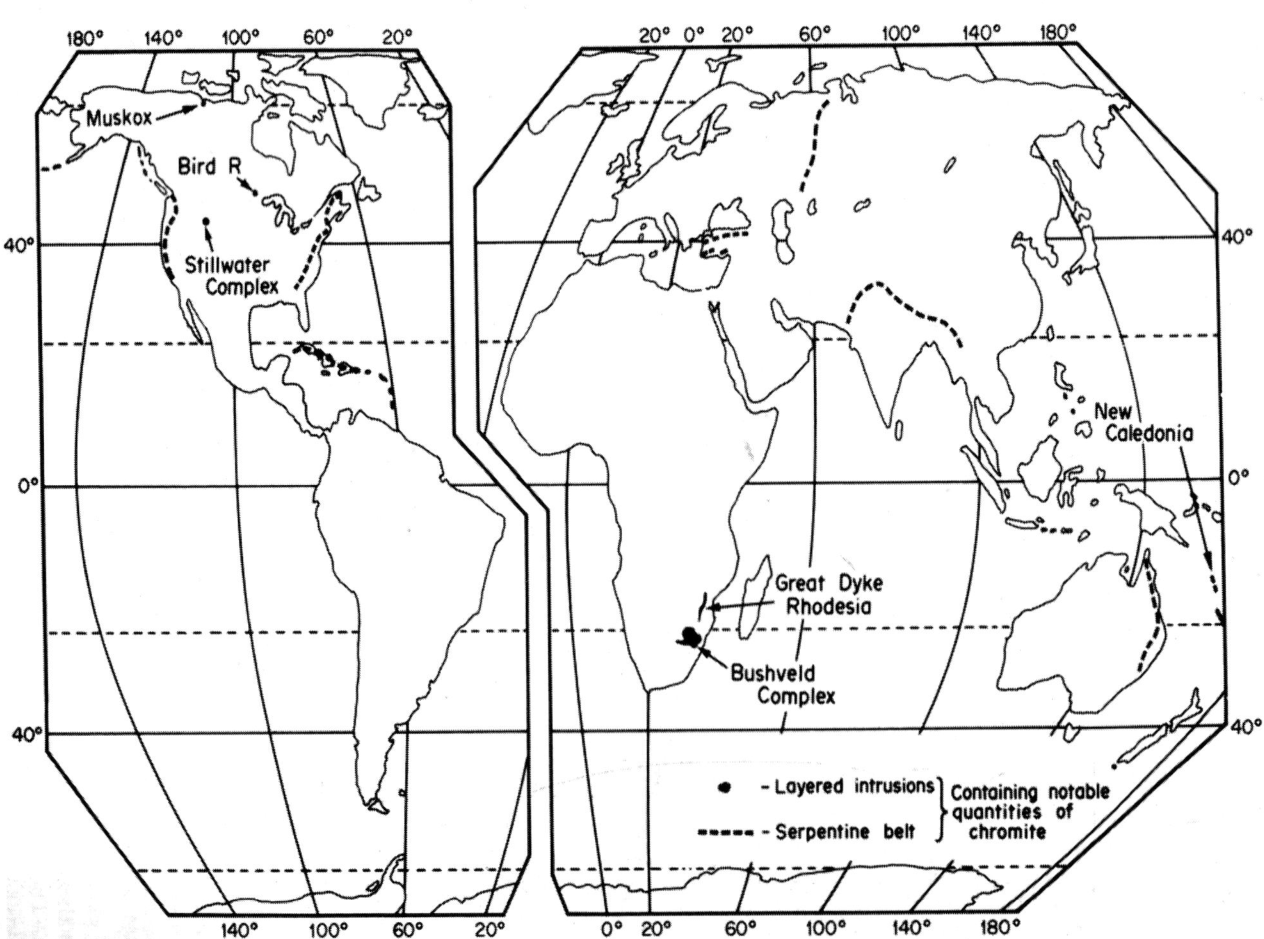

Fig. 1 Distribution of principal chromite-bearing layered intrusions and alpine-type serpentine belts.

(V) Metal Concentrations

(1) **CHROMITE**: It occurs abundantly in both environments, virtually entirely as chromite.

(2) **NICKEL**: It occurs in both, but the concentration in ores in lower than that of the chromite of chromite deposits. Its usual abundance in alpine-type intrusions is about 0.2 to 0.3 percent and rather less than this in layered intrusions. For the most part it occurs as "silicate nickel in olivine [(Mg, Fe, Ni_2) SiO_4)] through minor amounts are known to occur as sulfide or arsenide.

(3) **THE PLATINOID METALS**: These also occur in both, usually native in the alpine-type intrusions and most frequently as sulfides, arsenides, and antimonides in the layered intrusions. The chromite of layered and alpine-type cntrusions constitutes the world's sole source of chromium metal, and so deposits of this group are of substantial economic importance.

(VI) Distribution of Chromite Concentrations

Layered intrusions are themselves of very sporadic and localized occurrence, and in addition only a few of them contain sufficient concentrations of chromium to be considered sources of chromium ore.

Out of 6 important layered intrusions of the world belonging to Precambrian and Paleozoic terrains, only one or two are located in Tertiary rocks including the famous Skaergaard intrusion of Greenland. Of these, only 3, the Bushveld Complex of the Transvaal (Fig. 2), the Great Dyke of Southern Rhodesia and the Stillwater Complex of Montana are known to contain really substantial concentrations of commercial chromite. Thus, the layered-type chromite concentrations are few and isolated but as source of chromium they are very important.

Chromite in alpine-type; however, is wide-spread and virtually ubiquitous. There is hardly an orogenic belt, particularly in Post-Precambrian terrains, devoid of "serpentine" and there is probably no known serpentine belt that does not have, some where along its length, chromite concentrations of sufficient size and richness to constitute ore (Stanton, 1972, p., 307). The largest known deposits are those of the ultamafic belts of the Ural Mountains of Russia, the Philippines and Trukey. Substantial deposits are well known in ultramafic belts in Cuba, New Caledonia, India, Pakistan, Eastern Europe (Yugoslavia) and Greece, and Brazil, Minor deposits are known in the ultramafic belts of California Oregon, the Maritime Provinces of the United States and Canada and the Paleozoic belts of eastern Australia.

(vii) Mode of Occurrence of Layered and Alpine-Type Intrusions and Chromite Ore

The layered intrusions show regularity of layers and group of layers of great vertical regularity and wide lateral extent; behaving as strata and interbedded with bands of mafic-ultramafic members such as dunites, peridotites, pyroxenites and harzburgites. The structure of such sequences is often referred as pseudo stratification. The thickness of chromite bands ranges from few inches to 50 feet or so and the individual bands often extend for thousands of meters and in some cases for kilometers showing persistent stratigraphy like that of ordinary sediments.

In contrast to the generally high degree of order in layered intrusions, the chromite concentrations of alpine-type occurrences display what must be almost the highest possible degree of disorder.

The characteristic feature of chromite occurrence in this category is of extremely wide range of grain size and concentration. From sparsely distributed fine grains to dense, coarse-grained concentrations, discontinuous bands to stumpy lenses, podiform, to pipe shaped bodies or quite grosteque shaped to massive ore bodies and layered disseminations to sack form bodies are observed to occur in this category.

(Viii) Genetic considerations

The great layered intrusions and their constituent layered chromites all appear to have been injected into comparatively stable tectonic environments. While all have suffered some folding, faulting or tilting, none has suffered extensive deformation and metamorphism, indicating, their emplacement in comparatively late in the tectonic history of the area concerned. Although the bulk composition of each intrusion as a whole seems to be gabbroic, the chromites are now associated with ultramafic layers. The original emplacement does not seem to have been grossly influenced by major faults.

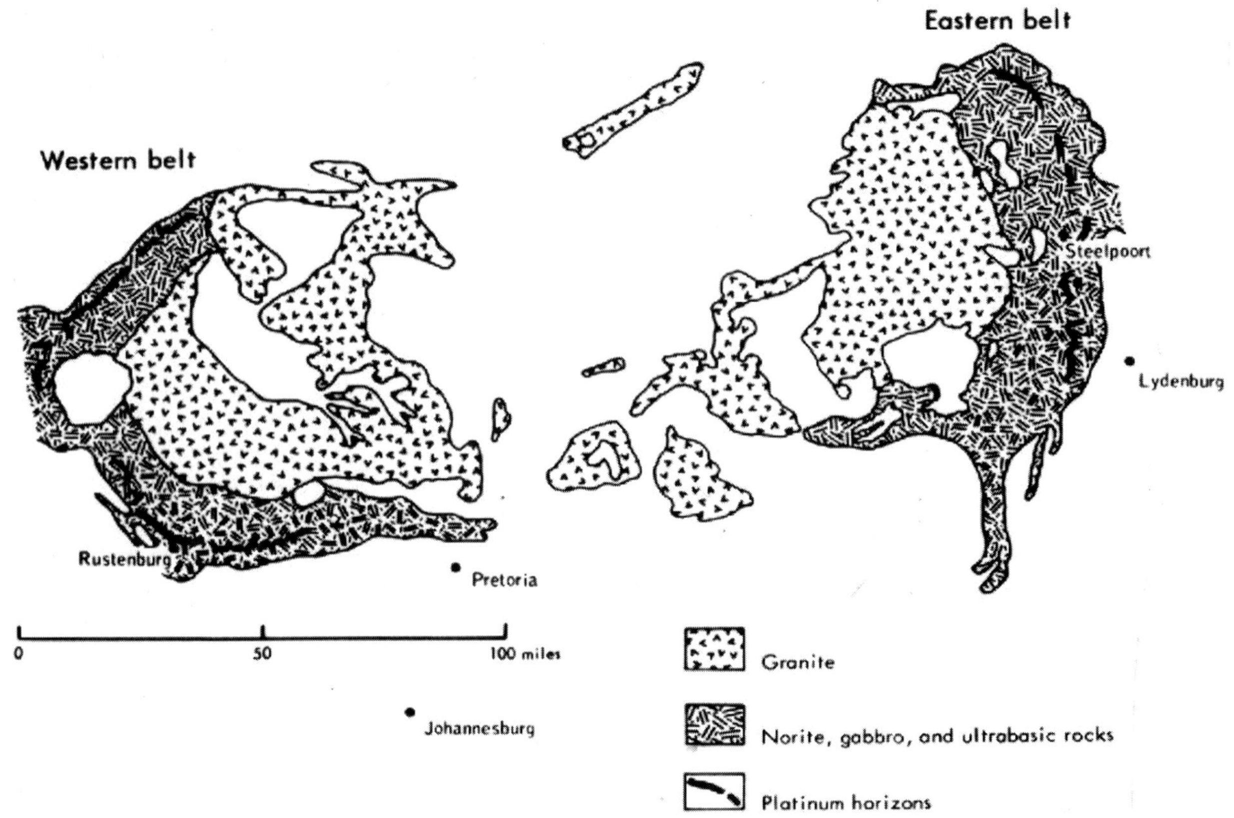

Fig. 2. Generalized geologic map showing distribution of the major rock types of the Bushveld igneous complex. Simplified after C.A. Cousins, *Engineering and Mining Journal,* vol. 160, no. 10, p. 114.

The alpine-type intrusions on the other hand have clearly been injected into highly unstable environment, in most cases into geosynclinal zones of volcanism, folding and faulting. The time of initial injection, too, is usually close to the most mobile period in the development of the geosyncline concerned. The bulk composition of the intrusions is peridotitic. The Chromite may be associated with a variety of rock compositions; the containing masses as a whole are quite definitely ultramafic. In addition a high degree of serpentinization is virtually ubiquitous in contrast to the layered intrusions, in which the serpentinization of magnesiam silicates is usually rather minor. 'Virtually all intrusions of alpine-type are clearly related to earlier faulting of major scale and most of them have been modified by faulting and squeezed and moved about by plastic flow.

(ix) Origin of Chromite Concentration

In the case of layered intrusions the mechanism appears to be on of fractional crystallization leading to the settling of successive crops of crystals on the progressively rising floor of the intrusive sheet concerned. According to Turner and Verhoogen (1960), the formation of stratification in layered intrusion is due to (1) development of large convection cells, with formation of crystals in the convicting magma as it moves across the top and down the sides of the chamber and sedimentation of the crystals as the magma moves across the floor; (2) multiple injection of magmas drawn periodically from a deeper magma as the differentiates; (3) change in mineral equilibrium boundaries accomparising changes in $H_2O_1O_2$, and other partial pressures, leading to the formation and precipitation of different minerals in alternate or cyclic fashion.

The formation of the alpine type intrusions, their structures and their various types of chromite concentration seems a vastly more complex problem or set of problems. The sharp differences from the layered intrusions in overall bulk composition, abundance of ultramafic types, olivine and pyroxene compositions, chromite compositions and textures, all argue against a common origin. It is interesting to note that in the environment existing in the cooling layered intrusions it was crystal faces that yielded energetically most economical solid-liquid interfaces during crystal growth. Whereas in the environments in the formation of alpine-type rocks it was the spherical shape that gave the lowest energy interfaces. Until recently there has been a very general belief that the alpine type rocks were of very fundamental nature probably mantle material pinched into and squeezed up, very deep faults. Lately, there has been the alternative suggestion, supported by a number of field and compositional features, that they are in the comparatively shallow differentiates of volcanic magma chamber, later serpentirized and squeezed into the great faults that had early localized the volcanism concerned. Whichever if either of these possibilities is the case, the deviation of the two class of rocks and their chromite, do seem to be different. In addition, whereas the layered intrusion occur virtually in their (pristine state apart from simple faulting and rifting, the apline-type intrusive are metamorphosed.

(x) Nickel Deposits

As a primary constituent Nickel is a very minor component of both layered and alpine-type intrusives. Some alpine-type masses are important resources of nickel but its concentration in this case is due to weathering and hence is secondary.

Most layered intrusions contain small concentrations of nickel, usually as sulfides and arsenides and usually fairly well confined to a few of the more mafic bands of the sequence. The sulfide assemblages are commonly pyrrohotite-pentlandite-chalcopyrite, generally with trace quantities of other nickel and copper minerals such as gersdorffite, cubanite, bornite and vallerite. Pyrite is sometimes as accessory.

The two best examples of nickel concentration in layered complexes are in Southern Africa the Bushveld complex (Fig. 2) and the great topolith intrusion of InsIzwa range in Cape Province.

By far the major part of the nickel of alpine-type intrusions is "silicate nickel" occurring as a minor constituent of olivine and serpentine derived from the latter. The abundance of Ni is usually of the order of 0.3 weight per cent of the whole rock and clearly far too low for the latter to be regarded as an ore at the present time. Residual deposits containing 1.5 to 2.5 percent nickel are often extensive. A minute amount of nickel may also occur as disseminated sulfide in alpine-type uframatic rocks – usually in pentlandite or heazlewoodite.

(xi) Platinoids and other precious metal deposits

Mafic and Ultramafic rocks of the present associations yield substantial proportions of the worlds supply of the platinoid metals and with them some gold and silver. Compared to Mafic-Ultramafic-Iron-Nickel-Copper Sulfide-Plationid Association the present association, although shows similarity of mineral assemblages to the former the distint difference lies in closely tied occurrence of platinoids in the present one which is not the case in the other associations. The Merensky Reef of Bushveld Complex demonstrates the above occurrence of platinoids. Cousins (1964, p 230) mentions that the concentration of the precious metals achieves maxima within or adjacent to the top and bottom of chromite bands. The platinium group minerals and their associations are ferroplatinum (platinum metal alloyed with varying amounts of iron), nickeliferous braggite [(Ni, pt Pd)S_2] cooperite (Pt AsS), laurite [Ru Os)S_2] stibiopalladinite (Pd_3 Sb) sperrylite, gold (alloyed with zinc), pentlandite[3], chalcopyrite, pyrrohotite, vallerite and chromite.

While small variations occur along the layer, the overall proportions of the platinoids and gold approximates to Pt. 60 per cent; Pd, 27 per cent; Ru, 5 percent; Rh, 2.7 per cent; Ir 0.7 per cent; OS, 0.6 per cent; Au, 4 percent. The chromite defining the layer is reported by Cousins, 1964, p. 231), as being a high-iron variety, Cr_2O_3, 40.5 per cent, total iron as Feo, 32.6 percent, Al_2O_3, 15.2 per cent; Mgo 9.7 per cent, Tio_2, 2.0 per cent; Cr/Fe, 1:1.

Similar concentrations of the platinoids have not been found in alpine-type intrusions, through platinoids are certainly well known to occur in ultramafic belts. The principal metals are platinum, osmium, and iridium and rather than occurring as arsenide-sulfide concentrations as in the layered intrusions, they are found native, in disseminated distribution. Economic concentrations are produced later by weathering and subsequent placer accumulation.

B. THE MAFIC-ULTRAMAFIC IRON-NICKEL-COPPER SULFIDE PLATINOID ASSOCIATION

1. Introduction

Although some deposits of this association show minor similarities to the small sulfide-platinoid occurrences in the layered intrusions of the association there are, however, divergences which are being given below.

(i) The ores of the present association are dominantly surfide-oxide and arsenide are usually very minor components.

(ii) The present ores occur neither in large conspicuously layered intrusions, nor in, alpine-type masses. The principal habitat is unbanded, essentially irregular gabbro-particularly norite gabbro-intrusions. A substantial proportion is norite gabbro-intrusions. A substantial propo-rtion is associated with ultramafic rocks, but these are neither bands of layered complexes nor intrusions of alpine-type. Some in deed may be basic to ultrabasic lavas.

(iii) While variable amounts of magnetite, titanomagnetite and ilmenite are common minor associations, chromite is conspicuous by its absence or near absence.

This group of deposits shows some variation in form and environment, and clear, if minor, differences in constitution, such deposits are familiarly known as sulfide-nickel ores and they have now been found on must of the continents.

2. Host rocks of the deposits

Most of the ore bodies occur in Precambrian rocks. A few are found in Paleozoic intrusions in Precambrian terrains and here another in post-Paleozoic. The prominent host rocks are norite-gabbro intrusives though other types-notably ultramafic rocks and their serpentinized equiva-lents are often involved. Among the adjacement rocks mafic pillow lavas and "green stones" (metamorphosed mafic lavas and associated pyroclastic rocks) are prominent, through again, other types including a verity of sedimentary rocks may be involved.

3. Distribution of deposits

Four major occurrences demonstrate the range of environment in which sulfide nickel deposits were formed. These are, (1) Sudbury, Ontario (Fig. 3), (2) Moak Lake-Setting Lake, Manitoba (The Thompson mine), (3) Lynn, Lake, Manitoba (Sheritt Gordon Mine and (4) Kambalda, Western Australia.

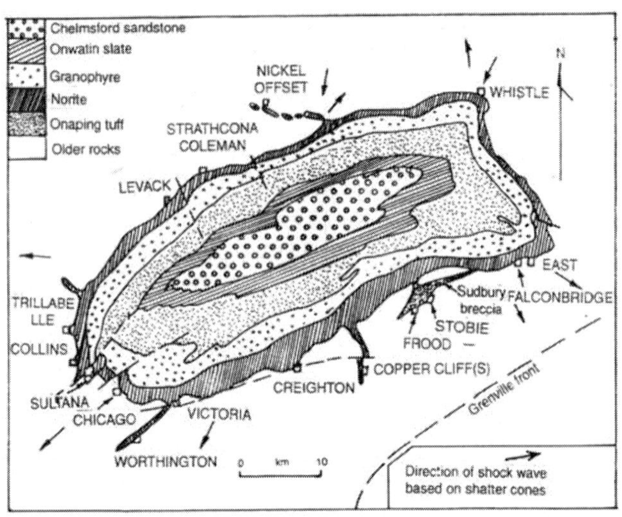

Fig. 3 Geological map of the Sudbury basin. (with modification after Philpotts, 1990)

By far the largest part of the worlds known resources, however, occur in Canada, whose production of sulfide nickel easily exceeds that of the rest of the world together. Other occurrences are in the Ungava district, in northern Quebec, Marbridge and Lorraine in Southern Quebec, Texmont and Gordon Lake in Ontaro and at Hope, British Columbia.

Several deposits occur in northern Russia, in westerly part of Kola Peninsual and north-western Siberia. The Australian deposits are in the vicinity of Kalgoorlie in the Southern part of Western Australia.

While norites are the host rocks of Sudbury deposit, the Thompon minle of Moak Lake – Setting Lake of Manitoba are located in the ultramafic rocks represented by peridotites, showing pseudomorphs of serpentine after olivine and orthopyroxene. The deposits of Lynn. Lake, Manitoba (Sheritt Gordon mines) lie in intrusive plugs composed of peridotite through norite to diorite. The principal ore bodies occur in "amphibolite" a metamorphosed gabbro rock'. The Kambalda deposits of western Australia, better referred as Kalgoorlie occur in sespenti-nized ultramafic rock mass, though whether this is a intrusive (a sill or sills), or a group of lavas or both is still not really clear (Stanton, 1972, p 336). According to Woodall and Travis (1969) the mass is a sill injected between two basalt flows, is a layered mass consisting of varying amounts of serpentine, amphibole, talc, chlorite, carbonate and biotite, together with accessory magnetite and very mirror chromite. The associated basalts show relict pillows are metamorphosed. The association of metasedimentary rocks and consideration that much of the ultrmafics were extrusive similar to high magnesian basalts found in the Tertiary to modern volcanics.

4. Ores of the deposits

The ore bodies of this association show their greatest similarity. The principal components are invariably pyrrhotite, pentlandite and chalcopyrite, commonly in that order of abundance. Pyrite and magnetite are frequently though not invariably-present as substantial accessories, as cubanite. A variety of Ni and Te arsenides and sulfarsenides often occur as minor constituents. The platinoid metals, particularly platinum and palladium may occur as important traces, usually as arsenides and bismuthides. Gold and silver occur native and as tellurides and selenides.

Pyrrhotite – is characteristically a dominant opaque mineral. This is almost invariably followed by pentlandite and then by the copper minerals, chieffy chalcopyrite, often with cubanite. Although Ni is almost always more abundant than Cu, Ni/Cu ratios are very variable. Ni/Cu ratios lie between about 2:1 and 3:1, are the most common.

The platinoids may be notable or insignificant in amount, and gold and silver while virtually ubiquitous, may vary substantially. In Sudbury deposit the major ore minerals are pyrrhotite, pentlandite, pyrite and cubanite and minor ore constituents are magnetite, ilmenite, magnetite –ilmenite intergrowths, pyrite, gersdorffite, niccolite, maucherite, heazlewoodite, bornite, vallerite, sphalerite, stannite, galena, parkerite, tetradymite, nafive-bismuth and bismuthinite. The precious metal constituents are native gold, native silver, sperrylite, michenerite (Pd Bi_2?), froodite (a Pb Bi_2), hessite and schapbachite. The mineralogy of Thompson mine is far less complex than at Subdury and spectacular assenide-platinoid and copper rich ores of Sudbury are lacking. Like the Thompson ores that of Kambalda has a high Ni/Cu ratio and contains principally of pyrrhotite, pentlandite and pyrite with minor chalcopyrite. The minor constituents of the banded ore are magnetite, magnetite- chromite intergrowths, altaite, galena and sphalerite.

(a) HOST ROCKS AND DISTRIBUTION OF INDIAN CHROMITE DEPOSITS:

1. **ORISSA**; 98% of Indian Production.

NAUSHAI, deposit in harzburgite (enstatite peridotite) or saxonite and enstatite-dillage peridotite (lherzolite) converted into serpentine and talc.

SUKINDA deposit located in enstatite-periodotite; dunite and orthopyroxenite **(2-7)** 2% of Indias production

2. **JHARKHAND** Jojohatu, Kimsiburu, Kittaburu, Roroburu Chitungburu deposits located in dunites, saxonites and pyroxenite-saxonite, largely serpentinized.

3. **MAHARASHTRA**: Kankauli and vagda deposits located in serpentine rocks.

4. **TAMILNADU**: Kondapalle: Serpentirized ultradbasic charnockites of the Chalk Hills. Deposits located in dunite, charnockites Siltamp-undi deposit located in pyroxenites and anorthosites.

5. **KARNATAKA**: Kalangavi Hill deposit located in ultrabsic rocks, largely converted into talc.

Nuggihalli schist belt – Number of deposits located in enstatite peridotite altered to talc and serpentine.

Jambur – Hulikere deposit located in talc-serpentine rocks.

Shinduvalli and deposits between Mysore and – Nanjangud located in ultrabasic rocks.

6. ANDHRA PRADESH,

7. MANIPUR

(b) NICKEL DEPOSITS OF INDIA:

1. **ORISSA**: The ores of Sukinda Ni deposits are located in dunite and orthopyroxenite in Kansa sector, Kumardah-Saruabil-Sukerangi Sector and Kaliapani, sector are represented by nickeliferrous limonite, serpentinous nickel ore, lateritic nickel ore siliceous nickeliferrous limonite and talcose nickeliferous limonite.

2. **JHARKHAND** Chromite ores association.

3. **KARNATAKA** Chromite ores association.

4. **NAGALAND** Chromite ores association.

(c) PLATINOID GROUP METALS: No Production from India

1. **ORISSA** Sukinda and Naushai Chromite deposits.

2. **TAMILNADU Chromite-Nickel Deposits** Sitampudi complex chromite Nickel deposits.

5. Genesis of the deposits

As the Sudbury ores have received great attention for a long time, hence, several theories have been put forward for the genesis of ores in this deposit.

1. Sulfide: Silicate Immiscibility and consequent magmatic differentiation in situ.
2. Magmatic differentiation again due to sulfide-silicate immiscibility at depth with subsequent injection of sulfides as magmas in their own right.
3. Hydrothermal activity with the deposition of the sulfides principally by replacement of rocks adjacent to the norite contact.
4. "Sulfurization" (i.e., the late addition of sulfur) to a cooling melt with consequent abstraction of iron, nickel, copper etc. as the sulfides.
5. Meteorite impact. The whole Sudbury intrusion, the Sudbury breccia, and the White water series are the result of the meteoritic impact, the sulfide ores representing original meteoritic material localized along the base of the crater. Alternatively, the impact of the meteorite may have triggered terrestrial magmatism which then yielded the sulfide concentrations.

The hyupothesis of differentiation in situ is based upon the experimental work of Vogt and his successors on the immiscibility of sulfides in silicates and upon the restriction of the sulfide concentrations to the sublayer and its derivatives.

The sulfide segregated from the main mass of norite during the cooling of its melt, after the latter intruded into the present position. The silicates separated in minor quantities from the developing sulfide mass.

The compositional differences in the various deposits of Sudbury region along with several other reasons such as simple segregation reprinting – injection and occurrence of sulfides at base of the intrusives as well as at the contract of chilled margins can not be explained by simple differentiation theory. The theory maintaining segregation of most of the sulfides from silicates at depth, with the injection of the two, as already separated phases, is hard to fault.

The hydrothermal replacement theory has no support as a major process at Shdbury. Except for the formation of some minor veins of hydrous silicates, carbonates and quartz the process can not be applied for the whole Sudbury mineralization.

Kullerud and Yoders (1963, p 215-218), 1964, pp 218-222) theory explaining that reaction between iron bearing silicates and sulfur result in the formation of pyrrhotite and/or pyrite and fyalite containing minor nickel, producing pentlandite in experiments can not be applied to Sudbury deposits as this theory can not account for the formation of the brecciated sulfides or those injected along chilled margins.

Dietz (1964) suggested that the Sudbury basin as a whole is an asteroid-impact structure. No evidences of meteorite impact theory are observed in the vicinity of other deposits of Sudbury region and it is difficult to believe that impact can generate magmatic activity such as the one occurred at Sudbury and hence, meteoritic impact is not of primary significance in the generation of iron nickel-copper sulfide ores as a class" (Stanton, 1972; p. 348). There are several other objections also.

The idea of sulfurization in the formation of ores encounters difficulties, particularly on constitutional grounds. The most critical difficulty lies in the relatively low magnetite content, and it is difficult to see how this could be accounted for in terms of the sulfurization hypothesis.

The only possible process which must have formed deposits of this category is magmatic segregation, in situ or at depth. Occurrence of sulfides as sheets between chilled margins of the intrusive and the intruded rock as in some Sudbury localities, and very high ratios of sulfides to silicate as at the Thompon mine, are two very substantial pieces of evidences, against formation entrirely by segregation in situ. These difficulties do not arise, how ever; segregation has taken place at least partly in some deeper, larger body of magma. In this case:

(1) Sulfide bearing silicate liquid and discrete sulfide liquid are injected together; segregation occurs and the sulfide concentrates at the base of the intrusions or along suitable interfaces.

(2) Silicate and sulfide liquids are injected – separately from time to time, both, however, following essentially the same conduits and coming to rest in the same locale.

Although one set of deposits – those of Sudbury appears to be related to layered gabbroic intrusions, it is very doubtful whether the ores of the present group have any general close tie with layered intrusions. Most of the deposits are associated with highly mafic to ultramafic rocks and there do not seem to be any felsic associates, such as are conspicous in the Bushveld and Still Water masses. As a class, too, deposits of the present group are usually associated with plugs and thin sheets.

One notable feature of the occurrence of these Ores is their remarkably frequent association with volcanic rocks, particularly pillow basalts. The environment usually looks to have been highly unstable. One suspects that emplacement of nickel bearing rocks may have been late stage tapping of some very deep source that had already supplied the volcanic rocks of the surrounding terrain. In this connection perhaps it should be kept in mind that some of the nickel bearing sheets may themselves be lavas rather than intrusions, the sulfides may have been extruded, as component of thick flows, rather than intruded.

It is suggested that the platinoids are preferentially associated with those ores showing higher copper nickel ratios. The ores associated with a higher proportion of gabbroic rocks show higher copper. Sudbury and Noril'sk both of norite association and relatively with high Cu/Ni rations are notable source of platinoids. Thompson and Kambalda both of ultramafic affiliation have conspicuously low Cu/Ni rations and both are also conspicuously low or lacking in platinoids.

The most remarkable constitutional feature of these iron – nickel – copper sulfide ores in their very low chromite content. We have already observed that many of the great layered mafic-intrusions and almost all alpine-type mafic intrusions contain substantial chromite. Many students of the mafic-ultramafic iron-nickel-sulfide ore type would consider the latter to be derived either from deeply differentiating layered intrusions or from ultramafic sources such as those that have provided alpine-type peridotites.

Jensen and Bateman (1979) considered the chromite deposits formation due to early magmatic (orthomagmatic and orthotectic process) in which segregation of early formed crystals by gravitative crystallization diffrentiation and injection of materials. Concentrated elsewhere by differentiation are included and the ore deposits are considered to have formed earlier than the rock silicates by sinking of ore minerals at the bottom of the magma chamber.

Magnetite, 'Ilmenite', chromite and some platinum deposits are considered late magmatic forming by residual liquid segregation and residual liquid injection at the close of the magmatic activity. The Ni-Cu sulfide-platinoid-Au-Ag deposits have formed by the process of immiscible liquid segregation at late magmatic stage.

C. COPPER-NICKEL DEPOSITS OF ULTRAMAFIC VOLCANIC ROCK ASSOCIATIONS: (ORE ASSOCIATION WITH KOMATIITES)

Copper-Nickel deposits of this association-represented by chalcopyrite-pentlandite-pyrrhotite ores are hosted by ultramafic high Mgo volcanic flows called komatiite in South Africa, Australia and Canada has led to a re-evaluation of basaltic volcanic Precambrian terrains. The ultramafic volcanic komatiites rocks are extrusive equivalents of peridoties, harzburgites and even dunites.

(1) Distribution of deposits

(a) **CANADA**: Alexo, the Marbridge, Near Timmins, Ontario.

(b) **SOUTH AFRICA**: Langmuir

(c) **ZIMBABWE**: Damba Prospect, Shangani Mines

(d) **AUSTRALIA**: Kambalda

(2) Mode of occurrence

All of above deposits involve lenses or sheets a few meters thick of pyrite-pyrrhotite chalcopyrite-pentlandite ore which are stratabound.

Within section of Komatiite flows. Kambalda deposit which is located in Archean rocks of Western Australia the gossan containing–nickeliferous limonite which lies at the lower contact of ultramafic Komatiites body that in turn lies between two metabasalts. Most of the ore is at the base of the komatiites in contact with metabasalt. At kambalda, the ore body occurs in a 3km thick sequence of Archean ultramafic-mafic volcanics in the Yilgara Craton. Post consolidation metamorphism thickned some ores and mobilized some into faults. The ore at Kambalda consists of both massive and disseminated sulfides; Canadian deposits are also associated with spinifex textrued komatiite flows including the classic outcrops in Munro Township. They occur in the lower ultramafic mafic platform basalts. The Langmuir Cu-N Sulfide mine was found in spinifex komatiites near Timmins, Ontario.

D. CHROMITE DEPOSITS ASSOCIATED WITH OPHIOLITES

(1) Mode of occurrence

Both ophiolites and Alpine periodotites involve the tectonic emplacement of slices of upper mantte and oceanic crust-mafic and ultra-mafic rocks formed at oceanic spreding centres into or onto continental crust. Ocean floor studies indicate that the lowerest layer 3 which is as much as 10 km thick is thought to be composed of vaguely layered medium to coarse grained, variably serpentinized dunites, peridotites, harz-burgites, gabbros, norites, troctolites and chromi-tites. "Steinmanns Trinity", the association that Steinmann mapped is rafts or slices of the 3 layers of the oceanic crust which are part of the assemblages that we now collectively call ophiolites.

Thus ophiolite complexes are layered assem-blages of mafic to ultramafic rocks with mantle – peridoties at the base, successively overlain by plutonic rocks (cumulate gabbro, norite, pyroxenite, metamorphic textured lherzolite, harzburgite and dunite), sheeted dykes, pillow laves and pelagic sediments, in other words, a slab of oceanic crust and upper most mantle.

Alpine complexes, along the Alps of southern Central Europe are important modes of occurrences of the ophiolite assemblages.

The chromite occurrence in ophiolites is atleast in 2 modes, as sacs or pod shaped masses in the tectonized periodotite upper mantle material or as discontinuous to quasi continuous layers in the highest temperature; most mafic basal cumulates of Layer 3. The former are typically smeared-out schlieren like lensoid bodies in dunite jackets; they can be seen to be part of a layered sequence. The layer 3 chromitities are normally more discernibly stratiforn, regular, laterally persistent, and with well preserved cumulate textures.

(2) Distribution of Chromite deposits

(i) Tiebaghi chromite deposit, Island of New Caledonia.

(3) Distribution of Nickel deposits

Ni ores associated with ophiolites are mostly – silicate – Ni constituting lateritic / saprolitic or garnierite variety produced by combined residual and chemical weathering process of concentration. Rare instances of small magmatic Ni $\pm$ Cu $\pm$ Pt – Group metals deposits / mineralization have been reported from the –

(i) Limassol Forest, of Troodos ophiolites Cyprus and (ii) Acoje from Zamboles ophiolites, Phillipines.

In Limassol Forest the ores are nikeliferous pyrrohotite which occur at the sheltered contact of gabbro and serpentinite. At stratigraphically higher zone, the mineral association, pyrrhotite-pentlandite-chalcopyrite, cubanite-vallerite, occasional troilite, Fe-Ni-Co arsenide, magnetite, chromite and rare ilmenite and rutile occurrence indicates early segregation of immiscible sulfides towards the end stage of formation of chromite deposits.

E. OPHIOLITE OF INDIA AND ASSOCIATED ORE DEPOSITS:

1. Distribution:

(a) The Indus Ophiolite Belt Ladakh (Tethys Belt)

(b) The Manipur-Nagaland Ophiolite Belt

(c) The Andman-Nikobar Ophiolite Belt

(d) The Shyok Ophiolite Belt Ladakh District (close to Indus Ophiolite Belt)

(e) The Malla Johar Ophiolite area of Himalayas.

2. Mineralization

Minor occurrences of chromite, magnesite clay, asbestos, talc and base metals have been recorded. No significant mineral deposit of any economic importance has so far been demarcated (Prasad, U., 1985, p. 13).

REFERENCES

Cousins, C.A; 1959: Generalized geologic map showing distribution of major rock types of the Bushveld Igneous Complex, engineering and Mining Journal, Vol. 160, No. 10, p. 114.

Cousins, C.A; 1964: The Platinum deposits of the Merensky Reef, in S.H. Haughton (ed.) "The Geology of some ore deposits in Southern Africa Vol. 11, pp 225-238. The Geol. Soc. South Africa, Johannesburg,

Dietz, R.S., 1964: Sudbury structure as an astrolabe, J. Geol. Vol. 72, pp. 412-434.

Jensen, M. & Bateman, A.M. 1981: Economic Mineral Deposits; John Wiley & Sons, Inc, New York, 593p.

Kullerud, and Yoder, H.S; 1963 and 1964: Sulfide-Silicate relations; Carnegie Inst. Wash. Ann. Rept. Dir, Gophys. Lab, Year Book 62 and 63, pp 215-218 and pp 218-222.

Mukherjee, A; 1999: Ore Genesis, A Holistic Approach; Allied Publishers Ltd., New Delhi, 657 p.

Philpots, A.R., 1990: Principles of Igneous and Metamorphic Petrology, Prentice Hall, Inc. Prentice Hall, India Pvt. Ltd., 1994, 498 p.

Prasad, U., 1985: Ophiolites of India, Rec. Geol. Surv. India, Vol. 115, pp 13-54.

Stanton, R.L.; 1972: Ore Petrology, McGraw-Hill Book Company, Inc. New York, 713 p.

Turner, F.J. and Verhoogen, J. 1960, "Igneous and Metamorphic Petrology, McGraw-Hill Book Company, New York, Indian Edition, 1987, 694 p.

Woodall, R. and Travis, G.A; 1969: The Kambalda Nickel Deposits, Western Australia; Ninth-Commonwealth Mining and Metallurgical Congress, London.

Wyllie, P.J., 1967: Ultramafic and Related rocks, John Wiley & Sons, Inc. New York.

Hon'ble Reviewers

1. **Prof. R.S. Sharma**
   Emeretus Professor
   Department of Geology
   Rajasthan University
   JAIPUR - 302 001
2. **Prof. K.K. Singh**
   Former Vice Chancellor VU
   702 Tech Apartments
   24 IP Ext. Patparganj
   NEW DELHI - 110 001

I-5

The Geological and Geochemical Environment of the major Gold Fields of South India

R.S.N. Sastry[1], K. Suryaprakash Rao[2], B.Srinivas[3] and A.Ravi Kumar[4]

1, 3, 4 Department of Applied Geochemistry, University College of Science, Osmania University, Hyderabad – 500 007

2. Professor of Applied Geochemistry (Retd.), Secretary, Indian Society of Applied Geochemists (ISAG), P.B. No. 706, Osmania University.

E-mail: rsnsastry@yahoo.co.in

Abstract

Archaean greenstone belts contribute, either directly, or indirectly, more than half of the world's gold production. In India, the entire gold production is exploited from the three greenstone belts of Kolar, Hutti and Ramagiri situated in the southern part of the Indian subcontinent.

Preliminary reconnaissance litho-geochimical surveys have been carried out in these three greenstone belts to understand the geochemical environment of the gold deposits & to establish a suite of pathfinder/ indicator elements useful in gold prospection.

Rock samples were collected from the Kolar, Hutti and Ramagiri mines and analysed for selected elements. Results indicate that tungsten as an element and as a mineral (Scheelite), arsenic, copper, silver zinc and to a lesser extent tin and fluorine can be applied as geochemical prospecting criteria for locating proximity to gold ore shoots. The geological and geochemical environment of gold deposits in the three greenstone belts have recognizable differences between them and therefore, the choice of pathfinder elements for each deposit are specialized and need to be selected based on the peculiarities of respected geochemical environment around the deposit.

Keywords: Gold deposits, Hutti, Kolar, Ramagiri, Geology, Pathfinder elements.

Introduction

The entire gold production of India is won from the three gold fields of Kolar, Hutti, and Ramagiri. Many investigators contributed to the understanding of these gold fields, few of them are Ananta Iyer (1967, 79, 80), Anhaeusser (1980, 81), Ghosh (1970), Giritharan et al (2001) Mukherjee (1985, 89), Natarajan 1986, 87), Narayanaswamy (1960), P. Nagabhushanam et al (2008) and Prabhakar Sangurmath (1994, 2001, 2003a and b, 2004, 2005, 2006), Prabhakar Sangurmath et al. (2001a and b), Roy (1982), Safanov (1980, 84), Suryaprakash Rao (1989a, b), Vasudev (1973), Ziauddin (1974)).

Most of the studies however have been oriented towards the genesis, petrography and geochemistry of the deposits. A comprehensive comparison to understand the different ore.

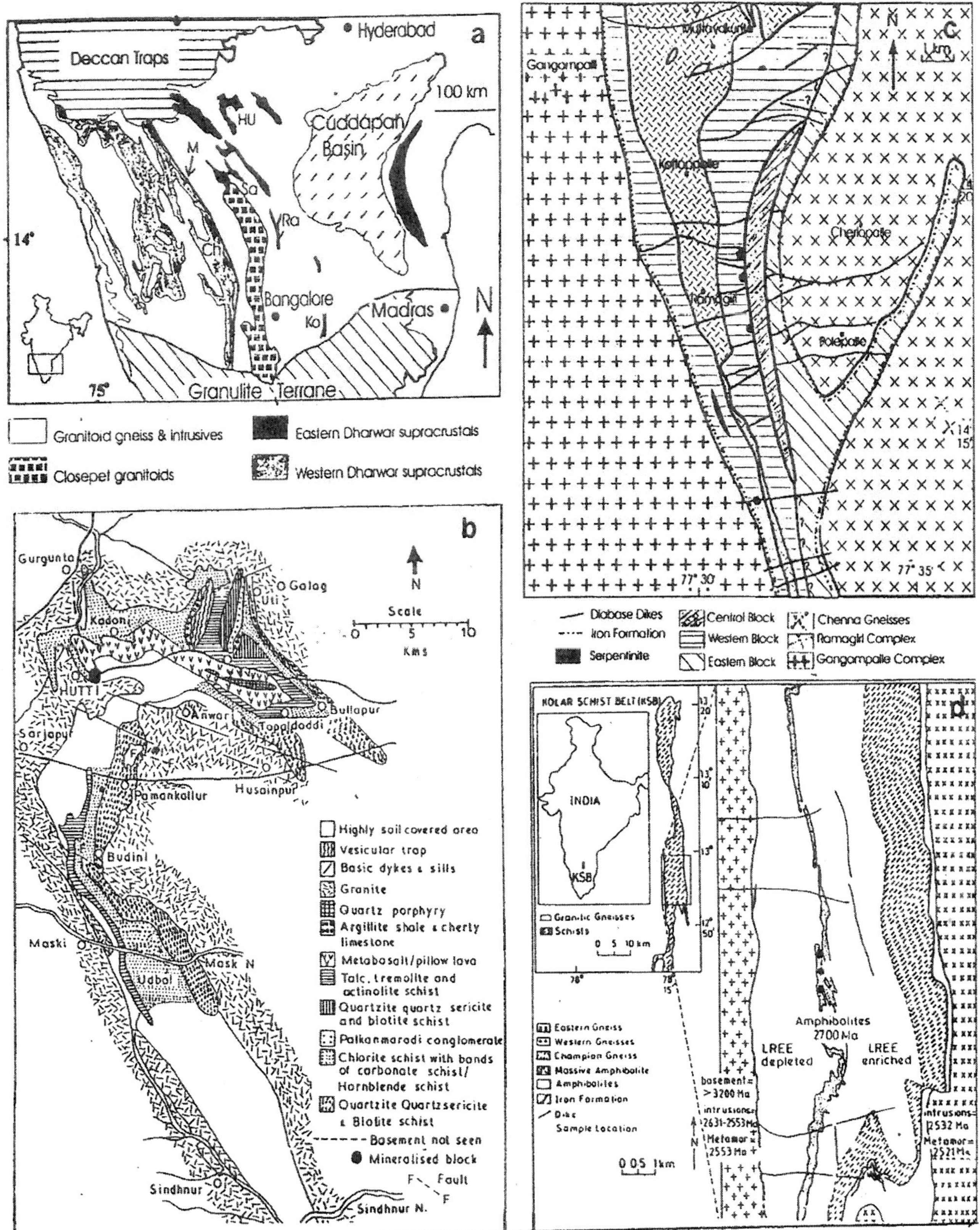

Fig. 1: (a) Geology of Southern India and Location of Kolar (Ko), Hutti (HU), and Ramagiri (Ra), (b) Geology of Hutti gold fields in Hutti Schist Belt, (c) Geology of Ramagiri Gold fields and (d) Geology of Kolar Gold Fields. (After T.S.Girithararan and V.Rajaramani, 2001).

environments and the trace element behaviour of these gold fields from an exploration point of view is attempted in this paper. Till recent times most of the exploration has been based on geological, geophysical and structural criteria in combination with searching for old workings. A systematic analysis of geochemical patterns for gold mineralization has not yet attempted in detail.

Preliminary investigations on the three gold fields of South India (Fig. 1a) have been carried out to understand the geochemical environment, and the behaviour of associated elements of gold distribution of trace elements across the auriferous veins, search for the most suitable pathfinder elements in the primary environment for locating potential ore zones

GEOLOGY AND MINERALIZATION

In southern India, a large part of the Precambrian shield area contains narrow bands of schistose rocks of the Dharwar system which are surrounded by gneisses and granite. The Dharwar rocks occur in wide areas in west near Chitradurga and are now recognized as of sedimentary origin. The present known gold fields seem to be limited to belts of a less sedimentary nature occurring in the east. These belts are elongated (50 to 100 Km) and narrow (1-5km) with predominantly NNW-SSE trends. The internal structure is fairly complicated by complex folding and shearing along longitudinal and transverse faults. A common feature is the presence of numerous cross-cutting dolerite dykes. The schistose rocks of these complex belts consist of the metamorphosed equivalents of mainly basic composition (greenstone). They have been subjected to alteration and now occur as phyllites, schistose chloritic metabasalts to pyroxene bearing amphibolite. Other components include former tuffs, and sediments of quartzitic, argillaceous, or carbonaceous composition, although the latter are rare.

There are several vein and placer gold occurrences in India. The primary gold deposits of India so far known today belong to the vein ore lode type. Kolar, Hutti and Ramagiri belong to this category. The salient features of gold deposits in the three schist belts of the eastern Dharwar craton (Giritharan, T.S. and Rajamani, V. (2001) are as follows.

	Kolar Schist Belt	**Ramagiri Schist Belt**	**Hutti Schist Belt**
Host rock	LREE enriched amphi- bolites and felsic schi- sts known as the Champion Gneiss.	LREE depleted to flat metabasalts represented by amphibolites.	Slightly LREE enriched to flat tholeiitic amphibolites
Wall rock altera- tion	Narrow zones of intense alteration with biotitization, albitization, sulfidation and carbonatization.	A wider zone (~ 200 mts) of not so intense alteration with sericitization, saussuritization, carbonatization and sulfidation of the host rock.	Narrow zones of intense alter- ation with bio- titization, albi- tization, carbonatization and sulfidation of the host schi- stose amphibolite.
Ore vein	Epigenetic Au-quartz carbonate vein with varying amounts of host rock fragments and minor sulfides.	A broad zone of chlorite-carbonate phyllites with sericitization and sulfidation.	Epigenetic Au-quartz-carbonate veins with varying amounts of host rock fragments and minor sulfides.
Mode of occur- rence	Fracture filled veins with disseminations of sulfides; gold occurring in native form in the quartz vein within the shear zones.	Fracture filling veins in the host actinolite-chlorite schists with minor sulfides. Gold occurs as free milling type within quartz vein and as inclusions in sulfides.	Fracture filling veins in the host amphibolite with minor sulfides. Gold occurs as both free milling gold in quartz and as inclusions in sulfides.
Sulfides	Pyrrhotite ± pyrite ± arsenopyrite.	Dominantly pyrite ± chalcopyrite.	Dominantly arsenopyrite ± pyrite ± pyrrhotite.

The gold mineralization at Kolar gold fields (Oorgaum) is situated in the easternmost belt of Bisanatham-Mallappakonda. The medium sized Ramagiri gold field is situated within the Penukonda schist belt. In the northern continuation of this belt fairly important gold mineralization is found at Hutti in the Raichur Doab. The origin of gold and sulphide mineralization in these belts is related most probably to the Precambrian lavas has been proposed by Anhaeusser (1975), Scherbakov (1967) and others for deposits in similar environments from other parts of the world. The geology and mineralization of individual gold deposits are described below in brief.

KOLAR GOLD FIELDS

The Kolar Gold Fields constitute an area of about 80 Km long and 3 to 4 km wide. The most important mine sites are located in the centre of this belt (Fig. 1d). The area is underlain by metamorphosed volcanic rocks, believed to have been extruded as basalt flows, into which sills and sheets of dolerite (and pyroxenite) are intruded. The environment of deposition seems to differ from that of Hutti and Ramagiri greenstone belts.

There are four mafic rock types in Kolar. The major part consists of schistose or massive fine grained amphibolite, representing basaltic and andesitic lava flows. Other types are granular and fibrous amphibolite, representing magma of doleritic composition. Some of the meta-basalts still retain pillow structure or are amygdaliod, despite the effect of later metamorphism There are no reports of meta-sedimentary components in the volcanic suites at Kolar.

Along the eastern margin of the Kolar schist belt, at Oorgaum, a zone of gneisses and auto-clastic conglomerate is exposed at the border with the peninsular gneisses. The Champion gneiss, containing quartz and pegmatites is considered by some) for instance de Smeeth, 1916), as the oldest part of the Dharwar system. In the western margin there is a highly schistose envelop around the mafic core of the belt and graphite containing quartz lodes have been detected here. Minor gold deposits occur in these marginal zones, but economically important deposits are confined to the central zone of the belt (Fig. 1d) which consists of metamorphosed basaltic lavas (amphibolites).

The major deposit in this belt is the champion quartz lode (sulphide poor). Other potential deposits are the Oriental and adjacent sulphide lodes with 5 to 15 percent combined sulphides, resulting in an ore not easily amenable for gold extraction. The major sulphide minerals are pyrite, pyrrhotite and arsenopyrite, sphalerite, and galena. Other minerals found in the veins or nearby in the wall-rock are cummingtonite, tremolite, tourmaline and occasionally scheelite. This observation opens up possibilities for application of multi-element geochemical exploration methods.

At Kolar, the paragenetic sequence seem to be the deposition of quartz followed by recrystallization of silicates (e.g., of diopside, biotite, and tremolite), followed by introduction and sulphides and finally mineralization of gold.

HUTTI GOLD FIELDS

The Hutti schist belt is about 80 km long and 3 to 25 km wide (Fig. 1b). The country rock consists of compact black fine grained rock with only few veinlets of quartz. It represents an altered metabasalt. Its major component is hornblende, minor components are epidote and occasionally chlorite. If not a simple basalt the original rock may have been a meta-spillite, pointing to submarine deposition as pillow lavas. The fine grained metabasalt also contains coarser grained granular amphbolite lenses, which could be remnants of coarser grained flows in the volcanic sequence. The incompatibility at the contact of coarser and finer flows might have acted as passage for later mineralizing fluids and this situation is common, though not universal.

Vasudev and Naganna (1973) have shown gold to be a later mineral in the sequence of paragenesis as it also occurs near calcite veinlets. Three generations of gold are delineated based on textural criteria. 1. Gold – pyrite association 2. Gold and arsenopyrite association and 3. Gold – Sphalerite association. It is also believed that the intense shearing of the host rocks produced schistose rocks which made the way for the mineralizing solutions. Silicification of the wall rock and an introduction of calcite also have occurred at about the same time.

RAMAGIRI GOLD FIELDS

The Ramagiri gold deposit is the only known gold mineralization in the centre of a 50 km long and bout 2 km wide schist belt striking almost N_S (Fig. 1c). Along a limited section of about 10 km gold mineralization of economic interest occurs in grey quartz veins which are localized in highly foliated schistose formation (phyllites) within a belt of essentially metavolcanics in the central part of a closed syncline. The metavolcanics are comprised of altered greenstones of basaltic and andesite composition. Lenticular bodies of meta-gabbro also occur. The greenstones probably represent sub-aqueous lava flows (Ghosh et al, 1970). In some places a volcanic textures and pillow structures and preserved (e.g., a few hundred meters to the west of Ramagiri mines).

The degree of metamorphism is lower than in the amphibolites found at Hutti and Kolar. This may indicate a different level of denudation. The centre of the belt is occupied by foliation and sub-vertical dips. The phyllite may represent a zone of desilicification of the greenstones as a possible source of quartz found in the gold reefs of the axial shear zones. Sericite-quartz schists and quartzites of possible sedimentary origin occur along the margins of the schist belt. The drag folds were produced with a northern plunge, and the position of the swarm of mineralized quartz veins and veinlets were determined due to the tectonic movement of the folded belt. The ore shoots are generally tongue shaped with the longest axis plunging north in an en-echelon pattern in dilation zones of folding. The gold-quartz deposition was result of a complex of geological events.

At Ramagiri, the gold mineralization is limited to the phyllite zone, ranging in width from a few to more than 150 m. Although gold mineralization is dispersed all over the phyllite zone, major concentrations occur in particular horizons. This is different from the setting at Kolar and Hutti where gold bearing veins occur in fracture zones at incompatible rock contacts finer and coarser grained amphibolite.

TRACE ELEMENT GEOCHEMISTRY

Lithological sampling has been carried out from Kolar, Hutti and Ramagiri gold mines across the strike of the gold bearing quartz veins and reefs. The samples have been analysed by XRF and Neutron activation analysis. Fluoride was determined by an Ion selective electrode method. The elements analysed are As, Sn, W, Ag, Cu, Zn and F. The analytical data together with sample location and description is presented in tables I, II and III.

In primary halos of epigenetic gold deposits in greenstone belts the elements most commonly accompanying gold, with the highest specificity as indicators, according to Boyle (1979) include, in their approximate order of importance:

In ore halos: SiO, S (mainly pyrite), Au, Ag, As, Sb, Te, Se, B, Bi, Hg, Mo, Sn, W, Cu, Zn, Pb, Co, and Tl.

Additionally, in wall rocks: F, Li, K, Rb, Ti, V and Cr.

Among the above elements, Au, Ag, As, Sn, W, Cu, Zn, f, Rb, Ti were analysed for the present investigations in addition to Se, Zr and P. Most of these elements are present in the Indian gold ore zones in the wall rocks.

In Kolar and Ramagiri (Table I&III) it is observed that As, W and Au occur in elevated amounts in contact of a principal gold bearing vein. The background value of Ag in the basic volcanic country rocks I very high ranging from 1.2 to 7.6 ppm. As per our earlier records Gold is present upto 1.2 ppm, 1 m away from the quartz veins. Its distribution is about normal for greenschist belts. The observed trends of Ag and Au are very similar to those of Yellowknife gold bearing greenstone.

As and W are found to occur in contact zone of mineralized veins. Sn concentration is slightly enhanced in the contact zone. Tungsten is evidently present in the Ramagiri deposits to some distance away from the contact. There seems to be less W in the samples of Oriental and Mundy lodes of Kolar gold fields. The fluorine content in contact zones of veins and in the surrounding country rocks is fairly irregular. However it decreases towards the major Ramagiri gold bearing quartz vein (Yappamanu mine).

The W, As, and Sn shows definite relation with contact zone. Similar is the case with Ag with respect to Kolar and Ramagiri gold fields and they form primary halos.

Rock samples from Hutti main gallery, 2nd level, were collected across the ore-bearing veins (from Oaklays to strike reef and further east). This includes closed spaced samples of wall-rock of gold-quartz veins but also intercalated rocks like amphibolite, meta-basalt and meta-spillites. Selected samples were analysed by AAS for elements As, Cu, and Ag. The results are presented in Table II. Arsenic is the most prominent indicator element. Its primary aureole does not exceed 10 to 20 meters. The dispersion trends of Cu, is fairly weak. Noteworthy is the high Cu content of 208 ppm in the volcanic component, intercalated in the metamorphosed sequence of basic rocks. The Ag content in contact rocks and normal country rock is high, but fairly irregular.

Table I: Trace elements in Oriental Lode,

KOLAR GOLD FIELDS							
S.N.	LOCATION	W	As	Sn	Cu	Zn	Ag
1	80 m west of lode	2	2	9	114	112	3.8
2	50 m west of lode	-	x	10	127	98	4,0
3	30 m west of lode	1	1	9	107	102	3.3
4	20 m west of lode	1	10	9	83	98	5.6
5	40 m east of lode	-	50	11	91	92	3.9
6	60 m east of lode	1	-	12	83	97	3.9

In Ramagiri the W, As, Sn shows additive halos where as the F shows Subtractive halo. In the case of Hutti gold mines the As shows additive halos and to some extent Ag shows Subtractive halos. In Kolar Gold fields Ag shows subtractive halos where as W and Sn shows additive halos. Though there are different geological and geochemical environments and trace element mobilities in Indian gold fields apart from Au we should look for As, W, Sn, Ag, and F.

Table II: Trace elements in the strike reef of Hutti Gold Mines

S.N.	LOCATION	As	Cu	Ag
1	1 m west of strike reef	3628	156	0.4
2	2.5 m west of strike reef	4040	132	0.8
3	5.0 m west of strike reef	828	84	1.2
4	10 m west of strike reef	10	160	1.2
5	20 m west of strike reef	26	120	0.8
6	30 m west of strike reef	2	24	0.8
7	50 m west of strike reef	5	156	4.0

Table III: Trace element data of Yappamanu Mine, Ramagiri Gold Fields

SAMPLE NO.	LOCATION	W	As	Sn	F
1	Ore Zone	35	105	10	55
2	1 m away from ore zone	20	30	13	95
3	5 m away from ore zone	20	22	11	60
4	10 m away from ore zone	17	35	7	75
5	15 m away from ore zone	10	1	5	550

DISCUSSIONS AND CONCLUSIONS

Gold mineralization in quartz reefs and veins occur mainly in the central part of elongated greenstone belts in South India. In the case of Ramagiri, Hutti and Kolar, economic mineralization is limited to relatively short section in the centre of the belts and not at the peripheries. The granitic intrusions in the greenstone belts act as heat engines supplying heat to the greenstone belts and thus cause the mobility of the elements and concentrate in respective receptacles. Thus the disseminated gold have been mobilized and concentrated in quartz bodies acting as receptacles.

The lithological environment and mineralogical composition of mineralization in the Precambrian greenstones of India are fairly similar to those of other greenstone belts as in Canada, Australia and Africa.

The deposits at Ramagiri exhibit a different geological environment (phyllite) than epidote-amphibolite wall rock at Hutti and Kolar. The surface targets for geochemical exploration for ore deposits are fairly elongated having extremely small width. Moreover, the metasomatic aureoles seem to be narrow and variable with respect to their geochemical pattern.

The accompanying major sulphides vary from mainly pyrite to arsenopyrite. Minor sphalerite occurs sporadically, chalcopyrite is occasionally found. The amount of total sulphides in the Au-bearing sections, ranges from practically nil to several percent (otherwise sulphides could constitute a safe and simple guide to gold).

The choice of suitable indicator elements for localizing gold quartz veins in the Indian greenstone belts seems to be limited. Significant and diagnostic value can be attributed to primary halos of As, Ag, Cu and to a lesser extent W, Zn, and perhaps F. Additionally, tungsten in the form of Scheelite mineral appears to be promising as prospecting criteria (Rao K S P et al., 1989).

The choice of pathfinder elements for each deposit is specialized based on geochemical environment in addition to geological environment.

Acknowledgements

The authors are grateful to Prof. R.D. Schuilling, Geochemist, Utrecht University, The Netherlands, Dr. M.G. Oosterom, Geochemist, Netherlands, and Dr. J.B. de Smeth, Geochemist (ITC, Enschde), The Netherlands for helping in sampling and analyzing the samples in Netherlands.

REFERENCES

Ananta Iyer, G.V. and Vasudeva Murthy, A.R. (1967): Geology, Geochemistry of Kolar Gold Gield and the origin of gold-quartz veins in Dhrwars of Mysore., Preceedings of upper mantle project, Hyderabad, PP. 231-238.

Ananta Iyer, G.V. and V.N.Vasudev (1979): Geochemisgtry of Archaean metavolcanics rocks of Kolar and Hutti Gold fields, Karnataka, India. Jour. Geol. Soc. India., vol. 20, pp.419-432.

Ananta Iyer, G.V., Vasudev, V.N., and Jayaram, S. (1980) Rare Earth geochemistry of metabasalts from Kolar and Hutti Gold fields, Karnataka Craton India, Jour. Geo. Soc. India, Vol. 2. PP. 603-608.

Anhaeusser, S.R., (1981): The relationship of mineral deposits to early crustal evolution. Eco. Geol. 75th anniversary, pp 42-62.

Anhaeusser, C.R., Fritz, K., Fyfe, W.S. and Fill, R.C.P., (1975): Gold in primitive volcanics. Chemical Geology, Vol. 16, pp. 129-135.

Ghosh, D.B., B.B.K. Sastry, Rao, A.J., and A.A.Rahim. (1970): Ore environment and ore genesis in Ramagiri Gold Fields, Andhra Pradesh, India. Eco. Geol. Vol. 65, pp. 801-814.

Giritharan, T.S. and Rajaramani, V. (2001): REE Geochemistry of ore zones in the Archaean auriferous schist belts of the Easter Dharwar craton, Proceedings of Indian Academy of Sciences (Earth and Planetary Sciences), Vol. 110, No. 2, June, 2001, pp 143-150.

Mukherjee M.M., (1989) Evolutionary aspects of gold mineralization in Southern part of Kolar schist belt., Indian Minerals, Vol.43., No. 5., pp. 347-360.

Mukherjee, M.M. and Natarajan, W.K. (1985): Exploration of gold in Kolar schist belt, India, in gold exploration and development. (U.N. Seminar on gold at Bangalore).

Natarajan, W.K., Shashidharan, K. (1987) : Economic Potentiality of new gold mineralization in South Kolar schist belt. Jour. Geol. Soc. India., Vol. 29., pp. 230-239.

Natarajan, W.W. and Shashidhran, K. (1986): Tectonically controlled mineralization in Chigaragunta area South Kolar Schist Belt, Chittor district, Andhra Pradesh, Jour. Geol. Soc. India., Vol.27. pp. 517-526.

Narayanaswamy, S., Ziauddin, M., and Ramachandra, A.V., (1960): Structural control and localization of gold bearing lodes, Kolar gold field, India., Eco. Geol., Vol. 55, pp. 1429-1459).

Nagabhushanam P., Prabhakar Sangurmath, M. L. Patil and B. S. Sukhija (2008): Ancient gold mining activity in the Hutti-Muski greenstone belt, Karnataka, India: Radiocarbon perspective, Current Science, Vol. 95, no. 3, 10 august 2008

Krishna Kumar N.and Prabakar Sangurmath (2001) : Heavy mineral assemblage of Mannucheeni-Maruda area, Wayanad, Nilambur Granulite terrain, South India. Quarterly Journal of the Geological Association and Research Centre Vol. 9 and No. 1, March 2001

Prabhakar Sangurmath (1994): Modelling of gold occurrences in the Archaean sediments of Gadag Schist belt, Karnataka, India. Quarterly Journal of the Geological Association and Research centre Vol.2, No.1.

Prabakar Sangurmath (2001a): Uti Gold deposit – Evolving scene Hutti-Muski Greenstone Belt, India, Geological Survey of India Special Publication No.58., pp289-292.

N. Krishna Kumar and Prabakar Sangurmath (2001b) : Heavy mineral assemblage of Mannucheeni-Maruda area, Wayanad, Nilambur Granulite terrain, South India. Quarterly Journal of the Geological Association and Research Centre Vol. 9 and No. 1, March 2001

Prabhakar B.C., Jayananda M., Mohamed Shareef and T. Kano (2009): Petrology and Geochemistry of Late Archaean Granitoids in theNorthern Part of Eastern Dharwar Craton, Southern India: Implications for Transitional Geodynamic Setting Journal Geological Society of India Vol.74, September 2009, pp.299-317

Prabhakar Sangurmath (2003): Detailed Exploration for Epigenetic Gold deposit: A Case History : Hira-Buddini Gold deposit. Hutti-Muski Schist Belt, India. 19th world mining congress 1-5 Nov, New Delhi, Mining in the 21st Century Quovadis.

Prabakar Sangurmath (2003): Amphibolites of the Hutti-Muski Supra Crustal Belt, India. Indian Journal of Geochemistry, Vol.18, pp 77-82.

Prabakar Sangurmath (2005): Geolocy and gold mineralization in Buddini Gold deposit – Hutti-Muski Greenstone Belt, Karnataka. Journal of Geological Society of India, Vol. 66, Nov. 2005, PP 552-560.

Prabakar Sangurmath (2004): Geochemistry of the Greenstone – Hosted Gold deposit of Buddini, Hutti-Maski Greenstone belt, India, Journal of Applied Geochemistry Vol.6, No.6, pp 235-244.

Prabakar Sangurmath (2006): Geochemical studies of the alteration zones of Buddini gold deposit in the Hutti-Muski greenstone belt, India. Journal of Applied Geochemistry, Vol. 8, No. 2A, July, 2006, pp458-470.

Nijalingappa R.and P.S. Sangoormath (2004): Metamorphic model for the Tuppadhur – Buddini Gold block in the Hutti-Maski Greenstone belt, Karnataka craton in the South Indian shield – Implications in Exploration. Journal of Economic Geololgy Vol.1, No.1, pp 71-64.

Surya Prakash Rao K, Sastry R S N, and Raju S V (1989 a) Scheelite as a prospecting tool for gold in the Ramagiri greenstone belt, Andhra Pradesh, India; *Journal of Geochemical Explo-ration.* **31** (307-317)

Roy., A., and Biswas, S.K., (1982): A note on the "Volcanic conglomerate/pyroclast horizon near Palkanmardi in the Hutti Schist Belt, Karnataka, Jour. Geol. Soc. India., Vol. 23., pp.365-478.

Safanov, Yu. G., Genkin, A.D., Vasudev, V.N., Krishna Rao and Ananta Iyer, G.V. (1984): Genetic features of gold ore deposit at Kolar Dharwar craton, India, Jour. Geol. Soc. India. Vol. 25. No.3.

Safanov, Yu, G., Radhakrishna, B.P., Krishna Rao, B., Vasudev, V.B., Krishnam Raju, Nosik, L.P., and Pashkov, Yb. (1980): Mineralogical and geochemical features of endogenic gold and copper deposits of South India. Jour. Geol. Soc. India. Vol. 21 pp. 365-478.

Suryaprakash Rao K., R.S.N. Sastry and S.V. Raju (1989b): Results of Geochemical soil survey in Ramagiri gold field, A.P., India. Transactions of the Institution of Mining and Metallurgy, Section B. Applied Earth Science. Transactions volume 98, January-April (1989), published by Institution of Mining and Metallurgy, 44 Portland place, London.

Hon'ble Reviewers

1. **Dr. Fareeduddin**
   Geological Survey of India
   VAsudha Bhawan
   BANGALORE - 560 078
2. **A. Sundramoorthy**
   Director General
   Geological Survey of India
   27, Jawaharlal Nehru Road
   KOLKATA - 700 016

I-6

A Study of Types of Iron Ores in Joda Block of Keonjhar, Orissa

S.H. Adil and P.S.Jena

Centre of Advance Study, Department of Applied Geology, Dr. H.S. Gour University, Sagar, India

Abstract

The BIF and its associates ore are belonging to the Keonjhar-Bonai synclorium which has been defined after the study. The Joda Iron Ore contains world-class high-purity hematite ore bodies hosted within banded iron formations. Hematite ores are always hosted by the Iron Formation, and ore formation resulted from a multistage, sequential removal of gangue minerals from the host, giving rise to residual concentration of iron. The first, hypogene, stage of ore formation removed silica only, leaving a thinned residue enriched in iron oxides, carbonates, magnesium silicates, and apatite, with no apparent change in the oxidation state of the iron minerals. The hard laminated ore has been formed in the second phase of supergene processes, where the deep burial upgrades the hydrous iron oxides to hematite. The massive ore is syngenetic in origin with BHJ. Soft laminated ores and biscuit ores were formed where further precipitation of iron was partial or absent. Blue dusts are formed by the complete leaching of silica. Lastly the canga ore are formed due to deposition. Canga ore are recemented products of Iron ore and laterite.

Keywords: Iron Ores, Keonjhar, Blue dust, Hematite, Laterite.

Introduction

Joda area is one of the main economical areas for its well known Iron Ore deposits. The main purpose of the study is to know the details about the geologic setting, and ore types of Joda Block. The Joda Iron ore contains several large high grade Iron-ore deposits. The three main high grade deposits of this area are Longalotta, Joda East and Thakurani reserve. This ore is having high Fe content upto 65%. All these ores are hosted by Precambrian BIF. In 1904 this area Iron ore deposit has been discovered.

Previous studies in this area have focused on the nature and origin of the BIF protolith. Dunn (1935) worked upon the origin of Iron Ore in Singhbhum area. J.A.Dunn worked upon the stratigraphy of south Singhbhum in (1940). Majumdar and Chakraborty studied the Stratigraphy and structure of the Precambrian Banded Iron Formation and chromite bearing ultramafic rocks of Sukinda valley, Orissa in (1980). Chaterjee and Mukherjee (1981) studied the structural set up of Malantoli Iron Ore deposits. Majumdar, Chakraborty and Bhattacharya studied the geochemistry of banded Iron Ore formation of Orissa (1982). Banerjee (1977) worked on the Precambrian banded iron formations and the manganese ores of the Singhbhum region, eastern India. Banerjee (1974) works on the stratigraphy and tectonic history of Iron Ore-bearing and associated rocks of Singhbhum and adjoining areas of Bihar and Orissa. Gutzmer, Mukhopadhyay and Buekes. Genesis of High-Grade Iron Ores of the Archean Iron Ore Group around Noamundi (2008). Venkatesh and Subrat Roy studied the mineralogy and geochemistry of Banded Iron formations and iron ores from eastern India with implications on their genesis in (2009). In this contribution, we aim to develop a metallogenetic model for the high-grade iron ore deposits of the Noamundi area. It is based on investigations at several large operating open-cast mines in the area, including the Longlotta, Murgabeda, Malangtoli, Mahaparvat, Jajang and Joda mines of TISCO. The ores mined in these operations can be broadly categorized into six major types: hard hematite ores, friable or soft hematite ores, Lateritic Ore, Shaly Ore, Canga Ore and Blue Dust. Nearly 65% of the ore those are mined in this area are Soft Laminated Iron Ore (SLIO). Nearly 25% ores are hard Iron ore.

Stratigraphy

The entire deposits are belonging to the Iron Ore group. Jones (1934) studied the iron-ore deposits of Bihar and Orissa. Acharya (2008) studies the Genesis of Banded Iron Formations and associated iron and manganese ore deposits of Orissa-Jharkhand region, India and its role as exploration guide. According to Acharya (1993) he has divided the Iron Ore group as western, eastern and southern Iron Ore group. These are well exposed along the eastern, western and southern perimeters of the Singhbhum granitoid. Acharya (1976) studied Iron Formation and iron ores of Orissa, their Stratigraphy and correlation. This group has similar lithostratigraphy comprising mafic and acid volcanic unit, shale and BIF with sub-ordinate banded cherts and carbonate beds. Three prominent provinces comprising BIFs such as Badampahar-Gorumahisani-Suleipat belt, Daitari-Tomka belt and Bonai-Keonjhar belt constitute the Iron Ore Supergroup of Odisha, which have been designated as BIF-I, BIF-II and BIF-III respectively.

An iron ore deposit occurs in all three green stone belts. Joda area deposits come under eastern Iron Ore group. The Archean-proterozoic banded Iron ore formation of Iron Ore Super Group is well developed in Keonjhar-Bonai belt. Keonjhar-Bonai synclorium is popularly known as Horse-Shoe synclorium and is the younger tier classification of BIF. It is known as BIF-III. Field investigations has been done in most of the areas like Murgabeda, Deojhar, Joda East, Jajang, Mahaparvat and Thakurani area which is coming under Bonai- Keonjhar belt in the Keonjhar district, Odisha and located in Survey of India toposheet numbers 73 G/5, 73 G/9, 73F/8, 73F/12. (22.06 degrees) 22°3' 35" North of the Equator and longitude (85.32 degrees) 85°19' 11" East of the Prime Meridian on the Map of the world. S.N.Acharya studies the structure and stratigraphy of this basin. A good amount of details on the infra-, intra- and supra- BIF nature of ore deposits have been known. BIF is considered as the litho marker in Iron Ore Super Group. Banded Hematite Quartzite (BHQ), Banded Hematite Jasper (BHJ), Banded Hematite Shale (BHS), Banded Shale, Banded Manganese Formation (BMnF), and Ferruginous Shale and Iron ore bodies. In BHQ and BHJ alternate bands of iron mineral and silica of varying thickness can be observed. At the base of BIF banded shale is present with different colors. The ferruginous shales younger to BIF occur discontinuously in the area. The iron mineral in the basin is predominately oxide facies. In this area hematite, martite and goethite can be observed. Hematite is mainly present. The mineral assemblages reveal primary sedimentary signatures and subsequently modified by digenesis and metamorphism. These iron ore can be formed mainly due to martyrisation, goethitisation and hematitisation. Hematite is mainly dominant ore present in this area. Chert, Jasper and Quartz are the silica minerals.

Local Geology of the Area

The Older Metamorphic Group, found in Champua (220 N, 850 40' E) in the central part of the basin, is the oldest (3200 Ma) rock formation and it consists of metasediments associated with basic intrusions (Sarkar 1977). Rocks of Older Metamorphic Group, comprising predominantly schists, forms the basement rocks. Hornblende schist and chloritic schist are exposed as patched in the southern part of the study area. Singhbhum granites are exposed at the eastern part of study area. The basal conglomerate which marks an erosional unconformity is designated as Raghunathpally conglomerate and contains pebbles of vein quartz, grey quartzite, and carbonaceous quartzite. Iron ore bodies, up to 3 km long along strike and several hundreds of meters wide (depending upon the dip of the strata), occur at various localities all along strike of the main BIF unit of the Lower Bonai Group (includes BIF-bearing supracrustals of Bonai–Keonjhar area which is co-relatable with the Noamundi Group of Banerji, 1974). The deposits appear strata bound and are mainly situated in the lower part of the iron formation immediately above the transition zone into underlying shale The Kiriburu and Meghatuburu deposits investigated during this study are located along the top of a steep mountain range comprised of iron formation that marks the overturned western limb of the Keonjhar-Bonai Synclorium. The Noamundi and Joda deposits, on the other hand, are located atop a less prominent range of hills defined by the same iron formation along the normal eastern fold limb of the Keonjhar-Bonai synclinorium. Shale, manganese and volcanics overlie the BIF-bearing horizons in this sequence. The thick and extensive mafic volcanic rocks flanking the Iron ore Group of rocks of the 'Horse Shoe Synclinorium' of Bonai-Keonjhar region, along its western and southern parts are designated as Bonai lava (Malongtoli lava). The volcanics/lavas are exposed over 800 sq. Km. area (Saha, 1994) to the west of Kendujhargarh and the north of Palalahara (21 28': 85 15'). The lavas are in general quartz-normative tholeiitic basalts – basaltic andesite – anderite in composition. The lavas are vesicular and the vesicles are filled up by quartz, chlorite – zeolite and chert. The lavas are, at places, intercalated with shale and are associated with ash beds and agglomeratic rocks. Close interbedded / intercalated relationship of lava and shale probably indicates a sub aqueous origin for the lava. An area of about 200 sq. Km. lying south of Jagannathpur at the Jharkhand – Orissa border and at the NE corner of Horse Shoe is occupied by Synclinorium composed of extensive flat lying lava flows – designated as Jagannathpur/Dangoaposi Lava, which are very weakly metamorphosed, locally basaltic rocks of having tholeiitic, andesitic in composition. The bulk composition of the basaltic rock is andesite tholeiite with differentiates of oligoclase andesites. The Kolhan Group is represented by an undeformed platformal sequence comprising basal polymictic conglomerate, fine to medium grained purple quartzite-sandstone, gritty quartzite with some shale slate, and minor limestone. This exposed well at the north-eastern part at the study area. The polymictic conglomerate at the base contains pebbles of banded iron formation, quartzite, and vein quartz and at places iron ore.

Ore Types of the area

The Hematite iron ore is the main ore which is used in commercial purpose because of its high Fe %. This hematite ore is mainly of six variety found in this area. They are HIO (Hard Iron Ore), SLIO (Soft Laminated Iron Ore), Canga Ore, Shaly Ore, Lateritic Ore and Blue Dust (Powder Ore). The recovery % is higher in Hard ore and there is very minor recovery % in case of the blue dust. There is lateritic ore; float ore are present over the hard ore body. Their thickness varies from 1-10m. There are joints and fractures can be noticed in between the ores, if we observe it clearly. These fractures are filled by the limonite and shale. In some cases due to more infilling of shale and limonite the recovery % of the ore decreases during processing.

Ore Types	Characteristic	Fe Content in %
Massive/ Hard	Often hard and compact, widely jointed	64-68
Soft Laminated Ore	Soft, porous, friable and often biscuit	62-65
Shaly Ore	Shaly texture and brittle	50-55
Lateritic Ore	Product of alteration of the former types, porous	56-58
Blue Dust	Fine grayish blue, occurs in bands and pockets	65-68
Canga Ore	Brecciated flat beds with subangular hematite boulders and embedded in lateritic matrix	58-60

Hard Iron Ore (HIO)

The most important belt is the classic Bonai-Keonjhar horse shoe shaped iron ore synclinorium of the Jamda- Koira valley, which occurs over the western fringe of Singhbhum craton, Keonjhar and Sundergarh districts. The elongated horse shoe shaped synclinorium trends NNE-SSW, open on NNE and is spread over an area about 60km X 25 km. The eastern limb features Thakurani, Joda east, Joribahal, Jiling-Langalotta, Jajang and Malangtoli. The western

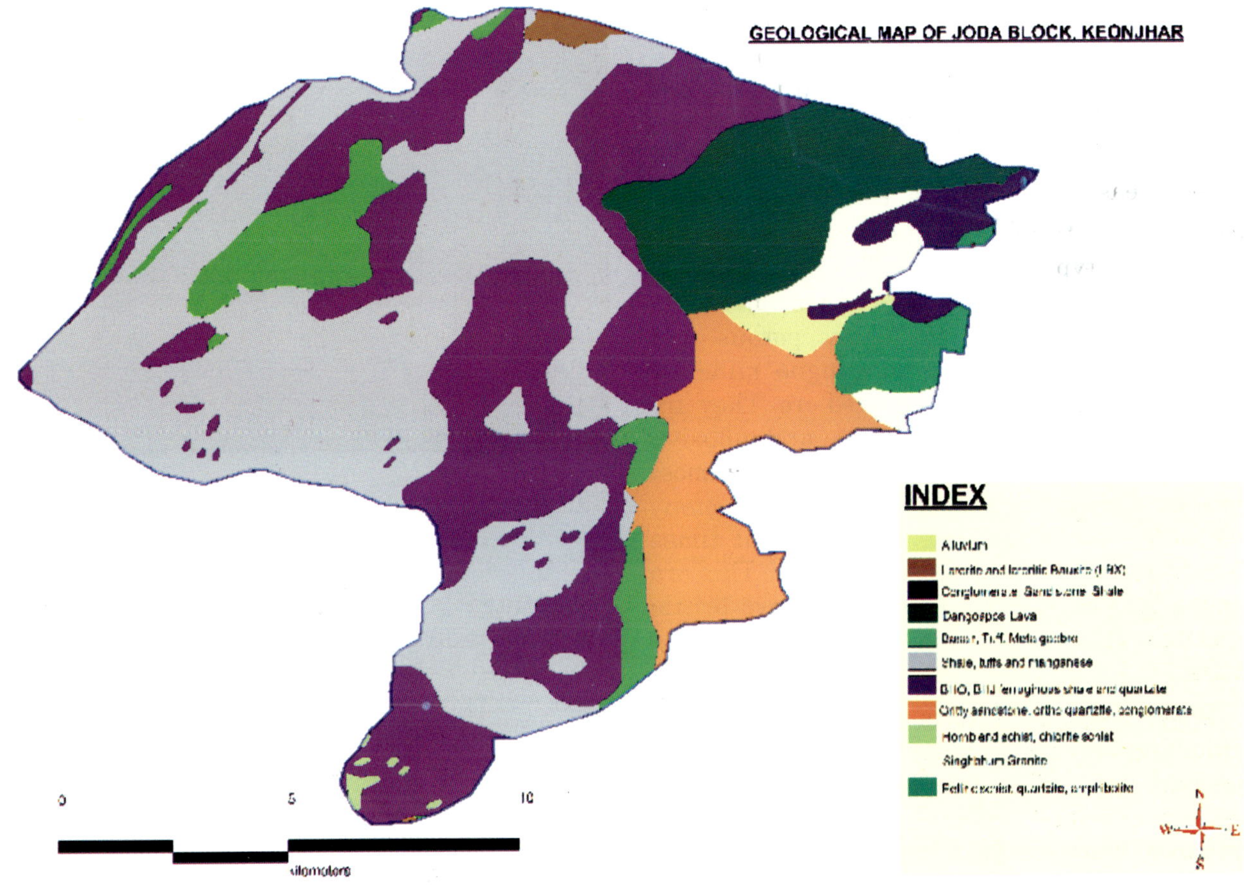

Geological Map of the study area (Joda Block). (After, Bhattacharya, 2010.

limb features Kiriburu, Bolani, Kalta, Barasua and Khandadar deposits. The top of the ridges of the synclinorium contain high grade ore. The hard iron ore is present at the top part of the ridges. The grade of the iron is very high. Hard iron ore colour is steel grey. In eastern limb of the horse shoe synclorium a large quantity of hard ore is exposed with less tripping ratio. It is characteristically porous (30% porosity on average), has a high lump yield and has low contaminants such as silica, phosphorus, aluminum, and alkalis. It consists essentially of randomly oriented fine-grained platy hematite and martite. Iron grade is high, in excess of 68 percent, and phosphorus is low, averaging less than 0.05 percent.

Hard Iron ore are abundant at the lower part of the Iron ore formation. The hard Iron Ore formed at the expense of the iron formation host rock by replacement of all mineral constituents, including quartz (chert), by hematite. The laminated iron ore, iron ore rich mesobands of the iron formation protolith have been transformed into dense, fine grained iron oxide and former cherty mesobands are replaced by a porous assemblage of fine-grained iron oxide. There is some magnetite present with the predominant hematite for which this ore is slightly magnetic in nature. These ores shows cross cutting contact with the bedding. This hard iron ore is strata bound deposit.

Soft Laminated Iron Ore

Soft Laminated Iron ore is formed by the supergene enrichment of oxidized and partly mineralized iron formation in the saprolite zone beneath a veneer of laterite and goethite. The saprolitic ore having higher porosity is mainly due to the leaching of the gangue components. The preservation of the bedding nature in this ore types suggests its saprolitic nature. This ore are present below the hard iron ore in most part. In few areas this ores are present below the laterite. The grade of this ore is Fe (62-65%).

In this type of ore, individual lamellae measure from a few millimeters to centimeters in thickness and the ore is highly porous and fragile. The principle ore and gangue minerals are same as that of hard laminated ore. They are relatively rich in goethite, kaolinite and gibbsite. The texture of soft laminated ore is almost similar to hard laminated ore but there are many voids between the lamellae, which are at times filled with secondary goethite and clay. This ore is more flaky in nature. This ore is more brittle in nature because of the very thin bedding nature.

Major deposits of this ore are present in Serajuddin, Jajang, Malangtoli, Mahaparvat and Bamebari area. Flaky-friable ores are brown to grey in colour and varies in characteristics from laminated, biscuity, flaky to friable, powdery in nature.

Shaly Ore

Shaly iron ore are present in most part of the area. This ore has no economic value. The grade range between 50-55% (Fe) content. This ore is very soft it is present below the friable ore unit. A very good exposure of shaly iron ore can be observed in Deojhar Iron Ore mines. This ore bearing zone has been affected by intense chemical weathering.

Canga Iron Ore

In near-surface environments, ores are recemented by the formation of goethite. Mechanical erosion of iron ores and host rocks is expressed by the accumulation of modern canga deposits. The canga ore has economic grade due to the higher % of hard iron ore boulders. Some canga iron ore grades can be nearly 60%. In this case the cementing material is also hematite ore. The low grade canga ore is due to the lateritic cementing material. This canga ore are present at downward or valley portion of the ore deposits and they are accumulated due to river action. In Murgabeda and Thakurani area this ores are predominantly present at the valley portion. Best exposure of canga ore can see just above the Murgabeda Temple inside Murgabeda Iron Ore Mines.

Lateritic Ore

Lateritic Iron ore are present on the top part of the ridges. This ore are low grade. There is not a constant grade in such deposits and grade variability can be observed. The thickness of laterite ore vary from 1-10m in many areas. Lateritic ores, known as soft ores are the products of lateralization of primary ores and hence their composition varies widely depending upon the nature of primary ore and degree of lateralization. They are usually vesicular or porous and observed as capping over ore bodies. The steel grey colored hard, massive, sometimes finely laminated ores found below lateritic ore, are characterized by their compact and homogenous nature. Goethite-lateritic ore is highly spongy and porous in nature. Goethite occurs as colloform bands and vein filling within the voids and cavities which is resulted due to the leaching out of preexisting minerals.

Blue Dust

Powdery ores are fine in nature and often closely associated with blue dust, ferruginous and shaly materials. Blue dust occurs as fine, steel grey colored powder with a few flaky ore pieces and generally found below hard massive ores in association with BHJ It occurs in limited extent at deeper levels of the mineralized zones. BHJ consisting of laminations of fine grained hematite and jasper vary in thickness from mere partings to a few centimeters. Blue dust is steel grey in colour and occurs as minor pockets and the patches which are randomly oriented along major fractures or joints or near the top of the ore horizon as thin but persistent beds. The blue dust is mostly a fine powdery material mainly composed of micro play hematite and martite. Presence of martite in blue dust indicates their origin by desilisification process from banded iron ore.

Conclusion

The Iron Ore formation of the Joda block belongs to the Keonjhar-Bonai Synclorium. The different types of Iron ore of this area are inticrately related with the BHJ. Hard ore, Soft Laminated, Goethite and Blue dust ore are formed due to the result of disilification process through the action of hydrothermal fluid. The hard laminated ore has been formed due to the supergene process, where the deep burial upgrades the hydrous iron oxides to hematite. Soft Laminated Iron ore/ Biscuit ore were formed where the precipitation of iron was partial or absent. The leached out space remains with the time and the ore becomes very fragile within lamina. Blue dusts are formed due to silica leaching from protore.

Reference

Acharya, S. (1976): Iron Formation and iron ores of Orissa, their Stratigraphy and correlation. Proc. Symp. Geol. etc. Ferrous and Ferroalloy Min., Bangalore, pp. 86-100.

Acharya, S. (2008) Genesis of Banded Iron Formations and associated iron and manganese ore deposits of Orissa-Jharkhand region, India and its role as exploration guide in 2008

Banerji, A.K. (1974): On the stratigraphy and tectonic history of Iron Ore-bearing and associated rocks of Singhbhum and adjoining areas of Bihar and Orissa. Jour. Geol. Soc. Ind., V.15, pp. 150-157.

Banerji, A.K. (1977): On the Precambrian banded iron formations and the manganese ores of the Singhbhum region, eastern India: ECONOMIC GEOLOGY, v. 72, p. 90-98.

Chatterjee, A. and Mukherjee, P. (1981): The structural set up of the Malangtoli iron ore deposit, Orissa. Jour. Geol. Soc. Ind.V. 22, pp 121-130.

Chakraborty, K.L., Chakraborty, T.L. and Majumdar, T. (1980): Stratigraphy and structure of the Precambrian Banded Iron Formation and chromite bearing ultramafic rocks of Sukinda valley, Orissa. Jour. Geol. Soc. Ind., V.21, pp. 398-404.

Dunn, J.A. (1935): The origin of iron ores in Singhbhum, India; *Econ. Geol.* 30 643–654.

Dunn, J.A. (1940): The stratigraphy of south Singhbhum, Mem. Geol. Surv. Ind., V.63, pp. 63(Pt.3).

Gutzmer, J., Mukhopadhyay, J., Beukes, N.J., Pack, A., Hayashi, K. and Sharp, Z.D. (2008), Genesis of High-Grade Iron Ores of the Archean Iron Ore Group around Noamundi, Inc. *Economic Geology,* v. 103, pp. 365–386

Jones, H.C. (1934): The iron-ore deposits of Bihar and Orissa. Mem. Geol. Surv. Ind., V. 63, Pt.2, pp. 167-202.

Majumder, T., Chakraborty, K. L., and Bhattacharyya, A. (1982): Geochemistry of banded iron formation of Orissa, India: Mineralium Deposit, v. 17, p. 107–118.

Mukhopadhyay, J., Buekes, J.N. and Gutzmer, J. (2006): Genesis of high grade Iron Ore of the Archean Iron Or Group around Noamundi, India.

Roy, S. and Venkatesh, A.S. (2009): Mineralogy and geochemistry of Banded Iron formation and Iron Ores from eastern India with implications on their genesis.

Saha, A.K. (1994): Crustal evolution of Singhbhum and north Orissa, Eastern India. Mem. Geol. Soc. Ind., V. 27.

Sarkar, S.N., Saha, A.K. and Miller, J.A. (1969): Geochronology of the Precambrian rocks of Singhbhum and adjacent regions, Eastern India. Geol. mag., 106, pp. 15-45.

Sarkar, S.N. and Saha, A.K. (1977): The present status of the Precambrian stratigraphy, tectonics and geochronology of Singhbhum-keonjhar-Mayur bhanj region, eastern India. Ind. Jour. Earth Sci. (Prof. S. Roy vol.), pp. 37-65.

Hon'ble Reviewers

1. **Prof. S. Acharya**
   Former Vice Chancellor, VU
   155, VIP Colony Nayapalli
   BHUBANESHWAR - 751 015
2. **Prof. S. Ibotombi**
   Department of Earth Science
   Manipur University
   IMPHAL - 795 003

I-7

Biogeochemical Prospecting of Pb-Zn Mineralization at and Around Zawar deposit, Using Floral species as Indicators

S.K. Trivedi and Kirti Chowdhary

Department of Geology J.N.V. University,Jodhpur, - 342 001-Rajasthan
*E-mail: skt.geology@gmail.com

Abstract

The use of Floral Species has been considered as a tool for Biogeochemical prospecting in the present study. Some plant species have high capacity to uptake metals in their organs among others, so they can use as a guide or good indicator plants for particular mineral. Biogeochemistry involves the analysis of the mineral content of the plant cover in order to detect mineralization. The Pb-Zn concentration in plant & soil has been evaluated among the different species namely: *Capparis serparia, Euphorbia tirucalli, Hibiscuss gibbsoni, Hyptis suaveolens, Pongamia pinneta, Leucena leucocephela, Sesamum indicum* and *Zenia elegans*

Keywords: biogeochemical prospecting; indicator plants; Pb-Zn deposit, Zawar deposit.

Introduction

Botanical methods of exploration for minerals incorporate two basically different techniques: "Geobotany" and "biogeochemistry". Geobotany is essentially a form of applied ecology, and involves the visual observation of the vegetation cover in order to detect mineralization by means of plant distribution, the presence of indicator plants or mutational or morphological changes induced by excesses of certain elements in the underlying strain (Carlisle, 1958).

Indicator geobotany has been regarded as that division of geobotany which is concerned with the study of the theoretical principles and practical applications of the vegetation cover and its component species as 'indicators' of the conditions of the environment. Indicator geobotany is very closely linked with plant ecology, physiology and biogeochemistry, on the one hand, and with the related geological sciences (Mining and Economic geology, soil science, terrain studies, hydrogeology and geochemistry) on the other.

Biogeochemical and geobotanical prospecting are some of the methods in which metal tolerating plants are used to indicate the presence of minerals (Brooks, et al., 1998b, Aery and Tiagi, 1990, Reveez, 2000, Adrino, 2001, Trivedi and Solanki, 2009). Biogeochemical method of prospecting involves chemical analysis of vegetation in order to detect mineralization in the underlying substrate (Goff et al., 1985). This method of prospecting involves chemical analysis of plants, as an alternative to soil analysis, in order to detect presence of mineralization beneath the earth surface (Brooks, 1972).

2.1 SOME WELL-KNOWN INDICATORS PLANTS

Table 1. Some plant indicators of mineral deposits

Species	Family	Locality
	Cobalt	
Crassula alba (L)	Crassulaceae	Zaire
Crotalaria cobalticola (U)	Leguminosae	Zaire
	Copper	
Crotalaria cobalticola (U)	Leguminosae	Zaire
Impatiens balsamina (L)	Balsaminaceae	India
	Lead	
Minuartia verna (L)	Caryophyllaceae	UK
Oligotrichum hercynicum	Barophyta	Alaska
	Uranium	
Astragalus pattersoni	Leguminosae	Western US
	Zinc	
Baptisia bracteata	Leguminosae	Wisconsin
Viola calaminaria	Violaceae	Belgium & Germany

(Source: Brooks R.R. "An Introduction to Geobotany in Mineral Exploration")

3. Study Area

The Zawar lies between latitude 24°18'48" and 24°22'48" and longitude 73°40' to 73°45'24" E to the south of Udaipur town. The general elevation of the area is 377.7 m above MSL. It is located about 40 km. south of Udaipur city.

The Zawar group of deposits is Proterozoic in age. The Aravalli Supergroup consists of shelf and deepwater facies of sediments deposited over the cratonized Archean rocks of Western India between 2500 to 1900 ± 80 Ma (Roy, 1988, 1991, Brooks, et al.).

The stratigraphic relationship is abundantly clear in sections around Udaipur and around Zawar mine area (Fig. 2). The succession overlies an Archaean granite-gnessic basement (Sarara inlier) exposed to South of Zawar area (Straczek and Srikantani, 1966). The major deposits of the Zawar belt area Balaria, Mochia, Baroi and Zawarmala. All of the metamorphic rocks of the Aravalli range including those of Delhi System have been involved in intense folding, giving rise to isoclinals or near isoclinals forms or initially affected by vertical tectonics i.e., rifting, producing a no. of faults and grabens. These rocks have been ruptured by major thrust fault (Roy et al., 1984, Paliwal, 1981, 1998).

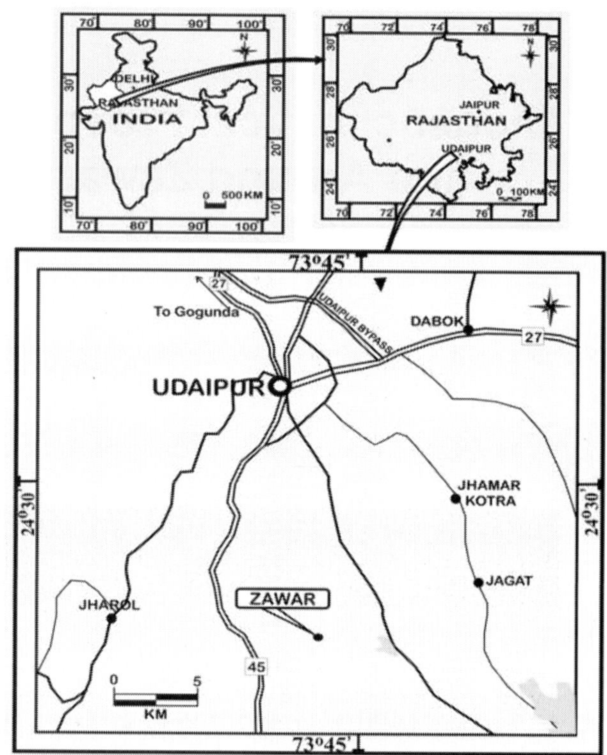

Fig. 1: Location Map of the study area, Zawar

Pb-Zn belt of Zawar provide a classic example of upgradation of ore mineral by tectonic deformation, accumulated in an intracontinental rift basin (Roy and Kataria, 1999). There are two types of mineralization in the Zawar mineralized belt one is carbonaceous siliciclastic turbidite hosted stratified ores resembling SEDEX-type (cf. Goodfellow et al., 1993) and the other is fault/fracture controlled replacement carbonate-quartz-sulphide veins or pure sulphide veins emplaced along fault planes or fractures (vein-type ores) within deep basinal dolomitic carbonate- sandstone turbidites.

The textural features of SEDEX-type ores attest to a synsedimentary/diagenetic origin for this type of mineralization. Talluri et.al. (2000) classified the Pb-Zn ores into three distinct modes: a) Stratiform or bedded ores hosted by black shale or carbonaceous phyllite, b) Vein type ores in open space filling cavities within dolomite, c) Massive ore of galena and sphalerite in large shear related quartz veins. The black shale hosted stratified ores resemble SEDEX type mineralization (cf. Goodfellow et.al., 1993) Mineralization is controlled by fractures and top of the surface of dolomite. The phyllite foliation is also mineralized.

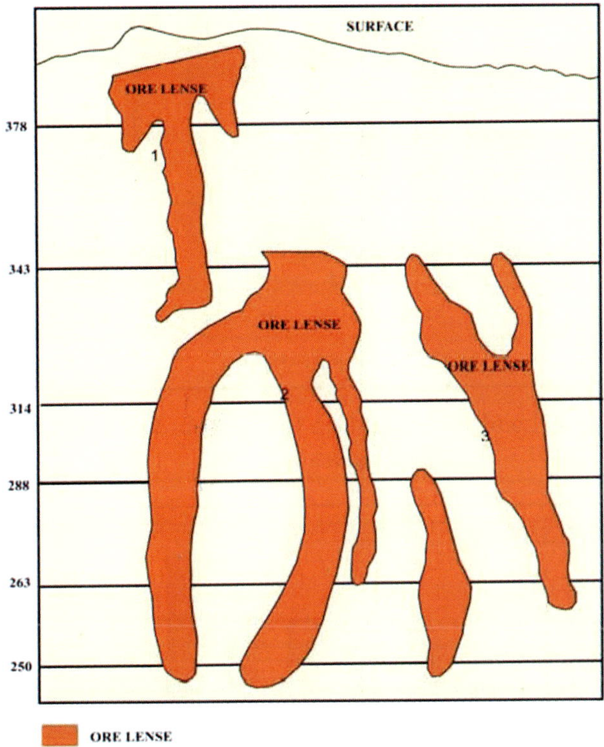

Fig. 2: Zawar mine section showing the shape of the ore body

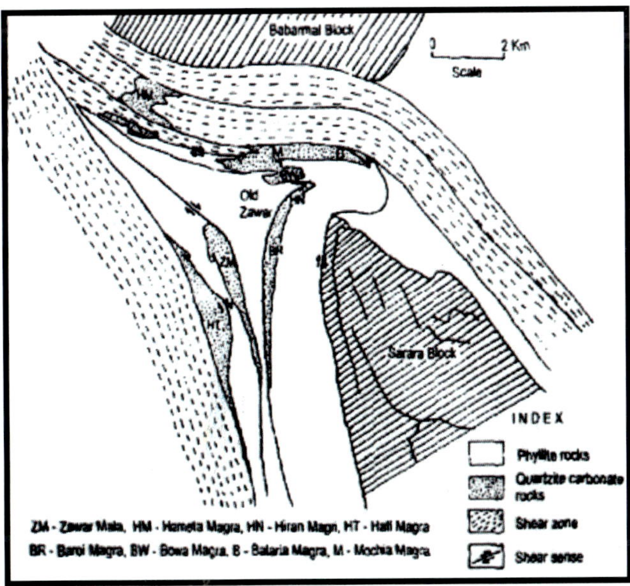

Fig. 3: Geological map based on aerial photograph showing major tectonic elements around Zawar mines (After Roy, 1995)

The ore minerals are Galena, Sphalerite, Minor amount of Molybdenum is also found. Pyrrohite, arsenopyrite, pyrite, pyrite, chalcopyrite and argentite minerals are also reported. Gangue minerals are Dolomite, Quartz etc.

4. Material and Methodology

Phytosociological studies carried out on these Zinc lead deposits revealed that several plant species mainly *Hyptis suaveolens, Capparis serparia, Pongamia pinneta, Leucena leucocephela, Zenia elegans, Hibiscuss gibbsoni, Euphorbia tirucalli, Sesamum indicum.* Plant and soil sample were collected from mineralized and non mineralized Zawar area both plant and soil sample were chemically analysed and digested on the dry weight ashing method of Greweling (1976). Final solution were atomised in an Atomic Absorption Spectrophotometer (AAS). The final results of plant and soil concentrations were obtined in ppm. The concentration of metal in soils and plants in normal and critical range as given by Alloway, 1990 is shown in tabular form (Table 2)

Table 2: Concentration of heavy metals in soils and plants (Alloway, 1990)

Elements	Normal range in soil (ppm)	Critical soil Concentration (ppm)	Normal range in plants (ppm)	Critical plant Concentration (ppm)
Cadmium	0.01-2	3-8	0.1-2.4	5-30
Copper	2-25	60-125	5-20	2-100
Manganese	20-10000	1500-3000	20-1000	300-500
Lead	2-300	100-400	0.2-20	30-300
Zinc	1-900	70-400	1-400	100-400

5. Data presentation and statistical results

Plants with high relative accumulations have the capacity to indicate anomalous metal concentrations where as species with low relative accumulation may be limited in the respect. This value varies within a species depending on the metal. In a preliminary survey, a number of plant species were tested for their elemental concentrations. The species having high constancy and fidelity values, i.e., the "Characteristic species", were selected along with certain other concentrators for further sampling and analyses work. Tables 2 and 3 contain highest data for elemental concentrations in the soils and the vegetation for Zawar area.

Table 3. Highest metals concentration (ppm) of Lead and Zinc in plant tissues in mineralized and non mineralized zone of Zawar

S. No.	Plant Name	Plant tissue	Name of Metals			
			Mineralized Zone		Non Mineralized Zone	
			Lead	Zinc	Lead	Zinc
1	*Hyptis*	Root	7.08	9.76	3.03	4.21
2	*suaveolens* (A1)	Stem	6.34	2.28	3.01	2.02
3	Herb	Leaves	7.54	5.76	2.18	2.01
4		Flower	2.36	4.83	1.03	1.01
5	*Capparis*	Root	8.10	8.89	3.35	3.37
6	*serparia* (A2)	Stem	7.10	3.08	3.03	1.14
7	Shrub	Leaves	5.06	7.08	2.14	3.56
8		Flower	3.08	4.50	1.15	2.07
9	*Pongamia*	Root	7.05	7.52	3.09	3.37
10	*pinneta* (A3)Tree	Stem	6.23	4.09	3.00	2.19
11		Leaves	7.52	7.74	1.01	2.10
12		Flower	3.05	7.75	1.09	2.12
13	*Leucena*	Root	8.56	6.22	4.29	2.19
14	*leucocephela* (A4)	Stem	7.36	5.61	3.09	2.11
15	Tree	Leaves	5.32	10.85	2.17	3.49
16		Flower	3.31	9.10	1.02	3.21
17	*Zenia*	Root	9.60	8.36	4.37	2.30
18	*elegans* (A5) Shrub	Stem	7.57	4.11	2.17	2.34
19		Leaves	3.57	6.67	1.28	3.24
21	*Hibiscuss*	Root	9.39	7.50	0.14	1.17
22	*gibbsoni* (A6) Herb	Stem	9.40	4.86	3.45	3.13
23		Leaves	6.42	6.77	3.4	3.17
24		Flower	5.44	5.50	2.1	2.13
25	*Euphorbia*	Root	9.33	10.07	1.89	2.01
26	*tirucalli* (A7) Shrub	Stem	7.89	5.95	3.57	4.41
27		Leaves	4.71	5.79	3.21	2.47
28		Flower	4.66	2.71	2.25	2.31
29	*Sesamum*	Root	4.39	4.39	1.27	1.19
30	*indicum* (A8)	Stem	4.36	4.36	1.10	1.07
31	Herb	Leaves	5.93	5.93	1.10	1.12
32		Flower	1.07	1.07	1.11	1.16

Abbreviations: **H** – Herb, **S**- Shrub, **T**- Tree

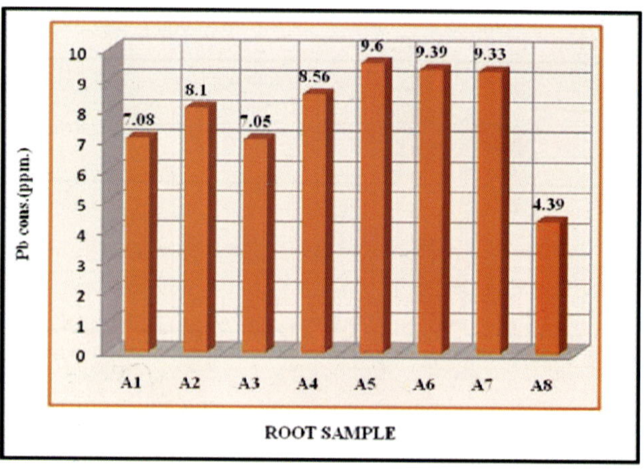

Fig.4: Lead concentration (in ppm) in roots of plant samples (A1-A8)

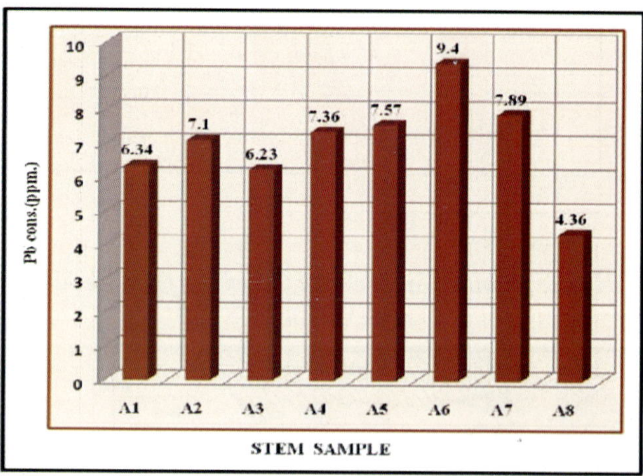

Fig.5: Lead concentration (in ppm) in stems of plant samples (A1-A8)

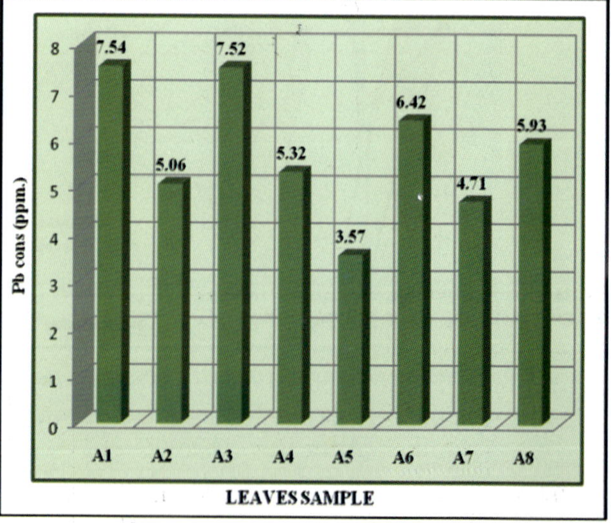

Fig. 6: Lead concentration (in ppm) in leaves of plant samples (A1-A8)

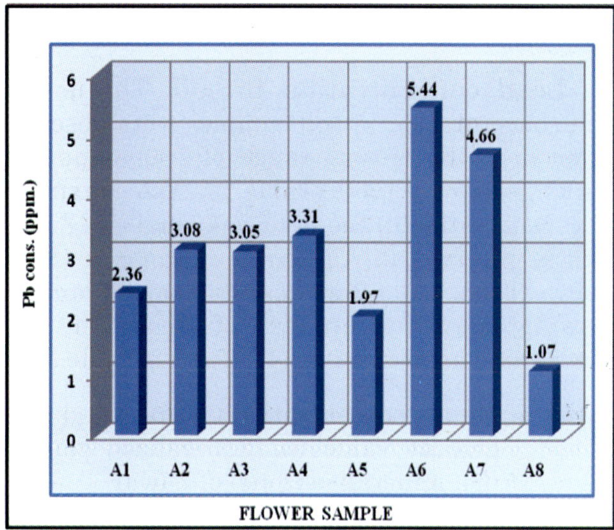

Fig. 7: Lead concentration (in ppm) in flowers of plant samples (A1-A8)

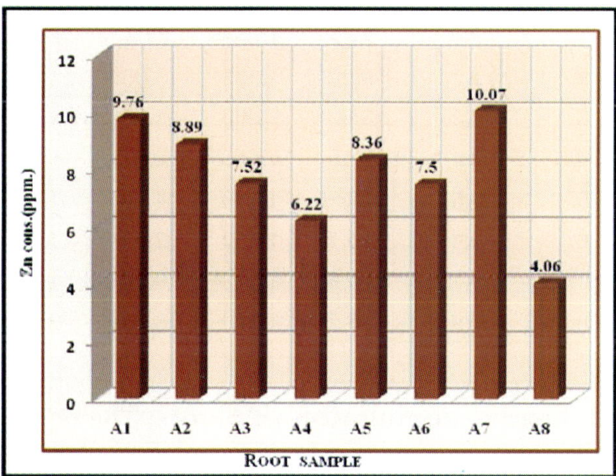

Fig. 8: Zinc concentration (in ppm) in roots of plant samples (A1-A8)

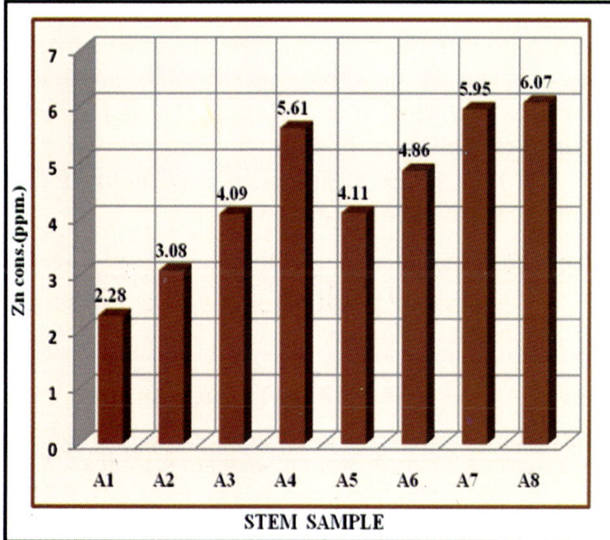

Fig. 9: Zinc concentration (in ppm) in stems of plant samples (A1-A8).

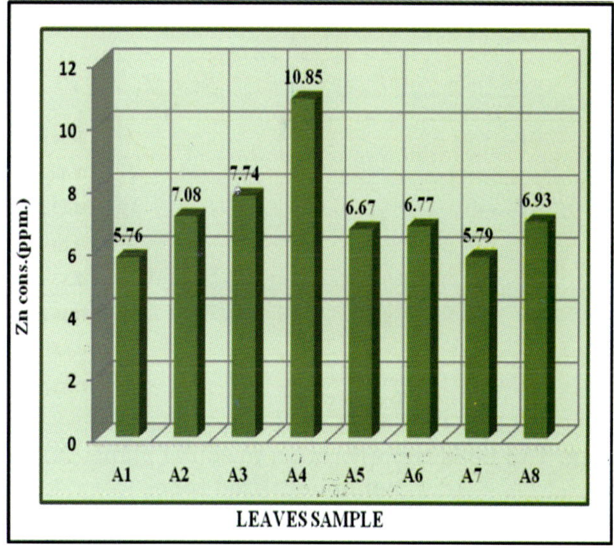

Fig. 10: Zinc concentration (in ppm) in leaves of plant samples (A1-A8)

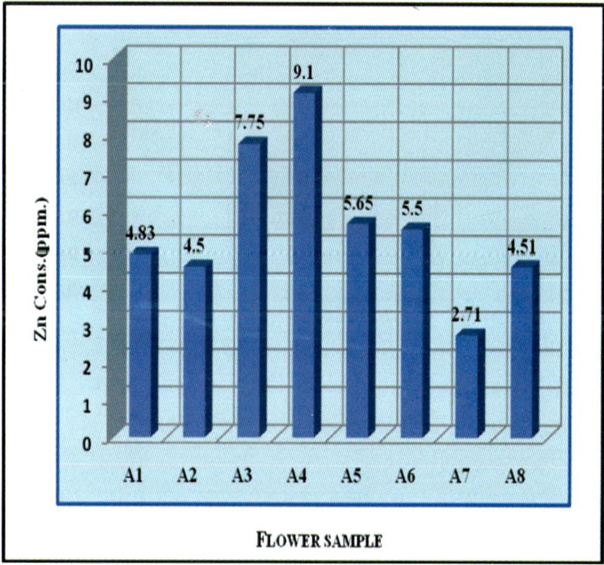

Fig. 11: Zinc concentration (in ppm) in flowers of plant samples (A1-A8)

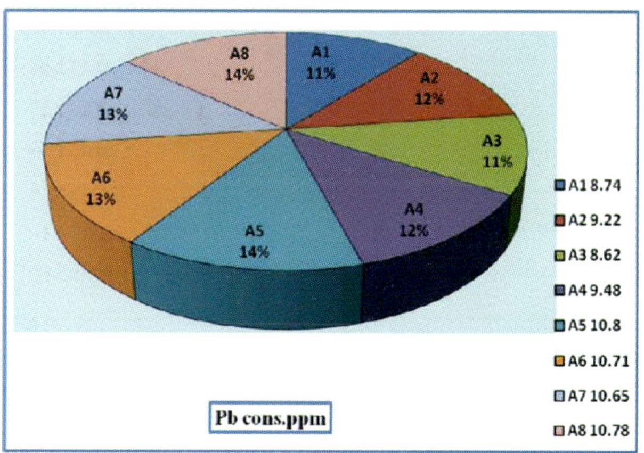

Fig. 12 : Lead concentration in soil of plants A1 to A8 in mineralized zone.

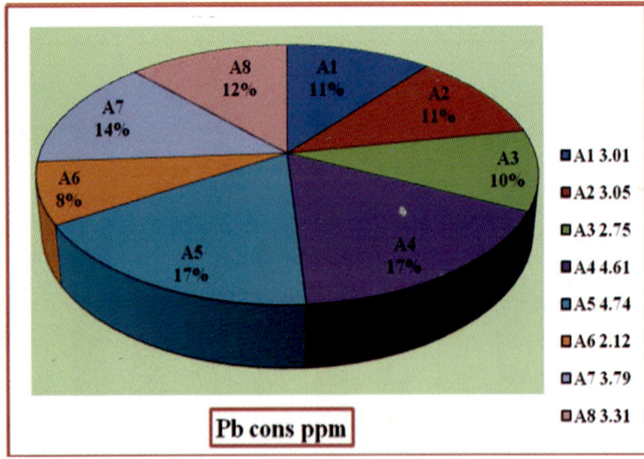

Fig. 13 : Lead concentration in soil of plants A1 to A8 in nonmineralized zone.

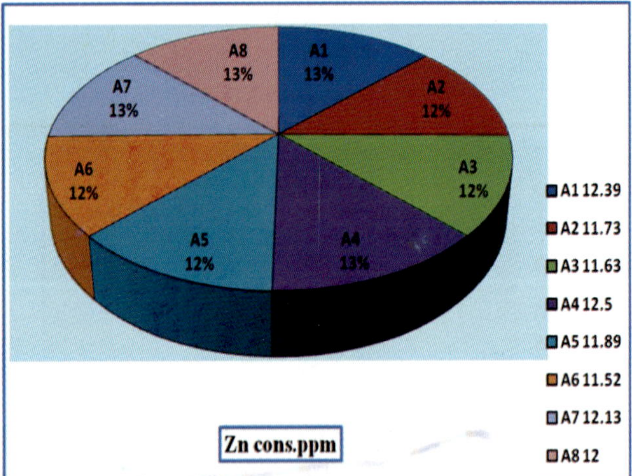

Fig. 15 : Zinc concentration in soil of plants A1 to A8 in mineralized zone.

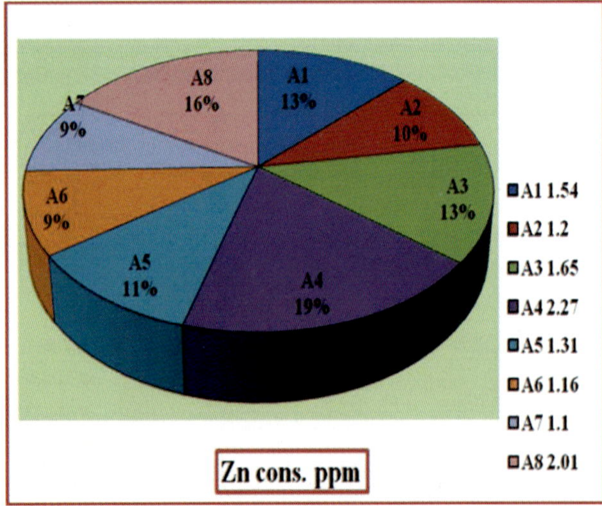

Fig. 16 : Zinc concentration in soil of plants A1 to A8 in nonmineralized zone.

This value varies within a species depending on the metal. Plant species with comparatively higher relative accumulation for zinc are A1 *(Hyptis suaveolens),* A7 *(Euphorbia tirucalli),* A4 *(Leucena leucocephela)* and A8 *(Sesamum indicum).*

Lead concentration in soil: The normal lead concentration in soil samples were observed lower than the critical range of 100-400ppm in plants, Alloway, 1990 (Table 2). The maximum Pb concentration in the field soil sample of Zawar area is observed in A5 (*Zenia elegans*) of (10.80 ppm) (Table 5). Significantly this plant is showing positive uptake of lead in non mineralized zone also i.e., (4.74 ppm) (Table 5).

Table 5: Metals concentration (ppm) in soil with highest value (in mineralized zone & non mineralizedzone) of Zawar

Serial No.	Name of Soil Sample	Mneralized zone		Non mineralized zone	
		Pb	Zn	Pb	Zn
1.	A1	8.74	12.39	3.01	1.54
2.	A2	9.22	11.73	3.05	1.20
3.	A3	8.62	11.63	2.75	1.65
4.	A4	9.48	12.50	4.61	2.27
5.	A5	10.80	11.89	4.74	1.31
6.	A6	10.71	11.52	2.12	1.16
7.	A7	10.65	12.13	3.79	1.10
8.	A8	10.78	12.00	3.31	2.01

Lead accumulation in plants: The normal lead concentration in plant samples is observed below then critical range (30-300ppm.) (Table 2) Alloway. In the study area the highest value of lead concentration in plant as found is (9.60 ppm) in root of A7 (*Zenia elegans*) (Table 6) in mineralized zone of Zawar. In the non mineralized zone of zawar the highest concentration of lead was observed for the root sample of A5 (*Zenia elegans)* of 4.37ppm (Table 7). Highest value of Lead (Pb) in plant samples of mineralized and nonmineralized area are given below-

Table 6: Highest Value (in ppm.) of Lead (Pb) metal in plant samples of mineralized area of Zawar.

Plant Sample	A1	A2	A3	A4	A5	A6	A7	A8
Value(ppm)	7.08	8.10	7.05	8.56	9.60	9.40	9.33	5.93
Plant organ	R	R	R	R	R	S	R	L
Order	A5 > A6 > A7 > A4 > A2 > A1 > A3 > A8							

Table 7: Highest Value (in ppm) of Lead (Pb) metal in plant samples of non mineralized area of Zawar

Plant Sample	A1	A2	A3	A4	A5	A6	A7	A8
Value(ppm)	3.03	3.56	3.09	4.29	4.37	3.45	3.57	1.11
Plant organ	R	L	R	R	R	R	R	R
Order	A5 > A4 > A7 > A2 > A6 > A3 > A1 > A8							

Zinc concentration in soil: Zinc concentrations were found higher than the other metals in soil samples. Normal range of zinc set of Alloway (1990) is 1-900 ppm and critical range is 70-400 ppm (Table 2). The highest value of zinc concentration in soil of Zawar area was found 12.50 ppm (Table 8) in A4 (*Leucena leucocephela*) for mineralized zone, where as for non mineralized it is 2.27 ppm (Table 9)

Zinc accumulation in plants: All the plant samples have zinc concentrations lower than the critical range on zinc in plants (1-900 ppm). The richest accumulation was found (10.85 ppm) in leaves of A4 (*Leucena leucocephela)* plant sample in mineralized zone (Table 5). In nonmineralised its highest in root of A7 (4.41ppm).

Table 8: Highest Value (in ppm.) of Zinc (Zn) metal in plant samples of mineralized area of Zawar

Plant Sample	A1	A2	A3	A4	A5	A6	A7	A8
Value(ppm)	9.76	8.89	7.75	10.85	8.36	9.49	10.07	6.93
Plant organ	R	R	F	L	R	R	R	L
Order	A4 > A7 > A1 > A6 > A2 > A5 > A3 > A8							

Table 9: Highest Value (in ppm) of Zinc (Zn) metal in plant samples of non mineralized area of Zawar

Plant Sample	A1	A2	A3	A4	A5	A6	A7	A8
Value(ppm)	4.21	3.37	3.37	3.49	3.24	3.17	4.41	1.16
Plant organ	R	R	L	L	L	S	R	L
Order	A7 > A1 > A4 > A2 ≥ A3 > A5 > A6 > A8							

Conclusion

The evaluation of plant metal concentration can be used to obtain information about specific plant behavior in biogeochemical prospecting and it reveals the metal distribution and their mobility. The analyzed geobotanical study is summarized below:

1. Different organs vary in their capacity to accumulate different elements. Roots and leaves accumulate much more Mn and Zn, then stem and flowers.

2. Generally an herb accumulates higher concentration of the elements than shrubs and trees.

3. The accumulation of Pb and Zn in mineralized zone except A5 (*Zenia elegans*), A1 (*Hyptis suaveolens*) to A8 (*Sesamum indicum*) accumulates higher concentration of Zn, while A5 *(Zenia elegans)* has been found higher enrichment of Pb then Zn in both zone, so it's considered as a purely Pb indicator in both zone as a herbs and Pb is absorbing in root organ whereas A1, A2, A3, A4, A6, A7, and A8 accumulates much concentration of Zn in mineralized zone except A5. Whilst A1, A3, A7 and A8 absorbing more Zn in both region.

A1, A3, A7, A8: Zn >> Pb (in both zone) & A2, A4, A6: Zn >> Pb (in mineralized zone).

A5: Pb >> Zn (in both zone) & A2, A4, A6: Pb << Zn (in non mineralized zone).

4. A2 *(Capparis serparia)*, A4 *(Leucena leucocephela)* and A6 *(Hibiscuss gibbsoni)* are working as Pb and Zn absorber in both regions. Conclusively they showing positive response toward enrichment of Zn in both region so they are considered as a purely Zn indicator plant.

5. A1 (*Hyptis suaveolens*) and A8 (*Sesamum indicum*) are herbs; A3 *(Pongamia pinneta),* is tree, while A7 *(Euphorbia tirucalli)* is shrubs. Mostly Zn is accumulating in root and leaves in these plants species.

Therefore, based on these biogeochemical analysis plants can be ideally used as possible application in exploration, mining reconnaissance surveys; to know metal status of an area and geological purposes.

References

Adriano, D.C.; 2001: "Trace elements in Terrestrial environments: Biogeochemistry, Bioavailability and Risks of metals", 2nd ed. Springer-Verlag, New York

Alloway, B.J., 1990: "Heavy metals in soils." Glasgow: Blackie Academic and Professional, 339 p.

Aery N.C., Tiagi Y.D., 1990: "The Significance of elemental ratios in Biogeochemistry" Jour. Trop. For., Vol.6, pp. 108-114.

Brooks, R.R. and Robinson, B.H., 1998: "The potential use of hyper accumulators and other plants for phytomining." in R.R. Brooks (eds.) 'Plants that hyper accumulate heavy metals: their role in Archaeology, Microbiology, Mineral Exploration, Phytomining and Phytoremediation.' CAB International, Wallingford, pp. 327-356.

Brooks, R.R., 1972: "Geobotany and biogeochemistry in mineral exploration" Harper and Row Publishers, New York, USA.

Carlisle, D. and Cleaveland, G.B., 1958: "Plants as a guide to mineralization." California Div. Mines Geol. Spec. Rept., pp. 50-31.

Greweling, T.; 1976: "Chemical analysis of plant tissue, Agronomy No. 6. Cornell University Agricultural Experiment Station, Ithaca, NY. Search Agriculture, Vol. 6, No. 8, pp. 1-35.

Goff, S., Brooks, R.R., Naidu, S.D. and Coppard, E.; 1985: "Delineation of potentially auriferous quartz reefs by analysis of the bark of *Finns radiata* (Monterey Pine)" Jour. Geochem. Explor., Vol.24, pp.273-280.

Goodfellow, W. D.; 1993: "Geological Association of Canada, Special Paper, No. 40, 201 – 252.

Paliwal, B.S., 1981: "Tectonic of Precambrian rocks southwest of Udaipur. Unpublished Ph.D. Thesis, University of Rajasthan, Jaipur, pp 1-121.

Paliwal, B.S., 1998: "The Aravalli fluvial sequence of Udaipur City, India: and example of well preserved sedimentary structures in complexly deformed Precambrian meta sediments." in: B.S. Paliwal (Ed.), The India Precambrian. Scientific Publishers (India), Jodhpur, pp.142-158

Reeves, R. D. and Baker, A.J.M., 2000: "Metal accumulating plants". In: Raskin, I. and Ensley, E.D. (eds). Phytoremediation of toxic metals: using plants to clean up the environment, Wiley, New York., pp.193-221

Roy, A.B., 1995: "Geometry and evolution of superposed folding in the Zawar lead-zinc mineralised belt, Rajasthan." Proc. Ind. Nat. Sci. Acad., Vol. 104, No.3. pp. 349-371. ISSN 0370-0046

Roy, A.B., 1988a: "Introduction in A.B. (ed.) Precambrian of the Aravalli mountain. Rajasthan Inadia, Mem Geol. Soc. Ind, 7, i-viii.

Roy, A.B. and Paliwal, B.S. and Bejarniya, B.R., 1984 : "The Aravalli rocks of an evolutionary model and metallogenic trend." Ind. Jour. Earth Sci., CEISM, Seminar, pp. 73-82

Roy, A. B., 1991: "Evolution of early Proterozoic Aravalli depositional Basin of India." In: Tandon, S.K., Pant, C.C. and Kashyap, S.M. (eds.) Sedimentary Bains of India: Tectonic context; Gyanodaya Prakashan, Nainital, pp. 1-11.

Roy, A. B. and Kataria, P., 1999: "Precambrian Geology of the Aravalli Mountain and neighbour-hood: Analytical update of recent studies; Proc. Sem. Geol. Raj., Status and Perspective MLS University, Udaipur, pp. 1–56.

Straczek, J.A., and Srikantan, B., 1966, The geology of Zawar zinc-lead area, Rajasthan, India: Memoirs of the Geological Survey of India, v. 92, 85 p.

Trivedi, S.K. and Solanki, K.K., 2009: "Present status and future trends of biological methods of prospecting for economic mineral deposits". in: Srivastava, K.L. (ed). Economic mineralization, Scientific Publishers (India), Jodhpur, INDIA, pp: 396-405.

Talluri, J. K. Pandalai, H. S. and Jadhav G. N., 2000: "Fluid chemistry and depositional mechanism of the epigenetic discordant ores of the Proterozoic carbonate-hosted, Zawarmala Pb-Zn deposit, Udaipur District, India." Vol. 95, pp. 1505-1525.

Hon'ble Reviewers

1. **Prof. Sakir Ali**
   Centre for Ad. Study in Geology
   Dr. Hari Singh Gaur University
   SAGAR - 470 003
2. **Dr. K.K. Sharma**
   Department of Geology
   Govt. College
   SIROHI - 307 011

I-8

On the Capacity Expansion in Malanjkhand Copper Deposit: Opportunity and Challenges

M.K. Shukla * and A.K. Shandilya **

* Hindustan Copper Limited, Malanjkhand, M.P., India
** Department of Geology, Government Dungar College Bikaner (Rajasthan)
*E-mail: shuklamohan55@gmail.com

Abstract

Huge gap in demand and supply of the copper metal has provided a great opportunity for continuance of ore production in Malanjkhand. Plans are afoot for inception of a highly mechanized underground mine with expanded capacity of milling and ore production from the present 2 million ton per annum to 5 million with an investment of Rs 1856.36 crore as approved by the Government of India. The underground mine will be within 340m to -60m RLs and the levels will be accessed through declines. Lodes will be worked along the strike through sub level open stoping method. Obtaining permissions and necessary approvals from the statutory authorities has so far proved to be quite time taking and challenging.

Huge production and processing of large quantities of ore will also pose a serious threat to ecology and environment not only in Malanjkhand but in the other parts of the country. It is imperative therefore that sustainable mining comes to existence through sustainable geological explorations taken up systematically and more than that, willingly.

Keywords: Malanjkhand Copper Deposit, Copper Economic, Mine Development, Baster Craton.

Introduction

Requirement of more metal is often responded by scaling up the mining operations for producing more ore of higher grade. This is often met by exploring larger rich areas for a new mine or developing the existing one-sideways or at depth. Malanjkhand is the large open pit copper mine in operation since 1978. Mining operations began by cutting the 2.6 km long Malanjkhand hill. The operations are now reaching an optimum depth of 340m RL. No more open pit mining is possible and if done so, large overburden will have to be removed for creating benches that will consume more and more surface area as a dumping ground at a huge cost. This is unlikely to be feasible and therefore all out efforts began for the inception of a highly mechanized underground mine with expanded capacity of milling and ore production from the present 2 million ton per annum to 5 million.

History

Malanjkhand is located at an altitude of 575m 90km north-east from the historical town of Balaghat, the district headquarter. Surface indications of the copper occurrences and old workings were first reported in year 1882 by the first deputy commissioner of the district,

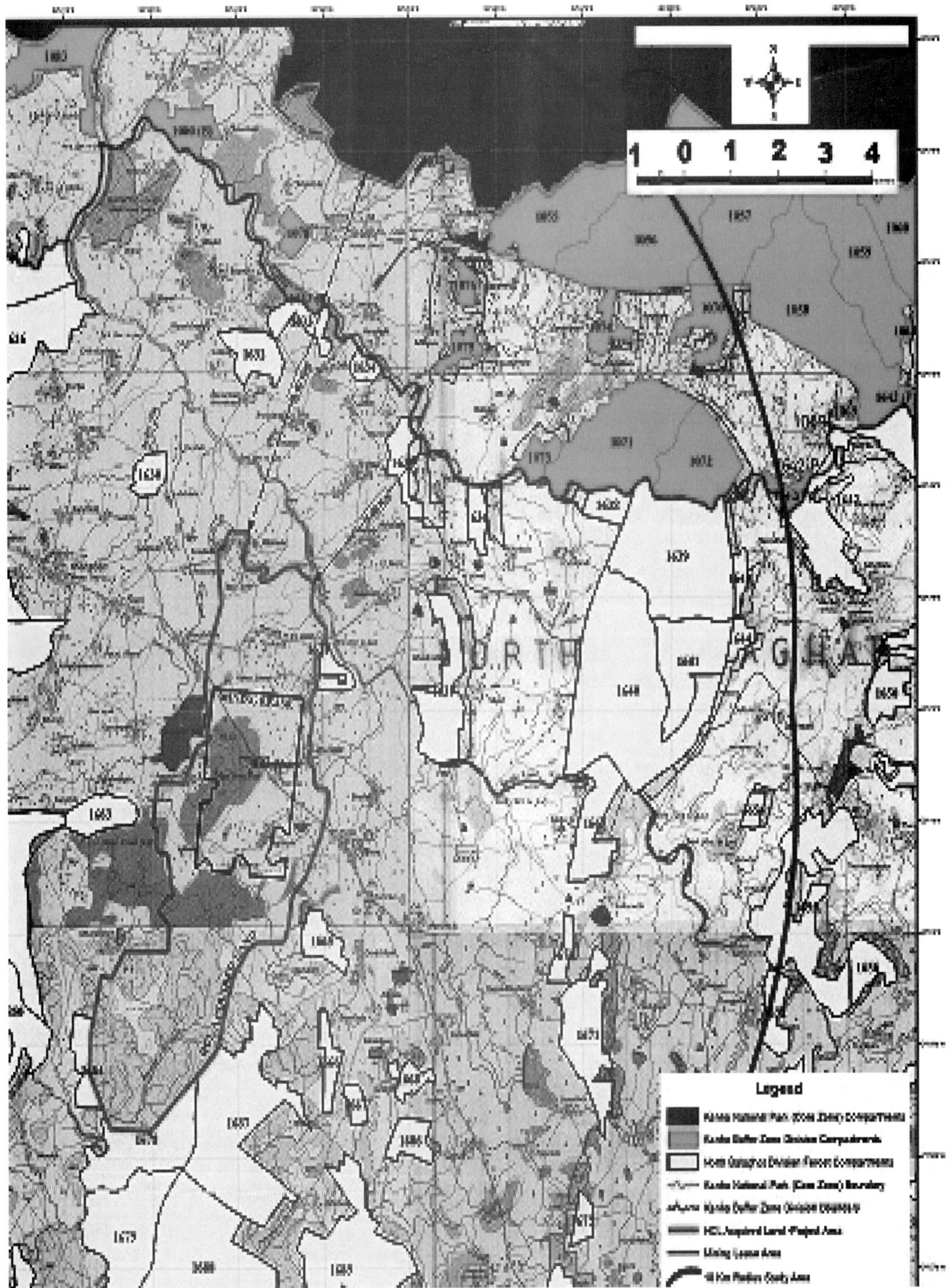

Location of Malanjkhand in Southern Vicinity of the Famous Kanha Tiger Reserve –A testimony that a Mining Industry and the Biodiversity can Co-exist in Harmony

Colonel Bloomfield and who brought the information to the notice of the GSI, a newly formed organization of the time. After the examination of the deposit by GSI's William King in year 1886, it was only in 1916, when M/s Glow Stanly first made an unsuccessful attempt to open the deposit followed by M/s Oke Brothers in 1937. Since then Malanjkhand had been a saga of the exploration expeditions by different agencies. Independence of Indian nation in 1947 gave a fillip to the mineral exploration which gave rise to a number of mineral and mineral based industries and plants in the country. Systematic exploration began in Malanjkhand in the year 1969. A little before this, the calm and quite deserts of Khetri Singhana in the western Indian state of Rajasthan had been witnessing the birth of a Government of India agency, the Hindustan Copper Limited (HCL), entrusted with the responsibility of the exploration, development and mining of the copper deposits of the area and in rest of the country. HCL was identified as the agency for the Malanjkhand also and a detailed project report was submitted by the year 1975. Project construction started in 1978 within an acquired area of 2016 after approval from the Government of India. Stripping operations began in 1979 with the commissioning of the first shovel. Beginning official production in July 1982, it was on 12 November 1982, when the then Prime Minister of India, Smt Indira Gandhi dedicated the project to the country.

This marked a beginning of a new era and a new civilization amidst the dense forests of Kanha which are famous since the time immemorial for tall trees, magnificent hills, glassy rivers, roaring tigers and panthers, chirping birds and galloping antelopes.

Geology

The ENE trending Son Narmada Lineament along the Narmada river cuts across the Indian Precambrian Shield. The Malanjkhand copper deposit is located in the southern part of the Central Indian Precambrian Shield which is bordered by Pranhita-Godavari valley and Son Mahanadi valley. The geology of the Malanjkhand area comprises the Basement complex and weakly metamorphosed sedimentary rocks of the Chilpighat series ranging from middle to upper Proterozoic ages.

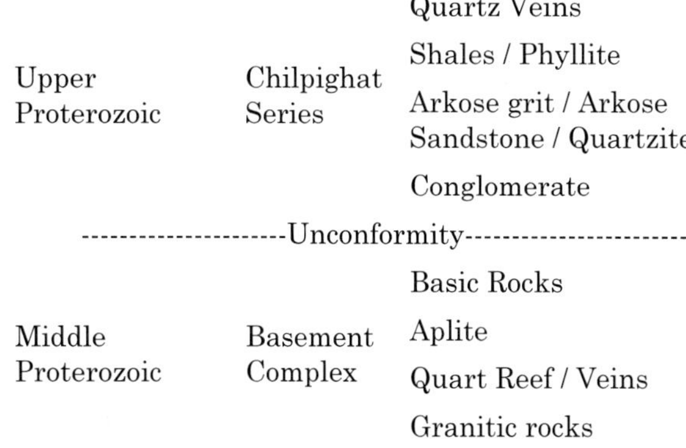

Upper Proterozoic	Chilpighat Series	Quartz Veins Shales / Phyllite Arkose grit / Arkose Sandstone / Quartzite Conglomerate
---Unconformity---		
Middle Proterozoic	Basement Complex	Basic Rocks Aplite Quart Reef / Veins Granitic rocks

Quartz reef is the principal host of mineralization and varies in composition from pure quartz to quartz feldspathic rock. Besides quartz reef, silicified altered granitic rocks are also seen hosting the sulphide mineralization.

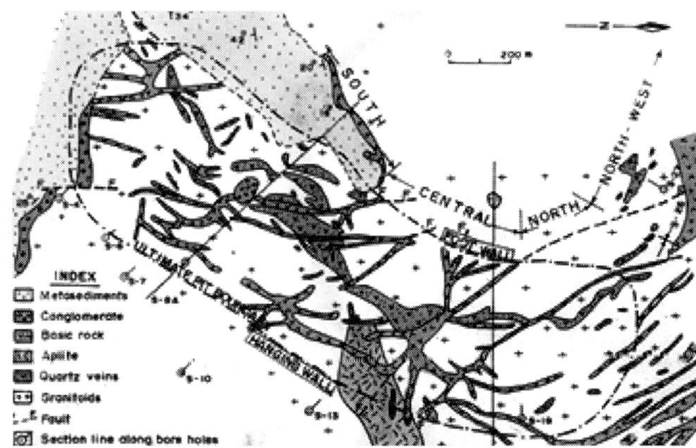

Geology in and around the Malanjkhand Copper Mine Open Pit

The reef was exposed on the erstwhile Malanjkhand hill ridges. It was positioned in an arcuate shape over a length of 2.6km in north-south direction dipping 60°-70°dueeast with an average width of 65 to 70 m. It was found emplaced along the fracture & fissure planes in the granitoids, subsequently sheared and fractured permitting sulphide rich hydrothermal fluids to precipitate giving rise to a rich copper deposit. This reef is being mined since 1982 and the open pit now exists in place of the reef.

Production

Beginning ore production and milling in 1982, the mine has so far produced and milled (31.03.2012) 57.11 million tonnes of 1.14% Cu grade ore that amounts to producing 2.41 million tonnes of copper concentrate at 27.07% Cu grade, that is, 603311 metal in concentrate. Accumulated tailings are estimated to be 54.88 million tonnes at 0.118% Cu grade. Waste rocks have been staked as the overburden dumps and are estimated to be 250 million tonnes.

Mining

Open pit mining, started in 1982 by cutting the then Malanjkhand hill that was then measured 652 m with respect to the mean seal level, has now gone to a depth of 400 m RL. Operated with 12 m high and 6 m wide benches, the mine is fully mechanized using rotary percussive drills, rope/hydraulic shovels, hydraulic excavators and dumpers. The pit will be deepened up to 340 m RL until 2017-18.

Huge gap in demand and supply of the copper metal provides a great opportunity for continuance of the ore production. A detailed project report was thus prepared that worked out the feasibility of expansion of the production capacity from the present 2 million tonne per annum to 5 million through the underground mining. A Rs 1856.36 crore investment proposal for the construction of the biggest underground mine in the country submitted to the Government of India, was approved on 30 September 2011.The shaft sinking and decline construction operations are likely to begin this year (2013) subject to obtaining statutory clearances from the government which are already under consideration.

The proposed underground mine will be accessed through declines and shafts. Basic design parameters will be as follows;

Open Pit Bottom	____	340m RL
Underground mining limits	____	340m to -60m RLs
<u>Mine Entries</u>		<u>*Level, Gradient, Dimension, Length*</u>
North Decline	____	532 to -60m RL at 1 in 7, 5.5 x 5 m, 4600m
South Decline	____	436 to -60m RL at 1 in 7, 5.5 x 5 m, 3850m
<u>Vertical Shafts</u>		<u>*Location, Depth, Collar, Bottom RL*</u>
Production Shaft	____	6350N 6500E, 695m, 585m, -110mRL
Service Shaft	____	6500N 6570E, 655m, 585m, -110mRL
North Ventilation Shaft	____	7320N 7145E, 633m, 573m, -60mRL
South Ventilation Shaft	____	4928N 5912E, 640m, 580m, -60mRL

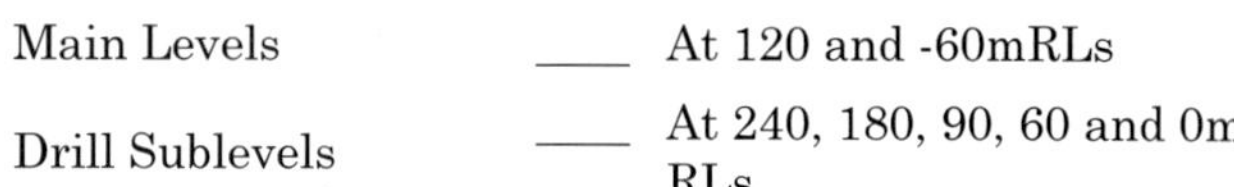

Main Levels	____	At 120 and -60mRLs
Drill Sublevels	____	At 240, 180, 90, 60 and 0m RLs

All the levels and sublevels will be developed from the declines. One hanging wall drive will be developed in the main levels for definition drilling. The proposed underground mine will feature the following:

i. An estimated copper ore reserve of 149 million tonnes with 1.34% Cu grade will be extracted between 340m RL and -60m RL whereas the present 340m RL will be the ultimate pit bottom of the ongoing open pit.

ii. There will be two extraction levels, one at 120m RL to draw ore between 120 and 340 and the other at -60m RL for drawing the ore between 120m RL and -60m RL.

iii. Lode will be worked along the strike through sub level open stoping method.

iv. Each stope will be worked through three sublevels at 55 to 60 m floor to floor vertical interval.

v. Basically it will be a trackless mining system using tyre mounted machines and conveyors.

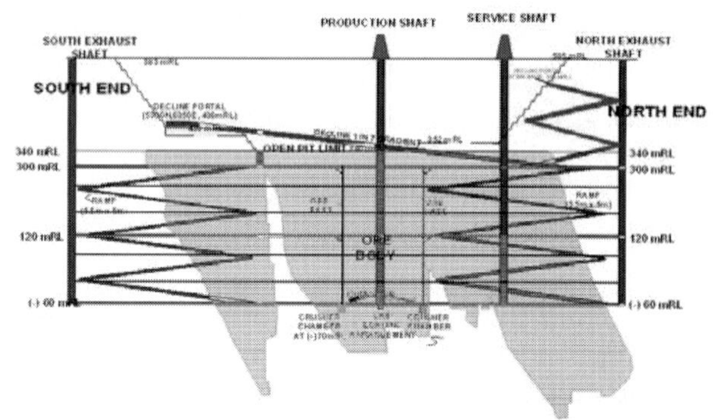

An Schematic Layout of the Proposed Underground Mine Under the Present Open Pit

vi. Each stope will be designed to produce 1.25 million tonnes of ore. Final dimension of the stope, however, will be ascertained only after definition drilling which will also help to determine the geotechnical properties of the hanging wall.

vii. At any given point of time, six stopes in the mine would be kept ready for production while six will be in the pipeline. This will always be crucial for achieving the target production.

Changes and Challenges

Demand and supply gap facilitates an opportune time for developing the deposit for underground. Already existing infrastructure, acquired land, mining lease, trained manpower and a fairly disciplined industrial culture which has been maturing since the last 35 years giving rise to a strong socio-economics in the area -are some of the ingredients which are favourably helping the switch over from the present open cast mining to the underground.

One of the biggest challenges, the mining industry has always faced, is fighting a number of statutes –the static factors that cannot be changed to the man's will and wish, but the man has to transform his own technology to fight out the statutes, so as to reach inside their womb and take out the desired metal bearing rocks, without inflicting injuries, or least possible if unavoidable, to the ecology and surrounding environment. Malanjkhand is a thirty five year old story of a pioneering work of copper mining of the sort and of a stark hinterland of agrarian agricultural communes, with poor socio economics, being slowly but steadily transformed to an awakened educated industrial society interacting with the outside world.

This was the first change put in to the motion by Hindustan Copper Limited, the public sector giant in late seventies that, alongside the strong foundation for a metal mining industry, has developed and matured Malanjkhand to a strong socio-economy. The stage is now set for the second change which aims to ensuring the longevity and sustainable growth. The Malanjkhand is now poised to take a giant leap from the present open cast to the modern technology-driven mechanized underground copper mine, matching to the world class standards. Plans are ambitious therefore. And the success will depend upon the right selection of technology, mobility of funds and the talent matching to the capital flow while gradually but steadfastly going into formulating the object oriented plans and working these plans without deviation and distraction within the given time frame and with class determination, misgivings notwithstanding, bringing production within the projected planned period, if not early.

Obtaining statutory clearances from the government and government agencies have so far proved to be the major challenges in implementation of the project. As per the grant of the ToR by the MoEF vide letter in March 2011 and stipulated conditions, a EIA/EMP study was conducted within the 10km radius drawn from the outer boundaries of the existing 479.9 ha mining lease. But the EIA/EMP study was not enough. Because of being located within the 10km linear distance from the internationally known Kanha National Park, the HCL had to undergo a cumbersome and highly difficult exercise of preparing a wildlife conservation plan that has successfully enumerated and overcome the objections of the forest and wildlife experts. Loss of wildlife habitats, pollution of air due to gas and dust, noise pollution, soil erosion, accidental fall of wild animals and increased dependence of people on forests are some of such impacts which have been properly addressed to and to the satisfaction of the authorities.

More than 17700 hectare area spread over 87 forest compartment was surveyed for enumerating and observing the presence of the wild animals and plants after seeking due permission from the forest officers in charge of the blocks. A large manpower was deployed for the purpose. Help of a retired IFS officer and scientists was also sought. According to this plan the Kanha National Park (KNP) was located 8.8 km north of the Malanjkhand Mining Lease within which the open cast mining activities are going on at present. But on discussions, the expert appraisal committee (EAC) asked for taking the 10km circle from the outer boundary of the 2016 ha project area and not the mining lease. This had forced the major overhaul of the study. All the permissions had to be sought again and additional forest compartments had to be taken up for the study. The number of the forest compartment thus rose from the earlier 87 to 114. The study area which was earlier a circle became an ellipse. As a result, the distance from the national park boundary reduced to 6.4 km (from the acquired 2016 hectare project area). The study area rose to 24925 hectare which really became challenging. Yet, this was completed working day and night and even on Sundays and holidays in war footing and a new revised study was submitted in a record time. All the above impacts and threat perceptions to the KNP's wildlife are largely reduced because of its being located at a safe distance from the Park area. The declining open cast activities along with the proposed gradual commencement of the underground mining will further reduce the threats, if still anticipated. The HCL in Malanjkhand is determined to protect not only the wildlife but will constantly work towards augmenting it and along with it, the forest cover. It is a known fact that the wildlife protection and augmentation in KNP and around has largely associated with the growth of the Malanjkhand breaking the myth that the mines and the wildlife can't exist together. KNP is better known in the last three decades forever rising population of tigers and panthers. Almost extinct hard ground Baragingha has been replanted and population has grown to a safe limit.

When presented before the EAC on 29 August 2012, environmental clearance (EC) has been agreed in principle. Final grant of approval is, however, awaited since then.

Another issue is obtaining clearance from the state forest department towards the 1808.54 acre of the forest land being held by the HCL since the year 1979. Despite efforts by the HCL since beginning, no denotification took place and therefore, the land transfer could not be notified. Although the forest land has been indicated 'transferred' in the forest maps, the forest department has insisted for completing the procedure since the proposal has not gone through the Forest (Conservation) Act 1980. The case is in consideration with the forest department and it is likely to be through in three month time.

In order to successfully bring about therequisite changes, the sector will need to provide lot of room to innovate, empowerment to the individual and an interface with opportunity to work. All these are the drivers for higher commitment for any individual to give greater performance. This is all the more important as the biggest challenge for the mining sector today is to access the right talent. There are not enough incentives to work in this industry located in deep interiors, considered hazardous with painful working conditions and without much entertainment that can be, anyway, compared with that in the urban settings. The mining industries are therefore introducing newer incentives to attract and retain the talent which so far have proved dangling carrots to attract and the engineering and earth science graduates are still shying away from this sector.

Over-enthusiastic militant lobby of the environmentalists is also a big challenge. This is being successfully met with by doing all that which is "plus" for the ecology and environment. Whereas in the other parts of the globe, the mines are being seen as the monsters, eating away the resources, the Malanjkhand copper mine has become a part and parcel of the local culture and the mining stories have found place in the local folklores.

The increasing investment for search of new ore horizons at depths is a difficult situation –a global problem. A mapper is considered to be a torch bearer for any geo-exploratory expedition. Once flourishing community of mappers within the earth science fraternity has greatly diminished with the overwhelming growth of white collar IT and office sectors. Whatever small number remained had to give in to the onslaught of the manpower reduction in the last two decades. The result is the geo-exploratory expeditions have suffered a great deal and the extra ore is being lost forever in the running mines without being uncovered.

Another issue is the need for lowering the cut-off grade of the ore under production. This will ensure the longevity of the deposit. Newer beneficiation technologies therefore will have to replace the existing ones. Old mine dumps and poor grade deposits will also have to be tested for economic recovery through biotechnological means. Bioleaching is an economic process of recovery of metal from ore. In our country where large reserves of poor grade ore are available, there is great scope for biotechnology for extraction of metal.

Waste rocks and copper ore tails (COT) will have to find uses through improved R&D efforts so that they find markets and equal amount of rock stripping elsewhere is saved. The stage will have to be kept in the ready mode to take on the new dimensions of the challenges on sustainably maintaining the ecology and environment. The mineral and metal community needs to tune itself to be at par with international technological trends in order to reap the full benefits.

Conclusion

The rate of mineral consumption has increased so high that the bulk of the minerals used by the mankind during the last 50,000 years have been used up only in the last 100 years. With this rate, more and more mineral wealth will be needed to sustain the industrial development in Malanjkhand. But we have now reached a critical phase in regard to exploration activities. The gradual diminishing chances of further discoveries of mineral wealth on the basis of surface showings need a reorientation of mineral exploration strategy. Huge production and processing of such quantity of ore will also pose a serious threat to ecology and environment. It is imperative therefore that sustainable mining comes to existence through sustainable geological explorations taken up systematically and more than that, willingly.

Acknowledgement

The authors thankfully acknowledge the management of the Hindustan Copper Limited for granting permission to present and publish this paper.

References

Malanjkhand Copper Mine, Modification of Scheme of Mining, 2012-13, Vo-I, published and presented in cooperation with the CIMFR, Dhanbad.

Published Reports of the HCL.

Wildlife Management Plan for Malanjkhand Copper Project in District Balaghat, Madhya Pradesh, India, published and presented in cooperation with the CIMFR, Dhanbad.

Hon'ble Reviewers

1. **Prof. R.K. Trivedi**
   Centre of Ad. Study in Geology
   Dr. Hari Singh Gaur University
   SAGAR - 470 003
2. **Prof. A.K. Sen**
   Department of Earth Science,
   Indian Inst. Technology Roorkee
   ROORKEE - 247 667

I-9

Zeolites Mineralization in Deccan Traps Rocks of Manmad Area, District Nasik, India

P.K. Jain[1] and D.C. Gupta[2]

1. School of Studies in Earth Science, Jiwaji University, Gwalior (M.P.)
2. Department of Earth Sciences, Barkatullah University, Bhopal (M.P.)
E-mail: dpkj_2005@yahoo.co.in

Abstract

The present study deals first time with the "Zeolitization" in the 366 meter thick Deccan Trap lava pile of Manmad, lies in between the Cartesian coordinates 20°10'16" – 20°14'20" N & 74°22'45"-74°30'00" E and covering approx. 120 sq.km. area. Total eighteen lava flows have been recognized on the basis of their physical characters in the field, out of 18 flows, 5 are showing 'compound' and rest 13 are shown by 'simple' nature in the field. All the flows are identified as 'pahoehoe' except one i.e. "aa" lava flow. Mineralogical, they are classified as basalt and olivine basalt.

Wide ranges of "zeolites group" of minerals are mineralized within the said lava pile. Zeolites are the "hydrous aluminum silicates" with predominantly Na and Ca cations. 'Six zeolites' are identified in the field and same are recognized by their optical properties namely – natrolite, thomsonite, heulandite, stilbite, chabazite and analcime. Structurally, they are classirfied in three sub groups – (i) the fibrous group (ii) the platy group (iii) the equant group and chemically as Na-zeolites, Ca – zeolites & both (Na – Ca zeolites).

Zeolites, generally occur in cavities, amygdales and fissures/pipe of basic volcanic rocks and they are crystallized at late stage in the hydrothermal environment. All the six recognized zeolites are well studied in the field and laboratory under thin section by the high precession petrological microscope first time for the area under investigations.

Key Words: Zeolitization, compound lava flow, Pahoehoe and Aa lava flow, Amygdales Hydrothermal, Deccan Traps.

Introduction

The authors carried out systematic mapping of the 366 meter thick Deccan Trap lava pile of Manmad area, which belongs to Nasik district of Maharashtra (Fig. 1). The investigation area lies in between the Cartesian coordinates 20°10'16" – 20°14'20" N & 74°22'45"-74°30'00" E and covering approx. 120 sq.km. area, falls in Survey of India toposheet no. 46 L / 8. Total 18 lava flaws of varied nature have been recognized in the field.

Materials and Methodology

Field investigations were carried out to mark the individual flow and their sequences in the field. All the flows have been separated by the distinct marker horizons in the form of RBH, GPB, pipe amygdales and green earth in the field.

Systematic sampling has been done along the travers, which were taken along and across the escarpment of the Manmad formation, Gupta et al. (2002). Identified zeolites in the field were further recognized by their optical properties in the laboratory. Field data and laboratory study were well incorporate to explain the zeolitization in the Manmad formation. They were also compared with their geochemistry to other worldwide distributed famous zeolites mineral chemistry.

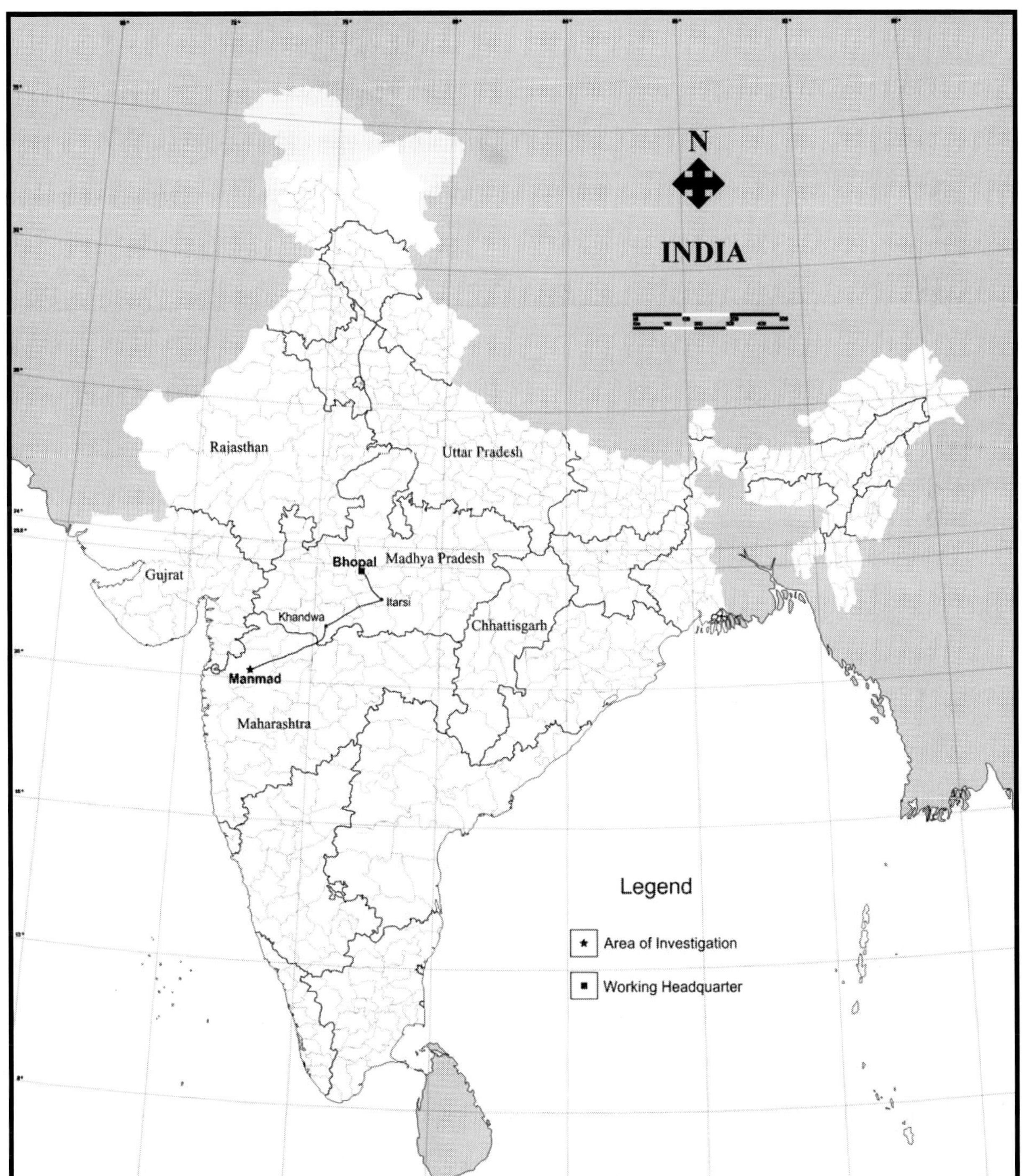

Fig. 1: Location Map of the Study Area

Geology of the Area

The Manmad volcanic rocks exhibits a thick lava flows sequence, constitute a part of the Deccan Trap Plateau. 18 lava flows have been observed and their individual thicknesses range from 5 to 60 meter. 5 flows are showing compound (Nicholas, 1933 and Walkar, 1969) and rest 13 flows are shown by simple nature in the field (Fig. 2). All the flows are identified in the field as "Pahoehoe" (Dutton, 1884) except one i.e. "Aa" (Morrison, 1985) type lava flow. These lava flows are melanocratic, hard and compact, but some flows are showing amygdaloidal and altered nature in the field. Classification of the Manmad formation and location of the zeolites within are shown in Table 1.

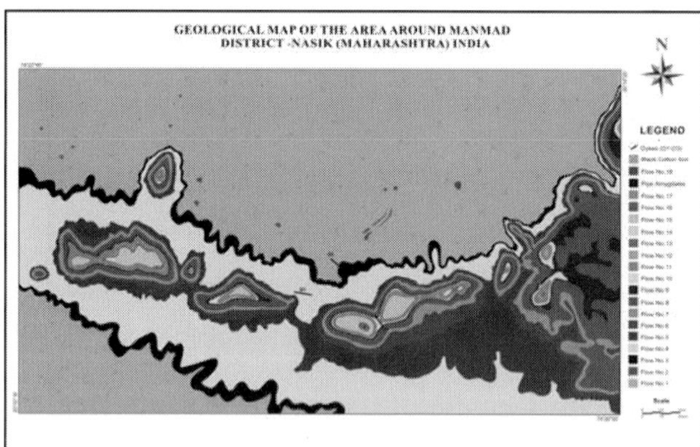

Fig. 2.

Petrographic variation

Generally most of the flows, when studied under microscope are composed mainly of plagioclase ranging from Ab_{49}-An_{51} to Ab_{32}-An_{68}, augite, olivine (Fo_{90}-Fa_{10}), ilmenite, magnetite and palagonite glass. The amygdaloidal flows consists of good amount of secondary minerals namely – Chlorite, Chlorophaite, calcite, quartz, zoisite, secondary glass and zeolites. Taxurally these flows are charcatrized by porphyritic (Glomarroporphyritic), nonporphyritic and sub-ophitic relationship. Mineralogical, they are classified as basalt and olivine basalt (Jain, 1995).

Zeolitization

Good amount of zeolites mineralization are seen in the amygdaloidal flows. 'Six zeolites' minerals of the zeolite group are identified in the field and well recognized in the laboratory under the high precision petrological microscope (Carl-Zeiss – Jenapol), namely (1) natrolite (2) thomsonite (3) heulandites (4) stilbite (5) chabazite and (6) analcime. (Jain, 1992).

Classification

(a) Based on structure (Morrison, 1985)

Structurally, they are classified in three sub groups:

(i) The fibrous group

(ii) The platy group

(iii) The equant group

(b) Based on Chemistry (Kerr, 1977)

Chemically, zeolites can also be grouped in three classes:

(i) Na - Zeolites

(ii) Ca - zeolites and

(iii) Na - Ca zeolites

Diagnostic properties of zeolites

(i) The fibrous group

Two minerals namely (a) Natrolite and (b) Thomsonite are fallen in this sub- group:

(a) Natrolite - A white color mineral composed of sodium hydrous aluminum silicate ($Na_2\ Al_2\ Si_3O_{10}.\ 2H_2O$), generally occurs as prismatic / radiating crystals (Fig. 3), showing uneven fracture and perfect one set of cleavage, Mohs scale hardness is 5.0 – 5.5, optically, colorless in thin section, it shows radiating aggregates (Fig. 4), parallel extinction along longitudinal section, moderate relief and 2V is varying from 61° to 63°.

Fig. 3: Field photograph shows Natrolite

Fig. 4: Photomicrograph shows Natrolite (X-Nicol, 10X)

Fig. 5: Field photograph shows Thomsonite

(b) Thomsonite - Sodium – calcium hydrous aluminum silicate ($NaCa_2\ Al_5\ Si_5O_{20}.\ 6H_2O$), snow white color mineral, occurring in masses of columnar radiating crystals (Fig. 5), shows uneven to sub conical fracture and one set perfect cleavage, hardness on Mohs scale is 5.0 – 5.5. Optically, colorless in thin section, shows fibrous aggregates (Fig. 6), parallel extinction, low relief and 2V is 50°, biaxial +ve sign.

(ii) The platy group

(a) Heulandites and (b) Stilbite are belonging in this sub - group:

Table 1. Classification of the Lava - flows

Flow (Nos.)	Thickness (Altitude in Meter)*	Number (Flow Unit)	Nomenclature and nature (Lava-Flow)	Zeolite (Mineralization)
18	11 (950-961)	Nil	Altered ferruginous lava-flow with olivine.	Stilbite
	05 (945-950)	-	---Red bole Horizon with pipe amygdales---	
17	15 (930-945)	-	Olivine basalt	-
16	10 (920-930)	-	Olivine rich basaltic lava flow --- Red bole Horizon---	-
15	60 (860-920)	Twenty one	Olivine bearing basaltic lava flow of compound nature	Natrolite
14	20 (840-860)	Nil	Basaltic lava flow --- Red bole Horizon---	Heulandite and Stilbite
13	15 (825-840)	Eight	Buff brown olivine bearing basaltic lava-flow of compound nature. --- Red bole horizon---	-
12	15 (810-825)	Eight	Basaltic lava-flow of compound nature	Thomsonite
11	35 (775-810)	Nil	Grey colored basaltic lava-flow	-
10	15 (760-775)	-	Picrite basalt of Ankaramite type	-
09	10 (750-760)	-	Olivine basalt ---Pipe amygdales---	Thomsonite and Stilbite
08	60 (690-750)	Twelve	Non-porphyritic pyroxene rich basaltic lava flow of compound nature --- Red bole Horizon---	2nd generation zeolite

07	20 (670-690)	Six	Grey colored basaltic lava flow of compound nature	-
06	15 (655-670)	Nil	Buff brown olivine basalt --- Red bole Horizon---	-
05	15 (640-655)	-	Olivine bearing basaltic lava flow ---Red bole Horizon---	Thomsonite, Stilbite and Chabazite
04	15 (625-640)	-	Olivine bearing doleritic lava flow	-
03	05 (620-625)	-	Buff brown amygdaloidal basaltic lava flow	-
02	20 (600-620)	-	Basaltic lava flow	Chabazite, Stilbite, Heulandite, Analcime and Natrolite
01	05 (595)	-	Pyroxene rich basaltic lava flow.	-

BASE NOT SEEN *(from MSL)

Table 2. Zeolites mineralization within the Manmad formation at a glance

S. N.	Name of Zeolite	Formula	Crystal System	Classification based on		Hardness Mohs	2V	Relief	Interference fig. and sign	Diagnostic properties
				Structure	Chemistry					
1	Natrolite	($Na_2 Al_2 Si_3O_{10}. 2H_2O$)	Orthorhombic	Fibrous Group	Na-Zeolite	5.0-5.5	61° to 63°	Moderate	Not obtained	Prismatic / Radiating aggregates, parallel extension along longitudinal section
2	Thomsonite	($Na Ca_2 Al_5 Si_5O_{20}. 6H_2O$)	Orthorhombic	Fibrous Group	Na-Ca Zeolite (Sodium rich)	5.0-5.5	50°	Low	Biaxial +ve	Columnar / Fibrous aggregate, parallel extinction
3	Heulandite	$Ca Na_2 Al_2 Si_7 O_{18}. 6H_2O$	Monoclinic	Platy Group	Na-Ca Zeolite (Calcium rich)	3.4-4.0	40° to 48°	Low	Biaxial +ve	Tabular crystals with vitreous lusture, extinction parallel to cleavage traces
4	Stilbite	$Na Ca_2 Al_5 Si_{13} O_{36}. 14H_2O$	Monoclinic	Platy Group	Na-Ca Zeolite (Sodium rich)	3.5-4.0	33°	Low	Biaxial –ve	Sheaf like aggregates at tabular crystals, wavy extinction
5	Chabazite	$Ca Na_2 Al_2 Si_4O_{12}. 6H_2O$	Hexagonal/ Trigonal	Equant Group	Na-Ca Zeolite (Calcium rich)	4.0-4.5	31°	Moderate	Biaxial +ve	Rhombohedral crystals with symmetrical extinction to crystal outline
6	Analcime / Analcite	$Na Al Si_2 O_6. H_2O$	Cubic/ Isometric	Equant Group	Na-Zeolite	4.5-5.0	Not obtained	Moderate	Not obtained	Trapezohedral crystals, two set cleavage at right angle

Table 3. Chemical Analyses of the Manmad Formation with their normative composition

Major oxides	Manmad Flows							
	F2	**F5**	**F8**	**F9**	**F12**	**F14**	**F15**	**F18**
SiO_2	48.92	48.00	49.02	48.72	49.40	48.25	49.34	50.25
Al_2O_3	14.00	14.19	13.75	13.60	14.50	14.00	13.80	13.65
TiO_2	1.77	2.10	1.89	1.95	2.02	1.98	1.90	1.96
Fe_2O_3	5.25	5.40	5.48	5.80	5.20	5.13	5.35	5.21
FeO	5.40	5.80	6.25	7.00	6.25	6.00	6.47	6.50
MnO	0.16	0.17	0.19	0.20	0.17	0.19	0.16	0.18
CaO	11.95	8.90	10.55	9.00	11.20	11.32	10.98	10.80
MgO	5.95	6.70	6.65	7.45	5.68	6.75	6.38	6.25
Na_2O	1.77	1.97	2.55	2.13	2.37	2.18	2.25	1.66
K_2O	0.50	0.59	0.51	0.56	0.76	0.48	0.61	0.60
P_2O_5	0.13	0.19	0.16	0.21	0.20	0.21	0.20	0.20
H_2O^+	1.21	1.76	1.15	0.90	1.06	1.20	0.70	0.65
H_2O^-	3.17	3.43	2.08	1.80	1.35	1.85	2.12	2.11
Total	100.18	99.20	100.23	99.32	100.16	99.59	100.26	100.02

Normative Composition

Major oxides	Manmad Flows							
	F2	**F5**	**F8**	**F9**	**F12**	**F14**	**F15**	**F18**
Q	7.95	7.91	4.27	6.14	6.30	4.36	6.53	9.73
Or	2.90	3.55	2.87	3.07	4.45	2.88	3.43	3.43
Ab	15.37	17.83	22.15	18.44	19.91	19.02	19.17	14.53
An	29.89	29.86	25.51	26.64	26.69	27.97	26.28	28.83
: Wo	12.96	6.79	11.49	8.20	11.72	12.27	11.79	10.61
Di : En	9.71	5.10	8.25	5.43	8.30	9.02	8.34	7.49
: Fs	1.93	0.85	2.18	2.05	2.38	2.05	2.42	2.17
Hy : En	5.85	12.76	8.87	13.63	5.90	8.51	7.93	8.52
: Fs	1.11	2.38	2.18	4.10	1.72	1.92	2.28	2.58
01	-	-	-	-	-	-	-	-
Mt	8.00	8.39	8.14	8.10	7.66	7.70	7.79	7.86
I1	3.49	4.21	3.76	3.74	3.80	3.94	3.71	3.90
Ap	0.35	0.36	0.35	0.35	0.35	0.35	0.35	0.35
Total	100.01	99.99	100.02	99.89	99.18	99.99	100.02	100.00

Table 4. Worldwide distributed famous Zeolites analyses, After Deer et al., 1992

	1	**2**	**3**	**4**	**5**
SiO_2	47.60	37.17	38.21	57.17	53.77
Al_2O_3	26.40	31.93	22.66	17.03	16.77
Fe_2O_3	0.51	-	0.10	0.04	0.08
MgO	-	-	0.04	0.02	0.03
BaO	0.09	-	0.90	0.31	-
SrO	0.02	-	5.27	0.19	-
CaO	0.10	13.98	6.57	7.13	7.75

Na_2O	15.41	4.00	0.48	1.39	1.81
K_2O	0.01	Tr.	3.89	0.73	0.07
H_2O^+	9.85	13.35	17.02	16.71	19.75
Total	99.99	100.43	99.56	100.72	100.03

Number of ions on the basis of framework of

	80 O	80 O	24 O	72 O	72 O
Si	24.17	19.88	7.05	26.70	26.26
Al	15.80	20.12	4.93	9.37	9.66
Fe^{3+}	0.20	-	0.01	-	0.03
Mg	-	-	0.01	0.01	0.02
Na	15.18	4.14	0.17	1.26	1.71
Ca	0.06	8.01	1.30	3.57	4.06
Sr	0.01	-	0.56	0.05	-
Ba	0.02	-	0.07	0.06	-
K	0.01	-	0.92	0.43	0.04
H_2O	16.58	23.80	13.20	26.02	32.17
Z	39.97	40.00	11.98	36.07	35.92

1. **Natrolite**, Schaffhausen, Germany (Alberti, A., Pongilsuppi, D. & Vezzalini, G., 1982, Neues Jahrb. Min., Abhdl., **143**, 231-48).
2. **Thomsonite**, radiating glassy prisms, Old Kilpatrick, Dunbartonshire, Scotland, UK (Hey, M.H., 1932, Min. Mag., **23**, 51-125).
3. **Chabazite**, Casal brunori, Rome, Italy (Passaglia, E., 1970, Amer. Min. **55**, 1278-301. Includes H_2O^- 4.42 percent).
4. **Heulandite**, cavity in Tertiary basalt, Faroes (Alberti, A., 1972, Tschemaks Min. Petr. Mitt., **18**, 129-46).
5. **Stilbite**, amygdaloidal Tertiary basalt, Flodigarry, N. Skye, UK (Passaglia E. et al., 1978, Bull. Min. **101**, 368-75).

(a) Heulandite - Sodium – calcium hydrous aluminum silicate (Ca Na_2 Al_2 Si_7 O_{18}. $6H_2O$). White, reddish brown, distinct tabular crystals (Fig. 7), vitreous lusture, uneven fracture, perfect one directional cleavage, Mohs hardness is 3.4-4.0, Optically, colorless in thin section, extinction parallel to cleavage traces (Fig. 8), biaxial +ve, low relief and 2V varies from 40°-48°.

Fig. 6: Photomicrograph shows Thomsonite (X-Nicol, 10X)

Fig. 7: Field photograph shows Heulandite

(b) Stilbite - Sodium – calcium hydrous aluminum silicate ($NaCa_2$ Al_5 Si_{13} O_{36}. $14H_2O$), white colour, occurs in sheaf like aggregates of tabular crystals (Fig. 9). Pearly lusture with uneven fracture, one directional cleavage, hardness 3.5-4.0. Optically, colorless in thin section, low relief, perfect one set cleavage, wavy extinction (**Fig. 10**), biaxial –ve and 2V is 33°.

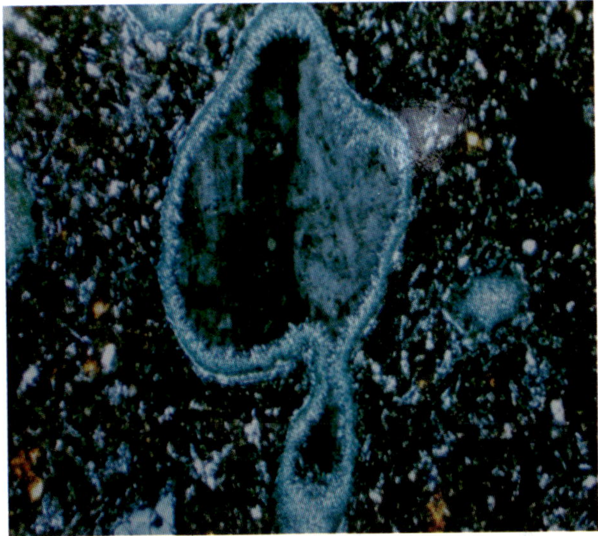

Fig. 8: Photomicrograph shows Heulandite (X-Nicol, 10X)

Fig. 9: Field photograph shows Stilbite

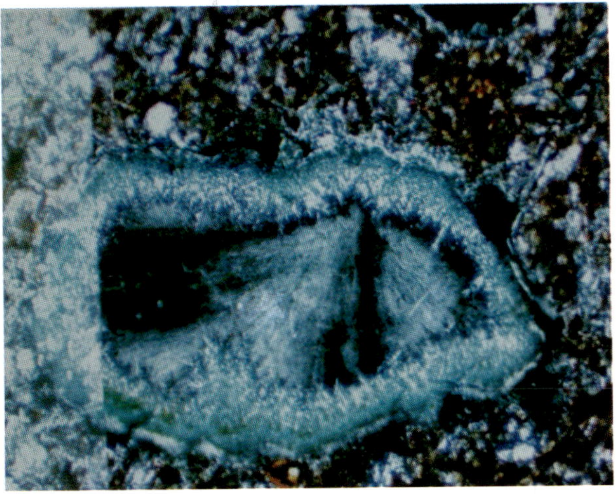

Fig. 10: Photomicrograph shows Stilbite (X-Nicol, 10X)

(iii) The equant group

(a) Chabazite and (b) analcine are shown the characters of this sub - group:

(a) Chabazite - Calcium – sodium hydrous aluminum silicates (Ca Na_2 Al_2 Si_4O_{12}. $6H_2O$), white, reddish colors, euhedral (Rhombo-hedral, Fig. 11) crystals, imperfect cleavage, vitreous lusture, hardness 4.0 - 4.5, Optically ,colorless in thin section, shows rhombohedral form (Fig. 12), relief moderate, extinction symmetrical to crystal outline, biaxial +ve and 2V is 31°.

Fig. 11: Field photograph shows Chabazite

Fig. 12: Photomicrograph shows Chabazite (X-Nicol, 10X)

(b) Analcime - Sodium, Hydrous aluminum silicates (Na Al Si_2 O_6. H_2O), it is also called as analcite. White color, occurs in equant crystals of trapezohedral (Fig. 13) cleavage imperfect, hardness 4.5 – 5.0. Optically colorless, relief moderate, cleavage two directional at right angle, some time it occurs in irregular ground mass (Fig. 14) also.

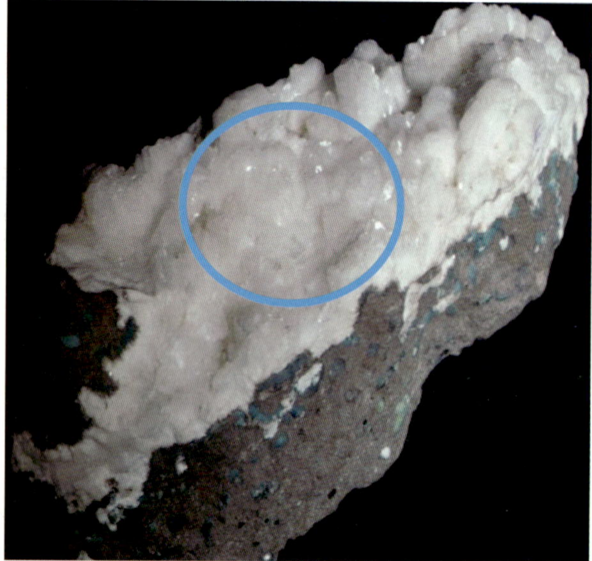

Fig. 13: Field photograph shows Analcime

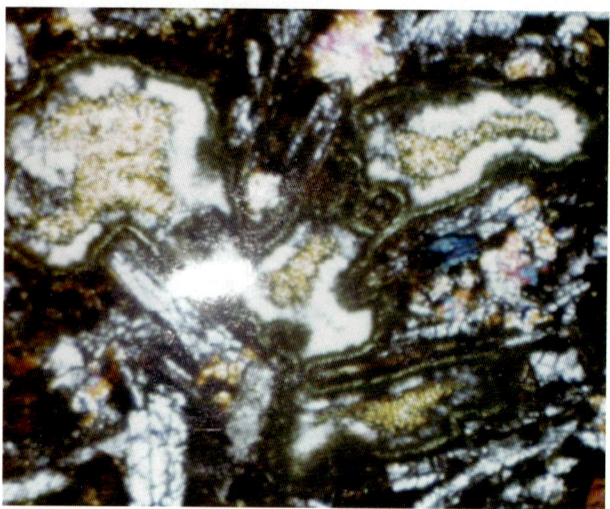

Fig. 14: Photomicrograph shows Analcime at the periphery of cavity (X-Nicol, 10 X)

Details of zeolites mineralization within the Manmad formation at a glance are shown in Table 2.

Geochemistry

Zeolites mineralization within the lava flows of the Manmad formation are seen only in eight flows out of eighteen under the microscope, namely flow nos. F02, F05, F08, F09, F12, F14, F15 and F18, their detailed whole rock chemical analyses (Jain 1992) are given in Table 3. World wide distributed and famous Zeolites group of minerals naming – natrolite, thomso-nite, chabazite, heulandite and stilbites and their chemical analyses are shown in Table 4 for geochemical correlation to the Manmad formation after the Deer et al. (1992).

Chemically classified all the zeolites of the study area, namely - Na, Ca and Na-Ca zeolites are shown well correlation with their geochemistry to the worldwide distributed famous zeolites mineral analyses.

Economic importance of the zeolites

White, red and yellow colors well crystallized zeolites are sometime used as gem stones. Fibrous, radiating and prismatic aggregates of crystals are well used as decorative stones; they are also known as decorative zeolites. Zeolites show peculiar powerful "base exchange" properties, Na-zeolites are originally used as water softeners. Dehydrated zeolites are used as "molecular sieves", have wide use in industry for the production of oxygen from air (Blackburn and Dennen, 1990). Sometime as absorbents and desiccants etc.

Discussion and conclusions

Zeolites are the hydrous aluminum silicates with predominantly Na and Ca cations, sometimes potassium and barium. In many ways, they are analogous in composition to the feldspar. The name zeolites is derived from the Greek words 'Zein' means **to boil** and 'lithos' means **stone**, that's why they are also known as "Boil Stone" (Read, 1984). Zeolites are the secondary minerals commonly occur in cavities, amygdales, vesicles and fissures in basic volcanic rocks, where they often form well developed crystals.

Rocks of the study area belong to basic volcanic, showing amygdales, cavities, vesicles and fissures (Fig. 15) in remarkable amount in the field (Fig. 16). Generally, vesicles / cavities are filled by secondary minerals in the form of zeolites. That is why amygdaloidal lava flows with pipe amygdales are seen in the field (Fig. 17). First generation zeolites are seen commonly in the field, the evidences of same are also found under the microscope in the lab. At few places, second generation zeolites formation evidences are seen, where plagioclase mineral is partly / fully converted in to the zeolites (Fig. 18). It is very interesting to note that, at a place two zeolite minerals of different group are seen in a cavity, namely- thomsonite and stilbite (Fig. 19), belongs of fibrous and platy sub-group respectively, showing different characters under the microscope.

Fig. 15: Field photograph shows cavitics, vesicles, fissures and Pipe amygdales in a lava flow

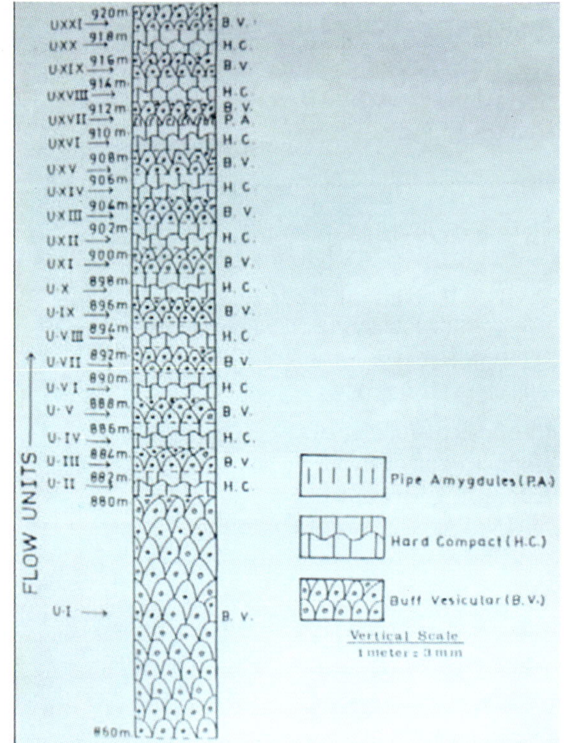

Fig. 16: Section graph shows alternate Compact and vesicular lava flow (Compound nature) with Pipe amygdales

Fig. 17: Field photograph shows Amygadaloidal Lava flow

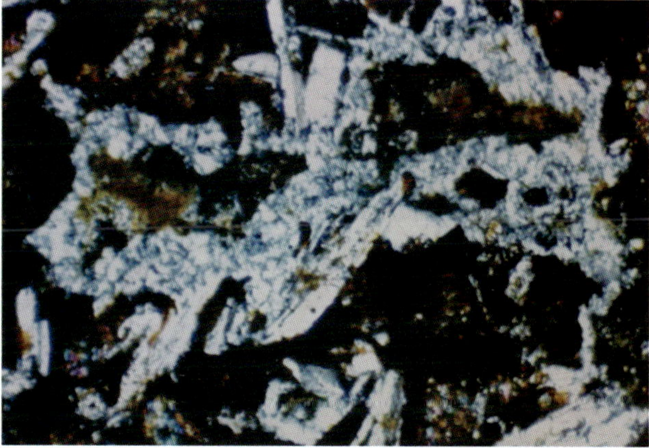

Fig. 18: Photomicrograph shows 2nd Generation Zeolitization (X- Nicol, 10 X)

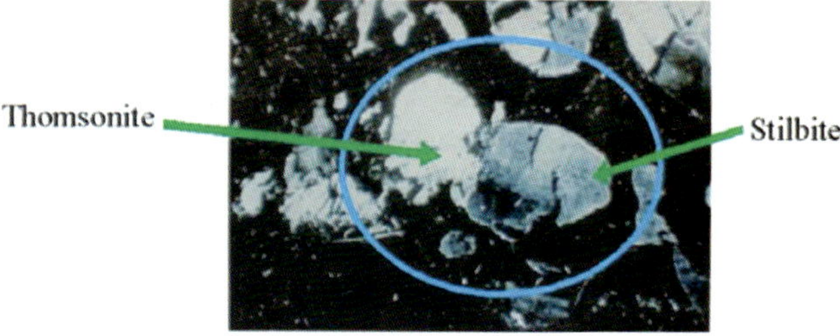

Fig. 19: Photomicrograph shows Stilbite and Thomsonite in a cavity (X- Nicol, 3.2 X)

Structurally, zeolite show aluminisilicate framework containing larger cavities and channels in comparison to some other analogues compositional minerals like feldspathoids and feldspars, because of their lesser specific gravity (Zeolite 2-2.3, feldspathoids 2.3-2.5 and feldspars 2.6 – 2.7). Zeolites are having some peculiar properties i.e. ion exchange and molecular absorption, which differentiates from the other analgous mineral groups by the presence of water molecules within the structural channels. They are relatively loosely bound to the framework and cations, to removed and replaced without disturbing the framework bonds. Cations (Na, Ca and K) replacement in the Zeolites can be expressed by these equations Ca $\leftrightarrow$2 (Na, K) and Ca Al $\leftrightarrow$ (Na, K) Si. Cation exchange capacity in general diminishes with loss of water content. Na is more mobile since it is monovalent. While Ca- zeolites absorb more water.

At the surface, zeolites are formed by the suitable materials like - volcanic glass, tuff, nepheline, feldspar, etc. in high pH conditions, resulting from the concentration of sodium/ calcium carbonate – bicarbonate by evaporation in an arid or semi-arid environment.

The known occurrences of zeolites are probably those which form well-developed crystals in the vesicles and amygdales of basalts. These zeolites are generally considered to be the result of late-stage fluids that permeated the basalts after their extrusion. Other famous occurrences of zeolites in basaltic rocks include those from the Deccan traps of India.

Field, laboratory and geochemical study of all the identified zeolite mineral of the Manmad formation, suggest that, they are the secondary minerals and crystallized in the hydrothermal environment at the late stage. Which are belonging of Deccan Trap lava pile of India.

Acknowledgement

The authors are wish to express their gratitude to Late Dr. M.G. Paithankar, outgoing Professor and Head and then Dr. R.N. Saxena, present Professor & Head, Department of Applied Geology, Barkatullah University, Bhopal for their kind cooperation and constant help during the entire work.

References

Blackburn W.H. and Dennin W.H., First Indian Reprint, 1990, Principal of mineralogy (ISBN-81-85392-53-6), Universal Book Stall, 5, Ansari Road, New Delhi - 110002 pp. 520-529.

Deer, W.A., Howie R.A. and Zussman J., 1992, Second edition. An introduction to the rock forming minerals, ELBS, educational Low priced book Scheme, Longman Scientific & Technical group (RE) Ltd., Harlow, England.

Dutton, C.E. 1884, Hawaiian Volcanics. B.U. S. Geol. Survey, 4th. Ann. Rept. Pp. 75-219.

Gupta D.C., Chatterjee A.C. and Jain P.K., 2002, Geochemical strategraphy of the Deccan Traps of Manmad, District – Nasik (Maharashtra) India, Bulletin of the Indian geologist association 35(1): 29-41, Chandigarh – 160014.

Jain P.K. 1992, Petrological study of the lava flows around Manmad, District – Nasik (Maharashtra), Unpublished Ph.D. thesis, Barkaullah University, Bhopal (M.P.), India pp. 52-96.

Jain P.K., 1995, A geological note on area around Manmad (Nasik) Maharashtra, Asian J. Ep. Sci (ISSN 0971 – 5444) Vol. 9, (1 & 2), pp. 19-23.

Kerr, P.F. 1977, Optical mineralogy, 4th edition, McGraw Hill Ind. USA, pp. 329-340.

Morrison W.G. 1985, A Dictionary of Geology, First edition, CBS Publishers & Distributors, 485, Jain Bhawan, Shahdra, Delhi – 110032 (India).

Nicholas R.L., 1983, Flow Units in basalt, Jour. Geol. 44, 617-630.

Read H.H., 1984, Rutleys elements of Minerology, Twenty Sixth edition, CBS Publishers & Distributors, 485, Jain Bhawan, Shahdra, Delhi – 110032 (India).

Walker G.P.L., 1969, Some observations and interpretations on the Deccan Traps, Unpub. Rep. Centre for Advance Study, Univ. of Saugar, 21 pp.

Hon'ble Reviewers

1. **Prof. T.K. Biswal**
   Department of Earth Science
   IIT Bombay, Powai
   MUMBAI - 400 076
2. **Prof. Santosh Kumar**
   Department of Geology
   Kumaun University
   NAINITAL - 263 002

I-10

Geological Characteristics and Genetic Model of Hydrothermal Breccia Fluorite Deposit in Malani Igneous Suite at Karada, Western Rajasthan, India

K.L. Shrivastava, Virendra Gaur and C.P. Dadhich

Department of Geology, J.N.V. University, Jodhpur-342005, India,
Email- klsgeology@yahoo.co.in

Abstract

The paper present investigations on the geological characteristics to suggest genetic model of the fluorite deposit at Karada and is the study of an economic deposit of considerable size in the Malani Igneous Suite (745±10Ma) after Jalera Khurd, the largest suite of anorogenic acid volcanics in India and the third largest acid volcanics of the globe. The fluorite deposit is situated (Latitude 72^0 14 N' and Longitude 25^0 15E') about 25 kilometer southwest of Bhinmal a tehsil headquarter in Jalore district, Rajasthan. India. Plume tectonics, circular intrusions and collapse caldera conditions are known for the regional geology. Bimodal volcanics of the Malani Igneous Suite (Neoproterozoic) at Karada is consisting of two mafic units and one felsic unit . The felsic unit is sandwitched between lower mafic unit I and upper mafic unit II. Lower mafic unit consist of mafic components of sub volcanic to volcanic nature and are of andesitic to trachytic composition with little variations. Middle felsic unit is only of porphyric rhyolite and further overlained by basaltic trachytic mafic components of mafic unit II. Volcanic breccia with fluorite is youngest among igneous components the host rocks. The main fluorite bodies are confined in volcanic breccia and having its penetration deep inside. The orientation of all mineralized veins is N 43^0 E-S43^{0}W with subvertical dip towards west. In both, Rekha and Santoshi mines, Fluorite is well filled in interspace available between the fragments of acidic and mafic volcanic rock apart from low grade fluorite fragments, finally constituting the volcanic breccia. In open spaces such as coatings, fillings and cocade structure, fluorites are very common and making this deposit a rich fluorite deposit. Detailed geochemical and petrological studies favour that the host rock magma must have been originated relatively at deeper parts of the earth. Crustal assimilation and fractional crystallization are the proposed mechanisms. The range of geological features displayed by fluorite breccia indicate that disruptive intrusion by a turbulent fluidized system was the dominant process, initially. The intrusive aspects of fluorite breccia show angular killas fragments encompassing a wide range of sizes all cemented by microcrystalline fluorite itself. This is followed by two episodes of collapse, a major and a minor; resulting in enclosure of early-formed fluorite fragments into later-formed fluorite matrix. These episodes were followed by pervasive silicification and metasomatic alterations. A thirteen stage genetic model is hence, proposed for the fluorite ore body, which combines the features of both, intrusive breccia and collapse breccia type hydrothermal deposit of volcanogenic origin.

Keywords: Malani Igneous Suite, Fluorite Mineralization, Hydrothermal Breccia, Volcanogenic Mineralization, Karada deposit.

Introduction

The Neo-Proterozoic Igneous activity in Northwestern India and adjoining part of Pakistan, named as Malani Igneous Suite is bimodal, although chiefly acidic, with extrusion and intrusion events to constitute a Large Igneous Provenance (LIP). Scattered out-crops of Malani cover an area of 50,000 sq. km. from south of Sirohi to north of Pokaran and east of Jodhpur to the edge of Thar desert of India where exposures are covered by sand dunes. Roy and Jakhar (2002) have recognized three stages of igneous activity on the basis of field evidences. Volcanics of the first stage are mostly basalt with occasional andesite or trachybasalts followed by a second stage of the voluminous outpouring of peralkaline and peraluminous rhyolite, basalt, dacite and trachyte flows covering the former. The third stage ceased with the outburst of ash flow deposits. The dominant felsic volcanics are rhyolites and rhyodacites spread over an area of about 31,000 km^2. The other rock types associated with the rhyolites are trachyte, dacite, pitchstone, welded tuff, vitric, lithic and crystal ash, ignimbrite, obsidian, pyroclastic slates, agglomerate, volcanic breccia and volcanic conglomerates. The majority of the acid volcanics are high potassic and a few are calcalkaline or low–potassic in composition. The Malani Large Igneous Provenance so far have not shown any mineralization of considerable size except Jalera Khurd Fluorite deposit (Shrivastava and Gaur, 2009); although the mineralization in Malani volcanic is well expected and pointed out by some earlier works (Kochchar, 1984, Shrivastava et al. 2000, Gaur, 2002). However, present investigation of fluorite deposit at Karada will be yet another important scientific/technical study of an economic deposit of considerable size in Malani volcanic, the largest suite of anorogenic acid volcanics in India. the present investigation includes field study, petrology and geochemistry of host rocks and mineralization to finally construct a genetic model of the karada fluorite deposit.

The Karada fluorite deposit (Latitude 72°14'N and Longitude 25°45'E) is situated in the form of two open cast mines. First, Rekha mine is five kilometer east and second, Santoshi mine is ten kilometer southeast of the village Karada which is situated 25 kilometer northwest from Raniwara, on Raniwara-Sanchore road and is part of Bhinmal tehsil, Jalore district, Rajasthan (Fig. 1a and 1b). The study area forms part of toposheet number 45 D/1 of Survey of India. The nearest Railway station is Raniwara of northern Railways of Indian Railways. The Karada thus is approachable by a 25 kilometer long metalled road from Bhinmal and further 5 kilometer by a cart road.

The terrain conditions vary widely, geomorphologically ranging from ergs or typical dune fields with interdunal plains to rocky and gravel pediments mainly. Older dunes are generally established with alluviated calcium carbonate horizon within it. Still flanks and crests of many are reactivated. Many active dunes of younger alluvium have also been formed in the study area. The rock outcrops are available as semi-circular hills, tors, inselbergs and scattered hummocks. Also at Karada fluorite deposit, some thirty percent of the area is blanketed by windblown sand and sand dunes, typical of desert.

Previous Work

There are many geological names have been given to this region. La Touche (1902) named the volcanics as "Malani series", whereas Coulson (1933) called them as "Malani System", Pascoe (1960) described them as "Malani granite and volcanic suite" and grouped them under Trans-Aravalli Vindhyan. Paliwal (1992) referred them as "Malani Series", Malani Volcanic Series", Malani beds; Murthy et al. (1961) grouped the rhyolites and associated granites as Malani Igneous Suite, Chore and Mohanty (1998) called these rocks as "Malani Igneous Province", while Bhushan and Khullar (1998) named them as "Malani Igneous Complex". The authors of this paper prefer to use the term 'Malani Igneous

Suite'. It is more as a conventional name, a noun, rather than in its true geological meaning.

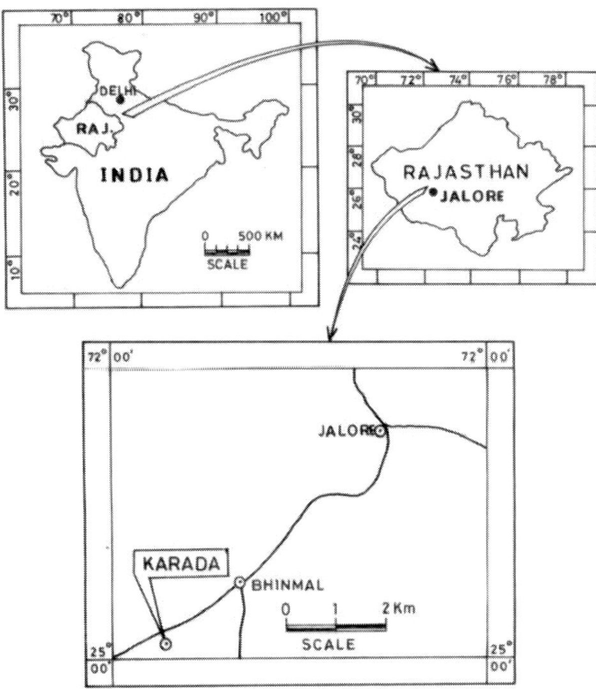

Fig. 1a. Location map of Karada Fluorite deposit.

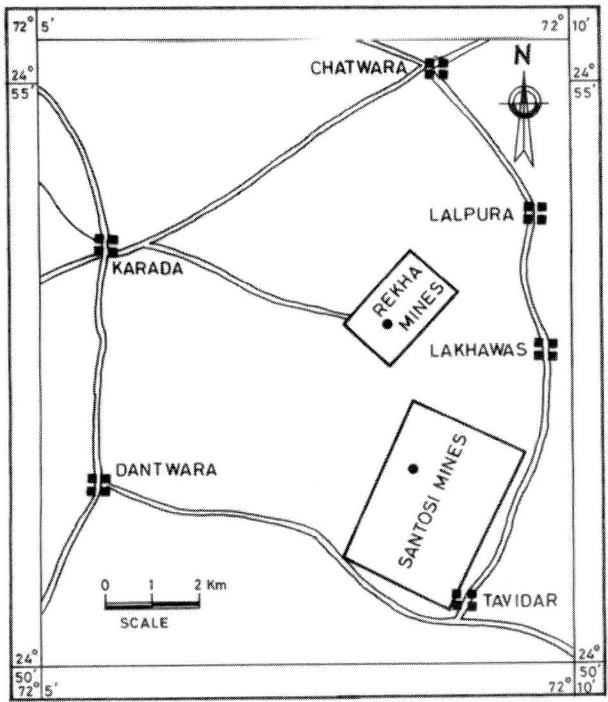

Fig. 1b. Location of Karada Fluorite Mine.

Several other attempts have been made to study different aspects of these rocks by La Touche, 1902; Heron, 1932; Pascoe, 1959; Kochhar, 1983, 1984, 1998; Murthy, 1979; Pareek, 1984; Bhushan, 1985, 1989, 1995, 2000; Vallinayagam, 1998; Vallinayagam and Kochhar, 1998; Srivastava, 1988; Srivastava et al., 1989; Rathore et al., 1996 a and b; Basu, 1982; Paliwal 1992, 1998, 1999 Pandit et al., 1999; Shrivastava et al. 2000; Paliwal and Rathore 2000 and Torsvik et al., 2001. La Touche (1902) discussed the physical, petrographic features and nature of eruption. Most of his field observations and interpretations continue to remain undisputed. Pascoe (1960) attempted a generalized classification based on previous data. His stratigraphic order however, did not include any basic volcanics, which are now firmly believed to prevail in many areas below the rhyolites. Murthy et al. (1961) studied the petrographic characters of sodic and peraluminous rhyolites of Barmer. The acid volcanics of the same area have also been investigated by Venkataraman et al. (1968), Krishnan (1968), Mukherjee (1962) and Mukherjee and Roy (1981). Pareek (1981a, b) attempted petrochemistry of Malani Igneous Suite (MIS). Kochhar (1984) and Eby and Kochhar (1990) studied the volcano-plutonic association and type of magmatism. Srivastava (op.cit.) elucidated the major and trace element geochemistry of felsic volcanics from Pali district. Maheshwari et al. (1996) described the crustal influence in the petrogenesis of rhyolites based on oxygen isotope studies. Dhar et al. (1996), on the basis of Sr, Pb and Nd isotope studies concluded that the Siwana alkaline magma for the granites as well as the rhyolites is mantle derived. Their Sr–isotopic data (Sr^{87}/Sr^{86} 0.7062 ± 0.0020) have indicated some crustal contamination and presence of Archaean crust in the area. Bidwai and Krishnamurthy (1997) illustrated some of the volcanic landforms from parts of the Siwana Caldera sequence. Barring some marginal attempts, a comprehensive genetic model pertaining to generation of both basic and acid magma, ascent of the fluids and emplacement of the lava piles and their physico-chemical condition of crystallization are yet to be clearly made out.

The Mineralized breccias of hydrothermal origin are known from various mineralized provinces in the world (Locke, 1926; Bryner, 1961; Kents, 1964; Johnston and Lowell, 1961; Llambias and Malvicini, 1969; Sawkins, 1969; Sillitoe and Sawkins, 1971; Fletcher, 1977; Knutson et al., 1979; Gultekin, 2005; Shrivastava and Gaur, 2009). Generally such bodies constitute one of the characteristic features of hydrothermal systems generated around plutons of acids to intermediate composition emplaced at shallow depths (<4-5 km) in the crust. The frequent occurrence of tourmaline-rich matrices in these intrusive breccias (Kents, 1964; Sillitoe and Sawkins, 1971; Fletcher, 1977; Grant et al., 1980) suggests that boron is strongly enriched in the residual phase during the final stages of crystallization of certain granodioritic and granitic plutons. Tourmaline breccia pipes are notably well developed in the copper-bearing porphyry instrusives in the Western Cordillera of the Andes (Kents, 1964; Sillitoe and Sawkins, 1971) and similar breccias are invariably found to be an important feature of the sub volcanic tin deposits of the Eastern Cordillera of the Central Andes in Bolivia (Grant et al., 1980). The Cornubian province of the Hercynian tourmaline within hydrothermal vein structures and as a major constituent of those granites and killas country hydrothermal solutions emanating from the granites (Dines, 1956; Exley, 1959; Charoy, 1979). In recent years it has been noted that more dynamic expressions of the activity of boron-rich fluids are also present in the Cornubian province suggesting that it is a chemical and mechanical sense there are certain similarities between the behaviors of the eruptives in the Cornubian region and those of the Bolivian tin best (Halls et al. 1977; Goode and Taylor, 1980). Sillitoe and Sawkins (1971) have given an account of certain tourmaline bearing breccia pipes in Chile and decided, on the basis of geological observation, that the process chiefly responsible for the formation of the Chilean pipes was one of chemical solution and in situ chemical brecciation (Sawkins, 1969) followed by collapse under the influence of gravitational forces and cementation by the residual boron-rich hydrothermal solution. An epithermal fluorite–barite–REE deposits of vein type in rhyolite–dacite–andesite–trachyte–basalt sequence at Kizilcaoren, Turkey have been evaluated by Gultekin (2005). The fluorite deposit at Jalera Khurd (Gaur, 2002; Shrivastava and Gaur, 2009) is the only economic deposit of considerable size in the Malani Igneous Suite. The fluorite deposit is situated one kilometre northwest of village Jalera Khurd, (Lat. 72°08'69"N and Long. 24°48'00"E) about 15 kilometer northwest from Raniwara, Bhinmal tehsil, Jalore district, Rajasthan, India. Bimodal volcanics of the Malani Igneous Suite (Neoproterozoic) consisting of lower porphyritic rhyolite and upper porphyritic trachyte are the host rocks. Out of three prominent Joint sets, at the fluorite deposit site, only one, named Joint I, has been mineralized. The main ore body constituting four veins of fragmental mass collectively constituting a tabular body, with its average thickness, length and depth 20m, 500m and 100m, respectively, within an elongated stockwork. The range of geological features displayed by fluorite breccia indicate that disruptive intrusion by a turbulent fluidized system. A twelve stage genetic model is suggested (Shrivastava and Gaur, 2009) for the fluorite ore body, which combines the features of both hydrothermal intrusive breccia and hydrothermal collapse breccia to form the economic deposit, after performing a detailed field study, petrological and geochemical study and details of the mineralization.

Regional Geology

The Neoproterozoic geology of Rajasthan as stated by Roy and Jakhar (2002), begins with opening of shallow rift basins to deposit non volcanic sediments described as the Sirohi Group. Before that, there are diorite intrusion in South Delhi Fold Belt (DFB) (ca. 1000 Ma, Volpe and Mac daugall, 1990) and exhumation of granulites around 1128 Ma (Fareeduddin and Kroner, 1998). West of DFB, the Sirohi Group allowed deposition of fine clastics and chemogenic sediments which were deformed and metamorphosed concomitantly with the emplacement of A-type granitoids (ca. 850 Ma) termed Erinpura Granites. Culmination of orogenic activities is marked by closing of Sirohi basins and concomitant emplacement of Erinpura Granites.

It was further followed by intrusion of a huge cratonic platform of Rajasthan Shelf, started geological activities in western Rajasthan with anorogenic, plume–related magmatism along with the development of ephemeral basins. The change of geological scenario from orogenic to typical platformal activities is observed at Sirohi, where an undeformed and unmetamorphosed volcano–sedimentary ensemble, termed the Sindreth Group, occurs unconformably overlying the deformed and metamorphosed sediments of the Sirohi Group. Similar relationships have also been observed in the hills of Diri, Gurapratap Singh and Manihari in the Pali District. The volcanic succession in these areas is dominantly bimodal representing both mafic and felsic volcanics. Associated with these volcanics are sedimentary conglomerates, grits and arkosic sandstones with minor shale beds. These dominantly volcanic sequences, which constitute linear N–S trending outcrops are the oldest beds of late–Neoproterozoic volcanic/plutonic & (minor) sedimentary formations of the Malani Group (Roy, 1998), popularly called 'Malani Igneous Suite', though not in strict geological sense.

The Neoproterozoic Malani rocks of western Rajasthan, occur as unconformity bounded sequence of dominantly bimodal volcanics with minor interbedded sediments and peralkaline meta–aluminous to peraluminous granites. Outcrops of rocks belonging to the Malani Igneous Suite (MIS) occur over widely scattered areas of western Rajasthan and Sindh (Pakistan).

Stratigraphically, Malani overlies Sirohi Group (ca. 835 Ma) unconformably. The upper age limit of the Malani is provided by the Marwar Supergroup (Pre–Palaeozoic, Pareek, 1984) which overlies unconformably the granites and lava flows of the Malani (745 Ma to 680 Ma).

Lithostratigraphy of MIS, according to Bhushan (2000) is : –

Phase III: Dykes of basic, acid, trachyte porphyry and andesite. Porphyry dykes, aplite and dioritic plugs.

Phase II: Intrusion of discordant peraluminous and peralkaline granites as plutons, ring dykes and bosses.

Phase I: Eruption of basic flows followed by voluminous acid flows, and culminating with eruption of ash flow deposits.

There are few uncertainties about Phase III of the lithostratigraphy of MIS. Roy and Jakhar (2002) questioned the characteristics of dykes of Phase III as there has also been KTB (Cretaceous–Tertiary Boundary) underplating activity in the region (Bhushan, 1984, Basu et al., 1993). Also there is possibility of post-granitic, volcanic activity (equivalent to Phase III). A second phase of flow post dating granites is inferred from the Rb–Sr isochron age of ultra–potassic rhyolite by Rathore et al. (1999).

There are evidences that the Malani area was reactivated during Tertiary times to produce a few younger volcanics. These are considered as manifestations of the Reunion plume (Basu et al. 1993, Roy, 2003). Such as those at Sarnu–Dandali in Barmer District (Basu et. al. 1993, Bhushan and Chandrashekar, 2002), Tavidar in Jalore District (Rathore, 1995) and Mundwara in Sirohi District (Subrahmanyan and Rao, 1972; Rathore et al. 1996). Some of the good reviews on Malani and associated rocks, that appeared in the literature are Pareek, 1984; Bhushan, 2000; Roy and Jakhar, 2002; Bhushan and Chandrashekhar, 2002; and Sharma, 2004.

The Geology of the area

Geological map of the area is presented in Fig. 2. And Geological cross sections of the same in Fig. 3. The local geology of the Karada deposit shows that the various lithounits available in the area, which have been shown as lithostratigraphic succession Table 1. As a matter of fact there are two mafic units and one felsic unit in the area. The felsic unit is sandwiched between lower mafic unit I and upper mafic unit II. Lower mafic unit I consist of mafic components of sub volcanic to volcanic nature and are of andesitic to trachytic composition with little variations. Middle felsic unit is only of porphyritic rhyolite and further over lained by basaltic trachytic mafic components of mafic unit II. Volcanic breccia with fluorite is youngest among igneous components and further covered by alluvium of Recent to Sub Recent time. This volcanic succession is resting over granite basement. Granite before meeting the volcanic succession shows thin bends of sedimentary chert. The stratigraphic succession have been summarized in Table 1.

The geological map of the area show that the Karada fluorite deposit is concentrated at two locations in the area, first one is called Rekha mine and is in the northern part, second one is called Santoshi mine and is in the southern part. In the Rekha mine there is one prominent vein having an uniform thickness of 10meter throughout its length and confined to the host volcanic breccia body. It is distinctly elongated in the N43^{0}E-S43^{0}W direction with subvertical dip towards west and penetrating deep inside the base of which has not been exposed upto a depth of 10 meter. There are two more veins having thickness of 10m and a length of 55m, parallel to the main vein and 200 meter apart from both sides, one towards east and another towards north. There is yet another vein available further 200m apart with moderate fluorite grade breccia only in the east. There is no such symmetrical vein available in the north or it must now have eroded totally because of nallah cuttings.

Table 1. Lithostratigraphy of Karada Fluorite Deposit

	Alluvium :	Sub Recent, stabilized and brown coloured 'older alluvium' is further overlained by blown sand and sand dunes of light cream colour and of Recent age. Alluvium is covering all the low level topography.
	Volcanic Breccia with Fluorite	Volcanic Breccia is consisting of sharp angular fragments of pre-existing rocks with dominating felsic components. Hence, are light coloured. The fragments are normally ranging in size from few millimeter to few centimeter.fragments are also low rank fluorites, cemented with new generation of fluorite. All fluoritized volcanic breccias are over grown by last phase of fluorite, filled along veins following the trend of the joints.
Mafic Unit-II	**Shoshonite :**	Typical basalt like rock with dark green melanocratic appearance present only as three small patches near Rekha mine. Smooth in appearance being fine grained.
	Trachyte II :	Well exposed at eastern and western parts of the ridge near village Tavidar. Olive coloured rock showing fine grained matrix and well oriented plagioclase phenocrysts in the direction of flow of lava.
Felsic Unit	**Rhyolite :**	Violetish pink rock sandwiched between two mafic units and extensively exposed in the northern part of the area. Rhyolite is constituting the main ridge being bold to weathering. Presence of plagioclase phenocrysts is making it porphyritic rhyolite.
	Latite :	Few spelly of erosion, are small patches of latite in the northern part of the area from west of Lalpura to Lakhawas village. Fine grained melanocratic rock with small plagioclase phenocrysts
	Trachye I :	Trachyte I is constituting inner most periphery of the semi-circular rock with low topography. Melanocratic olive green rock, weathered to dark grey. Flow oriented phenocrysts are embedded in fine matrix.
Mafnic Unit-I	**Trachyandesite :**	Dark grey melenocratic, crystalline rock with visible medium grained texture and partially oriented albite laths making trachytic texture. Constituting centre of the semi-circular ridge.
	Autobrecciated Chert	Thinly laminated, milky white to grey coloured bands showing auto brecciation and penecontemporaneous deformation.
	Unclassified : **Granite (basement?)**	Coarse to medium grained holocrystalline, typical grey coloured granite. Possibly serving as basements. Small exposure near village Tavidar.

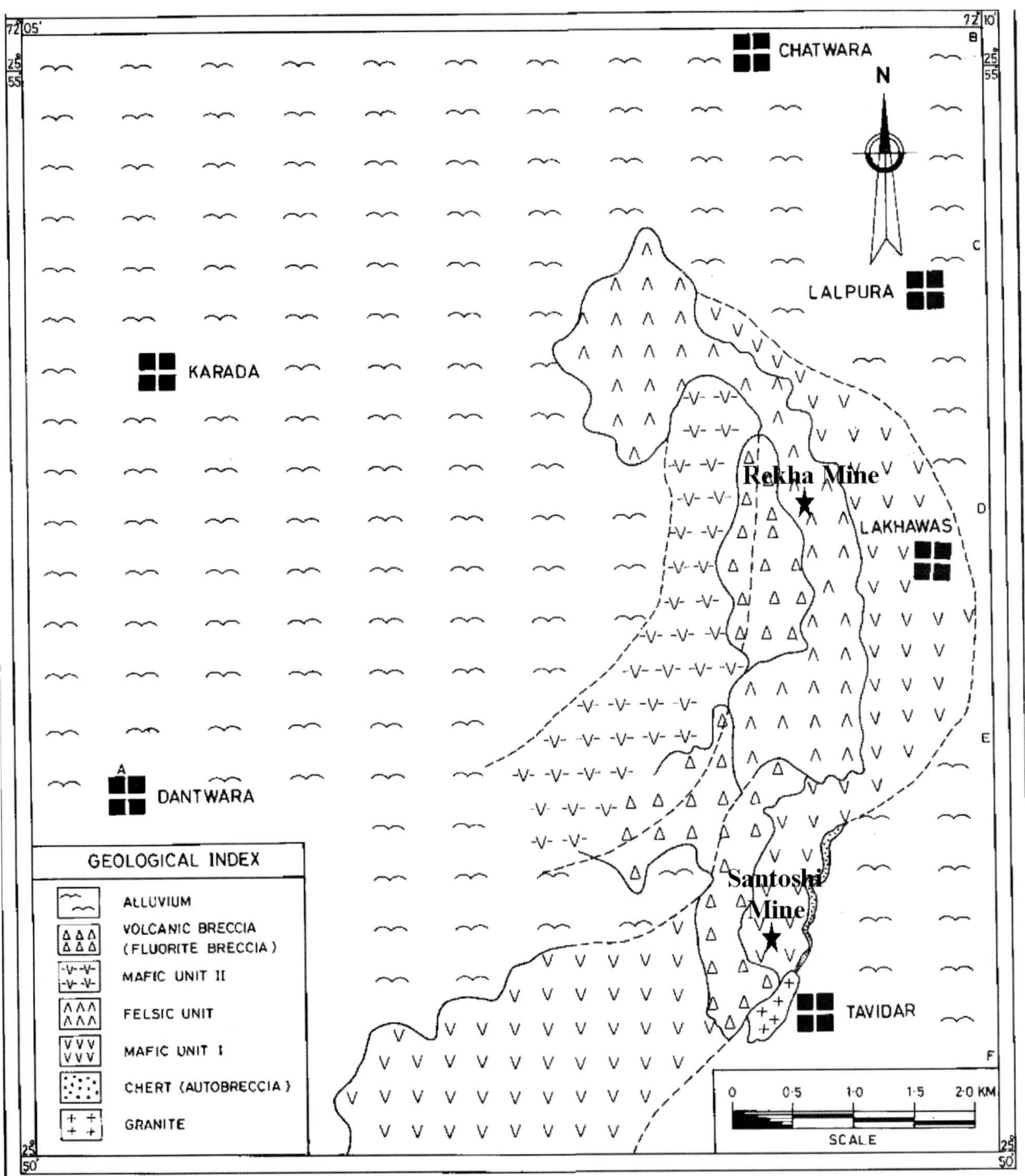

Fig. 2. Geological map of Fluorite Deposit at Karada, District Jalore, Rajasthan.

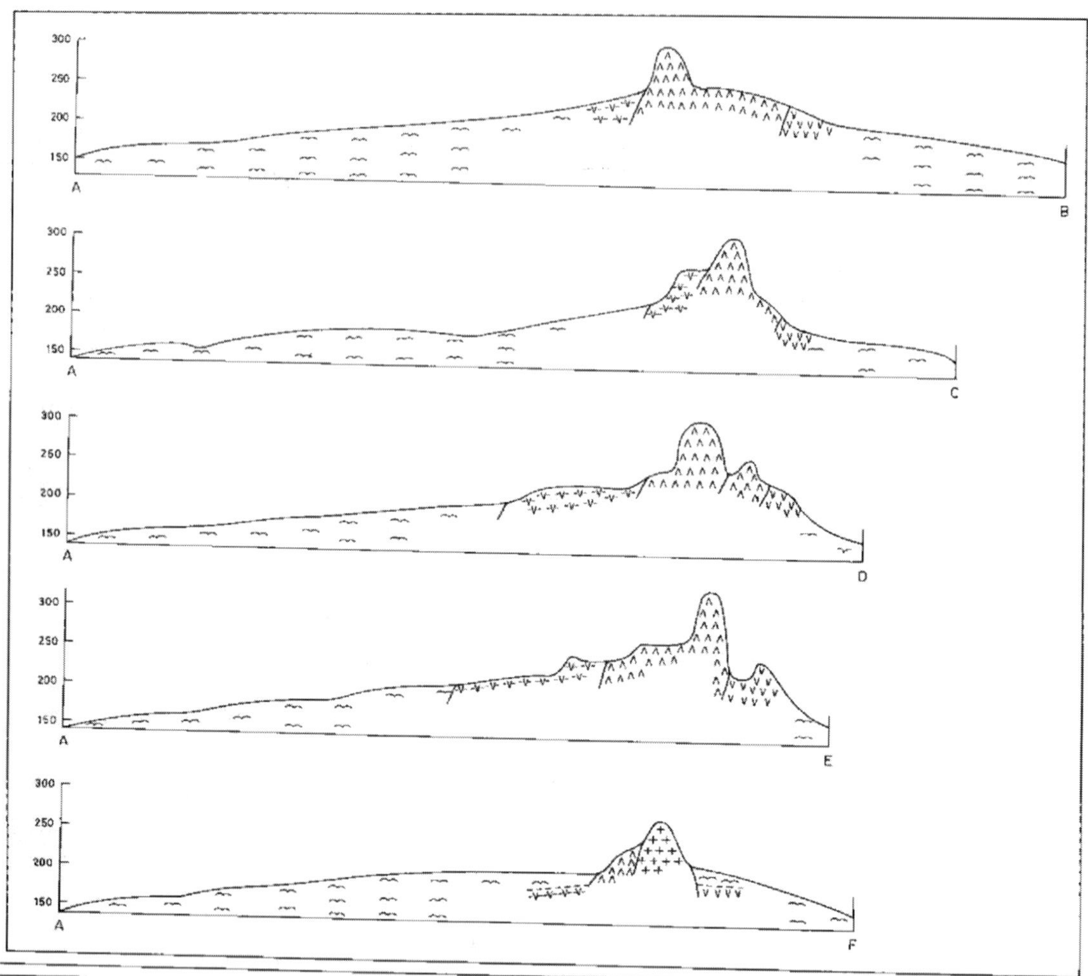

Fig.. 3. Geological Cross Section along various Profiles (Vertical Scale in Meters, Symboles and Profiles as per Geological Map)

The Santoshi mine is present in the south of the Rekha Mine. There is one vein with a thickness of 10m and length of 200m well filled with fluorite and volcanic breccia and is also well coated with 'dog tooth' and 'gem' varieties of fluorite. It is interesting to note that the orientation of this vein is also N43^{0}E-S43^{0}W with 85^{0} dip towards north.

The orientation of veins has strictly followed the fillings only of the first set of joints, named here as Joint I. This joint has the same orientation and dipping subvertically towards northwest .

As a rule, all the fluorite veins follow the trend of prominent Joint I showing N43°E-S43°W strike and dip 85°N. Though mineralization in the area is joint controlled, it has occurred mostly along the first set and not the later, Joint II and Joint III sets. However, all of the joints have fragmented the main mineralized body and host rock. Apparently, no post mineralization tectonic have been deserved to further fragment the mass. There is no movement of blocks or crushing of the available fluorite and the host.. Multi-generation fragmentation is well recorded at both the mines. Nonetheless the size of fragments is larger in the Santoshi mine than the Rekha mine.

As the second and third sets of joints are unaffected by the mineralization, there is no bifurcation of the mineralized veins (or the ore body, in general). It is clear from the plan of the breccia veins constituting stockwork that the shape of the ore body has been closely controlled only by the first set of joints, the formation of which presumably immediately pre-dates the emplacement of the breccia body itself. Trends of mineralized veins within the stockwork are almost sub-parallel to parallel, suggesting joint formation by tensional forces. Qualling apart of rock perpendicular to the trend of the joint (mineralized one) is further supported by lack of crushing along veins.

Petrology of Host Rocks

Unclassified Granite: There is only one small exposure of granite is available near Tavidar village. The major part of this granite, hence is unexposed. It is presumption of the author that this granite is serving as basement for the rest of the rocks available in the study area. This granite has been very well classified as 'unclass-ified' granite in the Malani terrain by the earlier workers (Bhusan and Chandrashekhar, 2002).

Autobrecciated Chert: A thin band of chert is observed and mapped (Fig. 2) only in the eastern part flank of the semi-circular ridge of karada area, near village Tavidar. The rock is thinly laminated, differentiated by colour banding, which otherwise is milky white. The chert band is showing penecontemporaneous deformations, many of which have resulted in to auto-brecciation.

Trachyandesite: Trachyandesite is present only in the south of the study area and west of Tavidar village, across the ridge. Seems, it constitutes a part of the centre of the semi-circular ridge. It has been exposed being deeply eroded by low order streams. In the handspeci-men, it is showing dark grey melenocratic appea-rance with visible medium grained crystalline grains. There is little orientation of the small laths of albite to justisfy the trachytic texture. Presence of albite (forty two percent), orthoclase (eighteen percent) and anorthite (eight percent) under petrological microscope is showing two clear generations of crystallization. The first generation mineral crop is coarse to medium grained and second generation is fine grained constituting matrix, and over all porphyric texture. There is presence of pyroxenes namely, hypersthene and dioposide Twelve and eight percent respectively), with little olivine, apatite, ilmenite and magnetite. Plagioclase displaying repeated and orthoclase simple twinning (Fig 5A). Pyroxenes showing 90° intersecting cleav-ages and third order pink, blue, orange colours of distinction. A single grain of olivine is surroun-ded by pyroxene (Fig 5A) illustrating 'corona' structure. Second generation mineral are not only of fine grained texture but has also surrounding the newly broken phenocrysts. Such broken phenocrysts are shooted and presented in photomicrograph (Fig 5B). It is indicating that second forceful explosion was accompanied with extrusion. Hence, plutonic to sub volcanic texture is preserved.

Trachyte I : Trachyte I is an intermediate component of the mafic unit I. It is constituting the eastern part of hill near Tavidar (Fig. 2). Light olive green is the colour of the rock in the field and in the hand specimen in which phenocrysts of feldspar are not only present but also oriented in the direction of the flow of the lava to make trachytic texture. Under petrolo-gical microscope (Fig 5C) phenocrysts of ortho-clase and albite are common and are embedded in the fine grained matrix also consting of quartz. This trachyte rock is infact, a quartz trachyte having some twenty four percent quartz. Albite and orthoclase are almost thirty three percent, each. Four percent diopside and five percent hypersthene are also showing their presence phenocrysts are also brocken with sharp angles (Fig 5C).

Latite : Latite rock is exposed as few small patches in the northern part of the study area, from west of Lalpura to Lakhawas village. While joining these patches in the map shows that these are remanents of one and the same flow unit. Under petrological microscope (Fig 5D), latite showing very fine matrix with phenocrysts. The mineral crop is consist of thirty percent plagioclase and only sixteen percent orthoclase, among feldspar. Diopside is the only dominating mineral among mafic ones, is almost thirty percent. Rest of the minerals are iron ore with rare grains of nepheline. All among the photo-micrographs shooted showing flow orientation of plagioclase laths. These laths are forming subophitic texture with penetration of plagioclase laths in pyroxene. The laths are also making spherulitic or radial arrangements (Fig 5D). Interestingly, one photomicrograph shooted to record digestion of a grain which is showing spinifex texture (Fig 5D).

Rhyolite: Rhyolite is the only rock representing felsic unit and is sandwhiched between two mafic units, named mafic unit I and mafic unit II. The Rhyolite is constituting the main ridge (Fig. 4A) being more bold to weathering, in comparison with the mafic units. Although, rhyolite shows consistent thikness in the study area, yet it is covering more surface area in the northern part. In general rhyolite are violetish pink in appearance when observed in the hand specimen. In the field too, the same colour is visible (Fig.5 A) rhyolite rock in both the mines, The Santoshi mine and Rekha mine. Hand specimen showing alteration by later solutions with iron oxides (Fig 4A) and to clays (Fig 5 F). Petrological studies reveal that orthoclase constitute some fifty percent of minerals with sub-ordinate albite and anorthite, which shows thirteen and five percent, respectively. Quartz is around twenty three percent. There is unusual presence of hypersthene and diopside with five and three percent, respectively. The texture too is typical of rhyolite. Phenocrysts of orthoclase are surrounded by fine grained plagioclase, orthoclase and glass, making porphyritic texture (Fig 5 E and F). One of the thin sections is showing spherulitic growth of acicular quartz. There are broken fragments of phenocrysts in the glassy and fine grained matrix (Fig 5 E and F) brocken fragments are with euhedral showing second phase of violent explosion and extrusion of rhyolites.

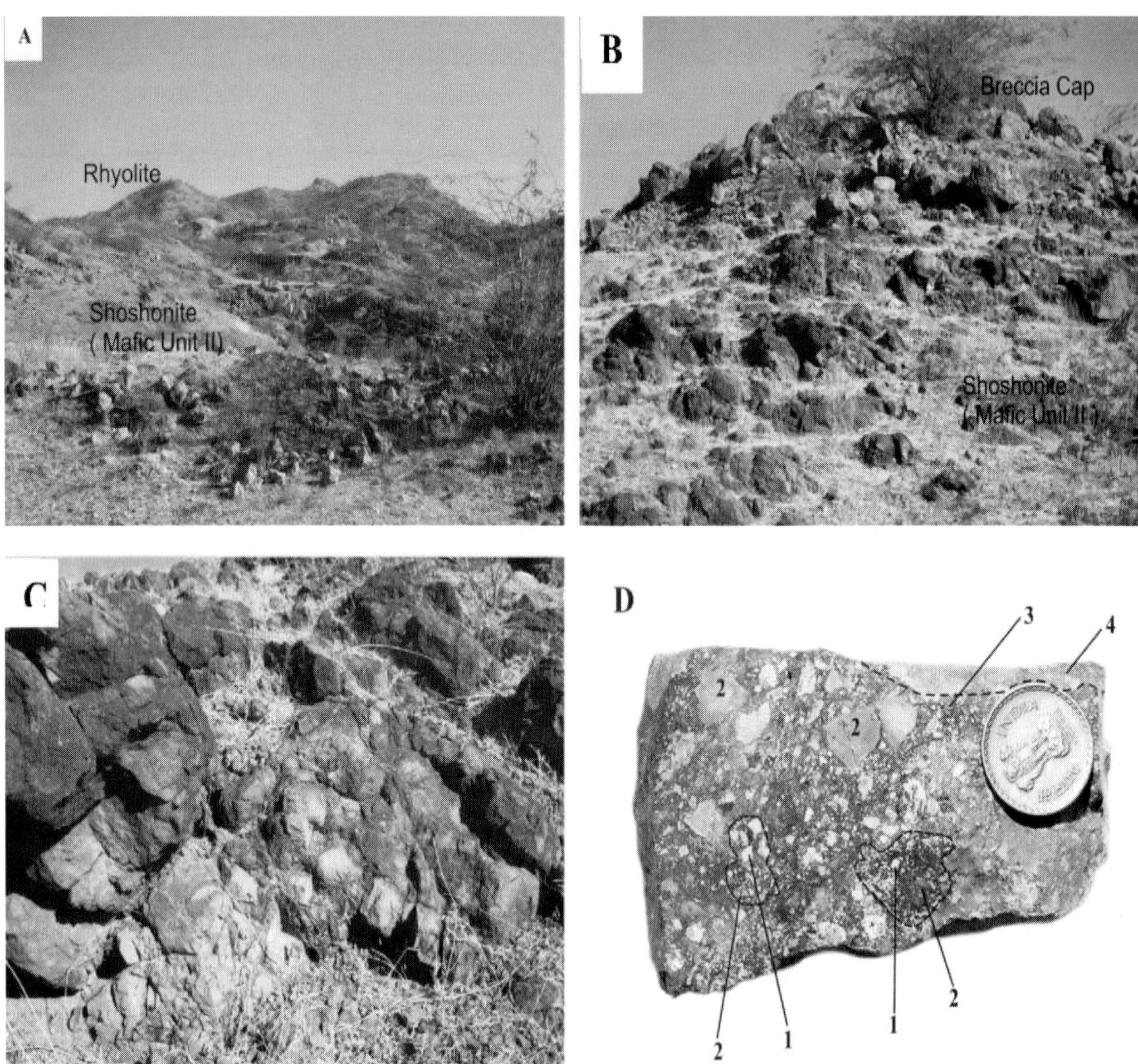

Fig. 4. (A) Field view of Rhyolite (at distal end) and shoshonite (mafic unit II) exposures at Rekha mine. **(B)** Boulders of shoshonite (mafic unit II) at Rekha mine showing pillows and exfoliation structures. Many pillows are close with each other to make compaction of the flow unit. Presence of pillows is showing sub-marine extrusion of lava. **(C)** Field view at Rekha mine showing clear relationship of volcanic breccia with shoshonite (mafic unit II) as cap rock, being later. Both the units have distinct colors even in the field view. **(D)** Volcanic breccia components are well demonstrated by a hand specimen various numbers are showing various generation of fragmentation, ultimately constituting the breccia. Fig. is showing four generation of fragments.

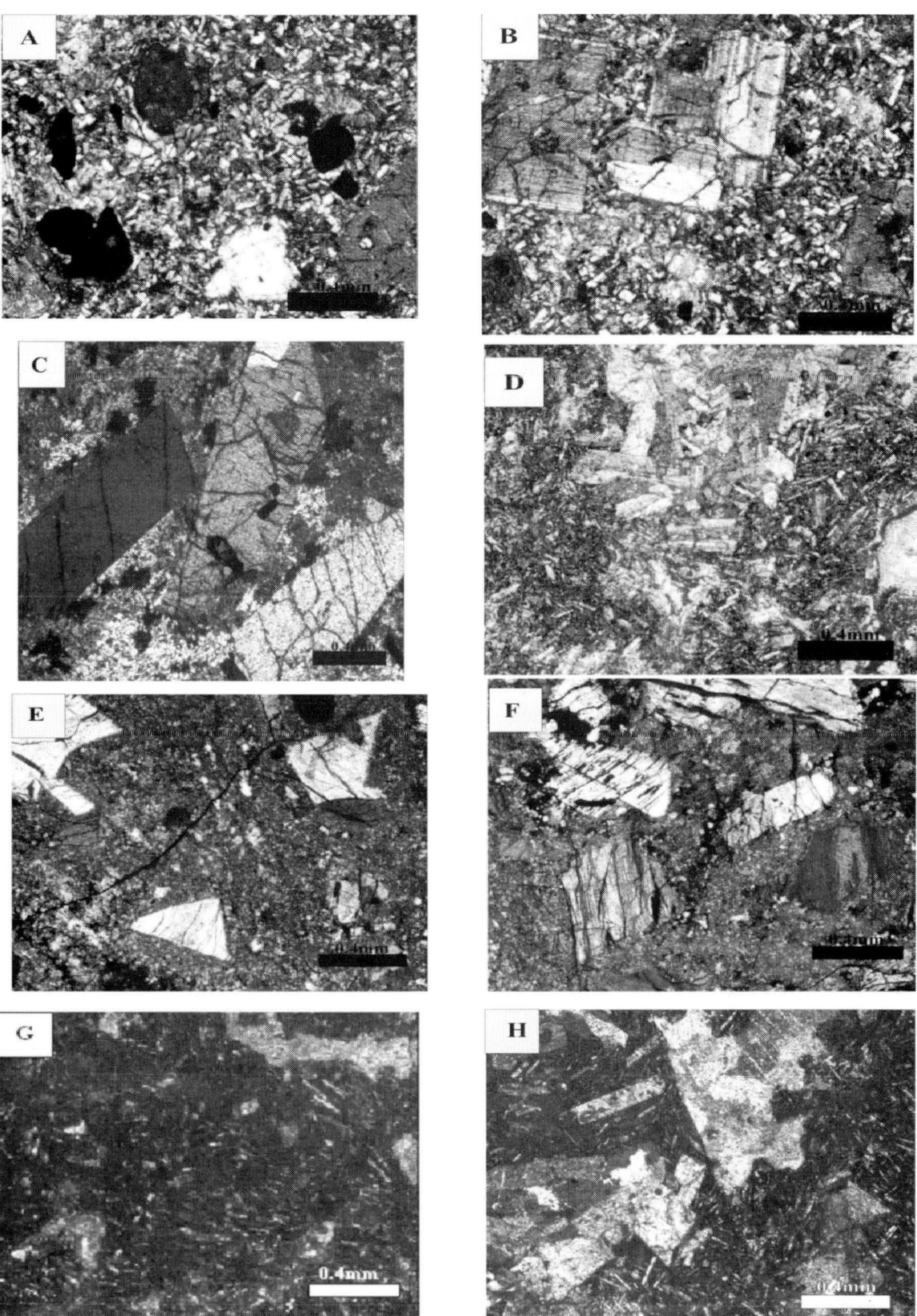

Fig. 5. (A). Photomicrographs of trachyandesite showing two clear generations of crystallizations of minerals. Medium grained crystals are embedded in the matrix of fine grains, showing earlier and later phases, respectively. There is presence of plagioclase, orthoclase, diopside, hypersethene and iron oxides. Plagioclases are with repeated twinning and orthoclases are with simple twinning. Pyroxenes with third order pink, blue, orange colours are distinct. Iron ores show isotropic and euhedral appearance. Plagioclase laths in matrix is showing partial flow orientation. **(B)** Photomicrographs of

trachyandesite, showing the earlier coarse crystals has broken due to second forceful extrusion. Both feldspars, plagioclase and orthoclase are better illustrated. Pyroxenes are present in both the generations. Plutonic to sub volcanic texture is preserved. **(C)** Photomicrograph of trachyte I showing phenocrysts of feldspars embedded in very fine matrix again of feldspars along with quartz (20%) and pyroxene (10%) with little iron ores. Coarse and euhedral feldspars are the phenocrysts. **(D)** Photomicrographs of latite showing very fine grained and dominant matrix consist of plagioclase (32%) and orthoclase (16%) with diopside (29%) *in toto*. Some nepheline and iron ores are constituting the rest of the mineral crop. Flow orientation of laths, sub ophitic texture and radial to spherulitic arrangement of feldspars are visible. **(E) and (F)** Photomicrographs of rhyolite showing Porphyritic texture which is well illustrated as phenocrysts of orthoclase and plagioclase (coarse to medium grained) well embedded in glassy and fine grained matrix of the same composition. Brocken fragments with sharp angles showing evidence of second phase of the extrusion. Rock in general showing fifty percent orthoclase with twenty percentage each of the plagioclase and quartz and rest, hypersthene and iron ore. **(G)** Photomicrographs of trachyte II samples showing well developed trachytic texture by orientation of lath minerals in the matrix to the direction of flow of lava. Plagioclase and orthoclase constituting forty and twenty percent, while diopside and olivine constituting twenty five and fifteen percent of minerals with the rest iron ores. Two generation crystallization is illustrated by porphyritic texture and forceful explosion by breaking of phenocryst. Later, metasomatic alteration of feldspar in to clay and intergranular filling of iron solutions is well illustrated. **(H)** Photomicrographs of shoshonite rock showing three feldspar, anorthite (15%), albite (25%) and orthoclase (15%). The mafic components consist of almost twelve percent olivine with two pyroxenes, twenty percent diopside and seven percent hypersthene. Texture is typical of basalt, with sub-ophitic texture. Fine laths of second generation of crystallization are oriented in the direction of flow of lava and sometimes making radial arrangements Phenocrysts are fragmented possibly due to forceful extrusion occur later.

Trachyte II : Trachyte II rock is exposed at eastern and western flank of the ridge near village Tavidar. One small exposure of the same rock is also observed and mapped at south-south-east of village Lakhawas. Trachyte II rock is light green in colour. This makes the trachyte II rock distinct than dark green coloured shoshonite. In the hand specimen trachytic texture is visible, as plagioclase laths are showing flow orientation. The trachytic texture has also well observed under peterological microscope (Fig 5 G). In thin section, there is presence of some forty percent plagioclase and twenty percent orthoclase, alongwith twenty five percent diopside and fifteen percent olivine. The rest of the minerals are isotropic iron minerals. The evidences favoring reactivation of the rock with forceful explosion is again observed, because out of two generation of crystallization, the first generation phenocrysts are broken (Fig 5 G). The rock have further suffered metasomatic alteration. Feldspar are greatly altered into clays (Fig 5 G), and iron solutions have managed their rooms in between the intergranular spaces (Fig 5 G).

Shoshonite: Shoshonite rock occur as three small dark green coloured exposures in the field (Fig 4 A and B). These exposures are in the inner part of the cauldron, hence, exposed only near Rekha fluorite mine with pillow structures (Fig 4 C). But probable extension of this rock unit is expected to constitute considerable surface exposures now concealed under thick cover of the alluvium. The hand specimen of the rock is also dark green with smooth appearance being very fine grained. Hence in hand specimen and in the field it shows typical basalt like appearance. The contact of shoshonite with trachyte II is sharp and dipping with high angle towards west. Under the peterological microscope, shoshonite shows the three feldspars, namely anorthite , albite and orthoclase with percentage of fifteen, twenty five and fifteen, respectively. The mafic components consist of olivine, diopside and hypersthene with percentage of twelve, twenty and seven, respectively. Very fine grained texture, typical of basalt is observed with some phenocrysts of feldspars (Fig 5 H). Infact, sub-ophitic texture is present when in the matrix the fine laths of plagioclase are inserted into the pyroxene. Some orientation of laths of plagioclase is also noted (Fig 5 H). The phenocryst of orthoclase and plagioclase are further fragmented during the course of second explosion as it is illustrated in Fig 5 H.

Volcanic Breccia with fluorite: Light coloured volcanic breccia (Fig 4 C) consisting of sharp angular fragments of the preexisting rocks, chiefly rhyolite and sub-ordinated by mafic components, high rank fluorite and iron oxide minerals. It infact, is indicating more than one phase of explosion and /or collapse with evidences of intrusion of fluortized breccia. The size of the constituting fragments ranges from few millimeter to centimeter almost in all part of the exposures specially at Rekha mine (Fig 4D). But size of the constituting fragments, suddenly increases from tens of centimeter to even a meter, at southwest of the Santoshi mine, near temple.

Alluvium : The alluvium large surface area, is of Recent to sub-Recent age. The lower component is older alluvium which is stabilized and shows dark cream to light brown colour. It is covering central part of the semi-circular ridge of the study area and serving as base for the blown sand. The blown sand is of Recent age. The blown sand is of light cream colour. In this field, lithostratigraphy and peterological set up, the Karada fluorite deposit is confined either as joint fills or as cement of the volcanic breccia.

Geochemistry of Host Rocks

All rocks and some of the pure fluorite samples and some fluorite samples inside the volcanic breccia have been analysed using XRF and ICP-MS and data have been presented in the form of Table 2a (for rocks) and 3 (for fluorite and host) for major oxides, trace element (Table 2b and 4) and rare earth elements (Table 2b and 5). It has been observed that the samples are rich in Zr, Sr, Rb, Zn and Nb as far as trace elements are concern. Among rare earth elements fluorites are rich in Ce, Pr, La and Nd.

To understand origin, evolution and economic potential, of the Fluorite deposit and associated rocks a detailed geochemical analysis and interpretation of the data have been added in the form of present chapter.

Result and Discussions

The major oxides determined includes SiO_2, Al_2O_3, Fe_2O_3, TiO_2, MnO, P_2O_5, CaO, MgO, Na_2O, K_2O. The trace elements determined includes, Ba, Cr, Ni, Cu, Zn, Th, Rb, U, Y, Zr and Nb. The rare earth elements includes La, Ce, Pr, Nd, Sm, Eu, Gd, Tb, Dy, Ho, Er, Tm, Yb and Lu. The results obtained for major oxides, various ratio and CIPW norm calculated after that are presented in Table 2a (Part I). Table 2b (Part II) is showing result obtained for trace elements and rare earth elements. The result and discussions are presented in the present chapter under following heads.

(i) Harkar variation diagrams
(ii) CIPW norms & Compositional diagram
(iii) Trace Elements and their distribution
(iv) Rare Earth Elements and their distribution
(v) Spider Diagrams

(i) Harkar Variation Diagram

The variation diagram was named after Alfred Harker. The diagram plots the wt% of SiO_2 along abscissa (x–axis) against the concentration of other oxides along the ordinate axis (y–axis). The diagram provide a better understanding of genesis of the rock which is afforded by the compositional variation within the body itself and between it and the other body that may be related in time and space. As the diagram plots the concentration values on variation basis, so that the trend or pattern in variation becomes more prominent to understand, thus it provide a better interpretation of the geochemical data.

Fig. 6 A to Fig. 6 B are showing Harkar variation diagrams for Karada fluorite deposit area. In these diagrams SiO_2 have been plotted against various oxide weight percentages. It is clear from the variation that no sets of points fits in any sort of pattern. It indicates that there must be no mixing of the two components (assimilation of rock or magma mixing).

Fig. 7 A to 7 B are showing Harker variation diagrams in these diagrams MgO plotted against various oxide.

(ii) CIPW Norms and Compositional Diagrams

CIPW norms have been calculated and presented in Table 2a. The rock have been divided in to three units, namely mafic unit I, felsic unit I and mafic unit II . In mafic unit *In toto* seventeen samples were selected for the chemical analysis to include thirteen from Rekha mines, four from Santoshi mine. Detailed compositions are presented in Table 2a and 2b and compositional diagrams are prepared and shown in Fig. 8 to Fig. 14 following work of Le Maitre (2002) that is Recommendations of the International Union of Geological Sciences sub–commission on the systematic of Igneous Rocks.

Table 2A. Geochemistry of Igneous Rocks of Karada Fluorite Deposit, District Jalore, Malani Igneous Suite India (Part I Major Oxides and CIPW Norms)

Sample Rock→	**MAFIC UNIT I**						**FELSIC UNIT I**				**MAFIC UNIT II**					
↓ Major oxide (Wt. %)	**KS 12 (Trachy- andesite)**	**KS 13 (Trachy- andesite)**	**KS 7 (Trachyte I)**	**KR 8 (Latite)**	**KR 9 (Latite)**	**AVG.**	**KR 2 (Rhyolite)**	**KR 3 (Rhyolite)**	**KS 4 (Rhyolite)**	**AVG.**	**KR 5 (Trachyte II)**	**KR 6 (Trachyte II)**	**KR 10 (Shoshonite)**	**KR 11 (Shoshonite)**	**KR 14 (Shoshonite)**	**AVG.**
SiO_2	58.64	58.47	72.52	47.06	48.06	56.95	69.62	70.6	69.7	69.97	50.33	49.33	45.12	49.2	54.23	49.64
Al_2O_3	14.18	14.22	12.4	13.58	13.49	13.57	13.41	12.42	14.12	13.31	16.51	15.22	13.81	13.94	14.5	14.79
Fe_2O_3	0.86	0.85	0.42	1.57	1.55	1.05	0.3	0.59	0.4	0.43	1.08	1.12	1.7	1.69	1.21	1.36
FeO	7	6.88	3.39	12.74	12.52	8.5	2.46	4.8	3.26	3.5	8.71	9.05	13.74	13.65	9.84	10.99
TiO_2	2.1	2.14	0.38	3.09	3.09	2.16	0.41	0.38	0.41	0.4	2.65	2.35	3.3	3.37	2.16	2.76
MnO	0.26	0.26	0.1	0.19	0.19	0.2	0.01	0.28	0.19	0.16	0.13	0.15	0.2	0.22	0.29	0.198
P_2O_5	1.01	0.99	0.04	0.35	0.35	0.54	0.03	0.03	0.04	0.033	0.28	0.25	0.34	0.35	0.15	0.27
CaO	4.88	5.2	0.87	9.1	8.76	5.76	4.58	0.9	0.71	2.06	8.14	11.13	10.2	5.41	7.05	8.38
MgO	3.15	3.11	0.23	6.92	6.57	3.99	0.32	0.52	0.36	0.4	5.91	5.74	7.24	6.78	3.24	5.78
Na_2O	4.92	4.9	4.57	2.68	2.67	3.94	2.06	0.51	1.94	1.5	2.97	2.77	2.77	2.96	4.44	3.18
K_2O	3.01	2.98	5.08	2.72	2.75	3.3	6.8	9.51	8.87	8.39	3.3	2.9	1.58	2.42	2.88	2.61
Total	100.01	100	100	100	100	100	100	100.54	100	100	100	100	100	99.99	100	100
Ratio																
Salic/Femic	2.48	2.45	8.52	1.08	1.15	3.13	7.29	7.71	12.02	9.00	1.87	13.05	0.95	1.32	1.88	3.81
$(Na_2O + K_2O)/ CaO$	1.277	1.192	8.23	0.44	0.461	2.32	1.29	6.83	9.96	6.02	0.57	0.379	0.337	0.759	0.812	0.57
K_2O/Na_2O	0.403	0.401	0.73	0.668	0.679	0.57	2.17	12.33	3.02	5.84	0.73	6.9	0.376	0.538	0.427	1.79
Q or F/L	0.06	0.059	0.36	0.09	0.043	0.12	0.35	0.38	0.278	0.336	0.024	0.103	0.125	----	0.008	0.06
							CIPW Norm :									
Quartz	4.17	4.28	24.34	----	----	6.55	23.06	24.5	20.1	22.55	----	----	----	----	----	-----
Anorthite	7.71	8	----	17	16.7	9.88	7.25	3.51	3.26	4.67	21.97	20.53	20.58	17.6	11.12	18.36
Albite	41.63	41.46	35.5	14.46	18.45	30.3	17.43	4.31	16.41	12.71	22.2	13.64	13.4	25.04	36.6	22.17
Orthoclase	17.79	17.61	31.53	16.07	16.25	19.85	40.19	56.2	52.42	49.6	19.5	17.14	9.34	14.3	17.02	15.46
Apatite	2.34	2.29	0.09	0.81	0.81	1.26	0.07	0.07	0.09	0.07	0.65	0.58	0.79	0.81	0.35	0.63
Iimenite	3.99	4.06	0.72	5.87	5.87	4.1	0.78	0.72	0.78	0.76	5.03	4.46	6.27	6.4	4.1	5.25
Magnetite	1.25	1.23	----	2.28	2.25	1.402	0.43	0.86	0.58	0.62	1.57	1.62	2.46	2.45	1.75	1.97
Nepheline	----	----	----	4.45	2.24	1.33	----	----	----	----	1.58	5.31	5.44	----	0.53	2.57
Diopside	8.23	9.41	3.56	21.41	20.27	12.57	8.51	0.66	----	3.05	13.59	27.19	23.03	5.77	19.28	17.77
Hypersthene	12.91	11.94	4.49	----	----	5.86	----	9.17	6.23	5.13	----	----	----	13.74	----	2.74
Corundum	----	-----	----	----	----	----	----	----	0.13	0.04	----	----	----	----	----	----
Wollastonite	----	------	----	----	----	----	2.27	----	----	0.75	----	----	----	----	----	----
Olivine	----	-----	----	17.66	17.16	6.96	----	----	----	----	13.91	9.53	18.69	13.88	9.24	13.05
Total	100	99.9	100	100	100	100	100	100.01	100	100	100	100	100	100	100	100

Table 2B. Geochemistry of Igneous Rocks of Karada Fluorite Deposit, District Jalore, Malani Igneous Suite. India (Part II Trace Element and Rare Earth Element)

Sample Rock→	MAFIC UNIT I						FELSIC UNIT I				MAFIC UNIT II					
↓ Trace Elements (ppm)	KS 12 Trachy-andesite	KS 13 Trachy-andesite	KS 7 Trachyte I	KR 8 Latite	KR 9 Latite	AVG.	KR 2 Rhyolite	KR 3 Rhyolite	KS 4 Rhyolite	AVG.	KR 5 Trachyte II	KR 6 Trachyte II	KR 10 Shoshonite	KR 11 Shoshonite	KR 14 Shoshonite	AVG.
Sc	16	15	4	37	34	21.2	8	7	4	6.33	21	22	36	37	19	27
Co	19	19	4	39	39	24	1	9	3	4.33	31	34	39	40	29	34.6
Ni	4.6	3.6	17.7	53.2	52.2	26.26	7.6	15.9	7.8	10.43	39.2	40.5	52.8	54.2	11.7	39.68
Cu	29.6	31	19.5	187.5	184.8	90.48	11.6	20.4	17.6	16.53	143.5	138.6	151.2	177	31.9	128.44
Zn	133.3	137.2	90.4	121.2	118.8	120.18	112.2	117.8	176	135.33	93.1	100	137.9	124	121	115.2
Ga	21.6	21.7	25.6	21.8	20.1	22.16	23.6	21.5	24.8	23.3	19.2	20	23.3	22.1	18.3	20.58
Pb	10.1	9.8	7.1	0.7	3	6.14	13.9	12.6	18.9	15.133	1.2	1.9	2.7	4.9	2.9	2.72
Th	8.4	6.4	31.6	1.3	1.2	9.78	25.8	25.6	26.5	25.96	3.3	1.2	0.6	0.3	3.9	1.86
Rb	71.9	69.8	193.5	141	141.6	123.56	208.8	321.4	276.3	268.83	145.1	120.5	87.2	129.9	101.1	116.76
U	1.8	1.8	2.2	3.6	3.3	2.54	3.4	2.5	3	2.96	2.8	2.7	2.5	4.4	2.5	2.98
Sr	599	594	14	348	343	379.6	40	50	24	38	428	395	347	300	176	329.2
Y	57.6	57.7	136	27.2	25.6	60.82	78.9	65.5	69.9	71.43	19.1	19.7	27.5	24.4	37.4	25.62
Zr	372	351	1165	199	199	457.2	920	882	890	897.33	150	132	205	202	290	195.8
Nb	68.6	68.4	130.8	27.5	27.7	64.6	146.8	112.8	142.2	133.93	27.4	23.8	28.5	29	55.8	32.9
Ba	---	----	----	----	----	---	---	----	---	---	---	---	---	----	1582	----
Rare Earth -Element (ppm)																
La	67.83	67.84	196.12	23.52	21.870	75.43	125.47	140.24	130.94	132.21	21.63	20.07	24.83	23.81	44.3	26.92
Ce	139.67	140.91	362.35	51.05	49.189	148.63	243.14	266.92	253.53	254.53	45.06	42.59	54.95	52.39	93.1	57.61
Pr	16.65	16.8	36.420	6.305	5.893	16.41	26.163	28.51	26.757	27.14	5.189	5.026	6.559	6.214	10.7	6.73
Nd	73.13	74.2	134.71	28.51	26.362	67.38	99.950	104.45	102.90	102.43	23.17	21.45	29.00	27.60	46.115	29.46
Sm	13.94	13.9	22.761	6.137	5.674	12.48	17.157	17.84	17.653	17.55	4.780	4.506	6.211	5.876	8.95	6.06
Eu	4.942	4.95	2.568	1.985	1.871	3.26	1.926	2.116	2.046	2.02	1.654	1.698	2.093	1.937	2.90	2.05
Gd	11.36	11.5	23.500	5.217	4.852	11.28	16.253	16.19	16.768	16.4	3.946	3.838	5.295	5.063	7.24	5.07
Tb	1.813	1.84	3.848	0.904	0.843	1.84	2.696	2.638	2.675	2.66	0.659	0.639	0.904	0.864	1.23	0.85
Dy	11.61	11.5	27.126	5.965	5.658	12.37	18.352	17.47	17.851	17.89	4.204	4.142	5.950	5.807	7.72	5.56
Ho	1.587	1.60	4.251	0.860	0.789	1.81	2.693	2.655	2.661	2.66	0.579	0.439	0.831	0.805	1.07	0.74
Er	4.510	4.486	12.832	2.386	2.239	5.29	8.355	8.297	7.915	8.18	1.575	1.392	2.304	2.242	3.02	2.1
Tm	0.626	0.63	1.878	0.344	0.317	0.75	1.299	1.228	1.216	1.24	0.221	0.167	0.321	0.321	0.42	0.29
Yb	5.253	5.32	15.793	2.898	2.641	6.38	11.333	11.49	10.603	11.14	1.808	1.617	2.700	2.621	3.53	2.45
Lu	0.771	0.77	2.229	0.416	0.389	0.91	1.626	1.602	1.591	1.6	0.255	0.252	0.386	0.380	0.110	0.27
Total	353.72	356.4	846.45	136.51	128.58	364.33	576.72	621.67	595.12	597.83	114.74	107.84	142.35	135.94	230.94	146.36

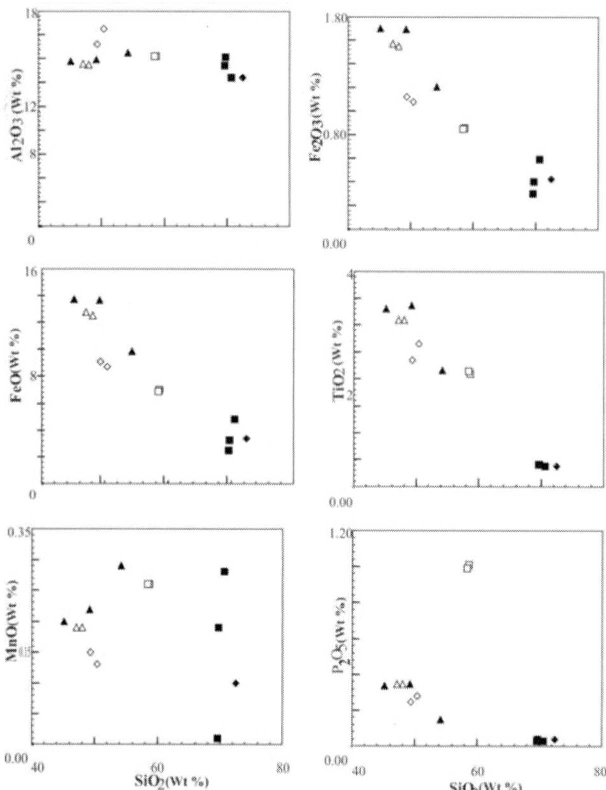

Fig. 6 A. Harker diagram displaying the correlation of major element oxides (wt%) with silica of Igneous rocks Rhyolite (■), Shoshonite (▲), Trachyte I (♦), Latite (p) , Trachyte II (u) and Trachyandesite (n) of Karada Fluorite deposit, District Jalore, Rajasthan.

Le Bas et al (1986) TAS Diagram

In a TAS (Total Alkali Silica) diagram Fig. 8 (Le Bas et al. 1986) a plot of Na_2O + K_2O wt% vs SiO_2 falls in the field of rhyolite, trachyte, trachyandesite, basaltic trachyandesite, trachy basalt and basalt.

Winchester and Floyd (1977) Diagram

A biaxial diagram of Zr/TiO_2 vs Nb/Y of Winchester and Floyd (1977) Fig. 9 shows that the plot falls under the field of trachyte, and alkali basalt.

Winchester and Floyd (1977) Diagram

Another biaxial diagram of Zr/TiO_2 vs SiO_2 of Winchester and Floyd (1977) Fig. 10 shows that the plot of the rock samples falls in sub-Albite andesite and rhyolite fields.

Jensen (1976) Diagram

Fig. 11 is a trilinear plot of Al_2O_3–(FeO + TiO_2) – MgO by Jensen (1976) shows that most of the rock samples falls in the field of Tholeiite series while one three samples belongs to high Fe- tholeiite series.

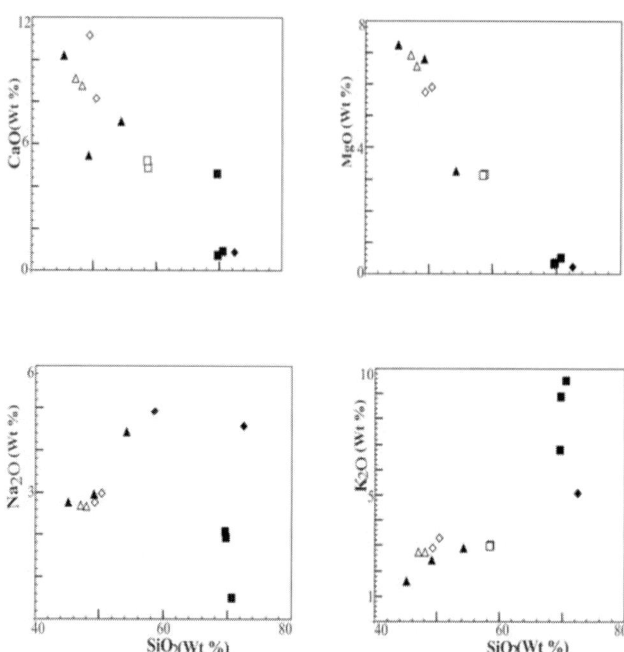

Fig. 6 B . Harker diagram displaying the correlation of major element oxides (wt%) with silica of Igneous rocks Rhyolite (■), Shoshonite(▲), Trachyte I (♦), Latite (p) , Trachyte II (u) and Trachyandesite (n) of Karada Fluorite deposit, District Jalore, Rajasthan

Maniar and Piccoli (1989) Diagram

Maniar and Piccoli (1989) presented a bilinear diagram of molar Al_2O_3/CaO+ Na_2O + K_2O vs Al_2O_3/Na_2O + K_2O. As shown in Fig. 12 all the rock samples falls in metaluminous fields only, except one sample which falls in peralkaline field.

Irvine and Barager (1971) Diagram

Subalkaline rocks have higher silica relative to alkalies and consequently generally have orthopyroxene and quartz in the norm and never feldaspathoids. Subalkaline rocks may be further divided into the tholeiitic suite and the calc-alkaline suite.the tholeiitic suite, comprised primarily of basalts but including more felsic rocks as well, shows in higher degree of iron enrichment on an AFM variation diagram and lower concentrations of aluminum. The calc-alkaline suite comprise the common volcanic basalt-andesite-dacite-rhyolite and plutonic gabbro-diorite-granodiorite-granite association, which contain combinations of olivine, Ca-rich clinopyroxene, Ca-poor orthopyroxene, honbland, biotite, feldspar, and quartz.

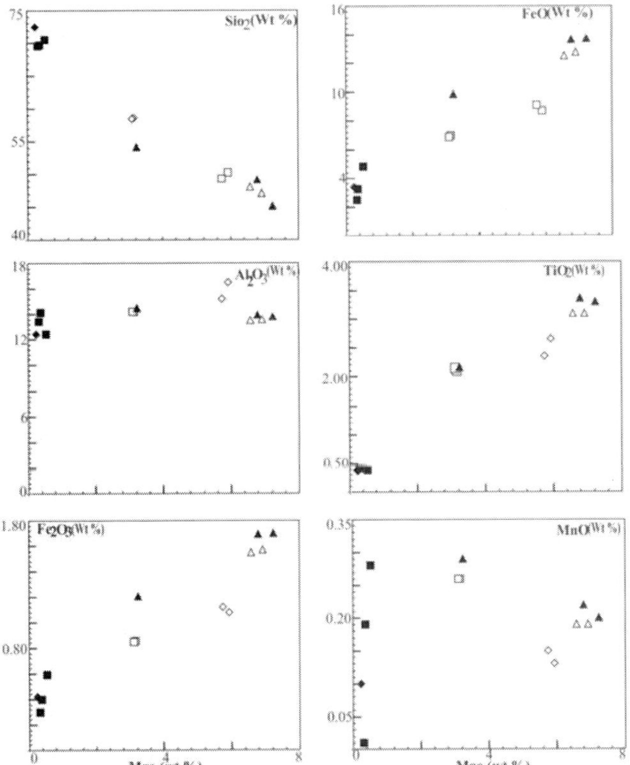

Fig. 7 A Harker diagram displaying the correlation of major element oxides (wt%) with Mgo of Igneous rocks Rhyolite (■), Shoshonite (▲), Trachyte I (♦), Latite (Δ) , Trachyte II (◊) and Trachyandesite (n) of Karada Fluorite deposit, District Jalore, Rajasthan.

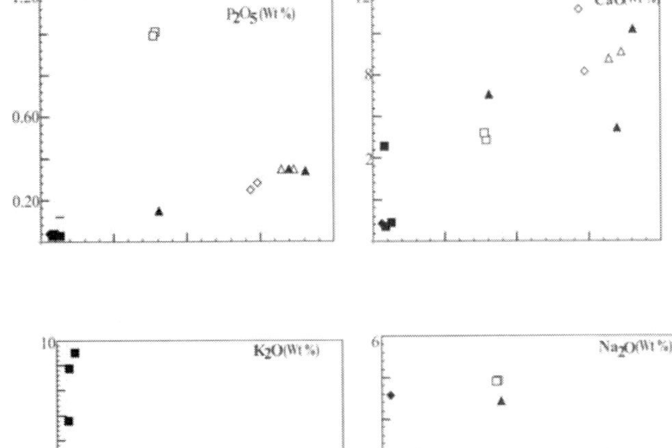

Fig. 7 B. Harker diagram displaying the correlation of major element oxides (wt%) with Mgo of Igneous rocks Rhyolite (■), Shoshonite(▲), Trachyte I (♦), Latite (Δ), Trachyte II (◊) and Trachyandesite (n) of Karada Fluorite deposit, District Jalore, Rajasthan.

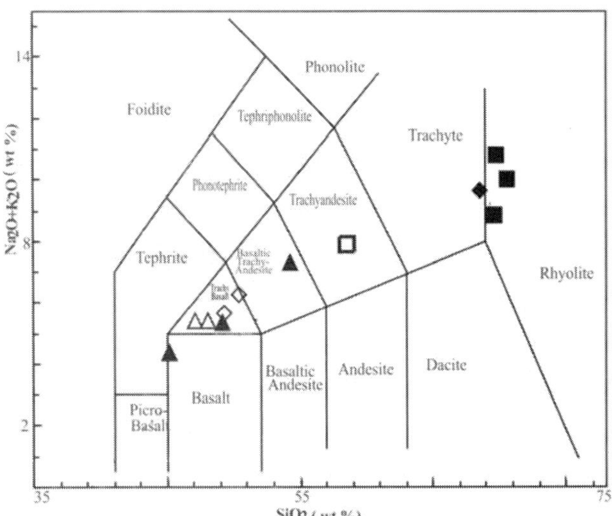

Fig. 8 Total Alkali Silica (TAS) diagram (Le Bas et al., 1986) showing plot for Igneous rocks Rhyolite (■), Shoshonite (▲), Trachyte I (Δ), Latite (p) , Trachyte II (◊) and Trachyandesite (n) of Karada Fluorite deposit, District Jalore, Rajasthan.

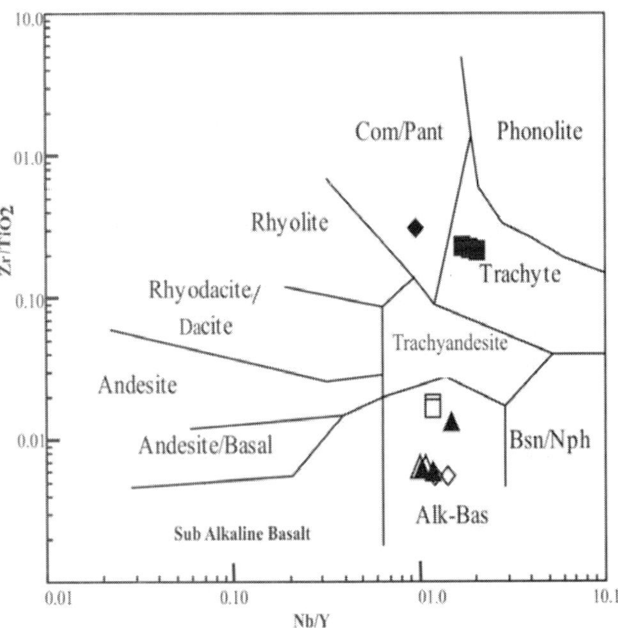

Fig. 9 Zr / TiO_2 - Nb / Y biaxial diagram (Winchester and Floyd, 1977) showing plot of Igneous rocks Rhyolite (■), Shoshonite (▲), Trachyte I (♦), Latite (Δ) , Trachyte II (◊) and Trachyandesite (n) of Karada Fluorite deposit, District Jalore, Rajasthan.

In Fig. 13, Irvine and Barager (1971), gave an trilinear diagram of $Na_2O + K_2O + MgO$–(FeO + 0.8998 Fe_2O_3) which shows a dividing line which separates the thoeliitic and calc–alkaline field. The rock samples area falls under tholeiitic field while rest of the rock samples belongs to calc–alkaline field.

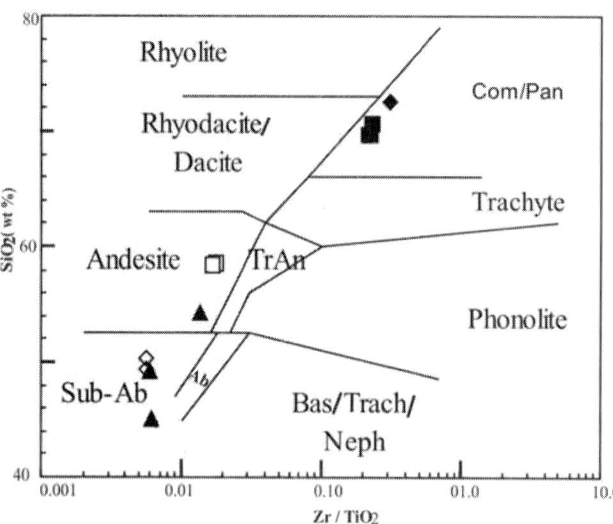

Fig. 10 Zr / TiO2 biaxial diagram (Winchester and Floyd, 1977) showing plot of Igneous rocks Rhyolite (n), Shoshonite(p), Trachyte I (u), Latite (p) , Trachyte II (u) and Trachyandesite (n) of Karada Fluorite deposit, District Jalore, Rajasthan.

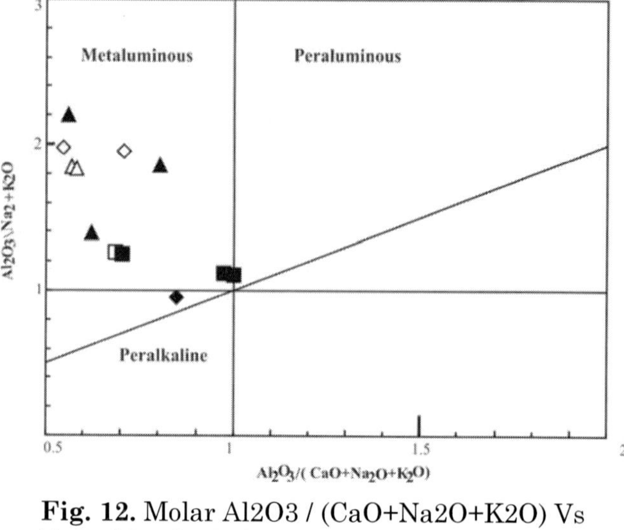

Fig. 12. Molar Al2O3 / (CaO+Na2O+K2O) Vs Al2O3(Na2O+K2O) diagram (after Mainar and Piccoli, 1989) showing plot of Igneous rocks Rhyolite (n), Shoshonite(p), Trachyte I (u), Latite (p) , Trachyte II (u) and Trachyandesite (n) of Karada Fluorite deposit, District Jalore, Rajasthan.

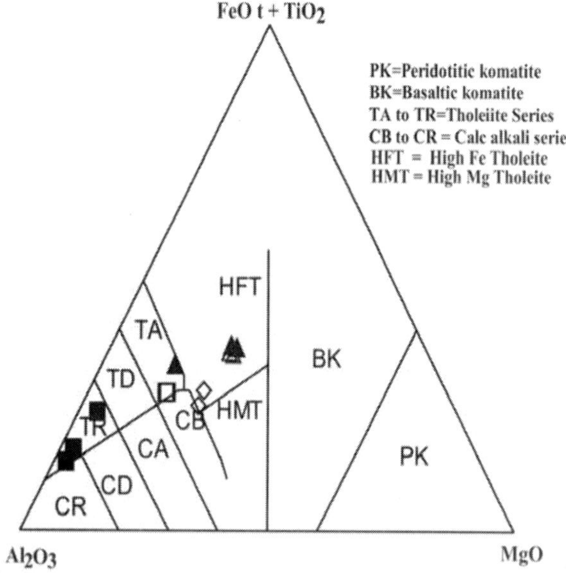

Fig. 11. Trilinear Diagram of Al2O3-(FeOt+TiO2)-Mgo (Jensen, 1976) showing plot of Igneous rocks Rhyolite (n), Shoshonite(p), Trachyte I (u), Latite (p) , Trachyte II (u) and Trachyandesite (n) of Karada Fluorite deposit, District Jalore, Rajasthan.

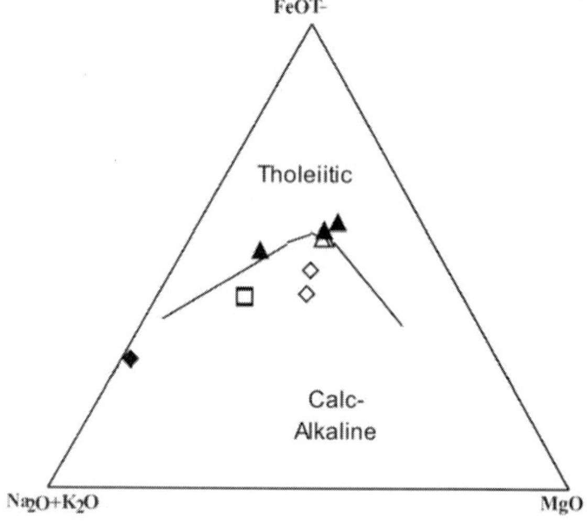

Fig.13 Trilinear diagram (Na2O + K2O)- (MgO)-(FeO+ 0.8998 Fe2O3) for dividing lines between the tholeiite and calc-alkaline Suites (Irvine and Barager, 1971) showing plot of Shoshonite(p), Trachyte I (u), Latite (p) , Trachyte II (u) and Trachyandesite (n) of Karada Fluorite deposit, District Jalore, Rajasthan.

Plagioclase–Olivine–Diopside Diagram

In Fig. 14 is illustrating a trilinear diagram of Plagioclase–Olivine–Diopside in which the rock samples data have been plotted. The rock samples at an average shows 22.17% Albite, 4.52% as normative Plagioclase13.05% Olivine and 17.77% normative Diopside.

Pearce et al. (1984) Diagram (Rb)- (Y+Nb)

In tectonomagmatic discrimination diagram Pearce et al.1984 (Rb)-Y+Nb showing plots (Fig. 15) of all samples are discriminately explaining with in Fig granite and two sample fall in volcanic arc granite.

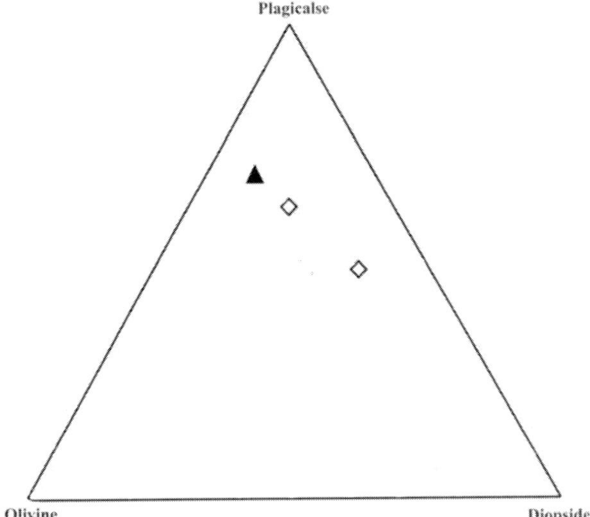

Fig. 14 Plagioclase- Olivine- Diopside diagram showing plot for Igneous rocks, Shoshonite(p), Latite (p) , Trachyte II (u) of Karada Fluorite deposit, District Jalore, Rajasthan.

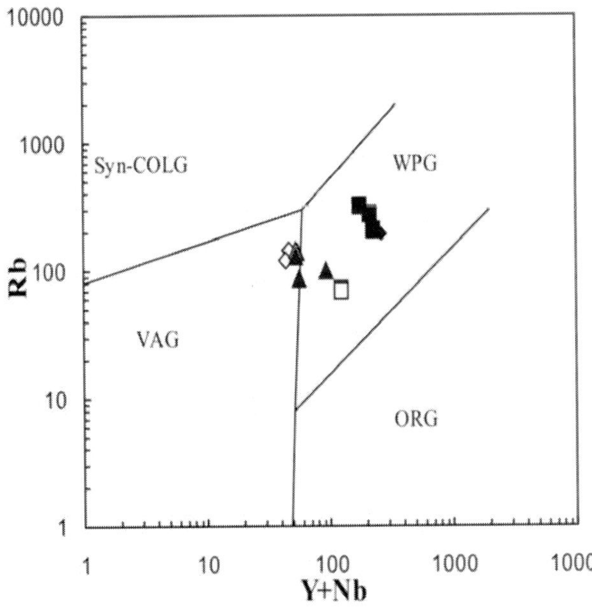

Fig. 15 Rb/Y+Nb biaxial diagram showing plot of Igneous Rocks of discriminating within Plate Granite (WPG), Syn collision (syn-COLG), Volcanic Arc granite (VAG), and orogenic Granite (ORG), (Pearce et al. 1984) showing plot of igneous Rocks Rhyolite (n), Shoshonite(p), Trachyte I (u), Latite (p) , Trachyte II (u) and Trachyandesite (n) of Karada Fluorite deposit, District Jalore, Rajasthan.

Pearce et al. (1984) Diagram (Nb-Y)

Another in Nb-Y biaxial tectonomagmatic discrimination diagram of Pearce et al.1984 showing plots of all samples are fall in within (Fig 15) granite and two samples fall in volcanic arc granite and syn- collision (Fig 16).

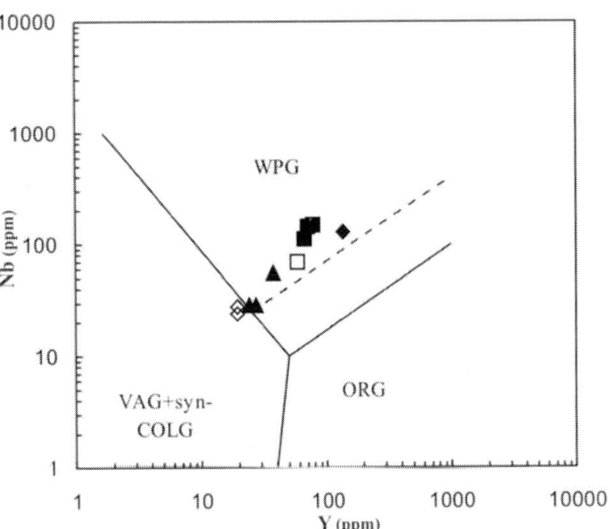

Fig. 16 Nb/Y biaxial diagram showing plot of Igneous Rocks of discriminating within PlateGranite (WPG), Syn collision (syn-COLG), Volcanic Arc granite (VAG), and orogenic Granite (ORG), (Pearce et al.1984)showing plot of igneous Rocks Rhyolite (n), Shoshonite(p), Trachyte I (u), Latite (p) , Trachyte II (u) and Trachyandesite (n) of Karada Fluorite deposit, District Jalore, Rajasthan.

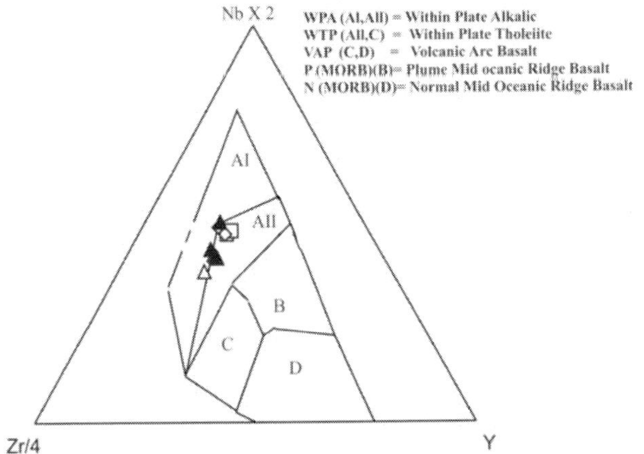

Fig. 17 (Nbx2)-(Zr/4)-Y tectonomagmatic discrimination diagram (Meschede,1986) showing plot for Igneous rocks, Shoshonite(p), Trachyte I (u), Latite (p) , Trachyte II (u) and Trachyandesite (n) of Karada Fluorite deposit, District Jalore, Rajasthan.

Meschede, (1986) Diagram

In Fig. 17 (Nb x 2)-(Zr/4)-Y tectonomagmatic discrimination diagram (Meschede, 1986) all the data of study area discriminately explaining with in Fig alkali nature.

Pearce and Norry, (1979) Diagram

In Zr-Zr/Y bilinear diagram (pearce and norry 1979) all samples showing in within Fig basalts nature (Fig. 18).

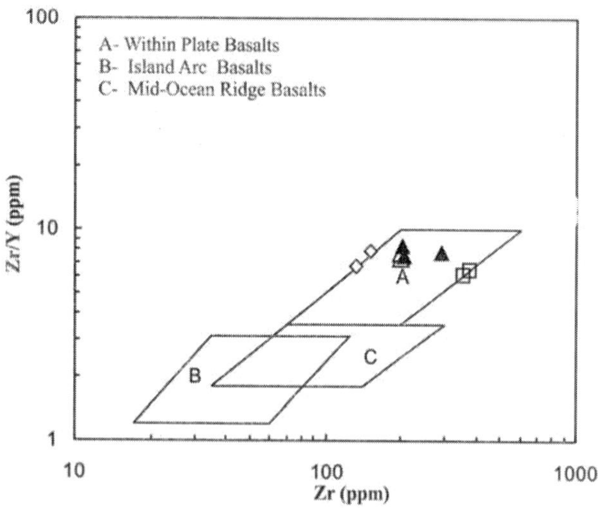

Fig. 18 Zr-Zr/Y bilinear diagram (Pearce and Norry,1979) showing plot for Igneous rocks, Shoshonite(p), Trachyte I (u), Latite (p) , Trachyte II (u) and Trachyandesite (n) of Karada Fluorite deposit, District Jalore, Rajasthan.

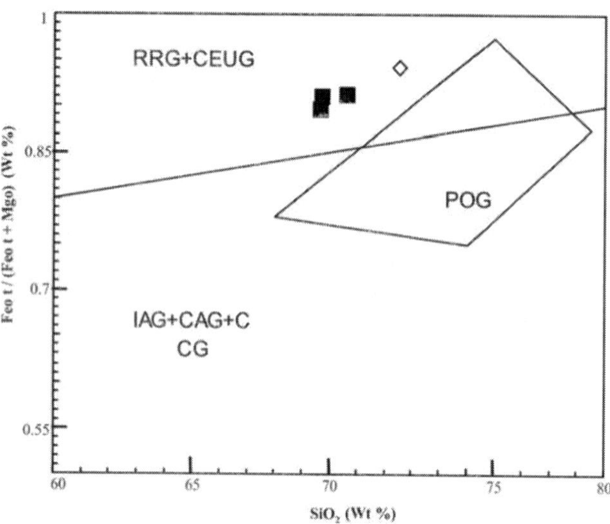

Fig. 19 Variation diagram of sio vs Feo t/ (Feo t + Mgo) after Maniar and Piccoli (1989). showing plot of Igneous rocks, Rhyolite (n), Trachyte (u) of Karada Fluorite deposit, District Jalore, Rajasthan. Abbreviations : IAG - Island Arc Granitoids, CCG - Continental Collision Granitoids, CAG - Continental Arc Granitoids, POG- Post-Orogenic Granitoids, RRG- Rift Related Granitods, CEUG- Continental Epeirogenic Uplift Granitoids.

Maniar and Piccoli (1989) Diagram

In the Feo^t (Feo^t+Mgo), tectonic discrimination diagram after Minar and Piccoli (1989) showing plot of study area samples of rhyolite and trachyte falls in Rift Related Granitoids and Continental Epeirogenic Uplift Granitoids (Fig.19).

Batchelor and Bowden (1985) Diagram

In the R_1-R_2 diagram multicationic factors proposed by Da La Roche et al. (1980) has been plotted to emphasis the classification and nature of the study area samples (Fig. 20) rather than tectonic environment (Batchelor and Bowden, 1985). In this diagram all of study area samples falls in late orogenic and pre Fig collision field.

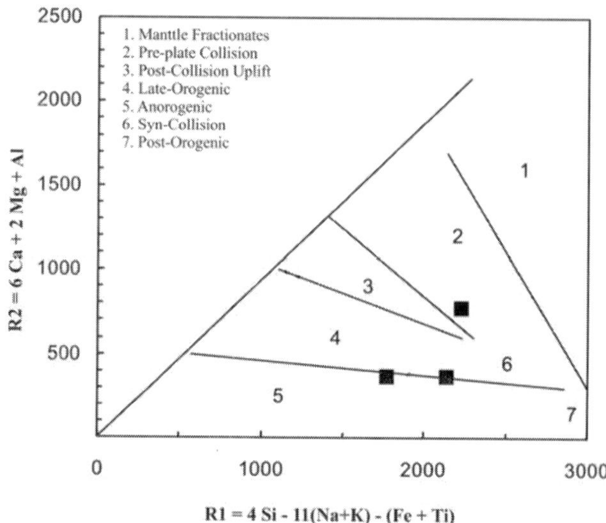

Fig. 20 R1-R2 Diagram of Batchelor and Bowden (1985) showing plot of Igneous rocks, Rhyolite (n), of Karada Fluorite deposit, District Jalore, Rajasthan.

(iii) Trace Elements and their distribution

Trace Element values of the study area are present in Table 4 for the fluorite and associated rocks. The trace elements determined includes Ba, Sc, Co, Ni, Cu, Zn, Ga, Pb, Th, Rb, U, Sr, Y, Zr and Nb. All the data are in ppm. As part of the general observations from the Table 4, it is clear that rocks are rich in Ba, Sr, Zr, Rb and Nb selectively. Also these rocks are showing relatively poor values for U, Sc, Pb Ni, and Th.

(iv) Rare Earth Elements and their distribution

Rare Earth Elements data for fluorite and associated host are present in Table 5. Rare earth elements are of great significance as partial melts derived from a source rock containing plagioclase will not completely dissolved during melting will contain relatively less Eu than other REE because K_D for Eu is larger. A similar "europium anomaly" could develop in residual melts from which plagioclase has been fractionated. K_D values involving garnet increase from light through heavy REE.

Therefore, a partial melt of a granetiferous source rock, or a residual melt from garnet fractionation, will have a high Ce/Yb ratio. These are the kinds of constraints that can be applied to evolutionary processes im magmatic systems using dispersed elements.

The Karada fluotite deposit rocks have been analysed for rare earth elements and data are presented in Table 5. The rare earth elements determined includes La, Ce, Pr, Nd, Sm, Eu, Gd, Tb, Dy, Ho, Er, Tm, Yb and Lu. All the values obtained are in ppm. It can be observed from the data that rocks are showing high values for Ce, Pr, La and Nd. Also, the low values for Lu, Tm, Ho and Tb rare earth elements.

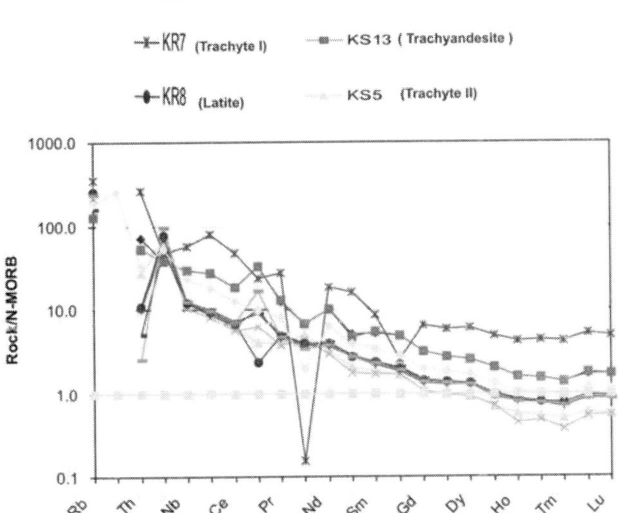

Fig. 21 N-MORB (Sun and Mc Donough, 1989) normalized spider diagram of host Igneous rocks of Karada Fluorite Deposit, District Jalore Rajasthan.

(v) Spider Diagrams

Most of the rocks of the study area are plotted as spider diagram Fig. 21 for rare earth elements with reference to the normalized Mid–Oceanic Ridge Basalts (normalization values are after Sun and Mc Donough; 1989). It shows depletion of Th and Nd remained prominent. In Fig. 22 E-type Mid–Oceanic Ridge Basalts (normalization values are after Sun and Mc Donough, 1989) it shows enrichment of P, Zr, Eu, Y Er, and depletion of Ce, Pr. In Fig. 23 sample plotted against primitive mantle (Sun and Mc Donough, 1989) normalized spider diagram and in Fig. 24 all samples data plotted against chondrite for normalized spider diagram (Sun and Mc Donough, 1989).

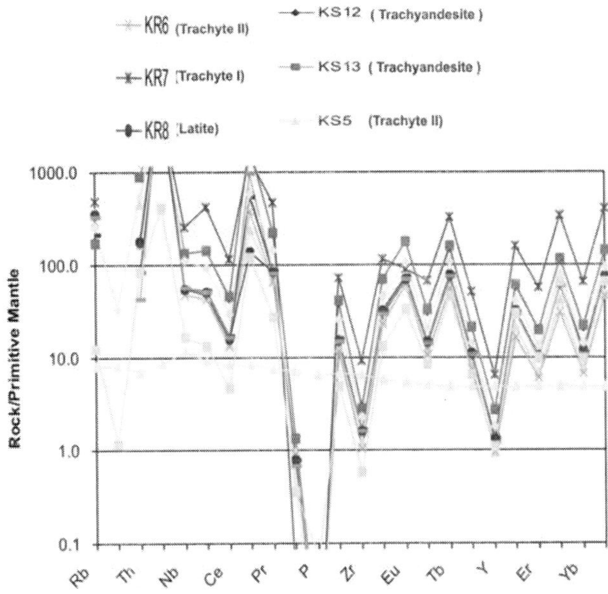

Fig. 22 E-Type MORB, (Sun and MC Donough 1989), normalized spider diagram of host Igneous rocks of Karada Fluorite Deposit, District Jalore, Rajasthan.

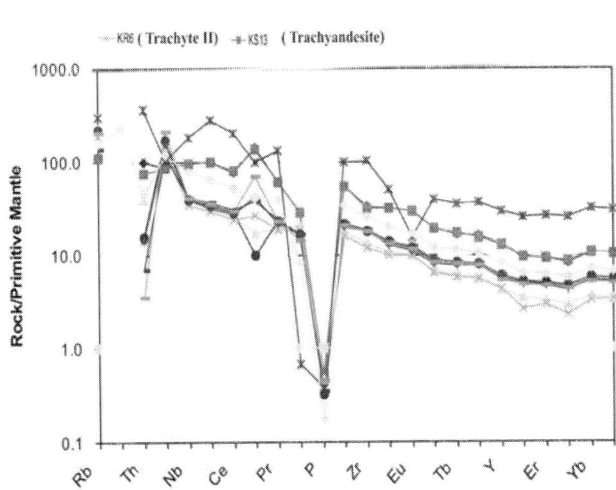

Fig 23 Primitive Mantel (Sun and Mc Donough,1989), normalized diagram for Igneous rocks of Karada Fluorite deposit, District Jalore, Rajasthan

This chapter is dealing with characteristic features of the fluorite mineralization in Rekha and Santoshi mines of Karada deposit as part first. The second section is explaining the possible mechanism of the formation and later modification in the deposit named as the genetic model. In both the sections various sub steps have been discussed.

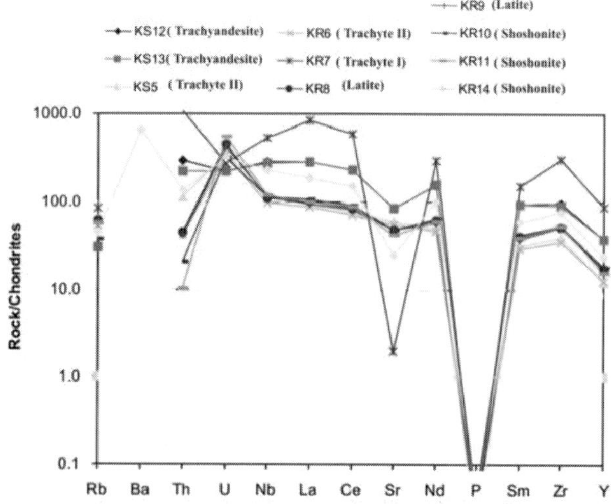

Fig. 24 Chondrite (Sun and McDonough, 1989) normalized spider diagram of Igneous rocks of Karada Fluorite deposit, District Jalore, Rajasthan

CHARACTERISTICS OF FLUORITE MINERALIZATION

1. The Mineralization

The Karada fluorite deposit is concentrated at two locations in the area, first one is called Rekha mine and is in the northern part, second one is called Santoshi mine and is in the southern part. In the Rekha mine there is one prominent vein having its uniform thickness of 10 meter almost uniform for a length of 100 meter (Fig 25 A). This main body is confined to volcanic breccia and having its penetration deep inside. The orientation of vein is N43^{0}E-S43^{0}W with subvertical dip towards north (Fig 25 A and B). Fluorite is well filled in interspace available between the fragments of acidic and mafic volcanic rock apart from low grade fluorite fragments finally constituting the volcanic breccia (Fig 25 F, G and H). In open spaces as coatings, fillings and forming cocade structures fluorites are very common and making this deposit as rich fluorite deposit (Fig 25 D and E).

There are two more veins available parallel to the main fluorite vein at Rekha mines. These are 200 meter a part from main vein both the sides, one towards east and another towards west. These are also having width of 10 meter each but there length is almost 55 meter each. There is not much 'gem' fluorite is available but only fluorite breccia (Fig 25 F, G and H) in volcanic breccia is present. There is yet another vein available further 200m apart with moderate fluorite grade breccia only in the east. There is no such symmetrical vein available in the west or it must now have eroded totally because of nallah cuttings.

Santoshi mine is present in the south of the Rekha mine area. There is one vein with a thickness of 10m and length of 200m well filled with fluorite and volcanic breccia and also well coated with 'dog tooth' 'gem' fluorite (Fig 25C and E). Unfortunately there are no such sympathetic veins present to this main fluorite vein as is the case in the Rekha mine. It is interesting to note that the orientation of this vein is also N43^{0}E-S43^0 W with 85° dip towards north.

The orientation of veins has strictly followed the fillings of first set of joints (Joint I). this joint has its orientation as N43^{o}E-S43^{o}W and dipping subvertically (85°) towards north (Fig 25 A & B)

As a rule, all the fluorite veins follow the trend of prominent Joint I showing N43°E-S43°W strike and dip 85°N. Though mineralization in the area is joint controlled, it has occurred mostly along the first set and not the later, Joint II and Joint III sets. However, all of them have greatly broken the main mineralized body. Apparently, no post mineralization tectonic observed in the fragmented mass. But along joints, there is neither shifting of the blocks nor crushing of the available minerals. Multigeneration fragmentation is well recorded at both the mines. Nonetheless the size of fragments is larger in the Santoshi mine than the Rekha mine.

As the second and third sets of joints are unaffected by the mineralization, there is no bifurcation of veins. It is clear from the plan of the breccia veins constituting stockwork that the shape of the ore body has been closely controlled only by the first set of joints, the formation of which presumably immediately pre-dates the emplacement of the breccia body itself. Trends of mineralized veins within the stockwork are almost sub-parallel to parallel, suggesting joint formation by tensional forces. Qualling apart of rock perpendicular to the trend of the joint (mineralized one) is further supported by lack of crushing along veins.

Some of the pure fluorite samples and some fluorite samples inside the volcanic breccia have been analysed using XRF and ICP-MS and data have been presented in the form of Table 3, 4 and 5 for major oxides, trace element and rare earth elements and rare earth elements. It has been observed that the samples are rich in Zr, Sr, Rb, Zn and Nb as far as trace elements are concern. Among rare earth elements fluorites are rich in Ce, Pr, La and Nd.

Fig. 25. (A) Abandoned part ofopen cast mine of fluorite at Santoshi mine showing orientation of N 43°E., illustrating joint control of the mineralization. **(B)**Mining of a vein of fluorite in Rekha mine controlled by a joint oriented N 43°E. **(C)** 'Dog tooth' form of 'gem' (high grade) fluorite attached on foot wall of mineralized vein on volcanic breccia oriented N 43°E following a joint at Santoshi mine. **(D)** Typical association of volcanic breccia showing off-white fragments of medium grade fluorite, over which translucent yellow cubes of fluorite are well grown, after getting rich hydrothermal solution as last phase of mineralization. This mine dump is observed at Rekha mine. **(E)** Well developed fluorite cubes from Rekha mine showing free growth from hydrothermal solution to develop all the faces. **(F)** A hand specimen of volcanic breccia showing a rich layer of fluorite (high grade, 'gem' variety) grown because of latest pneumato hydrothermal action. **(G)** A volcanic breccia constituent medium grade fluorite fragments (MF*l*), having various sizes (few mm to few cms) and angular nature. A forceful explosion, even after formation of the medium grade fluorite is proved. **(H)** Typical volcanic breccia having fragments of low grade fluorite (LFℓ) (off-white colour). The size of fragments varying. Close observation show that low grade fluorite fragments are showing 'Zig-Saw-Puzzle' pattern and a phase of iron metasomatism.

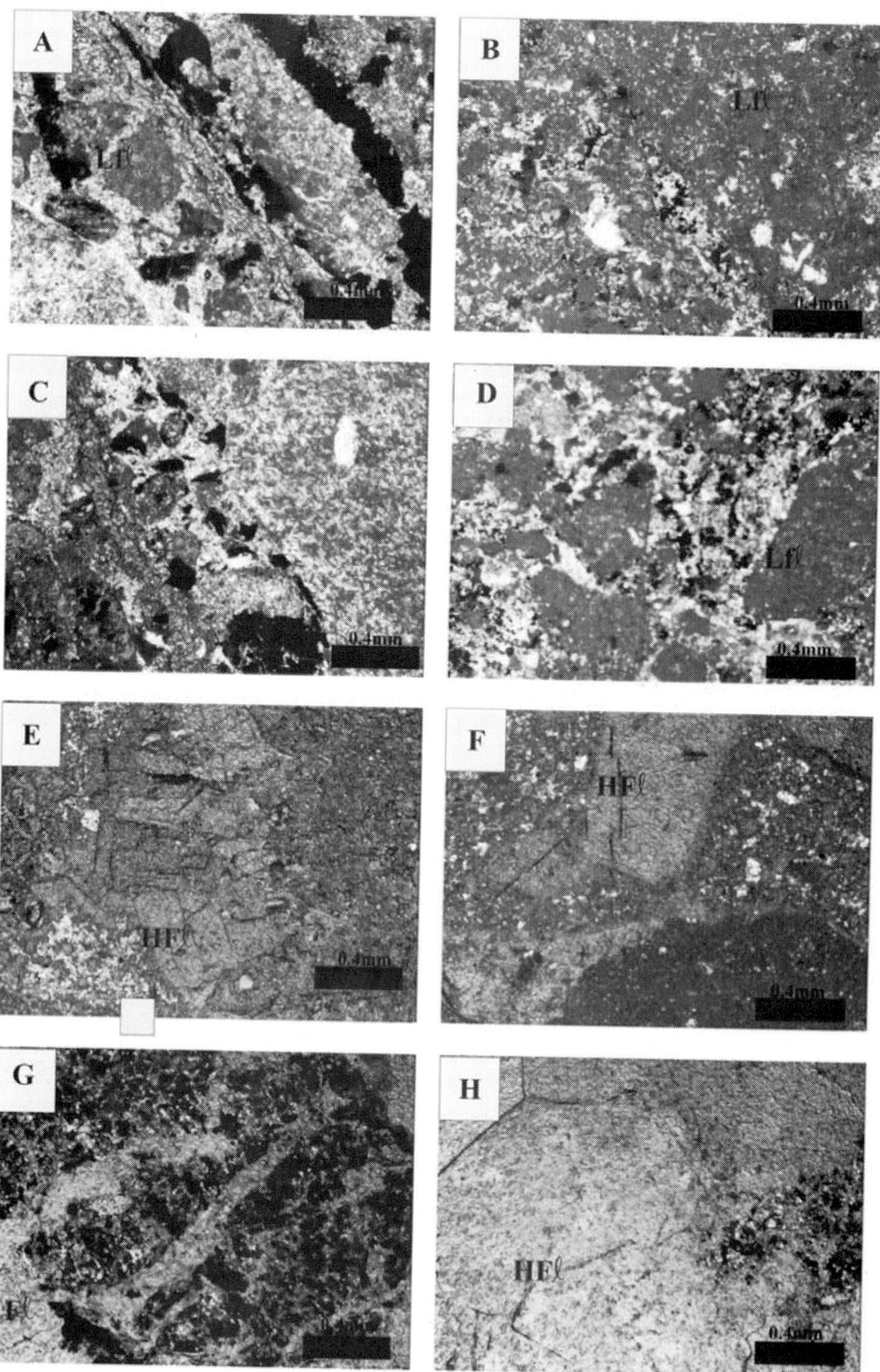

Fig. 26. (A-D) Photomicrograph illustrating fluorite of low grade (LFℓ) are one of the constituent of the volcanic breccia. Fluorite (gray) fragment are assembled with fragments of rhyolitic composition (light coloured) along with iron ore of isotropic nature. The matrix of fragments is again showing rhyolitic composition with very fine minerals and glass showing flow structures. **(E-F)** Photomicrograph illustrating high grade fluorite (HFℓ) fragments constituting volcanic breccia as one of the components. The matrix is of fine grained minerals and glass (rhyolite). **(G)** Photomicrograph demonstrating fluorite with its advancing front showing veins penetrating deep inside the iron minerals as part of metasomatic replacement, indicating that the last rich phase of fluorite formation was even later than iron maetasomatism. **(H)** Photomicrograph showing well developed fluorite cubes with free development of high grade fluorite (HFℓ) has occupied all the available spaces in the last phase

Table 3. Geochemistry of Fluorite bearing Volcanic Breccia, Karada Fluorite Deposit, District Jalore, Malani Igneous Suite, India. (Part I Major Oxides and ratio)

Sample No. → **↓ Major oxide**	**KR1**	**KR 2**	**KR 15**	**KR 16**	**KR 17**	**Avg.**
SiO_2	58.89	69.62	66.86	1.67	13.27	42.06
Al_2O_3	9.99	13.41	13.03	0.43	1.45	7.66
Fe_2O_3	0.27	0.3	0.67	0.04	0.05	0.26
FeO	2.21	2.46	5.4	0.32	0.42	2.16
TiO_2	0.22	0.41	1.13	0.04	0.07	0.37
MnO	0.12	0.01	0.34	0.02	0.02	0.1
P_2O_5	0.04	0.03	0.3	0.12	0.07	0.11
CaO	20.55	4.58	1.51	97.14	84.39	41.63
MgO	1.38	0.32	1.77	0.20	0.19	0.77
Na_2O	0.26	2.06	0.2	----	0.04	0.64
K_2O	5.7	6.8	8.8	0.04	0.03	4.27
Ratio						
Salic/Femic	1.11	7.29	5.11	--	--	4.50
$(Na_2O + K_2O)/ CaO$	0.13	1.29	3.598	--	--	1.67
K_2O/Na_2O	0.19	2.17	29.25	--	--	10.53
Q or F/L	6.07	0.35	0.393	--	--	2.27

Table 4. Geochemistry of Fluorite bearing Volcanic Breccia, Karada Fluorite Deposit District Jalore, Malani Igneous Suite, India. (Part II Trace Element in ppm)

Sample No. → **↓ Trace Elements (ppm)**	**KR1**	**KR 2**	**KR 15**	**KR 16**	**KR 17**	**Avg.**
Sc	15	8	5	34	32	18.8
Co	10	1	15	---	---	8.66
Ni	4.5	7.6	14.1	20	13.6	11.96
Cu	39.8	11.6	59	68	72.4	50.16
Zn	121	112.2	141.5	60	38.9	94.72
Ga	15.5	23.6	11.6	18	1	13.94
Pb	15.9	13.9	15.5	70	9.5	24.96
Th	18.5	25.8	8.5	2.9	0.9	11.32
Rb	199.6	208.8	298.9	3	17	145.46
U	1.3	3.4	3.6	3.2	3.1	2.92
Sr	159	40	38	619	503	271.8
Y	69.7	78.9	41.1	122	11	64.54
Zr	691	920	323	28	46	401.6
Nb	144.3	146.8	59.3	6.9	8.5	73.16
Ba	----	----	985	--	--	--

Table 5. Geochemistry of Fluorite bearing Volcanic Breccia, Karada Fluorite Deposit District Jalore, Malani Igneous Suite, India. (Part III Rare Earth Element in ppm)

Sample No. → **↓** **Rare Earth Element (ppm)**	**KR1**	**KR 2**	**KR 15**	**KR 16**	**KR 17**	**Avg.**
La	91.234	125.479	49.05	52.5	5.598	64.77
Ce	193.889	243.141	100.94	110.08	11.47	131.9
Pr	19.483	26.163	11.31	13.9	1.396	72.25
Nd	74.443	99.950	46.74	61.8	5.679	57.72
Sm	13.244	17.157	8.825	16.5	1.406	11.42
Eu	1.551	1.926	2.720	7.25	0.480	2.78
Gd	11.911	16.253	7.148	16.31	1.372	10.59
Tb	1.977	2.696	1.198	2.821	0.275	1.79
Dy	13.524	18.352	7.914	15.78	1.816	11.47
Ho	1.976	2.693	1.110	1.70	0.223	1.54
Er	5.642	8.355	3.236	3.62	0.559	4.28
Tm	0.847	1.299	0.481	0.37	0.071	0.61
Yb	7.737	11.333	4.088	2.48	0.564	5.24
Lu	1.110	1.626	0.604	0.30	0.080	0.74

2. Importance of Global Tectonics

The global tectonic the nature and origin of magmas of different compositions and their role in the genesis of mineral deposits is a field of general interest (Abdel Rahman and Elkibbi, 2001; Drury et al. 2001; Kabeto et al. 2001; Mocek, 2001; Okay et al. 2001; Phillips and Woods, 2001; Scambelluri and Philippot, 2001; Seghedi et al. 2001; Sheppard et al. 2001; Spath et. al. 2001; Waight et al. 2001; Wang et al. 2001 and Zhu et al. 2001). Role of magma chambers in the generation of felsic and basaltic magmas, crustal assimilation, contamination and their fractional crystallization, remain subject of studies by many workers (Aguirre Diaz, 2001; Blake and Bruno, 2001; Borisova et al. 2001; Kabeto et al. 2001; Kuritani 2001, Lindsay et al. 2001; Rabinowicz et al. 2001; Riley et al. 2001; Stevenson et al. 2001; Tatsumi 2001 and Verma 2001).

The parent magmas for the rhyolites must have depths and fractionated during the ascent to surface and assimilated crustal fragments. The emplacement of the magma must have been forceful and led to the development of regional

Caldera having tens of kilometer diameter. Occurrence of similar Caldera structures elsewhere have been reported by many (Aquirre Diaz, 2001; Chabiron, et al. 2001; Fiske, et al. 2001; Lindsay, et al. 2001; Riller, et al. 2001 and Verma, 2001). Occurrences of xenoliths from deeper parts of crust (or from mantle?) have also been discussed by a few (Andersen and Neumann, 2001; Chabiron et al. 2001; Vandenkerkhof and Thiery, 2001). Presence of CO_2 and H_2O in the magma and their possible role in the mineralization, have been studied by Diamond (2001) Edmonds et al. (2001) Evans et al. (2001) and Tamic et al. (2001).

The Karada fluorite deposit is located in parts of the outer circular ring fractures developed during collapse of caldera, in this region of Malani Igneous Suite. These fractures show only linear nature, locally. It is also expected that because of collapse of caldera which is deeper towards west, gases (like fluorine), migrated alongwith magma (basaltic) during petrogenesis and mineralization, the latter more towards east. Hence, resulted in the form of Rekha and Santoshi deposits of the Karada fluorite deposit.

3. Fluorite Breccia: Composition and Texture

Mineralization mostly occurs as filling of open spaces of joint I with permissive to forceful penetration. Texture suggesting that the filling of cavities, and to a lesser extent, replacement by digestion of fragments of host breccia certainly formed earlier. Open spaces along the joint I plane have provided channel ways for fluorite breccia and mineralizing solutions and sites for the deposition of the fluorite.

The characteristics of fluorite breccia are the following :

(i) Compositionally homogeneous to heterogeneous (basaltic to rhyolitic and fluorite fragments cemented into fluorite it self.

(ii) Clasts are angular to sub-angular fragments

(iii) Closed framework of angular clasts.

(iv) Preservation potential is moderate.

(v) Disorganized fluorite breccia

(vi) 'Jigsaw Puzzle' fit of clasts is preserved, where there has been no redistribution from site of fragmentation by turbulent mixing.

(vii) Recognition potential is moderate to good.

(viii) Massive to diffusely layered fluorite mass and its fragments.

(ix) The breccia is typical of large scale sector (up to hundreds of meters) generally known to form during collapse

Composition and texture of the fragments formed in the main stage breccia range from angular to sub-angular shapes and also high proportion of comminuted host rock fragments in the matrix of fluorite itself is evident. The matrix of breccia stage has formed chiefly by a process of precipitation into the spaces dynamically opened during explosive vaporization of the fluorite and fluorine-rich fluid during decompression (Fig 25 A, B and C). The central fluorite fill has also formed partly by replacement in a second episode of dilation fluid intrusion. The volume of metasomatically replaced material contributes much more to the substance of the structure than does the central fill. Such fractures are typical fractures, which were formed in a stage prior to the main episode of brecciation.

The volume ratio of fragments to matrix have been observed and the ratio of killas fragments to host rock fragments in various parts of the mineralized stockwork breccia have been noted. The general conclusion which can be drawn from these observations is that the lateral veins of breccia consisting predominantly of host fragments and that for the such parts the volume proportion of fragments lies between 75 to 35 percent. It defines the various grades of fluorite.

It is inferred that the eroded roof of the intrusion should possibly lie within 200 m vertically above the highest point now observed by projecting the contacts along reasonable lines, based on measured dips at the present level of exposure. The abundance of killas fragments in the deposit even indicate that the projected contact must lies even closure.

The breccia consists essentially host rock fragments and fluorite and lesser amounts of quartz. The fragments are cemented by fine grained fluorite and quartz (Fig 25 G and H). Within the population of host rock fragments which all carry either lesser or greater amount of fluorite, there is a large range of textures. For example, the degree of rounding of fragments is quite variable and a good percentage of the constituting fragments of the variable sizes (Fig 25 G) ranging from very large to moderate, retains an angular form convincingly.

The main breccia shows that within the population of the host rock fragments, there are many textural types of fluorite available. Fluorites themselves often carry still smaller bites of fluorite, mostly very fine grained and sometimes with lesser amounts of banded type and small segments of host within an aphanitic matrix.

Alteration of boundaries, moderate to intense are possibly the characteristic features for the identification of the killas fragments. This alteration has become intense in some cases. The rather uniform distribution of killas fragments within the breccia shows that considerable differential movement of fragment took place during the formation of mineralized bodies.

The hydraulic fracturing and disruption evident on the contact of the mineralized body indicates this body must have formed from a fluid under high pressure. As a consequence of this at lower levels of exposure, these fragments appear to have moved laterally or downward over almost two hundred meters.

4. Pneumato-Hydraulic Role

There are good number of field and microscopic evidences to show that a fluid system in subsurface must have developed a pressure possibly exceeding the tensile strength of the enclosing (host) rock and the minimum principal stress component, then that fluid must have extended its capacity resulting to propagate the fracture and fracture system, to open it in a tensile manner. Such fracture propagation and openings are always expected to lead a kind of the weaker forces produced narrow openings in the brittle rocks. It is, therefore, for this reason that above-mentioned host rocks carry numerous stringers and veinlets of fluorite in them. Such fracture-filled veinlets are better studied in rhyolite and under microscope (Fig 26, A-H). They are usually parallel to one another and are seen traversing the host rocks or the earlier formed minerals. These veins are characterized by nearly matching walls, though complete matching is not found.

Bastin has observed that "Perfect matching is not always found, even in undoubted fracture fillings, since one wall may have moved in the direction inclined to the plane of the section available for study, and fragments torn from the wall in the fracturing may also have moved out of that plane".

The rock is essentially shattered in tensile fashion by the overpressed fluids working through the host rock. Fracturing caused by this mechanism favour that the resultant fracture need not be involved in shear motion and have the breccia fragments and the associated wall rocks need not be pervasively foliated or sheared (Phillips, 1972). This fluid-induced fracturing and brecciation, called Pneumato-hydraulic fracturing, have effectively operated at Karada deposit

The manner of successive depositions of two or more mineral starting from the walls of the openings give crustified veins. The veins of this type provide information on paragenesis and nature of mineralizing solutions. The minerals is present next to the vein wall are the earliest formed followed by the minerals towards the center of the cavity. The crustification in all the cases was found to be of symmetrical type i.e., nature and sequence of minerals on both the walls is similar. Calcite grew along the length of the vein while crystallization of quartz commenced from the walls only, thus producing comb quartz. Not in the main fluorite veins but in other small veins, fluorite is followed by quartz crystals with their long axis perpendicular to the vein wall thus producing comb structures. In these cases, however, it is rare that successive mineral crusts have deposited in perfect parallelism for any appreciable distance. It is possibly because of the initial irregularity of the surface of deposition or even rate of growth of crystals.

The quartz crystallized last and completely closed the cavity on all the sides .The quartz crystals are oriented normal to the torn pieces. This shows that fluorite which was deposited along the walls was broken before solutions started depositing quartz. Hence, the quartz crystallized afterwards is seen to line the fluorite at the wall. In this such portion crystallization of quartz commenced at two surfaces, one next to fluorite at the wall and second around the broken up pieces.

The fragments which have been roll back in the deposit are present in a systematic pattern of deposition. The successive depositions around a rock fragment have been resulting in cockade structure, ring structure or crusted kernels.

Various shapes and sizes of rock fragments form nuclei of mineralization around which successive bands were formed. The phenomenon is reverse of what is seen in circular cavities, where the mineralization starts at wall and proceeds towards the centre of the cavity. In contrast, the mineralization in cockade ores starts around the rock fragments and proceeds outward from the centre thus producing convex fractures. The earlier formed quartz crystals are small but the later ones are much larger. These depressions show polygonal or square outline around a corresponding zoning typical of quartz.

The pneumato-hydrothermal fluid systems, in volcanic conditions, have enormous potential for causing hydraulic fracturing and brecciation. Though they have been reported from several places but association mineralization is not common. The hydraulic fracturing and brecciation showing angular, blocky to splintery fragments could be confused with quench-fragmented hyaloclastic deposits. Pneumatic-hydraulic breccias are characterized by (i) variably altered clasts (ii) angular to splintery clasts and (iii) 'jig saw puzzle' fit of clasts where little disturbance or transportation of clast has occurred. The Pneumato-hydraulic fracture breccia, in the outcrops, is distinguishable by their cross-cutting character and confinement to a fracture zone. The Karada fluorite deposit ofcourse shows both, the cross-cutting characters to produce stockwork and confinement of the mineralization along the fracture zone developed and oriented following N43°E-N43°W direction of joint I.

The thickness of the hydraulic fracture zone at Karada has been observed almost 20 metre. It is in the order with the concordant view that such zones may vary from several centimetres to many metres (Cas and Wright, 1987, p. 421). Karada fluorite breccia also shows incipiently brecciated zones. The breccia clasts show 'jigsaw puzzle' like fit. In the mineralized stockwork deposit in the field, the angular clasts show their movement relative to each other. Some of them have rotated significantly and even transported a little through the system, leading to heterogeneous mixture of clast lithologies and various generations of fluorite. Some small scale occurrences of the process have also been observed.

The hydraulic fracturing has also produced sites for considerable hydrothermal alterations. These alteration processes, have produced significant weakening of the rock. Such zones may preferentially absorb iron and other stains (Fig. 26 G) and become suitable site for subsequent hydrothermal action. The third and the last phase of fluorite formation has also occurred in the fractured brecciated sites and has cemented fragments of low grade and medium grade earlier formed fluorites, of pneumatic and hydraulic action.

The features observed in the ore breccia could best be explained in terms of the 'fluidization' model first applied to geological process by Reynolds (1954). Fluidized systems are those in which a dynamic state of suspension of solid particles by a flow of gas or vapour prevails, although a condensed liquid of lower density than the suspended fragments may also fluidize a system. Reynolds (1954) pointed out that the solid and gas (vapour) systems are characterized by a condition of turbulent flow within which great agitation and mixing of suspended fragments occurs. These results in abraded and well-rounded rock fragments, which may not have been appreciably transported away from their source and which, upon the exhaustion of the system, settle with lack of grading. Considerable amounts of kinetic energy would be absorbed by the multiple impacts of one fragment against another and it would be natural to expect that such systems should contain a significant proportion of well rounded fragments, the cause of which would be the physical process of attritive rounding by impact. The degree to which such rounding developed would depend upon the durability of the rocks, the viscosity of the suspending fluid phase, the kinetic energy of the system, and the duration of the fluidized condition. The possible corrosive chemical effects of the suspending fluid would also have to be taken into account (Sawkins, 1969; Sillitoe and Sawkins, 1971).

The mineralization of fluorite breccia at Karada does not show high population of well rounded fragments that might be expected if mechanical attritive effects had been of long duration Examination of the contacts of the well rounded fragments also shows that the rounded shape is product of concentric exfoliation, rather than of attrition. Evidence of the operation of this process is shown in which the concentrically spalled layers are apparently frozen in the initial stages of disaggregation within the matrix. The explanation chosen for this phenomenon is that of explosive rupture following an instantaneous drop in confining pressure. For this reason, it is termed 'decompressive shock exfoliation'. It is evidently this process, which is responsible for the population of larger spheroidal fragments within the breccia body.

A number of textural features observed at Wheal Remfry by Sillitoe and Sawkins (1971) have similarities with the Karada deposit. It is considered by Allman-ward et al. (1982) that the principal cause of disruption was the propagation of tensile fractures initiated during the escape of the boron-rich fluid from confinement at depth and that the disorientation of fragments depended more upon the turbulent passage of the fluids as they passed down a pressure gradient guided by the fracture system, than upon passive collapse under the influence of gravity into space created by chemical solution. The breccia thus displays features suggesting that, at the time of its formation, it consisted of a suspension of solid rock fragments in a stream of boron-rich fluid within which differential movements took place in response to the differences in hydraulic equivalence of fragments of different mass, shape, and density at different levels within the system. The fluid component was, no doubt, a boron-rich hydrous vapour carrying significant amounts of silicon, aluminium, and alkalis (Pichavant, 1979).

5. Collapse Fluorite Breccia

The presence of killas mixed within the fragments at all levels of the body is conclusive evidence of the turbulent mixing of fragments within a fluidized system at the time of formation. The impression conveyed by many of the textures of partial disaggregation is that the dynamic processes were frozen in situ by the rapid crystallization of the supporting matrix of fluorite. This inference is supported by the fact that the fluorite matrix is essentially microcrystalline and may have crystallized nearly instantaneously by a process of solute dumping as the fluids responded to a rapid change in physical conditions, in all probability a catastrophic fall in confining pressure. Although the breccia was certainly initiated as a fluidized suspension it is suggested that this condition was of comparatively short duration, in geological terms, perhaps, instantaneous.

6. Intrusive Breccia

The dynamic intrusive process of the fluorite breccia is different. The fluorite breccia is predominantly made up of angular killas fragments encompassing a wide range of sizes packed together and cemented again by high-grade fluorite. The constituent killas fragments are either partially or fully fluoritized. The matrix is made up chiefly of fluorite, but quartz and crystal-lined vugs are commonly observed in the larger interstices between fragments.

It is important to note here that predominance of killas fragments in the mineralized body some 200-300 m below the probable position of the porphyritic rhyolite (with granite at still lower parts) contact taken together with the purely siliceous and locally vuggy character of the matrix suggest an origin by collapse. Thus, this stage of breccia is, therefore, designated as fluorite collapse breccia

7. Relationship of Both Breccia

The fluorite collapse breccia evidently formed at a later stage the intrusive breccia. The matrix shows the same textural geometry as the fluidized matrix of the main stage. The comminuted rock fragments and all the larger clasts have suffered complete fluoritization so that the rock shows much less visual contrast between clasts and matrix and has a generally grey appearance. In the inner part of this elongated halo of fluoritization the host rock clasts are completely pseudomorphed by fluorite within which felted patches of fluorite and cavities lined with small terminated fluorite crystals mark the site of pre-existing feldspar phenocrysts. Towards the outer limits of the fluoritized zone some fragments, which have suffered only partial fluoritization display distinctive alteration haloes.

The precise definition of contacts between the main breccia and the collapse breccia is not made easily in the field because of the marked effects produced by the zone of fluoritization which evidently spread from the centre (in the elongated form) of collapse. There can, however, be little doubt that the collapse breccia formed, and was subject to the effects of pervasive fluoritization after the emplacement of the main ore body as a fluidized mass. Clearly the later stages of the formation of the breccia body were marked by collapse of the killas roof rocks into space created at a lower level by processes of physical displacement or chemical dissolution caused during the earlier stage of explosive release of fluids rich in fluorie. At the same time the composition of the active hydrothermal fluids changed from being fluorine-rich to being silica-rich.

The range of geological features displayed by the different parts of the Karada fluorite breccia body show that the process of disruptive intrusion by a turbulent fluidized system was the dominant process at the initial stages of formation. This was followed by a stage of collapse and that this took place in at least two episodes, one major and one minor. Resultantly, the early formed fluorite in the form of angular fragment included into the laterformed rich fluorite matrix. It was followed by pervasive silicification. The fluorite ore body, therefore, combines the features of both hydrothermal intrusive breccias and hydrothermal collapse breccias. The precise role of chemical processes in producing the brecciation of the rocks is difficult to assess but the balance of evidence suggests that hydraulic and mechanical forces were probably dominant.

8. Metasomatism

(a) Potash metasomatism

Potash metasomatism is main resultant of hydrothermal action of fluid and gases at Karada fluorite deposit and also in the nearby areas. Mineralogically, this is characterized by microclinization due to the partial or complete replacement of sodium by potassium in the feldspars, accompanied by haematitic reddening. There is development of new mica in the compositional range annite to ferrous siderophyllite. Chemically, there is an increase in K and a concomitant loss of Na.

(b) Acid metasomatism

Acid metasomatism (hydrogen ion metasomatism) resulted into progressive breakdown of rhyolitic and trachytic minerals in response to change K^+/H^+ ratio in the fluids. Perthite or microcline feldspar is destroyed and replaced by siderophyllite/ zinnwaldite and accompanied by fluorite and sericite. Chemically, hydrogen ion metamorphism is characterized by a marked decrease in potash and alumina due to breakdown of feldspar and an increase of Fe_2O_3 and SiO_2 content.

(c) Silica metasomatism

Silica metasomatism resulting in quartz vugs and replacement of early-formed minerals, especially in trachytes. The process of silicification is responsible in increasing total silica balance in the host rocks.

(d) Argillic alteration

Argillic alteration have been involved in the formation of clays at the expense of feldspars. As the main ore body is dipping somewhat westward, this alteration has accumulated kaoline at the hanging wall of the ore main body, at its centre.

The hydrothermal system responsible for kaolinization was clearly later than breccia formation (fluoritization). Towards the northwest of the main veins, (hanging wall of fluorite mineralization), the host rocks are completely kaolinized. Except in extreme fluoritized and silicified rocks, the breccia body retained permeability sufficient to allow access for solutions for kaolinization of feldspars.

(e) Hematitic alteration

Hematitic alteration is prominently seen in the extreme northeast end of the deposits at the lower reaches the host rocks of the stockwork. Within these areas there are fractures surrounded by oxidized iron oxides. This relationship is also not uncommon under microscopic observations for hematite and other iron oxides. It is probable that the haematization reflects a process of flushing of iron from the kaolinized parts of the system or from intersecting fracture systems but the precise relationship between the two processes requires further study before a coherent explanation can be given.

The fluorite breccia body itself posses no cogenetic mineralizations of economic significance although little barite have been noted.

THE GENETIC MODEL

Considering the geological evidences including the field relationships observed at

Karada, and the specimens, a genetic model is proposed for explaining the phenomenon of the rapid channeled escape of the fluorine-rich fluids from the apical part of the igneous rocks. Burnham (1979) has presented a model describing the emplacement and solidification of bodies of hydrous acid magma. This model has also been found satisfactory in the case of tourmaline rich breccia in the western part of the St Austell, England (Allman Ward et al. 1982); and Jalera Khurd in MIS (Shrivastava and Gaur, 2009). This type of model is considered to form a good basis for explaining the petrogenesis and mineralization events at Karada. There are as many as 13 distinct stages have been recognized in this chapter.

Presence of gases like fluorine, boron, CO_2 and H_2O in the magma and their vital role in the resultant processes including possible mineralization have been studied by Diamond (2001), Edmonds et al. (2001), Evans et al. (2001) and Tamic et al. (2001). Absence of gases and dry melt in Qz-Ab-Or system has also been evaluated by Hess et al. (2001). It is known that the presence of boron in significant concentration in the system also leads to enhanced solubility of H_2O in the silicate melt and the solidus temperature suffers a significant depression so that the hydrous phase may coexist with silicate melt at temperatures of the order of 575-600°C (Chorlton and Martin, 1978; Pichavant, 1979). It shows that boron plays a very important role in decreasing the miscibility gap between coexisting silicate and hydrous phases in water-saturated magmas and also lowers the temperatures at which silicate melts can survive in coexistence with the hydrous fraction. Whether fluorine play a similar role or not is still in question awaiting future experimental studies.

It is clear from geochemical and textural mineralogical considerations that the emplacement of the main stage of the fluorite breccia body took place as a consequence of disruption and fluidization by a fluorine rich fluid which escaped down a pressure gradient guided by a set of joint, subvertical in nature, in the apical part of the host rocks.

From further consideration of the fluorite-rich facies of the host rocks (two feldspar) and the ubiquitous occurrence of fluorite and or fluorine within the mineralized hydrothermal structures of Karada, it is clear that the fluorine-rich fluid was generated by differentiation and crystallization of the parent magma of the host rocks from subvolcanic to volcanic levels.

1. Evolution of Carapace (above the magma chamber)

The first stage of the model involves the rise of the magma under the influence of telluric pressure differences to a level in the crust between 5 ± 2 km depth under confining pressures of 2 ± 1 kbar. Charoy (1979) has estimated that around 4 ± 1 weight percent of initial water content of magma, the granite will cross the saturated solidus boundary at a depth almost of three kilometres (Harris et al., 1970). A solid crystalline crust will develop at the top of the magma chamber as soon as solidus-liquidus barrier is crossed (Burnham, 1979) seals the upward progress of the rest of the magma melt (740°C ± 40°C) temporarily, though the blocked pressure get released before development of the much solid carapace.

2. Crystallization and differentiation

The next stage is represented by the crystallization and differentiation in confinement begins within the rest of the magma beneath the solid crystalline carapace above the halted magma. It is assumed that before the development of solid crystalline envelope above the magma chamber, all the volatiles generated during solidification may have escaped into the surrounding country rocks. Differentiation due to progressive crystallization in the static magma would now take place leading to a saturated condition with respect to the volatile aqueous component. The latter exsolves to form a separate phase, and rise to get concentrated in the apex of the elongated balloon. Given the experimental results obtained by Pichavant (1979), a hydrous phase enriched in fluorine could possibly coexist with the 600°C silicate melt in the top of such balloon.

3. Formation of Joints

The crystallizing magma would fail at its top by fracturing the newly formed crust when the internal pressure achieved a value to the minimum principal stress and the tensile strength value of the newly-formed envelope of the rocks. This is a critical condition for the enlargement of the effective volume of the system by tensile fracturing. Presuming that this process in the case of Karada took place within a field of regional tectonic stress, the principal mechanical effect of increasing internal overpressure would be the creation of sets of subvertical tensile joints with an orientation normal to the trajectory of principal stress in the brittle envelope recently developed at the top of the host rocks of the fluorite deposit. This model is proposed following the similar sequence of stages explained by Phillips (1973).

4. Filling of Fluorite in the Joints

With the development of joints as described in stage iv the internal overpressure built up is relieved temporarily, as they enable the escape of portions of the hydrous fluids and gases from the elongated apical zone. At Karada, apart from the main central mineralized set of veins, there are poorly mineralized veins available, parallel to the main veins at both sides, at Rekha mine (Fig 25 A). This is attributed to the continued progressive buildup of internal overpressure arising from further crystallization of magma leading to the and the critical conditions once again followed by renewed fracturing producing the outer mineralized veins.

5. Origin of collapse breccia

Interestingly in the next stage, because of increase of the internal pressure, the above hard newly formed envelope would fail by fracturing. In fact, from the saturated liquidus boundary to the saturated solidus boundary change at this stage is represented by water saturated melt crystals and the hydrous phase. Following the opinion of Pichavant (1979), it would also allow the partition of fluorine in favour of hydrous fraction in a ratio of 2:1.

It is known that the volume covered by a unit mass of crystals and hydrous phase is always greater than the same unit mass of water saturated silicate melt. Thus, when the crystallization proceeds, the volume expansion cannot be accommodated by progressive plastic deformation and expansion owing to the effects of firstly, the wall rocks, secondly, the rigidity of the hard apex (recently crystallized) and lastly the confining lithostatic pressure. It is now clear that as the resultant of these factors, during crystallization and cooling, the internal pressure in the apex of the magmatic body will keep increasing gradually.

A possible answer to adjust this additional internal pressure could be a certain finite compression of the contiguous magma system. But it is too little to absorb the internal additional pressure generated during the crystallization and cooling.

According to Burnham (1979), within pockets of water saturated magma undergoing crystallization, an internal pressure of almost five kilobar could be generated. If so, at depths corresponding to internal pressures of one kilobar, the carapace developed at the apex of the body would fail much earlier before pressures of this magnitude were achieved, and produce the first set of collapse and brecciation, a scenario promoting formation of both fluorite crystals and the host rocks.

6. Enrichment of mineralization

In course of time, this repetition of the sequence of failure, renewed pressure build up in the apex, result in over pressure so large to increase number of fractures which will finally connect fractures open already and connected to the surface. This will sharply result in an instantaneous drop in the confining pressure. Corresponding to the depth of initiation, this value must almost be from lithostatic value to hydrostatic value. By this time, the fluids would have boiled and a large volume of vapour would have been generated. This turbulent fluidized system with explosive energy would be channeled down the controlled joint system within which this sudden change will lead to deposition of a large quantities of fluorite, silica and iron to finally produce the complicated mineral relationship now observed in the mineralized body.

7. Plume action (Bouyant or Pneumatic)

The last stage can also be explained following a classical theory of 'buoyant plumes' suggested by Phillips and Woods (2001). CO_2 is relatively insoluble in basaltic magma at low crustal pressures, and therefore, exists as a gas phase in the form of bubbles in shallow crustal reservoirs. Over time, these bubbles may separate gravitationally from the magma in the chamber. As a result, any fresh magma which recharges the chamber from the depth will be rich in such bubbles, creating a turbulent bubble plume within the chamber. A relatively deep bubbly layer of magma may then form below a thin foam layer at the roof. As their eruption proceeds, there will be continuing influx at the base of the chamber, followed by a gradual increase of the bubble content of the bubbly layer. This may ultimately lead to a transition from lava flow activity to one more explosion. The foam layer at the top of the chamber may provide a flux for the continual outgassing from the flanks of the volcano.

8. Origin of Intrusive Breccia

The emplacement of breccia evidently marked the climax of the sequence of events in which fluorine-rich fluids escaped from the apical zone of the crystallizing magma. Though the conditions necessary for the creation of such brecciated bodies are reported elsewhere e.g., local development in Cornubian province (Goode and Taylor, 1980) or regional as in Yukon, Canada (Thorkelson et al. 2001) nowhere they are so well developed or dramatically exposed as at Karada deposit.

9. Vigorous expansion

In the model proposed for the generation of the breccia, there is no implicit need for the breccia itself to have reached surface. But, there are indications that it has been emplaced in the roots of a volcanic system at Karada. The mechanism proposed does, however require that an instantaneous change in the confining pressure of the system took place and that explosive expansion of the hydrous vapour could occur. Thus, the vapour under hydrostatic confining conditions probably escaped into the meteoric system, having dispersed its energy during the process of fluidization, hydraulic fracturing, and expansion in the guiding structures closest to the focus of internal overpressure generated in the elongated apex of the host rocks.

10. Formation of Stockwork

The mechanism explained in the ninth Stage, operating under the influence of the prevailing tectonic stress field would explain the extended paragenetic and structural sequence observed in the pre-breccia fracture systems. Progressive stages in the relief of internal stresses by fracturing resulted in the development of intersecting sets of fractures changed in response to the jacking or wedging effect produced as the veins were filled under pressure. The interplay of this effect combined with slight changes in the regional stress field could explain in the principal sets of fractures (Joint I), which are developed in the in both the mines, Rekha and Santoshi in the Karada fluorite deposit.

11. Cyclic Repetition of Processes

The same processes possibly have been repeated atleast thrice to produce two generations of fluorite fragment in the breccia (Fig 26 A-D). The first set of fragments is one which is enclosed by later second set, and still later, the third set is represented, chiefly by fine grained fluorite which is either cementing the grains of earlier fluorites and/or host rocks or making somewhat layers of the fluorite itself.

12. Formation of 'Gem' fluorite

The 'Gem' fluorites are pure, well crystallized in open space fluorites, many times shows dog tooth structures. These varieties remain present in the main body of Rekha mine and also in the Santoshi mine. The very last more pneumatic action than fluids must have resulted in the formation of 'Gem' fluorite. This gaseous activity represents the last but richest phase of mineralization (Fig 26 H).

13. Metasomatism and alteration

Post-mineralization metasomatism and alterations are present as kaolinization, and haematitization. The central body of fluorite breccia showing good deposition of kaoline at the hanging wall, originated by metasomatic fluid action at the expense of feldspars. Pyrite generally represent the last activity of the volcanic action. It remain present at Jalera Khurd area of study (Shrivastava and Gaur, 2009). Interestingly it remain totally absent in the Karada area. It has a coat of hematite, showing iron oxide as a result of still later action of metasomatism (Fig 26 G). Rusty colours of iron oxides have also penetrated deep inside the host rock and the fluorite breccia and preferencial occurrence is at the contact of the two.

Conclusion

The Karada fluorite deposit is an excellent example of volcanogenic pnematohydrothermal intrusive and collapse breccia deposit of fluorite. The range of Geological features displayed by fluorite breccia shows that the process of disruptive intrusion by a turbulent fluidized system was the dominant process, initially. The intrusive aspects of fluorite breccia show angular killas fragments encompassing a wide range of sizes packed together and cemented again by microcrystalline fluorite itself. This was followed by a stage of collapse atleast in few episodes some major, and other minor. It resulted in inclusions of early formed fluorite fragments into later-formed rich fluorite matrix. In the last, there is intense pneumatic action, dominant over fluids to form rich fluorite crystals called 'gem' fluorites. Such fluorites are available as coating over volcanic and fluorite bearing volcanic breccia and also as cavity fillings. These episodes were further followed by pervasive silicification and metasomatic alterations. The genetic model suggested is comprise of thirteen stages. The fluorite ore body, combines the features of both hydrothermal intrusives breccia and hydrothermal collapse breccia to form the economic deposit.

To conclude the study, unfolding these events can now be summarized into the following stages as part of the geological history of the deposits. (1) Generation of magma deep inside the Earth, possibly at the upper mantle part and further migration upward by a plume associated with the regional hot spot activity at the site of subduction to form major episodes of Malani Igneous Suite (745 $\pm$ 10). (2) Crustal assimilation, contamination and fractional crystallization to produce the basement granite. (3) Forceful injection of the original basaltic magma from the magma chamber, showing initiation with violent extrusion to produce mafic unit I. (4) Extrusion of felsic magma to constitute rhyollite. (5) Second phase of extrusion of mafic pulse to produce mafic unit II. (6) Presence of fluorine, at all the stages of igneous activity (producing all mafic Unit I, felsic unit and mafic unit II) have been confirmed by the presence of two feldspars together, in all the acidic rocks. (7) Development of pneumato-hydraulic system to produce a pressure which might have exceeded the tensile strength of the host rock and the minimum principle stress component. Then the fluid must have extended its capacity to propogate the fracture system to open it in tensile manner. (8) Intrusive breccia of fluorite deposited alongwith very fine cementing fluorite, through joint plane I. Process of disruptive intrusion by a turbulent fluidized system was the dominant process at this stage. Microcrystalline fluorite matrix may have crystallized nearly instantaneously by the process of solute dumping as the fluids responded to rapid change in physical conditions, in all probability a catastro-phic fall in confining pressure. (9) Collapse breccia formed at this stage. The process of crystallization and differentiation in confinement begins at this stage within the remaining magma beneath the solidified corapace of a static elongated cupola. Differentiation due to progressive crystallization in the static magma become saturated with respect to the volatile aqueous component, which exsolving to rise, concentrate and form fluorite in the epical parts of cupola will collapse due to sudden release of the pressure. This process has been repeated atleast few times to produce few episodes of collapse breccia formation. This inference is on the basis of the fact that the early formed fluorite in the form of angular fragment occur into the later formed rich fluorite matrix. This phase of activity have produced 'dogtooth' and 'gem' varieties of fluorine. Chiefly it has resulted as coating over the fluorite and volcanic breccia and also over the hanging and footwall of the veins. In this stage gaseous action remained most active over fluids. (10) The active hydrothermal fluids, at this stage changed from being fluorine rich to silica rich, and is responsible to fill the vugs and producing crustification and comb structures. (11) This stage is also of metasomatism, namely kaolinization. At this stage fluorite breccia body retained a permeability sufficient enough to allow access by the kaolinizing solutions to feldspars in the host rock fragments. (12) A well known last phase to conclude all igneous activity in the analogues regions (Shrivastava and Gaur 2009) is also present at Karada to form hematite a minor process of hematitic alterations which have produced distinctive reddening.

The proposed genetic model of fluorite mineralization in Karada can serve as guide for the future extension of this deposit or new discovery in similar geological setup. It is evident in the present study that the Cauldron subsidence regions are possible sites of such mineralization as promising indicators. At Karada, the genesis of intrusive fluorite breccia and associated hydrothermal fluid show very forceful penetration. The prime cause for the same seems most likely, a subsidence of a regional cauldron in the region. Hence, it is logical to suggest that very similar geological circumstances may have produced equally rich fluorite deposits in peripheral regions of the cauldron subsidence. In Malani Igneous Suite and similar rock at distal places in east Africa, Tosham (Haryana) and Kirana hills (Pakistan) have shown many circular structures with collapse of Cauldron. Thus, such peripheral regions deserve to be explored for the mineralization and may prove a potential zone. Further, there a structural control of fluorite mineralization of the Karada area. The fluorite veins have formed only along Joint I. Its strike N43°E-S43°W and dip 85° towards north. It is suggested that this direction is the prime guide locally, for the possible extension of the Rekha and Santoshi mines of Karada area. Also, it is known experimentally that availability of two alkali feldspars together in an igneous rock like rhyolite in Karada, serves as an guide to suggest that the rock must have been formed in the presence of rich gasses especially fluorine. Hence, it is suggested that this petrological guide can also lead to the discovery of some new deposits.

References

Abdel Rahman, A.F.M., Elkibbi, M.M., (2001): "Anorogenic Magmatism: Chemical evolution of the Mount El-Sibai A-type complex (Egypt), and implications for the origin of within-Fig felsic magmas". Geol. Mag., Vol. 138, No. 1, pp. 67-85.

Aguirre Diaz, G.J., (2001): "Recurrent magma mingling in successive ignimbrites from Amealco Caldera, Central, Mexico". Bull. Volcano., Vol. 63, No. 4, pp. 238-251.

Allman-Ward, P., Halls, C., Rankin, A., and Bristow, C.M., (1982): "An Intrusive Hydrothermal Breccia Body at Wheal Remfry in the Western Part of the St Austell Granite pluton, Cornwall, England". In "Metallization Associated with Acid Magmatism" (Ed. A.M. Evans), John Wiley and Sons Ltd., pp. 1-28.

Andersen, T., Neumann, E.R., (2001) : "Fluid inclusions in mantle xenoliths". Lithos, Vol. 55, No. 1-4, pp. 301-320.

Ashiya, I.D.; (1989) : "Geochemistry of the volcanic rocks to the West of Barmer City". Unpublished Ph.D. thesis, University of Rajasthan, Jaipur, 106p.

Aswathanarayan, U.; (1964) : "Age determination of rocks and geochronology of India". Brochure XXII Int. Geol. Cong., New Delhi, 23.

Basu, A.R., Renne, P.R., Dasgupta, D.K., Teichmann, F., and Poreda, R.J.; (1993) : "Early and late alkali igneous pulses and a high ^{3}He plume for Deccan flood basalts". Science, Vol. 261, pp. 902 – 906.

Basu, S.K., 1982 : "Phases of acid igneous activity in Malani Suite of rocks around Jhunjhunu, Rajasthan". Rec. Geol. Surv. India, Vol. 12(7), pp. 89-94.

Batchelor, R.A.., and Bowden, p; (1985): "Petrogenetic interpretation of granitoid rock series using multicationic parameters". Chem. Geol., Vol.48,pp. 43-55.

Bhushan, S.K. and Chandrashekhran, V.; (2002) : "Geology and Geochemistry of the magmatic rocks of Malani Igneous Suite and Tertiary Alkaline Province of western Rajasthan". Mem. Geol. Surv. Ind. Vol. 126, 181 p.

Bhushan, S.K. and Chittora, V.K.; (1999) : "Late Proterozoic bimodal volcanic assemblage of Siwana subsidence structure, Western Rajasthan, India". Jour. Geol. Soc. Ind Vol. 53, pp. 433 – 452.

Bhushan, S.K. and Mohanty, M.; (1988) : "Mechanics of intrusion and geochemistry of alkaline granites from Siwana, Barmer District, Rajasthan", Ind. Jour. Earth Sci., Vol. 15, pp. 103 – 115.

Bhushan, S.K., 1989 : "Mineral Chemistry and Petrogenetic aspects of Malani Volcanics, Western Rajasthan". Ind. Min., Vol. 43, No. 3-4, pp. 325-338.

Bhushan, S.K., and Khullar, V.K.; (1998) : "Geochemistry and tectonic significance of dyke swarm in Malani Igneous Complex around Sankra, District Jaisalmer, Rajasthan". In : B.S. Paliwal (ed), The Indian Precambrian. Scientific Publishers (India), pp. 482 – 491.

Bhushan, S.K., and Kullar, V.K.; (1998) : "Geochemistry and tectonic significance of dyke swarm in Malani Igneous Complex around Sankara District Jaisalmer, Rajasthan. In : B.S. Paliwal (ed.), The Indian Precambrian Scientific Publication (India), pp. 482–491.

Bhushan, S.K.; (1973) : "The Malani rhyolite. Indian Minerals". Geol. Surv. Ind. Vol. 30, No. 2, pp. 99–100.

Bhushan, S.K.; (1984) : "Classification of Malani Igneous Suite. Symposium on Three decades of developments in Petrology, Mineralogy and Petrochemistry in India". Sp. Publ. Geol. Surv. Ind Vol. 12, pp. 199 – 205.

Bhushan, S.K.; (1985) : "The Malani volcanism in western Rajasthan", Ind, Jour. Earth Sci., Vol. 12 No. (1), pp. 58 – 71.

Bhushan, S.K.; (1991) : "Granitoids of Malani Igneous Complex, Western Rajasthan". Ind. Jour. Ear Sci., Vol. 18, pp. 184 – 194.

Bhushan, S.K.; (1995) : "Late Proterozoic continental growth Implication from Geochemistry of acid magmata events of west Indian craton". Mem. Geol. Soc. Ind. Vol. 34, pp. 339 – 353.

Bhushan, S.K.; (2000): "Malani Rhyolites – a review". Gond. Res., Vol. 3, pp. 65 – 77.

Bidwai, R. and Krishnamurthy, P.; (1997) : "Possible lava channels and tubes from the Siwana caldera sequence, Siwana, Barmer district, Rajasthan". Jour. Geol. Soc. Ind. Vol. 50 (2), pp. 103 – 115.

Blake, S., Bruno, B.C., (2001): "Modelling the emplacement of compound lava flows". Earth and Planetary Science Letters, Vol. 184, No. 1, pp. 181-197.

Borisova, A.V., Belyatsky, B.V., Portnyagin, M.V., Sushchevskaya, N.M., (2001) : "Petrogenesis of olivine-phyric basalts from the Aphanasey Nikitin Rise : Evidence for contamination by cratonic lower continental crust". Jour. Petrol., vol. 42, No. 2, pp. 277-319.

Bryner, L., (1961) : "Breccia and Pebble Columns associated with epigenetic ore deposits" Econ. Geol., Vol. 56, pp. 488-508.

Burnham, W.C., (1979) : "Magmas and hydrothermal fluids", in 'Geochemistry of Hydrothermal Ore deposits', H.L. Barnes (Ed.) Wiley, New York, pp. 71-136.

Cas R.A.F. and Wright J.V.; (1987) : "Volcanic successions : Modern and Ancient". Allen and Unwin. London. 528 p.

Chabiron, A., Alyoshin, A.P., Chuney, M., Deloule, E., Goluber, V.N., Velitchkin, V.I., Poty, B., (2001) : "Geochemistry of the rhyolitic magmas from the Streltsovka caldera (Transbaikalia, Russia) : a melt inclusion study" Chem. Geol., Vol. 175, No. 3-4, pp. 273-290.

Chandrasekaran, V.; (1987) : "Geochemistry of basic, acid and alkaline intrusives and extrusives of Sarnu–Dandali area, Barmer district, Rajasthan. Ph.D. Thesis, M.L. Sukhadia University, Udaipur, p. 108.

Charoy, B., (1979) : "Definition et importance des phenomenes deuteriques et des fluides associes dans les granites, consequences metallogeniques", Sci. Terre., Mem. No. 37.

Chittora, V.K. and Bhushan, S.K.; (1993) : "Volcanic stratigraphy and Geochemistry of Malani Igneous Suite around Siwana, Barmer district". Rep. (unpublished) Geol. Surv. Ind. (F.S. 1992–93).

Chittora, V.K., Maheshwari, A. and Coltori, M.; (1998) : "Flow Stratigraphy and Geochemical variants of Bhadrajan volcanics, Jalore district, Rajasthan : Implications for a Heterogeneous Protolith". Jour. Geol. Soc. Ind. Vol. 52, pp. 433–441.

Chore, S.A. and Mohanty, M., 1998 : "Stratigraphy and tectonic setting of the Trans-Aravalli Neoproterozoic Volcanosedimentary sequences in Rajasthan". Jour. Geol. Soc. India, Vol. 51, pp. 57-68.

Chorlton, L.B., and Martin, R.F., (1978) : "The effect of boron on the granite solidus", Can. Mineral. Vol. 16 pp. 239-244.

Coulson, A.I.; (1933) : "Geology of Sirohi State". Mem. Geol. Surv. Ind. Vol. 63, No. 1, pp. 1–166.

Crawford, A.R. and Compston, W.; (1970) : "The age of the Vindhyan system of Peninsular India". Quart. Jour. Geol. Soc. (London), Vol. 125, pp. 351 – 372.

Crawford, A.R.; (1970) : "The Precambrian geochronology of Rajasthan and Bundelkhand, Northern India". Cana. Jour. Ear Sci., Vol. 7, pp. 91 – 110.

Das Gupta, S.K. and Chandra, M.; (1978) : "Tectonic element of west Rajasthan shelf and their stratigraphy". Proc. Golden Jubilee Symp. Quart. Jour. Geol. Min. Met. Soc. Ind., (Calcutta), Vol. 50, pp. 1 – 16.

Debon, F. and Le Fort, P.; (1983): "A chemical mineralogical classification of common plutonic rocks and associations". Trans. of Roy. Soc. Edinburg. Earth Sci. Vol. 73, pp. 135 – 149.

Dhar, S., Frei, R. Kramers, J.D., Nagler, T.F. and Kochhar, N.; (1996) : "Sr, Pb and Nd isotope studies and their bearing on the petrogenesis of the Jalore and Siwana Complexes, Rajasthan, India". Jour. Geol. Soc. Ind., Vol. 48, pp. 151 – 160.

Diamond, L.W., (2001) : "Review of the systematics of CO_2-H_2O fluid inclusions". Lithos, Vol. 55, No. 1-4, pp. 69-99.

Dines, H.G., 1956 : "The Metalliferous Mining region of South-west England". Mem. Geol. Surv. Gt. Britain.

Drury, M.R., Van Roermund, H.L.M., Carswell, D.A., DeSmet, J.H., VandenBerg, A.P., Vlaar, N.J., (2001) : "Emplacement of deep upper-mantle rocks into cratonic lithosphere by convection and diapiric upwelling". Jour. Petrol., Vol. 42, No. 1, pp. 131-140.

Eby, G.N. and Kochhar, N.; (1990) : "Geochemistry and Petrogenesis of the Malani Igneous Suite, north peninsular India". Jour. Geol. Soc. Ind., Vol. 36(2), pp. 109 – 130.

Edmonds, M., Pyle, D., Oppenheimer, C., (2001) : "A model for degassing at the Soubriere Hills volcano,

Montserrat, West Indies, based on geochemical data". Earth and Planetary Science Letters, Vol. 186, No. 2, pp. 159-173.

Evans, W.C., Sarey, M.L., Kennedy, B.M., Stonestrom, D.A., Rogie, J.D., Shuster, D.L., (2001) : "High CO_2 emissions through porous media : transport mechanism and implications for flux measurement and fractionation". Chem. Geol. Vol. 177, No. 1-2, pp. 15-29.

Exley, C.S., 1959 : "Magmatic differentiation and alteration in the St. Austell granite". Jour. Geol. Soc. Lond., Vol. 114, pp. 197-230.

Fareeduddin and Kroner, A.; (1998) : "Single zircon age constraints on the evolution of Rajasthan granulite". In : B.S. Paliwal (ed.), The Indian Precambrian. Sci. Pub. (India), Jodhpur, pp. 547 – 556.

Fiske, R.S., Naka, J., Iizasa, K., Yuasa, M., Klaus, A., (2001) : "Submarine silicic caldera at the front of the Izu-Bonin arc, Japan : Voluminous sea floor eruptions of rhyolite pumice". Geol. Soc. Am. Bull., Vol. 113, No. 7, 813-824.

Fletcher, C.J.N., 1977 : "The geology, mineralization and alteration of Ilkwang Mines, Republic of Korea A Cu-W-bearing tourmaline breccia pipe". Econ. Geol., Vol. 72, pp. 753-768.

Gaur, V.; (2002): "Geological Investigations on Gnesis and Characteristics of Fluorite Mineralization around Jalera Khurd, District Jalor, Rajasthan in Malani Volcanics". 129 p., Ph.D. Thesis, J.N.V.University, Jodhpur.

Goode, A.J.J., and Taylor, R.T., (1980) : "Intrusive and pneumatolytic breccias in south-west England". Inst. Geol. Sci. Rep. No. 80/2.

Grant, J.N., Halls, C., Sheppard, M.F., and Avila, W., 1980 : "Evolution of the porphyry tin deposits of Bolivia". Min. Geol. (Japan), Special issue, vol. 8, pp. 151-173.

Gultekin, A.H. 2005 : "Epithermal Kizilcaoren fluorite – Barite – REE deposit, Eskisehir, Turkey : Preliminary Results". Jour. Geol. Soc. Ind. Vol. 66, pp. 105 – 116.

Halls, C., Rankin, A., Ferriday, I.L., Blain, C.F., Bristow, C.M. and Gronow, C., 1977: "A tourmalinized hydrothermal intrusion breccia at Wheal Remfry in the western part of the St. Austell pluton, Cornwall". Trans. Inst. Min. Metall., Vol. 86(B), 100-163.

Harker, A.; (1909): "The natural history of igneous rocks". London, Huen, 383 P.

Harris, P.E., Kennedy, W.Q., and Scarfe, C.M., (1970) : "Volcanism versus plutonism - the effect of chemical composition". In G. Newall and N. Rast (Eds.) Mechanism of Igneous Intrusion : Geol. Jour., Special issue, Vol. 2, pp. 187-200.

Heron, A.M., 1932 : "The Vindhyans of Western Rajputana". Rec. Geol. Surv. India, Vol. 65, No. 4, pp. 457-489.

Hess, K.U., Dingwell, D.B., Gennaro, C., Mincione, V., (2001) : "Viscosity-temperature behaviour of dry melt in the Qz-Ab-Or system". Chem. Geol., Vol. 174, No. 1-3, pp. 133-142.

Irvine, T.N., and Bragar, W.R.A.; (1971) : "A guide to the chemical classification of the common volatile rocks Cana. Jour. of Earth Science". Vol. 8, pp. 523 – 548.

Jain, R.B., Miglani, T.S., Kumar, S., Swarnkar, B.M. and Singh, R.; (1996) : "Rare metal and rare earth rich peralkaline, agpaitic granitoid dykes of Siwana Ring Complex, District Barmer, Rajasthan". Curr. Sci., Vol. 70 (9), pp. 854 – 861.

Jensen, L.S.; (1976): "A new cation plot for classifying subalkalic volcanic rocks, Ontario Division of Mines", M.P. 66, 22 p.

Johnston, W.P., and Lowell, J.D., 1961 : "Geology and Origin of Mineralized breccia pipes in Copper Basin, Arizona". Econ. Geol., Vol. 56, pp. 916 940.

Kabeto, K., Sawada, Y., Iizumi, S., Wakatsuki, T., (2001) : "Mantle sources and magma-crust interactions in volcanic rocks from the northern Kenya rift : geochemical evidence". Lithos, Vol. 56, No. 2-3, pp. 111-139.

Kents, P., 1964 : "Special breccias associated with hydrothermal developments in the Andes". Econ. Geol., Vol. 59, pp. 1551-1563.

Khan, M.S., Raza, M. and Safdar–E–Azam, M.; (2004): "Bombolai continental pillow Lavas (Neoproterozoic) from Trans–Aravalli Region", Pali District, Rajasthan and their Tectonic significance". Jour. Geol. Soc. Ind. Vol. 64 No. 6 pp. 803 – 812.

Khanna, P.P., Saini, N.K., Mukherjee, P.K. and Purohit, K.K.; (2009) : " An appraisal of ICP-MS technique for determination pf REEs : long term QC assessment of Silicate Rock Analysis". Himalaya Geol. Vol. 21(1), pp. 95-99.

Knutson, J., Ferguson, J., Roberts, W.M.B., Donnelly, T.H., and Lambert, I.B., 1979 : "Petrogenesis of the Copper-bearing breccia pipes, Redbank, Northern Territory, Australia". Econ. Geol., Vol. 74, pp. 814-826.

Kochhar, N. (1984) : "Malani Igneous Suite : Hot spot magmatism and cratonisation of the northern part of the Indian shield". Jour. Soc. Ind. Vol. 25, No. 3, pp. 155–161.

Kochhar, N. and Dhar, S.; (2000) : Comments on "Rb–Sr Isotope dating of Neoproterozoic (Malani Group) magmatism from Southwest Rajasthan, India : Evidence of younger Pan–African evetby $^{40}Ar/Ar^{39}$ studies". Gond. Res. Vol. 3, pp. 119.

Kochhar, N., 1983 : "Tusham ring Complex, Bhiwani, India". Proc. Indian Nat. Sci. Acad., No. 4, pp. 459-490.

Kochhar, N., 1998 : "Malani Igneous Suites of rocks". Jour. Geol. Soc. India, Vol. 51, pp. 120.

Kochhar, N.; (2006) : "Structure and tectonics of Kutchch basin and earth quakes". Curr. Sci. Vol. 91 No. 12 pp. 1598–1599.

Krishnan, T.K., 1968 : "Morphology of zircons from Jasai granite, Barmer district, Rajasthan". Geol. Surv. Ind., Misc. Pub. 9, pp. 70-76.

Kumar, A., Nirmal Charan, S., Gopalan, K., Macdougall, J.D.; (1998): "A long–lived enriched mantle source for two Proterozoic Carbonatite Complexes from Tamil Nadu, Southern India. Geochim. Cosmochim. Acta. 62, pp. 515 – 523.

Kuritani, T., 2001 : "Replenishment of a mafic magma in a zoned felsic magma chamber beneath Rishiri volcano, Japan". Bull. Volcano., Vol. 62, No. 8, pp. 533-548.

La Touche, T.H.D., (1902): "Geology of western Rajasthan". Mem. Geol. Surv. Ind. Vol. 35 (1), pp. 1–116.

Le Bas, M.J., Le Maitre, R.W., Streckeisen, A. and Zanettin, B., 1986 : "A chemical classification of volcanic rocks based on the total alkali-silica diagram". Jour. Petrol., Vol. 27, pp. 745-750.

Le Maitre, R.W. (Ed.); (2002): "A classification of Igneous rocks and glossary of terms". Recommendations of the IUGS subcommission on the systematics of Igneous rocks". Blackwell Oxford. 193 p. (Halo–I).

Lindsay, J.M., Schmitt, A.K., Trumbull, R.B., DeSilva, S.L., Siebel, W., Emmermann, R., (2001) : "Magmatic evolution of the La Pacana Caldera system, Central Andes, Chile : Compositional variation of two cogenetic, large-volume felsic ignimbrites". Jour. Petrol., Vol. 42, No. 3, pp. 459-486.

Llambias, E.J., and Malvicini, L., 1969 : "The geology and genesis of the Bi-Cu mineralised breccia-pipe, San Francisco de los Andes., San Juan, Argentina". Econ. Geol., Vol. 64, pp. 271-286.

Locke, A., 1926 : "The Formation of certain ore bodies by mineralization stoping". Econ. Geol., Vol. 21, pp. 431-453.

Maheshwari, A., Coltori, M. and Sial, A.N.; (1996) : "Petrogenesis and tectonic setting of Malani rhyolites : Evidence by trace elements and oxygen isotopic composition". Curr. Sci., Vol. 68 (9), pp. 954–56.

Maheshwari, A., Coltori, M., Sial, A.N., Mariano, G.; (1996) : "Crustal influence in the petrogenesis of the Malani rhyolite, south western Rajasthan : combined trace element and oxygen isotope constraints". Jour. Geol. Soc. Ind, Vol. 7, pp. 611 – 619.

Maheshwari, A.; (1983) : "Geochemistry of volcanic rocks of Gurapratap Singh and Diri, Pali district, Rajasthan". Unpublished Ph.D. thesis, University of Rajasthan, Jaipur, 99 p.

Maniar, P.D. and Piccoli, P.M.; (1989): "Tectonic discrimination of granitoids". Geol. Soc. Am. Bul., Vol. 101, pp. 635 – 643.

Meschede, M.; (1986): "A method of discriminating between different types of mid oceanic ridge basalts and continental theoliites with the Nb–Zr–Y diagram". Chem. Geol. 56, pp. 207 – 218.

Middlemost, E.A.K.; (1985): "Naming materials in the magma/Igneous rock system". Earth Sci. Rev. Vol. 37, pp. 215 – 224.

Miyashiro, A.; (1974): "Volcanic rock series in island arcs and continental margins". Am. Jour. Sci. Vol. 274, pp. 523 – 548.

Mocek, B., (2001) : "Geochemical evidence for arc-type volcanism in the Aegean Sea : the blueschist unit of Siphons, Cyclades (Greece)". Lithos, Vol. 57, No. 4, pp. 263-289.

Mukherjee, A.B. and Roy, A.; (1981) : "Cooling conditions of the high level Precambrian granite of Siwana : Evidence of experimental melting behaviour and the sodic amphibole–pyroxene reaction". Ind Jour. Ear Sci., Vol. 8(2), pp. 99–108.

Mukherjee, A.B., 1962 : "Pantellaritic assamblages from western Rajasthan". Quart. Jour. Geol. Min. and Metall. Soc. India, Vol. 34, No. 2-3, pp. 149.

Mullen, E.D.; (1983): "$MnO/TiO_2/P_2O_5$: "a minor element discrimination for basaltic rocks of oceanic environment and its implications for petrogenesis". Earth. Planet. Sci. Lett. 62, pp. 53 – 62.

Murthy, M.V.N., 1979: "Some aspects of mineralization associated with acid magmatism in Rajasthan. Proc. Workshop MAWAM Geol. Surv. India Spel. Publ., Vol. 13, pp. 15-24.

Murthy, M.V.N., Siddiqui, H.N. and Venkataraman, P.K.; (1961) : "Riebeckite–aegirine rhyolite and granites in the Malani suite of igneous rocks, western Rajasthan". Ind Min., Vol. 15, pp. 427 – 428.

Myron, G. Best.; (1968): "Igneous and metamorphic petrology" CBS publisher (India), pp.360.

Okay, A.I., Tansel, I., Tuysuz, O., (2001) : "Obduction, subduction and collision as reflected in the upper cretaceous - Lower Eocene sedimentary record of Western Turkey". Geol. Maga., Vol. 138, No. 2, pp. 117-142.

Paliwal, B.S. and Rathore, P.S.; (2000) : "Neoproterozoic volcanics and sedimentaries of Jodhpur : A Reappraisal in Tectonomagmatism, Geochemistry and Metamorphism of Precambrian terrains" (Eds. K.C. Gyani and P. Kataria) MLS Univ. Udaipur. pp. 75–96.

Paliwal, B.S., 1992 : "Tectonics of the Post-Aravali mountain building activity and its bearing on the accumulation of sediments along the western flank of the Aravalli Range, Rajasthan, India". In : Geology in south Asia-I Proceedings of GEOSAS-I Islamabad, Pakistan, R. Ahmed and A.M. Seikh (eds.), February 23-27, 1992. Hydrocarbon Development Institute of Pakistan, 1994, pp. 52-60.

Paliwal, B.S., 1998 : "Felsic volcanics interlayered with sediments of the Marwar Supergroup at Chhoti Khatu". Jour. Geol. Soc. India, Vol. 52 pp. 81-86.

Paliwal, B.S., 1999 : "Geology of the region to the west of Aravalli mountain : some observations". In : Geological Evolution of Northeastern India. B.S. Paliwal (ed.) Scientific Publishers (India), Jodhpur, pp. 387-414.

Pandit, M.K. and Amar Deep, (1994) : "Geological frame work of Sankara dyke swarm forming a part of Malani suite of rocks in western Rajasthan". Curr. Sci., Vol. 67, pp. 1015 – 1018.

Pandit, M.K. and Amar Deep, (1997) : "Geochemical evolution of the Late–Proterozoic Sankra Suite of rocks, western India". In : N.P., Wijayaanda, P.G. Cooray, and P. Mosley, (eds.), Geology in South Asia–II, Geological Survey and Mines Bureau, Sri Lanka, Professional Paper, 7, pp. 85 – 91.

Pandit, M.K., Shekhawat, L.S., Ferreira, V.P., Sial, A.N. and Bohra, S.K., 1999 : "Trondhjemite and granodiorite assemblages from west of Barmer : probable basement for Malani magmatism in western India". Jour. Geol. Soc. India, Vol. 53 pp. 89-96.

Pareek, H.S., 1981(a): "Petochemistry and Petrogenesis of the Malani Igneous Suite, India". Geol. Soc. America Bull. Vol. 92, Part I, pp. 67-70.

Pareek, H.S., 1981(b): "Petrochemistry and Petrogenesis of the Malani Igneous Suite, India". Geol. Soc. America Bull., Vol. 92, Part II, pp. 206-273.

Pareek, H.S.; (1984) : "Pre quarternary Geology and Mineral resources of northwestern Rajasthan". Mem. Geol. Surv. Ind. Vol. 115, 99 p.Pascoe, E.H.; (1960) : "Manual of the Geology of India and Burma". Geol. Surv Ind., Vol. 2, pp. 553–560.

Pareek, H.S.; (1984) : "Pre–Quaternary geology and mineral resources of northwestern Rajasthan". Mem. Geol. Surv. Ind, 115, 99 p.

Pascoe, E.H., 1959 : "Manual of the Geology of India and Burma". Mem. Geol. Surv. India, Vol. 2 pp. 485-1343.

Pascoe, E.H., 1960 : "Manual of the geology of India and Burma", Geol. Surv. Ind., Vol. 2, pp. 553-560.

Pearce, J.A. and J.R. Cann.; (1973): "Tectonic setting of basic volcanic rocks determined using trace element analysis". Earth planet. Sci., Lett. 19, pp. 290 – 300.

Pearce, J.A. and Norry, M.I.; (1979): "Petrogenetic implications of Ti, Zr, Y and Nb variations in volcanic rocks". Contrib. Miner. and Petro., Vol. 69, pp. 33 – 47.

Pearce, J.A., Niger, B.W.H. and Tindle, G.; (1984): "Trace element discrimination diagrams for the tectonic interpretation of granitic rocks". Jour. Petrol. Vol. 24 (4), pp. 956 – 83.

Pearce, J.A.; (1975): "Basalt geochemistry used to investigate past tectonic environments on Cyprus". Tectonophysics, Vol. 25, pp. 41-67.

Perace, T.H., B.E. Gorman and T.C. Birkett; (1975) : "The TiO_2–K_2O–P_2O_5 diagram : a method of discriminating between oceanic and non oceanic basalts". Earth Planet". Sci. Lett. 24, pp. 419 – 26.

Perace, T.H., B.E. Gorman and T.C. Birkett; (1977) : The relationship between major element chemistry and tectonic environment of basic and intermediate volcanic rocks. Earth planet. Sci. Lett. 36, pp. 121 – 32.

Phillips, J.C., Woods, A.W., (2001) : "Bubble plumes generated during recharge of basaltic magma reservoirs". Earth and Planet. Sci. Lett., Vol. 186, No. 2, pp. 297-309.

Phillips, W.J., (1972) : "Hydraulic Fracturing and Mineralisation". Jour. Geol. Soc. Lond., Vol. 128, pp. 337-359.

Phillips, W.J., (1973) : "Mechanical effects of retrograde boiling and its probable importance in the formation of some porphyry ore deposits".Trans. Inst. Min. Metall., Vol. 82(B), pp. 90-98.

Pichavant, M., (1979) : "Etude experimantale a haute temperature et 1 kbar du role du bore dans quelques systemes silicates. Interet Petrologique et metallogenique". Ph.D. Thesis, Institute National Polytechnique de Lorraine, CRPG-ENSG.

Rabinowicz, M., Genthon, P., Ceuleneer, G., Hillairet, M., (2001) : Compaction in a mantle mush with high melt concentrations and the generation of magma chambers". Earth and Planet. Sci. Lett., Vol. 188, No. 3-4, pp. 313-328.

Rathore, S.S., Venkatesan, T.R. and Srivastava, R.K.; (1996 b) : "Rb–Sr and Ar–Ar systematics of Malani volcanic rocks of southwest Rajasthan : Evidence for a younger post–crystalline thermal event". Proc. Indian Acad. Sci. (Earth. Planet. Sci.), Vol. 105, pp. 131 – 141.

Rathore, S.S., Venkatesan, T.R. and Srivastava, R.K.; (1999) : "Rb–Sr isotope dating of Neoproterozoic (Malani Group) magmatism from southwest Rajasthan", India : evidence of younger Pan–

African thermal event by Ar^{40}–Ar^{39} studies. Gond. Res., Vol. 2(2), pp. 271 – 281.

Rathore, S.S., Venkatesan, T.R., and Srivastava, R.K.; (1996 a) : "Mundwara Alkali Igneous Complex, Rajasthan, India : chronology and Sr isotopic characteristics". Jour. Geol. Soc. Ind, Vol. 48, pp. 517 – 528.

Rathore, S.S.; (1995) : "Geochronological Studies of Malani Volcanics and associated Igneous rocks of southwest Rajasthan, India, implications to crustal evolution". Unpublished Ph.D. thesis, M.S. University, Baroda, Vadodara, 175 p.

Reynolds, D.L., (1954) : "Fluidisation as a geological Process and its bearing on the problem of intrusive granites". Am. Jour. Sci., Vol. 252, pp. 577-614.

Riley, T.R., Leat, P.T., Pankhurst, R.J., Harris, C., (2001): "Origins of large volume rhyolitic volcanism in the Antarctic Penninsula and Patagonia by crustal melting". Jour. Petrol., Vol. 42, No. 6, pp. 1043-1065.

Riller, U., Petrinovic, I., Ramelow, J., Strecker, M., Oncken, O., (2001): "Late Cenozoic tectonism, Collapse Caldera and Figau formation in the Central Andes". Earth and Planet. Sci. Lett., Vol. 188, No. 3-4, pp. 299-311.

Roy, A.B. and Jakhar, S.R.; (2002): "Geology of Rajasthan (Northwest India) : Precambrian to Recent", Sci. Pub. Jodhpur. 421 p.

Roy, A.B. and Kataria, P., Upadhyaya, R. and Sharma, B.L.; (1995) : "Dyke rocks in the Precambrian crust of the Aravalli Mountain, Rajasthan". Mem. Geol. Soc. Ind., Vol. 33, pp. 169 – 182.

Roy, A.B. and Sharma, K. K., (1999) : "Geology of the region around Sirohi Town, western Rajasthan – a story of Neoproterozoic evolution of the Aravalli crust. In B.S. Paliwal (ed), Geological Evolution of Northwestern India". Scientific Publishers (India), Jodhpur, pp. 19 – 33.

Roy, A.B.; (1998) : "Malani Igneous Suite of rocks (correspondence)". Jour. Geol. Ind., Vol. 51, No. 4. pp. 556 – 557.

Roy, A.B.; (2003) : "Geological and Geophysical Manifestation of the Reunion Plume – Indian Lithosphere Interaction – Evidence from Northwest, India". Gond. Res. Vol. 6, No. 3, pp. 487–500.

Saini, N.K., Mukherjee, P.K., P.K., Rathi, M.S., Khanna, P.P. and Purohit, K.K.; (1998) : " A new geochemical reference sample of granite (DH-G) from Dalhousie, himachal Himalaya". Jour. Geol. Soc.Ind., Vol.52,pp.603-606.

Sawkins, F.J., (1969) :"Chemical brecciation, an unrecognized mechanism for breccia formation ?" Econ. Geol., Vol. 64, pp. 613-617.

Sayyed, Y.; (2007): " Geological investigation on search of mineralization in carbonatite-Nephelinite, alkaline suite of Sanpa-danddli area, Disrtict Barmer, Rajasthan". 147p., Ph.D. Thesis, J.N.V.University, Jodhpur.

Scambelluri, M., Philippot, P., 2001 : "Deep fluids in subduction zones". Lithos, Vol. 55, No. 1-4, pp. 213-227.

Seghedi, L., Downes, H., Pecskay, Z., Thirlwall, M.F., Szakacs, A., Prychodko, M., Mattey, D., (2001) : "Magmagenesis in a subduction-related post-collisional volcanic arc segment : the Ukrainian Carpathians". Lithos, Vol. 57, No. 4, pp. 237-262.

Sharma, K. K., (1996) : "Stratigraphy, Structure and tectonic evolution of the metasediments and associated rocks of the Sirohi region, southwestern Rajasthan". Unpublished Ph.D. thesis, Mohanlal Sukhadia University, Udaipur, 103 p.

Sharma, K.K., (2005) : Comments on paper "Bombolai Continental Pillow Lava (Neoproterozoic) from Trans – Aravalli region, Pali district Rajasthan and their tectonic significance "Jour. Geol. Soc. Ind. Vol. 64 No. 6 pp. 803 – 812.

Sharma, K.K.; (2004): "The Neoproterozoic Malani Magmatism of the northwestern Indian Shield : Implications for crust building processes". Proc. Ind. Acad. Sci. (Earth Planet. Sci.) Vol. 113, No. 4, pp. 795–807.

Sheppard, S., Griffin, T.J., Tyler, I.M., Page, R.W., (2001) : "High and Low K granites and adakites at a Palaeoproterozoic Fig boundary in northwestern Australia". Jour. Geol. Soc., Vol. 153, No. 3, pp. 547-560.

Shrivastava, K.L. and Gaur, V.; (2009): " Volcanogenic Pneumato- Hydrothermal intrusive and collapse Breccia Deposits of Fluorite at Jalera Khurd in Malani Igneous Suite, Northwest India". In 'Economic Mineralization', Editor K.L. Shrivastava, scientific publisher (India), pp. 108-138.

Shrivastava, K.L., Gaur, V. and Gaur. S., (2000) : "A Search of possible target Loci of Mineralization in Malani Volcanic Rocks, Northwestern India" (Abstract). International Association of volcanology and Chemistry of the Earth's Interior. General Assembly at Bali, Indonesia (July 18-22, 2000).

Sillitoe, R.H., and Sawkins, F.J., (1971) : "Geologic, Mineralogic and Fluid inclusion studies relating to the origin of copper-bearing tourmaline breccia pipes, Chile". Econ. Geol., Vol. 66, pp. 1028-1041.

Singh, A. K. and Vallinayagam, G.; (2006) : "High heat production granites from the piplun and Kundal Areas, Malani Igneous Suite, Western Rajasthan". Jour. Geol. Soc. Ind. Vol. 68, pp. 585–588.

Singh, A.K.; (2006) : "Dolerite Dykes of Kundal Area in Neoproterozoic Malani Igneous Suite, Western Rajasthan". WIHG, Northeast unit, Itanagar, Geol. Soc. Ind. Vol. 68, pp. 695–704.

Singh, S.P.; (1983) : "Pillow lavas from Delhi Supergroup near Bambolai, Pali district, Rajasthan". Jour. Geol. Soc. Ind, Vol. 24, pp. 206 – 211.

Sinha–Roy, S., Malhotra, G. and Guha, D.B.; (1995) : "A transect across Rajasthan Precambrian Terrain in relation to geology, tectonics and crustal evolution of south central Rajasthan. In : S. Sinha–Roy and K.R. Gupta (eds.), Continental crust of northwestern and central India". Mem. Geol. Soc. Ind. Vol. 31, pp. 63 – 90.

Spath, A., LeRoex, A.P., OpiyoAkech, N., (2001) : "Plume-lithosphere interaction and the origin of continental rift-related alkaline volcanism the chyulu Hills volcanic province, southern Kenya". Jour. Petrol., Vol. 42, No. 4 pp. 765-785.

Srivastava, R.K., Maheshwari, A. and Upadhaya, R.; (1989) : "Geochemistry of Felsic volcanics from Gurupratap Singh and Dhiri, Pali district, Rajasthan. I major elements. II Trace elements". Geol. Soc. India. Vol. 34, pp. 467–486 and pp. 617–631.

Srivastava, R.K.; (1988) : "Alkaline and peralkaline rocks of Rajasthan". Mem. Geol. Soc. Ind., Vol. 15, 24 p.

Srivastava, R.K.; (1988) : "Magmatism in the Aravalli Mountain range and its environs". In : A.B. Roy (ed.), Precambrian of the Aravalli Mountain, Rajasthan, India. Mem. Geol. Soc. Ind., Vol. 7, pp. 78 – 93.

Stevenson, R.J., Dingwell, D.B., Bagdassarov, N.S., Manley, C.R., (2001) : "Measurement and implication of "effective" viscosity for rhyolite flowemplacement". Bull. Volcano., Vol. 63, No. 4, pp. 227-237.

Subrahmanyan, N.P. and Rao, G.V.U.; (1972) : "Age of the Mundwara Igneous Complex, Rajasthan". Curr. Sci., Vol. 41, pp. 388 – 391.

Sun, S.S., and McDonough, W.F.; (1989) : " Chemical and isotopic systematic of oceanic basalts : implication of mantle composition and process. In A.D.Saundres and M.J. Norry (Eds.), magmatism in the ocean basin". Geol. Sco. London Spec. Pub., Vol.42,pp.313-345.

Tamic, N., Behrens, H., Holtz, F., (2001) : "The solubility of H_2O and CO2 in rhyolitic melts in equilibrium with a mixed CO_2-H_2O fluid phase". Chem. Geol., Vol. 174, No. 1-3, pp. 333-347.

Tatsumi, Y., (2001) : "Geochemical modeling of partial melting of subducting sediments and subsequent melt-mentle interaction : Generation of high - Mg andesites in the Setouchi volcanic belt, southwest Japan". Geol., Vol. 29, No. 4, pp. 323-326.

Thorkelson, D.J., Mortensen, J.K., Creaser, R.A., Davidson, G.J. and Abbott, J.G., 2001b - Early Proterozoic magmatism in Yukon, Canada: constraints on the evolution of northwestern Laurentia; *Canadian Journal of Earth Science,* v. 38, pp. 1479-1494.

Thorkelson, D.J., Mortensen, J.K., Davidson, G.J., Creaser, R.A., Perez, W.A. and Abbott, J.G., 2001a – Early Mesoproterozoic intrusive breccias in Yukon, Canada: The role of hydrothermal systems in reconstructions of North America and Australia; *PrecambrianResearch,* v. 111, pp. 31-55.

Torsvik, T.H., Carter, L.M., Ashwal, L.D., Bhushan, S.K., Pandit, M.K., Jantveit, B., 2001 : "Rodinia refined or obscured palaeomagnetism of the Malani Igneous Suite (NW India)". Precam. Res. Vol. 108, No. 3-4, pp. 319-333.

Vallinayagam, G. and Kochhar, N.; (1998) : "Geochemical characterization and petrogenesis of A–type granites and the associated acid volcanics of Siwana ring complex, North Peninsular India. In B.S. Paliwal(ed.) The Indian Precam. Sci. Pub.(India), Jodhpur, pp. 460–481.

Vallinayagam, G., 1998 : "Rare metal bearing and barren distinction of alkali granites, Piplun area, Siwana Ring Complex, Rajasthan, India : petrological and geochemical evidence". In : The Indian Precambrian, B.S. Paliwal (Ed.), Scientific Publishers (India), Jodhpur, pp. 492-501.

Vandenkerkhof, A. and Thiery, R., (2001) : "Carbonic inclusions". Lithos, Vol. 55, No. 1-4, pp. 49-68.

Venkataraman, P.L., Ray, S.M., Vishwanathan, T.V. and Murthy, M.V.N.; (1968) : "Comagmatic relationships among rhyolites and high level granites of the Malani suite, Rajasthan, India". Misc. Publ., Geol. Surv. Ind, Vol. 8, pp. 93 – 105.

Verma, S.P., (2001) : "Geochemical and Sr-Nd-Pb isotopic evidence for a combined assimilation and fractional crystallisation process for volcanic rocks from the Huichapan Caldera, Hidalgo, Mexico". Lithos, Vol. 56, No. 2-3, pp. 141-164.

Volpe, A.M. and Macdaugall, J.D.; (1990) : "Geochemistry and isotopic characteristic of mafic (Phulad ophiolite) and related rocks in Delhi supergroup, Rajasthan. India : Implications for rifting in the Proterozoic". Precamb. Res. No. 48, pp. 167 – 191.

Waight, T.E., Maas, R., and Nicholls, I.A., (2001) : "Geochemical investigations of microgranitoid enclaves in the S-type cowra granodiorite, Lachlan Fold Belt, SE Australia" : Lithos. Vol. 56. pp. 165-186.

Wang, J.H., Yin, A., Harrison, T.M., Grove, M., Zhang, Y.Q. and Xie, G.H., (2001) : "A tectonic model for cenozoic igneous activities in the eastern Indo-Asian Collision Zone". Earth and Planet. Sci. Lett., Vol. 188, No. 1-2, pp. 123-133.

Whalen, J.A., Currie, K.I., and Chappe, B.W.; (1987) : "A type granites – geochemical characteristics, discrimination and petrogenesis". Cont. Miner. Petrol., Vol. 95, No. 4, pp. 407 – 419.

Winchester, J.A. and Floyd, P.A.; (1977) : "Geochemical discrimination of different magma series and their differentiation products using immobile elements". Chem. Geol., Vol. 20, pp. 325 – 343.

Wood, D.A., J.L. Joron and M. Trevil; (1979) : A reappraisal of the use of trace element to classify and discriminate between magma series erupted in different tectonic settings. Earth Planet. Sci. Lett. 45, pp. 326-36.

Yadav, H.S.; (1991) : "Geochemistry of Jalore granites and associated rhyolites. Unpublished Ph.D. thesis, University of Rajasthan", Jaipur. 147p.

Zhu, Y.F. Sun, S.H., Gu, L.B., Ogasawara, Y., Jiang, N. and Honma, H., (2001) : "Permian Volcanism in the Mongolian Orogenic Zone, North east China". Geol. Mag., Vol. 138, No. 2, pp. 101-115.

Hon'ble Reviewers

1. **Prof. Ernst Hegner**
   Department of Earth Science
   University of Munich
   D80333 MUNCHEN
   GERMANY
2. **Prof. Naresh Kochchar**
   Centre of Advance Study
   Department of Geology
   Panjab University
   CHANDIGARH
3. **Dr. S.K. Bhushan**
   Former Dy D.G.
   Geol. Surv. India
   MPCL
   HOSPET - 583 203

I-11

Wollastonite Deposit at Belka Pahar: A Significant Georesource of Rajasthan

Kamal Kant Sharma

Government Postgraduate College, Sirohi 307 001 (Rajasthan)
E-mail: sharmasirohi@yahoo.com

Abstract

Wollastonite is one of the significant ceramic mineral produced from Rajasthan. The Wollastonite deposit is located at the tri junction of Sirohi, Pali and Udaipur district. The Belka Pahar Wollastonite deposit occurs near Khera Uprala, Khera Tarda and Khila village. The Wollastonite deposit forms lenses and pockets in the calc silicate rocks of Ajabgarh Group of south Delhi fold belt. The granitic intrusion at 850±25 Ma, in the region caused development of wollastonite and associated calcite mineral deposits.

Wollastonite is substantially used by ceramic industry as filler. Some other uses of wollastonite are in ceramic floor and wall tiles, marine wallboard, paint, plastic and in refractory liners in steel mills, and as a partial replacement for short-fibre asbestos in certain applications, such as brake-lining. Technical improvements in filler properties in plastic and rubber have been made in recent years.

Keywords: Wollastonite, Belka Pahar deposit, Ceramic, Refrectory, Skarn.

Introduction

Wollastonite ($CaSiO_3$) was named after the British mineralogist and chemist William Hyde Wollastone (1766-1826). The mineral occur as an aggregate of white bladed or needle like crystals with a vitreous to pearly lusture. In the year 1968 wollastonite was first reported from Khila village of Sirohi district in Rajasthan. Presently, entire production of wollastonite is being done by M/s Wolkem Private Limited from Belka Pahar mines (Fig 1, 2, 3).

The quality of wollastonite is expressed by accicularity or the aspect ratio i.e. ratio between length and width of a crystal. The 1.3 to 1.5 low range of aspect ratio is used for ceramic, metallurgical, flux and filler work. The higher aspect ratio of Wollastonite in the of 1:15 to 1:20 range is considered as semi-fibrous replacement of asbestos and used in the strengthening of plastic and resin.

Wollastonite is used as filler in a number of products like- ceramics, plastics, paint, resins, adhesive, bounded, abrasives wall boards, insulating, materials and replaces asbestos in various of industries. The low thermal expansion makes Wollastonite for reducing shrinkage in ceramic industries and useful for electrical insulating devises such as high tension circuit barkers. The wollastonite is also used in the manufacture of ceramic bounded abrasives, wheel stones, emery paper and industrial pumice. In the fiber glass industry, wollastonite is useful in manufacture of plywood substituting for alpha cellulose.

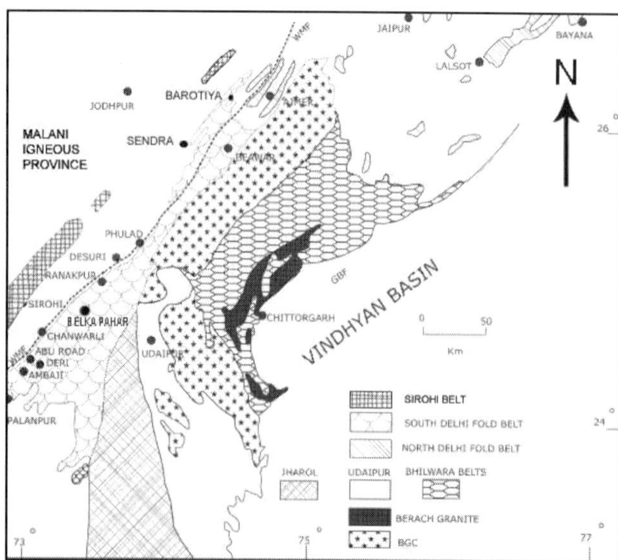

Fig- 1 Generalized Geological Map of Southern Aravalli Mountain Range (Modified after Gupta et al., 1980)

Fig-2 Location of Belka Pahar Mines

Fig-3 Google earth view of Belka Pahar Mines

Geological Setting

The Archaean basement of Aravalli craton in northwestern Indian shield is described as Banded Gneissic Complex (BGC) comprising granitoids, tonalite-trondhjemite-granodiorite gneisses, amphibolites, minor ultramafics and other meta-sediments (Table I, II). The presence of mafic-ultramafic enclaves alongwith granitoids is interpreted as granite-greenstone like evolution of BGC (Sinha Roy, 1988). The Palaeoproterozoic Aravalli rift basin evolved in ensialic setting after cratonisation of BGC continental part (Roy et al. 1988). The rocks outcropping from Delhi to Palanpur and northern Gujarat are described as part of Alwar and Ajabgarh series of Delhi System (Heron, 1953). Gupta et al. (1980) considered metasediments and associated rocks of Sirohi, Sindreth, and Punagarh region as part of Delhi Supergroup and extended its time framework from 2000 Ma to 740 Ma. The Delhi Supergroup (Table-I, II) was further subdivided into Gogunda, Kumbhalgarh, Sirohi and Punagarh Groups. On the basis of the ages of intrusive granites (Choudhary et al. 1984) in south Delhi rocks, Sinha-Roy (1988) suggested diachronous status for Delhi Fold belt and classified into South Delhi Fold Belt (SDFB) and North Delhi Fold Belt (NDFB).

Belka Pahar Wollastonite Deposit

Belka Pahar Wollastonite deposit is situated in the calc silicate rocks. Coulson (1933) describes these rocks as part of Ajabgarh Series of Delhi System (Table I). Gupta et al. (1980) described Gogunda group (Alwar Series, Heron, 1953) and Kumbhalgarh group (Ajabgarh Series of Heron, 1953) from SDFB. The Kumbhalgarh Group rocks are biotite schists, calc schists and calc-gneisses and hosts Belka Pahar Wollastonite deposit. The ultramafic and mafic amphibolites, referred to as Phulad Ophiolite Suite (Gupta et al., 1980), are interfolded with rocks of Kumbhalgarh group. The rocks of the Phulad Ophiolite Suite (POS) outcrops approximately 300 km-long and 20 to 50 km-wide linear belt along Phulad Shear Zone (Fig. 2). The gabbro-diorite intrusion along SDFB is dated at 1012 Ma (Volpe and Macdougall, 1990), this date almost coincides with the granitoids from Sendra region (Pandit et al., 2003). This date marks beginning of Neoproterozoic crustal building process in SDFB. Chaudhary et al. (1984) reported various granitoids at 850 Ma age from Erinpura, Pali, Ranakpur and Sai area. Chakranarayan et al., (1986) also reported 825±25 intrusive granites from Belka Pahar area, situated in SDFB. These granitoids show intrusive relationship with metasediments of Sirohi Group and Delhi Supergroup.

Volpe and Macdougall (1990) also recorded younger Sm-Nd mineral isochron for meta-gabbro, diorite and quartz diorite of Ranakpur to ages of 838±36; 835±43; 791±43 Ma respectively. These isotopic ages are interpreted as younger metamorphic equilibrium in the region. Deb et al. (2001) also reported zircon U–Pb age of 836+7/-5 Ma for the Siwaya granite gneiss situated south of Belka Pahar region.

Table I. Stratigraphic succession of the Sirohi state (Coulson, 1933)

Post-Tertiary and Recent	Wind-blown sand, talus, alluvium,and conglomerates,etc.
Post-Malani basic intrusives	Dolerites
Malani system (?Purana)	Idar granite, porphyries, rhyolites and dellenites.
Post-Erinpura-granite but pre-Malani basic intrusives	Dolerites, epidorites, gabbros, pyroxenites, picrites, basalts and sodalite-syenites.
Post-Delhi but pre-Malani granitic intrusives	Erinpura granite and accompanying quartz-reefs, pegmatites, and aplites.
Delhi and post-Delhi, but pre-Erinpura granite basic Intrusives amphibolites, hornblende- and actinolite schists.	Dolerites, epidiorites,
	3.Limestones and calc-rocks, with associated basic rocks.
Delhi System-Ajabgarh series	2.Mica-schists and thin quartzites, with abundant associated basic rocks.
	1.Quartzites
Unconformity	
Aravalli System	(Archaean) Mica-schists, phyllites, shales, crystalline limestones, quartzites, with ?contemporaneous tuffs and lavas.

Table II. The Precambrian Stratigraphy of Rajasthan

Heron, 1953		**Gupta et al.1980**		**Roy and Jakhar, 2002**
			Marwar Supergroup	Marwar Supergroup
Malani Series			Malani Suite	Malani Group
Erinpura granite		Proterozoic-III	Erinpura granite	=Sindreth Group 780-680 Ma
			and gneiss	Granite of Erinpura, Pali, etc 835 Ma
			Vindhyan Supergroup	Sirohi Group
				Intrusion of gabbro, diorite 1000 Ma
			Punagarh Group	
			=Sindreth Group	
		Delhi	Sirohi Group	
		Supergroup	= Sendra, Ambaji	Synorogenic granite 1400Ma
	Ajabgrh	2000-740 Ma	granite ca. 800 Ma	
Delhi System	Series	Proterozoic-II		Delhi Supergroup
			Kumbhalgarh Group	
	Alwar Series		=Ajabgarh Group	
	Railo Series		Gogunda Group	
			=Alwar Group	
				Darwal granite 1850 Ma
		Aravalli Supergroup 2500-2000 Ma		
Aravalli System		Proterozoic-I		Aravalli Supergroup
Banded Gneissic Complex		Bhilwara Supergroup		Untala, Ahar, Berach granitoids 2500 Ma
(BGC)		Archaean		Mewar gneiss (Heron's BGC) 2600-3300 Ma

Gupta et al. (1980; 1997) described this deposit as part of Todgarh Formation of Kumbhalgarh Group. The Kumbhalgarh Group is described as part of Delhi Supergroup and forms a linear belt from Pisangan (Ajmer Distt.) to Kherbrahma (Gujarat) over a distance of approximately 300 km. The Todgarh formation is outcropping between Sirohi Road to Gogunda Road section. The cal-gneisses and calc schists forms dominant part of this formation. This formation is affected by thermal metamorphism due to granitic intrusions of 825±25 Ma. This resulted into various minerals like- actinolite, garnet, Wollastonite, calcite etc. Belka Pahar Wollastonite deposit formed as contact aureole around intruded granite within impure marble and cal gneisses near Khera Uprala and Khera Tarla villages. A small occurrence is also situated near Khila village. The deposit occurs along one km strike length. The bedding in calc metasediments show north-south trend with 40°-75° dip towards east. The deposit forms ellipsoidal body of approximately 400 meter in length. The Wollastonite occurs with calcite in the deposit. Besides this, other skarn minerals like diopside and garnet developed due to thermal effect of granitic intrusion.

Structure and properties of Wollastonite

Natural wollastonite is an acicular calcium silicate mineral that occurs in triclinic and monoclinic varieties; both the varieties are very difficult to distinguish from one another. Generally it occur as triclinic, the unit cell parameters of wollastonite are: a = 0.79, b = 0.73 and c = 0.71 nm; a = 90°02', ß = 95°22' and y = 103°26' (Deer et al., 1978; Bauer et al., 1994).

Wollastonite consists of chains of indefinite length containing three Si04 tetrahedra per unit cell. The tetrahedral are joined apex to apex, and one is orientated with an edge parallel to the axis of the chain. These chains are paired; slight offsetting produces the different structural forms of the mineral. (Deer et al., 1978).

Chemical and physical properties from Bauer et al. (1994), unless otherwise specified

(a) Description: Triclinic crystals

(b) Form and habit: Bladed crystal masses; acicular

(c) Colour: White when pure; may be grey, pale green, yellowish brown with impurities

(d) Hardness: 4.5-5 on Mohs' scale

(e) Density: 2.87-3.09

(j) Cleavage: (100) perfect; (001) good; (102) good

(g) Melting-point: 1540 °C

(h) Inversion temperature: Pseudowollastonite (high-temperature polymorph) at 1120°C

Wollastonite is inert chemically; however, it can be decomposed in concentrated hydrochloric acid. Some wollastonite will fluoresce under ultraviolet light, with colours ranging from pink-orange to yellow-orange and, more rarely, bluish green. In addition, wollastonite may show phosphorescence. Wollastonite occurs in coarse-bladed masses. Because of its unique cleavage properties, wollastonite breaks down during crushing and grinding into lath-like or needle-shaped particles (fibers) of varying acicularity. This particle morphology imparts high strength and is therefore of considerable importance in many applications.

The acicularity of particles is defined by their length: width ratio or length: diameter ratio (aspect ratio). High-aspect ratio products with ratios of 15 : 1 to 20 : 1 are used as functional filers in the reinforcement of thermoplastic, thermoset polymer compounds and as a replacement for asbestos (Bauer et al., 1994).Wollastonite has a theoretical composition of 48.3% CaO and 51.7% Si02, although aluminium, iron, magnesium, manganese, potassium or sodium may partially substitute for calcium (Virta, 1995).

Mining and Processing

The mineral wollastonite is mined and processed by Wolkem India Ltd, Udaipur, India. This is ISO 9001:2000 as well as ISO 14001 certified. The Wolkem was established in 1972. This is one of the world's largest producer of mineral wollastonite and India's largest producer of calcite and wet ground calcium carbonate. The mines of Wolkem India Ltd. are the purest deposit of Wollastonite in the world, which will sustain their yields well into the future.

The mine has been developed along the slope of the hills and plunge of the lensoidal mineral body. Open cast method mining method is adopted here. Calcite and wallostonite are being mined simultaneously. The lenses have been developed into benches (Fig. 3). Wollastonite is being mined by opencast semi-mechanised method. The run-of-mine is selectively hand sorted to the size 30 cm to 50 cm to remove the associated minerals, such as calcite, diopside, garnet, quartz and iron. Wollastonite, thus separated, is then crushed to various sizes at two crushing plants near Sirohi railway station, having a total capacity of 80,000 tpy by Wolkem India Ltd. Principal commercial grades produced are: White Kemolit (S1 to S5) and off-white Kemolit (H1 to H5 and LG 25) which are milled products in the size range of 100 to 500 mesh. Besides, micronised products being marketed are Wolkron (1008, 1010, 1015, 1020, 1025 and 10825) in the low aspect ratio and Kemolit 1025 and 1020 in the high aspect ratio. In addition, specialty products and surface modified products are also marketed as Kemolit and Fillex, respectively. Wolkem Industries Ltd, a wollastonite mining and processing company, meets 20% of global requirements. Investigations carried out on a sample sent by the party at Central Building Research Institute (CBRI) revealed that it could substitute chrysotile asbestos to a little extent in the manufacture of cement products. The Belka Pahar deposit, Sirohi district, Rajasthan, produces wollastonite of the various grades as shown in Table-III. Department of Mines & Geology, Udaipur, Mineral wise Report (Major minerals) for the year 2010-11 indicates that the major production of wollastonite is from Sirohi (Table-IV and V). The number of mine workers engaged during the year were 260. (Table-IV).

Table-III. Belka Pahar Wollastonite deposit grades

Constituent	Grade-I	Grade-II	Grade-III
CaO	48.16%	47.62%	46.20%
SiO_2	46.12%	47.42%	48.00%
Al_2O_3	0.23%	0.60%	0.76%
Fe_2O_3	0.49%	0.40%	0.44%
MgO	0.20%	0.40%	0.30%
L.O.I.	1.51%	2.37%	2.71%

Table-IV. Wollastonite Production 2010-11 (Source-DMG, Rajasthan)

S. No.	Office	Leases (No.)	Area (in Hector)	Production (Tons)	Sale Value (`)	Revenue (`)	Employment (Nos.)
1	ME Ajmer	6	64.400	1,740	1392000	649280	15
2	ME Sirohi	1	49.48	122667	115306980	6451123	260
3	ME Sojat City	1	4	0	22000	22000	7
4	ME Udaipur	2	101.8				
Total		10	219.680	124407	116720980	7122403	282

Table V. Production of Wollastonite, 2008-09 to 2010-11 (Qty in tonnes; value in `'000)

	2008-09		2009-10		2010-11	
State	Qty.	Value	Qty.	Value	Qty.	Value
Rajasthan	111581	125957	132385	111930	182600	150093
Other State	-	-	-	-	-	-
India (Total)	111581	125957	132385	111930	182600	150093

(Source-Indian Minerals Yearbook 2011 (Part- II) 50th Edition WOLLASTONITE,), INDIAN BUREAU OF MINES, NAGPUR – 440 102)

Table –VI. World Production of Wollastonite India (Total) Source: World Mineral Production, (In tonnes)

Country	2008	2009	2010
China	325000	300000	300000
Finland	15600	16000	16000
India	111581	132385	182600
Mexico	46844	29728	46548
USA	90000	65000	67000

Production and Market Status

Presently, India is world's second largest producer of wollastonite after China (Table VI). The Belka Pahar deposit is in position to meet the domestic requirements of the ceramic industry as well as export demand. There is an increasing demand for wollastonite in the international markets, especially in ceramic and plastic industries and in construction activities. The domestic demand of wollastonite is estimated at 132 thousand tonnes by 2011-12, which may go upto 203 thousand tonnes by 2016-17 at 9% growth rate, as per the Sub- Group Report for 12th Plan Period.

Presently, the existing mines are in a position to meet the domestic requirements of the industry as well as export demand. The largest market for wollastonite in the world is ceramics followed by asbestos substitution, metallurgy and paints. Wollastonite is marketed under two grades, viz. high aspect ratio wollastonite and powdered (milled) wollastonite. The former type relies mainly on physical accicularity while the latter one depends on the chemical composition of the mineral. Recently, wollastonite demand is increased in the international markets, particularly in ceramic and plastic industries. The wollastonite is mined and export by only a few countries, looking to this, there is a scope for increasing the exports from India.

Acknowledgement

I am sincerely thankful to Prof. K.L. Srivastava for encouragement and necessary help to prepare this manuscript.

References

Bauer, R.R., Copeland, J.R. and Santini, K. (1994) Wollastonite. In: Carr, D.D., ed., Industrial Minerals and Rocks, 6th Ed., Littleton, CO, Society for Mining, Metallurgy, and Exploration, pp. 1119-1128.

Chakranaryan, A.B., Power, K.B., Pande, K., and Gopalan, K. (1986) Rb-Sr age of the Belka Pahar granite, Sirohi district, Rajasthan. J. Geol. Soc. India 28(4) pp. 325-327.

Choudhary, A. K., Gopalan K. and Anjaneya Sastry C. (1984) Present status of the geochronology of the Precambrian rocks of Rajasthan; Tectonophysics **105** pp. 131-140

Coulson, A.L. (1933) The Geology of Sirohi State, Rajputana. Mem. Geol Surv., Vol. 63 part-I, p. 166.

Deb, M., Thorpe, R.I., Krstic, D., Corfu, F., and Davis, D.W. (2001) Zircon U-Pb and galena Pb isotope evidence for an approximate 1.0 Ga terrane constituting the western margin of the Aravalli-Delhi orogenic belt, northwestern India. Precam. Res. 108, pp. 195-213.

Deer, W.A., Howie, R.A. and Zussman, J. (1978) Single-chain silicates. In: Rock FormingMinerals, 2nd Ed., London, Longman, p. 547-563

Gupta, S.N., Arora, Y.K., Mathur, R.K., Iqbaluddin, Prasad, B., Sahai, T.N. and Sharma, S.B. (1980) Lithostratigraphic map of the Aravalli region. Geol. Surv. India, Calcutta.

Gupta, S.N., Arora, Y.K., Mathur, R.K., Iqbaluddin, Prasad, B., Sahai, T.N. Sharma, and S.B. (1997) The Precambrain Geology of the Aravalli region, Southern Rajasthan and Northeastern Gujarat. Geol. Mem. Surv. India-123, Calcutta, pp.227.

Heron, A.M. (1953) The Geology of Central Rajputana. Mem. Geol. Surv. India, v.79, pp.389.

Roy, A B, Paliwal, B. S., Shekhawat, S. S., Nagori, D. K, Golani, P. R. and Bejarniya, B. R. (1988) Stratigraphy of the Aravalli Supergroup in the type area; In: Precambrian of the Aravalli Mountain, Rajasthan, India (ed.) Roy A B; Geol. Soc. India Memoir. **7** pp. 121–138.

Roy, A. B. and Jakhar, S. R. (2002) Geology of Rajasthan (Northwest India) Precambrian to recent; Scientific Publishers (India) Box 91, Jodhpur 421.

Sinha-Roy, S. (1988) Proterozoic Wilson cycles in Rajasthan; Mem. Geol. Soc. Ind. **7** pp. 95-107.

Virta, R.L. (1995) Mineral Industry Surveys: Wollastonite - Annual Review - 1994, Washington DC, United States Department of the Interior, Bureau of Mines.

Volpe, A.M, and Macdougall, J.D. (1990) Geochemistry and isotopic characteristics of mafic (Phulad Ophilite) and related rocks in the Delhi Super-group, Rajasthan, India: Implications for rifting in the Proterozoic; Precambrian Research 48 pp. 167-191.

Hon'ble Reviewers

1. **Prof. M.K. Pandya**
   Emeritus Professor
   Department of Geology
   MLS University
   UDAIPUR - 313 001
2. **Dr. Virendra Gaur**
   Department of Geology
   JNV University
   JODHPUR - 342 005

I-12

Multiple Deformation of Bar Conglomerate around Mineralized Bar-Birantiya-Giri Section of Delhi Supergroup, Pali, Rajasthan

Beena Tripathi, Rahul Paliwal and Bahula Verma

Department of Geology, J.N.V. University, Jodhpur-324 005
E-mail: beenagrabo@gmail.com

Abstract

The timing of events within Bar Conglomerate Horizon was critical to the style of deformation that occurred. An important aspect appears to be that the pebbles of Bar Conglomerate was first deposited with no gravity driven deformation. The pebbles exhibited a greater resistance suffered during second deforming event appear in the form of shear strain. The pebbles get deformed and during this stage about 80 % of the pebbles changed their shape from spherical to fish shaped. This style of deformation can be tracked regionally and has also been identified outside the investigated area i.e. from Phatakhera to Raira khurd. The evidence indicates that the deformational sequence penetrate to at least upto third deforming event, where the pebbles of the Bar Conglomerate Horizon get deformed and folded symmetrically in the form of anticline and syncline.

Strain in deformed rocks show extreme heterogeneous behavior. The heterogeneity arises in several ways and on a range of different scales, from ductility contrast between grains to regional structures, the presence of planar structures i. e. faults, folds, crenulations and variations in rocks is the result of the different level strains. The fourth event of deformation resulted in the Bar formation present in the form of diagonal shear zones along the pegmatite veins at Giri Dam as well as near Ramdev temple which gave rise to zonal crenulations. i.e. F_4 fold structure. Pegmatite veins are intruded at so many places, along the main strike of the horizon. Thick and thin exposures of pegmatite veins are present in fringes, and are post tectonic.

Keywords: Bar Conglomerate, Gravity Driven Strain, ductility contrast, Delhi Supergroup.

Introduction

The Precambrian rocks of the Bar area have been studied in detail in some parts leading to explain of multiple phases of deformation Raja Rao et al. (1971); Murty & Bhargava (1972); Sen (1980); Dasgupta et al. (1989); Ramakrishnan et al. (2008); Shrivastava D (1985), but the importance and quantitative analysis of the Delhi Supergroup was first realized after the publication of the classic memoirs of Heron A.M. (1953). He recognized three main stratigraphic units around Bar area; these units are the Banded gneissic complex, Delhi system and post Delhi intrusive granites. The contact between the Banded gneissic complex and the Delhi system of rocks has been strongly modified due to movement. Heron (1953) recognized the conglomerate horizon as the basal part of the Alwar series which crops out intermittently for a strike length of about 35 km. from Giri to Kalakot through Bar and noted that these rocks grade upwards into biotite-schist. About 80% of the pebbles undergone a flattering type of deformation. He further applied to analyse the various fold structures observed in this region. The present paper combines the structural studied of the Bar conglomerate and associated rocks at the southwest to northeastern margins around Bar, Barotiya and Giri (Fig. 1). About 47 sq. km area has been mapped to the scale of 1: 25,000 in general. Apart from these Google earth were also used for studying regional lineament of the Delhi fold belt.

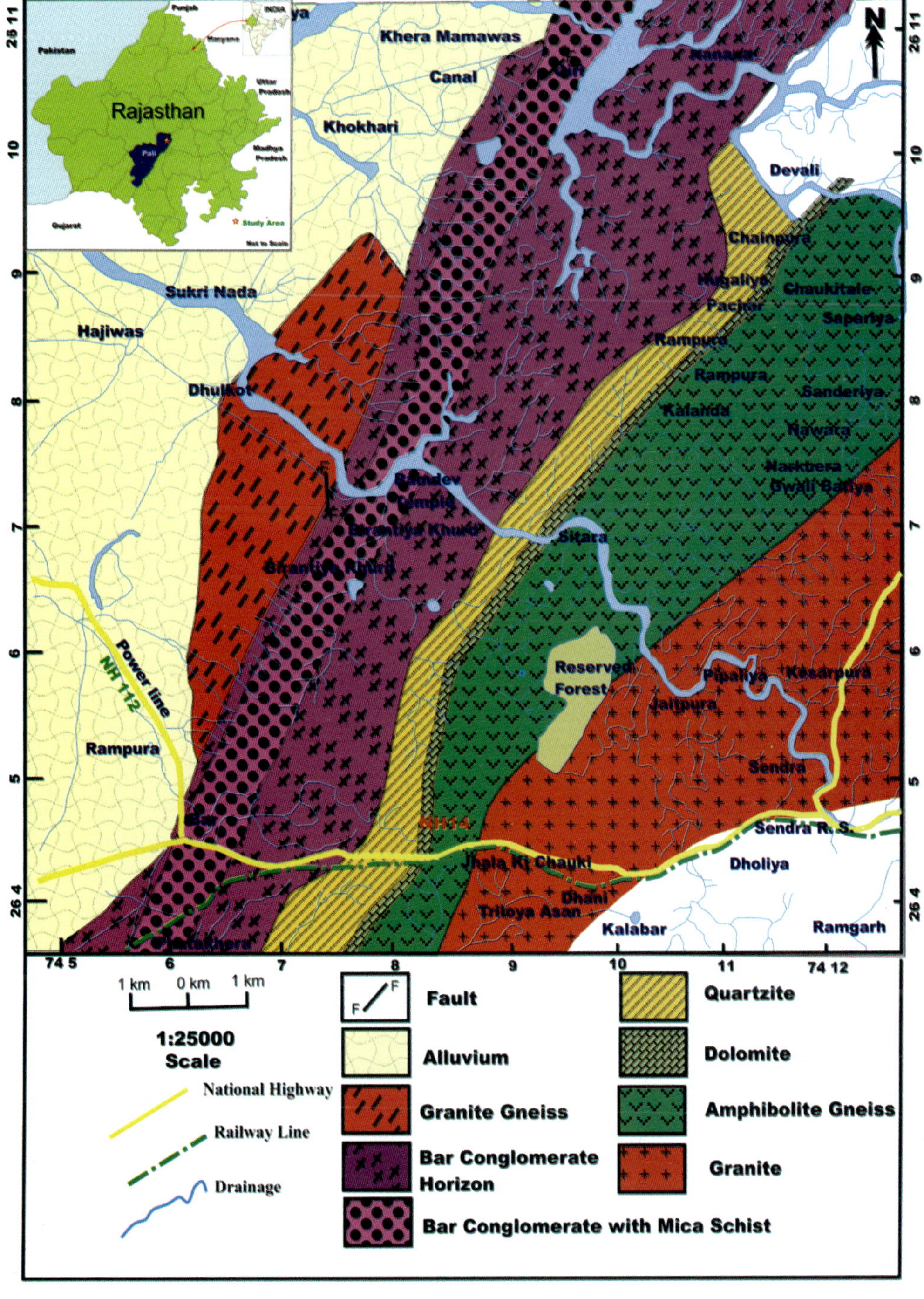

Table 1: General stratigraphy given by different authors

Heron (1953)	Gangopadhyay & Lahiri (1983)	Present Author	
Intrusive granite	Jetpur Granitic Intrusion	Intrusive granite	
Sendra Complex	Mica schists with bands of calc-schist	Sendra Formation	Calc gneiss
			With intercalated quartzite
			Metamorphosed limestone
			Calc amphibolite gneiss
			Foliated quartzite
			Calc gneiss
Nandana Crystalline Lst.	Nandana Crystalline Limestone	Dolomite	
Barotiya Sequence	Malni-ki-Bara tremolite-calcschist	Barotiya Formation Bar Conglo-merate horizon	Quartzite schist
			Calc amphibolite schist
			Kyanite Schist
			Staurolite schist
	Karmdoribora quartzite		Garnetiferous mica schist
	Bar garnetiferous schist		Bar conglomerate schist
	Bar semipelitic schist and conglomerate		Quartzofeldspathic mica schist and conglomerate
Unconformity	Unconformity	Unconformity	
Banded Gneissic Complex	Gneiss	Granitic gneiss	

GEOLOGICAL SETTING

The rocks of Bar area in Pali district, Rajasthan belongs to Delhi Supergroup. The Barotiya–Sendra basin in western Rajasthan belongs to sub-geosyncline of South Delhi Fold Belt. The generalized stratigraphic sequence in the area is tabulated below. The Banded Gneissic Complex, Barotiya Sequence and Sendra Complex of Heron (1953) have been termed as Granitic gneisses, Barotiya and Sendra formations respectively in the present study. Bar conglomerate horizon is a lower most unit of the Barotiya Formation and it is about 30 km long linear belt (Fig. 1) which is running from Southwest to Northeast and lies between granitic gneisses and garnetiferous mica schist of Barotiya Formation. Both the formations are intruded at several places by pegmatite veins. The rocks can be subdivided into three broad groups; granitic, schistose (foliated) and gneissose. The granitic and schistose types generally occur along the contact of Granitic gneiss and Bar conglomerate horizon while the gneissose type occurs at the northeastern part of the investigated area. Heron used a broad term i.e. 'Composite gneiss' to include rock types showing gneissose texture at Sendra such as amphibolite gneiss and calc-silicate gneiss. The Sendra Formation also includes quartzite, dolomite and garnetiferous mica schist intruded by thick and thin dykes ranging between50 cm to 2m. These dykes are quite competent and medium to dark in colour. However, slickensides and striations which are green in colour are well developed at the contact of the dyke indicates the planes of movement. This lineation represents prominent direction of movement in NNE-SSW and NE-SW direction. Boudinage structures have been observed in the quartz veins in schists of Barotiya Formation on the Phatakhera – Megarda road. Necking in boudins is quite prominent (Fig. h) and at Megarda village few quartz veins are also distorted due to later deformation (Fig. J). Three sets of fracture plane in dolomite at Jhala-ki-choki and in granite at Jhala-ki-choki – Sendra road (NH-14) are giving indication that the rocks are competent but not susceptible to normal shocks.

Fig. A Flattened pebble imbedded along S^2 schistosity near Ramdev temple. **Fig. B** Deformed pebble showing syn deformational effect with post deposional matrix inside the deformed pebble near Barotiya khurd village. **Fig. C** Photomicrograph of strained quartzite pebble showing two sets of micro strained shear planes cutting each other and distorted by simple shear to produce a parallelogram. 10 X **Fig. D** Crenulations are developed on the limbs of F_2 folds due to later deformative forces near Khejara village, south of Bar market. **Fig. E** Photomicrograph of crenulations showing fracture plane and alternate bands of mica and quartzofeldspathic material. 10 X **Fig. F** Crenulations are developed along with pegmatite vein near Barotiya khurd village. **Fig. G** Boudins are developed in quartz vein showing necking near Phatakhera. **Fig. H** Ptygmatic folding in quartz vein showing deformation by later deformative forces near Megarda village.

DEFORMATION IN CONGLOMERATE PEBBLE

Conglomerate belt of Bar (N26°05′: E74°06′) Birantiya (N26° 7': E74° 8') and Giri (N26° 11': E74°10') running from South to North is a narrow, elongated, intra-formational horizon situated in between granitic gneisses and garnetiferous mica schist of Bar conglomerate horizon, in the state of Rajasthan. The belt is intruded at several places by pegmatite veins. The pebbles of this belt occur not in a random arrangement but they are aligned along the main schistosity of the study area. The pebble size varies from about less than 2 cm to more than 1m. The general trend of the pebbles is NNE to SSW and NNW to SSE. Almost a complete sequence of deformation in Bar conglomerate horizon exposed in the present investigated area at Bar, Barotiya and Giri villages. The pebbles are deformed at different level and the results are discussed in this paper. The analysis reveals that the pebbles subjected to three event of deformation. Under the first deformational event the pebbles suffered slight elongation to spherical in the NW-SE direction (Fig. a). In the subsequent deformational event of pebbles got flattened. There were two flattening stages observed during second deformational event and interestingly both directions are not right angle to each other. In first stage of second deformational event, the pebbles get flattened along NE-SW direction that is along the strike direction of the belt while in second stage of second event the pebbles flattened along NW-SE direction that is along the general strike of the Aravalli belt (Fig. b). The main strain factor was directed vertically to sub vertically and second strain factor was directed nearly right angle to the previous factor. During this event syn-deformational shearing and post-deposional effect in the pebbles are clearly observed near Barotiya khurd village (Fig c). There is one more principal stress event acted across the main strike of the belt, pebbles get deformed and folded to attain a F_3 fold pattern. Mainly four types of stress factors are seen which deformed conglomerates of Delhi Supergroup rocks in mainly three events. All the four stresses were further applied to analyse the various type of fold structures observed in the region. Two major deformative forces brought in second event, which are responsible for deformation of conglomerate pebbles and elongation and flatting in pebbles takes place (Fig. d). Two set of lineations are clearly observed along the margins of the pebbles which were formed due to two different stress factors in second event of deformation in the second and third stage. F_3 folds are formed when pebbles get deformed and folded by third principal stress phase in the fourth stage. The fourth event of deformation resulted in the Bar formation present in the form of diagonal shear zones along the pegmatite veins at Giri Dam as well as near Ramdev temple which gave rise to crenulations. i.e. F_4 fold structure (Fig. e & f). Pegmatite veins are intruded at so many places, along the main strike of the horizon. Thick and thin exposures of pegmatite veins are present in fringes, and are post tectonic. The lithostratigraphic contact of the BGC and overlying Barotiya Formation is suggested to be placed at the contact of Bar conglomerate horizon with mica schist representing the typical unconformable plane. The beginning of deformation in the Bar conglomerate horizon, suggests to be marked with the first appearance of the undeformed pebble in association with quartzofeldspathic mica schist (Fig. a). Incidently it coincides with the appearance of deformed pebbles in next two events, where stretching and folding in pebbles take place respectively.

DEFORMATIONAL HISTORY

The end products of different tectonic events are well documented in the tectonites of the area. The rocks display structures of different generations on all scales. The study of mesoscopic and microscopic structures of Delhi Supergroup rocks exposed around Bar reveals that the area has undergone multiple deformation. Four major events of deformation have been recognized and are interrelated in space and time as successive events of Aravalli orogeny. The deformational history of the area has been worked out by analyzing foliation and lineation of different generations. Folds of different generations are identified on the basis of orientation of the axial elements and relationship with axial plane cleavage, crenulation cleavage, intersection lineation and mineral lineation.

FIRST EVENT OF DEFORMATION (D_1)

The mesoscopic folds F1 of first event of deformation are observed at Megarda and very thin chert bands with micritic dolomite suggests precipitation of silica along with the carbonates. Whereas, the profuse development of chert nodules within magnesite and dolomite, is outcome of diagenesis. Associations of talc with magnesite-silica, dolomite-silica or magnesite-dolomite-silica indicate that the formation of talc is related to the presence of magnesite/ dolomite + silica. Occurrence of the recrystallized grains of magnesite and dolomite adjacent to talc flakes points to their recrystallization during talc formation. The textural features of the talc-magnesite-dolomite- silica assemblage at Rema support the reaction (2) above, between magnesite and silica.

References

Ahmad, T., Harris, N., Bickle, M., Chapman, H., Bunbury, J., and Prince, C., (2000) Isotopic constraints on the structural relationships between the Lesser Himalayan Series and the High Himalayan Crystalline Series, Garwhal Himalaya: Geological Society of America Bulletin, v. 112, pp. 467–477.

Bhattacharya, A.K., Sengupta, H.P. and Mukherjee, A. (1985) Origin of upper Precambrian talc/ soapstone deposit at Chandak, district Pithoragarh, U.P. Current Trends in Geology,

Celerier, J., Harrison, T.M., Webb, A.A.G. and Yin A. (2009) The Kumaun and Garhwal Lesser Himalaya, India: Part 1. Structure and Straitigraphy. Geol. Soc. Amer. Bull., v. 121, pp. 1262-1280.

Gaur, C.S., Bagati, T.N. and Nautiyal, S.P. (1979) Magnesite deposits of Bagoli, Dist. Chamoli, Garhwal Himalaya: A preliminary exploration account. Him. Geol., v.9, pp.744-772.

Krauskopf, F.K. (1963) Factors governing the concentration of thirteen rare metals in seawater. In: The geochemistry of lithogenesis. Collected articles on the geochemistry of rock formation. Foreign Language Publishing House, Moscow, pp.294-338.

Nath, M. and Wakhaloo, G.L. (1962) A note on the magnesite deposits of Almora district, U. P. Indian Minerals, v.16, pp.116-123.

Nautiyal, S. P. (1953) The reconnaissance geological report of a part of the copper belt Kumaun Himalayas, Almora District, U. P. Rec. Geol. Surv. India, v.89, pp.341-358.

Miller, C., Klotzli, U., Frank, W., Thoni, M., and Grasemann, B., 2000, Proterozoic crustal evolution in the NW Himalaya (India) as recorded by circa 1.80 Ga mafic and 1.84 Ga granitic magmatism: Precambrian Research, v. 103, pp. 191–206,

Shandilya, A.K., Gahlot, S.K. and Sriravtave, N. (2009). Geological characteristics and Genesis of the Soapstone deposit around Sarada Inlier, Rajasthan. In: K.L. Shrivastava (Ed.), Economic Mineralization. Scientific Publishers, Jodhpur, pp.228-236.

Sharma, Rajesh, Joshi, Prabha and Pant, P.D. (2008) The role of fluids in the formation of talc deposits of Rema area, Kumaun Lesser Himalaya. Jour. Geol. Soc. India, v. 73, pp.237-248.

Sharma, Rajesh, Joshi, Prabha and Verma, Preeti (2010) Genetic issue of some of the non metallic minerals in Lesser Himalaya. International applied geological congress, Mashad Branch, Iran, pp.1938-40.

Srivastave Neeraj. (2007). Exploration of Soapstone Deposit around Devpura, District Udaipur, Rajasthan. . Unpublised Ph.D. Thesis, Bikaner University, Bikaner pp.106-207.

Srivastava, P., and Mitra, G., 1994, Thrust geometries and deep structure of the outer and lesser Himalaya, Kumaun and Garhwal (India): Implications for evolution of the Himalayan fold-and-thrust belt: Tectonics, v. 13, pp. 89–109,

Valdiya, K.S. (1968) Origin of the magnesite deposits of southern Pithoragarh, Kumaun Himalaya. Econ. Geol., v.63, pp.924-934.

Valdiya, K.S. (1979) An outline of the structural set up of the Kumaun Himalaya. Jour. Geol. Soc. India, v.20, pp.145-157.

Valdiya, K.S. (1980). Geology of Kumaun Lesser Himalaya. Wadia Institute Himalayan Geology, Dehra Dun, 288 p.

Valdiya, K.S. (1980). Geology of Kumaun Lesser Himalaya. Wadia Institute Himalayan Geology, Dehra Dun, 133-34 p.

Valdiya, K. S. (1995) Proterozoic sedimentation and Pan- African geodynamic development in the Himalaya. Precambrian Res., v. 74, pp.

Hon'ble Reviwers

1. **Prof. Arun Shandilya**
   Chairman
   Department of Appl. Geology
   Dr. H.S. University
   SAGAR - 470 003
2. **Prof. Devesh Walia**
   Department of Env. Studies
   North-Eastern Hill University
   Mawkynron
   SHILLONG 793 002

I-14

Gemstone Potential of Pegmatites in Southwestern Part of the Great Mica Belt, Jharkhand

Pankaj K. Srivastava and Surjeet Singh

Department of Geology, University of Jammu. J & K INDIA-180006.
Email: pankajsrivastava.ju@gmail.com

Abstract

India is a homeland of many gemstones. The gemstone mineralization associated with the Great mica belt of Jharkhand is one such region where a number of coloured precious and semi-precious stones have been reported.

The mica and gem minerals bearing zoned and unzoned pegmatite veins of varying dimension and shapes are emplaced in the meta-sediments and granites of the belt. The main precious and semi precious minerals present in the southwestern part of the mica belt are moonstone, bytownite, apatite and almandine. The mineralogical details of these minerals are reported in the present paper in order to characterize the available gemstones in the southwestern part of Great mica belt.

Keywords: Great Mica Belt, Hazaribagh Pegmatite, Gemstones, Garnet, Apatite.

Introduction

The Great Mica Belt, previously known as Bihar Mica Belt, extends to a distance of 160 km, having an average width of 25 kms, from Gurpa in Gaya district (Bihar) in the west through Nawada, Koderma, Hazaribag, Giridih and Jamui (Jharkhand) as well as Bhagalpur (Bihar) in the east. Its maximum width is about 40 km at Koderma-Hazaribagh-Nawada area. The whole belt encompasses roughly around 4,000 sq. km. area and lies on the northern fringe of Chotanagpur plateau. It runs in an east-northeast to west-southwest direction. The major part of the Great Mica Belt is located in Jharkhand state.

Numbers of pegmatite veins have intruded the mica schist and granites in this belt. Besides the occurrence good quality mica, the belt also hosts a number of precious and semi precious gemstones such as moonstone, indicolite, amazonite, citrine, almandine, cat's eye and bytownite, and some rare metals, viz. beryl, lepidolite, columbite-tantalite, lithium, refractory uranium ore-minerals and REE minerals. The pegmatites intruded in the southwestern part of the Great Mica Belt and exposed in the Kodarma district have moonstone, bytownite, apatite, tourmaline and almandine as main precious and semi precious minerals. Though quarrying of these minerals is going on these areas, no scientific details about the characterization and genesis of these gemstones are available in the published literature. In the present paper an endeavour has been made to mineralogically characterize the available gemstones in the southwestern part of Great mica belt.

20. Sinha-Roy (1988). Precambrian crustal interaction in Rajasthan, NW India, India Jour. Earth Sci., CISM Volume, 84-91.
21. Singh, S.P. (1984). Fluvial sedimentation of the Proterozoic Alwar Group in the Lalgarh graben, northwestern India, Sediment. Geol., 39, 95-119.
22. Hobbs, B.E., Mean, W.D. and Williams, P.F. An outline of structural geology.

Hon'ble Reviewers

1. **Prof. V.C. Thakur**
   Former Director
   Wadia Inst. Himalayan Geology
   DEHRADUN - 248 001
2. **Dr. Meenal Mishra**
   Department of Geology
   Indira Gandhi Open University
   NEW DELHI - 110 001

Genetic Implication in Talc Mineralisation and Causes low dipping talc veins in Rema-Kanda belt, Bageshwer district, Uttaranchal

Neeraj Srivastave[1] and A.K. Shandilya[2]

[1]Sr. Manager, A.S.D.C.P. Ltd.; Golcha Associated Group, Udaipur (Raj.)
[2]Department of Geology, Maharaja Ganga Singh University, Bikaner (Raj.)
E-mail: neeraj_geologist@yahoo.com

Abstract

Himalayas is a unexplored treasure of mineral wealth. The economic minerals are located mainly in the Lesser Himalaya. Talc is one of the most widespread mineral deposit located at several places in which deposits of Rema-Kanda belt of the Bageshwar district is very significant in the term of genesis and structural point of view. Talc mineralization in Rema area is restricted to the Mg-bearing carbonate lithounits of the Deoban Formation. The rocks of the study area show an antiformal structure, wherein talc bearing Deoban Formation is exposed in the middle part, after the erosion of overlying Berinag Formation. Talc is usually developed along the boundaries of coarse-grained magnesite. Sporadic development of talc is also seen within the recrystallized crystalline mosaic of dolomite grains wherein a coarsening in the granularity is noticed along the talc. Occurrence of the recrystallized grains of magnesite and dolomite adjacent to talc flakes points to their recrystallization during talc formation. Whereas, the recurrent presence of dolomite along with talc and silica indicate that conversion of dolomite to talc, at least to a large extent, can not be suggested for Rema. It is inferred therefore that magnesite and quartz acted as the main reactant minerals; consequently either magnesite or quartz is present with talc while one of these reactant minerals has been consumed.

Talc of this area having low dip as comparision to the deposition of Aravalli region even after same mineralisation age. The reason behind this difference is the thrust known as Deoban Thrust. There is striking difference in the trends and style of folding of the Chakrata and Deoban formation in the Bangaon area, justifying the placement of a thrust between the two.

Keywords: Talc Mineralization, Talc Genesis, Deoban Formation.

Introduction

Out of the four broad geotectonic divisions of Himalaya viz. Lower Himalaya, Lesser Himalaya, Central Himalaya and the Tethys Himalaya, much of the economic minerals are located in the Lesser Himalaya which is largely formed of the Precambrian elements of the Indian shield (Valdiya, 1995). The non metallic minerals in Himalaya are particularly hosted within the sedimentary sequences. Present discussion briefly focuses on talc from the sedimentary sequences of Lesser Himalaya emphasizing their formation and evolution.

The talc deposits of Rema are located 44 km NE of the Bageshwar in the Kumaun Inner Lesser Himalaya, wherein workable talc deposits occur within the siliceous magnesium bearing carbonates of Deoban Formation. Talc mineralization is intersected between depths of 10 to 40 m, and in this hilly terrain is easily accessible by open pit method of mining. Two views have been proposed for the origin of talc in Lesser Himalayan magnesite: (i) hydrothermal replacement (Nautiyal, 1953; Nath and Wakhaloo, 1962 and Gaur et al. 1979) and (ii) low grade metamorphism of siliceous carbonates (Valdiya, 1968; Bhattacharya et al.1985).

We present here petrography, geochemistry and fluid inclusions microthermometry of Rema talc deposit in order to understand the origin of carbonate hosted talc deposits and the causes of low dipping veins.

Regional Geology

The Lesser Himalaya, demarcated by the Main Boundary Thrust (MBT) in the south, and Main Central Thrust (MCT) in the north, represents a distinct lithotectonic unit of the Himalaya. This is broadly divisible into Inner Lesser Himalaya and the Outer Lesser Himalaya, having basement of the crystallines representing Indian cratonic rocks. The Lesser Himalayan sequences were deposited from Paleoproterozoic to Cambrian age in marine conditions on the passive margin setting of Indian continent (Valdiya, 1980; Srivastava and Mitra, 1994; Ahmad, 2000, Miller 2000; Celerier, 2009). The Inner Lesser Himalaya sediments are the lower flysh and quartz arenites with basic volcanics, and the upper carbonate shale assemblages. The 1800 ma old Berinag Formation constitute the lower most sequence of the Inner Lesser Himalaya (Miller, 2000), overlain by Chakrata and Rautgara Formation, Neoproterozoic Deoban Formation- mainly the carbonates & above them Mandhali Formation consisting slate carbonates.

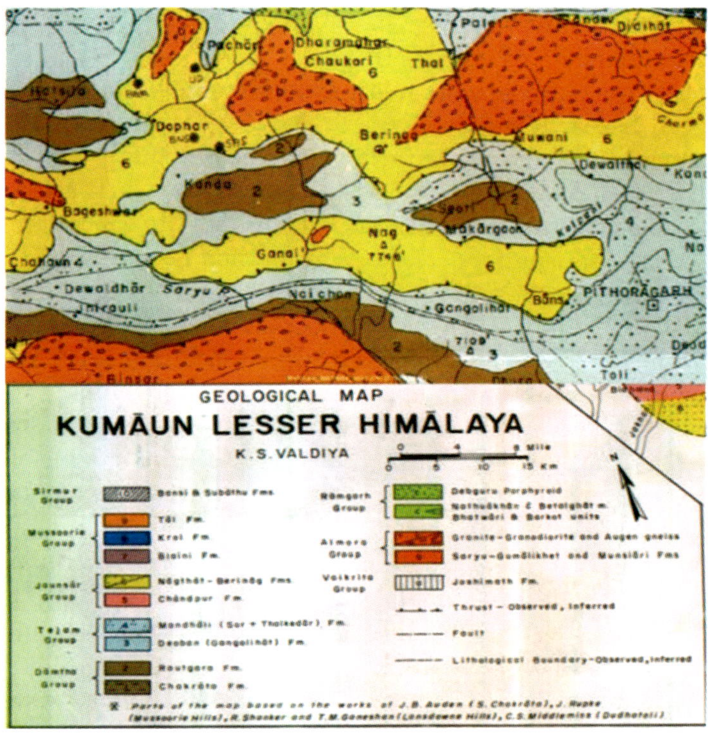

Fig. 1. (a) Location of the study area, (b) Regional geological map of Inner Lesser Kumaun Himalaya, (c) Geological map of the Rema area

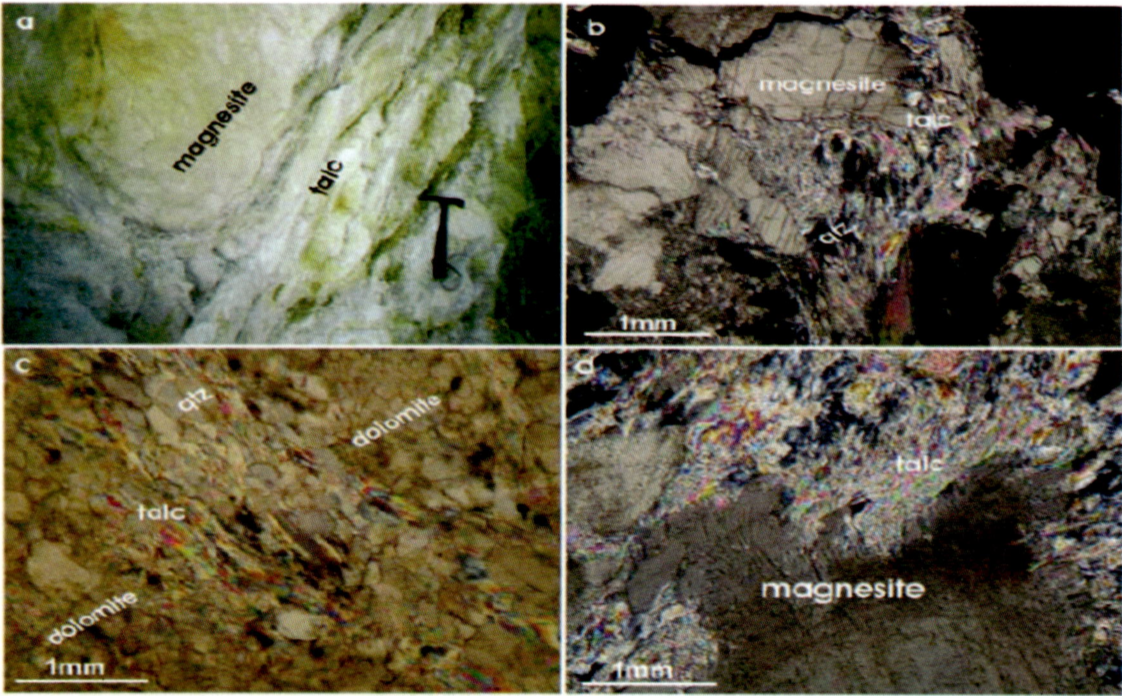

Fig. 2. (a) Field photograph of talc pocket within magnesite, (b) photomicrograph representing reaction between magnesite and quartz to form talc, (c) photomicrograph of talc developed within dolomite at the expense of magnesite, (d) highly corroded magnesite grain encircled by talc flakes, quartz has been consumed.

The outer Lesser Himalaya is folded sequence of sedimentary rocks comprising 850 ma turbidites of Chandpur Formation, overlain successively by Nagthat quartzites, Blaini conglomerates, Neoproterozoic Krol Formation which is largely calcareous, and above it Cambrian age clastic sediments of the Tal Formation. A number of the klippen of the crystalline rocks present in Lesser Himalaya are result of the thrusting of basement rocks over the sedimentary succession. Present focus is on the magnesite-talc which is hosted within the carbonates of the Deoban Formation. Talc mineralization in Rema area is restricted to the Mg-bearing carbonate lithounits of the Deoban Formation, which is a part of autochthonous Tejam Group and is overlain by allochthonous Berinag Formation (Fig. 1). The rocks of the study area show an antiformal structure, wherein talc bearing Deoban Formation is exposed in the middle part, after the erosion of overlying Berinag Formation. The talc and the magnesite mineralization are confined to the dolomite litho units in the uppermost part of the Deoban Formation. This formation is a thick succession of stromatolite bearing dolomite and limestone. Presence of clinochlore with talc in magnesite and dolomite suggest greenschist facies metamorphism of the host rocks. The upper limit of the Deoban Formation is marked by a sharp thrust contact with Berinag Formation (Valdiya, 1979; Valdiya and Pant, 1981), whereas base of the Deoban Formation is not exposed here.

Talc Mineralisation

Sizable talc deposits are also found at various places in the Deoban Formation, hosted within high Mg carbonates. Their mining by open pit method is operational particularly in Kumaun Himalaya. Talc occurs as irregular patches or pockets in the carbonate host rocks, showing close association with magnesite and restricted with dolomite. Bands, intercalations and pockets of talc also occur at the contact of dolomite and magnesite. The magnesite, quartz and talc are intimate partner in an assemblage, which shows development of talc at the expense of magnesite and silica. The fluid inclusion microthermometry was carried out at the fluid inclusion laboratory of Wadia Institute of Himalayan Geology, Dehradun, and using calibrated LINKAM THMSG 600 stage fitted onto the Nikon E600 microscope. The geochemical signatures in magnesite and talc do not invoke any external flux of Mg carrying fluid in talc formation

(Sharma et al., 2008), this is also substantiated by the fluid inclusions data of dolomite and magnesite suggesting within basin processes. The talc mineralization is confined to the magnesite bands and uncommonly to the dolomite in the upper part of the Deoban Formation. Lenses and pockets of pure talc are present within large magnesite or in the associated dolomite bodies. Rhythmic alteration of dolomite, magnesite and talc is seen at some places. Talc is generally found in intergranular spaces, in small clusters, lenses and pockets within coarse sparry magnesite. Talc also occurs as vermicular intergrowth and as thin layers along the foliation planes in the border zones of magnesite bodies (Fig.2a). Lenses of talc are generally 0.5 to 3 m thick and their strike lenght varies from 45 to 80 m. The massive to schistose talc mineralization mainly consists of talc + magnesite ± dolomite ± clinochlore. Pure talc is white, fine- grained and fibrous whereas compact- massive and the foliated varieties are marked by the presence of basal cleavage. Magnesite associated with talc is moderately crystalline, white in colour characterized by lustrous grains and soapy feeling. Thin films of talc developed around the stromatolitic nodules are quite often present within the host rock. A distinct relation between talc mineralization and the structural features like faults, folds or joints is not discernable in the area, but it is sporadically confined to the beddings within the magnesite. Dolomite associated with talc is fine to moderately crystalline, grey in colour, and varies in grain size. It is traversed by very fine network of calcite veins.

Petrography

Under microscope a close association of talc and magnesite is observed, though only minor talc is found associated with dolomite. Assemblage of highly corroded magnesite grains, fibrous chalcedony and well-developed talc flakes is commonly seen.

Talc is usually developed along the boundaries of coarse-grained magnesite (Fig.2b). Generally the grains of the host magnesite are recrystallized in the vicinity of talc, showing sutured grain boundaries and interlocking grain mosaic. Sporadic development of talc is also seen within the recrystallized crystalline mosaic of dolomite grains wherein a coarsening in the granularity is noticed along the talc (Fig.2c).

Cherty laminations are also present within the laminated dolomite. Quartz as a reactant mineral is seldom identified in the matrix, except for its local occurrences in which talc is developed within the cherty layers in dolomite (Fig. 2c). Talc flakes are aligned parallel to the schistosity of the dolomite. Calcite adjacent to the talc or representing a part of talc-magnesite- dolomite-silica assemblage is not found. Talc also contains numerous euhedral to subhedral pyrite grains, which are often limonitized. Study of number of thin sections of the host carbonate rocks as well as talc elucidate that highly corroded magnesite grains are present as remnants within the flakes of talc, and either magnesite or silica is found with talc (Fig. 2d).

Talc Genesis

In the talc deposit of Rema area, host magnesite and dolomite show number of sedimentary features for intertidal environment. There is prolific development of columnar as well as planar stromatolites in the study area. Presence of very thin chert bands with micritic dolomite suggests precipitation of silica along with the carbonates. Whereas, the profuse development of chert nodules within magnesite and dolomite, is outcome of diagenesis. Associations of talc with magnesite-silica, dolomite-silica or magnesite- dolomite-silica indicate that the formation of talc is related to the presence of magnesite/dolomite+silica. Occurrence of the recrystallized grains of magnesite and dolomite adjacent to talc flakes points to their recrystallization during talc formation. Further evidences such as (i) the variable proportion of talc with magnesite and silica and (ii) common absence of either silica or magnesite in the close vicinity of talc advocate that the magnesite and silica were mainly consumed in the development of talc. Whereas, the recurrent presence of dolomite along with talc and silica indicate that conversion of dolomite to talc, at least to a large extent, can not be suggested for Rema. It is inferred therefore that magnesite and quartz acted as the main reactant minerals; consequently either magnesite or quartz is present with talc while one of these reactant minerals has been consumed. We know that the formation of talc from magnesite/ dolomite in the presence of silica can be represented by the following reactions.

$$3CaMg(CO_3)_2+4SiO_2+H_2O \rightarrow Mg_3Si_4O_{10}(OH)_2+3CaCO_3+3CO_2$$

Dolomite Quartz Talc Calcite

$$3MgCO_3 + 4SiO_2 + H_2O \rightarrow Mg_3Si_4O_{10} (OH)_2 + 3CO_2 \quad (2)$$

Magnesite Quartz Talc

Although dolomite and sometimes talcose dolomite is found in association with talc, but calcite is absent in the talc- dolomite-silica assemblage. Conversely, if dolomite+silica produce talc, calcite should be present as a by-product. Hence, textural features of the talc-magnesite-dolomite- silica assemblage at Rema support the reaction (2) above, between magnesite and silica. Near identical major and minor elements in talc and host dolomite/magnesite, point that the talc is derived from these carbonates. However, very low CaO also indicate that talc was developed mainly at the expense of magnesite while un-reacted dolomite was left out. Such inference is also supported by many of the major and minor elements like MnO, CaO, Na_2O, TiO_2, P_2O_5, Ni, Co, Rb, U, in magnesite and talc, wherein these elements are near similar, and suggest that talc and magnesite shared the evolutionary process. A very low amount of Sr can also be correlated with the decreased CaO content in the talc. The high vanadium content of talc although can be considered for the possible vanadium migration from overlying amphibolites interbedded with Berinag quartzite, but the basic intrusive derive source of the higher vanadium is ruled out in view of very low concentration of Ni, Cr and Co contents. As these talc deposits are hosted within the marine stromatolitic carbonate vanadium might be derived from marine algae. It is known that some algae are capable of accumulating vanadium (Krauskopf, 1963). Higher Al content in pure talc samples and inverse relation of Si and Al in talc layers is possibly because of the presence of clinochlore in this assemblage, whereas pyrite associated with talc resulted in the higher average value of Fe_2O_3.

Reason behind Low Dipping Talc Deposits

Talc of this area having low dip as comparision to the deposition of Aravalli region even after same mineralisation age. The reason behind this difference is the thrust known as Deoban Thrust. There is striking difference in the trends and style of folding of the Chakrata and Deoban formation in the Bangaon area, justifying the placement of a thrust between the two.

It must be emphasized that at Marrow's Neck it is clearly reverse fault which has thrown the undulatorily folded file of the Deoban, with dip varying from 20^0 to 45^0 (or even 10^0) against the uniformly 45^0 NNE dipping Chakrata rocks. Drag folding, abrupt steepening and wedging with shattering is a common feature of this fault zone. This is clearly visible in the road of Lakhamandal. This fault has been named Marrow's Neck by Ram P. Sharma (privately circulated unpublished report, 1972). The same fault passes into a thrust eastward at Manjhgaon where tightly anticlinally folded steep depping (60^0NNE) Chakrata is overlain by 20^0-25^0 ENE dipping dolomites of Deoban formation in which talc vein exist. The evidences of trust is also seen in some of the talc veins in the form of minor fault lines.

Conclusion

Talc mineralization in Rema area is restricted to the Mg-bearing carbonate lithounits of the Deoban Formation, which is a part of autochthonous Tejam Group and is overlain by allochthonous Berinag Formation (Fig. 1). The rocks of the study area show an antiformal structure, wherein talc bearing Deoban Formation is exposed in the middle part, after the erosion of overlying Berinag Formation. The talc and the magnesite mineralization are confined to the dolomite litho units in the uppermost part of the Deoban Formation. Massive talc deposits are also found at various places in the Deoban Formation, hosted within high Mg carbonates. Their mining by open pit method is operational particularly in Kumaun Himalaya. Talc occurs as irregular patches or pockets in the carbonate host rocks, showing close association with magnesite and restricted with dolomite. Bands, intercalations and pockets of talc also occur at the contact of dolomite and magnesite. Lenses of talc are generally 0.5 to 3 m thick and their strike lenght varies from 45 to 80 m. In the talc deposit of Rema area, host magnesite and dolomite show number of sedimentary features for intertidal environment. Presence of very thin chert bands with micritic dolomite suggests precipitation of silica along with the carbonates. Whereas, the profuse development of chert nodules within magnesite and dolomite, is outcome of diagenesis. Associations of talc with magnesite-silica, dolomite-silica or magnesite-dolomite-silica indicate that the formation of talc is related to the presence of magnesite/ dolomite + silica. Occurrence of the recrystallized grains of magnesite and dolomite adjacent to talc flakes points to their recrystallization during talc formation. The textural features of the talc-magnesite-dolomite- silica assemblage at Rema support the reaction (2) above, between magnesite and silica.

References

Ahmad, T., Harris, N., Bickle, M., Chapman, H., Bunbury, J., and Prince, C., (2000) Isotopic constraints on the structural relationships between the Lesser Himalayan Series and the High Himalayan Crystalline Series, Garwhal Himalaya: Geological Society of America Bulletin, v. 112, pp. 467–477.

Bhattacharya, A.K., Sengupta, H.P. and Mukherjee, A. (1985) Origin of upper Precambrian talc/ soapstone deposit at Chandak, district Pithoragarh, U.P. Current Trends in Geology,

Celerier, J., Harrison, T.M., Webb, A.A.G. and Yin A. (2009) The Kumaun and Garhwal Lesser Himalaya, India: Part 1. Structure and Straitigraphy. Geol. Soc. Amer. Bull., v. 121, pp. 1262-1280.

Gaur, C.S., Bagati, T.N. and Nautiyal, S.P. (1979) Magnesite deposits of Bagoli, Dist. Chamoli, Garhwal Himalaya: A preliminary exploration account. Him. Geol., v.9, pp.744-772.

Krauskopf, F.K. (1963) Factors governing the concentration of thirteen rare metals in seawater. In: The geochemistry of lithogenesis. Collected articles on the geochemistry of rock formation. Foreign Language Publishing House, Moscow, pp.294-338.

Nath, M. and Wakhaloo, G.L. (1962) A note on the magnesite deposits of Almora district, U. P. Indian Minerals, v.16, pp.116-123.

Nautiyal, S. P. (1953) The reconnaissance geological report of a part of the copper belt Kumaun Himalayas, Almora District, U. P. Rec. Geol. Surv. India, v.89, pp.341-358.

Miller, C., Klotzli, U., Frank, W., Thoni, M., and Grasemann, B., 2000, Proterozoic crustal evolution in the NW Himalaya (India) as recorded by circa 1.80 Ga mafic and 1.84 Ga granitic magmatism: Precambrian Research, v. 103, pp. 191–206,

Shandilya, A.K., Gahlot, S.K. and Sriravtave, N. (2009). Geological characteristics and Genesis of the Soapstone deposit around Sarada Inlier, Rajasthan. In: K.L. Shrivastava (Ed.), Economic Mineralization. Scientific Publishers, Jodhpur, pp.228-236.

Sharma, Rajesh, Joshi, Prabha and Pant, P.D. (2008) The role of fluids in the formation of talc deposits of Rema area, Kumaun Lesser Himalaya. Jour. Geol. Soc. India, v. 73, pp.237-248.

Sharma, Rajesh, Joshi, Prabha and Verma, Preeti (2010) Genetic issue of some of the non metallic minerals in Lesser Himalaya. International applied geological congress, Mashad Branch, Iran, pp.1938-40.

Srivastave Neeraj. (2007). Exploration of Soapstone Deposit around Devpura, District Udaipur, Rajasthan. . Unpublised Ph.D. Thesis, Bikaner University, Bikaner pp.106-207.

Srivastava, P., and Mitra, G., 1994, Thrust geometries and deep structure of the outer and lesser Himalaya, Kumaun and Garhwal (India): Implications for evolution of the Himalayan fold-and-thrust belt: Tectonics, v. 13, pp. 89–109,

Valdiya, K.S. (1968) Origin of the magnesite deposits of southern Pithoragarh, Kumaun Himalaya. Econ. Geol., v.63, pp.924-934.

Valdiya, K.S. (1979) An outline of the structural set up of the Kumaun Himalaya. Jour. Geol. Soc. India, v.20, pp.145-157.

Valdiya, K.S. (1980). Geology of Kumaun Lesser Himalaya. Wadia Institute Himalayan Geology, Dehra Dun, 288 p.

Valdiya, K.S. (1980). Geology of Kumaun Lesser Himalaya. Wadia Institute Himalayan Geology, Dehra Dun, 133-34 p.

Valdiya, K. S. (1995) Proterozoic sedimentation and Pan- African geodynamic development in the Himalaya. Precambrian Res., v. 74, pp.

Hon'ble Reviwers

1. **Prof. Arun Shandilya**
   Chairman
   Department of Appl. Geology
   Dr. H.S. University
   SAGAR - 470 003
2. **Prof. Devesh Walia**
   Department of Env. Studies
   North-Eastern Hill University
   Mawkynron
   SHILLONG 793 002

I-14

Gemstone Potential of Pegmatites in Southwestern Part of the Great Mica Belt, Jharkhand

Pankaj K. Srivastava and Surjeet Singh

Department of Geology, University of Jammu. J & K INDIA-180006.
Email: pankajsrivastava.ju@gmail.com

Abstract

India is a homeland of many gemstones. The gemstone mineralization associated with the Great mica belt of Jharkhand is one such region where a number of coloured precious and semi-precious stones have been reported.

The mica and gem minerals bearing zoned and unzoned pegmatite veins of varying dimension and shapes are emplaced in the meta-sediments and granites of the belt. The main precious and semi precious minerals present in the southwestern part of the mica belt are moonstone, bytownite, apatite and almandine. The mineralogical details of these minerals are reported in the present paper in order to characterize the available gemstones in the southwestern part of Great mica belt.

Keywords: Great Mica Belt, Hazaribagh Pegmatite, Gemstones, Garnet, Apatite.

Introduction

The Great Mica Belt, previously known as Bihar Mica Belt, extends to a distance of 160 km, having an average width of 25 kms, from Gurpa in Gaya district (Bihar) in the west through Nawada, Koderma, Hazaribag, Giridih and Jamui (Jharkhand) as well as Bhagalpur (Bihar) in the east. Its maximum width is about 40 km at Koderma-Hazaribagh-Nawada area. The whole belt encompasses roughly around 4,000 sq. km. area and lies on the northern fringe of Chotan-agpur plateau. It runs in an east-northeast to west-southwest direction. The major part of the Great Mica Belt is located in Jharkhand state.

Numbers of pegmatite veins have intruded the mica schist and granites in this belt. Besides the occurrence good quality mica, the belt also hosts a number of precious and semi precious gemstones such as moonstone, indicolite, amazonite, citrine, almandine, cat's eye and bytownite, and some rare metals, viz. beryl, lepidolite, columbite-tantalite, lithium, refractory uranium ore-minerals and REE minerals. The pegmatites intruded in the southwestern part of the Great Mica Belt and exposed in the Kodarma district have moonstone, bytownite, apatite, tourmaline and almandine as main precious and semi precious minerals. Though quarrying of these minerals is going on these areas, no scientific details about the characterization and genesis of these gemstones are available in the published literature. In the present paper an endeavour has been made to mineralogically characterize the available gemstones in the southwestern part of Great mica belt.

Geological Setting

The Great Mica Belt has a geological setting distinct from extensive migmatitic granite-gneiss belts of the Chotanagpur Granitic Gneissic Complex (Mahadevan, 2002). It is characterized by large phaccoliths, often domical, plutons emplaced into relatively more open-folded schistose formations and "supracrustals". The belt trends roughly ENE-WSW.

The metasedimentary rocks are the main lithounits exposed in the area (Fig 1). They are dominantly represented by mica-schist and garnet-sillimanite-mica-schist. They are interbedded with prominent bands of hornblende-schists, micaceous quartzites and relatively minor calc-silicate granulites and local conglomerates. At places amphibolites are also exposed. The mineral assemblage associated with the metamorphites of the GMB has witnessed metamorphic conditions equivalent to an upper amphibolite facies (Bhattacharya,1986). The granitic bodies emplaced into the schistose formations are the next dominant geological formation. The granites vary from equigranular to weekly to strongly foliated varieties. Both the types consist of varying proportions of quartz, microcline perthite, plagioclase (oligoclase-andesine) and subordinate amounts of biotite and green hornblende. Allanite, epidote, fluorite, apatite, zircon, garnet, magnetite, illmenite and tourmaline occur as the main accessories. These granites are rimmed by various migmatitic zones. A number of mica pegmatite veins of different generations have intruded these rocks.

Mahadevan (2002) designated the formations of the Mica belt as a separate stratigraphic unit and classified it as lower Phulwaria Formation, middle Dhab Formation, and the upper most poorly developed Kakolat Formation. Presences of a conglomerate arenaceous sequence separate the Mica Belt from the Chotanagpur Granitic Gneissic Complex (CGGC). The proto-crust or the base of the CGGC has been inferred from the study of metapelitic clasts (K-feldspar- silliminite- muscovite - biotite schist, at Inderwa, Koderma) set in a conglomerate of granodioritic matrix at the base of the Bihar Mica Belt (BMB), a sub- basin of CGGC, showing basement-cover relationship (Mukherjee and Ghose, 1998, Ghose, 2009).

Saha et al (1987), Bhattacharaya (1988) and Ramchandran et al., (1994) have discussed the structural evolution of the Mica Belt. According to these workers three phases of deformation are observed in the GMB. In the first phase two interfering folds, F_{1a} and F_{1b} produced hook shaped pattern of folds. The second phase of folding is more open and has an E-W trending axial plane and plunge to the east at steep angles. This resulted in the development of a prominent axial plane schistosity S_2. This period is marks the synkinaematic anorthosite and granitic rocks (Ramchandran et al, 1994). The third and final phase of deformation (F3), which is relatively weak and less extensive, has produced NNE-SSW to NNW-SSE axial trends in some of the area (Mahadevan, 2002).

Pegmatites of the Southwestern part of Great Mica Belt

Pegmatite veins of varying dimension and shapes are emplaced in the metasediments, amphibolite and granites in the southern part of the belt. They are emplaced along the joints, foliations, bedding planes, slip planes, cores and limbs of folds. These pegmatite veins vary in length from few cm to about 500 m with varying width from a few cm to 5m. For mining purpose the common dimensions of the pegmatites are ranging from 50 to 300 m in length and 1 to 5 m in width. The shape of the pegmatites varies from one pegmatite to other, more commonly they form long narrow bodies thinning out at either end and also at depth. Some of the veins are lenticular in shape having a maximum thickness at central portion. Some of the pegmatites have arcuate to sinuous shapes.

The pegmatites are both zoned and unzoned in nature. In zoned pegmatites the massive white quartz and occasionally microcline perthite forms the core which is surrounded by intermediated zone of blocky perthite. The wall zone generally consists of an intergrowth of quartz, microcline, plagioclase, and muscovite. The border zone is normally very thin in such pegmatites. The zoned pegmatites are mostly economic in nature and contain variety of minerals including gem stones and rare metals. A variety of other minerals besides quartz, feldspar and mica are present in these pegmatites. These minerals include beryl, garnet and tourmaline. Some of the pegamtites from the Great Mica Belt have been reported to contain many rare metals and radioactive minerals like pitchblende, triplite, illmenite, zircon, columbite, tourbernite and graftonite (Tipper, 1919; Murkherjee and Parsad, 1958).

Ramachandran et al. (1994) suggested different age for the pegmatites of the belt. According to him the age of mica pegmatite is 1392-1200 Ma. The ages of different minerals like uraninite, monazite, allanite and lapidolite from different pegmatites of the belt vary from 1050 Ma -950 Ma. (Ramachandran et al., 1994). The formation of mica pegmatites has been suggested to be controlled by lithology as well as structure (Mahadevan and Maithani, 1967).

Mineralogical Characteristics of the Gemstones from the area

Precious and semi-precious varieties of some minerals have been found in the pegmatites of sourthern part of Great Mica Belt. These mainly include moonstone, bytownite, apatite, hossenite and indicolite. Two main localities from where these minerals are reported here are Bandarchua and Lokai-Indarwa in Kodarma district (Shan, 2009).

The mineralogy and gemological characteristics of these gemstones has been studied under various instruments at gemological Institute of India, Mumbai. The physical and optical properties of the gem minerals are summarized in Table 2. The refractive index is calculated using immersion method. The unit cell parameters of the gem minerals are obtained by using X-Ray Diffractometer (X'Calibur Single Crystal XRD with CCD camera). For the run of the samples the generator setting of 50 KV and 30 mA current is used. The MoKα radiations were used for the XRD. The mineralogical characteristics of these gemstones are reported here.

Moonstone

The name Moonstone is not confined to a single species, it is used to describe an optical effect which is likened to the reflected shine produced by moonstone. It is normally characterized by a milky white to bluish shimmering sheen effect that seems to float across the convex surface of the stone as it is turned and moved.

Moonstone's delicate beauty and its long heritage make it perhaps the most familiar gem quality member of the feldspar group. In this belt, this gemstone having export quality is concentrated in Lokai Inderwa area. The moonstone is a variety of feldspar which shows sheen of blue colour, caused by the reflection of light from the internal structure of alternate layers of albite and orthoclase feldspars When light falls between these thin, flat layers, it scatters in many directions producing the phenomenon called adularescence. Though the moonstone can come in different colours, the deep blue, which is present in this area (Fig.2A), is the most popular and thus most expensive.

The moonstone from the area is identified as oligoclase and labradorite. It is translucent colourless variety of the plagioclase which shows cleavages almost at 90 degree with some iron oxide stained surfaces. Under microscope the mineral shows development of crypto-perthite and also show characteristic very fine albite-pericline tartan twinning (Fig.2B). The X-ray data is utilized to calculate the unit cell parameters suggesting a= 8.187 Å, b= 8.698 Å and c= 9.364 Å (Table 3).

Bytownite is other variety of feldspar which is also present in the area. It is mostly concentrated in the intermediate zone of the pegmatite veins of Lokai-Indarwa. It is a variety of plagioclase in which exsolution takes place. It leads to physical optical effects of iridescence and chatoyancy. The mineral in this area shows zoned iridescence (Fig 2C) and hence is expensive. This has a bulk composition of An_{70} to An_{90} and shows huttenlocher intergrowth. XRD of the mineral suggest it to have similar cell dimensions as specified for bytownite.

Apatite

Apatite occurs as euhedral crystals (Fig 2D). It is another gemstone found in the Lokai-Indarwa area. It is found associated with the feldspar in the field. The colour of apatite from this area shows a transparent type and is used as a gem variety. But one thing which was noticed in the field was that the size of the columnar needles was very small, only few places the size was big enough to be used as gemstone.

Under microscope, it shows moderate relief and low birefringence with refractive index varying from 1.63 to 1.64. The blue colour with its typical optical properties and calculated cell parameters (Table-3) confirms this variety as apatite.

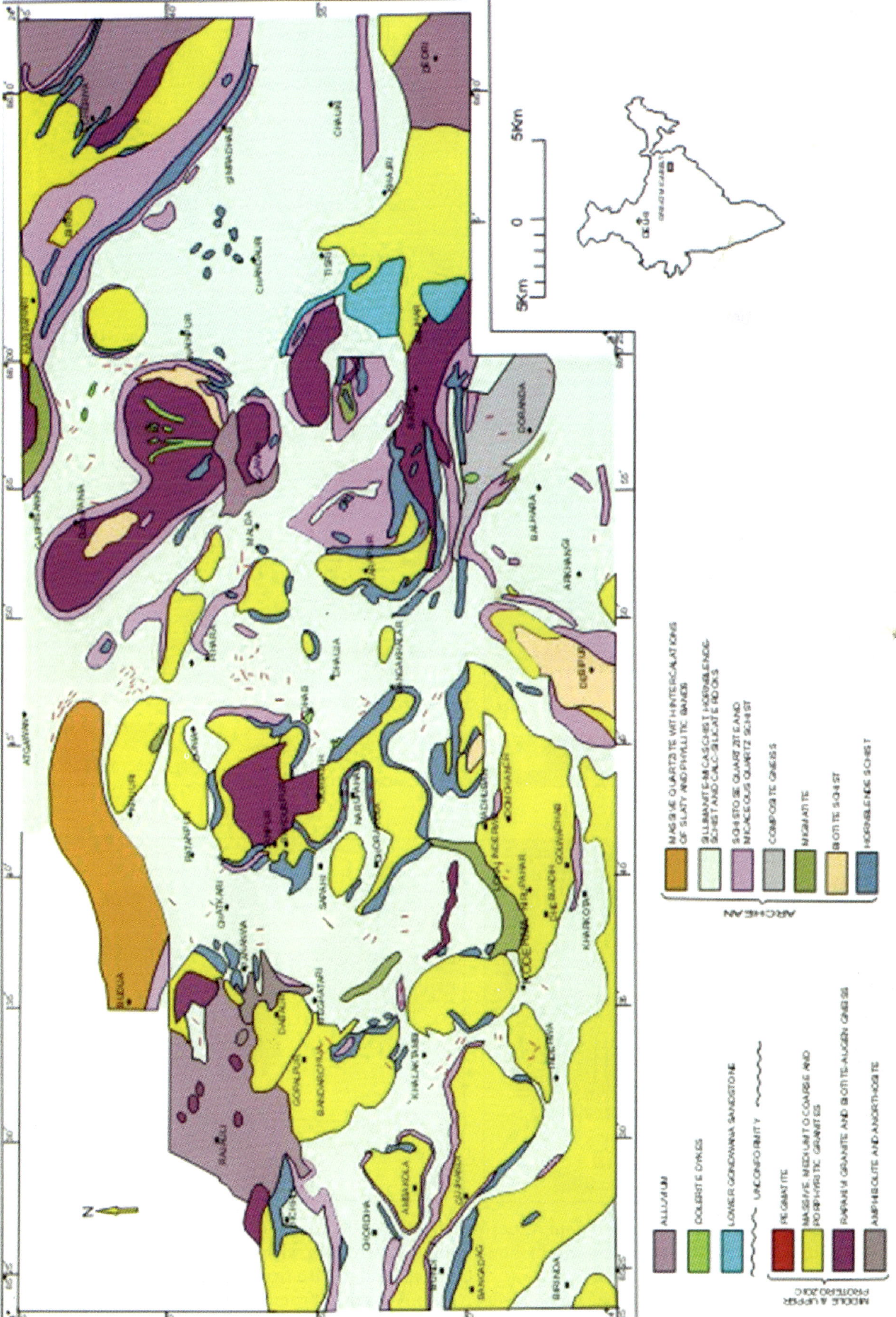

Fig.1. Geological map of parts of Great Mica Belt (after Ramachandran et al 1994).

Fig. 2: Photographs showing different gem minerals from GMB. (A) moonstone showing adularescence sheen of blue color, (B) photomicrograph showing tartan twinning within moonstone; (C) Zoned iridescence bytownite; (D) crystals of apatite with feldspar and biotite from pegmatite veins of Lokai-Indarwa area; (E) pockets of Almandine embedded within mica-schist from Bandarchua area; (F) Crystals of tourmaline separated from pegmatite veins of Bandarchua area.

Table 1: Geological Succession of Great Mica Belt (after Ramachandran et. al. (1998)

AGE	ROCK TYPE
Recent	Alluvium
Permo-carboniferous	Gondwana sediments
	------------------------------Unconformity------------- -----------------------------
	Dolerite dykes
	Rapakivi granite and pegmatites
	Biotite augen series, coarse porphyritic granite gneisses and pegmatites
	Medium grained massive granites and Pegmatites
	Amphibolites and anorthosites
Proterozoic	Massive quartzite with phyllitic and slaty intercalations
	Sillimanite-muscovite schist, calcsilicate & hornblende schist
	Schistose quartzite and quartz-mica-schist
	Hornblende schist, garnetiferous biotite-schist
	Migmatites and composite gneisses
	---------------------------Unconformity--------------------------------------
Archaean	Chotanagpur granite gneiss

Table 2: Mineralogical Charactersitics of different Gemstones

Gemstone	Moonstone	Apatite	Tourmaline	Garnet	Bytownite
Crystal System	Triclinic	Hexagonal	Trigonal	Cubic	Monoclinic
Colour	Bluish adularescence of variable intensity	Bluish - green	Black	Dark reddish	Bluish
Habit	Tabular	Columnar	Columnar	Dodecahedron	Tabular
Cleavage	Perfect (Two set)	Imperfect	Perfect	Poor	Perfect (Two set)
Fracture	Uneven	Conchoidal	Uneven	Conchoidal	Conchoidal
Hardness	6	5	7.3	7.5	6.4
Optic nature	Biaxial (-ve)	Uniaxial (-ve)	Uniaxial (-ve)	-	Biaxial (+/-ve)
Refractive Index	1.550 to 1.551	1.63 to 1.64	1.625 to 0.035	1.755 to 1.756	1.565 to 1.585
Birefringence	0.008	0.003	0.022-0.040	-	0.010-0.011
Specific Gravity	2.61	3.21	3.10-3.25	4.146	2.74-2.76
Lusture	Vitreous	Vitreous	Vitreous	Viterous	Viterous

Table 3: Unit Cell Dimensions for Gem Minerals from SW GMB

Gemstone	Origin	Unit-Cell dimensions						
		a	b	c	α	β	γ	V (in $Å^3$)
Moonstone	Pegmatite	8.187 Å	8.698 Å	9.364 Å	89.81°	82.76°	82.76°	658.42
Garnet	Mica- schist	11.557 Å	11.581 Å	11.549 Å	90.16°	89.64°	89.88°	1545.68
Apatite	Pegmatite	9.334 Å	9.371 Å	6.890 Å	90.12°	89.80°	119.81°	522.9
Tourmaline	Pegmatite	10.423 Å	16.042 Å	13.272 Å	90.20°	106.51°	89.67°	2127.63
Bytownite	Pegmatite	8.17 Å	12.85 Å	13.16 Å	93.5°	116.016°	90.833°	1239.15

Tourmaline

Tourmaline is present at Bandarchua and is black in colour (Fig. 2E). This variety is magnesium rich. Black opaque stones are also often cut and worn as jewellery. At places green and blue tourmaline (Indicolite) are also observed. Under the microscope, tourmaline does not show any zoning which makes it lower in the gem quality. The XRD suggest that the cell parameters of tourmaline is a=10.423 Å, b= 16.042 Å, c= 13.272 Å (Table 3).

Almandine

Almandine Stone is a variety of Pyralspite series of garnet group. Pyralspite normally forms as trapezohedrons, but in case of almandine, it is noticed that rhombic dodecahedral form is predominant. Almandine is mainly found associated with schist at Banderchua area (Fig.2F)

Almandine the commonest mineral in the garnet group is typically occurring in schists and gneisses, formed due to regional metamorphism of largely clay rich sediments and rocks. (Jackson B. 2006).

Conclusion

The southwestern part of Great Mica Belt is endowed with a gem quality feldspar mineralogically represented by moonstone and bytownite; Apatite and garnet (almandine). The microscopy and XRD details of these minerals confirm their gemological quality. This proves beyond doubt the gemological potentiality of the pegmatite from the area. A detailed exploration for the gemstones is recommended for the area.

REFRENCES

Bhattacharyya, B.P., 1986. Lithological characteristics of the metasedimentary country rocks as guides to mica exploration in Bihar mica belt. Geol. Surv. Ind Spec. Publi. No 18, pp 183-196.

Bhattacharyya, B.P., 1988. Sequence of deformation, metamorphism and igneous intrusions in Bihar mica belt. Mem. Geol. Soc. India. No. 8, pp.113-126.

Ghose, N.C. 2009. Discussion on Rb-Sr geochronology and petrogenesis of granitoids from the Chhotanagpur granite gneiss complex at Raikera-Kunkuri region, Central India. Jour. Geol. Soc. India, v.74, pp. 754-756.

Jackson, B. 2006. Gems, Their, sources, description and identification. Sixth Edition, 2006 pp 195-237.

Mahadevan, T. M. and Maithani, J. P. B. 1967. Geology and petrology of mica pegmatites in parts of Bihar mica belt. Mem. Geol. Surv. Ind., Vol. 93, pp 1-114.

Mahadevan, T. M., 2002. Geology of Bihar and Jharkhand. Geol. Surv. Ind., Bangalore, 291 p.

Mukherjee, B and Prasad, U., 1958. X-ray study of rare phosphate minerals. Nature, Vol. 182, pp. 472-473

Mukherjee, D. and Ghose, N.C. (1998) Conglomerate at the base of Bihar Mica Belt metasediments, Koderma district, Bihar and its stratigraphic significance. National Seminar on Advancement in Geological Sciences in Bihar, Geol. Surv. India, Patna, pp.15-16.

Ramachandran, S., Srivastava, P. K., Raminaidu, Ch. and Nateshwara Rao, B. 1994. Geological, structural, and geochemical studies in the Bihar mica belt, eastern India, with special reference to mode of emplacement of rare metal and other pegmatites. Explo. and Res. Atomic Minerals. Vol. 7 pp, 77-95.

Saha, A.K., Sarkar, S.S and Rej, S.S. 1987. Petrochemical evolution of the Bihar mica belt granites, eastern India. Indian Jour. Earth Sci., Vol. 14 (1), pp. 22-45.

Shan, S. S., 2009. Fluid inclusion study of gemstone bearing pegmatites from Bandarchua and Lokai-Indarwa areas of Great Mica Belt of Jharkhand (Unpublished M. Phil dissertation). University of Jammu. 66 p.

Tipper, G. H., 1919. Pitchblende, monazite and other minerals from Pichhli, Gaya district, Bihar and Orissa. Rec. Geol. Surv. Ind., Vol. 50, pt. 4 pp. 255-262.

Hon'ble Reviewer

1. **Prof. V.V. Nageshvara Rao**
   Department of Geology
   Andhra University
   VISAKHAPATNAM - 530 003
2. **Dr. Devesh Khandelwal**
   Department of Geology
   Dungar College,
   MGS University
   BIKANER - 334 003

I-15

Geochronological Studies in the Eastern Ghats Mobile Belt, India

D. Prakash, Deepak and P. Chandra Singh

Centre of Advanced study in Geology, Banaras Hindu University, Varanasi - 221 005
E-mail: snlal2010@gmail.com

Abstract

The Eastern Ghats Mobile Belt (EGMB) of India displays number of fault-bounded blocks in its northern margin whereas western margin is marked by a shear zone. Recent studies suggest that EGMB consists of several smaller blocks each having their own characteristics in respect of structural and metamorphic evolution. The isotopic data on the EGMB rocks are scanty and the data also lack proper correlation with tectonic and thermal events. A review of numerous isotopic-geochronological studies is given in the present paper.

Keywords: Geochronology, Eastern Ghat Mobile Belt, High Grade Metamorphism, Granitization.

Introduction

EGMB stretches between 15°30' and 20°51' N latitude and 80° to 85° E longitude occurring all along the east coast of India. EGMB covers a length of about 900 km starting from Ongole, with a minimum width of 30 km gradually increasing towards north to attain maximum width of 200 km in Ganyam area of Cuttack district and finally diffusing around Angul (Fig. 1). EGMB, on the Indian subcontinent occupying a key position and having therefore received considerable attention from petrologists over the last decade (Sarkar and Paul, 1998; Dasgupta and Sengupta, 2003; Lal, 2003; Dasgupta et al. 2008, Mukhopadhyay and Basak, 2009 and Dasgupta et al., 2012).

The old name of EGMB is Mahendragiri named after the peak of Mahendragiri (1501 m) which stands conspicuously above the coastal plain, about 80 km SW of Birhampur in Orissa. The average elevation of EGMB is about 1,100 m. This mountainous tract creates some important water shed area from which several small rivers originate. There are a few caverns in the lime-stone formations, such as Borra caves near Visakhapatnam and Gupteswar on western part of the EGMB.

The EGMB is typical Indian example of high-grade mobile belt besides other classical mobile belts of the world such as Limpopo Belt and Grenville Belt. The mobile belts, defined as "younger linear metamorphic belts, are charac-terized by high-grade metamorphism, granitis-ation and often with transcurrent dislocation which tend to surround the cratonic nuclei of shield area (Anhaeusser et al., 1969)." This definition gained wide acceptance but some aspects are still disputed. Later on Windley (1973) recognized two types of mobile belts such as high-grade and low-grade. The high-grade mobile belts correspond to the definition of Anhaeusser et al, (1969) whereas low-grade mobile belts contain volcanic and sediments of the type of Labrador trough and Wopmay orogen. According to Sarkar and Paul (1998) the special features, which suggest the EGMB to be mobile belt, are:

(a) Curvilinear deposition
(b) High-grade milieu (granulite-amphibolite facies)
(c) Contrasting tectonometamorphic entity in relation to adjacent cratonic zones
(d) Protracted tectonothermal history
(e) Presence of significant volumes of supracrustal rocks across the entire orogeny
(f) Evidence of extensive granitisation and abundance of migmatitic/grantitic rocks
(g) Abundant plutonic igneous activity covering a wide array of magma types

The rocks of the EGMB are highly deformed and metamorphosed crustal segment of Indian Precambrian terrain. The shape of EGMB is arcuate and separated from Eastern Dharwar and Bastar Craton in the West by eastward dipping NNE-SSW trending thrust. This thrust has been confirmed by the deep seismic studies. EGMB thrust was apparently active in Middle to Late Proterozoic times and is called as Eastern Ghats Front. This front is marked by the discontinuous array of sedimentary basins (Kale & Phansalkar, 1991). The marginal parts of these basins with mobile belt are deformed into folded belts.

EGMB is mostly composed of charnockite and khondalite. Khondalite are metamorphosed sedimentary rocks giving rise to smooth, rounded hills, which are capped with alluvium. Charnockite are associated with granitic intrusions. The Visakhapatnam hills and other surrounding hills are true mountain ranges of Alpine type which were formed during Achaean / Early Proterozoic and uplifted in Late Proterozoic. The metamorphic rocks of EGMB display gradual increase of metamorphism from Green schist to Granulite facies from SW to NE in the belt.

STRATIGRAPHY: PREVIOUS WORK

The stratigraphy of EGMB has been extensively studied by various workers from time to time. Earlier workers such as Blanford (1858), Foote (1870, 1880, 1883a, b) and King (1886) have done preliminary work in eastern part of the Indian shield. King (886) coined the terms Bezwada gneiss and Visakhapatnam gneiss for garnetiferous gneisses of Vijayawada and Visakhapatnam granulites respectively. Holland (1990) first coined the term charnockite for hypersthene-granite of Chennai and charnockite series for differentiated igneous suite of hypersthene-bearing rocks.

With the advent of 20th century, Walker (1902) coined the term khondalite for garnet-sillimanite gneiss with or without graphite and gave the following stratigraphy:

Archaean {
- Khondalite and Associated Rocks
- Charnockite Series
- Granitoid Series
- Crystalline Complexes

Later on Fermor (1936) divided the Indian shield broadly into two divisions.

Indian Shield {
- Charnock ite Region
- Non Charnock ite Region

After one decade, Krishnan (1950) and Pascoe (1950) proposed the following stratigraphy for Archaeans of EGMB :

Igneous {
- Charnock ite Series
- Gneissose Granite

Sedimentary {
- Calc Gneisses
- Khondalite

Summarizing the work of GSI geologists and other workers in fifties and sixties, Murthy et al., (1971) proposed the following stratigraphy.

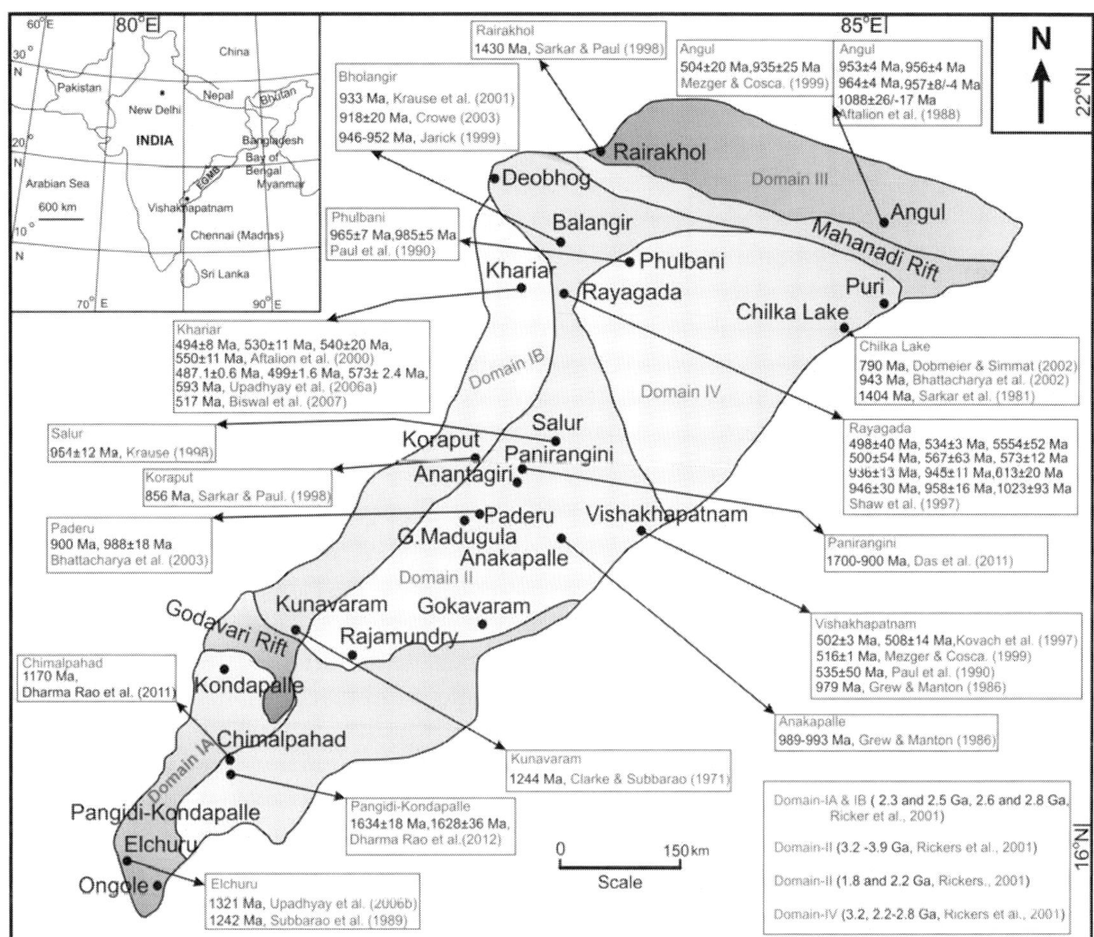

Figure 1: Geochronological ages for various events in different regions of EGMB

	Anorthosites and gabbroic Intrusives	Anorthosite, Pink and grey granite, porphyritic granulites and gneisses, Augen gneiss
	Migmatite Formation	Migmatite, Garnetiferous gneisses and leptynite
Eastern Ghats Group	Charnockite Formation	Acid, intermediate, ultrabasic and basic-charnockite, Meta-anorthosite and eclogites
	Khondalite Formation	Garnet-sllimanite-K-feldspar gneisses and schists (Khondalite), Garnetiferous quartzite, Calc silicate rock, cordierite gneiss and Magnetite-quartz-hypersthene-granite rocks
Sialic Basement not exposed		

Later on Benerji (1990) synthesized the available information from systematic geological mapping by GSI and other workers and proposed the following stratigraphy.

Late Proterozoic 900-500 Ma	Porphyritic gneiss, Granite and Dolerite
Middle Proterozoic 1600-900 Ma	Anorthosite, Quartz-magnetite granulite, Nepheline-syenite, Gabbro and Norite
Archaean to Early Proterozoic	Migmatite and biotite gneisses, Charnockite suite, Khondalite Group

The chronostratigraphy of the EGMB, using updated radiometric data (Table 1), may be summarized below:

Intrusives	535 Granite, Vishakhapatnam 600 K-Ar Alkaline gabbro	Pan-African metamorphism amphibolite-granulite facies 500-550 Ma- M-III
	956 Granite, Angul, Orissa	Granulite facies metamorphic event, 1000 Ma –M-II

Age	Lithology	Event
	1265 Alkaline igneous complex Kunavaram	
	1400 Anothosite, Chilka Lake	
	CHARNOCKITE ASSOCIATION	
	This includes charnockite and enderbite (± garnet), basic granulite (±garnet), ultrabasic granulite, meta-anorthosite And meta-layered igneous complex.	
>2600 – 1600 Ma	**KHONDALITE ASSOCIATION**	
	Khondalite (sillimanite), garnet-leptynite, quartzite, calc-silicate rocks, cordierite-sapphirine-bearing granulite and banded magnetite-quartz-hypersthene-garnet rocks (meta-BIF). These represent shallow water self facies, orthoquartizite-shale-carbonate evaporate association.	Granulite facies metamorphic event – MI Early Proterozoic/ Late Achaean?

STRUCTURE

The regional tends of EGMB are NE-SW and NNE-SSW direction (Murthy et al., 1971; Narayanaswami, 1975). Greenschists and Amphibolite facies rocks of southern part of EGMB are characterized by swerving of foliation tend from NNW-SSE/NW-SE to NNE-SSW/NE-SW. This swerving of foliation trend is found in Nellore Schist Belt from S to NE direction (Vasudevan & Rao, 1975). Sriramdas & Rao (1979) and Natrajan & Nanda (1981) suggested two phases of deformation in Visakhapatnam. The strike of foliation is NE-SW which dips towards SE and F_1 folds are isoclinal plunging NE and overturned to NW in the North of Visakhapatnam. This trend gradually changes to E-W to WNW-ESE with steep southerly dip which indicate a major isoclinal fold overturned to North. According to Sriramdas and Rao (1979), this major F_1 fold is overturning from NE-SW to E-W due to superimposed open F2 cross folds along the NW SE or NNW-SSE axis with southerly or south easterly plunge.

EGMB bears imprints of multiphase deformations. One school of thought advocates for three major episodes of folding and metamorphic fabrics (Sarkar et al., 1981; Halden et al., 1982; Bhattacharya et al., 1994; Chetty & Murthy, 1994), whereas the other school argues for four phases of deformation in which first three are dominated by folding and last one by shearing and fracturing (Natrajan & Nanda, 1981; Sarkar et al., 1981; Halden et al., 1982; Park & Dash, 1984; Sarkar & Paul, 1998). The first episode of folding is represented by rootless folds of bedding in metamorphites, with a strong and pervasive metamorphic foliation developed parallel to the axial planes of these folds. Major folds are indicated by hook – shaped folds on a map scale produced by coaxial refolding of earlier fold by later folding but this feature is not commonly present (Bhattacharya et at., 1994; Chetty & Murthy, 1994; Bhattacharya, 1996). The present attitudes of these minor earlier folds (F_1) are reclined. The most pervasive foliation, a gneissic foliation in the khondalite, corresponds to earlier (F_1) fold axial traces, which is broadly syntectonic with granulite facies metamorphism. Rootless and intrafolial earlier folds are also observed in the bands of charnockite and mafic granulite (Bhattacharya et al., 1994; Bhatta-charya, 1996).

Second episode of deformation is represented by commonly overturned to isoclinal folds on a regional scale. Axial planes of F_2 folds are parallel to F_1 axial planes. The S_2 foliation, axial planar to F_2 folds, is developed in most of the lithologies. Several small-scale shear zones marked by quartzite, mylonite and pseudotachylite have developed parallel to S_2. Steep plunge of S_2 is specially associated with the shear zones and their down dip stretching lineation. Thus, the steep plunge of F_2 folds is a consequence of rotation of fold axes by progressive simple shear (Bhattacharya, 1989).

The third episode of deformation is represented by open and upright folds with sub horizontal axes. The attitude of these folds is highly diverse. Diversity of F_3 axial plane foliation, particularly in Eastern Sector (Bhattacharya, 1994), is evidently related to the melt-dominated rheology of migmatites. A few large-scale strike-slip shear zones could be recognized from the lineament maps (Chetty & Murthy, 1994; Bhattacharya, 1996) and arc correlated with the third episode of folding and associated S_3 foliation (Bhattacharya, 1996).

Brittle ductile to brittle conjugate shear zones trending WNW and NNE represent manifestation of the last tectonic (fourth deformation) in which shearing and fracturing dominated (Sarkar et al., 1981; Halden et al., 1982; and Sarkar et al., 1989). Other workers demonstrated that the shear foliation related to thrusting/shearing transects the earlier S_1 and S_2 gneissic fabrics (Bhadra et al. 2004; Gupta et al. 2000; Das et al. 2008). The main structural element in the rocks of EGMB is a planar fabric (gneissosity) defined by the alignment of the platy minerals like flattened quartz, garnet, sillimanite, graphite etc. The parallelism between the foliation and lithological layering is related to isoclinal folding.

The field relations in the EGMB point to intense deformation of the terrain, apparently before, during and after metamorphism. This is grouped with close intermingling of granulite and khondalite suites.

GEOPHYSICAL ANOMALIES

The EGMB is made-up of metamorphosed rocks and constitute major Precambrian belt of Peninsular India. These rocks retain the signature of all major tectonic and metamorphic events.

The charnockite and Iron Ore Series of EGMB Group are magnetic in nature which has been found by the magnetic survey. These surveys were conducted with a vertical magnetometer of torsion type and help to bore out the subsurface geology and structure. These surveys show a highly faulted and shallow basement of the charnockite rocks.

The palaeomagnetic data of EGMB shows that the Virtual Geomagnetic Pole (VGP) position of charnockite are similar in different parts of EGMB and this VGP position also corresponds to the charnockite of Southern granulite Terrain (Lakhmipatiraju, 1998). The similarity of VGP position in charnockite shows that khondalite-charnockite suites magnetized during the earlier episode of regional metamorphism around 2600 Ma (Perraju et al., 1979). The magnetic activity in Cuddapah basin probably influenced the EGMB formation and remagnetized the iron ore. The tectonothermal events of this area are dated around 900 Ma (Paul & Sarkar, 1994). The charnockite are probably remobilized towards the end of EGMB orogeny. The palaeomagnetic data from EGMB provide evidence for the continual magmatic activity. The tentative magneto-stratigraphy given by Lakhimipatiraju (1998) is as follows:

AGE (Ma)	Events
900	Remagnetisation of iron ore due to the influence of magmatic activity in the Cuddapah basin
1000	Remobilization of charnockites emplacement and remobilization of iron ore
2600	Magnetism of the charnockite-khondalite of entire EGMB

The gravity evidence of EGMB is totally different from the eastern margin of Cuddapah basin and it exhibits a strong positive character. The gravity discontinuity and -40 mgal contour, characteristic of the western boundary of granulite belt of EGMB, continuous (with a few interruptions) up to the southern tip of the India and shows a smooth bend NW to merge with the Western Ghats granulite belt (Subrahmanyam & Verma, 1986). The NE-SW trend of EGMB also follows the gravity contour (Drury & Holt, 1980). The gravity signature of the EGMB boundary is positive and negative anomaly pair. This bimodality in gravity field mainly arises when two crustal blocks of normal and higher density get juxtaposed against each other during anomalies across structural boundaries as in Canadian Shield.

A gravity survey (Subramanyam & Verma, 1986) across this boundary shows a characteristic high gravity anomaly from the schist belt of the Bastar Craton. This gravity contrast may simply indicate high density crust and /or thickened crust underneath the EGMB (Subbra Rao, 1994). The high gravity over the khondalites and low gravity over the migmatites is in conformity with the geological observations by Nanda and Pati (1993).

The crescent shape of high gravity anomaly zone appears to be similar to the shapes of rift units described by Rosendahl (1987). These units are generally bounded by listric faults and movement along these faults may result in exhumation of the deeper layers of the crust (Lister & Davies, 1988). The mechanism may explain the presence of high-grade metamorphic rocks on ground surface in EGMB. Recently Kumar et al. (2007) and Ramesh et al. (2010) obtained geophysical data across the EGMB suggest that westerly dipping seismic high structures at a depth of 160–200 km beneath the Dharwar–Bastar Craton.

Table 1: Radiometric dating of rocks of the granulite facies terrain of Eastern Ghats Mobile Belt

S. No.	Area	Method	Rock type	Radiometric dates (Ma)	Author (References)
			(I). Charnockites, Khondalites, & Gneisses		
1.	Vishakha-patnam	(i) Sm-Nd, Model age, whole rock (ii) Rb/Sr-whole rock (iii) U-Pb date of zircon and sphene	(i) Basic granulites, Charnockites (ii) Garnet-biotite hypershthene gneiss, Khondalites (iii) Zircon from apatite magnetite veins	(i) 2860, 2600 & 2350 (ii) 2695, 2482, 2129 (iii) 516 ± 1	(i) Paul et al. (1990) (ii) Perraju et al. (1979) (iii) Mezger & Cosca (1999)
2.	Vishakha-patnam	(i) U-Pb (ii) U-Pb age in zircon	(i) Allanite from charnockite (ii) Charnockite-Zircon	(i) 1900 & 2000 (ii) 979	(i) Rao and Babu (1978) (ii) Grew & Manton (1986)
3.	Vishakha-patnam	Pb-Pb whole rock	Charnockite & basic granulites	1176	Paul et al. (1990)
4.	Puri	(i) U-Pb whole rock (ii) Rb-Sr whole rock (iii) Rb/Sr	(i) Detrital monazite from khondalite (ii) Khondalites (iii) Garnet leptynite	(i) 2600 (ii) 3090 (iii) 1211	(i) Vinagradov et al. (1964) (ii) & (iii) Perraju et al. (1979)
5.	Anakapalle	U-Th-Pb	Allanite in charnockite	1585	Aswathanarayan (1956)
6.	Anakapalle	(i) U-Pb, & Th-Pb (ii) U-Pb, lower intercept on Concordia	(i) Perrierite from sapphirine granulite (ii) Zirocn from charnockites	(i) 989 & 993 (ii) 979 & 500	Grew & Manton (1986)
7.	Angul	(i) Pb-Pb concordia-discocordia plot & Rb/Sr (ii) & (iii) U-Pb concordia-discocordia plot	(i) Zircon from augen gneiss (ii) Zircon from leptynites (iii) Monazite from Augen gneiss leptynites & grey gneiss	(i) 1111-1094 & 1080 (ii) 1114-1088 &1037 (iii) 973-953	Aftalion et al. (1988)
8.	Angul	(i) U-Pb Concordia lower intercept (ii) U-Pb Concortdia upper intercept	Calc silicate rock-Titanite	(i) 504 ± 20 (ii) 935 ± 25	Mezger & Cosca (1999)
9.	Chilka Lake	(i) Rb-Sr whole rock (ii) Pb-Pb zircon	(i) Xenoliths of khondalite in anothosite (ii) Patchy Charnockite	(i) 1388±22 (ii) 943	(i) Sarkar et al. (1981) (ii) Bhattacharya et al. (2002)
10.	Phulbani	U-Pb concordia-discocordia plot	Zircon & Monazite from Charnockite	985-965	Paul et al. (1990)
11.	Rayagada	(i) Rb-Sr whole rock isochron (ii) Sm-Nd whole rock isochron	Charnockite, Enderbites	(i) 958 ± 16 (ii) 1023 ± 93	Shaw et al. (1997)
12.	Kondapalle	K-Ar	Biotite from Khondalite	500	Grew & Manton (1986)
13.	Srikakulam	R/Sr	Garnet biotite gneiss inclusion in garnet gneiss	1369-1354	Perraju et al. (1979)
14.	Godawari Graben (South)	Sm-Nd Model ages	(i) Orthogneisses (ii) Metasediments	(i) 2300 & 2500 (ii) 2600 & 2800	Rickers et al. (2001)
15.	Panirangini	U-Pb zircon SHRIMP II	Monazite grains from orthopyroxene & sapphirine porphyroblsts	1700- 900	Das et al. (2011)
16.	Paderu	$^{207}Pb/^{206}Pb$ concordant U-Pb upper intercept	Sapphirine granulite-Monazite	900 988 ± 18	Bhattacharya et al. (2003)

			(II). Granite		
17.	Vishakha-patnam	(i) Sm/Nd model ages (ii) Sm/Nd	(i) Intrusive granites (ii) Garnet & Porphyritic granite	(i) 2100 (ii) 535	(i) & (ii) Paul et al. (1990)
18.	Angul	(i) Pb-Pb Concordia-discordia plot (ii) Rb/ Sr	(i) Monazite from pink granite (ii) Muscovite from late pegmatites	(i) 965-956 (ii) 854	(i) Aftalion et al. (1984) (ii) Halden et al. (1982)
19.	Nellore	(i) Rb/Sr W.R. (ii) U-Pb Zircon	(i) Biotite & Intrusive granites (ii) Ophiolite Emplacement	(i) 1615-999 (ii) 1330	(i) Gupta et al. (1984) (ii) Dharma Rao et al. (2011a)
20.	Srikakulam	Rb/Sr isochron	Garnet gneiss, garnet biotite gneiss xenoliths & porphyritic granite	1000 & 816 ±40	Perraju et al. (1979)
21.	Khariar Alkalinc complex	$^{207}Pb/^{206}Pb$ age of sphene	Nepheline Syenite- Titanite	593	Upadhyay et al. (2006a)
22.	Angole Domain	Monazite U-Pb-Th	Granite emplacement	1650 and 1550	Simmat & Raith (2008)
			(III). Anothosite, Alkaline Igneous Complex & Norite		
23.	Chilka lake	(i) Rb/Sr Isochron whole rock (ii) K-Ar	(i) Anorthosite (ii) Plagioclase from norite dyke into anorthosite	(i) 1400±89 (ii) 848±16	Sarkar et al. (1981)
24.	Kunnavaram	Rb-Sr whole rock isochron	Alkaline igneous pluton	1244	Clarke & Subbarao (1971)
25.	Chilka Lake	Th-U-Pb EPMA dating of monazite	Emplacement of Anorthosite	790	Dobmeier & Simmat (2002)
26.	Elchuru	SHRIMP, TIMS, U-Pb date of zircon	Alkaline pluton emplacement	1321	Upadhyay et al. (2006b)
27.	Khariar Alkaline complex	SHRIMP U-Pb on zircon	Alkaline igneous pluton	517	Biswal et al. (2007)
28.	Chimalpahad	Sm-Nd whole rock	Anorthosite Emplacement	1170	Dharma Rao et al. (2011b)
29.	Pangidi–Kondapalle Complex	(i) $^{207}Pb/^{206}Pb$ zircon ages for anorthosite of Pangidi (ii) $^{207}Pb/^{206}Pb$ zircon ages for anorthosite of Kondapalle	Anorthosites and Gabbronorites emplacement	(i) 1634 ± 18 (ii) 1628± 36 & 1693± 110	Dharma Rao et al. (2012)

Metamorphic Facies

The metamorphic rocks of EGMB range from green schist to granulite facies. Northern part of the belt is composed of granulite facies whereas southern portion, Nellore schist belt, is dominated by low-grade rocks (green schist facies). The low-grade rocks are chlorite schist, phyllite, quartzite, metavolcanics (Vasudevan & Rao, 1975). Nellore part of Nellore schist belt is characterized by amphibolite facies rocks which are exposed near Khammam.

Vasudevan and Rao (1975) found oligomictic conglomerate or fault separating the green schist and amphibolite facies rocks. A sharp break in metamorphic grade presumably relates to an unconformity at the base of younger green schists facies of Nellore schist belt.

The granulite facies rocks from EGMB are further divided into Khondalite Formation, Charnockite Formation and migmatite rocks (Murthy et al., 1971). These charnockite and mafic granulite recorded garnet-pyroxene-plagioclase-quartz assemblage showing high pressure and high temperature from few places (Lal et al., 1987; Mohan et al., 1997). P-T estimates show both isobaric cooling and isothermal decompression (Sengupta et al., 1990; Dasgupta et al., 1991, Sen et al., 1995; Mohan, 1995). Khondalites and Mg-A1 rich granulites present in this terrain show sapphirine-kornerupine-sillimanite-orthopyroxene assemblage (Grew, 1982; Narayan et al., 1994; Dasgupta, 1995; Divakar Rao et al., 1998; Charan et al., 1998) suggesting the P-T conditions around 800-850°C and 6-10 kbar. The near peak P-T condition (inferred) from Mg-A1 rich granulites using geothermobarometer and petrogenetic grid constraints reflect ultra-high temperature metamorphic (c. 950°C, 9 kbar) in the parts of EGMB.

GEOCHRONOLOGY

Based on the field observations and petrologic studies in the EGMB, three major deformational and metamorphic events have been identified by different workers. During the earliest granulite facies metamorphism (MI), the supracrustals were deformed under ultra-high temperature metamorphism as evidenced by sapphirine – spinel – quartz bearing paragenesis found in various domains within the EGMB. Later, the EGMB intruded by voluminous enderbitic charnockites that underwent deformation again under granulite facies metamorphism (MII). This is a dominant regional – scale granulite facies tectonothermal event (MII) widely recognized in the EGMB. Subsequent to MII, vast granitic and the anorthosite intrusion occurred, showing signatures of a metamorphic overprint (MIII) and related deformation exhibiting tentional fissures. The high precision geochronological data on EGMB are quite scarce.

More recently Mezger and Cosca (1999) have identified two segments in EGMB separated by the Godavari Rift. The northern segment is dominated by 1000 and 500 Ma events, whereas the southern segment indicates a major thermal event around 1600 Ma.

The mineral ages reported by Mezger and Cosca (1999) occurred at ca. 960 Ma. Several workers have also shown that the last major granulite facies episode (MII) in the northern segment of EGMB occurred at ca. 960 Ma – 1000 Ma (Grew & Manton, 1986; Aftalion et al., 1988; Paul et al., 1990; Mezger et al., 1996; Shaw et al., 1997). This granulite facies event (MII) that occurred late during global Grenvillian orogency ca. 1000 Ma in the northern sector is characterized by steeply decompressive retrograde P-T path (Grew & Manton, 1986; Sengupta et al., 1999; Bose et al., 2000 and the references therein). Thus MII was responsible for substantial uplift and exhumation of the deep-seated granulites at ca. 960 Ma – 1000 Ma.

The late Grenvillian orogenic event is coeval in the EGMB and Eastern Antarctica. In this way, MII metamorphism in EGMB has strikingly similar tectonothermal history with the Rayner complex exposed along the Mc Robertson Land and Kemp coast of East Antarctica.

However, the timing of earlier ultra-high temperature metamorphism (MI) preserved locally in Mg-Al rich granulites remains undermined. Mezger and Cosca (1999) have argued that this UHT metamorphism could either be a separate thermal event prior to the Grenvillian orogeny ca. 960 Ma or it could also be an early interval of this same tectonothermal event. It is quite possible that the UHT metamorphism corresponds to much older metamorphic age.

It is quite possible that this UHT metamorphism is at least Mid- Proterozoic or even an older event. The absence of older zircon ages may also be due to the general lack of U-Pb zircon date for most parts of the EGMB. The U-Pb cooling ages of 1672± 4 M (monazite) and ~ 1600 Ma (allanite) reported recently by Mezger & Cosca (1999) provide the first unequivocal documentation of Pre-Grenvillian granulite facies metamorphism in the EGMB. Shaw et al. (1997) have identified a 1400 Ma event through Sm-Nd systematics from Rayagada in EGMB.

Some older ages such as 3100 Ma (Rb-Sr W.R. in khondalite from Puri, Vinogradov et al., 1964), 2900 Ma (Sm-Nd W.R. in charnockites and basic granulites, Paul et al., 1990), 2580 Ma – 2600 Ma (U-Pb zircon, Vinogradov et al., 1964), 2600 Ma (Rb-Sr W.R. in charnockite, Paul et al., 1990) and 1858 Ma-2258 Ma (Pb-alpha in allanite, Rao et al., 1980) have been reported in literature from different area of EGMB. These ages indicate that orogenic belt in EGMB contain material of Achaean age that was reworked and mixed with more juvenile material during Proterozoic orogenic cycles (cf. Mezger & Cosca, 1999). Hence a further characterization of these ages becomes necessary.

It is now clear in recent years that the last major thermal overprint of amphibolite facies occurred during Pan-African orogeny at ca. 500 Ma-550 Ma that led to resetting of Ar-Ar systematics in hornblende, resetting of U-Pb systematics in allanite and sphene and neocrystallization of zircon in pegmatites (Mezger et al., 1996; Mezger & Cosca, 1999; Shaw et al., 1997; Kovach et al., 1997). So far there is no evidence in favour of widespread granulite facies metamorphism during this period, as opposed to that recorded in the Prydz Bay (Hensen & Zhou, 1995). This is also the time for the major granulite facies event in southern Indian granulite belt (c.f. Hariss et al., 1994; Unnikrishnan-Warrier et al., 1995; Mohan & Jaynanda, 1999) and Sri Lanka (e.g. Kroner & Williams, 1993; Kriegsman, 1994). Both in southern Indian granulite belt and Sri Lanka there is sufficient supporting evidence for a major orogeny that led to the regional-scale high-grade metamorphism. The lower P-T conditions extracted regional-scale high-grade metamorphism. The lower P-T conditions extracted from Mg-Al rocks may have been achieved during this Pan-African event.

A new Sm-Nd garnet W.R. internal isochron age (546 Ma) for garnetiferous leucosome in charnockite from Kantakapalli (Osanai et al., 1999) and the data presented by Mezger & Cosca (1999) provide strong evidence for a thermal overprint in EGMB at ca. 500 Ma – 550 Ma (MII). Besides these dates, Shaw et al. (1997) also gave zircon SHRIMP age of ca. 2747 Ma from metapelites in the central part of EGMB while Mezger & Cosca (1999) reported 863 Ma discordant U-Pb age of a titanite for the Rajamundry area. Significance of these haphazard dates is still not well understood.

It may be stressed here that for a reliable correlation of the currently available geochronological information with these three tectonothermal events which have influenced EGMB, modern systematic isotopic dating in detail will have to be rigorously pursued.

Acknowledgments

We are thankful to Head of the Department of Geology, Banaras Hindu University for providing the working facilities.

References

Aftalion, M., Bowes, D. R., Dash, B. & Dempster, T.J. (1988). Late Proterozoic charnockites in Orissa, India: U-Pb and Rb-Sr isotopic study. *Journal of Geology* **96**, 663-676.

Aftalion, M., Van breemen, O. & Bowes, D.R. (1984). Age constraints on basement of the Midland Valley of Scotland. *Transactions of the Royal Society of Edinburgh: Earth Sciences* **75**, 53-64.

Anhaeusser, C.R., Mason, R., Viljoen, M.J. & Viljoien, R.P. (1969). A reappraisal of some aspects of Precambrian shield geology. *Bulletin Geological Society of America* **80**, 2175-2200.

Aswathanarayana, U. (1956). Absolute age of the Archaean orogenic cycles of India. *American Journal of Science* **254**, 19-31.

Banerji, P.K. (1990). On the Geology of the eastern ghats of Orissa and Andhra Pradesh, India. In: S.M. Naqvi (ed.) precanbrian continental crust and its mineral resources, 391-407.

Bhadra, S., Gupta, S. & Banerjee, M. (2004). Structural evolution across the Eastern Ghats Mobile Belt-Baster Craton boundary, India: Hot over cold thrusting in an ancient collision zone. *Journal of Structural Geology* **26**, 233-245.

Bhattacharya, S. (1989). Ductile shear zone in Purulia, West Bengal. *Indian Journal of Geology* **61** (3), 172-178.

Bhattacharya, S. (1994). Boundary relations of Eastern Ghats granulite belts: problems and prospects. Abstract, *Workshop on Eastern Ghats Mobile Belts, Vishakhapatnam*, 12-13.

Bhattacharya, S. (1996). Eastern Ghats granulite terrain of India: An overview. *Journal of Southeast of Asian Earth Science* **14** (3-4), 165-174.

Bhattacharya, S., Deomurari, M.P. & Teixeira, W. (2002). Grenvillian thermal event and remnant charnockite:Isotopic evidence from the Chilka Lake granulite-migmatite suite in the Eastern Ghats belt, India. *Proceedings Indian Acadamy of Science (Earth Planetary Science)* **111** (4), 391-399.

Bhattacharya, S., Kar, R., Teixeira, W. & Basei, M. (2003). High-temperature crustal anatexis in a clockwise P-T-t path: isotopic evidence from a granulite-granitoid suite in the Eastern Ghats belt, India. *Journal of the Geological Society London* **160**, 39-46.

Bhattacharya, S., Sen, S.K. & Acharya, A. (1994). The structural setting of the Chilka Lake granulite-migmatite-anorthosite suite with emphasis on the time relation of charnockites. *Precambrian Research* **66**, 393-409.

Biswal, T.K., Waele, B.D. & Ahuja, H. (2007).Timing and dynamics of the juxtaposition of the Eastern Ghats Mobile Belt against the Bhandara Craton, India: A structural and zircon U-Pb SHRIMP study of the fold-thrust belt and associated nepheline syenite plutons.*Tectonics* **26**, 1-21.

Blanford, H.F. (1958). Pressure-temperature time path and a tectonic model for the evolution of granulites. *Journal of Geology* **95**, 617-632.

Bose, S., Fukuoka, M., Sengupta, P. & Dasgupta, S. (2000). Evolution of high Mg-Al granulites from Sunkarametta, Eastern Ghats, India: evidence for a lower crustal heating-cooling trajectory. *Journal of Metamorphic Geology* **18**, 223–240.

Charan, S.N., Subba Rao, M.V., Divakara, Rao, V., Balaram, V. & Govil, P.K. (1998). Prograde-retrograde mineralogy and geochemical variation trends in the charnockites of Gampori region, Vishakhapatnam district, Andhra Pradesh. *Geological Survey of India special publication* **44**, 220-231.

Chetty, T.R.K. & Murthy, D.S.N. (1994). Collision tectonics in the Eastern Ghats Mobile Belt: Mesoscopic to satellite scale structural observations. *Terra Nova* **6**, 72-81.

Clark, G.S. & Subbarao, K.V. (1971). Rb-Sr isotopic age of the Kunavaram Series, a group of alkaline rocks from India. Canadian Journal of Earth Science 8, 1597-1602.

Crowe, W.A., Nash, C.R., Harris, L.B., Leeming, P.M. & Rankin, L.R. (2003). The geology of the Rengali Province: implications for the tectonic development of northern Orissa, India. *Journal of Asian Earth Science* **21**, 697-710.

Das, K., Bose, S., Karmakar, S., Dunkley, D.J. & Dasgupta, S. (2011).Multiple tectonometamorphic imprints in the lower crust: first evidence of ca. 950Ma (zircon U-Pb SHRIMP) compressional reworking of UHT aluminous granulites from the Eastern Ghats Belt, India. *Geological Journal* **46**, 217-239.

Das, S., Nasipuri, P., Bhattacharya, A. & Swaminathan, S. (2008). The thrust-contact between the Eastern Ghats Belt and the adjoining Bastar craton (Eastern India): evidence from mafic granulites and tectonic implications. *Precambrian Research* **160**, 70-85.

Dasgupta, S. & Sengupta, P. (2003). Indo-Antarctic correlation: a perspective from the Eastern Ghats Granulite Belt, India, in: Yoshida, M., Windley, B.F., Dasgupta, S. (Eds.), Proterozoic East Gondwana: Supercontinent Assembly and Breakup. *Geological Society, London, Special Publications* **206**, 131-143.

Dasgupta, S. (1995). Pressure-temperature evolutionary history of Eastern Ghats granulite province: Recent advances and some thoughts. In: India and Antarctica during the Precambrian, Yoshida, M. and Santosh, M. (Ed.) *Geological Society of India* **34,** 101-110.

Dasgupta, S., Bose, S. & Das, K. (2012). Tectonic evolution of the Eastern Ghats Belt, India. *Precambrian Research,* http://dx.doi.org/10.1016/j.precamres.2012.04.005.

Dasgupta, S., Raith, M. & Sarkar, S. (eds) (2008). New perspective in the study of the Precambrian continental crust of India: An integrated sedimentologic, isotopic, tectonometamorphic and seismological appraisal. *Precambrian Research* **162**, 1-316.

Dasgupta, S., Sengupta, P., Guha, D. & Fukuoka, M. (1991). A refined garnet–biotite Fe–Mg exchange geothermometer and its application in amphibolites and granulites. *Contributions to Mineralogy and Petrology* **109**, 130–137.

Dharma Rao, C.V., Santosh, M. & Dong, Y. (2012). U–Pb zircon chronology of the Pangidi–Kondapalle layered intrusion, Eastern Ghats belt, India: constraints on Mesoproterozoic arc magmatism in a convergent margin setting. *Journal of Asian Earth Sciences* **49**, 362–375.

Dharma Rao, C.V., Santosh, M. & Wu, Y.B. (2011a). Mesoproterozoic ophiolitic melange from the SE periphery of the Indian plate: U-Pb zircon ages and tectonic implications. *Gondwana Research* **19**, 384-401.

Dharma Rao, C.V., Windley, B.F. & Choudhary, A.K. (2011b). The Chimalpahad anorthosite Complex and associated basaltic amphibolites, Nellore Schist Belt, India: magma chamber and roof of a Proterozoic island arc. *Journal of Asian Earth Sciences* **40**, 1027–1043.

Divakar Rao, V., Murthy, N.N. & Charan, S.N. (1998). Chemical constraints in modeling the evolution of Eastern Ghats Granulite Belt. *Geological Society of India Special Publication* **44**, 103-116.

Dobmeier, C. & Simmat, R. (2002). Post-Grenvilian transpression in the Chilka Lake area, Eastern Ghats Belt-implications for the geological evolution of peninsular India. *Precambrian Research* **113**, 243–268.

Dury, S.A. & Holt, R.W. (1980). The tectonic framework of the South Indian Craton: A reconnaissance involving LANDSAT imagery. *Tectonophysics* **65**, T1-T15.

Fermor, L.L. (1936). An attempt at the correlation of ancient schistose formation of Peninsular India. *Memoir Geological Survey of India* **70** (1), 1-52.

Foote, R.B. (1870). Geology of the neighborhood of Madras. Rec. *Geological Survey of India* **3** (1), 11-17.

Foote, R.B. (1980). On the geological structure of Eastern Ghats from latitude 15 northwards to Masulipatnam. *Memoir Geological Survey of India* **16** (1), 1-107.

Foote, R.B. (1983a). On the geology of Madurai and Tinnevalley districts. *Memoir Geological Survey of India* **29** (1), 1-103.

Foote, R.B. (1983b). Geology of South Travencore. *Records Geological Survey of India* **16** (1), 20-25.

Grew, E.S. & Manton, W.I. (1986). A new correlation of sapphirine granulites in the Indo-Antarctic metamorphic terrain: late Proterozoic dates from the Eastern Ghats Province of India. *Precambrian Research* **33**, 123-137.

Grew, E.S. (1982). Sapphirine, kornerupine and sillimanite+orthopyroxene in the charnockite region of south India. *Journal of Petrology* **21**, 39-68.

Gupta, J.W., Pandey, B.K., Chabria, T., Banerjee, D.C. & Jayaram, J.M.V. (1984). Rb-Sr geochronological studies on the granite of Vineekonda and Kanagiri Prakasam district, Andhra Pradesh, India. *Precambrian Research* **26**, 105-109.

Gupta, S., Bhattacharya, A., Raith, M. and Nanda, J.K. (2000). Contrasting pressure-temperature-deformation history acrossa vestigial craton-mobile belt boundary: the western marginof Eastern Ghats belt at Deobhog, India. *The Journal of Metamorphic Geology* **18**, 683-697.

Halden, N.M., Bowes, D.R. & Dash, B. (1982). Structural evolution of migmatites in a granulite facies terrane: Precambrian crystalline complex of Angul, Orissa, India. *Transaction Royal Society of Edinburg, Earth Science* **73**, 109-118.

Harris, N.B.W., Santhosh, M.and Taylor, P.N. (1994). Crustal evolution in south India: constrant from Nd isotope. *The Journal of Geology* **102**, 139-150.

Hensen, B.J. & Zhou, B. (1995). A Pan-African granulite facies metamorphic episode in Prydz Bay, Antarctica: Evidence from Sm-Nd garnet dating. *Australian Journal of Earth Science* **42,** 249-258.

Holland, T.H. (1990). The Charnockite series: A group of Archaean hypersthene rocks in Penninsular India. *Memoir Geological Survey of India,* **28**(2).

Jarick, J. (1999). Die thermotektonometamorhe Entwicklung des Eastern Ghats Belt, Indien – ein Test der SWEAT-Hypothese. Unpublished Ph.D. Thesis, Johann Wolfgang Goethe-Universität.

Kale, V.S. & Phansalkar, V.G. (1991). Purana basins of Penninsular India-An overview. *Basin Research* **3**, 209-220.

King, W. (1886). Geological sketch of Vishakhapatnam district, Madras. *Rec. Geological Survey of India* **19** (3), 143-156.

Kovach, V.P., Salnikova, E.B., Kotov, A.B., Yakovleva, S.Z. & Rao, A.T. (1997). Pan-African U-Pb zircon age from apatite magnetite veins of Eastern Ghats granulite belt, India. *Journal of the Geological Society of India* **50**, 421-424.

Krause, O. (1998). Die petrogenetische Bedeutung der porphyrischen Granitoide f¨ur die Krustenentwicklung des Eastern Ghats Belt (Indien). Ph.D. Thesis, Universit¨at Bonn, Germany.

Krause, O., Dobmeier, C., Raith, M.M. & Mezger, K. (2001). Age of emplacement of massif-type anorthosites in the Eastern Ghats Belt, India: constraints from U–Pb zircon dating and structural studies. *Precambrian Research* **109**, 25–38.

Kriegsman, L.M. (1994). Evidence for a fold nappe in the high grade basement of central Sri Lanka: Terrane assembly in Pan-African lower crust. *Precambrian Research* **4**, 178-187.

Krishnan, M.S. (1950). Bezwada gneiss, khondalite and leptynite. *Rec. Geological Survey of India* **81** (2), 315-320.

Kroner, A. & Williams, I.S.M. (1993). Age in the high grade rocks of Sri Lanka. *Journal of Geology* **101**, 513-521.

Kumar, K.V., Frost, C.D., Frost. B.R. & Chamberlain. B.R. (2007). The himakurthi, Errakonda, and Uppaladu plutons Eastern Ghat belts, India: anunusual association of tholeiitic and alkaline magmatism. *Lithos* **97**, 30-57.

Lakhmipatiraju, A. (1998). Review of paleomagnetic results from the Eastern Ghats Belt of Andhra Pradesh. *Geological Survey of India Special Publication* **44**, 95-102.

Lal, R.K. (2003). Metamorphic evolution of granulites from southern Indian Shield. In: Milestones in petrology and future perspectives. Mohan, A. (ed.), *Memoir Geological Society of India, Bangalore* **52**, 61–108.

Lal, R.K., Ackermand, D. & Upadhyay, H. (1987). P-T-X relationships deduced from corona textures in sapphirine-spinel-quartz assemblages from Paderu, southern India. *Journal of Petrology* **28**, 1139–1168.

Lister, G.S. & Davies, G.A. (1988). The origin of metamorphic core complexes and detachment faults formed during Tertiary continental extension in Northern Colorado Region, USA. *Journal of Structural Geology* **11**, 65-94.

Mezger, K. & Cosca, M.A. (1999). The thermal history of the Eastern Ghats Belt (India), as revealed by U-Pb and 40Ar/39Ar dating of metamorphic and magmatic minerals: implications for the SWEAT correlation. *Precambrian Research* **94**, 251-271.

Mezger, K., Cosca, M.A. & Raith, M. (1996). Thermal history of the Eastern Ghats Belt (India) deduced from U-Pb and Ar-Ar dating of metamorphic minerals. *The Journal of Conference Abstracts* **1,** 401.

Mohan, A. & Jayananda, M. (1999). Metamorphism and Isotopic Evolution of Granulites of Southern India: Reference to Neoproterozoic crustal evolution. *Gondwana Research* **2** (2), 251-262.

Mohan, A. (1995). Reaction history of sapphirine granulites and conditions of granulite facies metamorphism in the Eastern Ghats Belt, India. *Abstract, Symposium on Proterozoic Events in East Gondawana, Vishakhapatnam.* JGCP **368**, 3-4.

Mohan, A., Tripathi, P. & Motoyashi, Y. (1997). Reaction history of sapphirine granulites and a decompressional P-T path in a granulite complex from the Eastern Ghats. *Proceedings Indian Academy of Sciences (Earth and Planetary Sciences)* **106**, 115-130.

Mukhopadhyay, D. & Basak, K. (2009). The Eastern Ghats Belt – A Polycyclic Granulite Terrain. *Journal of the Geological Society of India* **73**, 489-518.

Murthy, M.V.N., Viswanathan, T.V. & Roy, C.S. (1971). The Eastern Ghats Group. Rec. *Geological Survey of India* **101**, 15-42.

Nanda, J.K. & Pati, U.C. (1993). Eastern Ghats of Southern Orissa with special reference to Petrochemistry of lithotypes. Workshop (in house, op. Orissa, GSI). In: Proceedings of Workshop on Eastern Ghats Mobile Belt, Special Publication (1998), *GSI special Publications*, 44.

Narayan, B.L., Rama Rao, P., Reddy, G.L.N., Balaram, V. & Govil, P.K. (1994). Rare earth element geochemistry of gabbro from Pedda Bailu area, Vishakhapatnam district. In: Eastern Ghats Terrain (Abs.) *Workshop on Eastern Ghats Mobile Belt, Vishakhapatnam, India.* **15-16**, 32

Narayanaswamy, S. (1975). Proposal for Charnockite-khondalite system in the Archaean Shield of Penninsular India. *Geological Survey of India Misc. Publication* **23**, 1-6.

Natrajan, V. & Nanda, J.K. (1981). Large scale basin and dome structure in the high grade metamorphics near Vishakhapatnam, South India. *Journal of the Geological Society of India* **22,** 62-54.

Osanai, Y., Hamamoto, T., Miyashita, Y., Yamasaki, T., Doyama, D., Ando, T., Saradhi, P. & Srinivas, S. (1999). Rb-Sr and Sm-Nd ages for potassium feldspar megacrystics charnockite pegmatite and garent-bearing charnockitic leucosome from the Central Eastern Ghats, India. *Gondwana Research Group Memoir* **5**, Eastern Ghats Granulites. Rao, A.T., Diwakara Rao, V. & Yoshida, M. (Eds.), 145-160.

Park, A.F. & Dash, B. (1984). Charnockite and related neosome development in the Eastern Ghats , Orissa, India. Petrographic evidence. *Phil. Royal Society of Edinburgh, Earth Science* **75**, 341-352.

Pascoe, E.H. (1950). A manual of Geology of India and Burma. *Govt. of India Publication*, 485.

Paul, D.K. & Sarkar, A. (1994). Geochronology of Eastern Ghats Precambrian Mobile Belt. Workshop on EGMB, Geological Survey of India.

Paul, D.K., Berman, T.R., Menaughton, N.J., Fletcher, I.R., Potts, P.J., Ramakrishnan, M. & Augustine, P.F. (1990). Archaean Proterozoic evolution of Indian charnockites, isotopic and geochemical evidences from granulites of the Eastern Ghat Belts. *Journal of Geology* **98**, 253-263.

Perraju, P., Kovach, A. & Sringer, E. (1979). Rubidium strontium age of some rocks from part of Eastern Ghats in Orissa and Andhra Pradesh. *Journal Geological Society of India* **20**, 290-296.

Ramesh, D.S, M.B. Bianchi, & S. Das Sharma (2010). Images of possible fossil collision structures beneaththe Eastern Ghats belt, India, from P and S receiver functions. *Lithosphere* **2**, 84-92.

Rao, A.J. & Babu, V.R.R.M. (1978). Allanite from Air Port hill charnockites, Vishakhapatnam, India. *American Mineralogist* **63**, 330-331.

Rao, T.C.S., Machado, X.T. & Murthy, K.S.R. (1980). Topographic features over the continental shelf off Visakhapatnam, Mahasagar. *Bulletin National Institute of Oceanography* **13**, 23-28.

Rickers, K., Mezger, K. & Raith, M.M. (2001). Evolution of the continental crust in the Proterozoic Eastern Ghats Belt, India and new constraints for Rodinia reconstruction: implications from Sm–Nd, Rb–Sr and Pb–Pb isotopes. *Precambrian Research* **112**, 183-212.

Rosendahl, B.R. (1987). Architecture of continental rifts with special reference to East Africa. *Annual Review Earth and Planetary Science* **15**, 446-503.

Sarkar, A. & Paul, D.K. (1998). Geochronology of the Eastern Ghats Precambrian mobile belt-a review. *Geological Survey of India (Special Publication)* **44**, 51-86.

Sarkar, A., Bhanumathi, L. & Balasubramanyam, M.N. (1981). Petrology, Geochemistry, and Geochronology of the Chilka lake igneous complex, Orissa State, India. *Lithos* **14**, 93-111.

Sarkar, A., Bhanumati, L. & Balasubramanyam, M. N. (1981). Petrology, geochemistry and geochronology of the Chilka Lake igneous complex, Orissa state, India. *Lithos* **14**, 93-110.

Sarkar, G., Barman, T.R. & Corfu, F. (1989). Timing of continental arc magmatism in northwest India: evidence from U–Pb geochronology. *The Journal of Geology* **97**, 607-612.

Sen, S.K., Bhattacharya, S. & Acharya, A. (1995). A multi-stage pressure–temperature record in the Chilka Lake granulites: the epitome of the metamorphic volution of Eastern Ghats, India. *Journal of Metamorphic Geology* **14**, 287-298.

Sengupta, P., Dasgupta, S., Bhattacharya, P.K., Fukuoka, M., Chakraborti, S. & Bhowmick, S, (1990). Petro-tectonic imprints in the sapphirine granulites from Anantagiri, Eastern Ghats Mobile Belt, India. *Journal Petrology* **31,** 971-996

Sengupta, P., Dasgupta, S., Raith, M., Bhui, U.K. & Ehl, J. (1999). Ultra-high temperature metamorphism of metapelitic granulites from Kondapalle, Eastern Ghats Belt: implications for the Indo-Antarctic correlation. *Journal of Petrology* **40**, 1065-1087.

Shaw, R.K., Arima, M., Kagami, H., Fanning, C.M., Shiraishi, K. & Motoyoshi, Y. (1997). Proterozoic events in the Eastern Ghats granulite belt, India: evidence from Rb–Sr, Sm–Nd systematics and SHRIMP dating. *Journal of Geology* **105**, 645-656.

Simmat, R. & Raith, M.M. (2008). U–Th–Pb monazite geochronometry of the Eastern Ghats Belt, India: Timing and spatial disposition of polymetamorphism. *Precambrian Research* **162**, 16-39.

Sriramdas, A. & Rao, A.T. (1979). Charnockites of Visakhapatnam, Andhra Pradesh. *Journal Geological Society of India* **20**, 512-517.

Subba Rao, D.V. (1994). Crustal structure beneath Eastern Ghats based on gravity data.Workshop on Eastern Ghats Mobile Belt, Visakhapatnam, India, pp. 24-25 (Abstract).

Subrahmanyamn C. & Verma, R.K. (1986). Gravity field, structure and tectonics of the Eastern Ghats. *Tecnophysics* **126**, 195-212.

Unnikrishnan-Warrier, C., Santosh, M. & Yoshida, M. (1995). First report of Pan-African Sm–Nd and Rb–Sr mineral ages from regional charnockites of southern India. *Geological Magazine* **132**, 253–260.

Upadhyay, D., Raith, M.M., Mezger, K. & Hammerschmidt, K. (2006b). Mesoproterozoic rift-related alkaline magmatism at Elchuru, Prakasam Alkaline Province, Se India. *Lithos* **89**, 447–477.

Upadhyay, D., Raith, M.M., Mezger, K., Bhattacharya, A. & Kinny, P.D. (2006a). Mesoproterozoic rifting and Pan-African continental collision in South-Eastern India: evidence from the Khariar alkaline complex. *Contributions to Mineralogy and Petrology* **151**, 434–456.

Vasudevan, D. & Rao, T. M. (1975). The high grade schistose rocks of the Nellore schist belt, Andhra Pradesh and their Geologic Evolution. *Indian Mineralogist,* **16,** 43-47.

Vinagradov, A., Tugarinov, A., Zhycov, C., Stapnikova, N., Bikinova, E. & Khorre, K. (1964). Geochronology of Indian Precambrian. *22nd Int. Geological Congo Report* **10**. 553-567.

Walker, T.L. (1902): Geology of Kalahandi state, Central provinces. *Memoirs of the Geological Survey of India* **33(3)**, 1-22.

Windley, B.F. (1973). Crustal development in the Precambrian. Philosophical Transactions of the Royal Society of London. Series A, Mathematical and Physical Sciences **273**, 321-341.

Hon'ble Reviewer

1. **Prof. S.P. Singh**
   Dean Department of Earth Science
   Bundelkhand University
   JHANSI - 284 001
2. **Prof. Madhumita Das**
   Head of the Department
   Department of Geology
   Utkal University
   BHUBANESHWAR - 751 001

I-16

The Pyrite – Pyrrhotite Deposit of Saladipura District Sikar, Rajasthan

R.G. Goswami

Geological/Mining Consultant, Saraswati Nagar, Basni, Jodhpur-342 005.
e-mail rgoswami2008@gmail.com

Abstract

Sulfide deposit containing Pyrite-Pyrrhotite occurs in Pre Cambrian rocks of Delhi system. Mineralization is like lenticular lens type. The main host rocks are quartzite and phyllite. These are intercalated with schists. Other minerals include Sphalerite, Chalcopyrite, Galena and Illmenite. Pyrite shows optical anisotropism. From geological evidence it is inferred that this is a stratabound sulfide deposit of volcanogenic origin. Subsequent deformation and metamorphism has altered the characters of original deposit and host rocks. Volcanogenic, yet strata bound genesis have been proposed.

Keywords: Pyrite, Pyrrhotite, Saladipura deposit, Fertilizer Industry, Delhi Supergroup

Introduction

The sulfide deposit of Saladipura consisting of Pyrite and Pyrrhotite orebodies is situated near the village Salapura in Sikar district. This deposit lies at the latitude 27°38'45" and longitude 75°31'32". This deposit was located after 2 gossan zones were discovered by Geological Survey of India during mapping on aerial photographs. Initial exploration work and geological studies were carried out by the same agency (Mukbtinath et al. 1968, Das gupta 1970). Recently attempts have been made to study the genetic aspects in detail (Ray-1974 Goswami 1975, Sarkar et al. 1980)

GENERAL GEOLOGY

The Saladipura orebody occurs in the rocks of Delhi System of precambrian age. This system has been divided into 2 parts lower and upper. In above area, rocks of upper part i.e. Ajabgarh series are exposed (Fig. 1).

The rock types are represented by metamorphosed argillaceous sediments. They consist of varieties of schists like biotite schist, chlorite schist, muscovite schist. These schists also contain minerals like garnet, stanrolite. kyanite and sillimanite. Different varieties of phyllite are also present. They include typical phyllite, carbon-phyllite. At places these are highly jointed and fractured phyllites is mainly associated with biotite, sericite and chlorite. Quartzite occurs throughout the area. They are mostly grey, fine grained, cherty, hard and compact. Other varieties of quartzite include thinly bedded flaggy quartzite, amphibole quartzite, argillaceous quartzite and phyllitic quartzite. Both phyllite and quartzite are found to be intercalated with schists.

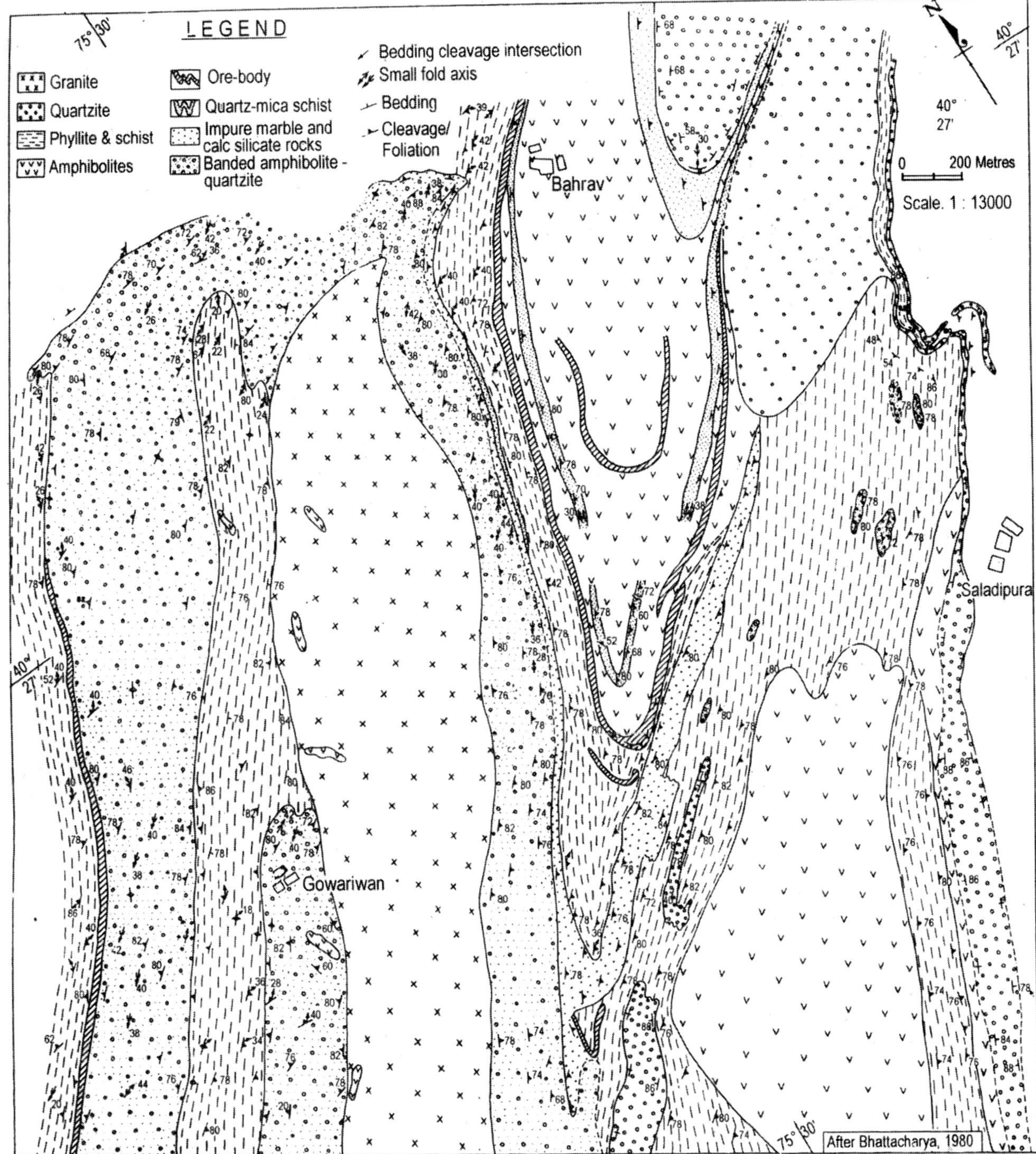

Fig. 1 Geological Map of the Area west of Saladipura (after Bhatacharya, 1980).

Other rock types of the area are amphibolites, granite, pegmatite, marble, calc silicate and calcgneisses. In certain places feldspars are kaolinised. Amphibolites appear to be related to mineralization as they are intimately associated with it. These are two types of amphibolites, the older and younger one. This is a dark coloured rock and is fine to coarse grain in texture. It is composed of actinolite tremolite, hornblende, plagioclase feldspar oligocloase to andesine, quartz, ilmenite, sphene, magnetite and chlorite.

The most conspicuous feature of this deposit is the presence of a V shaped gossan zone which has a strike length of 7.5 kms. This is composed of limonite, lepidocrocite, geothite lateritic limonite. A manganese dioxide band and limonitic jasper is also present. This has a thickness of about 2 to 2.5 meters. The colour of gossan varies from yellowish to brownish yellow, and dark brown to black and maroon. Limonite box works do not show regular pattern. The cavities are sub-rounded, avoid, rectangular, triangular as well as rhombic in shape. Study of box reveals that this is a result of oxidation of pyrite and pyrrhotite in part. The pyrite and pyrrhotite ore body is 30 to 50 meters below the gossan.

STRUCTURE AND METAMORPHISM

Structure of Saladipura area is very complicated. Large number of doubly plunging folds, both anticlines and synclines are present. Their regional strike is ranging from NE-SW to NNE-SSW. 3 sets of folds have been recognised in this area. Where intensity of deformation is greater rocks have been faulted (both normal and reverse).

Field and mineragraphic evidence shows that the ore body and host rocks have undergone intense metamorphisms. Sulfide minerals clearly show the effect of metamorphism. Development of granoblastic or coarse grained texture is a typical feature of metamorphosed sulfides. Metamorphosed sulfide also have the tendency to develop triple point junctions (Stanton 1964). Fig 3 shows coarse grain pyrrhotite having triple point junctions. Presence of minerals like staurolite, andalusite, garnet and kyanite in phyllites is an indication that retrograde metamorphism has taken place. It is probable that pyrrhotite has developed during metamorphism through breakdown of Pyrite. According to Das Gupta (1970) most of the silicate minerals in gangue fraction were originally iron rich but were de ironised and this iron was used in formation of this deposit. It is true that the existing gangue minerals are comparatively poor in iron content but iron so released has been utilised in the formation of Pyrrhotite. This mineral forms due to desulfurization of Pyrite. It has been proved that sulfur so released is not removed from the system but has interacted with iron bearing gangue minerals (Kanhaira et al. 1964, Thompson 1972). This reaction leads to iron removal from silicates and formation of Pyrrhotite. Pyrrhotite can also form from breakdown of hematite and pyrite. Hematite is very frequently associated with with pyrite in sulfide assemblage and coexists with it. In the present deposit hematite is totally absent and this can be attributed to the fact that hematite did exist with pyrite but during metamorphism it broke down and gave rise to Pyrrhotite.

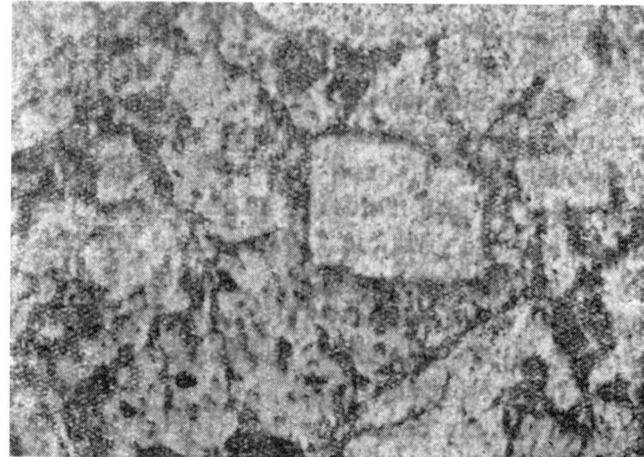

Fig. 2 Pyrite and other associated minerals.

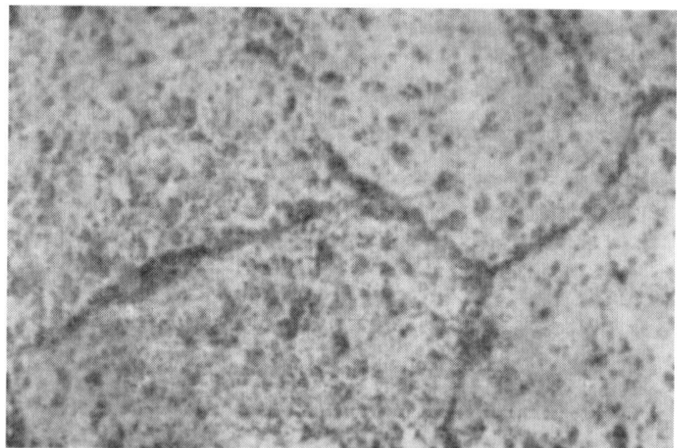

Fig. 3 Pyrite showing triple-Junction points

MINERALIZATION AND GENESIS

Mineralization occurs in quartzite and phyllite as lenticular zones of different grades. Mineralization varies from disseminated patches to rich and massive ore shoots. Pyrite is the chief sulfur bearing mineral and occurs in pyritohedral, octahedral and cubic forms. This mineral is followed by Pyrrohtite. Both Pyrite and Pyrrhotite occur in the form of bands as well as coarse grained aggregates. Other minerals found along with these two are Sphalerite. Chalcopyrite and Galena. Pyrrhotite mostly occupy the interstistcal space between Pyrite grains, Sphalerite occurs as thin veins. Some times Pyrrhotite is seen to occur as veins intercalated with pyrite. Ilmenite is found as gangue mineral.

Fig 2 shows pyrite and other associated minerals. Another important feature of the ore body is that hanging wall and footwall contacts of the ore body are sharp. Besides in places Pyrite shows conformable relations with the host rocks.

Pyrite exhibits unusual property of optical anistropism.

Research studies have suggested (Ray 1974, Sarkar et al 1980) that Saladipura deposit is of syngenetic sedimentary in origin. But comparison of different geological features of ores and host rocks with other sulfide deposits has led this author to believe that this deposit is a strata-bound sulfide deposit of volcanogenic origin. Subsequent metamorphism and deformation has chanced the original nature of sulfides as well as host rocks. Presence of iron bearing formations chest, manganiferrous shales, graphite bearing rocks, siltstones and pre-dominance of argillites over chemical precipitates are common features of volcanogenic sulfides (Hutchinson 1973) of proterozoic age. These features can also be observed in Saladipura deposit.

RESOURCE POTENTIAL

This area was given to GOI company M/S. P.P.C.L. Ltd. for conducting mining operations. This Co. got the exploration report prepared by M/S. RioTinto Zinc Ltd. As per the exploration data this deposit has 110 M.T. reserves of Pyrite-Phyrrhotite with average 22% Sulphur content. Mining was carried out for few years. The deposit is suitable for preparing Sulphuric acid and fertilizers.

ACKNOWLEDGEMENT

The author is grateful to Dr. M.K. Pandya and Mr. N.C. Jain for providing necessary facilities for this work. Grateful thanks are due to Dr. P.S. Ranawat for taking photomicrographs.

REFERENCES

Bhattacharyya, P. K., 1980: "Metamorphic reactions in static and dynamic environments-a petrographic illustration from the metampelites in the Khetri Copper Belt, Rajasthan, India. N. Jb. Miner. Abh., 139 (2), pp. 191-204.

Das Gupta, S.P. 1970 Sulfide deposits of Saladipura, Khetri Copper belt, Rajasthan, India. Eco. Geo 65 : 331-339

Goswami, R.G. 1975 A study of Saladipura Pyrite-Pyrrhotite deposit, district Sikar, Rajasthan. Unpub. M.Sc. (Tech.) Project dissertation, Univ. of Raj. P. 76,

Kanheira, K., Banno. S. Nitshida, K. 1964, Sulfide and oxide minerals. in some metamorphic terrains in Japan. *Japanese Jour. Geology. Geography* 15: 175-191

Hutchinson. R.W. 1973. Volcanogenic sulfide Deposits and their metallogenic significance. Eco. Geo 68(8) : 1223-1246

Muktinath, Natrajan, KWK, Mathur, A.L. 1968. A preliminary note on the pyrite pyrrhotite deposit near Saladipura, Sikar district, Rajasthan. Jour. Mines Metals & Fuels 16 (ii) 397-400.

Ray S.K. 1974. Structural history of the Saladipura Pyrite-Pyrrhotite deposit and associated rocks. Khetri Copper Belt, Rajasthan J. Geo. Soc. Ind. 15 (3) : 227238

Sarkar S.C., Bhattacharyya, P.K., Mukherjee, A.D.C. 1980. Evolution of the sulfide ores of Saladi-pura. Rajasthan India. Eco. Geo. 75 : 1152-1167

Stanton, R.L. 1964. Mineral interfaces in stratiform ores. Trans. Inst'n. Min. Met. 74 (2) : 45-79

Thompson, J.B. Jr. 1972. Oxides and Sulfides in regional metamorphism of pelitic schists. Internal Geol. Cong. 24th, Montreal, 10; 27-35.

Hon'ble Reviewers

1. **Prof. S.N. Mohapatra**
   Head
   School of Studies in Earth Science
   Jiwaji University
   GWALIOR - 474 001
2. **Prof. R.N. Sahu**
   Head
   Department of Geoinformatics
   Barkatullah University
   BHOPAL - 462 026

I-17

Mineralogy, Fluid Inclusion and Geothermobarometery of Root Zone of Almora Nappe in Lesser Himalaya, Kumaun India

S.N. Lal, Alpana Hyanki and Manisha Sanguri

Centre of Advanced Study in Geology, Kumaun University, Nainital, Uttarakhand.
E-mail: snlalgeo@gmail.com

Abstract

The controversy of the nappe rocks (Almora Nappe) of the Kumaun Lesser Himalaya to have their root zone in the Higher Himalaya. In the present paper we have tried to compare mineralogical data, geothermobarometric data and some fluid inclusion petrography data to testify the hypothesis whether Central Crystallines are root zone of Almora Crystallines or not. This investigation has observed that there is similarity in mineralogical and geothermobarometric data in both study areas, which is supportive of existence of Central Crystallines as a root zone of Almora Nappe, which was folded and thrusted later along. Occurrence of similar kind fluid inclusions in both Almora Nappe and Central Crystalline are suggestive of preexistence of root zone. Presence of CO_2 and Halite inclusions in Central Crystallines indicate that they were generated at greater depth while the absence of halite inclusions in Almora Crystallines may be due to fluid requilibriation during thrusting and Nappe formation. This similarity in mineralogy, geothermobaro-metric results and fluid inclusion petrography are more supportive of existence of Central Crystallines root zone for Almora Nappe.

Keywords: Central Crystallines (CC), Almora Nappe, Kumaun Himalaya, Root Zone, Mineralogy, Geothermobarometry, Fluid Inclusions

Introduction

Himalaya the "Paradise of Geologists" is the largest, highest and the youngest of all the mountain ranges on the earth. This magnificent orogeny on the earth has begun nearly 55 Ma ago, when Indian continent rammed into Tibet. The collision of two continents resulted into instance thrusting, faulting, volcanism and metamorphism. The compressional stress started in Paleocene about 55 Ma ago is still playing a vital role in rise up earthquakes, hot springs, neotectonic movements etc. Only after geological investigations for over a century, the enigmatic nature of the Himalayan geology has posted more problems than it has solved. On the basis of their large scale regional tectonic boundaries, the Himalaya has been divided into four lithotectonic and physiographic zones (Valdiya 1980); Siwalik or Outer Himalaya which is bounded by Main

Boundary Thrust (MBT) and Himalaya Frontal Fault (HFF) in north and south limits respectively and comprises 9500 meter thick fresh water sedimentary pile of molasse sediments of Miocene to Upper Pleistocene. This zone is also known as piedmont zone or fore land basin of India; Lesser or Sub Himalaya, consists of metasedimentaries of early to late Proterozoic, which are covered by crystalline nappe and klippe of Higher Himalaya (Gansser, 1964). Lesser Himalaya is bounded by Main Boundary Thrust (MBT) and Main Central Thrust (MCT) in south and north respectively; Higher or Great Himalaya, this zone is a 15-20 km thick pile of polymetamorphosed crystalline rocks showing metamorphic sequence from green schist to amphibolites facies. This region is bounded by Trans Himadri Fault (THF) in north and Main Central Thrust (MCT) in the south; Tethys or Trans Himalaya which occupies nearly 40 km wide zone of Trans Himalaya. This zone is characterized by 10km thick marine sediments of Precambrian to lower Eocene. Tethys Himalayan zone is sharply defined by a thrust called Indus-Tsangpo Suture Zone (ITSZ) in north and THF/ thrust in south. The ITSZ contains ophiolites, flysch and molasses of back arc basin (Fig: 1 A).

The Munsiari Thrust, which Heim and Gansser called up the Main Central Thrust, separates the crystallines of the Munsiari Formation at the base of the Central Crystalline zone of the Great Himalaya. According to Valdiya (1980), it is the root zone of the Almora Thrust and its extension defining the base of klippe. However, Gansser (1964) does not think that the Munsiari Thrust is the root zone of the Almora Thrust. Whereas the Almora Nappe is one of the largest nappe in Lesser Himalaya and is a large thrust sheet of solid rock that has moved a long distance about in kilometer or more at a low angle over the underlying rocks either by thrusting or recumbent folding. In Almora Nappe greenschist to amphibolite facies of rocks are exposed, the root of these crystallines is supposed from the base of Higher Himalaya, the Munsiari Formation (Heim and Gasser 1939, Gansser, 1964; Valdiya, 1980). In the present paper an attempt is made to compare mineralogical data, geothermobarometric data and some fluid inclusion's petrographical data to testify the hypothesis whether Central Crystallines are root zone of Almora Crystallines or not.

Geological Setting

The first study area lies between the Munsiari Thrust and the Main Central Thrust and forms a part of the Munsiari Formation of Main Central Thrust Zone (Latitude 30°4'- 30°8' N, Longitude 79°51'- 79°58' E) of the Central Crystallines of the Kumaun Himalaya (Loharkhet- Dhakuri- Khati- Phurkiya) (Fig: 1B, 1C). Valdiya et al. (1999) divided the Central Crystalline zones into two parts which is separated by a thrust plane, designated as the Vaikrita Thrust and recognized as the real Main Central Thrust. The lower tectonic unit consisting of metamorphic rocks exhibiting green schist to lower amphibolite facies is named Munsiari Formation. The thrust demarcating its base has been renamed as the Munsiari thrust in place of Main Central Thrust by Heim and Gansser (1939). The upper lithotectonic unit consisting of higher grade metamorphic rocks has been designated as Vaikrita group. The rocks occurring in the study area document evidence of at least three phases of deformation. The sandwiched Munsiari rocks have also suffered cataclastic deformation and mylonitization due to movement along the delimiting thrusts. The present study area comprises of chlorite-mica-schists, garnet-mica schists, garnet-staurolite-mica schists, kyanite-sillimanite mica schists with interbedded quartzites, granitic gneisses with the lenses of amphibolites. At places graphitic material is also noticed in Munsiari Formation.

The second study area Almora Crystalline lies about 40 km. ENE of Almora and bounded by long. 80° 0'- 79° 45' E and latt. 29° 45'- 29° 30' N (Jageshwar-Naichun, Jageshwar-Danya & Suyalbari-Naichun transects) (Fig: 1B, 1D, 1E). Almora Nappe in Uttarakhand, India is a Lesser Himalayan representative of the Himalayan Metamorphic Belt that was tectonically transported over the MCT from Higher Himalaya. The rock type exposed along the northern & southern margins are highly mylonitized and shows pattern of polyphase deformation & metamorphism. The degree of mylonitization decreases towards the central part where the rocks eventually grade into unmylonitized metamorphics. The north & south boundaries of Almora Nappe are demarcated by the North Ramgarh Thrust & South Ramgarh Thrust (Joshi, 1999), which extend to the western

Nepal. The schist and gneisses of the Saryu Formation of Almora group exposed upsection in the central part of the nappe, unlike its base, escaped the effects of mylonitization, and have preserved the pre-shear zone structures. Tectonic history of the Almora Nappe is preserved in three distinguishable structural regimes, viz. the pre-shear zone structures, the syn-shear zone structures and the post-shear zone structures with respect to the basal shear zone of the Almora Nappe. The present study area comprises of mica-schists, garnet-mica schists interbedded graphitic schists and quartzites, granitic gneisses and thickly bedded lenses of amphibolites.

Mineralogy

Coexisting minerals from the assemblages of Garnet mica schist from the representative areas of Almora Crystallines and Central Crystalline area have been analysed by electron microprobe technique (Tables 1-4). X_{Mg} of biotite in Almora Crystallines is from 0.43 to 0.45 and in Central Crystallines ranges from 0.55 to 0.56. TiO_2 content in biotite ranges from 1.29 to 1.86 in Almora Crystallines and in 1.3 to 1.34 in Central Crystallines. Al^{IV} content of biotite lies in between 2.3 to 2.33 in Almora Crystallines and 2.41 to 3.02 in Central Crystallines. Ti content of biotite in Almora Crystallines from 0.147 to 0.214 while Ti content in Central Crystallines is 0.14 to 0.15 while Ti content of muscovite in Almora Crystallines is from 0.037 to 0.044 while Ti content in Central Crystallines is increased from 0.03 to 0.85. X_{Mg} in muscovite from Almora Crystallines ranges in between 0.15 to 0.231 and in Central Crystallines it is 0.58 to 0.9. In the both study area garnet is basically almandine rich in garnet-mica schists. In Almora Crystallines 73.67 to 77.47 almandine, 3.77 to 3.96 pyrope, 17.3 to 17.11 grossularite and 1.63 to 5.06 spessertine mole% while in Central Crystallines garnet consists of 75.59 to 78.45 almandine, 5.89 to 7.26 pyrope, 7.29 to 10.55 grossularite and 6.98 to 7.95 spessertine mole%. The X_{Mg} of garnets in Almora Crystallines ranges in between 0.046 to 0.051 and 0.072 to 0.084 in Central Crystallines. The X_{Ca} ratio of plagioclase is 0.2 to 0.22 in Almora Crystallines and 0.225 to 0.23 in Central Crystallines. While the X_{Na} ratio of plagioclase is 0.77 to 0.79 in Almora Crystallines and 0.76 to 0.764 in Central Crystallines.

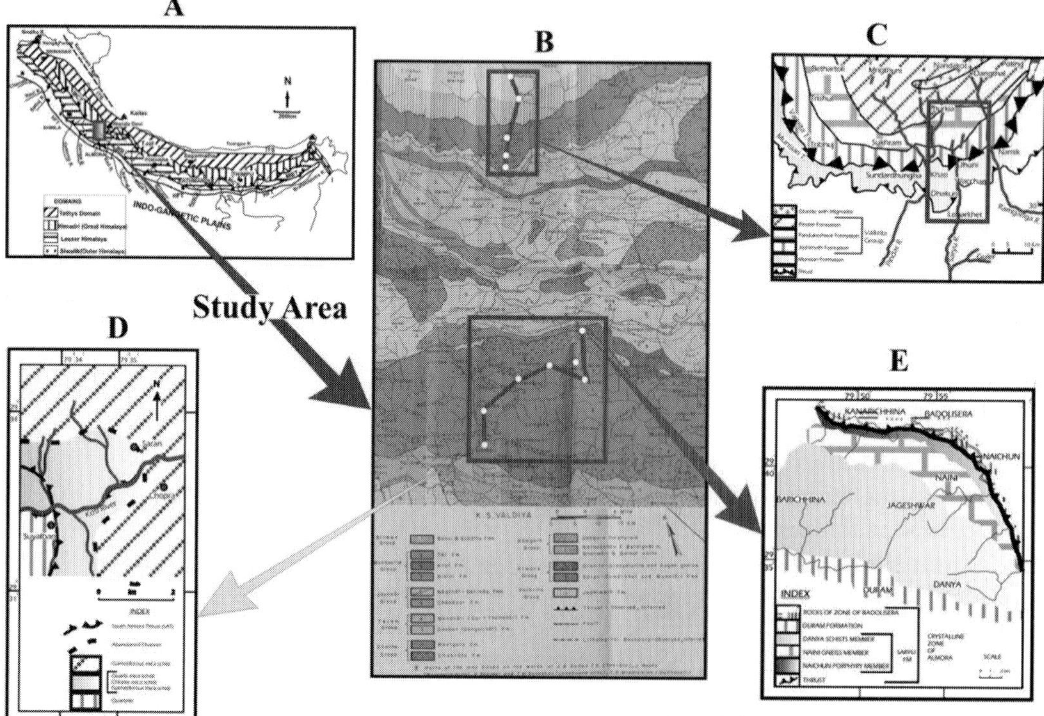

Fig. 1 A. Tectonic division map of Himalaya. (After Valdiya, 1980). **Fig: 1 B.** Geological & Location map of Lesser Himalaya (After Valdiya, 1980). **Fig: 1 C.** Location map of the Central Crystallines (After Valdiya, 1980). **Fig: 1 D.** Location map of the area around South Almora Thrust (SAT), Almora Crystallines. (After Pant and Agarwal, 2003). **Fig: 1 E.** Location map of the area around North Almora Thrust (NAT) Almora Crystallines. (After Misra and Kumar, 1972).

Table 1. Representative Microprobe analysis and Structural Formulae of Biotite Based on 22 oxygen) from Almora Crystallines and Central Crystallines.

BIOTITE				
OXIDES	**Almora Crystallines**		**Central Crystallines**	
SiO_2	37.49	37.26	36.34	36.27
TiO_2	1.29	1.86	1.34	1.30
Al_2O_3	17.96	16.94	17.83	17.75
Cr_2O_3	0.03	0.01	0.28	0.28
FeO	19.96	20.4	16.78	16.39
MnO	0.08	0.06	0.08	0
MgO	9.23	8.98	11.59	11.66
CaO	0.01	0	0	0.06
Na_2O	0.21	0.2	0.19	0.20
K_2O	9.19	9.17	8.70	8.21
Total	**95.45**	**94.88**	**93.13**	**92.12**
		CATIONS		
Si	5.67	5.692	4.98	5.59
Ti	0.147	0.214	0.14	0.15
Al	3.202	3.051	2.88	3.22
Cr	0.004	0.001	0.03	0.03
Fe^{2+}	2.524	2.606	1.92	2.11
Mn	0.01	0.008	0.01	0
Mg	2.08	2.044	2.37	2.68
Ca	0.002	0	0	0.01
Na	0.062	0.059	0.05	0.06
K	1.773	1.787	1.52	1.61
Total	**15.474**	**15.462**	**13.9**	**15.46**
Al^{IV}	2.33	2.308	3.02	2.41
Al^{VI}	0.872	0.743	0.14	0.81
X_{Fe}	0.548	0.56	0.447	0.44
X_{Mg}	0.45	0.43	0.55	0.56
X_{Ca}	0.001	0	0	0.005
X_{Na}	0.025	0.31	0.31	0.03
X_K	0.965	0.968	0.968	0.958
No. Oxygen	**22**	**22**	**22**	**22**

Table 2. Representative Microprobe analysis and Structural Formulae of Muscovite Based on 22 oxygen) from Almora Crystallines and Central Crystallines.

MUSCOVITE				
OXIDES	**Almora Crystallines**		**Central Crystallines**	
SiO_2	47.24	48.27	45.65	47.53
TiO_2	0.44	0.37	0.27	0.96
Al_2O_3	33.96	33.37	36.67	35.39
Cr_2O_3	0.03	0.18	0.07	0.05
FeO	1	0.95	0.82	1.21
MnO	0	0.03	0.04	0.03
MgO	0.1	0.16	0.43	0.92
CaO	0.02	0	0.03	0.00
Na_2O	1.48	1.42	2.11	1.16
K_2O	9.11	9.05	7.81	9.34
Total	**93.38**	**93.8**	**93.9**	**96.59**
		CATIONS		
Si	6.35	6.445	6.09	5.77
Ti	0.044	0.037	0.03	0.85
Al	5.382	5.253	5.76	5.06
Cr	0.003	0.019	0.01	0.01
Fe^{2+}	0.112	0.106	0.09	0.12
Mn	0	0.003	0.01	0.01
Mg	0.02	0.032	0.09	0.17
Ca	0.003	0	0.01	0
Na	0.386	0.368	0.55	0.27
K	1.562	1.542	1.33	1.45
Total	**13.862**	**13.805**	**13.97**	**13.71**
X_{Fe}	0.848	0.768	0.5	0.413
X_{Mg}	0.15	0.231	0.9	0.58
X_{Ca}	0.001	0	0.005	0
X_{Na}	0.197	0.192	0.29	0.15
X_K	0.8	0.807	0.703	0.843
No. Oxygen	22	22	22	22

Table 3. Representative Microprobe analysis and Structural Formulae of Garnet (Based on 24 oxygen) from Almora Crystallines and Central Crystallines.

GARNET				
OXIDES	Almora Crystallines		Central Crystallines	
SiO_2	37.77	38.06	37.32	35.37
TiO_2	0.13	0.05	0.11	0.09
Al_2O_3	20.71	21.07	20.92	21.27
Cr_2O_3	0.11	0	0.12	0.12
FeO	35.8	35.85	34.30	34.86
MnO	2.43	0.75	3.56	3.06
MgO	1.08	0.98	1.50	1.81
CaO	6.56	6.18	3.74	2.53
Na_2O	0.05	0.03	0.09	0.07
K_2O	0.01	0.02	0	0
Total	**104.65**	**102.99**	**101.66**	**99.18**
CATIONS				
Si	5.914	6	5.972	5.82
Ti	0.015	0.006	0.013	0.011
Al	3.823	3.916	3.947	4.126
Cr	0.014	0	0.015	0.016
Fe^{2+}	4.688	4.726	4.591	4.797
Mn	0.322	0.1	0.483	0.427
Mg	0.252	0.23	0.358	0.444
Ca	1.101	1.044	0.641	0.446
Na	0.015	0.009	0.028	0.022
K	0.002	0.004	0	0
Total	**16.146**	**16.035**	**16.048**	**16.109**
Al^{IV}	2.086	2	2.028	2.18
Al^{VI}	1.737	1.916	0.027	1.946
X_{Fe}	0.947	0.953	0.927	0.915
X_{Mg}	0.051	0.046	0.072	0.084
X_{Ca}	0.984	0.987	0.958	0.952
X_{Na}	0.013	0.008	0.041	0.047
X_K	0.001	0.003	0	0
Almandine	73.67	77.47	75.59	78.45
Pyrope	3.96	3.77	5.89	7.26
Grossularite	17.3	17.11	10.55	7.29
Spessertine	5.06	1.63	7.95	6.98
No. Oxygen	24	24	24	24

Table 4. Representative Microprobe analysis and Structural Formulae of Plagioclase (Based on 8 oxygen) from Almora Crystallines and Central Crystallines.

PLAGIOCLASE				
OXIDES	Almora Crystallines		Central Crystallines	
SiO_2	62.32	63.31	63.06	63.01
TiO_2	0.02	0.04	0.08	0.03
Al_2O_3	23.01	22.57	23.66	23.65
Cr_2O_3	0.02	0	0.01	0.00
FeO	0.16	0.02	0.00	0.04
MnO	0	0	0.00	0.06
MgO	0.01	0	0.00	0.00
CaO	4.77	4.46	4.89	4.62
Na_2O	9.17	9.73	8.92	9.11
K_2O	0.11	0.09	0.04	0.23
Total	**99.59**	**100.22**	**100.66**	**100.75**
CATIONS				
Si	2.776	2.8	2.77	2.77
Ti	0.001	0.001	0.01	0.01
Al	1.208	1.177	1.23	1.23
Cr	0.001	0	0	0
Fe^{2+}	0.006	0.001	0	0
Mn	0	0	0	0
Mg	0.001	0	0	0
Ca	0.228	0.211	0.23	0.23
Na	0.792	0.834	0.76	0.78
K	0.006	0.005	0.01	0.01
Total	**5.019**	**5.029**	**5.01**	**5.03**
X_{Fe}	0.857	1	0	0
X_{Mg}	0.142	0	0	0
X_{Ca}	0.22	0.2	0.23	0.225
X_{Na}	0.771	0.794	0.76	0.764
X_K	0.005	0.004	0.01	0.009
No. Oxygen	8	8	8	8

Geothermobarometry

On the basis of various geothermobarometric data of Almora Crystallines and Central Crystallines estimated pressure and temperature values were summarized (Table 5) as follows:

For Garnet-Biotite thermometry garnet-biotite pairs were chosen for the purpose. In few cases, the composition of biotite inclusions in garnet and composition of garnet at spots which are at few places microns away from the inclusion have taken. In some places a lower limit of peak temperature used. The estimated temperature by using the model of Holdaway (2000) is 461 to 495°C for Almora Crystallines and 498 to 508°C for Central Crystallines and according to Perchuk and Laverenteva (1983) temperature is 500 to 532°C for Almora Crystallines and 528 to 537°C for Central Crystallines; average temperature by Perchuk et al. (1985) is 446 to 473°C for Almora Crystallines and 472 to 486°C for Central Crystallines. According to Bhattacharya et al. (1992) the estimated temperature is 408 to 445°C for Almora Crystallines and 455 to 461°C for Central Crystallines and by Dwivedi et al. (2007) is 425 to 452°C for Almora Crystallines and 465°C for Central Crystallines. Temperature estimates calculated by Kaneko and Miyano (2004) with Fe3+ Garnet- Biotite geother-mometry temperature is 441 to 472°C for Almora Crystallines and 406 to 416°C for Central Crystallines. The lower temperature estimates in some cases may be attributed to requilibrium during cooling following the thermal peak of metamorphism.

Empirically calibrated geothermometers of Lal (1991) in which Ti content of biotite has taken gives temperature range from 581 to 591°C for Almora Crystallines and 583 to 561°C for Central Crystallines.

Average Pressure estimates through Garnet-Biotite- Plagioclase- Quartz Barometry calculated by Wu et al. (2004) garnet-mica schists is 5.6 to 5.9 for Almora Crystallines and 5.8 to 6.5 kbar for Central Crystallines.

Average P-T method has been used for the pelitic rocks (Garnet- Biotite- Plagioclase- Quartz Geothermobarometry) of Central Crystallines of Kumaun Himalaya by using the computer programme THERMOCALC (Holland and Powell, 1998) and its embedded thermodynamic properties for the end member and solutions the average P-T estimates are 7.4 to 7.8 kbar/ 558 to 572°C for Almora Crystallines and 7.3 to 7.9 kbar/ 520 to 527°C for Central Crystallines. The computer programme TWQ_ V 2.3 by Berman (2007) gives the temperature and pressure values about 450 to 490°C /6.5 to 7.6 kbar for Almora Crystallines and 475°C/ 6.7 kbar for Central Crystallines.

Table 5. P-T Conditions of Metamorphism in Garnet mica schists (Pressure in Kbars and Temperature in °C) from Almora Crystallines and Central Crystallines

S. No.	Reference	Models/ Methods	T & P	Almora Crystallines		Central Crystallines	
1	Perchuk et al. (1985)	Garnet- Biotite Geothermometry	T °C	446	473	486	472
2	Perchuk and Laverenteva (1983)	-- do--	T °C	500	532	528	537
3	Bhattacharya et al. (1992)	-- do--	T °C	408	445	455	461
4	Lal (1991)	P-T Calc, Ti content of Biotite	T_B °C	591	581	583	561
5	Kaneko and Miyano with Fe^{+3} (2004)	Garnet- Biotite Geothermometry	T °C	441	472	406	416
6	Holdaway (2000)	-- do--	T °C	461	495	508	498
7	Dwivedi et al. (2007)	Garnet- Biotite- Plagioclase- Quartz Geothermometry	T °C	425	452	465	465
8	Wu et al. (2004)	Garnet- Biotite- Plagioclase- Quartz Geobarometry	P Kbar	5.9	5.6	5.8	6.5
9	Holland and Powell (1998)	THERMOCALC, Garnet- Biotite- Plagioclase- Quartz Geothermobarometry	T °C & P Kbar	558 7.8	572 7.4	527 7.9	520 7.3
10	Berman (2007)	TWQ _2.3, Garnet- Biotite- Plagioclase- Quartz Geothermobarometry	T °C & P Kbar	450 6.5	490 7.6	475 6.7	475 6.7

Fluid Inclusion Study

In Central Crystallines fluid inclusions trapped in minerals are generally 10 to 50μm in diameter. Quartz present in garnet mica schist is the most common mineral for the fluid inclusions study. The doubly polished wafers of the selected samples with thickness of 0.4 to 0.8 mm have taken for fluid inclusion's petrography. The prepared sections were mounted on glass slides by using 'Canada balsam' because it makes easy to remove wafers for thermometry.

The fluid inclusion study of the metapelites of Central Crystallines (MCT zone) (Fig: 2B) and Almora Crystallines (Fig: 2A) is carried out to understand the nature of fluids and their evolution with respect to the metamorphism to correlate and to proof that Munsiari Thrust is the root of the Almora Thrust.

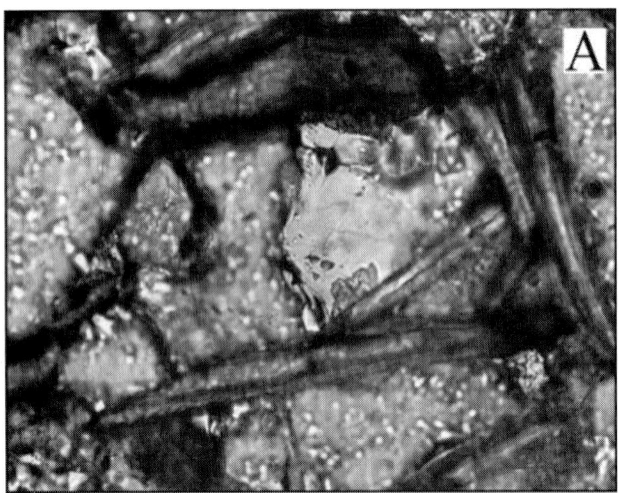

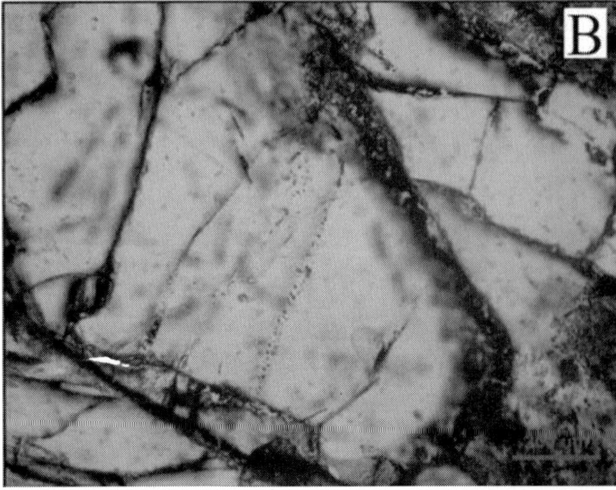

Fig. 2 A. CO_2 rich Fluid Inclusions in a group from Almora Crystallines. **Fig: 2 B.** CO_2 rich Fluid Inclusions in trails from Central Crystallines.

The observed fluid inclusions in Garnet mica schists of Central Crystallines are as follows:

Type (1): Monophase liquid inclusion (H_2O): Commonly filled up with carbonic fluid and less in amount. Mostly < 10 micron in size and sub-rounded or elliptical shaped. They are usually isolated, some were in trails. Various shapes like C- shaped inclusion is observed, which could be an example of exhumation of this mineral. Stretched within inclusions are very common, such shapes may form due to the pressure from the surrounding environment. Type (2): Monophase vapour inclusion (CO_2): Such inclusions were carbonic in nature entirely filled up with a low density vapour. They could be primary in nature. Their size varies <5- 10 microns. Type (3): Liquid rich Biphase inclusion (H_2O+ V): these are of various shapes like rounded, elliptical, cubical or irregular. Some stretched inclusions also observed which is due to the surrounding pressure. Many moving bubbles also observed, it occupies 30- 40 % volume of the inclusion. Type (4): Vapour rich Biphase inclusion (H_2O+ CO_2): Very less in amount. Vapour covers almost 80% area in such inclusions. Type (5): Biphase Immiscible Inclusions (H2O+ CO_2v+ CO_2g): The inclusion contains both liquid & vapour phase. Some immiscible inclusions i.e. bubble inside the bubble is present in a large number. The bubble occupies 70- 80% volume of the inclusion. Type (6): Multiphase Inclusions (H_2O+ CO_2+ NaCl): The inclusions filled with liquid and surrounded with a vapour phase, it may be of carbonic nature and a solid phase of NaCl (halite) crystals also observed. These crystals have perfect crystal boundary. Moving bubbles are also very common.

The observed fluid inclusions in Garnet mica schists of Almora Crystallines are as follows:

Type (1): Monophase liquid inclusion (H_2O): These inclusions filled up with carbonic fluid and are enriched in amount. Mostly < 5 micron in size and have different kind of shape like rounded, sub- rounded, elliptical, elongated. They are usually in groups of 10-20 inclusions, few are isolated or within trails. Few C- shaped monophase inclusions are also observed, which could be an example of exhumation of this mineral. Leaked cavity, stretched inclusion and necking are very common; such shaped inclusions indicate the pressure direction and conditions from the surrounding environment. Type (2): Monophase vapour inclusion (CO_2): Such inclusions were carbonic in nature entirely filled up with a low density vapour. They may be primary in nature. Their size is about 5 microns or is larger than 5 microns. Type (3): Liquid rich Biphase inclusion (H_2O+V): These are very common and abundant type of inclusions and mainly rounded, surrounded, elliptical, cubical or irregular in shape. Various stretched inclusions are observed which the resultant of the surrounding pressure. Many moving bubbles are also observed within it. Commonly it's L: V ratio is 75: 25 volume %. Type (4): Vapour rich Biphase inclusion (H_2O+ CO_2): Such inclusions are less in amount and have surrounded or elliptical shape. Type (5): Biphase Immiscible Inclusions (H_2O+ CO_2v+ CO_2g): These inclusions contain both liquid & vapour phase. Few immiscible inclusions are also observed. The bubble occupies 20- 30% volume of the inclusion.

On the basis of the distribution pattern of observed inclusions in quartz belongs to Garnet mica schist the entrapped inclusions suggests that inclusions present along growth planes and occur in isolated or random distribution are considered as primary in origin (Roedder, 1984). Those inclusions which present in clusters or in group or have a doubtful origin, they may be primary to secondary in origin; the inclusions present within the trails are secondary and are trapped along the cleavage of the host mineral; inclusions trapped in the healed planes/ microcracks are secondary and late in origin; the monophase carbonic inclusions and the biphase carbonic-aqueous inclusions in quartz are primary in origin. They represent the earliest fluid preserved in the metapelites, and are likely to be a pre garnet fluid and are helpful to understand prograde metamorphism of the metapelites. The observed fluid inclusions in Garnet mica schists of Central Crystallines and Almora Crystallines are showing almost same kind of fluid nature i.e. carbonic fluid.

Conclusion

It is not easy to directly support the commonly argued concept of the nappe rocks of the Lesser Himalaya to have the Higher Himalaya as their root zone. On the basis of above investigation it has been observed that there is similarity in mineralogical and geothermobarometric data in both study areas, which is supportive of existence of Central Crystallines as a root zone of Almora Nappe, which was folded and thrusted later along. Occurrence of similar kind fluid inclusions in both Almora Nappe and Central Crystalline are suggestive of preexistence of root zone. Presence of CO_2 and Halite inclusions in Central Crystallines indicate that they were generated at greater depth while the absence of halite inclusions in Almora Crystallines may be due to fluid reequilibriation during thrusting and Nappe formation. However further more advanced and precise studies like is needed to conclude its origin.

Acknowledgements

Authors are thankful to Head, and coordinator CAS, Department of Geology, Kumaun University, Nainital. UCOST is also acknowledged for financial assistance. Authors are grateful to the Director, WIHG for encouragements and providing necessary facilities.

References

Berman, R.G. (2007) 'TWQ' computer programme version 2.32.

Bhattacharya, A., Mohanty, L., Maji, A., Sen, S.K., and Raith, M. (1992). Non ideal mixing in the phlogopite-annite binary: Constraints from experimental data on Mg- Fe partitioning and a reformulation of the biotite-garnet geothermometer. Contrib. Min. Petrol. v.111, pp.87-93.

Dwivedi, S.B., Lalnunmawia, J., Vanthangliana V. and Lal, S.N. (2007). A refined garnet-biotite thermometer: constraints from empirically derived margules parameters for garnet (ΔWca & ΔWMn) and biotite (ΔWAl & ΔWTi). The Indian Mineralogist, v.41, pp.97-111.

Heim, A. and Gansser, A. (1939). Central Himalaya. Soc. Helv. Sci. Nat., Zurich, v.73, pp.1-245.

Gansser, A. (1964). Geology of Himalaya, Interscience Publishers. London, pp.273.

Holdaway, M.J. (2000). Application of new experimental and garnet Margules data to the garnet-biotite geothermometer. American Mineralogist, v.85, pp.881-892.

Holland, T.J.B. and Powell, R. (1998). An internally consistent thermodynamic dataset for phase of petrological interest. Journal of Metamorphic Geology, v.16, pp.309-343.

Joshi, M. (1999). Evolution of the basal shear zone of the Almora Nappe, Kumaun Himalaya. Mem. Gondwana Res. Group. Mem.6, v.6, pp.69-80.

Kaneko, K. and Miyano, T. (2004) Recalibration of mutually consistent garnet-biotite and garnet cordierite geothermometers, Lithos, v.73, pp.255-269.

Lal, R.K. (1991). Empirical calibrations of the Ti-contents of biotite and muscovite for geothermometry and their application to low to high grade metamorphic rocks. Third Indo-Soviet Symposium on experimental Mineralogy and Petrology, New Delhi, pp.18.

Mishra, R. C. and Kumar, S. (1972). Geology of the Danya area, Distt. Almora, U. P. Proc. Ind. National Sci. Acad., v.38, pp.161-166.

Pant, P.D. and Agarwal, P. (2003). Geological and geomorphological investigation of the Kakrighat-Chopra area: Implications to neotectonic movements of the South Almora Thrust (SAT), Kumaun Himalaya, Uttarakhand. Unpublished M.Sc. dissertation, pp.10.

Perchuk, L.L., and Laverent Eva, I.V. (1983). Experimental investigation of exchange equilibria in the system cordierite- garnet- biotite. In: saxena, S. K. (Ed.), Kinetics and Equilibrium in Mineral Reactions, Springer- Verlag, New York, pp.199-239.

Perchuk, L.L., Aranovich, L.Y., Podlesskii, R.K., Laverentiva, G., Kitsseol, V. Y., Kersakov, L.P., and Perdnikov, K.V. (1985). Precambrian granulites of the Aldam shield Eastern Siberia, USSR J. Meta. Geol., v.3, pp.265-310.

Roedder, E. (1984). Fluid inclusions: Reviews in mineralogy, Mineralogical Society of America, v.12, pp.646.

Valdiya, K.S. (1980). Structural setting of the Kumaun Lesser Himalaya, Geology of Kumaun Lesser Himalaya, pp.94-104.

Valdiya, K.S. (1980). Geology of Kumaun Lesser Himalaya, Wadia Institute of Himalayan Geology, Dehradun, pp.191.

Valdiya, K.S., Pal, S.K., Tara Chandra, Bhakuni, S.S. and Upadhyay, RC. (1999). Tectonic and Lithological characterization of Himadri (Great Himalaya) between Kali and Yamuna rivers, Central Himalaya. Himalayan Geology, v.20 (2), pp.1-17.

Wu, C., Zhang, J. and Liu- Dong, R. (2004). Empirical garnet- biotite-plagioclase-quartz (GBPQ) geobarometry in medium-to high-grade metape-lites. Jour. of Petrology, DOI: 10.1093/ petrology/egh038, v.45, pp.1907-1921.

Hon'ble reviwer

1. **Prof. Hari B. Srivastava**
   Head of the Department
   Centre of Advance Study in Gelology
   Banaras Hindu University
   VARANASI - 221 005
2. **Dr. Beena Tripathi**
   Department of Geology
   JNV University
   JODHPUR - 342 005

I-18

Geological Characteristics of Mandai Phosphorite, Barmar Basin, Westeren Rajasthan, India

V.S. Parihar*, S.L. Nama, Snehlata Mathur and S. C. Mathur

Department of Geology, Jai Narain Vyas University, Jodhpur – 342005, India.
Email: geoparihar@gmail.com

Abstract

The Mandai Phosphorite belonging to the Mandai Formation of Middle to Late Palaeocene age cropsout at and around Mandai village (26°22': 71°.08'); district Jaisalmer, western Rajasthan, India. The phosphorite bed of Mandai with an average one meter thickness has estimated reserves of 12 million tones of 10 per cent P_2O_5 contents. It is a potential source of phosphorite for future in the western Rajasthan. Detailed geological analysis of phosphorite bed reveal that intraclasts of internal moulds of bivalve (steinkerns), bivalve hash, and bivalve fragments are the dominant fraction with other orthochemical and allochemical sediments as such it is termed as phosphatic bivalve bed. Detailed field and petrographic studies suggest that the Mandai phosphorite is outcome of complex display of biogenic – diagenetic – reworking process in the realm of high energy, near coastal shallow marine environment.

Keywords: Barmer Basin, Phosphogenesis, Mandai Phosphorite, Bivale Bed, Steinkerns, Coastal Environment.

Introduction

Phosphorite is a single unusual lithofacies, received a good deal of attention of geoscientists since long time. It is one of the most extensively studied sedimentary facies, both because of its commercial value as a fertilizer material as well as its importance to sedimentology. It is a complex chemogenic-biogenic lithofacies, sedimentation of which is said to be related to the evolution of atmosphere, hydrosphere, lithosphere and biosphere (c.f. Cook and Shergold, 1986, Chauhan, 1989; Follomi, 1990, Glenn et. al., 1994; Rao et. al., 2000 and Mathur, 2002). The phosphorite facies of the Mandai Formation of the Barmer Basin came into lime light recently to the sceintific world after reporting of phosphatic moulds and cast of the bivalves (Parihar, 2009). In view of the significance of phosphorite facies of the Mandai Formation as future potential as low grade phosphorite, author's have attempted to study by covering basic information on geology, mode of occurrence, microfacies types, their biogenic framework elements along with phosphogenesis aspects.

Geological Framework

The Mandai phosphorite is located about 15 km south -west of Fatehgarh, which is situated on Fatehgarh –Jhinjinyali tar Road (Fig. 1).

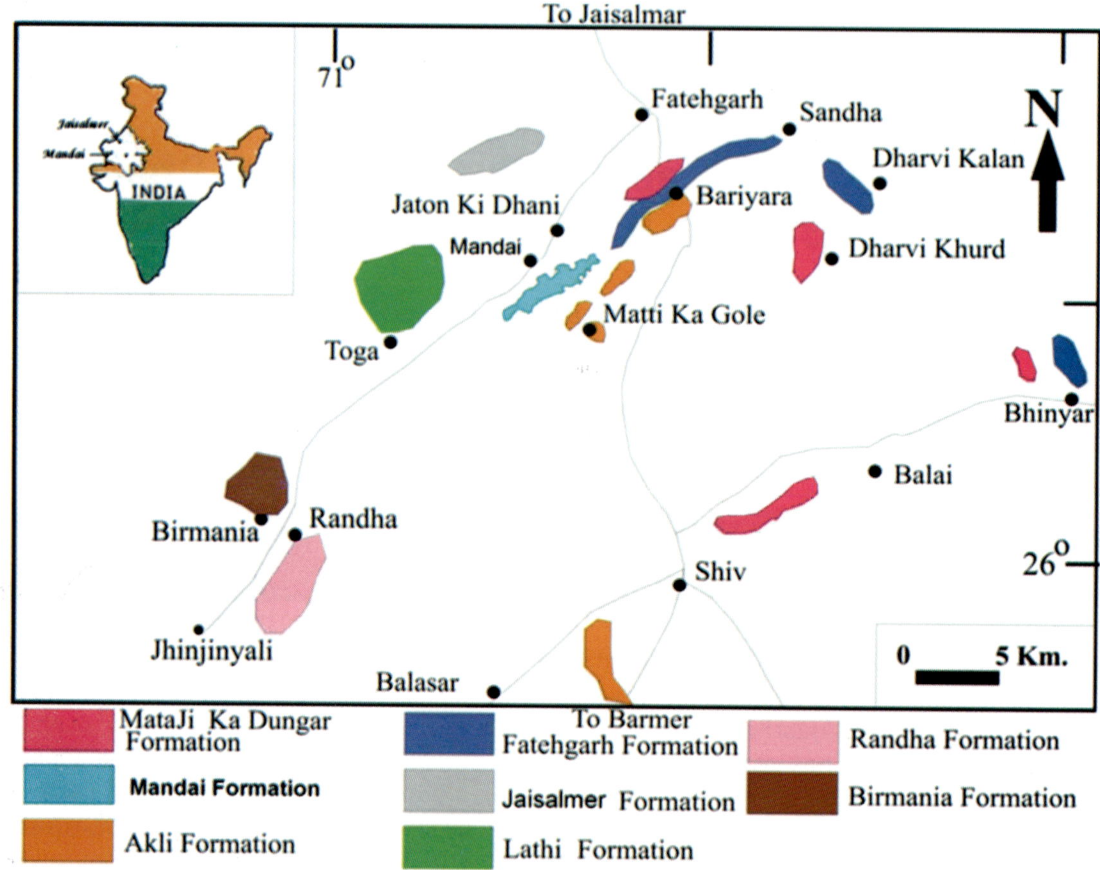

Fig. 1 Location map of the study area.

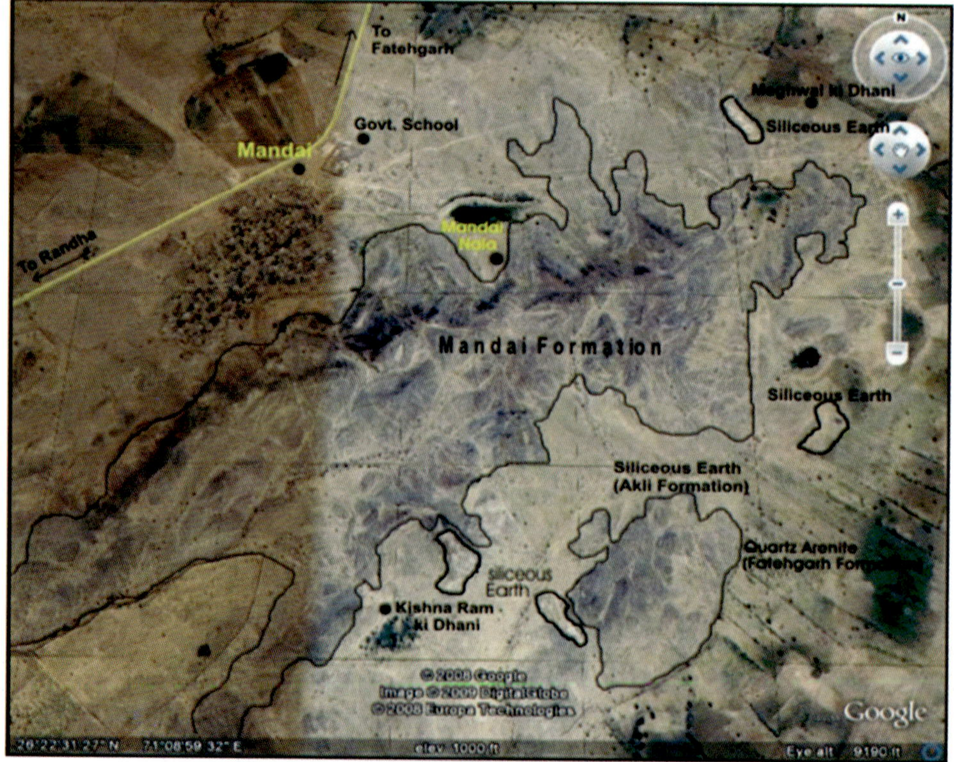

Fig. 2 Interpretative geological map of outcrops of Mandai Formation (Based on field relationship and Landsat imagery).

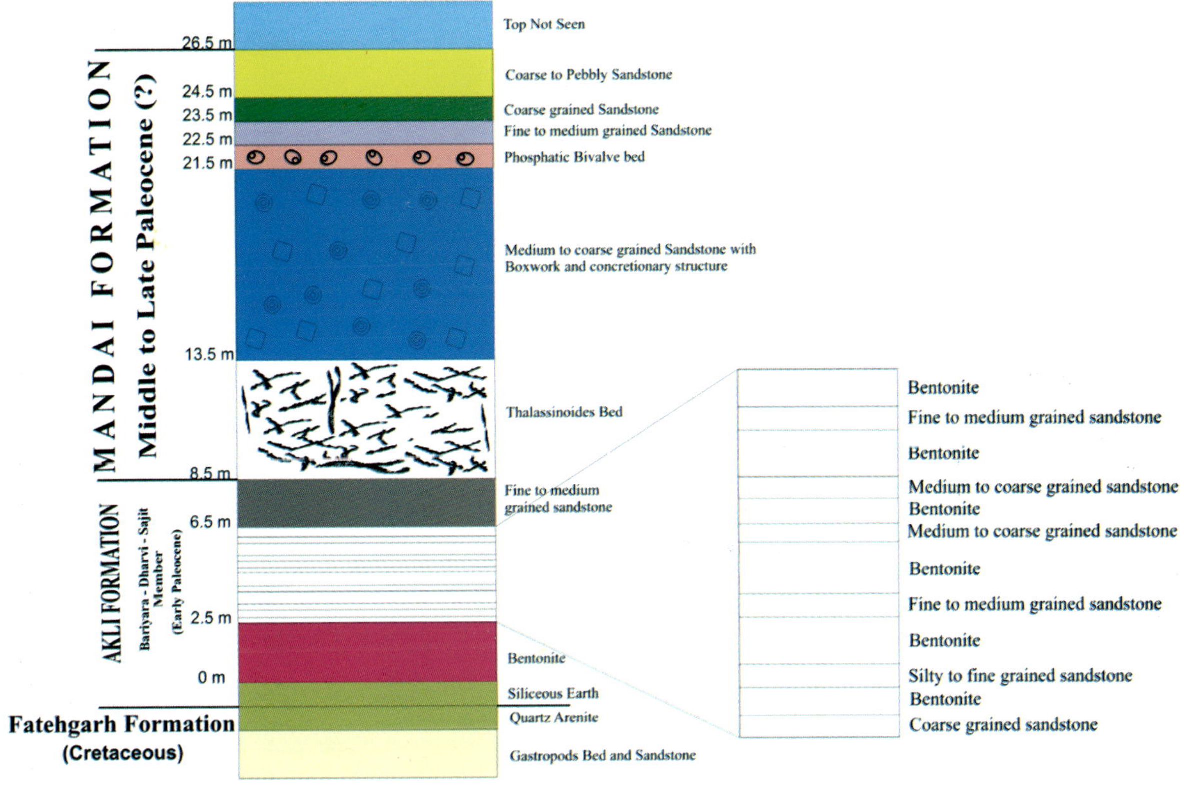

Fig. 3 Generalized lithostratigraphic section of the Mandai Formation measured at Mandai Nala Section.

The Mandai Phosphorite belonging to the Mandai Formation forms a single consistent horizon stretching for a strike length of about 4 kms. Its thickness varies from less than a meter to two meters with an average thickness of one meter. The P_2O_5 content of phosphate bed is 10 percent. The total estimated reserves are 12 million tones (Muktinath, 1967; Pareek, 1984; Roy and Jakhar, 2002). The main exposures of the phosphorite horizon are occurring mainly 0.5 km south of the Mandai village. The good outcrops are also available at 300 metres south and southwest of the Megwalo ki Dhani and about 0.7 km northwest of the Kishna Ram Ki Dhani respectively (Fig. 2).

On the basis of detailed geological studies, Parihar, (2009) named the Mandai Phosphorite bed as Phosphatic bivalve bed as per general practice for Cretaceous Phosphorite in the middle- East deposits (Follomi, 1990) and bone bed in Fatehgarh Formation in the Barmer Basin (Mathur et. al., 2006). The Field relationship indicates that it is sandwiched between fine to medium grained sandstone and Thalassinoides bed through out the study area. In Mandai Nala Section quartz arenite of Fatehgarh Formation (Cretaceous) is exposed at the base followed by repeated beds of sandstone and bentonitic clay of Akli Formation (Early Palaeocene) which in tern followed by sequences of the Mandai Formation (Fig. 3). The underlying and overlying relationship of Mandai Formation is very clear at Mandai village (Plate 1A), the field ralationship and whole section of Mandai Formation can be studied at Mandai Nala Section (Fig. 2 Plate 1B). The Mandai Formation starts with a consistantly thick Thalassinoides bed which contains number of trace fossils vis: *Thalassinoides* sp., *Ophiomorpha* sp. *Burrow Tubes*, *Halymenidium* sp. and *Rhazocorallium* sp.. As such it marks the unconformity between Akli Formation and Mandai Formation. The Thalassinoides bed is conformably overlain by a thick medium to coarse grained sandstone followed by a phosphatic bivalve bed. The fossils contents of these horizons showing Middle to Late Palaeocene affinity (Parihar, 2009). The phosphatic bivalve bed is conformably overlain by a coarsening upward siliciclastic sequence which starts with siltstone at the base followed by fine to medium grained sandstone, medium to coarse sandstone and pebbly to conglomeratic sandstone at the top (Parihar, 2009).

Phosphatic Bivalve bed

Field and Megascopic Characters

The phosphatic bivalve bed is brown to yellow in colour and megascopically it is medium grained in nature. This phosphorite horizon can be distinguished from the rest of the beds by its surficial azur blue and yellow colour (Plate 1C). This character has been of a great help in identifying, trenching and sampling this horizon. However, it is noted that there are considerable variation in composition, thickness and overall nature of the phosphorite horizon of the Mandai area. The interesting feature in phosphorite horizon can be observed by the presence of phosphatic intraclasts, internal moulds of bivalves (steinkerns), fragments of bivalve shells and bivalve hash (c.f. Dickinson and Wallace, 2009) with in a matrix of sediments that includes silt and medium grained sand of quartz grains (Plate 2A and 2B). Internal moulds and shell material are actually of Heterodont bivalves. The matrix supported beds are poorly to moderately sorted and massive in nature. Some elongated bivalve moulds do show preferential orientation to the horizontal, but generally no structures or bedding is observed accept at some places, where it forms lenticular beds.

Petrography

Petrographic attribute of the bivalve bed indicate that it is mainly composed of monocrystalline quartz grains which are cemented chiefly by phosphate cement, less commonly with iron oxide, and carbonate cement. Quartz as a chief ingredient of this lithounit, represented by 0.03 to 1.8 mm size fraction suggesting that it is dominantly a medium grained sandstone with some silt size grains and sporadic coarse quartz grains largely forming the matrix of the rock (Plate 2C). The bivalve intraclasts includes bivalve hash; phosphatic and non-phosphatic internal moulds of bivalve (steinkerns); broken shaells of bivalve also form major part of the rock. These framework elements are ranging in size from few mm to even cm or more and are cemented by phosphate, iron oxide and calcite cement. Among extraclasts, beside quartz grains mainly include heavy minerals i.e. well rounded grains of tourmaline, zircon and rutile.

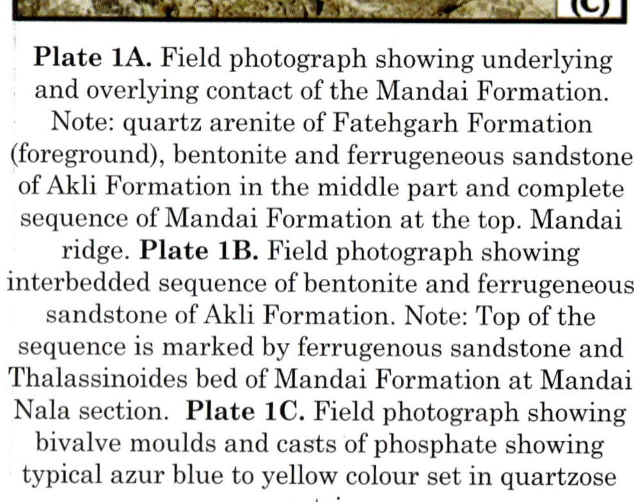

Plate 1A. Field photograph showing underlying and overlying contact of the Mandai Formation. Note: quartz arenite of Fatehgarh Formation (foreground), bentonite and ferrugeneous sandstone of Akli Formation in the middle part and complete sequence of Mandai Formation at the top. Mandai ridge. **Plate 1B.** Field photograph showing interbedded sequence of bentonite and ferrugeneous sandstone of Akli Formation. Note: Top of the sequence is marked by ferrugenous sandstone and Thalassinoides bed of Mandai Formation at Mandai Nala section. **Plate 1C.** Field photograph showing bivalve moulds and casts of phosphate showing typical azur blue to yellow colour set in quartzose matrix.

The petrographic studies show that the quartz grains are highly corroded by the phosphate. One such quartz grains can be seen in the center of the photo (Plate 2C) in which due to replacement by phosphate, skeletal quartz grain is left behind. Generally, phosphate cement form rims and envelops around the quartz grains (Plate 2C). Some time due to increasing amount of phosphate cement the quartz detritus seems to be float in the phosphate groundmass cement (Plate 2D). Beside phosphate cement there is occurrence of microsparitic to sparitic carbonate cement around the quartz grains (Plate 2C).

The phosphatic cement around intraclasts consists of microcrystalline collophane, replacing carbonate mud (Plate 2C). The petrographic studies indicate that phosphatic and non-phosphatic intraclasts of bivalve internal moulds comprised five different types of bivalve component.

(a) Bivalves remains mainly composed of carbonate shells and their infilling material are entirely composed of carbonate (Plate 2E and 2F).

(b) Bivalve remains consist of carbonate shells filled up by intraclasts comprises of quartz with carbonate cement (Plate 3A).

(c) Bivalve remains composed of partially phosphatized shells filled up partly with carbonate and partly with phosphate (Plate 3B and 3C).

(d) Bivalve remains consist almost fully of phosphatized shells and are filled up by phosphate and there by formed composite phosphate mould (Plate 3E).

(e) Bivalve remains composed dominantly of iron oxide shells and is filled up by intraclasts of quartz dominantly cemented by iron oxide and there by formed composite moulds of ironoxide (Plate 2F).

These components are either with complete shells or broken shells embedded in a matrix of fine to medium grained quartz matrix and are cemented mainly by sparitic carbonate and phosphate cement. As such they can be termed as bioclastic limestone and or biosparite (c.f. Folk, 1962). Further, it is noted that the bivalve component (a) is purely composed of carbonate shells of sparitic nature. Chambers of valves of bivalves are infilled by micrite to microsparite calcite often stained with iron oxide giving brownish appearance. The infilling material is devoid of any detritus, indicating their primary nature. The surrounding matrix of bivalve remains comprises mainly angular to sub-angular quartz grains of silt and sand size along with skeletal debris which are cemented by sparry calcite. The bivalve component (b) is consisting of shells of sparry calcite. The infilling material and matrix are of the same material, comprising silt to sand size, angular to subangular quartz grains, shale fragments with sparry calcite matrix, suggest dissolution followed by syngenetic deposition of both (Plate 2F). The shells of the bivalve component (c) are consisting of sparry calcite. The shells are normally unaffected by replacement process, only few fractures are filled up by phosphate (Plate 3B). The carbonates infilling (primary) are corroded chiefly at the periphery of the shell. The constituent quartz grains are also highly corroded and replaced by collophane. It is evident from their smaller size in term of strings, blebs and relict structures (Plate 2D). The bivalve shells of component (d) are almost dominantly composed of collophane. The shells of bivalve display fibrous franolitic composition of grey colour. The infilling material is mainly cryptocrystalline collophane of dark grey to black colour there by formed internal moulds of phosphate or stienkers of Dickinson and Wallace, (2009). The small amount of unreplaced relict part of calcite and quartz are often present (Plate 3E). The surrounding material of bivalve moulds are mainly made up of angular to sub-rounded sand size of quartz with small amount of microsparitic to sparitic calcite and phosphate cement. The nature of infilling collophane and cement phosphate are similar, indicating same stages of phosphate diagenesis.

The similar types of phosphate internal moulds of gastropodsand bivalves (steinkerns) fron Gippsland Basin of SE Australia (Dickinson and Wallace, 2009), gastropods moulds from Negav Phosphorite of Israel (Soudry and Lewy, 1988; Abed, 1990) and gastropods moulds from phosphorite of the Fatehgarh Formation (Gour, 2005) have been only the reports from the world.

Cement Phosphate

As mentioned earlier, the cement phosphates present with bivalve bed are mainly occur in three forms viz. rim, interstitial and groundmass cement. It has been observed that when the amount of phosphate in a rock is too small, i.e. 1 to 5 percent, occur with in a primary intraparticle and interparticle porosity. When amount of cement phosphate increases it forms thin rims (2-

5 μm) around quartz grains and bioclasts (Plate 2C). The increasing amounts of cement phosphate occur as groundmass cement. Normally, phosphate cement appears to be composed of cryptocrystalline collophane (Plate 2D).

The common occurrence of collophane rim and evelops around phosphate clasts, rounded quartz and bioclasts suggest that it must have formed under moderately agitated water condition. A corollary can be drawn from the processes of formation of carbonate oncollites (See Wilson, 1975; Flugel, 1982). Accordingly, it appears that when quartz grains and bioclasts move back and forth in shallow near coastal environmrnt by moderately strong currents over substrates covered with a thin layer of phosphate mud (microsphorite), they picked up small amount of phosphate around them to form rim and evelops around the clasts.

Calcite Cement

Calcite is ubiquously present in the phosphorite of the area. Three generations of carbonate minerals are evident:

(i) Micritic calcite or aragonite occurs as primary carbonate forming the shells of bivalve and skeletal debris (Plate 3A).

(ii) Sparry calcite forming the infilling material of bivalve remains and as cementing material for bioclasts and detritus (Plate 3B).

(iii) Microsparitic calcite occurring in associations with all the varieties of phosphate of the area as pervading fronts and mosaics.

Initially, bivalve shells, and skeletal material mainly composed of aragonite and or calcite of primary and fine grained in nature, which have been subsequently phosphatized as discussed above.

Iron Oxide Cement

Iron oxide cement similar to calcite occurs in almost all types of phosphate. They are occurs as thin rim and coating of hematite, isolated patches and as pore filling cement (Plate 3C).

Depositional Environment

The lenticular geometry of the phosphorite beds of Mandai Formation with their common association with moderatly sorted quartz detritus, Heterodonts shells, bivalve hash, and shell fragments and their cement nature indicates deposition in a high energy, shallow to near coastal areas of the Mandai Sea with finer material being winnowed away.

Phosphogenesis

Scanning of Litrature reveal that most of the concepts of phosphogenesis swing between unorthodox approach and uniformitarian concepts. At times, one thinks that the problem of phosphogenesis is finally settled but the incoming literature (c.f. Glenn, et. al., 1994; Mathur, 2002; 2008, Gour, 2005, Parihar, 2009 and Dickinson and Wallace, 2009), elucidates that the processes of "P" fixation in sediments to form phosphorite is much more complicated. None the less, the investigations so far carried out on the phosphate sediments have settled many important problems. The concepts concerning the intricacies of phosphorite research carried out so far mainly hinges around the following phosphogenic models:

(i) Direct precipitation model (Kozakov, 1937; Mckelvey et al., 1953; Sheldon, 1964, 1980; Freas and Riggs, 1965; Bushinski, 1969; Riggs, 1979b; Lucas and Prevote, 1982, 1984; Glenn et al., 1994).

(ii) Calcium – carbonate replacement model (Cayeus, 1939; Enigh, 1958; Ames, 1959; Rooney and Kerr, 1967; Crok, 1976; Howard and Hough, 1979; Chauhan, 1979, 1980; Cook and Shergold, 1986 b; Heggie et al., 1990; Nathan et al., 1990; Rao et al., 2000).

(iii) Primary biogenic precipitation model (See McConnel, 1973; Riggs, 1979 a; Sheldon, 1981; Cook and Shergold, 1986 b; Rao et al., 2000).

(iv) Biogenic – diagenetic - reworking model (See Shrakov, 1960, 1962; Bushinski, 1966, 1969; Baturin, 1982; Cook, 1984; Cook and Shergold, 1986 b; Mathur, 1987; Mathur and Kumar, 2000, Follmi, et. al., 2008).

In view of the field characters and petrography of the phosphorite of the area, it is suggested that they are outcome of complex multistage processes of direct primary precipitation (microsphorite layers); calcium carbonate intraclast replacement by phosphate followed biogenic –diagenetic and reworking mechanism to form bivalve bed of Mandai Formation.

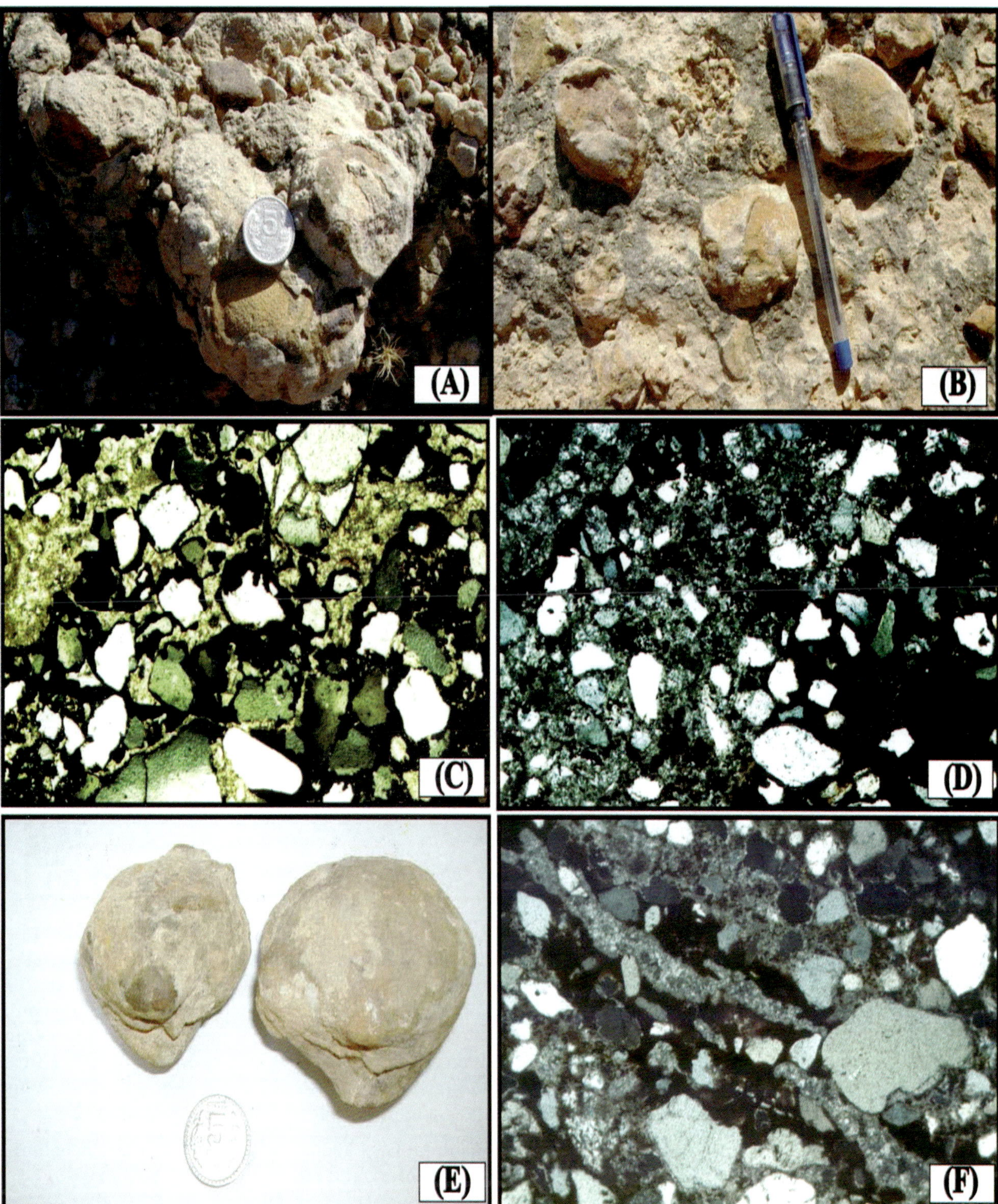

Plate 2A and B - Bivalve bed showing well to ill preserved bivalve shell. They are Hetrodonts bivalve shell of carbonate set in the silt to sand size quartzose matrix. **Plate 2C.** Photomicrograph showing of phosphatic intraclasts (bivalve fragments) with partial to total phosphatization of the infilling sediments. Note : Thin rim and interstitial phosphate cement around quartz grains. Crossed Nicols 5x. **Plate 2D.** Photomicrograph of bivalve bed showing increasing amount of phosphate cement in which quartz grains are highly corroded and floating in the phosphate groundmass cement. Crossed Nicols 10x. **Plate 2E.** Hand specimens of bivalve remains showing carbonate shell. **Plate 2F.** Photomicrograph of bivalve moulds of carbonate in which infilling material is dominantly composed of carbonate. Crossed Nicols 10x.

Plate 3A. Photomicrograph of internal moulds of bivalve showing partially phosphatized shell filled up partly by carbonate and phosphate minerals. Crossed Nicols 5x. **Plate 3B.** Photomicrograph showing infilling material of internal moulds of bivalve in which iron oxide cement replaces carbonate and phosphate cement. Crossed Nicols 5x. **Plate 3C.** Photomicrograph of bivalve remains showing composite phosphate moulds. Crossed Nicols 5x. **Plate 3D.** Field photograph showing iron oxide cement replaces the bivalve shell completely with its infilling matrix and converted it into ferrugenized bivalve moulds. **Plate 3E.** Shells of bivalve of calcite showing fractures filling of phosphate. **Plate 3F.** Field photograph showing phosphatized internal moulds (steinkerns) of bivalves which are set in the siliciclastic matrix.

References

Abed, A.M., and Fakhouri K., 1990. Role of microbial processes in the genesis of Jordanian upper Cretaceous phosphorite. Geol. Soc. London Spe. Pub., v. 52, 193-203.

Ames, L.L., 1959. The genesis of carbonate apatities. Economic Geology, v. 54, 829-841.

Baturin, G.N., 1982. Phosphorite of the seafloor origin, Composition and distribution. Elsevier New York, p. 343.

Bushinki, G.I., 1966. The origin of marine phosphorite : lithology and mineral resources. v. 3, 292-311.

Bushinki, G.I., 1969. Old phosphorites of Asia and Genesis. ISR. Prog. Sc. Transl. Jerusalem, p. 266.

Cayeus, L., 1939. Existences de nombrcuses bacteries dans les phosphates sedimentation de tout age – consequences C.R. Acad. Sci., Paris, 203, 1198-1200.

Chauhan, D.S., 1979. Phosphorite-bearing stromatolites of the Pre-Cambrian. Aravalli phosphorite deposits of the Udaipur region: their environmental significance and genesis of phosphorite. Precambrian Research, 8, 95-126.

Chauhan, D.S., 1980. Stromatolites of Nimachamata and Bargaon phosphorite deposits and genesis of rock phosphate. Proceedings workshop on stromatolites. Geology Survey of India, Miscellaneous Publication, 44, 314-329.

Cook, P.J., 1976. Sedimentary phosphates deposits. In: Wolf, K.H. (Ed.) Handbook of strata bound and stratiform ore deposits' Vol., New York, Elseveir 8, 525-535.

Cook, P.J., and Shergold, J.H., 1984. Phosphorus – phosphorites and skeletal evolution at P/C boundary. Nature, 308, 231-236.

Cook, P.T., and Shergold, J.H., 1986a. Proterozoic and Cambrian phosphorites an introduction In Cook P.T. and Shergold J.N. (Ed.), world phosphorites deposits. Cambridge University Press London, 1, 1-8.

Cook, P.T., and Shergold, J.H., 1986b. Proterozoic and Cambrian phosphorites – Nature and origin, In P.T. Cook and J.N. Shergold (Ed.) world phosphorites deposits. Cambridge University Press London, v. 1, 368-386.

Dickinson, J.A. and Wallace, M.W., 2009. Phosphate-rich deposits associated with the Mio-Pliocene unconformity in south-east Australia. Sedimentology (2009), Volume 56, No. 2, pp. 547-565.

Emigh, G.D., 1967. Petrology and origin of phosphorite in Hale, L.A. (Ed.) Anatomy of Western phosphate fields, Inter mountain Assoc. Geol., Salt lake city, Utah. 193-214.

Flugel, E., 1982. Microfacies Analysis of Limestone New York Springer Verlag, 633.

Follmi, K.B., 1990. Condensation and phosphogenesis of Helvetic Mid- Cretaceous (Northern Tethyan Margin). Sp. Pub. Geol. Soc. London, 52, 237-252.

Follmi, K.B., Gertsch, B., Renevey, J.P., Kaenel, De. E. and Stille, P., 2008. Stratigraphy and Sedimentology of phosphate-rich sediments in Malta and South –Eastern Sicily (latest Oligocene to early Late Miocene). Sedimentology (2008), Volume 5S, No. 4, pp. 1029-1051.

Folks, R.L., 1980. Petrology of sedimentary rocks. Hemphils, Austin, Texas, p. 170.

Freas, D.H., Riggs, S.R., 1965. Stratigraphy and sedimentation of phosphorite in central Florida phosphate districts. Am. Inst. Min. Eng. 65 H 84, 1-13.

Glenn, C.R., Follmi, K.B., Riggs, S.R., Baturin G.N., K.A., Trappe, J., Abed, A.M., Galli- Oliver, C., Garrison, R.E., Ilyin, A.,Jehl, C., Rohrlich, V., Sadaqah, R., Schidlowski, M., Seldom, R.E., and Stegmund, H., 1994. Phosphorus ans Phosphorites: Sedimentology and environment of formation of Eclogae Geol. Helv., v. 87, 747-784.

Gour, S.D., 2005. Geological Investigation of Phosphorite and associated Lithofacies of Sandha – Bhiyar area of Barmer District, Western Rajasthan, India. Unpublished Ph.D. thesis, J. N. Vyas Univ., Jodhpur, p.179.

Heggie, D.T., Skyring, G.W., O' Brien, G. W., Remenns, C., Herezeg, A., Moriarty, D.J.W., Burnet, W.C. and Milness, A.R., 1990. Organic carbon cycling and modern phosphate formation on the east Australian continental margin an overview: In: Nothalt, A.J.G. and Jarvis, I. (Ed.) phosphorite research and development by Geol. Soc. London. Spec. Publ., v. 5, 87-117.

Howard, P.F. and Hough, M.J., 1979. On the geochemistry and origin of the D. Tree, Wonarah and Sherrin Creek phosphorite deposits of the Georgina Basin. northern Australia Economic Geology, v. 74, 260-284.

Kozakov, A.V., 1937. The phosphate facies and the genesis of phosphorite in Geological investigation of Agricultural ores. Leningrad Sci. Inst. Fert., And Insect of fungicides Trans., 142, 95-113.

Lucas, J., and Prevote, L., 1982. The synthesis of Apatite by bacterial activity mechanism. In phosphorite In : Lucas, J., and Prevote, L., (Ed.) Sci. Geol. Mem., v. 77, 83-92.

Mathur. S. C., 1987. Geological Investigation of Birmania Phosphorite, western Rajasthan University of Jodhpur p.120 (Unpublished Ph.D. Thesis).

Mathur, S.K., 2002. Geological investigation of phosphorite and associated lithofacies of Fatehgarh area, western Rajasthan, India. Unpub. Ph.D. thesis. J.N. Vyas Univ, Jodhpur, p. 147.

Mathur, S.C., S.K. Mathur and R.S. Loyal, 2006. First Report of microvertebrate assemblage from Cretaceous Fatehgarh Formation. Geological Society of India, v. 67, No. 6.

Mathur S.C., 2008. Cretaceous phosphorite facies of the Barmer Basin, Western Rajasthan, India, Significance of its fauna and its magnetic framework elements. (Abstract), 33 IGC, 2008.

Mc. Connel, D., 1973. Apatite, its Crystal chemistry, Mineralogy, Utilization and Geologic and Biologic occurrences. Springer New York, p.111.

Mc. Kelvey, V.E., Swanson, R.W. and Sheldon R.P., 1953. The Permian phosphorite deposits of western united states 19th session of the Int. Conr. Algiers Section, v. 11, 45-64.

Muktinath, 1967. Phosphate deposits of Rajasthan Indian Min., v. 21, 83-101.

Muktinath, 1969. Phosphate deposits of Rajasthan Indian Min., v. 23, 29-42.

Nathan, Y., Soudry, D. and Avigur, A., 1990. Geological signification of carbonate substitution in apatite: Israel's Phosphorites as on example. In Nothalt A.J.G. and Jarvis I. (Ed.). Phosphorite Research and development Geological Society, London, Special publication 52, 179-191.

Parihar, 2009. Geological investigations of phosphorite and associated lithofacies of Mandai area, district Jaisalmer, Western Rajasthan India. Unpublished Theisis 178p.

Pareek, H.S., 1984. The Quaternary geology and mineral resources of North-western Rajasthan. Memmo. Geo. Sur. India 115, p.1-99.

Rao, Purnachandra, K., Mohan Rao and Raju D.S.N., 2000. Quaternary Phosphorite from the continental margin of Chennai, Southeast India. Journal of sedimentary Research, v. 70(5), 1197-1209.

Riggs, S.R., 1979b. Phosphorite Sedimentation in Florida: A Model of Phosphogenesis. Economic Geology, v. 74, 285-314. Riggs, S.R., 1979a. Petrology of the Tertiary phosphorite system of the Florida, economic Geology. 74, 195-220.

Roy, A.B. and Jakhar, S., 2002. Geology of Rajasthan (North West India) Precambrian to Recent. Scientific Publishers, (India), Jodhpur, p. 421.

Ronney, T.P., and Kerr P.F., 1967. Mineralogic nature and origin of phosphorite Beaufort country North Caralina. Geol. Soc. America Bull, v. 78, 731-748.

Sheldon, R.P., 1964. Palaeolatitudinal and paleogeographic distribution of phosphorite. Prof. Paper U.S. Geol. Surv., 501 C, 106-113.

Sheldon, R.P., 1980. Episodicity of phosphorite deposition and deep ocean circulation a hypothesis in marine phosphorites Betor Y.K. Soc. (Ed.) Geochemistry, Occurrence, Genesis Econ. Paleontal. Mineral Spec. Pub., v. 29, 239-247.

Sheldon, R. P., 1981. Ancient Marine phosphorite. Ann. Rev. earth planet Science. 251-284.

Soudry, D. and Lewyz. 1988. Microbiolly influenced formation of phosphate nodules and Mega fossils mould (Negav Southern Israel) Palaeogeogr. palaeoclimatol. v. 64, 15-34.

Wilson, J.L., 1975. Carbonate Facies in Geological History Springer New York. p 471.

Hon'ble reviewer

1. **Prof. V.C. Tiwari**
   Senior Scientists
   Wadia Institute of Himalayan Geology
   DEHRADUN - 248 001
2. **Prof. Viqar Husain**
   Department of Geology
   Karanchi University
   KARANCHI (Pakistan)

I-19

An Economic Analysis of Ball Clay Mining in Bikaner

Sharma Shishir*, Sharma Rajni and Singhal Parul****

*Department of Geology, Govt. Dungar College, Bikaner
** Department of Economics, Govt. M.S.College for Girls, Bikaner
E-mail: shishirsharma5@gmail.com

Abstract

The Bikaner district is gifted with huge deposits of Ball clay, gypsum, limestone, and lignite. Ball clay is the chief raw material for the ceramic industry. Ball clay is a plastic sedimentary clay. It imparts high green strength and good workability to pottery bodies. The name 'Ball clay' derives from the original method of production which consisted of cutting out the clay in open pits into cubes or balls. The Bikaner district alone accounts for more than 90% of Rajasthan's total Ball clay production. There are 71 (59.17%) leases of Ball clay in Bikaner district. Ball Clay has a significant position in the economy of Bikaner and has vast potential use in ceramics, cement industry, refractories, cosmetics and rubber industry. Data of last 10 years of ball clay production proves its vitality in Bikaner. The maximum clay production comes from Bikaner. From the mineral point of view Bikaner district can be called the clay district.

On the basis of available data regarding Ball Clay of Bikaner, a Time Series Analysis related with Production Revenue and Sale value of ball clay has been carried out. Their trend values are calculated by adopting ordinary least square technique (OLS technique). A sharp decline in the contribution in revenue from ball clay had been observed. The reasons behind the poor performance are :i. Unsystematic and un-organized mining, ii. Lacking of proper mechanization on the mines, iii. using obsolete technology in the mining of ball clay, iv. Lack of development of ceramic industries at local level in Bikaner and lack of ceramic education. The status of Ball Clay mining in Bikaner as revealed by statistical analysis is quite prosperous and has a good potential to flourish the economy of the district. But despite the large deposits of clay, the economy of Bikaner is not getting as significant profit as it must be. Above mentioned factors are responsible for the deterioration in production, revenue and employment potentials of clay.

Keywords: Ball Clay, Bikaner Basin, Clay Mineral Economics, Ceramic Industry

Introduction

The Bikaner district is largely gripped by Great Indian Desert of Thar and falls into arid and semi-arid climatic zone and has common features of harsh climate, erratic rain fall, extreme temperature variation, constant water shortage and frequent drought. But despite of these adverse conditions, the district is gifted with huge deposits of non-metallic minerals including Ball clay, gypsum, bajari, limestone, and lignite and masonry stones. Ball clay is the chief raw material for the ceramic industry. Therefore, the present work is concentrated on various aspects of economic analysis of Clay Mineral.

Ball Clay

"Most hard rocks when they have undergone the prolonged action of the natural agencies which cause decay in the upper portion of earth's crust, yield some of the fine grained material, we call clay (J.Allen Howe, 1914).

The word clay indicates a material which is generally white creamish in colour and have fine size of particles. On the basis of the physical and refractory properties, the clay can be classified into three categories-china clay, ball clay and fire clay (Bateman, A.M, 1996,).

Ball clay is a plastic sedimentary clay. It imparts high green strength and good workability to pottery bodies. The name 'Ball clay' derives from the original method of production which consisted of cutting out the clay in open pits into cubes or balls. The primary mineral phase is Kaolinite. Micaceous minerals and quartz are the dominent minerals present as impurities along with minor amount of tourmaline, feldspar, chlorite, montmorillonite, interstratified clay minerals, various titanium compounds, siderite, pyrites, limonite and haematite.

Ball clays have no definite chemical composition and in that respect they do not differ greatly from china clays However, it often contains a large proportion of silica and is more plastic. More than 850/0 of the particles are finer than 1 micron.

Distribution of Clay Mineral

The Bikaner district alone accounts for more than 90% of Rajasthan's total Ball clay production. There are 71 (59.17%) leases of Ball clay in Bikaner district. The clay is fine grained, plastic, and free from grits, whitish grey in colour with occasional iron stains. Ball Clay deposits are found in following places:

Khari Charnan, Guda, Kotri, Mudh, Kolayat (2 MT), Inda Ka Bala (2MT), Madhogarh (0.10 MT), Nal (.46MT), Guda (2 MT.)Palana (2 MT), Barsingsar (15 MT), Deh and Guda (DMGR,2008).

Table 1: Bikaner Mineral: Number of Leases of Ball Clay (1998-99 to 2007-08)

Years	1998-99	1999-00	2000-01	2001-02	2002-03	2003-04	2004-05	2005-06	2006-07	2007-08
No.	78	78	78	72	73	72	73	74	74	71
Total	164	120	115	114	110	105	107	110	113	120
%	47.56	65.00	67.83	63.16	66.36	68.57	68.22	67.27	65.49	59.17

Source: DMGR

Table 2: Bikaner Mineral : Mining Area of Ball Clay (1998-99 to 2007-08)

Years	1998-99	1999-00	2000-01	2001-02	2002-03	2003-04	2004-05	2005-06	2006-07	2007-08
Area	7819.35	7819.35	7819.35	6993.33	6998.33	6873.46	6392.34	5758.98	5272.41	5012.82
Total	14464.71	13910.54	13863.9	13981.89	14326.39	12147.87	11480.9	11473.08	11962.49	12117.91
%	54.06	56.21	56.40	50.02	48.85	56.58	55.68	50.20	44.07	41.37

Source: DMGR

Table 3: Bikaner Mineral :Production in Thousand Tonnes of Ball Clay (1998-99 to 2007-08)

Years	1998-99	1999-00	2000-01	2001-02	2002-03	2003-04	2004-05	2005-06	2006-07	2007-08
Production	492.097	658.899	698.426	704.879	792.820	878.542	590.000	320.000	584.500	892.930
Total	1888.772	2260.313	2670.998	2905.984	3348.525	3789.955	3651.382	3187.767	4398.035	6030.461
%	26.05	29.15	26.15	24.26	23.68	23.18	16.16	10.04	13.29	14.81

Source: DMGR

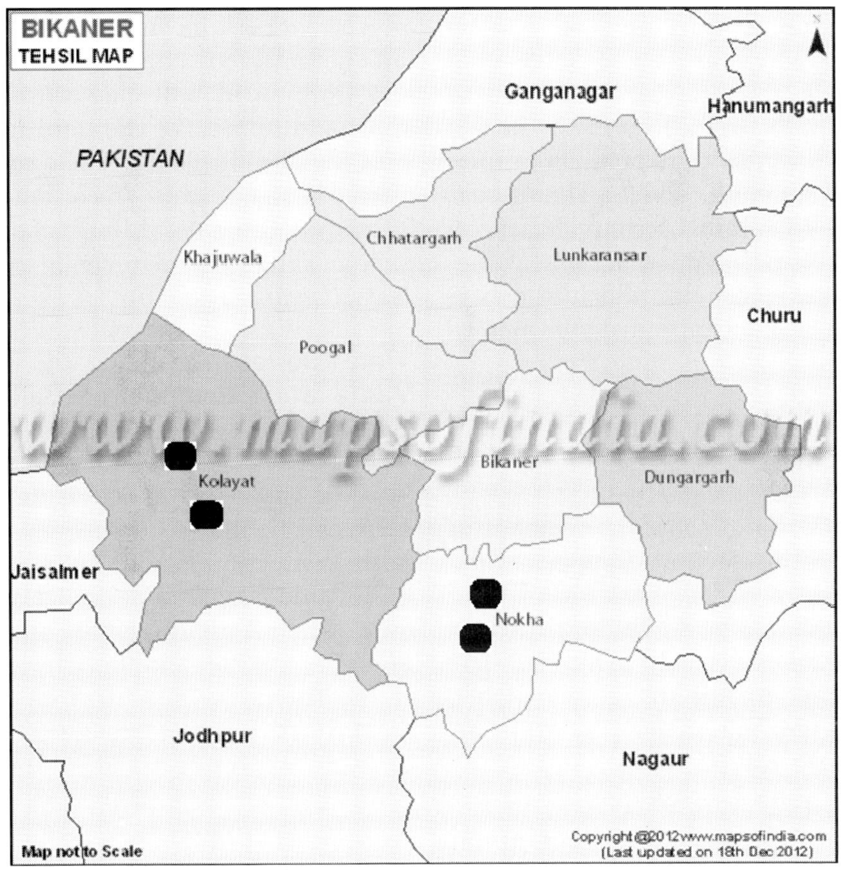

Fig. 1: Tehsil map of Bikaner showing the Ball Clay Mining Areas in Bikaner District

Table 4: Bikaner Mineral :Sale Value in Lac Rupees of Ball Clay (1998-99 to 2007-08)

Years	1998-99	1999-00	2000-01	2001-02	2002-03	2003-04	2004-05	2005-06	2006-07	2007-08
Sale Value	885.770	1015.324	1170.838	1409.750	1585.640	1781.00	1181.00	640.000	1753.000	4911.115
Total	3010.67	3552.092	4033.44	3966.95	4439.66	5331.01	3433.22	4600.705	10407.563	18572.986
%	29.42	28.58	29.03	35.54	35.72	33.41	34.40	13.91	16.84	26.44

Source: DMGR

Table 5: Bikaner Mineral: Revenue in Thousand Rupees of Ball Clay (1998-99 to 2007-08)

Years	1998-99	1999-00	2000-01	2001-02	2002-03	2003-04	2004-05	2005-06	2006-07	2007-08
Revenue	10515.00	12812.25	14326.524	16692.470	17642.000	17240.256	17157.830	13444.686	17562.322	22617.812
Total	36922.83	38275.75	52240.59	52277.64	57047	72712.78	71981.19	64342.52	109039.44	123834.03
%	28.48	33.47	27.42	31.93	30.93	23.71	23.84	20.90	16.11	18.26

Source: DMGR

Table 6: Bikaner Mineral: Employment by Ball Clay (1998-99 to 2007-08)

Years	1998-99	1999-00	2000-01	2001-02	2002-03	2003-04	2004-05	2005-06	2006-07	2007-08
Employment	504	442	370	378	380	459	250	125	250	575
Total	3341	3622	4235	4239	3175	3722	3789	2935	1770	2726
%	15.09	12.20	8.74	8.92	11.97	12.33	6.60	4.26	14.12	21.09

Source: DMGR

Although status regarding all these parameters of ball clay has shown increasing trend over the period of ten years (1998-99 to 2007-08) despite having less values in the period of 2004-2006 which was due to the ban on mining by the honorable supreme Court. The actual impact of this ban on mining activity has been reflected in the value of production, sale value, revenue and employment field, which shows a sharp decline in the above mentioned parameters in the year 2006-07.

Statistics and Time Series Analysis of Ball clay mining in Bikaner

The Table 1 to Table 6 show the Number of leases, Mining area, Production, Revenue, Sale Value and Employment in Bikaner in Ball Clay and their percentage wise contribution in these parameters. These Tables clearly point out that Ball clay is the chief contributor in the above mentioned parameters in major minerals in Bikaner from 1998-99 to 2007-08.

Ball Clay has a significant position in the economy of Bikaner and has vast potential use in ceramics, cement industry, refractories, cosmetics and rubber industry. Data of last 10 years of ball clay production proves its vitality in Bikaner. The maximum clay production comes from Bikaner. From the mineral point of view Bikaner district can be called the clay district. Out of 85 major mining leases in this district 71 are for ball clay (Table 1). Total mining area in 2007-08 is 5012.82 hectare which was 7819.35 hectare in 1998-99 (Table 2). This declination in mining areas and mining leases also reflects in production too. In 1998-99, the percentage contribution of ball clay was 26.05 % (492.097 thousand tonnes) of the total mineral production of Bikaner (Table 3). This contribution in the year 2007-08 was recorded 14.81 % only. However, the sale value increased more than 5 times (Table 4) and revenue just doubled in the last 10 years (Table 5).

Time Series Analysis of Production, Sale Value and Revenue of Ball clay mining in Bikaner

On the basis of these data regarding Ball Clay of Bikaner, a Time Series Analysis related with Production (Table 3) Revenue (Table 5) and Sale value (Table 4) of ball clay has been carried out. Their trend values are calculated by adopting OLS technique. The results are depicted in the Table 7:

Trend equation: $Y = a+bt$

Where,

a = intercept

b = coefficient of trends

t = time

Significance of coefficient is indicated by 't' value as * at 5 % level.

Table 7 presents the results of least square trend equations of time series data of ten years regarding production, revenue and sale value of Ball clay. 't' value is marked by *, at 5% level of significance.

Table 7: Results of Time Series Analysis of Production, Revenue and Sale Value of Ball Clay in Bikaner.

Minerals	Trend equation of Production	Trend equation of Revenue	Trend equation of Sale Value
Ball clay	661.308+5.671t	16001.22+840.99t	1633.34+231.799t
't' value	.278	3.275*	2.014*

*, significant at 5% level of significance.

Table 8.: Least Square Trend Values in Production of Ball Clay in Bikaner from 1998-99 to 2007-08 $Y = 661.308+5.671t$

Years	Prodction in Thousand Tonnes	a+bt – Yc
1998-99	492.097	635.7805
1999-2000	658.899	641.4515
2000-01	698.426	647.1225
2001-02	704.879	652.7935

2002-03	792.82	658.4645
2003-04	878.542	664.1355
2004-05	590	669.8065
2005-06	320	675.4775
2006-07	584.5	681.1485
2007-08	892.93	686.8195

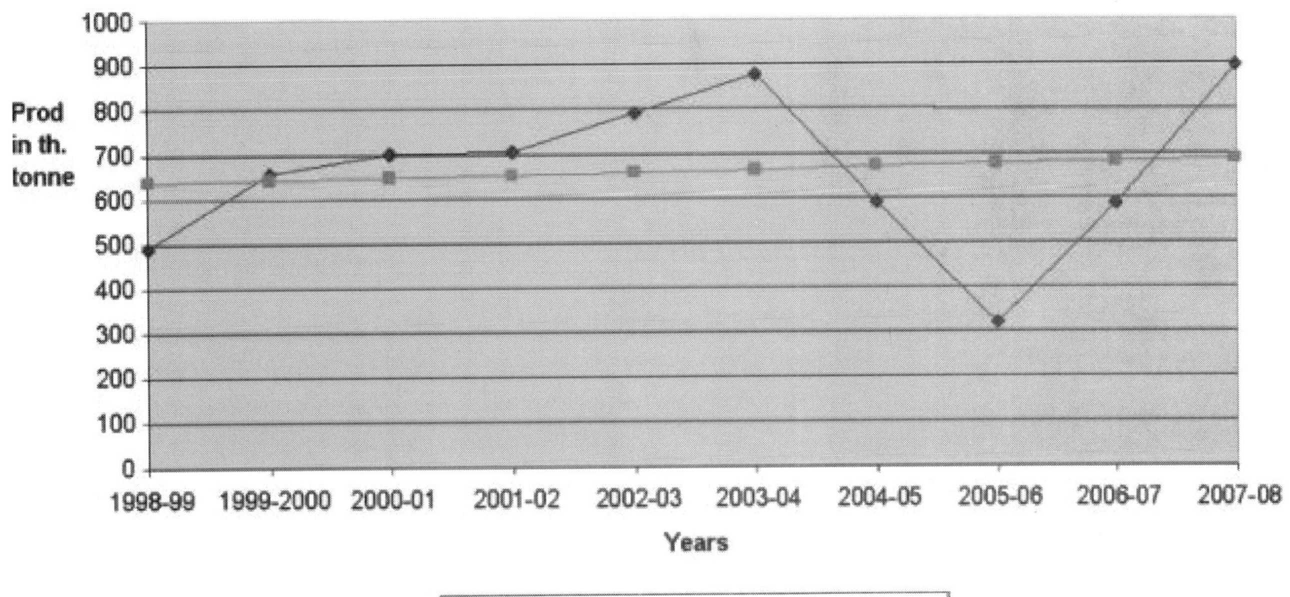

Fig: 2: Least Square Trend in Production of Ball Clay in Bikaner (1998-99 to 2007-08)
Origin year- 2003.5 a+bt = 661.308+5.671t

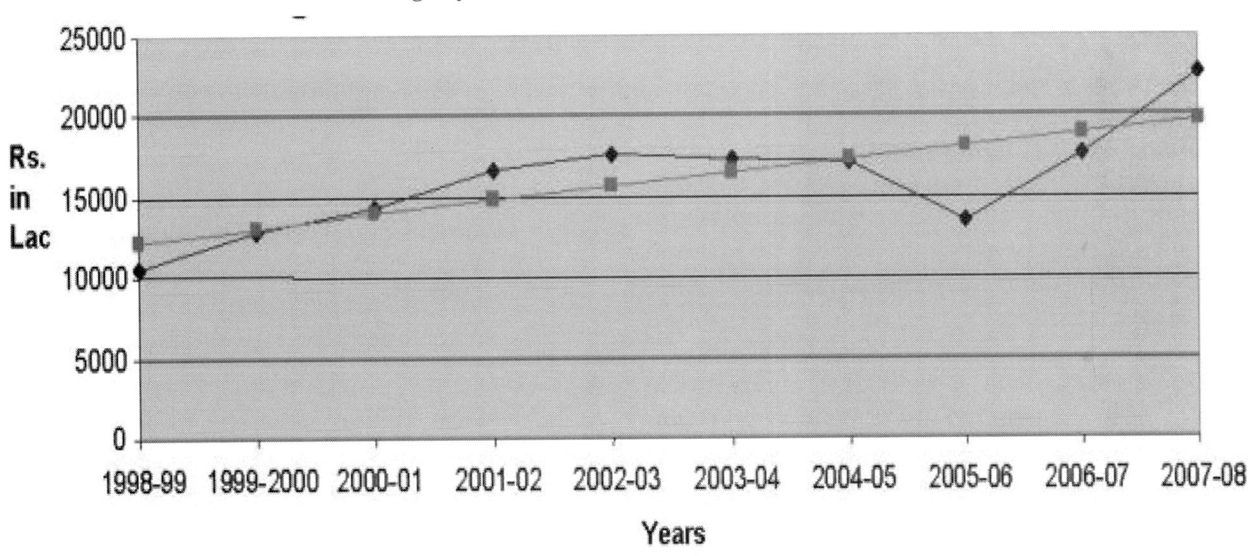

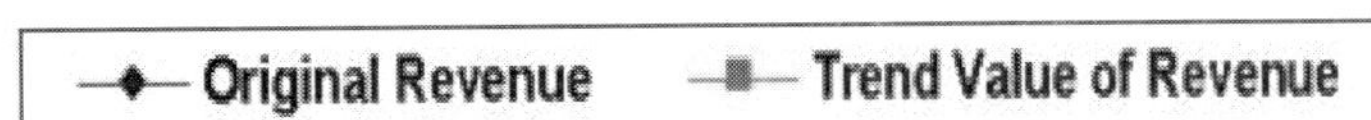

Fig. 3: Least Square Trend in Revenue from Ball Clay in Bikaner (1998-99 to 2007-08)
Origin Year- 2003.5 a+bt = 16001.22+840.99t

Table 9: Least Square Trend Values in Revenue of Ball Clay in Bikaner from 1998-99 to 2007-08 Y = 16001.22+840.99t

Years	Revenue in Lac Rs.	a+bt = Yc
1998-99	10515	12216.768
1999-2000	12812.25	13057.758
2000-01	14326.53	13898.748
2001-02	16692.47	14739.738
2002-03	17642	15580.728
2003-04	17240.25	16421.718
2004-05	17158.83	17262.708
2005-06	13444.68	18103.698
2006-07	17562.32	18944.688
2007-08	22617.82	19785.678

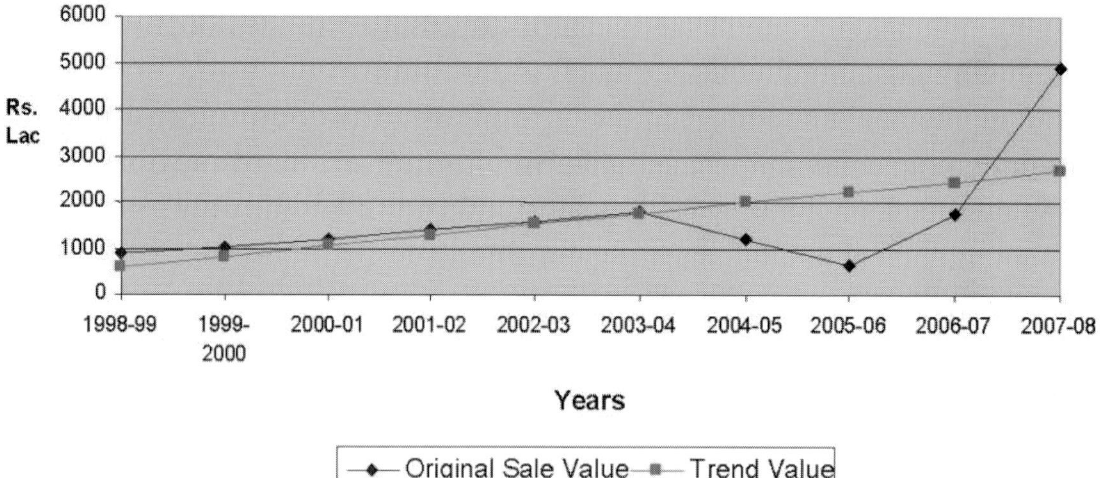

Fig: 4: Least Square Trend in Sale Value of Ball Clay in Bikaner (1998-99 to 2007-08)

Under major mineral production, ball clay is the principal mineral found in Bikaner District. In case of ball clay the rise in production is not much significant. Least square trend in revenue from Ball Clay has positive coefficient indicating their rising trend over the time period of ten years. As far as sale value is concerned, there is a significant rise in sale value of Ball clay. On the basis of these estimated least square equation (Table 7) trend value has been calculated of the Ball Clay mentioned in Table 8 to Table 10 and their respective graphical presentation has been depicted in Fig. 2 to Fig.4.

From Table 8 to Table 10 least square trend in production, revenue and sale value of ball clay of Bikaner has been estimated. The rate of production of ball clay is very slow despite maximum production in Rajasthan. Ball clay with 840.99 thousand rupees during last 10 years is the third highest revenue earner mineral of Bikaner. The sale value has also shown a rising trend.

Table 10: Least Square Trend Values in Sale Value of Ball Clay in Bikaner from 1998-99 to 2007-08
Y = 1633.34+231.799t

Years	Sale Value	a+bt = Yc
1998-99	885.77	590.2445
1999-2000	1015.324	822.0435
2000-01	1170.838	1053.8425
2001-02	1409.75	1285.6415
2002-03	1585.64	1517.4405
2003-04	1781	1749.2395
2004-05	1181	1981.0385
2005-06	640	2212.8375
2006-07	1753	2444.6365
2007-08	4911.115	2676.4355

A sharp decline in the contribution in revenue from ball clay had been observed. The reasons behind the poor performance are:

- Unsystematic and un-organized mining of ball clay in Bikaner.
- Due to small lease area, the mine owners are unable to carry out the proper mechanization on the mines.
- Most of the mine owners are stuck to obsolete technology in the mining of ball clay.
- Lack of development of ceramic industries at local level in Bikaner due to irregular unavailability of power, entrepreneurship and ceramic education.

Conclusion and Recommendations

In brief the result of least square trend values reveals some important aspects regarding Ball Clay Mining in Bikaner. The production of ball clay is not increasing at the significant rate (Table 3) despite the maximum leases among major minerals in North-Western Rajasthan (Table 1).

The factors responsible for that can be highlighted as:

1. There is an unsystematic and un-organized mining of ball clay in Bikaner. With a shortage of labour availability in mining areas and upsurge trend in labour wages the mine owners are unable to coordinate the demand and supply of ball clays, secondly, the mine owners are unable to compensate this labour shortage through incorporation of automatic mining due to small mining lease area.
2. Lack of proper marketing of mining product according to the demand of buyers. Grading of clay according to its uses was not proper in Bikaner because of the lack of Ceramic testing laboratory in the years of study. Recently the ceramic testing facility has been developed in Bikaner but the proper functioning of this lab is yet to establish.
3. Ball Clay mining, although, is facing the problem of water shortage for setting up improved mining system which utilizes water (wet mining) but the other systems for automatic production in clay mining can be modified to suit the mining environment in Bikaner.
4. Along with the mechanisation and technological development for higher production in ceramic industries, the demand for ball clay with quality consistency is increasing at faster rate. But there is a dearth of testing and beneficiation facilities at mine sites which are the vital factor to sustain the demand for ball clay.
5. Local ceramic industries could not compete with Indian ceramic industries as their products were inferior in quality. Lack of awareness among the youth about the mining engineering/ceramic engineering.
6. The Ball Clay Mining sector is facing a shortage of skilled and trained people.

The status of Ball Clay mining in Bikaner as revealed by statistics in the above mentioned tables is quite prosperous and has a good potential to flourish the economy of the district. But despite the large deposits of clay, the economy of Bikaner is not getting as significant profit as it must be. Above mentioned factors are responsible for the deterioration in production, revenue and employment potentials of clay.

In Ball clay mining the testing and beneficiation facilities should be provided at mine site to fulfill the demand for superior quality of clay by ceramic industries.

The establishment of a testing and common facilities centre at Bikaner, would directly help the ceramic industry, as a facility of this nature does not exist in the region. The product testing and common facilities centre would be used for carrying out various types test on ceramic products.

Due to inferior quality of finished product, the local ceramic manufactures cannot compete with the branded ceramic manufacturing units. Hence there is less demand for local product at local level which ultimately affects the production.

From the above analysis, this is clear that a mine owner can make his mine profitable by adjusting various cost variants in a planned and systematic way so that the profit enhances optimally.

References

Arogyaswamy, R.N.P. (1996): Mining Geology, 4th edition, Oxford & IBH Co. Pvt. Ltd. Kolkata Part 12th pp.629-631.

Bateman, M. Alan (1996): Economic Mineral Deposits, CBS, New Delhi, pp. 684- 687.

Chatterjee, K.K. (1991): An introduction to Mineral Economics, Prentice Hall, Kolkatta, pp. 25-26.

DMG, Rajasthan, Mineral Statistics (2008) pp. 33-45.

Howe, A.J. (1914): A Hand Book to the Collection of Kaolin, China Clay and China Stone in the Museum of Practical Geology, Jermyn Street, London, S.W.pp.189-193.

Mineral Policy (1994): Department of Mining and Geology, Govt. of Raj. in Rajasthan Mineral Bulletin vol. 18, pp-27.

Sharma Rajni (2010): An Economic Analysis of Mineral Resources of North-West Rajasthan(with special reference to Bikaner). Unpublished Ph.D. Thesis, pp.214

Shishir Sharma & Suman Sagar (2002): Mineral Resources of Rajasthan Desert in Advances in Resource Management of the Indian Desert Eds. BBS Kapoor et al. pp.1-18.

Sinha, R.K. and Sharma, N.L. (1988): Mineral Economics (4th ed.) Pub. Oxford IBH Publishing Co. Pvt.Ltd.-New Delhi, pp. 305-310.

Hon'ble reviewer

1. **Prof. M.K. Pandit**
   Department of Geology
   Rajasthan University
   JAIPUR - 302 001
2. **Dr. P.K. Jain**
   Dy. Mineral Economist
   Indian Bureau of Mines
   Indira Bhawan
   NAGPUR - 440 001

I-20

Palaeoecology of the Lakadong Limestone of Mawsmai Area, Cherrapunji, Meghalaya-Micropalaeontological Evidence

Yengkhom Raghumani Singh and Naorem Reshma Devi

Department of Earth Sciences, Manipur University, Imphal-795 003
* E-mail: yengmani@gmail.com

Abstract

The Palaeocene Lakadong limestone exposed in the eastern corner of Meghalaya particularly East Khasi Hill, NE India, comprises a rich assemblage of calcareous algae along with abundant larger benthic foraminiferal. In the present paper, twelve species belonging to ten genera of coralline red algae and twenty five species belonging to seventeen genera of the larger benthic foraminiferal are recorded from the sample of nearby the entrance Mawsmai cave of Meghalaya, India. On the basis of foraminiferal-algal fossils, the palaeoenvironmental interpretation of the Lakadong limestone suggests shallow marine environment of normal salinity and calm water condition at the time of deposition of sediments in the study area.

Keywords: Mawsmai cave, Lakadong Limestone, Calcareous algae, Foraminifera, Shallow marine.

Introduction

The state of Meghalaya is located in north eastern India and bounded by the country of Bangladesh and state of Assam, and characterized by the Shillong Plateau which includes Garo, Khasi, Jaintia and Mikir hills. The Mesozoic and Lower Tertiary successions are well exposed in this region. The Lower Tertiary successions of Sylhet Formation are well deposited in the eastern corner of Meghalaya particularly East Khasi Hill. The present study area of Mawsmai village which is situated at East Khasi Hill district near Cherrapunji, Meghalaya is a part of Lakadong Limestone. They bear a numerous of microfossil of calcareous algae and larger foraminiferal, which are dominant in this region.

Oldham (1859) reported the presence of Eocene beds with *Nummulitic* fossils from Khasi Hills. Evans (1932) recorded the occurrence of “Sylhet limestone stage” from Garo Hills, near Therria Ghat and Cherrapunji area of Khasi Hills. The pioneer workers on micropalaeontology of the Lakadong Formation were Pandey (1981), Misra *et al.* (2002), Gogoi *et al.* (2003), Matsumaru and Jauhri (2003), Sarma (2006), Matsumaru and Sarma (2010). The coeval successions of the Lakadong limestone in Meghalaya, India have been worked for their larger foraminiferal and calcareous algae, among others, by Jauhri (1988), Kalita and Gogoi (2003,

2006), Jauhri *et al.* (2006), Gogoi *et al.* (2009). However, few microfossils have been documented in this region. In this present study, larger foraminiferal assemblages with calcareous algae from the Lakadong limestone exposed in the Mawsmai area have been studied in detail and used for palaeoenvironmental interpretation.

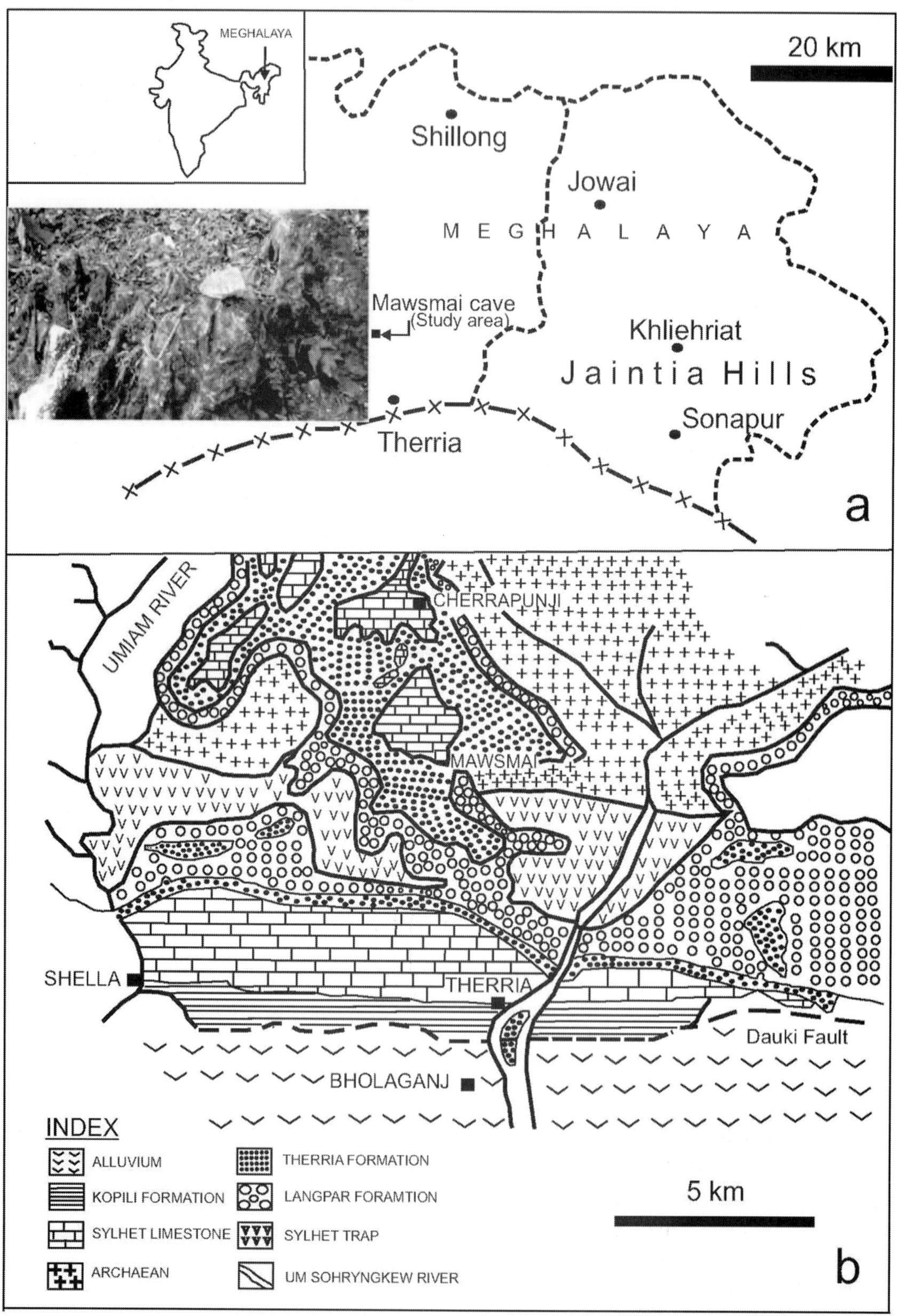

Fig. 1: (a). Location map of the study area with photograph. (b). Geological map of the area (after Ghosh 1940; Garg and Jain, 1995)

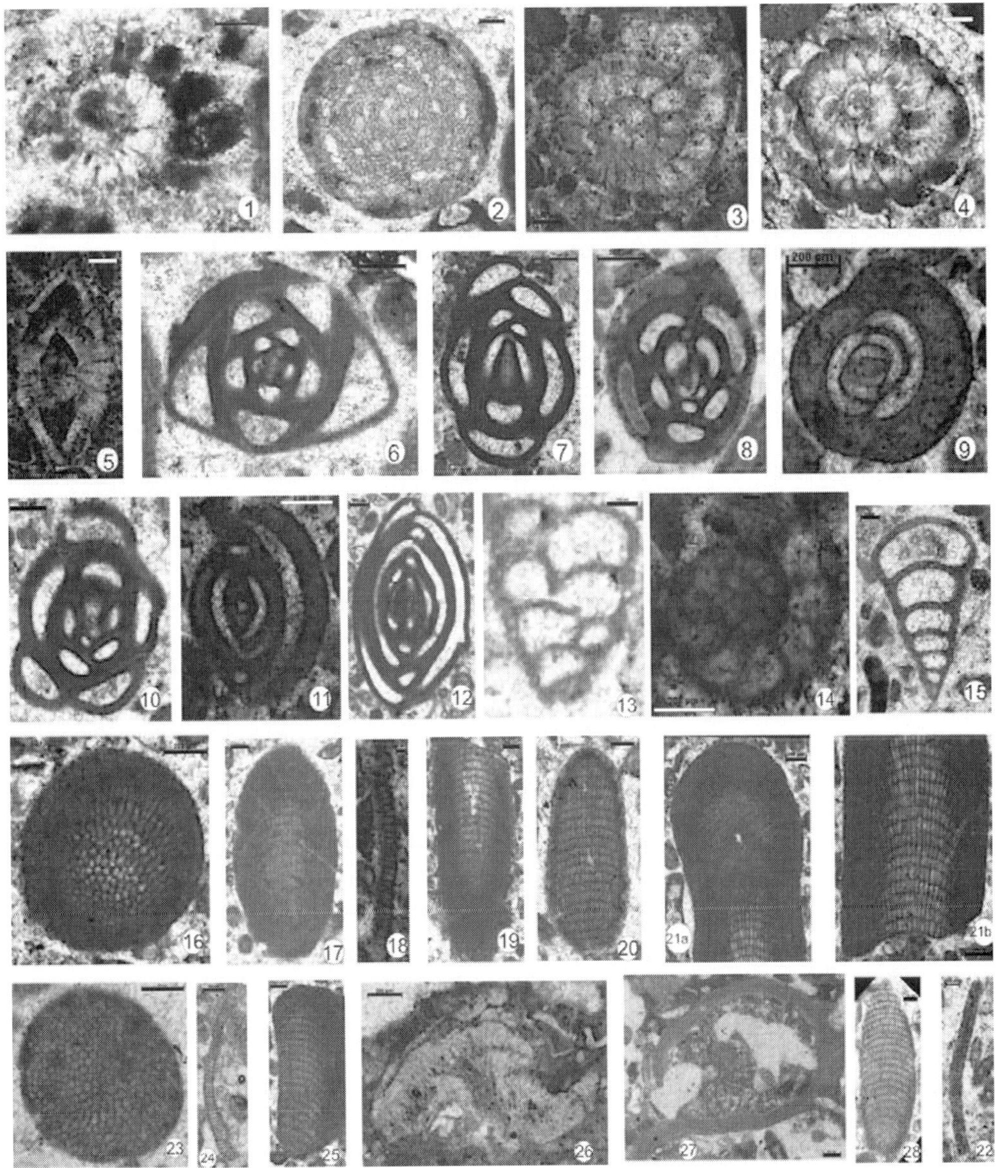

Plate-1: Scale bar represents 200µm.

Fig. 1, 4, 5. *Miscellanae juliettae* (lepping, 1988). Fig. 2. *Alveolina (Glomalveolina) primaeva* (Reichel, 1937). Fig. 3. *Miscellanea miscella* (d' Archiac & Haime, 1853). Fig. 6. *Alveolina* sp. Fig. 7. 10. *Quinqueloculina* sp. Fig. 8. *Quinqueloculina seminulum* (Linne, 1958). Fig. 9. *Alveolina subpyrenaica* (Leymerie). Fig. 11. *Spiroloculina* sp. Fig. 12. ? *Periloculina* sp. Fig. 13. *Textularia barretti* (Jones and Parker, 1867) (100µm). Fig. 14. *Eponides* sp. Fig. 15. *Textularia* sp.? Fig. 16, 23. *Sporolithon* sp. Fig. 17, 19, 25. *Amphiroa* sp. Fig. 18. *Lithoporella melobesioides* (Foslie) Foslie (60µm). Fig. 20, 28. *Corallina* sp. Fig. 21a. Upper part of Coralline algae *Lithophyllum* sp. Fig. 21b. Lower part of Coralline algae *Lithophyllum* sp. Fig. 22. *Distichoplax biserialis* (Diet) Pia (100µm). Fig. 24. *Lithoporella melobesioides* (Foslie) Foslie. Fig. 26. ? *Epiphyton* sp. Fig. 27. *Lithothamnion* sp.

MATERIAL AND METHODS

In this study, sampling was confined to rocks exposed nearby the entrance of the Mawsmai cave (GPS: N 25^{0}18' 48.6" and E 91^{0}43'46.1") shown in figure 1. Five samples of the Lakadong Limestone were collected from the same locality of the exposed limestone. The observation is based on the study of larger foraminiferal and calcareous algae in thin sections. The thin sections are prepared following normal procedure of thin section preparation. The primary data and figured material are housed in the collections of the Department of Earth Sciences, Manipur University, Imphal (India).

GEOLOGY OF THE STUDY AREA

Geologically the State of Meghalaya is also known as the "Meghalaya Plateau". East-West trending Khasi (Central Meghalaya), Jaintia (Eastern Meghalaya) and Garo (Western Meghalaya) Hill ranges are occupying the central part of Meghalaya State from approximately the watershed for Brahmaputra basin in north and Meghna basin in south. It consists mainly of Archean rock formations. These rock formations contain rich deposits of valuable minerals like coal, limestone, uranium and sillimanite. Meghalaya has many rivers. Most of these are rain fed and are therefore seasonal. In the southern Khasi Hills Region, where the Mawsmai cave exists, rivers have created deep gorges and several beautiful waterfalls.

Table 1: Stratigraphic succession in and around Cherrapunji

Formation	Member	Lithology	Age
Shella Formation	Lakadong sandstone	Fine to medium grained coal bearing sandstone	Thanetian (Late Palaeocene)
	Lakadong Limestone	Dark grey to gray coloured hard and compact algal and foraminiferal rich limestone	Thanetian (Late Palaeocene
	Therria Sandstone	Hard, compact, coarse to medium grained sandstone	Selandian (Early Palaeocene)
Langpar	---------	Calcareous shale, sandstone, limestone	Late Maestrichtian to Danian (Early Palaeocene)

The geology of the study area (Mawsmai Cave) and its adjoining areas is characterized by its diverse rock types which have been shown in the geological map (fig.1 A). The Jaintia Group is subdivided into two broad divisions, i.e. the Shella Formation and Kopili Formation. The Shella Formation is the lowermost litho-unit of the Jaintia Group comprising three alternating sandstone and limestone members, viz. Lakadong Limestone, Umlatdoh Limestone and Prang Limestone respectively, in chronological order. The pinching out of some of the members and interfingering of limestone beds with sandstone and shale made them regionally impersistent. The present study area is found to a regionally persistent horizon in east and western Khasi Hills and has been equated with the Lakadong Limestone of Cherrapunjee area. The general stratigraphic succession of in and around Cherrapunji area is given in table 1.

CHECKLIST OF FORAMINIFERA AND CALCAREOUS ALGAE

The Lower Tertiary rocks of the Shella Formation is represented by the members of Therria sandstone, Lakadong limestone and sandstone, Umladoh limestone, Narpuh sandstone and Prang limestone in the ascending order of stratigraphy. The present larger foraminifera and calcareous algae were recovered from the Lakadong limestone exposed at the entrance gate of Mawsmai cave.

The morphology and classification of foraminiferal and calcareous algae which are recorded from the sample of the study area are given as checklist in this paper. The classification given by Lobelich and Tappan (1964) in Treatise on Invertebrate Palaeontology is followed for the identification of different genera and species of foraminifers based on the morphology. On the other hand, the identification of Calcareous algae is followed by Wray (1977) and based on morphology. Twelve species belonging to ten genera of coralline red algae have been recorded from the present study area. On the other hand, twenty-five species belonging to seventeen genera of the larger benthic foraminifers are also recorded from the same sample. Besides this, a number of unidentifiable foraminifer and calcareous algae have been recovered these rocks. The selected species of calcareous algae and foraminifera are shown in plate 1. The detailed description of species is not included in the present text. Their checklist is given below:

Benthic Foraminifera

Alvoelina (*Glomalveolina*) *primaeva* (Reichel, 1937)

Alveolina (*Glomalveolina*) aff. *levis (*Hottinger, 1960*)*

Alveolina sp.

Alveolina subpyrenaica (Leymerie)

Biloculinites paleocenica (Rahagi, 1983)
Clavulina sp.
Discocyclina sp.
Eponides sp.
Idalina sp.
Miliolid sp.
Miscellanea juliettae (Leppig, 1988)
Miscellanea yvettae (Leppig, 1988)
*Miscellanea miscella (*d'Archiac and Haime, 1853)
Operculina sp.
Orbitosiphon punjabensis (Davies)-Tan, 1939
Orbitosiphon sp.
Pellatispira madaraszi (Von Hantken)
? Periloculina sp.
*Quinqueloculina seminulum (*Linne, 1958)
Quinqueloculina sp.
Ranikothalia sindensis (Davis) Caudri, 1944
Spiroloculina sp.
Textularia barretti (Jones and Parker, 1867)
Textularia sp.?
Triloculina sp.

Calcareous Algae

The non-geniculate coralline represented by genera *Lithophyllum, Mesophyllum, Distichoplax, Sporolithon, Lithoporella* and geniculate corallines are represented by the genera *Corallina, Amphiroa* and *Jania.*

Amphiroa sp.
Corallina typical (Ishijima)
Corallina sp.
Distichoplax biseralis
Distichoplax raoi
? Epiphyton sp.
Jania sp.
Lithoporella melobesiodes
Lithophyllum sp.
Lithothamnion sp.
Mesophyllum meghalayensis
Sporolithon sp.

PALAEOECOLOGY

To interpret the palaeoecological conditions during the Late Palaeocene Lakadong Limestone depositional episode a qualitative and quantitative distributional pattern larger foraminifera and calcareous algae (Fig. 1) were studied. The carbonate builds-up of the Lakadong Formation are characterised by poorly sorted, bioclastic lime-mud with modest representation of organic boundstones in the upper part. The increasing supply of clastic material to the basin towards the end deposition of the Lakadong Limestone inhibited the thriving of benthic foraminifers. Presence of mud free water in basin of deposition and predominance of larger foraminifera and calcareous algae in the Lakadong Limestone assemblage are indicative of low energy warm shallow water environment of deposition.

The exposed limestone of the study area has yielded rich assemblages of larger foraminiferl such as *Alvoelina* (*Glomalveolina*) *primaeva, Alveolina* (*Glomalveolina*) aff. *Levis, Alveolina subpyrenaica* (Leymerie), *Biloculinites paleocenica, Clavulina* sp., *Discocyclina* sp., *Eponides* sp., *Idalina* sp., *Miliolid* sp., *Miscellanea juliettae, M. Yvettae, M. Miscella, Operculina* sp., *Orbitosiphon punjabensis, Orbitosiphon* sp., *Pellatispira madaraszi, ? Periloculina* sp., *Quinqueloculina seminulum, Quinqueloculina* sp., *Ranikothalia sindensis, Spiroloculina* sp., *Textularia barrette, Textularia* sp.?, *Triloculina* sp. and the calcareous algae is represented by *Amphiroa* sp. *Corallina typical, Corallina* sp., *Distichoplax biseralis, D. Raoi, ? Epiphyton* sp., *Jania* sp., *Lithoporella melobesiodes, Lithophyllum* sp., *Lithothamnion* sp., *Mesophyllum meghalayensis, Sporolithon* sp.

In recent seas, this kind of assemblage is common in shallow water tropical environment of carbonate deposition. The presents of benthic foraminifera species like *Miliolid* indicate a typical of a shoal-reef environment with depths between twelve and twenty metre or even less (Arni, 1965) whereas *corallina* (geniculate coralline) and *Jania* and *amphiroa* (geniculate coralline) association is indicative of back-reef environment with low water energy and depths between twenty and forty meters in deeper Inner-shelf. The present of benthic foraminifera species like *Miscellanea juliettae* and *Sporolithon* sp. (non-geniculate coralline) suggests that the Lakadong Limestone was deposited in shallow shelf areas of carbonate sedimentation with influx of material from both fore-reef and back-reef environments with water condition. Moreover the presents of coralline algal genus such as *Lithophyllum, Mesophyllum* predict shallow marine environment with moderate to high energy open shelf environment.

From the above observation, it can be suggested that both the back-reef and fore-reef environment was prevalent during the deposition of the Lakadong Limestone unit. As it can be concluded that these limestones are reef deposits.

ACKNOWLEDGEMENTS

The authors are grateful to Prof. Soibam Ibotombi, Head of Department of Earth Sciences, Manipur University for encouragement and laboratory facilities. The Author (YRS) acknowledges University Grant Commission for financial assistance (Project No. 41-1024/2012, SR). We are also thankful to Shri Mr. Venus Gurumayum, Project Assistant and Shri Ksh. Attamajit Singh, Research Scholar for help during preparation of thin section.

References

Evans, P. 1932. Explanatory notes to accompany a table showing the Tertiary succession in Assam. *Transctions Mining and Geological Institute of India*, 27: 155-260.

Gogoi, B., Borgohain, R. and Kalita, K.D. 2003. Petrography, palaeontology and depositional environment of Lakadong Limestone of Mawsynram area, Meghalaya. *Journal of Indian Association of Sedimentologists,* **22 (1&2):** 133-149.

Gogoi, B., Kalita, K.D., Garg, R. and Borgohain, R. 2009. Foraminiferal biostratigraphy and palaeoenvironment of the Lakadong limestone of the Mawsynram area, South Shillong Plateau, Meghalaya. *Journal of the Palaeontological Society of India,* **54(2)**: 209-224.

Jauhri, A.K. 1988. Observations on the foraminiferal fauna of basal part of the Sylhet limestone Group (Lakadong Limestone) from Therria Ghat, South Shillong Plateau, Meghalaya, p. 34-42. In: *Palaeocene of India* (Ed. H. K. Maheshwari), Indian Association of Palynologists Publication, 4, Lucknow.

Jauhri, A.K. , Misra, P. K., Kishore, S and Singh S.K. 2006. Larger Foraminiferal and Calcareous algal facies in the Lakadong Formation of the South Shillong Plateau. NE India. *Journal of the Palaeontological Society of India,* **51(2)**:51-61.

Kalita, K.D. and Gogoi, B. 2003. Foraminifera from Late Palaeocene to Early Eocene of Mawsynram area, Meghalaya. *Gondwana Geological Society Magazine,* **18(1)**: 59-66.

Kalita, K.D. and Gogoi, B. 2006. First report of coralline algae in the Late Palaeocene Lakadong limestones of Mawsynram area, Meghalaya, NE India. *Journal Geological Society of India,* **68 (5)**: 779-783.

Matsumaru, K. and Jauhri, A.K. 2003. Lakadongia, a new Orbitoidal foraminiferal genus from then Thanetian (Palaeocene) of Meghalaya, NE India. *Micropalaeontology,* **49**: 277-291.

Matsumaru, K. and Sarma, A. 2010. Larger foraminiferal biostratigraphy of the Lower Tertiary of Jaintia Hills, Meghalaya, NE India. *Micropalaeontology,* **56 (6)**: 539-565.

Misra, P.K., Jauhri, A. K., Kishore, S. and Singh, S.K. 2002. Calcareous algal from the Lakadong Formation of the South Shillong Plateau, NE India. *Rev. Paleobiol.,* **21(2):** 717-734.

Oldham, T. 1859. On geological structure of a part Khasi Hills. *Memoirs Geological Society of India,* **1**: 99-210.

Pal, A.K. and Dutta, S.K. 1979. A study of fossil algae from Sylhet Limestone Formation of Meghalaya and Mikir Hills. *Geophytology,* **9(1-2)**: 144-154.

Pandey, J. 1981. Palaeocene smaller foraminifera of Um Sohryngkew River, Meghalaya, p. 70-152. In: *Proceedings of the VII Indian Colloquium of Micropalaeontology & Stratigraphy* (Ed. Rashid, D. A.), Madras, India.

Sarma, A. and Ghosh, A.K. 2006. A new record of calcareous algae from Shell Formation (Jaintia Group) of South Jaintia Hills, Meghalaya, India. *Current Science,* **90(9)**: 1276-1281.

Tiwari ,V.C., Kumar, K., Lokho and Siddaiah, N.S. 2012. Lakadong limestone: Palaeocene-Eocene boundary carbonates sedimentation in Meghalaya, North-eastern India. *Current Science,* **98 (10)**: 88-95.

Wray, J.L. 1977. Calcareous algae. *Elsevier Sceintific publishing company.* 1: 185.

Hon'ble reviewer

1. **Prof. P.K. Kathal**
   Dean
   Centre of Ad. Study in Geology
   Dr. H.S. Gaur University
   SAGAR - 470 003
2. **Dr. Rajneesh V. Khilnani**
   Director
   Hydrogeo Survey P. Ltd
   C-103, Shastri Nagar
   JODHPUR - 342 003

I-21

An Appraisal of Sandstone Deposits of Marwar Supergroup in Jodhpur and Nagaur Districts, Rajashan, India

Arun Vyas

P.G. Department of Geology, Government Bangur College, Didwana – 341 303, Rajasthan
E mail: avyasgeo@yahoo.com

Abstract

Rajasthan, the renowned and largest building stone producing province of India, is a veritable storehouse of building and decorative stone such as sandstone, marble and granite. Sandstone is mainly constituted by sand grains of 2 m. m. to 120 m. m. size; cemented by silica, iron, lime and various argillaceous and arenaceous matters. It is exploited for various types of building material.

Rajasthan is producing about 90% of sandstone in the country used for building purposes and as road metal along with quartzite. The intensive quarrying activities in Rajasthan are in the districts of Bharatpur, Sawai Madhopur, Karauli, Tonk, Bundi, Jhalawar, Kota, Bhilwara, Chittorgarh, Jodhpur, Nagaur and Bikaner. The red and blue Lower Bhander Sandstones which are exposed over a large area of about 5,000 sq km in south-eastern Rajasthan, covering Kota, Bundi, Bhilwara, Chittorgarh districts is quarried at a number of places, important ones being Bijolia, Barisal, Navanagar, Banio-ka-Talab in Bhilwara district and Dabi, Budhpura, Umarthan in Bundi district. The splittable sandstone areas are Bhanpura, Rajpura, Kasara, Chobe-ki-Guwari, Mokanpura, Berda and Bhakri in Karauli district. Other important areas of sandstone quarrying in Dholpur, Bharatpur and Sawai Madhopur districts are Sar Mathura, Bari Baseri, Hindon, Sapatra, Masalpur, Roopwas and Band-Barolla .

The sandstone deposits of the Marwar Supergroup occur at Fidusar, Soorsagar, Balesar, Mandore, Balsamand, Keru, Barli, Dechu, Dhelana, Lohawat and Osiyan in Jodhpur district; and Ladnun, Harsolav, Khatu and Khajwana in Nagaur district. Yellow and Rainbow sandstone of Khatu area in Nagaur district is famous for their quality in the world. It is being extensively used as a principal building material as well as decorative stone, which beautify the township in the major parts of western Rajasthan and Parliament and President House in New Delhi.

Keywords: Building stone, Marwar Supergroup, Soorsagar, Khajwana, Sandstone Deposit.

Introduction

Sandstone i.e. Sandy rocks consisting of small grains, mostly of quartz, cemented by iron oxide, calcium carbonate, clayey material etc. is used as decorative stone in flooring, roofing, facing, architectural work like carving, monuments etc. It is used in acid resistant buildings and buildings near sea shore. Industries based on it include cutting and polishing industries, carving work, artician work (cottage industries). Due to its various properties like regular bedding, uniform grain size, suitable nature, durability and resistance to eroding agencies; sandstone is extensively used for building purposes (D.M.G. Rajasthan, 2004). Sandstone distributed in 16 districts of the State (Table.1). Sandstone produced in the country is being marketed as Vindhyan Red, Rainbow, Teak, Modak, Bundi, Bansi Pink, Mandana, Dholpur Cream, etc. The sandstone may occur as massive, thick, non-splittable bands or thin beds or layers that can be split by applying slight pressure.

The occurrences of sandstone in India are spread over Andhra Pradesh, Assam, Bihar, Gujarat, Haryana, Madhya Pradesh, Meghalaya, Mizoram, Karnataka, Odisha, Punjab, Rajasthan, Uttar Pradesh, Tamil Nadu and West Bengal. The resource estimation has not been considered important because of its abundance and easy availability. Hence, there is no comprehensive inventory of sandstone. However, the Centre for Development of Stones (CDOS), a Government of Rajasthan undertaking has reported estimated reserves of sandstone at over 1,000 million tonnes in the country. Huge deposits of sandstone in Rajasthan are associated with Vindhyan and Trans-Aravalli Formations, exposed over an area of nearly 35,000 sq km covering districts of Dholpur, Bharatpur, Karauli, Sawai Madhopur, Tonk, Bundi, Jhalawar, Kota, Bhilwara and Chittorgarh. It is also found scattered in the districts of western desert plane. Splittable sandstone deposits are confined to an area of 16,000 sq km out of which 10,000 sq km lies in eastern and south-eastern Rajasthan and 6,000 sq km in western Rajasthan (I.B.M., 2012).

GEOLOGICAL SETTING

The Vindhyans, particularly the Upper Vindhyans i.e. Kaimur and Bhander of the trans-Aravalli region (now popularly known as the Marwar Supergroup in the area) rest unconformable on highly weathered and rugged surface of the Malani Igneous Suite (Blanford, 1877; Oldham, 1886,1888; La Touche, 1902; Heron, 1932; Pascoe, 1959; Pareek, 1981a, 1984; Chauhan et.al, 1991; Das Gupta and Bulgauda, 1994; Bhusan and Khullar, 1998; and Kochhar, 1998) producing excellent quality of sandstone (Vyas and Paliwal, 2003).

Jodhpur sandstone rests unconformably over a highly rugged basement of Malani Rhyolite (745±10 Ma) and granite (approximately 600 ± 70 Ma.) (Crawford and Compston 1970, Sharma et, al, 1975). It is considered to form the basal sequence of the Marwar Supergroup, designated as Jodhpur Group followed by sequence of carbonate lithofacies and siliciclastic evaporates facies named as Bilara Group and Nagaur Group respectively. This entire succession of Siliciclastic carbonates and evaporites facies which together represent the Marwar Supergroup is considered to be deposited in a NNW - SSE trending wide and far flung basin called as Marwar basin (Pareek 1981 a, 1984) (Figure - 1) (Table 2). This basin starts form the vicinity of Jodhpur city and continues subsurfacially further northwestward up to the Salt Range of Pakistan. Accordingly the rocks of the Marwar Supergroup are equated with the Cambrian succession of Saline Series of the Salt Range. Jodhpur sandstone resting unconformably over the Rhyolite basement exposed along the Jodhpur-Pokaran section represents the western margin of the Marwar basin (Cf. Narayanan 1971, Datta 1983, Pareek 1984).

SANDSTONE OF THE MARWAR SUPERGROUP

JODHPUR REGION

In Jodhpur region, sandstone mines around Soorsagar, Fedusar and Kaliberi areas (Figure - 2) situated about 12 km. from Jodhpur city. Soorsagar (North latitude 26^0 19', East longitude 73^0 01') is famous for sandstone quarries. At Kaliberi and Fedusar, sandstone is mined out manually. The total thickness of sandstone strata is about 20 mts. Besides these sandstone is also quarried from Daizer, Ganganna, Khetasar, Bhopalgarh, Rohila Kalan and Khurd, Kherapa, Lavera Kalan and Malunga.

Table 1. Sandstone Resources of Rajasthan (After D. M. G. Rajasthan, 2004)

Mineral	Distribution	Area of Leases (Hect.)	No. of Leases (2003-04)	Production in thousand tones (2003-04)
SAND STONE	Baran, Bharatpur, Bhilwara, Bikaner, Bundi, Chittorgarh, Churu, Dhaulpur, Jaisalmer, Jhalawar, Jodhpur, Karauli, Kota, Nagaur, Pali and Tonk.	270713.44	1199	7337.608

Table 2. The Stratigraphic Succession of the Marwar Supergroup (After Pareek, 1984)

Nagaur Group (75-500 mt.)	Tunklian sandstone	Sandstone, gritty sandstone and pebbly sandstone
	Nagaur sandstone	Brick-red sandstone with greenish clayey blotches
	Pondalo Dolomite	Dolomite, dolomitic limestone, cherty dolomite, stromatolitic limestone and dolomite
Bilara Group (100-300 mt.)	Gotan Limestone	Limestone with bands of chert and dolomite, Stromatolitic limestone
= Hanseran		
Evaporite Group	Dhanapa Dolomite	Dolomite, dolomitic limestone, chert and Cherty dolomite
	Girbhakar Sandstone	Brick-red siltstone, shale and sandstone
Jodhpur Group (125-240 mt.)	Sonia sandstone	Maroon siltstone and shale, creamish sandstone, salt pseudomorphs shales, banded chert and jasper with dolomite
	Pokaran Boulder Bed	Pebbles, cobbles, boulders in maroon/red clayey and silty matrix
	Precambrian basement rocks	

The Marwar Supergroup of rocks in the Jodhpur area is represented by the Basal Conglomerate and Jodhpur Group of rocks. Pareek (1981 a, 1984) divided the Jodhpur Group in to Sonia sandstone, shales and Girbhakar sandstone. On the other hand, Chauhan et. al. (1991) divided the Jodhpur Group into Lower Facies, Middle Facies Upper Facies. Paliwal (1984, 1992, 1998, and 1999) divided the Jodhpur Group into Basal Conglomerate, Sonia Formation and Girbhakar Formation.

In the Jodhpur region the Marwar Supergroup of rocks has been observed to comprise (1) The Basal Conglomerate and (2) Jodhpur Group. The letter includes (a) Lower Sonia Formation and Upper Sonia Formation separated by 'ash beds' and (b) Girbhakar Formation with clean washed conglomerates at the base grading in to pebbly sandstone. The Basal Conglomerate unconformably overlies the older Rhyolite flows of Malani. It occurs in the form of lenticular bodies at the base of Jodhpur Group. Exposures of the Basal Conglomerate are seen to the east of Rajbagh near of Bhomiaji Temple in Pratap Nagar area, inside Machia Safari Park and in a quarry section at Kankani area. The lower Sonia Formation overlies the Basal Conglomerate; it starts with sandstone traceable near Kailana Lake and Chopasni areas. The Lower Sonia Formation is being termed as the Lower Facies by Chauhan et.al. (1991); showing deltaic environment of deposition. Rocks of the Lower Sonia Formation are exposed at Chopasni, Masuria, Umaid Palace and Ratanada areas. The Upper Sonia Formation comprising pink to light gray Shaly sandstone with large number of mega ripples followed by medium grained thin-bedded pink sandstone with profuse development of ripple marks. Thin-bedded sandstone is followed by thick-bedded medium grained, cream-colored sandstone without ripples. Rocks are exposed in Soorsagar – Fedusar quarries, Balsamand – Mandore and Takhatgarh Fort areas. The Upper Sonia Formation represents the Middle Facies of Chauhan et. al. (1991) and denotes a beach environments of deposition. The sequence of Girbhakar Formation starts with a clean washed conglomerate grading into pebbly sandstone. The unit shows large scale cross bedding, particularly of Festoon type. The conglomerate is followed by thick-bedded brown to pink colored medium grained sandstone with the rare presence of cross bedding. The sandstone is also interlayered with shale's. The rocks of this unit are exposed at Wireless Hill opposite to the Mehrangarh Fort, Radar Hill in Kailana area, Takhatgarh Fort in Machia safari Park and at the top of Mandore and Motisar Bhakar. The Girbhakar Formation of the study area represents the Upper Facies of Chauhan et. al. (1991) denoting a braided fluvial environment of deposition. Paliwal (1992) suggested that the siliciclastic rocks of the Jodhpur Group and Calcareous rocks of Bilara/ Hanseran Group are of Proterozoic age (Neoproterozoic), whereas the siliciclastic and evaporitic sediments of the Nagaur Group belong to the Cambrian period (Paliwal and Rathore, 2000).

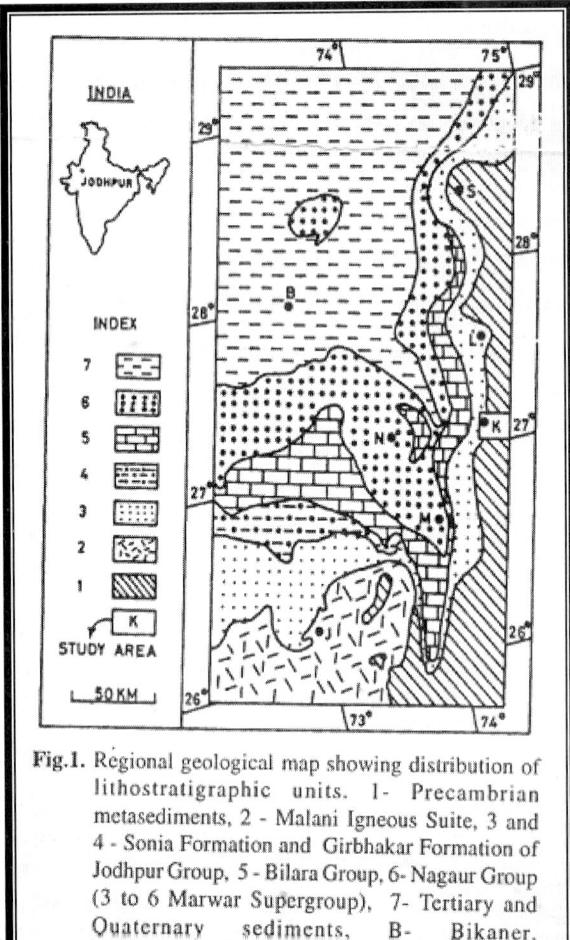

Fig.1. Regional geological map showing distribution of lithostratigraphic units. 1- Precambrian metasediments, 2 - Malani Igneous Suite, 3 and 4 - Sonia Formation and Girbhakar Formation of Jodhpur Group, 5 - Bilara Group, 6- Nagaur Group (3 to 6 Marwar Supergroup), 7- Tertiary and Quaternary sediments, B- Bikaner, S- Sardarshahar, L- Ladnun, N- Nagaur, K- Khatu, M- Merta, J- Jodhpur.

Figure- 1. Regional geological map of study area (After Paliwal, 1998)

Figure- 2. Exposures of sandstone and mining querry of sandstone mine around Jodhpur city

The lower Palaeozoic Jodhpur sandstone, representing the siliciclastic rocks of the Marwar Supergroup, is exposed in the vicinity of Jodhpur. It comprises a package of shales and sandstones; and is divisible into three distinct lithofacies viz. the lower deltaic, the middle beach and the upper fluvial facies. The middle facie consisting of a fine to medium grained sandstone of uniform nature. This sandstone facies shows prolific development of wave ripples, cross bedding, planar bedding etc. Geological examination of the outcrops reveals that it is only the middle horizon of the sandstone that is being quarried for building stone. Obviously, this sandstone of the middle horizon ought to be of specific nature. It has been observed that the sandstone of lower horizon is composed of interbedded sequence of lenticular bodies of sandy shale, Shaly sandstone and medium to coarse grained sandstone of variable dimensions, where as the sandstone of the upper horizon comprises mainly medium to coarse grained sandstone and very coarse grained sandstone with intercalation of pebbly to conglomeratic sandstone. It is therefore implicit that both of these varieties of sandstone do not make good quality of building stone. Unlike the above two varieties, the sandstone of the middle horizon is medium to fine grained regularly bedded and possesses fascinating light pinkish yellow to pinkish brown colors. It easily splits into regular slabs and can be shaped into suitable blocks and bricks, the material that makes buildings of the region beautiful and attractive (Vyas and Paliwal, 2003).

In an earlier study (Chauhan et al., 1991) the geological attributes of the sandstone of the lower, middle and upper horizons have been evaluated. It was concluded that these horizon sandstone represent three distinct sedimentary facies, which are designated successively as A, B and C lithofacies and these facies are the products of deltaic, beach and fluvial processes respectively (Chauhan and B. Ram, 1999) (Figure – 3). The middle horizon of the Jodhpur sandstone i.e. the 'B' lithofacies is characterized by reddish brown, grayish brown, pinkish brown, creamy pink- to- pink colored, fine to medium grained sandstone. These color changes are systematic from base upward and depend upon the nature and amount of iron oxide present. The basal units of sandstone are reddish brown, grayish brown, the middle pinkish brown and the upper creamy pink-to–pink in color. The grain size also increases, though slightly, from base upwards. The maximum thickness of sandstone of 'B' lithofacies is approximately 20 m. Petrographic study suggests that the sandstone is mainly composed of rounded to well rounded grains of monocrystalline quartz. A small fraction of polycrystalline quartz is also present. Authigenic quartz developed as partial overgrowths constitutes the main cementing material. Tourmaline occurs as well rounded grains within the range of fine sand. From the overall predominance of quartzose sand (98%) the sandstone of 'B' lithofacies is a typical quartz arenite (Chauhan and Ram, 1999).

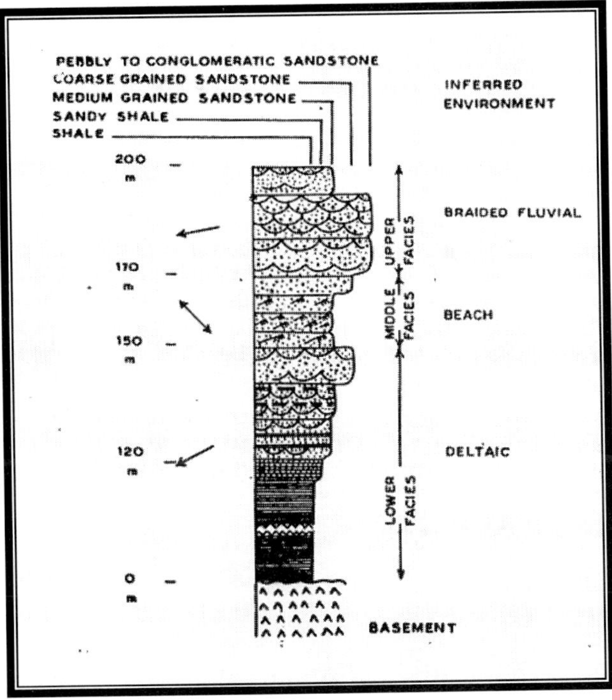

Figure- 3. Composite stratigraphic sections of Jodhpur group of rocks in the vicinity of Jodhpur city (After, Chauhan and Ram, 1999)

NAGAUR REGION

The Geological succession in Nagaur District is represented by the Archaean basement, Aravalli Supergroup, Middle Proterozoic Delhi Supergroup as the older sequences followed by Erinpura Granites, Malani volcanics, Marwar Supergroup and Tertiary rocks. The Aeolian sand, kankars, clays etc of Quaternary age are the youngest formations of the district. Rocks of Marwar Supergroup, which overlies rocks of Delhi Supergroup, occupy more than 50% area of the district. These are unfossiliferous marine sequences and have been classified in to Jodhpur, Bilara and Nagaur groups representing arenaceous, calcareous and arenaceous-argillaceous facies of sedimentary rocks. Jodhpur group represented by Sonia formation comprising of shale, sandstone, banded chert and carbonates, while the Girbhakar sandstone has alternate grit and pebble layer. Bilara group consists of calcareous facies represented by Dhanapa dolomite, Gotan limestone, and Pondalo dolomite. Hansera Evaporite sequence (containing potash minerals, anhydrite, clays, dolomite, Magnesite and gluconite) is equivalent to Bilara group. Nagaur group unconfirmably lies over Bilara group comprising of pebbles of dolomite, chert, sandstone, siltstone, clay stone, calcareous clay and gritty to pebbly sandstone. This group has a peculiarity to have an Evaporite sequence-gypsum in Nagaur area (Vyas and Paliwal, 2001; Gour and Vyas, 2007). Sandstone of Nagaur group belonging to Marwar Supergroup is of lower Cambrian age; occur as Tunklian sandstone and Nagaur sandstone.

Rocks of Jodhpur sandstone are exploited at Khatu and Ladnun. The Khatu Teak and Khatu Rainbow are being exported. Mining at present in Khatu area is being done below 4-5 meter thick cover of sand. Ladnun sandstone is being used for masonry purpose. Sandstone of Nagaur group is quarried from Harsolav and Nagaur (Rajasthan Mineral Bulletin, 2001; Gour and Vyas, 2007). Sandstone is also quarried from Khajwana, Kacharas, Danta, Bhilawas, Bhuwana, Chapari, Baldu, Pandolai, Nimbi Jodha, Sunari, Hirawati, Manda Basni, Lodsar, Chandrai, Idwa and Ratau. Two horizons of medium grained sandstone with color bands referred as "Layered Chromatographical Structures" lies at Bari Khatu. It is fine grained quartzose freestone. Color bands of different shades give a fascinating look and add to the decorative quantity of sandstones (Paliwal, 1976; Paliwal et.al., 2003). Felsic volcanics have been observed around Chhoti khatu; interlayered with sediments of the Marwar Supergroup (Figure – 4). They occur sandwiched between the underlying Sonia Formation and the overlying Girbhakar Formation of the Jodhpur Group. Stratigraphic column at Chhoti khatu is shown in Figure – 5 (Paliwal, 1998).

Figure- 4. Felsic volcanics and sandstone at Chhoti Khatu, Nagaur district

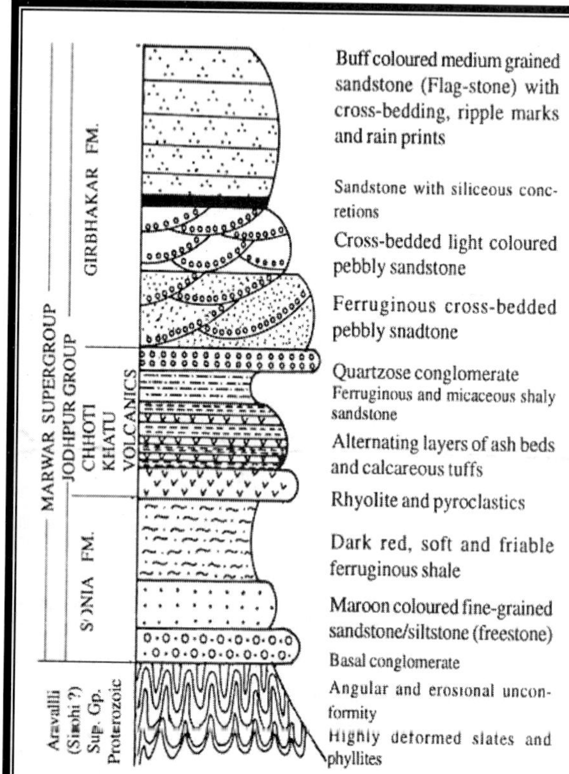

Figure- 5. Geological succession of rocks exposed at Chhoti Khatu (After Paliwal, 1998)

DISCUSSION

Rajasthan is the largest Sandstone producing state of Country. It is a sedimentary rock capable of desirable carving and suitable for decorative purpose and is used to make slabs, pillars, and nowadays as tiles in building constructions. Sandstone deposits occur in splittable and non- splittable (massive) varieties. Sandstone deposits are located in Baran, Bharatpur, Bhilwara, Bundi, Bikaner, Chittaurgarh, Churu, Dhaulpur, Jaisalmer, Jhalawar, Jodhpur, Karauli, Kota, Nagaur, Pali and Tonk districts. Its deposits in the State belong to Bhander Group (Vindhyan Supergroup), Jodhpur and Nagaur Groups (Marwar Supergroup), and are also of Tertiary age. The sandstone deposits of the Marwar Supergroup occur at Fidusar, Soorsagar, Balesar, Mandore, Balsamand, Keru, Barli, Dechu, Dhelana, Lohawat and Osiyan in Jodhpur district; Ladnun, Harsolav, Khatu, Kacharas (Figure – 6) and Khajwana in Nagaur district (Vyas and Paliwal, 2003, 2004; Vyas, A., 2009; D.M.G., Rajasthan, 2004). Sandstone of Jodhpur and Nagaur districts are marketed in the trade names like- Jodhpur-Pink, Jodhpur-Red, Rainbow, Teak Wood etc.

Figure- 6. Exposures of sandstone and mining querry of sandstone mine at Kacharas, Nagaur district

Mining of sandstone is generally done manually by using hammers and chisels of various shapes. The hard non-splittable sandstone is drilled and blasted to expose the underlying splittable sandstone. The majority of quarry owners produce hand dressed slabs and tiles in different thicknesses. But, in the present export market, machine-cut tiles are in demand.

Presently, the simple edge cutting machines with single or double cutters are used for getting machine-cut tiles. The further requirement of tiles in 10 to 12 mm thickness with one side natural and other calibrated has resulted in establishing cutting and polishing units of sandstone. In Rajasthan, there are four units engaged in the production of polished sandstone tiles measuring 30 x 30 x 1 cm and 40 x 40 x 1.2 cm. The sandstone is also exported with natural, honed and polish-finishes. BIS has prescribed IS: 3622-1977 (First Revision, reaffirmed 2003) as the specifications for sandstone slabs and tiles. In 2010-11, the exports of sandstone decreased to 0.29 million tonnes (Value of Rs. 3, 00, 36, 88000) from 0.63 million tonnes (Value of Rs. 4, 38, 77, 90000) in the previous year. The U.K. was the leading buyer in 2010-11 with 42% of the total value of sandstone exported followed by Belgium with 5% (I.B.M., 2012). Looking to the increasing demand of sandstones the more mechanized and systematic scientific approach for exploitation of such reserves is need of the hour.

Acknowledgement

Thanks are due to Indian Bureau of Mines, Nagpur and the Department of Mines and Geology, Rajasthan. I am sincerely thankful to my family members Sh. I. P. Vyas, Smt. Susheela, Mrs. Anita, Pallavi, Komal and Abhinav for their kind support. Last but not least I am also thankful to the editors - Prof. K. L. Shrivastava, Department of Geology, J. N. V. University, Jodhpur, Prof. Arun Kumar, Department of Earth Science, Manipur University, Imphal and Prof. A. K. Sinha, Jaipur for providing me the opportunity to contribute a paper in the volume entitled 'Georesources'- a festschrift for Professor Virendra Krishan Verma, of Department of Geology, Delhi University. Special thanks to anonymous Reviewers for their critical review and valuable suggestions.

REFERENCES

Bhusan, S.K. and Khullar, V.K. (1998) Geochemistry and tectonic significance of dyke swarm in Malani igneous complex around sankara, district Jaisalmer, Rajasthan. In B.S. Paliwal (ed.) The India Precambrian scientific publisher (India, Jodhpur, pp. 482-491)

Blanford, W.T. (1877) Geological notes on the Great Indian Desert between Sindh and Rajasthan. Res. Geol. Surv. India, 10 (1), pp.154.

Chauhan, D. S., Dubey, J.C., and Ram, B. (1991) Geological analysis of part of Nagaur basin in the vicinity of Jodhpur city. In: Tandon, S.K., Pant, C.C., Casshyap, S.M., (Eds.) Sedimentary basins of India: Tectonic context, Gyanodaya Prakashan, Nainital.

Chauhan D.S. and Ram B. (1999) Ripple marks & synthesis of beach sequence: a study of early Palaeozoic Jodhpur sandstone, Western Rajasthan. In B.S. Paliwal (ed.) geological evolution of Northeastern India. Scientific publisher (India), Jodhpur, pp. 129-147.

Crawford, A.R. and Compston, W. (1970) The age of Vindhyans system of peninsular India. Quartz. J. Geol. Soc. India, V. 125, pp. 351 –371.

Dasgupta, V. and Bulgauda, S.S. (1994) An overview of the geology and hydrocarbon occurrence in the western part of Bikaner-Nagaur basin. Ind. Petrol. Geology V. 3 (1) pp. 1-17.

Datta, A.K. (1983) Geological evolution and hydrocarbon prospect of Rajasthan basin. Petroleum Asia Jour., 6: 93-100.

D.M.G., Rajasthan. (2004) Mineral directory of Rajasthan. Directorate of Mines and Geology, Rajasthan, Udaipur Publication.193 p.

Gaur, C.P. and Vyas, A. (2007). Sustainable development of Mineral resources of Nagaur district, Rajasthan, India. In: Avdich, P.C. and Bhu, H. (eds.) Proceedings of the National seminar" Emerging trends of Research in Geology with special reference to Northwestern India."24-25 January, 2007, organized by M.L.Sukhadia University, Udaipur.pp.191-199.

Heron, A.M. (1932) The Vindhyans of western Rajputana, Rec. Geol. Surv. India, V. 65 (4), pp. 457-489.

I.B.M. (2012). Indian Minerals Yearbook (2011) Part-II. 50th Edition (Advance Release) Government of India, Ministry of Mines, Indian Bureau of Mines, Indira Bhavan, Civil Lines, Nagpur, October 2012. p. 72-1 to 13.

Kochhar, N. (1998) Malani igneous site of rocks. J. Geol. Soc. India, V. 51, pp. 120.

La Touche, T.H.D. (1902) Geology of western Rajputana. Mem. Geol. Surv. India, V. 35 (1)

Narayanan K. (1971) Problems of the stratigraphy of the Rajasthan Shelf. Proc. Symp. Problems of Arid zone of India, Jodhpur, 92-100.

Oldham, R.D. (1886) Preliminary notes of the geology of northern Jaisalmer. Rec. Geo. Surv. India, V. 19 (3), pp. 157-160.

Oldham, R.D. (1888) Memorandum on the results of an exploration of Jaisalmer with a view to the discovery of coal. Rec. Geo. Surv. India, V. 21 (3), pp. 30-33.

Paliwal, B. S. (1976) Note on the layered chromatographical structures in the Vindhyan sandstone of Badi Khatu, district, Rajasthan, India. Bull Ind Geol Assoc, V9 (1), 17-22.

Paliwal, B.S. (1984) Evolution of the Trans – Aravalli, Vindhyans of Rajasthan, India Proceeding 17th international geological congress Aug. 4-14, 1984. Moscow, USSR. Sec. C. 04.1.1 (Abstract).

Paliwal, B.S. (1992) Tectonics of post Aravalli mountain building activity and it's bearing on the accumulation of sediments along the western flank of the Aravalli range, Rajasthan, India. In: (Rajasthan) 341301 Ahmed, and A.M. Sheikh, (eds.), Geology in south Asia–1 proceeding of GEOSAS – I, Islamabad, Pakistan, February 23-27, 1992. Hydrocarbon development Institute of Pakistan, 1994 pp. 52-60.

Paliwal, B.S. (1998) Felsic Volcanic interlayered with sediments of the Marwar Supergroup at Chhoti Khatu. J. Geol. Soc. India, V. 52 pp. 81-86.

Paliwal, B.S. (1999) Geology of the region to the west of Aravalli mountain: Some observations In: B.S. Paliwal (ed.) Geological evolution of Northwestern India. India. Scientific publishers (India), Jodhpur, pp. 387- 414.

Paliwal, B.S. and Rathore P.S. (2000) Neoproterozoic volcanics and sedimentaries of Jodhpur: A Reappraisal. In: K.C. Gyani and P. Katariya (eds.). Tectonicmagmatism, Geochemistry and metamorphism of Precambrian Terrains. M.L. Sukhadia University, Udaipur publication. Pp. 75-96.

Paliwal, B.S., Pareek, U.S. and Vyas, A. (1997) Structural framework of the Makrana marble deposit and it's bearing on the Precambrian stratigraphy of Rajasthan. In: Wijayananda, N.P., Cooray, P.G. and Mosley, P. (Eds.), Geology in South Asia-II, Geological Survey and Mines Bureau, Sri Lanka, Professional Paper-7, PP. 123-136.

Paliwal B. S., Yadav, S.K., Rathi, G. and Bhaghela, A. (2003) Sandstone with layered chromatographical structures: A prospective decorative stone. Sandstone spectrum –03 Seminar organized by department of mines and geology, Govt. of Rajasthan, Udaipur. April 18-20, 2003. Souvenir Volume – 1. pp. 16-25.

Pareek, H.S. (1981 a) Basin configuration and sedimentary stratigraphy of western Rajasthan. Jour. Geol. Soc. India, 22: 517-527.

Pareek, H.S. (1984) Pre – quaternary Geology and Mineral Resources of Northwestern Rajasthan. Geol. Surv. India Mem. 115: 1-95.

Pascoe, E.H. (1959) Manual of the geology of India and Burma. Geol. Surv. India, V.2. pp. 485-1343.

Rajasthan Mineral Bulletin. (2001) Geology and Mineral Resources of Nagaur District. R.M.B., Department of Mines and Geology, Udaipur, Vol.22, No. 1&2, Jan-June, 2001, 5-17.

Sharma, K.K., Nagpal, K.K. and Jhingran, A.G. (1975) Fission track geochronology of early Precambrian of Rajasthan, Indian. Chayanica Geological, 1: 196.

Vyas, A. and Paliwal, B.S. (2001) Hydrogeological studies of the Nagaur District, Rajasthan, India. In: H. Bhu and V. Agrawal (Eds) Groundwater Resources-Evaluation and Management. I.G.C. – 2000 volume –1. Proceeding of the National seminar with the XII – I.G.C., Feb 8-12, 2000. Published by Department of Geology, M.L. Sukhadia University, Udaipur, India. 233-243.

Vyas, A. and Paliwal, B.S. (2003) Jodhpur sandstone: The crowning glory of building marvel. Sandstone spectrum –03 Seminar organized by department of mines and geology, Govt. of Rajasthan, Udaipur. April 18-20, 2003. Souvenir Volume – 2. pp. 15-20.

Vyas, A. and Paliwal, B.S. (2004) Mineral resources of Rajasthan, India- A developing scenario. Proceedings of the National Seminar on Mineral Exploration and Resources Survey. Base metals, Strategic Minerals, Rare Earths. December 19 and 20, 2001. Special Publication No. – 72. Published by Geological Survey of India, Jaipur. 2004. pp. 13-26.

Vyas, A. (2009) An overview of the Mineral Resources of Rajasthan, India. In: K. L. Shrivastava (Ed.). Economic Mineralization, Scientific Publishers (India), Jodhpur. pp.352-368.

Hon'ble reviwer

1. **Prof. D.C. Gupta**
   Pro Vice Chancellor
   Barkatullah University
   BHOPAL - 462 026
2. **Prof. Sunil Parihar**
   Head
   Department of Mining Engg.
   JNV University
   JODHPUR - 342 001

I-22

Geoeconomical and Geotechnical Evaluation During Subsequent Phases of Mineral Exploration

S.K. Gupta

Department of Applied Geology, Dr. Harisingh Gour Vishwavidyalaya, Sagar- 472 003
E-mail: skgupta@gmail.com

Abstract

The prospecting and exploration of mineral deposit primarily reveals 'what to search', 'where to seek', 'how to seek' and 'how to evaluate deposits of metallic ores, industrial minerals and rocks and fuel mineral resources. The main objective is to determine the industrial value of a mineral deposit which is done by its geoeconomic and geotechnical evaluation. There are many factors and indexes of such evaluation which are studied in details during the different phases of mineral exploration, depending upon the objectives, needs and funds available. As the exploration work advances the indexes become more and more clear and the accuracy of evaluation increases proportionately. In this paper the role of geotechno-economic evaluation at different phases of exploration and their relation are briefly discussed.

Keywords: Exploration programme, Mine development, Reconnaissance, Technological innovations, Mineral Depletion, Mineral Depot Evaluation.

Introduction

Minerals are basic raw materials for metallurgical industries and find wide application in several other industrial complexes. As the economy of a country is intimately linked with the programme of mineral development, a rational approach to their exploration and production is vital for mobilising internal resources and strengthening the country's economy. It is now universally accepted that the development of mineral resources is the biggest generator of economic growth. In developing countries, the exploration and utilization of mineral resources will greatly determine the scope and place of economic growth (Faruqi, 1994). Prospecting and exploration for minerals are undertaken in a logical time sequence. The target is to find useful mineral deposits and appraise their suitability for future evaluation. The discovery, study and exploration of a mineral deposit is a long-term procedure, which begins with the establishment of prospecting criteria and indications in the general and preliminary prospecting stages and ends with the depletion of the deposit (Harkin, 1973). Geoeconomical evaluation deals with the determination of the industrial importance of a mineral deposit during the period of construction, and the geological, geographical, economic and geotechnical conditions based on which the further exploration and extraction programme need to be planned (Karunakaran, 1971). It is therefore, the results of technical and economic calculations express the technoeconomical possibility of mining and extraction of metal and the economic indexes decide the mining of the mineral deposit.

Rapid development of minerals in India to meet the indigenous requirements and export demands has necessitated progressive diversification and intensification of mineral exploration. During such a rapid expansion in mineral exploration expertise becomes insufficient. Different norms of exploration and appraisal are adopted, leading to widely different interpretation. The situation can be partly remedied by adopting a general acceptable pattern of exploration and evaluation for mineral deposits. Standard methods of exploration and presentation of the data help in a realistic assessment of reliance that could be placed on the results of the exploration (GSI manual, 1965). This helps appropriate decisions, as otherwise, presumptions mark the real potential and subsequent operational/ developmental/ production schedules are greatly upset.

A conceptual approach to mineral exploration can be based on the fact that different types of mineral deposits are identified with specific geological parameters, which may be related to the age, structures, petromineralogy and geochemistry of plutonic and volcanic rocks, to the lithostratigraphy of volcanosedimentary associations, to the proximity of clastic and nonclastic rocks to shore lines, island arcs, or reef complexes, for example, at the time of their deposition; and to lineament representing present and past plate boundaries and other megastructures (Lacy, 1973). Various geological environments can be studied at the local, regional and global scales.

Mineral raw material and energy resources have always played key role in the global economy. In 1960's for the first time, the threat of a scarcity of mineral raw materials and energy resources became apparent. It was not until 1967 that the decrease in crude oil reserves caused by exploitation exceeded the increase in new oil reserves verified by exploration. The volatile situation in the Middle East accelerated the advent of the raw material and energy crises which, later became a gloomy reality and which contributed to the exponential inflation and devaluation of currencies. The crises also had its origins in the fact that world resources will not indefinitely be able to cover the 8-10% growth of the world economy. Even if the resources could stand it – abundant and cheap energy will enable utilization of sub marginal and very poor ores – the problem remains open whether nature itself could withstand it for the future of the world, discoveries of new methods for mining, recovery and processing of mineral raw materials have been realized (Kuzvart and Bohmer, 1986).

Exploration is a broad term covering all the complex investigations which are basically initiated in search of mineral deposit. The nature and magnitude of the problems occur during exploration depend upon the geological and geographical conditions (Ballurkar and Borisovich, 1977). Prospecting is a highly expensive and tremendously risky affair and almost all mining companies spend over 15-20% of their annual cash flow in prospecting and exploration. Apart from the geological risk of prospecting there are other kinds of risks that influence the theme and decision to begin, to continue or to abandon an exploration effort: political risk (excessive taxation, disastrous restrictions on mining), technologic risk (unsurpassable problems in mine development, operation, processing), and market risk (market price fall).

All mineral exploration is based on the premise that an eventual discovery will result in a mine that makes the effort worthwhile (Peters, 1987). Exploration programmes have no universal pattern, but they have some common features that are generalized in figure (1). Three major stages can be identified: programme design, reconnaissance, and detail. A fourth stage, prospect evaluation, is a part of the overall programmes. Prospecting and exploration work is generally carried out in different phases. It depends upon the main objectives and funds available for this purpose. However, in a systematic programme of geoexploration this work can be broadly carried out in four successive stages viz., preliminary prospecting, detailed prospecting, preliminary exploration and the detailed exploration (Figure 2). These stages in certain conditions overlap or merge with each other (Ballurkar and Borisovich, (1977). As work flows from one stage to the next, a great many difficult field decisions must be made. The results in rejections and revisions as shown in the "feed back" paths, data from rejected targets, reconnaissance areas and entire programmes are collected in the data bank (Fig. 1) for future use. The data bank is important because programmes laid to rest presently may be resurrected in future when budgets, personnel, concepts, of objectives have changed. The information from rejected targets must be preserved.

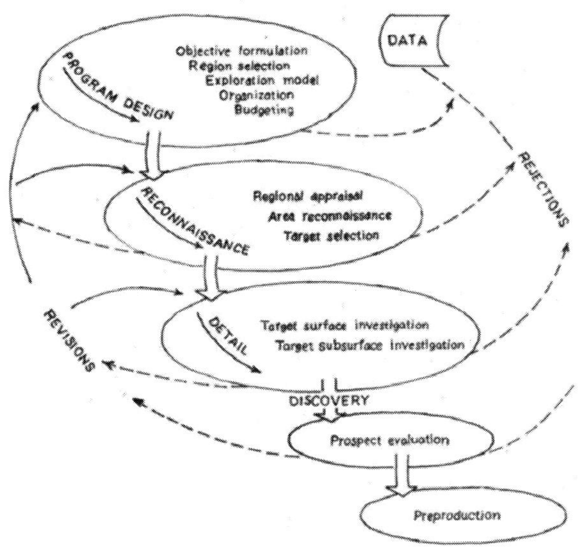

Fig. 1. Generalised regional exploration sequence. (After William C. Peters, 1987)

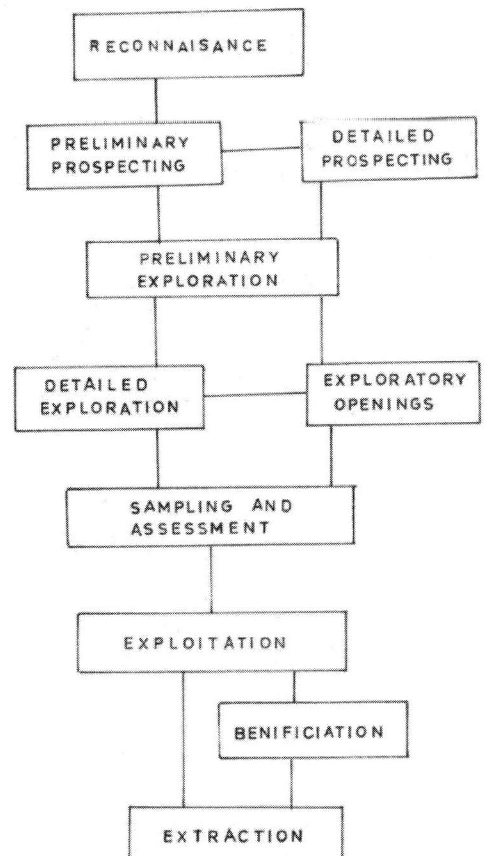

Fig. 2. Geoexploration programme (After Ballurkar, 1977)

Mineral exploration objectives are sensitive to the entire interplay of markets, geography, science and technology (Peters, 1987), Market conditions – time, place and unit value of minerals – affect the intensity of exploration and the material sought. Geographic factors have a bearing on where to explore, trend in politics, access to new transportation routes, new open-door economic policy, draws exploration from other areas where ore bodies are being depleted and mineral development has been discouraged or rejected. Geologic science and the technology of exploration provide springboards to new field investigation. Mining technology may take on a new aspect. Exploration will follow suit. Development work at a mine reaches a record depth; other deep exploration targets in other mines are reassed.

In the preliminary stages of exploration after geological reconnaissance and thorough field investigations, portions of the tract can be classified into following broad ground categories for further detailed probing on specific or target zones:

Ground Categories	Ground on which there are known ore bodies or promising indications.
	Ground on which favourable structural conditions are known to exist.
	Ground on which no favourable condition are known to exist.
	Ground on which favourable conditions are believed to be absent.

A distribution has to be made between mere absence of positive indications and actual presence of negative indications. The data obtained during this stage could be utilized in demarcating the areas and to bring out whether any geophysical and geochemical anomalies can be obtained to locate areas of interest (McKinstry, 1960). The results achived at the end of each stage of exploration are carefully scrutinized and if they satisfy certain standard requirements the subsequent stage of work is undertaken. In some countries, particularly developed ones, standard norms have been prepared to simplify this analytical work. Based on the final results of all these stages the reserve calculation of the deposit and its geoeconomic evaluation are made (Gupta, 2009). The main conditions of workability are the quality of the mineral and the geological and mining conditions of the mineral deposit, based on which the demarcation of workable zone (balance zone) from non- workable zone is made. This is generally done by contouring. The economical base for calculation of conditions is the minimum cost of production of unit volume of the final product which is calculated as:

$$C + EK = \text{minimum}$$

Where C= Cost price of exploration mining, beneficiation, dressing and extraction.

E= Coefficient of efficiency of capital investment which is called as: E=1/T

where T is the time required to recover the capital investment.

K= Capital investment to produce one unit of the final product.

This C+EK = minimum, is the main index for determination of the workable zone. The percentage of content on the boundaries of the workable zones should be such that it should be able to repay the cost of all the establishment of the mining enterprise. This aspect has been discussed in details by Buchover (1966) and Kajdan and Kobachidze (1968). The main objective of estimating the reserves is to ensure the amortization of the investment and a certain profit. There are different systems of classification of reserves such as the three fold classification of I - Proved, II - Probable, and III – Prospective or possible which was adopted in UK and some common wealth countries. In USA these classes are referred to as I – Measured, II – Indicated, and III – Inferred. However in USSR, the State Committee of Mineral Reserves has recommended the four fold classification as follows (Kreiter, 1968):

A – Designated for production, planning and mine projection.

B - For estimating mining investment and for planning the development of deposit.

C_1- For long term development plans of industry and for projecting detailed exploration.

C_2 - For planning further prospecting.

Figure 3 represents schematic classification of the reserve of a steep ore vein showing these four classes during the exploration work and data available of each of these classes. The classification of the reserves depends on the following factors:

1. The reliability of the estimated reserve
2. Technological properties of the mineral
3. Mining Conditions

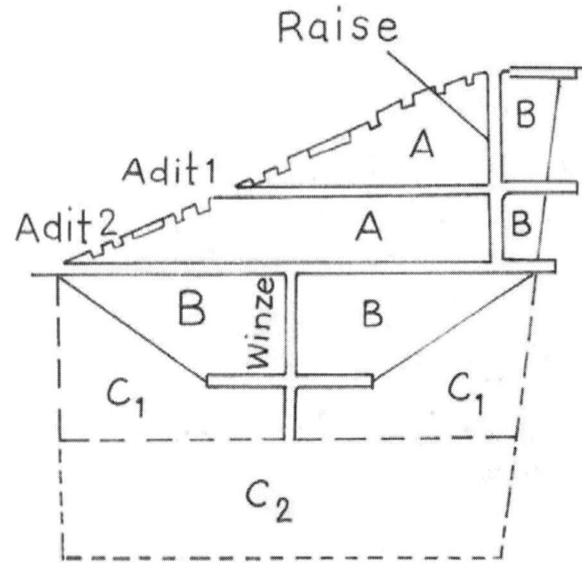

Fig. 3. Schematic classification of the reserves of a high-dipping ore vein (Longitudinal projection on to the vertical plane after Kreiter V.M. 1968)

Geoeconomic evaluation during subsequent stages of exploration

During preliminary prospecting the indexes of geoeconomic evaluation are the prognosis of the reserve and the approximate value of the metal content in the deposit. These two define the natural value of the mineral deposit which can be derived from:

$$V = \sum_{i=1}^{n} Ai.Bi.pi.qi.ri$$

Where:

V = value of deposit in money.

A = area of prospected deposit.

B = wholesale price of the metal.

$p.\ q.\ r.$ = the indexes of extraction of the metal while mining and extracting, dressing and beneficiation.

n = number of metals present in the deposit.

But in some cases the data received during the first stage, especially in complex geological conditions, is not sufficient to estimate the prognostic reserves. In this stage the geological value of the deposit can be determined based on the anomalies obtained by different surveys. This geological value is the base for the next stage which is the detailed prospecting or post reconnaissance phase. The aim of this stage is the estimation of industrial type of deposit for which it is necessary to make conditions to calculate reserves in class C_2 and prognosis reserves. The important characteristics are the mineral composition and technological properties of the minerals of the deposit. From geological and geophysical data and from some underground workings (drilling and mining) it is possible to decide the methodology of exploitation.

After receiving results in this stage it becomes possible to make technoeconomic speculation (TES) in which approximately stated the geoeconomic evaluation of the mineral deposit determines the scope of further exploration. During describing TES the utility of indexes is important aspect which characterise the natural value of the deposit such as:

1. Reserves of the deposit
2. Tenor of the mineral
3. Approximate natural value in terms of money
4. Approximate yearwise output of mining enterprise
5. Output of commodity production
6. Approximate cost price of mining

In this stage it is necessary to estimate capital investment for establishment of the mining enterprise (surface construction, shafts, drives, major equipment, etc.). The deposit investigated so far if gives positive value will be advised to be taken for the next preliminary exploration stage or detail phase. The two stages of exploration normally take a period of 5-7 years and it is necessary to calculate TES for future with this angle (7 years of changing prices and technological advancement). The indexes of TES should be calculated for future period in which the mining enterprise will be established. Therefore, geoeconomic evaluation in this stage has a task to carry out industrial evaluation of deposit based on which exploration work can be achieved at minimum time and cost.

The main task of the next stage of preliminary exploration would be the finding out the exact dimensions of the deposit, average quality of the ore, main geotechnological features of its dressing and beneficiation and the mining conditions. During this stage, the exploration is done in such a way that it would be possible to classify the whole ore body into C_2 and some portion or portions in C_1 and the reserves can be calculated accordingly. At the end of this stage it is necessary to prepare a technoeconomical report (TER) which should enable to decide whether to go for the further detailed exploration. The main geoeconomic indexes established here are the natural and economic factors which influence the choice of principles, technological and economic decisions of mining and extraction.

During the evaluation of TER not only the geological conditions are taken into account but aspects like economical, geographical, geotechnological, ore dressing and beneficiation, demand in and outside the country etc. are also considered. All these factors influence the geoeconomic evaluation of the deposit. Technoeconomical report should contain necessarily the following divisions:

1. Geographical conditions of the deposit
2. Geological situation:
   (i) Geological structure of the ore field
   (ii) Hydrological features
   (iii) Main feature of geoexploration
   (iv) Sampling
   (v) Reserves of the deposit and tenor of ore
   (vi) The significance of the deposit in that ore field and in the country
3. Geotechnology of mining and extraction:
   (i) Geotechnological conditions of mining
   (ii) Choice of mining method
   (iii) Choice of extraction method
   (iv) Technology of ore dressing
4. Economical division:
   (i) Yearwise output of mining enterprise
   (ii) Commodity production in volume, in terms of money
   (iii) Cost price of mining, dressing and beneficiation
   (iv) Profit and profitability of the deposit
   (v) Capital investment of establishment for the mining enterprice and its efficiency
   (vi) Technoeconomical indexes of mining of the deposit

5. Conclusions

The results of geoeconomic evaluation of the deposit at the end of preliminary exploitation are calculated by two methods (Kreiter, 1968), (i) by simple analogy (ii) by technical and economical analysis. In some cases while doing analogy method some feature of the deposit are not known and hence the comparison becomes difficult. Therefore, in such cases that part of the work is carried out to fill up the gap. Although this method (analogy) is cheaper and quicker, the author prefers the analysis method as it gives more correct picture of the ore body. If positive evaluation values are received after the preliminary exploration the deposit is taken for the detailed exploration.

The main task of detailed exploration is to finalise the geoeconomic evaluation of the deposit and handover the deposit for exploitation. During this stage the conditions prescribed include:

1. Reserve calculation

2. Division into industrial types of mineral raw material

3. Identification and demarcation of workable and non- workable zone.

4. Precise contouring and structural mapping of the ore body

The proportion of A, B and C categories depends on the complexity of geology of the deposit, size and morphological peculiarities and the distribution of variability of the content of the metal.

Based on the results of detailed exploration, the complete industrial evaluation of the deposit is made and along with the project for mining enterprise is also prepared. The indexes of geoeconomical evaluation during this stage are the same as in the case of preliminary exploration. But the difference is in the earlier stage the evaluation is made in general for the entire ore body where as during the detailed exploration evaluation is done for both, the entire ore body and different parts of it. The value of each index is derived by the detailed calculation. Besides, the expected workability of the labourers, necessary number of workers, commodity productions etc. may also be taken into consideration.

Geoexploration in and around mining enterprise

The main task here is to find out in details the structure of the ore body, and additional reserves, if any, in the peripheries of the mines, at depth between levels and in the surrounding area to continue the existence of the mining enterprise and to increase the efficiency of capital investment. During this stage geotechnoligcal aspects like the strength of the rocks in different sections of the mine, hydrogeological features etc., are studied in details, the balance reserves and the output are also considered and the extention programme of the mine is carefully planned. The indexes of this are same as in the case of exploration. The planning goes with the objective of expanding or reconstructing the enterprise. But in any case geoeconomic evaluation in mining exploration is most accurate than in other stage because the indexes here are most accurately available.

CONCLUSION

It is clear from the above discussions that there is a continuous increase in the number of indexes during successive stages of mineral exploration. Besides, during each subsequent stage there is also a gradual increase in the quantity of data of these indexes as they become more and more definite. Therefore, the geoeconomic evaluation can be best done while exploring in and around existing mines as the indexes here would be the most precise. Since the task of evaluation is more often to be undertaken in different stages of mineral exploration, the efficiency of evaluator lies in the best possible evaluation with the available indexes. The selection of indexes and methods of investigations should be such that with minimum expenditure and, time maximum required data should be obtained. A careful study of the worked out and working deposits along with an intelligent forecast of the situation, can certainly provide some valuable guidelines for the geoeconomic evaluation during different stages of exploration.

ACKNOWLEDGEMENT

The author is very much grateful to Prof. Bhagwan Das, retired Professor, Centre of Advance Studies in Applied Geology, Dr.

Harisingh Gour University, Sagar and Dr. V.K. Kanchan, retired Dy. D. G. Geological Survey of India, for some useful discussions and suggestions.

REFERENCES

Ballurkar, A. and Borisovich, V.T. (1977). Profit and profitability in geoexploration. Seminar on 'Exploration Geophysics in India', Osmania Universiy, Hydrabad, pp. 12-16.

Buchover, N.A. (1966). Principles of geoeconomic evaluation and its role in geoexploration (in Russia). Razvedka Ochrana, Nedra Publishers, USSR, no. 6, pp. 34-40.

Faruqui, F. A. (1994). The export oriented development of mineral resources and mineral based industries. 2nd Int. Conference, Geol. Sur. Pakistan, abstract vol. p.1-2.

Geological Survey of India (1975). Manual of mineral exploration. Misc. publication no. 33, Geol. Soc. India, pp. 3-18.

Gupta, S.K. (2009). Recent trends in exploration and mineral development and mitigating environmental impact . Int. Journ. of Ecology, Environment Conservation, vol. 15, no. 2, pp. 393-396.

Harkin, D. A. (1976). Systematic mineral exploration. Nat. Resource forum, vol. 1, no. 1, pp. 29-38.

Kajdan, A. B. and Kobachidze, L. P. (1968). Geoeconomic evaluation of mineral deposit (in Russia). Geologia in Razvedka, no. 12, pp. 147-158.

Karunakarn, C. (1971). Some problems in the exploration of non-ferrous mineral deposits in southern region by the Geological Survey of India. Journ.1, Gu. vol. VIII, pp. 29-38.

Kreiter, V. M. (1968). Geological prospecting and exploration. Mir. Pub. Moscow, USSR, pp. 129-136.

Kuzvart, M. and Bohmer, M. (1986). Development in economic geology, 21: prospecting and exploration of mineral deposits. Elsevir pub., New York, pp. 17-26.

Lacy, W. C. (1973). Exploration philosophy and future trends. Western Australia Conference, Australian Institute of Min. and Met., pp. 17-25.

McKinstry, H. E. (1960). Field exploration and mining geology. Asian Pub. House, Ist Indian ed., Bombay, pp. 407-427.

Peters, W.C. (1987). Exploration and mining geology. 2nd ed. John Wiley and Sons, New York, pp. 518-546.

Hon'ble reviwer

1. **Prof. M.P. Singh**
   Pro Vice Chancellor
   Lucknow University
   LUCKNOW - 226 007

2. **Dr. N.K. Dutta**
   Director General (Retd.)
   Geological Survey of India
   9, Utkarsh Nagar
   NAGPUR 440 001

I-23

Legislative Provisions for Corporate Social Responsibility (CSR) for Mining in India

P.K. Jain

Indian Bureau of Mines, Aashiyana, 46 Manish Society, Bornala, Katol Road, Nagpur - 440 013
E-mail pkjain3661@gmail.com

Abstract

Mining is an important segment of the Indian economy and has become necessary to social wellbeing. Since independence, there has been a phenomenal growth in mineral production in term of quantity and value. To great degree, minerals, forests and tribal are concentrated in same geographic areas. In present scenario to avoid conflict between mining and community it in necessary to pay adequate compensation and involve local community in mining activity. Currently, CSR practices are dictated by guidelines notified by the ministry of Corporate Affairs in December 2009.

Mines and Mineral (Regulation & Development) Act was amended in 1994 to be known as Mines and Mineral (Development & Regulation) Act to give prominence to development than regulation. In present paper an attempt has been made to analyse the various legislative developments like MMDR Act, 1957, Samatha Judgment, 1997, Hoda Committee, 2006, National Mineral Policy, 2008, MMDR Bill, 2011, to highlight the issue of fulfillment of CSR in mining.

Keywords: Corporate Social Responsibility, Mining Laws, MNDR Act, National Mineral Policy.

Introduction

The term "corporate social responsibility" came into common use in the late 1960s and early 1970s after many multinational corporations formed the term stakeholder, meaning those on whom an organization's activities have an impact. In India, the CSR concept promoted by Mahatma Gandhi during 1930s is well known which states the role of family-run-businesses conducting social and economic activities. World Business Council for Sustainable Development defines Corporate Social Responsibility (CSR) as "The continuing commitment by business to behave ethically and contribute to economic development while improving the quality of life of the workforce and their families as well as of the local community and society at large." World Business Council for Sustainable Development defines Corporate Social Responsibility (CSR) as "The continuing commitment by business to behave ethically and contribute to economic development while improving the quality of life of the workforce and their families as well as of the local community and society at large."

Currently, CSR practices are dictated by guidelines notified by the ministry of Corporate

Affairs in December 2009[1] then refined. These are, however, voluntaryand partnering in the endeavor of national development. Under these, business entities formulate their own CSR policy, approved by the company's board. The guidelines state that companies "should" allocate specific amounts in their budgets for CSR activities. "This amount may be related to profits after tax, cost of planned CSR activities or any other suitable parameter," the guidelines state.[2] Subsequently the National Voluntary Guidelines on Social, Environmental and Economic Responsibilities of Business (NVGs) have been released by the Ministry of Corporate Affairs in 2011[3].

The Mines and Minerals (Development and Regulation) Act, 1948 (MMDR) was the first legal framework for regulation and development of mines in independent Indiaenacted in 1957 by the Parliament. Two set of rules were made under the Act – the Mineral Concession Rules, 1960 (MCR) and the Mineral Conservation and Development Rules (MCDR), 1988.

The MMDR Act has been amended four times since 1957. First amendment came in 1972 which enhanced government control over mining. The second amendment in 1986 increased the Central Government control on mining and introduced the concept of an approved mining plan. In order to attract investments, the procedure for granting mineral concessions was simplified by an amendment in 1994. Third amendment also introduced the concept of a Large Area Prospecting Licence (LAPL). The fourth amendment to the MMDR came in 1999 after a committee reviewed the Act. This amendment comprised a number of changes – introduction of Reconnaissance Permit (RP) prior to prospecting licence (PL), delegation of powers to the state government, liberalization of area restrictions, etc.

The National Mineral Policy, 1993 highlighted the issue of social and economic upliftment of the local population including tribal population in mineral bearing areas. Following the mid-term appraisal of the Tenth Five-year plan, the Planning Commission constituted a High Level Committee under Shri Anwarul Hoda, (then Member, Planning Commission) to recommend changes in the Mineral policy 1993 and laws to address issues ranging from non-transparency in allocation of mineral resources, illegal mining to the social and environmental impacts of mining.

The Hoda Committee report was published in July 2006 with wide ranging recommendations like institutionalizing a Sustainable Development Framework (SDF) to address social, economic and environmental issues arising out of mining. Based on the Hoda Committee's report, the government brought out with a new National Mineral Policy in 2008 (NMP, 2008). Consequently, the Ministry of Mines framed a new Mines and Minerals (Development & Regulation) Bill to replace the existing MMDR Act, 1957.

The draft of the Mines and Minerals (Development and Regulation) Bill, 2010 was put in the public domain in June 2010. After this, a Group of Ministers (GoM) was constituted under Shri Pranab Mukherjee, then Minister of Finance, to consider the draft and give its recommendations on various issues including the sharing of the profits from mining with the local community.

The bill approved by Cabinet was introduced in Lok Sabha on 12th December 2011, which has been referred to the Parliamentary Standing Committee on Coal and Steel headed by Shri Kalyan Banerjee, Member of Parliament on 5th January, 2012. The suggestions were sought from stakeholders and state governments. The Parliamentary Committee is expected to finalise its report on the Mining Bill.

National Mineral Policy, 1993

As per Para 7.10 of National Mineral Policy, 1993 (NMP), the mineral deposits generally occur in remote and backward areas with poor infrastructural facilities which often inhibit their optimum development. Mineral bearing areas are also often inhabited by tribal population and exploitation of mineral resources has not always contributed adequately to their development. Contribution of mineral development to overall regional development has also not always been commensurate with huge investment in large mining projects. A major thrust needs to be given for development of infrastructural facilities in mineral bearing areas following an integrated approach for mineral development, regional

---

[1] http://www.mca.gov.in/Ministry/latestnews/CSR_Voluntary_Guidelines_24dec2009.pdf

[2] http://www.indiacsr.in/en/?p=3040

[3] http://www.giz.de/de/downloads/giz2012-corporate-social-responsibility-india-en.pdf

development and also social and economic upliftment of the local population including tribal population.

SamathaJudgment[4]

In 1997, the Supreme Court of India ruled that mining in area specified in Schedule V of the Constitution areas should not be allowed without the participation of the local people.

The judges laid out certain duties for any entity that mined in Schedule areas: viz. 20% of net profits to be set aside as a permanent fund for the establishment and maintenance of water resources, schools, hospitals, sanitation and transport facilities, reforestation and maintenance of ecology.

It was suggested to the state government to organize co-operative societies solely composed of schedule tribes, to undertake mining operation in those areas.

In the last few years, various state governments have been seeking clarification on these issues. The implementations of these provisions in the judgment remain an impasse and are yet to be implemented.

Hoda Committee Report, 2006[5]

Local Communities and Mining Activities

As per Hoda Committee report, land is often used without the consent of the indigenous people. Mining companies should act as if consent to gain access to land is required even when the law of the land does not require this. In making decisions to acquire land for mining, the cultural circumstances of the local people and loss of access to common resources should be kept in mind. Where resettlement takes place, companies need to ensure that living standards are not diminished, that community and socialites are preserved, and that they provide fair compensation for loss of assets and economic opportunity. Responsibility for ensuring the long-term wellbeing of resettled communities needs to be defined and monitored. Environmental mining and other interests should be considered along with the interests of the poor and politically marginalizedpeople in these areas.

The report further states that Mineral development can bring benefits to the local level. However, the relationship between mining companies and local communities has a legacy of abuse and mistrust. Widespread community demands for direct and sustained benefits from mineral wealth are a relatively recent phenomenon. The initiatives, which have started emerging, require that the focus should be on improving the capacity of the local government and other local institutions to provide the benefits from mining instead of the companies themselves seeking to take on governmental functions. Non-governmental organizations (NGOs) and other civil society groups can act as mediator to enable communities to join as partner with companies and governments in development. Supporting local businesses, preferential procurement policies towards local suppliers and distributors, employment of locals and skills training are important means of benefiting local communities and building human and financial resources. A broader role in community health programmes and participation in social infrastructure such as schools and hospitals are some of the other areas in which mining companies need to become active.

National Mineral Policy, 2008[6]

The National Mineral Policy, 2008 (NMP) states that preference for mining small deposits in Scheduled areas are to be given to Scheduled tribes formed as co-operatives.

As per Para 7.11 of NMP, 2008 mining operations often involve acquisition of land held by individuals including thosebelonging to the weaker sections. In all such cases asocial impact assessment will be undertakento ensure that suitable Relief and Rehabilitation packages are evolved. While compensation isgenerally paid to the owner for his acquired land, rehabilitation of affected persons in the form ofsubstitute land, land for housing and jobs is not always adequate. Appropriate compensation willform an important aspect of the Sustainable Development Framework mentioned in Para 2.3 and7.10 of NMP 2008. In so far as indigenous (tribal) populations are concerned the Framework shallincorporate

[4] Mines, Minerals & People
http://www.mmpindia.org/Samata_Judgement.htm

[5] Ministry of Mines website mines.nic.in

[6] Ministry of Mines website mines.nic.in

models of stakeholder's interest in the mining operation, especially insituations where the weaker sections like the local tribal populations are likely to bedeprived of their means of livelihood as a result of the mining operation.

In areas in which minerals occur and which are inhabited by tribal communities and weaker sections it is imperative to recognize resettlement and rehabilitation issues as intrinsic to the development process of the affected zone. Thus all measures proposed to be taken will be formulated with the active participation of the affected persons, rather than externally imposed. A careful assessment of the economic, environmental and social impact on the affected persons will be made. A mechanism will be evolved which would actually improve the living standards of the affected population and ensure for them a sustainable income above the poverty line. Forthis purpose, all the provisions of the National Rehabilitation and Resettlement Policy or anyrevised Policy or Statute that may come into operation, will be followed.

MM (D&R) Bill, 2011[7]

The various provisions under MM (D&R) Bill, 2011for social responsibility in mining areas are given below:

Sub-clause 7 of Clause 6 of the MM (D&R) Bill, allows state government to make provision for 'preferential' grant of mineral concession to co-operative society of Schedule Tribes in the areas specified in Schedule V and VI of Constitution for area as specified under Sub-clause 6 of Clause 6. These are areas with small deposits in isolated patches, which are not suitable for scientific mining. Instead of giving mining lease for each isolated small deposit, a lease for a cluster of such deposits will be granted in favour of a co-operative society of the Scheduled Tribes in the Scheduled areas.

Thus, although there is a move to depict that some sort of preference is being given to people of V and VI schedule areas, it is essentially up to the state government to notify such 'preference'.

Point (k) of Sub-clause 1 of clause 21 states that the prospecting licence holder has to pay compensation, as notified, to the person holding occupation rights of the surface of land.

There is also a provision under which the prospecting licence holder may have to pay compensation for damage to land as prescribed in the licence vide Point (a) under Sub-clause 2 of Clause 21.

Point (m) of Sub-clause 1 under clause 24 states that the lease holder has to pay compensation, as specified in sub clause (2) of clause 43, to the person holding occupation, usufruct or traditional rights of the surface of land.

Under clause 26, sub-clause 3, a Corporate Social Responsibility (CSR) document has to be attached with a mining plan. This shall comprise of a scheme for annual expenditure by the lessee on socio-economic activities in and around the mine area for the benefit of the host populations in the *panchayats* adjoining the lease area and for enabling and facilitating self employment opportunities, for such populations.

As per clause 43 (1) in respect of land in which minerals vest in the Government, the holder of a non-exclusive reconnaissance licence, high technology reconnaissance cum exploration licence or prospecting licence shall be liable to pay, to every person or family holding occupation or usufruct or traditional rights of the surface of the land over which the licence has been granted, such reasonable annual compensation as may be mutually agreed between the holder of such licence and such persons or in the absence of such agreement, which may be determined by an officer appointed, by notification, by the State Government in this behalf in such manner as may be prescribed by the State Government:

Provided that such amount shall be determined before commencement of operations and paid in advance each year, in such manner as may be prescribed by the State Government.

Profit sharing concept has been introduced for the first time in mining law in India under Sub-clause 2 of clause 43 of this Bill. A lease holder is to pay annually to the District Mineral Foundation (DMF), as specified in clause 56, an amount equal to the royalty paid during the financial year (except coal and lignite) and twenty six per cent of profit aftertax or a sum equivalent to in case of coal and lignite, whichever is higher.

Under Sub-clause 3 of clause 43, the lease holder (if a company) is also to allot at least one shareother than cash to each person of the family

---

[7] Ministry of Mines website mines.nic.in

affected by mining related operations. These shares areto be non-transferrable.

Sub-clause 5 of clause 43 makes provision for the leaseholder to provide employment and or other assistance as per the Rehabilitation and Resettlement policy of the state government topeople/family holding usufruct, occupation or traditional surface rights of the land over which thelease has been granted.

Under Sub-clause 7 of Clause 43, after the termination of a concession, the state government is to assess damages to the land, if any, and determine the compensation amount payable by the licencee or lease holder. This compensation is to be paid to person holding occupation or usufruct or traditional rights of the surface of the land and they are to be consulted inthe process of deciding the compensation.

As per Point (b) under Sub-clause 8 of clause 43 of the Bill the state government may initiate proceedings against lease holders if they fail to make the payments to the DMF as specified in Sub-clause2. Sub-clause 9 clarifies cases of no records or disputes on whether a person/family holds usufruct, occupation or traditional surface rights. For this purpose, the Bill lays down a procedurewhere the Collector of the district, in consultation with the *gram sabha/gram panchayat/* district council, shall take such decision.

Sub-clause 10 of clause 43 lays down the responsibility of identifying affected people with the state. As per point (b) under the sub-clause, the state government is to identify the directly or indirectly affected families by the mining operations, before the mining operations begin. The state governmentis also to ensure that monetary benefits are distributed directly or indirectly, to the affected people. The amount payable to the affected people may be decided based on the extentto which they are affected. This amount, on an average daily basis, is to be not less than at leastthe daily amount entitled to a person under the Mahatma Gandhi National Rural Employment Guarantee Act, 2005 (MGNREGA).

As per point (c)the State Government shall ensure that monetary benefits are distributed through a mechanism.

Under Sub-clause 2 of Clause 46, the Central Government is to form a National Sustainable Development Framework (NSDF) in consultation with the state governments. The main function ofthe NSDF is to facilitate and ensure scientific development and exploration of minerals, protection of environment and prevention and control of pollution. Sub-clause 4 lists down the important components of the NSDF like mitigation measures for adverse impact on environment and people, system of public disclosure of environmental parameters, consultative mechanisms for stakeholder interactions, develop indicators of sustainable development, etc. NSDF will also have broad criteria beyond which mining may not be deemed sufficiently sustainable and/orscientifically manageable.

Sub-clause 3 of clause 46 makes provision for the formation of a State Sustainable Development Framework (SSDF). The same can be formed only after prior approval of the central government.

Provisions for taking action in case of non-compliance

When operations are not carried out in line with the mineral concessions granted, the state government may under Sub-clause 1 of clause 12, issue a show cause notice to the concession holder. The state may forfeit the security and may suspend, curtail or revoke the licence or lease.

The government shall not permit the transfer of mining lease under certain conditions laid down in Sub-clause 5 of clause 18. These include fragmentation or unscientific mining, not in the interest ofmineral development and against national interest.

The Indian Bureau of Mines (IBM) may issue direction to a Reconnaissance Licence (RL) holder toensure compliance with conditions laid down in the RL under Sub-clause 3 of clause 19.

Under sub-clause 1, clause 47, the State Government may, in the interest of systematic development of mineral deposits, conservation of minerals, scientific mining, sustainable development and protection of the environment, issue directions to the owner, agent, mining engineer geologist or manager of a mine.

Sub-clause 2 of clause 49 says that if the Indian Bureau of Mines (IBM) or the Atomic Mineral Directorate (AMD) or the State Directorate is of the opinion that any mine or part of it poses a threat to conservation of minerals or the environment, then it may suspend operations. The same is to be done through a written order and the operations are to remain suspended till the order is complied with.

Clauses 104 to 107 define the offenses as recognised by the Bill. These may relate to carrying out exploration/mining without a licence, failure to implement the final mine closure plan, dis-obedience pertaining to any government/ authority ordersor any other contravention of the Bill.

Rights of Local Governing Body

Notification of public lands for inviting applications to bid for prospecting licence, large area prospecting licence or mining lease is to be done in consultation with the *gram sabha* or district council in V and VI schedule areas of Constitution of India according to Sub-clause 9 of clause 13. In non-schedule areas, the district *panchayats* are to be consulted.

As per Sub-clause 11 of clause 13, the *gram sabha*or the district council is to be consulted before granting mineral concession for minor minerals in a V or VI schedule area of Constitution of India.

Under the provision of Sub-clause 5 of clause 32, the concerned *panchayats*are to be consulted bythe IBM or the AMD or the State Directorate before approving or disapproving the progressive mineclosure plan. This is to be done within a period of ninety days from receipt of the plan.

Sub-clause 8 of Clause 32 specifies that the final mine closure plan be based on the planned land usefor the lease area after its closure. For deciding the planned land use, the concerned *panchayats* are to be consulted as the central government may prescribe.

The concerned *panchayats* are also to be consulted for suggesting modifications to the mine closure plan before approving it as under Sub-clause 10 of clause 32. This is to be carried out within aperiod of one year.

Various provisions for consultation with local governing bodies like *gram sabha*/district council/ *panchayats* before granting the concessions and on mine closure have been provided in the Bill, 2011. However, what this consultation means and howit will be conducted has not been defined. What will be relationship between consultation and consent hasnot been explored.

Mandatory CSR spending for coal mining firms

The Union government may make it mandatory for coal mining companies to spend a part of their net profit on the welfare of local people affected by the activities. The move, a part of the competitive bidding guidelines being framed by the coal ministry, is in line with the ongoing wave of policy decisions aimed at integrating land losers in development projects within the broader "inclusive" economic agenda.

If it goes through, the move would add to the controversy brewing on the similar proposal to ask mining companies to share 26 per cent of their profits with the local populace and make them stakeholders in projects. The policy on competitive bidding is getting framed. It is possible to make spending on corporate social responsibility (CSR) by companies mandatory in the bidding policy.

Presently, Coal India Ltd, the state-owned near-monopoly under the coal ministry and the largest producer of the dry fuel in the world, already spends around five per cent of its retained profits on CSR activities,[8]

Funding

The guidelines issued by the Government stipulate the following range of expenditure on CSR activities by different categories of public enterprises:

	Net Profit (previous year)	Range of funds in a fiscal year
I)	Less than Rs 100 crore	3% -- 5%
II)	Rs 100 crore to Rs 500 crore	2% -- 3%
III)	Rs 500 crore and above	0.5% --2%

The these fund has to be spent for Education, Water Supply and Sanitation,Human Health and Family Welfare, Animal / Livestock Care, Environment,Construction / Development of Infrastructure in Backward Areas, Social Empowerment, Promotion of Sports, Art & Culture and Participating in the Development and Welfare Programmes of the Govt, etc.

Amount allocated towards periphery development till 2009-10 (including special projects) by National Aluminium Company

[8]http://www.indiacsr.in/en/?p=3040

Limited (NALCO) at Damanjodi-Rs. 48.88 cr, at Angul-Rs. 61.10cr and other areas- Rs. 26.89 cr (TOTAL - Rs. 136.87 cr). Soon after the Super Cyclone in Orissa in the year 1999, Nalco has constructed 197 Primary Schools-cum-Cyclone Shelters, under Prime Minister's National Relief Fund Scheme. Tata Steel spends 5-7 per cent of its profit after tax on several CSR initiatives. The Vedanta Group spent Rs 230 crore on CSR when its net profit was Rs 13,130 crore, or 1.75% of its net profit. Ambuja Cement spent Rs 32.5 crore (2.6% of profit), the Ultra Tech Cement spent Rs 18.2 crore (0.8% of profit) under CSR activities.

Conclusion

The SamataJudgment in 1997 provided a semblance of hope for the tribal communities. Mining could no longer remain an economic activity to further the profits of a few and oblivious of the local communities. It has to seek consent of communities direct governments to be enabler of better development of the mining regions, make local communities partners in development of mines and alleged activities. Present MM (D&R) Act, 1957 enabled some form of a regulatory regime. The Constitutional Amendment empowering the local self governments and its extension in Scheduled Areas, which includes the expansion of the concerns on Human and Ecological right through various legislations and conventions. The various provision of MM (D&R) Bill, 2011 will improve the effectiveness of mining sector towards fulfillment of social responsibility for community in general and mining affected people in particular.

Notes: The views expressed in this paper are the author's own and not necessarily those of the Indian Bureau of Mines, for whom he is working.

References

http://www.mca.gov.in/Ministry/latestnews/CSR_Voluntary_Guidelines_24dec2009.pdf

http://www.indiacsr.in/en/?p=3040

http://www.giz.de/de/downloads/giz2012-corporate-social-responsibility-india-en.pdf

Mines, Minerals & People http://www.mmpindia.org/Samata_Judgement.htm

Ministry of Mines website mines.nic.in

Ministry of Mines website mines.nic.in

Ministry of Mines website mines.nic.in

http://www.indiacsr.in/en/?p=3040

http://articles.economictimes.indiatimes.com/2012-12-20/news/35933659_1_csr-initiatives-net-profit-corporate-social-responsibility

Hon'ble reviwer

1. **Dr. S.B.S. Chauhan**
   Former Jt. Advisor
   Planning Commission
   D6, 6021/6, Vasant Kunj
   NEW DELHI - 110 070
2. **Prof. R.N. Sharma**
   Vice Chancellor
   IASE University
   SARDARSAHR - 331 403

I-24

Geology and Mineral Resources of Meghalaya

Devesh Walia[1], and Dr Shailendra Chaudhary[2]

1. Department of Environment Studies, North Eastern Hill University, Shillong, Meghalaya
2. Department of Science and Technology, NEC, Shillong, Meghalaya
E-mail: deveshwa@gmail.com

Abstract

The Precambrian basement of Meghalaya, a remnant of the northeasterly extension of the Indian Peninsula, is a segment which has undergone multiple phases of deformation leading to folding and fracturing of rocks. The Shillong plateau is a region of basement tectonothermal reactivation. Along the eastern boundary of the plateau the Precambrian rocks disappear beneath folded sediments of the Assam area. The plateau standing as a watershed between the Surma Valley of Bangladesh on the south and the Brahmaputra Valley on the north is dissected by sevberal rivers and a network of their tributaries and lateral streams.

The magnificent gorges scooped out by the rivers in the southern Khasi and Jaintia Hills are result of massive headward erosion along joints in the gneissic granite rocks. The limstone covered country over the southern Garo, Khasi and Jaintia hills represent a typical karst topography. The concentric drainage pattern revealed from the study of aerial photographs, suggest that these areas represent "migmatite dome". It may be presumed that the radial to semicircular fractures, developed during the formation of such migmatite dome', have later become the repository of the Shillong quartzite.

The Shillong plateau manifests as a geomorphic arch A number of linear fractures on the regional scale are present in the Shillong plateau. The earliest structural trend have been E-W Subsequently the trend transposed to NE-SW probably accentuated by the intrusion of porphyritic granite plutons. However, some N-S and NW-SE faults and fractures are also reported from the plateau, which have affected the major E-W trending Dawki fault system and are still active. The important faults so far identified in the plateau, are Dauki fault, Raiban fault, Dapsi fault etc. The Raiban lineament (fault) is a southern dipping reverse fault having E-W trend and defines the contact between Sylhet Traps and the crystallines in the north. Along southern fringe of Shillong Plateau, the predominant structural lineament is the Dauki fault. In the western part of the plateau in the Garo hills, the E-W lineaments are represented by monoclines and faults among which the Dapsi reverse fault at the base of Tura range is spectacular. The Tura range, a horst, is bounded to the south by the Dapsi fault (up-thrust). In Garo hills, there is a ample evidence of reactivation activity along a number of E-W, N-S and NW-SE basement fault throughout, the Tertiary period.

The tectonic setup of the Meghalaya plays a very important role for developing the physiography, drainage pattern geomorphology and mineral resources localisation. Though no promising deposits of any metallic minerals have so far been located in the State. It is endowed with some of the important non-metallic minerals like lime stone, coal, clay, glass sand, quartz and silimanite. Recent find of Uraniferous arenaceous conglomerates makes the region much more prosperous and affluent.

Keywords: .Khasi and Jayantia hills, Shillong Plateau, Meghalaya Minerals, Non-Metallic Minerals.

Introduction

The state of Meghalaya comprising the Garo, Khasi and Jaintia hills districts which previously formed part of Assam, came into existence on 19th January, 1972. The state covers an area of approximately 23,000 sq km bordered to its northwest, north, east and southeast by the State of Assam and to the south and south west by Bangladesh. The surface elevation range from 150 m to 1950 m above msl. The State presents a picturesque landscape of plateau, ravines, brooks, rivulets, lakes, rapids and waterfalls. The Shillong is the capital of the state and is located at an elevation of around 1800 m in the central part of the Khasi Hills.

The region experiences tropical monsoon climate and the average annual rainfall is around 205 cm and the maximum annual average of 1143 cm being recorded around Cherrpunjee and Mawsynram.

Though no promising deposits of any metallic minerals have so far been located in the State, it is endowed with some of the important non-metallic minerals like lime-stone, coal, clay, glass sand, quartz and sillimanite.

The Meghalaya plateau (MP) in the northeast India region is a part of the Indian shield, is tectonically very active due to its collision with the Tibetan landmass in the North and the Shan Tenasserim block in the east, that is separated out from the peninsular shield and moved to the east by about 300 km along the Dawki fault. The gigantic E-W trending Dawki fault separates the MP to the north and the Bengal basin to the south. The mighty Brahmaputra river, on the other hand, separates the MP from the Himalaya to the north. The E–W segment of the river at the northern boundary of the MP is named Brahmaputra fault. The NW–SE trending Kopili lineament named Kopili fault separates MP from the Mikir massif. The region is jawed between the two arcs, the Himalayan arc to the north and the Indo-Burmese arc to the east. The region produced two great earthquakes ($M > 8.0$) and about 20 large earthquakes ($8.0 > M > 7.0$) since 1897. The Shillong plateau was the source area for the 1897 great earthquake of M 8.7, and numerous earthquakes due to its nearness to the Alpine Himalayan folded mountain chain, Assam-Arakan-Yoma thrust belt and Dawki Thrust. The tectonic setup of the Meghalaya plays a very important role for development of physiography, drainage pattern, geomorphology and mineral resources localisation.

Physiographically, Meghalaya represents a remnant of an ancient plateau of Pre-Cambrian Indian Peninsular shield, block uplifted to its present height. The plateau standing as a watershed between the Surma Valley of Bangladesh on the south and the Brahmaputra valley on the north is dissected by several rivers and a network of their tributaries and lateral streams. The present physiographic configuration of the plateau was attained through different geologic events as indicated by the polycyclic erosion surfaces at various levels.

The drainage pattern in the region represents a most spectacular feature revealing extraordinary straight courses of the rivers and streams, evidently along joints and faults. The magnificent gorges scooped out by the rivers in the southern Khasi and Jaintia Hills are result of massive head-ward erosion along joints in the gneissic granite rocks. The lime-stone covered area over the southern Garo; Khasi and Jaintia hills represent typical karst topography.

The earliest geological reference on the region was made by T. Oldham (1859). The systematic geological mapping of the region was subsequently carried out by H. B. Medlicott (1869), Godwin Austin (1869), La Touche (1883, 1887), C.S. Fox (1923), P.N. Mukherjee (1938-39) and others which has led to the establishment of the stratigraphic sequence of different rock suites of the region.

General Geology and Stratigraphy

The State of Meghalaya is occupied by rocks ranging from Archean to Quaternary period. It is also referred as 'geology in a nutshell' with rocks from almost all the metallogenic provinces and epochs. The main lithounits of the plateau are- Archaean – Proterozoic Gneissic Complex, Proterozoic Shillong Group of rocks with basic-ultrabasic intrusive, Neo Proterozoic- Lower Palaeozoic porphyritic granite plutons, small outcrops of Lower Gondwana sedimentary rocks at its western edge, Cretaceous volcanic rocks represented by Sylhet Trap, Upper Cretaceous Alkaline-Ultramafic-Carbonatite Complex, Cretaceous-Tertiary shelf sediments and Pleistocene to Recent fluvial sediments.

Stratigraphic succession of Meghalaya (after, GSI, 2009)

Age	Group Name	Formation (Thickness)	Lithology
Holocene	Newer Alluvium (Thickness not known)	Unclassified	Sand, silt and clay
Pleistocene	Older Alluvium (Thickness not known)	Unclassified	Sand, clay, pebble, gravel and boulder deposit
		...Unconformity...	
Mio-Pliocene		**Dupi Tila Formation (1050 m)**	Mottled clay, feldpathic sandstone and conglomerate
		...Unconformity / Disconformity...	
Oligo-Miocene	**Garo Group**	**Chengapara Formation (700 m)**	Coarse sandstone, siltstone, clay and marl
		Baghmara Formation (530 m)	Coarse feldpathic sandstone, pebble, conglomerate, clay, silty clay with a fossiliferous limestone horizon at the top
		Simsang Formation (1150 m)	Siltstone and fine sandstone and altermations of siltstone – mudstone
Eocene-Oligocene	**Barail group**	....	Coarse sandstone, shale, carbonaceous shale with streaks and minor lenses of coal
Palaeocene-Eocene	**Jaintia Group**	**Kopili Formation (50 m)**	Shale, sandstone, marl and coal
		Shella Formation (600 m)	Alternation of sandstone, limestone
		Langpar Formation (100 m)	Calcareous shale, sandstone, limestone
Upper Cretaceous	**Khasi Group**	**Mahadek Formation (150 m)**	Arkosic sandstone (often Glauconitic & Uraniferous)
		Conglomerate (25 m)	Conglomerate
		Jadukata Formation (140 m)	Conglomerate / sandstone
		...Unconformity...	
Cretaceous	**Alkaline Ultramafic-Carbonatite Complex of Sung**	....	Pyroxenite – Serpentinite with abundant development of melilite pyroxene rock, ijolite, syenite and carbonatite

Age	Group	Formation	Lithology
……………Unconformity……………			
Cretaceous	**Sylhet Trap (600 m)**		Basalt, alkali basalt, rhyolite and acid tuff
……………Unconformity……………			
Carboniferous to Permian	**Lower Gondwana**	**Karharbari Formation**	Very coarse to coarse grained sandstone with conglomerate lense, siltstone, shale, carbonaceous shale and coal
		Telchir Formation	Basalt tillite, with sandstone bands, siltstone and shale
……………Unconformity……………			
Neo Proterozoic – Early Palaeozoic		**Granite Plutons:** Kyrdem Granite Pluton (479±26 Ma) Nongpoh Granite (550±15 Ma) Mylliem Granite (607±13 Ma) South Khasi Granite (690±26 Ma)	Porphyritic coarse granite, Pegmatite, aplite / quartz vein traversed by epidiorite, dolerite and basalt dykes
……………Intrusive contact……………			
Proterozoic	**Khasi Basic-Ultrabasic intrusive (Khasi greenstone)**	…………………	Epidiorite, dolerite amphibolites and pyroxenite dykes and sills
Palaeo-meso-Proterozoic	**Shillong Group**	…………………	Quartzite, phyllite, quartz-sericite schist, conglomerate
……………Unconformity……………			
Archean (?) Proterozoic	**Meghalaya Gneissic Complex**		Biotite gneiss, biotite horn-blende gneiss, granite gneiss, mica schist, sillimanite-quartz schist, biotite-granulite-amphibolite, pyroxene granulite, gabbro and diorite

(A) GNEISSIC COMPLEX:

The Gneissic Complex mainly consists of two different units- Gneissic complex proper containing abundant granitic gneiss of amphibolite facies, and the non-porphyritic granitoid rocks. The Gneissic complex along with some other associated rocks containing diopsidic clinopyroxene and sillimanite-garnet-potash-feldspar-quartz gneiss without muscovite forms a common assemblage as observed in the MP.

However, among the Nonporphyritic Granitoid rock, the dioritic, granodioritic and granitic types of rocks are more common. The dioritic types of rocks are characterised by decussate texture and dominance of biotite in an aggregates of biotite, hornblende, sphene, apatite, zircon, epidote-zoisite, minor plagiclase and quartz. On the other hand, the granodioric types of rocks are classified as plutonic migmatite. These granitoid rocks occur in the core of the gneissic terrain which had developed during the most advanced stage of granitisation process. But none of the mobilised grnitoid intrudes into the Shillong Group of rocks contemplating that their upper age limit is pre-Shillong Group (Mazumder, 1976).

(B) SHILLONG GROUP

In the eastern, western and central part of the plateau, the Proterozoic NE-SW trending interior sag basin over the floor of Gneissic complex resulted in to the deposition of the Shillong Group of rocks with an unconformity defined by a basal conglomerate horizon under a continental environment in a tensional stress. These metasedimentary Shillong Group of rocks have been exposed in the plateau. In the northern part of the plateau, these rocks can be correlated with "Iron Ore Series" due to the presence of magnetite-quartzite and hematite-quartzite in it, with some difference in lithology.

The metasedimentary rocks of Shillong Group can be grouped into two broad lithofacies, the lower argillaceous unit in which metamorphism has reached upto "Garnet Zone" (argillaceous unit is represented by phyllites, siltstones and shale and often form thick piles in which intercalating bands/lenses of quartzite are common), and an upper arenaceous unit of low grade metamorphism or un-metamorphosed in some places.

The Shillong Group comprises mostly of clastic sequence of quartzite, phyllite, subordinate quartz- sericite schist with basal and intraformational conglomerate. The basal conglomerate invariably occurs over the basement rock, whereas the intraformational conglomerate occurs as discontinuous bands/lenses in the quartzite's at some localities. Among the quartzite, the ferruginous quartzites are dominant and, in some places where conglomerate is absent, it forms the base of Shillong Group.

These NE-SW trending lithopackage of Shillong Group of rocks traversed by acidic (granites with pegmatites and quartz), basic and ultrabasic (Khasi basic intrusive) intrusive at places.

(C) KHASI BASIC-ULTRABASIC INTRUSIVES:

The Khasi basic-ultrabasic intrusive (also known as Khasi greenstone) is an important intrusive phase occurring mostly as sills, dykes in the various subfacies of Shillong Group. The Khasi basic-ultrabasic intrusive have also intruded the basement Gneissic Complex in the west-central part of the plateau.

These intrusive rocks comprising of epidiorite, metagabbro, metadolerite represents an episodes of basic magmatism between the two major phases of granitic activities. The emplacement of basic magma was prior to the onset of metamorphism in the Shillong Group. The distribution of these intrusive rocks parallel to the strike of the enclosing Shillong Group of rocks indicates that they were either pre-fold sills or emplaced in a tensional zones of extensional regime related to the folding of the host rocks.

(D) GRANITE PLUTONS:

The entire Precambrian regime of the Meghalaya plateau has a number of syn to late tectonic plutons. Most of them occur within the gneissic complex. South Khasi batholith (690±26 Ma), Mylliem pluton (607±13 Ma), Kyrdem pluton (479±26 Ma) and Nongpoh batholith (550±15 Ma) represents a few significant porphyritic granite of the entire central Meghalaya plateau, broadly becoming younger in age from SW to NE. Amongst these, the South Khasi, Nongpoh and Kalong batholith have intruded into both the gneissic complex and Shillong Group of rocks. The presence of discordant xenolith within the plutons and the straight margins of some plutons like Nongpoh, South Khasi batholith etc. suggest a magmatic origin of these granitic plutons and their emplacement is largely controlled by the pre-existing fractures (Nandi, 2001).

Amongst the different plutons, some of them are discussed below:

(i) Kyrdem Granite Pluton (479±26 Ma)

This granitc pluton occurs as a small body in the northeastern part of Shillong and has intruded in the metasedimentary Shillong Group of rocks. Mainly comprising of two major variants- the dominant one is coarse grained porphyritic granite represented by grey and pink K-feldspar bearing variety and other one is fine grained granite which mostly occurs within the porphyritic granite as anastomizing vein.

(ii) Nongpoh Granite (550±26Ma)

The very coarse grained, porphyritic Nongpoh Granite occur as a discordant intrusive, show a sharp contact with the Basement Gneissic Complex/Shillong Group of rocks. Within the Nongpoh Granite, xenoliths of country rocks are visible, which are traversed by thin granitic vein along with the country rock.

(iii) Mylliem Granite (607±13 Ma)

The Mylliem Granite is a discordant body that intrudes into the Shillong Group of rocks. These are coarse to very coarse grained granite comprising of phenocrysts of microcline-microperthite in a groundmass of plagioclase, quartz, K-feldspar and biotite with accessories of apatite, zircon, sphene, epidote, muscovite and sericite. Within the Mylliem Granite, xenolith of quartzite and thin veins of pegmatite, apatite and quartz are visible.

(iv) South Khasi Granite (690 ± 19 Ma)

The South Khasi Granite comprising of porphyritic granite and fine grained homophonous pink granite has intruded in both the Basement Gneissic Complex and the metasediments of Shillong Group. The porphyritic granite variants consisting of coarse megacrysts of K-feldspar in a groundmass of fine grained K-feldspar, quartz, orthoclase and biotite, whereas the homophonous pink granite comprises of orthoclase, quartz, plagioclase and biotite. Within the pluton, numerous xenoliths, enclaves of amphibolities, diorite and granite gneiss are also present.

(E) LOWER GONDWANA GROUP

Most of the Gondwana rocks are concealed under the thick cover of Brahmaputra alluvium. However, an isolated N-S trending linear outcrop of Lower Gondwana rocks has been recognized at Singrimari near the western margin of Meghalaya Plataeu in Garo Hills. De and Boral (1978) classified these sediments as Talchir and Karhabari Formation.

Talchir Formation

The Gondwana sequence in the area starts with the boulder conglomerate comprising of pebble to large cobble size clasts of quartz, quartzite, gneisses and pegmatite's embedded in a silty matrix. These boulder conglomerates are unsorted, devoid of stratification and directional fabric. An outcrop of about 10 m thick alternating sequence of shale and conglomerate is observed in Boldamiri area.

Karharbari Formation

This overlying sequence is well exposed in the Singrimari area which exposes a thin layer of sandstone and carbonaceous shale with a thick layer of conglomerate over it. The brownish red to dark grey, highly micaceous shale contains well preserved plant impression, whereas, the conglomerate shows brownish grey to greenish grey color with clasts of vein quartz embedded within the sandy matrix.

(F) SYLHET TRAPS

The Sylhet Traps are represented by plateau (flood) basalts exposed in a narrow 80 km long and 4 km wide E-W belt along the southern margin Meghalaya plateau. These apparently overlie the eroded Precambrian basement complex and are overlain non-conformably by the upper Cretaceous-Eocene sediments. In general, the E-W trending Raibah fault limits the northern boundary of the Trap in which they are in contact with the gneisses and granites at Theriaghat section. The exposed limit of Sylhet Trap to the south is marked by the prominent E-W trending Dawki Fault system.

The sylhet trap mainly comprises of basalt with minor volume of alkali basalt (nepheline tephrite), rhyolite and acid tuff. The basaltic flows show a various texture like non-porphyritic, micro-porphyritic or megaporphyritic. The alkali basalt occur locally as two distinct flows in Umiew and Dwara gorge, mostly comprising phenocryst of diopside and hornblende in a groundmass of nepheline, aegirine-augite, aegirin, plagioclase, magnetite and felspathoid minerals. The acid volcanics occurs as intrusion within the basalt. The close association of rhyolites with the acid pyroclastic and the nature of the acid tuffs grading into the rhyolites shows that the rhyolites are really ignimbrites occurring as dyke.

(G) ALKALINE-ULTRAMAFIC-CARBONATITE COMPLEX OF SUNG VALLEY

The Alkaline-Ultramafic-Carbonatite Complex of Sung consisting of mainly peridotite, pyroxenite, serpentinite, uncompahgrite, alkaline rocks, biotite-alkali feldspar rock and carbonatite, occurs as intrusive body within the Shillong Group of rocks. Within the oval shaped complex, the pyroxenite is most abundant rock type which act as a host for emplacement of all the rock types except serpentinite. The central part of the complex is occupied by serpentinite which is surrounded by the pyroxenite rim except to the east where it is dirrectly in contact with quartzite. Carbonatites are occurs as small lenses and dykes in the southern half of the complex whereas, the northern part is mainly occupied by large oval shaped bodies and small dykes of uncompahgrite.

It is seen that near the vicinity of the intrusive body there are no evidence of contact metamorphism. However, a narrow zone of fenite and a brecciated admixed zone of fenite and quartzite have developed at places particularly along the contact in which these intrusive bodies sharply cuts the steeply dipping NE-SW and E-W trending quartzite and phyllite of Shillong Group.

(H) KHASI GROUP

The Cretaceous Khasi Group of rocks shows distinct arenaceous facies. The oldest Jadukata Formation consisting of conglomerate and sandstone (with characteristic zone fossil Inoceramus) unconformably overlying the Sylhet trap and occurs to the south of Raibah fault. On the other hand, in the north of Raibah fault, a younger conglomerate horizon of 25 m thickness has designated as Basal Conglomerate Formation. This Basal Conglomerate Formation is overlain by Mahadek formation consisting of coarse arkosic and glauconitic with thickness varying from 150-400 m.

(I) JAINTIA GROUP

The Langpar Formation of the Jaintia Group consisting of calcareous shale, sandstone and sandy limestone of Paleocene age overlies the Mahadek Formation. The contact between these two formations signifies the Cretaceous-Tertiary boundary. The Eocene Shella Formation consisting of three sandstone and three limestone members overlie the Langpar Formation. From bottom to top, these are- Theria sandstone (25m), Lakadong Limestone (200m), Lakadong sandstone (25 m), Umladoh limestone (55 m), Nurpuh sandstone (20 m) and Prang limestone (210 m). The Kopili Formation (150-800 m), representing the uppermost unit of Eocene Sequence of Meghalaya is comprised of thinly laminated shale's with phosphatic nodules, sandstone, limestone and marls, it conformably overlies the rocks of Shella Formation.

(J) BARAIL GROUP

The Oligocene sequence of shelf facies, around 1000 m in thickness, consisting of fairly coarse grained sandstone, shale, carbonaceous shale with streaks and minor seams of coal has been named as Barail Group in the east Khasi, Jaintia and Mikir Hill sectors. The Simsang/ Kherapara Formation in Garo Hills and Chokpotgiri Formation in the southern Khasi Hills are the local litho stratigraphic equivalent of Barail Group.

(K) GARO GROUP

The Simsang Formation of Garo Group consist of a cycle of massif cross-bedded sandstone alternating with siltstone-sandstone unit, conformably overlies the Kopili Formation without any break of sedimentation. In the eastern part of the plateau, the Baghmara Formation conformably overlies the Simsang Formation consist of irregular lenticular beds of coarse feldspathic sandstone with minor mudstone streaks, pebble conglomerate, massive clay and silty clay beds. This Formation is conformably underlying by the Chengapara Formation which consist of fine grained micaceous sandstone, siltstone and clays with a few thin marly beds at its base. From Bhagmara to few kilometers west of Barenapara, a fossiliferous calcarcous siltstone horizon of Acquitanian-Burdigalian can be delineated (Pascoe, 1964)

(L) DUPITILA FORMATION:

The Dupitila Formation representing mainly deltaic to fluviatile facies consists of alternations of coarse feldspathic sandstone with pebbly lenses containing vein quartz and mottled sandy clay.

(M) QUATERNARY AND RECENT DEPOSITS

The older Alluvium deposits consist of assorted pebble beds with coarse, loose sand and brownish clay forms a spectacular flat-topped low hillocks and mounds with red soil cover. The outcrop of such deposits overlie Tertiary rocks, which occur at various elevation levels along southern and western boarder of Garo Hills, southern fringe of Khasi Hills as well as northern fringe of Garo Hills.

The Recent (younger) Alluvium consisting of fine silty sand and light to dark grayish clay occurs in the river valleys on northern foot hill region of Garo and Khasi Hills, along the western border of Garo Hills and southern foot hill region of Garo and Khasi Hills.

Tectonic evolution

The prominent NE-SW structural trend of the Precambrian basement complex of Meghalaya which is a remnant of the northeasterly extension of the Indian Peninsula, suggest that these structures had developed in a NW-SE stress field and has undergone multiple phases of deformation leading to folding and fracturing of rocks. During the Proterozoic time, in the east and central part of the plateau, a NE-SW interior sag basin got developed which leads to the deposition of the Shillong Group of rocks under a continental environment in a tensional stress regime. Later on, these Shillong Group of rocks uplifted and folded, accompanied by low grade metamorphism and granitic (Mylliem granites) and basic/ultrabasic (Khasi greenstone) intrusions.

The post Precambrian landmass experienced peneplanation till the Upper Jurassic times which results in the formation of flat level surfaces preserved over the plateau till today. During the Late Jurassic-Early Cretaceous time, the eruption of Sylhet Trap through the E-W trending Raibah Fault had taken place in the southern block of the plateau. The Raibah fault is a southerly dipping reverse fault along which the southern block had subsided resulting into the marine invasion and deposition of Cretaceous sediments over this rapidly sinking block. The rate of subsidence gradually slowed down and attains a stable shelf condition during the Palaeocene-Eocene times in which foraminiferal formation of Jaintia Groups were deposited. Soon after this time, the E-W trending Dawki fault system became active in which the southern block of each fault segment had down-faulted relative to its northern block. Thus, the sedimentation pattern along the various segment of Dawki fault system was greatly influenced by the tectonic activity throughout the Tertiary time.

During Miocene, sedimentation continued uninterruptly over the southwestern part of Garo Hills and southern fringe of the Khasi Hills. The Jaintia block uplifted and remained as a landmass. However, at the end of Miocene, the major upliftment of the plateau results in the formation of land block shallow water lacustrine basins along southern fringe of Khasi and Garo Hills where deposition of Plio-Pleistocene Dupi Tila Formation had taken place. Along the old river valley, in the southern border of the Khasi and Jaintia hill and south and western parts of Garo hills, the deposition of Holocene fluviatile sediments has taken place.

Structure

The main structural elements of the basement gneissic complex are foliation, axes of folds and intrusives bodies which are predominantly E-W in the south and west of Nongpoh area and, NE-SW over rest of the area. These rocks of gneissic complex are different from that of Shillong Group of rocks which are mostly broad open folds with a few steeply dipping zones due to faulting. Murthy et. al. (1967), described three generations of deformations within the gneissic complex. The first one is represented by an E-W trending isoclinals to broad recumbent (or even reclined) which was superposed by a tight isoclinals fold having axial planes trending ENE to NE. The third phase is characterized by broad N-S warps.

In the central and eastern part of the plateau, a NE-SW trending intracratonic depression was developed in middle Proterozoic time (Kumar et al. 1990) for the deposition of molasse type of sediments of the Shillong Group. This structural trend of Shillong Group of rocks persistent all over the area and variation of this trend is discernible at places near the contact of intruded granite plutons. Within these rocks, current bedding and ripple marks are well preserved in the sandstone. Pene-contemporaneous structures, channel fills and sand lenses are also common in the siltstone. The NE-SW trending Tyrsad - Barapani Shear zone for which Porphyritic granitic intrusion has taken place within this intracratonic basin is an another important structural element.

The southern margin of Meghalaya plateau is marked by E-W trending Dawki fault system comprises of at least four E-W running normal faults with occasional reversal. These separate the plateau from the Sylhet plains of Bangladesh lying to the south. The Raibah fault is also E-W trending reverse fault dipping towards south. It separates the Sylhet trap which might have erupted during the Jurassic period from the Precambrian basement in the north.

In Garo Hills area (western part of the plateau), the WNW trending Dapsi reverse fault which runs at the base of Tura range demarcates the boundary between the Precambrian basement and self sediments. In the northern edge of the plateau, there are also some E-W faults among which, the Brahmaputra faults are most spectacular. Toward east, the NW-SE trending Kopili fault separates the plateau from Mikir Hills, is tectonically active till now as evidenced by intense seismic activity along this fault (Das Gupta and Nandy, 1995)

There are many N-S trending fault which runs across the western and central part of the plateau. The western boundary of Meghalaya Plateau is marked by Dhubri fault or Jamuna fault (Nandy, 2001). In the central part of plateau, there are Kulsi fault and Dhudnoi fault which traverse the plateau from Sylhet plain to the Brahmaputra in the north.

The emplacement of Precambrian plutons and later basic intrusive are mostly govern by these existing fracture planes which trend mainly in E-W to NE-SW direction. Another distinctive feature of the Shillong Plateau is their rectilinear drainage pattern controlled by these structures.

Mineral Resources of Meghalaya

The rocks of Meghalaya represent most of the metallogenic epochs and have deposits of major, minor, strategic and critical minerals. Four principal mineral deposits like limestone, coal, sillimanite and clay occurs in different rock formation of Meghalaya plateau. Besides these, occurrence of a number of mineral deposits like base metal, quartz, feldspar, corundum, glass sand, iron ore, pyrite, bauxite, rock phosphate has also been reported. The carbonaceous quartzite is traversed by vein quartz, near the Tyrsad, is the host rock for gold mineralization. The following paragraph describes the mineral deposits of Meghalaya, their host rocks and associates with the estimated deposits (Indian Mineral Year book (2007) and GSI Misc Pub (2009).

Limestone

In Jaintia, khasi and Garo hills of Meghalaya plateau, large deposits of limestone containing high calcium occurs in the Tertiary rocks. These high calcium limestone are suitable for cement and chemical industries. The largest deposits of limestone are found to be occur in Shella Formation of Eocene age located in and around the Litang River Valley area of Jaintia Hill.

Coal

Within the Meghalaya plateau, deposits of Coal are mainly confined to the Tertiary sediments of Garo, Khasi and Jaintia hill district and in the Lower Gondwana sediments of West Garo Hills district (exposed in the Singrimari basin). The coalfields of Khasi hills are located mainly in the southern periphery of the Meghalaya Plateau. Coal deposits in the Jaintia Hill occurs in the Tura sandstone of Eocene age.

Sillimanite

The deposits of sillimanite occur in Sonapahar area of West Khasi Hill as massive lensoid bodies of sillimanite-corrundum. They are contained mostly in Quartz-sillimanite schist and in a few cases in biotite-sillimanite-quartz schist and the surficial float concentrations are derived from these bodies due to prolonged and intensive weathering. The total reserve is 55807,000 tones.

Base metals

The base metal are occur in Umpirtha area within the sequence of high grade Archean gneisses comprising of granite and granite gneiss, migmatised biotite gneiss with amphibolites and micaceous quartz-sillimanite schist. The ore bearing rocks are lensoidal in shape and the total estimated reserve of Pb and Zn Ore is about 30,000 tones.

Feldspar

Deposits of feldspars are mainly located near Tura, Anegiri, Santhagiri, Chisakgiri and at certain other places. In Garo Hills, it occurs as thick pegmatite veins within the Archaean metemorphics including granite gneisses. The big augen shaped grains of orthoclase are used in glass industry

Glass sand

The white sandstone containing angular and pure well sorted grains of quartz which may be used in the manufacture of glass wares occurs in the Cherrapunjee and Laitryngew area. The inferred reserve is 406,400 tones in Laitryngew area. Some irregular beds of sandstone in the Lower and Middle Sylhet Sandstone (Jaintia Group) and the quartzites and sandstones of Shillong Group has also yield glass sand and is being used for making bottle glass.

Kaoline/clay

The Kaoline clay occurs in and around Smit, Latiyngkot of East Khasi Hills, Thandlaskein, Shaongpaung, Mulieh and Mynsngat of Jaintia hills and Darugiri in East Garo Hills districts. The Fireclay has been found to occur near Jowai and also around Lyrnai. It is used in the refractory, brick and pottery industry. The estimated reserves of Kaoline and Fireclay are about 88875,000 and 10999,000 tones respectively.

Dimensional Stone

Deposits of multicoloured granite have been located around Nongpoh in Ri-Bhoi district Mylliem in East Khasi Hills district and Nongstoin in West Khasi Hills district, Mendipathar and Songsakin East Garo Hills districts. The presence of the rifts and grains (primary joint planes) with more feldspar and less micacious minerals increases its value and eases the mining as slabs and blocks.

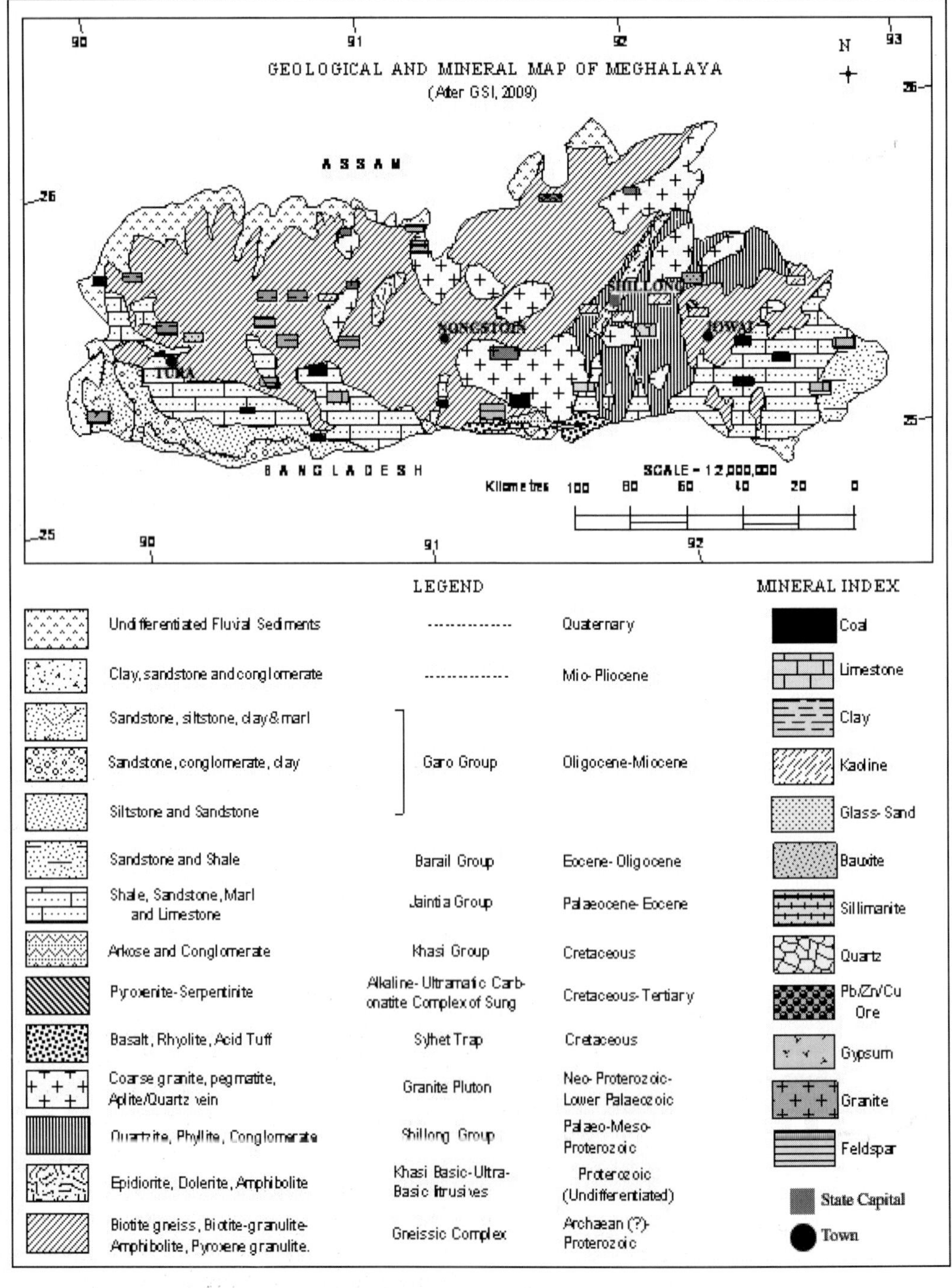

Table 1: Minerals resources and reserves of Meghalaya (Indian Mineral Year Book 2007)

	Recoverable reserves as on 01-04-2005			
	Proved	**Probable**	**Possible**	**Total**
China clay ('000 tonnes)	1890	5786	81199	88875
Copper Ore ('000 tonnes)	-	880	-	880
Metal ('ooo tones)	-	9	-	9
Feldspar (tones)	-	-	37449	37449
Fireclay	-	-	10999	10999
Apatite (tones)	-	-	1300000	1300000
Lead Zinc Ore	-	880	-	880
Lead (Metal) ('000 tonnes)	-	6	-	6
Zinc (Metal) ('000 tonnes)	-	16.5	-	16.5
Limestone ('000 tonnes)	459600	2588804	12662369	15896773
Coal (in million tonnes)	117.83	40.89	300.71	459.43
Quartz/silica sand ('000 tonnes)	-	176	6906	7083
Sillimanite (tones)	-	-	55807	55807
Silver Ore (tones)	-	880000	-	880000
Metal (tones)	-	19.80	-	19.80

The value of mineral production in Meghalaya in 2006-2007 is estimated at Rs.721 lakhs.

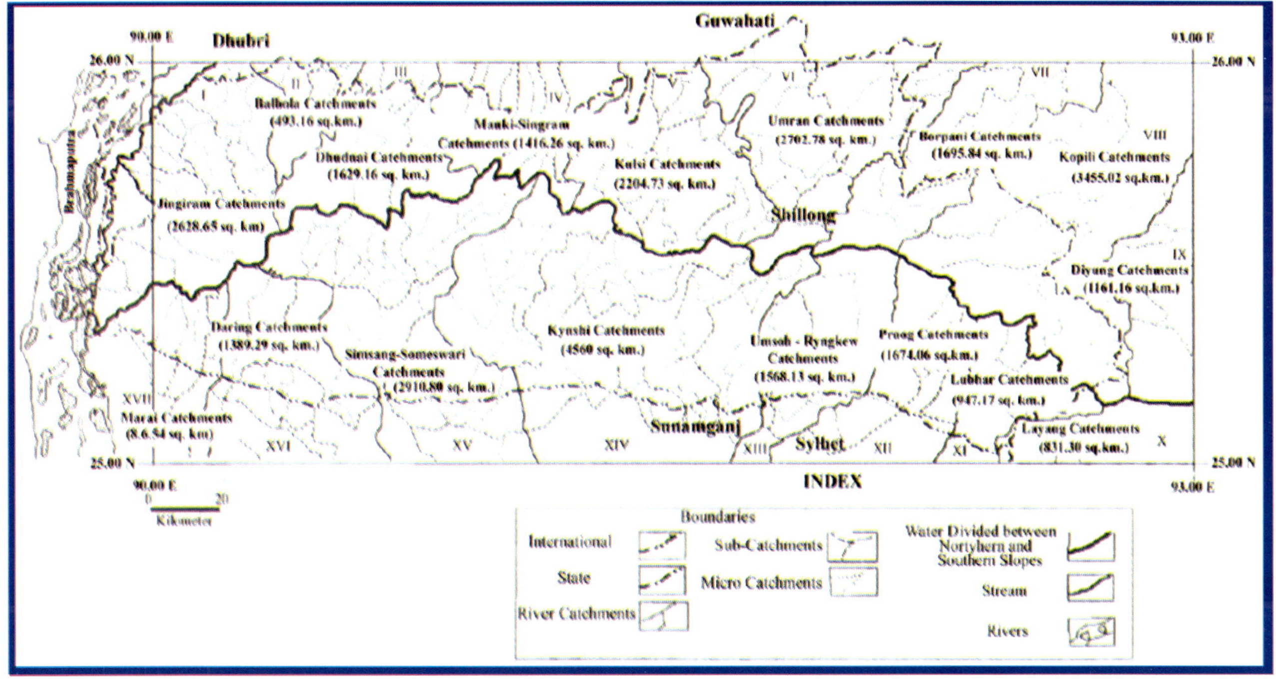

References

Das Gupta, Sujit and Nandy, D.R. (1995) Geological framework of the Indo-Burmese convergent margin with special reference to ophiolite emplacement. Indian Jour. Geol., Vol. 67, No. 2, pp. 120-125.

Das, P.K., Singha, L.J.C. and Sharma, A; (2001): Stratigraphy and sedimentation of the Lower Tertiary sediments around Sonapur-Lumshnong area, jaintia Hills, Meghalaya, India. Himalayan geology, Vol. 22 (2), pp. 55-60.

De, A.K. and Boral, M.C. (1977-78): A note on the Singrimari (Hallidayguni) area, Garo Hills, Meghalaya. Rep. F.S. 1977-78.

Geological Survey of India Misc. Pub. (2009), No. 30, Part IV, Vol. 2 (ii), Meghalaya.

Godwin Austin, H.H. (1869): Notes to accompany a geological map of the Khasi Hills. Jour. Asiatic Soc. Bengal, Vol. 38, Pt. 2.

Gokarn, S.G., Gupta, G., Walia, D., S.S. Sanabam and Nitu Hazarika (2008) Deep geoelectric structure over the Lower Brahmaputra valley and Shillong Plateau, NE India using magnetotellurics. Geophys. J. Int. Vol. 173, pp. 92–104.

Kumar, S; Bahuguna, R; Sengupta, B and Kaul, R. (1990): Geological environment of the sandstone type Uranium deposit, Domiasiat area, West Khasi Hill district, Meghalaya. Exploration and Research for Atomic Minerals, V. 8, No-1, pp. 3-14.

La Touche, T.H.D. (1983): Notes on traverse through Khasi Jaintia and North Cachar Hills. Rec.G.S.I. Vol. 10, pp. 198-203.

La Touche, T.H.D. (1989): Report on Cherrapunji coal fields. Rec. G.S.I., Vol. 22, Pt. 3, pp. 167-171.

Mazumdar, S.K. (1976): A summary for the Precambrian Geology of the Khasi Hills, Meghalaya. G.S.I. Misc. Pub. No. 23, Pt.-II, pp: 311-339.

Medlicott, H.B. (1868): Geological sketch of Shillong plateau in North Eastern Bengal. Mem. G.S.I., Vol.7, Pt. 1.

Mitra, S.K. (1998): Structural history of the shilling Group around Sohiong, East Khasi Hills, Meghalaya. Ind. J. Geol., Vol. 70 (1 & 2), pp. 123-131.

Murthy, M.V.N.; Mazumdar, S.K. and Bhaumic, N. (1976): Significance on tectonic trends in the geological evolution of the Meghalaya uplands since the Precambrian. G.S.I. Misc. Pub. No. 22, Part-II, pp: 471-484.

Nambiar, A.R. (2007): Early cretaceous Lamprophyre Dykes from Nongchram fault Zone, Meghalaya, northeastern India. J. Geol. Soc. Of india, Vol. 69, pp.-652.

Nandy, D.R. (1983): The Eastern Himalaya and the Indo-Buraman orogen in relation to the Indian plate movement. Symposium Geol. And Min. Resources of North Eastern Himalayas, Shillong. (1976). G.S.I. Misc. Pub. 43, pp 153-159.

Nandy, D.R.; (2001): Geodynamics of North Eastern India and the Adjoining Region. Abc publication.

Oldham, T. (1958): On the Geological structure of a portion of the Khasi Hills, Bengal. Mem. G.S.I. Vol. 1. Pt. 2.270.

Pascoe, E.H. (1973): A manual of the Geology of India and Burma, 3, 3rd Ed.

Walia, D., Lyngdoh, A. C., Saxena, A. (2009). Seismotectonic zones demarcation in the Shillong plateau using the micro earthquakes and the radon emanation rate. *Acta geophysica DOI:10.2478/ S11600-010-001-58.*

Hon'ble reviwer

1. **Dr. Vikas Mohan**
   Dy. Gen. Manager
   Reliance Cement Pvt. Ltd.
   C-Scheme
   JAIPUR - 302 001
2. **Dr. Sishir Sharma**
   Department of Geology
   Dungar College
   MGS University
   BIKANER - 334 003

Part - II:

GEO-ENERGY RESOURCES

Uranium exploration, resources and production in India

A.K. Sarangi

Gen. Manager (Corporate Planning), Uranium Corporation of India Ltd. Jaduguda, India
E-mail: aksarangi@gmail.com

Abstract

Uranium is the basic raw material for the nuclear power plants. It is ubiquitous in nature occurring in varied concentration in almost all type of rocks. Though the properties of uranium are favorable for formation of uranium deposits in a number of geological settings, major deposits are rare and have been found so far in a few countries. Only a small part of the land mass in Indian subcontinent is assumed to be geologically favourable for hosting uranium deposits. The exploration activities pursued over six decades in India, has brought to light several small to medium deposits in different geological settings. Some of the geological provinces of the country known for uranium mineralisation are the Proterozoic meta-sediments of Singhbhum shear zone (Jharkhand), phosphatic siliceous dolostone in south-western part of Cuddapah basin (Andhra Pradesh), the Cretaceous sediments of Mahadek basin (Meghalaya), major fractures / faults in limestone and granite of Bhima basin (Karnataka) and Delhi Supergroup of rocks (Rajasthan). Indian uranium deposits are generally of low grade. Development of uranium resources in India started in 1967-68 at Jaduguda in Singhbhum Shear Zone (Jharkhand). In line with the need of uranium, the uranium industry of the country has scaled commanding heights with new production centers and up-gradation of technology. Present uranium production activities are confined to Singhbhum Shear Zone and Cuddapah basin. The uranium mining and processing technologies in India have come a long way from conventional system progressively absorbing global expertise and adapting cutting-edge technology through radical innovations.

Keywords: Uranium, Nuclear Energy, Mineralisation, Exploration, Mining and Processing.

Introduction

Uranium is known for its high concentration of energy and is the essential raw material for the nuclear power plants. It has been playing a significant role in meeting the energy demand in the world [1, 2]. Today, about 13.5% of world energy requirement is met by nuclear power. The annual requirement of uranium (U) for presently operating 439 nuclear reactors with 375 GWe installed capacity is about 65000 tonnes[3] . With the concern over long-term security of fossil fuel and global recognition of nuclear power as a source of secure, clean and affordable source of energy, there is a general awakening of awareness on use of uranium leading to interest in its exploration by many countries.

Uranium is a common element on the earth crust with average concentration of 2 - 3 parts per million (ppm). It is as common as zinc or tin and is the constituent of many rocks. The geochemical properties of uranium, particularly its solubility, transportation and precipitation lead to the formation of ore deposits in a variety of geological settings. In spite of the widespread distribution, occurrences / deposits of uranium are found predominating in certain 'metallogenic provinces' only [4]. These deposits belong to more than one age and it has been studied that specific types are restricted to distinct epochs, exhibiting the time-bound phenomena.

There are several upheavals in the history of uranium exploration and production due to several geo-political events. However, the industry has been witnessing an upsurge of activities since 2003. These include increased production from currently operating mines as well as exploration and exploitation of new resources. Presently, about 48 countries in the world are involved in uranium exploration. Total known recoverable uranium resources in the world is estimated to be 5.3 million tonnes [3]. Average grade of uranium deposits in the world show wide variation ranging from 200 ppm to 2,10,000 ppm. During the year 2012, total of 58,314 t of uranium has been produced by 20 countries and more than 90% of this production is contributed by only eight countries - Kazakhstan, Canada, Australia, Niger, Namibia, Russian Federation, Uzbekistan and USA[5].

Uranium exploration:

The period after the discovery of fission power of the uranium atom and the Manhattan Project during the 2nd world war (1940s) besides other clandestine projects represent the first major effort to discover uranium deposits. The exploration programmers during this period (1945 to 1958) were primarily military driven so as to have access to the metal of fission power. The second cycle was during 1974 to 1983, driven by civil nuclear power programmes outlined by many countries and in the context of a perception that there may be shortage of uranium towards world energy need. The exploration efforts were mainly influenced by demand and supply, accentuated further by 'technology denial' or secrecy regimes. With greater understanding in geological processes and environments of formation of uranium deposits, a significant increase in activities has been observed in recent years. Concept based approaches, application of newer technology, deployment of modern precision equipment etc have given boost to the uranium exploration techniques all over the world.

While forming the uranium exploration strategy, fundamental parameters for consideration are the physical properties of uranium, geological settings, age, host rock characteristics, process of mineralisation, subsequent process of remobilisation etc. A basic exploration strategy follows a general sequence of localizing radioactivity anomalies followed by detailed ground surveys, drilling and logging.

The exploration is a multidisciplinary effort involving integrated application of different techniques - geological, geophysical and geochemical around a model developed around geological considerations. The desired geological activities in the exploration programme start with appraisal of the area's uranium potential, geological mapping and revisions of historical work. Certain geophysical investigations like airborne / car-borne/ ground surveys rely on the radioactivity of uranium deposits and their radioactive halos, while others rely on physical contrasts in electrical resistivity, conductivity and induction as a means of highlighting favourable geological aspects. Other geophysical methods such as gravity, magnetic and seismic surveys etc provide scope for investigation of deep subsurface structures and are valid aids in obtaining geological information in two and three dimensions.[6] Geochemical exploration includes collection and analyses of rock, soil and water samples.

Regional scale evaluation of geological terrain generally start by remote sensing surveys, making use of reflectance spectra to identify certain mineral assemblages. A variety of platforms are available to acquire such data, ranging from high-orbit, satellite-based systems to fixed wing aircraft and ground-based portable systems.

Uranium exploration in India:

India made a humble beginning of its inspiring atomic energy programme with the formation of Atomic Energy Commission in 1948. Consequent to this development, it was felt that the country must have indigenous resources of basic raw materials such as uranium, thorium etc. A group called Rare Metal Survey Unit (later on renamed as Atomic Minerals Directorate for Exploration and Research) was formed by Govt. of India to locate good uranium deposits in the country. These formative years were very daunting due to paucity of literature on the geology and geochemistry of uranium, unavailability of Geiger Muller counters and other physical assay equipments, unavailability of uranium standards and trained man power. However, with enthusiasm to create India's indigenous uranium resource, the emphasis of search was laid on the existing mineral belts and geologically favourable areas of the sub-continent. Interestingly, association of uranium in Shinkolobwe mine at Katanga copper belt in the then Belgian Congo and a few Cu-U mineralisation association in other parts of the world were known to many exploration geologists. On this analogy, known copper and gold provinces of the country were extensively investigated. The maiden effort in the Singhbhum copper belt (Singhbhum Shear Zone) in the then Bihar (presently in Jharkhand) became a major success story since a uranium province with over 50 anomalies were discovered in early 1950s. During this period, uranium mineralisation was also located in Precambrian black shales at Umra in Rajasthan. However, the lenses at Umra were found very discontinuous and exploration was discontinued.[7]

Since then, uranium exploration concepts and practices have undergone many changes with time due to increased knowledge on geology of the terrain and global development on exploration tools. Knowledge base on Precambrian geology of India, concept of plate tectonics as a major process to shape the earth governing the process of crust and mountain building, understanding of formation of uranium deposits and their spatial distribution etc have helped in widening the exploration programme in different basins of India.

Apart from the Singhbhum Shear Zone, potential for uranium mineralization has been recognized in thirteen Proterozoic basins for unconformity related and strata-bound deposits and Cretaceous Mahadek sediments of Meghalaya for sandstone type mineralisation. A score of new discoveries of different types such as stratabound dolostone hosted mineralisation around Tummalapalle in Andhra Pradesh, unconformity proximal type around Lambapur in Andhra Pradesh, fracture controlled mineralisation at Gogi in Karnataka, vein type mineralisation around Rohil in Rajasthan, Quartz-Pebble-Conglomerate (QPC) type in Karnataka, Jharkhand and Odisha, shear controlled vein type lenses at Bodal and Bhandaritola in Chattishgarh, sandstone hosted mineralisation in Mahadek basin of Meghalaya etc have already been established.[8]. The developments of high resolution airborne systems, availability of light weight, sensitive field spectrometers and total count rate meter, high resolution and deep penetrating systems of electrical, electromagnetic, magnetotellurics, airborne gradiometry systems etc are helping for rapid regional assessments. It is expected that future investigations adapting novel exploration strategies along with improvements of techniques focused towards deep-seated uranium mineralization shall enhance the uranium resource base of the country.

Uranium resources

Uranium resources of the world are very unevenly distributed, both in space and time. Though almost all types of deposits[9] have been found in different geological settings, the large-tonnage, high-grade ones are found in major uranium producing countries like Canada and Australia. Of the total 5,327,200 tonnes of recoverable uranium resources, 52% is known to occur in Australia, Kazakhstan and Russia[3]. The Proterozoic period spanning over 40% of earth's history accounts for 68% of world uranium resources. Bulk of this is related to unconformity related deposits (Athabasca basin in Canada and South Alligator basin in Australia) and some vein type deposits. This predominance can be related to numerous first order changes or features either singly or in combination that have directly or indirectly influenced the metallogeny of many elements including uranium[10]. The Phanerozoic period, dominated by sandstone type of deposits constitutes about 43% of the total deposits discovered so far and contains about 18% of the world uranium resources. They are formed in continental fluvial or marginal sedimentary environment (Western Cordillera region in USA, Central Asia and Niger). The genesis of these deposits is primarily linked with sedimentation, hydrological-geochemical regime and geotectonic evolution of the basin [11].

Uranium resources of India

Uranium exploration activities in India carried out by AMD since 1951, has established about 1,60,000 tonnes of U in different geological basins of India. Of these, Andhra Pradesh accounts for 51%, Jharkhand 29%, Meghalaya 11%, Rajasthan and Karnataka 4% each and remaining in other states. Unfortunately, large uranium deposits are not yet found in any part of India. Indian uranium deposits are of medium size and the country has a modest uranium resource mostly confined to following major types.

a. Shear-controlled vein type deposits:

i. Singhbhum Shear Zone (SSZ), Jharkhand

ii. Rohil- Ghateshwar shear zone, Rajasthan

b. Sandstone type deposits

i. Mahadek sedimentary basin, Meghalaya

c. Strata-bound uranium deposits

i. Vempalle Formation, Cuddapah Basin , Andhra Pradesh

d. Unconformity related uranium deposits

i. Srisailam sub-basin of the Cuddapah basin, Andhra Pradesh

e. Fracture controlled uranium mineralisation

i. Gogi area of the Bhima Basin in Karnataka

I. Uranium resources in Proterozoic basins of India

64% of the Indian continent, characterised by Proterozoic rocks (exposed and buried) is represented by the Singhbhum shear zone (craton-mobile belt shear zone) and thirteen other basins including the Purana and their peripheral parts. Of all the basins, spectacular resource findings have been established in only Singhbhum shear zone in Jharkhand and the Cuddapah basin in Andhra Pradesh. In Singhbhum shear zone, most part of uranium resources are of vein type, whereas in the Cuddapah basin, two different types - unconformity related (*sensu lato*) and strata-bound types have been located in two different geological settings.

A. Singhbhum shear zone

The Singhbhum shear zone (SSZ), also known as Singhbhum Thrust Belt in Precambrian Eastern Indian Shield extends over a length of 200 km trending NW-SE in the form of an arc covering southern part of Jharkhand and northern part of Odisha. The shear zone is a site of acid and basic volcanism and hydrothermal metasomatic activities[12]. The pioneering exploration work of geologists had brought to light the first significant uranium mineralisation at Jaduguda in April 1951. The systematic exploration programme during subsequent years reported about forty two uranium occurrences in this area. (Fig. 1) The exploration activities, intensified with deployment of additional resources, established ten small to medium sized deposits. From east to west within a stretch of 70 km, the deposits are Bagjata, Kanyaluka, Jaduguda, Bhatin, Nimdih, Narwapahar, Garadih, Turamdih, Banduhurang and Mohuldih.

The principal lithological units in these deposits are meta-sediments exhibiting different level of chloritisation, biotitisation, sericitisation and feldspathisation etc (varying grade of metamorphism)[13]. Uranium lodes extend as thin veins from surface following the general trend of the schistosity with fairly uniform persistence both along strike and dip [14,15,16]. Ore bodies are vein-like sub-vertical in the eastern sector (Bagjata and Jaduguda) whereas in the western sector (Turamdih and Banduhurang) they tend to be stratiform to strata-bound[15, 17]. (Fig. 2)

B. Cuddapah basin

The crescent shaped Cuddapah basin in southern part of Indian sub-continent is known for variety of economic minerals. Exploration for unconformity related, large-tonnage, high grade-uranium deposits began in 1980s that led to the discovery of two different types, previously unknown. These include the unconformity proximal type in the basement granites below the Kurnool Group of sediments at Lambapur and Peddagattu[18] and the stratabound type, dolostone-hosted minralisation at Tummalapalle[19], establishing it as an important uranium province of the country. (Fig. 3)

The northern side of the basin comprises dominantly of arenaceous and argillaceous rocks of Middle Proterozoic period overlain by Upper

Proterozoic calcareous formations. Close to the unconformity (basement granite and its overlying Srisailam quartzites) in the Srisailam outlier, cluster of uranium deposits have been found around Lambapur, Peddagattu, Kuppunur and Chitrial.

In the south western margin of the Cuddapah basin, uranium deposits occur in phosphatic siliceous dolostone of Middle Proterozoic Vempalle formation. Within the thick pile of carbonate rocks, the mineralized horizon is sandwiched between lower massive dolostone and upper shale. Mineralisation extends over 60 km length, but the mineabale lodes are persistent over 6.6 km length at Tummalapalle-Rachakantapalle sector near Pulivendla. Recent exploration findings in the area have established large tonnage of low grade uranium mineralisation extending over a stretch of 60 km upto a depth of 800 m leading to significant increase in uranium inventory of the country.

C. Bhima basin

The Bhima basin in the southern Indian plateau towards NW of Cuddapah basin has emerged as an important uranium district of the country. The basin predominantly comprises of carbonates with minor argillites and arenites. A number of fault zones have been identified in the basin, trends varying between E-W to NW-SE. Uranium mineralisation at Gogi has been discovered almost at the middle of East – West trending Kurlagere – Gogi – Gundahalli fault along the southern margin of Bhima basin. Uranium mineralisation at Gogi is mainly associated with two litho units e.g. brecciated limestone and sheared granite[20]. (Fig. 4)

D. Other Proterozoic uranium provinces and deposits

In addition to the Proterozoic uranium deposits / metallogeny described so far, there are a number of significant metallogenic provinces both in the Peninsular shield and the Central crystallines of Lesser Himalaya (believed to be the part of the northern extension of the peninsular shield). They host smaller deposits, often in contiguous areas.

i. Bodal - Bhandaritola in Rajnandgaon district of Chattishgarh occurring in Nandgaon group of Dongargarh supergroup of Late Archean - Early Proterozoic age

ii. Jajwal in Sarguja district of Chattisgarh occurring in central Sarguja crystalline of Lower - Middlle Proterozoic age

iii. Rohil - Ghateswar area in Sikar district of Rajasthan occurring in biotite schists and albites of Aravalli supergroup of Lower Proterozoic age

In addition, a number of smaller deposits (<500 t U) and significant occurrences belonging to the Proterozoic period have also been discovered, evaluated and reported from both the Peninsular shield and the Himalayan orogen [21, 22].

II. Uranium deposits in Phanerozoic basins of India

The Phanerozoic basins of India mainly comprise – Mahadek formations in Meghalya, Gondwana basins in Central India and Siwaliks in the foothills of Himalayas. Of all three basins, significant sandstone type uranium occurrences have been located only in the Cretaceous sediments in Mahadek basin of Meghalaya.

A. Mahadek basin

Significant uranium resources have been established in the Mahadek sediments of Cretaceous period in the southern part of Shillong plateau. A large uranium deposit with tabular to lensoidal orebody[23] has been planned for development at Kyelleng-Pyndengsohiong (formerly known as Domiasiat) in lower Mahadek sediments. The discovery of Wahkhyn deposit, located about 10 km west of Kyelleng-Pyndengsohiong under similar geological settings has substantially added to the uranium inventory of this region (Fig. 5). Other known uranium occurrences in this region are Wahkut and Lostoin.

III. Other Phanerozoic uranium provinces and deposits

A few low grade, small tonnage uranium occurrences have been located in the Siwalik basin in foothills of the Himalayas[24]. However, at present, because of their small size and low grade, they do not hold any promise for commercial scale mining operations. Discovery of uranium shows in the Gondawana sandstones around Motur near Betul in Madhya Pradesh encouraged to undertake exploration in Gondwana basins during the early 1970s. However, subsequent exploration has not shown any encouraging mineralisation[21].

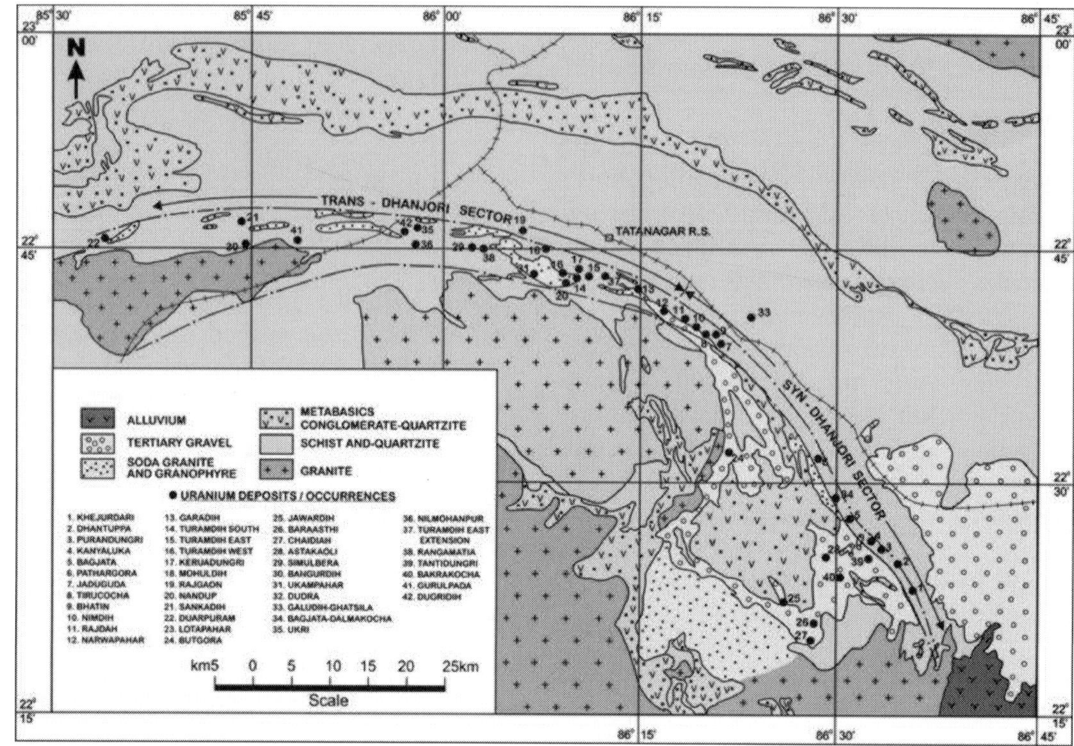

Fig. 1: Uranium occurrences identified in Singhbhum shear zone and adjoining areas. [28]

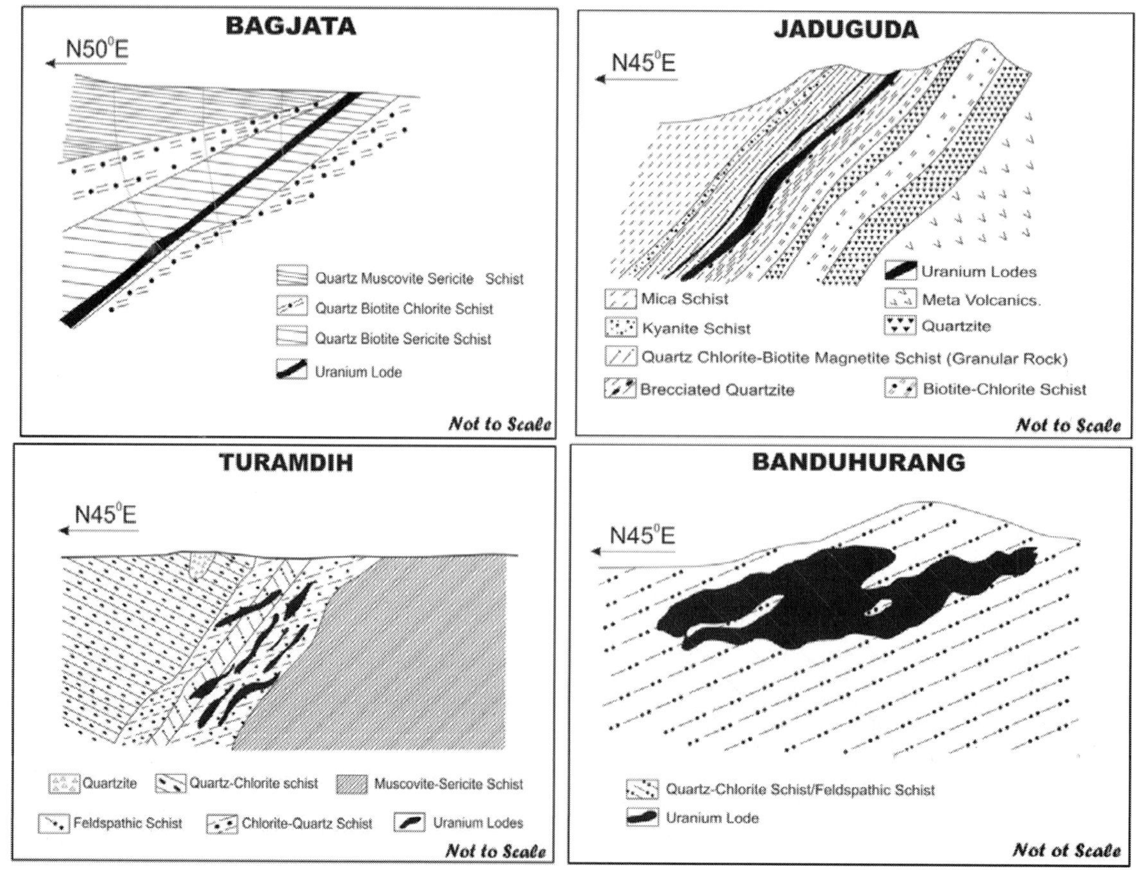

Fig. 2: Geological section showing orebody at Bagjata, Jaduguda, Turamdih and Banduhurang[10].

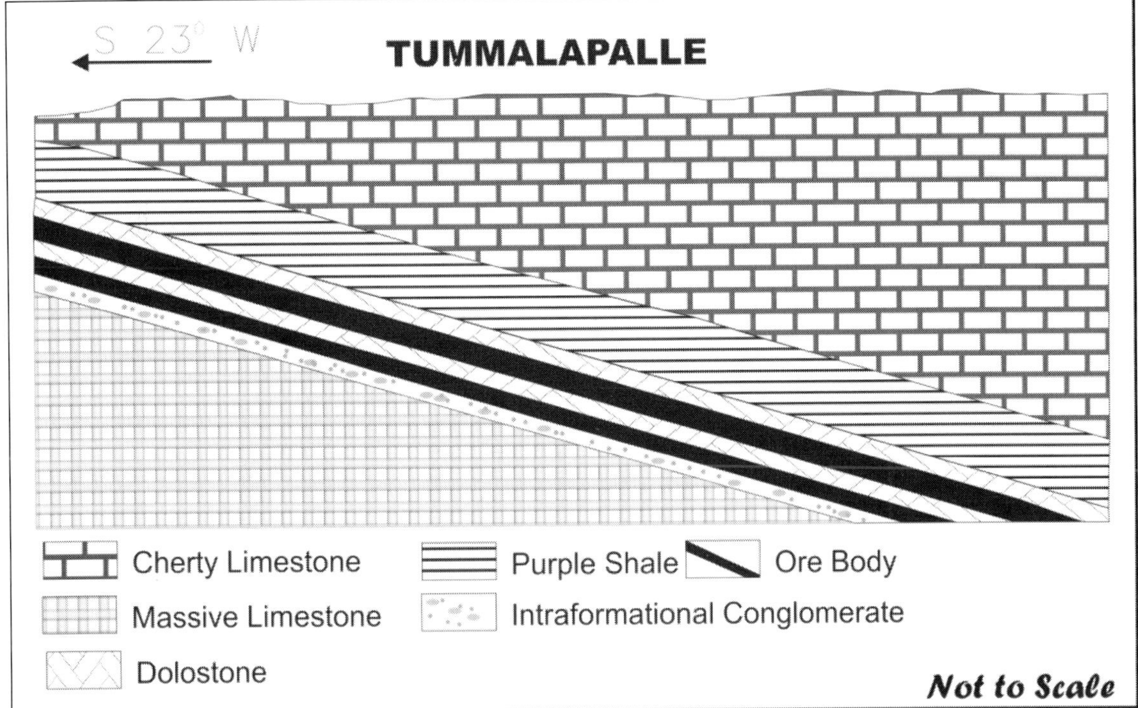

Fig. 3: Geological section showing orebody at Tummalapalle.[10]

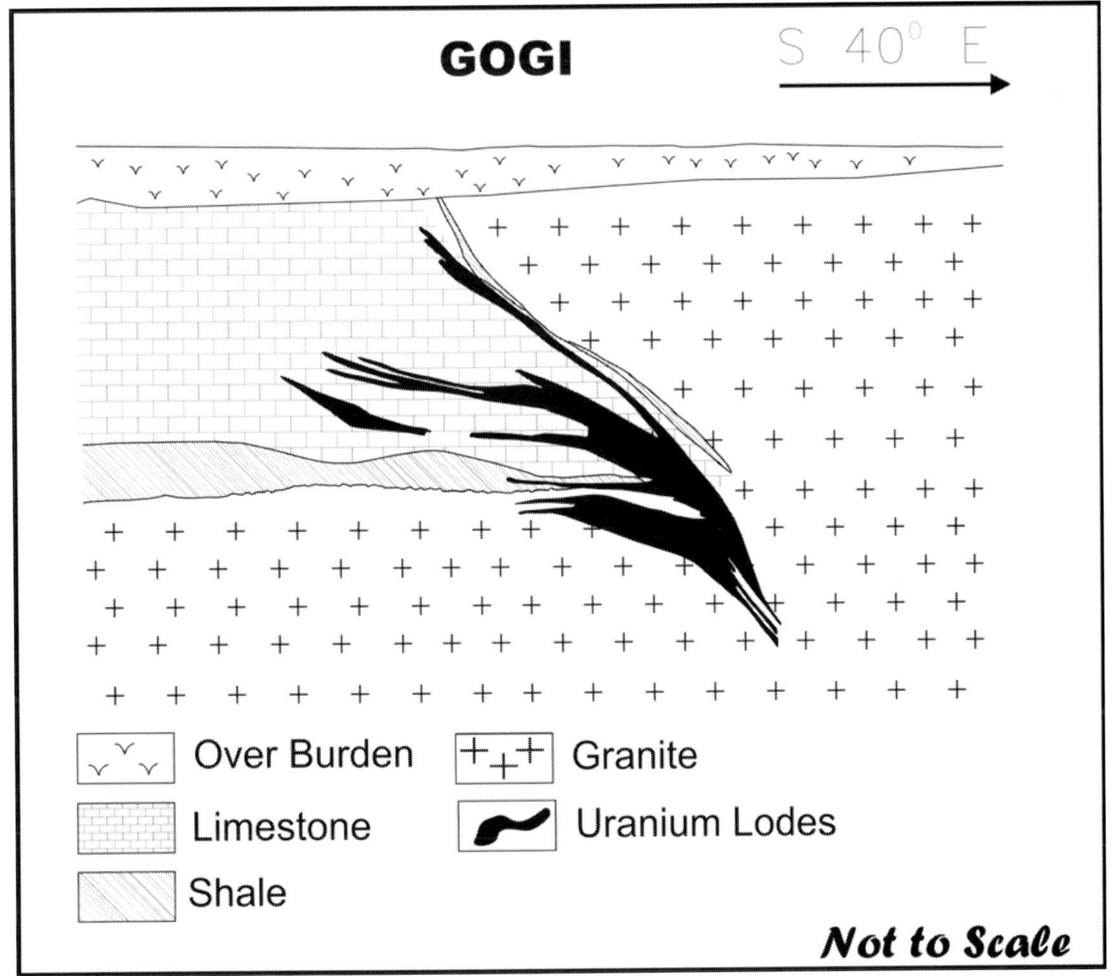

Fig. 4: Geological section showing orebody at Gogi Deposit.

N 35° E
DOMIASIAT
Not to Scale
S 56° E
WAHKYN
Not to Scale
INDEX
Oxidised Sandstone
Reduced Sandstone
Basement Granite
Ore Body
Over Burden
Conglomerate

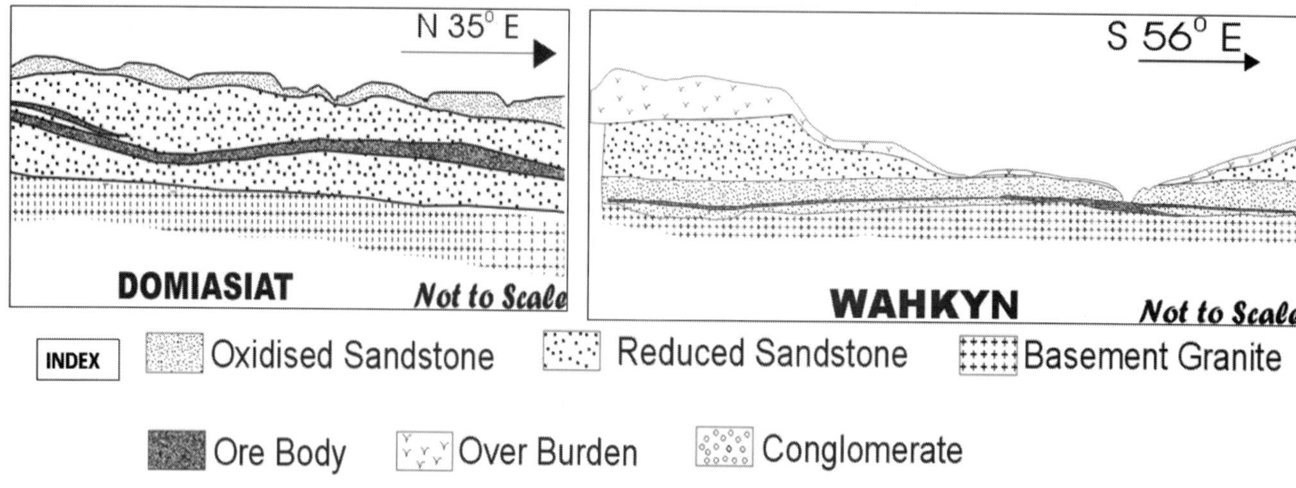

Fig. 5: Geological sections showing orebody at Domiasiat and Wahkyn.[10]

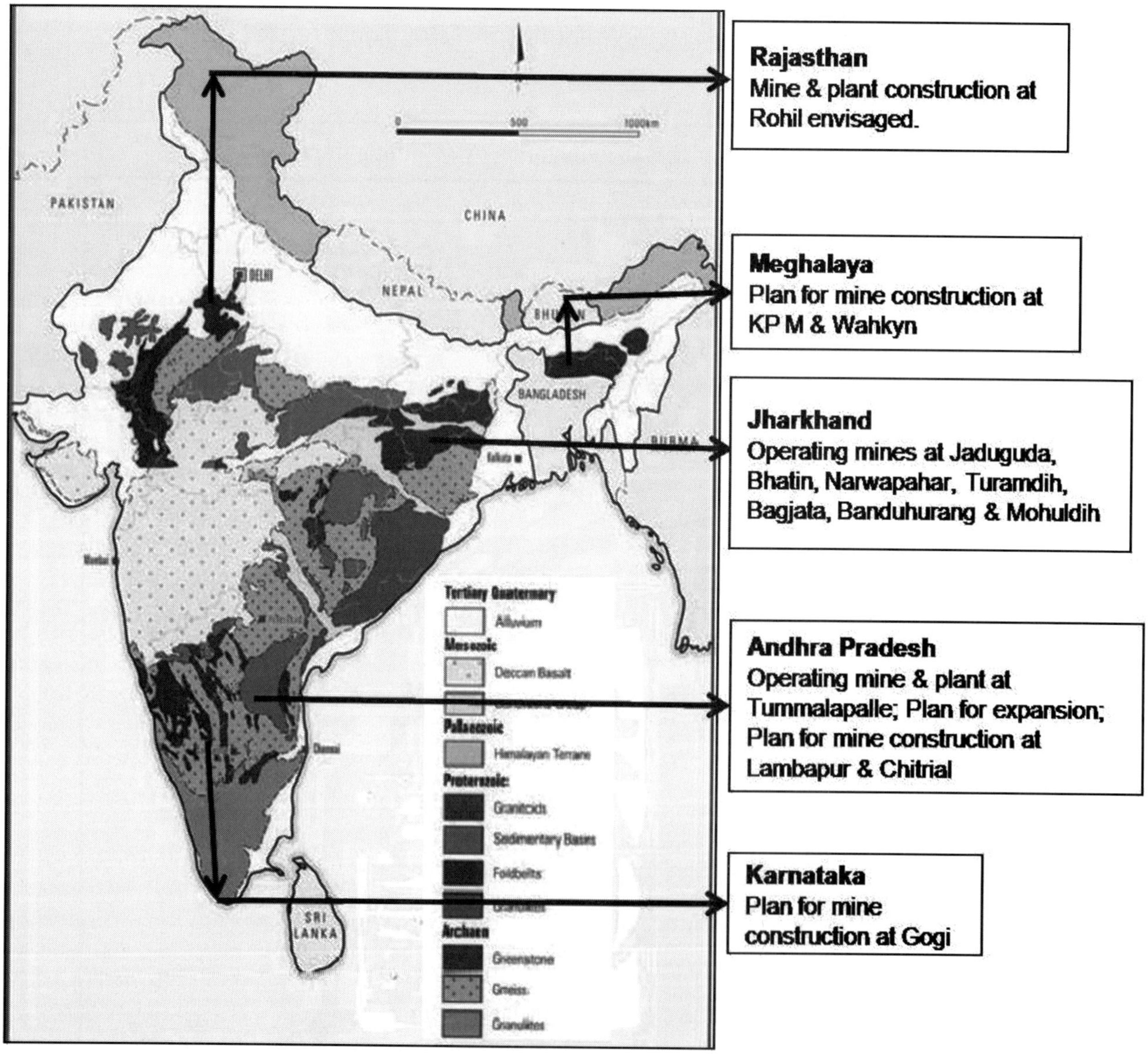

PAKISTAN
CHINA
NEPAL
DELHI
BANGLADESH
BURMA
SRI LANKA
Tertiary Quaternary
Alluvium
Mesozoic
Deccan Basalt
Palaeozoic
Himalayan Terrane
Proterozoic
Granitoids
Sedimentary Basins
Foldbelts
Archaen
Greenstone
Gneiss
Granulites
Rajasthan
Mine & plant construction at Rohil envisaged.
Meghalaya
Plan for mine construction at KP M & Wahkyn
Jharkhand
Operating mines at Jaduguda, Bhatin, Narwapahar, Turamdih, Bagjata, Banduhurang & Mohuldih
Andhra Pradesh
Operating mine & plant at Tummalapalle; Plan for expansion; Plan for mine construction at Lambapur & Chitrial
Karnataka
Plan for mine construction at Gogi

Fig. 6: Map showing operating, planned and potential uranium production centers of the country

Uranium resources vs. reserves

The geological in-situ resources identified / established at different deposits are sometimes never available for commercial mining operations. Several techno-economic factors are looked into on regular basis for evaluating the exploitable in-situ reserves considering factors like mining, processing, environmental, legal and social aspects etc. The resources established in some areas are sometimes not accessible for uranium mining due to environmental considerations, adverse public perception on uranium mining and sensitive administrative / political issues. As the above factors are of dynamic nature, the resources identified in these areas are re-looked / reassessed with every change in above scenarios to establish the reserves [25].

Uranium production centers of India

The uranium production in India to cater to the indigenous need made a formal beginning with the formation of Uranium Corporation of India Ltd (UCIL) in 1967 under Department of Atomic Energy. The operation was launched with the commissioning of an underground mine and ore processing plant at Jaduguda (1968) in Jharkhand (the then Bihar). Later, in line with the requirement of uranium, five new underground mines, one opencast mine and a processing plant have been commissioned in Jharkhand. Recently, a large underground mine and process plant have also been commissioned in Andhra Pradesh [26]

A. Jharkhand

In the Singhbhum shear zone, UCIL is now operating six underground mines and one openpit mine. All these operating uranium mines are located within 25 km from Jaduguda. Jaduguda underground mine with 1000 tpd production capacity was commissioned in 1968. The mine was initially explored with the development of a few adits as entry into the orebody. Regular mining operations started with the sinking of a fully lined vertical shaft of 5 m diameter. This was followed by commissioning of Bhatin mine in 1986, Narwapahar mine in 1995, Turamdih mine in 2003, Bagjata mine in 2008 and Mohuldih mine in 2013. New mines at Narwapahar, Turamdih, Bagjata and Mohuldih adopt decline method of entry and use of track less equipment. Vertical shaft provides access to deeper levels for ore and men & material hoisting. Ore extraction in underground uranium mines of UCIL is generally carried out by horizontal cut-and-fill method of stoping with site-specific modifications and introduction of newer equipment with a view for safety, improved ore recovery and less dilution. Banduhurang, the first opencast uranium mine of the country was commissioned in 2009.

Uranium ore produced from these mines are processed in two central plants located at Jaduguda and Turamdih. Jaduguda plant commissioned in 1968 had undergone expansion in phases to treat the ore of Bhatin, Narwapahar and Bagjata. Turamdih plant was commissioned in 2009 to treat the ore of Turamdih and Banduhurang mines. Turamdih plant is now under expansion and the ore from Mohuldih mine will be fed to this plant. Both the plants in Jharkhand follow conventional way of processing of uranium ore i.e. through acid leaching route because of siliceous nature of host rock. Over the years, many improvements have been introduced in the processing technology with utmost consideration on maximising the recovery, minimising the discharge of effluents and maximising the recovery of by-products.

B. Andhra Pradesh

In the SW part of Cuddapah basin at Tummalapalle, a large modern underground mine and process plant have been commissioned recently with state-of-art technology. Mine entry at Tummalapalle is through three declines along the apparent dip of the orebody and the central decline with 9^0 gradient has been equipped with conveyor for ore transport. Other two parallel declines are used as service path for movement of men and materials. The process technology at Tummalapalle is based on alkali leaching (under pressure) because of carbonate nature of host rock [27]

With recent findings and expansion of uranium resource base around Tummalapalle, plans for expansion of mining and milling operations are already in place. On successful implementation of alkali leaching technology, additional production centers with new mines and plants are envisaged in this region.

C. Karnataka

Uranium mineralisation hosted in limestone and granite at Gogi in Karnataka has been planned for development. A new underground mine and process plant (based on alkali leaching route) have been envisaged.

D. Potential uranium production centers

With recent findings of uranium mineralisation in different basins of the country, efforts are being made to expand the production base by constructing new mines and process plants. Mahadek basin in Megahalaya, northern part of Cuddapah basin and Delhi fold belt in Rajasthan are some of the areas under active consideration. The findings around Tummala-palle also hold the promise for expansion of existing operation and construction of new units.

The operating, planned and potential uranium production centers of the country are shown in Fig. 6.

Conclusion

Indian uranium deposits are of limited tonnage and low grade. Greater consideration towards nuclear power in the energy basket of the country preferably through indigenous resources demand large resources to be explored and evaluated for development. This, in turn has led to focus on advanced techniques in uranium exploration. Technology up-gradation in the fields of geophysics, 3D modeling, stable isotope geology, deviation controlled drilling etc are some of the domains drawing greater attention. Known uranium bearing areas are being assessed with greater intensity and unknown areas are being looked on the basis of concept based approaches. Proterozoic and Phanerozoic sedimentary basins contiguous to fertile granitoid-rich provenances are being prioritised as potential targets for exploration.

Uranium production of the country has also recorded a significant rise in recent years through setting up of new units and expansion of operating mines and plants. Use of integrated software to quickly establish the configuration of ore body, standard modules of mine layout and methods with flexibility to accommodate minor variations in site-specific geology etc have helped to achieve early commissioning of the mines. Technology up-gradation through use of more energy efficient electro-hydraulic equipment, improved underground ventilation system, strata control measures, bulk ore assaying system with automatic grade estimation etc are some of the unique developments associated with uranium mining activities in India.

However, the activities are sometimes hindered because of negative public perceptions on uranium mining often leading to social and legal problems in new construction sites. In the recent years, though nuclear energy is gradually becoming an acceptable option, anti-nuclear forces are still active confusing the society. This has already taken a heavy toll on the time schedule of some of the new uranium projects.

Development of uranium deposits of the country is inevitable for the sustained supply of fuel to the indigenous nuclear power programme. Apart from supplying the raw material for nuclear fuel, the uranium industry of India has been contributing towards development of infra-structure, mining technology and employment opportunity in the neigbourhood. Global developments in nuclear co-operation are also providing a newer opportunity to broaden the production base.

References

1. International Atomic Energy Agency, (1999), Sustainable Development – A role for Nuclear Power, IAEA, Vienna, 36 P.
2. Ott, G., (2000), The Global energy context – chances and challenges for the 21st century, Proc. Intern. Symp. Vienna, pp. 25-32.
3. International Atomic Energy Agency, (2011), URANIUM 2011: Resources, Production and Demands, IAEA and NEA, Vienna. pp 15-23.
4. Stanton, R.L., (1972), Ore Petrology. McGraw Hill, New York, Xviii, 714 P.
5. WNA, http://www.world-nuclear.org. (Accessed on 20th July 2013)
6. International Atomic Energy Agency, (2013), Advances In Airborne and Ground Geophysical Methods for Uranium Exploration, IAEA, Vienna, 58 P.
7. Krishnamurthy, P., (2006), "The Saga of Singhbhum Uranium Discoveries and Exploration", Min., Geol. and Met. Inst. India, Trans. Vol. 102, No. 1 & 2, pp.151-169.
8. Sarangi, A. K. and Yadav D S. (2011) "Uranium Resources of India and Potential for Development" 17th convention of the Indian Geological Congress

and International Conference on "New Paradigms of Exploration and Sustainable Mineral Development: Vision 2050 (NPESMD 2011)" Nov. 2011 at ISM Dhanbad, pp.99-110.

9. Dahlkamp, F.J., (1993), Uranium Deposits, Springer Verlag, Berlin, 460 P.
10. Sarangi, A. K., and Krishnamurthy, P., (2008), Uranium metallogeny with special reference to Indian deposits, Trans. Min., Geol. and Met. Inst. India, Vol. 104, No. 1 & 2, pp.19-54.
11. Adams, S.S., and Saucier, A.E., (1981), Geology and recognition criteria for uraniferous humate deposits, Grants Min. Reg. New Mexico. US-DOE, GJBX-2(81), 225 P.
12. Dunn, J.A., (1937), The Mineral deposits of Eastern Singhbhum and surrounding areas, Mem. Geol. Surv. India, Vol. 69, Pt. 1, 279 P.
13. Sarangi, A. K., and Singh, Bijay (2010) "Uranium Resources of Jharkhand: Genesis, Exploration and Exploitation" International Journal of Earth Science and Engineering ISSN 0974-5904, Vol. 03, No. 06- Spl issue, November 2010, pp.01-10.
14. Sarkar, S.C., (1984), Geology and ore mineralisation of the Singhbhum Copper-Uranium Belt, Eastern India, Jadavpur University, Calcutta, 263 P.
15. Mahadevan, T.M., (1988), Characterization and Genesis of Singhbhum Uranium Provinces India: Reorganization of Uranium Provinces IAEA Tech. Comm. meet. Sept. 18-20, London, pp 337-369.
16. Sinha, K.K., Das, A.K., Sinha, R.M., Upadhyay, L.D., Pandey, P., and Shah, V.L., (1990), Uranium and associated copper-nickel-molybdenum mineralization in the Singhbhum Shear Zone, Bihar, India: Present status and Exploration Strategy. Expln. Res. Atom. Min. Vol. 3, pp. 27-43.
17. Pandey, P., Kumar, P., and Upadhyay, L.D., (1994), Uranium deposits of Turamdih-Nandup area, Singhbhum district, Bihar, and their spatial relationship, Exploration and research for Atomic minerals, Vol. 7, pp. 1-13.
18. Sinha, R.M., Shrivastava, V.K., Sarma, G.V.G., and Parthasarthy, T.N., (1995), Geological favourability for unconformity related uranium deposits in northern part of Cuddapah basin, evidences from Lambapur uranium occurrence, Andhra Pradesh, India, Expl. Res. At. Min., Vol. 8, pp. 111-126.
19. Rai, A.K., Zakaulla, S., Jeyagopal, A.V., Vasudeva Rao, M., Nagabhushana, J.C., and Varadaraju, H.N., (2002), Uranium mineralisation in the south western part of the Cuddapah basin, Andhra Pradesh, Expl. Res. At. Min., Vol. 14, pp. 79-94.
20. Pandit, S.A., Natrajan, V., and Dhanaraju, R., (2002), Exploration for uranium in Bhima basin in parts of Karnataka, India, Expl. Res. At. Min., Vol. 14, pp. 59-78.
21. Saraswat, A.C., Krishnamurthy, P., and Mahadevan. T.M., (1988), Uranium provinces of the Indian subcontinent and surroundings, In: Uranium Deposits in Asia and the Pacific, Geology and Exploration, IAEA, TC-543/4, Vienna, pp. 39-57.
22. Singh, R., (1994), constraints and prospects of uranium exploration in Himalaya, Expl. Res. At. Min., Vol. 7, pp. 51-67.
23. Sen, D.B., Sachan, A.S., Padhi, A.K., and Mathur, S.K., (2002), Uranium exploration in the Cretaceous Mahadek sediments of the Meghalaya Plateau, Expl. Res. At. Min., Vol. 14, pp. 29-58.
24. Swarnkar, B.M., Umamaheshwar, K., Srinivasan, S., and Kothari, P.K., (2002), Exploration for sandstone-type uranium mineralisation in the Siwaliks of northwestern Himalaya, India, Expl. Res. At. Min., Vol. 14, pp. 01-27.
25. Gupta, R. and Sarangi, A. K. (2010) "Overview of Indian uranium production scenario in coming decades" International Seminar on Asian Nuclear Prospects (ANUP-2010), 11th to 13th October '10., Energy Procedia, Volume 7, 2011, pp. 146-152.
26. Gupta, R. and Sarangi, A. K. (2005); "Emerging Trend of Uranium Mining: The Indian Scenario", IAEA proceeding series on "Uranium production and raw materials for the nuclear fuel cycle-Supply and demand, economics, the environment and energy security", IAEA – CN – 128, pp. 47-56.
27. Gupta, R. and Sarangi, A.K. (2008); "A new approach to mining and processing of a low grade uranium deposit at Tummalapalle, Andhra Pradesh, India" IAEA Technical Meeting on "Uranium Small-Scale and Special Mining and Processing Technologies" Vienna during 19th – 22nd June 2007.
28. Bhola, K.L., Rama Rao, Y.N., Suri Sastry, C., and Mehta, N.R., (1966), Uranium mineralization in Singhbhum thrust belt, Bihar, India. Eco. Geol. Vol. 61, pp. 162-173.

Hon'ble reviewer

1. **Dr. S. Viswanathan**
   Former Director AMD
   B-203, United Avenue Apt.
   7-1-29, Ameerpet
   HYDERABAD - 500 001
2. **Dr. D.K. Sinha**
   Scientist `H'
   Atomic Mineral Directorate
   Begampet
   HYDERABAD - 500 001

II-26

Determination of High Heat Production to Evaluate Radioactive Mineralization potential of Granites of Mungeria Ring Complex of Malani Igneous Suite, Rajasthan, India

K.L. Shrivastava, Anju Mathur, Virendra Gaur, Tina Prajapat, C.P. Dadhich, Mamta Chauhan and Beena Tripathi

Department of Geology, Jai Narain Vyas University, Jodhpur-342005 India
E-mail : klsgeology@yahoo.co.in

Abstract

In Malani Igneous Suite of North western India, Mungeria Ring Complex (MRC) is relatively a small circular structure with around10 km diameter and also little studied. Form village Mungeria (N29°05'75": E71°00' 57") at its core it shows atleast three peripherial rocks. Core is consisting of sub volcanic andesitic-basaltic component followed by pink and grey granites toward periphery. The paper is reporting the average heat generation value of these granites, which comes as 7.09 HGU (heat generating unit). It is much higher than the average (3.8 HGU) known for the continental crust. The concentration of thorium is also higher then uranium and potassium. The radioelement concentration (Ur) varies from 12.19 to 28.14 in MRC. It clearly indicates a 'hot crust', hence favourble for the formation of mineralization of HFS elements, Nb, Ce, REE, U and Th, need yet to be explored.

Keywords: High heat producing (HHP) granite, Uranium deposits, Malani Igneous Suite (MIS), Mungeria Ring Complex (MRC), Large Igneous Provence (LIP)..

Introduction

High heat production (HHP) granites (Kochhar, 1989; Sharma, 1994; Shrivastava et al. 2013) are those evolved calc-alkaline granites that have comparatively higher content of Th, U, K and total REE and responsible for nearly half of the crustal heat flow (Plant, 1985; Morgan and Sass, 1984 and Morgan, 1985; Srivastava, 2003) through radiogenic decay of isotopes of Th, U and K, of which uranium is the dominant heat-production component. Therefore, the granites with nearly four times or more uranium than the general abundance [3.5 ppm: (Clark, 1990), 4 ppm: (Rogers and Adams, 1969)] are considered as HHP granites. Such granites act as 'heat engine' and prolongs the circulation of ore bearing hydrothermal fluids (Fehn et al., 1978; Kimberley, 1978), which may ultimately lead to mineralization.

The crystal evolution has occurred in the first stage, involved upward vertical migration of magma produced by partial melting of pyrolitic composition of the mantle, but once these igneous rocks had been differentiated and emplaced, forming the ocean lithosphere, the horizontal motion became operative as the lithosphere was carried away from the ridge. In the second stage island arc igneous rocks have been formed above the sinking oceanic lithospheric slabs. The resistance offered to the horizontal motion at the base of the plates and the friction caused by the subducting slab, could have been less in the Archaean due to the lower viscosity of the upper mantle, maintained by the elevated temperature (Roy, 2009; Patwardhan, 2010; Bhandari, 2009; and Shrivastava et al. In press). The heat required for greater partial melting of the upper mantle was largely provided by the exothermic reactions involved in the radioactive decay of unstable isotopes of elements, especially of U, Th and K. The heat generated by such decay progressively decreased with time, ever since the Earth became a part of the planetary system. Roughly twice as much heat is estimated to have been generated in the first two billion years (4.5-2.5 Ga), as has been generated since then till present (Valdia, 2010). This heat was responsible for the highly dynamic stage of the Earth, in the Archaean times, when basalt generation from the mantle (vertical movement) and its further transformation to felsic and silicic continental crust (horizontal movement, subduction and remelting) proceeded at rates which far exceed the present rate of plate divergence and subduction. It is well known, yet interesting that Malani Igneous Suite shown crustal evolution as late as during 745 Ma to 680 Ma. In the present paper granites of the Mungeria Ring Complex (MRC) have been studied for the evalution of their high heat production value, hence possibility of economic mineralization therein.

GEOLOGY AND PETROLOGY

The proposed study area Mungeria Ring Complex is at relative small circular structure with a diameter almost of 10 km, near village Mungeria (N 26° 05′ 75″:E 71° 00′ 57″) of Shiv tehsil of Barmer district of Rajasthan state. Mungeria is approachable from Shiv which is situated 75 km north of Barmer on Barmer-Jaisalmer highway. Shiv is also approachable from Jaisalmer on the same highway with the same distance, as well. The proposed area shows the occurrence of both, granites and basaltic rocks. These granites are part of the igneous activities of the Malani Igneous Suite (745 Ma-680 Ma) (Roy et al. 2012, Bhushan and Chandrasekaran, 2002; Kochhar, 2012). The field relationship of these granites favour its Malani age.

There are three kinds of rocks available in the area namely, mafic components, pink coloured granite and grey coloured granite. As it has been interpreted in the present study, the pink granite is metasomatised product of the grey granite itself, hence; it has been treated as younger rock in comparison with the grey granite which of course is the parent. The mafic lithounits are, basaltic-andesite and basalt. There are some tungs of aplite, a metasomatised product of pink granite in mafic components giving appearance of felsic dykes. Hence the stratigraphy of the area established in the present study is showing the following succession- komatiitic-basaltic andesite, grey granite, pink granite and felsic dykes and alluvium of the desert. Both pink and grey granite which appear different in the field, at least because of colors, show very similar petrology under microscope. It is observed that there is presence of equal amount of all three essential minerals namely, quartz, orthoclase and plagioclase. Some euhedral isotropic iron ores (magnetite and/or ilmenite) are also present in very little amount. The overall texture is holocrystalline, ranging in grain size from medium to coarse grained. Generally orthoclase is coarse grained. Alteration features in feldspar are common and have invariable altered into clays. These features show metasomatic activity. The interesting and dominating texture in these rocks is graphic texture. Various sizes and shapes of quartz especially cruciform are present. Many times there are presences of cauliflower type graphic texture. Sometimes very big and sometimes small ones 'Myremekite' is also present, showing vermicular growth of quartz and sodic plagioclase adjacent to potash feldspar. Each of which may suffer little subsequent exsolution. The process responsible for such solid solution structure essentially seems to be secondary.

GRANITE SAMPLE AND METHODOLOGY

Nine granite samples were collected from different locations to represent the whole granite bodies in the area. In the laboratory, the rock samples were crushed to make a fine powder of 53 micron or smaller size, before coning and quartering of each of the samples. Major oxides have been determined using XRF and rest (including minor, trace and rare earth elements) with the help of ICP (MS). Both the analysis have been performed in the laboratories of the Wadia Institute of Himalayan Geology, Dehradun, India.

Energy-dispersive X-ray Fluorescence has been used with isotopic excitation source ^{109}Cd/30 mCi and Si/Li detector system, 30mm^2 x 3mm, having resolution for 5.9 KeV radiations of 200 eV. It was coupled to series 90nMCA Canberra industries. The configuration of XRF system is of SIEMENS SRS 3000 sequential X-ray Spectrometer with end window Rh X-ray tube. For major oxides analysis pressed powder pellets have been used. Matrix correction procedures followed by Lucas-Tooth and Pyne (1964). For major oxides, the operating conditions include no filter, vacuum path, 20/40 kV and for trace elements, no filter, vacuum path, 55/60 kV have been used for the nominal time analysis 300 sec and 100 sec for major oxides and trace elements respectively. Normal minimum detection limit is almost 5ppm.

The configuration of Inductively Coupled Plasma mass Spectroscopy System is of Perkin Elmer SCIEX ELAN DRC-e, the easy sample introduction and quick analysis of ICP technology with accurate and low detection limit of Mass Spectrometer. This instrument is capable of trace multi element analysis, often at part per trillion levels. It is similar to mass spectroscopy that is highly sensitive and capable of determination of range of metal and several non-metals at concentration below one part in 10^{12}. It is based on Coupling Plasma as a method of producing ions (ionization) with mass spectrometer as a method of separating and detecting the ions.

RESULT AND DISCUSSION

The heat generation values were calculated (Table 1) from the measured abundance of U, Th and K using the relation of Birch (1954): A (μW m^{-3}) = 0.01*ρ(9.69*Cu* + 3.58 *Ck* + 2.65 *Cth*), where A is heat production and ρ is density in gm/cm^{-3} which is 2.7 gm/cm^{-3} for granites. The *Cu* and *Cth* are concentration of U and Th in ppm respectively, whereas *Ck* is concentration of K in wt%. The heat production (μWm^{-3}) unit may be converted into heat generation unit (1HGU = 10^{-3} cal cm^{-3} sec^{-1}) using the value of 1HGU = 0.42 μWm^{-3} obtained for the Gansboden granite gneiss at the Guspisbach heat flow site of Central Alps of Switzerland (Rybach et al., 1977). The radioelement concentration (Ur) is calculated following Ashwal et al. (1987); radioelements Ur equivalents: 1ppm of U in equilibrium (1 ppm eU) = 1 Ur; 1ppm of Th equilibrium (1ppm eTh) =0.5 Ur; 1 wt% of K = 2Ur.

Table 1. Radioelements, heat production and radioactive heat generation data of granites of Mungeria Ring Complex

Sample	U (ppm)	Th (ppm)	K (%)	Th/U	HP (μWm^{-3})	Ur	Heat Generation (in HGU) due to			Total HGU
							U	Th	K	
P1	3.5	46.7	3.18	13.34	4.56	12.19	2.18	7.95	0.73	10.86
P2	4	11.4	3.66	2.85	2.21	17.02	2.49	1.94	0.84	5.26
P3	4.5	30.4	4.22	6.75	3.76	28.14	2.80	5.18	0.97	8.95
P4	3.9	19.7	3.72	5.05	2.78	21.19	2.42	3.35	0.85	6.62
P5	3.4	10.6	4.55	3.11	2.08	17.8	2.11	1.80	1.04	4.95
G6	3.4	24.4	4.59	7.17	3.07	24.78	2.11	4.15	1.05	7.31
G7	4.2	28.7	4.56	6.83	3.59	27.67	2.61	4.89	1.04	8.55
G8	3.7	11.6	4.48	3.13	2.23	18.46	2.30	1.98	1.03	5.31
G9	3.5	16.6	4.59	4.74	2.54	20.98	2.18	2.82	1.05	6.05
Av.	3.79	22.23	4.17	5.88	2.98	20.91	2.35	3.87	0.96	7.09

The enrichment of U in granites (3.4-4.2 ppm) (Table 1) appears to be due to fractional differentiation that is indicated by the increase in K content of these rocks. The Ur values in granites vary from 12.19-28.14. The distribution of U and Th for granites is shown in Fig. 1 and Fig. 2. It exhibits positive correlation. Granites in the upper parts of the continental crust usually have 3 to 4 ppm of U and 10 to 15 ppm of Th with 3.7 as Th/U ratio (Wasserburg, 1964). The result obtain have comparable concentration of U and Th with the alkaline granite of Corsican magmatic province and A-type granite and rhyolite of Northwestern Ontario (Canada). Low Th/U ratio in few samples possibly explaining a fractionation of phase like Th-Orthosilicate and Zircon. These granitoids have very high content of radioelements, radioelements concentration (Ur) and heat production units hence identified as high heat producing (HHP) acidic rocks (Pagel, 1982; Plant et al., 1985).

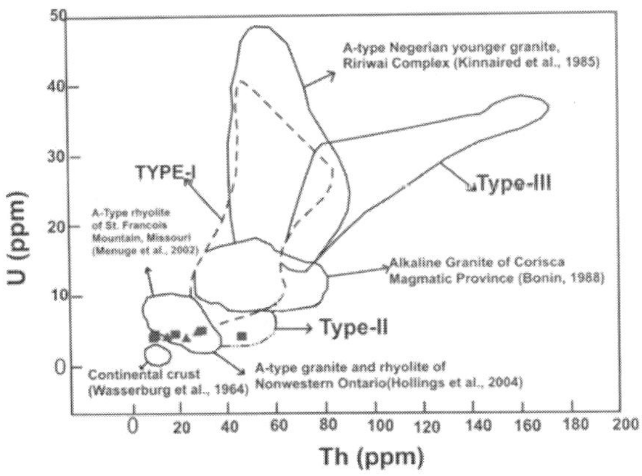

Fig. 1. Plots Th vs U for Mungeria granites. In plot, the field of continental Crust (Wesserburg, 1964), Alkaline granite of Corsican magmatic Province (Bonin, 1988), A-type Nigerian Younger granite, Ririwai Complex (Kinnaird et al., 1985a), Anorogenic A-type rhyolite of St. Francois Mountains, Missouri (Menuge et al., 2002) and A-type granite and rhyolite of Northern Ontario (Hollings et. al., 2004) are shown for comparison. The pink and grey granites are represents by squares and triangles, respectively and hence, Mungeria is showing A-type granites (of type II category).

In Fig. 3 Th/U ratio vs SiO_2 diagram (after Taylor and McLennan; 1985) further confirm Type II category of A-Type granite. In the K-U-Th ternary diagrams (Fig. 4), the concentration of elements fall near the Th-apex, indicating high content of Th in the Samples, and hence the heat generation of Th is much higher than that by U and K. Pink granite have a higher heat productivity compared to Grey granite.

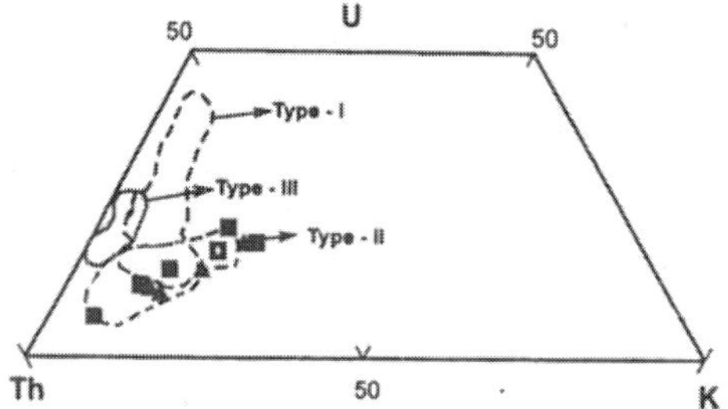

Fig. 2. Plots Th vs U for Mungeria granites. In plot, the field of continental Crust (Wesserburg, 1964), Alkaline granite of Corsican magmatic Province (Bonin, 1988), A-type Nigerian Younger granite, Ririwai Complex (Kinnaird et. al., 1985a), Anorogenic A-type rhyolite of St. Francois Mountains, Missouri (Menuge et. al., 2002) and A-type granite and rhyolite of Northern Ontario (Hollings et. al., 2004) are shown for comparison. The pink and grey granites are represents by squares and triangles, respectively and hence, Mungeria is showing A-type granites (of Type II category).

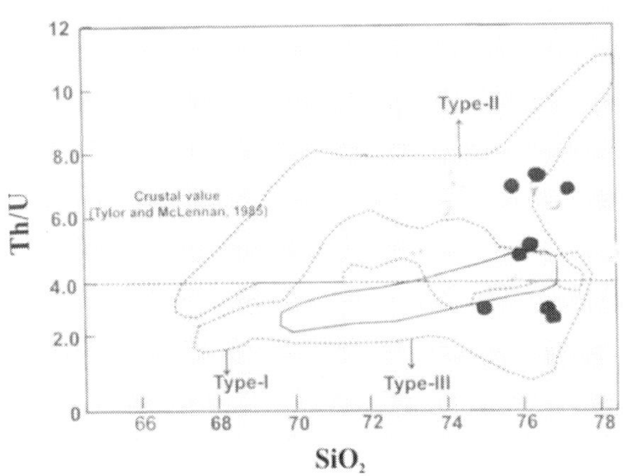

Fig. 3. Plot of Mungeria granites in Th/U vs SiO_2 (after Taylor and McLennan, 1985)

The average total heat generation value is 7.73 HGU for Pink granites, and 6.81 HGU for Grey granites. This is much higher than the average value of 3.8 HGU known for the continental crust (Taylor, 1964) and the value of 8.3 HGU obtained for the Peninsular India (Dhana Raju et al., 1983). Since the heat generation value of 7 HGU is taken as the boundary between the 'hot crust' and 'cold crust', the present data (>7HGU) support that Mungeria Granites belongs to the 'hot crust' category. It is also a well- known fact that the presence of long-lived isotopes of radioelement (U, Th and K) causes heat generation in the crust, which plays a significant role in continental heat flow. Bhushan and Chittora (1999) suggest that the continued influx of mafic magma is necessary to maintain the felsic magma reservoir for the local anomalously high heat flow. The generation of felsic magma is likely to form in a system of substantial dimension, driven by continued addition of heat from below as an ascending basic magma (Smith and Shaw, 1975). The felsic magma might have generated due to partial melting of crustal rocks at elevated temperature under the influence of a rising mantle plume. The present radioactive heat generation data suggest a possible linear relationship between the surface heat flow and the crustal heat generation in the Mungeria granites.

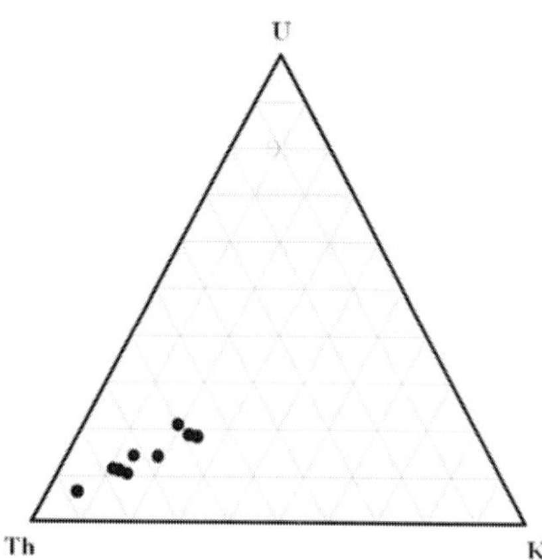

Fig. 4. Plot of Mungeria granites in Th-U-K ternary diagram of Taylor and Mclennan, (1685)

CONCLUSION

The genetic relationship between mineralization and granitoids indicate that U and rare earth elements have higher abundance in more felsic, alkalic and late stage members of igneous differentiation series rich in silica and potash than in mafic early stage lithologies (Kimberley, 1978; Boyle, 1982; Dahlkamp, 1993). The average value of total heat generation in MRC is 7.33 HGU for pink granite and for grey granite it is 6.81 HGU. These values are much higher than the average value of 3.8 HGU for the continental crust. Reactivations of basement and hydrothermal activities are always produced their impact on enrichment of radioelement and other incompatible HFS elements in these highly fractionated and differentiated granitoids. The increased heat flow triggers fluid convection of sufficient magnitude and helps further enrichment of these elements to develop an ore deposit (Fehn et al. 1978; Brown et al. 1980; Prajapat, 2011). In view of highly labile nature of uranium, the granitoids of MRC it is proposed that it can be a potential source for further enrichment leading to formation of economic deposit under favourable condition at suitable loci as evidenced by Nb, Ce, REE, U and Th mineralization in HHP Saudi Arabian granites (Jackson et al., 1985).

REFERENCES

Ashwal, L.D., Morgan, P., Kelly, S.A. and Preicival, G.A. (1987). Heat Production in an Archean Crustal Profile and Implcations for Heat Floe and Mobilization of Heat Producing Elements. Ear. Plan. Sci. Let., vol. 85, pp. 439-450.

Bhandari, N. (2009). Mineral Exploration of the Moon in Economic Mineralization. Ed. K.L.Shrivastava, Sci. Pub. (India), pp.47-52.

Bhushn, S.K. and Chandrasekharan,V. (2002). Geology and Geochemistry of rocks of the Malani Igneous Suite & Tertiary Alkaline Province of Western Rajasthan. Mem. Geol. Surv. Ind., No. 126, 1-181.

Bhushan, S.K. and Chittora, K.; (1999). Late Proterozoic Bimodal Volcanic Assemblage of Siwana Subsidence Structure, Western Rajasthan, India. Jour. Geol. Soc. India., vol. 53, pp. 433–452.

Birch, F. (1954). Heat from Radioactivity in: Nuclear Geology, Ed. H. Faul, John Wiley and Sons., pp. 148-174.

Bonin, B.; 1988: "Peralkaline granites in Corsica: some petrological and geochemical constraints". Rendiconti Della Society Italiana Di Mineralgia E Petrologia, Vol. 43 , No (2), pp.281-306.

Boyle, R.W. (1982). Geochemical prospecting for uranium & thorium Deposits. Elsevier, Amsterdam, 498p.

Brown, G.C., Cassidy, J., Oxburgh, E.R., Plant, J., Sabine, P.A. and Watson, J.V. (1980). Basement heat flow and metalliferous mineralization in England and Wales. Nature., vol. 288, No. 5792, pp. 657-659.

Clark, A.H. and 10 others (1990). Geological and Geochronological Construction on the Metallogenic Evolution of the Andes of Southestern Peru. Eco. Geol., vol. 85, pp. 1520-1583.

Dahlkamp, F.J. (1993). Uranium Ore Deposits. Springer-Verlag, Berlin. 460P.

Dhana Raju, R., Singh, J.P. and Bhattacharya, T. K. (1983). A Preliminary study on Radioactive Heat Generation of Some Precambrian Granitic Rocks of India. Indian Jour. Earth Sci., vol.10, No. 2, pp.112-119.

Fehn, U., Cathies, L.M. and Holland, H.D. (1978). Hydrithermal Convection and Uranium Deposits in Abnormally Radioactive plutons. Econ. Geol., vol. 73, pp. 1556-66.

Hollings, P., Fralick, P. and Kissin, S.; 2004: "Geochemistry and geodynamic implications of the Mesoproterozoic English Bay granite- rhyolite complex,

Northeastern Ontario". Can. Jour. Ear. Sci., Vol. 41, pp. 1329-1338.

Jackson, N.J., Drysdall, A.R. and Stoeser, D.B. (1985). Alkali Granites Related Nb-zr-REE-U-Th Mineralization in the Arabian Schield in High Heat Production (HHP) Granites, hydrothermal Circulation and Ore Genesis. Inst. Min. and Metall. London. pp. 479-488

Kimberley, M.M. (1978). Uranium deposits, their mineralogy, and origin. Min. Assoc. Can. Short Course Handbook, No. 3, 523p.

Kinnaird, J. A., Batchelor, R. A., Whittely, J. E. and Mackenzie, A. B.; 1985b: "Geochemistry, mineralization and hydrothermal alteration of the Nigerian high heat producing". In: High heat producing (HHP) granites, hydrothermal alteration and Ore genesis". Inst. of Mining and Metallurgy, London. pp. 169-199.

Kochhar, N. (2012). The Malani Supercontinent: Siberia Mangolia Kazakhistan Connections during Late Proterozoic. Section 18.3 Int. Geol. Cong. Brisbane, 207.

Kochhar, N. (1989). High Heat Producing Granites of the Malani Igneous Suite, Northern Peninsular India. Ind. Min., vol. 43, pp. 339-346.

Menuge, J.F., Brewer, T.S. and Seeger, C.M.; 2002 "Petrogenesis of Metaluminous A-type rhyolite from the St. Francois Mountains, Missouri & Mesoproterozoic evolution of the Southern Laurentian margin". Precam. Res., Vol. 113, pp. 269-291.

Morgan, P. (1985). Crustal Radiogenic Heat Production and the Selective Survival of Ancient Continental Crust. Proc. 15th Lunar Planet. Sci. Conf., Geophys. Res., vol. 90, pp. C561-C570.

Morgan, P. and Sass, J. H. (1984), Thermal Regime of the Continental Lithosphere. Jour. Geodyn. vol. 1, pp. 143-166.

Pagel, M. (1982). The Mineralogy and Geochemistry of Uranium, Thorium and Rare Earth Elements in Two Radioactive Granites of the Vosges, France. Min. Mag., vol. 46, pp. 149-161.

Patwardhan, A.M. (2010). The Dynamic Earth System, II ed. PHI Learning Pvt. Ltd., 422p.

Plant, J.A., O'Brien, C., Terney, J. and Hurdley, J. (1985). Geochemical Criteria for the Recognition of Heat Production Granites in High Heat Production Granite, Hydrothermal Alteration and Ore Genesis. Institute of Mining and Metallurgy, London, pp. 263-286.

Prajapat, T. (2011). Geological Characterization of post-Delhi Granites around Jodhpur and its bearing to Evaluate their Economic Potential. Unpublished Ph. D. Thesis, Jai Narain Vyas Univ., Jodhpur. 161 p.

Rybach, L., Werner, D., Mueller, S. and Berset, G. (1977). Heat Flow, Heat Production and Crustal Dynamics in the Central Alps, Switzerland. Tectonophysics, vol. 41, pp. 113-126.

Rogers, J.J.W. and Adams, J.A.S. (1969). Thorium in: Handbook of Geochemistry. Ed. K.H. Wedepohl, Springer- Verlag, Berlin, pp. 11-14.

Roy, A. B., Kroner, A., Rathore, S., Laul, V. and Purohit, R. (2012). Tectono-Metamorphic and Geochronological studies from Sandmata Complex, Northwest Indian Schield: Implications on Exhumation of Late Palaeoproterozoic Granulites in an Achaean-early Palaeoproterozoic Granite-Gneiss Terrance. Jour. Geol. Soc. Ind., vol.79, pp. 323-334.

Roy, A.B. (2009). Precambrian Crustal Evolution and Metallogeny in Economic Mieralization. Ed. K.L. Shrivastava, Sci. Pub. (India), pp. 26-35.

Sharma, R.S. (1994). High Heat Production Granites of JhunJhunu area, Rajasthan, India. Bull. Ind. Geol. Assoc., vol. 27, pp. 55-61.

Shrivastava, K.L., Tina Prajapat, Anju Mathur, Virendra Gaur, C.P. Dadhich, Mamta Chauhan and Beena Tripathi (2013). "High radioactive heat producing economically potential granites around Jodhpur city, Malani Igneous Suite, Northwestern India." Jour Appl. Geochem. Vol. 15 No. 4 pp. 410-418.

Srivastava, P.K. (2003). High Heat Producing Granites of Degana, Rajasthan, India. Jour. Geochem., vol.18, pp. 149 155.

Smith, R. L., and Shaw, H. R. (1975). Igneous Related Thermal System Assessment of Geothermal Resources of United States'. In: D.E. Whiten and D.C. Williams, (Eds.), U.S. Geol. Surv., Washington, D.C. Circ. 726, pp. 58-83.

Taylor, S. R. (1964). Trace Element Abundance and the Chondritic Earth Model. Geochem. Cosmochem. Acta, vol. 28, pp. 1989-1999.

Taylor, S.R. and McLennan, S.M.; 1985: 'The Continent Crust: Its Composition and Evolution'. Blackwell, Oxford, 312 p.

Wesserburg, G.G. (1964). Relative Contribution of U, Th and K to Heat Production in the Earth's Crust. Science. vol. 143, pp. 465-467.

Valdia, S. (2010). The making of India Geodynamic Evolution. Mc Millan Pub., 816p.

Hon'ble reviewer

1. **Prof. Mike Katz**
   Key Centre for Mines
   University of New South Wales
   SYDNEY - NSW 205
2. **Dr. A.K. Sarangi**
   Gen. Manager
   Uranium Corporation India Ltd.
   JADUGUDA (Jharkhand).

II-27

Scanning Electron Microscopic Studies of Radioactive and other Minerals of Motur and Barakar Sediments, Satpura-Gondwana Basin, Central India

Anil M. Pophare, B.H. Raut, I.K. Sheikh, P.K. Sinha[1] and Rahul Banerjee[2]

Department of Geology, RTM Nagpur University, Nagpur- 440001
Atomic Minerals Directorate for Exploration and Research, Nagpur[1] and Hyderabad[2]
*E-mail: apophare@gmail.com

Abstract

In the present study, backscatter Scanning Electron Microscopy (SEM) technique is applied for qualitative analysis and characterization of the fine clastic sediments of Barakar and Motur formations. These studies revealed that the preservation of large amount of detrital clay minerals implies absence of reworking of the sediments in a turbulent environment. Kaolinite reflects acidic environment and derivation from weathered granite and analogous Ca-poor igneous and metamorphic rocks. Kaolinite forms both in oxidizing and reducing environments, in presence of organic matter such as humic complex and organic acid. Montmorillonite is the alteration product of mostly basic rocks, where stable alteration minerals formed in alkaline conditions due to availability of Mg and Ca. Illite is relatively stable alteration product of mostly acidic rocks due to intense weathering in humid climate and mostly confined to shale. Chlorite is derived from aggradations of less organized sheet minerals, degradation of pre-existing ferromagnesian minerals and by crystallization from dilute solutions. In present samples, montmorillonite show high ion base-exchange and lie on the basal plane surface, whereas kaolinite has low ion base-exchange capacity. The morphology of monazite and xenotime suggests short transportation and derivation from nearby granitic source rocks. The clear and uneroded edges of the zircon reveal short transportation and first cycle of weathering and sedimentation.

Keywords: SEM studies, Monazite, Xenotime, Zircon, Barakar and Motur sediments, Satpura-Gondwana basin, India.

Introduction

Clay minerals are generated by weathering of alumino-silicates and formed in the early or late diagenetic depositional environment (Pettijhon 1980). Clay matrix is easily and rapidly winnowed from clastic aggregates by small amount of turbulence in flow. Small but significant quantities of clay can be trapped with sand during deposition in fluvial environment. Large amount of clay may possibly be trapped during deposition from dense turbidity current and still larger amount may be present in mud flow that implies absence or reworking of sediments in turbulent environment (Blatt 1972). The clay found in sandstones as the essential constituent of matrix and argillaceous rock fragments include all major groups of clay (Pettijhon *et al* 1972). Clay minerals are also commonly present as authigenic or diagenetic clay that has formed within sandstones and includes pore-filling kaolinite, pore-lining chlorite, pore-bridging illite (Wilson and Pittman 1977). The characteristic of clay deposits depends on source rock, weathering process, transportation and depositional environment.

Monazite (Ce La Th)PO_4, Xenotime (YPO_4) and Zircon ($ZrSiO_4$) are common accessory minerals in magmatic and metamorphic rocks and contains significant amount of radiogenic Th-U and non-radiogenic Pb (Williams *et al* 1999). These minerals occur as heavy minerals in clastic sediments being derived from magmatic and metamorphic source rocks. The SEM analysis of clastic sediments has proved to be an effective and useful tool in deciphering REE orthophosphates like monazite and xenotime along with zircon as detrital heavy minerals in clastic sediments.

The interlayered clay, carbonaceous material and shale bands are confined to the Motur and Barakar formations of the Gondwana Supergroup in the Satpura-Gondwana basin, Betul District, Madhya Pradesh. The fine clastic material comprising clay, carbonaceous materials occur in the form of bands, streaks, laminations in the Barakar and Motur sandstones and comprises of kaolinite, montmorillonite, illite and chlorite along with detrital monazite, xenotime and zircon grains. Monazite, xenotime and zircon are important accessory heavy minerals in the sediments and host REE, radiogenic Th and U that are useful for geochronology and geothermometry studies. The objective of present study is characterizations of sediments and radiogenic heavy minerals using SEM technique.

Geological Framework

Satpura-Gondwana basin is an intracratonic basin in which the sediments of Gondwana Supergroup were deposited over crystalline basement (Crookshank 1936). It is the western most basin located in peninsular region of central India and lies at the junction of Pranhita-Godavari and Narmada-Son-Damodar valleys. The Satpura-Gondwana basin covers an area about 12,000 km^2 of Permo-Triassic and Cretaceous sediments (Peters and Singh 2001). The basin is 200 km long in the ENE-WSW direction and about 60 km wide. Its longer side is marked by ENE-WSW trending Son-Narmada fault and Tapti North fault (Biswas 1999, 2003). The Satpura-Gondwana basin hosts ~5 km thick siliciclastic succession and preserves a stratigraphic record ranging in age from Permian to Cretaceous (Crookshank 1936). In this basin, Gondwana sequence of late Carboniferous to early Cretaceous period comprising of both lower and upper Gondwana formations of Talchir, Barakar, Motur, Bijori, Pachmarhi, Matkuli (Denwa and Bagra) and Jabalpur are exposed (Table 1). The study area (77° 50':78° 10' and 22° 05':22° 20') comprises of lower Gondwana formations that includes Talchir, Barakar Motur and Bijori. These formations have general E-W strike and dip approximately ranging between 8°-10°.

Table.1: Generalised stratigraphic succession of Satpura-Gondwana basin (after Peter and Singh 2001)

Age	Group	Formation	Lithology
Late Cretaceous	Deccan Trap		Basalt
Early Cretaceous to Late Jurassic	Gondwana Supergroup – UpperUnit	Jabalpur	Sandstone, conglomerate, clay
Rhaetic	Gondwana Supergroup – Middle Unit	Matkuli	Sandstone, conglomerate, clay
Carnian	Gondwana Supergroup – Middle Unit	Pachmarhi	Sandstone, pebbly sandstone, sandstone with minor clay
Middle to Late Triassic	Gondwana Supergroup	Bijori	Gritty, coarse sandstone, minor conglomerate, carbonaceous sandstone, carbonaceous siltstone and clay
Latest to Late Permian	Gondwana Supergroup – Lower Unit	Motur	Sandstone with clay pockets
Late Permian	Gondwana Supergroup – Lower Unit	Barakar	Sandstone, coal cyclotherm, carbonaceous clay
Early Permian	Gondwana Supergroup – Lower Unit	Talchir	Diamictites, sandstone, shale
Precambrian Undifferentiated Basements			

Talchir Formation is the lowermost litho-stratigraphic unit of the Gondwana Supergroup that unconformably overlies the Precambrian basement, and consists of diamictic light green sandstone, olive green needle shale and rhythmites. This formation underlies Barakar Formation without any depositional breaks. The Barakar Formation differentiated from Talchir through abrupt lithogical variation and consists of sandstone, coal and carbonaceous shale and clay. Motur Formation succeeding Barakar Formation comprising red clay, feldspathic sandstone and grey shale with thin coal laminae and underlies Bijori Formation. The Bijori Formation consists of sandstone, conglomerate, carbonaceous sandstone, carbonaceous siltstone and clay (Fig. 1).

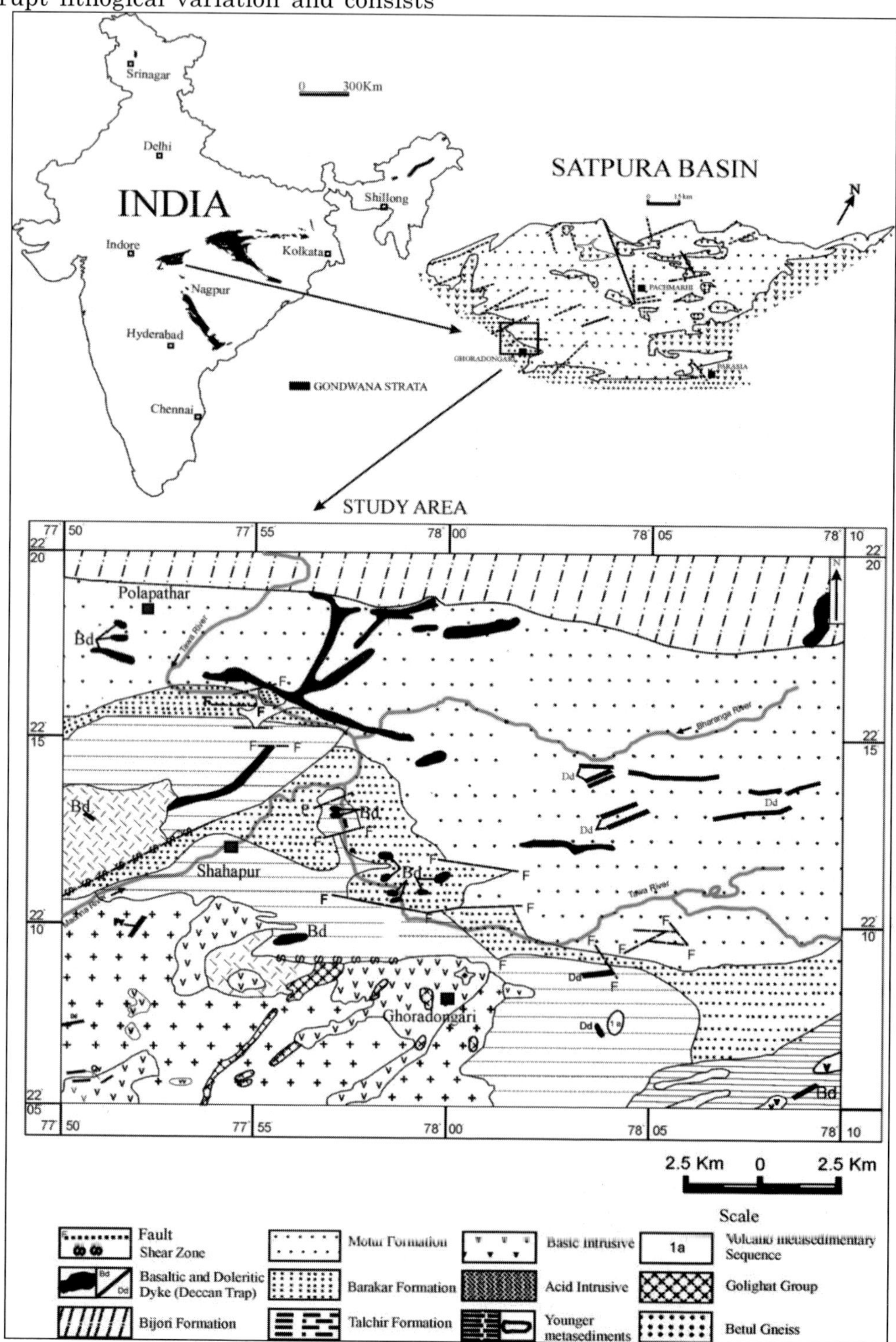

Fig. 1: Geological map of the study area, Satpura-Gondwana basin, Betul district, Madhya Pradesh

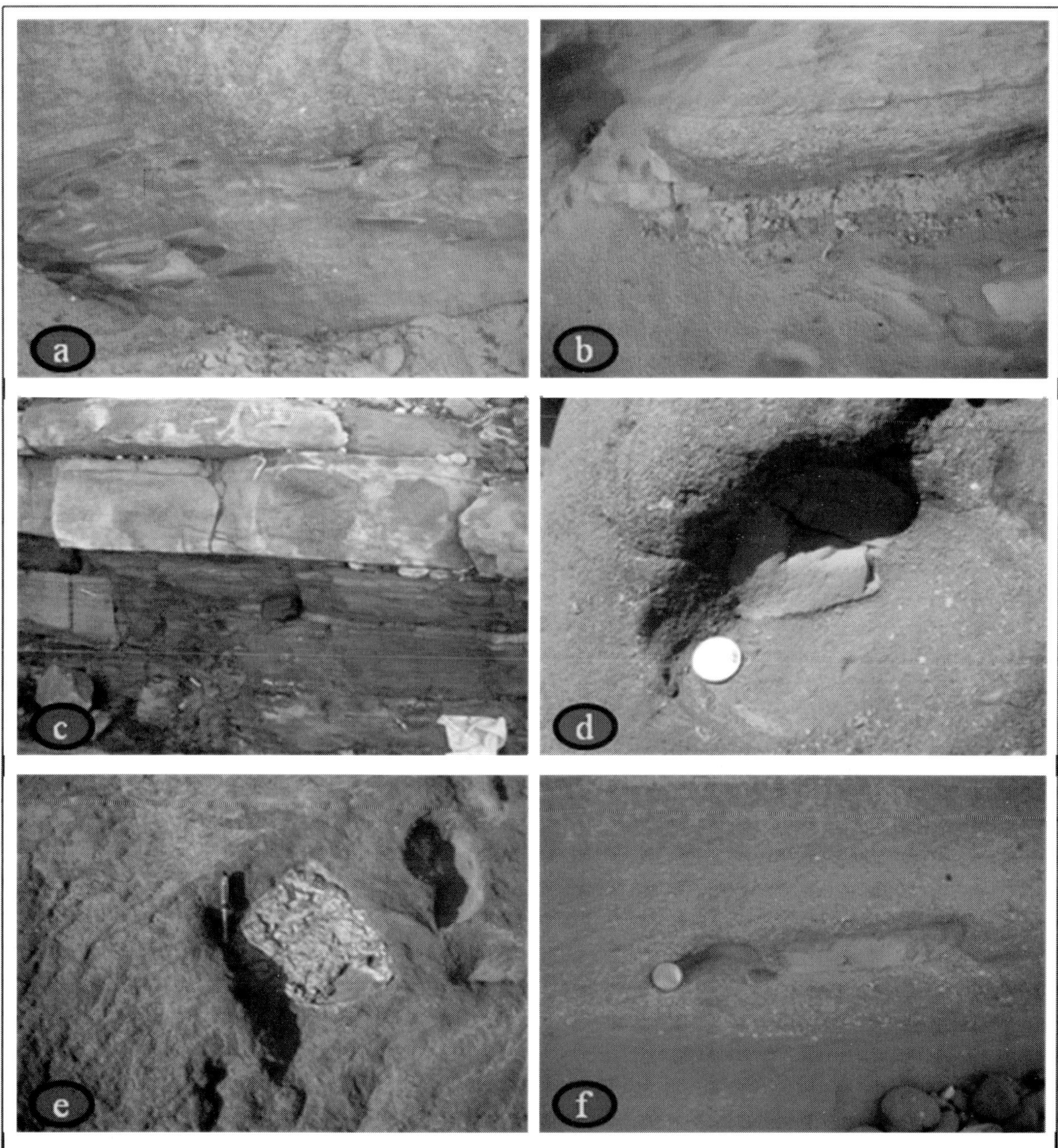

Fig. 2: Field photographs showing: (a) the patches and lenses of the clay material observed within the Barakar sandstone; (b) Interbedded clay layer with Barakar sandstone; (c) Carbonaceous shale with Barakar sandstone; (d) and (e) Clay balls in Motur sandstone; (f) Clay lenses within Motur sandstone

In the study area, shale bands, carbonaceous material as coal bands, streaks, lamellae and laminations inter-bedded with sandstone are mostly associated with the Barakar Formation (Fig. 2). The Barakar sandstone is whitish buff coloured, medium to coarse grained and cross laminated. The intercalated fine laminated clay layer along with sand and silt layer are observed within these sandstone beds. Clay patches of nearly 4 to 9 cm width along with carbonaceous lamellae are generally concentrated along cross bedding planes. Lenses and streaks of carbonaceous material are also observed in the Motur sandstone (Fig. 2). Weathered pyritic nodules also occur in the Barakar and Motur sandstones. Fine laminations of carbonaceous material are part of the upper member of Talchir Formation.

Methodology

The optical microscopic study cannot be applied to clay mineral because of their very fine character. The clay minerals in sandstone and clay were quantified using SEM technique. The crushed powder and polished sections of the sandstone and clay were used for SEM analyses. The scanning electron micrograph of polished and disaggregated clay material recorded on Hitachi-S 3400N SEM attached with EMAX energy dispersion spectrometer (EDS) at National Geophysical Research Institute (NGRI), Hyderabad. The scanning electron microscopy allows rapid examination of specimen at magnification between 20x to about tens of thousands. EDS allow qualitative elemental analysis of elements. The determination of a single clay mineral versus clay mineral matrix is difficult to identify. For precision a single mineral aggregate was used in SEM analysis. Secondary electron (SE) and backscatter electron (BSE) are two main images obtained with SEM. Backscatter electron images are used in the present study. As in the backscatter electron images, the grey level is associated with chemical contrast and give quality contrast between grains of different compositions, crystal orientations on flat well polished specimens.

Backscatter Imaging Mode in Scanning Electron Microscope

When an electron beam strikes a solid specimen about 70 to 80% of energy is expended within the interaction volume of the specimen. The remainder is scattered back out of the specimen by elastic collision with atoms of the specimen as backscatter electron which retains a high proportion of energy of the incident electron beam (Fig. 3) (White *et al* 1984).

The degree of back scattering is measured in terms of the backscatter electron coefficient, define as under -

$$\psi = \frac{\text{Number of electron scatter out}}{\text{Number of incident beam}}$$

An area of specimen composed of elements with higher atomic number emits more backscatter signals and thus appear brighter in image (atomic number z contrast). The dependency of backscattering coefficient on average atomic number allow many mineral phases to be distinguished on the basis of difference in grey level when a polished specimen is examined under SEM. Quartz and feldspar appears relatively darker than minerals containing heavy elements such as pyrite or illeminite (Pye and Krinsley 1984). BEM image gives quality contrast between grains of different composition, different crystal orientation on flat, well polished specimens (Lloyd and Hall 1981).

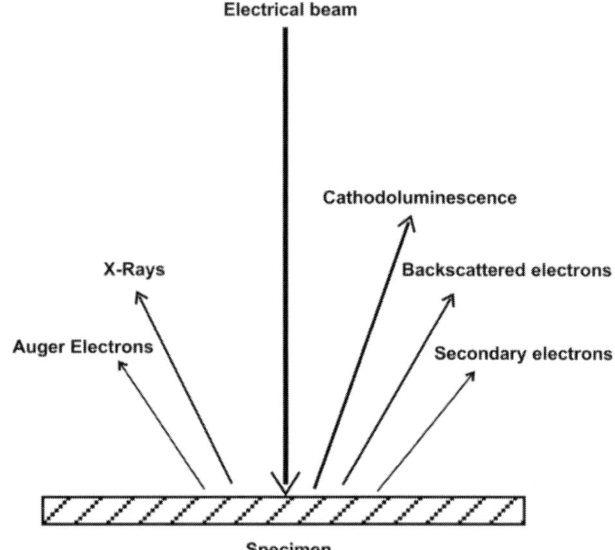

Fig. 3: Schematic diagram of different imaging and analytical method in Hitachi S3400N SEM

Clay Minerals and Their Characteristics

Clay mineral structure is based on composite layer built from components with tetrahedrally and octahedrally co-ordinated cations. Chemical composition may vary according to the extent of replacement of Si, Al and Mg by other cations, nature and quality of inter-layered cations and water content. The clay minerals vary in their dehydration and breakdown characteristic and decomposition product. These minerals also differ in their cation exchange properties as per the nature of their inter-layer cations and residual surface charge (Deer *et al* 1966).

Kaolinite group consist of kaolinite, dickite, nacrite, halloysite, meta-halloysite, nauxite and allophone minerals. The kaolinite structure is asymmetrical with two layer structure. One layer consists of gibbsite having octahedral co-ordination and other of tetrahedral coordination, which is the di-octahedral analog of septechlorite. Kaolinite clay is the fine grained alteration product of feldspars and other alumino-silicates. Excessive leaching in humid climate may form aluminous laterite (diaspore, gibbsite) at the expense of kaolin clay.

Montmorillonite (Smectite) group consists of montmorillonite, beidellite, nantrolite, saponite and hectorite. The montmorillonite structure is a three layer arrangement of one "brucite" or "gibbsite" octahedral layer enclosed between two inward pointing Si_4O_{10} layers of tetrahedral co-ordination. Montmorillonite clay is swelling clay. Interlayered cations and water molecules are required to neutralize cation substitution in both tetrahedral and octahedral sites.

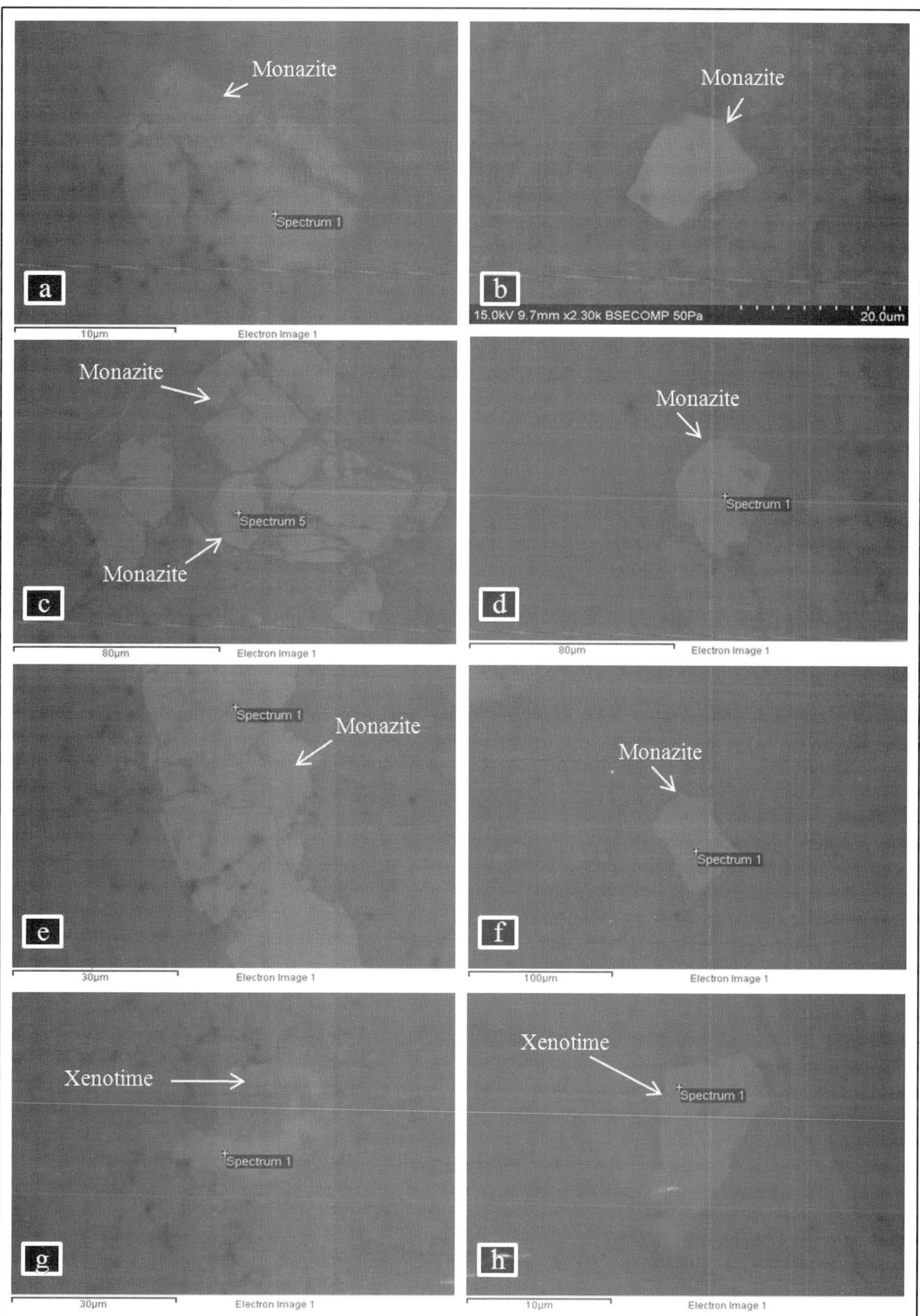

Fig. 4: (a-f). Monazite, (g and h) Xenotime in clay aggregates and along periphery of carbonaceous lamellae in Barakar and Motur sediments

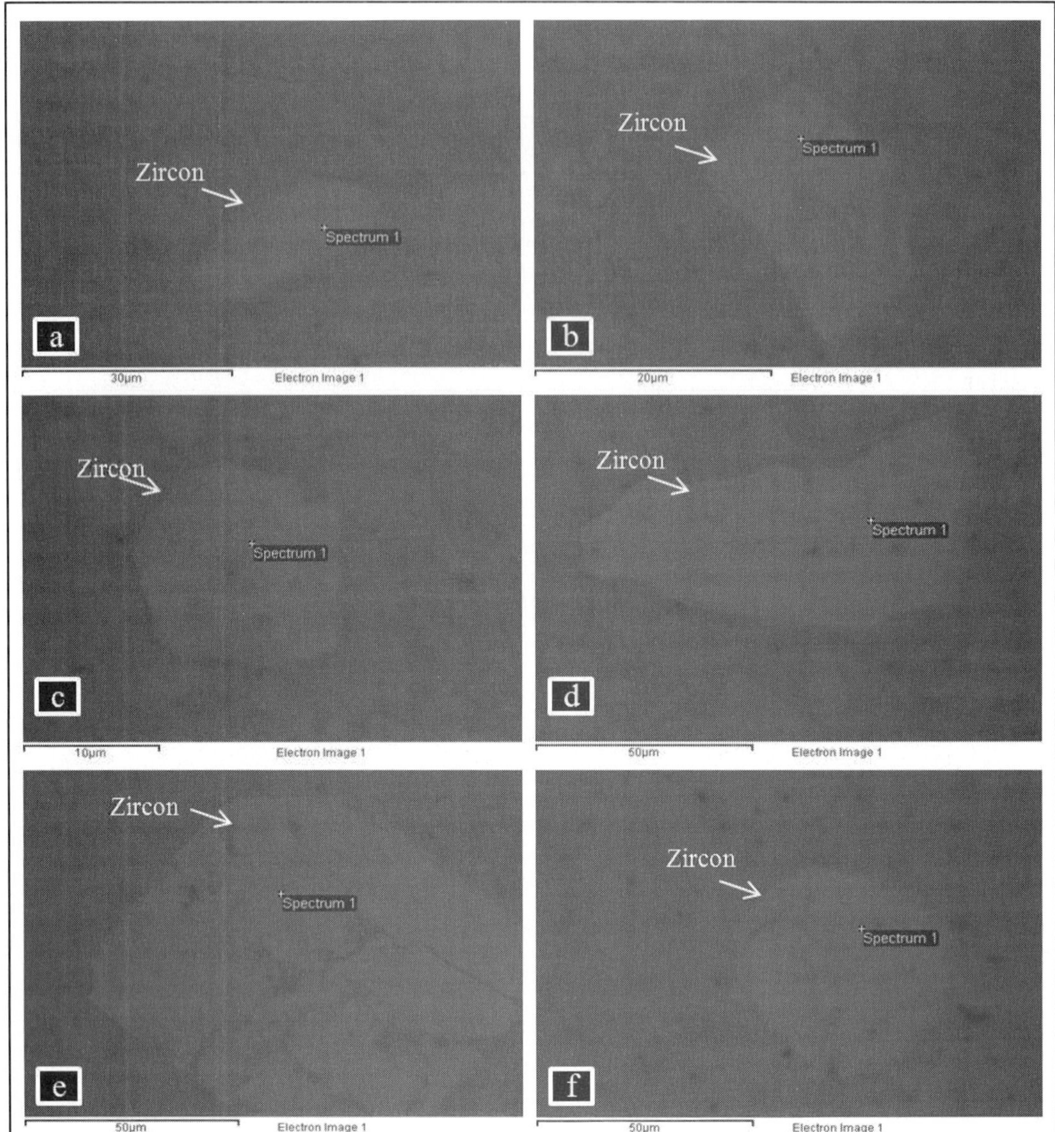

Fig. 5 (a-f). Angular and prismatic zircon mineral in Barakar and Motur sediments

Illite group display the typical three layer structure of mica, an octahedral layer sandwiched between tetrahedral layers with intersheet cations required to balance the substitution of Al^{3+} for Si^{4+}. The illite clay is relatively stable alteration product, largely of acidic rock type. Intense leaching in humid climates may yield Kaolin clay, Al-hydroxide or montmorillonite.

Vermiculite group represent wide range of composition and is closely related to both montmorillonite and chlorite. Vermiculite structure is again the three layer mica structure with interlayer cations balancing charge deficiency largely in tetrahedral sites. Interlayer cations are largely Mg^{2+}sandwiched between two water molecule layers to yield an interlayer configuration much like the "Brucite" layer of chlorite. Rapid heating vapourises the water molecule causing vermiculite layer to expand. Mixed layer associated with chlorite, montmorillonite or mica is common.

Polygorskite and sepiolite are fibrous clay. The structure of polygorskite contains amphibole like chains (Si_4O_{11}) elongated on c-axis and side by side along b-axis with tetrahedral of adjoining chain pointing alternately up and down. The structures of sepiolite differ only in the width of the "Talc" ribbon, which are wider by one silicon tetrahedron.

Glauconite [K Mg (Fe, Al)$_3$ Si_6O_{18} $3H_2O$] is usually in granule rarely as 0.1 mm long. It is supposed to be amorphous but actually composed of fine lamellae (Winchell 1951). Limited occurrence, colour and tropical pellet form usually distinguish glauconite. Limonite and goethite is common weathering product of glauconite. It often appears as mixed layered with expandable montmorillonite and become montmorillonite with loss of potassium (K). Glauconite characteristically occurs as rounded, digenetic pellets or granule in detrital sediments of marine origin (greensand). It may be intimately associated with detrital quartz, feldspar and mica. Glauconite suggests slow deposition and may be related to the action of sulphate-reducing bacteria and decaying organic matter on volcanic glass. Biotite and many other silicates may represent parent material for glauconite formation and K may be derived from the sea.

Charcterisation of Sediemnts by SEM Analysis

Clay minerals in Barakar and Motur sediments mostly comprises of kaolinite, montmorillonite, illite, and chlorite along with detrital monazite, xenotime, zircon grains and interlayer cation like La, Ce, Yb, Nd, Zn and Ba *etc.*

In the SEM analyses, monazite is very fine grained, flat and elongate exhibiting parting and micro-cleavages, commonly occur as separate and aggregate mineral in clay matrix and along carbonaceous lamellae in sandstone. The physical attributes of monazite grain reveals detrital mode of origin (Fig. 4a, 4b, 4c, 4d, 4e and 4f). The monazite grains show parting and cleavage (Fig. 4c and 4e). The angular xenotime grains embedded into the clay matrix and at periphery of carbonaceous lamellae (Fig. 4g and 4h).

Zircon grains mostly exhibit angular crystal outline and characteristically abundant in the finest grade size in sediments. Detrital zircon grain occurs as sub-rounded and prismatic grains. Sub-angular to sub-rounded, short prismatic zircon with uneroded corner reveals short transportation (Fig. 5a, 5b, 5c and 5d). The cluster of zircon grains and monazite minerals are interspaced between quartz and feldspar and along carbonaceous lamellae. Dark smudge and pseudo crack on zircon grains may be an effect of metamictization (Fig. 5e and 5f).

Conclusions

The clay minerals have number of characteristic and their structure is based on composite layer built from components with tetrahedrally and octahedrally co-ordinate cation. Chemical composition may vary according to the extent of replacement of Si, Al and Mg by other cations, the nature and quality of inter-layered cation and water content. The clay minerals vary in their dehydration and break-down characteristic and in their decomposition product and they also differ in their cation exchange properties according to the nature of their inter-layer cations and residual surface charge (Deer *et al* 1966). Clay mineral has base-exchange capacity characterized by their property of carrying cation (anions) that may exchange for other cation (anion) present in aqueous solute on which was brought into contact with clay minerals. The montmorillonite, kaolinite and illite show presence of La, Ce, Nd, Ba cations. The fine grained clastic material in Barakar and Motur sediments consist of mona-zite, xenotime and zircon as heavy constituent. The physical attribute of monazite and zircon response to detrital mode may derive from granite, granite gneiss and pegmatitic rocks.

The SEM studies revealed that the preservation of large amount of detrital clay minerals implies absence of reworking of the sediments in a turbulent environment. Kaolinite reflects acidic environment and derivation from weathered granite and analogous Ca-poor igneous and metamorphic rocks. Montmorillonite is the alteration product of basic rocks, where stable alteration minerals formed in alkaline conditions due to availability of Mg and Ca. Illite is relatively stable alteration product of acidic rocks due to intense weathering in humid climate and mostly confined to shale. Chlorite is derived from aggradations of less organized sheet minerals, degradation of pre-existing ferromagnesian minerals and by crystallization from dilute solutions. The morphology of monazite and xenotime suggests short transportation and derivation from nearby granitic source rocks and first cycle of weathering and sedimentation.

References

Biswas, S.K. (1999) A review on the evolution of rift basin in India during Gondwana with special reference to western Indian basin and their hydrocarbon prospects. In: Sahni, A. and Loyal, R.S. (eds), 'Gondwana assembly: new issue and prospective' Proc. Indian National Sci. Acad. Special issue, pp.261-283.

Biswas, S.K. (2003) Regional tectonic framework of the Pranhita Godavari Basin, India. Jour. Asian Earth Sci., Vol. 21, pp.1-9.

Blatt, H., Middleton, G. and Murray R. (1972) Origin of sedimentary rocks. Prentice-Hall, Inc., pp. 241-252.

Crookshank, H. (1936) The Geology of the northern slopes of Satpura between Mohand and Sher rivers. Geol. Surv. India, Memoir 66 (2), pp.217-218.

Deer W.A., Howie, R.A., Zussman, J. (1966) An introduction to rock forming mineral. Longman Group Ltd., London. pp. 250-274.

Lloyd, G.E and Hall, M.G. (1981) Application of scanning electron microscopy to study of deformed rocks.Tectophysics, Vol.78, pp. 687-698.

Peters, J. and Singh, S. K. (2001) Satpura basin an example of pre-rift, syn-rift and post-rift Gondwana sedimentation in India. Jour. Geol. Soc. India, Vol. 57, pp. 309-320.

Pettijhon, F.J. (1984) Sedimentary Rocks.harper and Rows, New York, CBS Publisher and Distrubutors, Delhi.

Pettijhon, F.J., Potter, P.E and Siever, R. (1972) Sand and Sandstone. Springer-verlag pub., pp. 610-618.

Pye, K. and Krinsley, D.H. (1983) Mudrock examined by backscatter electron microscopy. Nature, Vol.301, pp.412-413.

White, S.H., Shaw, F.H. and Huggett, J.M. (1984) The use of back-scattered electron imaging for the petrographic study of sandstones and shales. Jour. Sedi. Petro., Vol. 54, No. 2, pp. 487-494.

Williams, M.L., Jercinovic, M.J. and Terry, M.P. (1999) Age mapping and dating of monazite on the electron microprobe: Deconvoluting multistage tectonic histories. Geology, Vol. 27, pp. 1023–1026.

Wilson, M.D and Pittman, E.D (1977) Authigenic clay in sandstone: recognition and influence on reservoir properties and paleoenvironmental analysis. Jour. Sedi. Petrol., Vol.47, pp. 3-31.

Winchell, A.N. (1951) Elements of optical mineralogy. Willey, John and Son publ, New York, Vol.2.

Hon'ble reviewer

1. **Prof. N.C. Pant**
   Departmnet of Geology
   Centre of Ad. Study
   Delhi University
   DeLHI - 110 007
2. **Dr. Istvan E. Sajo**
   Senior Scientist
   Chemical Research Centre
   Hungarian Academy of Science
   BUDAPEST (Hungary)

II-28

Tectonic evolution of Andaman Inner Forearc basin and its implication in hydrocarbon exploration

S.D. Dogra* and Anil Sood**

*General Manager, ONGC, 7th Floor Scope Complex, Lodhi Road, New Delhi 110 003

** Group General Manager-HOI- GEOPIC, KDMIPE Campus, Dehradun-248 195

E-mail: susheeldogra@gmail.com

Abstract

Based on the newly acquired 3D seismic data and integration of this data with existing well data and 2D seismic data in the inner forearc part of Andaman Oceanic Island Arc system, two phases of evolution, accretionary prism phase (upper Cretaceous to Oligocene) and forearc phase (Early Miocene to Recent) have been proposed for theAndaman Inner Forearc basin situated on SE Asian region of Eurasian plate.

In the accretionary prism phase, most of the sediments are trapped in the trench and are involved in the process of accretion onto the overriding plate which eventually leads to an outerarc formation towards the close of Oligocene. This phase is characterised by complex thrusting and folding towards the trench side and normal faultingand associated deformation towards the volcanic front side responsible for the large structural trap formation in the area. In this phase, a thick prism of sediment containing good source rock (Eocene) and potential reservoir rock (Oligocene turbidite) has been identified in the area around present day deepwater. This source and reservoir combination constitute the first speculative petroleum sysyem, i.e, Paleogene-Paleogene (!) in the basin with the critical moment for oil towards the close of Mid Miocene and make the paleogene as the prime target. These sediments have been deposited in the deep water setting in the form of pelagite and turbidite with input from Sibumasu continental mass for the lower part of prism (upper Cretaceous to Eocene) and from Bengal fan and Irrawady delta for the upper part of prism (Oligocene). The study envisages short distance and vertical migration in this phase. The migration remains a critical risk for the propects in this phase.

In the forearc phase, most of the sediments are trapped in the arc-trench gap. This phase has been divided into two stages, i.e, ponded fill stage (Miocene) and sag fill stage (Post-Miocene).

Ponded fill stage is inundated by a number of tectonic events and dominantly by a turbidite fill and mudflow deposits.The ponded fill stage is started with an extensional event leading to half graben formation on sedimentary sequence of previous phase and compartmentalisation of forearc basin. This was followed by a quiescent period when carbonate sedimentation in shallow water to outer shelf has taken place and then two compressional events, one towards close of Mid Miocene and other towards the close of Miocene leading to inverted structures and thin skin thrust-fold belt. In this stage, two important reservoirs (carbonate and clastic) in Miocene and one potential source rock have been deposited. With the reservoirs and source rocks of this phase and previous phase, two more petroleum system, i.e., Paleogene-Neogene (.) and Neogene-Neogene (!) have been identified. The study also envisages short distance and vertical migration in this stage. The Neogene source rock has just entered in the early phase of oil window in the area of study. Therefore, the Mid Miocene carbonate with charge from Paleogene source rock is considered as the primary target and Mid Miocene clastics are considered as the secondary target in the area with reservoir as well as charge as the critical risk.

Sagfill stage is characterised by the passive fill of the basin with pelagite and has little hydrocarbon potential.

Andaman forearc basin has hydrocarbon potential in Paleogene and Miocene sequences with hydrocarbon charge as the critical risk in accretionary prism phase and both charge and reservoiras the critical risk in fore arc phase. Prospects overlying the kitchen holds more promises than the prospect away from kitchen. Probable kitchen area in the forearc is located around the present day deepwater area. Therefore, prospects close to the present day deep water are worth probing.

Although the viability of the fore arc area as a hydrocarbon province remains dependent upon the factors that can be confirmed only by drilling well, the shear size of the structural traps, envisaged stratigraphy and disposition within the basinal framework are sufficient to mean that the area VIII is perhaps one of the most attractive candidates for exploration of thermogenic play in the Andaman fore arc irrespective of first well results in the area.

Keywords: Hydrocarbon Exploration, Andaman, Inner Forearc Basin, 3D Sesmic Study, Tectonics Indian Ocean.

Introduction

The Andaman basin is situated towards southeastern part of Bay of Bengal around Andaman-Nicobar chain of islands (between 6^{0}N to 14^{0}N Lat and 91^{0}E to 94^{0}E Long.) and is a part of Oceanic Island Arc system that extends from Sumatra (Indonesia) Islands in the south to Myanmar (Burma) in the north (Figure-1). The basin has six major geotectonic elements like, as identified from west to east, Trench, Inner Slope, The Outer Fore Arc, The Inner Fore Arc Basin, Volcanic Arc and Back Arc, each with different hydrocarbon potential.

In the Andaman basin, eleven prospects have been drilled, nine in the Outer Fore Arc High and one each on Inner Slope and Volcanic Arc. First prospect (A-9) drilled on the Outer Fore Arc area is having a very small hydrocarbon accumulation in Middle Miocene carbonate. Rest of the prospects were dry. All these prospects are situated in the shallow water part in the Andaman Sea. After the exploration setback in shallow water opportunities in Inner Slope, the Outer Fore Arc and Volcanic Arc area, the focus has now shifted to deep water frontier areas, the Inner Fore Arc Basin and Back Arc area, of Andaman Sea. ONGC has acquired a good quality of 3D seismic data in different campaigns (I, VII, III, VIII, IV, IX, V & VI) along the Inner Fore Arc Basin and Back Arc area. The present study has made use of these high quality seismic data of VIIIth campaign in particular and other seismic campaign in general, situated in the inner fore arc area, and demonstrates the structural prospectivity of the Inner Fore Arc Basinin the framework of its tectono-sedimentary evolution and source rock maturity modeling. The present study has also assessed the hydrocarbon prospectivity of the Inner Fore Arc basin on the basis of structural and stratigraphic analogy.

Structural Interpretation

The study area is a part of Andaman Inner Volcanic Arc, which is a asymmetric basin deepening towards the accretionary prism in the

West and shallowing towards Volcanic arc in the East (Figure-2). As per the present study the Inner Fore Arc is separated from the accretionary prism by a near vertical fault active during Neogene time after the deposition of Early Middle Miocene carbonate but gradually rises towards Volcanic arc. The Inner Fore Arc is dominated by extensional faults both longitudinal and transverse in nature. The transverse extensional faults compartmentalize the Inner Fore Arc basin into number of segment (Figure-3) with differential subsidence where as the longitudinal faults are of enechelon (left stepping) nature brought out for the first time in the present study. However, earlier study indicates that the adjoining geotectonic elements like the accretionary prism dominated by strike slip and thrust faulting (Ananthanarayan et al., 1981) back arc setting by strike slip and normal faulting and volcanic arc mainly by normal faulting.

Interpretation of 3D seismic data (horizon mapping, horizon slice of dip azimuth cube and coherence cube and time slices) in the study area brings out large structural prospects at Oligocene and Middle Miocene levels and two fault systems, i.e. Paleogene extensional fault system (NW- SE, N-S & NE-SW direction) and Neogene (Upper Miocene) compressional fault system oriented in NE-SW direction (Figure-4). There are imprint of structural inversion along Paleogene fault systems due to the compressional force active in the Miocene period. Neogene compression is thin skin in nature and has resulted into a south verging fold-thrust belt (Figure-5).

In Salinsub basin in Myanmar in the north and Timor Basin in the south of study area, two fault system i.e. extensional in older section and compressional in younger section are considered as the structural analogue.

Stratigraphic Interpretation

In the absence of well control in the inner fore arc area, it is difficult to provide a detailed stratigraphy of the area. However, based on the seismic data, the stratigraphy of the area has been divided into two mega sequences such as Paleogene and Neogene mega sequences. The base of the strong reflector corresponding to Early Middle Miocene carbonate base demarcates the Paleogene from Neogene in the area (Figure-6).

Paleogene Mega sequence: Paleogene is having low frequency, high amplitude, and discontinuous seismic reflection character. This undifferentiated Paleogene may be equivalent to Baratang Formation (upper Cretaceous to Eocene in age) and Port Blair Formation (Oligocene). This unit is quite thick and having reservoir and source rock potential in the drilled wells in the adjoining accretionary prism and inner slope basin (Figure-7). However, the base of this sequence is not seen in the drilled section. Similarly the Paleogene base is not clearly seen in the Inner Fore Arc area in the seismic data. It makes the interpreters in wrongly estimating the thickness of the Paleogene in the inner fore arc area. However, the present study assumes a large thickness of the Paleogene (Figure-8) of about of about 2200 to 6500min the study area and also expects reservoir and source rock potential. Paleogene sediments mainly consist of pelagites and turbidites.

The Paleogene analogue in the adjoining inner fore arc basins (Figure-9) indicates thick Paleogene (about 10 km thick in Saline subbasin in Myanmar in the north)

Neogene Mega sequence: Neogene is composed of Early Middle Miocene at the bottom followed by Mid Miocene MTC (Mass Transport Complex), Upper Miocene clastic and Post Miocene clastic unit. Neogene sediments consist mostly of deep water turbidities and pelagites. The Early Middle Miocene carbonate is about 80-880 m thick and is correlatable with the carbonate in the outer fore arc (gas bearing) and also with volcanic arc area. The MTC Unit (Middle Miocene) is characterized by transparent seismic character and is 400 to 1600 m thick in the study area. Post Miocene unit is not of interest from hydrocarbon point of view in the study area and appears to be passive fill of pelagic deposits with sub marine fan complex toward top. Proposed stratigraphy of the area is presented in Figure-10.

Reservoir Depositional Model

Well and seismic data indicate the presence of reservoirs at three stratigraphic levels (Paleogene clastics, Early Middle Miocene carbonate and Mid Miocene clastics). While constructing the regional model, the study has considered the regional geology of the Myanmar and Sumatran forearc basins and existing plate reconstruction (Daly et al., 1991, Curray, 1991) of Indian plate.

Paleogene clastic reservoirs: For building aclastic dispersal model, the Paleogene section has been divided into Baratang and Portblair Formations. Baratang Formation has been further subdivided into two sequences, one of Upper Cretaceous to Palaeocene age and the other of Eocene age.

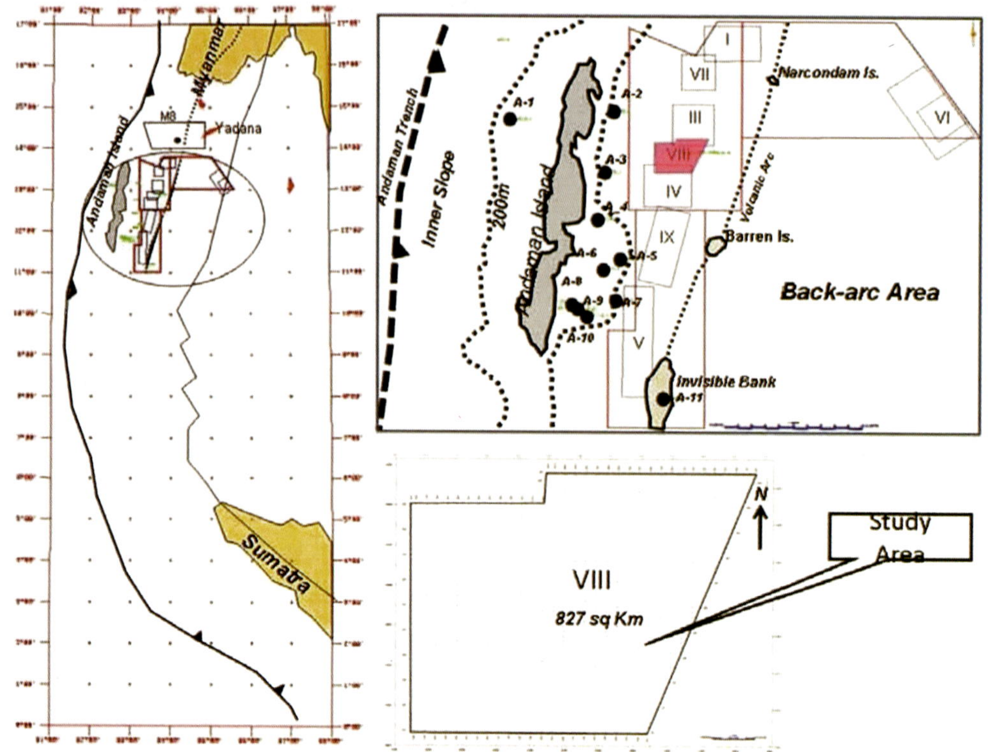

Fig. 1. Location of study area with respect to Andaman Island Arc System.

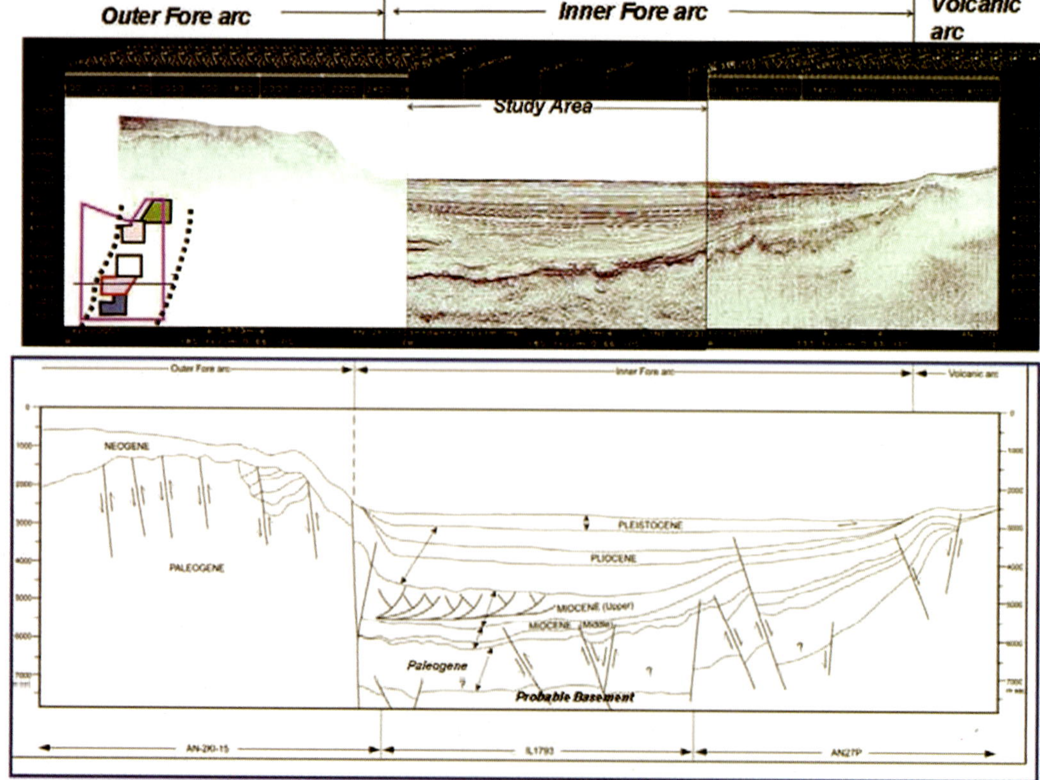

Fig. 2. Assymetric Andaman Inner Fore Arc Basin

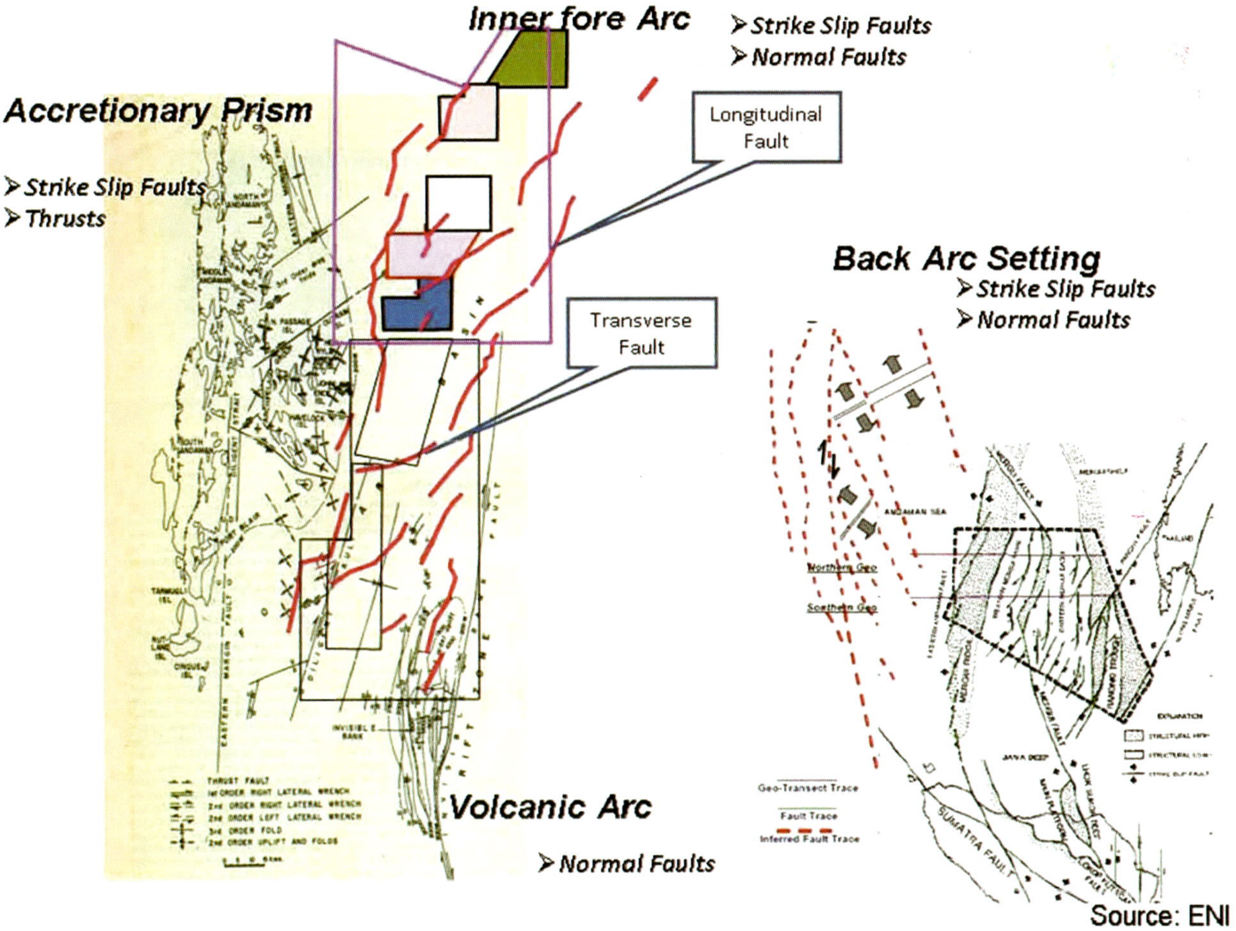

Fig. 3. Major Fault Trends in different geotectonic elements of Andaman Region

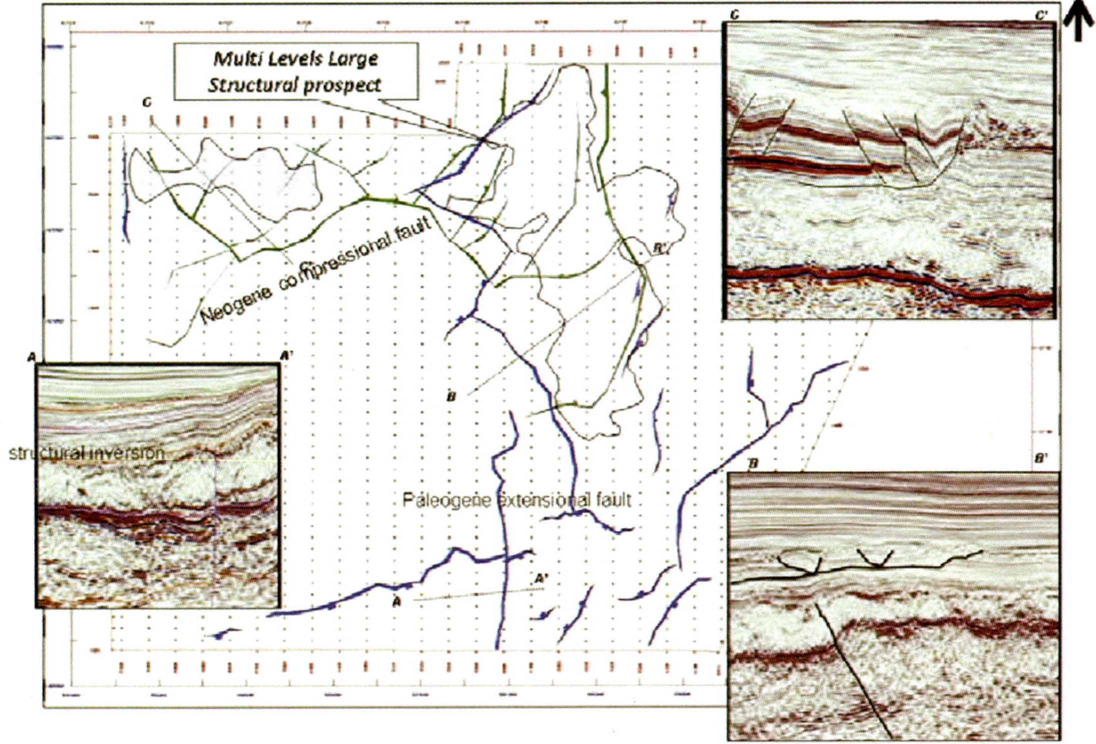

Fig. 4. Structural prospect and Fault Pattern in the Study area.

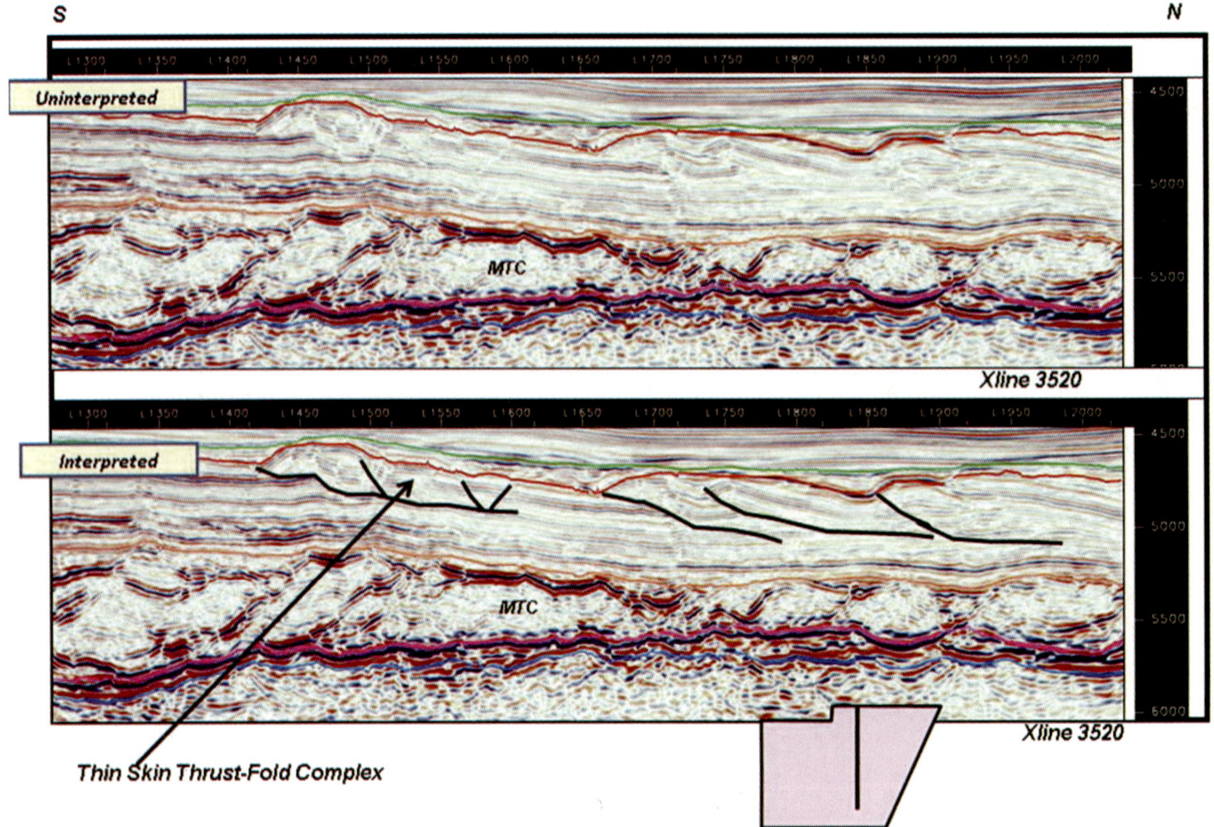

Fig. 5. Close Up-Compression in Miocene (Thin Skin Thrust-Fold Complex)

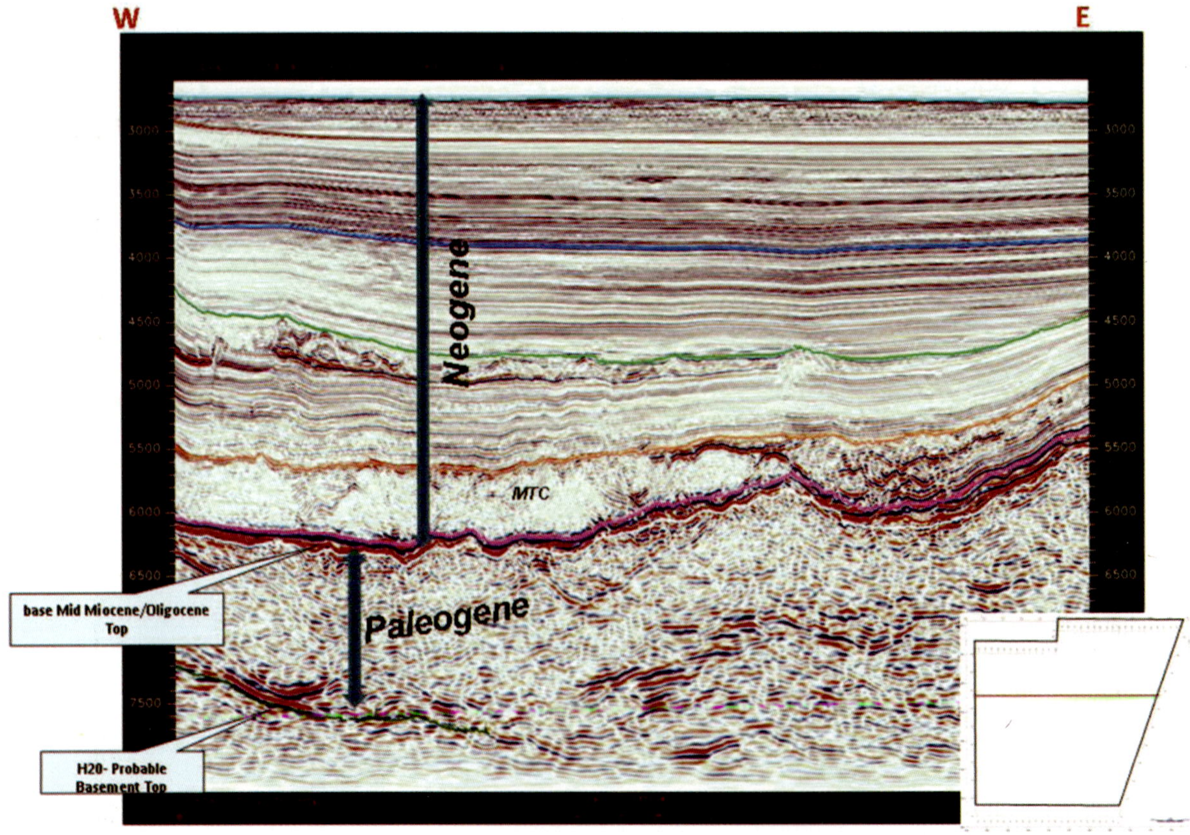

Fig. 6. Two fold stratigraphy of Inner Fore Arc Area

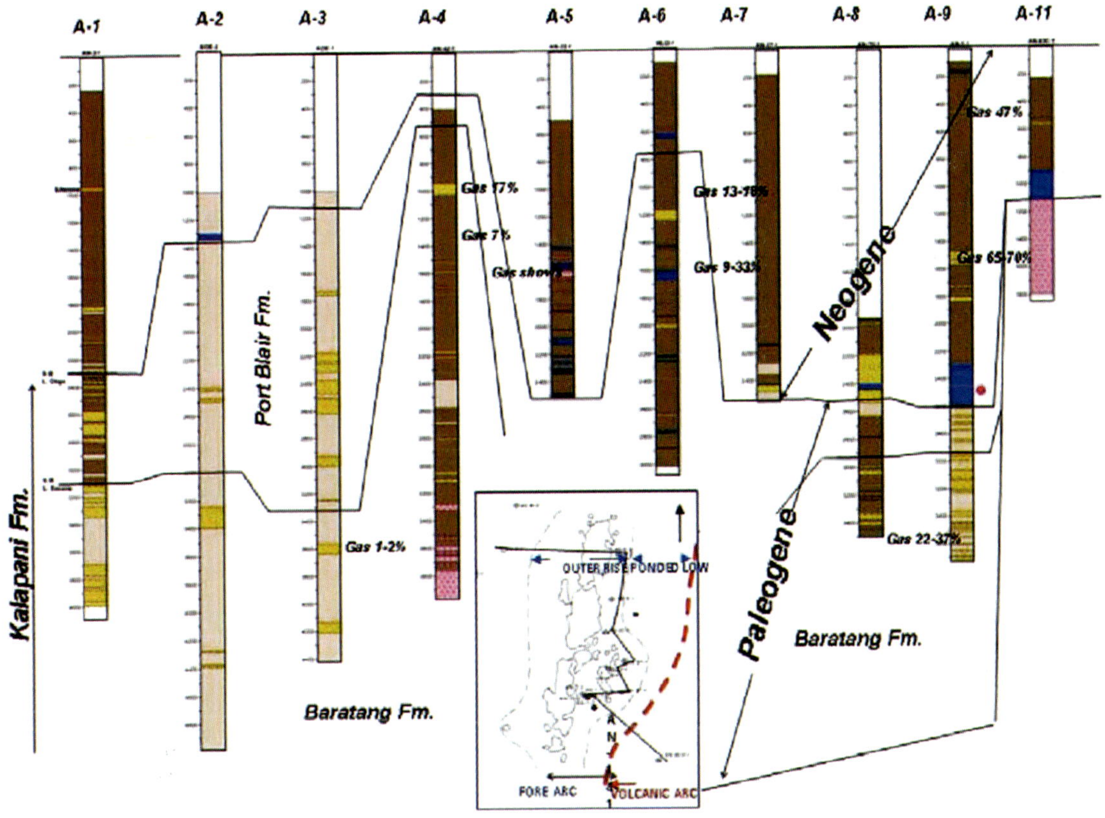

Fig. 7. Paleogene & Neogene Stratigraphy in Drilled wells.

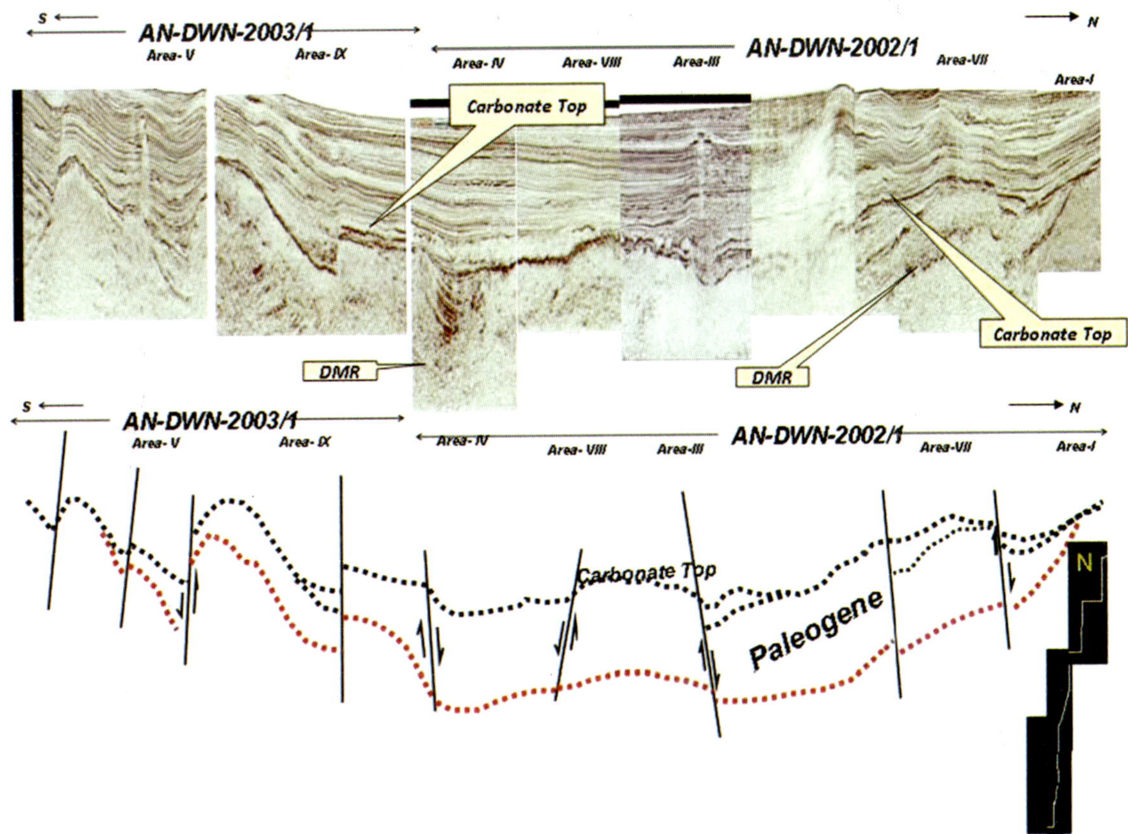

Fig. 8. Paleogene section in Inner Fore-arc area

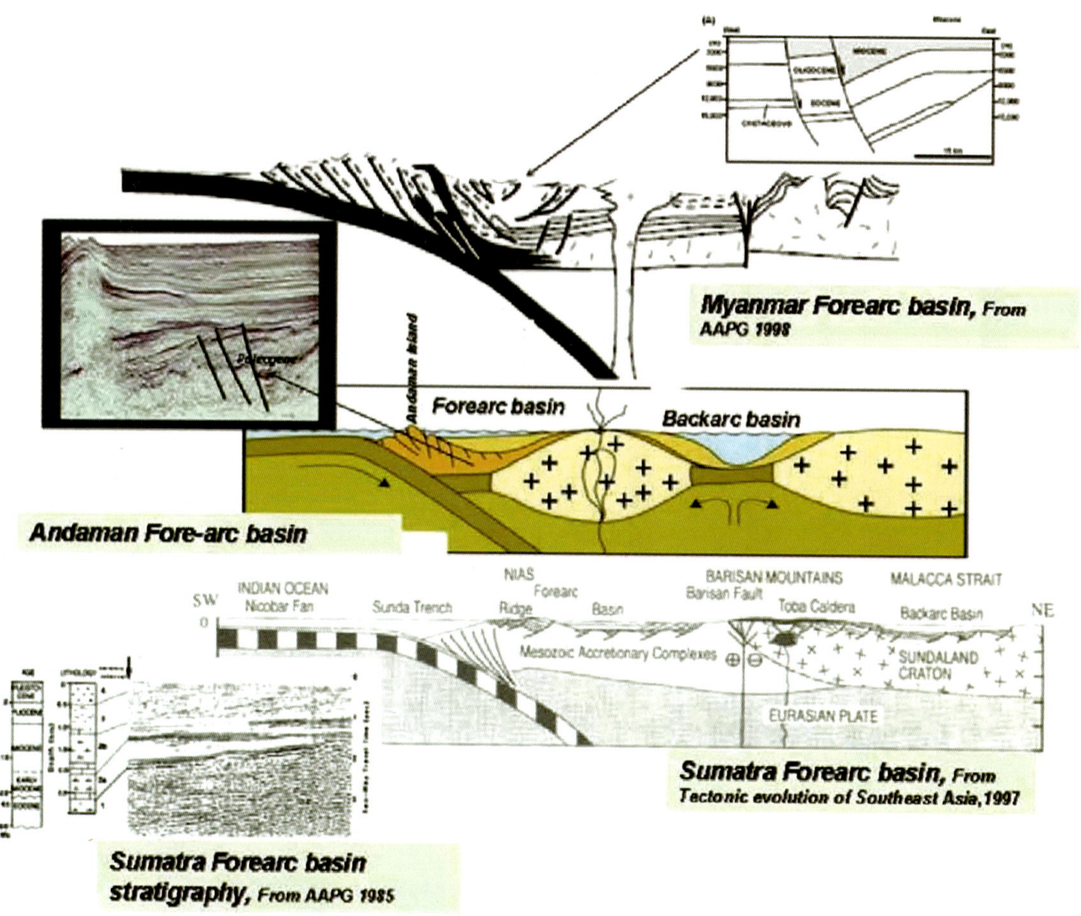

Fig. 9. Paleogene Analogue in Adjoining Fore-arc basin

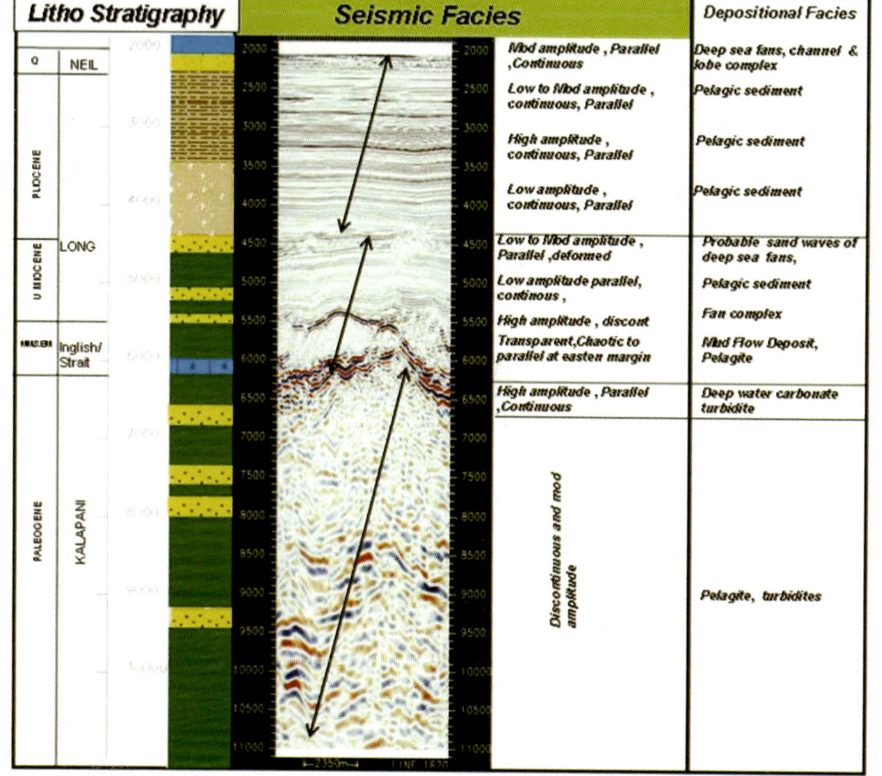

Fig. 10. Proposed Stratigraphy of Area

As per the earlier work, depositional set up for Baratang varies from shallow to deep water and provenance from Volcanic Arc/Sibumasu continental mass/Bengal and Nicobar fan. However, the existing plate reconstruction at the end of Palaeocene indicates that the Andaman area is a part of SE Asian plate and close to subduction zone between Indo-Australian plate and SE Asian plate of larger Eurasian plate. The above setting suggests, deep water set up for Upper Cretaceous-Palaeocene sequence of Baratang Formation with clastic input from North and NE (SE Asian plate, Figure-11a).

The existing plate reconstruction at the end of Eocene indicates the bending of subduction zone with north-south orientation adjacent to Andaman region. During Eocene, the Andaman region is a part of overriding plate and close to trench. Indo-Burmese range north of Andaman and an island chain came into existence during this time. The above setting, therefore, suggests deep water set up for Eocene sequence of Baratang Formation with clastic input from North(Indo-Burmese range) and East (island arc system) (Figure-11b).

Earlier work on sediment dispersal model for Portblair Formation indicates various depositional set up from Shallow water/Shallow to Deep Water /Deep Water & varying provenance as Bengal fan /Irrawaddy delta /NE /Bengal fan and Indo-Burmese range / Sediment from Himalayan.

However, as per the existing plate reconstruction at the end of Oligocene, the Andaman area is still situated close to subduction zone. Proto-Irrawaddy delta in north of Andaman and Bengal fan in NW had come into existence during this time. This overall association suggests a deep water set up for Oligocene sequence (Portblair Formation) with clastic input in the form of turbidite from North (Proto-Irrawaddy delta) for inner forearc and NW (Bengal Fan) for outer forearc area (Figure-11c).

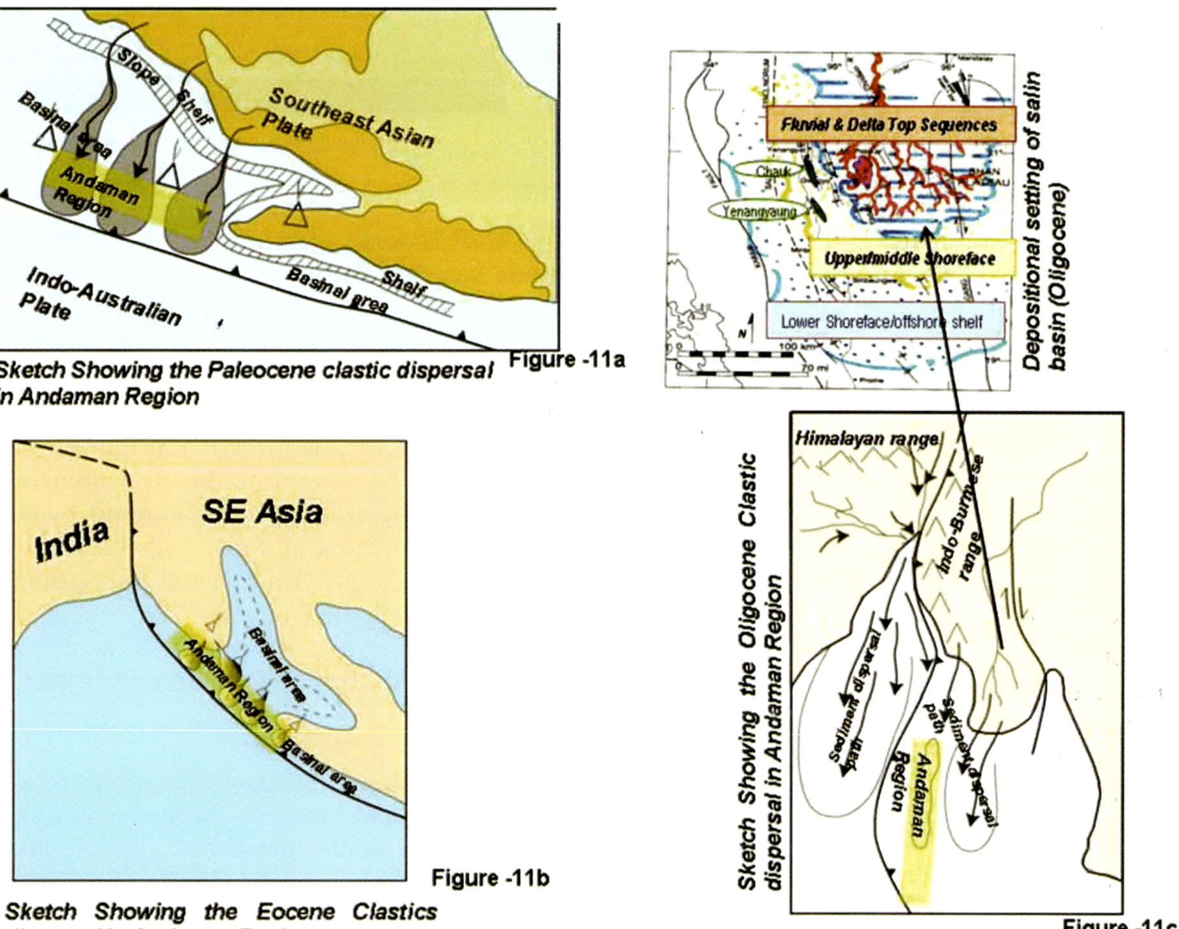

Fig. 11. Clastic dispersal pattern in Paleogene

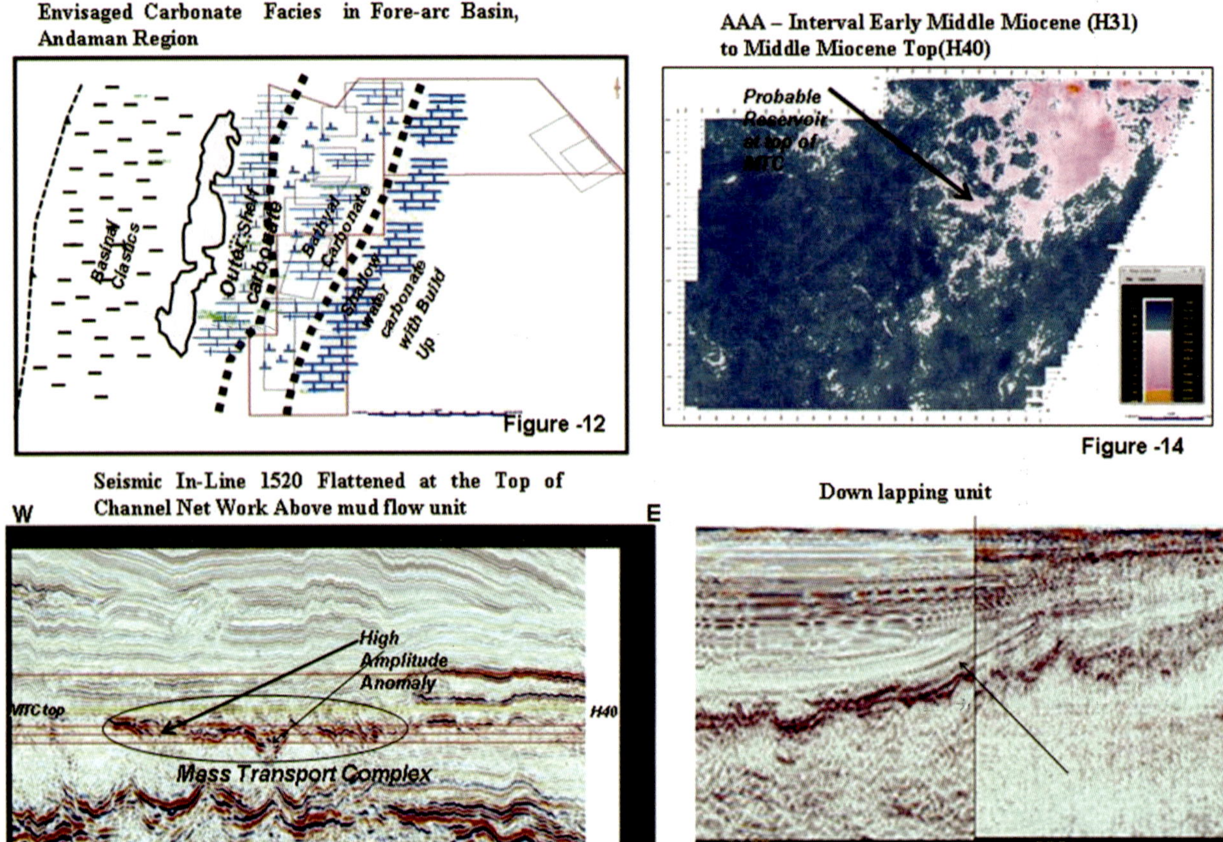

Fig. 12-15

Early Mid Miocene carbonate reservoir: Drilled well data show that there are two areas (A-9 and A-11) having good carbonate development. In other areas, the carbonate is represented by shale or shale with thin limestone bands. As per paleontological study this carbonate is of Early Middle Miocene age and paleobathymetry varying from middle to outer shelf to bathyal except for A-11carbonate which is of inner shelf to middle shelf.

This Early Middle Miocene carbonate has been encountered in exploratory well in the M-8 exploration block adjacent to our seismic campaign area I & Yadana field in North (Figure-1). Based on the analogues of adjacent area and geoscientific data, two types of carbonate facies (shallow water platform with build ups and deep water carbonate with off platform build ups) over a ramp profile have been envisaged in the forearc area of Andaman region (Fig.12). Isopach of Early Mid Miocene carbonate/or its equivalent facies indicates the presence of a thick carbona te /or its equivalent facies in the range of 100-600m.

Mid Miocene clastic reservoir: A high amplitude seismic anomaly towards the top of MTC unit is envisaged in the area (Figure-13). This may be a reservoir unit in the area. AAA attribute in the interval of MTC has brought the area of high amplitude unit in the form of a fan with input from NE direction (Figure-14). This reservoir is envisaged to be a submarine fan complex. This unit is down lapping over MTC probably indicating a force regression corresponding to late FSST/a normal regression unit corresponding to LST (Figure-15).

Tectonic evolution of Andaman Inner Fore Arc

The Andaman island arc system evolves from late Cretaceous to Recent through well known plate scale processes like subduction, magmatic intrusion and back arc spreading, leading to the development of various sub basins, viz. trench basin, inner slope basin, fore arc basin (outer fore arc and inner fore arc), intra volcanic arc basin and back arc basin. The present study has analyzed the development of inner fore arc basin and proposesa two phases of evolution, accretionary prism phase (Upper Cretaceous to Oligocene) and forearc phase (Early Miocene to Recent) on the basis of contrasting structural and stratigraphic characteristics of the basin.

The accretionaryprism phase: In incipient stage of Andaman island arc system, the subduction of Indo-Australian oceanic plate below Eurasian oceanic plate leads to the accretion of deepwater trench sediments as a prism (accretionary prism complex) onto the overriding Eurasian oceanic plate which eventually led to the formation of outer fore arc close to Oligocene. This initial period of growth of a thick accretionary prism qualify as a distinct phase (the accretionary prism phase) in the development history of inner fore arc basin. During this phase, the snout of lower plate does not subduct to such a depth to form a sub areal mountainous chain of volcanic arc of tholeiitic and basaltic type (Coleman, 1978). This accre-tionary prism forms the basal tectono-stratigraphic unit of Neogene fore arc basin. The Sumatran fore arc has a Mesozoic accretionary complex underlying the fore arc sequence (Figure-9).

In the inner fore arc of Andaman Island arc system, accretionary prism is of Paleogene age and is the first major event in the development of Andaman inner fore arc basin. The structure of this phase is characterized by complex thrusting and folding towards the trench side and perhaps much simpler extensional faulting and related deformation away from trench (Figure-9). There is high structural prospectivity in this phase. The Paleogene megasequence in this phase is having good reservoir being deposited in deep water setting in the form of submarine fan complex with input form SE Asian continental mass and Indo-Burmese range. This is also having a major source rock (Paleogene). During this phase the sediments are mostly land derived from the overriding plate and travel right upto the trench.

The forearc phase: In the fore arc phase, a structurally and stratigraphically different sequence overlies the accretionary prism. This sequence is thought to have been developed in the space between outer arc and volcanic arc after their full development. The period from the full development of outer arc and volcanic arc constitutes as the forearc phase. During this phase, the snout of lower plate has subducted to such depth to form a sub areal mountainous chain of volcanic arc of and esitic type. This phase has been divided into two stages (Figure-16) i.e., ponded fill stage (Miocene) and sag fill stage (Post-Miocene) based on the presence of a major unconformity at top of Miocene, nature of sediment fill and contrasting deformation style.

Ponded fill stage is characterized by a few tectonic events and dominantly consists of turbidite fill and mudflow deposits. The ponded fill stage has started with an extensional event leading (faults extending NW-SE, N-S & NE-SW direction) to formation of half graben filled with deep water clasticson sedimentary sequence of previous phase and east west compartmentalization of fore arc basin in lower middle Miocene. This was followed by a quiescent period when carbonate sedimentation with reservoir potential in shallow water to outer shelf has taken place and followed by compressional events, towards the close of Miocene leading to a thin skin thrust-fold belt (Figure-16). The post Early Middle Miocene sediments are mainly deep water clastics with a MTC having source rock potential at the bottom and reservoir (prograding unit) potential at the top.

The Sag fill stage during post Miocene is characterized by passive filling with mainly pelagite showing marine onlap.

Petroleum System

A preliminary study has been made to identify the elements in proved petroleum system and occurrence of other petroleum system if any.

Source rock: Source rock study of six wells (A-8, A-9, A-1, A-7, A-5, and A-4) indicates existence of two source rocks (Paleogene and Neogene) in the area. Paleogene source rock is generally having type-III kerogen with TOC values varying from 1.4-3.6% and HI values from 96-465 mg HC/g TOC. Neogene source rock generally having type-III kerogen with TOC values varying from 1.4-1.6% and HI values of 150 mg HC/g TOC.

Reservoirs: Reservoirs at three stratigraphic levels are anticipated. Paleogene coarser clastic is potential reservoir in the area and is distributed within Baratang Formation (lower unit of Paleogene) and Portblair Formation (upper unit of Paleogene). These upper cretaceous to Paleogene coarser clastics (sands) have been tested in well A-1 showing poor reservoir character. However, these reservoirs are primary target in the Myanmar inner fore arc basin. The upper Oligocene being a global regression, good coarser clastics (sands) reservoirs in the upper part of Oligocene (Portblair Formation) having most of the oil fields in inner fore arc setting in Myanmar are anticipated to form potential reservoir in the study area.

Early Middle Miocene carbonate is another potential reservoir in the area and is well developed in A-9 area and A-11. The porosity of this reservoir varies from 7-14% in A-9 area to 25% in A-11. This reservoir is having gas accumulation in A-1.

The Mid Miocene clastic is another reservoir corresponds to a seismic anomaly at the top of MTC unit.

Seal: Since the Andaman fore arc basin has evolved under deep water set up in its most of depositional history, the stratigraphy of the area is dominated by finer clastics. The 80% of the stratigraphic column of drilled wells are finer clastics. Therefore, there is no dearth of top seal in the area.

Generation-Migration-Accumulation: Petromod-II 2D application has been used for the 2D genetic modelling. A cross section of thirteen layers with two source rock, three reservoirs and corresponding seals are considered as the input for the G-M modelling (Figure-17).

For constraining heat flow values, vitrinite reflectance data and present day temperature data of well A-3 located on western side of block in outer fore arc was used. Petromod-II 1D application has been used for estimating the thermal history.

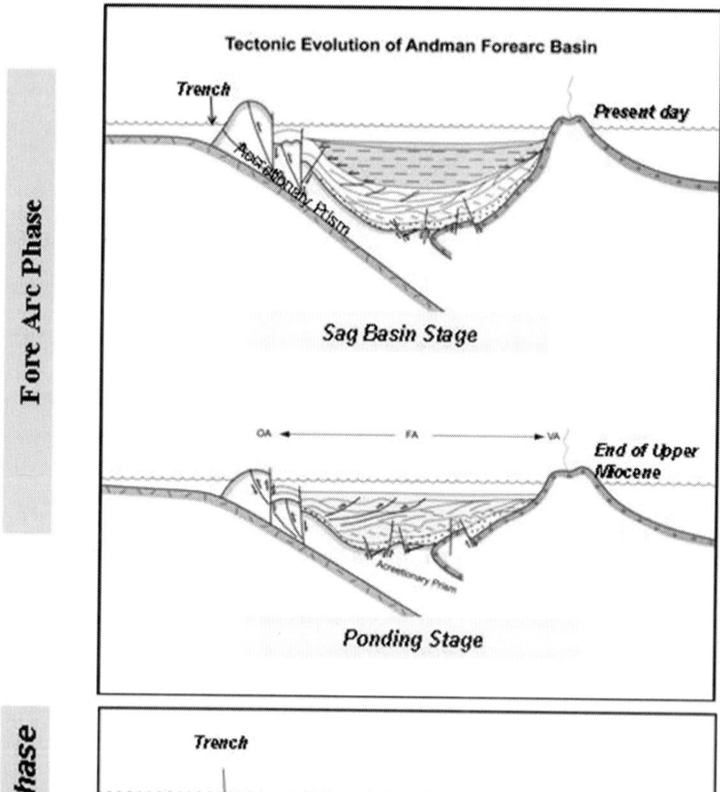

Fig. 16. Tectonic Evolution of Andaman Fore Arc Basin

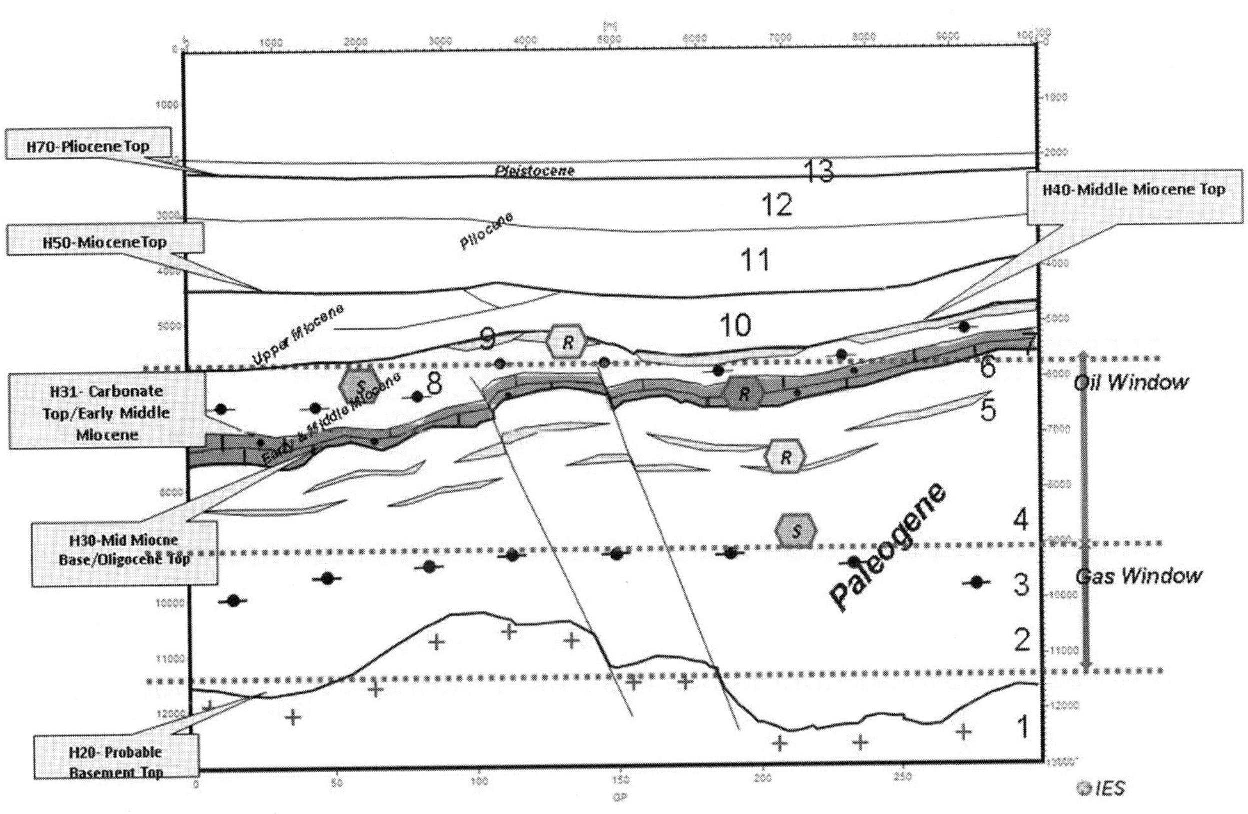

Fig. 17. Cross section of petroleum System Elements.

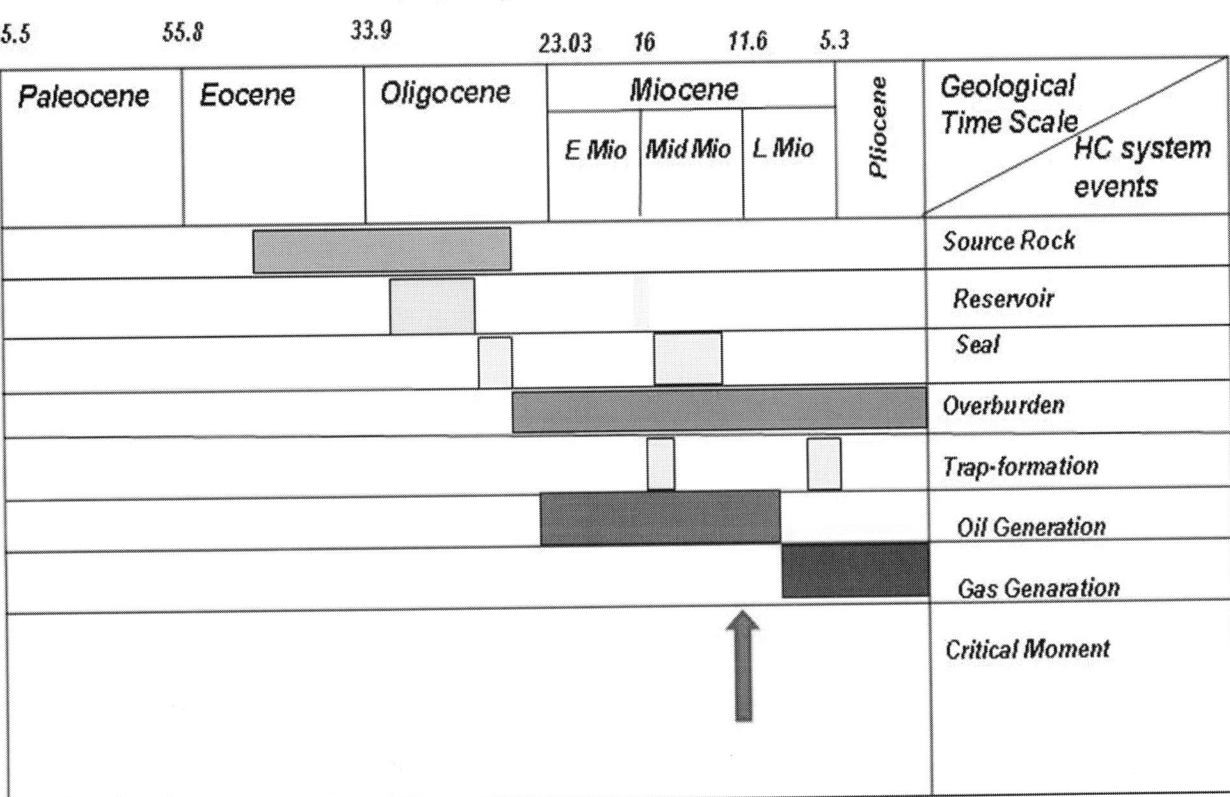

Fig. 18. Palaeogene Petroleum System Event Chart

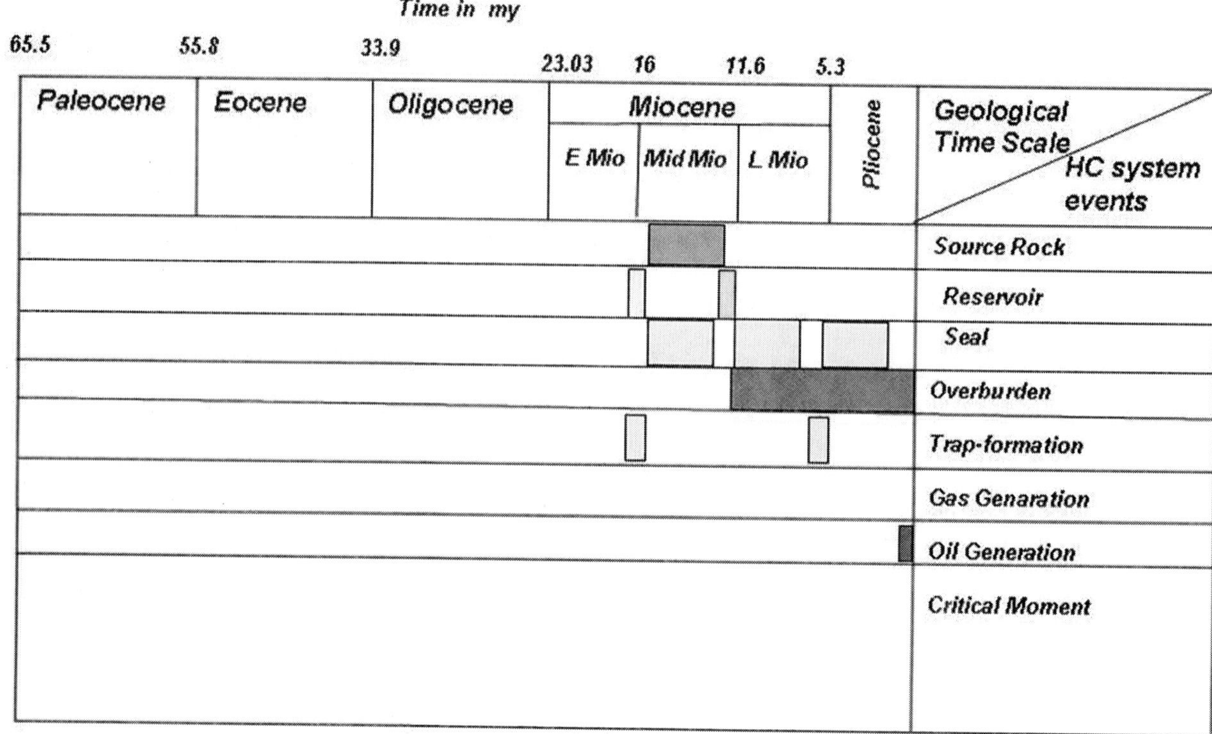

Neogene Petroleum System Event Chart

Fig. 19. Neogene Petroleum System Event Chart

The estimated geothermal gradient for Andaman deepwater area is in the range 1.5°–2°C/100m. Heat flow values 55mW/m^2 are used for period 40MYbp to 15Mybp, than reduced to 50mW/m^2 at 10 Mybp to 35 mW/m^2 at present. Thermal history for the period 40-65 Mybp is assumed to be same as at time 40 Mybp. Hence, heat Flow value of 55mW/m^2 during this period appears to be reasonable as this is close to the global average for forearc setting.

With the above input in the Petromod, it is observed that the middle part of Paleogene source rock has entered in onset of hydrocarbon generation at 23-26 Mybp at the corresponding paleo depth of source rock 3500-3600m. Later, it entered in peak oil stage during the period 12-14 Mybp with the corresponding paleo depth of source rock 5600-6100m. It has entered in the wet gas stage from 6-11 Mybp. The present day transformation ratio for this layer varies between 72-91%.

The upper source rock layer of Middle Miocene age has attained maturity for generation of hydrocarbons at places and present day depths are around 6500-6800m. The present day transformation ratio of this layer varies between 6-11%.

On the basis of above study, two petroleum system event charts one for Paleogene and another for Neogene have been prepared (Figure-18 & 19). Paleogene petroleum system event chart indicates that Paleogene source rock is likely to charge Paleogene reservoirs and Middle Miocene reservoir.

Neogene petroleum system indicates that in some part Neogene source rock has entered in early stage of oil window and might have entered in the peak stage in other part of fore arc basin.

The present study therefore indicates existence of three petroleum system such as

Paleogene-Paleogene (!), Paleogene-Neogene (.) and Neogene-Neogene(!)

Conclusions

1. Two phases of evolution in inner forearc basin, accretionary prism phase (Upper Cretaceous to Oligocene) and forearc phase (Early Miocene to Recent) have been proposed.
2. In the accretionary prism phase, most of the sediments are trapped in the trench and are involved in the process of accretion onto the

overriding plate which eventually leads to an outer arc formation towards the close of Oligocene. This phase is characterised by complex thrusting and folding towards the trench side and simpler normal faulting and associated deformation away from trench responsible for the trap formation in the area. Large structural prospectivity exists in this phase. A thick Paleogene section with a combination of source and reservoir rock is envisaged around area VIII, IV, III and VII. Input for lower part of Paleogene (Baratang Formation) is mainly from south East Asian land mass with subordinate amount from volcanic arc. Input for upper part of Paleogene (Portblair Formation) is mainly from proto-Irrawaddy delta in the fore arc area and from Bengal fan in the outer arc area.

3. Fore arc phase is divided into two stages namely Ponded fill stage (Miocene) and Sag fill stage (Post Miocene) based on a major unconformity at top of Miocene, nature of sediment fill and contrasting deformation style. Two primary reservoirs in Middle Miocene are anticipated. For Middle Miocene carbonate two depositional setting corresponding to two structural units, shallow water in volcanic arc and deepwater in Fore arc have been envisaged. Shallow water carbonate over the volcanic arc is extensive from Sumatra to Myanmar offshore area as evidenced by the presence of shallow water carbonate in Sumatra, AN-11(Andaman) and Yadana field in Myanmar. This carbonate is considered as the main source for redeposited deep water carbonate in fore arc. The Middle Miocene clastic reservoir is envisaged as a submarine fan complex. This phase is also associated with number of structural prospect.
4. Two probable source rocks, one in Paleogene and other in Neogene (Middle Miocene) are encountered in the drilled wells in the adjoining outer arc area. Paleogene and Neogene source rocks are of Type III with average TOC of 2% and HI of 200mgHC/gTOC.
5. 2D migration and generation model indicate existence of three petroleum systems, one known and two speculative i.e Paleogene-Neogene(.), Paleogene-Paleogene(!) and Neogene-Neogene(!).
6. Paleogene source rock has entered in peak oil stage during the period 12-14 Mybp. The upper source rock layer of Middle Miocene age has just attained maturity for generation of hydrocarbons at places.
7. Hydrocarbon charge by short distance migration has been invoked in the basin with long distance migration as the critical risk factor. Therefore, prospects overlying the kitchen hold better promises than the prospect away from kitchen. Probable kitchen area in the fore arc is located around the areas-IV, VIII and III.
8. Although the viability of the fore arc area as a hydrocarbon province remains dependent upon the factors that can be confirmed only by drilling well, the sheer size of the structural traps, envisaged stratigraphy and disposition within the basinal framework are sufficient to mean that the area VIII is perhaps one of the most attractive candidates for exploration of thermogenic play in the Andaman fore arc irrespective of first well results in the area.

Acknowledgments

The authors are indebted to Shri D.K. Pande, the than Director (Exploration) for assigning the G&G project in area VIII of Andaman deep waters to CEC-OG, New Delhi. The present paper is an outcome of this study. The authors are thankful to Shri S.R. Mangotra, DGM(C) for his help rendered in 2D genetic modelling.

References

Ananthanarayan P V, C M Ramdev, K V S Murti, Andaman Island Arc Wok shop on Geological Interpretation of Geophysical data, 1981, KDMIPE, ONGC.

ColemanPatrick J.,A Model of Arc Development, with Doubts and Questions: an Editorial offering, Bull. Aus. Soc. Explor. Geophys.V. 9. # 3. August 1978.

Curray Geoseph P., Geological History of Bengal Geosync line, Jour Assoc Expl Geophys., October 1991, V. XII, No. 4, p.209-219.

Daly M C, M A Cooper, I Wilson, D G Smith, B G D Hooper, Cenozoic plate tectonics and basin evolution in Indonesia, Marine and Petroleum Geology, 1991, Vol 8 February.

Moore Desiree Beaudry and Gregory E, Seismic stratigraphy and Cenozoic evolution of West Sumatra Fore Arc Basin Bull AAPG V 69 No 5 (May 1985) pp742-759

Pivinik D A, J Nahm, R S Tucker, G O Smith, K Nyein, M Nyunt, P H Maung, Polyphase Deformation in a Fore-Arc/Back-Arc Basin, Salin Sub basin, Myanmar (Burma), AAPG Bulletin, V. 82, No. 10 (October 1998), p.1837-1856.

Tectonic evolution of Southeast Asia, 1997.

Hon'ble reviewer

1. **Prof. K.S. Valdiya**
   Padam Shree
   Jawahar Lal Nehru Centre
   Ad. Scientific Research
   Jakkur
   BANGALORE - 560 064
2. **Prof. V.P. Dimri**
   Distinguish Scientist
   National Geophysical Res. Institute
   Uppal Raod
   HYDERABAD - 500 007

II-29

Palynofacies Proxies and Hydrocarbon Potential of Tertiary Rocks of Himalayan Foreland Basin, Northwest Himalaya, India

N.N. Dogra[1], O.P. Thakur[1] and Y. Raghumani Singh[2]

[1]Department of Geology, Kurukshetra University, Kurukshetra-136 119, India
2 Department of Earth Science, Manipur University, Imphal, Manipur - 795 003
E-mail:ndogra@rediffmail.com

Abstract

Palynofacies analysis acts as a potential technique for hydrocarbon prospecting and source rock evaluation studies. Fossil spore-pollen coloration and type and frequency of palynofacies are used to assess the thermal maturation level the sedimentary organic matter is subjected to and hydrocarbon source rock potential. In palynofacies spectrum, the category of gas-prone material includes brown, translucent, structured organic matter. The most important constituent of oil-prone material is amorphous organic matter present in the sedimentary rocks. The present paper is aimed at delineating the hydrocarbon source rock potential of Tertiary rocks of Himalayan Foreland basin based on palynofacies characterization work undertaken by the authors of this paper and published record of palynofacies and hydrocarbon source rock evaluation studies on these Tertiary rock sequences. The palynofacies maturation data based on quality, quantity and Thermal Alteration Index (TAI values) for majority of these sequences suggest a moderate to good source potential for hydrocarbons.

Keywords: Palyno facies, Hydrocarbon Exploration, Himalayan Foreland BAsin, Tertiary Rocks.

Introduction

The frontal thrust-fold belt of Himalaya is an example of compressional basin which has attracted the attention for hydrocarbon prospecting and exploration. The assessment of generation of hydrocarbons depends upon quality of organic matter in source rock and its degree of thermal alteration. It is now well understood that future trends in hydrocarbons exploration require integration of palynofacies data with the other geoscientific data to play an important role in cost effective prospecting and exploration of hydrocarbons, which warrant preparation of palynofacies map for all the 26 sedimentary basins of India. In the present context palynofacies analysis studies are focused in the Himalayan Foreland basin of frontal thrust-fold belt of Himalaya.

Himalayan foreland basin is subdivided into sub-basins by basement ridges and spurs (Raiverman et al., 1983). In the Subathu sub-basin, the foreland basin sediments consist of Subathu, Dagshai and Kasauli formations and the Neogene Siwalik Group (Gansser, 1964; Bhatia, 1982). The Subathu Formation is composed of shallow marine facies and consists predominantly of green mudstone, splintery shales with subordinate sandstones and limestones (Mathur, 1978). Uncommon red beds are interbedded in Simla Hills succession (Mathur, 1979). This basin started with a shallow marine environment eventually changed to estuarine and deltaic facies with time. By Mid-Miocene continental sedimentation marked by fluvial environment dominated the scene and this set-up has continued to the recent with minor modifications. The geological set-up of the Tertiary rocks of northwest Himalaya depicting the outcrop pattern of various lithounits is shown in Figure 1.

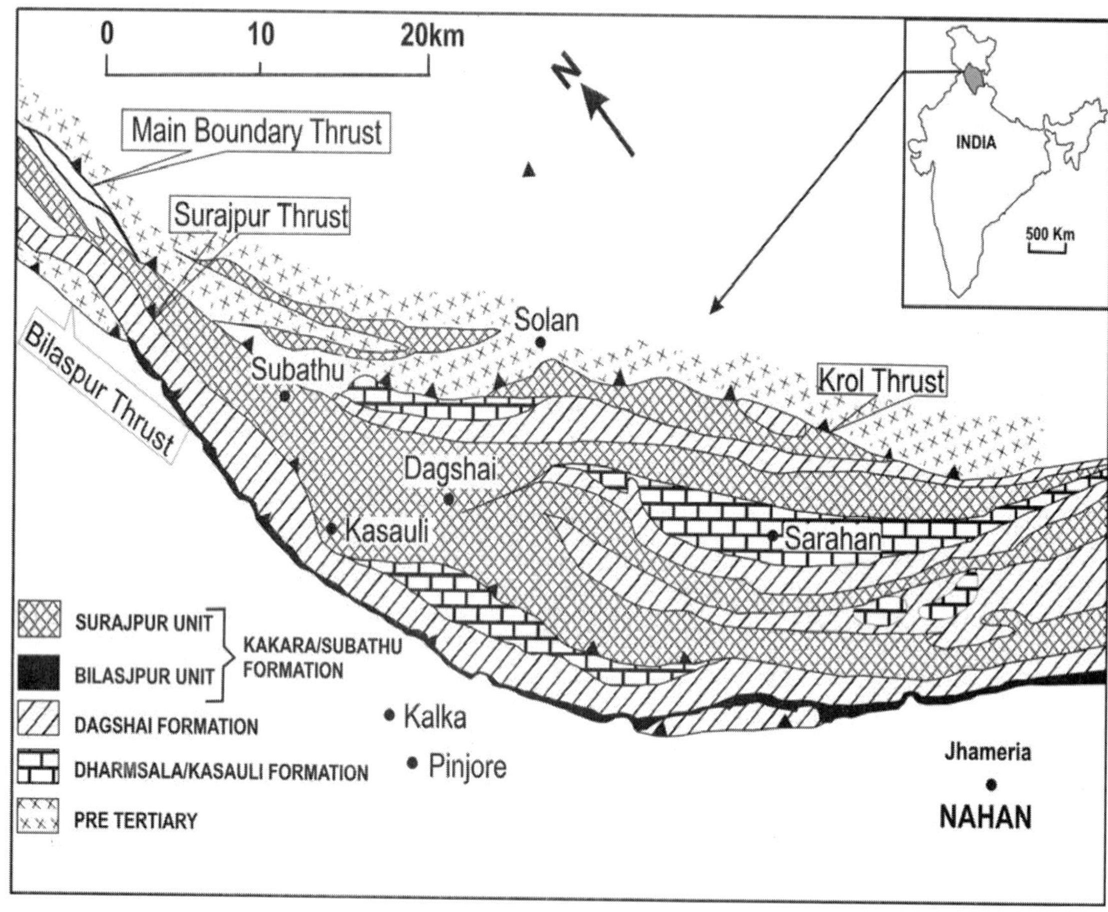

Fig. 1. Geological map of Tertiary rocks of NW-Himalaya depicting various Tertiary lithounits (After Thakur & Dogra, 2011)

Table 1. Exinite Colour, Thermal Alteration Index and Thermal Facies (Based on Staplin, 1977)

EXINITE COLOR	TAI	THERMAL FACIES	PRODUCT
Colourless-Pale Yellow-Yellowish Green	1.0 or Less		
Yellow-Light Brownish-Yellow	1.0^{+}-2.0	Immature	Mainly Biogenic Methane
Light Brown to Medium Brown	2.0^{+}-2.5	Transitional	Early stages of Oil generation
Brown	2.5-3.0	Mature	Oil generation active production
Dark Brown	3.0^{+}-3.5	Transitional	Condensate and Wet Gas
Brown Black-Black	3.5-5.0	Metamorphosed	Dry Gas or Barren

Table 2. Hydrocarbon Source rock Potential of Tertiary rocks from different localities of Northwest Himalaya.

Workers	Localities	Formations	Hydrocarbon Potential
Berry (1989)	Sediments exposed in Simla Hills from three energy sequences viz., Dharampur, Kumarhatti and Makreri	Subathu Formation	Sapropelic organic matter constituted the dominant type in Darampur and Kmarhatti energy sequence, while Makreri is dominated by humic matter and TAI values (2.5-2.75) suggested that the sediments are within the hydrocarbon generation phase.
Berry (1994)	Subsurface sequences of Nurpur-A and Suruinsar-A well of Himachal Pradesh and Jammu & Kashmir areas respectively	The Upper Dharamsala/ Upper Murree and Lower Siwalik sediments	Poor source potential for hydrocarbons
		Lower Dharamsala/Lower Murree and Subathu Formations	Good source potential for hydrocarbons.
Misra and Pundir (1994)	Subsurface sequence encountered in the well Jawalamukhi-B	Subathu to Middle Siwalik sediments	TAI (2.75-3) and the type of organic matter recorded suggested very good source potential for Palaeocene to Lower Eocene sediments (Subathu-Lower Dharamsala?). TAI (2.5-2.75) recorded in Lower Dharamsala-Middle Siwalik sediments are rated immature for generation of hydrocarbons.
Berry *et al.* (1998)	Batalghat-Bhararighat and Samog-Malhoti of Bilaspur area in Himachal Pradesh	Subathu Formation	Thermal alteration index value for the samples is 3.00. The facies recorded from these traverses are sapropelic-inertinitic and the source potential for hydrocarbon is moderate type.
Thakur & Dogra (2011)	Marhighat area of Sirmaur district, Himachal Pradesh	Subathu Formation	The lower and middle part of the Marhighat succession is dominant in amorphous organic matter whereas the upper part is dominant in black debris (charcoal). The lower and middle part of the sequence has sapropelic organic matter facies with adequate maturity (TAI 3.0) to generate liquid hydrocarbons, while the upper part of the sequence has sapropelic humic–charcoal facies with TAI 3.0, which is considered to be moderate to good for gaseous hydrocarbon generation.

PALYNOFACIES ANALYSIS AND HYDROCARBON SOURCE ROCK EVALUATION

The term Palynofacies was introduced by the French geologist Combaz in 1964. Palynofacies studies examine the preservation of Particulate Organic Matter and palynomorphs in sedimentary rocks. Palynofacies analyses have been discussed at length by Batten (1981, 1982), and Tyson (1993, 1995) among others.

Source Rocks:

Sediments that have been able to generate petroleum or which may have the potential to do so are called Source rocks. Organic rich shales, mudstones and very fine grained carbonates deposited in oxygen deficient conditions are considered excellent source sediments.

Types of Organic Matter:

The qualitative aspects such as type of organic matter and its facies are considered very important while evaluating source rock potential. These organic matter types must be distinguished and identified, for different types of organic matter have different hydrocarbon potentials and products.

Various laboratories use their own classification of organic matter. The important contributions on such classifications are those of Bujak *et* al., 1977; Burgess, 1974; Tissot *et al.*, 1974; Masron & Pocock, 1981; Venkatachala, 1981, 1984; Hart, 1986; Pocock *et al.*, 1987; Batten, 1996 and Tyson, 1994. The classification of dispersed organic matter proposed by Masron & Pocock, 1981 is briefly outlined as under:

Structured terrestrial organic matter: This organic matter comprises unaltered cellular remains of leaf, root and stem tissues. It contributes gaseous hydrocarbon.

Spore and Pollen: Spore and pollen are known to be rich in lipids, which contribute mostly to liquid hydrocarbons.

Biodegraded terrestrial organic matter: This matter is resulted by the microbial and bacterial biodegradation of structured terrestrial organic matter. It ranges from semi-amorphous to amorphous and is considered to possess enhanced hydrocarbon source potential than structured terrestrial. This category is mostly associated with deltas, bogs, swamps and such other environments.

Charcoal (Black debris): This black matter is the product of oxidation of structured materials. It is composed of almost pure carbon mixed with some mineral matter. It has no hydrocarbon source rock potential.

Fungal remains: Fungal remains are the major agents of biodegradation of organic matter eventually converting it to an amorphous mass.

Resin: Dried sap of many gymnospermous and a few angiospermous plants, in addition to the true resins and latexes are included in this category. Resin is considered good source of liquid hydrocarbons. It is found abundantly in fluvio-lacustrine, deltaic, estuarine and near shore deposits and commonly associated with rich plant debris.

Amorphous organic matter: This category is the final visible structureless product of biodegradation. In general, dark amber to dark brown amorphous matter is assumed to have been derived from higher vascular plants. Many algal material and certain lower plants produce a lighter amorphous product. According to Pocock *et al.*, 1987, the amorphous matter can be divided broadly into two types: 1) Amorphous matter of pale yellow to deep amber colour resulting from the aerobic bacterial activity 2) Amorphous matter of a neutral grey to dark brown colour formed by the action of anaerobic (reducing) bacteria.

Biodegraded aqueous organic matter: This category includes mostly altered algal remains sourced from marine material. Aqueous matter is easily convertible into variously textured amorphous mass and forms excellent source material for hydrocarbons.

Structured aqueous organic matter: It includes mostly algal material belonging mostly to shallow marine to marine environments. The category is important from both source rock character and environmental point of view.

The different types of organic matter identified in the sedimentary sequence are broadly classified into: Humic and Sapropelic facies from the hydrocarbon source point of view (Masron and Pocock, 1981). Humic organic matter is gas prone and it is land or terrestrially derived and is easily recognizable. Sapropelic organic matter is derived both from marine and terrestrial sources and is oil prone.

Thermal Maturation:

Thermal maturation of source facies is one important aspect to assess the hydrocarbon potential of source rock. Maturity is a measure of time-temperature history experienced by the organic matter within a sedimentary sequence and can precisely be defined as the process that generates the full spectrum of hydrocarbons as a result of 'thermal alteration. The visual assessment of Thermal Alteration Index (TAI) values based on the colour of the spores-pollen, cuticles, dinoflagellate cysts and sapropelic palynodebris has proved to be a rapid and fruitful technique to measure thermal alteration level of the organic matter. Generally these colour changes are from colourless to yellow to brown to black. The thermal alteration Scales practiced in various laboratories are different, but 1-5 scale of Staplin (1977) is one of the preferred scale for TAI estimation. Immature, mature, transitional, metamorphosed and thermal facies can be demarcated according to this scheme as shown in Table 1.

Sample Preparation for Palynofacies Analysis:

Standard maceration technique (Faegri and Iversen 1989) for palynofacies involves digestion with hydrochloric acid and hydrofluoric acid. The nitric acid treatment is avoided to get unoxidised residue. The macerate is thoroughly washed and smeared on zero no.cover glass and mounted on slides using polyvinyl alcohol and Canada-balsam. The quantitative estimation of the organic matter is made by visual estimation and counting. Thermal Alteration Index (TAI) values are assessed on the 1–5 scale of Staplin (1977).

HYDROCARBON SOURCE ROCK POTENTIAL OF TERTIARY ROCKS

As regards the detailed palynofacies analysis of Siwalik and Pre-Siwalik Tertiary rocks of the foreland basin is concerned, much emphasis has not yet been given by the earlier workers. However, a few published records on palynofacies and hydrocarbon source rock evaluation studies of the Tertiary rocks (Table.2.) are those of Berry (1989, 1994), Misra and Pundir (1994), Berry et al. (1998) and Thakur & Dogra (2011).

Berry (1989) carried out palynofacies and thermal alteration studies on Subathu sediments exposed in Simla Hills from three energy sequences viz., Dharampur, Kumarhatti and Makreri exposed in Talheri Ki Nadi, Nora Khondal Chakli Khud, Barog Kumarhatti, Koshalia River and Kasauli-Dharampur traverses. Sapropelic organic matter constituted the dominant type in Darampur and Kumarhatti energy sequence, while Makreri is dominated by humic matter and TAI values (2.5-2.75) suggesting thereby that the sediments are within the hydrocarbon generation phase.

Berry (1994) carried out palynofacies studies on Subathu-Lower Siwalik subsurface sequences of Nurpur-A and Suruinsar-A well of Himachal Pradesh and Jammu & Kashmir areas respectively. The palynofacies maturation studies based on quality, quantity and TAI values suggested that the Lower Dharamsala/Lower Murree and Subathu Formations contain good source potential for hydrocarbons. The Upper Dharamsala/Upper Murree and Lower Siwalik sediments have poor source potential for hydrocarbons.

Misra and Pundir (1994) carried out studies on palynofacies and Thermal Alteration Index (TAI) on subsurface sequence encountered in the well Jawalamukhi-B for source rock evaluation of Subathu to Middle Siwalik sediments. They have recorded twelve humic, sapropelic and mixed facies zones in the sequence. Biodegraded organic matter is the first dominant organic matter followed by charcoal, amorphous organic matter and wood/cuticle. TAI (2.75-3) and the type of organic matter recorded between 6727-5298m, suggested very good source potential for Palaeocene to Lower Eocene sediments (Subathu-Lower Dharamsala ?). TAI (2.5-2.75) recorded between 928-5298m in Middle Eocene to Middle Miocene (Lower Dharamsala-Lower Siwalik) . Middle Siwalik sediments are rated immature for generation of hydrocarbons.

Berry *et al.* (1998) also carried out the studies for hydrocarbon potential of Subathu sediments. Out of twenty one traverses, two traverses i.e., Batalghat-Bhararighat and Samog-Malhoti of Bilaspur area in Himachal Pradesh were studied. The total organic matter recorded was poor to moderate, however a few samples were having good concentration of organic matter. The facies are dominant of amorphous organic matter followed by biodegraded organic matter. Thermal alteration index value for the samples is 3.00. The facies recorded from these traverses are sapropelic-inertinitic and the source potential for hydrocarbon is moderate type.

Thakur and Dogra (2011) evaluated hydrocarbon source rock potential of the Subathu Formation exposed at Marhighat on Sarahan–Narag road in Sirmaur district of Himachal Pradesh on the basis of palynofacies analysis and thermal alteration index (TAI) values. The lower and middle part of the Marhighat succession is dominant in amorphous organic matter whereas the upper part is dominant in black debris (charcoal). The lower and middle part of the sequence has sapropelic organic matter facies with adequate maturity (TAI 3.0) to generate liquid hydrocarbons, while the upper part of the sequence has sapropelic humic–charcoal facies with TAI 3.0, which is considered to be moderate to good for gaseous hydrocarbon generation. The palynofacies analysis of Subathu Formation along Kalka–Simla section also indicate the dominance of amorphous organic matter with TAI value ranging from 2.75-3 which indicate moderate to good source potential for hydrocarbons.

CONCLUSIONS

Aspect of palynofacies study concerning the entire range of organic matter has gained importance to understand the type of organic matter contained in the sediments and the degree of thermal alteration achieved to predict the hydrocarbon source quality. With this technique it has been possible to predict occurrence of oil and gas possibilities in various sedimentary basins which are under exploration. The present work and published records on Tertiary rocks of NW- Himalaya suggest a moderte to good source potential for hydrocarbons on the basis of palynofacies maturation studies based on quality, quantity and TAI values from Subathu, Upper Dharamsala, Upper Muree and Siwalik sediments of Himachal Pradesh and Jammu and Kashmir

REFERENCES

Batten, D.J., 1981, Palynofacies, organic maturation and source potential for petroleum. In Organic Maturation Studies and Fossil Fuel Exploration Brooks, J. (Ed.): Academic Press, London, p. 201-224.

Batten, D.J., 1982, Palynofacies, palaeoenvironments and petroleum: Journal of Micropalaeontology, v. 1, p. 107-114.

Batten D.J., 1996, Palynofacies and palaeoenvironmental interpretation. In Palynology: Principles and Application, Jansonius J. and McGregor D.C. (Eds.): American Association of Stratigraphic Palynologists Foundation, v. 3, p.1011-1064.

Berry, C.M., 1989, Source rock palynology of Subathu sediments of Simla Hills: Geophytology, v. 19, p. 140-146.

Berry, C.M., 1994, Hydrocarbon potential of Siwlik and Pre-Siwalik sediments in Himachal Pradesh and Jammu-Kashmir areas - A Palynological perspective: Him. Geol., v. 15, p. 321-328.

Berry, C.M., Pundeer, B.S. and Mukherjee, B.K., 1998, Hydrocarbon potential of Subathu sediments in Jammu-Kashmir, Himachal Pradesh and Uttar Pradesh-A palynological perspective: Bull. ONGC, v. 35, p. 1-16.

Bhatia, S.B., 1982, Facies, fauna and flora in the lower tertiary formations of northwestern Himalayas: a synthesis: Palaeontological Society of India Spec. Publ., v.1, p. 8-20.

Bujak, J.P., Barss, M.S. and Williams, G.L., 1977, Offshore east Canada's organic type and color and hydrocarbon potential, part I: Oil & Gas Journal, v. 75, p. 198-202.

Burgess, J.D., 1974, Microscopic examination of kerogen (dispersed organic matter) in petroleum exploration, in R.R. Dutcher, P.A. Hacquebard, J.M. Schopf, and J.A. Simon, eds., Carbonaceous materials as indicators of metamorphism: GSA Special Paper, v. 153, p. 19-30.

Combaz, A., 1964, Les palynofacies: Rev. Micropaleont. V. 7, p. 205-218.

Faegri, K. and Iversen, J., 1989, Textbook of Pollen Analysis: New York, Wiley, pp.1–328.

Gansser, A., 1964, Geology of the Himalayas: Interscience Publishers, London, p. 273.

Hart, G.F. 1986, Origin and classification of organic matter in clastic systems: Palynology, v. 10, p. 1-23.

Masran, T.C. and Pocock, S.A.J., 1981, The classification of plant-derived particulate organic matter in sedimentary rocks. In Brooks, J. (Ed.), Organic Maturation Studies and Fossil Fuel Exploration: Academic Press, London, p. 145–175.

Mathur, N.S., 1978, Biostratigraphical aspects of the Subathu Formation, Kumaun Himalaya: Recent Res. Geol., v. 5, p. 96-112.

Mathur, N.S., 1979, Palaeoecology of the Subathu Formation, Kumaun Himalaya: Bull. Indian Geol. Soc., v. 1-2, p. 81-90.

Misra, C.M. and Pundeer, B.S., 1994, Palynofacies, maturation and source rock potential in the well Jawalamukhi-B, Himalayan foothills; In Proc. Second Seminar on Petroliferous Basins of India (eds) Biswas S K et al: (Dehradun, India: Indian Petroleum Publishers), v. 3, p. 161–171.

Pocock, S.A.J., Vasanthy, G. and Venkatachala, B.S., 1987, Introduction to the study of particulate organic materials and ecological perspectives: Journal of Palynology, v. 23-24, p. 167-188.

Raiverman,V., Kunte, S.V., and Mukherje, A., 1983, Basin geometry, cenozoic sedimentation and hydrocarbon prospects in North Western Himalaya and Indo-Gangetic plains: Petrol. Asia J., v. 6 p. 67-92.

Staplin, F.L., 1977, Interpretation of thermal history from colour of particulate organic matter -A review: Palynology 1 9–18.

Thakur, O.P. and Dogra, N.N., 2011, Palynofacies characterization for hydrocarbon source rock evaluation in the Subathu Formation of Marhighat, Sirmaur district, Himachal Pradesh: Journal of Earth System Science, v.120, p. 933–938.

Tissot, B., Durand, B., Espitalie, J. and Combaz, A., 1974, Influence of nature and diagenesis of organic matter in formation of petroleum: AAPG Bulletin, v. 58, p. 499-506.

Tyson, R.V., 1993, Palynofacies analysis. In Applied Micropaleontology, Jenkins, D.G. (Ed.): Kluwer Academic Publishers. The Netherlands, Amsterdam, p. 153–191.

Tyson, R. V., 1994, Sedimentary organic matter: organic facies and palynofacies: New York, Chapman and Hall, p.615.

Tyson, R.V., 1995, Sedimentary Organic Matter. Organic facies and palynofacies: Chapman and Hall, Londons, 615 p.

Venkatachala, B.S., 1981, Hydrocarbon source rock evaluation- A new Palynological approach: Petrol. Asia Journal, v.4, p. 80-93.

Venkatachala, B.S., 1984, Finely divided organic matter, its origin and significance as a hydrocarbon source material: Bull. Oil and Natural Gas Commission, v. 21, p. 23-45.

Hon'ble reviewer

1. **Dr. S.R. Jakhar**
   Department of Geology
   JNV University
   JODHPUR - 342 003
2. **Dr. Narendra Singh**
   Senior Petroleum Geologist
   SML
   ETHIOPIA

II-30

Gas Hydrates: An Emerging Energy Resource in India

Virendra Gaur, Ankit Ojha, Deva Ram, Seema Sharma, V.S. Parihar and K.L. Shrivastava

Department of Geology, Jai Narain Vyas University, Jodhpur-342005 INDIA
Email- klsgeology@yahoo.co.in

Abstract

In recent years, naturally occurring gas hydrates have attracted major interest globally as possible energy resource of the future with a reserve twice to the total amount of fossil fuel resources of the world. Infact, gas hydrate is a solid substance composed of water and low molecular weight hydrocarbons, mainly methane, and is known as calthrate. The paper is presenting a brief account of geology of gas hydrates with special emphasis on Indian prospects. Geoscientific investigations to infer more proxies related to gas hydrates before drilling of the hydrate bearing sediments have been identified in two areas of Indian offshore, namely, Kerala-Konkan basin in the west coast and Krishna-Godavari basin in the east coast and third, Andaman offshore basin. Initially, intensive exploration by GoI, National Companies like, GAIL and ONGC, in collaboration with laboratories of NGRI and NIO have given a promising response.

Keywords: Gas Hydrates, Calthrate, Nonconventional Energy Resource, Krishna-Godavari Basin, Kerala-Konkan Basin, Andaman Offshore Basin.

Introduction

Gas-hydrate is a solid substance composed of water and low molecular weight hydrocarbon (mainly methane) and is known as calthrate. It occurs worldwide in the oceanic and polar sediments where temperature is low enough and pressure is sufficiently high to crystallize the methane into gas-hydrates. The study of gas-hydrates has attracted the attention of the scientific community worldwide because of their widespread occurrences and potential energy content. Methane stored within and trapped below the hydrated sediments has a large energy potential. It is estimated to be twice the amount of total fossil fuel energy reserves of world.

In recent years the topics of naturally occurring gas hydrates have attracted major interest worldwide due to the fact that they may play a dominant role as possible energy resources in the future. Prior to this natural gas hydrates were mainly viewed as a source of operational problems in gas processing and transportation equipment. The world-wide carbonic gas-hydrates is estimated to be 10,000 X 1015g, which is double the carbon content in total fossil fuel (crude oil, natural gas and coal) reserves of the world (Kvenvolden, 1998). Among all renewable (solar, wind, wave, hydro, nuclear, hydrogen, bioprocess, geothermal, etc,) and nonconventional (gas shale, tight gas shale, coal bed methane, basin centered gas, gas-hydrates, etc,) energy resources, gas hydrates are regarded as one of the most suitable candidates for cleaner energy resources in this century. The paper is giving a brief account of geology of gas-hydrates with special emphasis on Indian prospects.

Previous work

Gas hydrates were recognized nearly two hundred years back by the British scientist Sir Humphery Davy, who discovered that, if a mixture of gas and water is exposed to high pressure and low temperature it turns into a solid compound called gas hydrate or calthrate (Henriet and Mienert, 1998). He found this by producing in 1810 an ice-like substance (chlorine hydrate) by effervesscing chlorine gas through water under elevated pressure. However, during the following century, gas hydrate was perceived as nothing more than a chemical oddity, and as such it did not receive much attention. In the 1930s, however, the oil and gas industry found that the formation of gas hydrate (or as called, dirty ice) clogged natural gas pipelines during cold weather. In the 1950s, the first two structures of gas hydrates were described. But it was not until the 1970s that Russian scientists, based upon theoretical models, postulated that there must be vast quantities of natural gas hydrate deposits on our planet (e.g. Vasil'ev et al., 1970; Trofimuk et al., 1973, 1975, 1979; Tucholke et al., 1977). This theory was confirmed when seafloor samples were recovered in the Black Sea by Russian vessels (Yefremova and Zhizhchenko, 1974), Black Ridge (Harrison and Curiale, 1982; Shipley and Didyk, 1982) and off Central America by the Deep Sea Drilling Project (Stoll et. al., 1971). It was also observed that methane was the predominant gas trapped in naturally occurring gas hydrates. Since then, many countries have set up national research programmes for the investigation of gas hydrate, owing to their geohazard and economic implications. The most important are run by Japan, Canada, the USA, Germany and India (Henriet and Mienert, 1998; Milkov et. al., 2002).

The first indication of presence of gas hydrates in Indian Offshore dates back to 1984 when Oil and Natural Gas Corporation Limited (ONGCL) (Chopra, 1984) identified BSR for the first time in Andaman Offshore. Indian efforts of study which have been presented here are by Gupta et al. (1998), Rao et al. (1998), Sain et al. (2000), Paull and Dillon (2001) and Gupta (2003). Exploratory assessment for the possible presence of gas hydrates in Indian Offshore areas by ONGCL and National Institute of Oceanography (NIO) sparked the interest in natural hydrates. The Indian Geophysical Union (IGU) conference held in 1995 at National Geophysical Research Institute (NGRI), Hyderabad, India deliberated the topic of Indian Gas Hydrates resources. Subsequently, Ministry of Petroleum & Natural Gas (MOP and NG), Government of India, established an expert committee to recommend the steps to be undertaken for exploration and development of gas hydrate resources in India. Based on the recommendations, a National Gas Hydrate programme (NGHP) was instituted in 1997. Simultaneously, ONGCL, GAIL and Directorate General of Hydrocarbons (DGH), (oil industry), and the national laboratories such as NIO and NGRI initiated in house R and D activities related to gas hydrates for mapping and evaluation of gas hydrate resources in Indian Offshore. To give further thrust to the gas hydrate related activities, MOP and NG reconstituted the NGHP in the year 2000. The ultimate goal of the programme was to achieve maximum self-sufficiency in hydrocarbons through tapping this non-conventional source of energy. The roadmap is in place to achieve the commercial production of gas hydrates.

Genesis of Gas hydrates

Hydrates can form in several ways. In the arctic; there is a growing belief that many hydrate accumulations represents pre-existing free gas accumulations that have been converted to hydrate by subsequent change in environmental conditions (onset of arctic climate postdated the migration of gas into shallow sandstone "traps". In the marine environment, hydrate is often considered to have formed from solution, as methane is generated by in-situ microbial processes to the point where the water becomes saturated with methane and hydrate growth begins. There is also a high likelihood that methane hydrate could accumulate in coarse grained marine sediments by the migration of gas from deeper, warmer zones, up through various pathways such as faults, and into water-bearing shallow sediments where it is then converted to methane hydrate.

The biogenic methane is the first way of formation of gas hydrates. Biogenic processes are capable of producing vast amounts methane; and are considered to be the dominant source of methane trapped in hydrate layers within shallow sea floor sediments. The common by-product of bacterial ingestion of organic matter can be explained as:-

$$(CH_2O)_{106} (NH_3)_{16} (H_2PO_4) \rightarrow 53CO_2 + 53CH_4 + 16NH_3 + H_2PO_4$$

Infact, it is the same process which produces methane in swamps, landfills, rice paddies, and the digestive tracts of mammals occur continually within burried sediments in geologic environments all around the globe.

The thermogenic methane is the second way of formation of gas hydrates, in which it is produced by the combined action of heat, pressure and time on buried organic material. In the geologic past, conditions have periodically recurred in which vast amounts of organic matter were preserved within the sediment of shallow, inland seas. Over time and with deep burial, these organic-rich source beds are literally pressure-cooked with the output being the production of large quantities of oil and natural gas. Along with the oil, the gas (largely methane, but also ethane, propane and other larger molecules) slowly migrates upwards due to its buoyancy or other regions, relative to water. If sufficient quantities reach the zone of hydrate stability, the gas will combine with local formation water to form hydrate.

In general, a combination of low temperature and high pressure is needed to support methane hydrate formation. Fig. 1 shows the combination of temperatures and pressures (the phase boundary) that marks the transition from a system of co-existing free methane gas and water/ice solid methane hydrate (Gupta et al., 1998, Chandra, 1998 and Gupta, 2003). When conditions move to the left across the boundary, hydrate formation will occur. Moving to the right across the boundary results in the dissociation (akin to melting) of the hydrate structure and the release of free water and methane.

The composition of both the water and the gas are critically important when fine-tuning predictions of the stability of gas hydrates in specific settings, apart from temperature and pressure. Experimental data collected thus far have included both fresh water and seawater. However, natural subsurface environments exhibit significant variations in formation water chemistry, and these changes create local shifts in the pressure/temperature phase boundary. Higher salinity restricts hydrate formation causing the phase boundary to shift to the left. Similarly, the presence of small amounts of other natural gases, such as carbon dioxide (CO_2), hydrogen sulfide (H_2S) and larger hydrocarbons such as ethane (C_2H_6), will increase the stability of the hydrate, shifting the curve to the right. As a result, hydrates that appear to be well above the base of hydrate stability, from pressure-temperature relationships, may actually be very close to the phase boundary due to local geochemical conditions. These local variations may be very common, as the act of forming hydrate, which extracts pure water from saline formation waters, can often lead to local, and potentially-significant increases in formation water salinity.

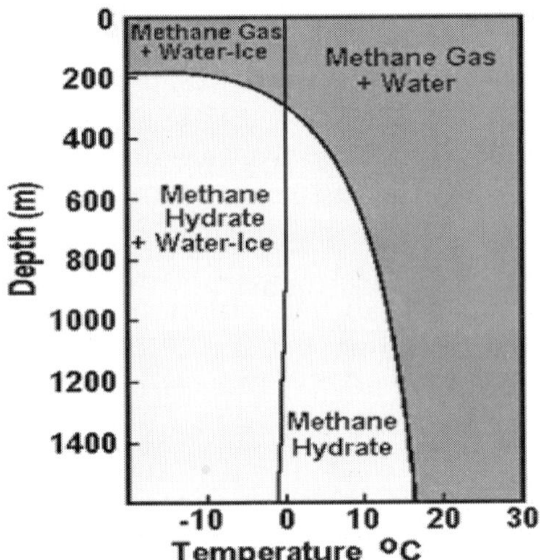

Figure 1: Methane Hydrate Phases

The phase diagram in Fig. 2 illustrates typical conditions in a region of arctic permafrost (with depth of permafrost assumed to be 600 meters) (Gupta et al., 1998, Chandra, 1998 and Gupta, 2003). The overlap of the phase boundary and temperature gradient indicates that the GHSZ should extend from a depth of approximately 200 meters to slightly more than 1,000 meters. Both the permafrost thickness and pressure/temperature gradients in the chart are examples and can vary with locality, so specially-tailored diagrams must be made before site-specific predictions of hydrate stability can be attempted. The phase diagram in Fig. 3 shows a typical situation on deep continental shelves (Gupta et al., 1998, Chandra, 1998 and Gupta,

2003). A seafloor depth of 1,200 meters is assumed. Temperature steadily decreases with water depth, reaching a minimum value near 0°C at the ocean bottom. Below the sea floor, temperatures steadily increase. In this setting, the top of the GHSZ occurs at roughly 400 meters-the base of the GHSZ is at 1500 meters. The hydrates will only accumulate in the sediments or as mounds on the seafloor over point sources of methane release. The final phase diagram in Fig. 4 illustrates why no hydrates are found in interior basins at sub-polar latitudes (Gupta et al., 1998, Chandra, 1998 and Gupta, 2003). At every depth (pressure), subsurface temperatures are too high for methane hydrate to be stable.

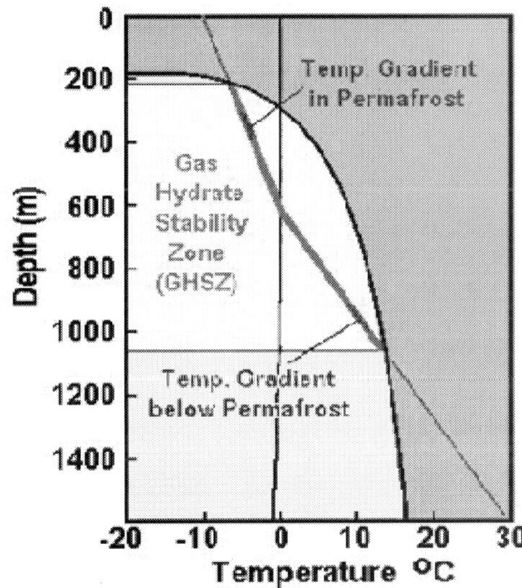

Figure 2: Specific Hydrate Stability for Arctic Permafrost

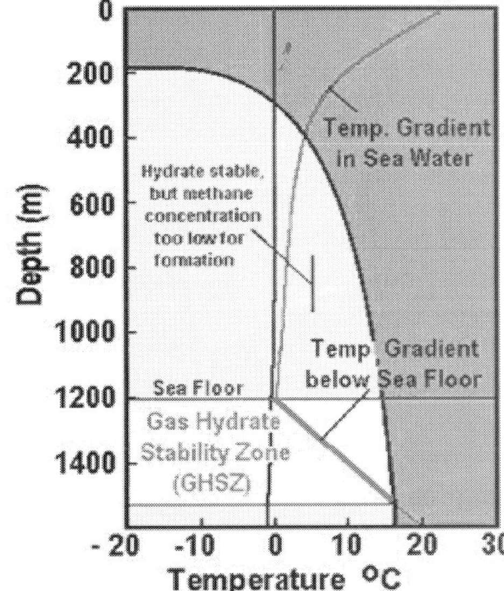

Figure 3: Typical occurrence of the gas hydrate on deep-water continental margin.

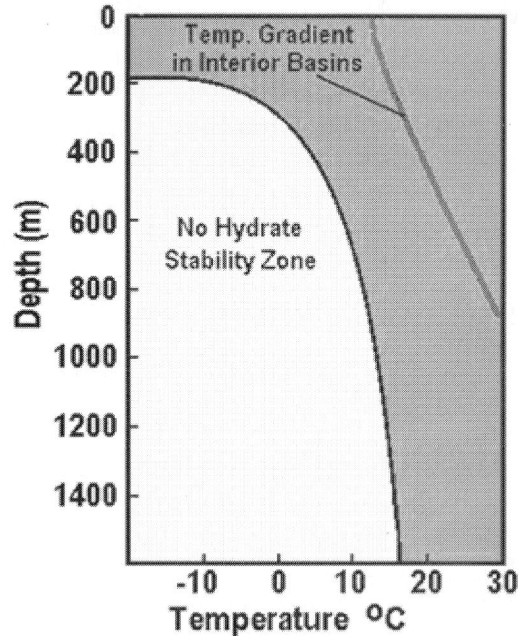

Figure 4: At sub-polar latitudes temp are too high for methane hydrate to be stable

Potential sites of Gas hydrates

Methane hydrates occur only in sediments characterized by well-known pressure and temperature conditions, in contrast to conventional natural gas. Meaning that exploration activities can be strictly limited to specific zones. The pressure-temperature conditions consistent with methane hydrate stability are widespread on earth. At pressure-temperature conditions outside the stability zone, methane is no longer bound in "methane ice" and exists only as free gas or gas dissolved in pore waters. An estimated 99% of worldwide gas hydrate occurs in ocean sediments, and appropriate temperature and pressure conditions predominate within the upper tens to hundreds of meters of seafloor sediments at water depths ranging from 300 to 500 m on the shallow end to greater than 4000m. In theory, methane hydrates are also stable on the seafloor and in the water column in large swaths of the world's oceans. Gas hydrates do not persist long in the water column, and sea floor gas hydrates are not significant as a resource.

Onshore, methane hydrates occur almost exclusively in areas with thick permafrost. The appropriate temperature and pressure conditions can occur over a zone that is typically several hundreds of meters thick and that encompasses the bottom part of the permafrost-bearing section and the top of the subpermafrost sedimentary section. Despite the worldwide occurrence of gas hydrate, the deposits are not ubiquitous within the gas hydrate stability zone. The key factor limiting gas hydrate formation in most locations is the lack of sufficient gas. A lack of free water in sediments can also limit gas hydrate formation, but this situation occurs almost exclusively in specific geologic settings (very low permeability formations) or possibly near the base of the stability zone. The amount of methane trapped in gas hydrates remains uncertain.

Recently, Boswell and Collett (2011) concluded that an estimate of 105 TCF of methane trapped in gas hydrates (gas-in-place or GIP) is a reasonable figure, but they strongly emphasize that the GIP number is not meaningful for resource-based studies. Most of the methane included in the GIP estimate is probably trapped in gas hydrates that are present in low saturations (<10% of pore space) in fine-grained, low permeability marine sediments. Such disseminated, low-saturation gas hydrate is nearly impossible to detect without drilling and is unlikely to ever produce commercial quantities of gas. From an energy perspective a more important figure is the amount of methane trapped in resource-grade deposits, estimated at 104 TCF in marine settings and several hundred TCF in permafrost-associated deposits (Boswell and Collett, 2011). This estimate of GIP in resource-grade methane hydrates is ~35% more than the 2010 estimate for global natural gas reseerves (~6600 TCF; EIA, 2010) and nearly 100 times greater than the annual global consumption of gas. There are as yet no proved reserves of gas hydrate since gas has never been produced from gas hydrate for more than a few days in research tests.

Prospects of Gas hydrates in Indian Offshore

Fig. 5 is illustrating major prospective areas of gas hydrates in India (Ramana et al, 2006). Indian studies which are adopted here are from the work of Gupta et. al. (1998), Rao et al. (1998), Sain et al. (2000), Paull and Dillon (2001) and Gupta (2003). The GAIL (India) Ltd., one of the National Companies in collaboration with the national laboratories such as NIO and NGRI carried out gas hydrates studies using about 4300/km of MCS data pertaining to Kerala-Konkan Basin in the west coast, and inferred BSRs which are strong, moderate and week. Based on the results of multidisciplinary studies carried out in collaboration with DGH, ONGCL, GAIL and NIO, three offshore areas i.e. Krishna-Godavari basin in the east coast, Andaman offshore and Kerala-Konkan basin in the west coast area were delineated for further geoscientific investigations to infer more proxies related to gas hydrate before drilling of the hydrate bearing sediments. The swath bathymetry, deep low digital side scan sonar images, Chirp Sonar, gravity, Magnetic, geological, geochemical and microbiological data were generated in two phases during 2002-2003 as a first step to understand the shallow geology of the areas. The gravity cores and water samples collected were analyzed for physical properties, gas chemistry and microbial signatures in the east and west coast areas.

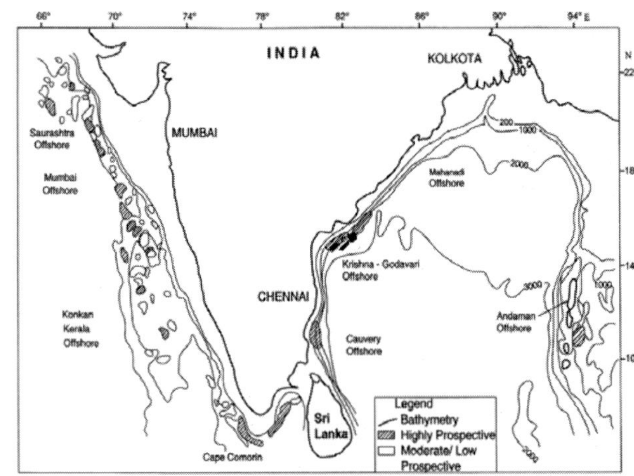

Figure 5. Map depicting prospective areas for gas hydrates along the Indian Continental margins (afterRamana et al., 2006)

Large amount of Multi-channel Seismic reflection data was acquired by oil industry for exploration of hydrocarbon in the Indian Deep-water basins, beyond 400m isobaths. These data were examined for BSRs, blnking above BSRs, polarity reversals etc (geophysical proxies of gas hydrates). A qualitative map prepared based on the inferred BSRs brought out a deepwater area of about 80,000 sq. km unto 3000m isobaths as favourable for gas hydrate occurrence (Chandra et al., 1998). Methodology for reprocessing of seismic data was evolved during 1997 to infer BSRs (gas hydrates) by ONGCL in collaboration with University of Victoria (UV), Canada. Subsequently several seismic line segments from the India offshore areas were reprocessed. Clear BSRs cutting across the sedimentary strata supported with interval velocity inversion in

Krishna-Godavari deep offshore Basin were observed. The swath bathymetry maps generated for Eastern and Western offshore areas depict the seafloor morphology. Based on the encouraging results of these initial studies, further reprocessing of the seismic data was carried out by ONGCL in a selected area in the Krishna-Godavari Basin. The studies performed in these three Indian offshore areas have been summarized as follows.

1. Krishna-Godavari Basin

Analysis and interpretation of the geophysical, geological, geochemical and microbial data have revealed the presence of several gas hydrate proxies in the east coast area (Gupta et al., 1998, Chandra, 1998 and Gupta, 2003). Two Digital side Scan Sonar images depict the presence of clusters of pockmarks and carbonate reef like features on seabed. The multi-frequency chrip Sonar and 3-5 KHz Sub-bottom profiler records depict the penetration upto 40-60 m below the seabed and several gas escape features and gas plumes unambiguously. Sometimes, the seabed is seen affected by gas up thrust and in some other cases the sub-bottom reflectors are affected by gas making. Blanking or disappearance of macro and micro bathymetry features more conspicuously. A WNW-ESE trending positive topography feature (sedimentary ridge) appears to be similar to Blake Ridge is a conspicuous geomorphic feature, which needs to be investigated thoroughly (Chandra, 1998). The crest of the ridge rises from the surrounding water depth of > 1400 to <900m. The width of the ridge about 10 km. in northwestern part of the study area, a fault controlled fan like geomorphic feature is also noted (Chandra, 1998).

Further study of 5 m long gravity cores reveal the variation of methane concentration in the top sediments between 0 and 2.27 nm, whereas the highest concentration of >158 nm was observed in the bottom sediments of one of the cores. The methane enrichment trend with the core depth was observed at 23 out of 76 sample locations. On the other hand SO_4^{2-} reduction trend with core depth was noticed at more than 48 out of 76 core locations. The Cl^- profiles do not depict much variation with depth. However, low concentration of Cl^- (~500NM) is observed in locations. The Organic Carbon (TOC) concentration (wt %) varies between 0.35 and 1.35 at middle level and a high concentration of > 2.5 were at one location. XRD and SEM studies indicate the presence of antigenic carbonates in subsurface layers, which might have formed in presence of the methane gas rising from the deeper layers. 'Ba' enrichment of nearly 5 to 6 times is observed in bottom of cores compared to core top. These elevated 'Ba' concentrations at the core bottom suggest Barium front phenomenon. The observed 'Ba front' serves as precursor to predict the occurrence of methane below the sulfate reduction zone. The 'Ba' enrichment trend has been noted at more than 25 locations. These observed trends of methane enrichment, sulfate reduction and Cl^- depletion with depth appear to be related to the processes of 'Sulfate Methane Interface' (SMI). The microbiological studies revealed presence of SRB_r, SRB_f and NRB in appreciable quantities at the bottom layers, probably a strong indicative of continuous supply of methane from the subsurface layers. Integration of geophysical, geological, geochemical and microbiological data results suggests that KG basin is highly conductive for gas hydrate occurrence. This belief is further strengthened by the presence of several gas hydrates layers through the recent drilling activity by private companies and ONGCL, and MWD logs.

2. Andaman Sea

The Directorate General of Hydrocarbons (DGH) carried out gas hydrates related studies in collaboration with Western Geophysical and LARGE of Russia, using MCS data that was acquired for hydrocarbon exploration in the Andaman deepwater area during the three different campaigns in 1996-97, 1999 and 2001 (Gupta et al., 1998, Chandra, 1998 and Gupta, 2003). The BSRs are seen cut across the upper Miocene to Middle Miocene strata, and broadly follow the seabed topography. The processed MCS data show the indications of free gas below. the BSRs and the corresponding seismic signatures resemble like the anticlinal closures. The BSRs appear to be serving as the cap for the underlying free gas. The BSRs were mapped over an area of about 3000 sq. km. Seven gas hydrate-bearing prospects have been identified below the seabed in the water depths of 850 m to 2000. high amplitude BSR zones are parallel to central halfgraben feature of the fore arc basin. The flat spot identified in few structural prospects beneath the BSRs, suggests possible gas water contact Pre-stack Genetic Algori-thm (GA) inversion, a full waveform inversion method, was carried out by the Western Geo-physical to confirm the seismic anomalies. The GA pre-stack inversion extracts Vp, VS, density and Poisson's Ratio (PR) from non-NMO corre-cted gathers were processed for AVO using pre-stack time migration and non-muted processing sequences. Results of post-stack hybrid inversion indicate that low PR anomalies generally associated with BSRs are not occurring, but localized anomalies were observed.

3. Kerala-Konkan Basin

Similar multidisciplinary studies were carried out in kerala-Konkan offshore basin in west coast (Gupta et al., 1998, Chandra, 1998 and Gupta, 2003). A positive NW-SW trending topography feature rising from 2400 m from surrounding seafloor at 3000 m in the central part of study area is noteworthy. The seafloor images obtained from deep low digital Side Scane Sonar system are devoid of pock marks and gas escape features and the multi-frequency Chrip Sonar and 3.5 KHz sub-bottom profiler data are devoid of gas escape features, excepting some micro features resembling as the gas escape feature. But, some seismic sections show the presence of BRS like features, and the analysis of core samples indicate the presence of methane enrichment, sulfate reduction and chloride depletion trends with core depth (Chandra, 1998). The geochemical and microbiological parameters show TOC increase at 10; $CaCO_3$ increase trend at 9, Ch_4 enrichment at 12, Cl^- depletion at 10 and SO_4^{2-} reduction at 11, SRB_r minimum count at 19 and SRB_f appreciate quantity at 10 locations respectively (Chandra, 1998).

Conclusions

The gas-hydrates and/or 'free gas' in the shallow sediments of outer continental margins have the most important implication as its economic aspects because a tremendous amount of energy is stored within and below the hydrated sediments. The energy content is estimated to be as large as twice the amount of total fossil fuel reserves. As the depth of BSR depends on pressure-temperature conditions, mapping of BSRs over the continental margins helps in preparing a heat flow map or to derive thermal structure without any well data. Nonetheless, geoscientific investigations to infer more proxies related to gas hydrates before drilling of the hydrate bearing sediments have been identified in two areas of Indian offshore, namely, Kerala-Konkan basin in the west coast and Krishna-Godavari basin in the east coast and third, Andaman offshore basin. There are several technical problems in extracting and producing gas from gas-hydrates at this moment. There is a strong need to prepare suitable strategies to assess the resource potential before exploitation. Initially, intensive exploration by GoI, National Companies like, GAIL and ONGC, in collaboration with laboratories of NGRI and NIO have given a promising response. A concerted effort is required to study the gas-hydrates in a scientific manner systematically, and efforts in this direction are on throughout the world.

References

Boswell, R. and Collett, T.S., 2011. Current perspectives on gas hydrate resources. Energy and Environmental Science, vol. 4, pp. 1206-1215

Energy Information Administration (EIA), 2010. Natural gas, in: International Energy Outlook 2010, U.S. Dept. of Energy, Washington DC. http://www.eia.doe.gov/oiaf/ieo/pdf/0484(2010).pdf

Harrison, W.E., and Curiale, J.A., 1982, Gas hydrates in sediments of holes 497 and 498A, Deep Sea Drilling Project Leg 67, J N Aubouin, J., von Huene, R., and others, Initial Reports of the Deep Sea Drilling Project, v. 67: Washington D.C., U.S. Government Printing Office, pp. 591-594.

Henriet, J.P. and Mienert, J., 1998: Gas hydrates: relevance to world margin stability and climate change. Geological Society Pub. House.

Chandra, Kuldeep, Singh, R. P. and Julka A. C., 1998: "Gas hydrate potential of Indian offshore area." 2nd Conference and Exposition on Petroleum Geophysics SPG 98, Chennai.

Gupta, H. K., 2003: "Has-hydrate- a Potential Source of Energy grom the Oceans" Bruun Memorial Lecture 2003 on Energy from the Sea. Intergov. Oceanog. Comm. (IOC).

Gupta, H.K., et al., 1998. Analysis of single channel seismic data from the continental margins of India for exploration of gas hydrates. NGRI Tech. Rep. No. NGRI-98- LITHOS-221.

Kvenvolden, K. A., (1993): "Gas hydrate as a potential energy resource a review of their methane content". In the Future of Energy Gases (eds. Howell, D. G. et. al), U. S. Geological Survey Professional Paper, vol. 1570, pp. 555-561.

Milkov, A. V. et.al., 2004: "Co-existance of gas hydrate, free gas, and brine within the regional gas hydrate stability zone at Hydrate Ridge (Oregon margin): evidence from prolonged degassing of a pressurized core." Earth Planet. Sci. Lett. Vol. 222, pp. 829-843.

Milkov, A. V., Sassen, R. and DeFreitas, D., (2002): "Gas hydrate in Gulf of Mexico". Applied Gas Hydrate Research Programme, Year III, Final Report, pp. 2-5.

NIO, (1997): "Gas Hydrate Resource map of India, prepared for GAIL (India) Limited, NIO/SP-25/97, p. 56.

Paull, C.K., and Dillon, W.P., 2001. Natural Gas Hydrates: Occurrences, Distribution, and Detection, Geophysical Monograph Series, 124.

Ramana, M. V., Ramprasad, T., Desa, M., Sathe, A. V., and Sethi, A. K., 2006: "Gas Hydrate-related proxies inferred from multidisciplinary investigations in the Indian offshore areas." Curr. Sci. Vol. 91, No. 2, pp. 183-189.

Rao, Hannumatha Y., Reddy, S. I., Khanna, R., Rao, T. G., Thakur, N. K., and Subrahmanyam, C., 1998. Potential distribution of methane hydrates along .the Indian continental margins. Current Science, vol.74, 466-468.

Sain, K., Minshull, T. A., Singh, S. C., and Hobbs, R. W., 2000. Evidence for a thick free gas layer beneath the bottom simulating reflector in the Makran accretionary prism. Marine Geology, Vol. 164, pp. 3-12.

Sethi, A. K., Sathe, A. V., and Ramana, M. V., 2004: "potential Natural gas Hydrates Resources in Indian offshore areas" in Gas Hydrates: Energy resources Potential and Associated Geologic Hazards." AAPG, HEDBERG Conference Sept. 12-16, 2004, Vancouver, BC, Canada.

Shiepley, T.H., Houston, M.H., Buller, R.T., Shaub, F.J., McMillen K.J., Ladd, J.W., Worzel, J.L., 1979. Seismic evidence for widespread possible gas hydrate horizons on continental slopes and margins. Am. Assoc. Pet. Geol. Bull. Vol. 63, pp. 2204-2213.

Shipley T, Didyk BM (1982) Occurence of methane hydrates offshore southern Mexico. In: Watkins JS, Moore (eds) Initial Reports DSDP 66. US Printing Office, Washington, DC, pp 547–555

Stoll, R. D., Ewing, J. I., and Bryang, M., 1971: Anomalous wave velocities in sediments containing gas hydrates. J. Geophys. Res., vol. 76, pp. 2090-2094.

Trofimuk A.A., Chersky N.V., Tsarev V.P., 1979: "Gas hydrates – new sources of hydrocarbons." *Nature*, N 1, pp.18-27, 1979.

Trofimuk A. A., Cherskiy N. V., Tsarev V.P. (1975): The biogenic methane resources in the oceans. Dokl. Acad. Nauk SSSR Vol. 225, pp. 936–943.

Trofimuk, A., Cherskiy, N, Tsarev, V., 1973. Accumulation of natural gases in zones of hydrate—formation in the hydrosphere. Doklady Akademii Nauk SSR 212, 931-934 (in Russian).

Tucholke, B. E., Bryang, M., and Ewingj, I. 1977: Gas-hydrate horizons detected in seismic-profiler data from the western North Atlantic. Amer. Assoc. Pet. Geol. Bull., vol. 61, pp. 698-707.

Vasil'ev V.G., Makogon Y.F., Trebin F.A. 1970: "Properties of natural gas found in Earth's crust in the solid state and forms a gas hydrate reservoir." Discoveries of the USSR, 1968-1969. *M., Nedra*, pp.5-18, 1970.

Yefremova, A.G. and Zhizhchenkb, O.P., 1974: Obnaruzheniye kristallgidradov gazov v osadkakh sovremennykh akvatoriy ["Occurrence of crystal hydrates of gases in the sediments of modern marine basins"] Doklady Akademii Nauk SSSR, Vol. 214, pp. 1179-1 181 (in Russian); Doklady-Earth Sci. Section 214 (1975), pp. 219-220 (in English).

Hon'ble reviewer

1. **Dr. Alok Tripathi**
   Chief Geologist
   Oil Natural Gas Commission Ltd.
   Cinnamara
   JORHAT - 785 704
2. **Dr. S.S. Deb**
   Chief Geoscientist
   Oil India Limited
   Basni
   JODHPUR - 3422 005

II-31

Palaeozoogeographic implications of the Jurassic Ostracods of Western India and Importance of ostracods in Petroleum Exploration

S.R. Jakhar

Department of Geology, MBM Engineering College, Jai Narain Vyas University, Jodhpur 342 011
E-mail: srjakhar@yahoo.com

Abstract

The palaeozoogeographic implications of the Jurassic ostracods of western India are discussed at length in the paper. So far as 131 species of the Jurassic ostracods from Kachchh, Gujarat and Jaisalmer, Rajasthan are published (Khosla *et al.*, 2009). Of these, 100 species are known from Kachchh and 73 species from Jaisalmer. As far as 58 species are showing affinity with other than their original geographical place, among them 46 species, are common between Gujarat and Rajasthan. 27 species occur in the equivalent horizons in Madagascar, Tanzania, Somalia and Saudi Arabia. The overall evidence based on distribution and correlation of ostracod taxa suggest that the Jurassic ostracods of the western India have strongest affinity with above named countries and these were very close to each other during the Jurassic time.

Besides biostratigraphic zonation, correlation and dating of rocks, ostracods are precious source of information for petroleum exploration. Their importance in petroleum exploration is discussed in the paper. Ostracods are used as a tool for determining palaeoenvironment of petroleum bearing strata, even today.

Keywords: Jurassic, Ostracods, Palaeozoogeography, Petroleum, Western India.

Introduction

The ostracods are small size animals but most are between 0.1 mm to 2 mm and belongs to class Crustacea of the phyllum Arthropoda. They consist of a shrimp-like organism, encased in a bean shaped, bivalve shell. Ostracods are still living today and their fossil record stretches back into the Cambrian period. They live in all aquatic environments, including fresh and brackish water. They are quite useful in palaeontological interpretation, particularly in recognizing important physical interface such as those between land and sea, sea and estuary, and sea and ocean. On global scale ostracods can be utilized to the prior juxtaposition of subsequently separated continental blocks and conversely, the recognition of prior separation of subsequently adjacent blocks (Whatley, 1988). All these features render them as ideal tools for many aspects of geological investigations viz. well correlation in petroleum exploration, stratigraphic correlation and palaeozoogeographic interpretations.

In so far as ostracods from the Jurassic beds of western India is concerned fairly good work has been done during past 5 to 6 decades. In western India mainly the Middle Jurassic beds are extensively developed in Kachchh District of Gujarat State and Jaisalmer District of Rajasthan State. In Kachchh District the beds crop out in three east-west trending anticlinal hill ranges and an isolated mass near Wagad (Fig. 1A). In Jaisalmer district the beds are exposed in a stretch surrounding the Jaisalmer Town (Fig. 1B). The ostracods of Jaisalmer are mainly reported from the Jaisalmer Formation of the Jurassic beds.

Previous work

The ostracods of the Jurassic beds of Kachchh have been described by a number of micropalaentologist. Lyubimova *et al.* (1960) have described, among others, eight new species from Khavda and Lodai localities. Subsequently, Guha (1977) recorded 18 species, including one new, from the late Jurassic and early Cretaceous subsurface succession of well No.2, Banni, Rann of Kachchh. Neale and Singh (1986) re-examined the ostracods fauna of the previously mentioned well No. 2, and reported the presence of a much richer fauna comprising 34 species. These included 19 new species. Khosla and Jakhar (1993) described two species of *Amicytheridea* - *Amicytheridea triangulata* Bate, 1975 and *Amicytheridea* sp. - from the Jurassic of the Habo Dome, Mainland Kachchh. In addition, they established a new subgenus *Paratrichordis* to accommodate four species including one new, *Trichordis triangula* Bate, 1975 (type species), *T. (Paratrichordis) devexa, T. (P.) grumosa,* and *T.(P.) parvicarinata* n. sp. The occurrences of the last three species have been reported from the Jurassic of Kachchh. Khosla *et al.* (1997) described 49 ostracod species from the Jurassic of the Habo Dome, in which two genera (*Batella* and *Habocythere*), 20 species and one subspecies are new. Khosla and Jakhar (1999) recorded 28 species from the Jurassic of the Jumara Dome. Of these, 25 species have been assigned to already described from the Chari Formation of the Habo Dome by Khosla *et al.* (1997), while three species have been left in open nomenclature. Khosla *et al.* (2003a) recorded 54 species from the type section of Khadir Formation, Khadir Island, Rann of Kachchh. Of these 30 species are assigned to the previously known taxa. Twenty-three species are left in open nomenclature and one species described as new. Khosla *et al.* (2003b) reported 57 species from the Khadir Formation, Bela Island, Rann of Kachchh. Of these, 31 species are assigned to the previously known taxa while twenty-six are left in open nomenclature. Khosla *et al.* (2003c) described three species of genus *Progonocythere* including two new species from Jurassic beds of Kachchh and Jaisalmer of western India. Khosla *et al.* (2004) reported 49 species from the Jurassic of the Jhura Dome, of these thirty-six have been assigned to previously known species, six as new species and seven left under open nomenclature. Khosla *et al.* (2005) described 74 species from the Jurassic beds of Khadir, Bela and Pachchham islands, Rann of Kachchh. Of these 27 have been described as new, seven species left in open nomenclature and forty species are assigned to already known taxa from Kachchh, Jaisalmer and elsewhere and Andreu *et al.* (2012) reported 71 species from the Middle and Upper Jurassic beds of Kachchh, of which 34 are previously published and thirty-seven left under open nomenclature.

In so far as the Jurassic ostracods of Jaisalmer is concerned, they were first described by Lyubimova *et al.* (1960) who recorded, among others, six species from the Jaisalmer Formation. Thereafter, Singh and Jai Krishna (1969), reported six species from the Jaisalmer Formation and all left in open nomenclature. Singh, S. N. and Kulshreshtha (1973) described a new genus - *Pokornya*, with *P. kuldharensis* as the type species – from the Kuldhar Member of the Jaisalmer Formation. Bhatia and Mannikeri (1977) recorded 12 species, along with charophytes, of fresh and brackish water ostracods from the Fort Member of the Jaisalmer Formation, of these 11 are left under open nomenclature. Kulshreshtha *et al.* (1985) recorded 34 species from the Kuldhar Member, Jaisalmer Formation including 12 new species. Mannikeri (1996) described two new genera *Jainiana* including two species and *Bhatiana* including three species from the Joyan, Fort and Badabag members of Jaisalmer Formation. Khosla *et al.* (2006) described 72 species from the Fort and Kuldhar Members of Jaisalmer Formation, of these thirteen are described as new, fourteen species are left in open nomenclature and forty-five assigned to previously known taxa. It will be beyond to the scope of this paper to write name of all these species here. The detail of all these species may be sought from the above mentioned respective papers. Very recently Khosla *et al.* (2009) compiled all published 131 species of western India in a form of atlas. These includes 29 species left in open nomenclature.

Important contributions to our understanding of Jurassic Ostracods from other countries bordering the Indian Ocean, with which western India ostracods have affinities, are by Grekoff (1963), Bischoff (1964), Dingle (in Dingle and Klinger, 1972), Bate (1975, 1976, 1977), Basha (1980), Dépêche (in Dépêche *et al.*, 1987), Rosenfeld *et al.* (1987, 1988), Rafara (1990), Rosenfeld and Honigstein (1991, 1998), Mette (2004a, 2004b, 1995, 1997), Mette and Geiger (2004a. 2004b, 2004c), Schudack and Schudack (2002) and Same (2008).

On the basis of distribution of ostracods Khosla *et al.* (2006) have recognized three biostratigraphical range zones in the middle Jurassic of Mainland Kachchh; five range zones in Northern Island belt and four range zones in Jaisalmer district (Table 1).

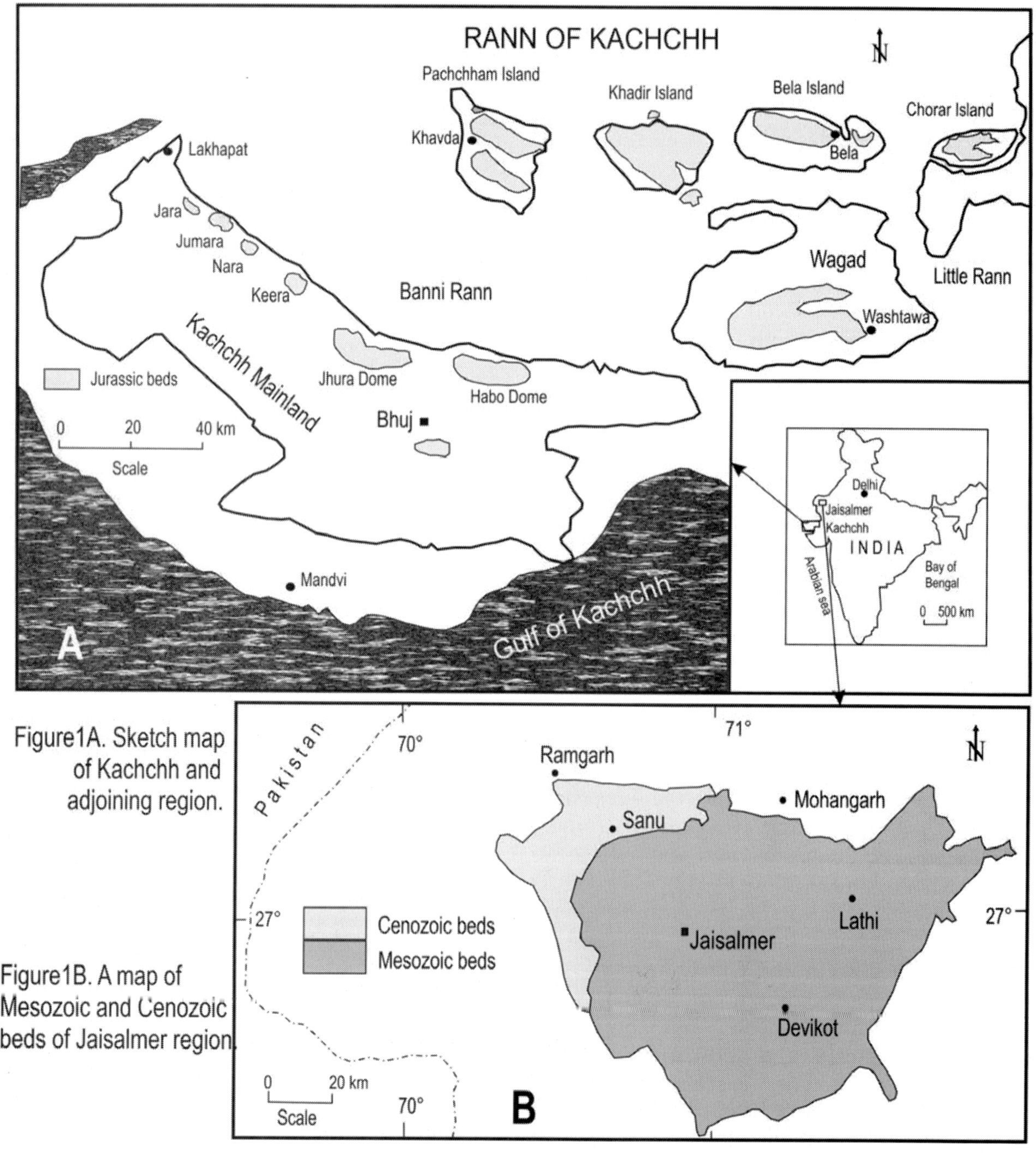

Figure1A. Sketch map of Kachchh and adjoining region.

Figure1B. A map of Mesozoic and Cenozoic beds of Jaisalmer region

Table 1. Correlation of Ostracod Biostratigraphical Zonation of Middle Jurassic of Mainland and Northern Island Belt of Kachchh and Jaisalmer districts, Western India (Khosla *et al.* 2006).

Age	Kachchh District		Jaisalmer District
	Mainland	**Northern Island Belt**	
			Majungaella perforata kachchhensis-Galliaecytheridea remota Interval Zone
middle-late Callovian	*Majungaella perforata kachchhensis-Galliaecytheridea remota* Concurrent Range Zone	*Majungaella perforata kachchhensis-Galliaecytheridea remota* Concurrent Range Zone	*Majungaella perforata kachchhensis-Galliaecytheridea remota* Concurrent Range Zone
	Majungaella perforata kachchhensis Partial Range Zone		
Callovian		*Fastigatocythere (Habocythere) mouwanaensis* Range Zone	
late Bathonian-early Callovian	*Progonocythere laeviscula* Range Zone	*Progonocythere laeviscula* Range Zone	*Majungaella perforata kachchhensis- Fastigatocythere (Batella) falcula* Concurrent Range Zone
Bathonian		*Cytheropteron micropunctata* Range Zone	
Bajocian-Bathonian		*Trichordis hadibhadangensis* Range Zone	*Trichordis hadibhadangensis* Range Zone

Palaeozoogeography

Most of the earlier workers have described ostracods affinity with Madagascar and south-east Africa. The reference of following publications viz. Grekoff (1963), Guha (1977), Bate (1975, 1977,) Dingle (1988) and Benjamin, (2008) may be made for further details. The palaeozoogeography of the marine ostracods of the Jurassic beds of Kachchh has been discussed at length by Khosla *et al.* (1997, 2005) and has shown strong affinity with Madagascar and lesser with Tanzani and Saudi Arabia.

Very recently Andreu *et al.* (2012) published a paper on palaeobiogeography of the Middle and Upper Jurassic beds of Kachchh and found that out of 16 species of ostracods reported from Bajocian of Indian localities only one species *Fastigatocythere malgachica* has been described from the Bajocian of Madagascar. In the Bathonian out of 20 species of India nine are in common between India and Madagascar. *Fastigatocythere triangulata* is present in the Upper Bathonian of Somalia, *Progonocythere laeviscula* in the Bathonian of Saudi-Arabia and Israel, and *Procytheridea ihopyensis* in the Bathonian of Morocco. In the Callovian, 23 species are described from India of which 17 are common with Madagascar, 4 species are common with Tanzani, 2 are common with Saudi Arabia and one with Somalia. In the Oxfordian, 10 published species reported, of which four are common with Madagascar, two with Tanzania, one with Somalia and Israel. In the Kimmeridgian-Tithonian, 6 published species are described in India, of which one species *Majungaella perforata* is common with Madagascar, Tanzania and Somalia. From Andreu *et al.* (2012), it is clear that most of the Bathonian and the Callovian common ostracods species are attributed to Indo-East African Province belonging to South Gondwana and after the Callovian they are rapidly declining between India, Madagascar, Tanzania, Somali, Saudi Arabia and Israel.

A compilation on the ostracods from the Jurassic beds of Kachchh, Gujarat and Jaisalmer, Rajasthan published by Khosla *et al.* (2009) indicates that Jurassic beds of western India comprise 131 species. Of these, 100 species are known from Kachchh and 73 species from Jaisalmer. Twenty-seven species occur in the equivalent horizons in Madagascar, Tanzania, Somalia and Saudi Arabia. Their distribution is given in table 2. A perusal of this table shows that as many as 46 species, are common to Rajasthan (for name of the species see table 2). Twenty-two species, *Acrocythere oculata*, *Cytherella obscura*, *Cytherelloidea ipis*, *C. paradifficila*, *Fastigatocythere* (*Amicytheridea*) *triangulata*, *Fastigatocythere* (*Batella*) *befotakaensis*, *F.*(*B.*) *falcula*, *Fastigatocythere* (*Habocythere*) *bicruciata*, *F.*(*H.*) *dorsoangulata*, *F.*(*H.*) *jakhari*, *F.*(*H.*) *juglandica malgachica*, *F.*(*H.*) *ventrisulcata* , *Fastigatocythere* (*Fastigatocythere*) *retusa*, *Majungaella perforeta*, *Neurocythere whatleyi*, *Procytheridea ihopyensis*, *Progonocythere laeviscula*, *P. tenuisulcata*, *Trichordis devexa*, *T. jaisalmerensis*, *T. parvicarinata* and *T. praetexta* are common to Madagascar. Six species, *Cytherella disjuncta*, *C. masuguluensis*, *Fastigatocythere* (*Amicytheridea*) *triangulata*, *Galliaecytheridea remota*, *Majungaella perforeta* and *Progonocythere tenuisulcata* are common to Tanzania. Two species, *Fastigatocythere* (*Amicytheridea*) *triangulata* and *Majungaella perforeta* are common to Somalia. Four species, *Fastigatocythere* (*Amicytheridea*) *triangulata*, *Mandelstamia banniensis*, *Procytheridea ihopyensis* and *Progonocythere laeviscula* are common to Saudi Arabia and two species, *Cytherelloidea badiensis* and *Progonocythere laeviscula* are common to Israel.

The overall evidence based on distribution and correlation of ostracod taxa suggest that the Jurassic ostracods of the western India have strongest affinity with those of the Madagascar and to a lesser extent with Tanzania, Somalia, Saudi Arabia and Israel (Fig. 2). This analysis is in conformity with the earlier findings of Khosla *et al.* (1997, 2005) and Andreu *et al.* (2012). The Jurassic ostracods of western India also strongly supports the theory of plate tectonics that continents of today, were once very close to each other.

Importance of Ostracods in Petroleum Exploration

The importance of ostracods as microfossils in high resolution biostratigraphic zonation, correlation, dating of rocks and interpretation of palaeoenvironment, especially of Jurassic to Pleistocene rocks is second only to the foraminifers. Several factors in the ecology of ostracods may be used in study of the palaeoenvironment of rocks of fresh water, brackish waters or marine water origin.

The researches of the last 60 years have resulted in selecting some of the ecological factors as tool in oil exploration mainly for identifying the source rocks for hydrocarbons in sedimentary basins. Oertli (1971) studied the data of some 25 oil fields and numerous well sections and, suggested some simple and rapid methods which does not need much knowledge of systematic as general aspects of the fauna are utilized. The basic approach is to determine the conditions of sedimentation based on shell characters and to interpret whether or not these were optimum for the generation of hydrocarbons?

Oertli *et al.* (1983) found that ostracods are related to possible hydrocarbon reservoirs. Marine ostracods faunas that are very rich in the number of individuals and species (and few foraminifera) are situated (in the Middle East) near grainstones or dolomitic reservoirs. Further they have mentioned that in Western Europe ostracods associations of the Purbeck and Wealden of England and northwest Germany, which permitted excellent zonations and proved to be of individual help in petroleum exploration, where in the latter country the Wealden is one of the best reservoir rocks.

The organic matter may be converted into hydrocarbon if the deposition of sediments takes place rapidly, in reducing aquatic conditions under the optimum pressure and temperature.

Some morphological characters of ostracods reflect such ecological factors:

(a) The carapace and valve ratio;

(b) The degree of pyritization of the carapace;

(c) The current index;

(d) The compaction ratio based on the percentage of deformed carapaces; and

(e) The preservation of carapace.

Table 1. Distribution of ostracods in Gondwanaland.

Sl. No.	Name of Species	India		Mada-gascar	Tanzania	Somalia	Saudi Arabia	Israel
		Gujarat	Rajasthan					
		Kachchh	Jaisalmer					
1	*Acrocythere oculata* Mette and Geiger	+	+	+				
2	*Bairdoppilata badiensis* (Khosla, Jakhar and Mohammed)	+	+					
3	*B. jhuraensis* (Khosla, Jakhar and Mohammed)	+	+					
4	*Bairdoppilata* sp. E	+	+					
5	*Citrella belaensis* Khosla, Manisha Kumari and Darwin Felix	+	+					
6	*Cytherella disjuncta* Lyubimova and Mohan	+	+		+			
7	*C. kalajarensis* Khosla and Jakhar	+	+					
8	*C. masuguluensis* Bate	+			+			
9	*C. obscura* Lyubimova and Mohan	+	+	+				
10	*Cytherelloidea badiensis* Khosla, Jakhar and Mohammed	+	+					+
11	*C. bhujensis* Khosla, Jakhar and Mohammed	+	+					
12	*C. dhrangensis* Khosla and Jakhar	+	+					
13	*C. ipis* Grekoff	+	+	+				
14	*C. paradifficila* Khosla, Jakhar and Mohammed	+	+	+				
15	*Cytheropteron devai* (Khosla, Jakhar and Mohammed)	+						
16	*C. micropunctata* Khosla, Darwin Felix and Manisha Kumari	+	+					
17	*Fastigatocythere (Amicytheridea) flebilis* Khosla and Darwin Felix	+	+					
18	*F. (A.) globosa* Mette and Geiger	+	+					
19	*F. (A.) indica* Khosla and Manisha Kumari	+	+					
20	*F. (A.) triangulata* (Bate)	+		+	+	+	+	
21	*Fastigatocythere (Batella) befotakaensis* (Grekoff)	+	+	+				
22	*F.(B.) clavata* (Khosla, Jakhar and Mohammed)	+	+					
23	*F.(B.) depressa* (Khosla, Jakhar and Mohammed)	+	+					
24	*F.(B.) falcula* (Grekoff)	+	+	+				
25	*Fastigatocythere (Habocythere) bicruciata* (Grekoff)	+	+	+				
26	*F.(H.) dorsoangulata* (Grekoff)	+		+				

Sl. No.	Name of Species	India		Mada-gascar	Tanzania	Somalia	Saudi Arabia	Israel
		Guiarat	Raiasthan					
		Kachchh	Jaisalmer					
27	*F.(H.) jakhari* Khosla, Darwin Felix and Manisha Kumari	+		+				
28	*F.(H.) juglandica malgachica* (Grekoff)	+		+				
29	*F.(H.) ventrisulcata* (Khosla, Jakhar and Mohammed)	+		+				
30	*Fastigatocythere (Fastigatocythere) retusa* (Grekoff)	+	+	+				
31	*Galliaecytheridea grekoffi* Khosla and Jakhar	+	+					
32	*G. lodraniensis Khosla and Manisha Kumari*	+	+					
33	*G. remota* (Grekoff)	+	+		+			
34	*Glabellacythere hussaini* Khosla, Darwin Felix and Manisha Kumari	+	+					
35	*G. mathuri* (Khosla, Jakhar and Mohammed)	+	+					
36	*Lophocythere vertipolycostata* Khosla and Manisha Kumari	+	+					
37	*Majungaella perforeta* Grekoff	+		+	+	+		
38	*M. perforeta kachchhensis* Khosla, Jakhar and Mohammed	+	+					
39	*Mandawacythere kachchhensis* Khosla, Manisha Kumari and Darwin Felix	+	+					
40	*Mandelstamia banniensis* (Neale and Singh)	+	+				+	
41	*M. biswasi* Khosla, Manisha Kumari and Darwin Felix	+	+					
42	*M. kachchhensis* Khosla, Darwin Felix and Manisha Kumari	+	+					
43	*Monoceratina mouwanaensis* Khosla and Manisha Kumari	+	+					
44	*Morkhovenicythereis rectangularis* Khosla, Manisha Kumari and Darwin Felix	+	+					
45	*Neurocythere denticulate* (Kulshreshtha, Singh and Tewari)	+	+					
46	*N. whatleyi* (Khosla and Jakhar)	+		+				
47	*Paracypris contermia* Lyubimova and Mohan	+	+					
48	*P. kachchhensis* Khosla and Darwin Felix	+	+					
49	*P. mohani* Khosla, Darwin Felix and Manisha Kumari	+	+					
50	*Procytheridea ihopyensis* Grekoff	+		+			+	
51	*Progonocythere jaisalmerensis* Khosla, Jakhar, Nagori and Darwin Felix	+	+					
52	*P. laeviscula* Lyubimova and Mohan	+	+	+			+	+
53	*P. tenuisulcata* (Mette and Geiger)	+		+	+			
54	*Trichordis devexa* (Grekoff)	+	+	+				
55	*T. hadibhadangensis* Khosla, Darwin Felix and Manisha Kumari	+	+					
56	*T. jaisalmerensis* (Kulshreshtha, Singh and Tewari)	+	+	+				
57	*T. parvicarinata* Khosla and Jakhar	+		+				
58	*T. praetexta* (Grekoff)	+	+	+				

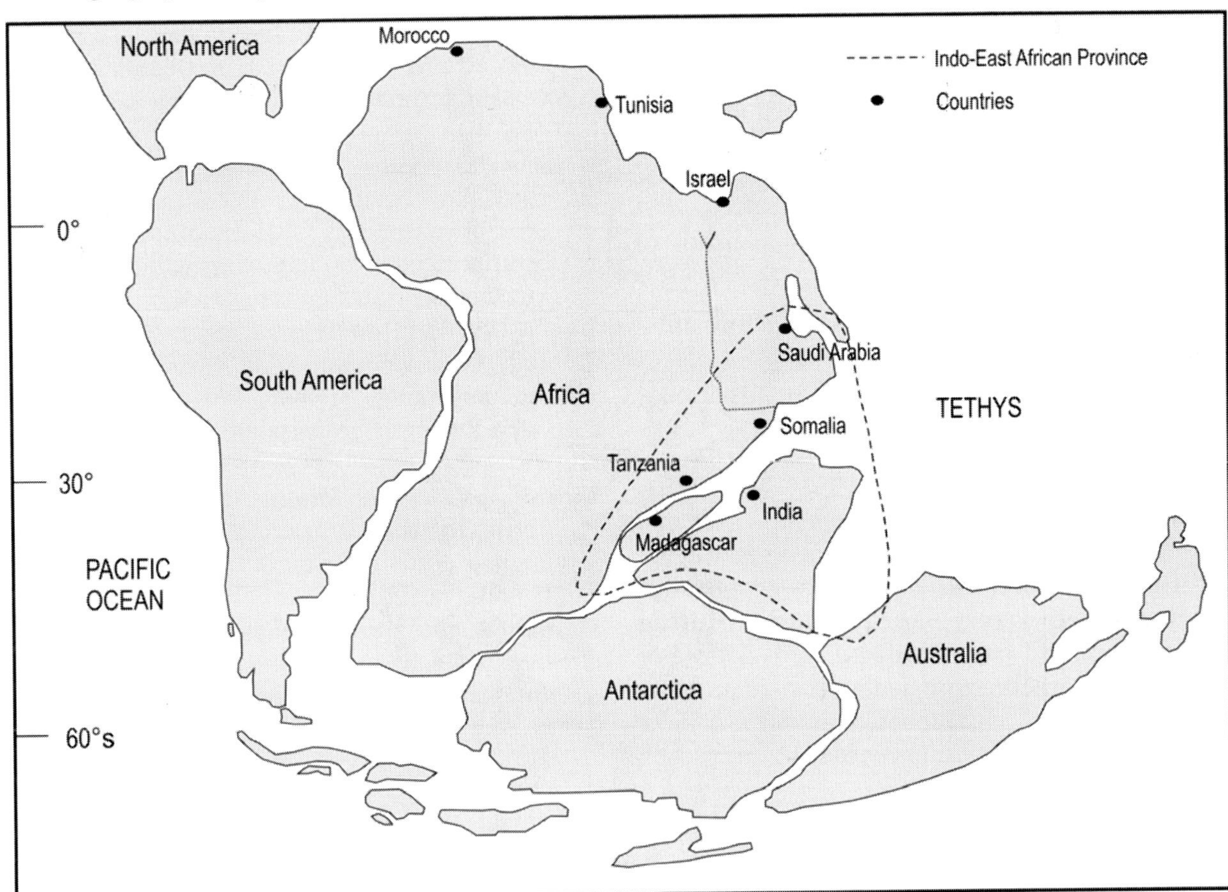

Figure 2. Palaeogeographic map of the Gondwanaland during Jurassic time (Based on Andreu *et al.* 2012)

Carapace and valve ratio

After the death of animal the adductor muscles of carapace disintegration and the valves tend to spring open during the transportation of ostracods. However, if they get buried earlier than opening of their valves the percentage of closed carapaces will be more than open valves. More is the number of closed carapaces higher will be the rate of sedimentation. Guha (1983) opined that the Eocene sediments of Cambay basin where 100% closed carapaces are found, were deposited rapidly while the Miocene sediments of Kachchh, here the percentage is only 80 indicate slower rate of sedimentation.

Degree of pyritization

The white or light brown color of carapace of the optimum oxygen condition changes into a shade of dark brown to black in the reducing conditions due to the accumulation of tiny pyrite crystals. This factor gives an access to determine the degree of reducing environment which is important for transformation of organic matter into hydrocarbons.

Current index

The current index is determined on the basis of percentage of adult and juvenile shells in a rich fauna. In an undisturbed fauna, both the adult and juvenile carapaces are expected. If the currents are active they may: (i) disturb the ratio of adult/juvenile carapaces; and (ii) create considerable aeration resulting into oxidation of organic matter. Usually a high current index, is not favorable for generation of petroleum. Higher valve of the index is met in the littoral, and low in the lagoonal assemblages.

Compaction ratio

Compaction of sediments is necessary for conversion of organic matter of sediments into hydrocarbons. This ratio may be determined directly on the basis of degree of deformation of carapaces or valves. It has been noticed that the thin carapaces easily get deformed in comparison to the thicker ones.

Preservation of carapace

In the poorly preserved specimens the protective organic coating is not disintegrated. This indicates the incomplete conversion of organic matter into hydrocarbons.

Some other important aspect, helpful in petroleum exploration are

(a) the relationships between normal and dwarfed forms;

(b) the relation between smooth and ornamented forms; and

(c) the diversity of species and their variation

Ostracod will always remain a precious source of information for petroleum exploration, even if other techniques (essentially geophysics) have undergone a very rapid evolution during last few years and have gained a high degree of precision in characterizing sediments. Even then characterizing of palaeonvironmental data remain one of the trump cards of ostracods research.

REFERENCES

Andreu, B., Colin, J.P., Singh, J. 2012. Middle and Upper Jurassic Ostracods from Western Kachchh, Gujarat, India: Biostratigraphy and paleobiogeography. Gondwana Research 22, 1110–1124.

Bate, R.H., 1975. Ostracods from Callovian to Tithonian sediments of Tanzania, East Africa. Bulletin of the British Museum (Natural History), Geology 26, 163–223.

Bate, R. H. 1976. New name for *Rhadinocythere* Bate, 1975. *Geol. Mag.,* 113 (5) : 489.

Bate, R. H. 1977. Upper Jurassic Ostracoda from Tanzania, East Africa. Actes VIe Colloq. Africain Micropal. Annales des Mines et de la Géologie, Tunis, 28 :163-183.

Basha, S. H., 1980. Ostracoda from the Jurassic System of Jordan, including a stratigraphical outline. Revista Española de Micropaleontología, vol. 12, no.2, pp. 231-254.

Benjamin S., 2008. Application of Ostracoda and Charophyta from the Late Jurassic to Early Cretaceous Tendaguru formation at Tendaguru, Tanzania (East Africa) — Biostratigraphy, palaeobiogeography and palaeoecology. Palaeogeography, Palaeoclimatology, Palaeoecology 264, 213–229.

Bhatia, S.B. and Mannikeri, M.S. 1977. Callovian Charophyta from Jaisalmer, western India. *Geol. Palaeont., Marburg* 11 : 187-196.

Bischoff, G. 1964. Ostracoden-studien im Libanon, 3: Die gattung *Cytherelloidea* im Oberen Jura und in der Unterkreide. Senckenbergiana Lethaea, 45 (1-4): 1-27.

Dépêche, F., Nindre, Y., Manivit, J., and Vaslet, D., 1987 Les ostracodes du Jurassique d'Arabie Saoudite centrale: Systématique, Répartition Stratigraphique et Paléogéographique. Geobios, Memoire spécial, no. 9, pp. 221-275.

Dingle, R.V. and Klinger, H.C., 1972. The stratigraphy and Ostracod fauna of the Upper Jurassic sediments from Brenton, in the Knysna Outlier, Cape Province. Transactions of the Royal Society of South Africa 40 (4), 279–298.

Dingle, R.V., 1988. Marine Ostracod distributions during the early breakup of Southern Gondwanaland. In: Hanai, T., Ikeya, N., Ishizaki, K. (Eds.), Evolutionary Biology of Ostracoda. Elsevier/ Kodansha, Amsterdam-Tokyo, pp. 841–854.

Grekoff, N. 1963. Contribution a l'étude des ostracodes du Mésozoique moyen (Bathonien-Valanginien) du Bassin de Majunga, Madagascar. Rev. Inst. Francais Pétrole et Annales des Combustibles Liquides, 18 (12) : 1709-1762.

Guha, D.K. 1977. On some Mesozoic Ostracoda from subcrops of Banni, Rann of Kutch, India. *Proc. VI Indian Colloq. Micropal. and Strat.* : 84-90.

Guha, D.K. 1983. Use of Mesozoic and Tertiary Indian Ostracoda in oil Exploration. In Maddocks, R. F. (ed.): Application of ostracoda, Proc. 8th Internat. Symp. on Ostracoda, July 26-29, 1982: 433-441, Houston (Texas).

Khosla S. C. and Jakhar S. R., 1993 : *Amicytheridea* Bate, 1975 and *Paratrichordis,* a new subgenus of ostracoda, from the Jurassic beds of Kachchh, India. K. G. McKenzie & P. J. Jones (eds.), Ostracodes in the Earth and Life Sciences, Proceeding of the 11th International Symposium on Ostracoda, Warrnambool / Victoria / Australia / 8-12 July 1991, A. A. Balkema / Rotterdam / Neatherland, pp.141 - 152.

Khosla S. C. and Jakhar S. R., 1999 : A note on the Ostracode fauna from the Jurassic beds of Jumara dome, Kachchh. Jour. Geol. Soc. India, vol. 54, pp. 43-49.

Khosla S. C., Jakhar S. R. and Mohammed M. H., 1997 : Ostracodes from the Jurassic beds of Habo Hill, Kachchh, Gujarat, India. Micropalaeontology, vol. 43(1), pp. 1-39.

Khosla S. C., Jakhar S. R., Nagori M. L., A. Darwin Felix and Manisha Kumari 2003a. Jurassic Ostracode Fauna from the Type Section of Khadir Formation, Kachchh and *Lophocythere vertipolycostata* n. sp. Jour. Geol. Soc. India, vol. 61(1), pp. 61-74.

Khosla S. C., Manisha Kumari, A. Darwin Felix, Jakhar S.R. and Nagori M. L., 2003b. The Jurassic Ostracode fauna of Bela Island, Rann of Kachchh, Gujarat. Current Science, vol. 85(3), pp. 249 - 251.

Khosla S. C., Jakhar S. R., Nagori M. L. and A. Darwin Felix, 2003c. Ostracod Genus *Progonocythere* from Jurassic Beds of Western India. Gond. Geol. Magz., Spl. Vol. 6, pp. 167-174.

Khosla S. C., Jakhar S. R. and Mohammed M. H., 2004. Ostracodes from the Jurassic beds of Jhura Hill, Kachchh, Gujarat. Jour. Geol. Soc. India, vol. 63(1), pp. 15-28.

Khosla S. C., Manisha Kumari, A. Darwin Felix, Jakhar S. R. and Nagori M. L., 2005. Middle Jurassic Ostracoda from the Northern Island Belt, Rann of Kachchh, Gujarat. Journal of the Paleontological Society of India, vol. 50 (1), pp. 17-64.

Khosla S. C., Jakhar S. R. Manisha Kumari, and S. Dubey, 2006. Middle Jurassic Ostracoda from the Jaisalmer Formation, Jaisalmer District, Rajasthan, India. Journal of the Paleontological Society of India, vol. 51 (1), pp. 1-31.

Khosla S. C., Jakhar S. R., and Manisha Kumari, 2009. Atlas of the Middle Jurassic Ostracods from western India. Journal of the Paleontological Society of India, Lucknow. 167p.

Lyubimova, P. S., Guha, D. K. and Mohan, M. 1960. On Ostracoda of Jurassic and Tertiary deposits from Kutch and Rajasthan (Jaisalmer), India. Bull. Geol. Min Metall. Soc. India, 22 : pp. 1-61.

Mannikeri, M.S., 1996. Two new ostracode genera *Jainiana* (Family: Progonocytheridea) and *Bhatiana* (Family: Trachyleberididae) from the Middle Jurassic of Jaisalmer, Western Rajasthan, India. Proc. XV Indian Colloq. Micropal. and Strat. Dehradun : 397-406.

Mette, W., 1995. Ostracods from the Middle Jurassic of southern Tunisia. Beringeria 16, 259–348.

Mette, W., 1997. Palaeoecology and palaeobiogeography of the Middle Jurassic ostracods of southern Tunisia. Palaeogeography, Palaeoclimatology, Palaeoecology 131 (1–2), 65–113.

Mette, W., 2004a. Deciphering and interpreting Jurassic diversity and distribution trends for marine ostracods in Gondwana, the role of biologic endogenous and exogenous processes. Beringeria 24, 3–35.

Mette, W., 2004b. Middle to Upper Jurassic sedimentary sequences and marine biota of the early Indian Ocean (Southwest Madagascar): some biostratigraphic, palaeoecologic and palaeobiogeographic conclusions. Journal of African Earth Sciences 38, 331–342.

Mette, W., Geiger, M., 2004a. Bajocian and Bathonian ostracods and depositional environments in Madagascar (Morondava Basin and southern Majunga Basin). Beringeria 34, 37–56.

Mette, W. and Geiger, M. 2004b. Taxonomy and palaeoenvironments of Callovian ostracoda from the Morondava Basin (south-west Madagascar), 34: 57-87.

Mette, W., Geiger, M., 2004c. Middle Oxfordian to early Kimmeridgian ostracoda and depositional environments of south-west Madagascar. Beringeria 24, 89–115.

Neale, J.W. and Singh, P. 1986. Jurassic Ostracoda from the Banni well no. 2, Kutch, India. *Rev. Esp. de Micropal.* 17 (3) : 347-372.

Oertli, H.J., 1971. The aspects of ostracodc fauna – a possible new tool in Petroleum Sedimentology. In Oertli, H. J. (ed.), Paléoécologie des Ostracodes, Bulletin du Centre des Researches Pau-SNPA, suppl. 5. pp. 137-151.

Oertli, H.J., Grosdidier, E and Fevre J. Le, 1983. Ostracods in petroleum exploration. In Maddocks, R. F. (ed.): Applications of ostracoda, Proc. 8th Internat. Symp. on Ostracoda, July 26-29, 1982: 19-34, Houston (Texas).

Rafara, A. H., 1990. Les ostracodes du Jurassique Moyen-Neocomien du Bassin de Majunga (Madagascar) - comparison avec les autres faunes Gondwaniennes. Geobios, no. 23, fasc. 4, pp. 415-443.

Rosenfeld, A., Gerry, E., and Honigstein, A., 1987. Jurassic ostracodes from Gebel Maghara, Sinai, Egypt. Revista Española de Micropaleontología, vol. 19, no. 2, pp. 251-280.

Rosenfeld, A., Honigstein, A., Gerry, E., Oertli, H. J. and Flexer, A.,1988. Early Jurassic Ostracodes from the Ardon Formation in Israel and Sinai. Geological Survey of Israel Current Research, vol. 6, pp. 50-55.

Rosenfeld, A., and Honigstein, A., 1991. Callovian-Oxfordina ostracodes from the Hamakhtesh Hagadol Section, Southern Israel. Revista Española de Micropaleontología, vol. 23, no. 3, pp. 133-148.

Rosenfeld, A., and Honigstein, A., 1998. Kimmeridgian ostracodes from the Haluza Formation in Israel. Revista Española de Micropaleontología 30 (2), 77–87.

Sames, B., 2008. Application of Ostracoda and Charophyta from the Late Jurassic to Early Cretaceous Tendaguru Formation at Tendaguru, Tanzania (East Africa). Biostratigraphy, palaeobiogeography and palaeoecology. Palaeogeography, Palaeoclimatology, Palaeoecology 264, 213–229.

Schudack, M.E., Schudack, U., 2002. Ostracods from the Middle Dinosaur Member of the Tendaguru Formation (Upper Jurassic of Tanzania). Neues

Jahrbuch für Geologie und Paläontologie, Monatscheft 6, 321–333.

Singh, S.N. and Jai Krishna, 1969. A preliminary note on the Mesozoic of Jaisalmer area, Rajasthan. Journal of the Palaeontological Society of India, vol. 12, pp. 41-43.

Singh, S.N. and Kulshreshtha, S.K., 1973. A new genus of Ostracoda from Jurassic of Rajasthan, India, pp. 184-187. In Singh, S.N. (ed.), Proceedings of the 2nd Indian Colloquium on Micropalaeontology and Stratigraphy, Lucknow.

Whatley, R.C., 1988. Ostracoda and palaeogeography, pp.103-123, figs 1-2. In Deckker, P.D., Colin, J.P., and Peypouquet, J.P. (eds.), Ostracoda in the Earth Sciences, Elsevier, The Netherlands.

Hon'ble reviewer

1. **Prof. G.R. Jakhar**
   Vice Chancellor
   MGS University
   BIKANER - 334 001
2. **Prof. S.C. Mathur**
   Chairman
   Department of Geology
   JNV University
   JODHPUR - 342 005

32

Petrographic Characteristics of Coal Seams of Kamptee Coalfield, Central India

Anil M. Pophare[1], C.P. Ramteke[1] and A.L.V. Prasad[2]

[1]Department of Geology, Rashtrasant Tukadoji Maharaj Nagpur University, Nagpur-440 001
[2]Central Institute of Mining and Fuel Research, Nagpur Unit, Nagpur-440 001
E-mail: apophare@gmail.com

Abstract

Petrographic characters and maceral content of coal determines the nature of coal, mode of deposition and coalification. The petrographic study of macerals from coal seams of Kamptee coalfield show strong inverse relationship between coal macerals of inertinite and vitrinite groups, which indicates common source material for both the macerals but different genetic tracts. The coal seams of this coalfield are rich in inertinite and relatively poor in vitrinite. The IV factor of 60.69 for coal seams of Kamptee coalfield is close to the values obtained for wet/dry forest swampy conditions of peat development. Thus, coal petrography and maceral studies reveals that the coal seams of Kamptee coalfield were deposited in fluvio-lacustrine environment with alternating oxic and anoxic conditions in wet/dry forest swampy condition during peat stage development.

Keywords: Coal petrography, Macerals, Depositional environment, Kamptee Coalfield.

Introduction

Coal is the main fossil fuel and source of heat and energy and accounts for ~ 80% of total energy generation in the world. India has a huge coal reserve of ~ 276 billion tonnes, spread in more than 60 coalfield of Gondwana and Tertiary basins. The Gondwana basins of Peninsular India constitute main coal repository of India. These basins have considerable similarity in terms of geographical distribution pattern and mode of evolution and occurs as rectilinear belts of failed rift origin that are controlled by Precambrian crustal structure such as Archean-Proterozoic cratonic sutures and mobile belts. The Gondwana basins of Peninsular India are located along the major river valleys, which include Damodar valley, Son-Narmada valley, Mahanadi valley, Pranhita-Godavari valley, Wardha valley and Rajmahal-Birbhum belt (Acharyya, 2000).

In central India, Gondwana Supergroup host coal deposits in Pench-Kanhan-Tawa valley coalfields, Kamptee coalfields, Wardha valley coalfields in Maharashtra and Sohagpur, Sonhat and Jatpura coalfields in Madhya Pradesh (CMPDIL, 1993). These coalfields contain mostly bituminous to sub-bituminous rank of coal, which occurs at a depth of about 300 to 1200 m (Acharyya, 2000; Dasgupta, 2006). Geophysical surveys carried out to ascertain the coal potential of eastern Maharashtra have revealed that Gondwana sediments were deposited along three axes. The principle axis is continuation of Godavari valley, a northen axis extends from Chimur to beyond Katol and a smaller southern axis that extends from Punwat to west of Warora (Ramkrishna *et al.*, 1999; GSI, 2008).

The Kamptee coalfield has potential coal reserves, where exploration for coal was carried out up to 300 m depth (Basu *et al.*, 1978). The estimated coal reserves proved up to the depth of 300 m are 0.24 billion tonnes (Dutt *et al.*, 2001; Dutt, 2007).

The Kamptee coalfield is located towards the north-central part of Godavari-Wardha valley (Fig. 1). This is one of the important coalfields of Central India, situated along Kanhan river and extends towards the north up to Saoner-Kelod areas over a length of ~ 45 km between Tekari and Saoner. This coalfield has an areal extent of ~ 136 Km^2 and bounded by latitude 21° 12': 21° 30" and longitude 78° 50': 79° 16" (Fig. 1).

The study area of present investigation lies towards the north and north-western part of Nagpur and located at a distance of about 15 -30 km from Nagpur city (Fig. 1). The study area is covered under the Survey of India Topographical Sheet Nos. 55 K/15 and 55 O/3, and includes the underground coal mines of Silewara (21° 17' 58.5" : 79° 07' 25.5"), AB-Incline (21° 18' 15": 79° 06' 27.4"), Pipla (21° 18' 56.6": 79° 04' 56.9"), Saoner WCL-1 (21° 21' 29.3": 78° 55' 48"), Saoner WCL-2 (21° 22' 0.2": 78° 55' 48") and Saoner WCL-3 (21° 21' 1.7": 78° 54' 30.7") (Fig. 1). During the course of present study, coal samples were collected by channel sampling from the coal seams of these underground coal mines of Western Coalfields Limited (a subsidiary of Coal India Limited). The study was concentrated on petrographic characterisation and maceral composition of coal seams of Kamptee coalfield of Central India to ascertain their environment of deposition.

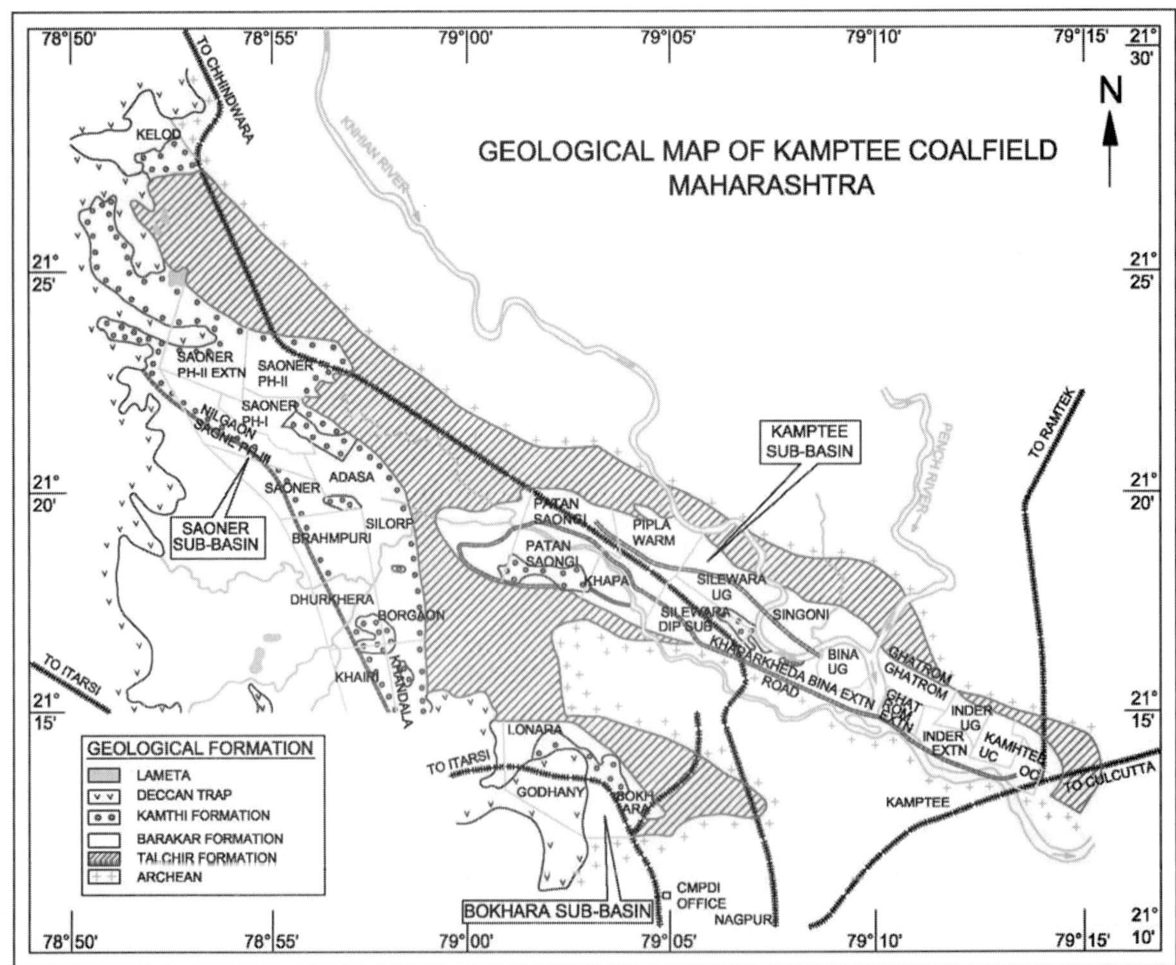

Fig. 1: Generalised geological map of Kamptee coalfield (after Gee and Khedekar, 1952; Shingte and Sah, 1993)

Table 1: Generalised stratigraphic succession of Kamptee coalfield

Age	Supergroup/ Group	Formation	Thickness (m)	Lithology
Recent		Alluvium	1.00-53	Black cotton soil and sandy soil with pebbles
Eocene to Upper Cretaceous	Deccan Trap	Basalt extrusive and intrusive	0.25- 63	Fine grained basaltic rock
Cretaceous		Lameta	0.83-26	Cherty limestone and sandstone
~~ Unconformity ~~~~~~~~~~~~~~~~~~~~				
Upper Permian	G o n d w a n a s	Kamthi	149	Brown colour, medium to coarse grained ferruginous sandstone and clay
~~			~~~~~~~~~~~~~~~Unconformity~~~~~~~~~~~~~~	
Middle Permian		Motur	1.22-300	Variegated clay, chloritic sandstone with occasional shale band
Lower Permian		Barakar	15.35-284	Fine to coarse grained sandstone (white to grey colour) with shale, carbonaceous shale, fireclay and coal seams
Upper Carbonaceous		Talchir	3.75-118	Greenish shale with siltstone, sandstone, basal conglomerate and absence of coal seams
~~ Unconformity ~~~~~~~~~~~~~~~~~~~~				
Archean			---	Gneisses and schist

Geology

The Kamptee coalfield has horse-shoe shaped configuration and aligned in the NW-SE direction in the drainage areas of Pench and Kanhan rivers (Fig. 1). The Kamptee coalfield comprises of an ascending succession of Talchir, Barakar, Motur, Kamthi and Lameta formations of Gondwana Supergroup (Table 1) (Gee and Khedekar, 1952; Raja Rao, 1982; Shingte and Sah, 1993; Deolasee, 1996; GSI, 2008).

Major portion of Kamptee coalfield along with coal measures are covered by younger sediments, Deccan traps and alluvium (Vaidy-anadhan and Ramakrishnan, 2008). In the study area, lower Gondwana rocks occur in three distinct patches separated by Deccan traps in large tract (Fig. 1), where isolated exposures of Kamthi Formation are found as small ridges or in valleys. The strike of these beds varies from NW-SE to WNW-ESE with dip towards SW to SSW. The amount of dip varies from nearly horizontal to as much as 45° (Dutta *et al.*, 1961).

The Kamptee coalfield has been divided into *three* sub-basins namely, Kamptee sub-basin, Saoner sub-basin and Bokhara sub-basin. These sub-basins are separated by metamorphic highs, exposing Talchir Formation in between the sub-basins, below the detrital cover (Fig. 1). The southern boundaries of Kamptee and Bokhara sub-basins are faulted. The Saoner sub-basin is mostly concealed below the Deccan traps (Gee and Khedekar, 1952; Shingte and Palker, 1974; Shingte and Krishnan, 1981; Shingte and Sah, 1987, 1993).

Coal Seams of Kamptee Coalfield

Barakar Formation is the main coal bearing entity in Kamptee coalfield and composed of buff to white coloured fine to medium grained sandstone, shales, clays, carbonaceous shale and coal seams (Gee and Khedekar, 1952; Dutt, 1953; Dutta *et al.*, 1961; Wadia, 1966; Krishnan, 1982; Shingte and Sah, 1993; Nandgaonkar, 1996).

There are total *five* workable coal seams in Kamptee coalfield designated as seam no. I to V in ascending order (Sharma and Ram, 1966; Shingte and Sah, 1993). All the coal seams are laterally continuous and younger seams are successively above the older one in the northern direction due to homoclinal nature of the strata. The coal seams have variable thickness and some seams split into two to three sections with sand-stone and shale partings and designated as bottom, middle and top (Table 2). All the coal seams are composed of inter banded carbona-ceous shale-shaleycoal-coal and at some places fire clays (Gee and Khedekar, 1952; Chandra and

Shingte, 1976). The inter seam sand bodies are of sheet, elongated (ribbon, dendroid, belt, shoestring) and wedge shaped, which are similar to the fluvial sand bodies described from various part of the world (Wadia, 1966; Krishnan, 1982; Pettijohn, 1984; Sengupta, 1996; Hota, 2000, 2008; Hota and Maejima, 2005).

The coal seams of Kamptee coalfield are mostly non-gassy seams of C and D grade and contain bituminous to sub-bituminous coal of non-coking nature (Chawdhury, 1971; Dutt *et al.*, 2001; Dutt, 2007). The generalised thickness of these coal seams in the entire Kamptee coalfield area is as under-

Coal Seam	Thickness (m)
V	6.88-11.88
IV	13.72-23.56
III	0.40-4.12
II	1.65-9.2
I	1.34-7.26

However, in the study area of coal mines under investigation, the coal seams have been designated as seam I, II, III, IV and V from bottom to top. The thickness of sampled coal seams is variable (Table 2). The inter seam partings have minimum and maximum thickness of about 0.4 - 5m, 2 - 40m, 0.71 - 17 m, 0.63 - 12.61 m and 1- 12.26 m, respectively (Gee and Khedekar, 1952; Shingte and Sah, 1993).

The Coal Macerals

Coal is a heterogeneous and complex substance that consists of various micro-components called macerals. Macerals can be defined as the physically discrete components of coal mass distinguishable under the microscope (Stopes, 1935) and may be differentiated on the basis of morphology, reflectance, size and polishing relief. The properties of maceral become more akin to one another as the rank of coal increases.

The International Committee for Coal and Organic Petrology (ICCP) and International Standard Organization (ISO) has recommended three groups of macerals viz. vitrinite, liptinite, (exinite) and inertinite. Whereas, classification developed in India include one more maceral group i.e. semifusinite (ICCP, 1971, 1994, 1998, 2001; Chaudhuri and Ghose, 1978; Chaudhuri et al., 1998; 2000; 2001).

Vitrinite macerals are derived from woody and panenchymatus tissue of roots, barks and leaves composed of cellulose and lignin (Pareek, 2004). It is derived from woody tissues with cell structure (tellinite) or without cell structure (collinite) and occurs as detrital fragments (vitrodetrinite) with gelinite that combine to form the groundmass vitrinite (collodetrinite). Gelinite probably form in cell lumens (corpogelinite) or migrate into fractures (ICCP, 1998; Taylor and Liu, 1998).

Liptinite macerals include cutinite (remains of outer protective cuticles of plants), sporinite (remains of spores), resinite (preserved resin bodies or migrated resin), alginite (preserved algae) and liptodetrinite (fragments of various liptinitie macerals) (Rimmer *et al.*, 2006).

Table 2: Sample location, thickness and depth of coal seams of Kamptee coalfield

Sample No	Coal Seams	Depth (m)	Thickness (m)	Type of Mining Activity	Type of Coal Samples Collected
CR-2	Pipala Seam-II	132.61	6.42	Underground	Channel Sampling
CR-10	AB- Incline Seam-IV (Top)	107.30	3.56	Underground	Channel Sampling
CR-8	Silewara Seam-II (Top)	152.03	4.80	Underground	Channel Sampling
CR-1	WCL-2 Seam-IV (Middle)	97.04	4.90	Underground	Channel Sampling
CR-4	WCL-2 Seam-IV (Middle)	97.04	4.90	Underground	Channel Sampling
CR-7	WCL-2 Seam -V	66.40	5.24	Underground	Channel Sampling
CR-9	WCL-2 Seam -V	66.40	5.24	Underground	Channel Sampling
CR-11	WCL-3 Seam-IV (Middle)	98.50	6.15	Underground	Channel Sampling
CR-12	WCL-3 Seam-IV (Bottom)	108.70	2.05	Underground	Channel Sampling
CR-13	WCL-3 Seam-V (Bottom)	48.20	5.31	Underground	Channel Sampling

The inertinite present in coal mostly represents woody tissue consisting of organic material that were oxidised at the time of deposition, either due to sub-aerial exposure, fungal and bacterial alteration or due to wild fire (Taylor and Liu, 1998; Scott, 2002; ICCP, 2001; Rimmer *et al.*, 2006).

The fusinite macerals are concentrated in the desert fractions and characterised by higher reflectance level. This may be due to forest fires, contemporaneous with peat swamp development and accumulation (Scott, 2002; Rimmer *et al.*, 2006).

Macerals of liptinite group react slowly in the early part of coal maturation and more rapidly during latter part. The inertinite macerals react vigorously in the early part of maturation history, whereas vitrinite alter more steadily throughout the rank of coal (Murchison, 1991; Chaudhuri *et al.*, 2000; Choudhury *et al.*, 2001, 2008a, 2008b).

The vitrinite and liptinite macerals are distinguished by their origin from various plant materials. Vitrinite is a coalification product of humic substances and mainly originate from lignin and cellulose of the cell walls. Exinites are derived from the hydrogen rich plant remains *e.g.* spores, pollens, resin, waxes, fats, *etc.* and are not the product of humifiable materials. Vitrinite and exinite are characterised by higher aromatic fraction and correspondingly higher oxygen content, whereas macerals of liptinite group are distinguished by comparatively high carbon and low hydrogen content along with increased levels of aromatization. Although inertinite macerals are derived from the same original plant substances as vitrinite, but they pass through a different primary decomposition stage. For instance, cell walls of wood can be transformed into vitrinite with cell structure by the process of humification and gelification, while forest fire or mouldering produces fusinite and semifusinte. During mouldering, aerobic bacteria and fungi take part in the reaction, producing hydrogen poor humic substances, which latter may become oxyfusinites and macrinite. The first process *i.e.* formation of vitrinite is termed as vitrinisation and second process is termed as fusinisation. In both the cases cellulose structure of plant tissues are visible up to the bituminous coal rank stage. When plant material is completely destroyed it may be fusinitized to form micrinite, which develops perhaps during coalification from liptinite bodies (Chaudhuri *et al.*, 2000; Choudhury *et al.*, 2001, 2008a, 2008b).

Changes in property of macerals are the result of exposure of macerals to various influences of natural environment that include rate of temperature rise, duration of exposure to elevated temperatures, oxidative conditions, overburden pressure and stresses. Inter-relationship of these influences is complicated, but normally rise of temperature has been regarded as dominant factor that causes coal reactions. These factors have equal significance in the regional coalification changes (Choudhury *et al.*, 2001, 2008a, 2008b).

Petrographic Characteristics of the Coal Seams

During the present work, total *seven* coal samples were selected (Table 2) for petrographic studies by preparing the pellets (polished blocks) using standard techniques (ICCP, 1971, 1994, 1998). The coal pellets (polished blocks) were studied for maceral and micro-lithotypes analysis using Leica MPV and DMRXP HC Polarized light microscopes at the Central Institute of Mining and Fuel Research (CIMFR), Digwadih Campus, Dhanbad, following ICCP (1971), ISO:7404 (Part-III, 1994; Part-V, 1994-10-1) and IS: 9127 (Part-I, 1992; Part-III, 2002; Part-IV, 2002; Part-V, 2004) standards.

The polished blocks of coal samples were studied under incident light with oil-immersion objective. The macerals and micro-lithotypes were measured using both white light and fluorescent light irradiation using the same system (Cloke and Lester, 1994; Choudhury *et al.*, 2001, 2007, 2008a, 2008b). The macerals of the entire three groups *i.e.* vitrinite, liptinite and inertinite are identified in coal seams of Kamptee coalfield. The representative microphotographs of various macerals occurring in the coal seams of the study area are shown in Figs. 2 to 9.

Vitrinite Group of Macerals

Petrographic studies of the coal seams from Kamptee coalfield indicate that vitrinite maceral is characterised by grey to white colour with their reflectance lies between darker liptinite and brighter inertinite (Figs. 2a, 2b, 2c, 2d, 2e and 2f) (Singh, 1998; Singh and Singh, 2001). The macerals of vitrinite group *i.e.* tellinite, collinite and vitrodetrinite are distinguished in white light on the basis of presence or absence of cell structure. Corpocollinite is a sub-maceral of vitrinite group (collinite). It is characterised by circular or oval shape structure and occurs as homogenous bodies or cell filling with colour and reflectance similar to other vitrinite macerals of the same coal (Fig. 3a) (Diessel, 1992; Singh, 1998). Corpocollinite has been reported to occur in very large sizes, measuring up to 300 micron (Singh, 1998).

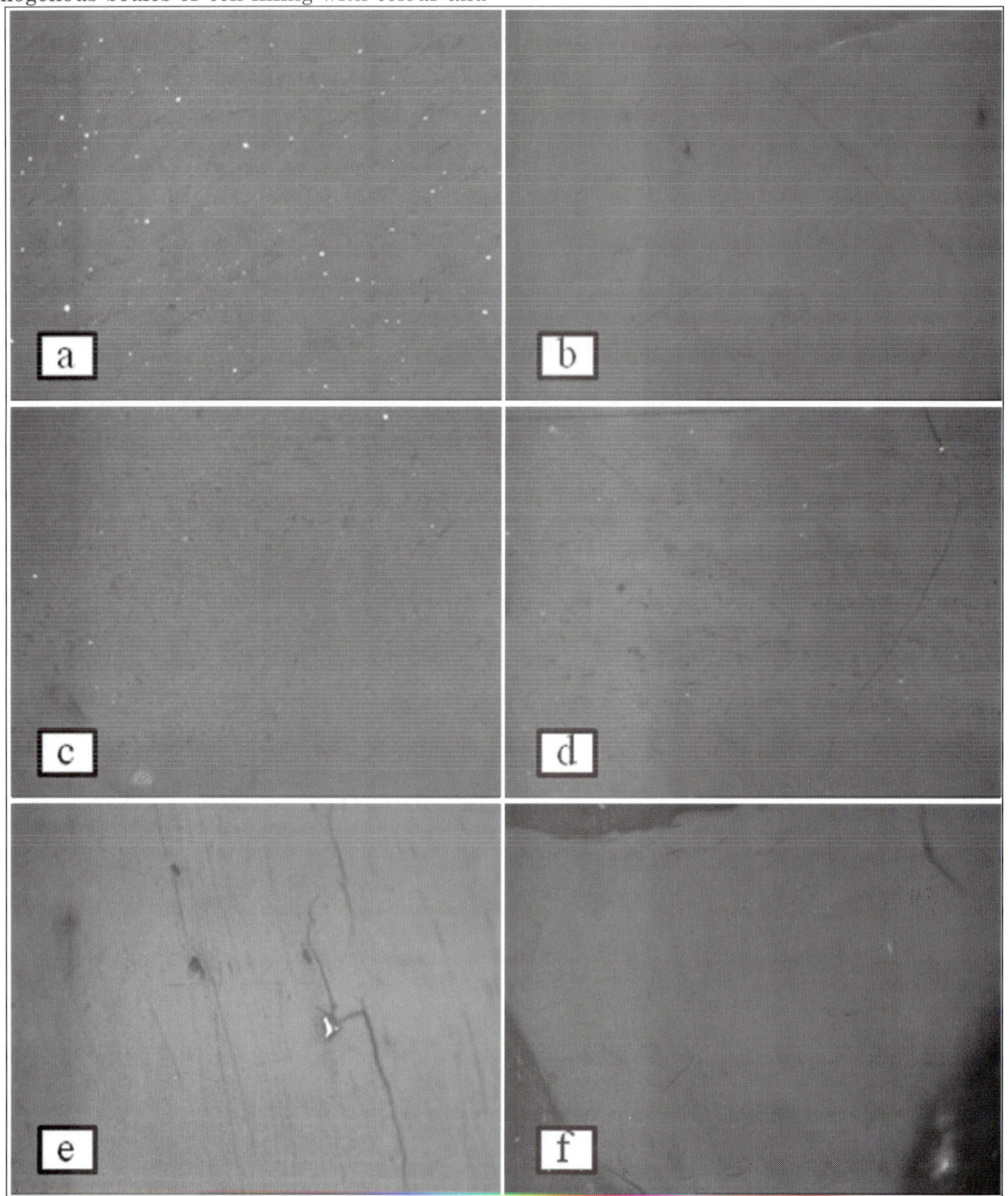

Fig. 2: Photomicrographs of coal macerals from coal seams of Kamptee coalfield (incident light with oil immersion objective; 50x) a) Vitrinite [AB-Incline mine, coal seam-IV (top)] b) Vitrinite [Pipla mine, coal seam-II] c) Vitrinite [Saoner WCL-2 mine, coal seam-IV (middle)] d) Vitrinite [Saoner WCL-2 mine, coal seam-V] e) Vitrinite [Saoner WCL-3 mine, coal seam-V (bottom)] f) Vitrinite [Saoner WCL-3 mine, coal seam-V (bottom)]

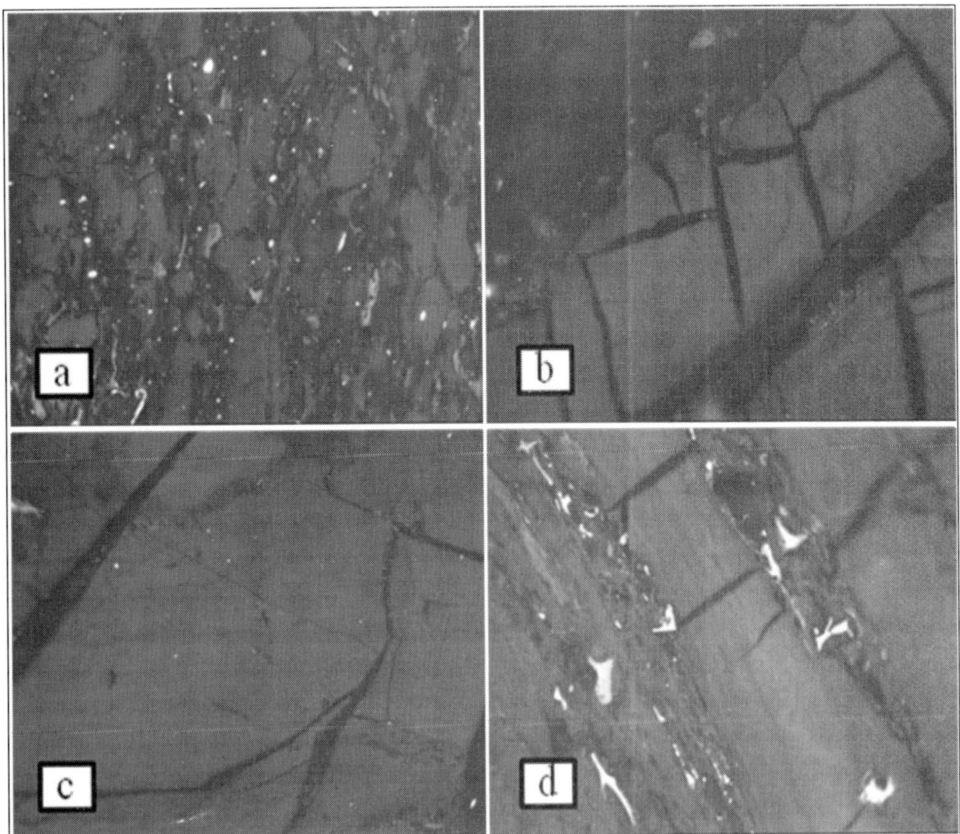

Fig. 3: Photomicrographs of coal macerals from coal seams of Kamptee coalfield (incident light with oil immersion objective; 50x) a) Corpocollinite [AB-Incline mine, coal seam-IV (top)] b) Oxidised vitrinite [Pipla mine, coal seam-II] c) Oxidised vitrinite [Saoner WCL-2 mine, coal seam-IV (middle)] d) Oxidised vitrinite [Saoner WCL-3 mine, coal seam-V (middle)]

Oxidised vitrinite is recorded in the coal seams of Kamptee coalfield at Pipla (seam-II), Saoner WCL-2 (seam-IV, middle) and Saoner WCl-3 (seam-V, middle) coal mines (Figs. 3b, 3c and 3d). Vitrinite is oxidised due to exposure to atmosphere leading to deterioration of quality (Singh and Singh, 2001; Pareek, 2004). In coal seams of Kamptee coalfield, vitrinite is associated with inertinite, liptinite and mineral matter (Figs. 4a, 4b, 4c, 4d, 4e and 4f).

In this group the identified macerals are fusinite, semifusinite, sclerotinite and inertodetrinite (Ahmed, 1998). Inertodetrinite represent broken fragments of inertinite components (Chandra and Taylor, 1982; Ahmed, 1998). Fusinite is characterised by presence of cell structure and higher reflectance. Occasionally fusinite bands without cell structure are also observed. In some cases cell structure is crushed and produces bogen structure. The bogen (bow) structures are developed when the brittle cell walls are broken into curved fragment either by overburden or tectonic pressure (Diessel, 1992; Ahmed, 1998). Fusinite and semifusinite occur in the form layer and lenses in which cell cavities (lumens) are either empty or may be filled with a wide range of substances including gelovitrinite, geloinertinite, resinous material and various minerals (Diessel, 1992; Pareek, 2004).

Inertinite Group of Macerals

In coal seams of Kamptee coalfield, the maceral fusinite occurs in the form of layer or lenses and is highly reflective (Figs. 5a, 5b and 5c). It has well defined cellular structures of parenchyma (wood) or sclerenchyma, with cell cavities displaying various sizes and forms. These cavities may be empty or filled with gelinite, exudetinite, clay, pyrite and mixture of organic and inorganic matter (Pareek, 2004). These coal samples show fusinite with bogen structure (Fig. 5c). Presence of bogen structure in fusinite, shining structure in the coal seams of the study area along with abnormalities in coal rank indicates tectonic disturbance in this region (*c.f.* Ahmed, 1998).

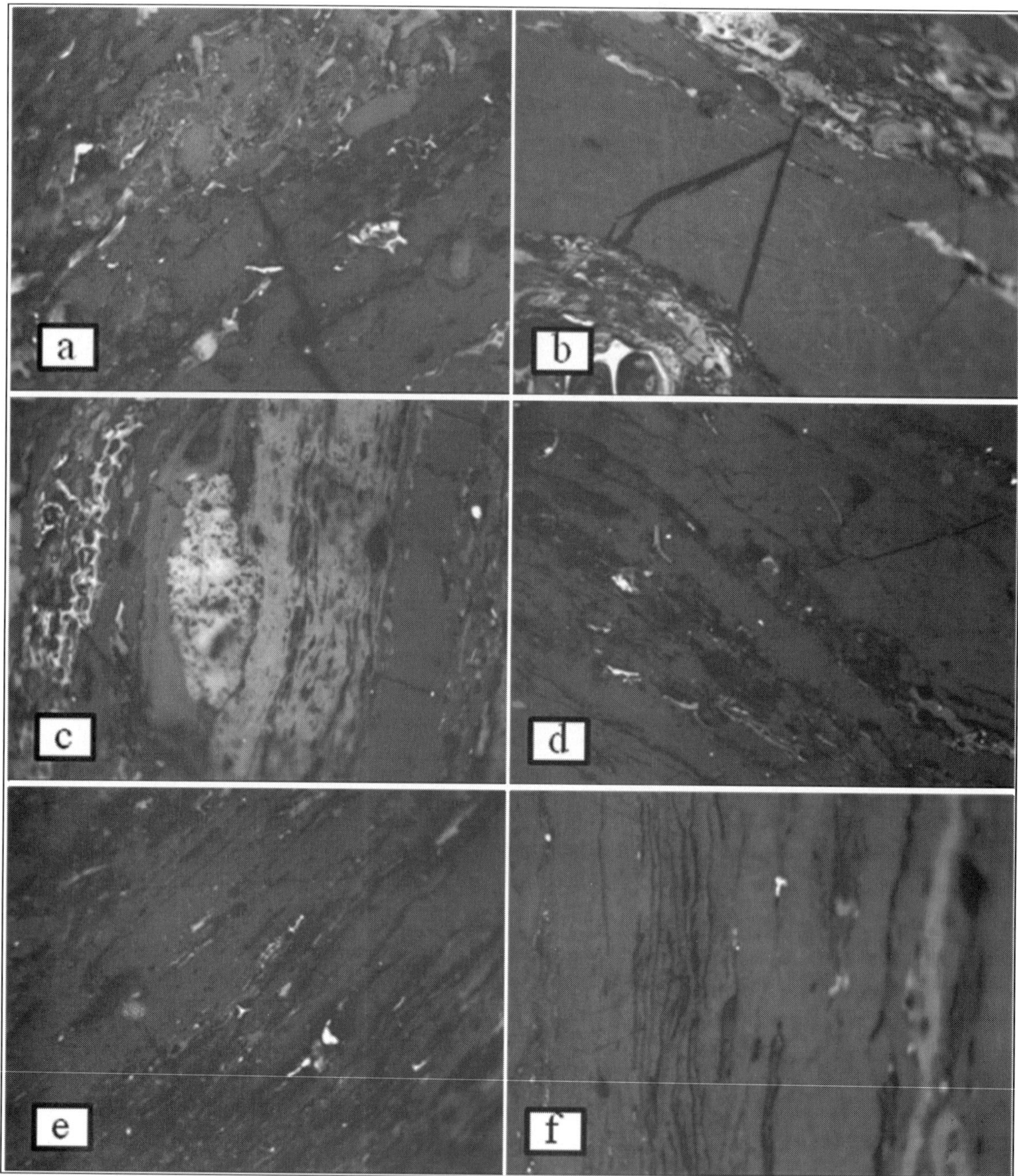

Fig. 4: Photomicrographs of coal macerals from coal seams of Kamptee coalfield (incident light with oil immersion objective; 50x) a) Vitrinite with inertinite [Saoner WCL-2 mine, coal seam-IV (middle)] b) Vitrinite and inertinite [Saoner WCL-2 mine, coal seam-IV] c) Vitrinite, inertinite with mineral matter [Saoner WCL-2, coal seam-IV (middle)] d) Vitrinite and liptinite [Saoner WCL-2 mine, seam-IV (middle)] e) Vitrinite and liptinite [Saoner WCL-2 mine, seam-V] f) Vitrinite and liptinite [Saoner WCL-2 mine, seam-IV (middle)]

Semifusinite also occur in the form of layer or lenses associated with fusinite and vitrinite (Figs. 5c, 5d and 5e). Semifusinite shows reflectance in between vitrinite and fusinite, where cell structure are less preserved (Ahmed, 1998; Pareek, 2004). Sclerotinite occur as oval, circular bodies with high reflectance and preserved either single cell or as number of cells.

Funginite is highly reflecting single and multi celled fungal spores in polished section. It occurs as single celled (spores), twin celled and multi celled spindle shaped (teleutospores) and multi celled round shaped (sclerotia). It also occurs in tabular form (hyphae) and sutured structure of thin tube (mycetia) (Figs. 6a, 6b and 6c) (Pareek, 2004).

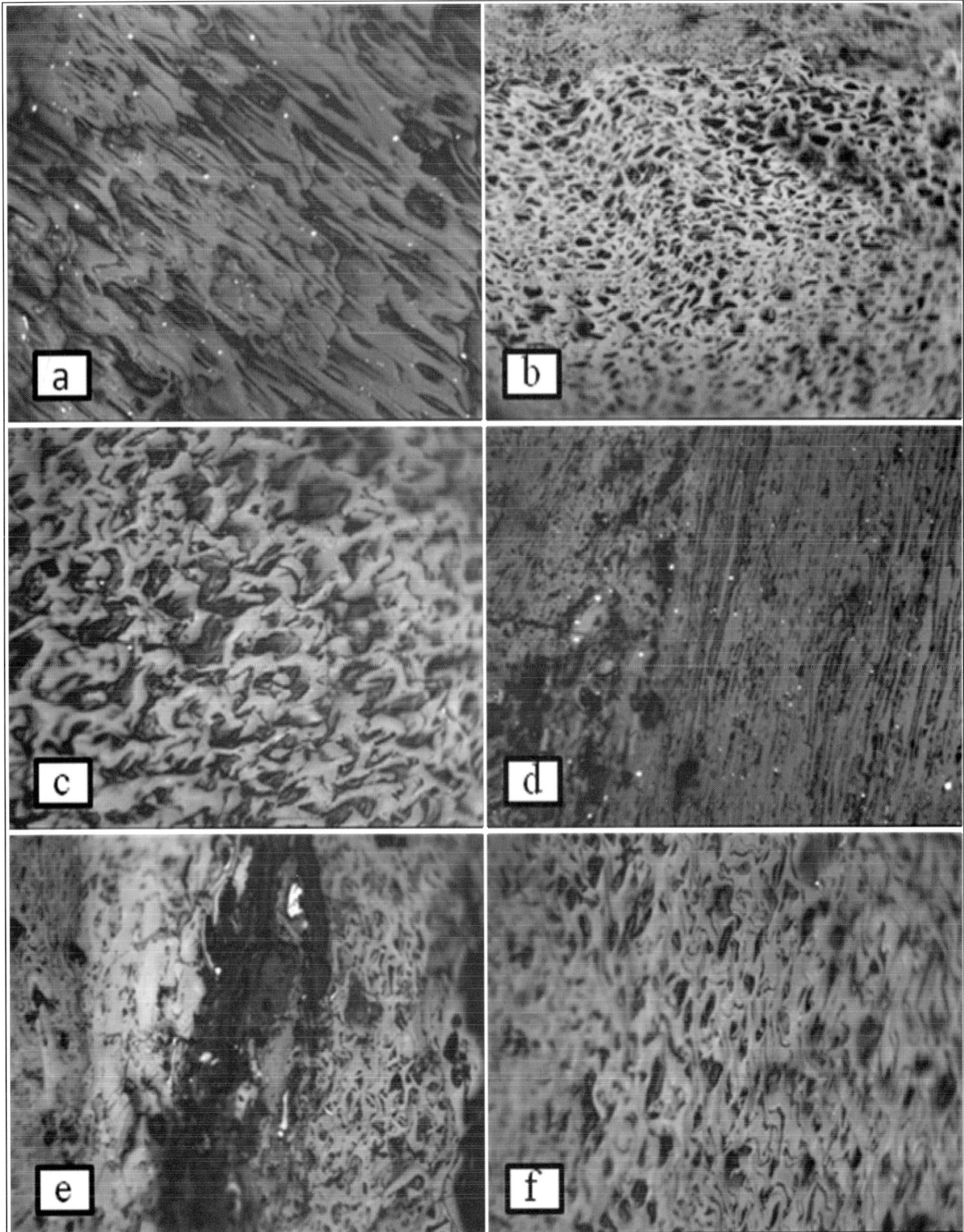

Fig. 5: Photomicrographs of coal macerals from coal seams of Kamptee coalfield (incident light with oil immersion objective; 50x) a) Fusinite [AB-Incline mine, coal seam-IV (top)] b) Fusinite [Pipla mine, seam-II] c) Fusinite with *Bogen Structure* [WCL-2, coal seam-IV (middle)] d) Semifusunite [AB-Incline mine, coal seam-IV (top)] e) Semifusinite and fusinite [Pipla mine, coal seam-II] f) Semifusinite and fusinite [WCL-3, coal seam-V (bottom)]

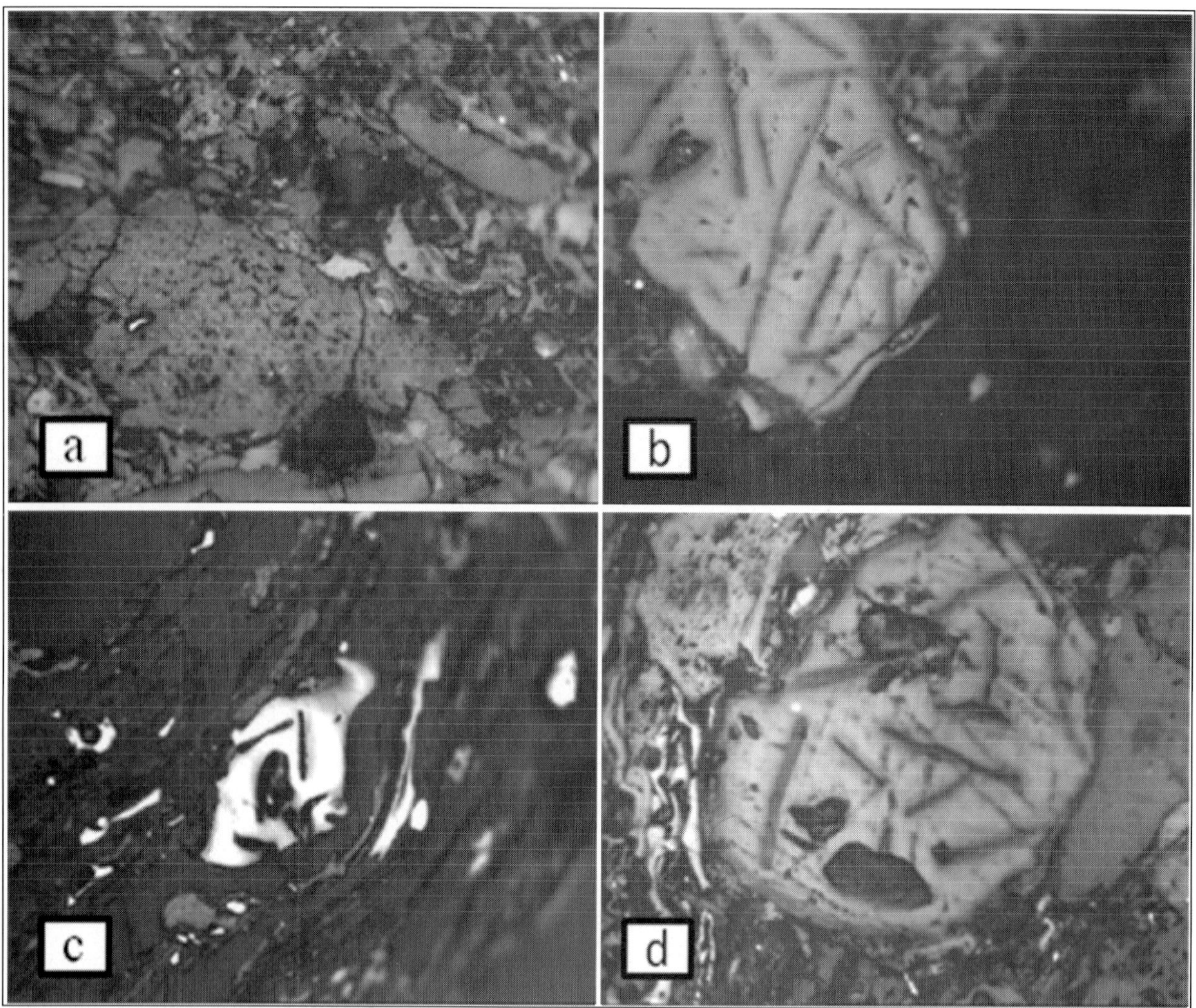

Fig. 6: Photomicrographs of coal macerals from coal seams of Kamptee coalfield (incident light with oil immersion objective; 50x) (a) Fuginite [Pipla mine, coal seam-II)] (b) Fuginite [Silewara mine, coal seam-II (top)] (c) Fuginite and vitrinite [Saoner WCL-2 mine, coal seam-IV (middle)] (d) Secretinite [Saoner WCL-2 mine, coal seam-V]

Secretinite is a macerals of inertinite, which shows round to oblong, non-circular forms with high reflectance (Fig. 6d). It is different from funginite by its non-cellular nature and from micrinite by round-equant form and high relief. Secretinite shows characteristic fracture, oxidised rim and notch (Fig. 6d). It is oxidised product of resin (Ahmed, 1998).

Tremacerite occurs in coal seams of Kamptee coalfield as lenticular bodies with high reflectivity or brightness and associated with funginite (Figs. 7a, 7b and 7c). At places, inertinite macerals are associated with inorganic mineral matter in coal of the study area (Figs. 8a, 8b).

Liptinite Group of Macerals

Liptinite group include the macerals of sporinite, resinite and cutinite (Chandra *et al.*, 2001; Pareek, 1970, 2004; Ahmed, 1998; Singh, 1998; Singh and Singh, 2001). The liptinite macerals in coal seams of Kamptee coalfield are normally few and mainly composed of spores and cuticles (Figs. 8c, 8d). Sporinite occur as long slender bodies having low reflectivity than vitrinite. Most of these macerals occur inside the vitrinite as miospores and identified as tenui-sporinite. Resinite is found as irregular and oval shaped bodies within vitrinite and is almost opaque in reflected light. Rarely observed cutinite occur as a thin long band with serrated edges and show very poor reflectivity (Ahmed, 1998).

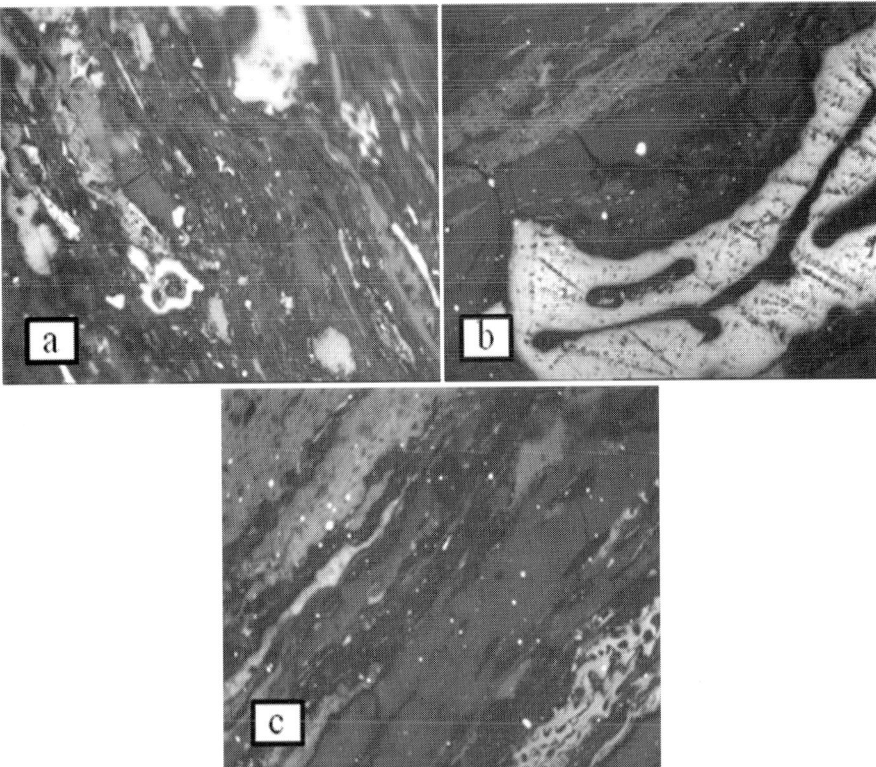

Fig. 7: Photomicrographs of coal macerals from coal seams of Kamptee coalfield (incident light with oil immersion objective; 50x) a) Trimacerite [Pipla mine, coal seam-II], b) Trimacerite and funginite [AB-Incline mine, coal seam-IV (top)], c) Trimacerite with mineral matter [AB-Incline, coal seam-IV (top)]

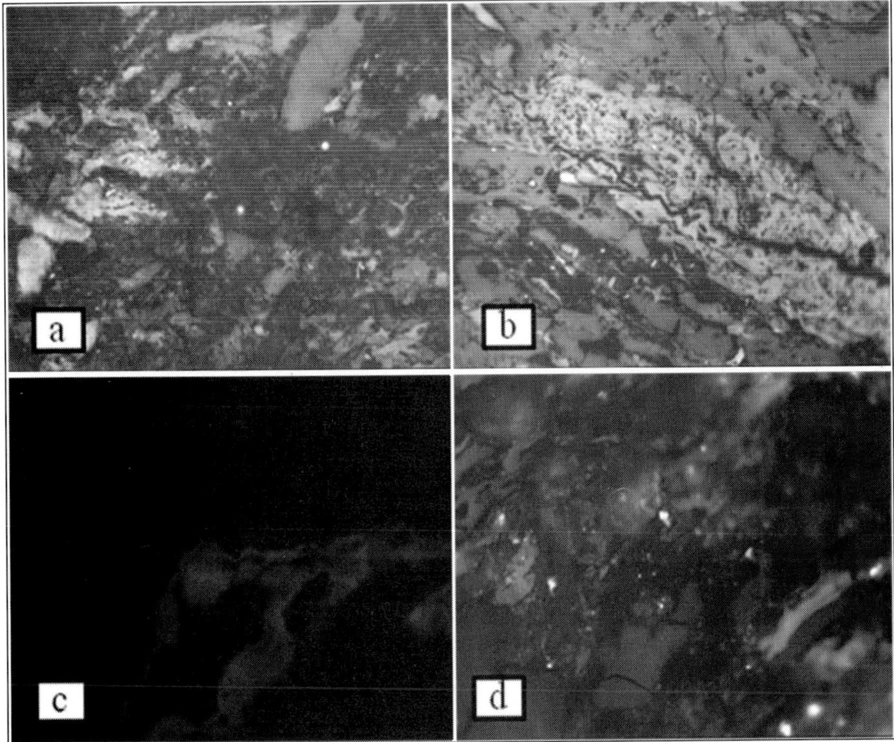

Fig. 8: Photomicrographs of coal macerals from coal seams of Kamptee coalfield (incident light with oil immersion objective; 50x) a) Inertinite and mineral matter [Saoner WCL-2 mine, coal seam-IV (middle)] b) Inertinite and mineral matter [Saoner WCL-2 mine, coal seam-V] c) Liptinite and mineral matter (fluorescent light) [Pipla, coal seam-II] d) Liptinite and mineral matter (normal light) [Pipla mine, coal seam-II]

Mineral matter

The coal seams of Kamptee coalfield are rich in mineral matter, which is dominantly represented by pyrite. These occur as nodular, disseminated grains and framboidal balls inside vitrinite, liptinite and inertinite maceral groups (Figs. 9a, 9b and 9c). Carbonate minerals are also seen within the mineral matter (Fig. 9d). Opaque minerals are observed with carbonate and pyrite (Fig. 9e).

Maceral Composition of the Coal Seams

Petrographic maceral modal analyses were carried out with an electronic point counter attached to the microscope stage for determination of maceral composition of the coal seams of Kamptee coalfield. Individual macerals are counted and finally computed to corresponding groups (Diessel, 1992; Choudhary et al., 2001). The quantitative modal analysis of various maceral groups is given in Table 3. Maceral categories counted include total vitrinite, total liptinite and inertinite macerals (fusinite, semifusinite and micrinite).

The maceral composition of coal samples from Kamptee coalfield shows that the vitrinite content varies from 13.5% to 40.5%, with an average value of 25.84%. Semi-vitrinite content varies from 0.1% to 0.4%, with an average value of 0.25%. Inertinite content ranges from 24.9% to 59.7%, with an average value of 41.16%, whereas the liptinite content varies from 8.4% to 18.4%, with an average value of 13.34% in these coal samples. The mineral matter content ranges from 11.9 to 25.5%, with an average value of 19.59% (Table 3).

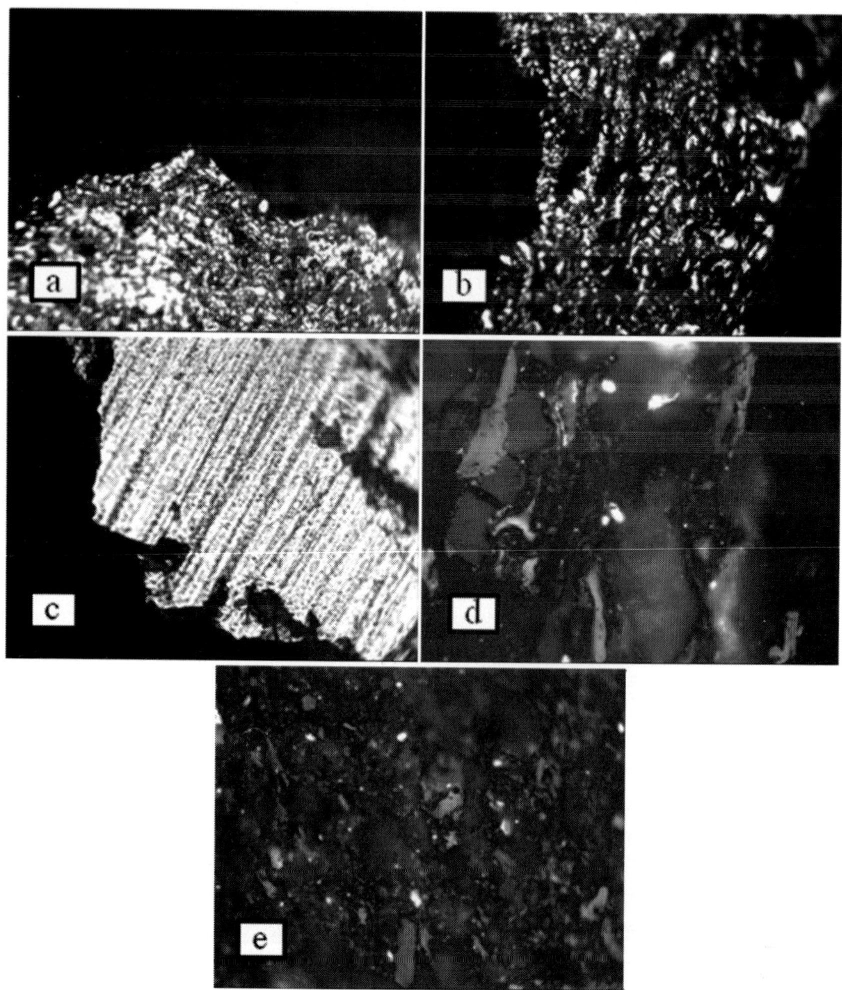

Fig. 9: Photomicrographs of mineral matter from coal seams of Kamptee coalfield (incident light with oil immersion objective; 50x) a) Pyrite [Pipla mine, coal seam-II], b) Pyrite [Silewara mine, coal seam-II (top], c) Pyrite [Saoner WCL-3 mine, coal seam-V (bottom)], d) Mineral matter and carbonate mineral [Saoner WCL-2 mine, coal seam-IV (middle)] e) Mineral matter [Saoner WCL-2 mine, coal seam-IV (middle)]

Table 3: Maceral content of coal (vol %) from coal seams of Kamptee coalfield

Sample No.	Vitrinite	Semi-Vitrinite	Liptinite	Inertinite	Vitrinite + Liptinite	Mineral Matter	IV Factor = Inertinite x 100 / (Inertinite +Vitrinite)
CR-2	13.5	---	8.4	52.6	21.9	25.5	79.57
CR-10	20.5	---	13.8	44.7	34.3	21	68.55
CR-8	16.1	---	12.3	59.7	28.4	11.9	78.75
CR-1	28.1	---	13.6	32.3	41.7	26	53.47
CR-4	27.8	---	11.1	47.7	38.9	13.4	63.17
CR-9	40.5	0.4	15.8	24.9	56.3	18.4	38.07
CR-11	34.4	0.1	18.4	26.2	52.8	20.9	43.23
Range	13.5 - 40.5	0.1 -0.4	8.4 - 18.4	24.9 - 59.7	21.9 - 56.3	11.9 - 25.5	38.07 - 79.57
Average	25.84	0.25	13.34	41.16	39.18	19.59	60.69

Table 4: Maceral content (vol %) on visible mineral matter free basis from coal seams of Kamptee coalfield

Sample No.	Vitrinite	Semi-Vitrinite	Liptinite	Inertinite	Vitrinite + Liptinite
CR-2	18.1	---	11.3	70.60	29.4
CR-10	25.9	---	17.5	56.60	43.4
CR-8	18.3	---	13.9	67.80	32.2
CR-1	37.9	---	18.4	43.70	56.3
CR-4	32	---	12.9	55.10	44.9
CR-9	49.6	0.5	19.4	30.50	69
CR-11	43.5	0.2	23.2	33.10	66.7
Range	18.1 - 49.6	0.2 -0.5	11.3 - 23.2	30.5 - 70.6	29.4 -69
Average	32.19	0.35	16.66	51.06	48.84

The macerals contents of coal seams of Kamptee coalfield on visible mineral matter free basis are also calculated, where moderate to low vitrinite content ranges from 18.1 to 49.6%, with an average value of 32.19% (Table 4). The semi-vitrinite content ranges from 0.2 to 0.5% with an average value of 0.35%. Moderate to low liptinite content ranges from 11.3 to 23.2% with an average value of 16.66%, whereas high to very high inertinite content ranges from 30.5 to 70.6% with an average value of 51.06% (Table 4).

Higher proportion of inertinite content in coal seams of Kamptee coalfield suggest that the coal deposition has taken place at a shallow depth under aerobic condition with frequent exposure of plant debris to the atmosphere (Mukherjee *et al.*, 1983; Chandra *et al.*, 2001; Chaudhuri *et al.*, 2000, 2001). Rotting process (corroded, crumbling, decay, decomposition and mouldering) seems to have played significant role in the peatification of coals from Kamptee coalfield, which may account for micro-banded nature of many of coal seams of the Kamptee coalfield. High inertinite bearing coals of this coalfield contain high moisture (4.2- 9.40%), which on evaporation renders these coals friable (Sander and Brookes, 1986; Chaudhuri *et al.*, 2000, 2001).

The coal seams of Kamptee coalfield has low vitrinite content and comparatively low hydrogen content that ranges from 4 to 5.2% (dmmf), with an average value of 4.57% (dmmf) along with high mineral matter. This indicates relatively high porosity and high moisture content, which result in lower heating value of coal seams of Kamptee coalfield as compared to Carboniferous coals of same maturity (Sander and Brookes, 1986; Chaudhuri *et al.*, 2000, 2001).

Coal Macerals and Environment of Deposition

Different coal macerals are formed from identical source material, depending upon the position of water table or groundwater conditions at the time of peat swamp development and coalification. The woody and herbaceous cell tissues (lignin and cellulose) forms macerals of the vitrinite group under anaerobic (high hydrogen content) and mostly reducing conditions. Whereas, the inertinite group macerals forms under aerobic (low hydrogen content) conditions, which were subjected to period of drying and oxidation during the early peat stage development (Mackowski, 1968; Diessel, 1992, 2006). Formation of vitrinite is termed as vitrinisation and formation of inertinite is termed as fusinisation. Different environmental conditions of deposition of peat swamp are responsible for formation of vitrinite and inertinite macerals (Murchison, 1991).

Numerous studies have been conducted to establish the relationship between coal macerals and depositional environment (*c.f.* Duguid, 1980; Diessel, 1982; Hunt *et al.*, 1986; Teichmuller, 1989). Diessel (1986, 1992) used two indices to define the palaeo-depositional environment of the peat-mire. First is the Tissue Preservation Index (TPI), which is the ratio of tissue derived structural macerals of vitrinite and inertinite groups to their tissue derived unstructured macerals. The TPI can be used to measure the degree of humification as well as proportion of wood that has been preserved in the peat-mire. The other index is Gelification Index (GI), which is the ratio between gelified macerals (*e.g.* vitrinite and macrinite) to fusinitised macerals (*e.g.* fusinite and semifusinite). The TPI-GI cross plot is used to decipher the depositional environment and maceral types.

Smyth (1984) used macerals in triangular diagram for comparison and identifying the depositional environment of coal during stage of peat swamp development *i.e.* genetic relation. Macerals concentration of vitrinite, liptinite and inertinite groups being used to indicate the depositional environment of the basin such as fluvial, lacustrine, upper deltaic, lower deltaic and lagoonal or brackish water.

The macerals of coal seams of Kamptee coalfield, trends from fluvial to lacustrine condition during peat swamp development as these coals contain high percentage of inertinite as compared to vitrinite and liptinite (Fig. 10) (Smyth, 1984; Singh and Singh, 1998). This indicates that these coal seams were deposited under fluvio-lacustrine environment. This can be further corroborated by model based on quantitative mineral-matter-macerals association, which suggest these coals have been originated from alternate oxic and anoxic condition (Fig. 11) (Akande *et al.*, 1992; Singh and Singh, 1998).

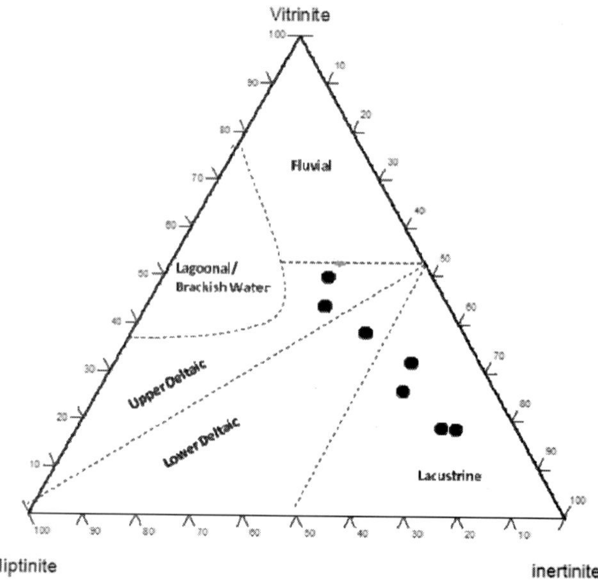

Fig. 10: Depositional environment of coal seams of Kamptee coalfield based on coal macerals (after Smyth, 1984)

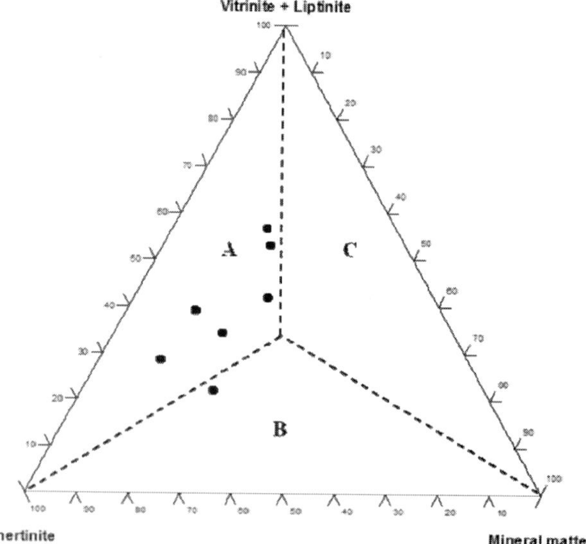

Fig. 11: Depositional condition of coal seams of Kamptee coalfield based on macerals and mineral matter (after Smyth, 1984) (A-Alternate oxic and anoxic condition; B- Oxic dry with sudden dry flooding; C- Wet with intermittent moderate to high flooding condition)

These observations are further corroborated by the independent study of IV factor (*c.f.* Warbrooke, 1987). The environment of deposition of coal seams of Kamptee coalfield is worked out from the study of IV factor (Inertinite + Vitrinite factor) as suggested by Warbrooke (1987). The IV factor for coal seams of Kamptee coalfield varies from 38.07 to 79.57 with an average value of 60.69 (Table 3), which is close to the values obtained for wet/dry forest swampy conditions of peat development (Warbrooke, 1987). This indicates that the deposition of coal seams of Kamptee coalfield was taken place in wet/ dry forest swampy condition during peat development.

Moreover, the coal seams of Kamptee coalfield have high inertinite content and low vitrinite content (Table 3), which indicates that aerobic condition of deposition prevailed during the early peat stage *i.e.* low water content or oxic condition of the peat swamp. The percentage of inertinite macerals increases with decreasing vitrinite content (Fig. 12). This strong inverse relationship between coal macerals of inertinite and vitrinite indicates a common source material of both the macerals but different genetic tracts. This relationship also suggests that under the condition of fusinisation more cell tissues are destroyed than during vitrinisation. This leads to the relative enrichment in the partially oxidized peat of spores, pollen, cuticles, resin bodies and other resistant non-tissue component, including inorganic mineral matter (Diessel, 1992).

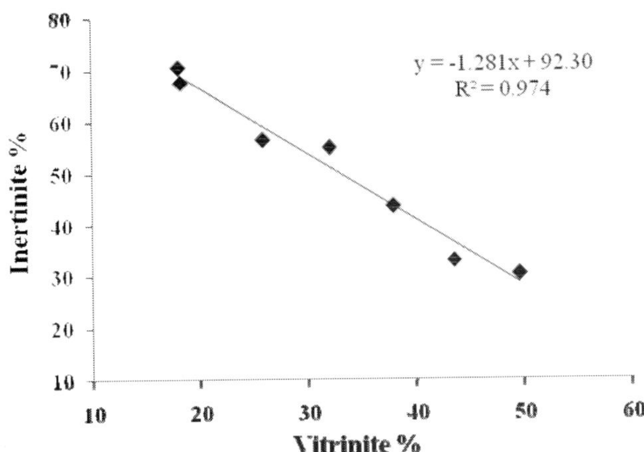

Fig. 12: Relationship between vitrinite and inertinite group macerals in coal seams of Kamptee coalfield

Discussion and Conclusions

The inertinite content (on mineral matter free basis) in coal seams of Kamptee coalfield is as high as 70.6%, which are equal to the average inertinite content in Gondwana coals of the world (70%). The maceral contents of coal from Kamptee coalfield lie in the field of Gondwana coals, which has high inertinite content as compared to high vitrinite containing Carboniferous coals. This indicates that flora and climate of the Kamptee coalfield coal seams are similar to Gondwana coals of other parts of the world. The high inertinite and low vitrinite content of the coal seams of Kamptee coalfield indicates that aerobic condition of deposition *i.e.* low water content or oxic condition prevailed during early peat stage development of the peat swamp and the coal from this basin have medium volatile bituminous rank.

Strong inverse relationship between coal macerals of inertinite and vitrinite groups is observed, which indicates a common source material of both the macerals but different genetic tracts. This relationship also suggests that more cell tissues are destroyed during fusinisation as compared to vitrinisation. This leads to relative enrichment of spores, pollen, cuticles, resin bodies and other resistant non-tissue component including inorganic mineral matter in partially oxidised peat.

Further, the macerals of coal seams of Kamptee coalfield, trends from fluvial to lacustrine condition during peat swamp development, which indicates that these coal seams were deposited under fluvio-lacustrine environment. The IV factor for coal seams of Kamptee coalfield varies from 38.07 to 79.57 with an average value of 60.69, which is close to the values obtained for wet/dry forest swampy conditions of peat development. This indicates that the deposition of coal seams of Kamptee coalfield was taken place in dry forest swampy condition during peat stage development.

Acknowledgements

Authors are thankful to Dr. Nandita Choudhury, Head of Department, Petrographic Division CIMFR, Dhanbad and Prabal Boral, Scientist, CIMFRI, Dhanbad for their help in petrographic studies, to Shri A. K. Chowdhury, Office-in-Charge, CIMFR, Nagpur Unit and Dr. Khanwalkar, Scientist, CIMFR, Nagpur Unit for providing analytical help to carry out this work.

References

Acharyya, S.K. (2000) Coal and lignite resources of India- An overview. Geological Society of India, 50p.

Ahmed, M. (1998) Petrogrphic characters of Eocene coals of Lairyngen coalfield, East Khasi Hills, as parameters for geo-environment, coal and organic petrology. Recent Researches in Economic Geology, Vol. 1, pp. 92-98.

Akande, S.O., Andreas, H. and Erdtmann, B.D. (1992) Upper Cretaceous and Tertiary coal from the southern Nigeria: composition, rank, depositional environments and their technological properties. Bulletin of Nigerian Association of Petroleum Explorationists, Vol. 7, pp. 26-38.

Basu, T.N., Chandra, T.K. and Shrivastava, B.B.P. (1978) Strategy of exploration in the concealed coalfield of the Kamptee in Central India. In: Proc. 2nd International Conference on 'Coal Exploration', Denver, U.S.A.

Chandra, D. and Taylor, G.H. (1982) Gondwana coals. In: Stach, E., Mackowsky, M.Th., Teichmuller, M., Taylor, G.H., Chandra, D. and Teichmuller R. 'Stach's Textbook of Coal Petrology', 3rd revised edition, Gebr. Borntraeger, Stuttgart, pp. 191-197.

Chandra, D., Singh, R.M. and Singh, M.P. (2001) Text book of coal (Indian Context). 1st edition, Tara Book Agency, Varanasi, 402p.

Chandra, T.K. and Shingte, S.K. (1976) Consolidated geological report on the area covered by Kanhan (Walni)-Pipala Incline, Kamptee Coalfield Maharashtra. Unpubl. Report, Central Mine Planning & Design Institute.

Chaudhari, S.G., Choudhury, N., Chatterjee, C.N. and Mukherjee, A.K. (2001) Application of main petrographic study in the prospecting of coal bed methane-An appraisal. In: Proc. 'Coal Bed Methane: Prospects and Potentialities', South Asian Association of Economic Geologist, pp. 110-116.

Chaudhuri, S.G. and Ghose, S. (1978) Preliminary study of relative semifusinite of Indian coking coals. Journal of Mines Metals Fuels, Vol. 26 (3), pp.137-141.

Chaudhuri, S.G., Chatterjee, C.N. and Choudhury, N. (1998) Application of coal petrology in the coal utilization processes. Resent Researches in Economic Geology, Vol. 1, pp.12-30.

Chaudhuri, S.G., Choudhury, N., Chatterjee, C.N. (2000) Role of coal petrography in combustion and pulverised coal-a review. Journal of Mining, Geological and Metallurgical Institute of India, pp. 67-80.

Chawdhury, A.N. (1971) Chemical properties of Indian coals. In: 'Coal Resources of India'. Geological Survey of India Memoir No. 88, pp.111 152.

Choudhury, N. Chaudhuri, S.G. and Chatterjee, C. N. (2001) Petrographic approach in studying maceral reactivity with special references to Indian coals. Transactions of Mining, Geological & Metallurgical Institute, Vol. 97(102), pp. 87-104.

Choudhury, N., Boral, P., Mitra, T. Adak, A.K., Choudhury, A. and Sarkar, P. (2007) Assessment of nature and distribution of inertinite in Indian coals for burning characteristic. International Journal of Coal Geology, Vol. 72, pp. 141-152.

Choudhury, N., Biswas, S., Sarkar, P., Kumar, M., Ghoshal, S., Mitra, J., Mukherjee, A. and Choudhury, A. (2008a) Influence of rank and macerals on the burnout behaviour pulverized Indian coal. International Journal of Coal Geology, Vol. 74, pp.145-153.

Choudhury, N., Mohanty, D., Boral, P., Kumar, S. and Hazra, S.K. (2008b) Microscopic evaluation of coal and coke for metallurgical usage. Current Science, Vol. 94, No.1, pp.74-81.

Cloke, M. and Lester, E. (1994) Characterization of coals for combustion using petrographic analysis-A review. Fuel, Vol. 73, pp. 315-320.

CMPDIL (1993) Coal Atlas of India on Pench-Kanhan-Tawa Valley coalfields. Central Mine Planning & Design Institute Ltd, pp.110-111.

Dasgupta, S.C. (2006) A prospective of exploration and resource evaluation of CBM in India vis-a-vis the role of the Geological Survey of India. Journal of Applied Geochemistry, Vol. 8 (2A), pp. 359-365.

Deolasee, C.B. (1996) Coal in Vidharbha: present status and exploration prospects. In: Mohbey, N.K., Gwalani, L.G. and Mohabey, D.M. (eds.) Golden Jubilee Symposium Volume, Department of Geology, Nagpur University, pp. 133-147.

Diessel, C.F.K. (1982) An appraisal of coal facies based on maceral characteristics. In: Mallet, C.W. (ed.) 'Coal resources-origin, exploration and utilization in Australia'. Australian Journal of Coal Geology, Vol. 4, pp. 474-484.

Diessel, C.F.K. (1986) The relationship between coal macerals and sedimentary environments in Australian coals. In: Proc. 'Sediments down-under', 12th International Sedimentological Congress, Canberra, 82p.

Diessel, C.F.K. (1992) Coal-bearing depositional systems. Springer Verlag Publ. Berlin, 721p.

Diessel, C.F.K. (2006) Utility of coal petrology for the sequence stratigraphic analysis. International Journal of Coal Geology, Vol.70, pp. 3-34.

Duguid, K. (1980) Semivitrinite in South African coals. World Coal, Vol. 6, pp.2-19.

Dutt, A.B. (1953) Coal its history and utilization. Indian Minerals, Vol.7 (2), pp. 69-79.

Dutt, A.B. (2007) Strategies for regional exploration for solid fuel and CBM- A Futuristic outlook. Gondwana Geological Magazine, Spl. Vol. No.9, pp.81-86.

Dutt, A.B., Mukhopadhyay, A. and Chakrabarti, N.C. (2001) Coal bed methane potential in Central Indian coalfields-possibilities and prospects. In: Proc. 'Coal Bed Methane: Prospects and Potentialities', South Asian Association of Economic Geologist, pp. 26-36.

Dutta, K.K., Narayanamurthy, B.R. and Shrivastava, A.K. (1961) A Note on Kamptee coalfield, Nagpur district, Maharashtra. Unpubl. Report, Geological Survey of India.

Gee E.R. and Khedekar, V.R. (1952) The Kamptee Coalfield, Nagpur district. Records of Geological Survey of India, Vol. 78, Part-4, pp. 407-418.

GSI (2008) Geology and Mineral Resources of Maharashtra. Miscellaneous Publ. No.30, Part-II, 2nd revised edition, Geological Survey of India, Central Region, Nagpur, 120p.

Hota, R.N. (2000) Quantitative stratigraphic correlation-a case study from two borehole section of Barakar Formation, Talchir coalfield, Orissa. Journal of Geological Society of India, Vol. 56, pp. 517-534.

Hota, R.N. (2008) Geometry and compositional variation of coal seam III of Damuda Group, Talchir Gondwana basin Orissa. Gondwana Geological Magazine, Vol. 23 (2), pp.103-109.

Hota, R.N. and Maejima, W. (2005) Geometry of the inter-seam sand bodies of Barakar Formation, Talchir Gondwana basin, Orissa. Journal of Indian Association of Sedimentology, Vol. 24, pp.15-26.

Hunt, J.W., Brakel, A.T. and Smyth, M. (1986) Origin and distribution of the Bayswater seam and correlatives in the Permian Sydney and Gunnedah Basins, Australia. Australian Journal of Coal Geology, Vol. 6, pp. 59-75.

ICCP (1971) International hand book of coal petrology. International Committee for Coal and Organic Petrology, CNRS, Paris, 2nd edition, 184p.

ICCP (1994) Vitrinite classification. International Committee for Coal and Organic Petrology, Aachean, pp. 1-24.

ICCP (1998) The new vitrinite classification. International Committee for Coal and Organic Petrology, (ICCP system 1994). Fuel, Vol. 77 (5), pp. 349-358.

ICCP (2001) The new inertinite classification. International Committee for Coal and Organic Petrology, (ICCP system 1994). Fuel, Vol.80 (4), pp.459-471.

IS: 9127 (Part-I, 1992) Method for the petrographic analysis of coal: Definition of terms relating to petrographic analysis of coal.

IS: 9127 (Part-III, 2002 / ISO: 7404-3, 1994) Method of petrographic analysis of coal: Method of determining maceral group composition.

IS: 9127 (Part-IV, 2001/ISO: 7404-3, 1988) Method of petrographic analysis of coal: Method of determining micro-lithotypes, carbominerite and minerite composition.

IS: 9127 (Part-V, 2004/ISO: 7404-3, 1994) Methods of the petrographic analysis of coal: Microscopic determination of the reflectance of vitrinite.

ISO: 7404 (Part-III, 1994-10-01) Methods for the petrographic analysis of bituminous coal and anthracite: Method of determining maceral group composition.

ISO: 7404 (Part-V, 1994-10-01) Methods for the petrographic analysis of bituminous coal and anthracite: Methods of determining microscopically the reflectance of vitrinite for petrographic analysis.

Krishnan, M.S. (1982) Geology of India and Burma. CBS Publ. and Distributors, Delhi 518p.

Mackowski, M.T. (1968) Mineral matter in coal. In: Murchison, D. and Well, T.S. (eds.) 'Coal and Coal Bearing Strata'. Olvier and Boyd, London, pp. 309-321.

Mukherjee, A.K., Alam, M.M., Chatterjee, S.B. and Choudhury, S.S. (1983) Depositional environmental and physico-chemical properties of lignite deposit of Palana basin, Rajasthan. Journal of Mines, Metals and Fuels, pp.21-29.

Murchison, D.J. (1991) Petrographic aspects of coal structure, reactivity of macerals in laboratory and natural environments. Fuel, Vol.70, pp. 296-315.

Nandgaonkar, R.R. (1996) Petrographic characterisation of coal from Wardha Valley Coalfields. In: Mohbey, N.K., Gwalani, L.G. and Mohabey, D.M. (eds.) Golden Jubilee Symposium Volume, Department of Geology, Nagpur University, pp. 149-155.

Pant, P.R. and Murty, B.V.S. (2004) Geophysical exploration for coal- the Indian scenario. Geological Society of India, Memoir No. 59, 310p.

Pareek, H.S. (1970) On the geology and correlation of coal seams of Pench-Kanhan-Upper Tawa Valley coalfields, dist. Chindwara, Madhya Pradesh. The Palaeobotonist, Vol.18, No.1, pp.95-102.

Pareek, H.S. (2004) Progress of coal petrology in India. Geological Society of India, Memoir No. 57, 161p.

Pettijohn, F.J. (1984) Sedimentary Rocks. 3rd Indian edition, CBS Publisher and Distributers, Delhi, 628p.

Raja Rao, C.S. (1882) Coalfield of India. Geological Survey of India, Bull. Series A, No. 45, Vol. II, 103p.

Ramakrishna, T.S., Rama Rao, M.S.V., Bhaskara Rao, K.V.S. and Punekar, D.Y. (1999) Geophysical study of Gondwana basin of eastern Maharashtra. Geological Survey of India, Spl. Publ. No.47.

Rimmer, S.M., Rowe, H.D., Taulbee, D.N. and Hower, J.C. (2006) Influence of macerals content on ò13 C and ò15N in the middle Pennsylvanian coal. Chemical Geology, Vol. 225, pp. 77-90.

Scott, A.C. (2002) Coal petrology and the origin of coal minerals: a way ahead? International Journal of Coal Geology, Vol.30, pp. 119-134.

Senders, G.J. and Brookes, G.F. (1986) Preparation of Gondwana coals-1, washability characteristics. International Journal of Coal Preparation, Vol. 3, pp.105-132.

Sengupta, S.M. (1996) Introduction to Sedimentology. Oxford and IBH Publ. Co-Pvt. Ltd., New Delhi, 305p.

Sharma, N.L. and Ram, K.S.V. (1966) Introduction to geology of coal and Indian coalfields. Oriental Publ., Jaipur, 183p.

Shingte S.K. and Palkar, S.G. (1974) Preliminary geological report on Inder-Godegaon sector of Kamptee coalfield Maharashtra. Unpubl. Report., Central Mine Planning & Design Institute Ltd.

Shingte, S. K. and Krishnan, S. (1981) Geological report on Kamptee colliery leasehold and eastern extension. Unpubl. Report. Central Mine Planning & Design Institute Ltd.

Shingte, S.K. and Sah, R. (1987) Geological report on Waghoda sub-block, Saoner phase-II, Kamptee coalfield, Maharashtra, Unpubl. Report. Central Mine Planning & Design Institute Ltd, Ranchi.

Shingte, S.K. and Sah, R. (1993) Kamptee coalfield present status of exploration and future potentiality. Gondwana Geological Magazine, Spl. Vol. 1 'Gondwanas of India', pp. 379-389.

Singh, A. and Singh, B.D. (2001) Petrology of Kanhan coals, Satpura Gondwana basin (India) vis-à-vis Coal Bed Methane. In; Proc. 'Coal Bed Methane-Prospects and Potentialities', South Asian Association of Economic Geologist, pp. 117-130.

Singh, M.P. (1998) Micropetrographic constituent of brown and hard coals: Recent developments. Recent Researchers in Economic Geology, Vol. 1, pp. 1-11.

Singh, M.P. and Singh, P.K. (1998) Environment of deposition of Karharbari coals of Saharjuri coalfield, Deoghar basin, Bihar: Application of coal petrography based model, coal and organic petrology. Recent Researches in Economic Geology, Vol.1, pp. 48-67.

Smyth, M. (1984) Coal micro-lithotypes related to sedimentary environment in the Cooper basin, Australia. In: Rahmani, R.A. and Flores, R.M. (eds) 'Sedimentology of coal and coal bearing sequences'. Society of Mining Engineering of AMIE, Illinois, pp.11-39.

Stopes, M.C. (1935) On petrology of banded bituminous coal. Fuel, Vol. 14, pp.4-13.

Taylor, G.H., Liu, S.Y. (1998) Micrinite- its nature, origin and significance. International Journal of Coal Geology, Vol. 14, pp. 29-46.

Teichmuller, M. (1989) The genesis of coal from the view point of coal petrology. In: Lyons, P. C. and Alpern, B. (eds) 'Peat and Coal: Origin, facies and Depositional models'. International Journal of Coal Geology, Vol. 12, pp. 1-87.

Vaidyanandan, R. and Ramakrishnan, M. (2008) Geology of India. Geological Society of India, Vol.2, pp. 615-661.

Wadia, D.N. (1966) Text Book of Geology of India. McMillan and Ltd. London (3rd edition, 536p.

Warbrooke, P.R. (1987) Depositional and chemical environments in the Permian coal forming swamps from the Newcastle area. In: Proc. 'Advance Study of Sydney Basin', Vol. 21, pp. 1-10.

Hon'ble reviewer

1. **Prof. T.K. Pandya**
   Chairman
   Department of Geology
   MLS University
   UDAIPUR - 313 001
2. **Prof. A.S. Sheoran**
   Department of Mining Engg.
   JNV University
   JODHPUR - 342 011

II-33

Palynoassemblage from Palaeocene – Early Eocene Lignite-Clay Sediments of the Barmer Basin, Western Rajasthan, India with Remarks on Palaeoenvironment

S. L. Nama[1], V. S. Parihar[2], Akashdeeep[2], Suibir Bera[3] and S. C. Mathur[2]

Department of Zoology[1] and Geology[2], Jai Narain Vyas University, Jodhpur – 342005 India
Department of Botany[3], University of Calcutta, Kolkata, India
Email: nama.shankerlal@gmail.com

Abstract

Pollen spectrum recovered from the Palaeocene – Eocene Clay - lignite facies of Akli Formation from Giral mine in Barmer Basin, Rajasthan, India consists of 69 taxa of plants remains. Among these, pteridophytes are represented by 13 taxa, gymnosperm by 2 taxa and angiosperm by 34 taxa. Fungal spores of 19 taxa are also encountered in good frequency. Two pollen assemblages i.e., GLI and GLII have been identified; the former is characterized by the dominance of fungal and pteridophytic members suggesting a tropical-subtropical humid forest cover where the biomass suffered severe biodegradation by microbes and the latter by the dominance of angiosperms especially coastal palms pointing towards a near shore palaeodepositional environment. The assemblages are closely comparable to the earlier known Palaeocene – Early Eocene palynoassemblages of Barmer basin and other sedimentary basins of India.

Key Words: Lignite Deposit, Barmer Basin, Palynoassemblage, Palaeocene-Eocene, Palaeoenvironment.

Introduction

The Barmer Basin, in the great Thar Desert of Rajasthan, represented the south-eastern most entity of the west Rajasthan shelf. The West Rajasthan shelf has been divided into the following broad tectonic zones from north to south, viz., (1) The Nagaur – Ganganagar basin, (2) The Pokran Nachna high, (3) The Jaisalmer basin, (4) The Barmer-Birmania-Nagar Parker high and Devikot high (5) The Barmer Basin (6) The Sarnu high (7) The Sanchor basin and (8) The Therad high (Dasgupta and Chandra, 1978 ; Mathur and Kumar, 2003 and Compton, 2009). The Barmer Basin is a narrow elongated rift Basin situated between the Cambay – Sanchore Basin and Jaisalmer Basin and showing continuity with the Cambay Graben of Gujrat (Das Gupta, 1977, Sisodia and Singh, 2000 and Mathur and Kumar, 2003). The Barmer Basin initiated in Early Cretaceous time with extensional rifting in the south-eastern Indus shelf followed by strike slip episode and fault reactivation which produce series of horst and graben structures on the pre-rift sediments of the Birmania and Lathi Formation (Dasgupta and Chandra, 1978 and Mathur and Kumar, 2003).

Tectonically, the Barmer Basin is a graben aligned roughly in a north-south direction (Dasgupta and Mahesh chandra, 1978). The Barmer Basin is bounded by Barmer fault zone in the west, by Bhiyar fault zone in the east, and Sarnu fault to the south-east. The northern boundary is marked by the Fatehgarh transverse fault with downthrow to the south. The ridge of Devikot high demarcates the Barmer Basin from the Jaisalmer basin in the north (Mathur and Kumar, 2003).

The Barmer Basin comprises sediments of Middle Jurassic to Early Eocene age of about 4 km thickness (Dasgupta, 1977). The sediments of basal Cretaceous age in this basin are named the Fatehgarh Formation, which is represented by fluvio-deltaic to shallow marine mixed siliciclastic, carbonated phosphorite facies. The phosphorite facies is highly fossiliferous and is correlated with global Late Cretaceous phosphogenic event (Mathur and Kumar 2003 and Mathur et. al., 2006). It is overlain by Akli Formation which comprises lignite – clay – bentonite sequences of Palaeocene to Early Eocene age.

The outcrops of Akli Formation are available at throughout the central part of the Barmer Basin (Fig. 1). The Giral lignite belonging to Akli is situated about 42km. North from Barmer town. The lignite horizons of Akli Formation are defined by a prominent reflector in seismic profile and as such, its extension all over the basin is documented by seismic surveys. A few deep boreholes drilled in the Barmer Basin intersected the lignite at various depths. The maximum thickness of lignite varies from 30-70m. The gas content of lignite seams of the Barmer Basin as evident from desorption studies, is 3 - 4 m^3/t within a 600-1200 m depth range. The lignite seams are often faulted or fractured, which may account for high permeability (Jain et. al., 1973). The Present investigation is undertekan to scan whole lignite – clay beds of the Giral mine (Fig. 2). Palynologically, we collected about 50 samples at close stratigraphyic interval for the present study. The present paper embodies detailed study of pollen spectrum recovered from the Giral mines to establish correlation with equvilant sequences in India along with climatic and palaeoenvironmental studies.

Geology of the Giral Mine

Thickness of the Akli Formation at Giral Lignite mine (26° 03'; 71° 15') is about 45 to 90 meters. The sequence starts with gluconitic sandstone at the base followed by repetitious carbonaceous clay and lignite beds. The top of the sequence represented by a 17.25 meters thick Bentonite horizone. Total 13 lignite seams are presents MS – 0 and MS – 1 and LS-01 to LS – 11. The medium grained greenish coloured gluconitic sandstone form the base of the sequence. It is friable sandstone showing almost horizonetal beds of 1-2 meters thickness. It is mainly composed of quartz and gluconite with argillaceous matrix. No underlying contact with gluconitic sandstone is seen and no overlying contact with upper most bentonite is seen at Giral mine section. However, to know the stratigraphic position we utilized the lithological correlation near Bariyara and Khatri Siliceous Earth mine where siliceous earth is directly overlain the sequence of the Fatehgarh Formation and Siliceous earth overlain the Akli sandstone respectively. The same contact can be seen at Hathi Singh Ki Dhani also. As such Akli Formation is post Late Cretaceous.

There are numerous thin bends of gypsum and selenite was observed in carbonaceous clay between lignite seam LS-01 and LS-02. These bends are parralel and also cut across the carbonaceous clay lamination showing evaporatic conditions.

The Carbonaceous clay presents below ignite seam (LS-06) contain numerous shelly and skeletal meterial, comprising bones, teeth, vetebrae, spicules spines and scales of fishes with calcareous biogenic debris (Rana et al., 2005 and Nama, 2007). The carbonaceous clay occuring between LS -06 and LS – 05 also contains microvertebrate and gastropode remains indicating Palaeocene to Early Eocene age (Rana et al., 2005 and Nama, 2007).

The prelimnary observations based on biota and stratigraphic correlations, it can be inferred that possible palaeocene – Eocene boundary and Palaeocene – Eocene thermal Maximum (PETM) can be marked in the stratigraphic sequence of Giral mine section (Fig. 2), for which the detail study is going on.

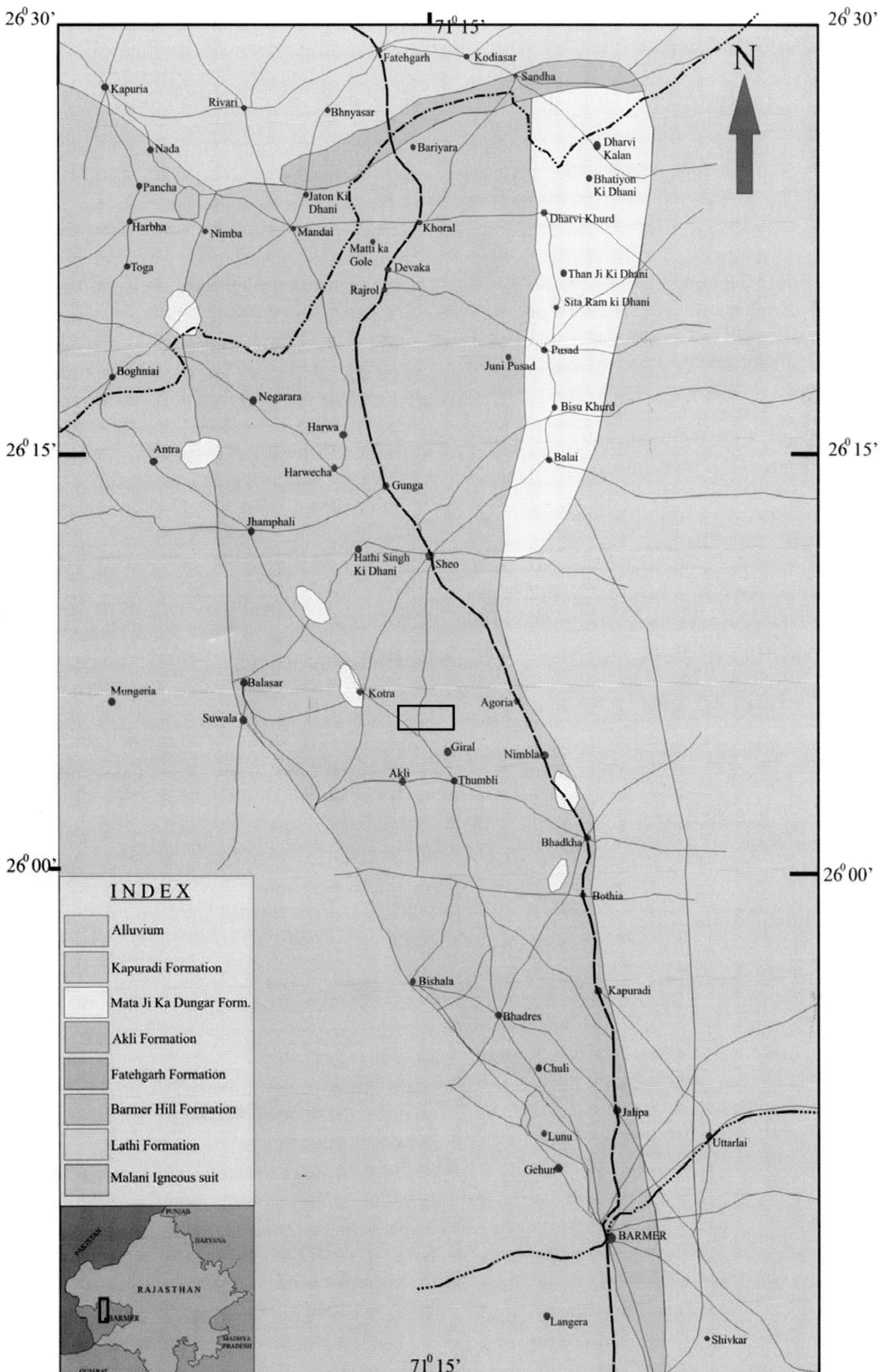

Figure 1: Interpretative geological map of the Barmer Basin (Mathur et. al., 2006)

Previous Palynological works

Tripathi (2002) carried out comprehensive study of palynofloras recorded from Giral lignite mine and those from bore-holes drilled near Kapurdi and Jalipa in Barmer district, with an aim to correlate sequences. Palynological assemblage from the lignite mine and that from the bore-hole MJ-4 are very similar as, both of these exhibit coastal elements as the dominant constituents. The assemblage is marked with profuse occurrence of monosulcate pollen assignable to different species of *Spinizonocolpites*. These forms possess an extended sulcus parallel to the polar axis and exhibit spinulate exine. This pollen, in all probabilities is related with modern brackish water Palm *Nypa*. Other significant monosulcate pollen taxa present in the assemblage are *Kapurdipollenites gemmatus* and *K. baculatus*. This pollen also possesses an extended sulcus but is distinctly different in exinal characters. Affinity of this pollen with those of *Nypa* can not be ruled out. Another interesting monosulcate form is *Retiverrumonosulcites barmerensis*. The sulcus in this pollen is small and is characteristically bordered with clavae of varying sizes. It is significant to note that these forms are confined to the Barmer Basin only.

Tripathi *et al.* (2003) have done palynological investigation as well as environmental interpretation on interbedded shales in lignites of Akli Formation which unconformably overlies Fatehgarh Sandstone and Siliceous Earth. The result shows the dominance of *Dandotiaspora, Lycopodiumsporites, Palmaepollenites, Spinizonocolpites, Palmidites, Psilastephanocolporites* and *Granustephanocolpites*, most of which are index fossils of Late Palaeocene. The assemblage lacks the early Eocene marker palynotaxa like *Tricolporopollis matanomadhensis, Meliapollis ramanujamii, M. pachydermis, M. simplex, Umbelliferoipollenites ovatus and Pellicieroipollis langenheimii*. The observed palynofloral assemblage thus indicates Middle to Late Palaeocene age to the sequence unconformably overlying Siliceous Earth.

A partial palynological assemblage from the Palaeocene – E. Eocene succession of Mankera-Tibba area, Jaisalmer was recorded by Lukose (1974). Naskar & Baksi (1978) recorded palynomorphs from Akli lignite of Barmer. Palynological work of subsurface Palaeocene-Eocene sediments near Kapurdi, Barmer district have been done by Tripathi (1995). Bose (1949, 1952), Jain *et al.* (1973) and Tripathi (1994) worked on the Barmer sandstone. Their assemblage shows abundance of angiosperms pollen followed by pteridophytic spores. Some of the significant species of the assemblage are *Cyathidites minor, Trilites tuberculiformis, Seniasporites verrucosus, Couperipollis wodehousei, Palmaepollenites nadhamuni, Proxapertites* spp., *Tricolpites reticulatus*.

The Palana sediments were initially investigated from Bikaner by Rao & Misra (1949), Rao & Vimal (1950, 1952) and Singh & Dogra (1988). Detailed palynological work on Palana sediments of Rajasthan was carried out by Sah & Kar (1974). They have reported 44 genera and 67 species. The assemblage is dominated by angiospermous pollen (70%) followed by pteridophytic spores (20%). Kar (1995b) described a rich palynological assemblage recovered from borehole K12, drilled by the M.E.C.L. at Kuchaur – Benia area in Bikaner District, Rajasthan. The assemblage is characterized by *Palmaepollenites kutchensis, Proxapertites operculatus, P. cursus, Tricolpites reticulatus*. Recently Kar & Sharma (2001) recovered the Late Palaeocene and Early Eocene palynoforms from the subsurface sediments of Bikaner district, Rajasthan and identified one Cenozone viz., *Piladiporocolpites caratinii* Cenozone.

Nama (2007) studied the Giral section and suggested stratigraphically up to the level of GI-9 the iron containing clay came from the land through erosion and weathering processes and were transported by river draining the restricted near coastal to lagoonal areas of the Barmer Basin initially, under the limited oxidation to high reducing conditions. Mathur et. al., (2009) studied bentonite facies from the top part of the Giral section and suggested genesis and basalt source for bentonite.

The present study deals with palynoassemblage analysis of complete lignite – clay sequence from Giral lignite mine of Barmer to reconstruct the vegetation vis-à-vis palaeoenvironment of the area during Palaeocene – Eocene time (Fig. 2).

8 lignite seems and interbeded clay beds were selected and samples were collected from the Giral lignite mine located about 48 km. from

the Barmer (Fig. 1)., Rajasthan, India. The lithostratigraphic section showing sample position is presented in Figure 2.

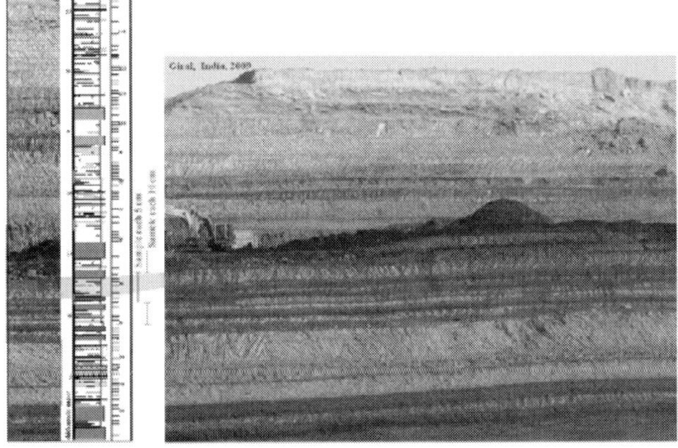

Fig. 2. Lithostratigraphic section showing interbedded sequence of lignite and clay from the Giral mine section of Akli Formation.

Method and Materials

Usual maceration technique has been adopted for recovery of palynomorphs with particular caution to avoid contamination. 10 gm. of each sample were treated with 10% aqueous solution of HCl to dissolve the carbonates, if any, washed thoroughly with distilled water followed by 40% HF treatment for 24 hrs. The HF free samples were then treated with conc. HNO_3 for few hours to few days depending on the maturation status of the samples. After thorough washing, samples were treated with 5-10% KOH, sieved, washed with distilled water and dried partially. The samples were then suspended in a heavy liquid of KI-CdI_2 mixture adjusted to a specific gravity of 2.3. To the supernatant liquid, 5 times distilled water is added and then few drops of 10% glacial acetic acid were added and kept for overnight, centrifuged, washed and finally preserved in distilled water or 50% glycerin. 5 slides with 22mm x 50mm coverglass were prepared and studied for each sample, scanned thoroughly under biological compound microscope. The best-preserved forms were photographed and studied through Image Analyzing Kit for detail morphographic character analysis including measurement. Each of the palynomorphs is stored in the hard disc of the computer. The prepared slides are stored in the repository of Palaeobotany-Palynology section, Department of Botany, University of Calcutta.

Observation

Microscopic analysis of the samples reveals the presence of rich palynoassemblage belonging to diverse forms of plants viz., algal remains (Plate I, Figs.1-3), fungal spores, hyphae, fruit bodies (Plate I, Figs. 4-33), pteridophytic spores (Plate II, Figs. 1-11), gymnosperms (Plate II, Figs. 12-13) and angiosperm pollen grains of both monocotyledonous and dicotyledonous affinity (Plate II, Fig. 14 ; Plate III, Figs. 1-14 ; Plate IV, Figs 1-29). Among the 69 palynotaxa, 13 taxa of pteridophytes, 34 of angiosperms, 2 of gymnosperms, 19 of fungi and only 1 species under 1 genus belonging to algal affinity. Out of 69 palyno texa are recovered in the present assemblage we are reporting 14 new palyno taxa from the Giral mine section. A checklist of recovered taxa is given in Table 2.

Table 2: Checklist of recovered palynotaxa

Algae	**Angiosperm**
cf. *Pterospermella* sp.	*Arecipites bellus*
Fungi	*Arecipites* sp.
Basidiosporites sp.	*Couperipollis brevispinosus*
Brachysporisporites sp.	*Couperipollis rarispinosus*
Ctenosporites sp.	*Ctenolophonidites* sp.
Dicellaesporites minutus	*Ghosiacolpites globatus*
Diporisporites anklesvarensis	*Lakiapollis ovatus*
Diporicellaesporites sp.	*Liliacidites* sp.
cf. *Desmidiospora* sp.	*Longapertites retipilatus*
Exesisporites sp.	*Longapertites* sp.
Fractisporonites sp.	*Palaeosantalaceaepites minutus*
Fusiformisporites sp.	*Palmaepollenites kutchensis*
Inapertisporites kedvesii	*Palmaepollenites* sp.
Inapertisporites sp.	*Palmidites naviculus*
Lacrimasporonites sp.	*Palmidites* sp.
Monoporisporites sp.	*Pellicieroipollis langhenheimii*
Multicellaesporites sp.	*Polybrevicolporites cephalus*
Phragmothyrites eocenica	*Proxapertites reticulatus*
Polyadosporites sp.	*Proxapertites* sp.
Staphlosporonites sp.	*Psiloschizosporis psilata*
Microthyriaceous germling	*Rhoipites* sp.
Pteridophyta	*Retimonosulcites ovatus*
Cyathidites minor	*Retitetrabrevicolporites* sp.
Cyathidites sp.	*Spinizonocolpites echinatus*
Dandotiaspora dilata	*Striacolporites striatus*

Dandotiaspora telonata
Dictyophyllidites granulatus
Foveosporites splendus
Laevigatosporites cognatus
Lygodiumsporites lakiensis
Lygodiumsporites sp.
Osmundacidites cephalus
Tricolpites reticulatus
Triporopollenites sp.
Verrumonosulcites foveolatus
cf. *Proteacidites* sp.
cf. *Tiliaepollenites* sp.
cf. Euphorbiaceae
cf. Rubiaceae
cf. *Polypodiaceaesporites* sp.
Todisporites kutchensis
Todisporites sp.

Gymnosperm

cf. *Taxodiaceaepollenites* sp.
cf. *Podocarpidites* sp.
cf. Verbanaceae
Phytolith (Panicoid)
Incertae sedis

Frequency of occurrence (%) of individual taxa has been presented in Table 3.

Table 3 : Frequency of occurrence (%) of individual taxa

Code	Name of taxa	Sample No.							
		LS 1	**LS 2**	**LS 3**	**LS4**	**LS5**	**LS6**	**LS7**	**LS8**
1	cf. *Pterospermella* sp.			2.1					
2	*Basidiosporites* sp.				8.2		4.9	5.9	4.8
3	*Brachysporisporites* sp.				4.8	4.5	3.5	4.2	4.5
4	*Ctenosporites* sp.			2.5	1.2				
5	*Dicellaesporites minutus*	3.80	3.40	3.5	2.1	5.5	5.8	6	6.5
6	*Diporisporites anklesvarensis*				2.5				
7	*Diporicellaesporites* sp.			1	4.7	5.1	6.5	7.4	7.8
8	cf. *Desmidiospora* sp.				3.2	1.5	3.5		
9	*Exesisporites* sp.	2.86	2.50	5.2	8.2	10.5	8.8	12.8	10.5
10	*Fractisporonites* sp.				5.6	4.8			
11	*Fusiformisporites* sp.				5.5	5.2	3.4	3.5	5.4
12	*Inapertisporites kedvesii*	6.50	9.80	4.2	10.4	14.7	14.5	15.8	16.9
13	*Inapertisporites* sp.	5.80	5.50	4.6	5.3	4.6	5.8	5.5	5.8
14	*Lacrimasporonites* sp.				4.7	3.8	3.9	4.2	5
15	*Monoporisporites* sp.			2.1	4	4.5	4.5		
16	*Multicellaesporites* sp.			5.5	6.8	9	8.5	10.3	10.5
17	*Phragmothyrites eocenica*	0.93			2.3				
18	*Polyadosporites* sp.			2.5	4.2	3.9	7.6	8.6	8.9
19	*Staphlosporonites* sp.			3.2	6.5	6.5	6.2	5.1	5.9
20	Microthyriaceous germling			1	2.6	5.8	5.9	5.2	7.5
21	*Cyathidites minor*	0.95		2.5					
22	*Cyathidites* sp.			2.8			1.2		
23	*Dandotiaspora dilata*			2.4			2.5		
24	*Dandotiaspora telonata*			2.2	2.2		1.5		
25	*Dictyophyllidites granulatus*		7.76						
26	*Foveosporites splendus*		2.58						
27	*Laevigatosporites cognatus*	0.95		4.9			1.5		
28	*Lygodiumsporites* sp.			2.1				3.4	
29	*Lygodiumsporites lakiensis*	6.67			1.5			2.1	
30	*Osmundacidites cephalus*	0.95							
31	*Polypodiaceaesporites* sp.		2.54						
32	*Todisporites kutchensis*		6.03						
33	*Todisporites* sp.	1.90							

Code	Name of taxa	Sample No.							
		LS 1	LS 2	LS 3	LS4	LS5	LS6	LS7	LS8
34	cf. *Podocarpidites* sp.			0.50					
35	cf. Taxodiaceae	0.50							
36	*Arecipites bellus*	0.95							
37	*Arecipites* sp.	0.95	6.03						
38	*Couperipollis brevispinosus*	0.95							
39	*Couperipollis rarispinosus*	1.90							
40	*Ctenolophonidites* sp.	3.10		4.5					
41	*Ghosiacolpites globatus*			2.67					
42	*Lakiapollis ovatus*	2.50							
43	*Liliacidites* sp.	2.20							
44	*Longapertites retipilatus*	3.00							
45	*Longapertites* sp.	8.20							
46	*Palaeosantalaceaepites minutus*	2.86	2.50	3.9					
47	*Palmaepollenites kutchensis*			2.26					
48	*Palmaepollenites* sp.	5.50	3.45	2					
49	*Palmidites naviculus*		4.31						
50	*Palmidites* sp.	4.25		3.2	2				
51	*Pellicieroipollis langhenheimii*	5.60	5.20						
52	*Polybrevicolporites cephalus*	4.50	4.50	3.4					
53	*Proxapertites reticulatus*	4.80							
54	*Proxapertites* sp.		9.50	3.5					
55	*Psiloschizosporis psilata*	0.95		1.79					
56	*Retimonosulcites ovatus*		3.45						
57	*Retitetrabrevicolporites* sp.			0.95					
58	*Rhoipites* sp.	1.30		6.7					
59	*Spinizonocolpites echinatus*	3.50	4.80	4.5					
60	*Striacolporites striatus*			1.79					
61	*cf. Proteacidites* sp.			1.57					
62	*Tricolpites reticulatus*	4.50	4.50	2.1	1.5				
63	*Triporopollenites* sp.	1.90							
64	*Verrumonosulcites foveolatus*	0.95							
65	cf. *Tiliaepollenites* sp.		2.70	0.8					
66	cf. Euphorbiaceae								
67	cf. Rubiaceae			1.67					
68	cf. Verbenaceae	3.83	5.80						
69	Incertae Sedis		3.15			10.1			
70	Phytolith (Panicoid)				0.5				

Two distict palynoassemblage zones viz., GL I & GL II (GL=Giral Lignite) have been suggested based on the basis of pollen spectrum represented in the studied samples (Fig. 3). GL I (LS 4,5,6,7,8) is characterized by high frequency of diverse types of fungal remains and GL II (LS 1, 2 and 3) (Fig. 2) by prevalence of angiosperm pollen grains belonging to the families Arecaceae, Araceae, Liliaceae, Anacardiaceae, Rhizophoraceae, Bombacaceae, caesalpiniaceae, Theaceae, Proteaceae etc.

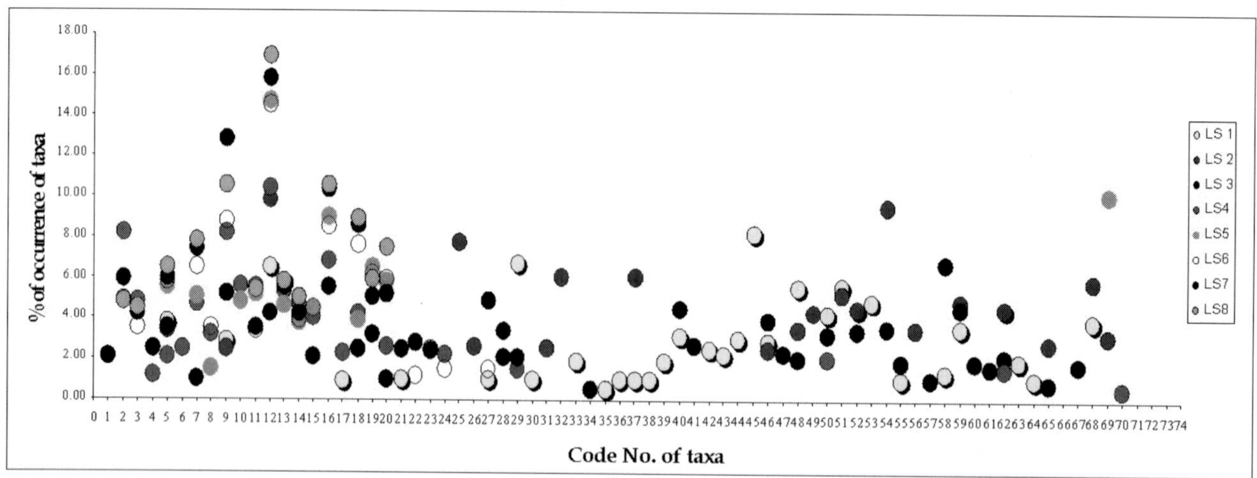

Figure 3 - Scatter diagram for distribution of taxa establishing two distinct zones : GL I (LS4, 5, 6, 7, 8) & GL II (LS1, 2, 3)

Environmental and climate Analysis

The palynomorphs recovered from the samples are presented in Table 4 along with their nearest Living Relatives, habit and climatic regime to which they grow.

Table 4 : Recovered palynomorphs from Palaeocene sediments of Barmer Basin, Rajasthan with their Nearest Living Relatives (NLR), habit form and environment consideration

Palynotaxa	Affinity	Habit	Environment / Climate
Cf. *Pterospermella* sp.	Algal cyst	Thallophyta	Brackish water-Marine
Basidiosporites sp.	Basidiomycetes	Thallophyta	Tropical, Humid
Brachysporisporites sp.	Fungi	Thallophyta	Tropical, Humid
Ctenosporites sp.	Fungi	Thallophyta	Tropical, Humid
Dicellaesporites minutus	Phragmosporae	Thallophyta	Tropical, Humid
Diporisporites anklesvarensis	Fungi	Thallophyta	Tropical, Humid
Diporicellaesporites sp.	Fungi	Thallophyta	Tropical, Humid
cf. *Desmidiospora* sp.	Fungi	Thallophyta	Tropical, Humid
Exesisporites sp.	Amerosporae	Thallophyta	Tropical, Humid
Fractisporonites sp.	Fungi	Thallophyta	Tropical, Humid
Fusiformisporites sp.	Fungi	Thallophyta	Tropical, Humid
Lacrimasporonites sp.	Fungi	Thallophyta	Tropical, Humid
Monoporisporites sp.	Fungi	Thallophyta	Tropical, Humid
Multicellaesporites sp.	Fungi	Thallophyta	Tropical, Humid
Phragmothyrites eocenica	Microthyriaceae	Thall0phyta	Tropical, Humid
Polyadosporites sp.	Fungi	Thallophyta	Tropical, Humid
Staphlosporonites sp.	Fungi	Thallophyta	Tropical, Humid
Microthiriaceous germling	Microthyriaceae	Thallophyta	Tropical, Humid
Cyathidites minor	Cyatheaceae	Herb	Tropical to Sub Tropical
Cyathidites sp.	Cyatheaceae	Herb	Tropical to Sub Tropical
Dandotiaspora dilata	Matoniaceae	Herb	Tropical to Sub Tropical
Dandotiaspora telonata	Matoniaceae	Herb	Tropical to Sub Tropical
Dictyophyllidites granulatus	Matoniaceae	Herb	Tropical to Sub Tropical
Dictyophyllidites sp.	Matoniaceae	Herb	Tropical to Sub Tropical

Palynotaxa	Affinity	Habit	Environment / Climate
Foveosporites splendus	Uncertain	------	------
Laevigatosporites cognatus	Polypodiaceae	Herb	Tropical to Sub Tropical
Lygodiumsporites sp.	Schizaeaceae	Herb	Tropical to Sub Tropical
Lygodiumsporites lakiensis	*Lygodium*, Schizaeaceae	Herb	Tropical to Sub Tropical
Osmundacidites cephalus	Osmundaceae	Herb	Tropical to Temperate
cf. *Polypodiaceaesporites* sp.	Polypodiaceae	Herb	Tropical to Sub Tropical
Todisporites kutchensis	Osmundaceae	Herb	Tropical to Temperate
Todisporites sp.	Osmundaceae	Herb	Tropical to Temperate
cf. *Taxodiaceaepollenites* sp.	Taxodiaceae	Tree	Tropical to sub tropical
cf. *Podocarpidites* sp.	Podocarpaceae	Tree	Tropical to sub tropical
Arecipites bellus	Arecaceae	Tree	Tropical to Sub Tropical
Arecipites sp.	Arecaceae	Tree	Tropical to Sub Tropical
Couperipollis brevispinosus	*Arenga*, Arecaceae	Tree	Tropical to Sub Tropical
Couperipollis rarispinosus	*Arenga*, Arecaceae	Tree	Tropical to Sub Tropical
Ctenolophonidites sp.	*Ctenololphon*	Herb	Tropical to Sub Tropical
Ghosiacolpites globatus	*Ctenololphon*	Herb	Tropical to Sub Tropical
Lakiapollis ovatus	*Cullenia, Durio* of Bombacaceae	Tree	Tropical to Sub Tropical
Liliacidites sp.	Liliaceae	Herb	Cosmopolitan
Longapertites retipilatus	Arecaceae	Tree	Tropical to Sub Tropical
Longapertites sp.	Arecaceae	Tree	Tropical to Sub Tropical
Palaeosantalaceaepites minutus	Santalaceae / Rhizophoraceae	Tree	Tropical to Sub Tropical, Coastal
Palmaepollenites kutchensis	Arecaceae	Tree	Tropical to Sub Tropical
Palmaepollenites sp.	Arecaceae	Tree	Tropical to Sub Tropical
Palmidites naviculus	Arecaceae	Tree	Tropical to Sub Tropical
Palmidites sp.	Arecaceae	Tree	Tropical to Sub Tropical
Pellicieroipollis langhenheimii	Theaceae	Shrub	Tropical to Sub Tropical
Polybrevicolporites cephalus	Uncertain	------	------
Proxapertites reticulatus	Araceae	Herb	Tropical to Sub Tropical
Proxapertites sp.	Araceae	Herb	Tropical to Sub Tropical
Psiloschizosporis psilata	Uncertain	------	------
Rhoipites sp.	Anacardiaceae	------	------
Retimonosulcites ellipticus	Uncertain	------	------
Spinizonocolpites echinatus	*Nypa*, Arecaceae	Tree	Tropical to Sub Tropical, Coastal
Striacolporites striatus	Caesalpiniaceae	Tree	Tropical to Sub Tropical
Retitetrabrevicolporites sp.	Uncertain	------	------
cf. *Tiliaepollenites* sp.	Tiliaceae	Uncertain	Tropical
cf. Proteacidites sp.	Proteaceae	Tree	Tropical to temperate
Tricolpites reticulatus	*Gunnera*, Gunnerraceae	Herb	Tropical
Triporopollenites sp.	Betulaceae	Tree	Temperate
Verrumonosulcites foveolatus	Arecaceae	Tree	Tropical to Sub Tropical
cf. Euphorbiaceae	Euphorbiaceae	Uncertain	Cosmopolitan
cf. Rubiaceae	Rubiaceae	Uncertain	Cosmopolitan
cf. Verbenaceae	Verbenaceae	Uncertain	Cosmopolitan

Discussion and Conclusion

The present palynoassemblage is compared to palynoflora recorded from Palaeocene sediments of peninsular and extra peninsular basins of India viz., Rajasthan (Bose 1949, 1952; Jain *et al.*, 1973; Kar and Sharma 2001; Lukose, 1974; Rao & Mishra 1949; Rao & Vimal 1950, 1952; Sah & Kar 1974; Singh & Dogra, 1988; Tripathi 1994, 1995, 2002; Tripathi *et al.* 2003), Kutch (Kar 1985; Sah & Kar 1969, 1970), Cambay (Koshal and Uniyal 1986), Bombay Offshore (Rawat *et al.* 1996), Krishna–Godavari (Rawat *et al.* 1991; Venkatachala and Sharma 1984), Cauvery (Venkatachala 1974; Venkatachala and Rawat 1972), Bengal (Baksi 1972; Baksi and Deb 1980; Deb 1976), Meghalaya (Kar and Kumar 1986; Kar *et al.* 1994; Mandal 1990; Saxena *et al.* 1996; Tripathi 1988; Tripathi & Singh 1984, 1985), Assam (Baksi 1974; Mehrotra and Sah 1982; Sah & Singh 1974; Salujha *et al.* 1972; Sah and Dutta 1966, 1968, 1974; Dutta and Sah 1970) and Western Himalaya (Bhandari *et al.* 1977; Mathur 1984). Rich fungal assemblage of present study is closely comparable to Early Tertiary sediments of India (Gupta, 2002) and abroad (Kalgutkar, 1997; Parson and Norris, 1999). Earlier, Tripathi (2002) reported palynoassembalges from the Akli Formation of same lignite mine. He suggested two zones in the studied sequences assignable to Late Palaeocene age; lower one is characterized by high frequency of dinoflagellate cysts along with low number of pteridophytic spores, whereas the upper part is distinctly dominated by angiosperm pollen having affinity with the families Arecaceae, Liliaceae, Caesalpiniaceae, Onagraceae etc. Palynoassemblage zone GL II of present study is closely comparable to the upper zone made by Tripathi (2002). GL I is closely comparable to Late Palaeocene to Early Eocene fungal remain dominated palynoassemblages of Iceberg Bay Formation, Eukeka Sound Group, Axel Heiberg Island, Northwest Territories, Canada (Kalgutkar, 1997), Caribou Hills Mackenzie Delta, northern Canada (Parson & Norris, 1999) and Sirmaur district, Himachal Pradesh, India (Gupta 2002). Therefore Palaeocene – Early Eocene age is suggested for both the zones GL I & GL II.

From environmental analysis (Table 4) an overall tropical-subtropical, humid Climatic condition is suggested for the presently studied lignite - clay deposition. Within this tropical humid environmental condition, GL II records pollen grains with mangrove plant affinity viz., *Spinizonocolpites Palaeosantalaceaepites* , other palms and also occurrence of algae - *Pterospermella* have gained much significance in environment analysis suggesting near shore influence in the area during Palaeocene - Early Eocene time. Frequent occurrence of fungal remains including epiphyllous fungi in GL I corroborates the development of a forest under tropical to subtropical climatic regime with high rate of precipitation which favored severe biodegradation of organic remains by microbes.

References

Baksi S. K. (1972). On the palynological biostratigraphy of Bengal Basin. in Ghosh A. K. *et al.* (Editors)-Proceedings of the seminar on Paleopalynology and indian Stratigraphy Calcutta 1971, Bot. Dept., Calcutta University : 188-206.

Baksi S. K. (1974). On Oligocene palynological biostratigraphy of Assam – Bengal Region, India. Symp. Strat. Palynol., B.S.I.P. Spl. Publ. No. 3 : 106-116.

Baksi S. K. and Deb U. (1980). Palynostratigraphic zonation of the Upper Cretaceous - Palaeogene sequence of Bengal basin, Geophytology, 10:199-224.

Bhandari L.L., Venkatachala B.S. and Singh P. (1977). Stratigraphy, palynology and paleontology of Ladakh Molasse group in the Kargil area. Proc. IV Colloq. Indian Micropalaeont. Strat.,: 127-133.

Bose M. N. (1949). Angiospermic remains from Barmer Sandstone. Curr. Sci., 18 : 246 – 247.

Bose M. N. (1952). Plant remains from Barmer district, Rajasthan. Journal of Scientific and Industrial Research, 11B : 85 – 190.

Compton Paul M. 2009. The geology of the Barmer Basin, Rajasthan, India, and the origins of its major oil reservoir, the Fatehgarh Formation. *Petroleum Geoscience.* Vol 15, No 2, May 2009 pp. 117 - 130

Dasgupta and Mahesh Chandra, 1978. Tectonic elements of the western Rajasthan self and their stratigraphy. Proc. Golden Jubilee. Sym. Geo. Min. Met. Soc. Ind. Calcutta. 50: 1-16.

Dasgupta S. K. (1977). The Stratigraphy of the West Rajasthan. Proc. 4th Colloq. Indian Micropalaeontol. Stratigr : 219-233.

Deb U. (1976). Upper Cretaceous – Eocene palynology and stratigraphy of Bengal – Meghalaya region. Unpublished Ph. D. thesis. J.U.

Dutta S. K. and Sah S. C. D. (1970). Palynostratigraphy of the Tertiary sedimentary formations of Assam : 5. Stratigraphy and palynology of South Shillong Plateau. Palaeontographica, 131B (1-4) : 1 – 62.

Gupta A. (2002). Algal / Fungal spores from Early Tertiary sediments of Sirmaur district, H. P., India. Tertiary Research 21(1-4) : 123 – 154.

Jain K. P., Kar R. K. and Sah S. C. D. (1973). A palynological assemblage from Barmer, Rajasthan; Geophytology. 3(2) : 150-165.

Kalgutkar, R. M. (1997). Fossil fungi from the Lower Tertiary Iceberg Bay Formation, Eukeka Sound Group, Axel Heiberg Island, Northwest Territories, Canada. Review Palaeobotany Palynology. 97 : 197-226.

Kar R. K. (1985). The fossil floras of Kachchh - IV. Tertiary palynostratigraphy. The Palaeobotanist, 34 : 1-279.

Kar R. K. (1995b). Some new spore pollen genera from Early Eocene sediments of Rajasthan. J. Palynol., 31 : 161 – 170.

Kar R.K. and Kumar M. (1986). Palaeocene palynostratigraphy of Meghalaya, India; Pollen et Spores, 28(2) : 177-217.

Kar R. K. and Sharma, P. (2001). Palynostratigraphy of Late Palaeocene and Early Eocene sediments of Rajasthan, India; Palaeontographica, Abt. B. 256 :123-157.

Kar R. K., Handique, G. K., Kavita C. K., Mandal J., Sarkar S., Kumar M. and Gupta A. (1994). Palynostratigraphical studies on subsurface Tertiary sediments in Upper Assam Basin, India. Palaeobotanist, 42(2) : 183-198.

Koshal V.N. and Uniyal S.N. (1986). Palynostratigraphy of the Cenozoic succession of Cambay Basin, Gujarat. Bull. Geol. Min. Metall. Soc. India, 54 : 208 – 226.

Lukose N.G. (1974). Palynology of the subsurface sediments of Manhera – Tibba Structure, Jainsalmer, Western Rajasthan, India. Palaeobotanist, 21(3) : 285 – 297.

Mandal J. (1990). Palynological investigation of Palaeocene sediments from Thanjinath, Meghalaya. Palaeobotanist, 37(3) : 324 – 330.

Mathur Y. K. (1984). Cenozoic palynofossils, vegetation, geology and climate of the North and NorthWestern subhimalayan region, India; The evolution of the East Asian environment vol. II Palaeobotany, Palaeozoology and Palaeoanthropology (ed. Robert Orr. Whyte. Centre of Asian Studies, University of Hong Kong), 2 : 504-551.

Mathur S.C.and Sudhir Kumar. 2003. Sedimentation in Barmer Basin. Jour. Geol. Soc. India. V.61, No.3. pp. 368-369.

Mathur, S.C., S.K. Mathur and R.S. Loyal. 2006. First Report of microvertebrate assemblage from Cretaceous Fatehgarh Formation. Geological Society of India. 67, 759-769.

Mathur S. C., Loyal, R. S., Nama S. L. and Negi, N. S. 2009. Geological Constraints on the Bentonitic facies of the Barmer Basin, Western Rajasthan, India. In Prof. K. L. Srivastva (Editor) – *"Economic Mineralization"*, Festschrift in Honour of Prof. B. S. Paliwal. Scientific Publishers, Jodhpur. 274-286.

Mehrotra N. C. and Sah S. C. D. (1982). Palynostratigraphic correlation of Mikir Formation in the type area. Geoscience Jour., 3(2) : 113 – 123.

Nama, S. L., 2007. Sedimentological and Palaeontological Investigation of Rocks of Akli Formation of Barmer Basin, Western Rajasthan, India. Unpublished Ph.D. thesis, Department of Geology, J. N. Vyas University, Jodhpur, p. 275.

Naskar, P. and Bakshi, S.K., 1978. Palynological investigation of Akli lignite, Rajasthan, India. Palaeobotanist, 25, 314-329.

Parsons M. G. and Norris G. (1999). Palaeogene fungi from the Caribou Hills Mackenzie Delta, northern Canada. Palaeontographica. Abt. B. 250 : 77-167.

Rana, R.S., Kumar K., Singh H. and Rose K.D., 2005. Lower vertebrates from the Late Palaeocene – Earliest Eocene Akli Formation, Giral Lignite Mine, Barmer District, western India. Curr. Sci., vol. 89, 9, 1606-1613.

Rawat M. S., Sharma K. D. and Juyal N. P. (1991). A standard palynozonation for Krishna – Godavari Basin. In: J. Pandey and V. Banerjee (Eds.), Conference on Integrated Exploration Research, Achievements and Perspectives, Dehra Dun, : 281-297.

Rawat R. S., Misra C. M., Sharma J. and Roy H. (1996a). Palynostratigraphy and Palaeoenvironment of Basal Clastic (Panna Formation) Sequence of Mahim area, Bombay Offshore. Proc. XVI Indian Colloq. Micropal. & Strat., Dehra Dun, : 657-662.

Rao K. P. and Misra S. S. (1949). An oil bearing alga from the Palana lignite (?Eocene) of Rajputana. Curr Sci., 18(10) : 380.

Rao A. R. and Vimal K. P. (1950). Plant microfossils from Palana lignite (Eocene), Bikaner. Curr. Sci., 19 : 82 – 84.

Rao A. R. and Vimal K. P. (1952). Tertiary pollen from lignites from Palana (Eocene) Bikaner. Curr. Sc. India, 18(6) : 595 – 601.

Sah S.C.D. and Dutta S. K. (1966) Palynostratigraphy of Tertiary sedimentary formations of Assam-1. Stratigraphical position of the Cherra Formation. Palaeobotanist, 15 : 72-86.

Sah S.C.D. and Dutta S.K. (1968). Palynostratigraphy of the Tertiary sedimentary formations of Assam – 2. Stratigraphic significance of spores and pollen in the Tertiary succession of Assam. Palaeobotanist, 16(2) : 177-195.

Sah S.C.D. and Dutta S.K. (1974). Palynostratigraphy of sedimentary formations of Assam –3. Biostratigraphic zonation Cherra Formation of South Shillong Plateau. Palaeobotanist. 21(1) : 42-47.

Sah S.C.D. and Kar R.K. (1969). Pteridophytic spores from the Laki Series of Kutch, Gujarat, India. J. Sen Memorial volume, Bot. Soc. Beng. Calcutta edited by Santapau, H. *et al.* : 109 – 122.

Sah S. C. D. and Kar R. K. (1970). Palynology of the Laki sediments in Kutch-3. Pollen from the bore holes around Jhulrai, Baranda and Panandhra; Palaeobotanist. 18(2) : 127-142.

Sah S.C.D. and Kar R.K. (1974). Palynology of the Tertiary sediments of Palana, Rajasthan. Palaeobotanist, 21(2) : 163 – 188.

Sah S.C.D. and Singh R.Y. (1974). Palynological biostratigraphy of Tura Formation in Type Area. Sym. Strat. Palynol. Spl. Publ. Birbal Sahni Inst. Palaeobot., Lucknow, 3 : 76 – 98.

Salujha S.K., Kindra G.S. and Rehman K. (1972). Palynology of the South Shillong Front. Part I : The Palaeogene of Garo Hills. Proc. Sem. Paleo-palynol. Indian Stratigr, Calcutta, 1971: 265–292.

Saxena R.K., Tripathi S.K.M. and Prasad V. (1996). Palynofloral investigation of the Tura Formation (Palaeocene) in Nongwal Bibra area, East Garo Hills, Meghalaya. Geophytology, 26(1) : 19 – 31.

Singh R.Y. & Dogra N.N. (1988). Palynological zonation of Palaeocene of India with special reference to western Rajasthan In: Maheshwari HK (ed) – Palaeocene of India : Proceedings of the Symposium on Palaeocene of India : Limits and subdivisions 1986 : 51-64. Indian Association of Palynostratigraphers, Lucknow.

Sisodia, M.S. and Singh, U.K., 2000. Depositional environment and hydrocarbon prospect of Barmer Basin NAFTA. Zagreb, Croatia. 51(9): 309-326.

Tripathi S. K. M. (1988). Age of the Therria Formation, Meghalaya, India, a review. Geophytology, 18(1) : 121 – 123.

Tripathi S.K.M. (1994). New angiosperm pollen from subsurface Early Palaeogene sediments of Barmer district, Rajasthan, India. Palaeobotanist, 42(2) : 61 – 65.

Tripathi S.K.M. (1995). Palynology of subsurface Palaeocene – Eocene sediments near Kapurdi, Barmer district, Rajasthan, India. Palaeobotanist, 43(1) : 45 – 53.

Tripathi S.K.M. (2002). Palynological study of Late Palaeocene sediments of Giral Lignite mine, Barmer, Rajasthan. Nat. Conf. Biodiversity- Past and Present, Lucknow: 110-111.

Tripathi S.K.M. and Singh H.P. (1984a). Two new pollen genera from the Lower Tertiary sediments of Meghalaya. Palaeobotanist, 32(2) : 153 – 157.

Tripathi S.K.M. and Singh H.P. (1984b). Palynostratigraphic zonation and correlation of the Jowai Sonapur Road Section (Palaeocene-Eocene), Meghalaya, India. Proc. V. Indian geophytol. Conf., Lucknow (1983), Spl. Publ. The Palaeobotanical Society, Lucknow : 316-328.

Tripathi S.K.M. and Singh H.P. (1985). Palynology of the Jaintia Group (Palaeocene - Eocene) exposed along Jowai – Sonapur road, Meghalaya, India – Part I. Systematic Palynology. Geophytology 15(2) : 164 – 187.

Tripathi S.K.M, Singh U. K. and Sisodia M. S. (2003). Palynological investigation and environmental interpretation on Akli Formation (Late Palaeocene) from Barmer Basin, Western Rajasthan, India. Palaeobotanist, 52 : 87 – 95.

Venkatachala B.S. (1974). Palynological zonation of the Mesozoic and Tertiary subsurface sediments in the Cauvery basin. Aspects and Appraisal of Paleobotany, Lucknow : 476 – 494.

Venkatachala B.S. and Rawat M.S. (1972). Palynology of the Tertiary sediments in the Cauvery basin, 1, Palaeocene-Eocene palynoflora from the subsurface; Proc. Seminar on Palaeo-palynology and Indian Stratigraphy 1971, Univ. of Calcutta : 292-335.

Venkatachala B.S. and Sharma K.D. (1984). Palynological zonation in subsurface sediments in Narasapur well no. 1, Godavari – Krishna Basin, India. Proc. X Indian Colloq. Micropal. & Strat., Poona : 445 – 466.

Hon'ble reviewer

1. **Prof. Patricia Vickers - Rich**
   Pro Vice Chancellor
   Monash University
   VIC 3800
   AUSTRALIA
2. **Dr. C.P. Sisodia**
   Dy. Director General (Retd.)
   Geological Survey of India
   94/38 Agarwal Farm
   Mansarovar
   JAIPUR - 302 001

II-34

Geology of Coal Bed Methane and its Potential in India

Virendra Gaur, Amit Kumar, Vijendra Chouhan, Deva Ram, Seema Sharma and K. L. Shrivastava

Department of Geology, Jai Narain Vyas University, Jodhpur-342005 INDIA
Email- klsgeology@yahoo.co.in

Abstract

The paper is presenting a brief account of origin, extraction mechanism and status of exploration of Coal Bed Methane as a geoenergy resource in India. In coalification process, Coal Bed Methane is a natural gas produced by bio-and thermogenic degradation of buried organic matter. If CBM can be tapped from coal beds economically that is more than ninety percent Methane, it can be used as non-conventional geoenergy resource. Globally, the extraction of CBM is in progress as per geological conditions prevailing at the resource site. Extraction of CBM is normally carried out using three types of wells. Vertical beds are traditionally used only in seams of shallow depth. Gob wells are used from the gob areas of any mine where roof is caved with overlying strata. Horizontal wells are used for greater depths to withstand with the higher pressure levels. Indian coals have been targeted for CBM potential evaluation category wise. The first category includes Jharia, Bokaro, Raniganj and North Karanpura coal fields. South Karanpura, Raniganj, Pench-Kanhanand and Sohagpur coal fields are of seccond category. The third category includes Talchir, Pranhita-Gadavari and Wardha coal fields. Tertiary coal/lignites belongs to the fourth and last category. As required, the understanding of Coal Bed Methane has greatly enhanced in recent years by way of exploration research in India. Present status of knowledge is assuring a bright future of CBM as geoenergy resource in this country.

Keywords: Coal Bed methane, Potential of CBM, Extraction of CBM, Jharia Coal field, Raniganj Coal field, Bokaro Coal field.

Introduction

During the process of coal formation, CBM is a natural gas produced by bio- and thermogenic degradation of buried plant materials. It is mixture of hydrocarbon gases dominantly methane (90%-95%), traces of ethane, propane and 2%-3% of carbon dioxide and nitrogen. If CBM is trapped from coal beds economically and used properly as a fuel; gives advantages: as an important source of non-conventional energy; with additional advantages to reduce greenhouse gases to minimize global warming impact, mining hazards and deforestation activities. These benefits have attracted geoscientists and technologists in recent years towards exploration and exploitation activities.

It has been estimated that there are over 909 billion tones of proven coal reserves worldwide and distributed on every continent in over 70 countries, with the biggest reserves in the USA, Russia, China and India. Therefore very good prospects for CBM availability exist in the country. All fossil fuels will eventually run out and it is essential to use them as efficiently as possible. While it is estimated that there is enough coal to last us 155 years; this could extend still further through a number of developments including; the discovery of new reserves through ongoing and improved exploration activities; advances in mining techniques, which will allow previously inaccessible reserves to be reached. Additionally, significant improvements continue to be made in how efficient coal is used so that more energy can be generated from each tone of coal produced. (Optima, 2005; WCI, 2005)

The work done in the field of CBM exploration by western countries and later Russia, U.K., Canada, South Africa and India have emphasized the exploration activity for estimation of CBM potentially. The noteworthy contribution in the field was made by Biswas (1995), Pandey (1996), Chandra (1997), Acharyya (1999), Chandra et al. (1999), Singh and Singh (199a,b), Mitra (1999), Banerjee and Gill (1999), Peters (2002), Guha (2002) Singh (2010) and Singh and Saxena (2010) and several others.

Origin of Coal Bed methane

Coal bed methane evolve during the transformation of the organic matter in the swamp, which later converts it into peat after burial under reducing condition. As temperature increases the peat converts into lignite followed by sub-bituminous, bituminous, low-volatile, medium-volatile, high volatile, anthracite and graphite. This process is known as coalification. During this process at early stage biogenic methane evolve, later thermogenic methane is formed. Much of the methane generated by the coalification process escapes to the surface or migrates into adjacent reservoir or other rocks, but a portion is trapped within the coal itself. In early stages of coalification, biogenic methane is generated as a by-product of bacterial respiration. Aerobic bacteria first metabolize any free oxygen left in the plant remains and surrounding sediments. In fresh water environments, methane production begins immediately after the oxygen is depleted (Rice and Claypool, 1981). Species of anaerobic bacteria then reduce carbon dioxide and produce methane through anaerobic respiration (Rice and Claypool, 1981).

Later, when a coal's temperature underground reaches about 122°F and after a sufficient amount of time, most of the biogenic methane has been generated, and about two-thirds of the original moisture has been expelled, the coal attains an approximate rank of sub-bituminous (Rightmire, 1984a). As the temperature increases above 122°F through increased burial or increased geothermal gradient, thermogenic processes begin and additional water, carbon dioxide, and nitrogen are generated as coalification proceeds to approximately the rank of high-volatile bituminous (Rightmire, 1984b). Maximum generation of carbon dioxide, with little methane generation occurs at about 210°F. Generation of thermogenic methane begins in the higher ranks of the high volatile bituminous coals, and at about 250°F, generation of methane exceeds generation of carbon dioxide. Maximum generation of methane from coal occurs at about 300°F. With even higher temperatures and higher rank coals, methane is still generated, but at somewhat lower volumes (Rightmire, 1984 a and b). Primarily adsorbed on or absorbed within micro pores of the coal. Coal bed methane is found associated with the coal/lignte beds. This is the product that evolves during the process of coalification generally more in quantity than the coal can hold. It is obvious that some of the gas escape in the surroundings yet remaining of it lie in the coal under hydraulic pressure as CBM.

Interestingly, coal acts as a source as well as a reservoir rock for CBM. Generally the CBM content increases with the rank of coal. Permeablity in coal is the result of naturally occurring fractures referred to as cleats. Generally the coal bed acts as an aquifer in view of its richness in water holding capacity in comparison to other conventional gas reservoir owing to the higher internal surface area. Methane is normally adsorbed in the internal surface of the coal and can not be easily detected by conventional gas-well drilling technology. To produce the gas from the coal reservoir, water must be pumped out, and suitable methods should be chalked out for the disposal of the water, depending upon the quality to avoid the environmental problems in and around the area.

As each basin has its own geological characteristics vis-a-vis coal characterization, hence to make analogy of one basin with others may not be fruitful and is rather often misleading. The first assessment of CBM resources can be adjudged by availability of the total coal resources.

In any virgin area, for the prospectivity of CBM, the acquisition of varied exploration and laboratory data are important viz., The total area of the project, the number of seams, their thicknesses, roof and floor characteristics along with the bands of each seam together with their time and space relationship in the area; proximate and ultimate analyses, rank of coal, geological structure, hydrological regime of the area geothermal gradient, reflactance, maceral composition, microlitho types, mineral matter as per International Commission for Coal and Organic Petrology (ICCP), fluorescence microscopy for ascertaining the micro-cleat under UV light, Gas composition and percentages, desorption characteristics, adsorption isotherms study, Permeability and porosity Cleat and fracture pattern, analysis of cleat- filled minerals, quality of adsorbed water etc. Besides, reservoir engineering and modeling, the rock mechanical testing etc. are also to be carried out for better assessment.

Extraction of Coal Bed Methane

Methane from unmined coal seams is recovered through drainage systems constructed by drilling a series of vertical or horizontal wells directly into the seam. Water must first be drawn from the coal seam in order to reduce pressure and release the methane from its adsorbed state on the surface of the coal and the surrounding rock strata. Once dewatering has taken place and the pressure has been reduced, the released methane can escape more easily to the surface via the wells.

Vertical wells- The choice of vertical or horizontal wells is dependent on the geology of the coal seam. In the case of seams at shallow depths, vertical wells have been traditionally used. These vertical systems often use layers of fracture wells, which drain the methane from fractures in the coal seam produced as result of the increased pressure created during the dewatering process. At these shallow depths, the combination of high permeability and low pressure make the vertical systems ideal as extra methane flow enhancement is not required and the structure of the vertical and fracture wells remains stable.

GOB WELLS- The designation "gob well" refers to the type of coalbed methane (CBM) recovery well that extracts methane from the gob areas of a mine after the mining has caved the overlying strata. Gob wells differ from vertical wells in the sense that they are normally drilled to a point 10 to 50 feet above the target seam prior to mining, but are operated only after mining fractures the strata around the wellbore. The methane emitted from the fractured strata then flows into the well and up to the surface. The flow rates are mainly controlled by the natural head created by the low-density methane gas or can be stimulated by blowers on the surface. an active gob well used to drain methane from the gob area of a mine and a second well drilled above the active mining section that will be placed in operation as the longwall moves under the well. Gob wells can recover 30% to 70% of methane emissions depending on geologic conditions and the number of gob wells within the panel. They have several disadvantages when compared to vertical wells. lack of consistently high gas content, relatively short production life, and application only to overlying strata.

HORIZONTAL WELLS- At greater depths, the structure of the vertical and fracture wells may not be able to withstand the higher pressure levels and extra flow enhancement may be required to produce the methane. This is often true in cases of Vertical Coal Bed Methane recovery due to the depths at which the coal is found. In these instances, horizontal drilling techniques may be used for increased accuracy and flexibility. Within these horizontal systems, flow enhancement techniques such as extra hydraulic fracturing - where water is pumped into the seam at high pressure - may be deployed to further facilitate the release of the methane from coals seams. Although horizontal systems can recover much higher volumes of methane from coal seams at extreme depths than a vertical system possibly could, recovery efficiency is relatively low and heavily dependent on the overall length of the drill through the coal seam. Horizontal systems are still in their infancy and over time there may be increased movement towards their use as the technologies mature and efficiencies are improved.

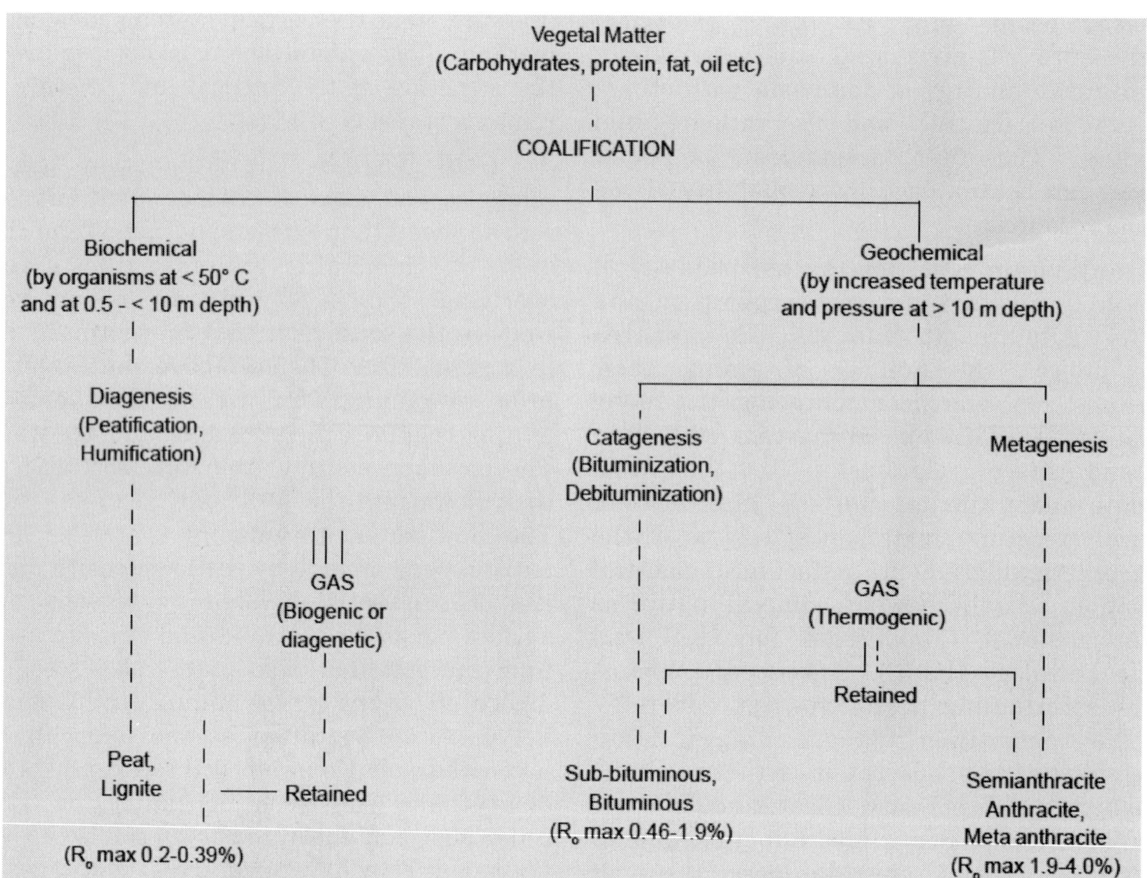

Fig. 1. Flow chart of Methane gas generation during coalification (after Singh and Singh, 1999b).

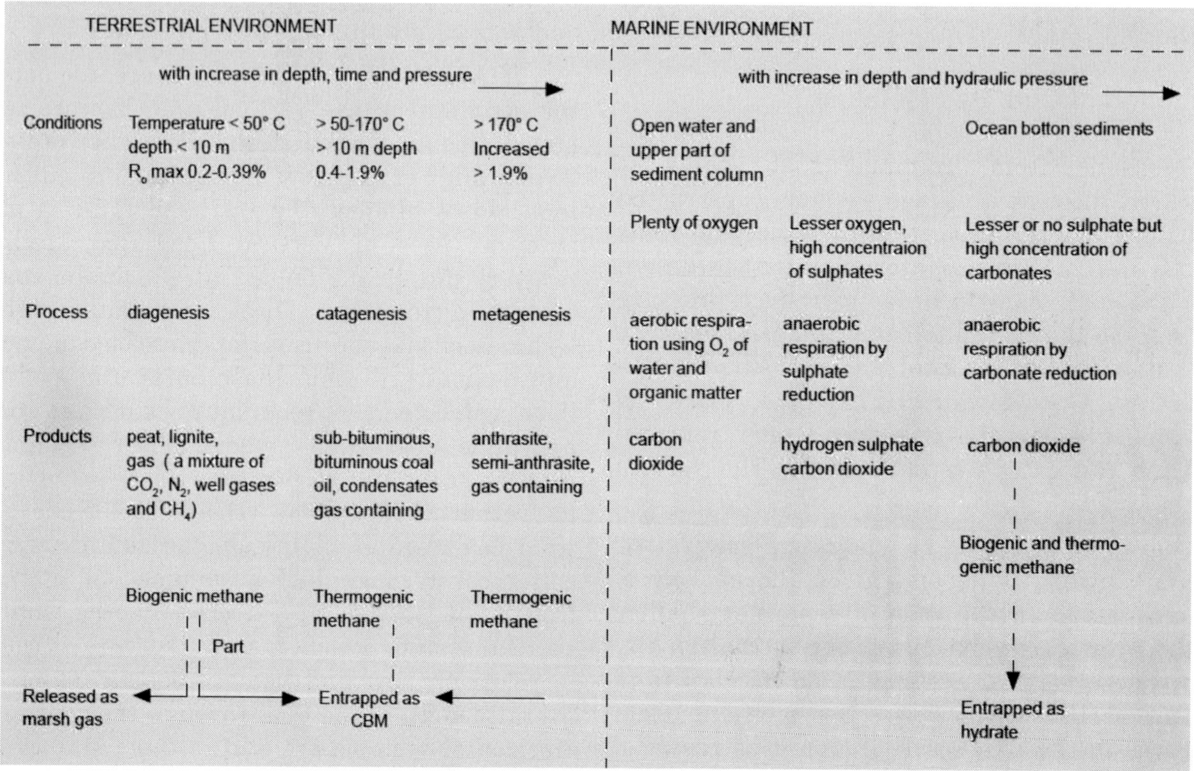

Fig. 2. Schematic representation of methane generation in terrestrial and marine environments (after Singh and Singh, 1999b).

Table 1. Categorization of Indian Coals for CBM Potential (after Peters et. al. 2001; Singh and Saxena, 2010)

Category I	Gondwana Coals ranking high volatile bituminous A and above.	Jharia, Bokaro, Raniganj and North Karanpura Coalfields.
Category II	Gondwana Coals ranking high volatile bituminous A and below.	South Karanpura, Raniganj, Pench-Kanhanand Sohagpur Coalfields .
Category III	Low ranking Gondwana Coals.	Talchir, Pranhita-Godavari Valley and Wardha Valley Coal field.
Category IV	Tertiary Coal / Lignite resources.	Cambay, Bikaner-Nagaur, Barmer, Assam-Arakan, Cauvery and Himalayan Foot Hills Basins.

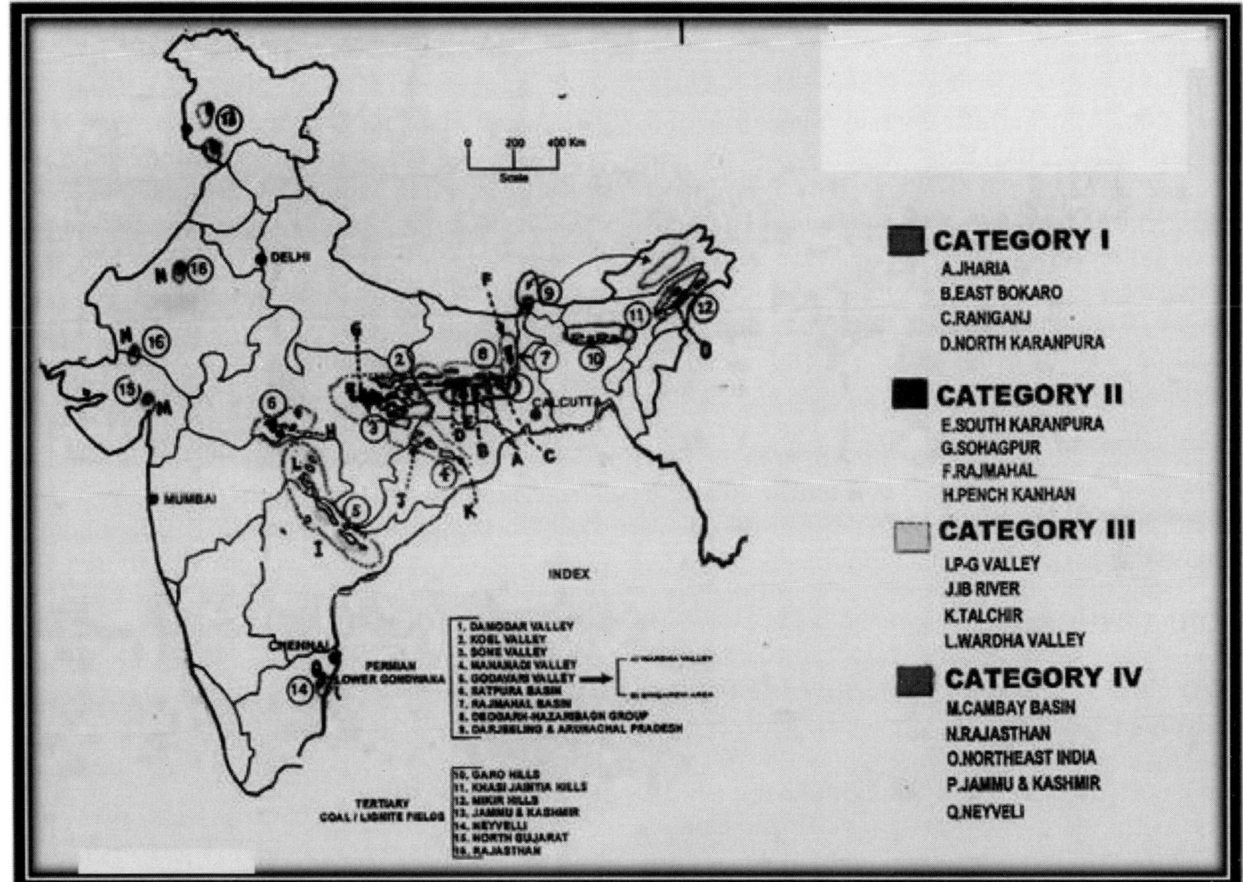

Fig. 3. Category of Indian Coals for CBM Potential (after Peters et. al., 2001)

INDIAN PROSPECTS, POTENTIAL AND EXPLORATION OF CBM

Recently, Singh and Saxena (2010) have given a brief account for Indian prospects, potential and exploration of CBM. The proceeding sections have been adopted from their great contribution.

(A) MADHYA PRADESH

(i) ***SINGRAULI MAIN BASIN CBM BLOCK***- Singrauli Coalfield is a composite basin composed of Moher sub-basin in the east and main sub-basin on the west has very large spread where the Barakar coal seams of Lower Gondwana sequence (Permian) are concealed below a variable cover of younger sediments. The deeply buried coal seams of the main basin offers good prospect for CBM exploration. The Barakar coal measures in the main sub-basin where the CBM block has been carved out, contains eight coal seams. The adsorption isotherm for coal occurring at 300-600m depth level show gas content of 8.3-9.7 m3/t at pressure of 49.79-66.96 atmosphere. The persistent development of moderately thick Barakar seams of high volatile bituminous B/C rank, with an average gas content of 4 m3/t with localized supply of heat from igneous intrusions, all in combination are likely to contribute to CBM gas generation and storage in seams. One CBM block viz. SR-CBM-2005/III covering an area of 330 sq km. is on offer for exploration and production of CBM. The CBM resource of the block is estimated at 31.0 BCM (1.1 TCF) (Singh and Saxena, 2010).

(ii) SOHAGPUR COAL FIELD- The northern part of Sohagpur coalfield has a complex history of dolerite intrusion, which may influence the generation of gas. The exploration in Nigwani-Bakeli area of Barakar coal measures below 300m cover of Barren Measure and Raniganj Formation. Five regional seams are developed in Barakar Formation. The cumulative coal thickness in the block varies from 3-10m and attain higher rank with low moisture and high carbon content. Gas seepage in nearby area also indicates storage of gas in deeper seams. The adsorption isotherm of seam-I (558m) at 47.31 atmospheric pressure give moisture equivalent gas content of 6 m^3/t. One CBM block viz. SP (North)- CBM-2005/III covering an area of 609 sq. km is on offer for exploration and production of CBM. The CBM resource of the block is estimated at 16.72 BCM (0.59 TCF).

(B) CHATTISHGARH

(i) MAND-RAIGARH COAL FIELD- The Mand-Raigarh Coalfield occupies the central part of the upper Mahanadi valley Gondwana belt and extends over a large stretch lying between Ib valley in the east and Korba and Hasdo-Anand Coalfields in the west and northwest. This coalfield displays a shallow synclinal structure with axis trending in NW-SE direction and the southern basin margin is demarcated by the prominent boundary fault. In the central part of the coalfield where Barakar coal measures lie at greater depth below the younger sediments, a CBM block has been demarcated. In general twelve coal seams are developed in Barakar Formation. The cumulative coal thickness in the CBM block varies from 30 to 35 m, the Coal thickness tends to increase towards south near Chhal area. The coals broadly correspond to high volatile bituminous 'B' to 'C' in rank. With the increase of depth of burial the coal seams are likely to store substantial amount of methane. Desorption studies of coals indicate that deep-seated coal seams in the CBM block hold at least 4.5-5.5 m3/t of gas if not higher. The hydrogeological sudies also reveal that the coal seams are under normal hydrostatic pressure, which will facilitate good sorption of gas on coal. Singh and Saxena (2010) have mentioned that one CBM block viz. MR-CBM-2005/III covering an area of 634 sq km is on offer for exploration and production of CBM. The CBM resource of the block is estimated at 119 BCM (4.2 TCF).

(ii) TATAPANI– RAMKOLA COAL FIELD- Tatapani-Ramkola Coalfield occurs in the structural junction between the Damodar valley and the Son valley Gondwana basin belts. The basin depicts structural and stratigraphic similarity with that of adjacent Gondwana basin of the Damodar valley whereas the pattern of coal formation and sandstone dominated cycles of Barakar coal measures are more characteristic of the Son valley belt. Exploration in Barkar gaon revealed the occurrence of twelve regional Barakar coal seams at a depth of 300-750 m. Adsorption isotherms of Coal from Barkagaon show a sorption capacity of 7.4-8.1 m3/t of gas at a pressure of 58.2- 86.7 atmospheric in the block. The coalification gradient in Tatapani-Ramkola coalfield (Singh and Saxena, 2010) is suggestive of significant methane generation in the deep-seated seams of the CBM blocks. One CBM block viz. TR-CBM-2005/III covering an area of 458 sq km is on offer for exploration and production of CBM. The CBM resource of the block is estimated at 53.78 BCM (1.9 TCF).

(C) JHARKHAND

(i) RAJMAHAL COALFIELD-Recently Singh and Saxena (2010) have given a brief account for Indian prospects, potential and exploration of CBM. The proceeding sections have been adopted from their great contribution. Rajmahal coal belt has several disconnected coal basins viz., Hura, Chuperbhita, Pachwara, Mahu-agari and Bramhani Coalfields along the western fringe of the Rajmahal hills age of Lower Gondwana period (L Permian). Thickness of coal seams (about 40-60m) below a variable cover of 300-600 m of younger strata with an average gas content of 5-6 m3/t. One CBM block viz. RM-CBM- 2005/III covering an area of 469 sq km is on offer for exploration and production of CBM.

The CBM resource of the block is estimated at 158 BCM (5.58 TCF) (Singh and Saxena, 2010).

(D) ANDHRA PRADESH

(i) SOUTHERN GODAVARI VALLEY COAL FIELD- The Godavari valley coalfield is the largest Gondwana basin covering an area of about 17500 sq km in Andhra Pradesh, India. It extends in NW-SE direction from the border of Maharashtra in the heart of central India to the east coastal region. The basin is filled by 6000 m of sediments of early Permian to Cretaceous age and is the major coalproducing area in south India. The Gondwana basin displays a typical rift structure with prominent boundary fault along its margin and is analogous to the petroliferous Cooper basin of Australia where large resource of gas has been estimated from Gondwana equivalent horizons. The coals of the Godavari CBM block belong to high-volatile bituminous B/C rank. Desorption studies indicate that coal seams in the 600-1000m depth range within the CBM block, may hold on an average 4-5 m3/t of gas. The CBM block has been delineated in the eastern side of the Kothagudem area and covers part of the main sub-basin across the Paloncha neck in the dip side of Anisetthipalli and Kothagudem blocks. The CBM block shows development of lower Gondwana sequence comprising coal-bearing Barakar Formation, Barren Measures and upper coal measures of Raniganj Formation. The Barakar Formation of Kothagudem sub-basin where a large part of the CBM block is located contains two coal seams. The cumulative coal thickness shows variation from 4.5 m to 26 m. One CBM block viz. KG (East)-CBM-2005/III covering an area of 750 sq.km. is on offer for exploration and production of CBM. The CBM resource of the block is estimated at 57.2 BCM (2.02 TCF) (Singh and Saxena, 2010).

(ii) NORTHERN GODAVARI VALLEY COAL FIELD- The Barakar Formation (Lr. Permian) of the Lower Gondwana sequence is the main coal-bearing unit. Presence of one coal seam has been recorded from Raniganj Formation (Up. Permian) also. The coal-bearing Barakar Formation is exposed along the northwestern margin of the basin where extensive mining operation is in progress from Dorli-Belampalli belt in the north through Somagudem-Indaram in the centre to Ramagundam belt further south. The dip side of these mining belts is considered to be the most potential target area for CBM exploitation. In these areas, as many as 10 coal seams of 0.6 to 18.2 m thickness have been established in Barakar Formation though 4 to 5 seams are more persistent. The coal seams are by and large high volatile bituminous 'B' in rank with vitrinite reflectance of 0.6 – 0.7% within 600m depth limit. There are, evidences of enhancement of rank of coal with greater depth of burial. Adsorption isotherm of Barakar coal seams and limited desorption data show that the coals are likely to store 4-5 m3/t of gas upto 1000 m depth, which may increase at greater depth of burial with better permeability due to intensive high fracture density in rift setting. Singh and Saxena, 2010have reported that one potential CBM block viz. G V(North)-CBM-2005/III covering an area of 386 sq km in the northwestern part of the Godavari Gondwana basin, is on offer for CBM exploitation. The gas in place resource of the CBM block is estimated at 29.65 BCM (1.05 TCF) within a depth range of 450 to 1500m.

(E) WEST BENGAL

(i) BIRBHUM COALFIELD- The Birbhum Coalfield southwesterly extension of Rajmahal coal-belt in West Bengal. Here, Barakar Formation contains four thick to very thick Coal seam zones has good quality Coal and the heat flow from 200 m thick volcanic suite to underlying coal measures all contribute to the CBM potentiality of the field. Desorption and adsorption isotherm data generated from the northern part of the field shows that coal seams often store 5-6m 3/t of gas. Singh and Saxena, 2010 have reported that one CBM block viz. BB-CBM-2005/III covering an area of 248 sq km is on offer for exploration and production of CBM. The CBM resource of the block is estimated at 50 BCM (1.77 TCF).

(F) RAJASTHAN

(i) BIKANER-NAGAUR AND BARMER-SANCHOR BASIN- In Rajasthan three districts Bikaner, Barmer and Nagaur has reserves of more than one billion tones of lignite deposits. The Barmer-Sanchor basin which lies in the Thar Desert of Rajasthan is a narrow elongated rift extending from Fatehgarh Fault in the north to Sanchor depression in the south and shows continuity with the Cambay graben of Gujrat

(Chouhan et. al., 2009). The sedimentary sequence in Barmer basin ranges in age from cretaceous to recent, having a maximum thickness upto 4 km. thick lignite horizon is reported to occur at variable depth over the length and breadth of the basin. The lignite occur in middle Eocene Tharad Formation and its equivalent Thumbli Formation in the basin. This lignite bearing sequence is homotaxial with the Kadi and Kalol Formation of Cambay basin. Lignite horizon attains its maximum thickness of about 60-70 m. the desorption tests indicate a gas content of 3-4 m^3/t of gas in 600-1200m depth.

Palana and Barsinhsar (Bikaner)- Both mines are situated about 17 km from Bikaner. CBM has yet to be explored.

Matasukh and Kasnau (Nagaur)- Matasukh and Kasnau mines are situated near villages Kasnau and Matasukh of Nagaur district (Tripathi and Shrivastava, 1990) in the central Rajasthan, which 42 km from district head quarter have commercial production is 12,00,000 MT per year. CBM has yet to be explored.

Two CBM blocks viz. BS (4)-CBM-2005/III in the western flank and BS (5)-CBM-2005/III in the eastern flank of Barmer rift, measuring an area of 1168 sq. km and 739 sq. km respectively have been carved out and are on offer for CBM exploration and production. The estimated CBM resources of BS (4)-CBM-2005/III and BS (5)-CBM-2005/III are 82.0 BCM (2.9TCF) and 38.0 BCM (1.34 TCF) respectively (Singh and Saxena, 2010).

CONCLUSION

The understanding of coal Bed Methane has greatly enhanced in recent years by ways of research to find the possible sites and intensive exploration of the resource. It is need of the hour specially when India is the third largest producer of coal and having sixth largest reserves of the coals. It is also true that reserves alone does not guarantee that CBM availability or exploibility. Therefore adequate understanding of reservoir fundamentals is on its way to be generated before taking up a site as a prospective target area for CBM exploration and production. The government has thus so far awarded 26 coal-bed methane blocks for exploration. These are spread over 13,600 sq. km. and the total methane resources in these blocks are estimated at 1,374 bcm, which translates into a production potential of around 38 mscmd. Present status of knowledge certainly assuring a bright future of CBM as resource in this country.

References

Acharyya, S.K. (1999) Indian Coal Basins as a Source of Coal Bed Methane- Prospects and Retrospect. *In:* Coal Bed Methane – Prospects and Potentialities. Proc. Vol. Int. Sem., S.A.A.E.G., pp.1-13.

Banerjee, M. and Gill, T. (1999) Coal Bed Methane industry. Proc. Vol. Int. Sem., S.A.A.E.G., pp.108-110. BISWAS, S.K. (1995) Prospects of coal bed methane in India. Indian Jour. Petroleum Geol., v.4, pp.1-23.

Biswal, S. K., (1995): "Prospects of Coal Bed Methane in India." Ind. Jour. Petro. Geol. vol. 4 pp. 1-23.

Chandra, K. (1997) Alternative hydrocarbon resources in the next millennium. Geohorizons, v.2, no.2.

Chandra, K., Peters, J. and Singh, S.K. (1999): Coal Bed Methane Exploration in India- An Overview. *In:* Coal Bed Methane –Prospects and Potentialities. Proc. Vol. Int. Sem., SAAEG, pp.63-70.

Chandra, D. Singh, R.M. and Singh, M. P. (2000): "A Textbook of Coal (Indian context)" Tara Book Agency, Varanasi, pp. 280-321.

Chauhan, M. Swadia, J. Prajapat, T. and Shrivastava K.L. (2009) "Petroleum Geology and Hydrocarbon Prospects in Barmer-Sanchor Basin, Northwest India" in Economic Mineralization 'Editor K. L. Shrivastava.' Sci. Pub. (Ind) pp. 433-441.

EIA., (2005): Data obtained from International Energy Annual 2003, U.S. Energy Information Administration, Washington, DC., Table posted June 13, 2005.

Guha, S. (2002): "Petrological and Geochemical environment of Coals and associated Sediments, S-E part of Ib river valley MCL Orissa. Unpublished Ph. D. Thesis, Vikram University, Ujjain.

Mitra, N.D. (1999) A few thoughts on coal bed methane prospects in India. Proc. Vol. Int. Sem. SAAEG, pp.14-15.

Optima, 2005, ECOAL Newsletter: v.1 No.1, 2005, World Coal Institute.

Pandey, S.K. (1996) Prospects of coal bed methane and its introduction in India. Minetech, vol. 17, pp 14-23.

Peters, J. (2002): "Evaluation of Coal Bed Methane Potential of Jharia basin, India." SPE paper No. 4, pp. 1-23.

Rice, D. D. and Claypool, G. E. (1981): "Generation, accumulation and resource potential of biogenic gas": American Association of Petroleum Geologists Bulletin, vol. 65, (1), pp. 5-25.

Rightmire, C.T. (1984a): "Coal bed methane resources", *in* Rightmire, C.T., Eddy, G.E. and Kirr, J.N., eds., Coal bed methane resources of the United States: American Association of Petroleum Geologists Studies in Geology Series No. 17, pp. 1-13.

Rightmire, C.T. (1984b): "Coalbed methane resources", *in* Rightmire, C.T., Eddy, G.E. and Kirr, J.N., eds., Coal bed methane resources of the United States: American Association of Petroleum Geologists Explorer, vol. 21, no.4, ps. 16, 18-20, 22-23.

Singh, A. and Singh, B.D. (1999a) Petrology of Kanhan coals, Satpura Gondwana Basin (India) vis-à-vis Coal Bed Methane. Proc. Vol., Int. Sem. SAAEG, pp.117-130.

Singh, A. and Singh, B.D. (1999b) Methane gas: An unconventional energy resource. Curr. Sci., vol.76, pp. 1546-1553.

Singh, K.N. (2010): Coal Bed Methane Potentiality-case studies from Umaria, Korba and Ib-valley Coals, Son-Mahanadi basin." Jour. Geol. Soc. Ind., vol. 76, pp. 33-39.

Singh, M.P. and Saxena, R. (2010): "Status of Coal Bed Methane Investigations in India" in "Glimpses of Geoscience Research in India,Indian Reports to IUGS 2004-2008, Eds. Singhvi, A.K., Bhattacharya, A., and Guha, S., Indian National Science Academy, WWW.iypeinsa.org/updates-09/geosc-rpt-1htm, pp. 229-238.

Tripathi, R. P. and Shrivastava, K. L. (1990): "Characterization of Merta Road Lignite and decompositional study of its 'Pyrite' using Mossbauer Spectroscopy." FUEL Vol. 69 pp. 928-930.

WCI, 2005, Coal Reserves: Optima, vol. 1, No.1, World Coal Institute WEC (2001): World Energy Council, "India: Extract from the Survey of Energy Resources, 2001".

Hon'ble reviewer

1. **Prof. Pankaj K. Srivastava**
   Dean
   Department of Geology
   Jammu University
   JAMMU - 180 006
2. **Prof. Anil M. Pophare**
   Head
   Department of Geology
   Nagpur University
   NAGPUR - 440 001

Part - III:

WATER RESOURCES

Multiparametric and Multidisciplinary Approach Aimed at Developing a Decision Support System for Groundwater

(A case study of Milan area in one of the most productive within southern Europe)

Maurizio Gorla, Emmanuele Cavalli and Roberto Simonetti

Water Balance and Geology Department, CAP Holding S.p.A. Assago, Milan, Italy
E-mail: m.gorla@tiscali.it

Abstract

This paper intends to depict the workflow needed to develop a new D.S.S., called I.A.P. (Infrastructural Aqueduct Plan), based on a robust multilayer analysis (M.L.A.) and the creation of a certain number of environmental, anthropic and infrastructural critical indexes. This kind of innovative IT approach can be considered a new managing tool to evaluate the sustainability of groundwater abstraction compared with the demographic and urban increase. The criticality indexes can be gathered into four groups: (1) geological-stratigraphical (permeability, effective porosity, isobaths and isopachs); (2) chemical (spatial and time-dependent distribution of physical-chemical parameters, major ionic species and pollutants); (3) infrastructural (water pipelines' state of health); (4) demographic (growth of population within the supplied municipalities). The cartographic output is an user-friendly summary map, or in alternative a certain number of single thematic maps. They represent an easy way to give a support in decision, dealing with sustainable management of groundwater resources. Nevertheless, only when we will complete the analysis on a hydrogeological basin scale, we could say that the method proposed was completely vindicated.

Keywords: Infrastructural Aqueduct Plan, Decision Support System, Groundwater, Multilayer Analysis, Criticality Indexes, Summary Map.

Introduction

CAP Holding S.p.A. is a publicly funded company, with local authority share holding, dealing with Integrated Water Management (public water supply, sewage network and treatment), so that it manages the network and treatment infrastructure, plans and undertakes infrastructure investment and long term maintenance, maps underground networks for more than 80 years withinthe Provinces of Milan and Monza/Brianza (Figure 1).

Its main roles are briefly resumed:

- 197 municipalitiesand 2 provinces are share holders serving for more than two millioncustomers;
- 800 boreholes owned;

- 154 underground and above ground reservoirs;
- more than 6800 km of public water supply network.

In 2012, CAP Holding's Board of Directors gave to Water Balance and Geology Department the task of developing a new D.S.S., called Infrastructural Aqueduct Plan (I.A.P.). Between May 2012 and May 2013, a pilot project has been elaborated, with the purpose of testing the work procedure created.

A decision support system (D.S.S.) is a computer-based information system supporting business within organizational management, operations, and planning levels, which may be rapidly changing and not easily specified in advance, in order to help decision making activities (Power, 2002, 2003 and 2004).

Within CAP, it has always been a certain consciousness regarding the importance of evaluating the cost/benefit ratio of a new infrastructural work. In the past few years, nevertheless, this approach was primarily based on a empiric experience of the technicians, that is to say on their deep knowledge of the territory supplied instead of using complicate IT tools.

The new multiparametric and multi-disciplinary approach is going to completely change the way of working and even planning the future infrastructures, aimed to get better the Aqueduct systems of the municipalities supplied by CAP Holding S.p.A., through a user-friendly interface.

So, the aim of this paper is to illustratethis new D.S.S. to manage the environmental, demographic and infrastructural criticalities, firstly based on a robust knowledge of the hydraulic, hydrogeological and hydrochemical features of the aquifer systems existing in the subsurface within the Province of Milan.

As a matter of fact, groundwater resources will play an increasingly strategic role in supplying fresh drinking water in the future, and we have to maintain a correct and sustainable withdrawal and a balanced management of these water resources.

Thus, CAP Holding S.p.A. is currently promoting R&D project, devoted to sustainable management not only for environmental characters, but even regarding infrastructural ones (water wells, pipelines, hubs, etc.).

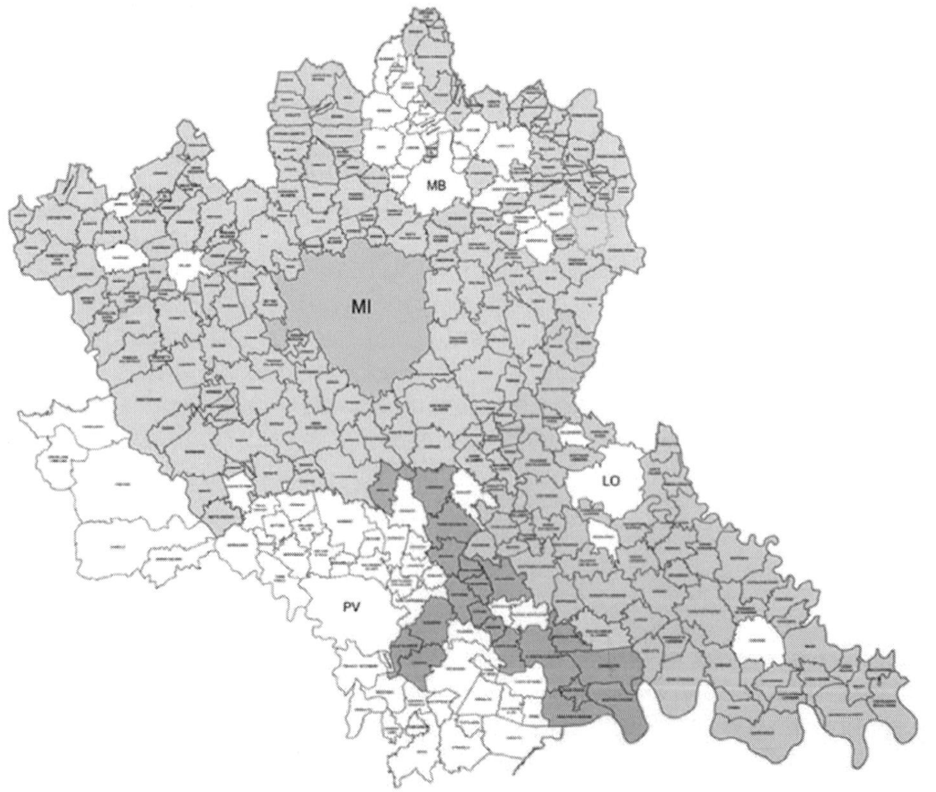

Figure 1. The CAP Holding S.p.A. domain.

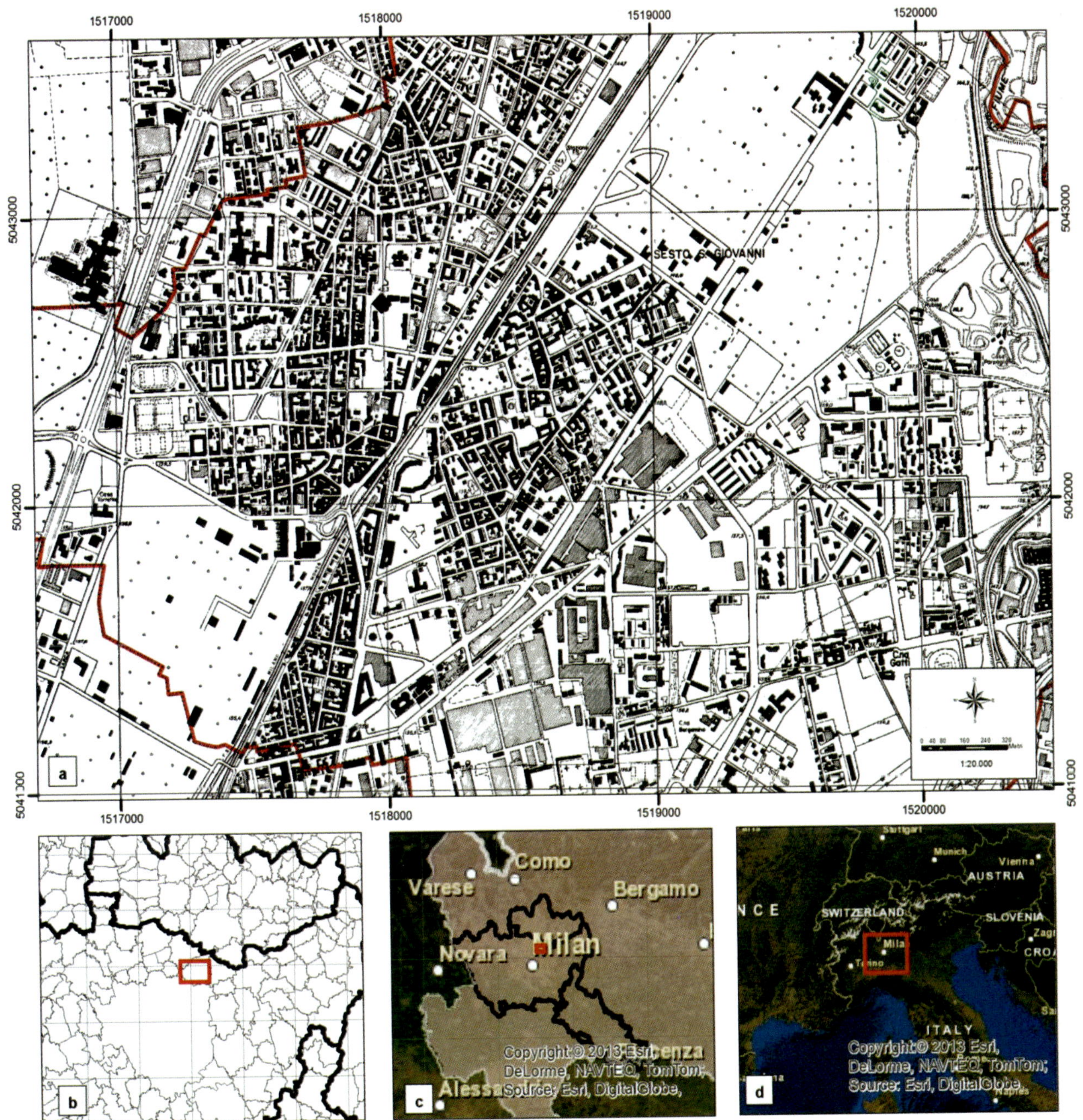

Figure 2. Area of study: developing a pilot project (CTR, RegioneLombardia, 1994; ESRI©, DeLorma©, NAVTEC©, TomTom©, 2013). Legend: a: local scale; b: province scale; c: region scale; d: nation scale.

2. AREA OF STUDY

The case study proposed in this paper is located north of Milan (Figure 2), covering an area of 190km^2, with the aim of analyzing the environmental, demographic and infrastructural features of 28 municipalities (Milan and Monza too).

This choice has been made firstly to test the D.S.S.within a very urbanized and highly densely populated territory (more than two million inhabitants), suffering from many problems related to the quality of groundwater distributed by the public Aqueduct.

These criticalities are mainly represented by heavy concentration of anthropic pollutants, both organic (TCE, PCE, etc.) and inorganic (nitrate, chromium, etc.), that requires an adequate treatment (granular activated carbon G.A.C. and reverse osmosis R.O.) of groundwater before its distribution.

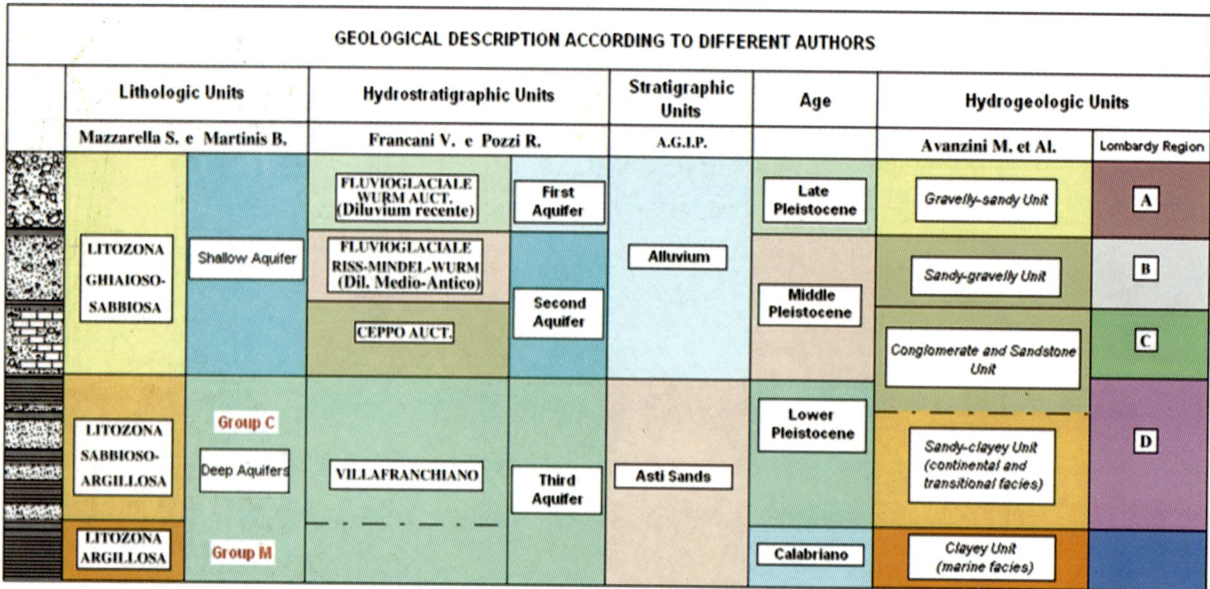

Figure 3. Stratigraphical sketch of the Milan Province quaternary deposits (Gorla, 2010).

3. GEOLOGICAL AND HYDROGEOLOGICAL SETTING

The first 200-300 m in depth show the following stratigraphical profile (CAP, 1995; Gorla, 2001 a & b); from top to bottom, we can distinguish these lithologic units (Figure 3):

- Clayey Unit (UnitàArgillosa). Composed of prevalent grey-blue clayand silt, with macro and microfossils (i.e Hyalinea Balthica and Ammonia beccarii) and thin cemented sandy levels. It derives from ashallow-deep marine palaeo-environment, referred to Calabriano age. This unit can be observed at up to 280-300 m depth.
- Sandy-Clayey Unit (Unità Sabbioso-argillosa). During the Lower Pleistocene, there was a phase of marine regression, which constituted lagoon, beach, palustrine and alluvial deposits. This unit, that represents thebottom of the traditional unconfined aquifer, shows a predominance of grey and yellow clay and silt (with frequent interposition of colour), blendedwith peat, and locally lenses of gravel, sand and conglomerates, which generate confined ground water reservoirs, recognizable within adepth range between 100 m and 160 (eastern sector)-200 m (western sector).
- Conglomerate and Sandstone Unit (Unità a Conglomerati e Arenarie – "Ceppo"). Made up of conglomerate and sandstone lithologies, of Lower Pleistocene age, passing laterally and southward to uncemented graveland sand. This unit can be found principally in the pedemontan region, atup to 50-100 m depth. Its maximum thickness is about 80 m. It derivesfrom a palaeo-environment of river channels.
- Sandy-Gravelly Unit (Unità Sabbioso-ghiaiosa). Composed of sand deposits, prevailing on the gravel ones, of Middle Pleistocene age. Locally, silt-clay strata and arenitic-conglomeratic lenses are recognizable. Thickness is 40-60 m. Within the city of Milan, this formation reaches adepth of 100-110 m.
- Gravelly-Sandy Unit (Unità Ghiaioso-sabbiosa). It indicates a fluvio glacial palaeo-environment, when streams were meandering throughout thealluvial plain during the Late Pleistocene - Olocene. The gravel fractiongenerally dominates the sandy one, and there are rare clayey lenses witha limited spatial extension. The thickness of this unit is about 20-40 m.

From the hydrogeological point of view, we can distinguish three primary aquifer systems (Figure 4):

- The first one, mostly in a phreatic, unconfined condition, which results from the coalescence of the two/ three lithologic/

hydrogeologic unit. The bottom of this aquifer can be encountered, on average, at 100 m beneath the ground level. It has a great yield, due to high transmissivity and saturated thickness, respectively of about $1 \div 2 \cdot 10^{-2}$ m^2/s or even greater ($1 \cdot 10^{-1}$m^2/s) and upto 70-80 m. The isopiezometric map clearly shows the flow pattern of this aquifer (Figure 5). The hydraulic gradient diminishes from north to south, varying from 1% within the upper partof the province and 1÷2‰ in its southern portion.

- The second and the third ones, are deep confined aquifers: the so called C and M aquifers (C stands for continental and M stands for marine environment). Both of them are confined aquifers, with transmissivity values mainly ranging from $1 \div 2 \cdot 10^{-3}$m^2/s to $8 \div 9 \cdot 10^{-3}$m^2/s.Deep aquifers are mostly wedge-shaped, with thicknesses varying from 5-6 m, southward, up to > 20 m, within the highland zone, in the northern part of the province.The flow pattern at provincial scale shows a principal flow direction from north to south, with a hydraulic gradient varying from 5÷6‰ upstream to 1÷2‰ downstream.

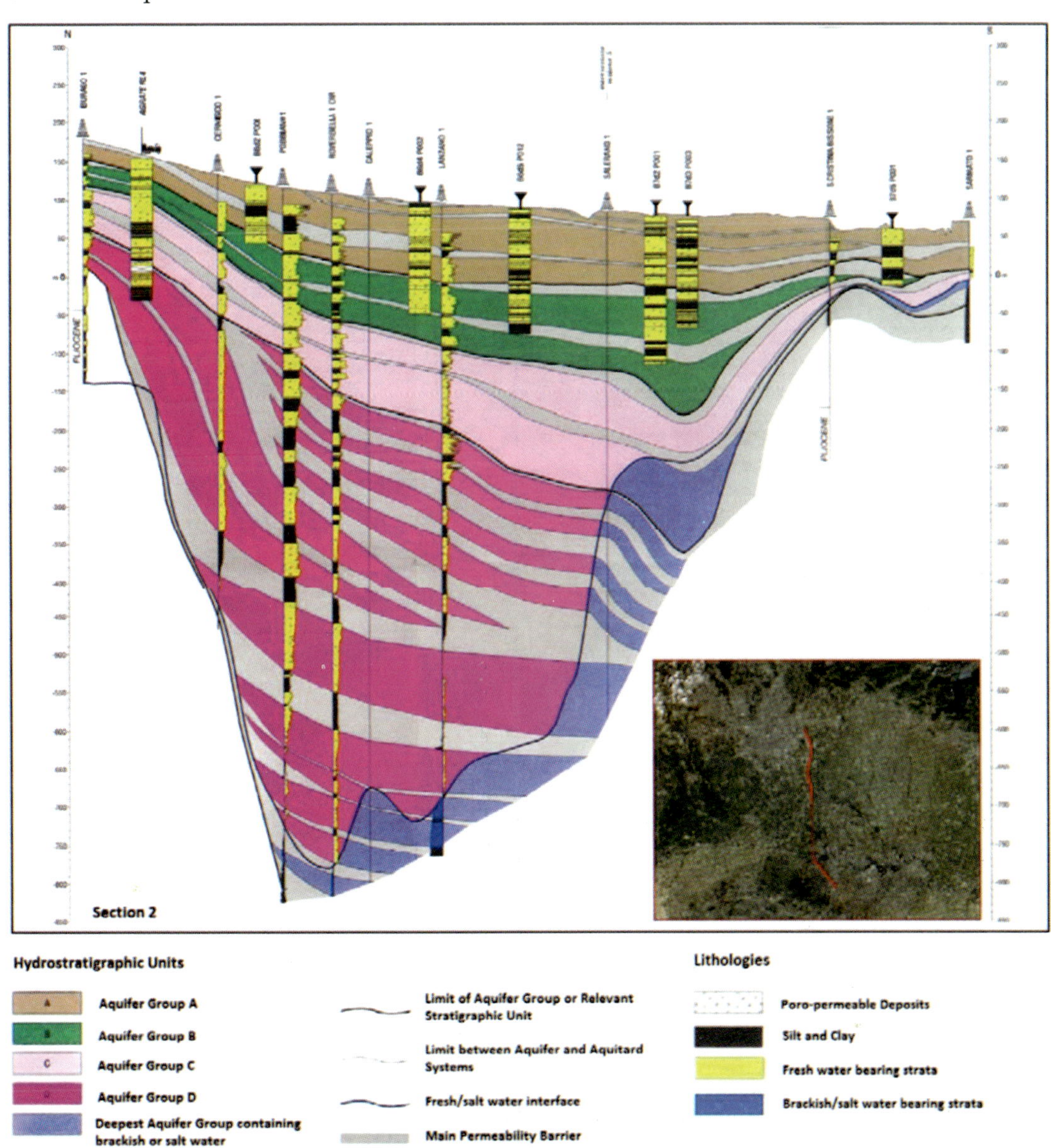

Figure 4. Cross section showing the main aquifer groups identified within Milan Province (from RegioneLombardia – ENI Divisione AGIP, 2002, modified).

4. METHODOLOGICAL BACKGROUND

The environmental analysis undertaken within the pilot area, located north of Milan, had to start from these methodological requirements (Figure 6):

- strong knowledge of anthropic and environmental features within the investigated area;
- development and implementation of relational & territorial databases (RDBMS)
- variables regionalization;
- quantitative and qualitative analysis of the parameters, even in a comparative way;
- experienced in discretion of cartographic output.

The final purpose of the work is to reach a multilevel informative comparison, based on ageostatistical (probabilistic) analysis (Cressie, 1990 and 1991), enabling the Department to produce a synthetic cartographic output.

Up until the recent past, all the work was done by means of empiricism, without using any geomodelling e GIS tools.

As a matter of fact, this kind of approach can't be used anymore, because the planning of complex infrastructures, like water wells, pipelines, etc., requires a 3D reconstruction of the geological and hydrogeological matrix, flow and transport of water and solutes, and even their time-dependent relationships and interferences with a lot of other demographic and infrastructural factors.

So, it's easy to understand that the whole project I.P.A. can be correctly managed only by means of a mighty informative architecture (hard ware + software), coupled with high geological, hydrogeological, geochemical and IT skills.

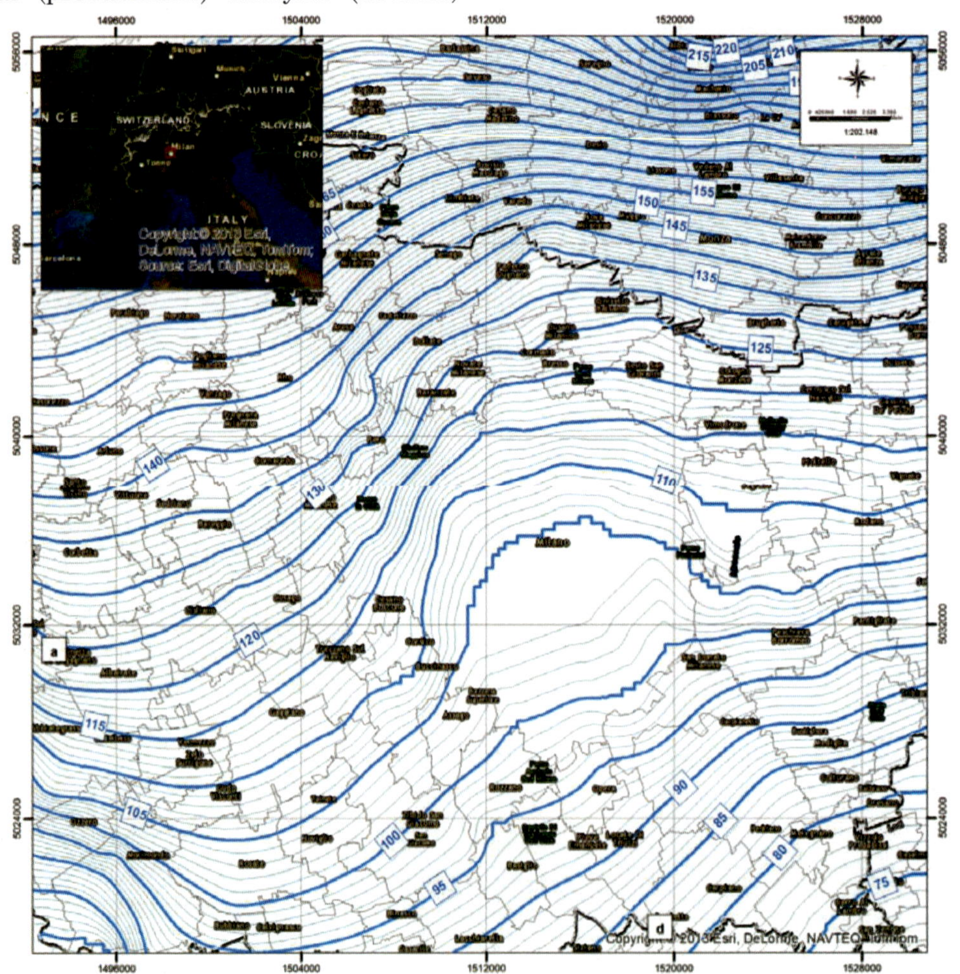

Figure 5. Interpolated groundwater head map of the main aquifer system identified within Milan Province. Hydraulic heads (blue numbers) are expressed in meters above mean sea level (from Provincia di Milano, 2009-2011, modified; and ESRI©, DeLorma©, NAVTEC©, TomTom©, 2013).

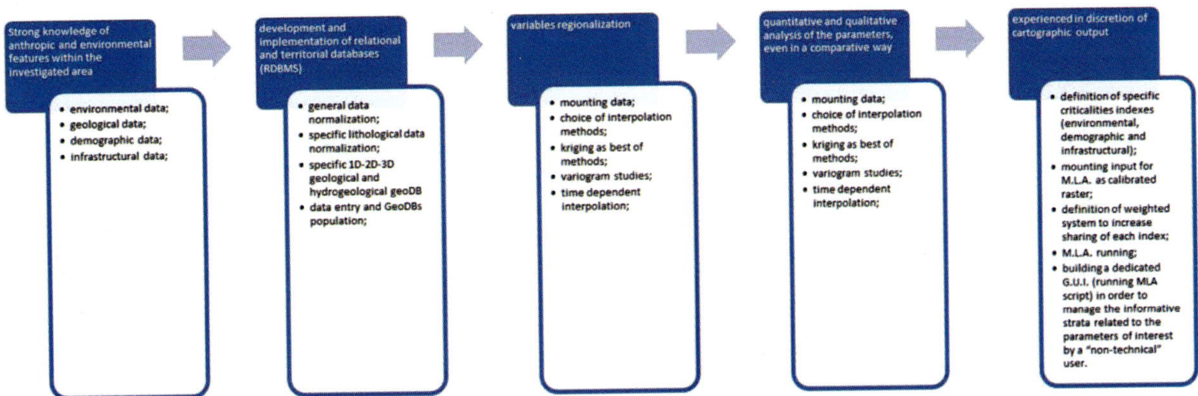

Figure 6. Flow chart of methodological background.

Developing the I.P.A. pilot project, required the *ad hoc* creation of a workflow procedure.

It can be summarized in the following steps:

- Data collections:
  - environmental data
    - chemical;
    - pollution data;
    - meteorological.
  - geological data
    - geophysical;
    - lithological;
    - hydrogeological.
  - demographic data
    - ISTAT;
    - trends.
  - infrastructural data
    - arcs and nodes of water-work net, as GeoDB features.
- Database (DB) and GeoDatabase (GeoDB):
  - general data normalization;
  - specific lithological data normalization;
  - single format rDB (from many different datasheet and DB);
  - 2D geoDB;
  - specific 1D-2D-3D geological and hydrogeological geoDB
  - data entry and GeoDBs population;
  - testing data and strength test.
- 3D heterogeneities interpolation:
  - 3D hydrogeological migration (from GIS to Geostatistical Software);
  - 3D simulations and interpolations;
  - mapping 2D hydrogeological heterogeneities (2D mapped output).
- 2D Spatial analysis:
  - mounting data;
  - choice of interpolation methods;
  - kriging as best of methods;
  - variogramstudies;
  - time dependent interpolation;
  - 2D mapped output;
  - calibration of the output coming out from the geostatistical analysis, just to properly manage numerical data.
- Multilayers Analysis (M.L.A.):
  - definition of specific criticalities indexes (environmental, demographic and infrastructural);
  - mounting input for M.L.A. as calibrated raster;
  - definition of weighted system to increase sharing of each index;
  - M.L.A. running;
  - building a dedicated G.U.I. (running MLA script) in order to manage the informative strata

related to the parameters of interest by a "non-technical" user.

5. INPUT DATA

The input data were very heterogeneous (hardcopy archive, digital, etc.), so as it needed a very hard homogenization work for the different implemented data-set, with an informative level variable from null to high.

The data-set available are the following ones:

- Stratigraphic (about 600 stratigraphic profile were analyzed)
- Geophysical borehole logs (about 12 geophysical logs were analyzed)
- Hydraulic heads (about14.000records and more than one hundred historic groundwater head graphs, since 70's)
- Hydrological
- Rain (a ten years historic report has been collected and analyzed)
- Chemical (ten years of laboratory analyses, equal to 12.000 records)
- Demographic(web source from the institute of national statistics)
- Water wells and pipelines data-set (about 200 wells and 1400 kmof water pipes are available).

The classes of analyzed data can be further grouped into:

- **geological-stratigraphical**: all the features related to aquifer's hydraulic conductivity, effective porosity, isobaths and isopachs of the three main aquifer groups surveyed at hydrogeological basin scale. The analysis takes into account even the hydraulic head profiles, when available;
- **chemical**: qualitative analysis of groundwater abstracted, due to the presence of various type of pollutants, and requiring a treatment to supply good drinking water;
- **infrastructural**: health state of water pipes network, identified both as a percentage growth of pipelines within the past years andas a number of maintenance operations;
- **demographical**: proportionality with the number of citizens within a single municipality. This prediction has funded on a ten years trend analysis.

6. MULTILAYER ANALYSIS

In order to manage properly the huge amount of environmental, demographic and infrastructural data collected, the best way to operate is undertaking a multilayer analysis M.L.A.

Making a M.L.A. means overlaying a certain number of raster images, intended as a form of numerical cartography, which depict the each specific character deriving from the analysis evaluated.

Measuring association between categorical maps in GIS typically involves categorical overlay, representation of overlay results as a contingency table, and statistical analysis of the latter with various integral measures of association, log-linear models and so on.

The determinacy analysis of spatial relationships, in terms of proportions, from overlaid maps is a very useful and viable alternative to classical statistical methods.

The raster data are data stored thanks to the creation of a regular grid: to every single cell (it's a pixel of desired dimensions) has assigned an alfa-numerical value representing an attribute.

Acting this way, the areas can be represented with groups of consecutive cells having the same value.

The values assigned to the cells can show both single natural or anthropic phenomenon (concentration of a pollutant in a water well, percentage of population increasing, etc.) and combination of multiple information through proper techniques of analysis (i.e. for every cell, the resultant, as a sum or a subtraction, between contaminant's concentration in a well and growth of population) or merely as a graphic attribute (grey hues or color).

Examples of M.L.A. carried through within the pilot area are the following ones:

- population density and groundwater quality;
- water networks implementation and demographic growth;

- plume localization and water treatment's economic investment;
- water demand – volume abstracted by water wells – aquifer productivity/permeability;
- effective porosity – isopachs – chemical profile of the reservoirs.

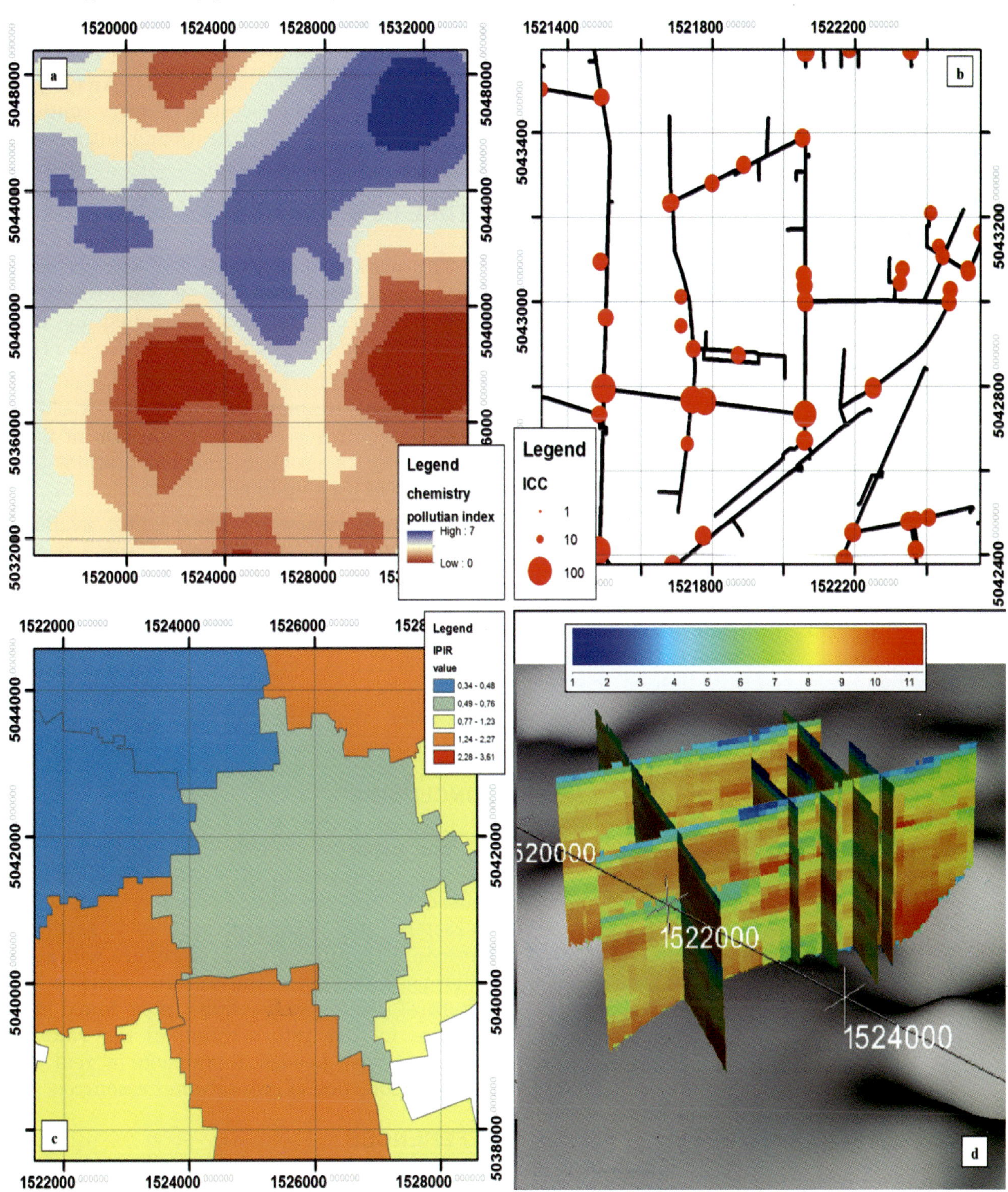

Figure 7. Typical M.L.A. output.
Legend: a: chemistry pollutant index; b: pipelines criticality index; c: demographic index; d: subsurface geology.

CASE STUDY AND OUTPUT

In 2012, CAP Holding's Board of Directors gave to Water Balance and Geology Departmentthe task of developing a new D.S.S., called Infrastructural Aqueduct Plan (I.A.P.). Between May 2012 and May 2013, a pilot project has been elaborated, with the purpose of testing the work procedure created.

A pilot area has been selected, located north of Milan and clustering 28 municipalities, very densely populated (more than two millions people live there).

The output has represented by thematic maps, in paper or digital format (raster, shapefile, kml., etc.), and each of them can be considered separately or showing the synthesis of multiple indexes, weighted according to different *w* factors.

This kind of approach allow us to make a certain number of customizable maps, focused on the analysis requirements.

So, after having created appropriate criticality indexes, and giving every of them a "balanced weight", we finally superimposed and summed/subtracted thematic maps, derived from the different kind of analyses, up to work out final synthetic map, supplying the interaction among them.

This effort is the essential base of the new D.S.S. here proposed.

This tool has intended not only for the optimal and sustainable management of groundwater resources, but even with a predictive scope (environmental emergency, urban growth decision, etc.).

The output created are the following ones (Figure 7):

- Underground geology
  - bottom isobaths of stratigraphic layers (Figure)
  - top isobaths of stratigraphic layers
  - isopachs of overburden aquifer strata
  - permeability.
- Network infrastructure

The output is areal, coinciding with the localization of the municipality within the water pipes have been posed.

- Demography

The output is polygonal and rasterized.

- Chemistry

Putting a specified weight *w* to every pollutant, chosen as a marker, and decreasing in direct proportion to the level of danger, processing data in a GIS environment and interpolating them with geostatistical tools (ArcGIS), it has been drawing the isocones of every contaminant and the whole concentrations multiplied by *w*.

FUTURE DEVELOPMENT

The workflow developed and applied to a pilot area of 190 km^2, will be widened to the whole hydrogeological basin supplied by CAP (Ticino-Adda and Po rivers area), during the time period 2013-2016.

We want to highlight that the workflow proposed requires a constant collection of data, followed by a robust implementation and analysis, real time updated.

A critical review of indexes now available and the creation of new more sophisticated indexes, but also the refinement of all the procedures regarding flow and transport models, will be targets of future evaluation.

Nevertheless, only when we will complete the analysis on a hydrogeological basin scale, we could say that the method proposed was completely vindicated.

CONCLUSIONS

The paper presented a case study dealing with the development of a new D.S.S., mainly focused on interactions between environmental characters and infrastructural and anthropic features. This kind of work procedure will allow to properly manage the future water infrastructures (water wells, hubs, pipe lines, etc.), and also to take into account all the valuable "point of view" needed to promote a real and efficient protection of groundwater resources.

REFERENCES

CAP Milano (1995). Indagine preliminare sull'uso sostenibile delle falde profonde nella Provincia di Milano. PubblicazioneCAP.

Cressie N.A.C. (1990). The origins ofkriging. Mathematical Geology, vol. 22.

Cressie N.A.C. (1991). Statistics for spatial data. Wiley, New York.

Gorla, M. (2001a). Acquiferi confinati profondi della Provincia di Milano: sviluppi delleconoscenze idrogeologiche e idrochimiche ai fini di un loro uso sostenibile. Parte 1 – Premessa e caratteristiche idrogeologiche. AcqueSotterranee, fascicolo n. 72. Geo-Graph, Segrate, pp 9-24.

Gorla, M. (2001b). Acquiferi confinati profondi della Provincia di Milano: sviluppi delleconoscenze idrogeologiche e idrochimiche ai fini di un loro uso sostenibile. Parte 2 – Caratteristiche geochımi che, inquinanti e conclusioni. Acque Sotterranee, fascicolo n. 73. Geo-Graph, Segrate, pp 9-27.

Gorla (2010). The age of deep aquifers in Milan Province: development of a new tritium – I.E.B. calibration curve. In Paliwal (Ed.), Global Ground water Resources and Management. Selected Papers from The 33rd International Geological Congress,General Symposium: Hydrogeology, Oslo (Norway) August 6-14, 2008, Scientific Publishers (India), Jodhpur, pp. 281-296.

Power D.J. (2002). Decision Support Systems: concepts and resources for managers. Westport, Greenwood/Quorum.

Power D.J. (2003). A brief history of Decision Support Systems. DSSResources.COM, World Wide Web, URL DSSResources.COM/history/dsshistory2.8.html, version 2.8.

Power D.J. (2004). Decision Support Systems: from the Past to the future. Proceedings of the 2004 Americas Conference on Information Systems, New York, pp. 2025-2031.

Provincia di Milano (2009-2011). Carte delle isopiezometriche e della soggiacenza della Provincia di Milano. Sistema Informativo Falda (S.I.F.), Milano.

Regione Lombardia (1994). Carta Tecnica Regionale.

Regione Lombardia – ENI Divisione AGIP (2002). Geologia degli acquiferi padani della Regione Lombardia. S.El.Ca., Florencc.

Hon'ble reviewer

1. **Prof. U. Aswathanarayan**
   Director MICWM
   B 16, Santishikha
   Somajiguda
   HYDERABAD - 500 082
2. **Prof. S.K. Tandon**
   Pro Vice Chancellor
   Delhi University
   DELHI - 110 007

III-36

Geochemical Characterization of Groundwater of Some Parts of Mahanadi River Basin, Cuttack district, Odisha, India

Rosalin Das[1], Madhumita Das[2] and Shreerup Goswami[3]

[1]Department of Geology, Banki Autonomous College, Banki-754008, Orissa
[2]Department of Geology, Utkal University, Vani Vihar, Bhubaneswar-751004, Orissa
[3]Department of Geology, Ravenshaw University, Cuttack-753003, Orissa
E-mail: madhumitadas_geo@rediffmail.com

Abstract

An evaluation on geochemical characterization of groundwater and its suitability for drinking, irrigation and industrial purpose was carried out in some parts of Mahanadi river basin, Cuttack district, Odisha. Altogether 53 groundwater samples were collected from open wells and bore-wells during the pre-monsoon and post-monsoon periods. The quality assessment is made through the estimation of Ca^{2+}, Mg^{2+}, Na^{+}, K^{+}, Cl^{-}, SO_4^{2-}, CO_3^{2-}, HCO_3^{-} , total hardness as $CaCO_3$, TDS, EC, and pH. Based on these analyses, parameters like sodium adsorption ratio, % sodium, residual sodium carbonate, non-carbonate hardness, potential salinity, magnesium ratio and permeability index were calculated. The groundwater falls under (Doneen's) class-I and excellent to good (Wilcox) zone indicating its suitability for irrigation use. According to Gibbs' ratio samples in both the seasons fall in the rock dominance field. Groundwater in the area is generally saline and an increasing problem of bicarbonate is marked. Various classifications show that present status of groundwater is better for irrigation purpose except for a few locations with a caution that it may deteriorate in near future. The overall quality of water in the study area is suitable for drinking, irrigation and industrial purpose.

Keywords: Water quality, Mahanadi river basin, Cuttack district, Groundwater Chemistry

Introduction

The study area of Mahanadi river basin of Cuttack district is well known as Athgarh formation in Indian stratigraphy, which belongs to Upper Gondwana Formation. It consists of a thick pile of fresh water fluviatile and lacustrine sediments. It covers an area of about 512.80 sq. km. The area is situated between 20° 15' and 20° 29'N Latitude and 85° 20' and 85° 45'E Longitude. It comes under survey of India toposheet Nos. 73H/7 and 73H/11 (Fig. 1). The study area comprises essentially a clastic succession of 700 m thick constituting conglom-erate; sandstone (argillaceous to ferruginous); mudstone; shale (purple, yellow and grey); and carbonaceous shale (Ball, 1877; Adyalkar, 1965; Das and Pandya, 1997, Goswami *et al.*, 2006; 2008). Mahanadi River forms the main drainage artery of the study area. It is flowing in the northern border of the study area from west to east direction. The different geomorphic features marked in the area are hills, pediments, buried pediments, flood plains, intermontane valley along with structural features such as lineament, fault, joint etc., which are the favourable sites for ground water exploration. The area enjoys humid tropical climate.

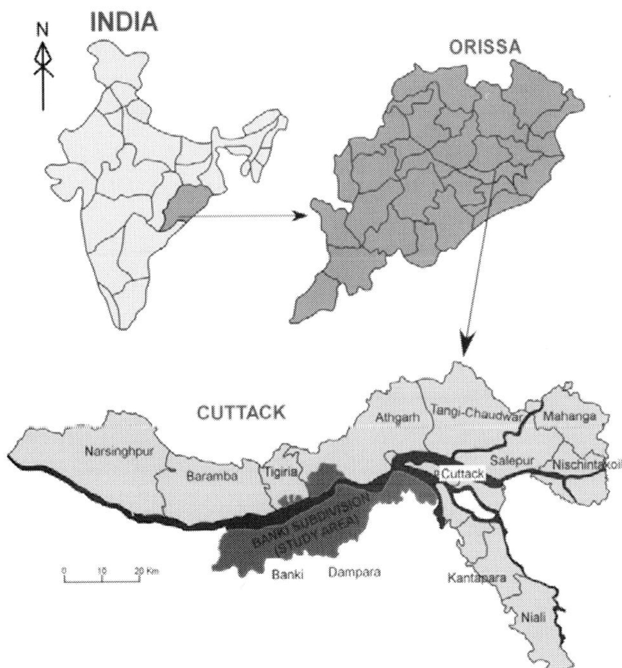

Fig. 1. Location Map of the Study Area

Irrigated agriculture is a major consumer of water, using 75 to 80% of available surface and groundwater resources. In the present study, various chemical parameters of groundwater were analysed to find out its suitability for drinking, irrigation and industrial purpose.

METHODOLOGY

In total, 53 water samples were collected during pre-monsoon and post-monsoon periods (2010) from some parts of Mahanadi river basin, Cuttack district (Banki sub-division). Out of 53 samples, 25 samples were collected from bore wells and 28 samples from dug wells. The samples were analysed for pH, electrical conductivity (EC), total dissolved solids (TDS), major cations and anions by adopting standard analytical procedures (Trivedi and Goel, 1984; Saxena, 1995; Gupta, 2004; Jaiswal, 2004; Gray, 2005; APHA, 2005). Electrical conductivity (EC) and pH were measured using digital meters immediately after sampling. Ca^{2+}, Mg^{2+}, HCO_3^-, CO_3^{2-}, Cl^- and TDS were analysed by volumetric titrations. Concentrations of Ca^{2+} and Mg^{2+} were estimated titrimetrically using 0.05N EDTA and 0.01N H_2SO_4 was used to determine the concentrations of HCO_3^- and CO_3^{2-}. $AgNO_3$ was used to estimate Cl^-. Flame photometer was used to measure Na^+ and K^+ ions. SO_4^{2-}, NO_3^- and F^- were determined by spectrophotometric techniques.

RESULTS AND DISCUSSION

The analytical results and computed values of water samples of study area are given in Tables 1a and 1b. The groundwater quality data interpretation for irrigation has been carried out as per guidelines given by Ayers (1977) and Christiansen *et al.* (1977).

Understanding the quality of groundwater with its temporal and seasonal variation is important because it determines the suitability for drinking, domestic and agricultural purposes. The pH of water is an indicator of its quality and geochemical equilibrium for solubility calculation (Hem, 1985). pH of the pre-monsoon samples are acidic and ranged from 5.25 to 6.91 with an average 6.08, while in the post-monsoon an increase is observed and ranged from 6.00 to 8.21 with an average of 7.31 (Tables 1a, 1b). The EC values ranged from 103 to 1985 µS/cm during pre-monsoon and 133 to 2250 µS/cm during post-monsoon period. Higher values are generally noticed in the southeastern part of the study area. The TDS values ranged from 77 mg/l to 1469 mg/l and 101 mg/l to 1687 mg/l during pre and post-monsoon periods respectively. Total alkalinity values ranged from 86 mg/l to 489 mg/l and 85 mg/l to 492 mg/l during pre-monsoon and post-monsoon. The concentrations of chloride ranged from 20.7 mg/l to 179.5 mg/l during pre-monsoon and 19.76 mg/l to 177.25 mg/l during post-monsoon period. The concentrations of fluoride ranged from 0.101 mg/l to 1.76 mg/l during pre-monsoon and 0.04 mg/l to 1.73 mg/l during post-monsoon period. Sulphate concentrations ranged from 0.27 mg/l to 127.5 mg/l in pre-monsoon and 0.062 mg/l to 125.75 mg/l during post-monsoon period. Nitrate values ranged from

0 mg/l to 40.76 mg/l and 0.01 mg/l to 43.97 mg/l during pre and post-monsoon periods respectively. The calcium concentration ranged from 26 mg/l to 127 mg/l and 20 mg/l to 120 mg/l during pre and post-monsoon periods respectively. The magnesium content of the water samples ranged from 6.09 to 38.21 mg/l during pre-monsoon and 6.48 mg/l to 37.65 mg/l during post-monsoon period. Sodium concentrations ranged from 1.7 mg/l to 88.3 mg/l during pre-monsoon and 9.76 mg/l to 97.0 mg/l during post-monsoon period. Potassium concentrations of the study area ranged from 0.68 mg/l to 83.3 mg/l and 1.0 mg/l to 77 mg/l during pre and post-monsoon periods respectively (Tables 1a, 1b). The bicarbonate value ranged from 94.7 mg/l to 437.5 mg/l and 103.7 mg/l to 451.4 mg/l during pre and post-monsoon periods respectively. Tables 1a and 1b explicitly demonstrate that most of the subsurface water of the study area is suitable for the drinking and domestic use with little exceptions.

Table 1a. Chemical Parameters of Groundwater (Pre-monsoon)

S N.	Location	Well Type	pH	EC μS/cm	TDS mg/l	Alka-linity	TH mg/l	CO_3^{2-} mg/l	HCO_3^- mg/l	CL^- mg/l	F^- mg/l	SO^{2-}_4 mg/l	NO^-_3 mg/l	Ca^{++} mg/l	Mg^{++} mg/l	Na^+ mg/l	K^+ mg/l	Na%	SAR	PS	PI	RSC	CR	MR
1	Kadalibadi	DW	6.15	376	282	122	183	0	326.00	49.70	0.8	32	15.5	45	17.01	26.1	6	34.109	4.69	1.74	50.11	1.69	0.317	38.89
2	Ostia	DW	5.86	314	236	153	215	0	289.10	37.61	0.8	34	22.1	65	12.6	40	10.3	39.32	6.42	1.06	48.47	0.46	0.304	24.3
3	Kanta-pahara	DW	6.01	875	657	97	157	0	436.20	34.70	1.1	37	21.4	43	11.9	19.3	13.7	37.54	3.68	1.37	54.16	4.02	0.2	31.31
4	Ananda-pur	DW	6.36	617	457	199	205	0	387.30	49.60	0.7	28	18.7	52	18.3	32.4	3.2	33.61	5.47	1.69	50.71	2.25	0.254	36.83
5	Ghoda-bara	TW	6.17	546	410	244	163	0	378.70	51.70	0.2	28	12.1	31	20.7	41.5	7.6	48.71	8.16	1.75	65.41	2.96	0.268	52.31
6	Gholapur	DW	5.98	289	217	149	172	0	297.60	67.50	0.1	25	11.2	47	13.1	39	15.1	47.37	7.12	2.16	56.76	1.45	0.404	31.49
7	Gobin-dapur	TW	6.06	103	77	127	211	0	202.40	26.81	0.4	20	4.35	65	11.7	34.6	12.4	37.99	5.59	0.97	43.86	-0.88	0.289	22.86
8	Tentulia	TW	6.53	493	370	255	206	0	152.60	33.70	0.9	23	7.08	51	19.2	47	6.9	43.43	7.93	1.19	10.54	-1.62	0.46	38.35
9	Bandala	DW	6.12	715	523	230	222	0	96.50	40.02	0.4	34	10.2	58	18.7	19.9	6.93	25.91	3.21	1.48	30.76	-2.85	0.94	34.76
10	Baunsaput	DW	6.91	729	547	145	205	0	94.70	44.70	0.2	28	12.5	56	15.8	32.1	6.75	35.09	5.35	1.55	40.24	-2.54	0.968	31.78
11	Ghasiput	TW	6.52	224	166	92	174	0	126.60	36.80	0.3	28	7.96	41	17.46	14.3	11.6	30.7	2.65	1.33	35.12	-1.42	0.636	41.26
12	Ghasiput	TW	5.86	198	149	118	145	0	157.30	20.91	0.2	15	7.84	43	8.94	10	6.03	23.58	1.96	0.74	36.39	-0.31	0.282	25.61
13	Ramac-handrapur	TW	6.06	176	131	143	166	0	163.20	47.42	0.2	15	3.76	36	18.62	10.3	27	40.57	1.97	1.5	35.53	-0.66	0.502	45.95
14	Karabara	DW	6.67	498	374	158	145	0	167.50	85.67	0.3	8.7	3.21	41	10.3	19.7	7.8	34.89	3.89	2.51	45.97	-0.15	0.77	29.31
15	Simiipur	DW	6.12	701	515	151	133	0	117.30	93.91	0.2	15	13.1	35	11.1	39	10.3	51.67	8.13	2.5	58.55	-0.74	1.26	34.21
16	Harirajpur	TW	6.89	610	458	216	130	0	206.10	109.60	0.2	26	12.6	34	11.25	46	37.5	25.1	9.79	3.36	66.5	0.78	0.87	35.77
17	Mantrisahi	TW	6.42	566	425	194	209	0	205.90	78.52	0.1	46	23.5	50	20.4	27	15	37.36	7.58	2.69	42.44	-0.81	0.75	40.19
18	Sisua	TW	6.72	923	693	237	126	0	307.30	107.30	1	25	12.7	34	9.97	54	42.1	28.21	11.5	3.04	73	2.52	0.57	32.54
19	Sahada-pada	DW	6.63	907	681	284	208	0	347.20	127.80	1	57	6.37	53	18.37	88.3	59	10.74	14.8	4.21	66.96	1.54	0.689	36.39
20	Charc-hika	TW	6.34	1086	815	289	329	2	345.00	101.72	0.4	46	4.92	85	28.4	84.7	83.3	44.73	14.2	3.35	52.13	-0.86	0.55	35.56
21	Kotad-wara	DW	5.67	1985	1469	127	303	0	294.60	117.32	0.3	41	5.76	91	18.4	55	7.1	36.76	7.31	3.74	43.89	-1.22	0.7	13.69
22	Ragadi	TW	6.02	986	740	129	286	0	163.70	101.70	0.1	25	5.04	100	8.76	47	8.6	46.96	6.37	3.14	38.38	-3.03	1.03	18.2
23	Singha-nathpur	DW	6.73	852	631	238	263	0	199.80	74.70	0.9	54	11.8	83	13.53	76	44	53.29	10.9	2.68	52.24	-1.16	0.8	24.18
24	Pakhala-khala	DW	6.51	598	449	155	164	0	176.80	25.71	0.5	28	2.45	39	16.02	63	27.3	58.76	12	1.02	64.64	0.45	0.36	29.6

25	Gobin-dapur	DW	6.02	235	177	95	159	0	135.00	29.92	0.9	41	12.6	47	9.97	26.3	15.4	37.23	7.6	1.27	45.52	-0.96	0.62	36.83
26	Damapada	DW	5.68	281	208	105	153	0	133.20	37.45	0.2	18	0.04	34	16.6	12.7	7.5	20.91	2.95	1.25	38.29	-0.89	0.533	44.98
27	Damapada	TW	6.24	792	594	147	169	0	135.00	36.82	0.7	24	0.86	40	16.91	29	0.68	53.99	5.61	1.3	47.54	-1.15	0.56	42.4
28	Rajib	DW	6.72	998	739	366	308	0	437.50	102.32	1	24	0	94	17.61	64	37.2	37.61	8.57	3.14	48.35	1.03	0.38	15.04
29	Kuspangi	DW	5.93	989	742	195	169	0	139.70	40.61	0.5	25	17.8	51	10.05	71	3.3	55.73	12.9	1.41	62.71	-1.08	0.59	37.09
30	Jatam-undia	TW	5.25	816	612	217	190	0	127.10	29.62	0.7	27	1.02	49	16.39	79	5.9	55.45	13.8	1.13	62.51	-1.72	0.55	36.78
31	Dolan-apur	TW	5.60	901	667	229	157	0	128.00	23.59	0.2	42	2.07	37	15.51	59.2	2.4	55.12	11.6	1.11	63.11	-1.03	0.59	52.02
32	Bilipada	TW	6.23	576	432	187	247	0	299.60	77.41	0.2	40	7.09	63	21.83	49	5.3	38.71	7.52	2.6	49.54	-0.03	0.49	51.29
33	Patapur	DW	6.42	788	591	207	466	0	269.70	89.62	0.7	50	8.24	127	36.23	25	4.6	14.5	2.77	3.05	22	-4.91	0.658	19.75
34	Gopalpur	TW	6.38	584	438	127	273	0	159.40	105.62	0.3	48	0.87	78	19	20.7	2.7	34.37	2.97	3.48	28.3	-2.84	1.24	25.19
35	Anuary	TW	6.49	695	522	239	272	0	193.90	102.80	0.2	33	0.51	83	15.61	56.7	30.1	45.82	8.08	3.24	45.47	-2.24	0.91	28.5
36	Gopa	TW	6.01	636	477	208	340	0	273.50	94.70	0.2	42	10.4	103	20.02	36.1	26.7	37.08	4.6	3.11	33.07	-2.31	0.647	23.85
37	Kurum-pada	TW	6.19	889	667	228	340	0	275.60	89.50	0.4	51	40.8	104	19.61	48.7	36.4	32.19	6.21	3.05	38.01	-2.25	0.64	20.55
38	Kurum-chaini	TW	6.29	862	638	296	295	0	103.60	77.90	0.5	53	36	91	16.31	43.4	9.76	42.38	7.91	2.75	41.21	-4.18	1.58	24.46
39	Kotapalla	DW	6.06	694	507	289	281	0	116.70	83.90	0.5	47	13.7	83	17.93	56	21.1	29.13	4.96	2.86	33.79	-3.7	1.42	43.13
40	Baideswara	DW	6.12	901	676	211	277	0	247.50	93.70	0.3	128	1.36	48	38.21	58	6.3	42.18	8.83	3.96	51.12	-1.48	1.06	41.61
41	Balabh-adrapur	DW	5.42	992	744	192	216	0	261.30	100.70	0.5	62	4.56	52	20.82	61	4.9	53.14	10.1	3.49	57.66	-0.02	0.786	38.48
42	Pasania	TW	5.73	968	717	160	264	0	204.70	138.70	0.3	53	0.09	73	19.68	43	21.6	33.21	6.32	4.46	42.24	-1.9	1.218	39.64
43	Pathuri-pada	DW	6.25	1003	753	182	370	0	209.00	179.50	0.2	52	1.76	100	29.07	31.7	3.1	25.25	3.95	5.6	28.71	-3.95	1.46	20.54
44	Brahma-pura	DW	6.69	794	588	173	218	0	261.70	69.89	0.2	55	0.08	61	15.72	28	11.9	31.143	4.52	2.55	42.19	-0.04	0.59	37.7
45	Hatimara	DW	6.03	699	525	205	225	0	194.30	36.60	0.8	64	3.62	53	22.34	21.5	6.7	25.5	3.5	1.7	36.59	-1.3	0.6	33.67
46	Dhansara	DW	5.72	618	464	110	190	0	155.50	34.60	1.3	18	5.96	49	16.31	17	4.3	24.06	2.98	1.16	35.79	-1.24	0.42	35.19
47	Talabasta	TW	6.15	279	210	107	152	0	108.60	30.40	0.8	1.1	0.98	34	16.21	1.7	3.7	10.29	0.34	0.87	23.34	-1.25	0.4	45.69
48	Talabasta	DW	6.49	742	557	129	234	0	143.30	22.90	1.8	0.3	0.17	65	17.38	16.5	4.06	22.99	2.57	0.66	54.96	-2.32	0.225	28.63
49	Arapur	TW	5.25	270	203	95	150	0	167.10	20.70	1.3	7.5	3.58	34	15.82	23	8.1	39.56	4.61	0.66	49.32	-0.26	0.22	27.04
50	Tulasipur	TW	6.07	436	327	174	97	0	202.60	22.70	0.9	27	0.06	26	7.63	26	9.61	46.58	6.34	0.93	67.47	1.39	0.29	58.97
51	Baraput	TW	6.13	872	646	489	262	0	211.60	145.70	0.7	34	0.99	65	24.15	18.6	3.32	19.74	2.79	4.46	30.77	-1.76	1.131	13.37
52	Baghei	DW	5.89	186	140	86	123	0	100.00	37.51	0.6	20	2.7	39	6.09	13.3	4.04	27.77	2.8	1.27	39.9	-0.81	0.73	20.41
53	Kalapa-thara	DW	6.21	476	358	279	178	0	142.50	56.72	0.5	39	5.02	48	14.04	20.9	0.96	26.05	3.75	2	39.55	-1.21	0.83	32.39

Table 1b. Chemical Parameters of Groundwater (Post-monsoon

S.N.	Location	Well Type	pH	EC µS/cm	TDS mg/l	Alka-linity	TH mg/l	CO_3^{2-} mg/l	HCO_3^- mg/l	CL^- mg/l	F^- mg/l	SO_4^- mg/l	NO_3^- mg/l	Ca^{++} mg/l	Mg^{++} mg/l	Na^+ mg/l	K^+ mg/l	Na%	SAR	PS	PI	RSC	CR	MR
1	Kadalibadi	DW	7.13	524	393	125	152	0	340.3	52.6	0.77	32	17.76	34	16.2	28	4	38.94	5.59	1.87	59.41	2.55	0.31	43.9
2	Ostia	DW	6.59	346	260	156	127	0	280.7	36.5	0.82	33	22.61	33	10.8	42	9.7	54.14	8.97	1.38	68.47	2.11	0.31	12.7
3	Kantapahara	DW	7.03	1660	1245	99	133	0	427	32.6	1.05	37	20.76	32	12.8	18.2	13	40.66	3.85	1.3	61.68	5.35	0.2	39.6
4	Anandapur	DW	7.27	789	584	207	152	0	366	46.8	0.51	28	19.6	34	16.3	32	2	40.35	6.38	1.62	62.16	2.96	0.26	5.56
5	Ghodabara	TW	6.98	645	484	246	154	0	366	52.8	0.19	29	13.62	28	20.3	48	6	52.79	9.77	1.79	69.71	2.93	0.28	54.4
6	Gholapur	DW	7.59	377	276	152	142	0	280.4	63.5	0.07	25	9.71	36	12.5	47	15	55.9	9.55	2.05	66.74	1.77	0.41	36.4

7	Gobindapur	TW	7.53	133	101	130	178	2	195.4	25.5	0.31	20	5.58	55	9.81	37	11	42.7	6.5	0.93	50.06	-0.28	0.29	22.8
8	Tentulia	TW	7.92	670	503	259	192	0	158.6	32.4	0.76	23	6.72	48	17.5	51	7.2	47.05	8.91	1.15	54.58	-1.24	0.43	37.5
9	Bandala	DW	7.63	849	637	232	190	0	104.7	38.6	0.32	33	12.81	46	18.3	25.6	8	34.32	4.51	2.25	39.85	-2.09	0.84	39.6
10	Baunsaput	DW	7.92	843	633	148	197	0	98.5	42.5	0.19	29	11.76	52	16.2	38	4.3	38.29	6.51	1.5	45.13	-0.64	0.9	33.9
11	Ghasiput	TW	7.71	377	264	95	140	0	115.9	35.5	0.26	28	8.08	32	14.6	17.2	11	37.45	3.56	1.29	43.83	-0.9	0.68	42.9
12	Ghasiput	TW	6.59	313	210	120	125	0	146.4	21.3	0.17	14	8.01	34	9.72	11.3	4.7	26.79	2.42	0.75	42.51	2.38	0.3	32
13	Ramacha-ndrapur	TW	7.05	307	230	148	137	0	158.6	46.8	0.15	15	3.21	28	16.2	19.2	23	48.83	4.08	1.48	50.13	-0.13	0.51	48.7
14	Karabara	DW	7.05	642	482	163	125	0	158.6	82.8	0.31	14	4.53	34	9.72	23.1	6.5	40.37	4.94	2.48	53.41	0.1	0.82	32
15	Simiipur	DW	7.83	752	564	159	116	0	122	92.7	0.15	16	12.71	32	8.47	46	9.7	57.91	10.2	2.78	65.96	-0.3	1.2	30.4
16	Harirajpur	TW	7.94	658	494	220	110	0	195.2	113	0.26	26	11.76	28	9.72	52	32	69.01	12	3.44	73.52	1	1.08	36.4
17	Mantrisahi	TW	7.03	852	631	198	183	0	195.2	76.3	0.08	49	24.6	43	18.7	32	12	41.82	5.79	2.66	49.32	-0.46	0.81	42.8
18	Sisua	TW	7.92	1031	774	241	134	0	280.6	105	0.1	24	13.62	36	10.4	67	43	70.32	13.9	3.2	73.84	1.94	0.61	33.1
19	Sahadapada	DW	8.02	976	732	287	195	0	337.1	125	0.36	56	7.05	52	16.2	97	65	70.53	16.7	4.11	70.05	1.63	0.69	34.1
20	Charchika	TW	7.52	1153	863	300	335	4	366	106	0.18	44	5.23	8.8	27.9	88	77	81.78	20.6	3.46	85.88	3.26	0.53	83.9
21	Kotadwara	DW	6.83	2250	1687	124	290	0	280.6	114	0.16	40	6.31	80	16.2	37	4	29.87	6.33	3.64	40.34	-0.73	0.72	25.1
22	Ragadi	TW	7.76	1004	753	136	210	0	158.6	98.2	0.09	24	4.86	55	9.85	45	5	43.53	7.9	3.02	52.42	-0.95	1.02	22.8
23	Singhanathpur	DW	7.79	897	673	241	227	0	195.2	76.8	0.63	52	12.81	70	12.5	81	42	59.86	12.6	2.72	58.09	-1.31	0.83	22.6
24	Pakhalakhala	DW	7.33	665	499	159	170	0	128.6	26.2	0.25	29	7381	44	14.6	59	29	60.03	10.9	1.04	59.82	-1.29	0.52	35.3
25	Gobindapur	DW	6.79	298	221	98	131	0	122	27.6	0.38	40	11.7	36	9.76	27.1	13	46.82	5.67	1.2	52.35	-0.6	0.66	30.8
26	Damapada	DW	7.62	347	269	110	125	0	134.2	35.5	0.04	22	0.09	26	14.6	17.8	6.1	37.06	3.95	1.23	50.32	-0.3	0.54	46.9
27	Damapada	TW	7.57	940	696	152	173	0	122	32.4	0.68	27	0.15	38	19	32.5	1	37.14	6.09	1.19	48.67	-1.46	0.6	45.1
28	Rajib	DW	8.21	1073	827	370	300	0	451.4	99.3	0.74	26	0.01	90	18.2	72	36	49.94	9.79	3.08	51.73	1.41	0.37	25
29	Kuspangi	DW	6	1002	762	198	146	0	158.6	37.8	0.33	26	20.61	44	8.76	67	2.3	56.99	13	1.34	43.19	-0.32	0.51	24.7
30	Jatamundia	TW	6.05	872	637	220	187	0	122	28.5	0.26	26	0.76	46	17.5	82	4.7	57.73	14.6	1.08	63.95	-1.73	0.54	38.5
31	Dolanapur	TW	7.17	941	707	232	131	0	124	24.7	0.18	4.7	3.21	28	14.6	63	1.7	60.27	13.6	0.75	70.17	-0.57	0.49	46.2
32	Bilipada	TW	7.03	856	689	184	234	0	317.2	54.6	0.09	40	7.65	58	21.5	51.2	3.2	40.62	8.12	1.96	52.79	0.6	0.32	38
33	Patapur	DW	7.69	832	616	208	421	0	276.3	86.8	0.27	47	9.81	112	34.3	28.1	2	11.06	3.29	2.94	25.64	-3.88	0.46	33.5
34	Gopalpur	TW	7.69	670	494	125	280	0	152.5	95.7	0.27	49	0.371	82	18.2	25.9	1.4	21.4	3.66	3.21	30.32	-3.09	1.2	26.8
35	Anuary	TW	6.78	748	561	242	254	0	195.2	99.7	0.2	35	0.671	78	14.3	62.1	31	50.22	9.14	3.17	49.27	-1.87	0.97	23.1
36	Gopa	TW	7.2	686	501	210	320	0	280.6	92.5	0.24	40	9.61	98	18.2	49	28	39.85	6.43	3.03	39.79	-1.79	0.59	23.5
37	Kurumpada	TW	7.89	917	686	230	320	0	280.6	88.6	0.38	49	43.97	98	18.2	53.6	33	42.69	7.03	3.01	41.42	-1.79	0.59	23.5
38	Kurumchaini	TW	7.23	891	670	300	295	0	122	76.6	0.32	52	36.7	89	17.5	61	7	38.97	8.36	2.7	43.02	-3.88	1.5	24.5
39	Kotapalla	DW	6.83	759	570	242	268	0	168	84.6	0.25	48	12.6	78	17.8	38.5	18	37.22	5.56	2.89	38.31	-2.61	1	27.4
40	Baideswara	DW	7.53	960	690	215	295	0	262.3	92.2	0.24	126	2.77	56	37.7	72	5.4	45.24	10.5	3.91	53.24	-1.59	0.99	52.6
41	Balabhadrapur	DW	7.25	1024	768	195	203	0	258.7	98.3	0.31	60	3.12	48	20.5	69	4.3	51.69	11.8	3.4	61.87	0.15	0.77	41.3
42	Pasania	TW	6.71	1162	860	163	261	0	198.8	136	0.22	49	0.981	72	19.5	51	21	43.85	7.54	4.36	45.66	-1.94	1.21	31
43	Pathuripada	DW	7.01	1071	704	185	425	0	225.7	177	0.27	51	0.78	120	30.4	35.4	1.3	19.61	5.23	5.53	27.14	-4.79	1.34	29.5
44	Brahmapura	DW	7.63	865	649	176	193	0	254.6	72.3	0.2	57	0.761	54	14.2	32.5	13	39.87	5.57	2.63	48.13	0.32	0.61	30.1
45	Hatimara	DW	7.15	758	554	206	207	0	186.6	34.1	0.65	62	2.56	48	21.1	26.7	5.4	31.71	4.54	1.6	42.12	-1.08	0.6	42
46	Dhansara	DW	7.03	622	468	112	126	0	147.3	2.53	0.91	16	4.81	32	11.1	19.6	3.2	26.53	4.22	1.08	50.58	-0.1	0.13	36.3
47	Talabasta	TW	6.53	268	211	105	110	0	123.1	28.4	1.73	0.1	0.652	20	14.6	11.3	2.3	28.22	2.72	0.8	48.8	-0.18	0.32	54.6
48	Talabasta	DW	6.89	336	215	132	165	0	139.4	24.7	0.95	0.1	0.55	34	19.5	17.6	3.9	28.67	3.4	0.7	41.35	-1.02	0.25	48.5
49	Arapur	TW	7.03	524	396	98	134	0	158.5	19.8	0.87	8.8	2.72	28	15.4	27.6	5	42.88	5.93	0.65	56.57	-0.07	0.23	47.6
50	Tulasipur	TW	7.26	680	510	176	87	0	215.7	20.4	0.53	27	0.86	24	6.48	26.7	7.1	52.58	6.84	0.86	72.36	1.81	0.26	30.6
51	Baraput	TW	7.87	941	669	492	281	0	208.6	143	0.29	32	1.59	76	22.2	19.8	1.7	17.97	2.82	4.3	29.03	-2.19	1.12	32.4
52	Baghei	DW	7.57	295	207	85	115	0	103.7	35.5	0.33	20	1.82	34	7.29	15.2	2.5	30.01	3.35	3.13	44.92	-0.6	0.68	26.1
53	Kalapathara	DW	8.09	528	382	284	155	0	138.5	58.5	0.37	38	3.61	42	12.1	23.6	1.5	31.7	4.54	2.04	45.53	-0.82	0.88	2.95

Table 2. Comparison of water quality of the study area with WHO and BIS 10500 standards

Para Meter	Range		WHO (1993)	IS 10500	
	Pre-monsoon	Post-monsoon		Highest Desirable	Maximum Permissible
pH	5.25-6.91	6-8.21	6.5-8.5	6.5	8.5
EC	103-1985	133-2250	400-2000		
TDS	77-1469	101-1687	500-1000	500	2000
Calcium	26-127	20-120	100-200	75	200
Magnesium	6.09-38.21	6.48-37.665	30-50	30	100
Sodium	1.7-88.3	11.3-97	20-1756	-	-
Potassium	0.68-83.3	1.0-77	10-12	-	-
Bicarbonate	94.7-437.5	98.5-451.4	-	200	600
Sulphate	0.27-127.5	0.062-125.75	25-250	200	400
Chloride	20.7-179.5	19.76- 177.25	25-600	250	1000
Nitrate	0-40.76	0.01-43.97	-	45	100
Fluoride	0.04-1.73	0.97-1.76	-	1.0	1.5
Alkalinity		85-492	-	200	600
Total Hardness	97-466	87-425	-	300	600

Table 3. Salinity levels of Groundwater samples of study area. (Ayers, 1977)

Salinity Zone (EC, μScm^{-1})	Effects	Sample Numbers (out of 53)			
		Pre-monsoon (Sample location)	%	Post-monsoon (Sample location)	%
0-750	No Problem	1-2, 4-17, 24-26, 32, 34-36, 45-50, 52-53	43%	1-2, 5-8, 11-14, 16 , 24-26, 34-36,46-50, 52-53	60%
750-2750	Increasing problem	3, 18-23, 27-31, 33, 37-44, 51	57%	3-4,9-10, 15,17-23,,27-33, 37-45,51	40%
Above 2750	Severe Problem	Nil	0	Nil	0

Drinking Suitability

To assess the suitability of groundwater for drinking purpose, the concentrations were compared with the WHO standards (2004) for drinking water, which shows that most of the surface and subsurface water of the study area are suitable for the drinking and domestic use with few exceptions, as most of the parameters are within the permissible limits (Table 2).

Hydrochemical Facies

In the present study, water quality data is analyzed through statistical distribution diagram such as Piper's trilinear diagram by plotting the concentrations of major cations and anions to gain better insight into the hydrochemical processes operating in the groundwater flow system. Piper's trilinear diagram reveals that water of the study area belongs to the Ca–Mg - HCO_3 facies (Fig. 2). The plot of the ground water samples of pre-monsoon data and post-monsoon data falls in the fields 4 and 5, which suggest that alkaline earth exceeds strong acids. Calcium and magnesium are major cations in the study area (about 75-90% of the total cations) and bicarbonate is major anion (about 70-85% of the total anions).

Agronomic Aspects

The major parameters have been grouped in two categories namely Salinity and Sodicity (Rhodes, 1972; Ayers and Westcot, 1985; WHO, 1989; Kuchanwar *et al.*, 1999), which may affect the soil, plant and human use directly or indirectly.

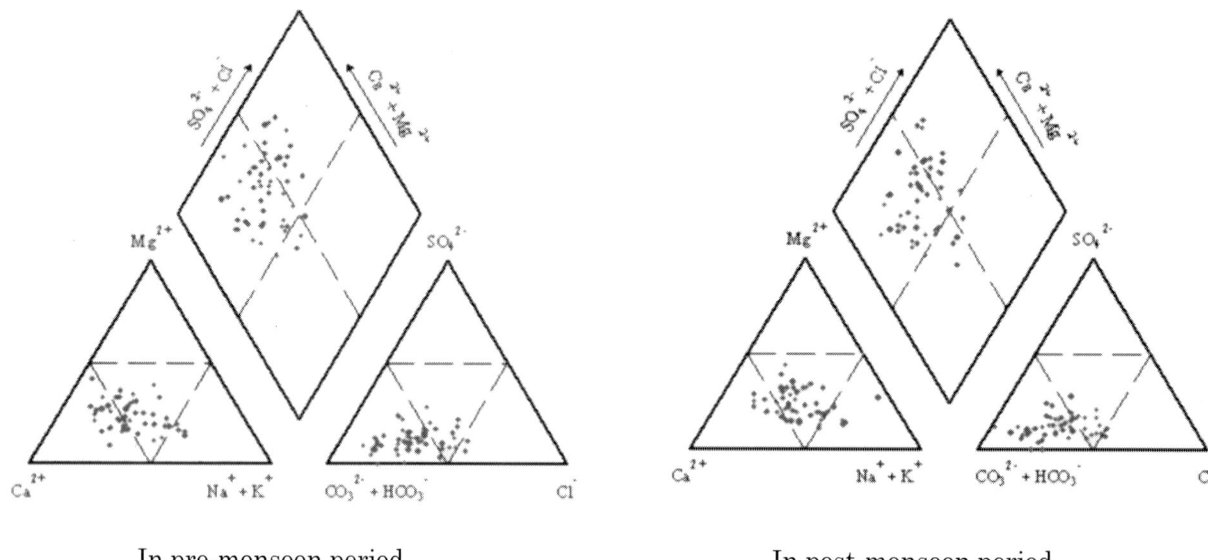

In pre-monsoon period In post-monsoon period

Fig. 2. Piper's trilinear diagram

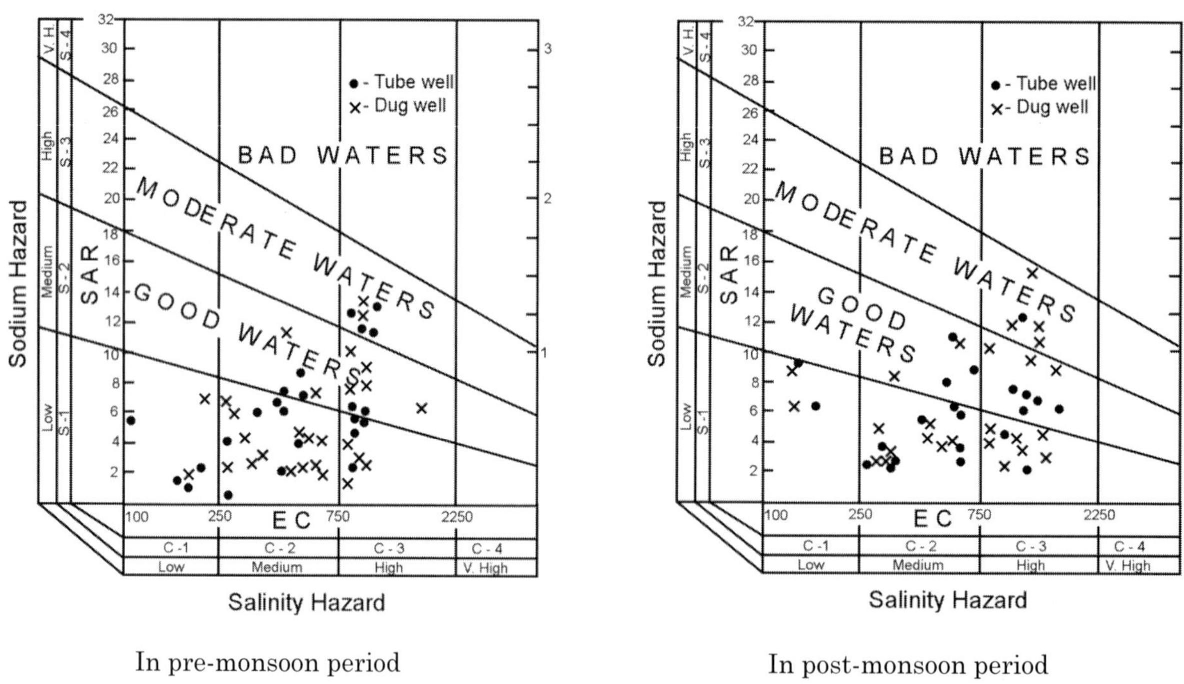

In pre-monsoon period In post-monsoon period

Fig. 3. U. S. Salinity Laboratory diagram showing classification of water samples for irrigation purpose

Salinity

The salinity usually reported as Electrical Conductance (EC) affects availability of water to crops. The EC varies from 103 to 1985 μS/cm in pre-monsoon and 133 to 2250 μS/cm in post-monsoon. 57% samples in pre-monsoon and 40% in post-monsoon periods fall under "increasing problem" category (Table 3). The EC values of other samples are excellent to good and good to permissible limit of Wilcox (1967).

Sodium Absorption Ratio (SAR)

The sodium or alkali hazard in groundwater for irrigation is determined by the absolute and relative concentration of cations and is expressed in terms of Sodium Absorption Ratio (SAR). There is a significant relationship between SAR values of irrigation water and the extent to which sodium is absorbed by the soil.

$$SAR = \frac{Na^{+}}{\sqrt{\left(Ca^{++} + Mg^{++}\right)/2}} \quad \text{(All ions in epm)}$$

Irrigation waters are classified based on sodium adsorption ratio (WHO, 1989). Based on the classification, 66% of the water samples of the area fall in excellent category, 22% samples fall in good category and 12% samples fall in medium category in pre-monsoon season and 58% of the samples fall in the excellent category, 26% fall in the good category and 16% in medium category in post-monsoon season. Therefore, none of the samples are of the poor category for irrigation in either season. A more detailed analysis, with respect to the irrigation suitability of the water was made by plotting the data on US Salinity Laboratory diagram (Fig. 3).

Magnesium Ratio

Generally calcium and magnesium maintain a state of equilibrium in water. More Mg^{2+} present in water will adversely affect the soil quality converting it to alkaline and decreases crop yields.

The magnesium ratio can be expressed as

$$MR = \frac{Mg^{++} \times 100}{Ca^{++} + Mg^{++}}$$

(All ions are expressed in epm)

The magnesium ratio of groundwater varies from 11.90 epm to 57.38 epm in pre-monsoon and 2.95 epm to 83.94 epm in post-monsoon (Tables 1a and 1b). Maximum value is marked in the sample location of Charchika. Except four locations, the magnesium ratio in almost all the investigated areas of both pre-monsoon and post-monsoon period is below 50, which depicts that the water can be used for agriculture purpose.

Residual Sodium Carbonate

Residual Sodium Carbonate (RSC) is defined as (CO_3^{2-}+HCO_3^-) – (Ca^{2+}+ Mg^{2+}), where all concentrations are expressed in epm. Water having excess of carbonate and bicarbonate concentration over the alkaline earth mainly of calcium and magnesium (beyond the permissible limit) affects agriculture unfavorably (Eaton, 1950; Richards, 1954). Table 4 shows the classification of water on the basis of RSC value, which demonstrates that in pre-monsoon 78% samples are safe and suitable for agricultural purposes; 11% samples are marginally suitable and the rest 11% are unsuitable for irrigation use. Where as in post-monsoon 85% samples are safe, 6% samples are marginally suitable and 6% are unsuitable.

Bicarbonate

Bicarbonate concentration of all the samples of both the seasons comes under increasing problem (Table 5). Bicarbonate ranges from 1.6 epm to 7.2 epm in the study area. Bicarbonate content more than 1 epm in the water is necessarily attributed from the biological activities of plant roots, from the oxidation of organic matter included in the soils and in the rock, and from various chemical reactions (Mandel and Shiftan, 1981).

Table 4. Residual Sodium Carbonate in groundwater

RSC (epm)	Water category	No. of Samples (out of 53)			
		Pre-monsoon		Post-monsoon	
		No. of wells	% of sample	No.of wells	% of sample
<1.25	Safe	41	78%	45	85%
1.25-2.5	Marginally	6	11%	5	6%
>2.5	Unsuitable	6	11%	3	6%

Table 5. Bicarbonate Concentration of Groundwater Samples (Mandel and Shiftan, 1981)

Sl. No.	Bicarbonate (epm)	Effects	No. of Samples (out of 53)	
			Pre-monsoon	Post-monsoon
1.	0-1.5	No Problem	Nil	Nil
2.	1.5-8.5	Increasing Problem	53	53
3.	>8.5	Severe Problem	Nil	Nil

Doneen's Permeablity Index

Permeability of the soil is influenced by the sodium content of the irrigation water. The permeability index (PI) is obtained by considering the ions (epm), which influence permeability (Domenico and Schwartz, 1990). Permeability Index is defined as:

$$PI = \frac{Na^{+} + \sqrt{HCO_{3}^{-}} \times 100}{Ca^{++} + Mg^{++} + Na^{+}}$$

The concentration of cations and anions are in epm.

The groundwater samples of the study area falls in class-I and class-II of Doneen chart (Fig. 4). 18 samples fall in class-II and 35 samples are plotted in class-I in pre- monsoon and in post monsoon period, 37 samples fall in class-I and 16 samples in class-II. It is inferred on the basis of the permeability index that the ground water of the study area is of good quality for irrigation purposes. The increased percentage of groundwater samples under class –I was due to dilution and subsequent lower values of permeability index.

Potential Soil Salinity (PS)

Doneen (1962) proposed a criterion based on the salinity of the irrigation water, which is an improvement over the U.S. soil salinity. The potential soil salinity (PS) is given by the concentration of chloride and half of the sulphate ions,

$$PS = Cl^{-} \text{ (epm)} + 1/2\ SO_4^{2-} \text{ (epm)}$$

The potential soil salinity varies from 0.66 epm to 5.60 epm in pre-monsoon and from 0.65 epm to 5.53 epm in post-monsoon periods. The whole study area has excellent to good class of PS value. Only one sample has good to injurious class of PS value. PS classification (Table 6) explicitly demonstrates that out of 53 samples, 52 samples come under excellent to good category and only one sample is classified as good to injurious for irrigation in both the seasons.

Table 6. Potential Soil Salinity Classification of Ground Water

Potential Soil Salinity	Class	Number of samples (out of 53)	
		Pre-monsoon	Post-monsoon
<5	Excellent to Good	52	52
5-10	Good to Injurious	1	1
>10	Injuries to Satisfactory	Nil	Nil

In pre-monsoon period

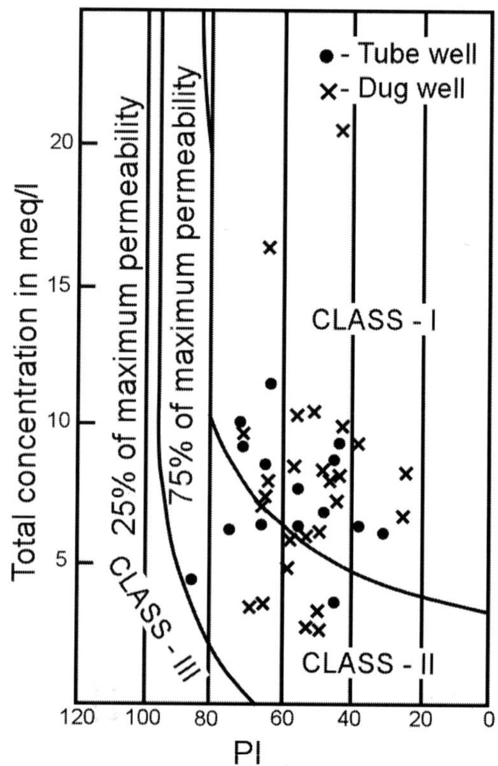

In post-monsoon period

Fig. 4. Classification of irrigation water based on Permeability Index (Doneen, 1962)

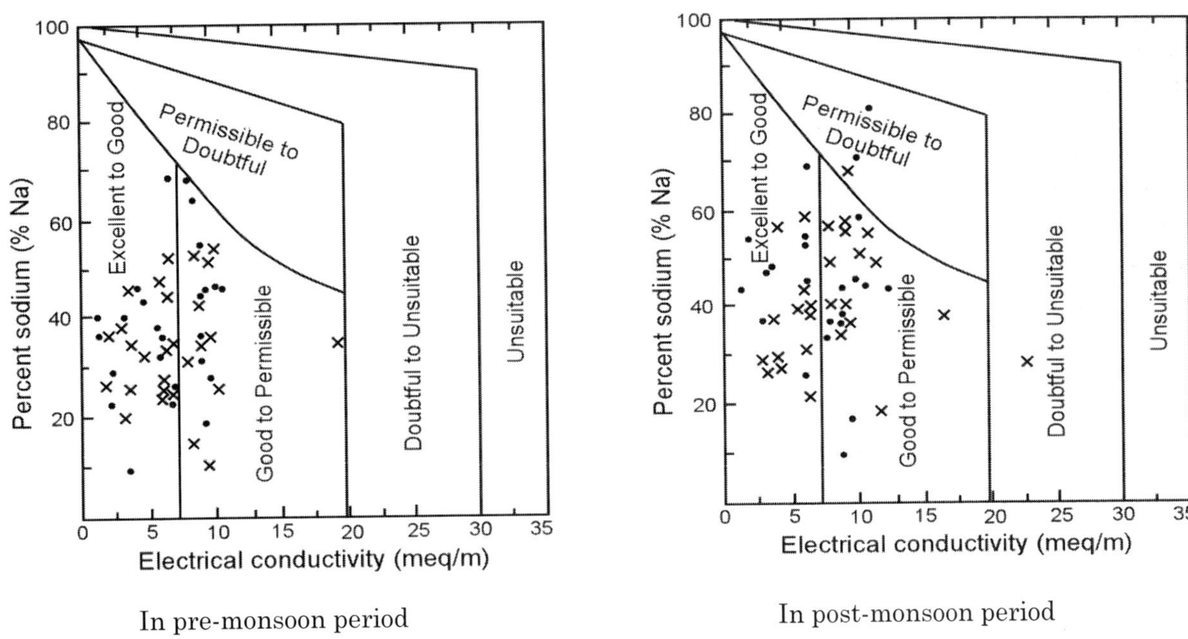

In pre-monsoon period In post-monsoon period

Fig. 5. Wilcox diagram for classification of groundwater based on EC and Na%

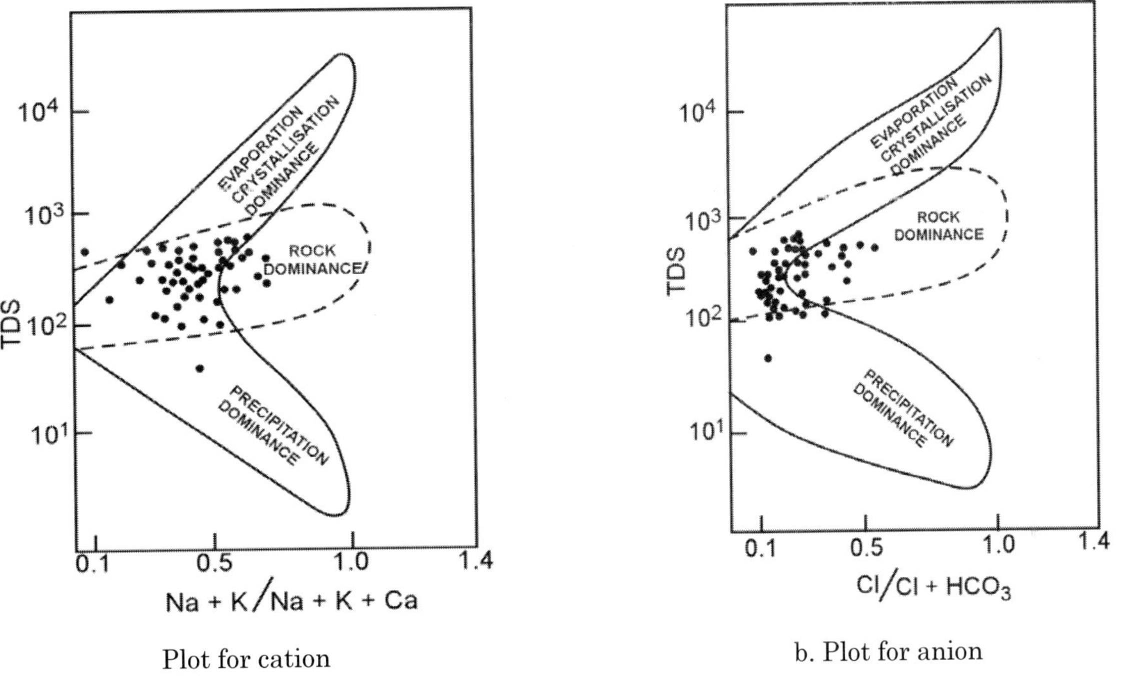

Plot for cation b. Plot for anion

Fig. 6. Gibb's plot for cation and anion

Wilcox diagram

Percentage of sodium content in natural water is an imperative parameter to assess its suitability for agricultural use. A maximum of 60% sodium in groundwater is allotted for agricultural purposes (Wilcox, 1948, 1967; USSL, 1954). Sodium percentage can be defined in terms of epm of the common cations (Wilcox, 1948).

$$Na\% = \frac{(Na^{+} + K^{+})100}{Ca^{++} + Mg^{++} + Na^{+} + K^{+}}$$

The concentrations of cations are in epm.

The sodium percentage (Na %) in the study area ranges from 10.29% to 58.16%. The highest percentage of sodium is found in the dug well water sample of Pakhalakhala (Sample No. 24). The minimum value of Na% is located in the tube well water sample of Talabasta (Sample No. 48).

Plotting the data of the study area on Wilcox diagram (Wilcox, 1948) relating to electrical conductivity and sodium percentage (Fig. 5) show different water classes for irrigation value and are presented in the Table 7. Excellent to good and good to permissible water can be used for the purpose of irrigation.

Table 7. Water classes for irrigation on the basis of Na%

Water class for irrigation	% of Na	Number of samples (out of 53)	
		Pre-monsoon	Post-monsoon
Excellent to	Up to 20	31	23
Good	20-40	22	25
Permissible to	40-60	Nil	3
Doubt full	60-80	Nil	1
Unsuitable	>80	Nil	Nil

USSL Diagram

In the U.S. Salinity Laboratory diagram (Fig. 3) based on Sodium Absorption Ratio (SAR) vis a vis specific conductance values (USSL, 1954), the two most significant parameters of sodium and salinity hazards determine the suitability of water for agricultural purpose. In pre-monsoon, 40% of water samples and in post-monsoon 30% of samples are plotted in C_2S_1 which reveal medium salinity and lower sodium water. 15% samples of pre-monsoon and 20% of post-monsoon are plotted in C_3S_1 plot which indicate moderate to high salinity and are suitable for irrigation purposes. Rest of the samples fall in C_1S_1, C_2S_2, C_3S_2, C_3S_3 field (Table 8). In pre-monsoon no samples and in post-monsoon only one sample is plotted in C_3S_4 field which is highly alkaline in nature and have high salinity (Table 8).

Industrial Suitability

For Industrial use water must be of much higher quality than water used for cooling. Quality requirements for industrial water may vary widely according to potential use. The purest of waters are required for the manufacturer of pharmaceuticals and paper, while for certain phases of mining industry, quality requirements are less exacting. The groundwater of the study area is suitable for industrial purposes, because all the samples show total hardness and TDS within the permissible limits in both pre and post-monsoon seasons.

Table 8. Ground water classification based on USSL diagram

Category	No. of samples		Water quality
	Pre-monsoon	Post-monsoon	
C_1S_1	6	4	Low salinity and lower alkali water
C_2S_1	21	16	Medium salinity and lower sodium water Good for medium permeable soil
C_3S_1	8	11	Moderate to high salinity and less alkaline water
C_2S_2	5	5	Moderately alkaline and medium salinity. Good for medium permeable and coarse grained permeable soil.
C_3S_2	7	9	Moderate to high salinity and moderate alkaline
C_3S_3	6	7	Highly alkaline and have moderate to high salinity
C_3S_4	-	1	Highly alkaline and saline

Table 9. Corrosivity Ratio of Ground Water

Corrosivity Ratio (CR)	Water category	Out of 53 water samples	
		Pre-monsoon	Post-monsoon
<1	Safe Zone	44	44
>1	Unsafe Zone	9	9

Corrosivity Ratio (CR)

The corrosivity ratio is the susceptibility of groundwater to corrosion. It is expressed as ratio of alkaline earths to saline salts in groundwater. The corrosivity ratio is defined by the formula,

$$\text{Corrosivity Ratio (CR)} = \frac{\dfrac{Cl}{35.5} + \dfrac{2(SO_4)}{96}}{\dfrac{2(HCO_3 + CO_3)}{100}}$$

where all the ions are expressed in ppm of groundwater

Ryner (1944), Raman (1985), Balasubramanian (1986), Sankar (1995) and Aravindan *et al.* (2004) have used the ratio to evaluate corrosive tendency of groundwater on metallic pipes of various regions. 83% of both pre and post-monsoon water samples have corrosivity ratio (CR) less than one (safe zone), while 17% of both pre and post-monsoon samples have (CR) more than one (unsafe zone) (Table 9). In the area, where groundwater has Corrosivity Ratio (CR) value more than 1, Polyvinyl Chloride (PVC) pipes should be used.

Gibb's diagram

Based on aquifer lithology, the mechanism controlling chemical relationship of groundwater had been studied by Gibb (1970) and Viswanathaiah *et al.* (1978). In the Gibb's diagram, three kinds of fields are recognized namely, precipitation dominance, evaporation dominance and rock-water dominance. In case of anion of the water samples, 45 samples are pointed towards rock dominance of the chemistry of groundwater in the region, which reflects the influence of the chemistry of aquifer lithology vis-à-vis groundwater. Rests of the samples are in the field of precipitation dominance (Fig. 6). Similarly, in case of cations of the water samples, 50 samples (out of 53 water samples) are plotted in the field of rock dominance (Fig. 6). Hence from both the diagrams of Gibb, it can be established that the groundwater of the study area is mainly dominated by the lithology of aquifer of the concerned region.

CONCLUSION

The present chemical analysis explicitly demonstrates that the groundwater of the study area is alkaline and weakly acidic in nature. Calcium and magnesium are the major cations and bicarbonate is the major anion in the study area. The higher concentration of the bicarbonate indicates chemical weathering of rocks. The groundwater belongs to Ca-Mg-HCO_3 facies. The Ca, Mg, HCO_3 contents depict the temporary hardness and alkalinity of the investigated water. Total hydro-geochemistry of the study area is dominated by alkaline earth and weak acids. Various parameters such as sodium absorption ratio (SAR), sodium percent (Na %), residual sodium carbonate (RSC), magnesium ratio, permeability index and potential soil salinity were estimated to identify the suitability of water for irrigation purpose. The groundwater falls under class-I for most of the sites as per classification of Doneen's permeability index and could be treated as good for irrigation. The Wilcox classification shows most of the samples come under good to permissible zone. The residual sodium carbonate values show that 78% of the water samples are under 'safe'. According to U.S. Salinity diagram, the majority of groundwater samples belong to C_2S_1 (medium salinity and low SAR). The quality of groundwater belongs to "good to permissible" for irrigation purpose. The groundwater samples of the study area falls in Class-I and Class-II as per the permeability index. The water has excellent to good potential soil salinity. Gibb's diagrams are plotted to infer water quality for industrial use. The chemical parameters indicate that the groundwater of the study area is also safe for industrial purpose. The diagrams of Gibb reveal that the groundwater of the study area is mainly dominated by the lithology of aquifer of the concerned region. In general, groundwater is suitable for domestic and irrigation purposes with few exceptions. The chemical parameters indicate that the groundwater of the study area is also safe for industrial purpose.

Acknowledgements

The authors are thankful to the Director, Ground Water Survey and Investigation (GWSI), Bhubaneswar for the chemical analysis. Prof. R.N. Hota is thanked for his help during the preparation of the manuscript. The authors are also thankful to Prof. B.C. Guru for critically going through our manuscript.

References

Adyalkar, P.G. (1965) Sedimentological study of Athgarh Sandstones. Proc. Natl. Inst. of Sc., v. 31, pp. 66-78.

APHA (2005) Standard methods for the examination of the water and waste water. APHA (American Public Health Association), AWWA, WPCF, 21st edition, p. 1134.

Aravindan, S., Manivel, M. and Chandra Sekhar, S.V.N. (2004) Groundwater quality in the hard rock area of the Gadilam River Basin, Tamilnadu. Geol. Soc. India, v. 63, pp. 625-635.

Ayers, R.S. and Westcot, D.W. (1985) Water quality for agriculture, FAO of the United Nations, Paper 19, Rome, Italy.

Ayers, R.S., (1977) Quality of Water for Irrigation., J. Irrigation and Drainage Div., ASCE, Vol.103, No.IR2, pp. 135-154.

Balasubramanian, A. (1986) Hydrogeological Investigations in the Tambraparani River Basin, Tamil Nadu, Unpublished Ph.D. Thesis, University of Mysore, 345p.

Ball, V. (1877) On the Athgarh Sandstones near Cuttack. Rec. Geol. Surv. Ind., v. 10 (2), pp. 63-68.

Christiansen, J.E., Olsen, E.C. and Willardson, L.S., (1977) Irrigation Water Quality Evaluation, J. irrigation and Drainage Div., ASCE, v. 103 (2) pp. 155-169.

DAS, R. and PANDYA, K.L. (1997) Palaeocurrent Patteern and provenance of a part of Gondwana Succession, Talcher Basin, Orissa. Jour. Geol. Soc. India, v. 50, pp. 425-433.

Domenico, P.A. and Schwartz, F.W., (1990) Physical and chemical Hydrogeology. John Wiley and Sons, New York. pp. 410-420.

Doneen, L.D. (1962) The influence of crop and soil on percolating water. Proc. 11961 Biennial conference on Groundwater Recharge, pp. 156-163.

Eaton, S.N. (1950) Significance of carbonate in irrigation water and soil. Science, v. 39, pp. 123-133.

Gibb, R. J. (1970) Mechanism controlling world water chemistry. Science, v. 170, pp. 1088-1090.

Goswami, S., Meena, K.L., Das, M. and Guru, B.C. (2008) Upper Gondwana Palynoflora of Mahanadi Master Basin, Orissa, India. Acta Palaeobotanica, v. 48 (2), pp. 171-181.

Goswami, S., Singh, K.J. and Chandra, S. (2006). Palaeobotany of Gondwana basins of Orissa, India: A bird's eye view. Journal of Asian Earth Sciences, v 28 (4-6), pp. 218-233.

Gray, N.F. (2005) Water Technology: An introduction for Environmental scientists and engineers. Elsevier, Amsterdam, pp. 1-645.

Gupta, P.K. (2004) Methods in Environmental Analysis: Water, Soil and Air. Agrobios Publication, Jodhpur, India, pp. 1-322.

Hem, J. D. (1985) Study and Interpretation of chemical characteristics of natural water, U.S.G.S. Water supply Publication. pp. 2254-2264.

ISI (Indian Standards Institute) (1995) Indian Standard Specification for drinking water. ISI New Delhi, IS 10 500.

Jaiswal, P.C. (2004) Soil, plant and water analysis. Kalyani Publishers, Ludhiana, India, pp 1-213.

Kuchanwar OD, Kale CK, Deshpande VP, and Dharmadhikari D.M. (1999) Irrigation water quality and farm management decisions. Water Sci Tech v. 40(2), pp. 97–103.

Mandel, S. and Shiftan, Z.L. (1981) Ground Water Resources Investigation and Development, Academic Press Inc., New York.

Raman, V. (1985) Impact of Corrosion in the conveyance and Distribution of water, Jour. of the I.W.W.A., v. 15(11), pp.115-121.

Rhodes J.D. (1972) Quality of water for irrigation. Soil Science, v. 113(4), pp.227–234.

Richards, L.A., (1954) Diagnosis and improvement of saline and alkaline soils U.S. Deptt. Agri. Hand Book No.60, p. 160.

Ryner, J.W. (1944) A new index for determining amount of calcium carbonate Scale Formed by water.Jour. Amer. Wat. Assn., v. 36, pp. 472-486.

Sankar, K. (1995) Hydrological studies in the Kanyakumari District, Tamil Nadu. Unpublished Ph.D. thesis,University of Madras, pp. 90-91.

Saxena, M.M. (1995) Environmental analysis of water, soil and air. Agro Botanical Publishers, India, pp. 1-112.

Trivedy, R.K. and Goel, P.K. (1984) Chemical and biological methods for water pollution studies. Env. Publ., Karad, India, pp. 1-215.

U.S.S.L. (1954). Diagnosis and Improvement of Saline and Alkaline soils. U.S. Salinity Laboratory (USSL), U.S. Dept. of Agriculture Hand Book, Washington D.C., v. 60, pp. 1-160.

Viswanathaiah, M.N., Sastri, J.C.V. and Rame Gowda, B. (1978) Groundwater of Karnatak. Indian Mineralogist, v. 19, pp. 65-69.

WHO (1989) Health Guidelines for the Use of Wastewater in Agriculture and aquaculture. Report of a WHO Scientific Group – Technical Report Series 778, WHO Geneva, pp. 74.

WHO (2004) Guidelines for Drinking Water Quality Recommendation, v. 1, 3rd edition, World Health Organization, Geneva.

Wilcox, L.V., (1948) The Quality of water for irrigation use. U.S. Dept.of Agriculture, Tech, Bull, Washington, D.C., p. 19.

Wilcox, L.V., (1967) Classification and use of irrigation water. USDA circ.969, Washington D.C., p. 19.

Hon'ble reviewer

1. **Prof. V.K.S. Dave**
   Professor Emeritus
   Department of Earth Science
   IITR, ROORKEE - 247 667
2. **Prof. Vibhuti Rao**
   Chairman
   Department of Geology
   Lucknow University
   LUCKNOW - 226 001

III-37

Interlinking of Rivers in India: Pros and Cons

Droupti Yadav and Richa Mishra

Centre for Advanced Study in Geology, University of Lucknow, Lucknow 226 007
E-mails: droupti.yadav@gmail.com

Abstract

India is a vast country with its population spread out all over, starting from mountainous region to Gangetic plane, deserts, plateau, coastal areas and southern hilly terrains. Depending upon its various physiographic regions, the population density varies in its different parts. A major role in this distribution is being played by the availability of fresh water as a natural resource from both underground as well as from surface water bodies. Rivers play a pivotal role in sustaining the human settlements including small and large cities. Apart from the major rivers like Ganga, Yamuna, Narmada, Godavari, Krishna and Kaveri, there are a large number of smaller rivers which cut across different regions of the country and are the main source of fresh water both for the agriculture as well as domestic requirements.

Lately, there has been a suggestion put forward by various government and private agencies that in order to enhanced the water availability, some of these rivers can be interlinked through a network of canal development. The present paper discusses the aspect of interlinking of rivers, both as an environmental issue as well as societal issues. It is suggested that looking in to the various issues of interlinking of the rivers, both positive and negative impacts exist and depending on the particular interlinking, success and failure can be judged. The pros and cons of interlinking of rivers can only be established once a well researched project is put into operation and executed considering several factors.

Keywords: River linking Projects, Inidan River Linking, Water Management water uses, Water ecology.

Introduction

Rapid urbanization and industrialization in India have lately started showing its impact on the earth system at an alarming rate. The maximum impact has been on the environment, particularly on the natural resources such as water, soil and air. Water, as a sustainable resource, is indeed one of the most vulnerable entities as the biosphere cannot sustain without its existence. As a matter of fact, majority of ancient civilizations of the world evolved and flourished by the river banks. Rivers are, therefore, an integral part not only of human existence but the very existence of life on our planet Earth.

Indian culture owes its supremacy in the world scenario on account of rivers which are the life and blood of the nation and worshiped as gods and goddesses all over the country. Role of rivers in our life, in national development, progress and prosperity, development of agriculture, science and technology or industry is needs to be discussed at length. India is a country with vast population with extremes of climate, topographic variability, varied types of soils, annual rainfall ranging from just a few cm to hundreds of cms. Some parts facing the havoc of floods and other parts are remaining thirsty for just a few drops of rain. It is therefore, a scheme for effective and efficient management of water resources was considered to be the need of the hour which includes the aspect of interlinking of rivers in order to enhance and optimize the river based water resources of the country.

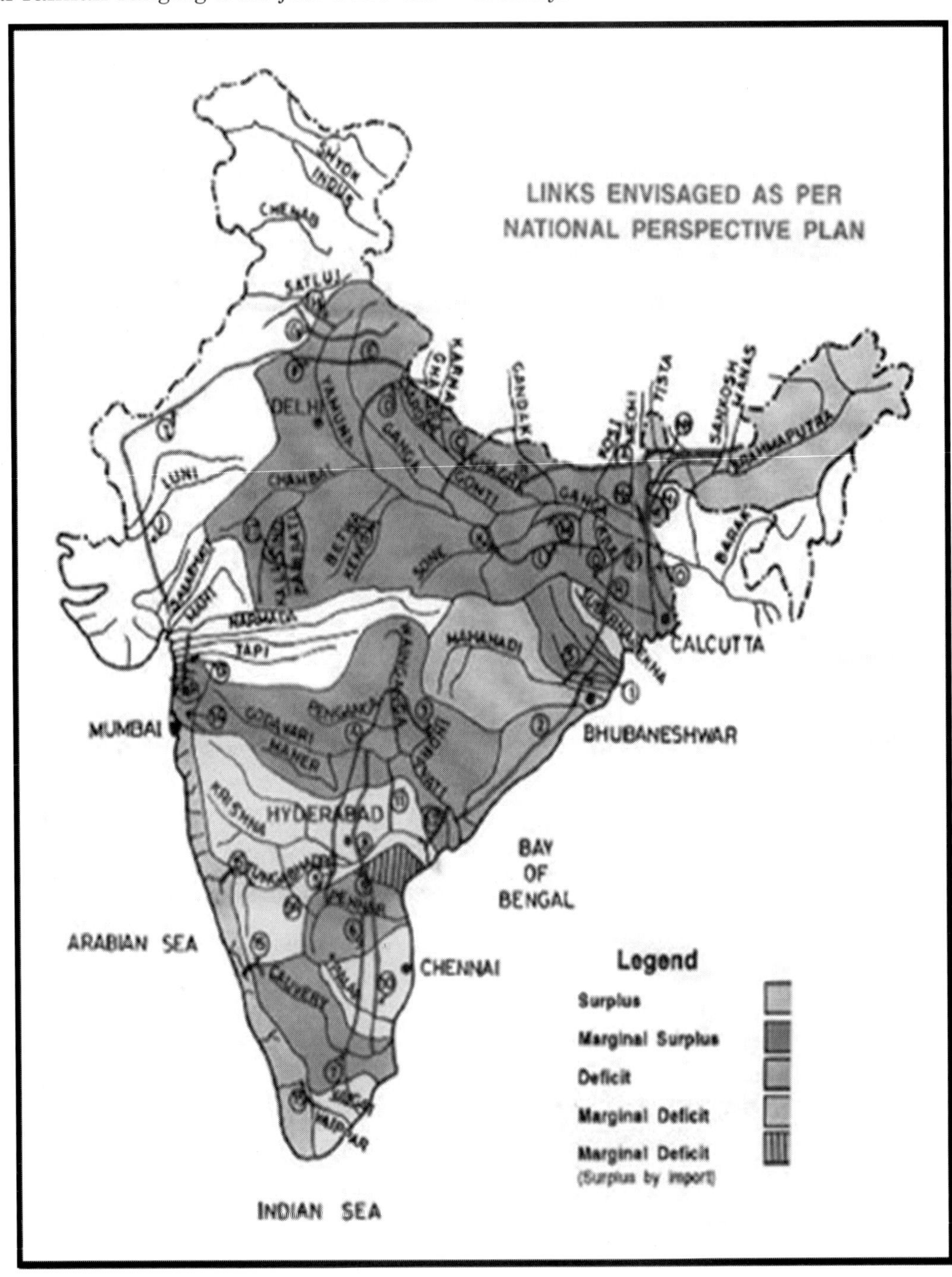

Fig. 1: Water deficit and water surplus regions of India
(**Web resource1:** http://chellaney.net/2012/03/)

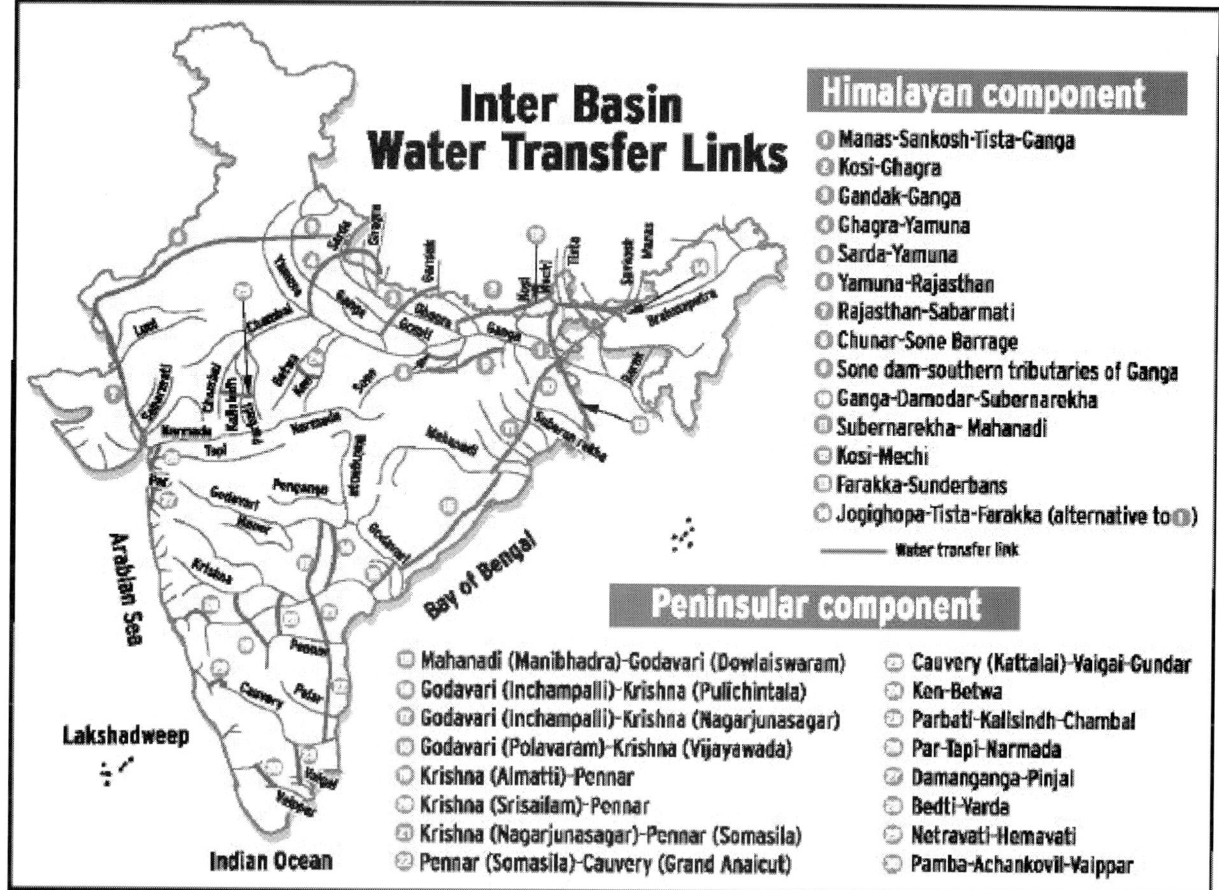

Fig. 2: Inter basin water transfer links
(**Web resource2:** http://www.thehindu.com/ opinion/op-ed/a-river-sutra-without%20links/article2954807.ece)

The main objective of interlinking of rivers is to achieve "water-balance" or to divert water from water sufficient regions to water deficient regions i.e. from areas with flood prone river systems to the areas that chronically remain in drought for long periods of time (Fig. 1). For peninsular India, the National Water Development Agency (NWDA) has considered 16 possible links in major river basins, including the Mahanadi, the Godavari, the Krishna, the Pennar, the Cauvery, the Vaigai, and the west-flowing rivers of Kerala, Karnataka, north of Mumbai, and south of Tapi, and the southern tributaries of the Yamuna, which were studied to establish sustainability (Fig. 2). However, in northern part of the country, the scenario is different as most of the areas get sufficient amount of water during rainy season (Monsoon). Nevertheless, this part being the most populated part of the country, requires also pure water and hence an appropriate river-stream management plan is also required from the point of view of national growth and development. The National Water Development Authority (NWDA) after a thorough study indicated that Himalayan rivers, especially, Brahmaputra and Ganga have exceedingly surplus quantum of water and hence, proposed transfer of water from these surplus basins to deficit basins in peninsular region (Vidyasagar Rao *et al.*, 2003; Reddy, 2003).

WATER DEPLETED REGION OF THE COUNTRY

A BACKGROUND OF THE INTERLINKING PROPOSAL IN INDIA

- 1972—Ganga Cauvery link proposed by Union Minister Dr. K.L. Rao.
- 1974—Garland Canal proposal by Captain Dinshaw J Dastur, a pilot. Both plans rejected due to technical infeasibility and huge costs.
- 1980—Ministry of Water Resources frames the National Perspective Plan (NPP) envisaging interbasin transfer.

- 1982—The National Water Development Agency (NWDA) set up to carry out pre-feasibility studies. These form the basis of the ILR plan.
- 1999—A national commission (NCIWRDP) set up to review NWDA reports concluded that it saw 'no imperative necessity for massive water transfers in the peninsular component' and that the Himalayan Component would require more detailed study.
- August 15, 2002—President Abdul Kalam mentions the need for river linking in his Independence- day speech; based on which senior advocate Ranjit Kumar filed a PIL in Supreme Court.
- October 2002—Supreme Court recommends that the government formulate a plan to link the major Indian rivers by the year 2012.
- December 2002—Govt. appointed a Task Force (TF) on Interlinking of rivers (ILR) led by Mr. Suresh Prabhu. The deadline was revised to 2016. (Lakra, *et al.* 2011)

India and Pakistan are locked in water war over the construction of power projects on Jhelum on their respective territories and so is the case between India and China. Snow-fed rivers like Ganga, Indus and Brahmaputra, which originate in the Himalayas and their tributaries, are perennial. They continue to flow throughout the year. During monsoons, they tend to overflow and inundate large flood prone areas and cause loss of life, livestock, crops and property. Rain-fed rivers like Luni in Rajasthan remain dry for most part of the year because the rainfall is scanty in that area resulting in drought like conditions. The volume of water in the west flowing rivers of Central Highlands, Narmada and Tapti is directly proportional to the amount of rainfall received during the monsoon season. Hence, there is always an element of uncertainty in availability of water.

Thirty Himalayan and peninsular rivers have been identified by the National Water Development Agency for such transfers. Fourteen links (route or line) between the Ganges-Brahmaputra-Meghna river basins have been proposed in the Himalayan zone, similarly in the peninsular zone, 16 rivers including Mahanadi, Krishna, Godavari, Pennar and Cauvery river basins have been chosen for the transfers (Shukla, 2009; Boddu, 2011).

ALTERNATIVE REMEDIES FOR WATER CONSERVATION

India's Interlinking of Rivers Plan (ILR) has been offered as a solution to the inequitable distribution of water resources around the country by redistributing excess water in flood-prone areas towards more arid areas. The proposed outcome of the plan is the creation of a flowing water link between every major river basin in the country.

The ILR would allegedly alleviate flooding in the Ganga and Brahmaputra river basins during monsoon season (June to September) by storing "surplus" water in reservoirs in the Himalayas and then piping this water towards more arid areas of the country throughout the year. Some alternative remedies are "Water grid" i. e. similar to the power grid so that water can be shared between the States without any friction, conjunctive use of groundwater and surface water with improving techniques of conventional water harvesting, decentralization of local rainwater harvesting, improving the efficiency of waste water management system and recharging groundwater Effective Techniques of Irrigation and Cloud Seeding (Prakasa Rao *et al.* 2010).

DISCUSSION

1. Some Benefits of interlinking of river i.e. increased Inland navigation i.e. River transports is not only cheaper but also a non-polluting transport alternative.
2. Interlinking of rivers will generate employment opportunities all over the country especially in agriculture sector, power, transport and construction works.
3. Migration from rural areas will come down which would reduce congestion in urban areas; decentralization of industries would be a natural phenomenon with the availability of water and power.
4. Cheap water irrigation potential and drinking water would achieve food security and self-sufficiency by increasing area of arable land, increased hydropower production which would increase electricity production in India, control floods because

floods are a major cause of mortality and property damage in our country.

5. National integration of some societal issues like economic growth, improved public life system with creeping improvement in other infrastructure sectors notably, roads, railways and power. New institutional models for infrastructure creation and management are likely to restore the country's confidence in its capacity to create and manage large infrastructure projects, agricultural diversification, rising energy costs and urbanization.
6. According to some scientists, monsoon rains come all over the country around the same time, hence interlinking of rivers may cause extensive flooding and large areas can get submerged during that time. It would also affect the cropping pattern of the region. Inter-basin water transfer could introduce alien / hostile species of plants causing ecological imbalance.
7. Other major problem associated with this ILR projects are ecological issues like deforestation, soil erosion and impact on aqua life.
8. Rehabilitation and resettlement of public due to interlinking of rivers would create public unrest and would not be an easy task.
9. The project may also triggered drought like conditions in certain areas for example the region of existing dam on Ken and Betwa rivers require additional water inputs during summers and have excessive water during the monsoons. The proposed link may act as a double edged sword as either of seasons, the impact may be negative.
10. On a geopolitical platform, the issue may trigger strained relationship with neighbouring countries like China, Pakistan, and Bangladesh.
11. Reverse impact on irrigation can increase salinity and water logging in arid areas and would actually make agricultural land less productive (Rai et al., 2005).
12. Loss of Biodiversity may result due to certain projects where two widely different regions get connected due to interlinking and would therefore, generate a new ecosystem. In this ecosystem the existing flora and fauna can get adversely affected and may result in loss of biodiversity.
13. Detailed scientific studies with relevant ground data of the rivers to be interlinked needs to be done by looking into the pros and cons of the post interlinking scenario in order to have the perspective about any major change that may be environmentally catastrophic. This is necessary because once the interlinking process is completed, the process cannot be reversed.
14. If such a link between rivers is found appropriate then sufficient funds should be kept for maintenance of the link on a regular basis otherwise the project would became a failure within few years of its commencement.
15. Previous projects of India and abroad where such interlinking has been proved to be successful should be considered as guiding line so that the interlinking can became a boon for the society and the nation. Any failure in the process of interlinking may become a bane with its serious environmental and social impact.
16. We can conclude that the interlinking of the rivers have both positive and negative impacts and depending on the particular interlinking, success and failure can only be judged once the projects is completed and is put in operation. However, a deep pre-emptive study needs to be done in order to know the future, whether it is a boon or a bane!

ACKNOWLEDGEMENTS

The authors extend their heartfelt thanks to the Head of the department of Geology, University of Lucknow for providing the working facilities of the department. They are also thankful to Prof. Vibhuti Rai, Ms. Purnima Sharma, Ms. Shashi Verma and Mr. Rohit Aggarwal for helping them in the present work. The authors are thankful to Prof. Madhu Tripathi of Zoology Department, University of Lucknow for inspiring them for the present and related research themes.

REFERENCES

Bandyopadhyaya J. and Perveen, S. 2003. The Interlinking of Indian Rivers: Some Questions on the Scientific, Economic and Environmental Dimensions of the Proposal. Paper presented at Seminar on Interlinking Indian Rivers: Bane or Boon? at IISWBM, Kolkata 17 June 2003, SOAS Water Issues Study Group, Occasional Paper No 60.

Boddu, M, Gaayam, T. and Annamdas, V. G. M. 2011. A Review on Inter Basin Transfer of Water" IPWE 2011, Proceedings of 4th International Perspective on Water Resources & the Environment, January 4-6, 2011, National University of Singapore (NUS), Singapore.

IWMI (International Water Management Institute) 2000. World water supply and demand 1995 to 2025. Draft report prepared for World Water Vision. Colombo, Sri Lanka.

Prakasa Rao, B.S., Vasudeva Rao, P.H.V. , Jaisankar, G., Amminedu, E., Satyakumar, M. and Koteswara Rao, P. 2010. Interlinking of river basins: A mega harvesting plan-a review, J. Ind. Geophys. Union, vol.14, no.1, pp.31-46.

Sarkar, W.S., Dubey, U.K., Sani, V.K. and Pandey, R. 2011. River inter linking in India: status, issues, prospects and implications on aquatic ecosystems and freshwater fish diversity. Review Fish Biol. Fisheries 21:463–479.

Rai, Vibhuti, Verma Anju and Rai Madhu 2005. Interlinking of river: Environmental impact in Reverse. ISEG news, Vol.2 No.2.pp-30.

Reddy, M.S., 2003. Linking of rivers in India-Retrospect and Prospect. Journal of applied Hydrology, 16 (4A), 14-30.

Shukla, S. K. 2009, "Indian river systems and pollution" In: Encyclopedia of Earth. Eds. Cutler J. Cleveland (Washington, D.C.: Environmental Information Coalition, National Council for Science and the Environment).

Thatte C. D. 2006. Augmentation of water resources through inter-basin water transfer (IBWT): A review of needs, plans and prospects in the light of the global situation. World Affairs, Vol. 10, No.4, pp. 38 -62.

Vidyasagar Rao, R., 2003. Inter- basin water transfer-A vital necessity but a distinct reality. Journal of applied Hydrology, 16 (4A), 1-13.

Web resource1: http://chellaney.net/2012/03/

Web resource2: http://www.thehindu.com/opinion/op-ed/a-river-sutra-without%20links/article2954807.ece.

Hon'ble reviewer

1. **Prof. P.K. Verma**
   Director General, MPCST
   Vigyan Bhawan
   BHOPAL - 462 003
2. **Dr. Ashutosh Shrivastava**
   Hydrogeo Scientist
   Rajasthan State Ground Water
   JAIPUR - 302 001

III-38

Fast Depleting Water Resources of Uttar Pradesh: An Overview

C. Prasad

489-Uphar Udyan-I, Jail Raod, Lucknow
E-mail: drchittaranjaprasad@gmail.com

Abstract

Uttar Pradesh is considered to be a state quite rich in its water resources, however, since 1990, the rapid decline of ground water level and shrinkage of great rivers like Ganga and Yamuna, are very alarming. Under such conditions the water resources may not be able to sustain the fast growing population and development processes. This depletion has nothing to do with the climate change, but is mainly caused by the uncontrolled extraction of ground water and wastage of surface waters.

For replenishment of this resource, efforts like rain-water harvesting and interlinking of rivers may help. The interlinking of rivers may be useful for flood control and may also check desertification.

Keywords: Groundwater, Water Resources, Uttar Pradesh, Water Table Depletion, Water Management.

Introduction

It is generally believed that Uttar Pradesh is sitting over the largest fresh water reservoir of world, viz the Ganga Yamuna plains, still the state is rapidly moving towards shortage in supply of water for human consumption, viz irrigation, industry and domestic usage. If adequate measures are not taken up, a "water crisis" will be there. The records show that water level in underground aquifers is rapidly declining. Also, the overground water level in most rivers of state are lowering significantly, since 1990. The irony is that while the eastern part of the state is flood prone, south western part is reeling under desertification. On the other hand the demand of water for human consumption is increasing rapidly on account of increasing population, urbanisation and the growing industries.

As per the available data, the declining trend of the groundwater level has affected about 2/3 of the state. In this context, while in 2000, the ground water level was 30-40 meters deep in 54 districts, now over 60 districts have below 40 meters deep level. Moreover, the surface level of most rivers including Ganga has gone down considerably. Many newspaper reports state that while at Kanpur, the Ganga has left behind the bathing ghats, in Allahabad, now the pilgrims can walk down to the confluence point or "Sangam". Deep borings are being done at many places which has resulted in declining yield of many wells. Also many deep wells in central U.P. are yielding toxic water with high F or Cd content. On the other hand, subsidence has occurred at many places in 2002 (Sinha 2005).

Contrary to the declining trend of ground-water level at many places in the eastern U.P. particularly in the command areas of the canals, water-logging conditions have emerged. May be because of faulty irrigation practices, about 36 lakh ha. of the once fertile land has turned to wasteland and to a high alkaline terrain.

GEOHYDROLOGICAL CONDITIONS

Studies show that, the state possesses a complex state of underground aquifers. U.P. is located over the Ganga alluvium which might be over 3000 meters thick at places. However in northern part near the Himalayan foot hills, the alluvium is mixed with colluvial material from piedmont fans. The aquifers are usually contained in generally lenticular sand bodies enclosed in clayey layers. There are gravels also at deeper level. Anantha Rao and Sahni (2005) have identified mainly five hydrogeologic zones as given in the following table:

Table 1: Hydrogeological zones in U.P.

Zones	Geologic Condition	Strata Content	Aquifer Condition
Bhabor	Piedmont fans	Gravel beds	Ground water level 8-40 m, poor
Terai	Adjacent to piedmont fans	Thin alluvium with colluvium	Level 10-20 m, moderate
Ganga plains	Central part of the Ganga basin	Thick alluvium sandy clayey layers	Good aquifers, with better yield
Marginal plains	The plains of Yamuna, Chambal etc	Thin alluvium colluviums etc	Erratic aquifer with little yield
Southern Plateau	Vindhyan and granitic terrane	Rocky areas with thin alluvium	Erratic aquifer with very low yield

Anantha Rao and Sahni have also identified 4 levels of aquifers as given in the table below:

Table - 2 : Levels of aquifers

Zones	Level in meters
I (Top)	0-150
II (Upper)	160-210
III (Middle)	250-360
IV (Lower)	380-700

Of the above given aquifer levels only the top level is presently being exploited. The potential of other layers is yet to be assessed fully.

DECLINE OF GOURND WATER LEVEL

The data published by U.P. Govt. (Sinha 2005), state the decline of ground water level since 1995. Following table shows availability and extraction of ground water in U.P.

Table 3 : Ground water availability in lakh ha. m. (2005)

Zones	Availability (Lakh ha. m)	Extraction	Remained
Eastern	29.42	13.58	15.84
Western	18.93	10.43	8.42
Central	27.75	18.43	9.32
Southern	4.72	1.35	3.37

The above table shows that while ground water conditions in eastern U.P. are normal, they are alarming in central and western U.P. However, the conditions in southern U.P. have become "critical".

Following table shows the decline of ground water level (1995 to 2005)

Table 4: Decline in ground water level since 1995

No. of blocks	Decline in meters/Year	Conditions
146	1-2	Normal
65	2-3	Alarming
29	3-4	Alarming
12	4-5	Critical
8	Over 5	Critical

The critical conditions exist in south-western U.P. whereas alarming conditions prevail in western and central U.P.

According to Prasad (2007), and some the newspaper reports, the extraction in urban areas with increase in population and growth of industries has made the decline of ground water level very alarming. Following table shows the decline since 2002 upto 2012.

The condition in the capital city, Lucknow shows that decline in ground level has become quite critical as shown in the following table:

Table - 5 : Decline of ground water level in cities

Cities	Decline in meter (2002-2007)	Decline in meter (2007-2012)
Lucknow	2.8	17.5
Kanpur	2.3	16.0
Agra	3.3	20.5
Meerut	4.0	23.0

Table - 6 : Ground water level in Lucknow

City	Level in 2007 (Depth)	Level in 2012 (Depth)
Lucknow	26.3 meters (western part)	43.8 meters (southern Part)

The decline is more in urban areas particularly in western central and southern U.P. However conditions in the eastern U.P. may be considered as normal with even water-logging at some places. This part of state is also floodprone.

CAUSES OF DECLINE

It is often stated that the decline in the ground water level may because of the ongoing climate change. However, a warming trend of temperature will cause more evaporation and consequent rainfall, Moreover, there will be ice ablation in the catchment area of U.P. rivers viz Uttarkhand and Nepal. In this context, following set of data may be interesting (Prasad 2011)

Table - 7 : Rise in annual temperature

Period	Temperature in hottest months	Temperature in coldest month
1940-1970	40.5°C	7.9°C
1970-2000	42.0°C	6.9°C

Above table shows little rise and fall in temperature.

However, the effects of warming are more significant in the catchment areas. It is interesting to note that while the Ganga and Yamuna are glacierfed the smaller rivers of central and eastern U.P. are mainly rainfed. The following table shows a significant decline in snow fall in the Himalaya.

The catchment area of Ganga-Yamuna and other rivers largely come under western Himalaya. The decline in snow fall will cause the glaciers to retreat. The glacial retreat has also been noticed in Gangotri glacier, since 1971 (Prasad and Naithani 2003). A large scale glacial retreat would normally cause floods, which already has affected the eastern and central U.P. but not the western U.P.

Table - 8 : Decline in snow fall in Himalaya

Zone of Himalaya	Decline in percentage	Source
Western	14	Down to earth (2007)
Eastern	25	

The newspaper reports have given some annual rain fall data since 1997 (Prasad 2011)

Table - 9 : Annual rain fall in cms

Year	Rain fall	Year	Rain fall
1997	104.5	2003	98.2
1998	69.1	2004	81.4
1999	67.7	2005	70.3
2000	122.7	2006	88.1
2001	81.2	2007	97.4
2002	76.9	2008	105.6 (estimated)

(Normal value 103.7)

The above table shows that, rainfall in the state has largely been deficient, except for 1997, 2000 and 2008.

However, the flooding during rainy reasons has been mainly in eastern and central U.P.

Table - 10: Major flood events in India (Prasad 2011)

Year	Areas affected	Probable cause
1998	Eastern U.P., Bihar	High rain fall
1999	Central and Eastern U.P.	High rain fall
2000	Orissa, A.P., Maharastra	Tropical cyclone
2001	Eastern U.P. Bihar, Assam	High rainfall
2002	Orissa, A.P., West Bangal	Tropical cyclone
2003	A.P., Maharastra	Tropical cyclone
2004	A.P., Tamilnadu	Tsunami
2005	Bihar, Assam	High rainfall
2006	Rajasthan & Gujrat	High rainfall

The above table shows that Central and Eastern U.P. and Bihar have frequently been flooded. Whereas Western U.P. has hardly been affected even when the ice has ablated in the western Himalaya. Therefore, climate change has hardly affected the depletion of water resources in U.P.

DISCUSSIONS

The available data and other observations indicate, that the main cause for depletion of water resources in U.P. may be the intense and uncontrolled extraction of groundwater through tube-wells, closely spaced in the urban and industrial areas. It is reported that in the Lucknow city alone, There are over 5000 tube-wells working continuously. Moreover, the surface water of rivers are being wasted in absence of any provision for conservation. The decline of ground water level is more in the western U.P. which is urbanized with more industrial plants. The decline has increased in central U.P. also, since 2002.

It is generally agreed that the ground water is recharged from surface sources only. The great rivers of U.P. and their tributaries carry huge quantum of water, which may recharge the subsurface water in the nearby areas. However the recharge is a slow process, whereas the depletion takes place rapidly (Chandra 2005). The Newspaper reports state that Ganga, the major river of western and central U.P. is carrying less water now, it has become shallower at Allahabad and Kanpur.

The deficient rainfall in the state can be indicated by the shrinkage of the rivers. However it may not be true for Ganga and Yamuna, as they are fed by glaciers, which are retreating. While the recession is effective since 1971 (Prasad and Naithani 2003), the melting of ice would have flooded western U.P. which is not being observed. On the other hand, the rivers of eastern U.P. which are rain fed mainly, are flooded frequently, despite of deficient rain fall.

In the opinion of the author, the main reason for the shrinkage of Ganga and Yamuna, may be the large river valley projects of Uttarkhand. The reservoir of Tehri dam is over 40 sq kms and has blocked enormous quantum of water. Moreover, about 22 river valley projects have been planned in the tributaries of Ganga and Yamuna in Uttarakhand many of them are storage structures, although experts have suggested only diversion structures in the Himalaya.

SUGGESTIONS FOR REPLENISHMENT

(i) Because of rapid growth of the cities, less open space is available, between the concrete jungles. Therefore little natural recharge of ground water is possible. Wherever, the new cities or residential areas are planned, a lot of open space should be left as parks, forests etc. The vegetational cover should be increased with new plantations.

(ii) Artificial recharge of ground water should be attempted by rainwater harvesting with roof top storage in urban areas and impounding reservoirs in rural areas.

(iii) Interlinking of rivers in the state can also be attempted, as the rivers in eastern U.P. are frequently flooded. The old forgotten concept of “garland canal”, may be revived and attempt to link Rapti saryu and Ganga-Yamuna may help in the replenishment of water resources and also the recharge of ground water in central and western U.P.

CONCLUSIONS

On the basis of above discussions, following conclusions can be arrived at

(i) The global warming vz climate change might have caused a general depletion of water resources, but as far as U.P. concerned, it’s effect has not been significant.

(ii) The depletion, as evidenced by the decline of the ground water level and shrinkage of rivers, is mainly because of overextraction of groundwater and unplanned management of surface waters.

(iii) The proliferation of river valley projects specially the storage structures in the adjoining Uttarakhand for power generation might have been responsible for the reduction of flow and shrinkage of the channels of Ganga and Yamuna.

(iv) Artificial recharging of ground water by rain water harvesting and interlinking of the rivers may help to check the depletion of water resources.

REFERENCES

Anantha Rao C., and Sahni MK. (2005)- Annual Bulletin of Geohydrology Deptt. U.P. Govt. 14-16

Chandra, K. (2005) Annual Bulletin of Geohydrology Deptt. U.P. Govt. 17-18

Prasad, C. (2007) Proceedings of conference on Real Estates, Institute of Engineers, India (Lucknow Chapter) 77-79

Prasad, C. (2011) Water Problems of U.P. Mountain resource management (eds,) Bisht, MPs and Pal, D. Tansmedia Srinagar (GWL) 52-57

Prasad, C. and Naithani, AK (2003)-"Glacier Morphometry – A case study of the Gangotri group of glaciers, Himalaya, India" J. Geol. Soc. of India, Bangalore 61, 325-334

Sinha R.S. (2005) – Lectures on water harvesting Geohydrology Deptt. U.P. Govt.

Down to Earth (2007) issues

Times of India (2008) issues.

Hon'ble reviewer

1. **Dr. D.K. Chaddha**
   Former Chairman
   Central Ground Water Board
   G-66, Vikaspuri
   NEW DELHI - 110 018
2. **Prof. L.P. Chaurasia**
   Department of Applied Geology
   Dr. H.S. Gaur University
   SAGAR - 470 003

III-39

A Study of Groundwater in the Middle Part of Ramganga Valley of Almora Crystalline, Kumaon Lessor Himalaya

Manisha Tripathi

Department of Geography, Kumaun University, Nainital, Uttarakhand
E-mail: manishatripathi.jnu@gmail.com

Abstract

Groundwater plays an important role in the Himalayan mountain system. The Quaternary geological and geomorphological study has been undertaken in the middle part of the intermontane Ramganga valley covering an area of about 200 Km^2 from Chaukhuitia to Bhikiasain in Ranikhet tehsil of Almora district. The Quaternary deposits uncomfortably overlie the Precambrian rocks belonging to Kumaun Super Groups. The study of satellite images topographic maps, real time photographs supported by ground truth survey reveals that the study area has a network of interlinked subsurface fractures. The features of neo-tectonic activities in the form of faults and lineaments have a definite control on the alignment of the Ramganga River (w), small streams and rivulets in the area. The paper aims to measure the hydrologicalas well as hydrogeomorphological behavior of the groundwater resource potential in different landscapes of seven micro-watersheds of the Western Ramganga River. The study of slopes, aspects lineament represents the hydrogeology of the area which helps in categorization of the landforms into different hydro-geomorphological classes representing the relationship of the geological structures vis-à-vis the groundwater occurrence.

Keywords: Groundwater, Hydrogeology, Hydrogeomorphology, Ramganga Valley, Almora Crystalline, Kumaon Lessor Himalaya.

Introduction

Groundwater hydrology in mountains is the study of hydrologic processes and interactions within mountain watersheds (Alford, 1985). The topographic factors are local relief; slope angle and aspect influence the timing, volume and spatial variability of water and energy in a mountainous terrain (Alford, 1985). Groundwater resources are subjected to change in space and time domains due to natural and anthropogenic causes. The study area exhibits diverse hydrogeological and hydrogeomorphological conditions where the groundwater regime is controlled mainly by topography and geology and some extent due to man induced landscape transformation. The Kumaun Lesser Himalayan terrain has highly rugged terrain marked with steep and high ridges and deep and narrow valleys, through which flow the western

Ramganga with its tributaries. The behavior of groundwater in the Kumaun lesser Himalayan region is highly complicated due to the occurrence of diversified geological formations with considerable lithological and chronological variations, complex tectonic framework, climatological dissimilarities and various hydrochemical conditions. As a result of the structurally controlled terrain with steep valley slopes, and the diverse lithologies, occurrence of springs is very common. Springs are the main source of groundwater for the local communities. While the interplay of all these factors affects the groundwater condition and nature of the springs.

The groundwater potential of an area is dependent upon the intrinsic characters of the rocks, soil, and surface hydrology and hydrogeomorphology. Rodda (1994) in his paper '*Mountains- A hydrological paradox or paradise*', although mountains are the source of the bulk of the water resources of the world, knowledge of mountain water resources is generally much extensive, reliable and precise then other physiographic regions. Klemes (1990) has made a similar statement: 'Hydrology of mountain areas is lagging behind many other areas of hydrological inquiry in proportion to its greater difficulty'.

Hydrogeomorphological study of groundwater incorporates relationship of geomo-rphic units with their groundwater potential as interpreted from the landform characteristics (nature of landform occurrence, lithology, structure, inter-relationship with other units etc.) as well as sub-surface geology. Geomorphological features influence the occurrence, storage and movement of groundwater and therefore the study of hydrogeomorphological parameters of this area are eminent in groundwater resources study.

METHODOLOGY

The methodology involves the study and identification of the ideal geo-hydrological conditions prevailing in the area within the regional geological setup with respect to the broad geological structures based on the satellite data, Survey of India maps at 1:50,000 scale followed by ground truth surveys and previous studies carried out in the study area (Figure 1). A total eleven springs have been considered for the spring hydrogeomorphological investigation.

Most of the springs selected for the study are unconfined in nature, and the hydrology of such types of springs is mainly influenced by rainfall and surrounding catchments area hydrogeology and geo-hydromorphology. Different hydrogeomorphic features mapping and other field mapping has also been employed for delineating and refining spring catchment area boundaries in the study area. The location of springs during field survey has verify in-situ conditions and provides additional data to further characterize the springs and their catchment area in the study area.

STUDY AREA

The study area constitutes an important part of the catchment area of the middle valley of western Ramganga River in Kumaun Lesser Himalayan terrain which forms a part of Almora district. It lies between 29° 45' N to 29° 53' N latitudes and 79° 15' E to 79° 23' E longitudes covering an area of ~97.8 Km^2. The study of satellite images topographic map, real time photographs supported by ground truth survey reveals that the study area has a network of interlinked subsurface fractures. The features of neo-tectonic activities in the form of faults and lineaments have a definite control on the alignment of rivers and streams in the area.

GEOLOGICAL SETTING

The area under investigation is a part of the Kumaun Lesser Himalaya consisting of several NW-SE trending longitudinal distinct geotectonic belt demarcated by faults and thrusts (Ghose, 1973). Out of these the metamorphic belt know as the Almora Crystalline and termed as the Almora Group (Ghose and Chakrabarti, 1972; Ghose *et al.* 1974)) occurs as a distinct and separate tectonic unit in the form of a NW-SE trending asymmetric synform and bounded by planes of tectonic discontinuity on both the sides. Folds related linear and planar structures; faults (thrusts) and joints are the major dystrophic structures observed in the Almora Crystalline tectonic unit.

The rock formations met within the area constitute two main tectonic units:

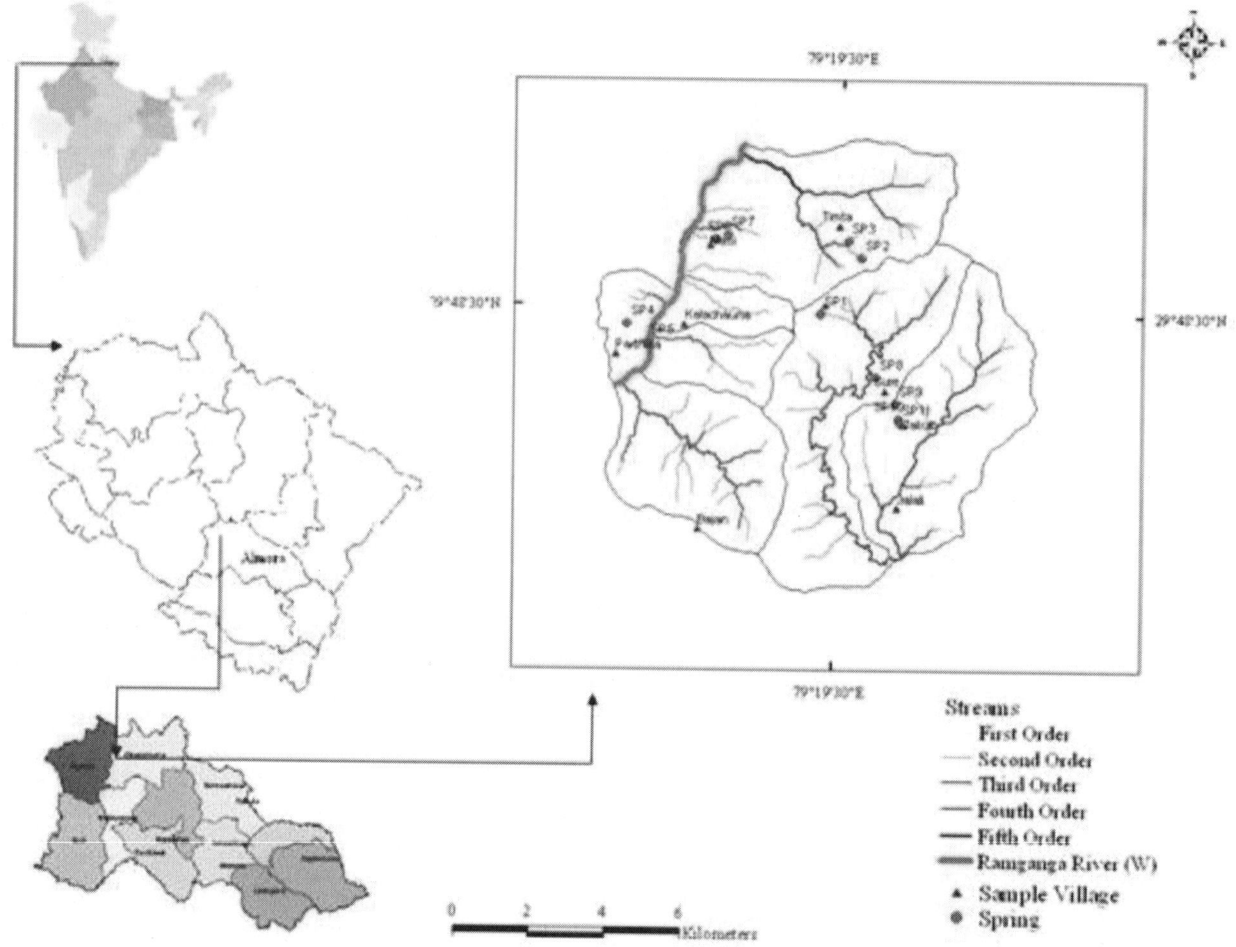

Figure 1. Reference map of the study area

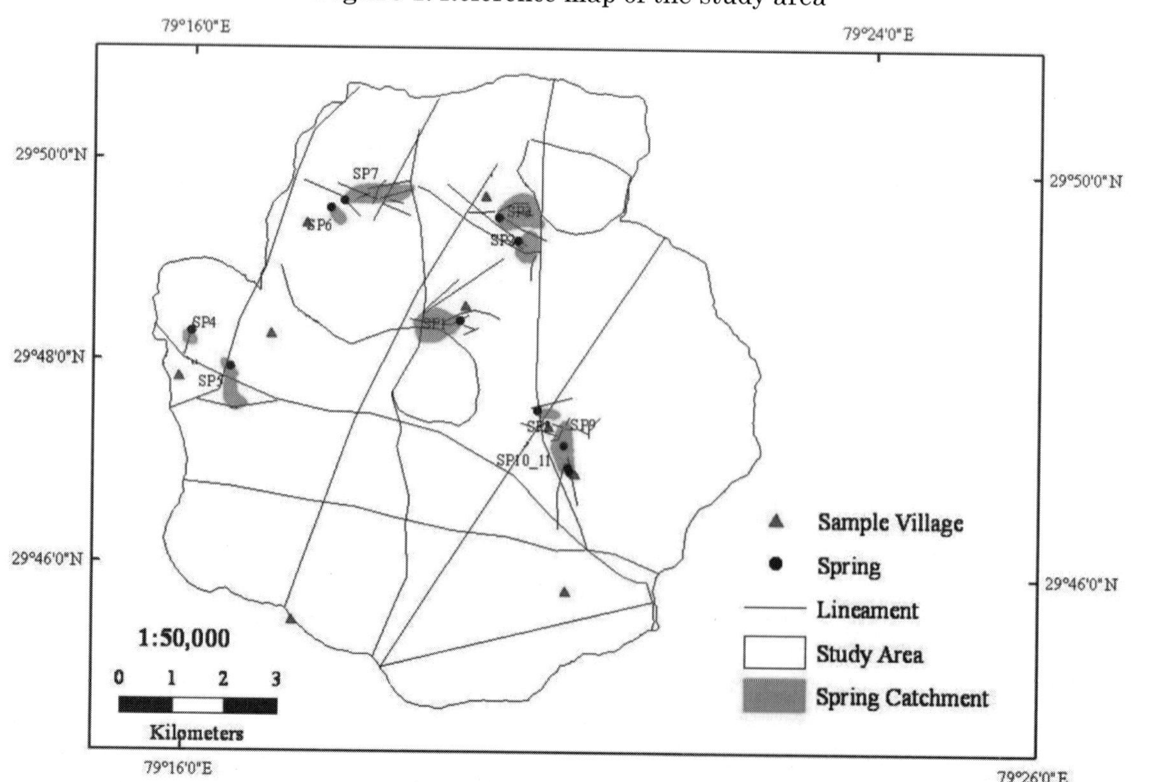

Figure 2. Major lineament and associated springs

i. Crystalline Formation (Almora–Dudatoli zone) chiefly made up of muscovite gneiss, garnet-mica schist flaggy quartzite, sericite schist and grit.

ii. Meta-Sedimentaries, comprising the Quartzite (quartzite-spilite etc.) and Calcareous (dolomite-limestone) Formations.

These are separated by the 'North Almora Thrust', which has been traced between Chaukhutia and Basarkhet, trending in NNW–SSE direction. The Crystalline Formation in the study area is designated as the 'Almora-Dudatoli Crystalline' by many scholars. The vast stretch of the Crystalline Formation, between Chukhutia and Masi form a major synclinal structure (Masi Syncline) with the axis running N50°W-S 50°E direction. It comprises mainly of muscovite gneiss, garnetiferous mica schist, flaggy and schistose quartzite, siliceous limestone with lenses of magnetite, sericite-quartz schist, chlorite schist, and schistose grit. These rocks are separated from those of the Quartzite and Calcareous Formations by the North Almora Thrust (NAT). The schistose grit with limestone quartzite and sericite-quartz and chlorite schist forms the base of the Almora-Dudatoli Crystalline in the study area. The rocks of the Calcareous and quartzite formations are highly affected by tectonic disturbances. There is no evidence of regional inversion in the area.

GROUNDWATER

The occurrence and movement of groundwater in the study area is governed by several factors such as, geology, structure, lithology, climatic conditions, topography, extent of fractures and lineaments, slope, drainage pattern, land-use/land cover and interrelationship between these factors. In hard rocks, especially crystalline rocks, topography and stage of geomorphic evolution give clues to the development of weathered layers. Attitude of rocks underlying a region provides information on the vertical and lateral extent of the aquifers and associated layers and helps in forecasting their depths at any location.

Groundwater resources in the study area occur in dynamic state and hence subjected to periodic changes. Personal communication with the local communities highlights the fact that there has been a noticeable change in the weather conditions during the last three to four decades in the area. Several landscapes of the Himalayan region are facing an acute shortage of drinking water, including the present studied landscapes of the Kumaun Lesser Himalaya. In such a situation one has to depend on alternative source, i.e., groundwater which moves out in the form of spring or seepage. Springs are the main source of groundwater for the area of study (Table 1).

Table 1. Recharge area characteristics of the springs

Spring	Elevation (m asl)	Spring type	Slope/Aspect
SP_1	1488	P/L/J	10-20° SE
SP_2	1481	P/Fr/J	10-30° WNW
SP_3	1363	P/L	10-30° NW
SP_4	958	P/Fr/J	10-20° ESW
SP_5	854	P/ACF	5-30° WNW
SP_6	993	P/RT	10-20° WNW
SP_7	1026	P/L	5-30° WSW
SP_8	1304	P/L/J	10-20° WNW
SP_9	1356	P/F	5-10° ESE
SP_{10}	1272	P/F/Fr/J	10-30° SE
SP_{11}	1246	P/F/Fr/J	30-40° SE

P-Perennial, L-Lineament, F-Fault, Fr-Fracture, J-Joint ACF-Alluvial-colluvial Fan, RT-River terrace

HYDROGEOLOGY

Geologically the study area lies in the Kumaun Lesser Himalayan region and constituted by the Crystalline Metamorphic rocks of Dudatoli Almora Group. It has been found that the groundwater conditions in the folded Almora Crystalline are controlled by structures reactivated by recurrent tectonic activities. These are responsible for secondary porosity of rocks of the recharge areas. There is a strong relationship in tectonic linears and recharge of groundwater through spring recharge, the major springs lying directly on or within a few meters of them (Table 1). Presence of several horizons of springs along valley slopes may indicate permeability stratification controlled by structure in the area. The features of neo-tectonic activities in the form of faults and lineaments have a definite control on the alignment of rivers and streams in the area. Lineaments and fractures that cut across the river and other structural features may prove to be good water potential zones, since they have more chances to divert water into them from channels they cut across. The main source of water in the area is Ramganga River. The Ramganga River with its very wide catchment area receives its flow from snow melt, rainfall and through effluent seepages. Since the slopes are high and hills are barren a major part of rainwater goes as surface runoff. Another part which percolates downward through cracks, fissures and joints in the rocks, appear on surface as springs and in the gadheras/gads (small rivulets) under favorable circumstances.

Bedrock Geology, Structure and Lithology

The area comprises quartzite and phyllite of Maithana Formation and granite, gneisses and garnetiferousmica schist and quartzite of Ranikhet Formation. These rocks possess negligible primary porosity but develop secondary porosity and permeability due to fracturing and weathering. Not only do these structures control the drainage pattern, but they also play an important role in promoting groundwater recharge and in the location of springs (Figure 2). Unconsolidated granular materials like gravels, cobbles, pebbles and sand deposits are potential aquifers and geological structures like synclinal folds, faults, joints and stratified rocks are potential zones for the movement and storage of groundwater. These types of geological composition act as a barrier to groundwater flow, which results in the origin of springs.

Folds, Faults, Fractures and Joints

The study of satellite images, topographic map, real time photographs supported by ground truth survey reveals that the study area has a network of interlinked subsurface fractures. The study area is cut by numerous faults, fractures and joints which are conspicuous in all rock formations. Antiformal folds develop intense fracturing at crests which is responsible for higher secondary porosity and permeability, and hence greater water yields. However, in the case of open synformal folding (Almora synformal fold axis passes through the centre of the study area) groundwater is more controlled by the dip of strata.

Lineament

It is well known that the fracture traces and lineaments are important in rocks where secondary permeability and porosity dominate and where intergranular characteristics combine with secondary opening influencing weathering and groundwater movement.

The most extensive and very clearly marked features are the major lineaments. The study area is criss-crossed by major and minor lineaments. Some of the major lineaments in study area are Ramganga lineament, Dwarahat lineament, South Almora Thrust and Almora-Syncline).

Some of these appear to extend laterally along their strike but being less clearly marked, has been mapped as probable major lineaments. At few places they occur parallel to each other and form en-echelon pattern. The minor lineaments are identified in structural and denudational hills, valley fill and valley wall zones, controlling the channel of streams and rivulets in the study area. Intersecting lineaments (Figure 2) are considered as good potential zones for groundwater targeting as they reflect high porosity and hydraulic conductivity of the underlying materials.

Spring Hydrology

Gentle slope with unconsolidated sediment and pine forest in upper reaches make the micro-watershed area of the springs. Sp_1 has been classified as contact springs Axis of a major syncline i.e. Almora Syncline passes through from this area. Sp_2 is classified as fracture spring as fracture zone trending SE-NW cuts across the rock sequence leading to emergence of Sp_2 along it. Sp_4 is a fracture/lineament controlled spring which appears at contacts between quartzite and phyllite.

Sp_5 spring is a perennial spring. Highly fractured phyllite with cultivated fields (River terraces) and alluvial-colluvial deposits are exposed around this (Sp_5) spring. Two springs i.e. Sp_4 and Sp_5 springs are located on the opposite sides of a valley running in the NNW-SW direction. Both the valley sides on which the springs are located are entirely made up of a dipping sequence of quartzite, and phyllite.Sp_6 and Sp_7 springs are perennial. The prominent fractures in highly weathered granitic gneiss and schist are controlling the movement of groundwater in these springs. Sp_8, Sp_9, Sp_{10}, Sp_{11} are fracture-joint controlled springs. These fracture springs are observed at places where the quartzites overlie the phyllite. Table1 indicates the number of springs selected and studied in each of the micro-watershed of the study area.

HYDROGEOMORPHOLOGY

The entire study area is manifested by hills and undulating surfaces. Four distinct geo-hydromorphologic units have been identified in the study area. These distinct geomorphic features are resulted from the complexity of geomorphic evolution. The distribution and extent of the geomorphic zones are varying in different geologic, lithologic and topographic, conditions and having a pronounced effect on groundwater occurrence in the study area. Various types of erosional and depositional landforms have been found in this area. The commonly developed landforms in crystalline rock terrains are structural hills, erosional valleys and valley fills. Numbers of colluvial deposits were reported at the higher reaches and at the lower elevation valley fill deposits are present. Colluvial-alluvial deposits, river terraces, flood plains and other alluvial features are good indicators of groundwater potential whereas the structural hills, rock cliffs and steeply sloping surfaces are generally poor in groundwater resources. In the spring cathment areas various types of erosional and depositional landforms have been investigated. They promote high overland flow and consequently have low infiltration of rainwater, particularly where vegetal cover is lacking or disturbed by various activities. In the upper reaches of micro-watersheds slope is very steep and soil cover is also thin. Some colluvial debris deposits have reported in the upper micro-basin area. In the low lying areas, various levels of Ramganga river terraces are present along the channel of the river. These can be terraced on both the sides of Ramganga River from Chaukhutia upto Naugaon.

Thus, the hydro-geomorphic study of the area reveals close relationship between groundwater condition and geomorphology of the area.

RESULT AND DISSCUSSION

Structures of rocks have a strong bearing on the movement and storage of water in the area. Because metamorphic rocks are exceedingly dense, they have virtually no primary porosity or permeability. The secondary permeability due to faults and fractures constitute the principal mechanism for groundwater flow in metamorphic rocks. In crystalline rocks, fracture zones and fracture intersections are generally the sites where groundwater occurs. It has been noted that the fracture systems are generally hydraulically connected with the overlying weathered saturated residium. The yield potential of the crystalline and metasedimentary rocks shows wide variations. In most of the granite/gneiss area, the weathered residium serves as an effective groundwater repository. The most obvious structural features that are important from the groundwater point of view are the lineaments. Lineaments and fractures that cut across the river and other structural features may prove to be good water potential zones, since they have more chances to divert water into them from channels they cut across. It has been found that the springs in the folded Almora Crystalline are controlled by structures reactivated by recurrent tectonic activities. These are responsible for secondary porosity of rocks of the recharge areas.

The rocks are highly folded, faulted, fractured and jointed throughout the catchments of the springs. Water is coming through the joints, of the detached rock mass body. The rocks are intruded by quartz-veins at various places. In some parts interbedded granite and gneisses of rocks has also been found. The quartzite is also showing the gritty nature. This sequence belongs to an Argillo-Arenaceous Facies and referred to as Ranikhet Formation (Sharma and Sinha, 1972). In the quartzite along some joints shear zones are also reported. Owing to the multiplicity of joints and sheared nature of the quartzite rocks, the rainwater percolates down and recharges the groundwater or perched water table of the catchment. These rocks have been very good aquifer for the water retention in the area. The quartzite is medium to fine grained and at some localities interbedded with thin bands of phyllites.

The hydro-geomorphic units in the area reveal close relationship between groundwater condition and geomorphology of the area. Upslope areas in general and gentle slopes in particular, are conducive to develop good hydraulic gradient, adequate recharge and quicker movement of groundwater. Colluvial and alluvial deposits retain and transmit significant amount of groundwater. The toe of talus cones, alluvial fan, middle to lower part of landslide, debris fan, cross section of various lineaments, fault, contact between pervious and impervious bed rocks, weathered zones, etc., are good location for the seepage zone and subsequently for groundwater occurrence.

CONCLUSIONS

Groundwater resources in the study area occur in dynamic state and hence subjected to periodic changes. In this sector of Kumaun Lesser Himalayan Mountains, the steep and irregular topography with its great diversity in hydrometeorology, geology, land-use and forest cover create a variety of groundwater recharge conditions in the form of spring recharge areas in which rock formation, altitude, slope, aspect, water contributing area, vegetation, biotic interference, etc. all vary. However, despite some excellent geological work, there is limited work on the hydrogeology of springs and other crucial groundwater resources of the area.

It is concluded that in this part of the Kumaun Himalaya groundwater recharge conditions are primarily controlled by geological structures, secondly land-use pattern and thirdly by the extent of vegetal cover. Human activities in the recharge area also bring about some quantitative and qualitative changes in the groundwater regime. Geomorphological features influence the occurrence, storage and movement of groundwater and therefore, the study of hydro-geomorphological parameters of an area are eminent in groundwater resources.

ACKNOWLEDGMENTS

The author gratefully acknowledges her Research Supervisor Prof. Milap Chand Sharma (CSRD, JNU, New Delhi), for the constant guidance, encouragement, advice and valuable inputs to improve this manuscript. Author would also like to acknowledge the help and support received from Institute of Himalayan Environmental Research and Education (Masi, Almora), during field visits. Finally the co-operation of the villagers is acknowledged without which this work would not be completed.

REFERENCES

Alford, D. (1985) 'Mountain hydrologic system'. Mountain research and development, Vol. 5, No 4, pp 349-363.

Ghose, A.; Chakraborti, H. (1972) 'Report on the metamorphism and stratigraphy of the area around Chaukhutia, Dwarahat, Someshwar, Khairna and Champawat, district Almora and Nainital U.P'. Unpub. Report Geol. Surv. Ind.

Ghose, A. (1973) 'Tectonic evolution of Lesser Himalayas of Kumaon U.P. Seminar on Geodynamics Himalayan Region, NGRI Hyderabad.

Ghosh, A.; Chakrabarty, B. and Singh, R.K. (1974) 'Structural and metamorphic history of the Almora Group, Kumaon Himalaya, Uttar Pradesh'. Him. Geol., Vol.4, pp 171-194.

Klemes, E. (1990) 'Forward'. L.Molnar (ed) 'Hydrology of mountainous areas'. IAHS Publication. No 190.

Rodda, J.C. (1994) 'Mountains-a hydrological paradox or paradise?. Hydrology Kleiner Einzugsgebiete, gedenkschrift Hans, M. Keler, Beitragezur hydrologie der schweiz, No. 35, pp.41-51.

Sharma, R.P. and Sinha, A.K. (1972) 'Stratigraphy and structure of Ranikhet area Kumaun Himalaya' Proc. Ind. Sci. Cong. 59th session part III, 176.

Hon'ble reviewer

1. **Prof. U.C. Singh**
   School of Earth Science
   Jiwaji University
   GWALIOR - 474 001
2. **Dr. Prakhar Kumar**
   Scientist H
   Atomic Mineral Directorate
   NAGPUR - 440 001

III-40

A Study of Ground Water and its Fluctuation in Raisar, Bikaner District, Rajasthan

A.K. Shandilya*, Sushma Jain** and Monika Agrawal**

*Department of Geology, **Department of Chemistry, Government Dungar College, Bikaner (Rajasthan)
*E-mail: arunshandilya@gmail.com

Abstract

Water is most essential commodity for human consumption and is one of the most important renewable resources, which must be prevented from deterioration in quality and quantity. In this study an attempt has been made to determine ground water fluctuation by the help of periodic ground water fluctuation maps and long term groundwater fluctuation map. The selected project area is 24 villages lie in Bikaner district which present in the arid zone of western Rajasthan .Hydro geological studies of the area were carried out to evaluate the water bearing properties present in the study area.

Keywords: Groundwater Management, Water Table Fluctuation, Raisar in Bikaner, Arid Zone

Introduction

Ground water is one of earth's most vital renewable and widely distributed resources as well as an important source of water supply throughout the world. The quality of water is a vital concern for mankind since it is directly linked with human welfare. In India, most of the population is dependent on groundwater as the only source of drinking water supply [1].Ground water is a valuable natural resource that is essential for human health, socio-economic development, and functioning of ecosystems [2]. In India severe water scarcity is becoming common in several parts of the country, especially in arid and semi-arid regions. Ground water is the major drinking water source of western Rajasthan, which is facing both the problems of quality and quantity of potable ground water.

The Bikaner district lies in arid zone of western Rajasthan. It covers an area of about 27,244 sq km and lies between North Latitudes 27^{0}11' to 29^{0}03' and East Longitudes 71^{0}54' to 74^{0}12'. The present research work covers an area around Village Raisar and falls between North Latitude 27^{0}50' to 28^{0}05' and East Latitude 73^{0}30' to 73^{0}45'[3]. The important villages located in the study area are Raisar, Naurangdesar, Bamblu, Gersar, Pemasar, Udasar, Bhinasar, Himtasar, Shivbari, and Ridmalsar etc. Here ground water is the main source of drinking water for local population because there is no seasonal and no perennial stream in this area. The quality of ground water resources of these villages vary naturally and widely depending on climate, season, geology of bedrock as well as anthropogenic activities[4]. The area is characterized by very low rainfall, high evapo-transpiration, lake of vegetation, high diurnal variation and prolonged summers and high velocity hot winds during summer and cool breeze in winters.[5]

The study area is the part of the Thar Desert where the water is the most precious thing. The desert is suffering from acute water needs of life in this desert are mainly fulfilled by the surface water sources (ponds and tanks – artificial R.C.C. tanks) and the ground water. The surface water sources of the region supplies water only for three to six month after rainy season. So that ground water is only source in this region, which is usually deep and its quality is variable. The occurrence, movement and origin of ground water depends upon various factors viz. geomorphic features, rainfall, climate, hydro geological characteristic and structure of the geological formations present in the area.

Hydrogeology

Hydro geological studies of the area were carried out to evaluate the water bearing properties present in the study area. There are total 24 key wells, which have been taken into consideration for well inventory.The selection of key wells depends on their good conditions, proper maintenance and easy accessibility. The key well have been selected are monitored and proper maintained by Public Health Engineering Department (PHED) and Groundwater Department (GWD), of Rajasthan and Some other private organizations (including farmers etc.). These wells have been plotted on Survey of India Topo sheet of 1:2,50,000 scales to prepare the base map. These wells and the base map have been used to prepare various kinds of hydro geological maps. The ground water data of Pre-monsoon (May-June) and post monsoon (Oct.-Nov.) of years 2010, 2011 and 2012 have been measured and recorded. While collecting ground water data some necessary help has been taken from Ground Water Department and PHED (Rajasthan). The well inventory data (depth to water table) of pre-monsoon and post-monsoon 2010, 2011 and 2012 is shown in Table No.1. The periodical fluctuations between years 2010-2011, 2011-2012 (short term) and 2010-2012 (long term) are shown in the Table No. 2, 4 & 6 respectively.

Ground Water Fluctuation Maps

The ground water fluctuation maps are prepared by calculating the difference in the water levels of two different periods for all the Key Wells of the study area. The difference is plotted on the locations of Key Wells and maps have been prepared by employing linear interpolation method.

The variation in the water table fluctuation depends upon the climatic factors, drainage, topography, geological conditions and anthropogenic (man induced) causes like presence of canal, application of irrigation, artificial recharge and ground water withdrawal etc. A decline in water table represents ground water abstraction in excess of increment showing negative fluctuation whereas ground water rise represents ground water increment in excess of ground water abstraction showing positive fluctuation.

The ground water level fluctuations can be studied mainly in two ways –

1. **Seasonal fluctuation** – This can be studied between two seasons.
2. **Periodic fluctuation** – This can be studied between a certain periods, which may be one, two or more years.

Use of fluctuation map

1. Delineating the area of positive and negative fluctuation.
2. In determining the areas of future recharge and extraction.
3. The overall withdrawal or recharge certain period can be determined.

Periodic fluctuation of study area is studied in two ways as follow–

1. Short Term (One year) 2010-2011, 2011-2012
2. Long Term (two year) 2010-2012

A. Short Term Fluctuation

The Short Term Fluctuation maps have been prepared for one year period on the basis of ground water data of pre monsoon period. The contour interval taken to prepare these maps is 1m. There are only two maps prepared; (1) 2010-

2011 and (2) 2011-2012.These maps are described below-

1. Periodic Fluctuation between Years 2010-2011

The Short Term Fluctuation map for 2010-2011 (Fig. 1) has been prepared by using ground water data presented in Table 1. The data (Table 1) and map (Fig. 1) shows a mix picture of positive and negative fluctuation and data shows that out of total 24 Key Wells, 15 Key Wells show negative fluctuation, whereas 9 Key Wells show positive fluctuation. The Table 2 is prepared on the basis of Table 1 shows that the maximum positive fluctuation +3.56m observed at village Pemasar (Key Well No. 7), and maximum negative fluctuation value is -5.26 m found at village Gadwala (Key Well No. 18).

Table 1: Periodic Water Level Fluctuation in 2010-2011

Key Well No.	Name of Village	Depth to Water 2010 (m)	Depth to Water - 2011 (m)	Fluctuation
1	Ranisar	71.32	70.62	0.7
2	Naurangdesar	86.67	85.3	1.37
3	Bamblu	68.19	68.32	-0.13
4	Gersar	59.94	60.4	-0.46
5	Nagasar	62.1	60.86	1.24
6	Bichwal	72.37	71.72	0.65
7	Pemasar	72.4	68.84	3.56
8	Udasar	82.99	82.65	0.34
9	Raisar	73.9	74.5	-0.6
10	Himtasar	72.5	73.56	-1.06
11	Ridmalsar	70.63	71.53	-0.9
12	Sheobari	54.68	55.36	-0.68
13	Bhinasar	65.7	64.52	1.18
14	Udramsar	71.53	72.9	-1.37
15	Ambasar	85.61	86.47	-0.86
16	Surdhana	73.7	74.26	-0.56
17	Kilchu	95.11	96.5	-1.39
18	Gadwala	93.3	98.56	-5.26
19	Napasar	74.96	75.2	-0.24
20	Sinthal	73.2	72.4	0.8
21	Mundsar	84.76	85.9	-1.14
22	Belasar	86.67	83.58	3.09
23	Tejrasar	91.1	91.21	-0.11
24	Gusainsar	93.65	96.26	-2.61
Average of Fluctuation				-0.19

Table 2: Short term periodic fluctuation of year 2010-2011

S. No.	KeyWell No.	Name of Village	Fluctuation	Average Water Level Fluctuation
1.	7	Pemasar	Max +ve 3.56 m	-0.19 m
2.	18	Gadwala	Max –ve 5.26 m	

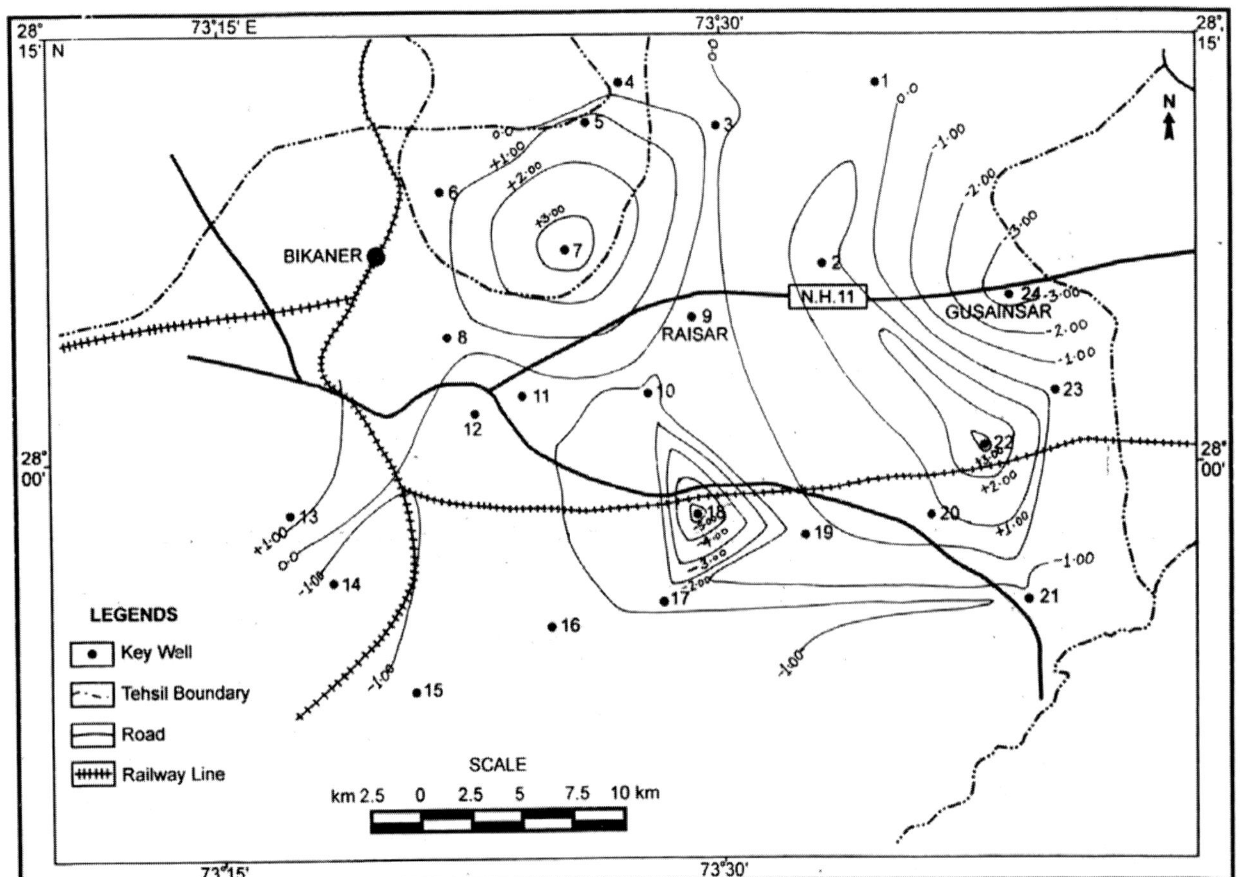

Fig. 1: Periodic Fluctuation Map of Pre Monsoon 2010-2011

Table 3: Periodic Water Level Fluctuation in 2011-2012

Key Well No.	Name of Village	Depth to Water 2010 (m)	Depth to Water 2011 (m)	Fluctuation
1	Ranisar	70.62	72.22	-1.6
2	Naurangdesar	85.3	87.36	-2.06
3	Bamblu	68.32	68.34	-0.02
4	Gersar	60.4	59.97	0.43
5	Nagasar	60.86	59.5	1.36
6	Bichwal	71.72	72.65	-0.93
7	Pemasar	68.84	70.83	-1.99
8	Udasar	82.65	82.62	0.03
9	Raisar	74.5	74.95	-0.45
10	Himtasar	73.56	72.73	0.83
11	Ridmalsar	71.53	71.96	-0.43
12	Sheobari	55.36	55.98	-0.62
13	Bhinasar	64.52	64.74	-0.22
14	Udramsar	72.0	73.05	-0.15
15	Ambasar	86.47	89.36	-2.89
16	Surdhana	74.26	75.55	-1.29
17	Kilchu	96.5	101.15	-4.65
18	Gadwala	98.56	97.32	1.24

19	Napasar	75.2	75.3	-0.1
20	Sinthal	72.4	71.98	0.42
21	Mundsar	85.9	86.45	-0.55
22	Belasar	83.58	82.48	1.1
23	Tejrasar	91.21	92.78	-1.57
24	Gusainsar	96.26	99.54	-3.28
Average of Fluctuation				-0.72

Table 4: Short term periodic fluctuation of year 2011-2012

S. No.	Well No.	Name of Village	Fluctuation	Average Water Level Fluctuation
1.	7	Nagasar	Max +ve 1.36 m	-0.73 m
2.	17	Kilchu	Max –ve 4.65 m	

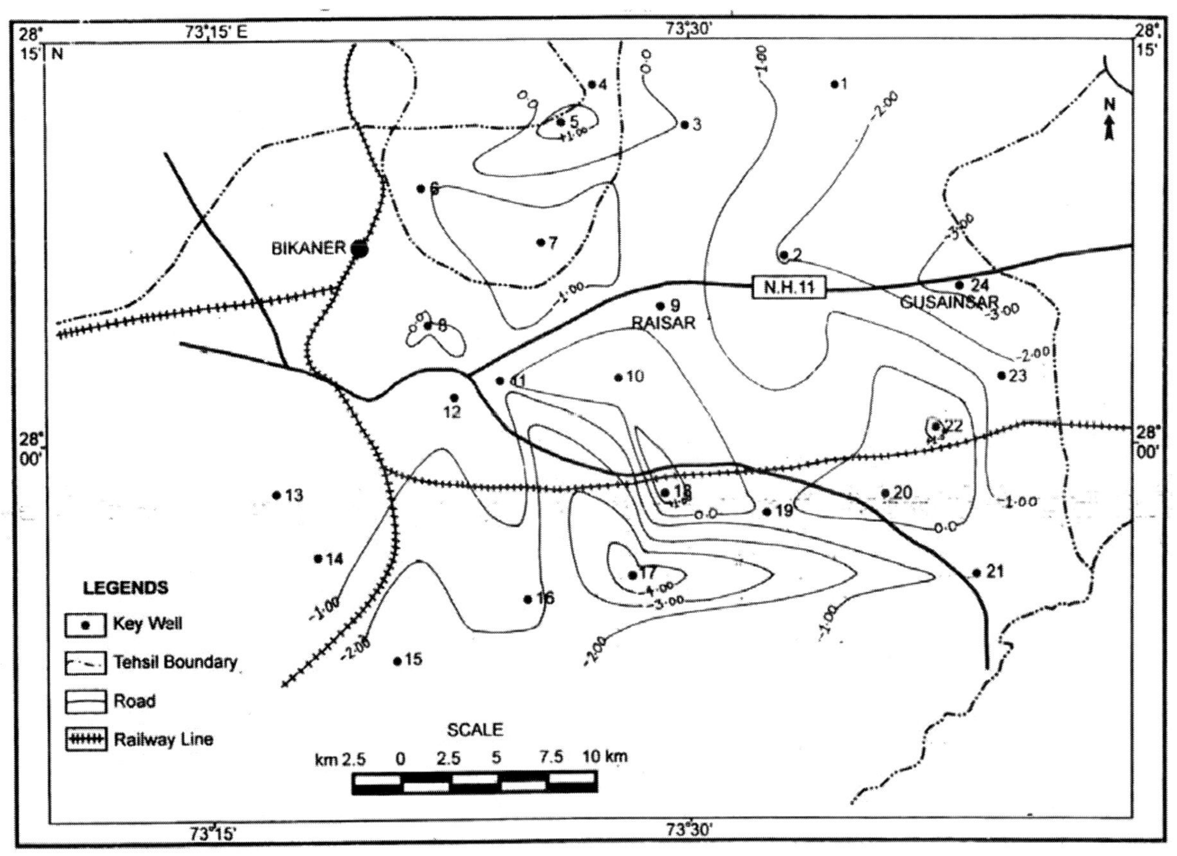

Fig. 2: Periodic Fluctuation Map of Pre Monsoon 2011-2012

The map (Fig. 1) shows the areas with positive fluctuation exist mainly at two patches. The first and the largest patch is located in the north western part which surround the villages Pemasar (Key Well No. 7), Nagasar (Key Well No.5) & Udasar (Key Well No.8). The second patch is located in the eastern part around villages Belasar (Key Well No. 22) and Naurangdesar (Key Well No. 2). The maximum positive fluctuation shows around villages Pemasar (Key Well No.7) and Belasar (Key Well No. 22), which is represented by a highest contour of 3m. The remaining area show negative fluctuation. The negative fluctuation present in southern part of the study area in around the villages Gadwala (Key Well No. 18) , Kilchu (Key Well No. 17) and Himatsar(Key Well No. 10), here the highest negative contour of -5m present around the village Gadwala (Key Well No. 18). The map (Fig. 1) shows that eastern part also have negative fluctuation around villages Gusainsar (Key Well No. 24) and Tejrasar (Key Well No.23), here the lowest contour of -3 m is presented around village Gusainsar (Key Well No.24). The average water level in this period in the study area is calculated -0.19 m shown in Table 2.

2. Periodic Fluctuation between Years 2011-2012

The periodic (yearly) Ground Water Fluctuation Map (Fig 2) for 2011-2012 is prepared on the basis of pre monsoon data of 2011 and 2012 shown in Table No.3. The table shows that out of total 24 Key Wells, only 7 Key Wells show positive fluctuation, while remaining Key Wells show negative fluctuation. The study of map reveals that majority of area are showing negative fluctuation. The Table No.4 is prepared on the basis of Table No. 3 shows that the maximum positive fluctuation +1.36m observed at village Nagasar (Key Well No.5) and the maximum negative fluctuation value is -4.65 m found at village Kilchu (Key Well No. 17).

This fluctuation map (Fig 2) shows a substantial increase in negative fluctuation area. There are only four small patches in the map, where the positive fluctuations are shown. First one is situated in the northern part of the study area around village Nagasar (Key Well No. 5) and village Gersar (Key Well No. 4).The second patch of positive fluctuation is situated in the central part around village Gadwala (Key Well No. 18), Ridmalsar (Key Well No. 11) and Himatsar (Key Well No. 10). The third patch of positive fluctuation is situated in the south eastern part around Belasar (Key Well No. 22) and Sinthal (Key Well No. 20). A small patch is also shown around village Udaser (Key Well No. 8). In this map remaining area shows negative fluctuation. The map shows the highest positive contour of +1m and highest negative contour of -4m. The average water level fluctuation calculated for 2011-2012 is -0.73 m shown in Table 4.

B. Long Term Fluctuation Map 2010 to 2012

The Long Term Ground water Fluctuation Map for period between 2010 to 2012 (Fig. No.3) is prepared on the basis of pre monsoon data of 2011 and 2012 shown in Table No.5. In comparison to Short Term Fluctuation Maps, the Long Term Fluctuation Maps gives a better picture of fluctuation with regard to the planning and management of ground water aquifers. The Table No.4.13 shows that out of total 24 key well, 18 key well shows negative fluctuation.

The Table No. 6 is prepared on the basis of the Table No. 5 shows that the maximum positive fluctuation 4.19 m is at village Belasar (Key Well No. 22) and the maximum negative fluctuation -6.04 is at village Kilchu (Key Well No. 17). The map (Fig No.3) of 2010-2012 shows that there is a mix picture of both positive and negative fluctuation. This map shows that the positive fluctuation in the northern part around village Nagasar (Key Well No. 5) and Pemasar (Key Well No. 7) which elongated up to east central part area around village Belasar (Key Well No.22) and Sinthal (Key Well No. 20). The area show more positive fluctuation in the east central part around village Belasar, which have a highest contour of +4m. The southern and eastern part of the area shows negative fluctuation. The maximum negative fluctuation in the southern part around village Kilchu (Key Well No. 17) which is represented by a lower contour of -6m. The area with negative fluctuation indicates more abstraction than the recharge. These areas has slightly better quality of the water hence its exploitation is more to fulfill the domestic and agriculture needs of water.

The average water level fluctuation calculated for 2010-2012 is -0.91 m shown in Table No.6.

Result and Conclusion

The study of fluctuation maps show that in periodic fluctuation map of year 2010-2011 the maximum positive fluctuation of +3.56m is observed in village Pemasar (Key Well No. 7) and maximum negative fluctuation of -5.26m is observed in village Gadwala (Key Well No. 18). The average fluctuation of this year is -0.19m.

The study of periodic fluctuation map of year 2011-2012 the maximum positive fluctuation of +1.36m is observed in village Nagasar (Key Well No. 5) and maximum negative fluctuation of -4.65m is observed in village Kilchu (Key Well No. 17). The average fluctuation of this year is -0.73m.

The long term fluctuation map of year 2010-2012 shows that the maximum positive fluctuation of +4.19m is shown in village Belasar (Key Well No. 22) and maximum negative fluctuation of -6.04m is observed in village Kilchu (Key Well No. 17). The average fluctuation of this year is -0.91m.

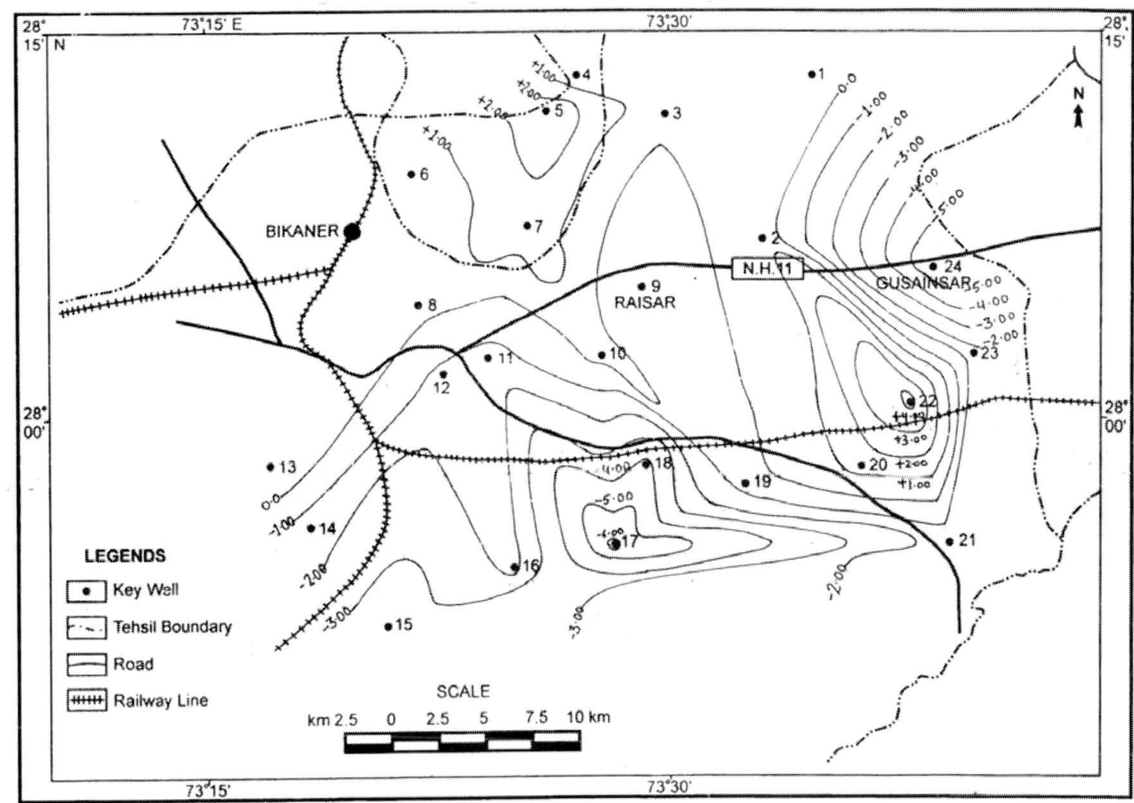

Fig. 3: Long Term Fluctuation Map of Pre Monsoon 2010-2012

Table 5: Long-Term Water Level Fluctuation in 2010-2012

Key Well No.	Name of Village	Depth to Water 2010(m)	Depth to Water 2011(m)	Fluctuation
1	Ranisar	71.32	72.22	-0.9
2	Naurangdesar	86.67	87.36	-0.69
3	Bamblu	68.19	68.34	-0.15
4	Gersar	59.94	59.97	-0.03
5	Nagasar	62.1	59.5	2.6
6	Bichwal	72.37	72.65	-0.28
7	Pemasar	72.4	70.83	1.57
8	Udasar	82.99	82.62	0.37
9	Raisar	73.9	74.95	-1.05
10	Himtasar	72.5	72.73	-0.23
11	Ridmalsar	70.63	71.96	-1.33
12	Sheobari	54.68	55.98	-1.3
13	Bhinasar	65.7	64.74	0.96
14	Udramsar	71.53	73.05	-1.52
15	Ambasar	85.61	89.36	-3.75
16	Surdhana	73.7	75.55	-1.85
17	Kilchu	95.11	101.15	-6.04
18	Gadwala	93.3	97.32	-4.02
19	Napasar	74.96	75.3	-0.34
20	Sinthal	73.2	71.98	1.22

21	Mundsar	84.76	86.45	-1.69
22	Belasar	86.67	82.48	4.19
23	Tejrasar	91.1	92.78	-1.68
24	Gusainsar	93.65	99.54	-5.89
Average of Fluctuation				-0.91

Table 6: Long term periodic fluctuation map 2010-2012

S.No.	Well No.	Name of Village	Fluctuation	Average Water Level Fluctuation
1.	22	Belasar	Max +ve 4.19 m	-0.91 m
2.	17	Kilchu	Max –ve 6.04 m	

The observation data shows that water level of this area is continuously decreases as the average value of negative fluctuation is increases from year 2010 to 2012. So it is necessary to continuous monitoring of ground water level along with quality study to minimize the chances of further deterioration of ground water.

Reference

1. Mahmood A, Kundu A (2005). India's demography in 2050: Size, structure and habitat. Discussion Paper, IWMI-TATA partners meet - 2005, Anand, Gujarat.
2. Humphreys WF (2009). Hydrogeology and groundwater ecology: Does each inform the other? Hydrogeol. J., 17(1): 5–21.
3. Survey of India-Catalog.'.
4.. Shandilya A. K. and Sharma S., Hydro- Meterological Analysis of Bikaner for Climatic Classification, Oikoassay, Vol. 17, No. 1 & 2 , pp 41 – 44.(2004)
5. Shandilya, A.K. Dubey, P. and Parihar, P. (2007) Groundwater Fluctuation around Nokha (Bikaner, Rajasthan). Oikoassay, Vol. 20, No. 1 & 2, pp. 23 - 28.

Hon'ble reviewer

1. **Prof. S.S. Thigale**
   Director, Symbiosis
   28 Sai Dhan
   Ashok Nagar
   PUNE - 411 007
2. **Prof. L. Elango**
   Department of Geology
   Anna University
   CHENNAI - 600 025

III-41

Water Sustainability in Urban Metropolis: A Study from Lucknow City, Uttar Pradesh, India

Vibhuti Rai and Purnima Sharma

Centre of Advance Study in Geology, University of Lucknow, Lucknow - 226 006
E-mail: vibhutirai@gmail.com

Abstract

In developing countries like India the urban centres are growing not only in size but also in its populations. One of the most important natural resources for such urban centres is the water, which has reached at a critical stage due to ever widening gap between the requirement and the availability. In order to minimize this gap, some conventional and non-conventional techniques are to be employed so that in the coming years at least minimal water requirement can be made available to the masses.

The present study has been conducted in Lucknow city on account of its untamed growth both in its size and population. Geographical Information System (GIS) based information on the sub-surface water-bodies, surface depressions and elevation modelling using satellite data and water table fluctuation data has been used to develop a model that can help in achieving the goal of water availability to all in the present time as well as in future. This model has been discussed in the light of data generated over the last several years.

Keywords: Water Resources, Geographical Information System (GIS), Water Management, Water Table Fluctuation.

Introduction

The water resources of Lucknow city has been exploited due to rapid urbanization and overpopulation. City dwellers have encroached the wetlands and water-bodies for residential and commercial purposes. These water-bodies and wetland are the natural resource of fresh water and also act as a recharging zone in rainfall. The encroachment of these water resources has caused depletion of aquifer system and waterlogging in the low-lying areas of the city during monsoon.

This paper attempts to highlight some management methodologies for conserving surface and sub-surface water resources of the city. With the help of Survey of India toposheets, Satellite data and field data a conceptual model is prepared which focus on sustainable use of surface and sub-surface water resources of Lucknow city.

STUDY AREA

Lucknow, the capital of Uttar Pradesh is one of the most popular cities of India, with many distinctive characteristics. The city is famous for its cultural heritage, gardens, buildings and monuments. Lucknow is also known as *city of palaces*, *city of culture* and *city of gardens.*

Lucknow is situated in between 26° 52' North and 80° 56' East on the banks of river Gomati that divides the city into two unequal halves. The southern half being larger and densely populated it is called as *Cis- Gomati Area* and the northern part is called as *Trans- Gomati Area.* It is located in the central Gangetic plain and forms a part of Sai-Gomati sub-basin bound by Sitapur, Raibareli, Barabanki and Unnao on North, East and West, respectively.

Administratively, the Lucknow district has been divided into 4 tehsils namely Lucknow with an area of 616.13 sq. km, Malihabad with an area of 471.84 sq. km, Bakshi-Ka-Talab with an area of 350.42 sq. km and Mohanlalganj with an area of 644.07 sq. km and 8 Community Development blocks. The district comprises a total rural area of 2168.50 sq. km. and 359.50 sq. km urban area, which is increasing rapidly.

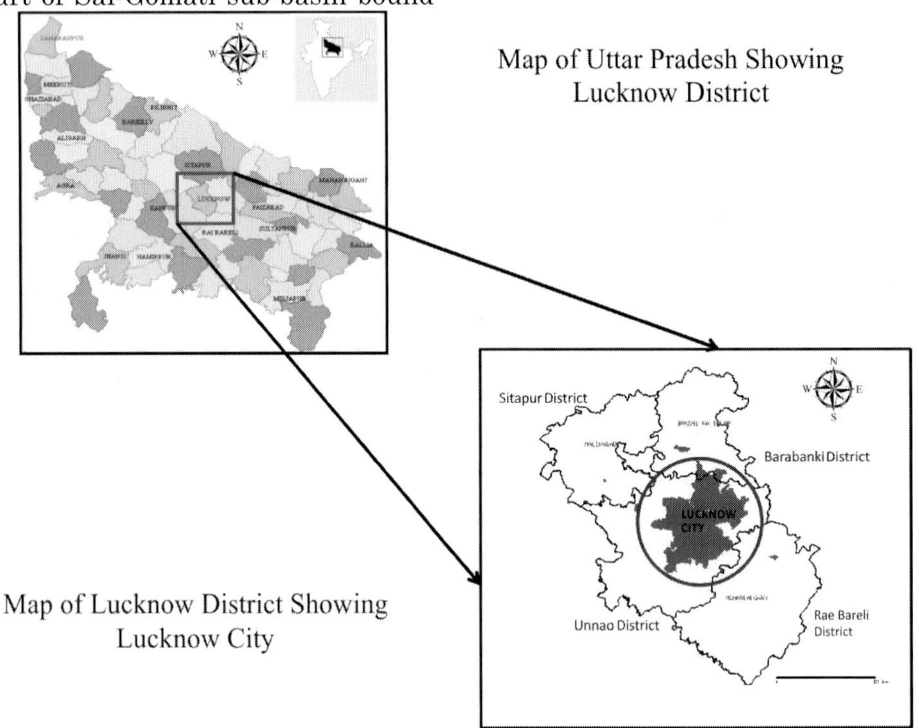

Fig . 1. Location Map of Study Area

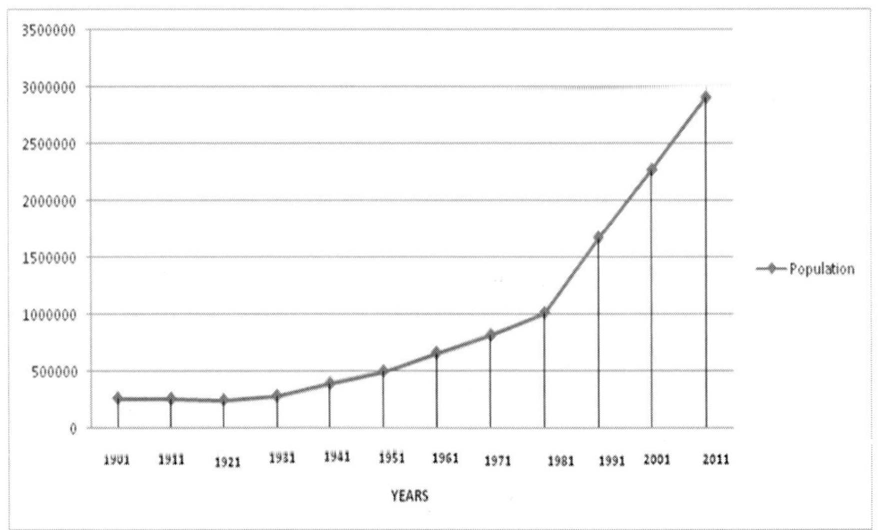

Fig .2. Graph Shows the trend of Population growth of Lucknow Metropolitan/Urban Agglomeration from 1901 to 2011 (Census of India 2001 and 2011).

PATTERN IN POPULATION GROWTH

As per data released by Govt. of India for Census 2011, Lucknow is an Urban Agglomeration coming under category of Million plus Urban Agglomeration/Metropolitan City. Lucknow city is governed by Municipal Corporation and is situated in Lucknow Urban Region.

The total population of Lucknow Urban Agglomeration (UA) /Metropolitan region is 2,901,474. The male population of which is 1,518,951 while female population is 1,382,523. The population of Lucknow city in 2011 is 2,815,601; of which male and female are 1,470,133 and 1,345,468 respectively.

The density of population of Lucknow (UA) has increased 35.8 per cent during the last decade. It shows an exponential growth pattern of population of Lucknow Metropolitan areas (Table 1 and Fig. 2).

Table 1. Population of Lucknow Metropolitan/ Urban Agglomeration from 1901 to 2011 (Census of India 2001 and 2011)

S.No	Years	Population of Lucknow UA/Metropolitan City
1	1901	256239
2	1911	252114
3	1921	240566
4	1931	274659
5	1941	387177
6	1951	496177
7	1961	655673
8	1971	813982
9	1981	1007604
10	1991	1669204
11	2001	2266933
12	2011	2901474

GROUND WATER CONDITION

Lucknow city is extended across both banks of the river Gomati, around 85% of the land area is situated on the Central Ganga alluvial plain. The area is enclosed by Quaternary alluvium. The clay particles, occasional kankar, and sand of various grades in different proportions are found in soil. The main aquifers of ground water in the city are found in the sandy horizons.

The area can be broadly divided as follows into two geologic units:

Younger alluvial plain- it lies all along the river Gomati and forms a wide flood plain.

Older alluvial plain- it is a higher level than younger alluvial plain along with uplands and as natural levies, paleo-channels and meander scars

The city, resting on a rich alluvial aquifer system of Central Ganga plain, is a glaring example of 'Hydrogeological Stress', where pressure on ground water for drinking water supplies has increased manifold and this has resulted into its continuous & excessive withdrawals. As such the aquifers from the dynamic zone have alarmingly depleted, leading to significant lowering of water table in different parts of the city.

Since last 2 decades, the city is witnessing widespread exploitation of ground water and such abstractions have reached to a point where the geoscientists of Central Ground Water Board have projected possible threat of land-subsidence in some prominent localities of the city in next 10-20 years (State water resource agency, Jal Sandesh, 2010).

In Lucknow city, as per study of Central Ground Water Board, 3 Aquifer Groups, each separated by 5 to 10 meter thick clays, are distinguishable (Table 2).

Table 2. Table showing three groups of aquifer in Lucknow city (CGWB Report, 2008)

S.N.	Aquifer Group	Depth range (meter below ground level, mbgl)
1.	First Aquifer Group	Water level to 67/155
2.	Second Aquifer Group	67/155 to 165/327
3.	Third Aquifer Group	165/327 to 485/542

The top Aquifer Group (< 150 mbgl) of the city is presently under High Stress. The granular zone of this dynamic aquifer is gradually drying up, causing almost irreversible damage to ground water domain of this capital city of the state (Fig 3).

The trend of ground water exploitation in Lucknow shows continuous rise in resource withdrawals. Construction of drinking water tube wells to meet water demand of this growing urban agglomerate had started in early 70's and by 1985 about 70 tube wells were operating. Now, this number of tube wells, under the control of Lucknow Jal Sansthan, has gone up as 470 or even more (almost 7 times increase).Apart from this, as the citizen's tendency to have an alternative & secured water supply, private tube well construction activity in the city has almost mushroomed, causing intensive & unregulated extractions which is yet to be assessed. Although, there have been no systematic ground water studies, but as per preliminary estimates, the gross ground water withdrawal in the city (extending over an area of 285 sq. km.) from both the municipal & private drinking water systems as well as from other sectors is very high and at present, this withdrawal may be tentatively taken as 550 million litres per day (MLD) (State water resource agency, Jal Sandesh, 2010).

SURFACE GEOMORPHOLOGY

Lucknow city is located in the Central Alluvial Plain of the Ganga Plain, in the region of the Ghaghara-Ganga interfluves (Khanna, 1992). In this area, the Gomati River exhibits meandering characteristics, which are highly distorted indicating Neo-tectonic activity in this region. The general slope of the city area is from North and Northwest to South and Southeast. The elevation of the city area varies from 103 to 130 m above mean sea level. This area exhibits three different geomorphic surfaces with their characteristic features, namely:

(i) Interfluve Plain or Upland Area (T2)

(i) Older Flood Plain (T1)

(iii) Younger Flood Plain (T0)

Interfluve Plain/ Upland Area (T2)

The Upland interfluve surface is also called as Older Alluvium or Bangar (Mukherji, 1963). It occurs between major drainage ways. It makes the high surface showing a variety of alluvial features, namely bhur ridges, abandoned channels, lakes, ponds and patches of alkaline soils. This surface exhibits kilometer scale undulations making high grounds and low grounds. It resulted in regions (basins) of centripetal drainages of several square kilometer areas. Within each of such basins a number of small and large ponds and lakes are present. There is no outflow from such basins. The ponds and lakes are usually associated with raised natural levees and alluvial ridges, which are often ancient sites of human settlements. These water bodies were formed in early to middle Holocene and provide good record of palaeo-climate and palaeo-vegetation. The sediments are sands and clays admixed with kankar in varying proportions. The soils are clayey to coarse loamy in nature. The occurrence of alkaline/Usar soils characterize this zone as poor drainage and poor vertical permeability. The T2 surface is about 15-20 m above the T1 surface.

Older Flood Plain (T1)

The surface is located within the Gomati River Valley, above the level of annual floods of the river. The older flood plain of Gomati in the city is present as detached pockets in the central (north of river) and south- eastern part (south of river). The characteristics landforms in this zone are abandoned channels, meander scars, meander cutoffs etc. The sediments are fine-grained sand with silt and occasional clay beds. The characteristic feature of this zone is absence of alkaline soil. Ravenous tracts can be seen at the junction of younger flood plain with the older flood plain, where occasionally thin kankar beds are seen indicating presence of older alluvium at depth.

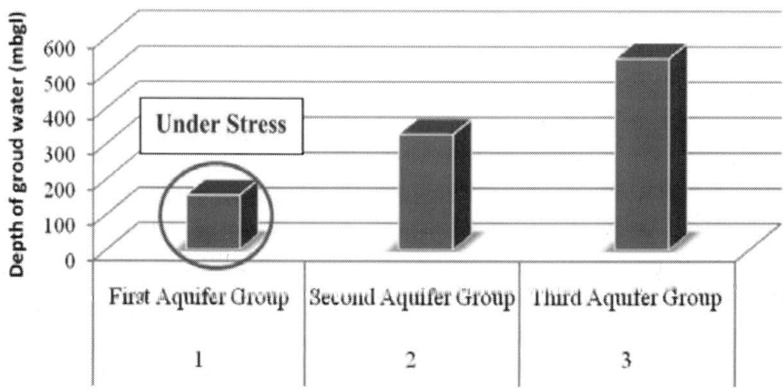

Fig .3. Histogram showing three groups of aquifer in Lucknow city (State water resource agency, Jal Sandesh, 2010)

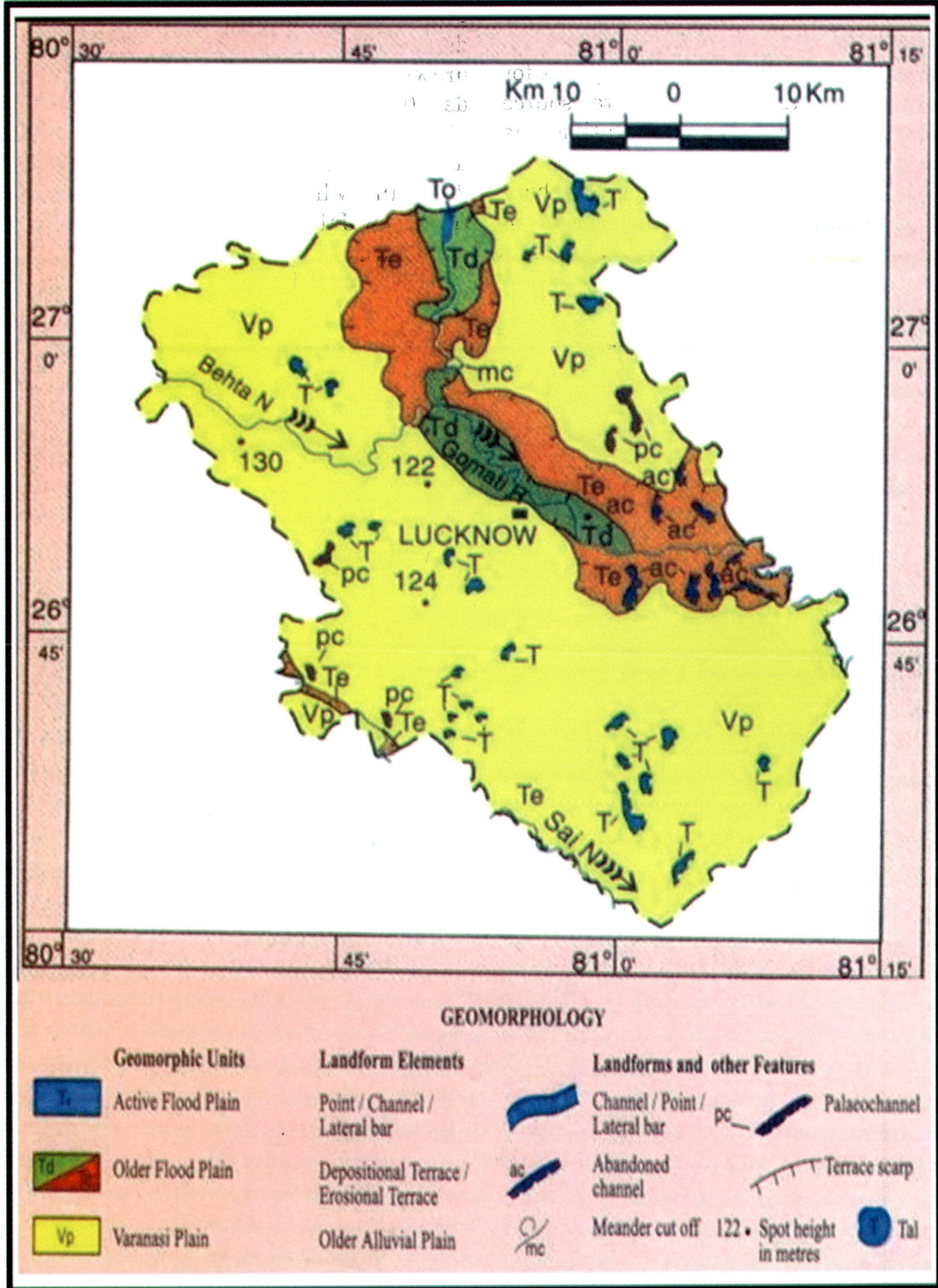

Fig. 4. Map showing the Geomorphological Condition of Lucknow District showing landforms and palaeochannels (District Resource Map, Geological Survey of India, 2001).

Younger Flood Plain (T0)

Surface makes the river channel and its active flood plain. The characteristics geomorphic forms present in this zone are point bars, channel bars, etc. It has an area of about 40 Sq. Km. within the city which has been restricted to about 10 Sq. Km. due to urban habitation and earthen embankments on either side of Gomati. The area gets periodically flooded and is widely cultivated, upstream of Gaughat. The sediments in younger flood plain belong to younger alluvium and are poor in calcareous matter (Fig 4).

WATER RESOURCES

The two sources of water are mainly use for water supply in city namely surface water source and ground water source. River Gomati is the only source of surface water supply where as deep tube wells, hand pumps and private bore wells are the main sources for extraction of groundwater in the city to fulfill the water demand of the population. The surface water drawn from river Gomati is 280 million liters per day (MLD) and the groundwater withdrawal is 170 MLD for domestic uses. The demand of drinking water is estimated to be 546 mld in 2011 and which increase upto766 mld in 2021 (Lucknow Jal Sansthan Report).

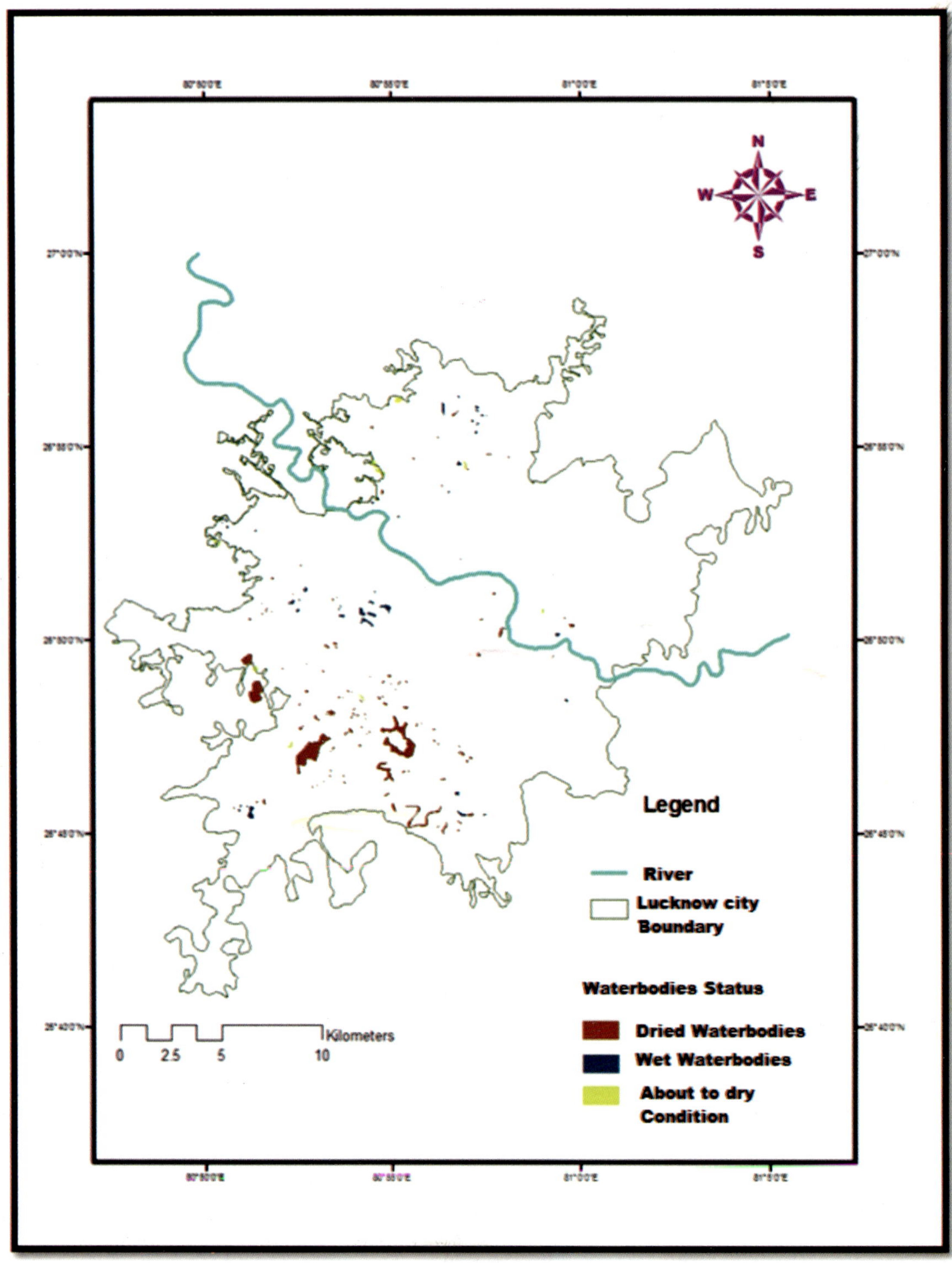

Fig. 5. Status of Water bodies in 1973 in Lucknow City

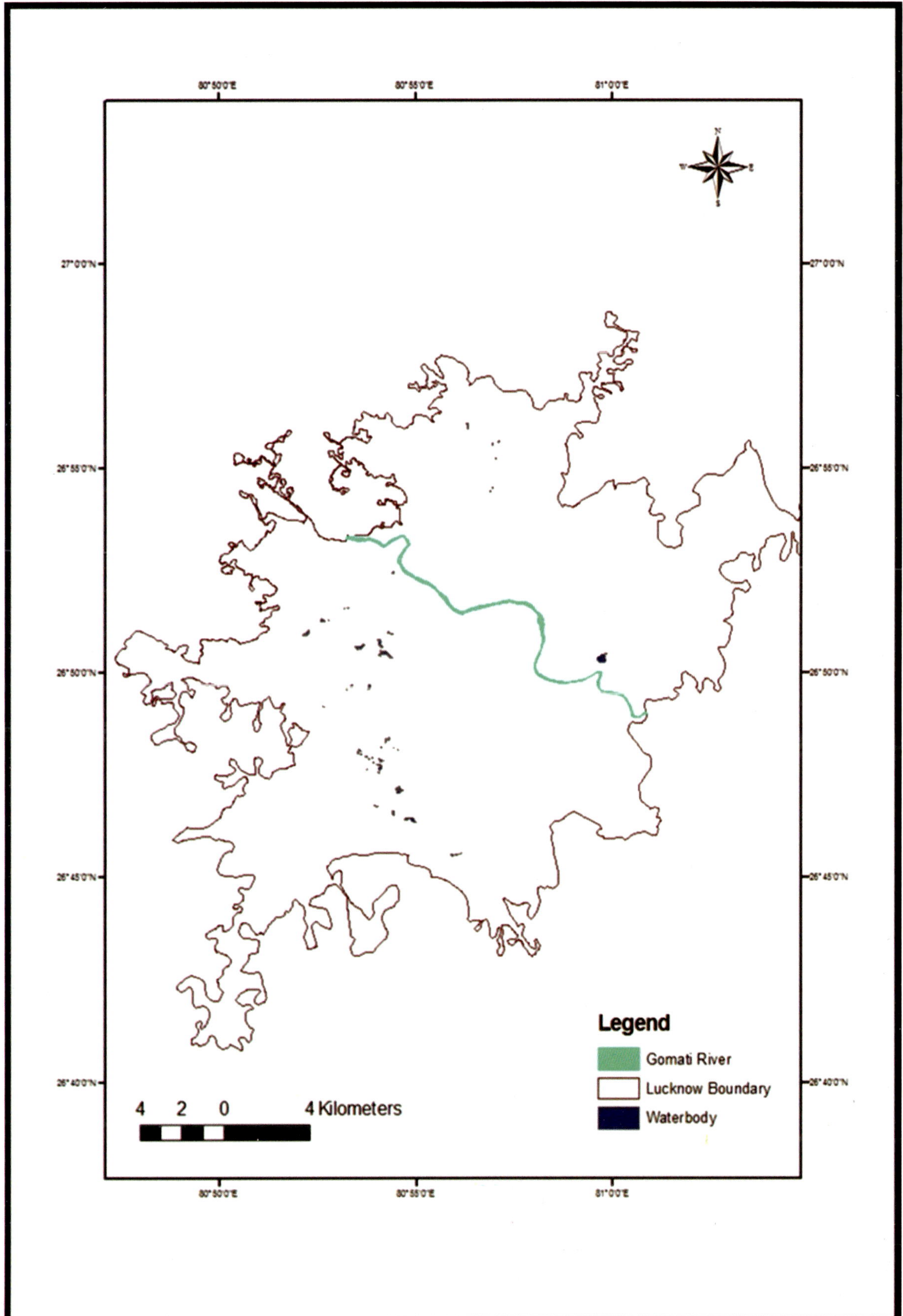

Fig .6. Status of Water bodies in 2003 in Lucknow City

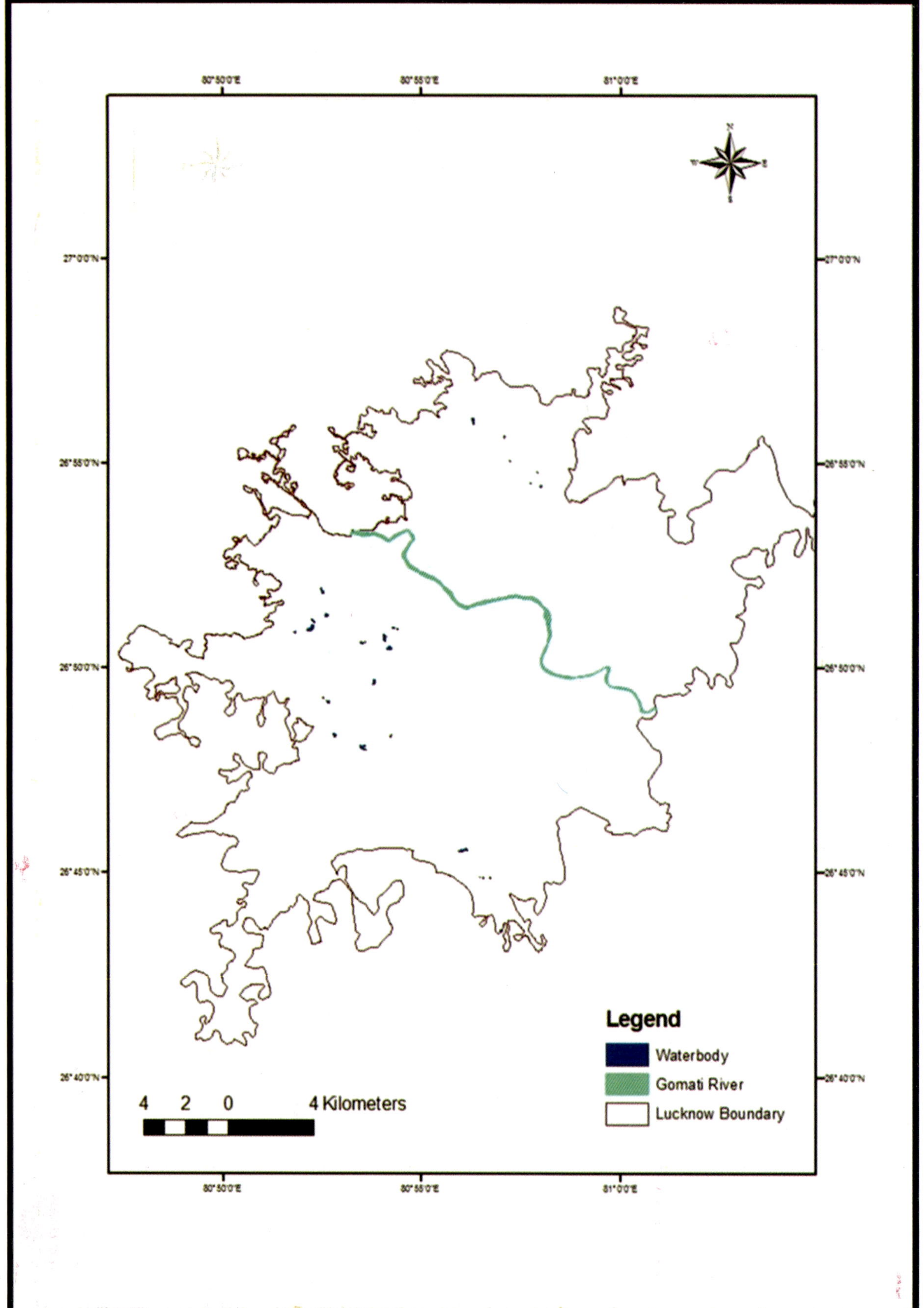

Fig .7. Status of Water bodies in 2012 in Lucknow City

Table .2. Drains and their discharge in Lucknow City (CPCB, 2002)

S. No	Name of Drains Cis- Gomati (Right Bank).	Discharge (MLD)	Name of Drains Trans-Gomati (Left Bank).	Discharge (MLD)
1	Nagariya Drain.	10.00	Maheshganj Drain.	4.570
2	Gaughat Drain.	1.0	Rooppur Drain.	0.656
3	Sarkata Drain.	34.57	Mohan Meakins Drain.	5.1875
4	Pata Drain.	18.54	Daliganj Drain. I	12.275
5	Drain to U/S of NER Bridge.	3.00	Daliganj Drain II	6.950
6	Drain to D/S of NER Bridge.	1.00	Arts College Drain.	8.890
7	Wazirganj Drain.	75.60	Hanuman Setu Drain.	0.825
8	Ghasiyari Mandi Drain.	19.53	Trans Gomti Pumping Station.	1.640
9	China Bazar Drain.	7.01	Kedarnath Road Drain.	3.400
10	La-Place Drain.	3.00	Nishatganj Bridge Drain.	1.700
11	Joppling Road Drain.	3.00	Babupurwa Drain.	0.65
12	CIS-Gomti Pumping Station drain	19.64	Kukrail Drain.	50.00
13	Gayasudin Haider Canal.	106.20		
14	Jiamau Drain.	3.00		
15	Lamartenier Drain.	1.00		
16	Last Drain U/S of Rly Bridge.	3.00		

Beside Gomati River, city has also various water bodies and canal system. According to the Land Records of the Lucknow Nagar Nigam, the city has 846 tanks and ponds -the majority of which are unidentifiable due to reclamation. Lucknow has a number of tanks that were constructed by the Nawabs and their chieftains. Some of these were ancient and famous as religious places and were renovated and developed further by them such as Suraj Kund, Bara Imambara Baoli, Raja Tikait Rai Ka Talab, Hussainabad Ka Talab, Chandganj Ka Talab, Moti Jheel etc (City development plan, 2006).

On the basis of Survey of India toposheets, satellite data and Google earth data the condition of water bodies demarcated in three different years, in 1973, 2003 and 2012.The result reveal that in 1973 there were more than 60 water bodies present in Lucknow city, but in 2003 the number has decreased as less than 50 and in 2012 it reduced to less than 40, not only the number but size of water bodies are also decreasing rapidly. Some major water bodies such as Jamuna Jheel, Moti Jheel, Rajaji puram jheel, Butler Jheel, Lakes in around Alambagh area etc has been eutrophied or encroached by city dwellers (Fig 5, 6 and 7).

WASTE WATER MANAGEMENT

Drainage systems are the major source of conducting waste water and also rain water.. There are 16 drains from Cis-Gomti and 12 drains from trans-Gomti area meeting river Gomti directly including the sewage pumping stations. It is estimated that 305.09 mld discharge generated from right bank of the river Gomati River, on the other hand 156.24 mld discharge from the left bank of the River Gomati (Table 2 and Fig 8).

PROPOSED MODEL

The Digital Elevation Model (Fig 10) of Lucknow City is prepared with the help of Shuttle Radar Topographic Mission (SRTM) data, which have downloaded from GLCF-ESDI site of 90 m resolution is used The image is registered and re-projected in Arc GIS and with the help of 3D Analyst tool of Arc-GIS the DEM is prepared It shows the elevation range of Lucknow city as lowest 98masl (Mean Average Sea Level) and highest elevation of 142 masl. The highly elevated areas are situated on the left bank of Gomati River. The lowest elevation from 90 masl to 106 masl is mostly observed along the basin and drainage of Gomati River.

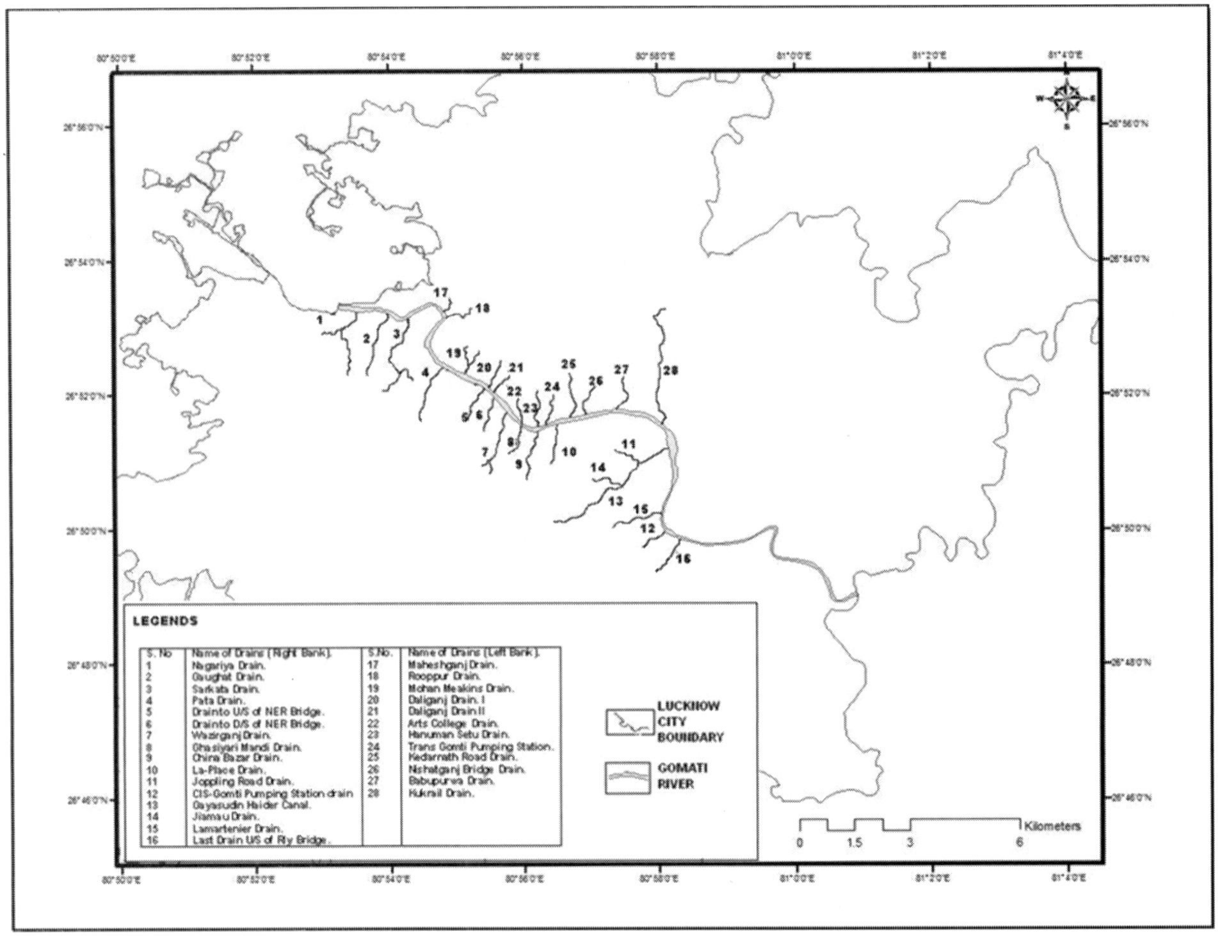

LEGENDS
S. No
Name of Drains (Right Bank).
1 Nagariya Drain.
2 Gaughat Drain.
3 Sarkata Drain.
4 Pata Drain.
5 Drainto U/S of NER Bridge.
6 Drainto D/S of NER Bridge.
7 Wazirganj Drain.
8 Ghasiyari Mandi Drain.
9 China Bazar Drain.
10 La-Place Drain.
11 Joppling Road Drain.
12 CIS-Gomti Pumping Station drain
13 Gayasudin Haider Canal.
14 Jiamau Drain.
15 Lamartenier Drain.
16 Last Drain U/S of Rly Bridge.
S.No.
Name of Drains (Left Bank).
17 Maheshganj Drain.
18 Rooppur Drain.
19 Mohan Meakins Drain.
20 Daliganj Drain. I
21 Daliganj Drain II
22 Arts College Drain.
23 Hanuman Setu Drain.
24 Trans Gomti Pumping Station.
25 Kedarnath Road Drain.
26 Nishatganj Bridge Drain.
27 Babupurwa Drain.
28 Kukrail Drain.
LUCKHOW CITY BOUNDARY
GOMATI RIVER
0 1.5 3 6
Kilometers

Fig .8. Map showing the networks of drains in the Lucknow City

HOTSPOTS
(where Ground Water Level is declining at an alarming rate)
Cantonment 1.01
Hazratganj-Narhi 1.06
Indira Nagar 1.10
Vikas Nagar 1.13
Gomti Nagar 1.69
0 0.5 1 1.5 2
Decline (m/ year)

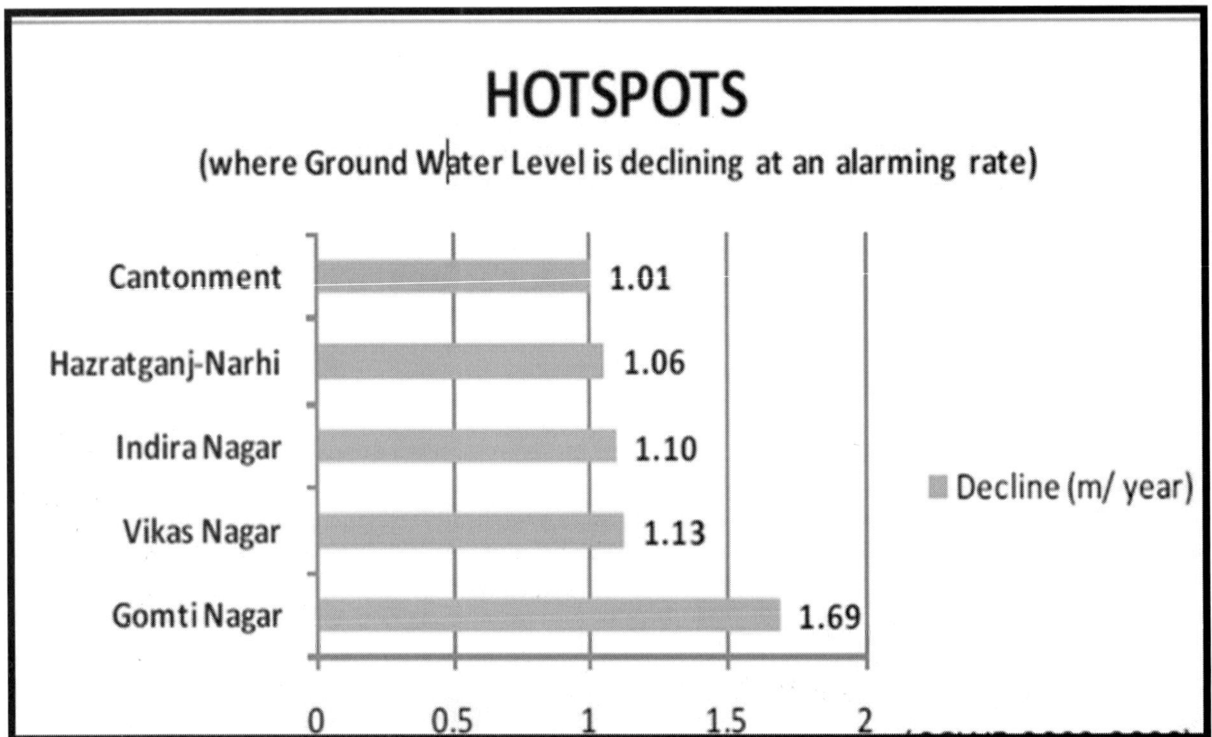

Fig.9. Ground water declining rate in different areas of Lucknow City (CGWB, 2010)

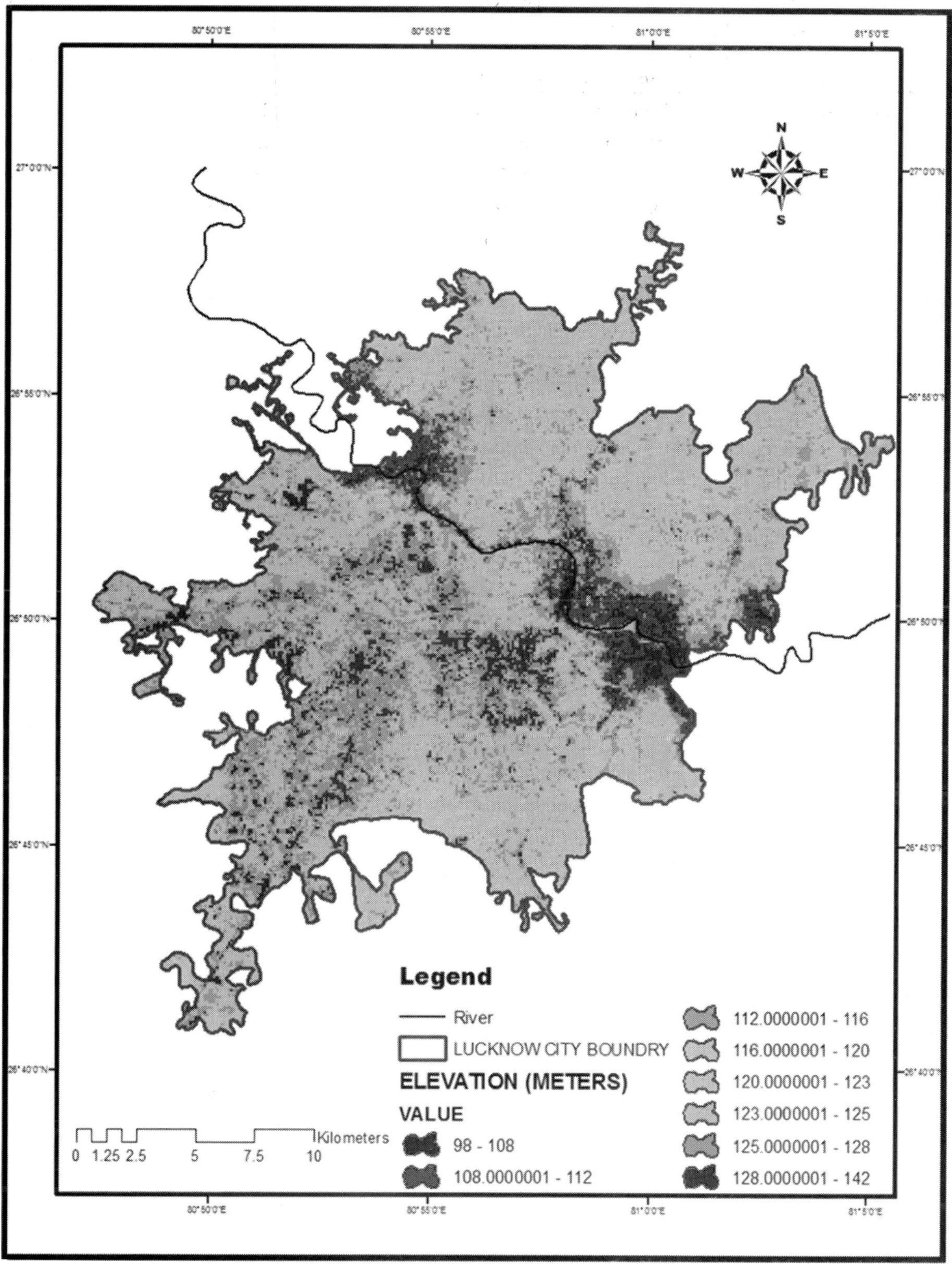

Fig .10. Digital Elevation Model of Lucknow city (SRTM Data)

The natural flow of water, low-lying areas and water bodies demarcated with the help of this surface elevation model. As the result of integration of Central Ground Water Board data and elevation data the areas of Gomti Nagar, Hazratganj-Narhi, Indira Nagar, Vikas Nagar and Cantonment are the marked as severely affected by declining of ground water level (Fig 9).

Table 3. Field Data of waterlogged areas in Lucknow city from North to south direction in years 2010, 2011 and 2012.

S.No	Name of the Location	S.No.	Name of the Location	S.No.	Name of the Location
1	Jankipuram Sec-I	28	Polytechnique Churaha	55	Mill Road Aishbagh
2	Sahara state to mariyon	29	Jalsansthan , Thakurganj	56	Mal Avenue near Haider canal
3	Sahara state	30	J.J. Bakers, Near IT crossing	57	Naka hindola
4	Jankipuram Sec-5/H	31	HAL , Faizabad Road	58	Talkatora Rajajipuram
5	IET campus sitapur road	32	Mery mount Restaurant near Vivekanand Puri	59	Shaheed nagar near Sadar
6	Tedhi pulia near petrol pump	33	Lekhraz market, Faizabad Road	60	Near Husainganj Police Station
7	Indiranagar Sec-14 near Power House	34	Kalyangiri temple, Thakurganj	61	Nano showroom, Charbagh
8	Khargpur Indiranagar	35	Ramnagar, Balaganj	62	Mavaiya crossing
9	Sector E Aliganj	36	Old Kalyanpuri, Thakurganj	63	City bus Station
11	Indiranagar Sec-16	37	Hari Bazar, Balaganj	64	KKC Railway Bridge
12	Indiranagar Sec-19	38	Old Hyderabad, kailash hostel	65	Mavaiya Railway Crossing
13	Indiranagar Sec-17	39	Nishatganj	66	Green Eco garden Rd
14	Kapoorthala Crossing	40	Shaktinagar, Nishatganj	67	Shantipuram colony Aishbagh
15	Indiranagar Sec-23	41	Slums near Kukrail Nala	68	Manyavar kanshiram Sthal
16	Mandir marg Mahanagar	42	New kalyanpuri, Thakurganj	69	Awadh Hospital chauraha
17	Channilal chauraha	43	Gari peer khan , Thakurganj	70	Phoenix mall Smarak rd
18	Indiranagar Sec-22	44	Nehru enclave Gomti nagar	71	Smarak rd behind Bodh vihar
19	Polytechnique overbridge end point	45	MG rd near Parivahan nigam	72	Barabirwa Alambagh
20	Mahanagar Civil	46	Vishwash Khand 2	73	Himalayan Colony, Telibagh
21	Opposite Vivekanand Hospital	47	Vishwash Khand	74	BSNL office Alambagh rd
22	Vivekanandpuri	48	IMS Gomti nagar	75	Sector H Alambagh
23	Chinhat	49	MG rd near Parivartan Chauraha	76	Hindnagar
24	Kamta	50	Dainik Jagran Chauraha	77	Purani chungi kanpur rd
25	Geeta guest house near Channilal chauraha	51	Near Taj Hotel, Gomti Nagar	78	Transport Nagar
26	Gol market Mahanagar	52	Naval Kishor marg Hazratganj	79	Mansarovar Yojna near Transport nagar
27	Nagaria, Thakurganj	53	Hazratganj Chauraha	80	Amausi Airport
		54	Lal Bahadur Shastri Marg		

DISCUSSION AND CONCLUSIONS

1. Water is one of the natural resources on which the biosphere is totally dependent. It is therefore that we must understand the significance of water in our day today life.
2. Fresh water on land is mainly responsible for controlling the metabolic activity of the biotic entities as well as for providing the medium for transportation of nutrients to each and every cell of it.
3. Despite being the most crucial resource for the biosphere for its existence, the anthrosphere has put extreme pressure on it.
4. In the course of industrialization, on one hand we have started exploiting water at uncontrollable limits, on the other hand,

discharging it in the form of effluents into the ground.

5. India is a country with its population growing at an alarming rate. In future as per the current estimate, is going to be double in the next twenty year or so. The existing water requirement would be increased at least two times from present level within a span of 20 years.

6. Lucknow being the capital city of the state of Uttar Pradesh, is one of the most populous and rapidly urbanizing cities. The city's expansion in both space and population needs to be addressed in terms of planning for the future, there by suggesting that we should be prepared not only in the present times but also for the future so that we are able to safeguard the interest of the younger generation.

7. Considering the minimal requirement of fresh water, it was calculated that the present demand of fresh water is 546 million liters per day (MLD) and according to the Lucknow Master Plan 2021 the demand would be increased as 766 MLD in 2021.

8. However, considering the present availability and future growth of fresh water demand with blasting population there seems a major gap between the government and the planners. As such there seems to be total apathy amongst the government planners to look into the future requirement of this crucial natural resource.

9. Every year, on an average, there has been steep decline in the water table depth in and around Lucknow city and more so the upper few strata (water table) have been totally squeezed out of its water.

10. As one of the major contributions of the present work, we have suggested to provide a viable solution by way of utilizing water resources from the existing system.

11. A survey of the Lucknow city using GIS technique and locating fresh water wells suggests that sustainability of water requires additional inputs by way of controlled utilization.

12. In this direction utilization of rain water during the monsoon period can be one of the very useful recharging resource that we have.

13. Identification of area with pathetic situation should be the starting point to rejuvenate the surface and sub-surface water conditions.

14. The excessive rains during the monsoon in June and July that causes pause in the day today life of population in general due to waterlogging in the low-lying areas can be channelized into ground water percolation system which can enhance the water recharge capacity of the already depleted water table.

15. For such a situation, sub-surface sediment and soil characteristics and their layering pattern can help us in identifying the best percolating sites which can be effectively used to dewater the waterlogged areas. This surplus water can be utilized by recharging of the ground water table after proper treatment.

16. The current study has clearly indicated the key waterlogged areas of the Lucknow metropolitan area. These are given in the Table 3. These areas should be considered for recharging points in the city areas.

17. In addition, surface water bodies like lakes and ponds, both natural and man-made should be revived. Rain water can feed these drenched out natural water bodies and as a result, such areas can eventually become major rain water harvesting centers.

18. Long distance water lifting by pumps from neighboring areas and districts such as Mohanlalganj and Mati villages situated in Barabanki District can be used for Lucknow metropolitan areas as in these areas, water table is at a much higher level.

19. Upgrading the green cover of the parks and open areas can effectively enhance the water conditions of Lucknow and similar metropolitan areas. Since water in plants is not only retained into the exposed part but also exists in sufficient amount in the root zone.

20. Care should be taken to avoid excessive usage of water during the summer period because water is at its critical level of

availability and at the same time water requirement is enhanced many fold.

21. In the light of above mention deductions based on our study, it can be concluded that water sustainability in urban metropolitan areas require better management and planning for making it as a perennial source, otherwise the already grim situation can become violent with lots of civil unrest and can trigger major health hazard.

22. Similar study should be initiated in other areas for proper management and better availability of this very prestigious resource- WATER.

ACKNOWLEDGEMENTS

We are grateful to the Head of the Department of Geology, University of Lucknow for providing us the laboratory facilities. We are also very thankful to Ms. Droupti Yadav and Mr. Rohit Agarwal for helping us in various steps.

REFERENCES

Anonymous, 2002: A Report on State of Environment of Lucknow, Central Pollution Control Board, Control of urban pollution series-52.

Anonymous, 2006: City Development Plan, 6p.

Anonymous. 2001: Census of India, Towns & Urban Agglomerations.

Anonymous. 2010: Jal Sandesh, State Water Resource Agency, Uttar Pradesh, Issue-3rd, 3p.

Anonymous. 2010: Central Ground Water Board Report

Anonymous. 2011: Census of India, http://www.census 2011.co.in/census/city/127-lucknow.html

Khanna, S.P., 1992: Hydrology of Central Ganga Plain, Uttar Pradesh Gangetic Plain Tera Incognita, 23-27 p.

Mukherji, A.B., 1963: Alluvial morphology of the upper Ganga-Yamuna doab, The Deccan Geographer- 2, 1–36p.

Hon'ble reviewer

1. **Trilochan Singh**
   Scientist (Retd.) WING
   97 Chand nagar
   NEW DELHI - 110 018
2. **Prof. Sreeup Goswami**
   Chairman
   Department of Geology
   Ravenshaw University
   CUTTACK - 753 003

Assessment of Geoenvironment on Water Quality: A Case Study of Jodhpur City, Rajasthan, India

Purnima Sharma[1], S.C. Mathur[2], Vibhuti Rai[1] and Shiv Singh Rathore[2]

[1] Centre of Advance Study in Geology, University of Lucknow, Lucknow
[2] Geology Department, J.N. Vyas University, Jodhpur
E-mail: vibhutirai@gmail.com

Abstract

Geoenvironment referes to the aspect of geosphere such as rock, soil, water and biotic entities that affect the environment. As one of the key environmental domains, the water constitutes a major component that gets affected due to its dynamic nature. Since water on earth exists in all of its three states, as snow/ice/water and vapour (solid, liquid and gas), it gets interacted with practically every geospheric entity. This interaction leaves behind a series of change in water quality and it remains as one of the most affected material due to geoenvironmental impact.

The industralization, natural and man-made pollution and the aspects of geoenvironment gradually affect the nature, which in turns affect the quality of surface and sub-surface water resources. The present study demonstrates the physico-chemical characteristic of the different surface and sub-surface water resources such as open wells, tube-wells, *Baowri*, *Jhalra* and ponds of Jodhpur district based on analytical data. This study deals with the various parameters of water such as pH, Total dissolved solids (TDS), Electrical Conductivity (EC), Heavy metals, namely- Zinc (Zn), Chromium (Cr), Cadmium (Cd), Lead (Pb), Copper (Cu) and Manganese (Mn) which describes the present condition of surface and sub-surface water characteristics of the Jodhpur city. A total of 40 samples have been analysed which are spread across the length and breadth of the Jodhpur city.

The findings reveal that some of the samples contained heavy metals beyond permissible limits prescribed by Bureau of Indian Standards (B.I.S), which can become a serious health hazard. Based on these results; it is recommended that any water resource in the study area should be tested before use for its potability and other domestic or industrial usages.

Keywords: Water Resources, Water Table Fluctuation, Water Chemistry, Water Management Geoenvironment.

Introduction

Water is elixir of life. Although, ¾ of earth is covered with water, yet the quantity of usable fresh water is very limited. Water being a very potent solvent, it dissolves minerals from the rock with which it comes in contact. Due to the rapid increase of population and industries, the environment and Geoenvironment, both are getting affected badly.

Jodhpur being one of the metro cities of Rajasthan, caters to a large population (over 1 Million).There is a limited amount of fresh water bodies in Jodhpur region. The rainy days are limited to maximum 15 days in a year, which is not sufficient enough to cater to the yearly need of the city dwellers. Ground water occurs under unconfined to semi-confined conditions in different rock groups of Jodhpur. These form the chief source of ground water in the district. In view of long term change in ground water regime, it is observed that over the time, a major part lying in central and eastern portions of the district show maximum decline. This significant decline is observed due to increased ground water draft. In some areas of Osian, Bhopalgarh and Bilara, water level decline of more than 25m has been noticed during last 10 years (District Ground Water Brochure, Jodhpur, 2008). Although lately, in the city area, due to impact of Indira Gandhi canal, water table has considerably improved and in some areas, waterlogging has started.

There is a large variation in chemical quality of surface and sub-surface water in the district depending on the characteristics of water bearing formation, movement of ground water, depth of water levels etc. Chemical contaminants with special reference to heavy metals pose a very serious health hazard. It is therefore that the present study has been taken up in order to know the quality of water that is being used for drinking, agriculture and industrial purposes.

STUDY AREA

Jodhpur district is situated between 25°51' 08" & 27°37' 09" North latitude and 71°48' 09" & 73°52' 06" East longitude covering geographical area of 22,850 sq km. This district comes under arid zone of the Rajasthan State. Jodhpur district is part of Jodhpur Division. The district is divided into 5 sub-divisions namely Jodhpur, Shergarh, Pipar City, Osian & Phalodi and comprises of 7 tehsils and 9 blocks (District Ground Water Brochure, Jodhpur, 2008) (Figure 1).

Jodhpur city is governed by Municipal Corporation which comes under Jodhpur Metropolitan Region. The Jodhpur city is located in the southern part of Jodhpur district. As per provisional reports of Census of India, population of Jodhpur in 2011 was 1,033,918; of which male and female are 544,057 and 489,861 respectively (Figure 3).

HYDROGEOLOGY OF DISTRICT

Ground water occurs under unconfined to semi-confined conditions in rocks of Delhi Super Group, Jodhpur sandstone, Bilara limestone, Nagaur sandstone, Lathi sandstone and unconsolidated sediments (valley fills and alluvium) of the desert region. These horizons forms the chief source of ground water in the district. Confined condition is also met sometimes at deeper levels in the north-western part of the district (District Ground Water Brochure, Jodhpur, 2008) (Figure 4).

WATER QUALITY ASSESSMENT

For demonstrating the chemical quality of water resources, various physico-chemical properties have been checked through several experiments.

Sampling methods

The water samples were collected from 40 different sites and source, among these 21 samples were collected from Open well, 5 from Baori, 5 from Tube well, 3 from Jhalra, 4 frpm Basement supply water, 1 from Talab and 1 sample from Tap water, supplied by Kaylana Lake. The samples were collected in distilled sampling water bottle. For heavy metal analysis, samples were preserved by dilute Nitric acid (Figure 5).

Method for analysis

The physico-chemical characteristics of the water samples were determined by standard methods and instruments according to APHA, 1988 (Table 1).

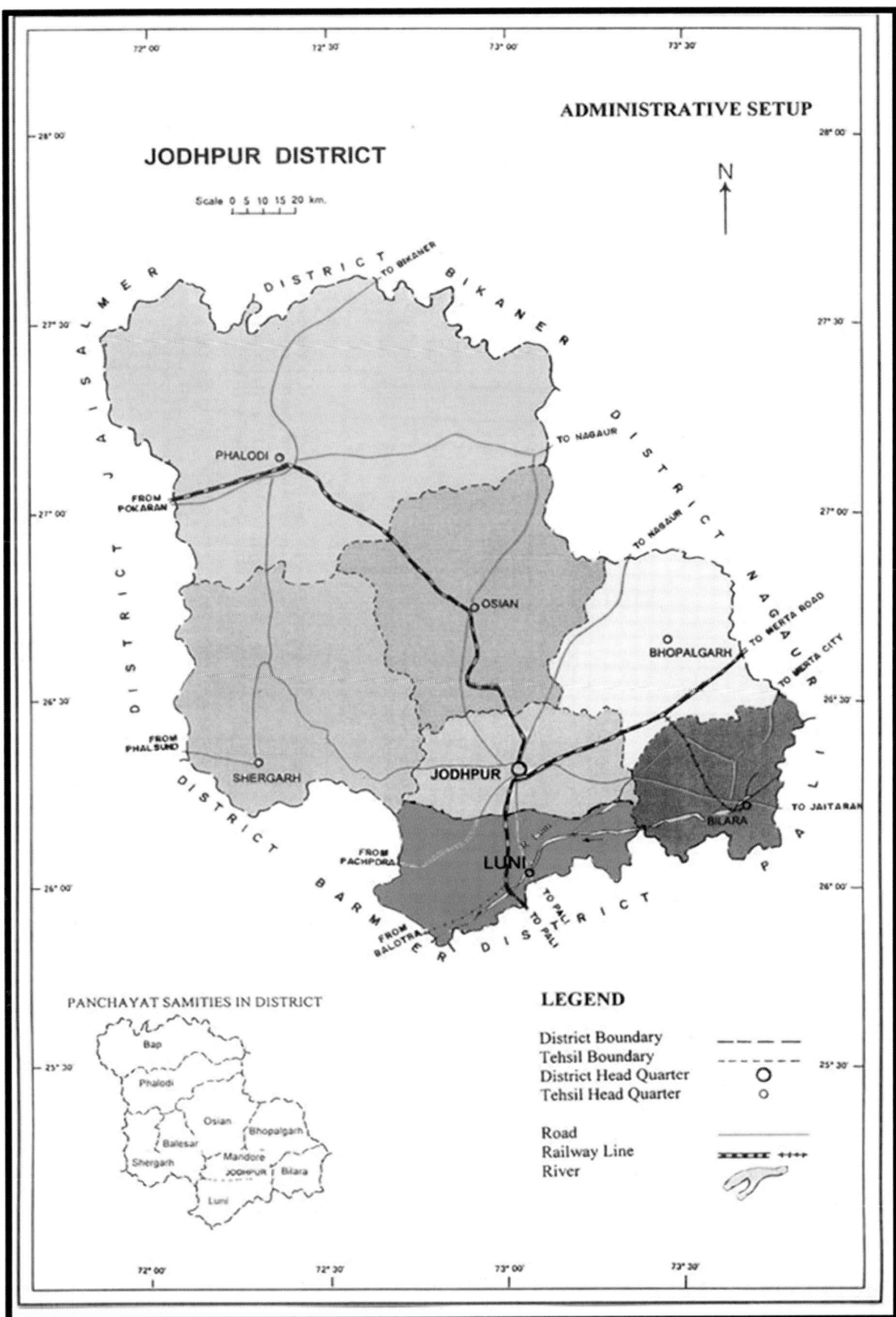

Fig. 1. Administrative boundaries of Tehsils of Jodhpur district (http://jodhpur.nic.in/map.html).

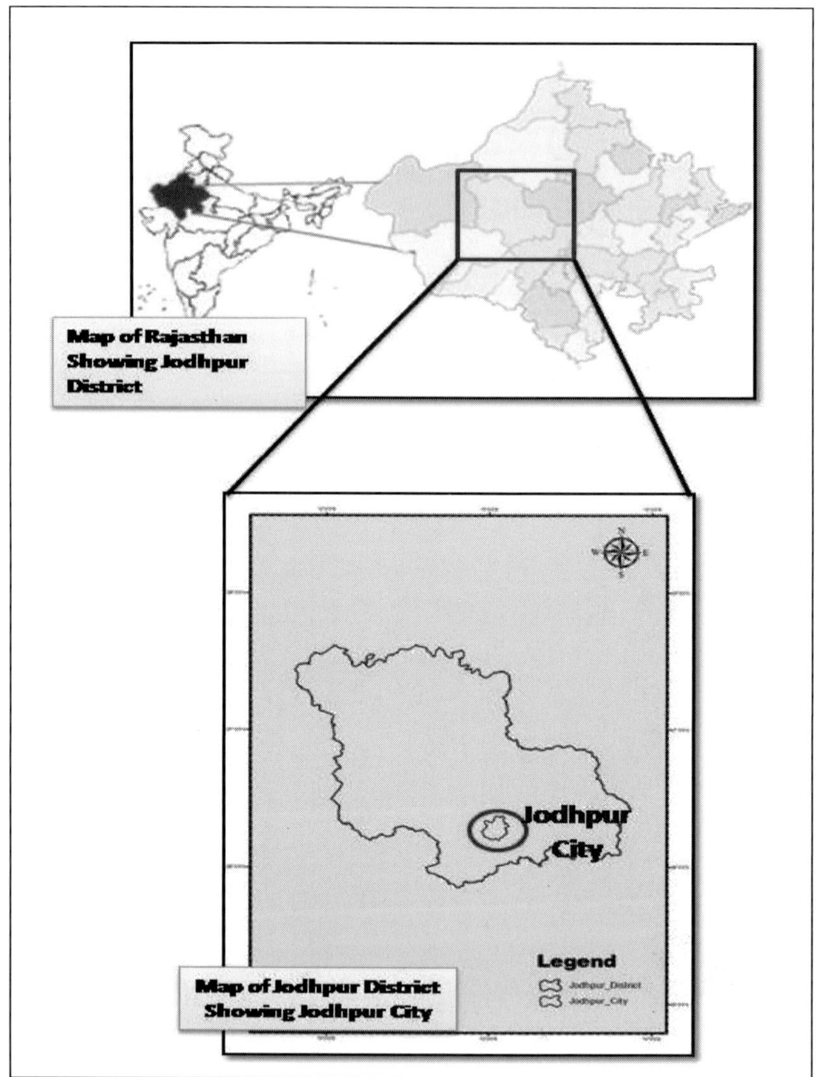

Fig. 3. Location of study area: Jodhpur city

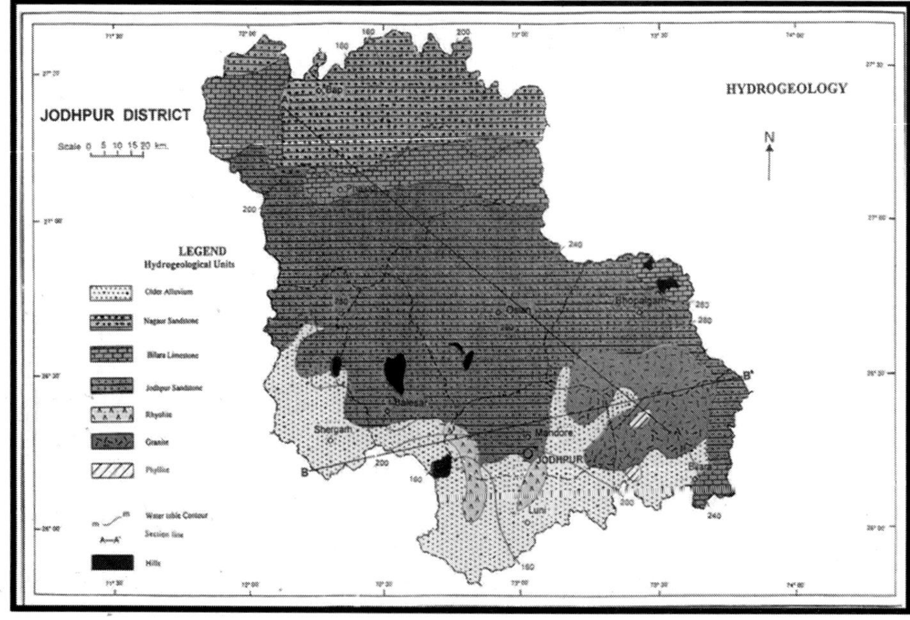

Fig.4. Map showing Hydrogeology of Jodhpur district (http://jodhpur.nic.in/map.html).

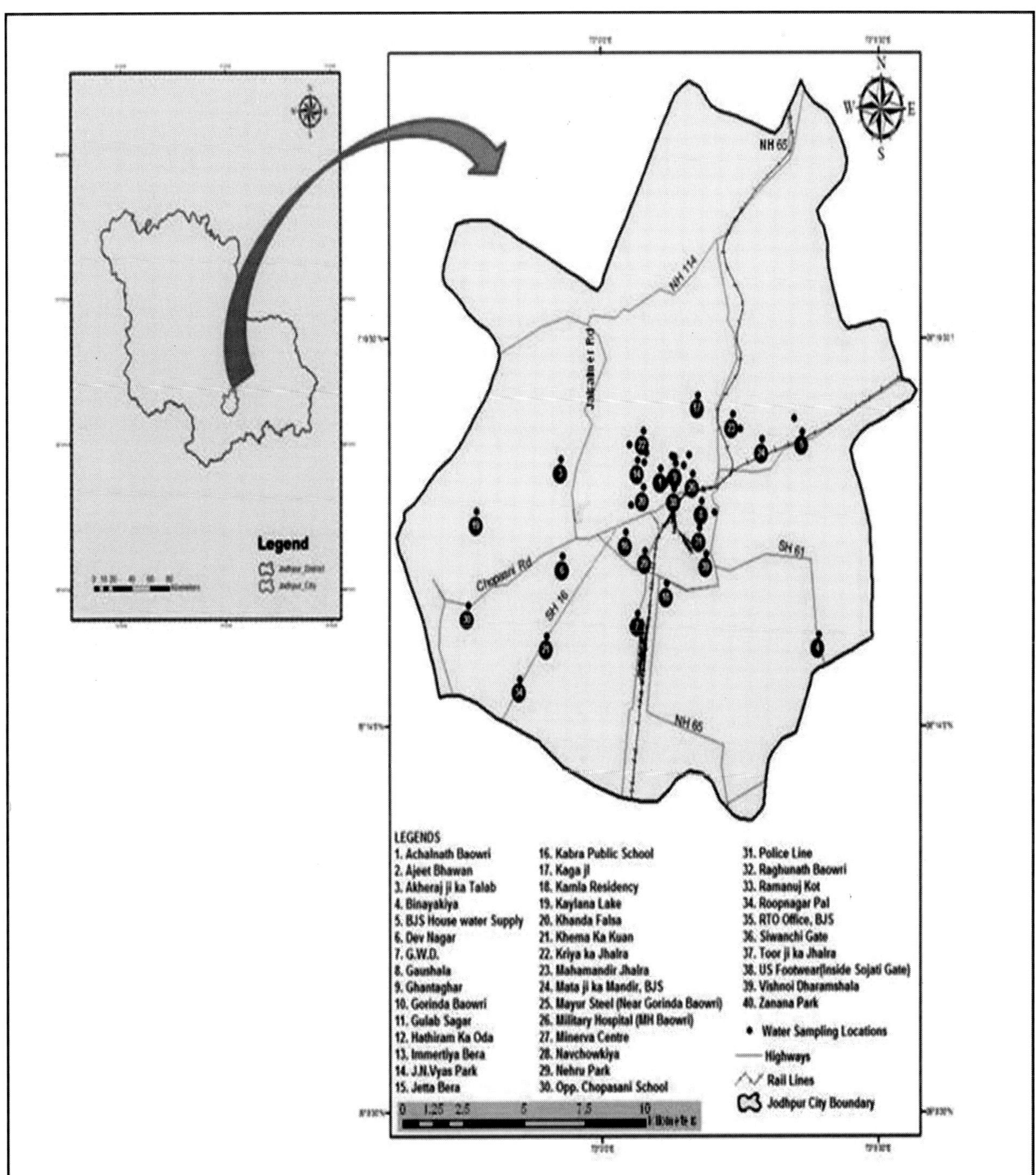

Fig. 5. Water sampling locations of the Jodhpur city

Table 1. Parameter, Method and Instrument used in the present study

S.N.	Name of Parameters	Name of Method	Instrument Name
1	pH Determination	Electrode method	Water Analysis Kit (ELICO PE 138)
2	Conductivity	Electrode Method	Water Analysis Kit (ELICO PE 138)
5	Total Dissolve Solid	Electrode Method	Water Analysis Kit (ELICO PE 138)
6	Trace Elements (Zn, Cr, , Mn, Cd, Cu,Pb)	Digestion method	Atomic Absorption Spectrophotometer (Perkin Elmer Analyst 700)

Table 3. Concentration of different chemical constituents and other parameters of water samples collected from different location of Jodhpur (O/W= Open Well, T/W= Tube Well, EC= Electric Conductivity, TDS= Total Dissolved Solid, Zn= Zinc, Pb= Lead, Cd=Cadmium, Cr=Chromium, Mn= Manganese, Cu=Copper, ppm= parts per million or mg/litre, BDL= Below Detection Limit).

S.No.	Name of the Location	Source	Latitude	Longitude	pH	EC (mhos/cm)	TDS (ppm)	Zn (ppm)	Pb (ppm)	Cd (ppm)	Cr (ppm)	Mn (ppm)	Cu (ppm)
1	Achalnath ji ki Baori	Baori	26.294	73.02	7.723	65.26	31.25	0.1803	0.0213	0.0561	0.3663	BDL	0.0006
2	Ajeet Bhawan	O/W	26.284	73.037	7.763	67.74	49.21	0.1191	0.0473	0.0468	0.5236	BDL	BDL
3	Akheraj ji ka Talab	O/W	26.297	72.985	7.459	146.9	68.15	0.1066	BDL	0.0822	0.4542	BDL	BDL
4	Binayakiya	O/W	26.255	73.072	7.906	129.6	62.01	0.1056	0.0434	0.1042	0.5326	0.0702	BDL
5	BJS House water supply (PS House)	Tap water (supply from Kaylana)	26.303	73.066	8.078	128.4	62.01	0.127	BDL	0.0544	0.5	BDL	BDL
6	Dev Nagar	T/W	26.274	72.985	7.487	122.3	49.84	0.1126	0.0046	0.1048	0.6054	BDL	BDL
7	G.W.D.	T/W	26.26	73.012	8.202	91.18	45.46	0.0252	BDL	0.1036	0.5098	BDL	BDL
8	Gaushala	O/W	26.286	73.033	7.426	60.32	46.03	0.0915	0.1121	0.0591	0.6372	BDL	0.0064
9	Ghantaghar	O/W	26.295	73.024	7.811	94.98	26.92	0.2265	0.0177	0.0513	0.3186	0.0237	BDL
10	Gorinda Baori	O/W	26.292	73.022	7.839	101.9	54.83	0.2583	0.0344	0.0706	0.3706	0.0185	BDL
11	Gulab Sagar	Talab	26.297	73.024	7.792	55.19	25.92	0.0824	0.0162	0.0726	0.3185	0.0118	BDL
12	Hathiram ka Oda	O/W	26.295	73.027	7.826	71.04	31.81	0.0838	BDL	0.0715	0.4265	BDL	BDL
13	Immertiya Bera	O/W	26.304	73.046	7.662	75.39	31.81	0.1077	0.0727	0.0298	0.5649	BDL	BDL
14	J.N. Vyas Park	O/W	26.296	73.012	7.276	164.2	97.63	0.2656	0.0064	0.1008	0.5842	0.069	BDL
15	Jetta Bera	O/W	26.298	73.015	7.313	73.08	49.84	0.1132	BDL	0.0782	0.4268	BDL	BDL
16	Kabra Public School (Sardarpura)	O/W	26.279	73.008	7.695	161.8	81.54	0.0928	0.045	0.0932	0.7044	BDL	BDL
17	Kaga ji	O/W	26.312	73.032	7.862	102.1	31.25	0.0862	BDL	0.0615	0.4539	BDL	BDL
18	Kamala Residency	Basement water supply	26.267	73.022	8.389	68.46	33.33	0.142	0.0466	0.0478	0.6566	BDL	BDL
19	Kaylana Lake (Canal raw water)	O/W	26.298	73.974	8.172	113.2	68.59	0.103	0.1582	0.089	0.3906	BDL	BDL
20	Khanda Falsa	O/W	26.29	73.014	7.523	92.22	34.78	0.109	0.0046	0.1052	0.52	BDL	BDL
21	Khema ka Kuan	O/W	26.255	72.982	7.709	65.02	49.84	0.1212	BDL	0.115	0.6082	BDL	BDL
22	Kriya ka Jhalra	Jhalra	26.303	73.014	7.512	260.5	164.1	0.217	0.0368	0.0548	0.689	0.0692	BDL
23	Mahamandir Jhalra	Jhalra	26.307	73.043	7.482	87.1	46.03	0.0345	0.036	0.0492	0.54	0.0482	BDL
24	Mataji ka Mandir, BJS	T/W	26.301	73.053	7.692	68.84	48.5	0.1086	0.0365	0.0642	0.5806	BDL	BDL
25	Mayur Steel (Near Gorinda Baori)	Basement water supply	26.29	73.021	8.7	195.5	48.73	0.2276	0.1222	0.1094	0.5226	0.011	0.052
26	Military Hospital (MH Baori)	Baori	26.293	73.047	7.748	80.95	35.02	0.0581	0.0452	0.1018	0.5895	BDL	BDL
27	Minerva Centre	Basement water supply	26.287	73.024	7.806	128.8	46.82	0.1796	0.041	0.0252	0.6112	BDL	BDL
28	Navchowkiya	O/W	26.296	73.014	7.352	71.57	33.33	0.326	BDL	0.0746	0.637	0.0326	BDL
29	Nehru Park	O/W	26.275	73.014	7.59	121.5	70.21	0.0724	0.0486	0.0768	0.773	BDL	BDL

30	Opp. Chopasani School	T/W	26.262	73.954	7.785	62.02	47.54	0.2062	0.0134	0.0688	0.3844	BDL	BDL
31	Police Line	Baori	26.28	73.032	7.87	105.5	49.21	0.0663	0.0366	0.0713	0.5518	BDL	BDL
32	Raghunath Baori	Baori	26.3	73.009	7.311	192.3	57.94	0.4016	0.0434	0.0444	0.6526	0.0048	BDL
33	Ramanuj Kot	O/W	26.297	73.029	7.663	91.52	46.3	0.1181	0.0392	0.0787	0.3396	0.0145	BDL
34	Roopnagar Pal	O/W	26.245	72.971	7.315	97.26	28.57	0.109	0.0724	0.1008	0.4812	BDL	BDL
35	RTO, Office, BJS	T/W	26.306	73.064	7.57	105.9	34.78	0.0681	0.0354	0.0207	0.5564	BDL	BDL
36	Siwanchi Gate	O/W	26.285	73.01	7.541	148.1	52.39	0.1672	0.0276	0.0545	0.424	BDL	BDL
37	Toor ji ka Jhalra	Jhalra	26.297	73.023	7.803	108.1	34.09	0.0649	BDL	0.0594	0.429	BDL	BDL
38	US Footwear (inside Sojati Gate)	Basement water supply	26.289	73.024	7.816	83.1	66.1	0.1224	BDL	0.0886	0.6234	BDL	BDL
39	Vishnoi Dharamshala	O/W	26.274	73.035	7.815	112.2	46.82	0.054	0.0281	0.0672	0.5839	BDL	BDL
40	Zanana Park	Baori	26.29	73.031	7.69	50.31	463.03	0.0708	0.402	0.0411	0.3678	BDL	BDL

All the parameters have been compared with the Bureau of Indian Standards (BIS), IS: 10500 of drinking water standards (Table 2).

Table 2. BIS standards of permissible and excessive limits of various parameters

Parameters	**Acceptable Limits**	**Permissible Limits**
pH	6.5-8.5	No Relaxation
TDS (ppm)	500	2000
Zinc (ppm)	5	15
Lead (ppm)	0.01	No Relaxation
Cadmium (ppm)	0.003	No Relaxation
Chromium (ppm)	0.05	No Relaxation
Mangnese (ppm)	0.1	0.3
Copper (ppm)	0.05	1.5

RESULTS AND DISCUSSION

The results obtained from analysis of water sample from different places of Jodhpur are given in Table 3.

1. pH

pH is a measure of the activity of hydrogen ions in a solution, resulting in its acidic or basic quality. In other words, pH is the negative Logarithm of the hydrogen ion concentration in an aqueous solution.

$$pH = -\log_{10} [H^+]$$

pH is measured on a logarithmic scale (Figure 6) that commonly ranges from 0 (acidic) to 14 (basic), with 7 being neutral (Poe, 2000).

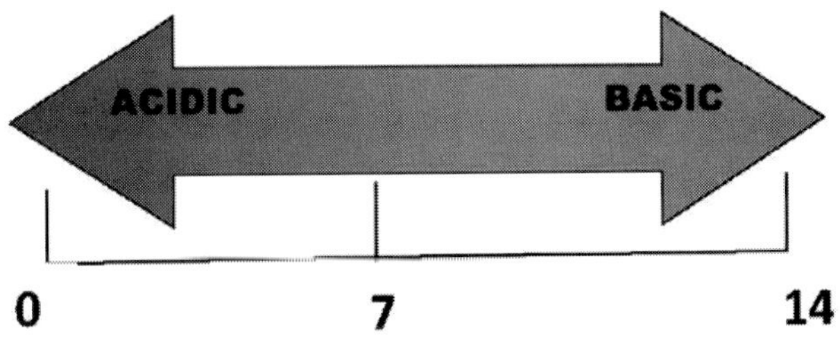

Fig. 6. pH Scale

The pH values of water samples ranges from 7.2-8.7 in Jodhpur regions. Almost all the samples show the pH concentration under the permissible limits defined by BIS, except some sample indicating slightly alkaline samples (Figure 7).

2. Electrical conductance (EC)

Electrical conductivity is the ability of substance to conduct an electrical current, measured in microsiemens per centimetre (mS/cm) or micromhos (μmhos). It is also termed as *conductivity*, or *specific electrical conductance* (Poe, 2000). It is well known that electrical conductance is a good measure of dissolved solids and excessive presence of sodium in water in not only unsafe for irrigation but also makes the soil uncultivable.

In the present investigation, the electrical conductivity of water samples varies from 50-260 mhos/cm. The U.S. salinity laboratory has classified ground water on the basis of electrical conductivity : upto 250 mhos/cm is excellent, 250-750 is good; 750-2250 is fair and > 2250 μ mhos/cm is poor. Based on this classification, all water samples except one sample belong to excellent category (Figure 8).

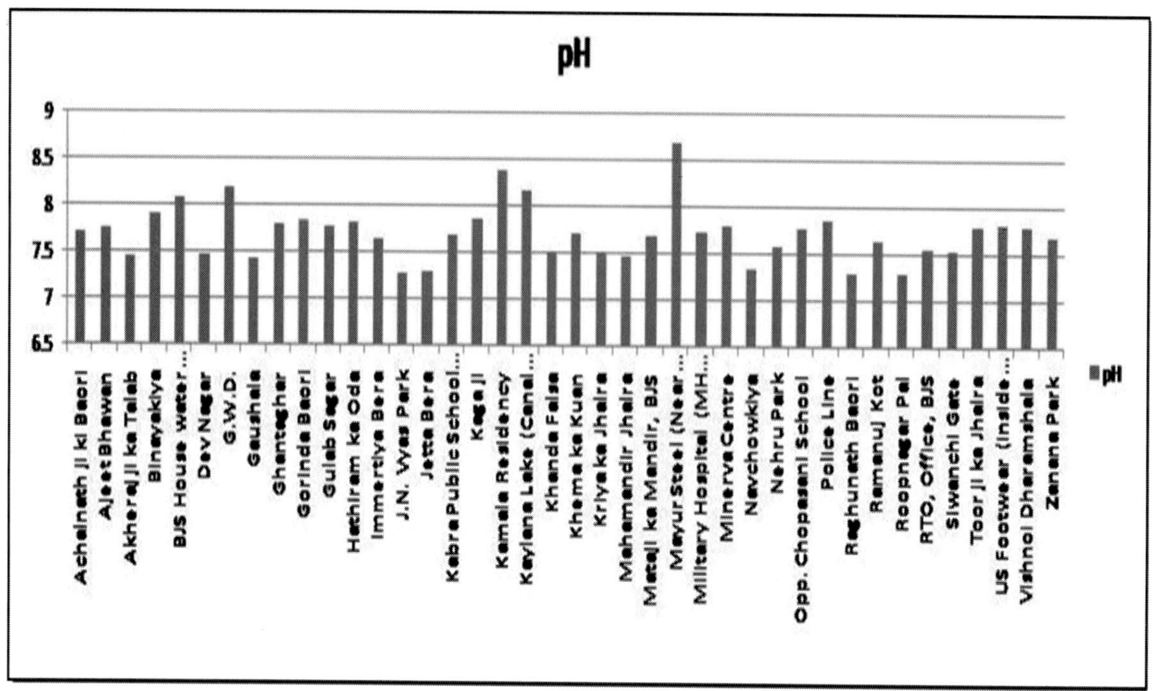

Fig. 7. Histogram showing the pH value of water samples

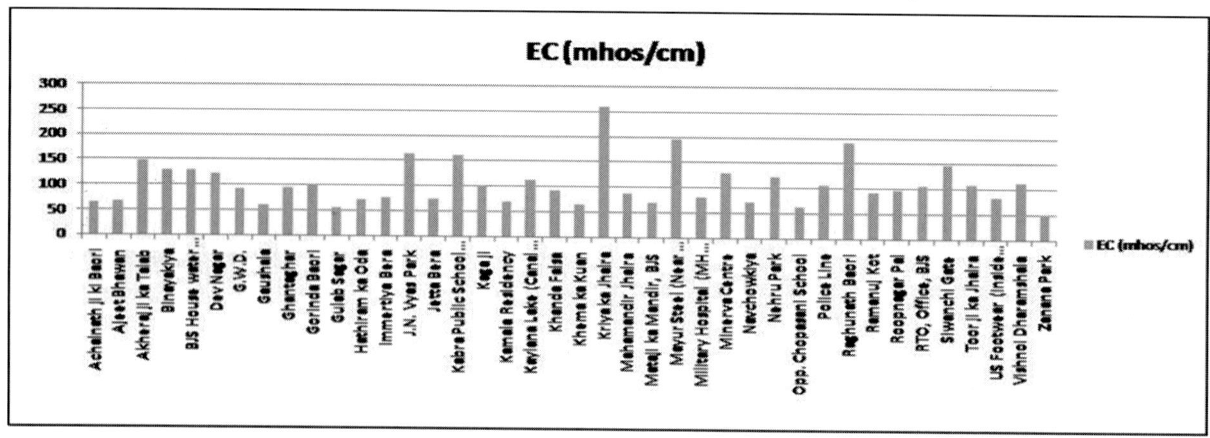

Fig. 8. Histogram showing EC value of water samples

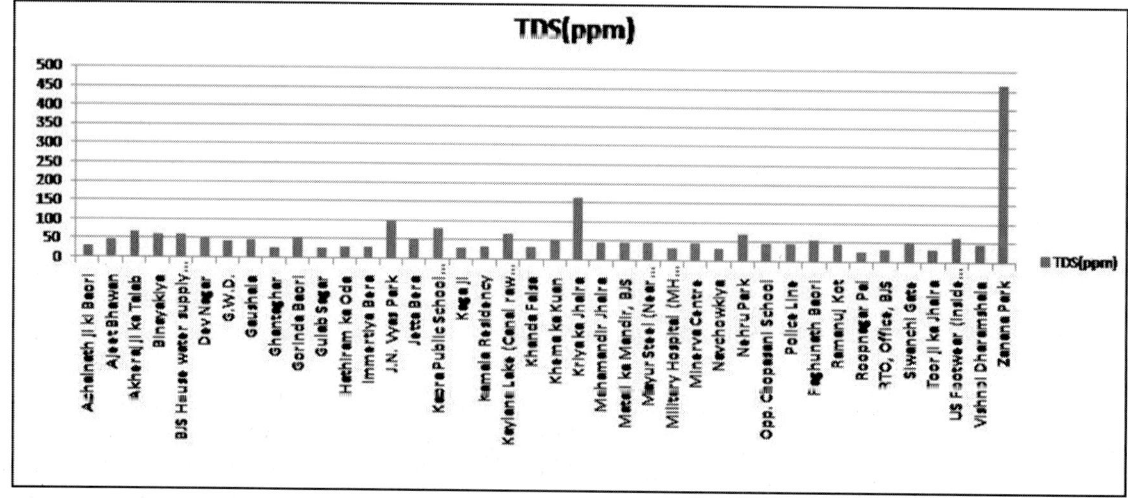

Fig. 9. Histogram shows the TDS value of water samples.

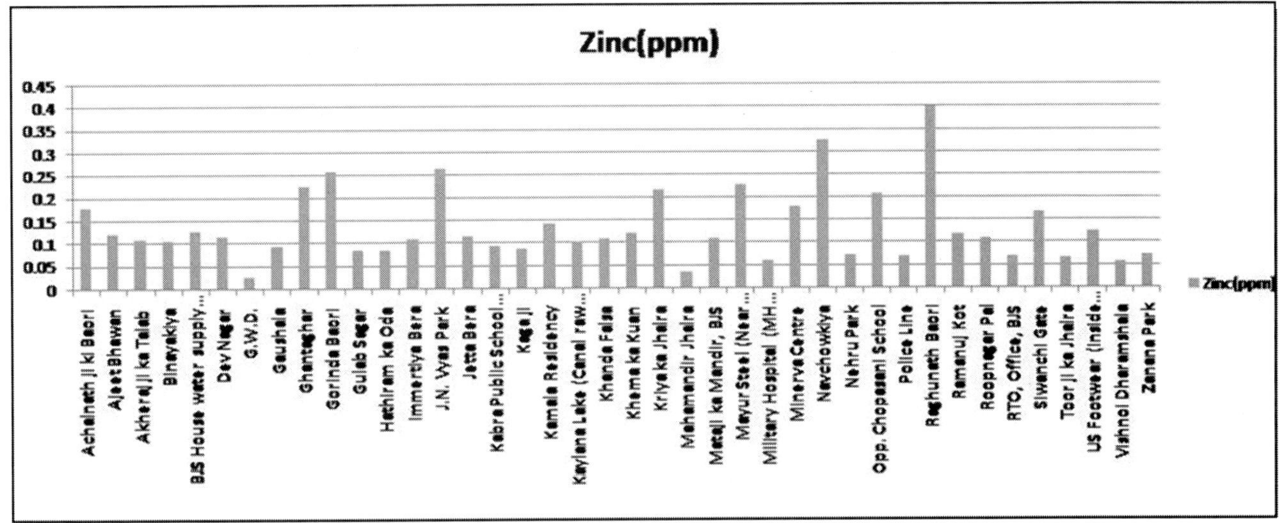

Fig. 10. Histogram showing Zinc Concentration in water samples

3. Total dissolved solids (TDS)

Total dissolved solid (TDS) is a measure, in milligrams per liter (mg/L or ppm), of the amount of dissolved materials in the water. Ions such as potassium, sodium, chloride, carbonate, sulfate, calcium, and magnesium all contribute to the dissolved solids in the water (Poe, 2000). TDS indicate the nature of water quality for salinity.

The water samples in the study fall in the range of 25-463 mg/L. According to BIS specification, all the water samples come under the permissible limit (Figure 9).

4. Zinc

Zinc is a very common substance that occurs naturally. Many foodstuffs contain certain concentrations of zinc. Drinking water also contains certain amounts of zinc. It is very important and essential micronutrient for human health. It is necessary element for physical and neurobiological growth of infants, children and adults. It also influences the activity of various enzymes and hormones. According to WHO (World Health Organization), the recommended dietary dose of Zinc for an adult male is 9.4 mg/day and 6.5 mg/day for female and the permissible limit of zinc in water is not more than 5 mg/liter.

Both deficiency and excess amount of zinc causes various health related problems in human-beings. Zinc deficiency is characterized by growth retardation, loss of appetite, and impaired immune function. In more severe cases, zinc deficiency causes hair loss, diarrhea, delayed sexual maturation, impotence, hypogonadism in males, and eye and skin lesions. Zinc toxicity can occur in both acute and chronic forms. Acute adverse effects of high zinc intake include nausea, vomiting, loss of appetite, abdominal cramps, diarrhea, and headaches (Dietary reference Intakes for Vitamin A, Vitamin K, Arsenic, Boron, Chromium, Copper, Iodine, Iron, Manganese, Molybdenum, Nickel, Silicon, Vanadium, and Zinc, 2001).

In present analysis, the Zinc concentration in water samples varies from 0.02-0.4 parts per million (ppm) (Figure 10). Although all water samples come under the permissible limit (5ppm) described by Bureau of Indian Standards for drinking water (BIS), yet this low to very low concentration of Zinc can also adversely affect the human health.

5. Chromium

Chromium compounds are present in water samplkes only in trace amounts. Chromium is found in two forms either in anionic trivalent or hexavalent. Chromium has a large influence upon drinking water quality. It cannot normally be found in groundwater and surface water in considerable concentrations. The concentration increases due to discharge of waste water into the surface water bodies and over ground through various industries. The estimated safe and adequate daily dietary intake range for chromium varies from 50-200 micrograms (National Research Council, Food and Nutrition Board, Recommended Dietary Allowances, 1989).

Among the two form of chromium, the trivalent form is not toxic but hexavalent form is very toxic for human health. Trivalent chromium is an essential trace element for humans. Together with insulin it removes glucose from blood, and it also plays a vital role in fat metabolism. Chromium deficits may enhance diabetes symptoms. Chromium can also be found in RNA.

Trivalent Chromium (III) toxicity is unlikely, at least when it is taken up through food and drinking water. It may even improve health, and cure neuropathy and encephalopathy. Hexavalent chromium (VI) is known for its negative health and environmental impact, and its extreme toxicity. It causes allergic and asthmatic reactions, is carcinogenic and is 1000 times as toxic as trivalent chromium. Health effects related to hexavalent chromium exposure include diarrhoea, stomach and intestinal bleedings, cramps, and liver and kidney damage. Hexavalent chromium is mutagenic. Toxic effects may be passed on to children through the placenta (www.lenntech.com).

In present analysis, the Chromium concentration in water samples varies from 0.3-0.7 ppm. According to BIS specification, all the 40 water samples show a higher concentration beyond the permissible limit (0.05 ppm). This value can be harmful for human health (Figure 11).

6. Manganese

Manganese compounds exist naturally in the environment as solid compounds in the soil and as dissolved particles in the water. Humans enhance manganese concentrations in the air by industrial activities and through burning fossil fuels. Manganese that derives from human sources can also enter surface water, groundwater and sewage water.

Manganese is an essential nutrient involved in the formation of bone besides its active role in amino acid, cholesterol, and carbohydrate metabolism. The recommended daily intake dose of manganese is 5 mg/day. The deficiency of manganese can also cause adverse health effects such as Skin problems, Changes of hair colour, Glucose intolerance, Neurological Symptoms etc.

On the other hand, the overdose of manganese severely affect the respiratory tract and brain. Symptoms of manganese poisoning are hallucinations, forgetfulness and nerve damage. Manganese can also cause Parkinson, lung embolism and bronchitis. When men are exposed to manganese for a longer period of time they may become impotent. A syndrome that is caused by manganese has symptoms such as schizophrenia, dullness, weak muscles, headaches and insomnia (www.lenntech.com).

The present analysis shows that the Manganese is found in only 11 samples, the concentration ranges from 0.004-0.07 ppm. According to BIS specification, all the water samples show the concentration below permissible limit (0.1ppm). The result shows that there is deficiency of manganese in the samples, which can be harmful for human health (Figure 12).

7. Cadmium

A limited concentration of Cadmium is found in water bodies naturally through weathering of rocks and the rest of the Cadmium is released through human activities, such as manufacturing plants and industries. Another important source of Cadmium emission is during the production of artificial phosphate fertilizers. Part of the Cadmium ends up in the soil after the fertilizer is applied on farmland and the rest of the Cadmium ends up in surface waters due to waste from fertilizer production plants (www.lenntech.com).

It is a by-product of Zinc production plant and one of the most toxic elements to which man can be exposed through inhalation, contaminated food and water. Once absorbed, Cd is efficiently retained in the human body, in which it gets accumulated for life. Cd is primarily toxic to the kidney, especially to the proximal tubular cells which is the main site for accumulation. Cd can also cause bone demineralization, either through direct bone damage or indirectly as a result of renal dysfunction (Bernard, 2008).

In the present investigation, the Cadmium concentration in water samples varies from 0.02-0.1 ppm. According to BIS specification, all the water samples show the higher concentration beyond the permissible limit, which can be harmful for human health (Figure 13) of Jodhpur residents.

8. Copper

Copper is a very common substance that occurs naturally in the environment and spreads through the environment in a natural way. On account of human activities such as disposal of waste water that is contaminated with copper to the water bodies such as ponds, Bawori, canals etc. there would be increased concentration of copper.

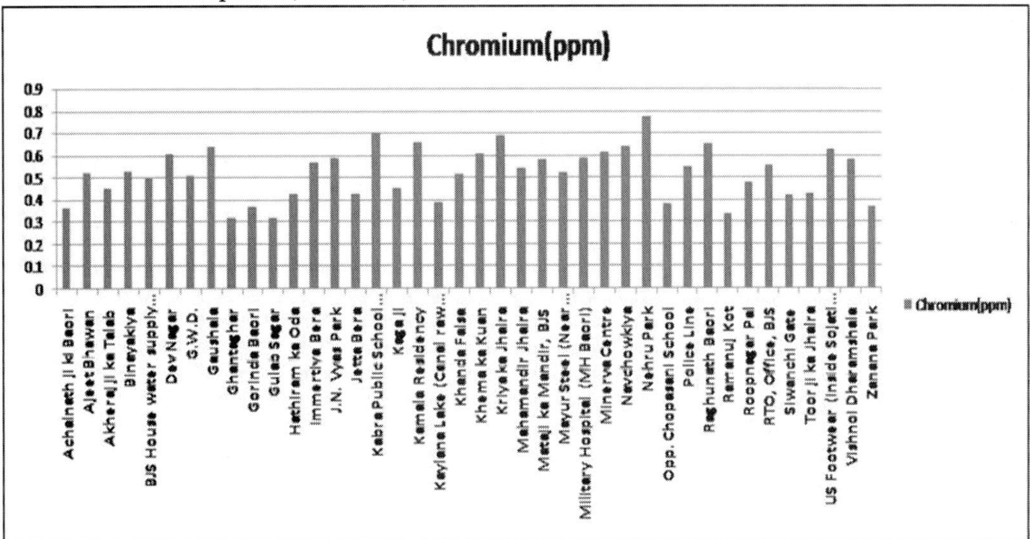

Fig. 11. Chromium concentration of water samples

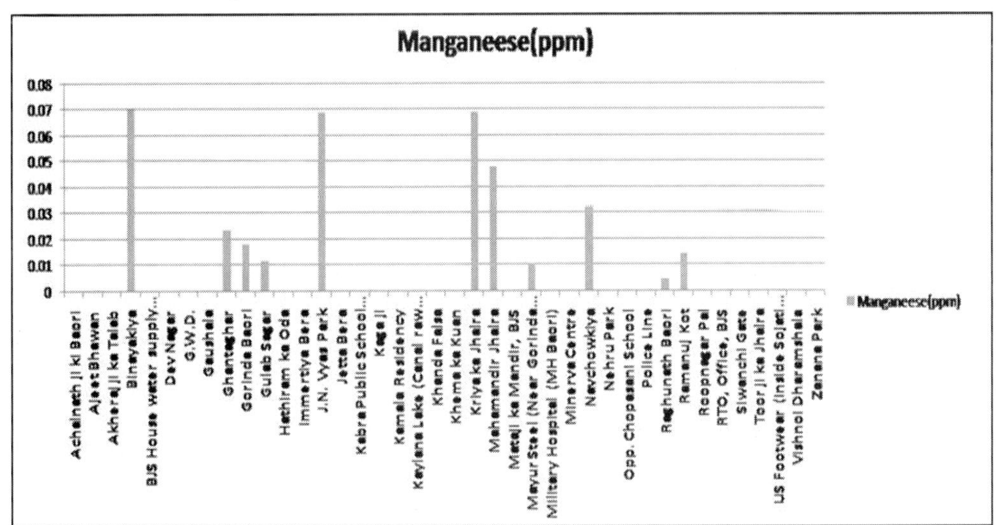

Fig. 12. Manganese concentration of water samples

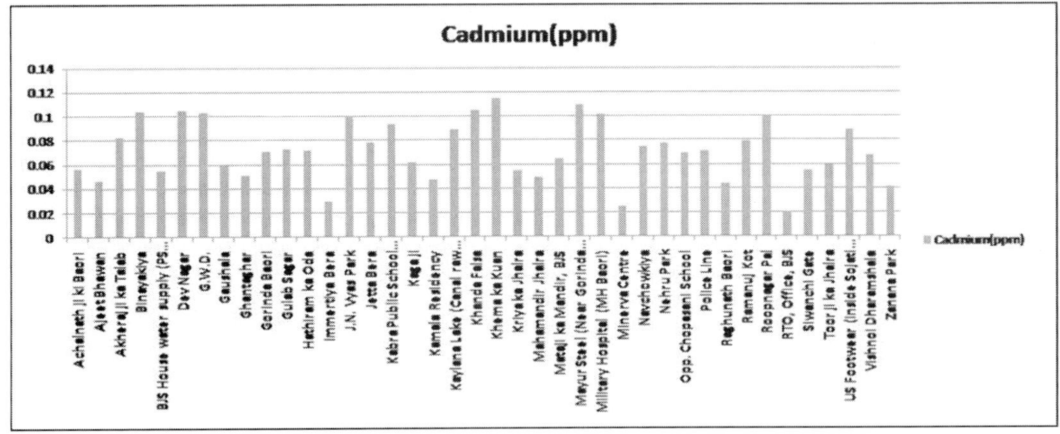

Fig. 13. Cadmium Concentration of water samples

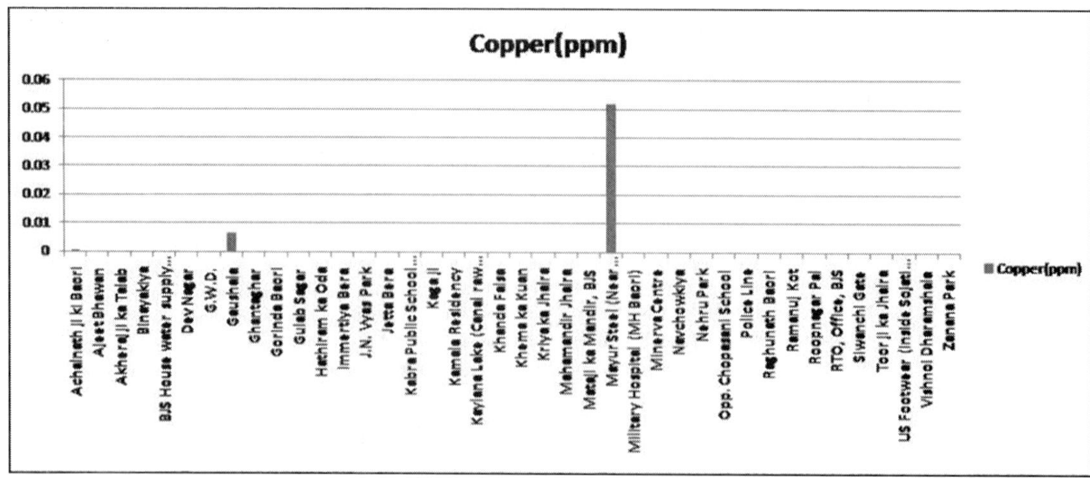

Fig. 14. Copper Concentration of water samples

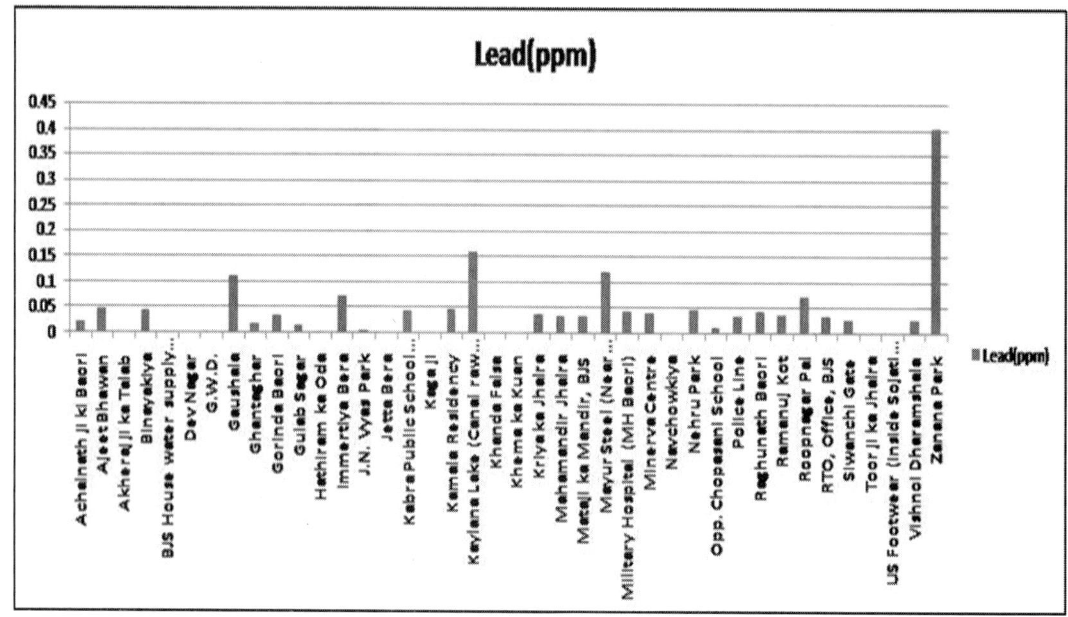

Fig. 15. Lead Concentration of water samples

Although Copper is a toxic heavy metal, yet it is essential for good health. The recommended daily dose of copper is 1.5 to 2.5 mg per day by WHO. Copper is incorporated into a variety of proteins and metallo-enzymes which perform essential metabolic functions; the micronutrient is necessary for the proper growth, development and maintenance of bone, connective tissue, brain, heart and many other body organs. Copper is involved in the formation of red blood cells, the absorption and utilization of iron, the metabolism of cholesterol and glucose and the synthesis and release of life sustaining proteins and enzymes. These enzymes in turn produce cellular energy and regulate nerve transmission, blood clotting and oxygen transport. However, exposure to higher doses can be harmful. Long-term exposure to copper dust can irritate our nose, mouth and eyes, and causes headaches, dizziness, nausea, and diarrhea (Agency for Toxic Substances and Disease Registry, 2004).

In the present study, the measurable Copper was found only in 3 sample out of 40 with a concentration range from 0.006- 0.05 ppm, which comes under permissible limit according to BIS standards (Figure 14). This result shows that all the samples sites are deficient in copper, which can be harmful for human health and should be a matter of concern for inhabitants of Jodhpur city.

9. Lead

Lead is one of the most commonly used metals in the world and like many other metals, is rarely found in its elemental form; rather, it is found in a variety of compounds, complexes and alloys. It occurs naturally in the environment. However, most lead concentrations that are found in the environment are a result of human activities. Due to the application of lead in gasoline, an unnatural lead-cycle has taken place. In automobile engines, lead is burned, so that lead salts (chlorines, bromines, oxides) originate.

These lead salts enter the environment through exhausts of automobiles. The larger particles will drop to the ground immediately and pollute soils or surface waters, the smaller particles will travel long distances through air and remain in the atmosphere. Part of this lead will fall back on earth when it is raining. Beside this, Lead enters in drinking water primarily as a result of corrosion or the wearing away of materials that are in the water supply system and household plumbing. These materials can include lead-based solder, brass and chrome plated faucets.

Lead fulfils no essential function in the human body, it can prove to be a serious health hazard if taken through food, air or water. It can cause several unwanted effects, such as disruption of the biosynthesis of haemoglobin and anaemia, a rise in blood pressure, Kidney damage, miscarriages and subtle abortions, disruption of nervous systems, brain damage, declined fertility of men through sperm damage, diminished learning abilities of children and behavioural disruptions of children, such as aggression, impulsive behavior and hyperact-ivity. Lead can enter a foetus through the placenta of the mother. Because of these issues, it can cause serious damage to the nervous system and the brains of unborn children (www.lenntech.com).

The present analysis shows that Lead is found in 30 samples, concentration varies from 0.004-0.4 ppm. Among these 30 samples, 27 samples show the values more than permissible limit (0.01ppm) define by BIS, which can be harmful for human health (Figure 15).

CONCLUSIONS

The present results reveal that the quality of water of the Jodhpur city is generally within the permissible limits of various parameters including concentration of metals barring Chromium, Cadmium and Lead which are found to be sufficiently high in several samples. These heavy metals show the concentration above the permissible limit as set by BIS, which can cause various adverse health impacts. However, some parameters like Zinc, Copper and Manganese are present in very low concentration, which can also create health problems as these heavy metals are essential for the proper development and metabolism of the human body.

It is therefore, recommended that the proper treatment of the water should be done before being consumed by population in general. Whether it is removal of certain toxic elements or addition of essential elements, both should be taken care of by the authorities in order to provide safe and useful drinking water. It is also suggested that a long term plan should be put into operation where continuous and regular monitoring of the quality of water is being done specially in pre- and post- monsoon seasons. Such a planned growth and development can provide better and healthy conditions for the residents of the Jodhpur city.

ACKNOWLEDGEMENTS

We are grateful to Prof. K.K. Agarwal, Head of the Department of Geology, University of Lucknow for providing us the laboratory facilities and to Prof. K.L. Shrivastava of the Department of Geology, J.N. Vyas University of Jodhpur for providing the working facilities of the department to carry out the present work. We are also very thankful to Ms. Droupti Yadav, Ms. Anamika Singh and Ms. Shashi Verma for helping us in various steps. The samples were collected by S. S. Rathore and the analysis was done by P. Sharma on the Perkin Elmer Analyst 700 Atomic Absorption Spectrophotometer at the Geology Department, University of Lucknow.

REFERENCES

Anonymous, 1988: APHA, 19th Edition, American Public Health Association AWWA, Water Pollution Control Federation, New York.

Anonymous, 1989: National Research Council, Food and Nutrition Board, Recommended Dietary Allowances, 10th Edition, National Academy Press, Washington, DC.

Anonymous, 2001: Dietary Reference Intakes for Vitamin A, Vitamin K, Arsenic, Boron, Chromium, Copper, Iodine, Iron, Manganese, Molybdenum,

Nickel, Silicon, Vanadium, and Zinc. A Report of the Panel on Micronutrients, Subcommittees on Upper Reference Levels of Nutrients and of Interpretation and Uses of Dietary Reference Intakes, and the Standing Committee on the Scientific Evaluation of Dietary Reference Intakes, Food and Nutrition Board, Institute of Medicine, National academy press, Washington D.C.

Anonymous, 2004: Agency for Toxic Substances and Disease Registry,Toxicological profile for Copper. Atlanta, GA: U.S. Department of Health and Human Services, Public Health Service.

Anonymous, 2008: District Ground Water Brochure Jodhpur, Central Ground Water Board, Ministry of Water Resources, Government of India, 2008, p-1, 11.

Anonymous, 2009: Bureau of Indian Standards, Draft Indian Standard, Drinking water- Specification, Second Revision of IS: 10500, ICS No. 13.060.20, p-9.

Bernard, A. 2008: Cadmium & its adverse effects on human health, Department of Public Health, Catholic University of Louvain, Belgium Received April 2, 2008 Indian J Med Res 128, October 2008, pp 557-564

Poe, K.F, 2000: Water Quality and Monitoring, Master Watershed Steward, p-3 and 4.

Web source: http://jodhpur.nic.in/map.html.

Web source: http://www.census2011.co.in/census/city/80-jodhpur.html

Web source: http://www.lenntech.com.

Hon'ble reviewer

1. **Prof. Hema Achythan**
   Chairperson
   Department of Geology
   Anna University
   CHENNAI - 600 001
2. **Dr. Mahesh Kumar**
   Vice President
   Lajarge Cement
   GURGAON 112 001

III-43

Micro-Watershed Morphometry of the Gunjari River Basin, District Shivpuri, Madhya Pradesh, India

Anita Sharma and U.C. Singh

School of Studies in Earth Science, Jiwaji University, Gwalior-474001, (M.P.)
E-mail: anitasharma.phd@gmail.com

Abstract

The morphometric analysis has been carried out for 5 micro-watersheds (A, B, C, D, E) of the Gunjari river basin to determine the basic causes of waterborne problems in the basin area and decentralized approach to develop watersheds at micro level. The quantitative analysis of morphometric parameters like linear, areal and relief reveal that the river basin is susceptible to hydrological processes like water logging, sheet erosion and flooding. At the micro-watershed scale, significant differences were observed and it was found that morphometric properties of micro-watersheds A, C, D, and E are likely to induce high magnitude of these problems compared to morphometric properties of micro-watershed B. These watersheds (A, C, D and E) have high values of bifurcation ratio in the catchment area of 2nd order streams and very low in the section of 3rd order streams with low drainage density, low stream frequency and high circularity ratio. Due to these characteristics, the surface runoff will not quickly be removed from the basin area alarming for the high possibilities of floods in the segments of 3rd order streams and high occurrence of sheet erosion in altitudinal zones. The segment of 1st order streams is also associated with the problem of water logging due to surface undulation and poor surface drainage. Area of trunk channel is subjected to very low runoff and very high potentiality of flooding. However, identified sites should be under taken for the soil conservation and water harvesting programs to minimize the loss of agricultural land and restricting to flood possibilities in the study area.

Key Words: Micro-Watershed, Geo-Morphometry, Soil Erosion, Soil Conservation, Gunjari River Basin.

Introduction

Morphometric study of a river basin describes the precise nature and configuration of the earth topography, its geographic relationship to the land mass and the characteristics of geomorphologic processes (Dinesh, 2008). Now a days the quantitative analysis of drainage system is an important aspect of watersheds (Strahler, 1964), because all hydrologic and geomorphic processes occur within the watersheds. Therefore, micro-watershed morphometry has become the important research unit for a detailed understanding and analysis of hydrological processes and elementary treatment of the fundamental properties of hydrological features. The Gunjari river basin is characterized by hydrologic and geomorphic problems like water logging, sheet erosion and it is liable to flooding so the quantitative study of micro-watersheds will not only help in unraveling the complications of basin morphometry but also in the improvement of river regime.

Drainage System

The region falls in the water resources region '2' (The Ganga river system). The region is subdivided into 4 basins and Shivpuri district falls into the Yamuna river basin [C(2C)]. The Gunjari river is the main water body in Kolaras Block which forms a small part of the Sindh river basin and constitutes catchment [3(2C3)] of the Yamuna river basin. It flows easterly in the investigated area. Most of the streams of the study area are of consequent type, their flow directions are guided by the situation of underlying rocks. Keeping in view of the physical and economic importance of the Gunjari river basin for the area, it has been selected for the detailed analysis. It has been sub-divided into 5 micro-watersheds namely A, B, C, D and E (Fig.1).

Study Area

The Gunjari river is one of the important left bank tributary of the Sindh river and originates from a place near Daharwara village and confluences with the Sindh river near Bharata village after flowing a distance of about 26 kms. Its total drainage area measures 153.7 sq km and lies between north latitudes 25°9'20" to 25°18'33" and east longitudes 77°31'30" to 77°40'18". It is falls in Survey of India toposheet no. 54G/11 and 54G/12. The study basin in the southern uplands of the Vindhyan plateau having undulating topography with hills, plateaus and plains with valleys are the main physiographic units. Climatically, the area comes in semi-arid zone. The yearly average rainfall of the study area is 711mm, with most of the rains falling from June to September. The maximum temperature of the area is 42°C in summer and minimum is 5°C in winter. The annual average potential evapo-transpiration for the area is high i.e. about 1396.6 mm. The Gunjari river and its all tributaries dry up during the cold and hot seasons. The area, characterized by tropical dry deciduous forest, is limited to hill tops and some upland of plateaus.

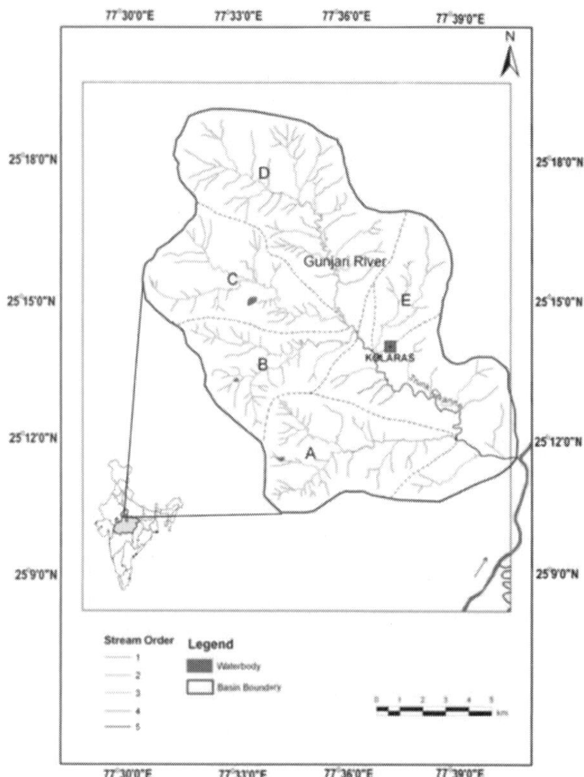

Fig. 1 Drainage Map of the Study Area

Geologically, the Gunjari river basin is a part of Madhya Bharat plateau. The rock units belong to the Upper Vindhyan Supergroup represented by Rewa Sandstones and Shales exposed in the whole study area. Basin forms a large part of alluvial plain of quaternary period. Mainly six types of geomorphic units are identified as denudational hills, residual hills, dissected plateaus, pediplains, alluvial plains and valley fills. The slope of residual hills and plateaus ranges from 10° to 15° keep steepness, whereas alluvial plains and valley fills possess very gentle to nearly level slopes ranging from 0° to 1°. The average mean height varies between 420m to 480m. Most of the area is composed of clay and silty clay soils.

Methodology

The morphometric analysis of the Gunjari river basin was prepared based on published topographic maps on a 1:50,000 scale. The drainage network of the basin was scanned and digitized as available on toposheets. The basin was divided into 5 micro-watersheds and morphometric analysis was carried out at micro level in Erdas Imagine 9.1 software. Basin parameters viz area, perimeter, length, stream lengths, stream orders were also calculated which were later used to calculate linear, areal and relief aspects of the basin. Drainage density map was prepared in Arc GIS 9.2 with input from the drainage pattern map and was classified into high, moderate and low. The drainage pattern map is then superimposed over the drainage density map and used to identify the site-specific mechanism of hydrological process (Fig. 2).

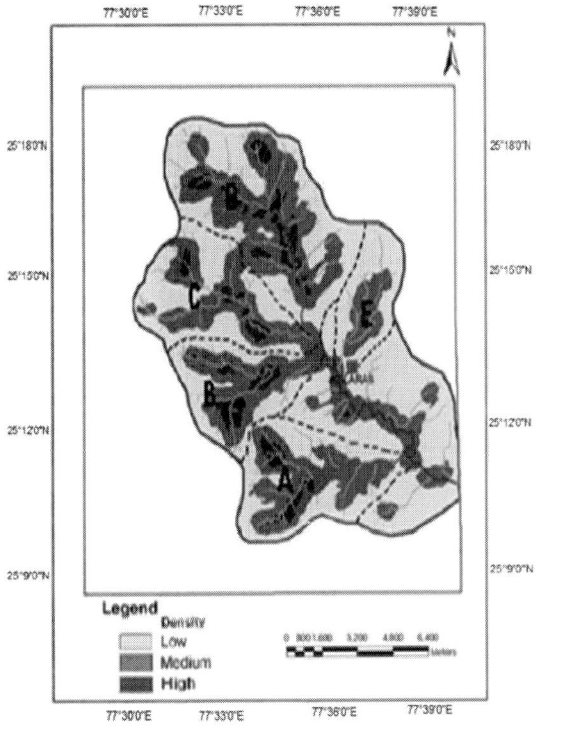

Fig. 2. Drainage Density Map of the Study Area

Drainage Pattern

Drainage network of the basin exhibits dendritic and sub-parallel patterns (Fig. 1) which indicate homogeneity in texture and somewhat structurally controlled. The drainage pattern in sandstones is controlled by joints and fractures while in shale meandering and dendritic patterns are present. In the basin under study, some of the second order streams show the sub-parallel type of drainage pattern mainly because of stepped topography (Baber, 2005).

Results and Discussion

Morphometric Parameters

Morphological parameters along with hydrological parameters are used to quantify and simulate the hydrological behavior of various catchments particularly ungauged catchments (Maidment, 1993; Chalam et al., 1969). The morphometric parameters were analysed and divided in three categories – linear, areal and relief parameters. The procedure used in deriving each of the variables is equally shown below.

Linear Parameters

In the present study, the linear properties were determined with respect to parameters like stream orders, stream numbers, stream length, basin length, bifurcation ratio, length of overland flow and sinuosity indices.

Stream Orders

Strahler's (1957) system of stream orders which is in fact a slightly modified version of Horton's (1945) was followed because of its simplicity. Accordingly, the smallest unbranched stream segment was designated the first order stream, the one formed by the merging of two such first order segments, the second order stream and so on (Subramanyan, 1974). In this ordering, the Gunjari river basin was found to be a fifth order stream (Fig.1) consisting of 5 micro-watersheds. Two micro-watersheds (C, E) are under third order, three micro-watersheds (A, B, D) are identified under fourth order, and only the trunk channel is under the fifth order (Table 2). In micro-watersheds A, C, D and E the 2nd order streams are dominating.

Stream Numbers (Nu)

The Gunjari river basin exhibits that the number of streams of a given order forms an inverse geometric sequence by decreasing systematically with increasing order and conforms to the Horton's (1945) law of stream numbers. The total number of streams in the Gunjari river basin is 261 (Table 2). The D micro-watershed has maximum first order streams (57) while minimum streams (11) have been noticed in micro-watershed E. The presence of the highest number of first order streams indicates the possibility of low permeability and poor infiltration of surface water in the said micro-watershed.

Stream Length (Ls)

The lengths of various stream segments have been measured order wise with the help of

Erdas Imagine software. The total length of all streams in the Gunjari river basin is 202.15 km. The D micro-watershed has maximum first order stream length i.e. 19.45 km indicating minimum loss of drainage water through infiltrations.

Basin Length (Lb)

According to Gregory and Walling (1973), the L is the longest length of the basin from the catchment to the point of confluence. Basin length is used to calculate drainage shape (Sreedevi et al., 2005). The basin length of the Gunjari river is 19.05 km while lengths of the five micro-watersheds are given in (Table 2).

Bifurcation Ratio (Rb)

According to Strahler (1964), the ratio of number of streams of a given order (Nu) to the number of streams of the next higher order (Nu+1) is termed as Rb. The mean Rb of the entire basin is 3.85, usually this value is common in the areas where geologic structures do not exercise a dominant influence on the drainage pattern. It will be seen that the 13 bifurcation ratios calculated for all micro-watersheds fall between 2 and 8 (Table 2). The highest values of Rb common for 2rd order streams in mostly all micro-watersheds are the results of large variation in frequencies between successive orders and indicate undulating surface.

Length of Overland Flow (Lg)

The length of overland flow is considered as a dominant hydrologic and morphometric factor and is the mean horizontal length of flow path from the water divide to this stream in a first order basin and is a measure of stream spacing and degree of dissection (Chorley, 1969). In the area under investigation, the length of overland flow value is 0.38 km/sq km for the entire river basin (Table 3). The values of Lg are 0.40, 0.32, 0.34, 0.39, and 0.35 for A, B, C, D and E micro-watersheds respectively. The Lg values of A, C, D and E micro-watersheds show that the rainwater has to run over more distance before getting concentrated in stream channels, which confirm the coarse texture of the micro-watersheds. The comparatively low value of Lg for micro-watershed B (0.32) indicates that the slope of the channels and surrounding land area is steeper as compared to other 4 micro-watersheds.

Sinuosity Indices (SI)

The Standard Sinuosity Index is calculated using SSI = CL/VL (Muller, 1968). The value of Standard Sinuosity Index (SSI) 1.0 is indicative of straight river course whereas value between 1.0 and 1.5 represents a sinuous course (Singh, 1998). The value of SSI for the Gunjari river basin is 1.3 indicating that the Gunjari river is in sinuous shape and flows in plain land.

Areal Parameters

By introducing a second dimension, the areal properties of a basin can be measured. The areal aspects of the drainage basin include the study of basin area, drainage density, stream frequency, form factor, circularity ratio, elongation ratio etc.

Basin Area (A)

The area of the drainage basin is one of the important parameters like the length of streams draining the basin. The area of a given order is defined as the total area projected on horizontal plane, contributing overland flow to the channel segments of the given order, which includes all tributaries of the lower order. The total area of the Gunjari river basin is found to be 153.7 sq km and areas of all micro-watersheds are given in Table 3.

Drainage Density (Dd)

According to Horton (1932), the drainage density (Dd) is defined as the length of streams per unit area divided by the area of the drainage basin. As the value of the drainage density of the basin under investigation is 1.31 km/sq km; therefore the texture of the basin can be classified as coarse. The values of drainage density for all micro-watersheds range between 1.4 and 1.52 (Table 3), which are low confirming that the area is having low relief and high resistance with permeable soil. The drainage density map (Fig. 2) which is grouped into three categories i.e. low, medium and high reveals high drainage density zone in the catchment area of 2nd order streams than the other streams associated with poor groundwater prospects.

Table 1: Formulae for Computation of Morphometric Parameters

Category	Parameters	Formulae	References
Linear	Stream orders		Strahler (1957)
	Stream numbers (Nu)		Horton (1945)
	Stream Length (Ls)		Horton (1945)
	Basin length (Lb)		Gregory and Walling (1973)
	Bifurcation ratio (Rb)	Rb = Nu/(Nu+1)	Strahler (1964)
	Length of overland flow (Lg)	Lg = A/2L	Chorley (1969)
	Standard Sinuosity Index (SSI)	SSI = CI/ VI	Muller (1968)
Areal	Basin Area (A)		Gregory and Walling (1973)
	Drainage density (Dd)	Dd = L/A	Horton (1932)
	Stream frequency (Fs)	Fs = Ns/A	Horton (1945)
	Form factor (F)	$F = A/L^2$	Horton (1932)
	Circularity ratio (Rc)	$Rc = 4\Pi A/P^2$	Miller (1953)
	Elongation ratio (Re)	$Re = 2\sqrt{A/\Pi}/L$	Schumm (1956)
Relief	Basin relief (R)	R = H-h	Hadley and Schumm (1961)
	Relief ratio (Rr)	Rr = R/L	Schumm (1956)
	Ruggedness number (Rn)	Rn = H* Dd	Strahler (1968)

Table 2: Linear Parameters of the Gunjari river Basin and its Micro- Watersheds

Watersheds	Stream order	Stream number (Nu)	Stream length (km)	Basin perimeter (km)	Basin length (km)	Bifurcation ratio (Rb)	Mean Bifurcation ratio (Rbm)
A	1 Order	44	12.1 km	22.3	8.35	4	3.8
	2 order	11	9.2			5.5	
	3 order	2	5			2	
	4 order	1	5.25				
B	1 order	36	10.65	20.1	6.35	5.1	3.4
	2 order	7	6.2			2.3	
	3 order	3	4.15			3	
	4 order	1	6.5				
C	1 order	33	17.05	20	8.05	5.5	5.7
	2 order	6	9.6			6	
	3 order	1	8				
D	1 order	57	19.45	27.8	10.4	3.4	4.5
	2 order	16	17.55			8	
	3 order	2	6			2	
	4 order	1	9.05				
E	1 order	19	9.2	16	6.5	3.8	4.4
	2 order	5	6.2			5	
	3 order	1	2.8				
Trunk channel	1 order	11	16.3	40.50	8.65		
	2 order	3	10				
	5 order	1	11.65				

	1 order	200	84.75			4.1	
Total	2 order	48	58.75			5.3	
(Gunjari river	3 order	9	26.2	54	19.05	3	3.8
basin)	4 Order	3	20.8			3	
	5 order	1	11.65			-	

Table 3: Areal Parameters of the Gunjari river Basin and its Micro-Watersheds

Watersheds	Basin area (sq km)	Drainage density (km/sq km)	Stream frequency (Fs)	Length of overland flow (km/sq km)	Form factor (Rf)	Elongation ratio (Rc)	Circularity ratio (Rc)
A	25.75	1.22	2.25	0.40	0.36	0.68	0.65
B	18	1.52	2.61	0.32	0.44	0.75	0.55
C	24	1.44	2.84	0.34	0.37	0.68	0.75
D	41.5	1.25	1.83	0.39	0.38	0.69	0.67
E	13	1.4	1.15	0.35	0.30	0.62	0.63
Trunk channel	31.5	1.20	1.50	0.41	0.41	0.72	0.25
Total (Gunjari river basin)	153.7	1.31	1.69	0.38	0.42	0.73	0.66

Table 4: Relief Parameters of the Gunjari river Basin and its Micro-Watersheds

Watersheds	Basin Relief (R) in m.	Relief Ratio (Rr)	Ruggedness No. (Rn)
A	40	0.0047	0.04
B	40	0.0062	0.06
C	40	0.0049	0.05
D	40	0.0038	0.05
E	20	0.0030	0.02
Trunk channel	20	0.0023	0.02
Gunjari river basin	60	0.0031	0.07

Stream Frequency (Fs)

The stream frequency (Fs) of the basin may be defined as the ratio between the total number of segments cumulated for all orders within a basin and the basin area (Horton, 1945). In this study, the stream frequency of 1.69/sq km for the whole river basin is an indication that the intensity of dissection in the area is very low. These very low values for the micro-watersheds D (1.8) and E (1.1) have very little effect on the extent to which the surface has been lowered by agents of denudation (Eze and Efiong, 2010).

Form Factor (F)

Form factor (Rf) is defined as the ratio of the basin area to the square of the basin length. Horton (1945) proposed this parameter to predict the flow intensity of the basin of a defined area. Form factor of the Gunjari river basin is 0.42 indicating moderate runoff in the basin. In addition, the form factor values range between 0.30 and 0.44 for all micro-watersheds (Table 3). Low form factors micro-watersheds A, C, D and E have flatter peak flows for longer duration, while the micro-watershed B with moderate value has moderate peak flow for moderate duration.

Circularity Ratio (Rc)

The circularity ratio (Rc) has been used as an areal aspect and is expressed as the ratio of basin area (A) to the area of a circle (Ac) having the same perimeter as the basin (Strahler, 1964). The Rc of the whole basin is 0.66, which indicates that the basin is sub-circular in shape and has the dendritic stage. The Rc values for five micro-watersheds are given in Table 3, which are in the elongated to sub-circular shape. The high values of circularity ratios for micro-watersheds A, C, D and E signify an early topographical old stage of the geomorphological cycle while micro-watershed B reveals mature topographical late stage of the geomorphological cycle.

Elongation Ratio (Re)

Elongation ratio (Re) is the ratio between the diameter of a circle with the same area as that of the basin (A) and maximum length (L) of the basin (Sreedevi et al., 2009). A circular basin appears more hasten in the discharge of run-off than that of an elongated basin (Singh and Singh, 1997). Re for the entire basin is estimated to be 0.73 which also confirms sub-circular shape of the basin. The Re values obtained for all micro-watersheds vary between 0.75 and 0.62 (Table 3), which indicate that the flow of water in these micro-watersheds is quickly distributed.

Relief Parameters

The relief aspects of the drainage basin include the study of basin relief, relief ratio, and ruggedness number.

Basin Relief (R)

The maximum height in the Gunjari river basin is 480m and the lowest is 420m. Therefore, the relief of the basin is 60m. The analysis reveals that the basin and all micro-watersheds have low to moderate relief values. The low values of micro-watershed E and area of trunk channel indicate low gravity of water flow, high infiltration and very low runoff condition in comparison to micro-watersheds A, B, C and D.

Relief Ratio (Rr)

When basin relief R is divided by the horizontal distance on which it is measured, there results a dimensionless ratio called relief ratio (Baber, 2005). It is an indicator of rates of erosion operating along the slope of the basin (Schumm, 1954). The relief ratio value obtained is 0.0031 for the Gunjari river basin, while that of the five micro-watersheds are shown in Table 4. Generally the relief ratio of the basin is low and is a characteristic feature of less resistant rocks and low relief of the area. In field it has been observed that there is a high degree of correlation between low stream channel slope and long distance between two stream valleys which bring out extensive sheet erosion in a short duration. These are characteristics of micro-watersheds A, C, D, E and in trunk channel indicating that the erosional work are more active in these watersheds.

Ruggedness Number (Rn)

Strahler's (1968) ruggedness number (Rn) is the product of the basin relief (R) and drainage density (Dd) and is useful combination of slope steepness with its length. It indicates the structural complexity of the terrain (Bagyaraj and Gurugnanam, 2011). Calculated accordingly, the Gunjari river basin has a ruggedness number of 0.078 (Table 4), indicating that the basin lies within low to moderate sloping terrain and there is some structural complexity. The analysis for all micro-watersheds shows that the slope of micro-watershed B is not only steep but long as well as having Rn value more than 0.05.

Conclusion

The quantitative analysis of morphometric parameters like linear, areal and relief has characterized that the drainage area of the Gunjari river basin is almost plain having some basement complexities. It has been noticed that the river basin is more undulating because of more variations in the values of bifurcation ratio between 2nd and 3rd order streams, show high altitudinal zones of 2nd order traverse, while 3rd, 4th and 5th order stream segments occur in comparatively flat lands. The basin is sub-circular in shape, quickly discharge water and highly susceptible to hydrologic processes like water logging, sheet erosion and flooding. However, some segments of the basin are more susceptible to these problems as due to surface undulation and poor surface drainage, 1st order streams have given rise to the problems of water logging and incrustation of soil. Due to the sloppy surface, the areas of 2nd order streams have undergone severe sheet erosion and because of the low values of bifurcation ratios in some parts of 3rd order streams are more liable to flooding. The mophometric characteristics at the micro-watershed scale reveal that A, C, D and E express very low relief values, sub-circular shapes with high variations in bifurcation ratios which implies that they are more prone to these environmental problems. Whereas micro-watershed B indicates medium runoff zone and less relief variations showing that this micro-watershed is more efficient in managing these problems. In the context of drainage area of trunk channel, it keeps very low morphometric values compared to all other micro-watersheds. These values imply high orthogonal distance of water divider and high probabilities of flooding in this segment of the basin. Efficient basin management techniques therefore call for reducing rates of catchment degradation in the basin. The segment of 2nd order streams in micro-watersheds A, C, D and E should be undertaken for the soil conservation programs whereas the adjoining area of 2nd order streams can be used for check dams construction for both the water harvesting and reducing the flood potential in these micro-watersheds.

Acknowledgements

The first author is very grateful for the research facilities and editorial supports from the Head of the Earth Science Department, Jiwaji University, Gwalior (M.P.). He significantly helped me in improving the quality of this work. I also thank Dr. S.N. Mohapatra for his constructive comments on this paper.

References

Baber, M. (2005). Hydrogeomorphology. Fundamentals application and techniques, New India Publishing Agency, pp. 110-261.

Bagyaraj, M. and Gurugnanam, B. (2011). Significance of Morphometry Studies, Soil Characteristics, Erosion Phenomena and Landform Processes Using Remote Sensing and GIS for Kodaikanal Hills, A Global Biodiversity Hotpot in Western Ghats, Dindigul District, Tamil Nadu, South India. Res. Jour. of Env. & Ear. Sci., **3**(3): 221-233.

Dinesh, S. (2008). Computation and characterization of basic morphometric measures of catchments extracted from digital elevation models. Journal of Applied Science Research, **4**(11): 1488-1495.

Chalam, B. N. S., Krishnaveni, M. and Karmegam, M. (1969). Correlation analysis of runoff with geographic parameters. J. Appl. Hydrology, **9**(3-4): 24-31.

Chorley, R. J. (1969). The drainage basin as the fundamental geomorphic unit. In water, Earth and Man (ed. By R.J. Chorley), Methuen, London, pp. 77-100.

Eze, Bassey and Efiong, J. (2010). Morphometric Parameters of the Calabar River Basin: Implication for Hydrologic Processes. Journal of Geography and Geology, **2**(1): 18-26.

Gregory, K. J. and Walling, D. E. (1973). Drainage basin form and process a geomophological approach. Arnold, London.

Hadley, R. F. and Schumm, S. A. (1961). Sediment sources and drainage basin characteristics in upper Cheyenne river basin. United State Geological Survey Water-Supply Paper, 1531-B, pp. 137-196.

Horton, R. E. (1932). Drainage basin characteristics. Trans. Am. Geophys. Union, **13:** 350-361.

Horton, R. E. (1945). Erosional development of stream and their drainage basin; hydrological approach to quantitative morphology, Geol. Soc. Mer. Bull., **56:** 275-370.

Maidment, D. R. (1993). Handbook of Applied Hydrology. McGraw hill Inc., New York.

Miller, V. C. (1953). A Quantitative Geomorphic Study of Drainage Basin Characteristics in the Clinch Mountain Area. NR Technical Report **3:** Va. And Tenn. Office Naval Research Project.

Muller, J. E. (1968). An introduction to hydraulic and topographic sinuosity index. Ann. Assoc. Amer. Geogr., **58**(2): 371-385.

Schumm, S. A. (1954). The relation of drainage basin relief to sediment loss. Assoc. Ind. Sci. Hydrol. Proc. Ass. Gen. Rome., **1**(36): 216-219.

Schumm, S. A. (1956). Evolution of drainage systems and slopes in Badlands at Perth Amboy. New Jersey. Nat. Geol. Soc. Am. Bull., **67:** 597-646.

Singh, S. and Singh, M. C. (1997). Morphometric analysis of Kanhar river basin. National Geographical J of India, **43**(1): 31-43.

Singh, S. (1998). Geomorphology. Prayag Pustak Bhawan, Allahabad, India, pp. 1-586.

Sreedevi, P. D., Subrahmanyam, K. and Ahmed, S. (2005). The significance of morphometric analysis for obtaining groundwater potential zones in a structurally controlled terrain. Environmental Geology, **47:** 412-420.

Sreedevi, P. D., Owais, S., Khan, H. H. and Ahmed, S. (2009). Morphometric Analysis of a Watershed of South India Using SRTM Data and GIS. Journal Geological Society of India, **73:** 543-552.

Strahler, A. N. (1957). Quantitative Analysis of Watershed Geomorphology. Trans. Am. Geophys. Un., **38:** 913.

Strahler, A. N. (1964). Quantitative Geomorphology of Drainage Basin and Channel Networks. In: Chow

VT (ed), Handbook of Applied Hydrology, McGraw Hill Book Co., New York, pp. 39-76.

Strahler, A. N. (1968). Quantitative geomorphology. In: Fairbridge, R.W. (ed). The Encyclopedia of geomorphology, Reinhold Book Crop. New York.

Subramanyam, V. (1974). A quantitative analysis of two drainage basins around Sagar. M.P., Proceedings INSA, **40:** 76-99.

Hon'ble reviewer

1. **Prof. Rameshwar Bali**
   Department of Geology
   Centre of Ad. Study
   University of Lucknow
   LUCKNOW - 226 007
2. **Prof. Amita Pande**
   Department of Geography
   DBS Campus
   Kumaon University
   NAINITAL - 263 001

III-44

Investigation of Geomorphic Development and Categorization of Small Watersheds in Central Himalaya

Anita Pande

Associate Professor, Department of Geography, Kumaun University, Nainital, Uttarakhand
E-mail: anni_pan@rediffmail.com

Abstract

Investigations of geomorphic development were done for small watersheds of Central Himalaya. Categorization of watersheds was done on the basis of i. Erosion stage ii. Critical-height zone iii. Erosion integral iv. Hypsometric integral. The watersheds are categorized under two categories on the basis of erosion stage, viz. equilibrium stage and inequilibrium stage; under four categories on the basis of critical-height-zone, viz. low altitude zone (1300-1500m), moderate altitude zone (1500-1700m), moderately high altitude zone (1700-1900m) and high altitude zone (1900-2100m); under three categories on the basis of Erosion integral, viz. low erosion integral (10%-30%), moderate erosion integral (30%-50%) and high erosion integral (50%-60%); under three categories on the basis of hypsometric integral, viz. low hypsometric integral (40%-60%), moderate hypsometric integral (60%-80%) and high hypsometric integral (80%-90%).

As far as erosion intensity of watersheds was concerned, 78% watersheds experienced their equilibrium stage while only 22% watersheds attained their inequilibrium stage of geomorphic development. For critical-height-zone of watersheds, maximum (about 78%) watersheds possess moderately high altitude while minimum watersheds fall under moderate (11%) and high (11%) altitude zone. Under erosion integral of watersheds, maximum watershed (about 78%) were confined to moderate erosion integral while minimum watersheds (22%) belong to high erosion integral. The hypsometric integral of watersheds possesses maximum watersheds (about 78%) under low hypsometric integral while minimum watersheds (22%) belong to moderate hypsometric integral.

Keywords: Central Himalaya, Small Watersheds, Watershed Management, Geomorphic Development, Hypsometric Integral, Erosion Integral

Introduction

The Himalayan environment is an outcome of adverse land use conditions due to its fragile behaviour associated with increasing anthropogenic interference. The varying altitude as well as slope condition significantly contribute in the determination of land use status of Himalayan terrain. Consequently the problem of erosion intensity and consequent slope instability as well as deteriorating ecological conditions need for an instant conservation and reorganization of land in this area.

THE STUDY AREA

The 'Jaigan Watershed' extends in between the latitudes of 29°47'25"N to 29°39'5"N and the longitudes of 79°41'45"E to 79°53'40"E. Administratively it lies in Bageshwar District of Uttarakhand. Encompassing an area of about 150 km^2, it is a part of the greatest Sarju-Kali drainage system of Kumaun Himalaya (Fig.1).

OBJECTIVE

The major objective of present study is to assess the geomorphic traits and land use status of the study area and to estimate the impact of land use upon the geomorphic development of Central Himalayan watersheds.

DATA BASE AND METHODOLOGY

Nine fourth order watersheds of Jaigan watershed was selected for the detailed geomorphic study in Himalayan terrain. The stream ordering was done using Strahler's (1964) stream segment method. Base map was prepared with the help of SOI topographical sheet nos.. 53 O/9, 53 O/10, 53 O/13 and 53 O/14. Area computation and length measurements were grid based. The stages of geomorphic development (Strahler, 1952) of watersheds were based on percentage hypsometric curve (Fig. 2, 3, 4, and 5).

ANALYSIS AND DISCUSSION

The physiographic characteristics of the erosional topography of small watersheds of Central Himalayan was examined under different variables viz. stage of erosion, critical-height-zone, erosion integral and hypsometric integral:

I. Stage of Erosion

The watersheds were categorized under two categories on the basis of stage of erosion; viz. equilibrium stage and inequilibrium stage. Out of all studied watersheds about 78% watersheds were experiencing their equilibrium stage while only about 22% watersheds attained their inequilibrium stage of geomorphic development (Fig. 6). Variability in erosion stage is noticed under different land use/land cover:

(a) Cultivated Land Scenario About 56% watersheds have undergone their equilibrium stage while about 44% are still experiencing their inequilibrium stage of geomorphic development.

(b) Forest Land Scenario About 67% watersheds have undergone their equilibrium stage while about 33% are still experiencing their inequilibrium stage of geomorphic development.

(c) Barren Land Scenario 100% watersheds of the study area are experiencing their inequilibrium stage of geomorphic development.

II. Critical-Height Zone

The watersheds were categorized under four categories on the basis of critical-height-zone; viz. low altitude zone (1300-1500 m), moderate altitude zone (1500-1700 m), moderately high altitude zone (1700-1900 m) and high altitude zone (1900-2000 m). About 45% watersheds possess critical-height-zone of 1700-1800 m above which erosion was found acute followed by critical- height zone of 1800-1900 m and this belongs to about 33% watersheds. The-critical-height zone of 1500-1600 m and 1900-2000 m which are infect the lowest and highest critical-height-zone categories respectively, possess only about 11% watersheds under each category (Fig. 7). Variability in critical-height-zone is noticed under different land use/land cover:

(a) Cultivated Land Scenario The critical-height-zone of 1400-1500 m and 1500-1600 m possess about 34% watersheds under each height-zone respectively while the critical-height-zone of 1300-1400 m, 1600-1700 m and 1700-1800 m represent about 11% watersheds under each zone respectively.

(b) Forest Land Scenario About 45% watersheds possess critical-height-zone of 1800-1900 m while critical-height-zone of 1700-1800 m and 1900-2000 m belongs to 22% watersheds under each zone. In case of forest land the lowest critical-height-zone i.e. 1500-1600 m governs only 11% watersheds of the study area.

(c) Barren Land Scenario Maximum watersheds i.e. about 40% possess the critical height-zone of 1500-1600 m while the rest critical-height-zone viz. 1300-1400 m, 1400-1500 m and 1700-1800 m are represented by about 20% watersheds under each height zone respectively.

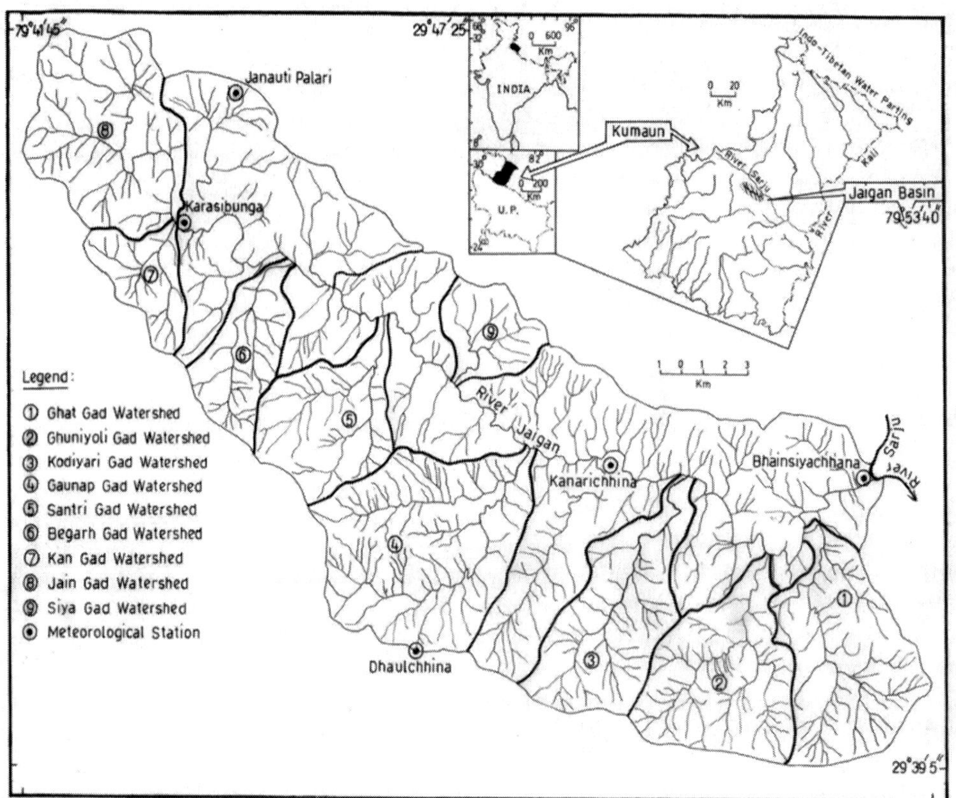

Fig. 1 Location Map

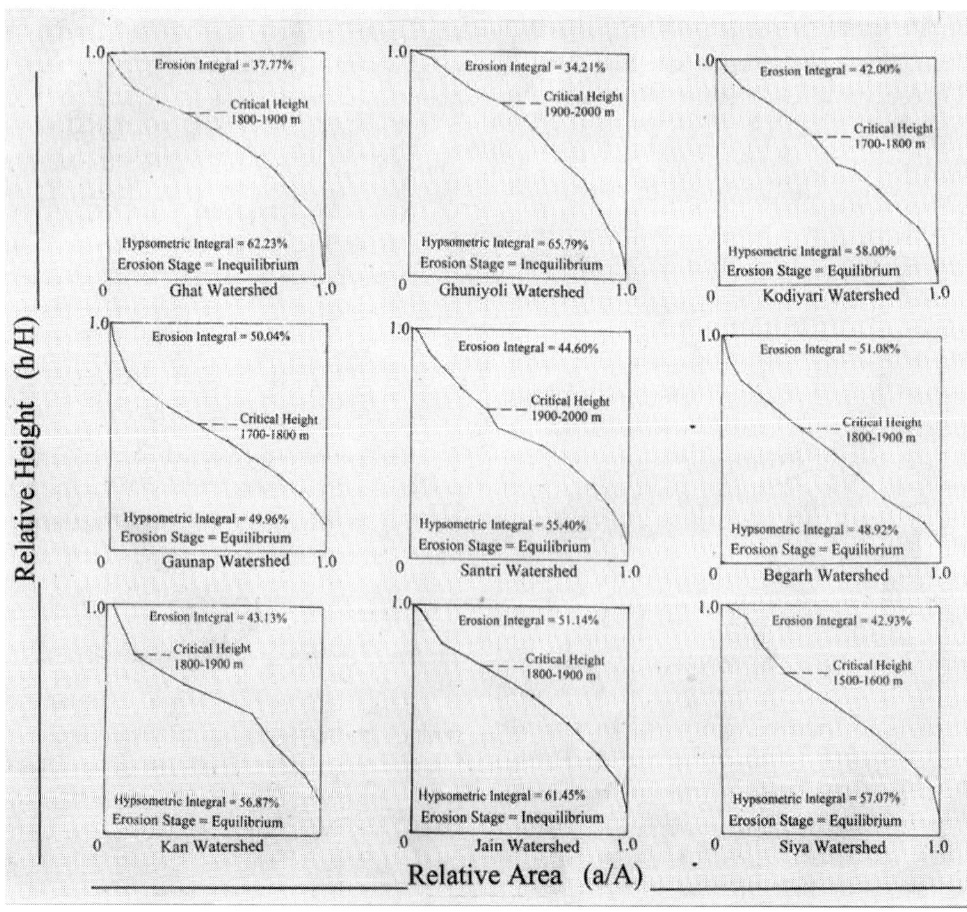

Fig. 2. Erosion Intensity in different Watersheds

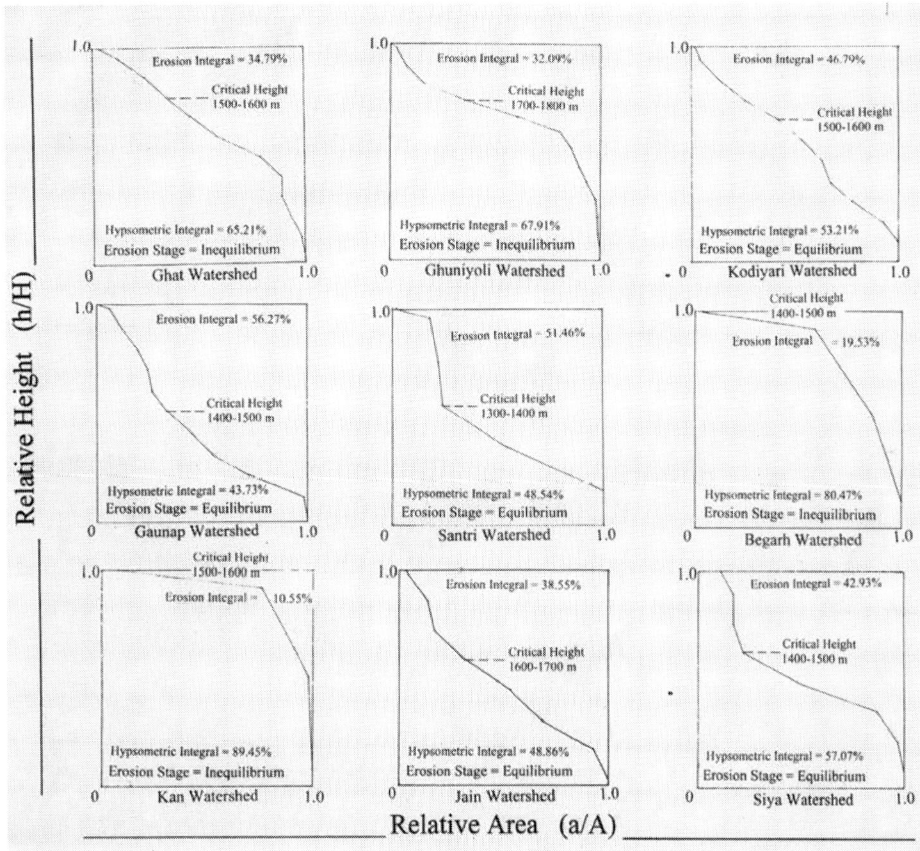

Fig. 3. Erosion Intensity in Cultivated Land under different Watersheds

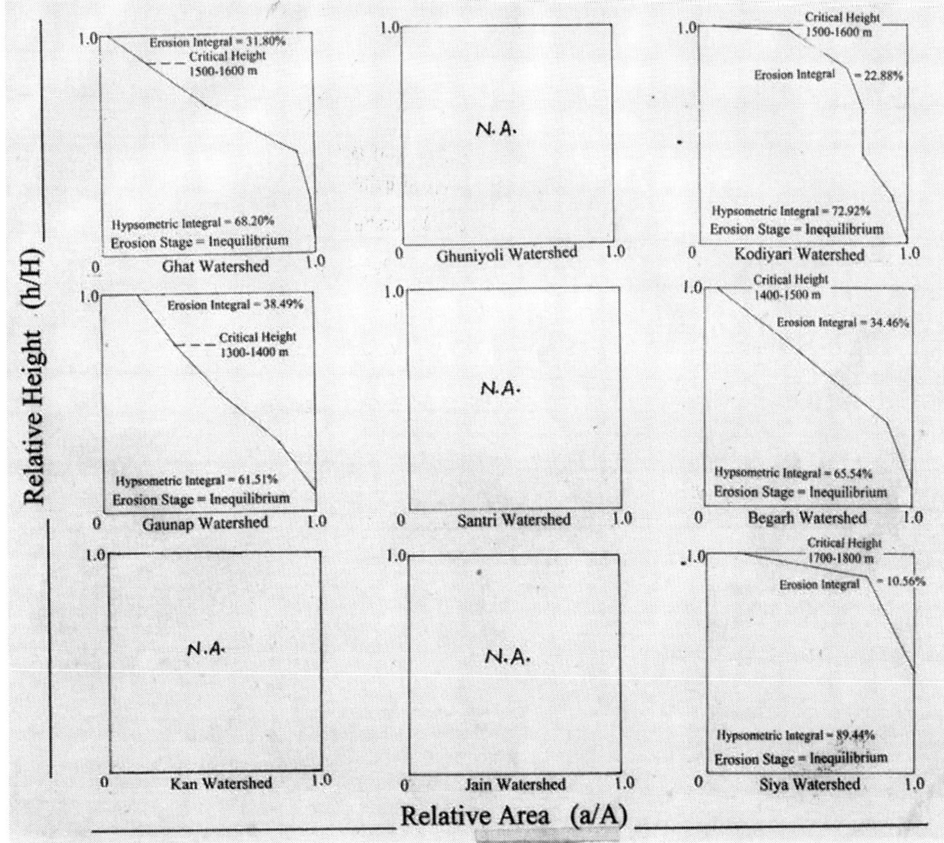

Fig. 4. Erosion Intensity in Barren/Rocky Land under different Watersheds

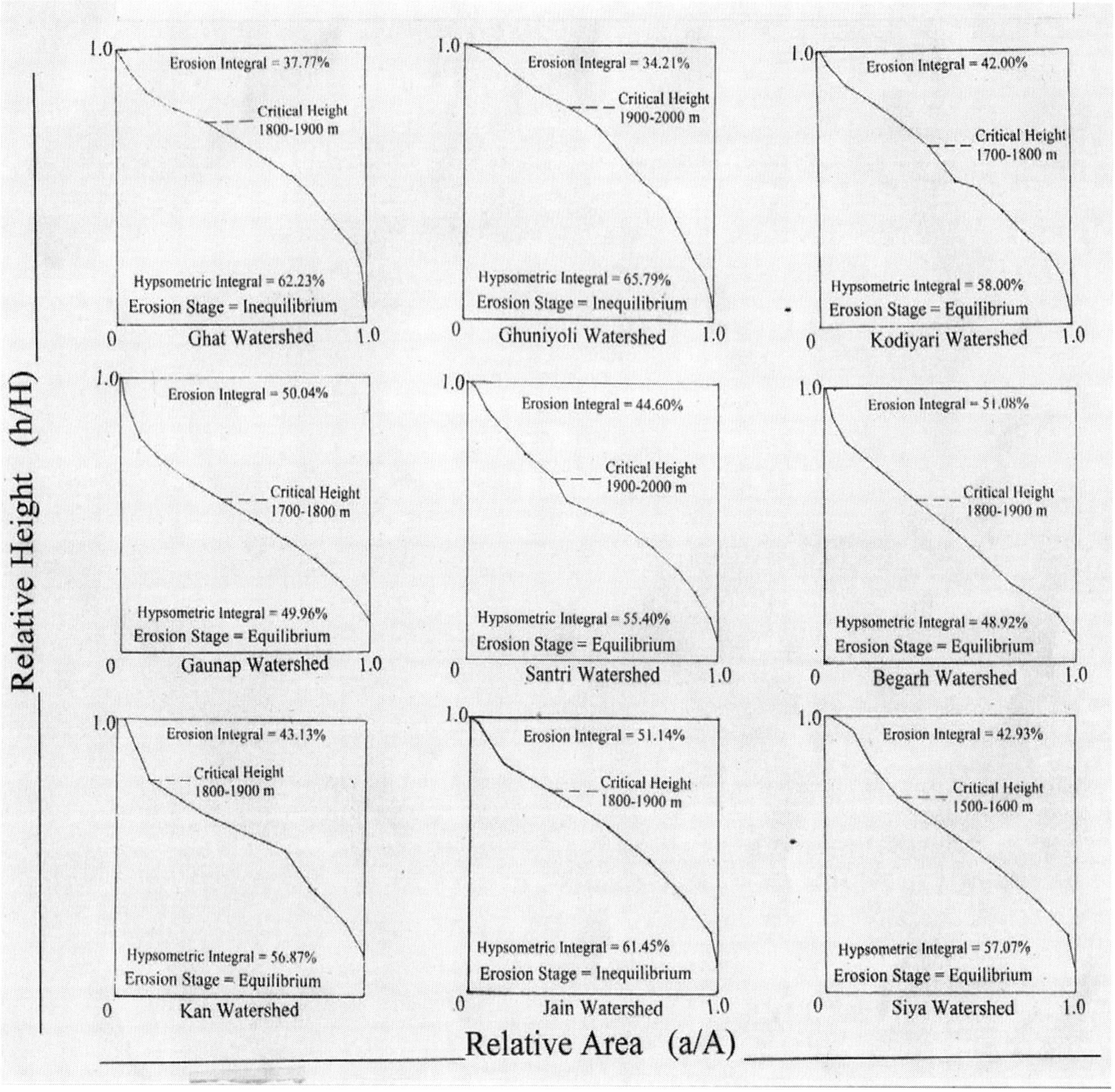

Relative Height (h/H)
Relative Area (a/A)
Erosion Integral = 37.77%
Critical Height 1800-1900 m
Hypsometric Integral = 62.23%
Erosion Stage = Inequilibrium
Ghat Watershed
Erosion Integral = 34.21%
Critical Height 1900-2000 m
Hypsometric Integral = 65.79%
Erosion Stage = Inequilibrium
Ghuniyoli Watershed
Erosion Integral = 42.00%
Critical Height 1700-1800 m
Hypsometric Integral = 58.00%
Erosion Stage = Equilibrium
Kodiyari Watershed
Erosion Integral = 50.04%
Critical Height 1700-1800 m
Hypsometric Integral = 49.96%
Erosion Stage = Equilibrium
Gaunap Watershed
Erosion Integral = 44.60%
Critical Height 1900-2000 m
Hypsometric Integral = 55.40%
Erosion Stage = Equilibrium
Santri Watershed
Erosion Integral = 51.08%
Critical Height 1800-1900 m
Hypsometric Integral = 48.92%
Erosion Stage = Equilibrium
Begarh Watershed
Erosion Integral = 43.13%
Critical Height 1800-1900 m
Hypsometric Integral = 56.87%
Erosion Stage = Equilibrium
Kan Watershed
Erosion Integral = 51.14%
Critical Height 1800-1900 m
Hypsometric Integral = 61.45%
Erosion Stage = Inequilibrium
Jain Watershed
Erosion Integral = 42.93%
Critical Height 1500-1600 m
Hypsometric Integral = 57.07%
Erosion Stage = Equilibrium
Siya Watershed

Fig. 5. Erosion Intensity in Forest Land under different Watersheds

Entire Watershed Scenario
22%
78%
Cultivated Land Scenario
44%
56%
Equilibrium Stage
Inequilibrium Stage
Forest Land Scenario
33%
67%
Barren Land Scenario
0%
100%

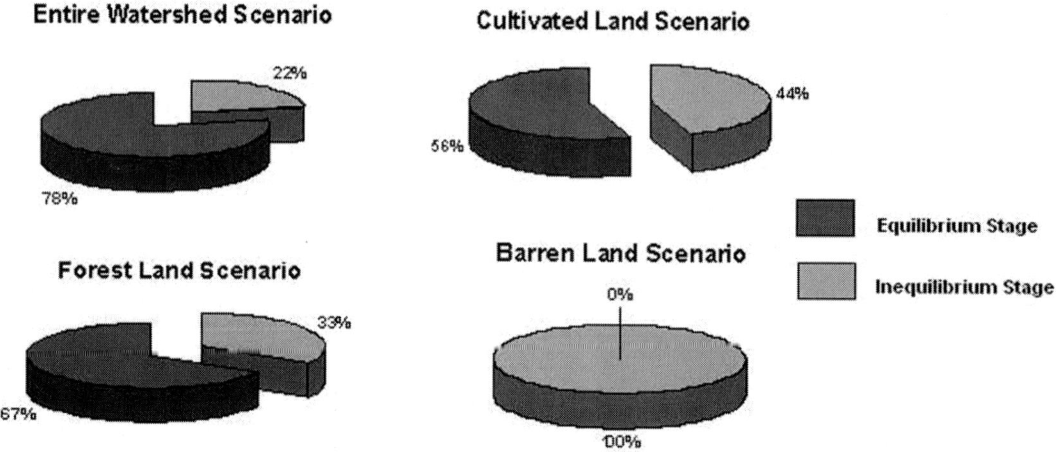

Fig.6. Categorization of Watersheds on the basis of Stage of Erosion

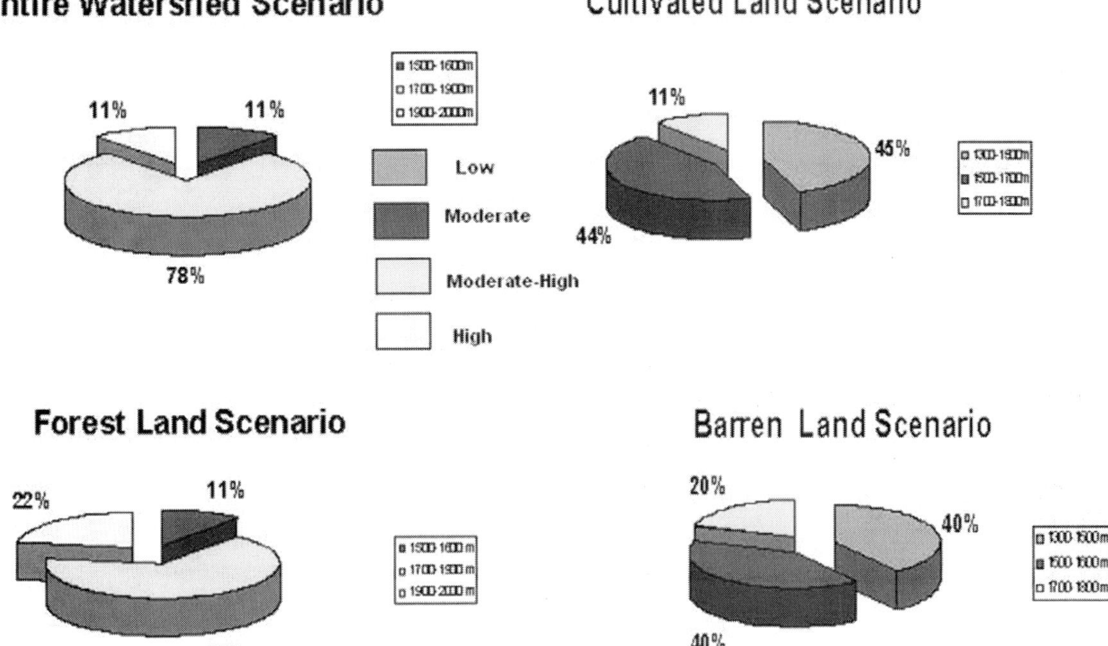

Fig. 7. Categorization of Watersheds on the basis of Critical-Height-Zone

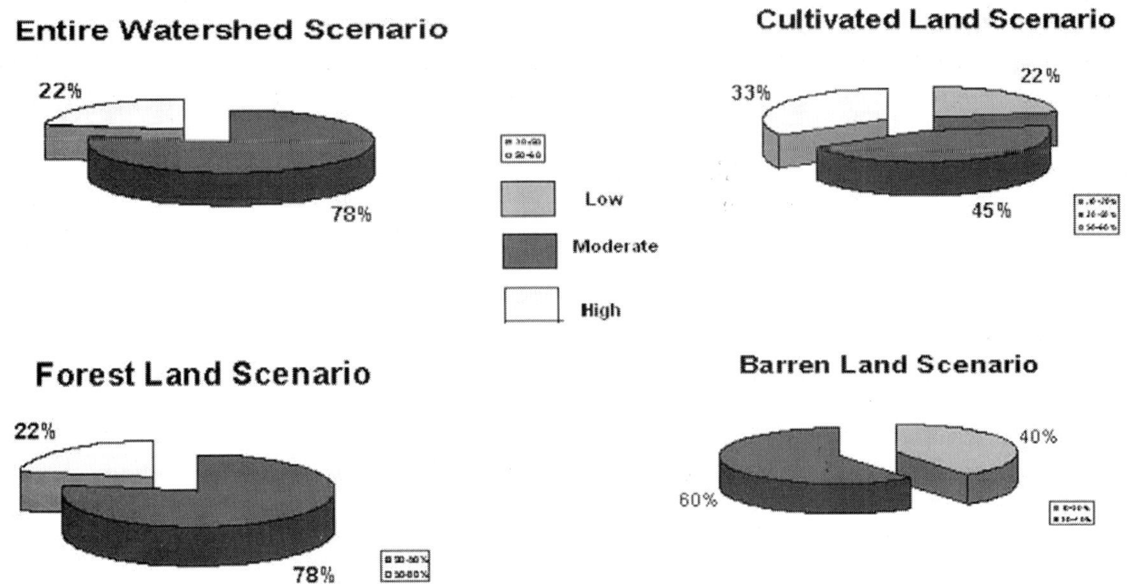

Fig. 8. Categorization of Watersheds on the basis of Erosion Integral

III. Erosion Integral

The watersheds were categorized under three categories on the basis of erosion integral or 'Intensely eroded area' due to erosion processes, viz. low erosion integral (10-30%), moderate erosion integral (30-50%) and high erosion integral (50-60%). About 56% watersheds possess erosion integral group of 40-50% while the erosion integral group of 30-40% and 50-60% possessed by about 22% watershed under each group respectively (Fig. 8).

Variability in erosion integral is noticed under different land use/land cover:

(a) Cultivated Land Scenario The highest erosion integral group i.e. above 50% (50-60%) covers about 33% watersheds while the erosion integral group of 10-20%, 30-40% and 40-50% are represented by about 22% watersheds under each group respectively.

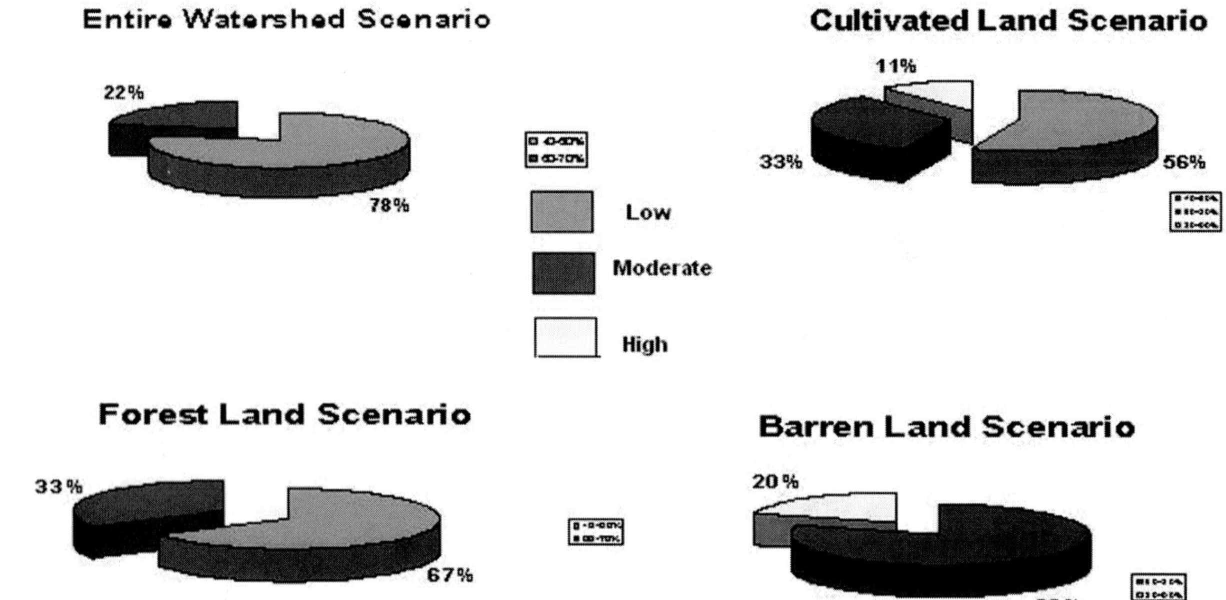

Fig. 9. Categorization of Watersheds on the basis of Hypsometric Integral

(b) Forest Land Scenario Maximum watersheds, i.e. about 45% fall under moderate erosion integral group i.e. 40-50%. The lowest erosion integral group has only about 33% watershed while highest erosion integral group belongs to about 22% watersheds.

(c) Barren Land Scenario About 60% watersheds represent 30-40% erosion integral while only 40% watersheds have 10-20% erosion integral or eroded area.

IV. Hypsometric Integral

The watersheds were categorized under three categories on the basis of hypsometric integral or 'sparsely eroded area' due to erosionary process; viz. low hypsometric integral (40-60%), moderate hypsometric integral (60-80%) and high hypsometric integral above 80%. About 78% watersheds possess low hypsometric integral while 22% watersheds belong to moderate hypsometric integral (Fig.9).

Variability in hypsometric integral is noticed under different land use/land cover:

(a) Cultivated Land Scenario About 34% watersheds fall under minimum hypsometric integral group i.e. below 50%. The hypsometric integral group of 50-60% and 60-70% represent about 22% watersheds under each category while group of 70-80% and 80-90% belong to about 11% watersheds under each group.

(b) Forest Land Scenario About 45% watersheds belong to the hypsometric integral group of 50-60%. About 33% watersheds represent the hypsometric integral group of above 60% and about 22% watersheds belong to the hypsometric integral group of below 50%.

(c) Barren Land Scenario Maximum watersheds about 60% have less affected area of 60-80% while hypsometric integral or less affected area i.e. 70-80% and above 80%, about 20% watersheds are observed under each category.

CONCLUSION

The studied physiographic characteristics of the erosional topography of small watersheds of Himalayan terrain can be better discussed under different criterion, viz. stage of erosion, critical-height-zone, erosion integral and hypsometric integral. High kinetic energy of Himalayan fluvial system has created acute erosion conditions within the watershed. The land use scenario of Himalayan terrain is proved major contributing factor differentially for erosion status consequently the Himalayan watersheds present varying degradation scenario under different land covers.

REFERENCES

Pande, Anita: 1998: Geomorphic Hazard Mapping in Jaigan Watershed. Project report submitted to

CSIR, New Delhi under Scientists' Pool Scheme. 159.

Strahler, A.N: 1952: Hypsometric (area-altitude) analysis of erosional topography. Bulletin of Geological Society of America, 63,117-1142.

Strahler, A.N: 1964: Quantitative geomorphology of drainage basins and channel networks: In Chow, V.T. (ed) Hand book of applied hydrology, Mc Grow Hill, New York, pp 4-11.

Hon'ble reviewer

1. **Prof. Rana Pratap**
   Department of Geography
   Magadh University
   BODHGAYA 824 234
2. **Prof. H.S. Sharma**
   Prof. Geography (Retd.)
   Rajasthan University,
   A-3 Santiniketan, Tonk Road,
   JAIPUR - 302 015

III-45

Morphometric Analysis of A Part of Shivna River in Mandsaur District, Madhya Pradesh, India

T. K. Pandya and Ranjana Vyas

Department of Geology, M.L. Sukhadia University, 51-Saraswati Marg, Udaipur-313001, Rajasthan, India
E-mail: vyasranjana@gmail.com

Abstract

Qualitative and quantitative geomorphologic analysis of a small drainage of Shivna River in Mandsaur district, Madhya Pradesh in India has been studied. The quantitative analyses of the morphometric characteristics of the basin include determination of linear and aerial aspects. The different parameters such as Stream pattern, bifurcation ratio, drainage density, stream frequency, form factor, circulatory ratio, elongation ratio, lemniscate ratio, basin relief, ruggedness number, relief ratio, constant of channel maintenance, slope of ground surface, infilteration index and texture ratio were determined for the study area. The main purpose of this work was to discover holistic stream properties from the measurement of various stream attributes.

Keywords: morphometric analysis, bifurcation ratio, drainage density, stream frequency, lemniscates ratio, ruggedness number.

Introduction

Mandsaur district is one of the areas, which is facing the situation of water scarcity during summer season, in spite of Shivna River drainage system. Mandsaur district fall in the Chambal sub-basin, is a part of Malwa plateau in western Madhya Pradesh. It is a leading district of the state for the cash crop production. The water requirements for agriculture and drinking mainly depend upon its groundwater potential. The depleting groundwater reserves, declining water level, increasing water demands and unexpected occurrence of brackish to saline groundwater at places give an alarming threat of groundwater drought situation in the District. Realizing the present situation of over-extraction of groundwater, it has been considered feasible to undertake a systematic hydrogeomorphological investigation in the vicinity of Mandsaur area.

The present investigation has been carried out in and near the Mandsaur town which is a part of Mandsaur district in Malwa region of Madhya Pradesh state. The study has been confined to longitude 75°0' E to 75°10' E and latitude 24°0' N to 24° 10' N lie on Survey of India, Toposheet no. 45 P/4.

2. MATERIALS AND METHODS

The base map for the study is digital topographic map on 1:50,000 scale. Using

Geomatica software the topographic sheet is georeferenced. After georeferencing, onscreen digitization process was carried out to extract the drainage network of the study area according to their stream order (Figure 2). The stream length, area and perimeter of the drainage basin have been measured with the help of measurement tool in the Geomatica software (Table 1).

Various geomorphic parameters such as stream pattern, bifurcation ratio (Rb), drainage density (Dd), stream frequency (Sf), form factor (F), circulatory ratio (Rc), elongation ratio (Re), lemniscate ratio (RL), basin relief (H), ruggedness number (HD), relief ratio (Rh), constant of channel maintenance (C), slope of ground surface, infilteration index and texture ratio were computed, forming the basis for the analysis of drainage basin.

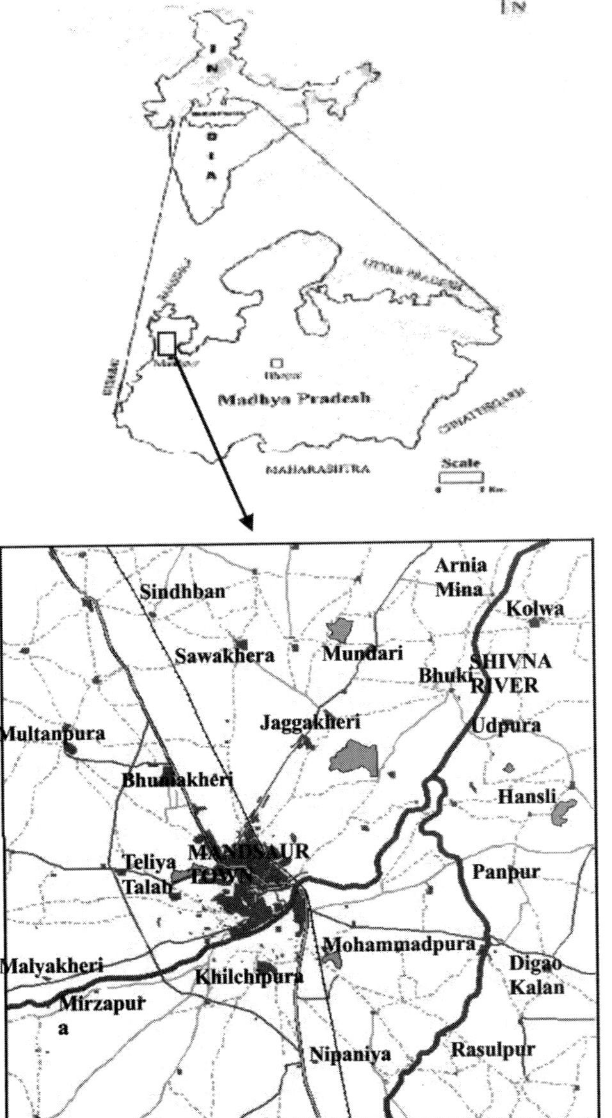

Figure 1: Location map of the study area

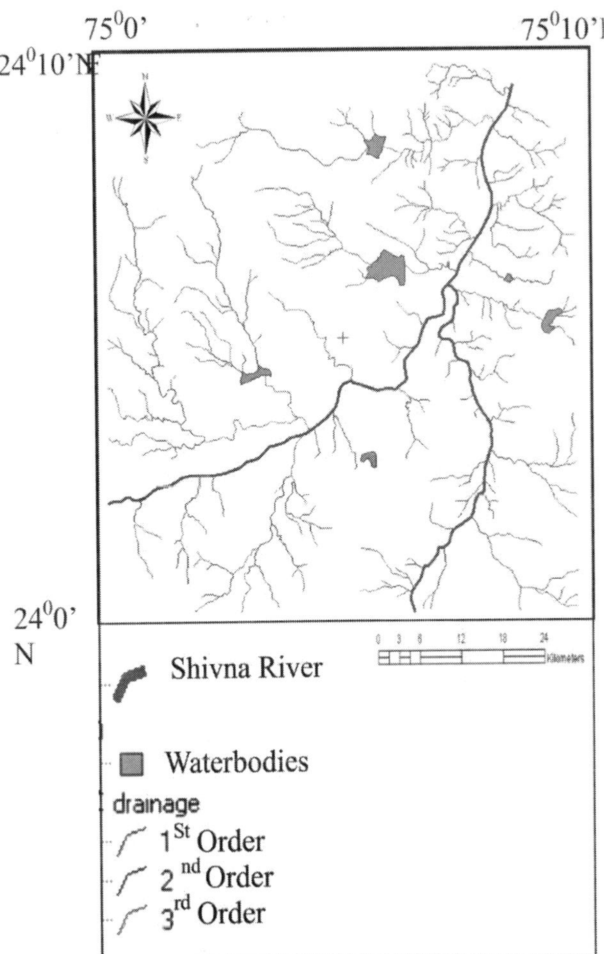

Figure 2: Drainage map of the Study Area

3. GENERAL GEOLOGY AND GEOMORPHOLOGY OF THE AREA

The area belongs to the Deccan Trap Volcanic Province ranging in age from the Upper Cretaceous to Eocene. The study area is a part of Malawa Plateau and locally known as "Malaw Trap". Three lava flows have been observed during field study. Alluvium is developed along the river course of Shivna and maximum part of the area is covered by black cotton soil.

The study area mainly consist of a few isolated low, flat-topped hills and mounds at places. The depositional and erosional landforms are developed in the area of present investigation. The differential hardness of rock masses is evidence of their resistance to weathering. The general topographical slope of the area is towards North and North-East. The height of study area from Mean Sea Level (MSL) is 462.805 meter.

4. MORPHOMETRIC ANALYSIS

Table 1: Determinations of Morphometric variables of drainage of Shivna River in Mandsaur, Madhya Pradesh.

S. N.	Morphometric Determinators	Basin
1	Number of first order stream	210
2	Number of second order stream	57
3	Number of third order stream	8
4	Number of fourth order stream	2
5	Total number of stream	277
6	Length of first order stream (km.)	206.01
7	Length of second order stream (km.)	83.50
8	Length of third order stream (km.)	18.34
9	Length of fourth order stream (km.)	37.84
10	Total length of stream (km.)	345.69
11	Length of basin (km.)	18.6
12	Basin Perimeter (km.)	72.49
13	Area of basin (Sq. km.)	275.29
14	Highest elevation within the basin (m.)	488
15	Lowest elevation within the basin (m.)	413

4.1. Stream Pattern (Order, Number and Length):

Horton's (1945) system and Strahler's (1952) methods are followed in the present test. The stream length has been measured with the help of measuremet tool in Geomatica software. The Shivna River has a length of 24.16 kilometer in the vicinity of the study area. All the streams have been categorized according to their respective orders and all the segments of order are connected to yield the number of the each order in the basin.

4.2 Bifurcation Ratio

The bifurcation ratio is denoted by symbol Rb. It can be expressed by the formula: $R_b = N_\mu$ / Nu+1, Where, R_b = Bifurcation Ratio, N_μ = Number of streams of particular order, $N_{\mu+1}$= Number of streams of next higher order

The bifurcation ratio of the basins of the area are given in Table 2.

The bifurcation ratio of Shivna River Basin varies from 3.68 to 7.12.

In the study area, the computed values of bifurcation ratio in the basin show a variation from 3.68 to 7.12, with an average of 5.4, which indicates that the drainage pattern is not distorted by geologic structures.

Table 2: Showing Bifurcation Ratio of Shivna River Basin of Study Area

	Stream orders	Stream numbers	Bifurcation ratio
Basin	I	210	3.68
	II	57	7.12
	III	8	4
	IV	2	-

4.3. Drainage Density

Drainage density is expressed by the symbol 'D_d' and has been defined by R.E. Horton (1945) as a ratio of total length of all stream segments in a given drainage basin to the total area of that basin and thus it can be derived as follows: $D_d = L_u / A$, Where, D_d = Drainage density, L_u = Total length of all stream segments of the basin in km. = 345.69 km., A = Total area of the basin in sq. km. = 275.29 sq. km. The average drainage density of the study area has been calculated to be 1.26, which indicate that the basin point out rather greater runoff and lesser infiltration.

4.4. Stream Frequency

Stream frequency is also known as the drainage frequency. The stream frequency is given by the formula: F_μ or $S_f = \Sigma N\mu / A$, Where, F_μ or S_f = Stream frequency, $\Sigma N\mu$ = Sum of all streams in a basin = 275, A = Total area of drainage basin in sq. km. = 275.29 sq. km. The stream frequency of the study area has been calculated as 0.99 sq. km.

The basin exists in low value (0.99) of stream frequency; it shows high permeability and low relief of the basin.

4.5. Form Factor

The form factor provides an area of the shape of the basin. It is expressed by the symbol 'F'. Formulae is $F = A / L_b^2$, Where, F = Form factor, A = Area of basin = 275.29 sq. km., L_b = Basin length = 18.6 km. The average form factor of the study area was found to be 0.8

4.6. Circulatory Ratio

The shape of basin changes with the order and with increasing size of drainage area. The larger basins would be more or less circular in case of region with homogenous rocks and structures. The circulatory ratio is usually one for the theoretical circular basin (Miller, op. cit). In the study area the average value is 0.657. Formula of circulatory ratio is: $R_c = 4\pi A / P^2$, Where, R_c = Circulatory ratio, A = Basin area = 275.29 sq. km., P = Basin perimeter = 72.49 km.

4.7. Elongation Ratio

It is also known as the basin elongation. It is generally expressed by the symbol 'R_e' or 'E'. Formulae is $R_e = 2/L_b \times \sqrt{A/\pi}$, Where, R_e = Elongation ratio, A = Area of basin = 275.29 sq. km., L_b = Length of basin = 18.6 km.

The average elongation ratio has been calculated for the study area as 1.00

4.8. Lemniscate Ratio

It determines the shape of the drainage basin. The value of lemniscate ratio calculated for the drainage basin of the study area as 0.314, which reveals that the drainage basin is elongated in shape. It is calculated by the formula: $R_L = L_b^2 / 4A$, Where, R_L = Lemniscate ratio, L_b = Length of basin = 18.6 km., A = Area of basin = 275.29 sq. km.

4.9. Basin Relief

It is expressed by the symbol 'H' or 'Hb'. The formula for basin relief is: $H = H_1 - H_2$, Where, H = Basin relief, H_1 = Highest point of basin = 488 m. , H_2 = Lowest point of basin = 413 m. Basin relief is found to be 75 meter or 0.75 km.

4.10 Ruggedness Number

It can be expressed by the following formula: $H_D = H \times D_d$, Where, H_D = Ruggedness number, H = Maximum basin relief = 0.75 km., D_d = Drainage density = 1.26

It is dimensionless unit, extremely high values ruggedness number occurs when both variables are large and the basin having high ruggedness number values are susceptible to soil erosion and sedimentation load. The ruggedness number has been calculated for the study area as 0.945

4.11 Relief Ratio

The effective measure of gradient aspect of the basin is the relief ratio. Symbole 'R_h' or 'R' generally denotes it. It measures the overall steepness of a drainage basin and reflects the intensity of erosion processes. Relief ratio can be expressed as: $R_h = H / L_b$, Where, R_h = Relief ratio, H = Basin relief = 0.75 km., L_b = Horizontal distance along longest dimension of basin = 18.6 km. The relief ratio has been calculated for the study area as **0.040**

4.12 Constant of Channel Maintenance

It is actually the slope of a linear regression of area on length: $C = 1/D_d$, Where, C = Constant of channel maintenance, D_d = Drainage Density = 1.26

Constant of channel maintenance has been calculated as 0.79. It means on the average 0.79 sq.km. of surface are required to maintain each kilometer of channel length.

4.13 Slope of Ground Surface

This brings into consideration the aspect of relief in drainage basin geometry: $S_g = H \times 2D_d$, Where, S_g = Ground Surface Slope, H = Basin Relief = 0.75, D_d = Drainage Density = 1.26. The slope of ground surface is calculated as S_g = 1.89. This method might be applied to a watershed as a means of assessing both slope steepness and orientation simultaneously.

4.14. Infilteration Index

The infiltration index is expressed by the formula: $I_f = D_d \times F\mu$

Where, I_f = Infiltration Index, D_d = Drainage Density = 1.26, $F\mu$ = Stream Frequency = 0.99. The value of infilteration index is calculated as 1.247 which shows that the infiltration-capacity drops off rapidly at first, then more slowly as it approaches during rainfall. The geomorphic significance of this decrease of infiltration capacity during rain is illustrated by the fact that if infiltration capacity remained constant at the high value it usually has at the beginning of the rain, then there would be little surface runoff or soil erosion.

4.16. Texture Ratio

For quantitative purposes, drainage density and stream frequency are needed, since two drainage nets with the same drainage densities may have quite different numbers and lengths of streams. Texture ratio is calculated by the formula: Texture ratio = $N\mu/P$, Where, $N\mu$ = Total number of streams = 275, P = Perimeter of basin = 72.49. Texture ratio = 3.793

5. RESULTS AND DISCUSSIONS

The bifurcation ratio of Shivna river drainage indicates that the geological structures do not distort the drainage pattern. The average drainage density is 1.26, which indicates that the sub basins point out rather greater runoff and lesser infiltration. The drainage length of overland flow is rather moderate, and indicates that before reaching to drainage channel, water travels to moderate distance. The basin exists in low value (0.99 sq. km.) of stream frequency; it shows high permeability and low relief of the basin and the average form factor is found to be 0.8 which indicates that the drainage has moderately high flow for shorter duration. The average value of circulatory ratio, elongation ratio and lemniscate ratio is calculated as 0.657, 1.00 and 0.314 respectively which reveal that the drainage basin is elongated in shape. Basin relief is found to be 75 meter. The ruggedness number has been calculated for the study area is 0.945. The relief ratio has been calculated for the study area as 0.040; it measures the overall steepness of a drainage basin and reflects the intensity of erosion processes. Constant of channel maintenance has been calculated as 0.79. It means on the average 0.79 sq.km. of surface are required to maintain each kilometer of channel length and the slope of ground surface is calculated as 1.89. The value of infilteration index is calculated as 1.247 which shows that the infiltration-capacity drops off rapidly at first, then more slowly as it approaches during rainfall. The geomorphic significance of this decrease of infiltration capacity during rain is illustrated by the fact that if infiltration capacity remained constant at the high value it usually has at the beginning of the rain, then there would be little surface runoff or soil erosion. All these analyses indicate that the drainage basin is more or less elongated with flat topography, less undistorted and greater runoff and less infiltration.

ACKNOWLEDGEMENT

The authors are very thankful to Prof. Vinod Aggarawal, Head of the Geology Department, Mohanlal Sukhadia University, Udaipur and all the faculty of the department for their support and guideline to prepare this research work.

References

[1] Clarle, J.J., Morphometry from maps In Dury, G.H. (Ed) 'Essays in Geomorphology, Heinmann, London, 1966, p. 235-274.

[2] Dhokarikar, B.G., Ground water Resource Development in basaltic rock terrain of Maharashtra. Water Industry Publ. Pune, 1991, 275 p.

[3] Joshi, V.N., A study on environmental geology of Mandsaur area Vikram University Ujjain, M.Sc. Thesis, 1977, 127 p.

[4] Kulhada, A., Hydrogeological Investigation of Mandsaur Township, Mandsaur District (M.P.). Unpubl. M.Phil. Thesis, Vikram University, Ujjain, 2001, 89 p.

[5] Kulshreshtha, V.,Dev, P., Groundwater quality evaluation for irrigation application in Mandsaur region, Madhya Pradesh, India, Jour. Nep. Geol. Soc., Vol. 34, 2007, 159-164p.

[6] Kulshreshtha, V., Geoinvernental Management Modelling of Ground water Resource of Shivna River Basin, Mandsaur district, Madhya Pradesh. Unpubl. Ph.D. Thesis, Vikram University, Ujjain, 2006, 194 p.

[7] Savingear, R.A.G., A technique of morphological mapping, Ann. Assoc. Amer. Geol., vol. 55, 1969, p.514-538.

[8] Horton, R. E., (1945). Erosional development of stream and the drainage basin, Hydrophysical approach to quantitative morphology, Geo. Soc. Amer Bull. V-56p 275-370

[9] Strahler, A.N. (1964). Quantitative Geomorphology drainage basin and channel Network. In V.T. Chow (Ed). Handbook of Applied Hydrology, Section 4-II, Mcgraw Hill Book Co., New York, p. 4.39-4.76

[10] Millel, V.C. (1953) Relation of quantitative Geomorphic study of drainage basin characteristics in the Clinan Mountain area, Verginia and Tenesse, Tech. Report-3 Office of Neval Research, Department of Geology, Columbia Univ. New York.

Hon'ble reviewer

1. **Prof. S.M. Bijukchhen**
   Central Department of Geology
   Tribhuvan University,
   KATHMANDU (NEPAL)
2. **Dr. D.R. Patel**
   Director
   Department of Geology and Mines
   RAIPUR - 492 001

III-46

Morphometric Analysis and Prioritization of Sub-Watersheds Using GIS and Remote Sensing Techniques for Babina Block of Jhansi District, Uttar Pradesh, India

Deepak Kumar Agarwal[1], Rameswar Bali[1] and Shiva Shankar, V.[2]

1 Centre of Advanced Study in Geology, University of Lucknow, Lucknow. 226007
2. Department of Disaster Management, Pondicherry University, Brookashbad campus, Port Blair. 744103
E-mail: rameshbali@rediffmail.com

Abstract

The region under investigation encompasses an area of 706.27 km^2 and lies between Latitudes 25° 07' N to 25° 31' N and Longitudes 78° 18' E to 78° 35' E. Babina, cantonment town in Babina Block of Jhansi district in the state of Uttar Pradesh, India, despite receiving fairly decent average rainfall (880mm), has uncertain and erratic dry and wet spells causing droughts and also floods. While carrying out the environmental impact assessment, the entire study area has been divided into 10 sub-watersheds named as SW01 to SW10 covering a geographical area ranging from 32.33 Km^2 to 167.3 Km^2 and four mini-watersheds covering an extent ranging from 14.34 Km^2 to 26.81 Km^2 and coded as MN01 to MN04. These have been subjected to prioritization based on morphometric analysis using state of art GIS and remote sensing techniques. Sub-watershed- SW01, SW04, SW05, SW06, SW07, SW09 and Mini-watershed-MN01, MN03, MN04 are having elongation ratio of 0.6 or more, which indicates that these watersheds are located in high relief and steep slope. Drainage density ranges from 2.09 km/km^2 to1.20 km/km^2 indicating the basin has highly permeable sub-surface geological units. This anomaly is due to diversity of slope, relief and structural conditions prevailing in this watershed. The compound parameter values are calculated and the watershed with the highest compound parameter is given the highest priority. The watersheds SW02, SW03, SW10 and MN02 have maximum compound parameters are likely to be subjected to maximum soil erosion and susceptible to natural hazards. Hence, these should be provided with immediate soil conservation measures.

Keywords: Morphometric Analysis, GIS, Prioritization, Remote Sensing, Sub-Watersheds.

Introduction

Land and water resources are limited and their wide utilization is imperative, especially for countries like India, where the population pressure is continuously increasingly. Drainage basins, catchments and sub-catchments are the fundamental units for administrative purposes to conserve natural resources. The watershed management concept recognizes the interrelationships among the linkages between uplands, low lands, land use, geomorphology, slope and soil. Soil and water conservation is the key issue in watershed management while demarcating watersheds. However, while considering watershed conservation work, it is not feasible to take the whole area at once. Thus the entire basin is divided into several smaller units, as sub-watersheds and mini-watersheds, by considering its drainage system.

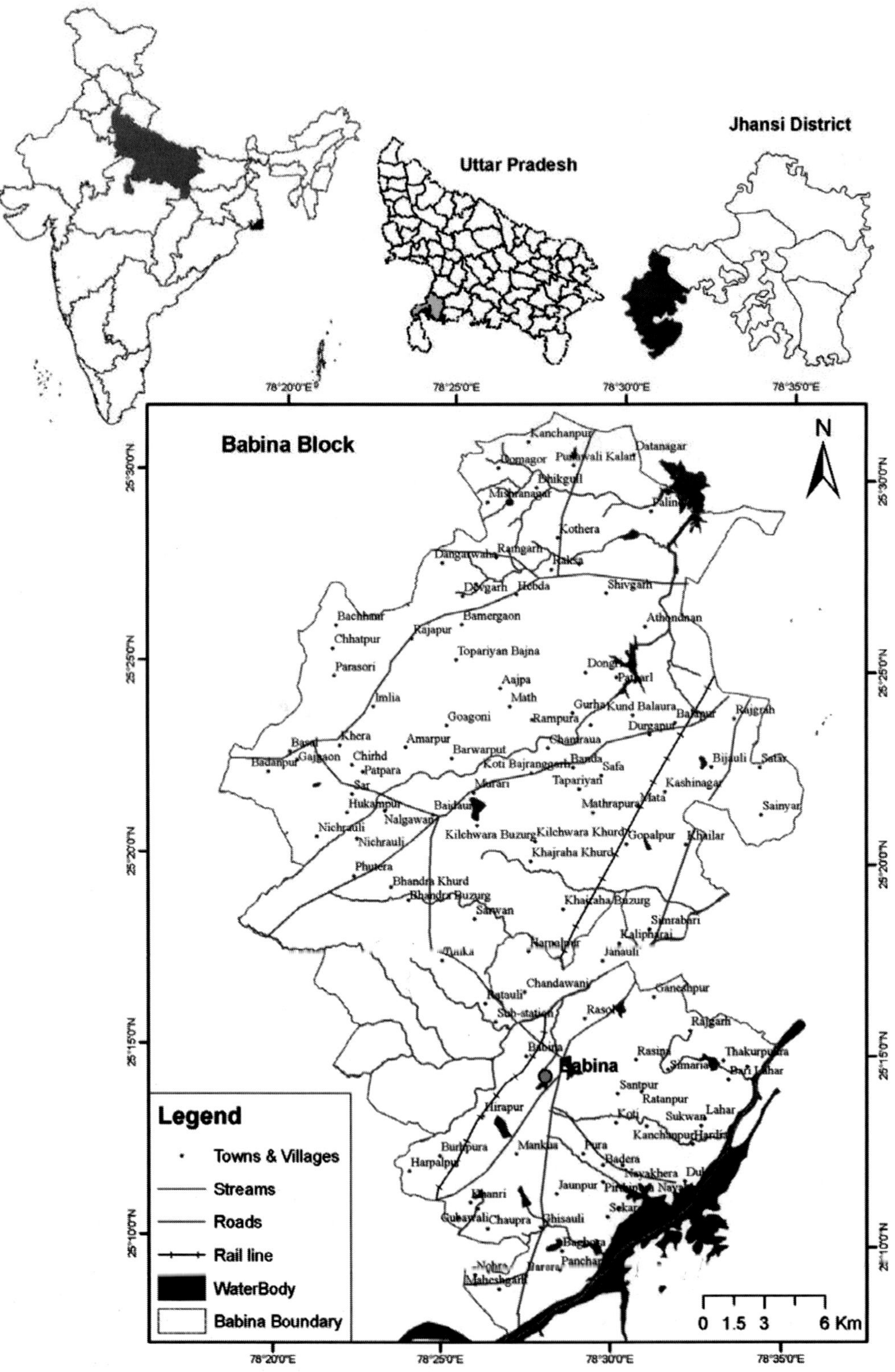

Fig. 1. Map showing the location of the study area

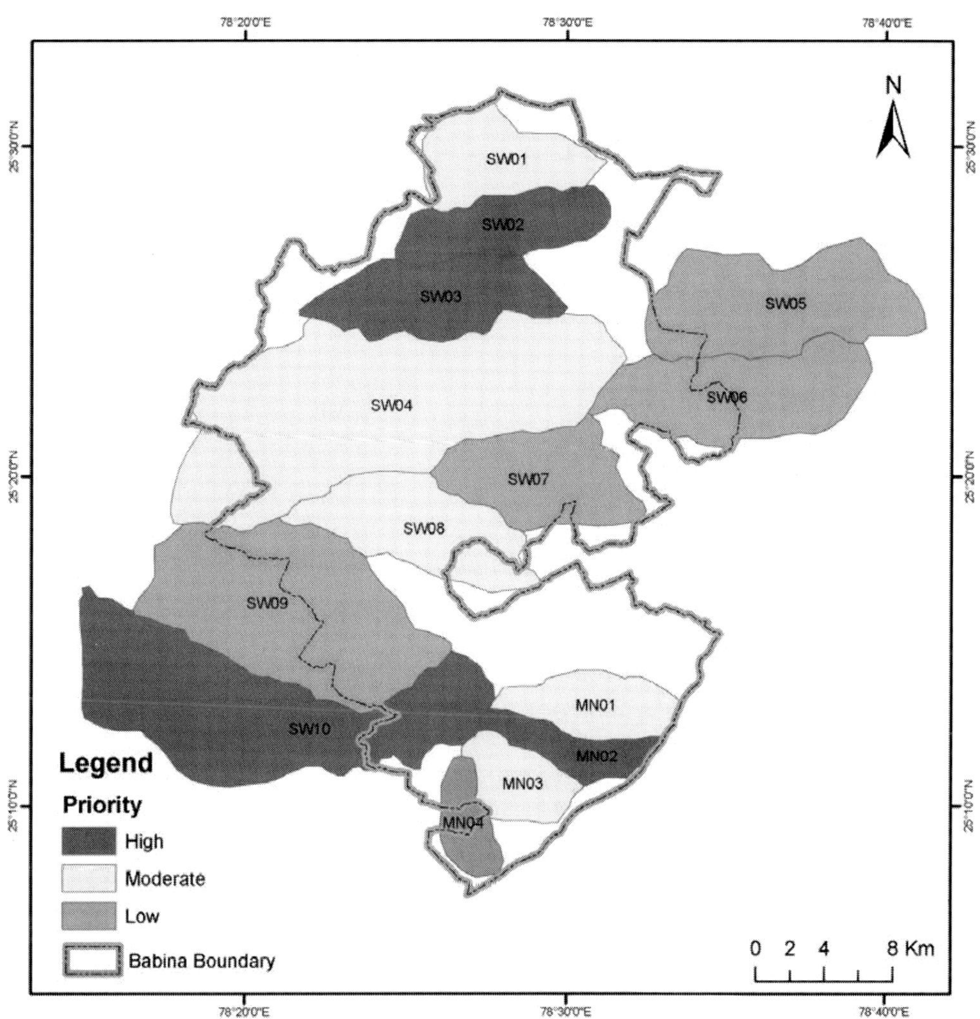

Fig. 2.: Watershed Priority Map of the area

Morphometric analysis is a significant tool for prioritization of sub-watersheds even without considering the soil map (Biswas et al., 1999). Such study requires measurement of the linear features and gradient of channel network contributing ground slopes of the drainage basin. GIS and remote sensing technique has been effectively used for the delineation and prioritization of soil erosion areas in the Sukhana lake catchment in the Siwalik hills and sub-watersheds of Pawagada area, Tumkur district, Karnataka respectively (Shrimali et al., 2001, Srinivasa et al., 2004). Similarly, Khan et al., 2001 carried out morphometric analysis of Guhiya basin and Chopra et al., 2005, for sub-watersheds of Gurdaspur district, Punjab, respectively for watershed prioritization. Study on check dam positioning by prioritization of micro-watersheds using the sediment yield index (SYI) model has been effectively been carried out by Nookaratnam et al., 2005. Morphometric analysis has also been used to delineate the glacial till covered overburden material contributing to the groundwater budget, to identify areas prone to flash floods as well as zones of active tectonics (Bali et al., 2011). In the present study, morphometric analysis and prioritization of sub-watersheds are carried out for Babina Block in Jhansi district of Uttra Pradesh, to understand the hydrologic response of the watershed.

Study Area

The Babina block of Yamuna basin, encompasses an area of 706.27 km^2 and lies between Latitudes 25° 07' N to 25° 31' N and Longitudes 78° 18' E to 78° 35' E. Babina is a cantonment town in Babina Block of Jhansi district in the state of Uttar Pradesh, India. BABINA is an acronym for "British Army Base in North Asia". This British base near Jhansi was created during British Raj. It has an average elevation of 303 meters.

The study area shows undulating topography. It is primarily dominated by the outcrops of Bundelkhand Gneissic Complex with conspicuous recent intrusive bodies like quartz reef, porphyries, tonalite, quartz reef with pyrophylite and dolerite dyke. Small pockets of amphibolite intrusions are also seen in the southern regions of the study area (Basu, 1986). The impervious surface and sub-surface lithological association does not facilitate water percolation resulting in the development of dendritic drainage surface run-off pattern.

The entire Babina block is well drained so that there are no areas that suffer water stagnation. The rivers are still active and grading their courses and adjusting themselves to the underlying structure (Venkatesan, 1973). Although the deepening process is restricted to the harder gneissic bed-rocks, the widening process has removed the upper soil layer and carved out deep ravines in the adjacent softer stratum. Samogar Nadi, Pahuj Nadi, Ghurari Nadi, Satar Nadi, Kharobaro nadi and Betwa River are few major streams in the study area. Pahuj Reservoir, Pahuj Tal, Manpur Tal, Bara Tal, Raja Tal and Ghisauli Tal are few prominent lakes which are well fed by the streams. Finally these streams flow into the Betwa River.

The climate of Babina Block of Jhansi District is characterized by a hot, dry summer and cold winters. The coldest months in the year are December and January in which the temperature goes as low as 10°C. From March onwards, the temperature begins to rise progressively and peaks in May during which temperature goes upto 45°C. The average rainfall of Babina is 880 mm. Despite fairly decent average rainfall of the districts, its uncertainty and erratic behavior leads to dry spells causing droughts and also floods in the Monsoon (CGWB, 2002).

Methodology

The Environmental Impact Assessment of the area has been carried out by synthesizing and integrating the data collected in the field and that interpreted using the remotely sensed data, ancillary data and published reports. Data collected from various sources has been trained and tailored to carry out the analysis and the inference requirement of the present study. This information has then finally been put in a Geographic Information System (GIS) platform and the output data is recorded in the form of various thematic maps. Prominent administrative features like village names, administrative boundary, bench marks, spot heights, drainage have been extracted from Survey of India toposheet nos. 54K/06, 54K/07, 54K/08, 54K/10, 54K/11 and 54K/12 in conjunction with the satellite imagery. The information has later been used in the preparation and validation of the base map entities. Ancillary data like rainfall, water level data, water quality data, etc are obtained from published Central Groundwater Board.

IRS LISS III Data has been utilized for extracting different thematic maps as well as for preparing the data for drainage morphometric analysis. Landsat enhanced thematic mapper + was utilized for geo-rectification of LISS-III data and delineation of geological and geomorphological features. ASTER (Advanced Spaceborne Thermal Emission and Reflection Radiometer) DEM Data has been employed for generation of Digital Elevation Model (DEM) and Digital Terrain model (DTM) which is of paramount importance for studying the characteristics of watershed like slope, slope-aspect, flow direction, topography, etc. Google earth data provides a seamless natural coloured composite data, which is quite useful in image classification and visual interpretation of the real world features and entities.

Following Horton, 1932, 1945; Strahler, 196; Schumm, 1956 and Miller, 1953, a number of morphometric parameters have been computed for the delineated watershed area. These include stream order, stream length, bifurcation ratio, drainage density, drainage frequency, relief ratio, elongation ratio, circularity ratio and compactness constant. Prioritization rating of all the ten sub-watersheds and four mini- watersheds of Babina watershed is carried out by calculating the compound parameter values. The sub and mini-watersheds with the lowest compound parameter value is given the highest priority.

Morphometric Analysis

Drainage pattern is characterized by irregular branching of tributaries in many directions with an angle less than 90°. The Babina watershed is divided into ten sub-watersheds (>30 sq km) coded as SW01 to SWS10 and four mini-watersheds (10-30 sq km), coded from MN01 to MN04 based on the Technical guidelines of Integrated Mission for Sustainable Development (NRSA, 1996). For prioritization of the sub-watersheds and mini- watersheds, following morphometric parameters have been computed.

Stream Order (Nu): Stream ordering refers to the determination of hierarchical position of stream within a drainage basin. Classification of streams based on the number and type of tributary junction has proven to be a useful indicator of stream size, discharge and drainage area. Ordering of stream begins from the fingertip tributaries, which do not have their own feeders (Strahler,1952). Two first order streams when join together, form second order stream just below the junction. Similarly, two second-order streams meet to make stream of third-order and this process continues till the trunk stream is given the highest order.Out of ten sub-watersheds, only sub-watershed number SW05 have 6th order stream, sub-watershed SW04 and sub-watershed SW07 have 5th order stream. Sub-watershed SW01 has the least stream order of 3. The details of all stream order of all sub-and mini- watershed are tabulated in Table 1. The variation in order and size of the sub-watersheds is largely due to physiographic and structural conditions of the region.

Table 1: Stream Order, Stream Length and Stream Length ratio of Babina Block

watershed Name	Stream Order					Total Count of all Order	Stream Length in Km					Total Stream Length	Stream Length Ratio				Mean RL
	I	II	III	IV	V		I	II	III	IV	V		II/I	III/II	IV/III	V/IV	
SW01	52	30	19	0	0	101	38.64	16.32	8.45	0	0	63.41	0.42	0.52	0	0	0.47
SW02	54	21	18	11	0	104	36.45	9.38	8.02	5.91	0	59.76	0.26	0.86	0.74	0	0.62
SW03	75	36	26	2	0	139	46.68	16.49	12.65	1.61	0	77.42	0.35	0.77	0.13	0	0.42
SW04	247	130	58	10	28	473	163.75	75.9	34.05	9.46	15.2	298.33	0.46	0.45	0.28	1.6	0.7
SW05	139	75	24	10	29	277	78.26	31.24	9.88	5.98	8.53	133.89	0.4	0.32	0.61	1.43	0.69
SW06	137	80	21	32	0	270	75.99	28.22	9.92	11.19	0	125.33	0.37	0.35	1.13	0	0.62
SW07	80	32	22	11	7	152	52.43	10.85	10.85	6.33	2.4	82.86	0.21	1	0.58	0.38	0.54
SW08	95	43	24	27	0	189	60.61	19.45	9.74	11.36	0	101.15	0.32	0.5	1.17	0	0.66
SW09	177	67	72	32	0	348	111.55	37.06	33.13	14.35	0	196.09	0.33	0.89	0.43	0	0.55
SW10	161	75	42	36	0	314	99.97	46.1	26.93	19.84	0	192.83	0.46	0.58	0.74	0	0.59
MN01	44	21	21	0	0	86	28.24	11.06	9.05	0	0	48.35	0.39	0.82	0	0	0.61
MN02	17	8	6	0	0	31	12.81	5.4	5.49	0	0	23.7	0.42	1.02	0	0	0.72
MN03	16	10	4	0	0	30	12.69	8.27	3.91	0	0	24.86	0.65	0.47	0	0	0.56
MN04	16	5	10	0	0	31	10.27	4.31	6.76	0	0	21.34	0.42	1.57	0	0	0.99

Table 2: Derived and shape morphometric parameters of Babina block

Watershed Name	Fs	Mean Rb	Ff	Re	Rc	Dd	Rt	Cc
SW01	2.83	1.66	0.4	0.71	0.65	1.78	3.85	1.24
SW02	3.22	1.79	0.27	0.58	0.55	1.85	3.82	1.35
SW03	3.08	5.49	0.24	0.55	0.52	1.71	4.21	1.39
SW04	2.83	2.57	0.28	0.6	0.52	1.78	7.43	1.39
SW05	3.89	1.93	0.31	0.63	0.57	1.88	7.01	1.32
SW06	4.51	2.06	0.28	0.6	0.55	2.09	7.27	1.35
SW07	3.33	1.88	0.36	0.68	0.64	1.82	5.09	1.25
SW08	3.86	1.63	0.25	0.56	0.53	2.06	5.57	1.37
SW09	3.25	1.94	0.41	0.72	0.69	1.83	7.88	1.2
SW10	2.66	1.7	0.26	0.58	0.4	1.63	5.13	1.59

MN01	3.21	1.55	0.28	0.6	0.61	1.8	3.67	1.28
MN02	1.77	1.73	0.21	0.52	0.42	1.35	1.35	1.55
MN03	1.45	2.05	0.49	0.79	0.73	1.2	1.59	1.17
MN04	2.16	1.85	0.3	0.62	0.64	1.49	1.84	1.25

Table 3: Basic morphometric parameters of Babina Block

Watershed Name	Basin Area (Sq Km)	Perimeter (Km)	Basin Length (Km)	Min. Height (m)	Max. Height (m)	Mean Relief	Basin Relief (H) in m	Relief Ratio (Rh)	RHO coeffi-cient
SW01	35.64	26.22	9.48	216	351	247.9	135	0.0142	0.28
SW02	32.33	27.25	11.03	219	348	251.5	129	0.0117	0.34
SW03	45.19	33.01	13.75	237	370	273.22	133	0.0097	0.08
SW04	167.3	63.62	24.43	236	377	283.28	141	0.0058	0.27
SW05	71.17	39.51	15.04	184	303	223.23	119	0.0079	0.36
SW06	59.91	37.16	14.61	181	332	232.33	151	0.0103	0.3
SW07	45.6	29.84	11.23	232	324	261.65	92	0.0082	0.29
SW08	49.01	33.96	14.13	244	311	279.02	67	0.0047	0.41
SW09	107.15	44.16	16.11	264	396	298.29	132	0.0082	0.28
SW10	118.04	61.23	21.21	260	399	307.27	139	0.0066	0.35
MN01	26.81	23.45	9.7	240	313	267.6	73	0.0075	0.39
MN02	17.53	23.01	9.1	243	340	271.16	97	0.0107	0.42
MN03	20.65	18.86	6.46	246	303	276.03	57	0.0088	0.27
MN04	14.34	16.83	6.89	254	297	276.1	43	0.0062	0.54

Table 4. Ranking for different sub- and mini-watersheds for Prioritization

watershed Name	Dd	Fs	Ff	Rc	Re	T	Cc	Rb	Compound Factor	Final Priority
SW01	8	8	2	2	2	9	9	9	6.125	4
SW02	4	6	7	5	7	10	6	7	6.5	3
SW03	9	7	10	8	10	8	3	1	7	2
SW04	7	9	6	9	6	2	2	2	5.375	6
SW05	3	2	4	4	4	4	7	5	4.125	8
SW06	1	1	5	6	5	3	5	3	3.625	9
SW07	6	4	3	3	3	7	8	6	5	7
SW08	2	3	9	7	9	5	4	10	6.125	4
SW09	5	5	1	1	1	1	10	4	3.5	10
SW10	10	10	8	10	8	6	1	8	7.625	1
MN01	1	1	3	3	3	1	2	4	2.25	3
MN02	3	3	4	4	4	4	1	3	3.25	1
MN03	4	4	1	1	1	3	4	1	2.375	2
MN04	2	2	2	2	2	2	3	2	2.125	4

Stream Length (Lu): It is the total length of streams of a particular order. The stream lengths for all sub-basins of different orders have been calculated from the streams digitized from topographical maps and satellite data products using ArcMap software. Horton's law of stream lengths supports the theory that geometrical similarity is preserved generally in the basins of increasing order (Strahler, 1964). The numbers of streams of various orders in the sub-and mini-watersheds have been enumerated in 1. It is noticed that the maximum length of stream segments is contributed by first order streams in all the basins.

Stream Length Ratio (Rl): Stream length ratio has been defined as the ratio of the mean length of the higher order to the next lower order of stream segment. Horton's law of stream length ratio states that mean stream length segments of each of the successive orders of a basin tends to approximate a direct geometric series with stream length increasing towards higher order of streams (Horton, 1932). The stream length ratio between different sub- and mini-watersheds showed an increasing trend in the length ratio from lower order to higher order indicating their mature geomorphic stage in sub-watersheds, with an exception in SW07, a change from one order to another order indicating the late youth stage of geomorphic development of streams in the inter basin area. The stream length ratios of all the sub- and mini-watersheds are tabulated in Table 1.

Stream frequency (Fs): Stream frequency (Fs) or channel frequency is the total number of stream segments of all orders per unit area (Horton, 1932):

$$Fs = \sum Nu / A;$$

where $\sum$ Nu = total number of stream segments of all orders, and A = area of the basin.The stream frequencies of all the sub- and mini-watersheds are mentioned in Table 2. The study revealed that the sub watersheds having large area under dense forest have low drainage frequency and the area having more agricultural land have high drainage frequency. High value of drainage frequency in SW06 produces more runoff in comparison to others.

Bifurcation Ratio (Rb): This parameter expresses the ratio of number of streams of a given order (Nu) to the number of stream segments of the higher order (Nu + 1) (Horton, 1945). In the study area, a maximum mean Rb of 5.49 (SW03) and a minimum Rb of 1.55 (MN01) has been measured. The mean Rb value varies from 1.5 to 3. The mean Rb of the sub- and mini-watershed has been computed in 2. Chow (1964) stated that, an Rb range from 3 to 5 for watersheds does not exercise a dominant influence of geologic structure on the drainage patterns. The bifurcation ratio values of Babina watershed suggests that geologic structure exercises a dominant influence on the drainage pattern as the study area has Rb between 1.5 and 2.5. This further suggests inhomogeneous lithology in the area (Huggett, 2003).

Form Factor (Ff): The Ff of a drainage basin is expressed as a ratio between the area of the basin (A) and the square of the basin length (L) (Horton, 1945), and is expressed as:

$$F_f = A/L^2$$

For a perfectly circular basin, the value of form factor is always less than 0.7854. Thus for smaller value of form factor, the basin tends to be elongated. The basin with high F_f have high peak flows of shorter duration, whereas elongated sub watershed with low F_f have lower peak flow of longer duration (Chopra et al. 2005). In Babina area sub- and mini- watershed have the form factor between 0.2 and 0.5 which indicates that the watersheds are elongated in shape. Elongated watershed suggests flatter peak-flow for longer duration. The values of all sub–and mini-watershed are shown in 2.

Elongation Ratio (Rc): Elongation ratio (Re) is the ratio between the diameter (D) of a circle of the same area as the drainage basin and basin length (L) (Schumm, 1956). It is calculated as:

$$Re = D/L = 1.128 \sqrt{A} / L$$

where A is the area of the basin.

In the present study area, the values range between 0.52 and 0.79 (2). Sub-watershed SW01, SW04, SW05, SW06, SW07, SW09 and Mini-watershed-MN01, MN03, MN04 are having elongation ratio of 0.6 or more, which indicates that these watersheds are located in high relief and steep slope. The values of 0.6 to 0.8 are associated with high relief and steep ground slope and are in accordance with Strahler, 1964.

Circulatory Ratio (Rc): The circulatory ratio ($R_c = 4 \pi A / P^2$, where A = area of the basin and P = perimeter of the basin), has been used as a quantitative measure to express the ratio of the basin area (A) to the area of a circle having the same perimeter as the basin (Miller 1953; Strahler 1964). The medium values (Rc between 0.4 and 0.7) of the ratio in the study area indicates that this is a mature stage of the drainage pattern with strongly elongated watershed and highly permeable homogenous geologic materials as reported by Miller in 1953 shows the Rc values of Sub- and Mini-watersheds.

Drainage density (Dd): Drainage density is measured as the total length of streams of all orders per unit area divided by the area of drainage basin. It is a measure of the degree of fluvial dissection and is influenced by numerous factors, among which resistance to weathering and erosion of rocks, permeability of rock formation, infiltration capacity of the land and climatic conditions rank high (Verstappen, 1983).

Dd of the entire Babina watershed ranges from 2.09 km/km^2 to 1.20 km/km^2 (Table 2). It gives the detailed information about the drainage density in sub- and mini-watershed. It is suggested that the low drainage density indicates the basin is highly permeable subsoil and thick vegetative cover (Nag, 1998). The type of rock also affects the drainage density. Generally, lower values of Dd tend to occur on granite, gneiss and schist regions. The study area has Bundelkhand Gneissic Complex with conspicuous recent intrusive bodies like quartz reef, porphyries, tonalite, quartz reef with pyrophylite and dolerite dyke.

Drainage Texture (Rt): It is the ratio between total numbers of stream segments of all orders to the perimeter of the basin (Horton, 1945). The drainage texture (Rt) is an expression of the relative channel spacing in a fluvial dissected terrain. It depends upon a number of natural factors such as climate, rainfall, vegetation, rock and soil type, infiltration capacity, relief and stage of development of a basin. Five different drainage textures are classified based on Drainage Texture. For drainage texture below or equal to 4.0 is coarse, between 4.0 and 10.0 is intermediate, between 10.0 and 15.0 is fine and above 15.0 is ultra-fine texture having badland topography. (Smith, 1950).

Coarse to intermediate drainage texture is observed in the study area. Drainage texture in the area is generally below 8 indicating that the area mainly has coarse to intermediate drainage texture presents the details of drainage texture of Sub- and Mini-watersheds of Babina watershed.

Compactness coefficient (Cc): It expresses the relationship of a hydrologic basin with that of a circular basin having the same area as the hydrologic basin. A circular basin is the most hazardous from the drainage stand point because it will yield the shortest time of concentration before the peak flow occurs in the basin. The values of Cc in the area under investigation ranges from 1.17 to 1.59 showing narrow variations across the Sub- and mini-watersheds and are as shown in Table 2. These Cc values indicate that the basin is elongated and non hazardous from the drainage point of view.

Relief Ratio (Rh): The relief ratio (Rh) for the different sub- and mini-watersheds of the study area has been calculated and has been found to range from 0.0142 to 0.0047 (3). The lower values for the SW08 sub-basins indicate the presence of less resistant rocks on the surface. Higher values indicate higher resistance to weathering and erosion. Sub-watershed SW01 has the highest relief ratio amongst all the sub and mini watershed in the area and it indicates the most resistant and least eroded area in the block.

RHO Coefficient (RHO): It is an important parameter that determines the relationship between the drainage density and the physiographic development of the basin, and allows the evaluation of the storage capacity of the drainage network (Horton 1945). It is influenced by climatic, geologic, biologic, geomorphic, and anthropogenic factors. The RHO coefficient ranges from 0.08 (SW03) to 0.54 (MN04) for the area under investigation, indicating that the study area is having low storage capacity during flood periods as the area is having impervious sub-surface geological units. The RHO coefficient of sub and mini-watersheds are as shown in Error! Reference source not found.

Prioritization of Sub-watershed and Mini-watershed

The resources-based approach is found to be realistic for watershed prioritization since it involves an integrated approach (Vittala et al., 2008). The present study puts forward an integrated approach for developing a preliminary prioritization of sub-watersheds in the study area. Hence, the entire area, divided into ten sub-watersheds four mini-watersheds has been taken-up for prioritization, using various morphometric parameters. Morphometric parameters coupled with integrated thematic map of drainage density and land use can help in decision making process for water resources management (Mishra and Nagarajan, 2010). During the course of the present study, knowledge-based weightage system has been adopted for sub-watershed prioritization based on the significant parameters and after carefully observing the field situation. The basis for assigning weightage to different themes was according to the relative importance to each parameter in the study area. Each of the morphometric parameters have been prioritized by assigning rank, based on the values calculated for each of them (Biswas et al., 1999; Thakkar and

Dhiman, 2007; Mishra and Nagarajan, 2010). Ranking of a parameter is done for individual sub-watershed amongst the group of ten sub-watersheds and similarly ranking of each mini watershed has been done amongst the group of four mini-watersheds. Ranking is based on the absolute values of parameters obtained for each sub-watershed and mini-watershed (Table 4).

The rank calculated for morphometric parameter of individual sub-watershed and mini-watershed is presented in Table 4 e.g Drainage Density among ten sub-watersheds in the area ranges from 1.63 (SW10) to 2.09 (SW06), then accordingly the Drainage Density rank for each sub watershed have been given values ranging from 1 (SW06) to 10 (SW10) in the decreasing order of Drainage Density value. Compounded parameter is then calculated by taking average of all parameters' rank for each sub-watershed and mini-watershed. This compounded parameter for each sub-watershed and mini-watershed is then ranked, in similar fashion as stated above, amongst the group sub-watershed and mini-watershed. The ranking thus obtained (4) are grouped into three class viz. High priority, Moderate priority and Low priority. In sub-watershed group, ranks 1-3 are given high priority, 4-6 moderate and 7-10 rank as low priority.

The sub-watershed SWS10 with a compound factor of 7.625 receives the highest priority followed by sub-watershed SW03. These sub-watersheds are of immediate concern. Erection of conservative structures like check dams, en-echelon dams, etc., at the confluence of streams downhill is the need of the hour for watershed development.

Error! Reference source not found.4 shows the priority of individual parameters measured within a sub-watershed or mini-watershed. The outcome has been depicted as a watershed priority map (). The priori-tization based on the compound ranking for the mini-watershed are; rank 1 as high priority, rank 2-3 as moderate priority and rank 4 as low priority. As per this MN02, with a compound factor of 3.25, is highest prioritized among all the mini watersheds in the area 4 shows the details of prioritization in the mini-watersheds of the study area. The same has been depicted in Fig. 2.

Conclusion

Watershed prioritization is one of the most important aspects of planning for implementation of its development and management practices. The recent study demonstrates the usefulness of GIS and remote sensing techniques for morphometric analysis and prioritization of the sub- and mini-watersheds of Babina block in Jhansi district of Uttar Pradesh, India. The morphometric characteristics of different sub- and mini-watersheds show their relative characteristics with respect to hydrologic response of the watershed. Morphometric parameters coupled with integrated thematic map of drainage density land use can help in decision making process for water resources management. Results of prioritization of watersheds show that sub-watersheds SW02, SW03 and SW10 and mini-watershed MN02are more susceptible to soil erosion. Therefore, immediate attention towards soil conservation measures is required in these sub- and mini-watersheds to preserve the land from further erosion and to alleviate natural hazards.

References

Bali, R, Agarwal, K.K., Nawaz Ali, S., Rastogi, S.K. and Krishna, K., 2012. Drainage morphometry of Himalayan Glacio-fluvial basin, India: Hydrologic and Neotectonic implications Environ. Earth Sci. v.66, pp.1163-1174.

Basu, A,X., 1986. Geology of parts of Bundelkhand Granite Massif, Central India. Rec.Geol. Surv. India, v.117(2). pp. 61-124.

Biswas, S., Sudhakar, S.and Desai, V.R., 1999. Prioritization of sub-watersheds based on morphometric analysis of drainage basin, district Midnapore, West Bengal. Jour. Indian Soc. Remote Sensing, v. 27(3). pp. 155–166.

Chopra, R., Dhiman, R., and Sharma, P. K., 2005, Morphometric analysis of subwatersheds in Gurdaspur District, Punjab using Remote Sensing and GIS techniques, Journal of Indian Society of Remote Sensing, v. 33(4), pp 531-539.

Chow, V. T., 1964. Handbook of Applied Hydrology: A Compendium of Wafer-Resources Technology. McGraw-Hill, New York. 1495 p.

Central Ground Water Board (CGWB), 2002. Hydrogeological framework and ground water development prospects in Jhansi district, UP. 103 p.

Horton, R.E., 1932. Drainage basin characteristics. Trans. Amer. Geophys. Union. v. 13. pp. 350-361.

Horton, R. E., 1945. Erosional development of streams and their drainage basins; Hydrological approach to quantitative morphology. Geological Society of American Bulletin. v. 56. pp. 275-370.

Hugget, R..J., 2003 Fundamentals of Geomorphology, Routledge, London. 401 p.

Khan, M.A., Gupta, V.P. and Moharana, P.C., 2001, Watershed prioritization using remote sensing and geographical information system: a case study from Guhiya, India, Journal of Arid Environments, v. 49, pp 465-475.

Miller, V. C., 1953. A quantitative geomorphic study of drainage basin characteristics in the clinch mountain area, Technical report 3, Department of Geology, Columbia University. pp1-30.

Mishra, S. and Nagarajan, R., 2010. Morphometric analysis and prioritization of subwatersheds using GIS and Remote Sensing techniques: a case study of Odisha, India. International Journal of Geomatics and Geosciences. v. 1(3). 501-510.

Nag, S.K., 1998, "Morphometric analysis using remote sensing techniques in the Chaka subbasin Purulia district, West Bengal", Jour. Indian Soc. Remote sensing, 2 v. 6, pp 69-76.

National Remote Sensing Agency (NRSA), 1996. Integrated Mission for Sustainable Development; Technical guidelines. NRSA/NMSD/TG: 1996. NRSA, Department of Space, Hyderabad, India

Nookaratnam, K., Srivastava, Y. K., Venkateswarao, V., Amminedu, E. and Murthy, K.S.R., 2005, Check dam positioning by prioritization of microwatersheds using SYI model and morphometric analysis – Remote sensing and GIS perspective, Journal of the Indian Society of Remote Sensing, v. 33(1), pp 2528.

Schumm, S. A., 1956. Evolution of drainage systems and slopes in Badlands at Perth Amboy, New Jersey. Geol. Soc. Amer. Bull. v. 67. pp.597-646.

Smith, K. G., 1950. Standards for grading texture of erosional topography. An. J. Sci. v. 248. 655-68.

Shrimali, S.S., Aggarwal, S. P. and Samra, J. S.,2001, Prioritizing erosion prone areas in hills using remote sensing and GIS – a case study of the Sukhna Lake catchment, Northern India, JAG, v. 3(1), pp 54-60.

Srinivasa, V. S., Govindaonah, S. and Home Gowda, H., 2004, Morphometric analysis of subwatersheds in the Pawagada area of Tumkur district South India using remote sensing and GIS techniques, Journal of Indian Society of Remote Sensing, v. 32(4) pp 351-362.

Strahler AN (1952) Hypsometic (area-altitude) analysis of erosional topography. Bull. Geol. Soc. Amer. v 63 pp. 1117–1142.

Strahler, A.N., 1964. Quantitative geomorphology of drainage basins and channel networks section 4-2, Handbook of Applied Hydrology. Ed. Ven te Chow, McGraw-Hill, New York. v. pp. 4-35.

Thakkar, K.A. and Dhiman S.D. 2007. Morphometric Analysis and Prioritization of Mini-watersheds in Mohr Watershed, Gujarat, using remote sensing and GIS Techniques. Jour. Indian Soc. Remote Sensing, v. 35(4). pp. 313–321.

Venkatesan, S., 1973. Hydrogeological conditions in Jhansi district, Uttar Pradesh. Central Ground Water Board. pp.10.

Verstappen, H.T., 1983. Applied Geomorphology: Geomorphological Surveys for Environmental Development. Elsevier, Amsterdam. pp.437.

Vittala, S.S., Govindaiah, S. and Home Gowda, H., 2008. Prioritization of sub-watersheds for sustainable development and management of natural resources: An integrated approach using remote sensing, GIS and socio-economic data. Curr. Sci., v. 95(3). pp. 345–354.

Hon'ble Reviewer

1. **Prof. S.C. Mathur**
   Chairman
   Dept. of Geology, J.N.V. University
   Jodhpur - 342 005
2. **Dr. P.S. Dhinwa**
   Sr. Scientist
   Space Appl. Centre
   ISRO
   Ahmedabad - 380 001

III-47

Application of Hydroseims for Prediction of Seismic Phenomenon, Delhi Region, India

D.K. Chadha

Global Hydrogeological Solutions, G-66, Vikas Puri, New Delhi-18, India
E-mail: devinderchadha27@gmail.com

Abstract

In the recent past, the occurrence of many earthquakes and the associated activities in the Indian continent has made the realization to promote further research on the various possible pre-cursors to predict the occurrence of earthquakes to protect life and property. Delhi is one of the seismic prone areas as it lies in Zone IV of the seismic zoning map of India which means that earthquakes of magnitudes in the range of 6-6.5 (Richter scale) can be expected to occur. It is now recognized that the water level changes form one of the best options to predict seismic events. In order to promote research in this field, a pilot project was taken up under which five piezometers were installed, two in the hard rock, one in the weathered rock and two in the alluvial areas up to the depth of about 110 m. The piezometers were fitted with the digital automatic water level recorders and telemetric system for transfer of data online and provision for downloading at site. The changes in fluctuation of water levels, though very small, were observed during the period of seismicity even of the low intensity earthquakes with epicenter within less than 60 km. During the period of study there had been occurrence of 19 low intensity earthquakes but some significant changes in the hydroseisms were observed in the piezometers. From the study, it seems apparent that the geological and structural set up, hydro geological framework, occurrence of regional aquifer system are the key parameters for design of the system in its ability to predict the occurrence of earthquakes based on hydrologic anomalies.

Keywords: Hydroseims, Earthquakes in Delhi, Hazards, Urban Planning, Water Table Fluctuation.

Introduction

Delhi region is considered as seismic prone area because of its proximity to the Himalayan region and highly faulted basement rock in and around the Delhi from where low intensity seismic activities are frequently observed. It is asserted that since there had been a marked cessation of seismic activity since 1994 but the stresses which were earlier being released periodically in the form of low intensity earthquakes, have now been stored stresses which will be released any time resulting in high intensity earthquakes.

The occurrence of earthquakes around the city is attributed to seven tectonic units and this peculiar geological and structural setting makes Delhi prone to seismic events. The tectonic activity in Himalayan region has in the past affected the Delhi region like the Chamoli earthquake of 1999, 280 km from Delhi, when a lot of buildings were damaged.

Over the years, scientists have employed various methods and instrumentation in their attempts to predict earthquakes. These include statistical probability, physical measurements, and geochemical observations, Fault-creep measurement, measuring changes in surface tilt, Strain measurement, ground water levels' fluctuations and observations of animal behavior. The interactions between earthquakes and hydrological processes is shown in Figure 1.

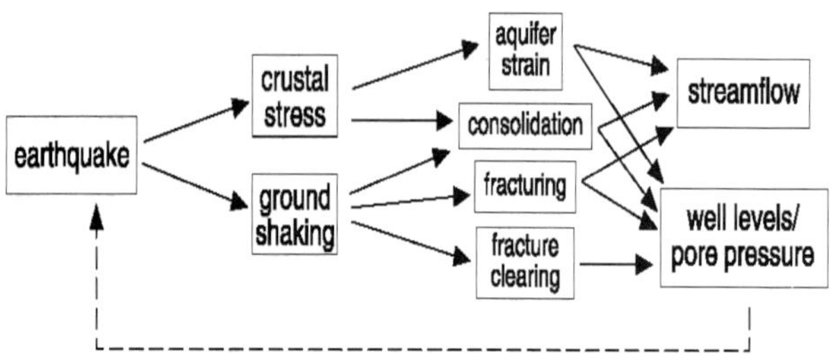

Figure 1: Interactions between earthquakes and hydrological processes.

GEOLOGY AND TECTONIC SETTINGS

In the study of precursors based on ground water level fluctuations, it is important to fully comprehend the regional and local geological and structural frame work, identify the different litho units & select an aquifer zone which is regionally extensive, not influenced by pumping of tube wells around the area and the intermittent rainfall during non-monsoon period. Geologically, the area comprises of quartzites of Delhi Super group (~1500 m.y) and Quaternary to Recent alluvial sediments (> 1.64 m.y). The quartzites are exposed on the surface along the 'Ridge' in the central and southern parts of the region. The contact between these quartzites and overlying unconsolidated sediments of Older and Newer alluvium is unconformable..The quartzites are weathered to varying degree from a few cm to more than 10m and fractures are observed at different depths which impart secondary permeability and constitute the water bearing strata within the quartzites.

On a regional scale, Delhi is bounded by the Indo-Gangetic alluvial plains in the north and east, Thar Desert in the west and by Aravali ranges in the south. The NNE – SSW trending Ridge marks the northern extension of the Aravali ranges of Rajasthan and forms a prominent physiographic feature in the region. The area is dissected by number of faults & fractures; the trend of which varies from NNE-SSW to ENE-WSW. The important faults west of the ridge area are Rajendra Nagar fault, MES Depot-East Patel Nagar fault, Anand Parbat-west Patel Nagar fault and Inderpuri fault. The notable faults east of ridge are Kishangarh fault, a WNW-ESE trending fault between Qutab Minar and Mehrauli and Lado Sarai fault. . The seven tectonic faults include Moradabad, Sohna, Aravali, a hidden fault in the Indo-Gangetic basin, Sonepat-Delhi-Sohna, the junction of the Aravali and Sohna faults and the Delhi-Haridwar ridge fault. The tectonic activity in Delhi is associated with the Delhi-Hardwar ridge of the Delhi Super Group, Aravalli-Delhi folding, Sohna, Mathura and Muradabad faults. The most seismic active zone of the area is located in the vicinity of the Delhi-Haridwar ridge, Lahore-Delhi ridge and axis of folding (Figure 2).

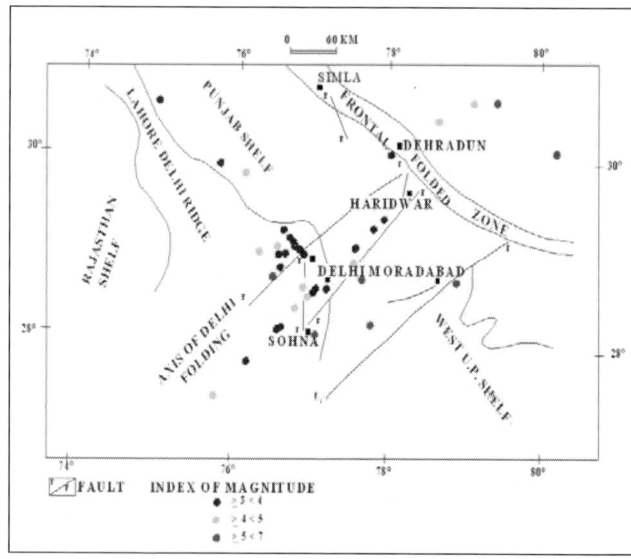

Figure 2: Tectonic map showing distribution of epicenters around Delhi

Subsurface Lithology

The rapid variations in sub surface lithology within the Newer and Older alluvium make the aquifer geometry of Delhi complicated and complex. In the Newer alluvium, the disposition of different sediments particularly the pervious layer constituting the unconfined aquifer is well delineated in the Yamuna flood plain area. In the older alluvium, the disposition of different lithological units is not well defined and they are heterogeneous in nature, making it difficult to identify the deep aquifer zones which are regionally extensive, both vertically and laterally. In the hard rock aquifer, the aquifers are defined by the presence of fractured zones at different depths. These fractured zones at places are locally well defined but not regionally extensive.

Site Selection and construction of piezometers

The selection of five sites for installation of piezometers is based on the geological, hydrogeological and sub-surface lithological data which was synthesized to comprehend the presence of different aquifer zones in the alluvial strata and configuration of bed rock and the fractured zones within it. The data was used to prepare 2-D sub-surface lithological sections to know the occurrence of regional and local aquifer zones and the depth to bed rock. The study of the depth to bedrock and its configuration in the present case is very important to know the regional extensiveness of this geological formation and the aquifer system thereof. The regional occurrence of the aquifer system for such a study is preferred as any earthquake event with epicenter within this geological formation or bearing shocks will show the first changes in the ground water levels. Considering the extensiveness of the Delhi quartzite and the aquifer system thereof, three piezometers sites out of five were selected in the Quartzite formation. The three sites i.e. at IMD station Kamla Nehru Ridge, Geology Department North Campus and CRRI Mathura road were further probed through deep resistivity survey to determine the thickness of the weathered mantle and the presence of fractured zones at different depths up to 120m depth.

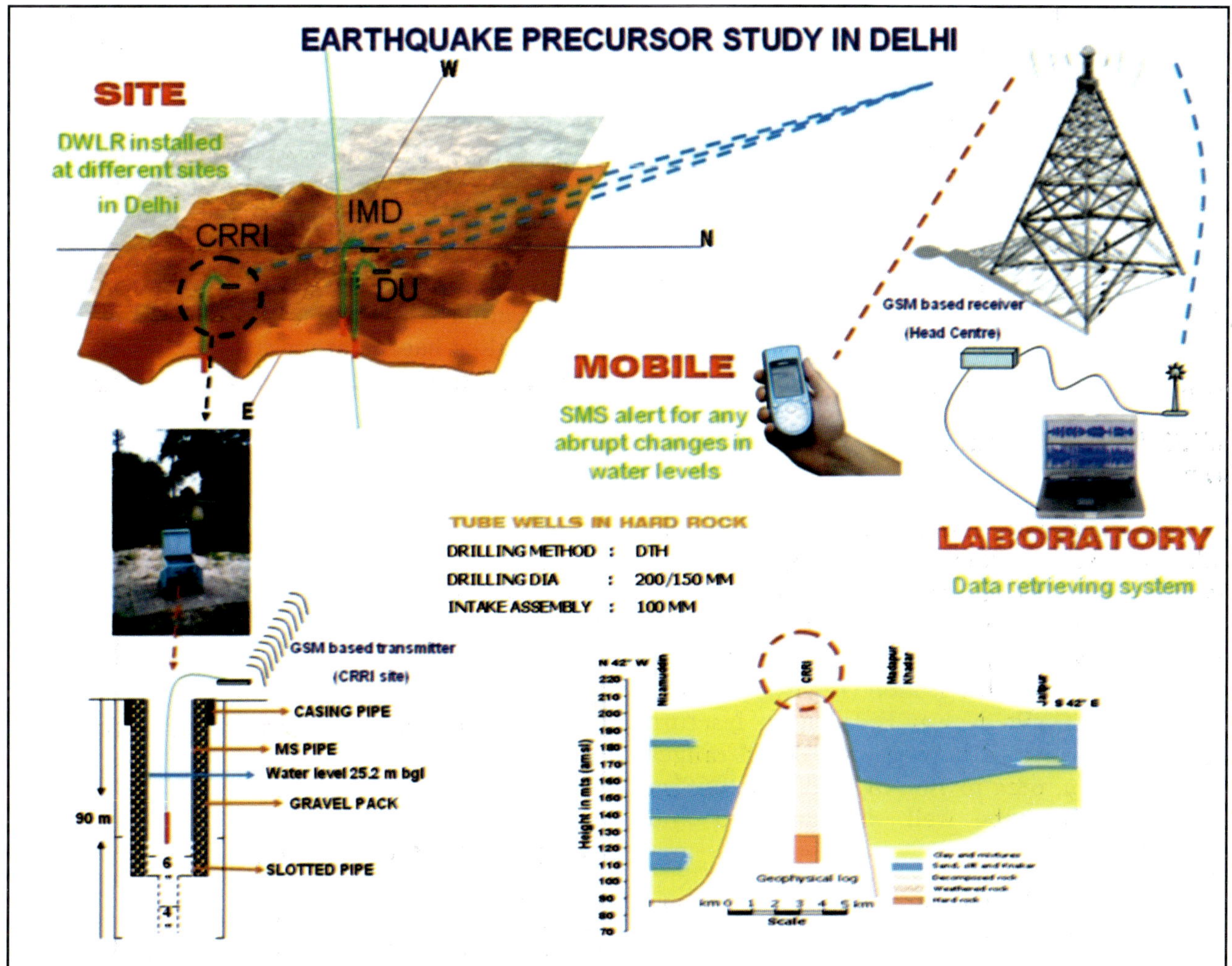

Figure 3: Earthquake Precursor study System

The deep fracture zones below 55m depths were selected for the design of piezometers in hard rock i.e. Delhi quartzites. Similarly two sites were identified in the alluvial sediments. The selection of the aquifer zone in the alluvial strata was such that the ground water table is under semi-confined/confined condition, the aquifer zones are deep seated below 65 m depth so that the pumping of the shallow tube wells influenced up to 50m depth will not possibly affect the ground water level of deep aquifer zones and influenced by non-monsoon intermittent rains. The piezometers thus established tapped different aquifer zones in both the quartzites and the alluvial sediments. The design of the tube wells in hard rock is very important and the piezometer (tube well) should be bottom open with a slotted assembly against the deep fractured zones. The tube well in alluvial area should gravel packed with proper slotted assembly tapping the deep aquifer zones with a bail plug at the bottom. All the five piezometers were fitted with automatic water level recorders (Troll 500) which records water levels, temperature and pressure. The water level data and other parameters were downloaded using rugged system or laptop with software for the purpose. The automatic water level recording system was also connected with telemetry system for on line/real time transfer of data using GSM mode.

A representative diagram of total earthquake precursor system is given in Figure 3.

The sub surface lithology of the five sites for the pilot study is described below:

PIEZOMETER 1: Kamla Nehru Ridge (IMD Complex)

The subsurface formation from Nizammuddin to Kamla Nehru Ridge extending to Madanpur Khadar shows that the configuration to bed rock is uniform though the gradient is comparatively steep, no undulation of the bed rock is observed in this section (Figure 4). In the bed rock at Kamla Nehru Ridge deep fracture zones were identified between the depth ranges of 33m to 46m, 68m to 76m and 106m to 110 m bgl accordingly tube well was designed and installed.

PIEZOMETER 2: Geology Dept Complex, Delhi University

The site is located about 2km east of the IMD station of Kamla Nehru Ridge and it falls in part of the ridge where Delhi Quartzite are exposed. The sub surface formation is of weathered quartzite, the thickness of which is about 35 meter. The tubewell was installed tapping the weathered zone and the fractured zone of the quartzite.

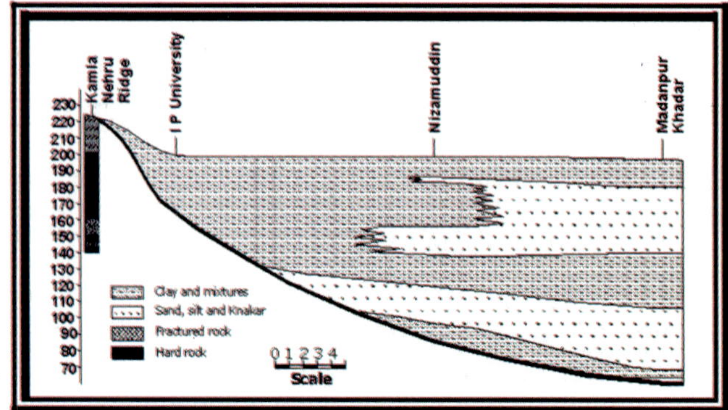

Figure 4: Cross section through Kamla Nehru Ridge - Madanpur Khadar

PIEZOMETER 3 : CRRI Complex

A regional section from Mandoli to Okhala shows that the bed rock is present at shallow depth and gradually slopes to different depths with different gradient towards Okhla (SW) and Mandoli (NE) (Figure 5). At CRRI the quartzites are exposed near to the surface & are highly fractured at different depth. The deeper fracture zones below 50 m depth have been identified for designing & construction of piezometer. The configuration of the bed rock gives the indication of the faulted basement rock.

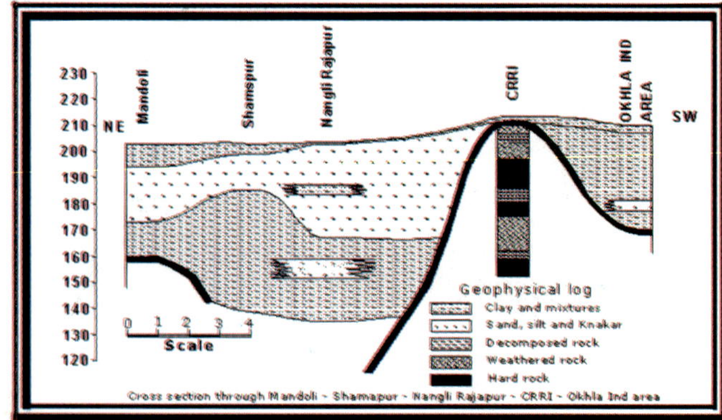

Figure 5: Cross section through Mandoli – Okhla Ind area

PIEZOMETER 4 & 5: Dwarka

Both the sites are located in the older alluvial plain. The groundwater is fresh at shallow depth but brackish at depth more than 30m bgl. A panel diagram showing the lithology in sector 5 and sector 20 where the piezometers have been installed is in shown in Figure 6. The sub-surface lithology is dominated by clay with kankar beds. The piezometers have been installed upto a depth of 110 meter tapping the aquifer zone occurring between the depth of 70-90m bgl.

The summary statement of the5 piezometer is given in Table 1.

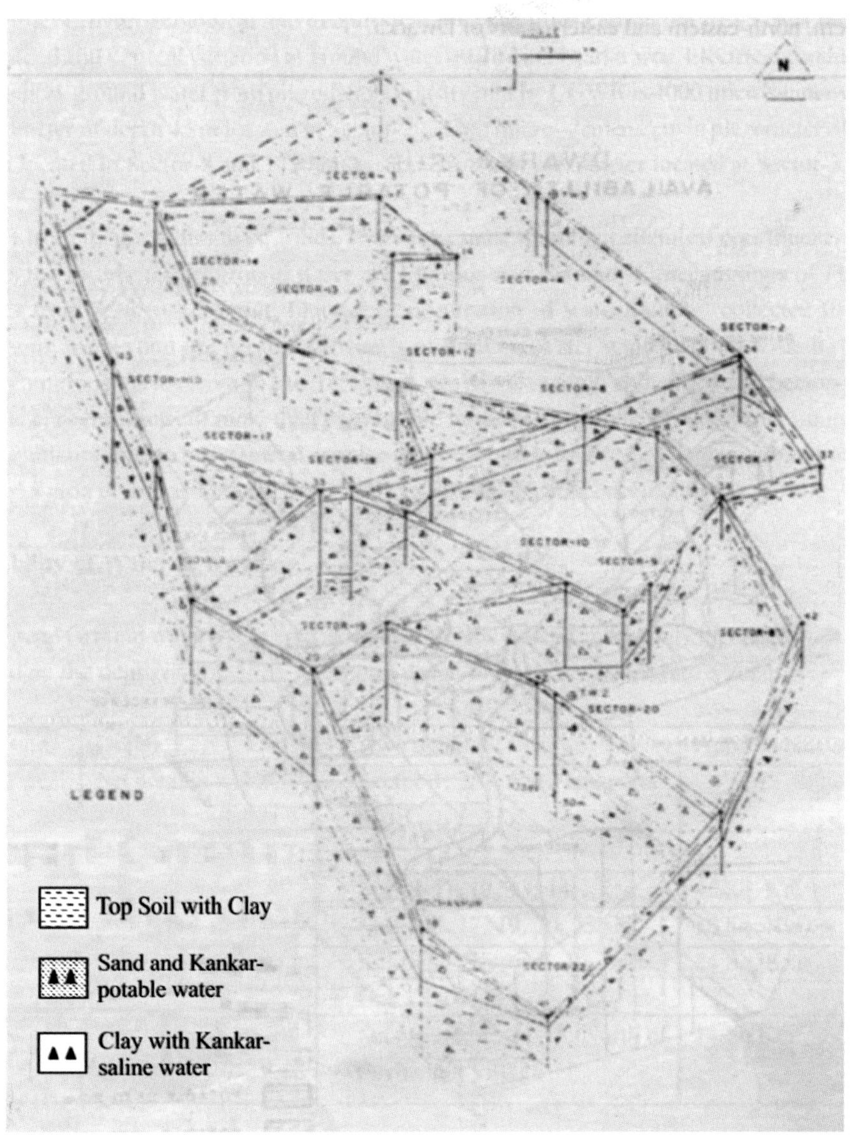

Figure 6: Panel Diagram

Table 1. Summary statement of the 5 piezometer

S.No.	Location	Rock Type	Depth to water level	runoff
1	Kamla Nehru Ridge		103m	9.8 m
2	CRRI campus		110m	24.2m
3	Dwarka Sector 5		115m	17.7m
4	Dwarka Sector 20		110m	8.7m
5	D.U		90m	3.6m

Data Analysis

It has been known that the response to change of water table in relation to earthquakes generally has two forms: oscillations and step-like changes (Wang, et al, 1983; Wang, et al, 1984). Oscillations are fluctuations of water surface in deep wells responding to the effects of large, distant earthquakes in a way that suggests a similarity to seismographs (Stearns, 1928). The remarkable feature of oscillations is the great amplitude of fluctuations, primarily due to surface waves from distant earthquakes. The amplitude and period of oscillation are related to the characteristics of a well-aquifer system (Eaton and Takasaki, 1959; Cooper et al., 1965). While the step like changes in confined aquifers within a few source dimensions of an earthquake epicenter can often be quantitatively accounted for as the poroelastic response to the earthquake static strain (Wakita, 1975). And strong ground shaking near the earthquake rupture may increase permeability of hard rock aquifers and, consequently, change ground water levels as step-like and pulse etc (Rojstaczer et al, 1995, Ohisma, 1996).

In the present research three cases were selected to show the unusual behavior of the water fluctuation level before and after Earthquake. In the first case given below in the Figure7, recorded at CRRI on 31st March, 2006 shows a steep decrease During the period of the project, most of the earthquakes recorded were of low intensity earthquakes (less than 4 magnitudes on Richter scale) and 19 in number & only 4 earthquakes were more than magnitude 4 (Table 2). In the present paper three cases are selected where the impacts of all these earthquakes before, during & after the events were studied to observe the change in water levels linked to earthquakes. The following pages give the details of the water table fluctuation vis-à-vis the different intensity of earthquakes, both observed from the piezometers installed in different aquifer systems of the hard rock & the alluvial areas.

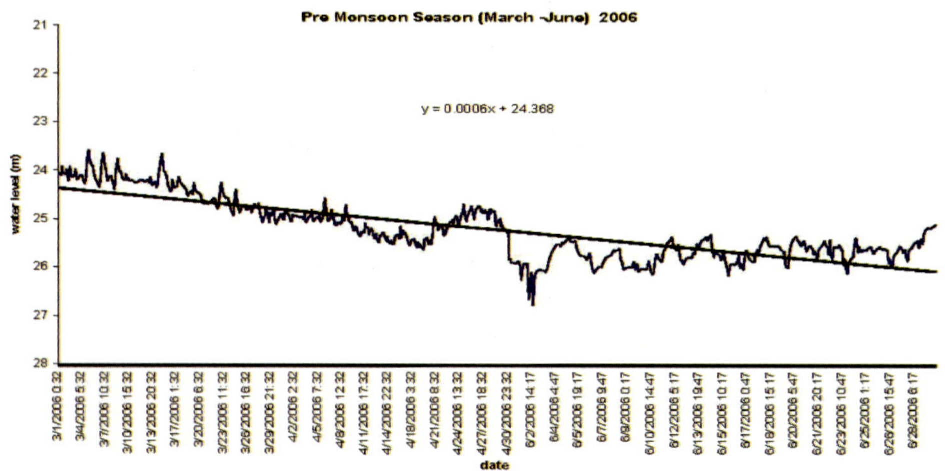

Figure 7: Pre-Monsoon Graph for CRRI 2006

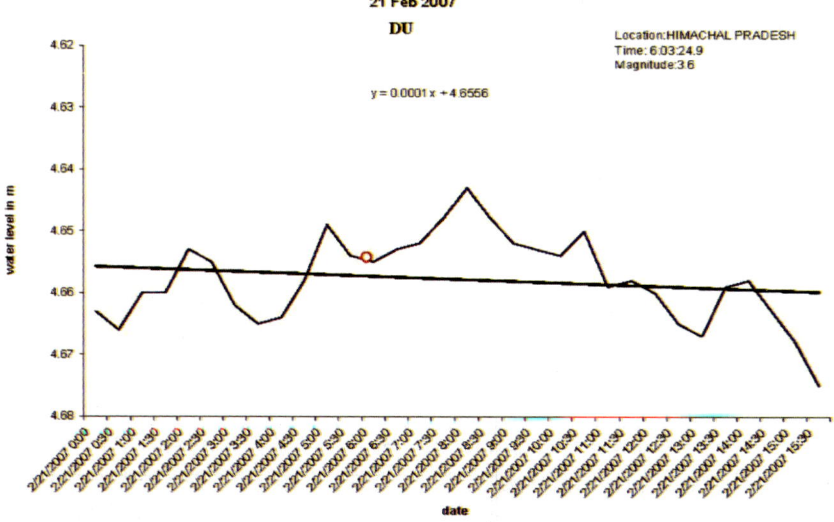

Figure 8: Graph Showing Water Level Fluctuation at DU on 21 Feb

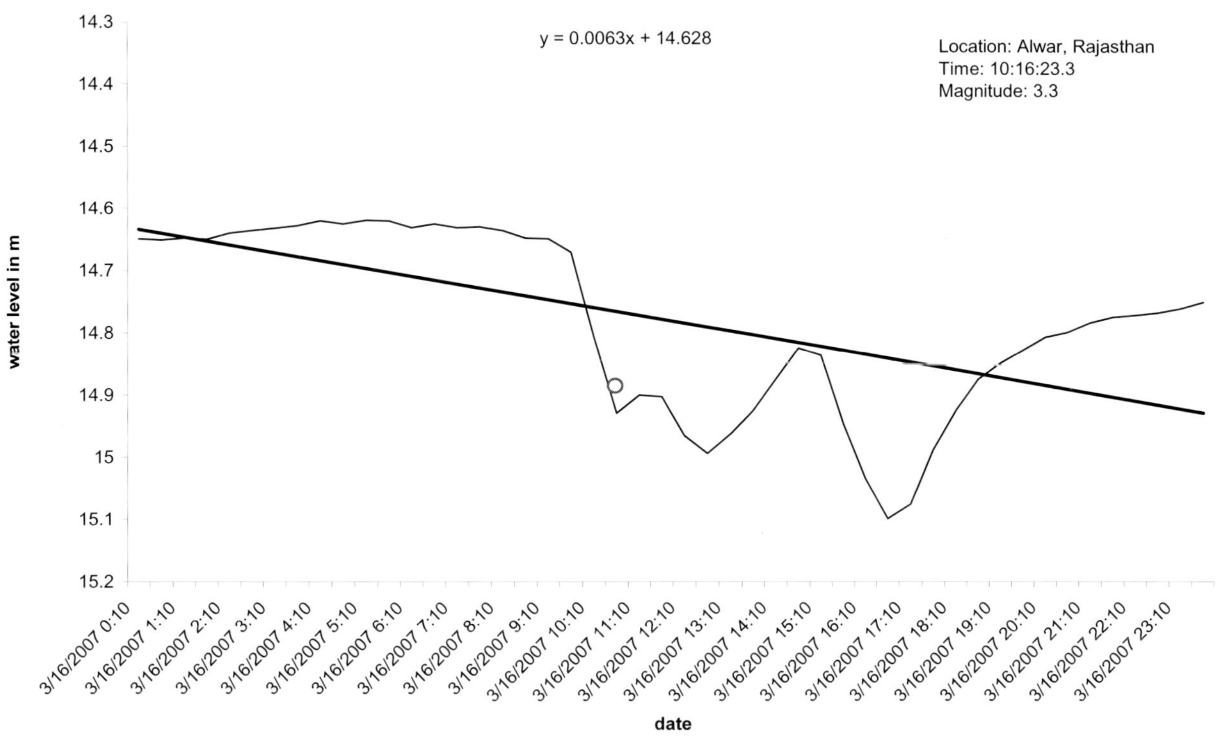

Figure 9: Graph Showing Water Level Fluctuation at CRRI on 16 March

27 March CRRI

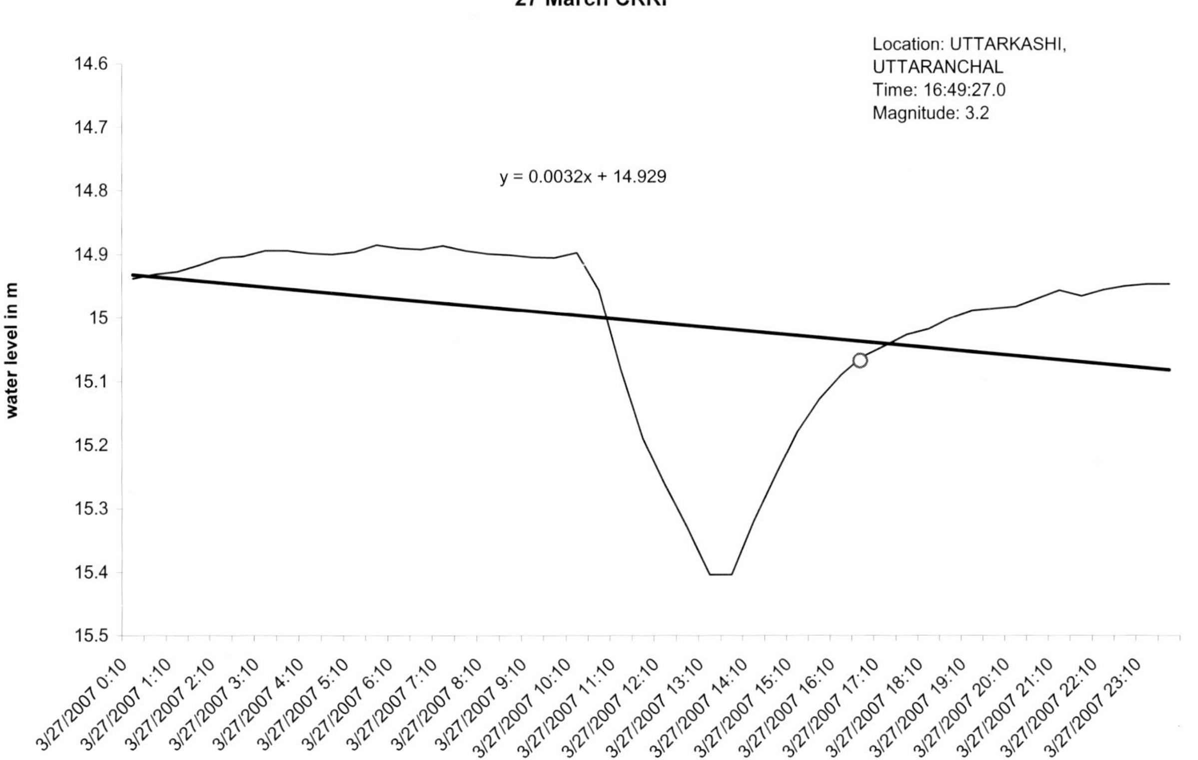

Figure 10: Water Level Fluctuation at CRRI on 27 March

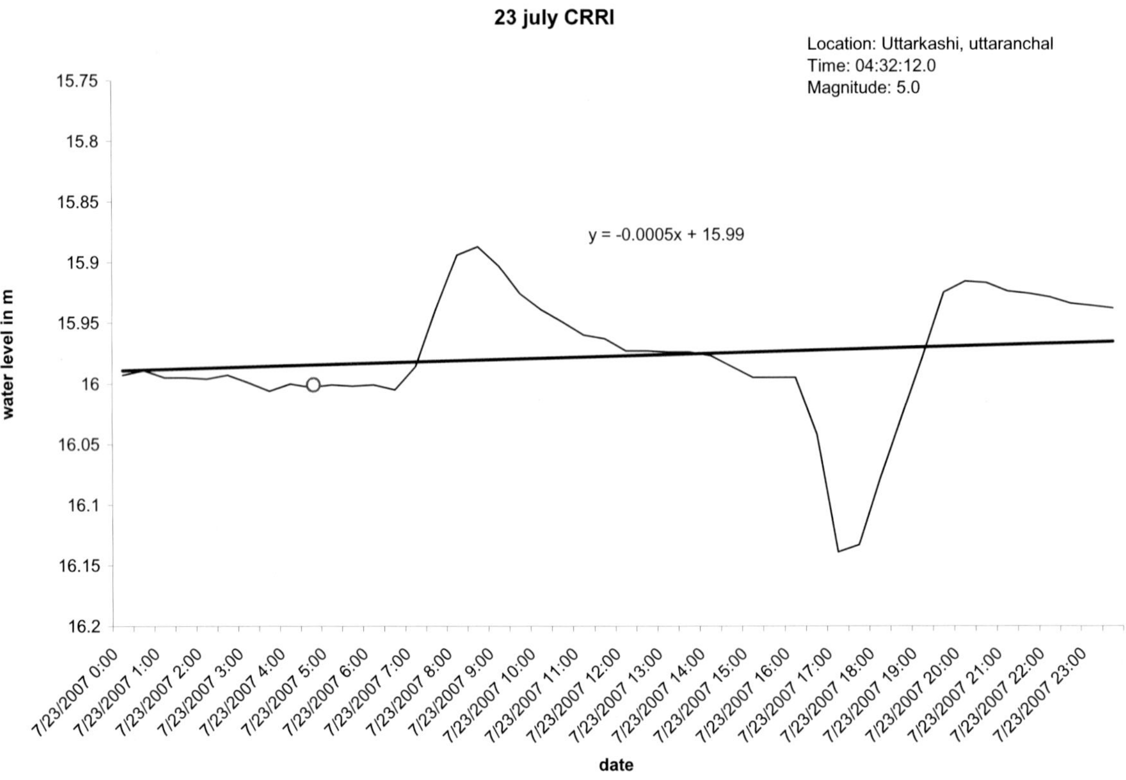

Figure 11: Graph Showing Water Level Fluctuation at CRRI on 23 July

Table 2. Table showing water level fluctuation prior to earthquake

Location	Origin	Depth (Km)	Magnitude	Epicenter Distance (Km)	Prior to Earthquake Water Level Fluctuation (m)
		31 March 2006			
CRRI	**Jhajjar, Haryana**	22.0	3.4	60.10	0.116
IMD	Jhajjar, Haryana	22.0	3.4	41.03	0.008
DU	Jhajjar, Haryana	22.0	3.4	55.21	0.039
		07 April 2006			
CRRI	**Sonipat Haryana**	11.5	3.0	51.99	0.040
IMD	Sonipat Haryana	11.5	3.0	47.84	0.016
DU	Sonipat Haryana	11.5	3.0	38.40	.048
		01 May 06			
CRRI	**Rohtak, Haryana**	10.0	3.6	74.58	0.124
IMD	Rohtak, Haryana	10.0	3.6	59.57	0.009
DU	Rohtak, Haryana	10.0	3.6	64.45	0.016
		07 May 06			
CRRI	**Jhajjar,Haryana**	20.2	4.1	60.10	0.302
IMD	Jhajjar, Haryana			41.03	0.085
DU	Jhajjar,Haryana	20.2	4.1	55.21	0.078
		15 May 06			
IMD	**Faridabad**	05.0	2.5	30.02	0.047
		20 July 06			
IMD	**Himachal Pradesh**	10	3.5	395.05	0.008

		29 Nov. 06			
CRRI	**Alwar, Rajasthan**	13.0	3.9	129.57	0.226
DU	Alwar, Rajasthan	13.0	3.9	137.98	0.016
		9 Dec 06			
CRRI	**Rohtak, Haryana**	06	2.8	74.58	0.675
DU	Rohtak, Haryana	06	2.8	64.45	0.015
		10 Dec 06			
CRRI	**Himachal Pradesh**	33	3.5	391.87	0.158
DU	Himachal Pradesh	33	3.5	379.57	0.035
		23 Dec 06			
CRRI	**Jaipur, Rajasthan**	10	3.7	234.52	0.027
DU	Jaipur, Rajasthan	10	3.7	241.30	0.008
		21 Feb 07			
CRRI	**Himachal Pradesh**	33.0	3.6	391.87	0.042
IMD	Himachal Pradesh	33.0	3.6	395.05	0.013
DU	Himachal Pradesh	33.0	3.6	379.57	0.013
		13 March 07			
CRRI	**Delhi**	10.1	2.3		0.023
IMD	Delhi	10.1	2.3		0.113
DU	Delhi	10.1	2.3		0.004
		16 March 07			
CRRI	**Alwar, Rajasthan**	10.0	3.3	129.57	0.030
IMD	Alwar, Rajasthan	10.0	3.3	119.58	0.032
DU	Alwar, Rajasthan	10.0	3.3	137.98	0.025
		3 April 07			
CRRI	**Rohtak, Haryana**	08.8	2.8	74.58	0.066
IMD	Rohtak, Haryana	08.8	2.8	59.57	0.103
		14 May 07			
CRRI	**Rohtak, Haryana**	05.0	3.2	74.58	0.70
IMD	Rohtak, Haryana	05.0	3.2	59.57	0.154
		15 Sept 07			
CRRI	**Rajasthan**	27.9	2.8	235.20	0.606
IMD	Rajasthan	27.9	2.8	217.19	0.046
Dwarka Sector 20	Rajasthan	27.9	2.8	218.88	0.038
		18 Oct 07			
CRRI	**Uttar Pradesh**	05.6	3.6	299.25	1.893
Dwarka Sector 5	Uttar Pradesh	05.6	3.6	319.93	0.019
Dwarka Sector 20	Uttar Pradesh	05.6	3.6	318.30	0.016
		20Nov 07			
Dwarka Sector 5	**Rajasthan-Haryana border**	01.2	3.3	55.48	0.132
Dwarka Sector 20	Rajasthan-Haryana border	01.2	3.3	57.21	0.36
		25 Nov 07			
Dwarka Sector 5	**Delhi-Haryana Border**	20.3	4.3	0	0.008
Dwarka Sector 20	Delhi-Haryana Border	20.3	4.3	0	0.023

In all the above mentioned earthquakes and the subsequent hydroseisms, different shapes and different amplitudes are attributed to different responses

The earthquake events dated 13 March, 3 April, 15 September and 3 November 2007 were of low intensities of 2.3, 2.8, 2.8 and 2.7 respectively. All the events though showed fluctuations & the highest value of water level was 700 millimeters while the lowest value of water level was 23 millimeters. It has been seen that, if the hydroseisms for 12 hours are interpreted, a sudden rise or fall is observed either before or after the event. The normal variations in ground water levels can be attributed to the lower magnitude of the earthquake and also to the epicenters of the respective events which are quite far from all the three water level recording sites. The event of 13 March 2007 had its epicenter at Delhi & the maximum fluctuation in the water level was 113 millimeters. Though the 15 September 2007 event was of 2.8 magnitudes & its epicenter was at Sikar, Rajasthan which is approximately at a distance of 230.56 kilometers from Delhi, showed a maximum fluctuation of 606 millimeters & minimum fluctuation of 38 millimeters in water level.

The earthquake events dated 21 February, 16 March, 27 March and 23 July 2007 were relatively higher intensity earthquakes of magnitude 3.6 (Figure 8), 3.3 (Figure 9), 3.2 (Figure 10) and 5.0 (Figure 11) respectively. It is observed that in all the hydroseisms for these events, water level jumps and drops regularly but on 27 March 2007 water level shows a significant rise of 0.377 meters or 377 millimeters. If the trends for 21 February 2007 event are noticed it is seen that the difference in the highest and lowest values of ground water levels at Delhi University, CRRI & IMD are nearly 30-40 millimeters respectively. The earthquake that took place on 16 March 2007 showed more or less same variations in the water levels ranging between 0.25 to 0.32 meters. The ten days plotting of data shows a maximum fluctuation of 522 millimeters in water level.

The event on 23 July 2007 was the highest intensity earthquake event of magnitude 5.0 on a Richter scale. The epicenter was Uttarkashi, Uttarakhand which is at an aerial distance of 267.95 kilometers from Delhi University, 263.07 kilometers from CRRI and 265.85 kilometers from IMD Station. At CRRI, the day readings showed a difference of 17 millimeters when the highest and lowest values were compared & 387 millimeters difference is seen over a period of 18-28 July. On the other hand, both the sites, Delhi University and IMD Station (12 millimeters) failed to show a steep rise and fall. The five sites are at different epicentral distances; hence show different sensitivities. Table 2 gives the list of the earthquakes recorded during March 2006 to March 2008, epicenter distances, depth of origin and the fluctuation in water levels.

SUMMARY OF RESULTS

During the period of implementation of the project, earthquakes having intensity of 2.3 to 5.2 were recorded with epicenters in Delhi, Rajasthan, and Haryana and nearby Himalayan areas in Himachal Pradesh and Uttaranchal. There had been only one earthquake with epicenter in Delhi with depth of origin at about 10 km and of intensity 2.3. The earthquakes with epicenters in Jhajjar, Rohtak and Rajasthan were of the intensity varying from 2.5 to 3.9 with depth of origin mostly between 10 km and 20 km whereas the depth of origin in Himachal Pradesh is about 33 km and intensity 3.5 and above.

The hydroseisms were studied in relation to 19 earthquakes recorded during the period of the research programme i.e. from 2006 and 2008 (refer Table 2). The changes in water levels, though very small, were observed of the earthquakes originating in Delhi and surrounding areas with epicenters in less than 100 km but the notable changes were observed for the following earthquakes:

- 7 May 06-Intensity of earthquake was 4.1 with epicenter at Jhajjar, Haryana & the maximum fluctuation in water was 0.302 meters. (Figure 12)
- 10 Dec 06- Intensity of earthquake was 3.5 with epicenter at Himachal & the maximum fluctuation in water level was 0.158 meters. (Figure 13)
- 14 May 07- Intensity of earthquake was 3.2 with epicenter at Rohtak, Haryana & the maximum fluctuation in water level was 0.70 meters (Figure 14)
- 15 Sept 07-Intensity of earthquake was 2.8 with epicenter at Rajasthan & the maximum fluctuation in water level was 0.606 meters. (Figure 15)
- 3 Nov 07- Intensity if earthquake was 3.3 with epicenter in Rajasthan and the maximum fluctuation in water level was 0.132 meters. (Figure 16)

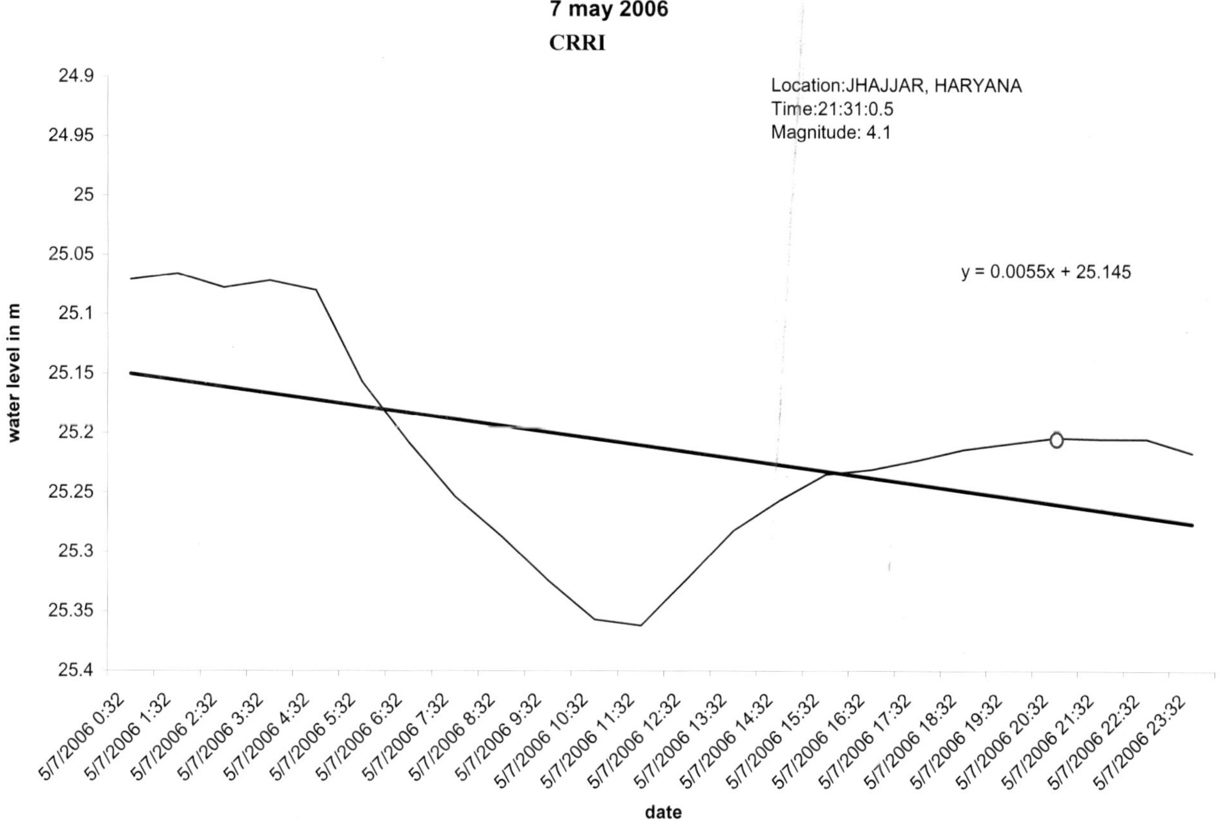

Figure 12: Graph Showing Water Level Fluctuation at CRRI on 7 May

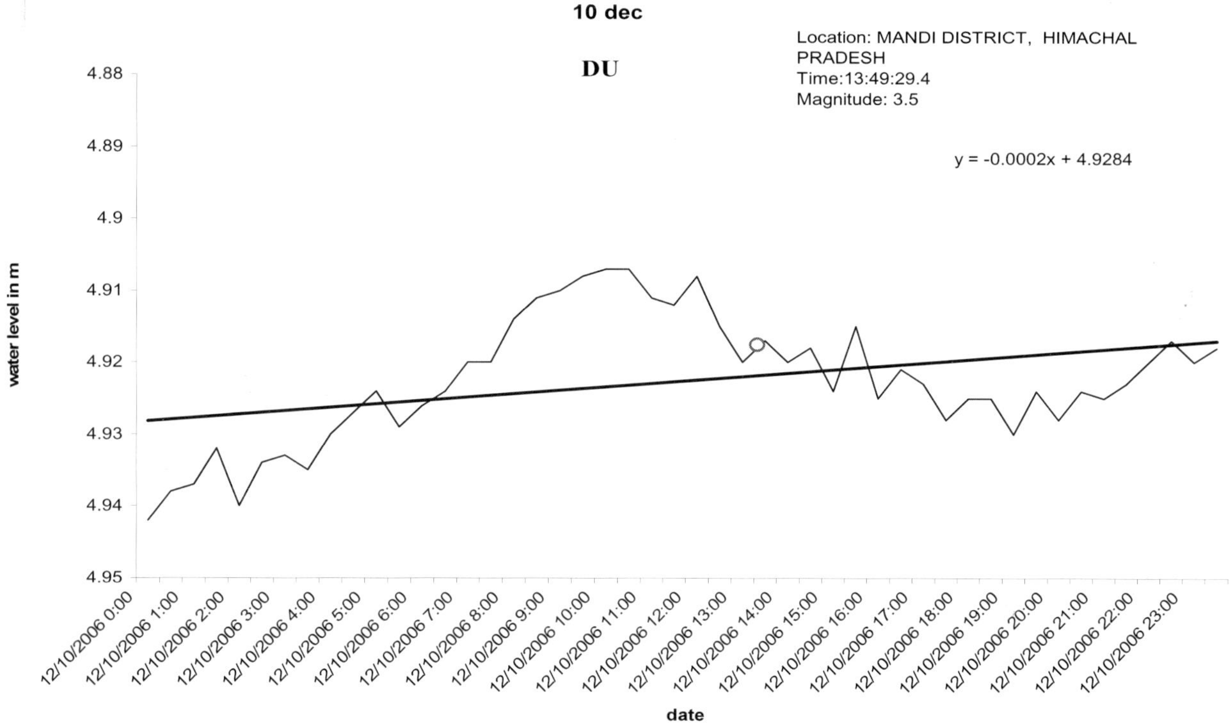

Figure 13: Graph Showing Water Level Fluctuation at DU on 10 Dec

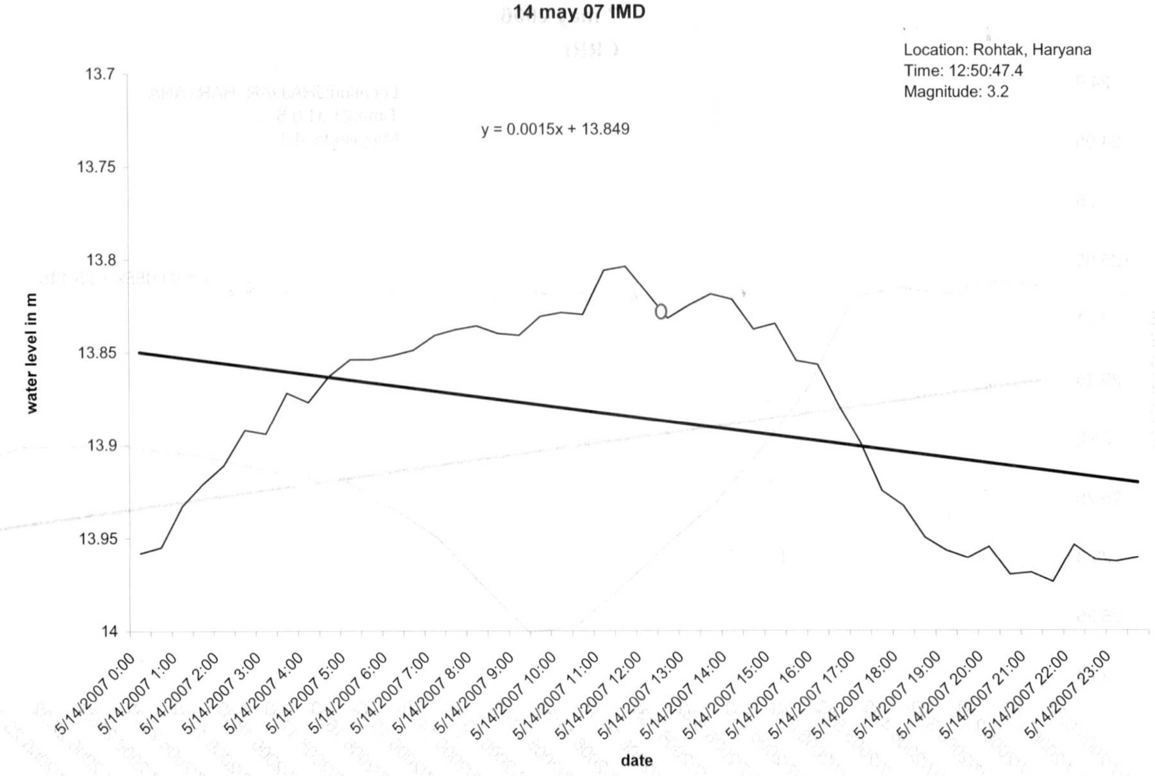

Figure 14: Graph Showing Water Level fluctuations at IMD on 14 May

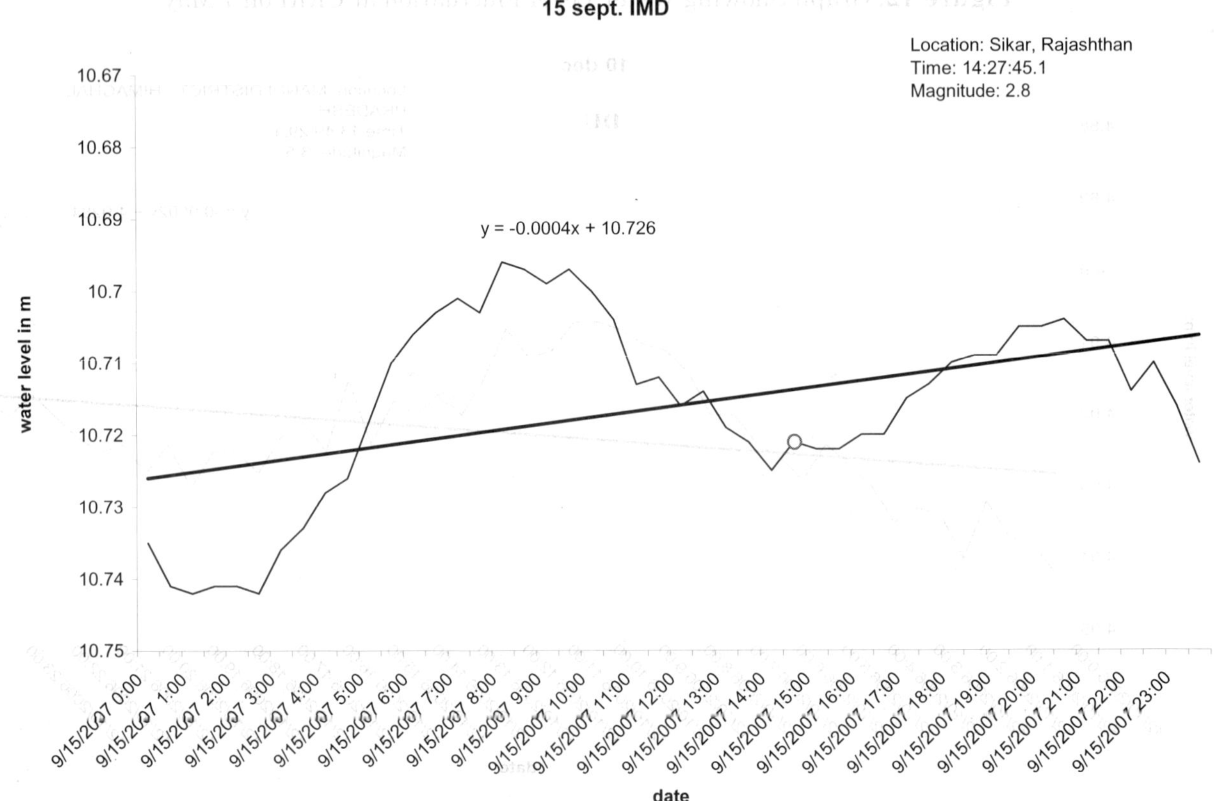

Figure 15: Graph Showing Water Level Fluctuation at IMD on 15 Sept

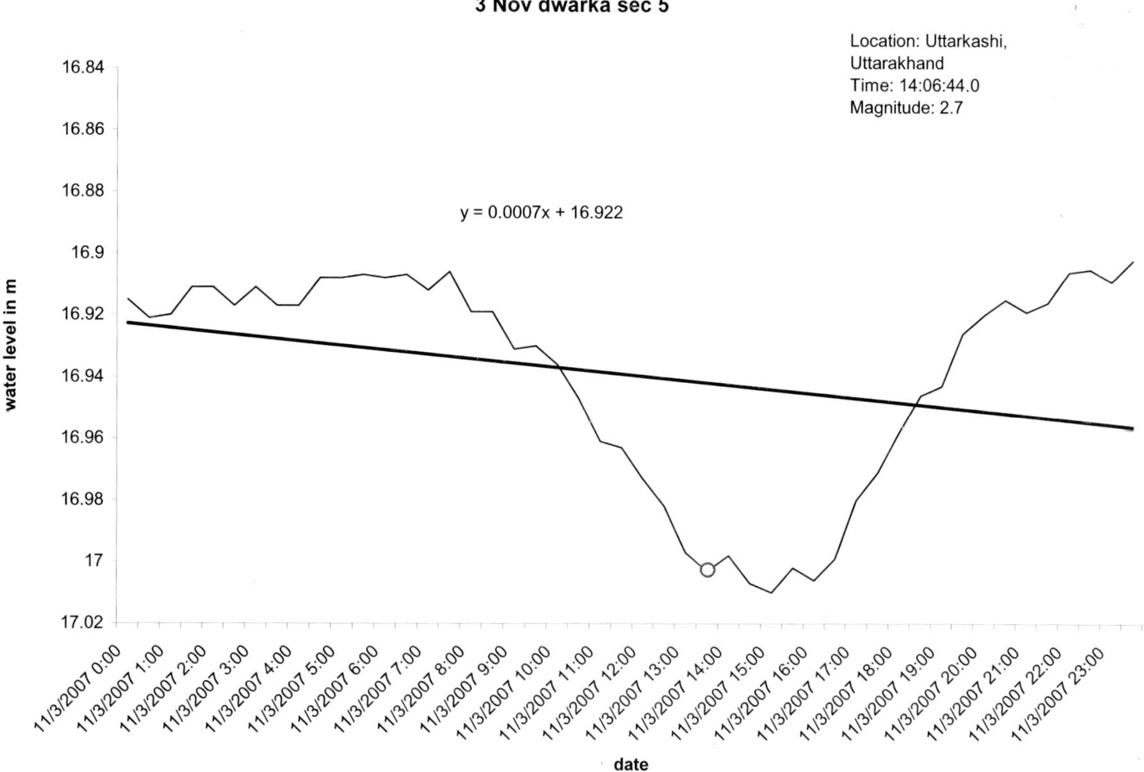

Figure 16: Graph Showing Water Level Fluctuation at Dwarka Sector 5 on 3 Nov

It is observed that the response to water levels seems to be related to the depth of occurrence, distance of epicenter from the recording stations, magnitude of earthquake and the behavior of aquifer system. The trend and inclination of water levels changes sharply i.e. abrupt rise or fall just before the earthquake and water table lines show different shapes. It is also observed that the water level fluctuations are not observed in all the piezometers for the same earthquake.

CONCLUSION

Delhi and the surrounding areas are considered prone to seismicity as it comprises of quartzite of Delhi Super group which are overlain with a marked unconformity by alluvial sediments of Quaternary to Recent period and traversed by number of faults. The notable faults east of ridge are Kishan garh fault, a WNW-ESE trending fault between Qutab Minar and Mehrauli and Lado Sarai fault. The seismicity in the Delhi region is attributable to 7 major tectonic features but the most active being Delhi - Haridwar ridge and Delhi- Lahore ridge. In the past between 1720 and 1996, Delhi has experienced six high intensity earthquakes of magnitude 6 and above besides a number of low intensity earthquakes. In view of its vulnerability to high intensity earthquakes, Delhi is placed in Zone IV of the Earthquake zonation map of India wherein earthquakes up to 6.5 on the Richter scale become a possibility. It is found essential to continue to work on precursor study to predict the earthquake events for taking necessary protection measures.

Groundwater in Delhi is being exploited to meet the domestic water demand and the tube well depths have already reached to 250m bgl, therefore the piezometers for the precursor study should be drilled before the depths so as to avoid the impact of tube well running o the piezometers. This will be in line with recent findings that the deeper the aquifer systems better the response of earthquake on water levels. In order to meet the R & D requirements of the pre-cursor studies based on ground water levels, there is further improvement in the automatic water level recording systems which not only record the water level but also the change in water quality, also a good precursor. The improvement in continuous multi data recording systems and transmission of real time data makes study of earth quake prediction on the behaviour of water levels vis-à-vis earthquakes more acceptable than the other pre-cursor systems.

For any future studies a number of deep piezometers should be installed in entire NCR area in a grid pattern, each grid measuring 100 to 120 sq.km. This is important in the view of the recent observation that the regionally extensive aquifer systems like the Delhi quartzite will be more responsive to earthquakes. The piezometers for measuring the fluctuation water level should also fitted with strain meter/tilt meter to relate it to the stresses built up during earth quake.

REFERENCES

Cooper HH & et al 1965: The response of well-aquifer systems to seismic waves. J.Geophys. Res. 70, pp 3915-3926.

Eaton, J.P. and Takasaki, K.J., 1959. Seismological interpretation of earthquake-induced water level fluctuation in wells. Bull. Seism. Soc. Amer. 49, pp 227-245.

GHS, 2006, Geophysicla survey for determination of bed rock configuration and aquifer geometry Delhi, Unpublished report.

Igarashi, G., Wakita, H., and Sato, T 1992: Precursory and co seismic anomalies in well water levels observed for the 2 February 1992, Tokyo Bay earthquake, Geoph. Res. Lett., 19, 15, pp 1583-1586.

Kissin, I.G., 1984. Hydrogeodynamic precursors in an earthquake prediction system. In: A.V. Nikolaev and I.G. Kissin (Editors), Hydrogeodynamic Earthquake Precursors. Nauka, Moscow, pp.3-30 (in Russian).

Kissin, I.G. and Grinevsky, A.O 1990: Main features of Hydrogeodynamic earthquake precursors, Tectonophysics, 178, pp 277-286.

Kovach, R.L., Wesson, R.L., Robinson, R. and Nur, A.,1975. Water-level fluctuation and earthquake on San-Andreas fault zone. Geology, 3 (8) pp 437-456

Merifield, P.M. and Lamar, D.L., 1981.Anomalous water-level changes and possible relation with earthquakes. Geophys. Res. Lett., 8 (5) pp 437-440.

Mukhopadhyay, S. & et al 2002: Seismic microzonation of Delhi for ground-shaking site effects, Current Science, 82, pp 877-881

Ohisma, H. & et al 1996: Groundwater fluctuations caused by the earthquake. Journal of the Japan Society of Engineering Geology 37, pp 351-358.

Stearns, H.T., 1928. Record of earthquake made by automatic recorders on wells in California. Bull. Seism. Soc. Amer. 18, pp 9-15

Rojstaczer, S., Wolf, S. and Michel, R., 1995. Permeability enhancement in the shallow crust as a cause of earthquake-induced hydrological changes. Nature 373, pp 237-239.

Wakita, H., 1975. Water wells as possible indicators of tectonic strain. Science 189, pp 553-555.

Wang, C.M & et al 1984: Main results of groundwater observation for 15 years and their applications in earthquake prediction in China.

Hon'ble Reviewers

1. **Dr. Harsha Gupta**
   President Geol. Soc. India
   NDMA Bhawan
   New Delhi - 110 021
2. **Dr. V.K. Gaur**
   Distinguished Scientist
   Nat. Geophy. Res. Inst.
   Uppal Road,
   Hyderabad - 500 007

Part - IV:

LAND RESOURCES, GEOHAZARDS AND ENVIRONMENT

IV-48

Airborne Suspended Particulate Matter and its Impact on Human Respiratory System – Mineralogical Study from Shahdara and Shahzada Bagh Industrial Areas in Delhi

Jitendra K. Nagar[1], J.P. Shrivastava[1] and Raj Kumar[2]

[1]Department of Geology, University of Delhi, Delhi-7
[2]Department of Respiratory Medicine, Vallabhbhai Patel Chest Institute, University of Delhi, Delhi-7
E-mail: jpshrivastava.du@gmail.com

Abstract

Airborne suspended particulate matter (SPM) is associated with frequent cases of respiratory illness as airborne particulates easily reach the deepest recesses of the lungs which damage respiratory system. Present results show that the concentration of indoor SPM is very high (~ 3 times higher than the standard values) in Shahdara and Shahzada Bagh industrial areas, mainly due to the presence of a large amount of industrial establishment. On the basis of relative abundances of mineral phases, SPM have been subdivided into three distinct groups, such as Group – A: Quartz - Carbonates Dominant Group, Group – B: Quartz - Mica Dominant Group and Group – C: Quartz - Feldspar Dominant Group. Relatively high proportion of five major mineral groups such as quartz, carbonates (calcite and dolomite), mica (illite and muscovite), feldspar, talc-chlorite (talc and chlorite) were found in the indoor suspended particulate matter. Illite, chlorite and talc are the clay minerals and all others are non - clay minerals. Present work revealed abundance of mineral phases in the indoor SPM is in the order of quartz > carbonates > Feldspar > mica > talc-chlorite. The concentration of major and trace elements observed in indoor SPM isin the order of Si > Al > Ca > Fe > Zn > Cu > Na > Mo > Pb > Mg > Cr > K > Mn > Ni > Co > Cd. The mean level of indoor SPM is high in the houses of Shahdara and Shahzada Bagh, where children had diagnosed to have asthma, rhinitis and upper respiratory tract infection. The high values for mean level of indoor SPM was associated with environmental factors. Quartz and mica contents were high in indoor SPM where children having asthma, rhinitis and upper respiratory ailments.

Keywords: Airborne Suspended Particulate Matter (SPM), Human Respiratory Diseases, Airborne Minerals, Airborne Elements.

Introduction

Airborne suspended particulate matter consists of a mixture of solid particles and liquid droplets suspended in the air for long periods of time, varying microscopic in size, composition, and origin (Dockery et al., 1993; Bascom et al., 1996 and Harrison et al., 1997). It refers to a stable mixture of particles suspended in a gas (Bascom et al., 1996). The size of airborne particle ranges over a wide scale from 0.005 μm to 100 μm in aerodynamic diameter, although the suspended portion is generally less than 40 μm. As per the United States Environmental Protection Agency (US EPA Report, 2003) the suspended particles fall into three basic modes. These are ultra fine (smaller than 0.1 μm in diameter), fine (between 0.1 and 2.5 μm) and coarse particles (larger particles than 2.5 μm). U.S. national health standards for ambient air quality is based on the mass concentration of "inhalable particles," defining only very small particles PM_{10} (with an aerodynamic diameter of less than 10 μm size) easily are inhaled into the lung. Fine-particulate air pollution includes particles with an aerodynamic diameter equal to or below 2.5 μm ($PM_{2.5}$), leading to great risk to human health because they can be breathed more deeply into the lungs and are more toxic than larger particles (Dockery et al., 1993). Suspended particulate matter is responsible for serious health problems such as asthma, bronchitis, rhinitis, irritation of lungs, pneumonia, decreased resistance to respiratory infection, chronic obstructive pulmonary disease (COPD), chronic cough, phlegm production, shortness of breath, cancer and allergy leading to premature death of human beings (Bascom et al., 1996; Horak et al. 2002; McConnell et al., 2003; Kan et al., 2004).

Airborne particulates reach ambient atmosphere and indoor air from a wide range of sources (Merefield and Stone, 1998). In general, concentrations of suspended particulate matter vary greatly from one to another area, depending on the nature and intensity of local sources. It also depends upon other factors such as topography, time of year, general weather conditions (Trindade et al., 1981; Wahab, 2003). Particulate matter in urban areas having several sources such as - (a) combustion of fossil fuels for space heating, power generation and automobile exhaust (b) incineration of refuse in apartment houses and municipal incinerators (c) industrial processes and (d) natural sources such as sea salt, volcanic eruption, wind blown dust, pollen grains and particles of soil (Kleinman et al., 1976).

Based on the origin, particulate matter is classified into two classes such as primary and secondary particles. Primary particles are released directly into the atmosphere from the source of generation such as dust from roads or black carbon from combustion sources (United States Environmental Protection Agency, 2004). They include metals, rocks and mineral particles from mining, quarrying, construction and industries. The primary sources also include sea spray, windblown dust, volcanic eruption, fossil fuel power stations, motor vehicles and factories flues (Panyacosit, 2000). Additionally, flyash from coal combustion and soot particles from other carbon based combustion processes such as from diesel fuel and wood burning add to the dust flux (Merefield and Stone, 1998). Secondary particles are formed in the atmosphere as a result of reactions that involve gases. Principal gases involved in the secondary particulate formation are sulphur dioxide (SO_2), nitrous oxide (NO_x), ammonia (NH_3) and volatile organic carbons (VOCs). They are the production of chemical transformation of SO_2, NO_2, HCl and ammonium compounds under atmospheric conditions (Panyacosit, 2000). It includes ammonium sulphate $(NH4)_2\ SO_4$) and gypsum ($CaSO_4$), formed by SO_2 emissions from the power plants and industrial facilities (Merefield and Stone, 1998). Secondary particulate include nitrates formed from NO_X emissions from power plants, automobiles and other combustion sources; chloride salts derived from sea spray; and carbon formed from organic gas emissions from automobiles and industrial facilities (Merefield and Stone, 1998; US EPA, 2004). However, mineralogical composition of indoor suspended particulate matter and their effects on children respiratory allergy in industrial areas have not been studied in India. Present work is aimed at the mineralogical study and distribution of indoor suspended particulate matter (SPM) and its effects on the respiratory illness in children (7-15 years age), living in close vicinity of the industrial areas of Delhi. Based on industrial units and high pollution level, Central Pollution Control Board identified Shahdara and Shahzada Bagh as industrial areas of Delhi (http://www.cpcb.nic.in/). Children are most vulnerable as they spend more time indoors and are exposed to indoor air pollutants. Appraisal of indoor air quality is one of the most important aspects in understanding the impact of particulate pollution on children health (Lee et al., 1997).

Study area

National Capital Territory of Delhi (latitude 28^0 24' 17" - 28^0 53' 00", longitude 76^0 45' 30" - 77^0 21' 30" at about 160 km south of Himalayas at an elevation of 216m a. s. l.) is spread over an area of 1,483 km^2, of which 783 km^2 is designated as rural and 700 km^2 urban (Fig. 1). It is situated on the bank of river Yamuna, almost in the centre of Northern India - located near Thar Desert in the west and south-west, central hot plains to the south, Gangetic plains to the east and cooler hilly regions to the north. The Delhi has a maximum length of 51.9 km and the maximum width of 48.48 km. The Yamuna flood plain and the Delhi ridge are the two prominent features in Delhi. The World Health Organization (1992) listed Delhi as world's fourth most polluted city with respect to particulate matter. It has been well recognized that in Delhi atmosphere the SPM is a major air pollutant originating mainly from vehicular and industrial emission besides numerous domestic coal burning units and three thermal power plants. Vehicular pollution contributes 67% of the total air pollution load (approximately 3,000 mt per day) in Delhi. Another 25% of air pollution is generated by industries and coal based thermal power plants. These power plants in Delhi generate 6000 Metric Tons of fly ash per day (http://en.wikipedia.org/wiki/Delhi). Increasing industries have been the major cause of concern for increasing airborne particulate pollution problem in the city. Generally low wind speed and temperature inversion conditions lead to an accumulation of airborne pollutants over the city, particularly in winter months, while in summer the situation is exacerbated by dust storm. The climate and weather play a crucial role in the distribution and dispersion of air pollutants (Balachandran et al., 2000). The atmosphere rarely exhibits a steady state even during short intervals of time; therefore frequent climatic changes have a bearing on ambient air quality in the region. The climate of Delhi is hot and humid in the summer and cold in the winters. The cold wave from the Himalayan region makes winters very chilly. The region has a tropical steppe climate (Srivastava and Jain, 2007). The Delhi region experiences a total annual rainfall between 700 mm to 800 mm with maximum rainfall occurring during the monsoon period (July-September). A regular pre-monsoon feature is the dust storm where westerly winds from the Thar Desert region of Rajasthan deposit large concentrations of suspended particulate matter into Delhi's atmosphere (Khillare et al., 2004). Extreme temperature ranges from -0.6 °C to 47 °C. The annual mean temperature is 25 °C; monthly mean temperatures range from 14 °C to 33 °C (Srivastava and Jain, 2007).

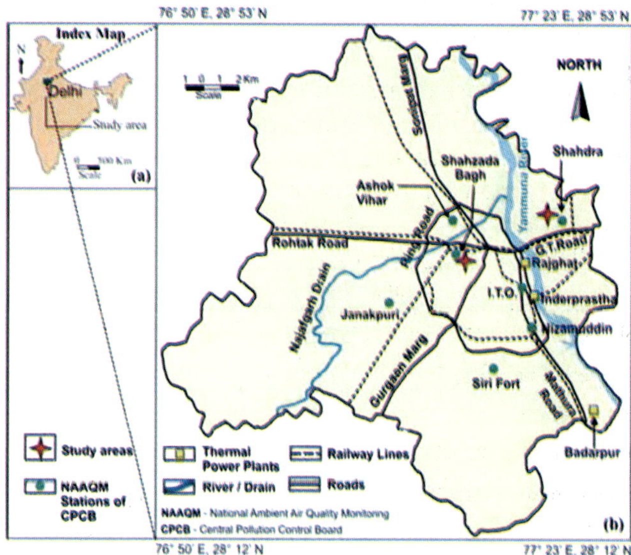

Fig. 1 Shows (a) Index map and (b) map of Delhi, where Shahzada Bagh and Shahdara industrial areas are located

Central Pollution Control Board (CPCB) has established the National Ambient Air Quality Monitoring (NAAQM) Network during 1984-85 at the national level. Within its purview, seven monitoring stations were established in Delhi (Fig. 1). The parameters monitored in these areas are suspended particulate matter (SPM) or total suspended particulates (TSP), PM_{10}, oxides of nitrogen (as NOx) and sulfur dioxide (SO_2) (http://www.cpcb.nic.in/). The annual average concentration of ambient suspended particulate matter (SPM) and PM_{10} remains well above the national standard (Table 2) set by CPCB for residential locations throughout the years (2004-2006). Based on classification of Central Pollution Control Board (http://www.cpcb.nic.in/), the main sites for the present investigations are Shahdara and Shahzada Bagh, recognized as industrial areas of Delhi (Fig. 1). Hundred houses with children aged 7-15 years from each socio-economic class were selected for the present investigations including their health check-up, where 25% houses of these houses selected for monitoring of indoor suspended particulate matter. Shahdara industrial area is located in the North - Eastern part of Delhi (Fig. 1). The surrounding area is open and flat with dense residential and industrial units. There are number of industrial units of steel industries, chemical units, pickling units, copper wire drawing units, plastic units, engineering units and other small scale industries located in Shahdara industrial area. Inderprastha and Rajghat thermal powers plant situated approximately 15 km distance away from the Shahdara area.

Table 1: Average annual concentrations of SPM and PM_{10} in ambient air as recorded in the national ambient air quality monitoring stations located in Delhi

Locations	SPM				PM_{10}			
	2004	**2005**	**2006**	**Std***	**2004**	**2005**	**2006**	**Std***
Shahdara	342.25	300.41	387.10	140	130.33	132.17	145.20	60
Shahzada Bagh	335.50	301.00	417.11	140	138.50	129.92	160.00	60
Ashok Vihar	311.25	313.33	478.50	140	128.25	108.00	196.00	60
Janakpuri	319.64	316.50	349.90	140	132.72	138.91	143.90	60
Nizamuddin	342.75	272.41	328.70	140	133.25	104.33	113.50	60
Delhi	357.36	335.12	387.62	140	145.72	140.29	152.33	60

Data source and Standard* = (CPCB, 1999)
(http://www.cpcb.nic.in/Air_quality_data.php)

Shahzada Bagh industrial area is located in North -West part of Delhi (Fig. 1). The surrounding area is open, with flat plateau consisting of a large number of industries in west and south west, including Lawrence road, Moti Nagar, DLF and Kirti Nagar industrial areas comprising large scale industrial units of steel industries, metal based industries and chemical plants. There are about 600-700 small iron-based industries using liquid fuel in Wazirpur industrial area. In extreme south a large number (about 2,000) of pickling and electroplating units are also located in the Anand Parbhat industrial area. A few stone crushers had also been situated, and these were in operation till June, 1992 at Anand Parbat area contributing atmospheric particulate load, but presently these are not in operation. Inderprastha and Rajghat power plant is situated approximately 20 km distance away from the Shahzada Bagh area. (Electronic version of the list of industries is available with JPS).

Fig. 2. Children participated in the medical camp at Shahdara

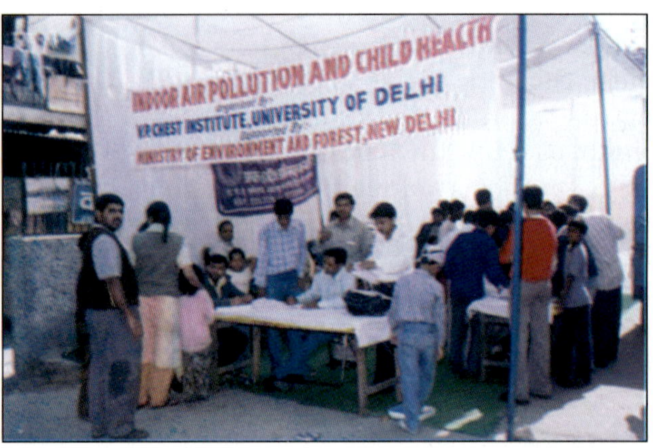

Fig. 3 Children participated in another medical camp at Shahzada Bagh

Fig. 4. PFT and medical examination of children carried out in medical camp organized in Shahdara

Materials and methods

Based on works of the American Thoracic Society (Ferris, 1978), British Medical Research Council (BMRC, 1960), a questionnaire was developed. Data regarding symptoms such as cough, phlegm production, shortness of breath, wheezing were recorded. For this, JKN and RK visited houses from first corner of the residential community to study all the children between 7-15 years. The questionnaire was administered at the house itself. The detailed questionnaire contained questions about a history of smoking in the family, type of cooking fuel used, ventilation at the cooking place, occupancy per room, and history of the respiratory symptoms (cough, phlegm production, shortness of breath, wheezing, common cold, throat congestion) of children. Examination of the child including pulmonary function test, was done at home. If they were not present in the house, a special medical camp was also organized in nearby area where examination and other tests were done under supervision of medical doctors (Fig. 2, 3 and 4).

Indoor suspended particulate matter level measured by using the Handy Air Sampler (Fig. 2) or Low Volume Sampler (APM 810, Envirotech Instrument Pvt. Ltd., New Delhi) with a flow rate of 1 liter/minute. Indoor SPM was collected on 25 mm Whatman glass microfiber filters. The filter papers were desiccated for 24 hours and weighted before sampling. Handy Air Sampler for indoor samples was placed in the center or corner of the room, with the inlet roughly 1 meter above the ground level, corresponding with the breathing height of the children. It was run continuously 6-8 hours of sampling period. At the end of the sampling period, the filter paper was removed. The filter papers were again desiccated for 24 hours then re-weighted to determine the mass of the particles collected. The particle concentration was measured by the weight gain of the filter divided by the volume of air samples using following equation.

$$C\,(\mu g/m^3) = \frac{(W2 - W1) \times 10^6}{T \times [(R1 + R2)/2]}$$

Where, C = Concentration of indoor SPM ($\mu g/m^3$), W1 = Initial weight of the filter paper in mg, W2 = Final weight of the filter paper in mg, R1 & R2 = Initial and final flow rate in Liter Per in Minute and T = Sampling time in minutes.

Medical and environmental data of Shahdara and Shahzada Bagh industrial areas were collected simultaneously. Details pertaining to present data is summarized and electronic version is obtainable from JPS. Total 2122 houses were surveyed in Shahdara and Shahzada Bagh industrial areas of Delhi. Children between 7 to 15 years were present in 1103 houses, but family members allowed for study in only 647 houses (Table 3). Thus, total 831children (59.7% male and 40.3% female) were studied in Shahdara and Shahzada Bagh industrial areas.

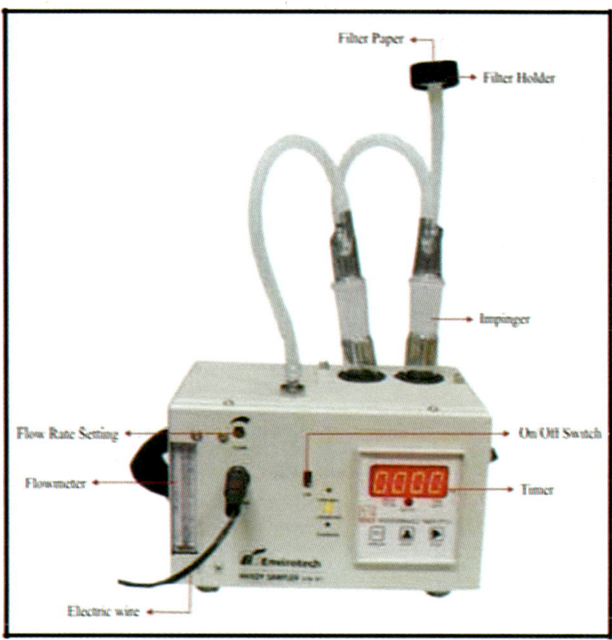

Fig. 5 Handy air sampler

Table 2: Details of house visit for the present study

Study area	House surveyed	Children found in houses	Houses allowed for study	Children examined
Shahdara	1467	640 (43.63%)	319 (21.8%)	394 (11.4%)
Shahzada Bagh	655	463 (70.69%)	328 (50.1%)	437 (12.6%)
Total	2122	1103 (51.97%)	647 (30.49%)	N=831

Distribution of indoor suspended particulate matter (SPM) was measured in Shahdara and Shahzada Bagh areas. Mean concentration level of indoor SPM was 1049.7±452.4 $\mu g/m^3$ (260 to 2380 $\mu g/m^3$) in Shahdara and 1108.7±509.8 $\mu g/m^3$ (310 to 2420 $\mu g/m^3$) in Shahzada Bagh (Table 2). The indoor SPM of Shahdara and shahzada Bagh mainly distributed between 500 to 1500 $\mu g/m^3$ (Fig. 2.6). The SPM level was very high in Shahdara and Shahzada Bagh areas and it was approximately 3 times higher than annual standard for industrial areas that was set up by Central Pollution Control Board (CPCB, 1999) and 7 times higher than the WHO standards. High concentration of SPM was mainly due to the presence of a large amount of industrial establishment and high density of road traffic in these areas. Large scale construction activities such as metro rail, roads, flyovers, buildings and special construction for 'Common Wealth Games 2010' may be also responsible for the high concentration level of particulate matter in Delhi. Cooking and smoking also play an important role to increase indoor SPM concentration in the houses. Most of the children show respiratory symptoms and suffering from respiratory diseases. It may be possible that children affected by high concentration of SPM, therefore it is necessary to understand geochemistry of SPM and find out relationship between respiratory diseases and SPM containing mineral phases and their major and trace elemental concentration.

Table 3. Concentration of the indoor suspended particulate matter

Study areas	Houses (N)	Concentration of indoor SPM (μg/m³)		
		Minimum	Maximum	Mean± SD
Shahdara	103	260	2380	1049.7±452.4
Shahzada Bagh	109	310	2420	1108.7±509.8
Total	212	260	2420	1080.0±482.4

X-Ray Diffraction studies

Mineralogical analysis of indoor SPM was carried out using X-Ray Diffraction (XRD). To obtain a uniform diffraction mount and to avoid the differential settling of particles, oriented mounts were prepared by filtering the SPM on a membrane filter and transferring it on a glass slide. Similar procedure was adopted for each sample. The XRD spectra were obtained by scanning 2θ between 5 to 90° with a scanning speed of 1 degree/minute using, PHILIPS X-ray diffractometer (X' Pert Model, PW/830/00) with Cu-Kά radiations. The mineral phases were identified by comparing the spectra with standard X-ray diffraction data of the minerals available in the JCPDA / IDD Files, Brown (1961), Tucker (1988) and Tan (2005). Quantitative estimation has been done by direct comparison of the intensity or peak area of identified minerals. Semi quantitative evaluations of the mineral assemblages were made on the integrated peak area basis. The reflections of an individual mineral on diffractogram was identified by the nature of the peak, d-spacing values and comparative study of the intensity rates from basal reflection. Indexing of the reflections was carried out by comparing the analytical value with the known calculated values available (Brown, 1961; Grim, 1968; Olphen and Fripiate, 1979; Tucker, 1988 and Tan, 2005). The intensity of diffraction pattern estimated either as peak height or preferable peak area of a mineral in the mixture is proportional to its concentration for quantitative analysis of minerals. The principal alternative to the addition of an internal standard is to use a mineral already present in the sample (e.g. quartz) as the internal standard (Hooton and Giorgetta, 1977 and Tucker, 1988).

$$W_{Min} / W_{Qt} = H_{Min} / H_{Qt} * I_{Min} / I_{Qt}$$

Where, W_{Min} = relative W % of mineral, W_{Qt}= reference Wt % of quartz, H_{Min} = relative value of peak height of mineral, H_{Qt} = reference value of peak height of quartz, I_{Min} = intensity of mineral and I_{Qt} = reference intensity of mineral.

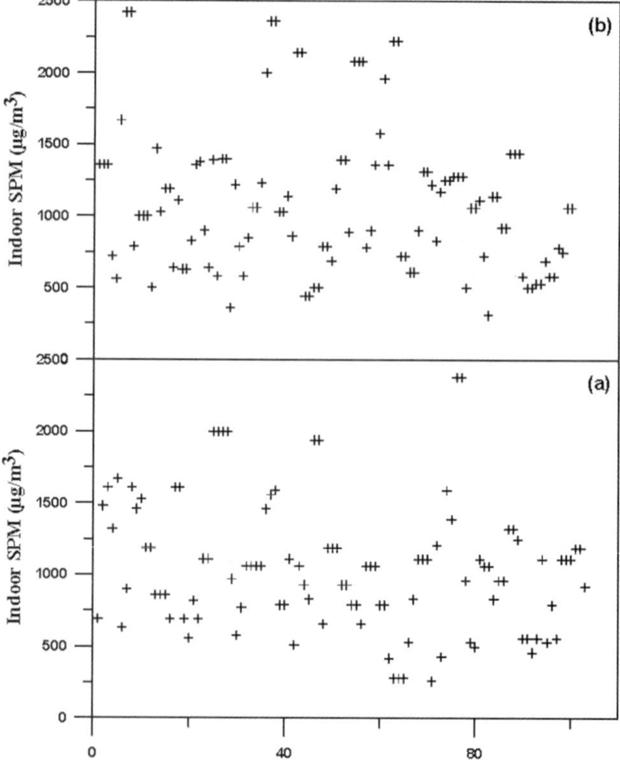

Fig. 6 Distribution of indoor SPM at (a) Shahdara and (b) Shahzada Bagh Industrial areas. Abbreviations: + = suspended particulate matter.

The identified minerals in the indoor suspended particulate matter of the Shahdara and Shahzada Bagh industrial areas contain quartz, calcite, dolomite, illite, muscovite, feldspar, talc and chlorite minerals (Fig. 7). Quartz was identified by series of basal (100 & 101) reflections in all the 40 samples. The intense basal (100) reflection at 20.67-20.92 Å, (101) reflection at 26.47-26.74 Å, (102) reflection at 39.27-39.57 Å, (112) reflection at 49.97-50.25 Å and (211) reflection at 59.75-60.12 Å regions were observed in case of quartz (Fig. 3.1). Calcite was identified by characteristics basal reflection viz: 104 reflections in the region 29.25-29.55 Å in the all 40 samples (Fig. 3.1). Dolomite was identified in all 40 samples except J_2 and J_{39} samples, by their (104) basal reflection at 30.76-31.13 Å and (030) reflection at 67.83-68.31 Å (Fig. 7). Illite was identified by the presence of basal reflections (002) at 8.70-8.83 Å (Fig.7). Muscovite was identified by their (003) basal reflections at 8.72-8.92 Å and (015) reflection at 45.32-45.58 Å (Fig. 7). Feldspar was identified by its characteristic basal reflection (002) in the region of 27.81-28.76 Å in all the samples collected from Shahdara and Shahzada Bagh industrial areas (Fig. 3.1). Chlorite was identified by their (002) basal reflection at 12.02-12.68 Å (Fig. 7). Talc identified by the presence of basal reflections (002) at 9.31-9.54 Å in indoor suspended particulate matter collected from the houses of Shahdara and Shahzada Bagh industrial areas (Fig. 7). Semi - quantitative estimation of these indentified minerals in SPM samples is given in Table 4.

Group A = Quartz Carbonate Dominant Group

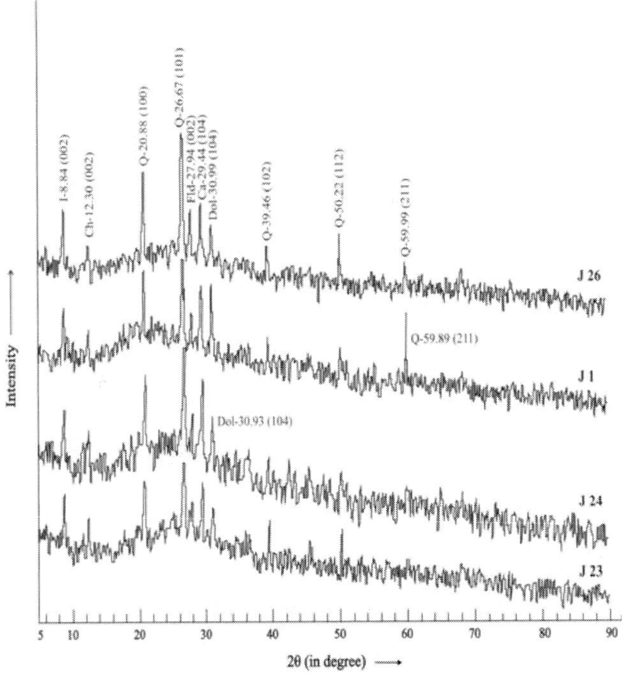

Fig. 7 (*continued*)

Group A = Quartz Carbonate Dominant Group

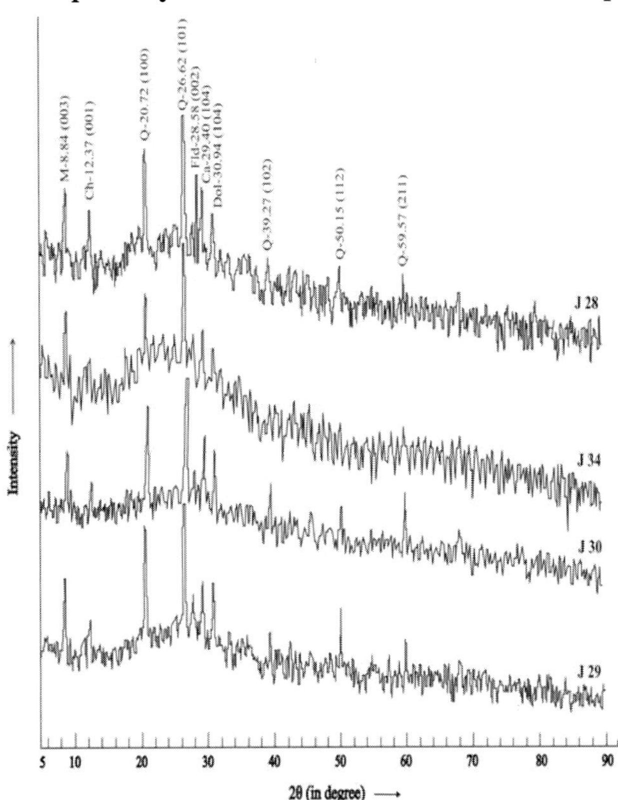

Fig. 7 (*continued*)

Group A = Quartz Carbonate Dominant Group

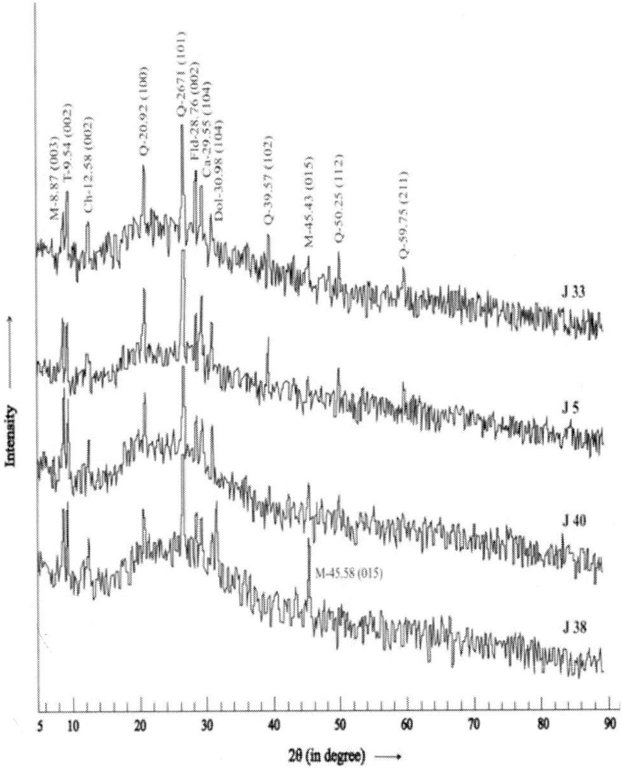

Fig. 7 (*continued*).

Group A = Quartz Carbonate Dominant Group

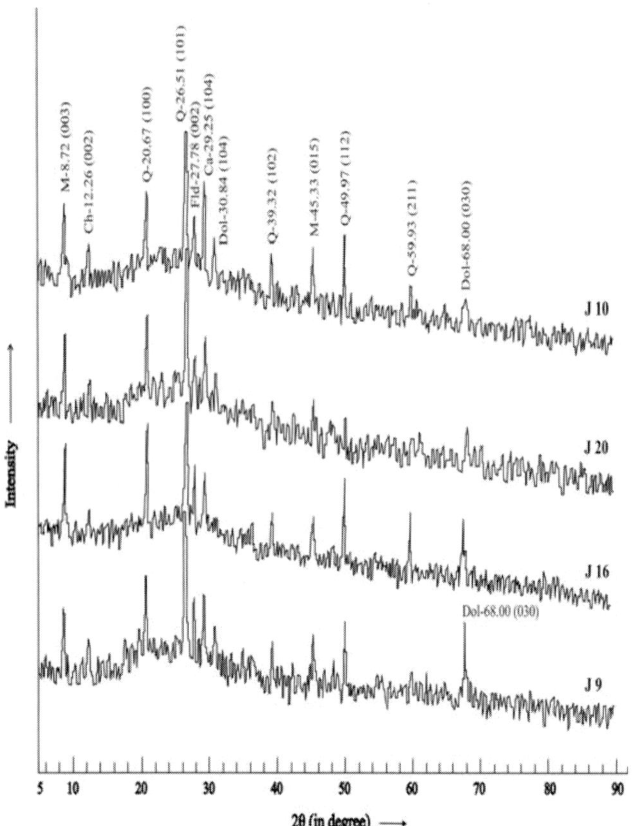

Fig. 7 (*continued*).

Group B = Quartz Mica Dominant Group

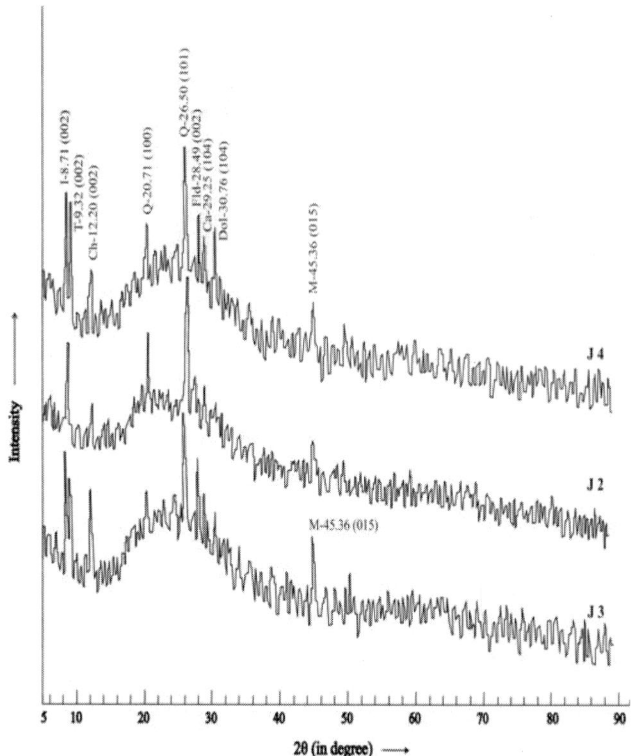

Fig. 7 (*continued*).

Group A = Quartz Carbonate Dominant Group

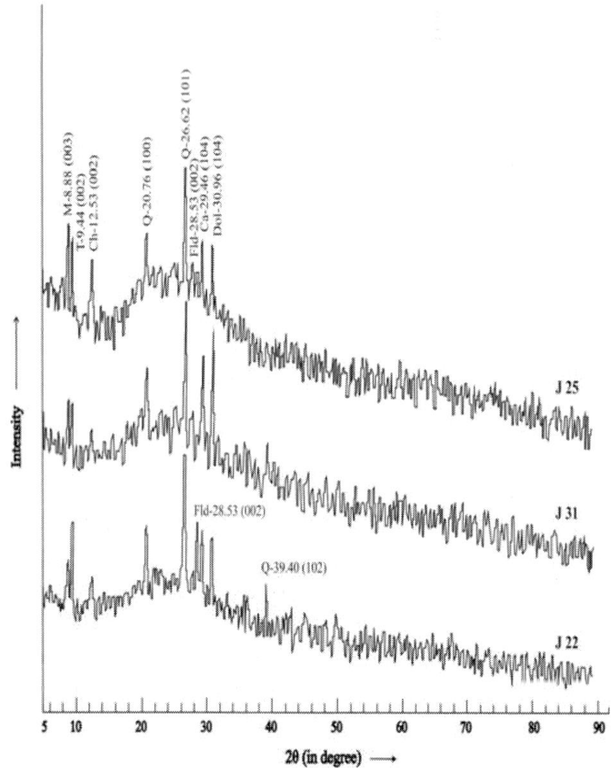

Fig. 7 (*continued*).

Group C = Quartz Feldspar Dominant Group

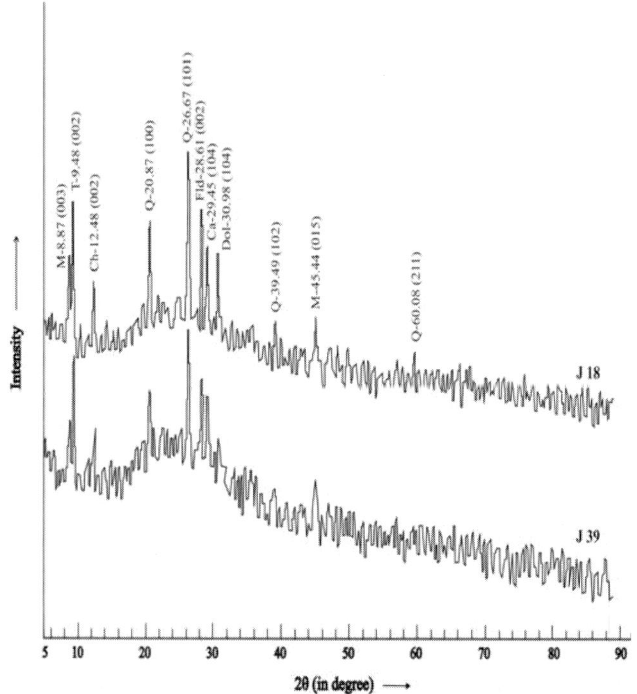

Fig. 7 (A-C): X-ray diffraction patterns of indoor suspended particulate matter from Shahdara and Shahzada Bagh industrial areas of Delhi. Abbreviations: Q = Quartz, I = Illite, M = Muscovite, Ca = Calcite, D = Dolomite, Fld = Feldspar, T = Talc, Ch = Chlorite.

Table 4 Semi-quantitative estimation of mineral phases present in indoor SPM

Samples	Mineral Phases (%)					Groups
	Quartz	Carbonates	Mica	Feldspar	Talc-Chlorite	
J1	58.58	31.40	3.10	4.69	2.23	
J5	59.45	23.48	3.69	9.69	3.69	
J9	66.80	14.03	9.75	7.62	1.80	
J10	60.79	19.82	11.00	6.20	2.19	
J16	75.02	11.40	7.35	6.23	-	
J20	69.78	14.79	6.63	6.98	1.81	
J22	52.55	24.54	-	15.92	6.99	
J23	73.02	17.28	3.23	4.77	2.70	
J24	67.29	23.49	3.16	4.78	1.28	
J25	45.44	29.16	8.20	5.35	11.85	A
J26	67.16	17.43	4.08	9.30	2.33	
J28	50.05	20.12	4.95	14.56	10.31	
J29	64.72	23.11	4.34	5.89	1.94	
J30	67.39	23.05	3.44	4.18	1.95	
J31	45.25	43.94	2.08	6.15	2.58	
J33	48.69	21.28	1.51	19.18	9.35	
J34	50.76	22.13	5.99	16.60	4.52	
J38	42.72	22.85	19.22	7.56	7.65	
J40	59.00	16.87	6.67	9.32	8.14	
J2	74.07	4.44	12.74	5.56	3.19	
J3	36.90	10.77	22.95	14.83	14.76	B
J4	44.62	14.41	19.10	12.14	9.73	
J18	40.36	19.81	5.45	22.40	11.99	C
J39	43.37	12.27	5.46	23.85	15.05	

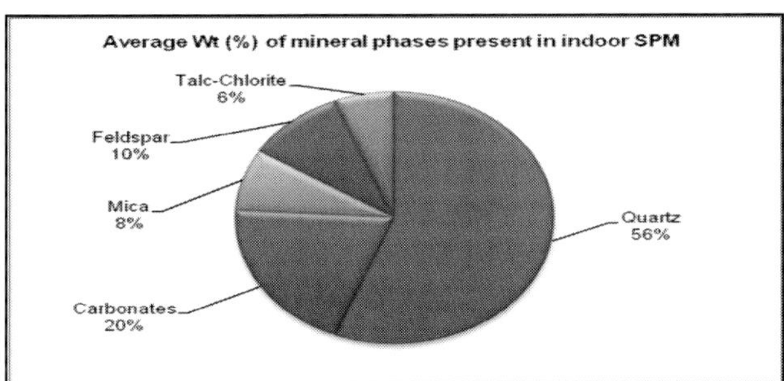

Fig. 8. Average mineral phases (wt. %) of indoor SPM of Shahdara and Shahzada Bagh areas

Discussion

The relatively high proportion of five major mineral groups such as quartz, carbonates (calcite and dolomite), mica (illite and muscovite), feldspar, talc-chlorite (talc and chlorite) were found in the indoor suspended particulate matter. On the basis of relative abundances of mineral phases present in the indoor suspended particulate matter of Shahdara and Shahzada Bagh areas, suspended particulate matter have been subdivided into three distinct groups, such as: Group - A: Quartz Carbonates predominant Group, Group - B: Quartz Mica predominant Group and Group - C: Quartz Feldspar predominant Group (Table 4). Illite, chlorite and talc are the major clay minerals which are present in the SPM whereas, other non - clay minerals such as quartz, calcite and feldspar are also present. Quartz is present in all the SPM samples of Shahdara and Shahzada Bagh areas. Its relative abundances is more than 50% in all the samples (Table 4) except J_{25}, J_{31}, J_{33} and J_{38} (Group - A), J_3 and J_4 (Group - B), J_{18} and J_{39} (Group - C). Carbonates were present in all the groups of SPM in Shahdara and Shahzada Bagh areas with relative abundance less than 25% except J_1, J_{25} and J_{31} samples. Mica was present in all the samples of SPM, except sample J_{22} with relative abundances less than 10% (Table 4). Feldspar was present in all the samples of indoor SPM in Shahdara and Shahzada Bagh industrial areas of Delhi. Its relative abundances were less than 10% in all samples except J_3, J_4, J_{18}, J_{22}, J_{28}, J_{32}, J_{33} and J_{34}. Talc - chlorite was present in all the samples of SPM except J16 with relative abundances less than 10% except J_3, J_{18}, J_{25}, J_{28} and J_{39} samples collected from the houses of Shahdara and Shahzada Bagh industrial areas of Delhi (Table 4).

The high amount of mineral phases present in the indoor SPM indicated that the identified minerals at both Shahdara and Shahzada Bagh sites were injected into the air possibly through wind erosion and anthropogenic sources such as industrial emission (steel industries, metals based industries, battery manufacturing, etc.), power plants (Inderprastha, Rajghat and Badarpur), vehicular traffic, construction activities (construction of metro rail, buildings, roads, flyovers, and special construction for Common Wealth Games 2010), burning of fossils fuels and others human related activities. The homogeneity in the identified mineral phases at both the study areas leading to the indoor environment carried particulate matter from cooking, smoking, dusting, outdoor sources and well mixed in air as a result of continuous deposition and uptake from the grounds. Air particulate matter with mineral phases existing in the air other than its normal constituents is hazardous to the health of children and adults, particularly those suffering from asthma (US EPA Report, 2003). It may also causes severeness in symptoms like coughing, wheezing, shortness of breath, chest discomfort and burning feeling in the lungs (US EPA Report, 2003). Relative abundances of quartz was found to be maximum than other minerals such as calcite, dolomite, illite, muscovite, talc and chlorite. Quantitative analysis (Table 5) revealed that quartz was the dominant constituent found in all the groups with its average weight percentage varied from 59.22%, 51.86% and 41.87% in group A, B and C respectively (Fig 7). In group A the second largest phase is of carbonates, having average weight percentage of 21.60% followed by feldspar (8.90%), mica (6.19%) and talc-chlorite (4.77%). In group B mica is the second largest phase, having weight percentage 18.26%, followed by feldspar (10.84%), carbonates (9.87%) and talc-chlorite (9.23%). In group C the second largest phase is the feldspar, having average weight percentage of 23.13%, followed by carbonates (16.04%), talc-chlorite (13.52%) and mica (5.46%) (Table 5).

Table 5: Average weight (%) of mineral phases present in SPM samples of all groups

Minerals	Group A	Group B	Group C
Quartz	59.22	51.86	41.87
Carbonates	21.60	9.87	16.04
Mica	6.19	18.26	5.46
Feldspar	8.90	10.84	23.13
Talc-Chlorite	4.77	9.23	13.52

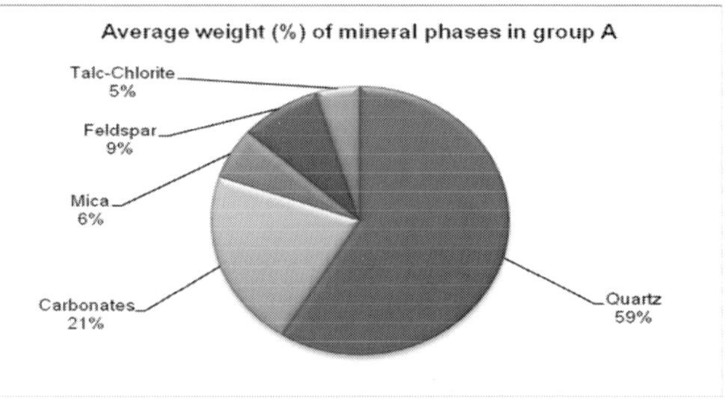

Fig. 9 Average weight percentage of mineral phases present in SPM samples of group A

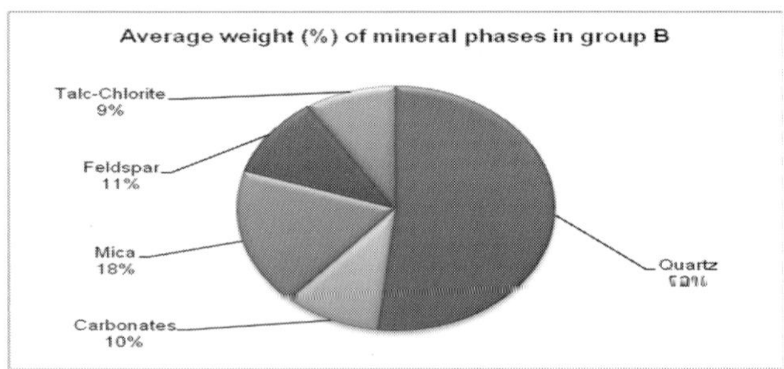

Fig. 10 Average weight percentage of mineral phases present in SPM samples of group B

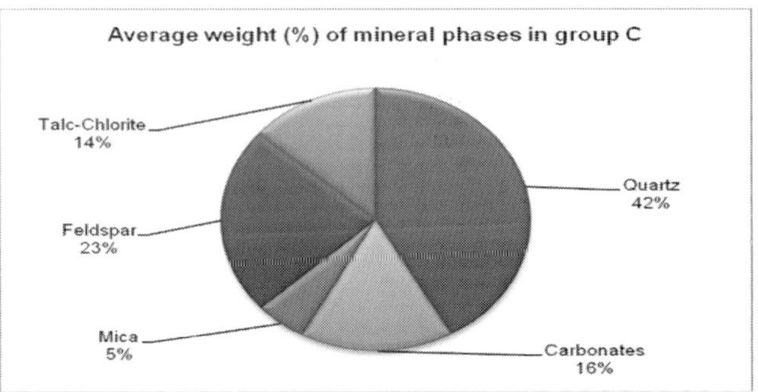

Fig. 11 Average weight percentage of mineral phases present in SPM samples of group C

Several studies (Crocker et al., 2002; Yadav and Rajamani, 2004; Wahab et al., 2005; Hussain et al., 2008) show occurrence of clay and non clay minerals in indoor and outdoor suspended particulate matter and supporting the present observation. Merefield et al. (1998) had undertaken analysis of an environmental dust collected from the south Wales opencast mine area, using a combination of X-ray diffraction and scanning electronic microscope (SEM), recognized 10 regularly occurring minerals including quartz, bassanite, halite, calcite, illite, chlorite, kaolinite, montmorillonite (smectite), dolomite and feldspar. Jones et al. (2002) characterized airborne particles collected within and proximal to an opencast coalmine, South Wales, U. K. and found minerals with relative percentage such as carbonates (12.9%), mica (7.3%), quartz (10.4%), clays (26.2%), siderite (8.6%), gypsum (10.0%), halite (0.8%), barytes (0.6%), pyrite (3.7%) and organics (13.9%) in PM_{10}. Reynolds et al. (2003) also investigated the toxicity of airborne dust generated by opencast coal mine, U. K. and found mean percentages of major minerals including carbonates (10.3%), mica (7.3%), quartz (10.6%), clays (27.3%), gypsum (10.0%), pyrite (3.6%) and siderite (9.0%) in airborne the particulates. Hussain et al. (2008) have investigated indoor suspended particulate matter as a source of air pollution in residential, commercial and industrial areas of Faisalabad city of Pakistan and showed that illite (73.8%) was the dominent constituent where the other minerals phases found as quartz (1.8%), albite (8.77%), calcite (7.08%), gypsum (0.5%) and talc (1.83%) in the indoor SPM collected from industrial areas. Mineralogy of atmoshpheric suspended particles in three indoor and one outdoor locations in the Oman have been discussed by Wahab et al. (2005). Sampling locations included a residential house located nearby the cement plant, refinery plant and Al-suwayq residential area. The minerogical analysis of these samples showed presence of quartz as the principal phase inside the house of Al-Suwayq residential area, whereas quartz, dolomite and gypsum were common phases outside the house. In the residential house nearby the cement plant, it was found that calcite, quartz, dolomite and goethite were the pricipal mineral phases, whereas the particles collected from the house nearby the refinery composed primarily of dolomite and calcite. Crocker et al. (2002) characterized $PM_{2.5}$ from rural midwestern United States sites and identified the mineral phases such as quartz, feldspar, clays (kaolinite, chlorite, biotite), mica (muscovite and illite), carbonates (calcite, dolomite, ankerite, siderite), iron-rich (pyrite, magnetite, ilmenite and hematite) and sulphur-rich anhydrite and gypsum in the $PM_{2.5}$. Ozaki et al. (2005) analysed of crystalline phases in airborne particulates by X-ray diffractometry and found that the quartz is present as the most predominent mineral folowed by gypsum, calcite and halite in airborne particulates. In India, Yadav and Rajamani, (2004) studied the aerosols (free fall, SPM and PM_{10}) of northwestern part of India adjoining the Thar Desert and investigated mineralogy and geochemistry of particles. Quartz is the dominant mineral followed by feldspar, mica, calcite, chlorite and plagioclase. Garnet, amphibole, titanite and zircon were some of the heavy minerals present in the suspended particulate matters and PM_{10}. These all studies have established that quartz, carbonates, mica, clays, feldspar and talc-chlorite are the most common mineral phases of suspended particulate matters of indoor and outdoor environments. These studies also support the present study and proved the presence of quartz, carbonates, mica, feldspar and talc-chlorite in indoor SPM of Shahdara and Shahzada Bagh areas of Delhi.

A limited work was carried out recently on indoor air quality (IAQ) in non-occupational settings. However, it is suggested that elevated concentration of some airborne contaminants have been encountered in the indoor environments and considered as serious health hazards, need appropriate steps to protect human beings. Indoor air quality has more affects on human health, chiefly due to large amount of time we spend indoors in modern times. Jenkins et al. (1992) have observed that urban people spend on an average 87% time in indoors and remaining 6% in outdoors. Persons (old, the very young and the children) spend their time indoors and are susceptible to the affects of pollutants. Women and children are the most vulnerable as they spend more time indoors and are exposed to smoke. Appraisal of indoor air quality is one of the most important aspects in understanding the impact of particulate pollution on human health (Lee et al., 1997). Indoor SPM concentrations depend on both indoor and outdoor sources and removal processes, such as air exchange or chemical reactions (Wadden and

Scheff, 1983; Phillip et al., 1993). The main sources of indoor particulate matter are cooking, tobacco smoking, kerosene heating, wood burning, re-suspension of particles by people and pets, dusting & vacuuming, showering, operation of humidifiers, electric motors and by transport from outdoors via leakage through the walls, windows, or the ventilation system (Kamens et al., 1991; Gulliver and Briggs, 2004; He et al., 2004). Re-suspension of particles by human movement contributed to the coarse mode of indoor particles (< 1 μm size). Secondary particles in the indoor environment are formed by gaseous pollutants through the process of gas-to-particle conversion.

It is well established that indoor air is simply an extension of ambient air (Jones et al., 2000). Air penetrates indoors through doors, windows, air conditioners. Assuming that there is no source of pollutant in the indoor, the entering air would be filtrated by whatever it passes through. In a humid or rainy day, outdoor pollutants tend to be washed out while, due to the relative constant humidity indoor, the air quality is much worse even without any indoor activity (Jones et al., 2000). Yocom (1982) pointed out that the indoor and outdoor relationships are complex due to interactions of various factors such as meteorological, indoor sources and sinks, pollutants depletions, filtrations and ventilations. Lee et al. (1997) studied air quality during the summer season in Korea and their field measurements showed that outdoor air contributed 50-100% of indoor pollution. Koponen et al. (2001) studied effects of outdoor air pollution on indoor air with emphasis on the ventilation system and found that indoor particles were mainly of outdoor origin and that the ventilation had a strong influence on indoor particle characteristics and concentrations. Moreover, indoors and outdoors airborne particulates differ in their types, characteristics, concentrations and sources (Repace and Lowrey, 1980; Allen and Miguel, 1995 and Jones et al., 2000).

Environmental tobacco smoke (ETS) is a major source of indoor air contaminants and produce a huge amount of particulate matter in the air. The U.S. Surgeon General asserted in his report on *Smoking and Health* that tobacco smoke can be a significant source of atmospheric pollution in enclosed areas (Department of Health, Education, and Welfare, 1979). Tobacco use including both the smoking and the nonsmoking forms of tobacco is common in India. The few studies (Behera and Malik, 1987; Kumar et al., 1997; Reddy and Gupta, 2004) on tobacco use in different population groups reported its prevalence from about 15% - 50% among men. Tobacco combustion in indoors contributes to concentrations of repairable particles, nicotine, polycyclic aromatic hydrocarbons, CO, acrolein, NO_2, and many other toxic substances (Reddy and Gupta, 2004). Measurements in bars, restaurants, airplanes (smoking section), buses, conference rooms, homes and offices indicate CO and particulate matter concentrations of 2 to 35 ppm and 10 to 1000 $\mu g/m^3$, respectively (Spengler and Sexton, 1983). Analysis of repairable particle samples collected in 80 homes over several years indicate that a "pack-a-day" smoker will contribute approximately 20 $\mu g/m^3$ to 24-hour indoor particle concentrations (Spengler and Sexton, 1983). In the presence of cigarette smoke, many normal nonsmokers experience eye and throat irritation, headache rhinitis and coughing; allergic person report wheezing, sneezing and nausea as well. Particularly acute symptoms may be found in infants, children, persons with cardiovascular or respiratory diseases (Pope et al., 1995).

Biomass fuel used for cooking which is another important source of indoor particulate matter and the use of wood and other forms of biomass as a cooking fuel is common in developing countries like India. In rural areas of India, biomass is used as the primary cooking fuel of households. Wood smoke contains hundreds of chemical compounds (Smith and Liu, 1994). Some of the components present in wood smoke which are concerned with the health include particles, polycyclic aromatic hydrocarbons and carbon monoxide (Samet et al., 1987). Exposure of air pollutants levels are usually much higher among women who tend to do most of the cooking (Behera et al., 1988) and among young children who stay indoors and who are often carried on their mother's back or lap while cooking (Albalak, 1997). A baseline survey conducted in two district of Zimbabwe found that women and young children spend an average of 5 hours per day in the kitchen area, where air pollution levels from biomass fuel combustion for cooking tend to be very high (Mika, 2002).

Boleij et al. (1989) measured domestic air pollution from biomass burning in Kenya and found that the indoor respirable particles were 10 times higher (mean value 1400 $\mu g/m^3$) than the recommended air quality standard. In another study in Kenya that was based on biomass combustion for cooking, the mean level of indoor SPM was 4898 μg/m³ (Ezzati et al., 2000). A similar study undertaken by Ezzati and Kammen (2001) in rural Kenya indicated that the SPM from biomass combustion was approximately 1000-2000 μg/m³. Reid et al. (1986) have reported indoor SPM concentration to the level of 2000 μg/m³ in the Nepal. Singh et al. (1990) have measured the concentrations of suspended particulate matter over Delhi. The mean values of the concentration of SPM for three months (November, December and January 1987-1988) in the Ashok Vihar, Shahzada Bag, Shahdara, Siri Fort, Janakpuri and Nizamuddin areas was 782.0, 789.9, 546.5, 461.5, 462.7 and 460.7 μg/m³, respectively. Study on concentration of suspended particulate matter (SPM) has been undertaken in Delhi by Goyal and Sidhartha (2002) and found that the monthly mean SPM concentration reached to its highest value (465.68 μg/m³) in November and to lowest value (150.07 μg/m³) in August. Quiterio et al. (2004) have assessed concentrations of total suspended particles and airborne particulate trace metals in the industrial district of Santa Cruz, Rio de Janeiro, Brazil, where the main source of pollution was from the local industries. The geometrical mean of total suspended particles (TSP) was found to be 87±40 μg/m³ which were higher than the Brazilian primary and secondary standards (80 and 60 μg/m³) of air quality. The major elements including Ca (23770 ng/m³), Mg (6643 ng/m³), K (48 ng/m³), Na (175 ng/m³), Al (5774 ng/m³), Fe (38903 ng/m³), Mn (1216 ng/m³) and trace elements such as Zn (2120 ng/m³), Cu (335 ng/m³), Cr (421 ng/m³), Co (1.1 ng/m³), Pb (101 ng/m³), Cd (0.9 ng/m³) were found high in the airborne suspended particles. The Mn, Cr and Ni` concentration levels were 2.4, 4.2 and 2.1 times much higher than the standard values of the elements set by United State Environmental Protection Agency (Agency for Toxic Substances and Disease Registry, 2002) in the air. The mean concentration of Cd was 1.6 times higher as compared to the values recommended by the European Community (European Commission, 2000). The mean value of Fe concentration was found 33 times higher that of the reported value (1185 ng/m³) in the Chicago city (Sweet et al., 1993). Wahab and Yaghi (2004) studied the total suspended dust and heavy metals emitted from an industrial workplace and further compared these concentrations with the nearby residential houses in Sohar Industrial Estate (SIE) in Oman. The indoor TSP and heavy metal concentrations were assessed from inside of workplace and nearby houses and analyzed for heavy metals such as Cr, Cu, Mn, Ni, Pb, and Zn. The mean concentrations of Cr, Cu, Mn, Ni, Pb and Zn observed in the TSP were 4.7, 5.0, 3.9, 3.9, 30.2 and 2060 ng/m³ in the workplace of industrial site and 3.0, 0.3, 2.9, 4.9, 3.2 and 748 ng/m³ in the indoor houses respectively. The mean values for all the heavy metals except Ni were found higher in the workplace of industrial site as compared to the houses. Dams and Jonge (1976) determined chemical composition of atmospheric aerosols at the Jungfraujoch in Switzerland (3752 m above sea level) and found that the geometric mean concentration of total SPM was 3.1 μg/m³. Concentrations of 31 trace elements such as Na, Mg, Al, K, Sc, Cr, V, Ti, Fe, Mn, Co, Cu, Zn, As, Se, Mo, Ag, Cd, Ir, Sb, Ba, Cs, La, Sm, W, Au, Hg, Pb, Cl, Br and I were determined in the suspended particulate matter enriched aerosols. Richter et al. (2007) have determined the chemical fractionation in the airborne particulate matter (PM_{10}) from the Cerrillos monitoring station in Santiago, Chile. Elemental concentrations determined in the PM_{10} on yearly basis (1997-2003) for Pb, As, Mo, Cd, Zn, Cr, V, Cu, Ba, Ca, Ti, Mn, Mg, Ni and Al. These data-sets were statistically analysed, using cluster analysis to identify groups of elements having similar behaviuor along with the time. Elements such as Cd, Cu, Pb, Ni, As and Mg show a clear decrease in concentration with time. In this study a great concern was that elements high toxicity such as Pb, Cd and As have a direct impact on the environment and health of the exposed population. Godoi et al. (2008) have assessed the chemical compositions and deposition in the human respiratory tract of inhalable particulate matter from lime industries in South Brazil. The elements, including Al, Si, S, Cl, K, Ca, Ti, Cr, Mn, Fe and Cu were identified in the respirable particulate matter. Five different classes of particulate matter have been detected based on the major element concentrations such, as aluminosilicate, Ca-Mg oxide, carbon-rich particles, mixtures of organic particles and Ca-Mg carbonate, soot and biogenic particles. The chemical composition and size distribution of the aerosol particles, deposition probabilities in the three main parts of the human respiratory system (tracheobronchial, pulmonary and lung) were estimated in this study. Winchester and Bi (1984) have analyzed fine and coarse aerosol compositions of Beijing, China. They studied interrelationships amongst the concentrations of 18 elements such as S, Cl, K, Ca, Mn, Ni, Cu, Zn, Pb, Al, Si, P, Ti, V, Cr, Fe, As and Br and found that the particle distribution of most of the elements was bimodal, with a gradual progression from mainly coarse mode of Ca, Ti, and Al to fine mode of S, As, and Br. The Pb, Zn and Cu occurred mainly in the fine particles, derived from the combustion of coal and other fuels, but not from gasoline. Fine particles of Cr, Mn and Fe concentrations were high and derived from the industrial air pollution. Coarse particles were apparently from local surface dust raised by vehicular traffic and were enriched in Pb, Zn and Cu. Querol, et al. (2007) studied source origin of trace elements in particulate matter from regional background, urban and industrial sites of the Spain. Industrial emissions from different forms of metallurgy (steel, stainless steel, copper, zinc), ceramic and petrochemical industries were evaluated. At industrial and traffic hotspots, average trace metal concentrations were high, exceeding rural background levels in cases of Cr, Mn, Cu, Zn, As, Sn, W, V, Ni, Cs and Pb. Steel production emissions were linked to high levels of Cr, Mn, Ni, Zn, Mo, Cd, Se and Sn (and probably Pb). Copper metallurgy areas showed high levels of As, Bi, Ga and Cu. Zinc metallurgy was characterized by high levels of Zn and Cd. Glazed ceramic production areas were linked to high levels of Zn, As, Se, Zr, Cs, Tl, Li, Co and Pb. High levels of Ni and V were tracers of petrochemical plants and fuel-oil combustion. Wahab et al. (2005) studied the mineralogy of atmospheric suspended dust in three indoor and one outdoor location in the Oman. Sampling locations included a residential house located nearby the cement plant, a residential house located nearby a refinery plant, and a residential house located at Al-Suwayq residential area. Indoor air was sampled from these three houses. Moreover, for the Al-Suwayq residential house, sampling was also taken outside the house for comparison. The dust samples were analyzed for their microanalysis characterization and their mineral contents as well. The microanalysis identified the elements (Na, Al, Si, K, Ca, Ba, Zn, Fe, Mg and S) present in the particles. Furthermore, the mineralogical analysis of the sample filters showed presence of quartz as the principal phase inside the house of Al-Suwayq, whereas quartz, dolomite, and gypsum were common mineral phases present outside of the house. In the residential house nearby the cement plant, it was found that calcite, quartz, dolomite and goethite were the principal phases, whereas the particles collected from the house nearby the refinery composed primarily of dolomite and calcite. However, at the cement factory the crushing and grinding of limestone have contributed significantly to the airborne dust load. Williamson (2001) have also studied the mineralogical composition of atmospheric airborne particulates in Russia, and identified minerals such as hematite, anglesite, zincite, pyrite, chalcopyrite, sphalerite and a range of other metal-bearing species in particulates, dominantly PM_{10} in close vicinity of the smelter and industries.

In Indian context, Srivastava and Jain (2007) investigated the suspended particulate matter in indoor environment at Jawaharlal Nehru University (JNU), New Delhi. The indoor SPM concentration values crossed the 500 μg/m^3 level and some times were greater than 1000 μg/m^3. SPM concentrations were much higher than the standards set up by the Central Pollution Control Board (CPCB), New Delhi. The AAS analysis carried out for the concentration of different metals such as Mg (2.07μg/m^3), Ca (2.65 μg/m^3), Cu (0.2 μg/m^3), Zn (2.22 μg/m^3), Cd (0.1 μg/m^3), Pb (1.5 μg/m^3), Cr (0.9 μg/m^3), Mn (0.3 μg/m^3), Fe (7.4 μg/m^3), Co (0.3 μg/m^3) and Ni (0.8 μg/m^3), showed significant amount in the indoor SPM. Furthermore, the chemical analysis revealed that the Ca, Mg and Fe metals have maximum concentration and Pb concentration was higher than the CPCB standard values for the sensitive area. Vijayanand et al. (2008) assessed the heavy metal contents in the ambient air of the Coimbatore city, Tamilnadu, India. Suspended particulate matter was collected at ten stations and analysed for the heavy metal contents. The concentrations of seven heavy metals (Zn, Fe, Cu, Pb, Ni, Cr and Cd) were estimated. The level of SPM was found to be either at permissible or non-permissible limit depending upon the category of the sampling station. At majority of sampling stations, concentrations of Zn were found to be maximum than other heavy metals. The order of average concentrations of heavy metals in Coimbatore atmospheric air was Zn > Fe > Cu > Pb > Cr >Ni

> Cd. The usage of Zn for protective coating on iron, steel etc. by the industries in Coimbatore city could be the major reason for the higher concentration of this heavy metal in this region. Thakur et al. (2004) investigated the distribution patterns and chemical composition of airborne dust particulates in different representative areas of Raipur in Chhatisgarh and found the high concentration of toxic metals such as Cr, Mn, Fe, Ni, Cu, Zn, Sb and Pb in the airborne dust particulates. Air quality and trace metal chemistry of different size fractions of aerosols in N-NW India were studied by Yadav and Rajamani (2006). Three different size fractions of aerosols - free fall (FF), SPM and PM_{10} as well as surface sediments were collected from four different locations situated along a dust transport path of nearly 600km in NW India starting from Bikaner in the Thar to Garhmuktesar through Jhunjhunu and Delhi and were studied for their heavy-metal chemistry. The aerosol load in the atmosphere increased to several orders of magnitude for all size ranges (FF = 21 g/m^2 /day; SPM = 10,702 $\mu g/m^3$ and PM_{10} = 2907 $\mu g/m^3$) during the summer and winter. Concentrations of trace elements (Ba, Sr, Ni, Cr, Cu, V, Zn and Pb) were analyzed in different size fractions of environmental aerosols. Multiplicity of sources for the heavy metals also studied. Ba and Pb are added largely by fossil fuel burning; Cu, Ni and Zn are contributed by various industries. Anthropogenically added heavy metals show maximum enrichment relative to the upper continental crustal component in finer fractions and especially in winter. This study suggested that the aerosol loading in the atmosphere and its heavy-metal chemistry is strongly influenced by the climatic regime and the presence of desert and industrialization in this area. Yadav and Rajamani (2004) also studied the mineralogy and geochemistry of three different size fractions of aerosols (Free Fall, SPM, PM_{10}) of northwestern part (Bikaner, Delhi and Jhunjhunu) of India. The mineral composition of all three types of aerosols studied by X-ray diffraction and electron microprobe methods showed presence of albite, calcite, dolomite, gamet, muscovite, plagioclase, titanite in the SPM samples collected from Bikaner. Albite, chlorite, muscovite, plagioclase, quartz, and titanite minerals identified in the free fall samples collected from Delhi region and amphibole, calcite, chlorite, feldspar, muscovite plagioclase, titanite was identified in the SPM samples collected from Jhunjhunu. Quartz is the dominant mineral followed by K-feldspar, mica, calcite, chlorite and plagioclase.

Human beings are exposed to heavy metals and other elements in airborne particulate matter, soils, water and food and in other materials which come in contact with the man. Certain heavy metals often occur in abundances, considerably above their background levels (US EPA, 1982). Anthropogenic processes in urban areas that increase rates of mobilization of these elements from their natural states (US EPA, 1982). These increased concentrations poses threat to the human health. Whether the immediate pathways to man are directly from inhalation of polluted air, from ingestion, or by other means; the overall sequence of transport links between the pollution sources of heavy metals and a human receptor generally include an atmospheric route as an important part. Consequently, the measurement of heavy metal concentrations in airborne suspended particulate matter is of global significance to human health (US EPA, 1982).

The biological effects of a particle is determined by physical and chemical nature of the particles itself, the physics of deposition, distribution in the respiratory tract (Fig 1), and the biologic events occurring in response to the particles. Toxicological approaches may focus the effects of a single agent or mixture of particles on the airways or alveoli of human beings may be acute and chronic in beginning (Bascom et al., 1996). The most important characteristics influencing the deposition of a particle in the respiratory system is based on the size. Models (ICRP, 1994; Berico et al., 1997) of inhaled particles deposition related to aerodynamic diameter size of particle to the site of deposition in the respiratory tract (Fig 1). Most inhaled particles which are > 5 μm in aerodynamic diameter in size (such as acid fog), deposit in the upper airways or larger lower airways of the respiratory system of human beings. Smaller particles are more likely to be deposited in the periphery of the lung. The sites of particles deposition is also influenced by hygroscopic growth of particles in the airways (Fig 1), the shape and dimensions of the airways and alveoli, ventilatory patterns (breathing rate and tidal volume), oral versus nasal breathing, and respiratory secretions (Brain and Valberg, 1979). Lung diseases can alter particle deposition by changing airway dimension or airflow patterns. Exercise increases oral breathing, bypassing the nasal passages, and increases total ventilation, increasing particle velocity and inertial impaction. Both these changes result in greater particle deposition in the lower airways (Brain and Valberg, 1979). Particles are cleared by several mechanisms. Those deposited in the anterior nasal cavity are expelled by sneezing or wiping, while the remainder of particles deposited in the nose is cleared posteriorly to the pharynx. Particles deposited in the trachea, bronchi, and bronchioles (Fig 1.1) rise on the mucociliary ladder to be expelled by coughing or swallowing (Harmsen et al., 1985). Particles deposited distal to the terminal bronchioles are largely cleared by lung macrophages and/or dissolution. The lung macrophages in turn will transport the ingested particles onto the mucociliary ladder, or into the lymphatic system (Harmsen et al., 1985). A small fraction of these deposited particles will migrate through alveolar tissue directly into the lymphatic circulation (Fig 1). When ultrafine particles are inhaled, i.e., those about 0.2 μm in diameter, highly increased interstitial access and acute inflammatory reaction have been demonstrated by lung parameters (Oberdorster et al., 1992). Epidemiological studies (Dockery et al., 1993; Bascom et al., 1996; Horak et al., 2002; McConnell et al., 2003) have established that the suspended particulate matter is associated with more frequent respiratory allergy because airborne particulates easily reach the deepest recesses of the lungs and damage. Particulates are especially harmful to people with lung diseases such as asthma, bronchitis, rhinitis and chronic obstructive pulmonary disease. Exposure to particulate matter can trigger asthma attacks and cause cough, wheezing, phlegm production, shortness of breath, common cold, throat congestion, and respiratory irritation in individuals with sensitive airways (US EPA, 1994; Bascom et al., 1996; Nagar et al., 2007). Zemp et al. (1999) in Switzerland found positive correlation between annual mean concentrations of total suspended particulates and PM_{10} and chronic phlegm production, chronic cough, breathlessness, wheezing, chest tightness and current asthma. Bronchitis was reported to be more prevalent among children with asthma living in communities with high content of ambient PM_{10} (particulate mass less than 10 μm in diameter) in combination with high sulfur dioxide and sulfate particulate matter (PM) in 24 cities of the eastern United States (Dockery et al., 1989; Dockery and Pope, 1994). Environmental health effect studies conducted between 1990-2000 has shown an association between exposure to particulate matter and increases in daily mortality and symptoms to certain illness in human beings (Dockery and Pope, 1994; Schwartz, 1994). The United States Environmental Protection Agency has reported that PM_{10} increased the level of respiratory illness in children exposed to levels of 40-58 $\mu g/m^3$ for extended periods (US EPA Report, 1993). Studies (US EPA Report, 1997) indicated that tens of thousands of elderly people die prematurely each year from exposure to fine parti-cles and showed that exposure to fine particles was associated with hospital admissions each year.

The effect of indoor and outdoor particulate level on respiratory health was examined in never smokers (40 to 69 years of age) residing in industrial, residential and suburban area of Beijing, China and diagnosed bronchitis, asthma, and respiratory symptoms including chronic cough, chronic phlegm, shortness of breathe (SOB), wheeze, and wheeze with SOB (Xu and Wang, 1993). Household with coal stoves had substantially higher indoor particulate levels than those with gas stoves. Chronic phlegm, wheeze, and wheeze with SOB were significantly higher in their indoor use of coal stoves for both cooking and heating (Xu and Wang, 1993). The particulate levels were highest in the industrial areas and lowest in the suburban areas. The total particulate exposure, represented by the sum of indoor and outdoor particulate exposure was significantly associated with increased prevalence of all the respiratory symptoms (Xu and Wang, 1993). This study also suggested that both indoor and outdoor particulate exposure may be important risk factors in the development of chronic obstructive pulmonary disease (COPD). Sivacoumar et al. (2001) have studied air pollution in the stone crushing industry and their association with human health in Chennai. A detailed air pollution survey was conducted at Pammal, 26 km to the southwest of Chennai. The particle size analysis indicates high percentage of finer particles with silica content, posing serious health problems to the people exposed for longer duration. The total silica dust and respirable particulate fraction in the work environment was significantly associated with respiratory problems and occupational health standards. Dusseldorp et al. (1995) have studied the association of

PM_{10} and airborne iron with respiratory health of adults living near an industrial area in the Wijkaan Zee, Netherland. During the study, the 24 hour average PM_{10} concentrations that ranged from 36 to 137 μg/m^3, while the concentration of iron, silicon, sodium and manganese ranging from zero to 6.95, 1.84, 12.02, and 0.37 μg/m^3 respectively. The steel industry was found contributing significantly to the PM_{10} concentrations, and especially to the iron and manganese concentrations in the air. A statistically significant decrease in peak expiratory flow (PEF) was found to be associated with increasing PM_{10} concentrations (Dusseldorp et al., 1995). It was concluded that PM_{10} was found to be associated with acute changes in the respiratory health and increases in respiratory symptoms. Airborne iron was also found to be associated with exacerbation of some respiratory symptoms. Dockery et al. (1993) have studied the association between air pollution and mortality in six cities of United States and observed statistically significant association between air pollution and mortality. Air pollution was positively associated with death from lung cancer and cardiopulmonary diseases but not death from other causes. Mortality was most strongly associated with air pollution with fine particulates, including sulfates. The result of this study suggested that fine-particulate air pollution, or a more complex pollution mixture associated with fine particulate matter, contributes to excess mortality in the cities of United States.

Conclusion

The concentration of indoor SPM is very high in Shahdara and Shahzada Bagh industrial areas and it was approximately 3 times higher than the standard values with respect to industrial areas setup by Central Pollution Control Board (CPCB Report, 1999) and 7 times higher than the WHO standards (Bascom et al., 1996). High concentration of SPM was mainly due to the presence of a large amount of industrial establishment and high density of road traffic in these areas. Mineralogical analysis of indoor SPM samples collected from Shahdara and Shahzada Bagh industrial areas showed presence of several mineral phases - quartz, calcite, dolomite, illite, muscovite, feldspar, talc and chlorite (Fig. 8). On the basis of relative abundances of mineral phases present in the indoor suspended particulate matter of Shahdara and Shahzada Bagh industrial areas, SPM have been subdivided into three distinct groups, such as Group – A: Quartz - Carbonates Dominant Group, Group – B: Quartz - Mica Dominant Group and Group – C: Quartz - Feldspar Dominant Group (Table 4). Relatively high proportion of five major mineral groups such as quartz, carbonates (calcite and dolomite), mica (illite and muscovite), feldspar, talc-chlorite (talc and chlorite) were found in the indoor suspended particulate matter. Illite, chlorite and talc are the clay minerals and all others are non - clay minerals. Over all average weight percentage of mineral phases present in indoor SPM samples was 56% of quartz, 20% of carbonates, 8% of mica, 10% of Feldspar and 6% of talc-chlorite and the decreasing order of average weight percentage of mineral phases in indoor SPM is quartz > carbonates > Feldspar > mica > talc-chlorite (Fig 3).

The major elements including Si, Al, Fe, Mg, Ca, Na, K, Mn and trace elements such as Cr, Co, Ni, Pb, Zn, Cu, Mo and Cd were identified in indoor suspended particulate matter. The decreasing order of mean concentration of major and trace elements in indoor SPM are Si > Al > Ca > Fe > Zn > Cu > Na > Mo > Pb > Mg > Cr > K > Mn > Ni > Co > Cd. Among these major and trace elements, Si was found to be the highest as corroborated by the abundance of quartz identified in indoor SPM, while Cd found to be lowest concentration in indoor SPM. The mean concentration level of all major and trace elements were extremely higher than the reported values (Table 4). Mean concentration level of Mn was 3 times higher than United State Environmental Protection Agency standards (Agency for Toxic Substances and Disease Registry Report, 2002) and 9 times higher than the WHO standard (WHO report, 2002). Mean concentration level of Pb was 2 times higher than the USEPA standard and 7 times higher than the WHO standard. Mean concentration level of Cr was 18 times higher than the US EPA and 1.6 times higher than the WHO standards. Mean concentration level of Cd was also 6 times higher than the USEPA standard and 8 times higher than the WHO standard (Table 4). Several studies (Dams and Jonge 1976; Winchester and Bi, 1984; Quiterio et al., 2004; Wahab and Yaghi, 2004; Richter et al., 2007; Godoi et al., 2008) are also supportive of the present study and determined the chemical composition of airborne particulate matter and found major and trace elements such as S, Cl, K, Ca, Na, Mn, Ni, Cu, Zn, Pb, Al, Si, P, Ti, V, Cr, Fe, As, Br in suspended particulate matter.

Moreover, the present study revealed that the identified minerals at both the selected sites (Shahdara and Shahzada Bagh) were injected into the air of Delhi city possible through wind erosion, re-suspended soil and road dust and anthropogenic sources. The homogeneity in the identified mineral phases in both the study areas leading to the indoor environment carried particulate matter from cooking, smoking, dusting, outdoor sources and well mixed in air as a result of continuous deposition and uptake from the grounds (topsoil). Air particulate matter with mineral phases and major and trace elements existing in the air other than its normal constituents is highly dangerous to the health of children and adults particularly those who are suffering from respiratory diseases including asthma, rhinitis and chronic obstruction pulmonary disease (USEPA Report, 2003). It also causes coughing, wheezing, shortness of breath, chest discomfort and burning sensation in the lungs (USEPA Report, 2003).

Study deals with investigation of respiratory symptoms and diseases in children (7 to 15 years) of Shahdara and Shahzada Bagh areas. Pulmonary function test of each child was carried out with electronic portable spirometer and the maximal expiratory flow volume curves were obtained as per the recommendations of American Thoracic Society (ATS, 1995). On the basis of health history, symptoms and health check-up including Pulmonary Function Test, Peak Expiratory Flow Rate and Chest X-ray, the respiratory diseases such as asthma, rhinitis and upper respiratory tract infection were diagnosed in the children. The results when correlated show respiratory symptoms in children including cough, phlegm production, shortness of breath and wheezing (Table 5.4). Children were diagnosed to have asthma, rhinitis and upper respiratory tract infection (Table 5.5). Children who are exposed to passive smoking or environmental tobacco smoke (ETS) have more rhinitis and upper respiratory tract infection. There was significantly more asthmatic, rhinitis and upper respiratory tract infection problem in the children who generally liquid petroleum gas for cooking compared to biomass fuel (coal, wood, kerosene, cow dung, etc.) for cooking.

Epidemiological studies (Dockery et al., 1993; Bascom et al., 1996; Horak et al., 2002; McConnell et al., 2003) have established that the suspended particulate matter is associated with more frequent respiratory allergy because airborne particulates easily reach the deepest recesses of the lungs and damage. Exposure to particulate matter can trigger asthma attacks and cause cough, wheezing, phlegm production, shortness of breath, common cold, throat congestion, and respiratory irritation in individuals with sensitive airways (USEPA Report, 1994; Bascom et al., 1996). The relationship between respiratory allergy in children and the mineral phases present in the indoor SPM with reference to their elemental composition have been established in the present study. Asthma in children was significantly associated with indoor SPM. The mean level of indoor SPM was high in the houses of Shahdara and Shahzada Bagh where children had diagnosed to have asthma, rhinitis and upper respiratory tract infection. The high values for mean level of indoor SPM was associated with environmental factors. Quartz and mica contents were high in indoor SPM where children having asthma, rhinitis and upper respiratory tract infections.

References

Agency for Toxic Substances and Disease Registry (ATSDR) (2002) Regulations and guidelines applicable. U.S. Department of health and human services, Public Health Service.

Albalak, R. (1997) Cultural practices and exposure to particulate pollution from indoor biomass cooking: Effects on respiratory health and nutritional status among the aymara Indians of the Bolivian Highlands (Ph.D. Thesis). Ann Arobar, Michigan: University of Michigan.

Allen, A.G. and Miguel, A.H. (1995) Indoor organic and inorganic pollutants: in-situ formation and dry deposition in south-eastern Brazil. Atmospheric Environment. v.29B, pp.3519-3526.

Balachandran, S., Meena, B.R. and Khillare, P.S. (2000) Particle size distribution and its elemental composition in the ambient air of Delhi. Environmental International, v.26, pp.49-54.

Bascom R. et al. (1996) Health effects of outdoor air pollution. American journal of respiratory and critical care medicine. v.153, pp.3-50.

Behera, D., and Malik, S.K. (1987) Chronic respiratory disease in Chandigarh teachers. Indian J chest Dis Allied Sci., v.29, pp.25-28.

Behera, D., Dash, S. and Malik, S.K. (1988) Blood carboxyhaemoglobin levels following acute exposure to smoke of biomass fuel. India J. Med. Res., v.88, pp.522-42.

Berico, M, luciani A and Formignani (1997) Atmospheric aerosol in an urban area – measurements of TSP and PM10 standards and pulmonary deposition assessments. Atmospheric Environment, v.31, pp.3659-3665.

Boleij, J.S.M., Ruigewaard, P. and Hoek, F. (1989) Domestic air pollution from biomass burning in Kenya. Atmospheric Environment, v.23, pp.1677-1681.

Brain, J.D. and Valberg P.A (1979) Deposition of aerosol in the respiratory tract. Am. Rev. Respir. Dis., v.120, pp.1325-1373.

Brown, G. (1961) The X-ray identification and crystal structure of clay minerals. (ed.) Mineral. Soc. London, 544p.

ckery, D.W., and Pope, C.A. (1994) Acute respiratory effects of particulate air pollution. Annu. Rev. Public Health, v.15, pp.107-132.

Crocker, C.R., Eylands, K.E., Mccollor, D.P., Helmowski, B.S., Benson, S.A. and Galbreath, K.C. (2002) Characterization of PM2.5 from rural Midwestern U.S. sites. Fuel Chemistry Division Preprints; v.47 (2), pp.632-633.

Dams, R. and Jonge, J.D. (1976) Chemical composition of Swiss aerosols from the Jungfraujoch. Atmospheric Environment, v.10, pp.1079-1084.

Dockery, D.W. (1989) Effect of inhalable particles on respiratory health of children. Am Rev Respir Dis, v.139, pp.587-594.

Dockery, D.W., Pope III, A.P., Xu, X., Spengler, J.D., Ware, J.H., Fay, M.E., Ferris, B.G., and Speizer, F.E. (1993) An association between air pollution and mortality in six U.S. cities. The New England Journal of Medicine. v.329(24), pp.1753-1759.

Dusseldorp, A., Kruize, H., Brunekreef, B., Hofschreuder, P., Meer, G. and Oudvorst, A.B. (1995) Associations of PM10 and airborne iron with respiratory health of adults living near a steal factory. Respir Crit Care Med., v.152, pp.932-1939.

Ezzati, M., Saleh, H. and Kammen, D.M. (2000) The contributions of emissions and spatial microenvironments to exposure indoor air pollution from biomass combustion in Kenya. Environment Health Perspectives. v.108, pp.833-839.

Ezzati, M. and Kammen, D.M. (2001) Indoor air pollution from biomass combustion and acute respiratory infections in Kenya: an exposure-response study. The Lancet v.358, pp.619-624.

European Commission, (2000) D G environmental working group on arsenic, cadmium and nickel compounds. Ambient air pollution by As, Cd and Ni compounds, Position Paper (Final Version).

Ferris, B.G. (1978) Epidemiology standardization project. Am Rev Respir Dis, v.118, pp.1-53.

Grim, R.E. (1968). Clay mineralogy. Mc Graw Hill book company, New York, 600.

Gulliver, J. and Briggs, D.J. (2004) Personal exposure to particulate air pollution in transport microenvironments. Atmospheric Environment, v.38, pp.1-8.

Godoi, R. H. M., Braga, D.M., Makarovska, Y., Alfoldy, B., Filho, M. A. S. C., Grieken, R. V. and Godoi, A. F. L. (2008) Inhable particulate matter from lime industries: Chemical composition and deposition in human respiratory tract. Atmospheric Environment, v.42, pp.7027-7033.

Goyal, P. and Sidhartha (2002) effect of winds on SO2 and SPM concentrations in Delhi. Atmospheric Environment, v.36, pp.2925-2930.

Harmsen, A.G., Muggenburg, B.A., Snipes, W.B. and Bice, D.E. (1985) The role of macrophages in particle translocation from lung to lymph nodes. Science, v.230, pp.1277-1280.

Harrison, R.M., Deacon, A.R., and Jones, M.R. (1997) Sources and processes affecting concentrations of PM10 and PM2.5 particulate matter in Birmingham (U.K.). Atmospheric Environment. v.31, pp.4103-4117.

He, C., Morawska, L., Hitchins, J. and Gilbert, D. (2004) Contribution from indoor sources to particle number and mass concentrations in residential houses. Atmospheric Environment. v.38, pp.3405-3415.

Hooton, D.H. and Giorgetta, N.E. (1977) Quantitative X-ray diffraction analysis by a direct calculation method. X-ray Spectrom, v.6(1), pp.2-5.

Horak Jr, F. et al (2002) Particulate matter and lung function growth in children: a 3-yr follow-up study in Australian Schoolchildren. Eur Respir J, v.19, pp.838-845.

Hussain, M.Y., Yousuf, M., Shahzad, I. and Imran, M. (2008) Investigating indoor suspended particulate matter as source of air pollution in residential, commercial and industrial areas of Faisalabad city. Pak. J. Agri. Sci. v.45 (1) pp.112-115.

ICRP (1994) Human respiratory tract model for radiological protection. ICRP Publication,66, Ann. ICRP, v.24, (1-3).

Jenkins, P.L., Phillips, T.J., Mulburg, J.M. and Hui, S.P. (1992) Activity patterns of Californians: use of and proximity to indoor pollutant sources. Atmospheric Environment. v.34, pp.2603-2612.

Jones, N.C., Thornton, C.A., Mark, D.A. and Harrison, R.M. (2000) Indoor/outdoor relationships of particulate matter in domestic homes with roadside, urban and rural locations. Atmospheric Environment. v.34, pp.2603-2612.

Jones, T., Blackmore, P., Leach, M., Berube, K., Sexton, K. and Richards, R. (2002) Characterization of airborne particles collected within and proximal to an opencast coalmine: South Wales, U. K. Env. Mont. Assess., v.75, pp.293-312.

Kamens, R., Lee, C., Wiener, R. and Leith, D. (1991) A study to characterize indoor particles in three non-smoking homes. Atmospheric Environment. v.25A, pp.939-948.

Khillare, P.S., Meena, B.R., Balachandran, S. (2004) Spatial and temporal variation of heavy metals in atmospheric aerosol of Delhi. Environmental Monitoring and Assessment. v.90, pp.1-21.

Kleinman, M.T., Kneip, T.J. and Eisenbud, M. (1976) Seasonal patterns of airborne particulate concentrations in New York City. Atmospheric Environment. v.10, pp.9-11.

Koponen, I.K., Asmi, A., Keronen, P., Puhto, K. and Kulmala, M. (2001) Indoor air measurement campaign in Helsinki Finland 1999 – The effect of outdoor air pollution on indoor air. Atmospheric Environment. v.35, pp.1465-1477.

Kumar, A., Mohan, U. and Jain, V.C. (1997) Influence of some socio-demographic factors on smoking status of academicians. Indian J. chest Dis. Allied Sci., v.39, pp.5-12.

Lee, H.S., Kang, B.W., Cheong, J.P. and Lee S.K. (1997) Relationship between indoor and outdoor air quality during the summer season in Korea. Atmospheric Environment. v.31, pp.1689-1693.

McConnell, R. et al (2003) Prospective study of air pollution and bronchitic symptoms in children with asthma. Respir. Crit. Care Med., v.168, pp.790-797.

Merefield, J.R. and Stone, I.M. (1998) Environmental Issues in Opencast Mining, In: Proceeding of the Seventh International Symposium on Mine Planning and Equipment Selection/calgary/ Canada: Mine Planning and Equipment Selection, edited by: Singhal R. K., pp.667-672.

Nagar, J. K., Shrivastava, J. P., Rajkumar, Chandra, U., Rathi, B., Rana, S.V.S. and Ahmad, K. (2007) Urban air pollution: A global environmental health problem – a review. Bull. Env. Sci. XXV (3rd issue): pp.201-223.

Oberdorster, G., Ferin, J., Gelein, R., Soderholm, S. and Finkelstein, J. (1992) Role of the alveolar macrophage in lung injury: studies with ultrafine particles. Environ. Health Perspect., v.97, 193-199.

Olphen and Fripate (eds.) (1979) Data handbook for clay materials and other non metallic minerals, Pergmon press London.

Ozaki, K., Hanatani, T. and Nakamura, T. (2005) Analysis of crystalline phases in airborne particulates by grazing incidence X-ray diffractometry. Analyst. v.130, pp.1059-1064.

Panyacosit, L. (2000) A review of particulate matter and health: Focus on developing countries, in International Institute for Applied Systems Analysis, Interin Report IR-00-005, Austria.

Phillip, J.L., Field, R., Goldstone, M., Reynolds, G.L., Lester, J.N. and Perry, R. (1993) Relationship between indoor and outdoor air quality in four naturally ventilated office in the United Kingdom. Atmospheric Environment. v.27A, pp.1743-1753.

Pope III, C.A., Thun, M.J., Namboodiri, M.M., Dockery, D.W., Evans, J.S., Speizer, F.E. and Heath, C.W. (1995) Particulate air pollution as a predictor of mortality in a prospective study of US adults, Am. J. Respir. Crit. Care Med. v.151, 669p.

Repace, J.L. and Lowrey, A.H. (1980) Indoor air pollution, tobacco smoke, and public health. Science, v.208, pp.464-472.

Reynolds, L., Jones, T. P., Berube, K. A., Wise, H. and Richards, R. (2003) Toxicity of airborne dust generated by opencast coal mining. Mineralogical Magazine, v.67 (2), pp.141-152.

Reddy, K.S., Gupta P.C. (2004) Prevalence of tobacco use. Report on tobacco control in India. New Delhi: Ministry of Health and Family Welfare, Govt. of India. pp.49-56.

Reid, H., Smith, K. and Sherchand, B. (1986) Indoor smoke exposures from traditional and improved cookstoves: comparisons among rural Nepali women. J Mountain Res. Dev., v.6, pp.293-304.

Richter, P., Grino, P., Ahumada, I. and Giordano, A. (2007) Total element concentration and chemical fractionation in airborne particulate matter from Santiago, Chile. Atmospheric Environment. v.41, pp.6729-6738.

Querol, X. et al (2007) Source origin of trace elements in particulate matter from regional background, urban and industrial sites of Spain. Atmospheric Environment. v.41, pp.7219-7231.

Srivastava, A. and Jain, V. K. (2007) a study to characterize the suspended particulate matter in an indoor environment in Delhi, India. Building and Environment, v.42, pp.2046-2052.

Spengler, J.D. and Sexton, K. (1983) Indoor air pollution: a public health perspective. Science. v.221(4605), pp.9-17.

Smith, K.R. and Liu, Y. (1994) Indoor air pollution in developing countries. In: Samet JM, editor. Lung biology in health and disease, v.74: Epidemiology of lung cancer. New York: Marcel Dekker. pp.151-184.

Samet, J.M., Marbury, M.C. and Spengler, J.D. (1987) Health effects and sources of indoor air pollution. Part I. Am. Rev. Respir. Dis., v.136, pp.1486-1508.

Schwartz, J. (1994) Particulate air pollution and chronic respiratory disease. Environmental Research, v.62, pp.7-13.

Singh, M.P., Goyal, P., Panwar, T.S., Aggarwal, P. and Nigam, S. (1990) Predicted and observed concentrations of SO_2, SPM and NOx over Delhi. Atmospheric Environment. v.24A, pp.783-788.

Sivacoumar, R., Jayabalou, R., Subrahmanyam, Y.V., Jothikumar, N. and Swarnalatha, S. (2001) Air pollution in stone crushing industry, and associated health effects. Indian J. Environ. Health, v.43, pp.169-173.

Sweet, C.W., Vermette, S.J. and Landsberg, S. (1993) Sources of toxic trace elements in urban air in Illinois. Environmental Science and Technology. v.27, pp.2502-2510.

Trindade, H.A., Pfeiffer, W.C., Londres, H. and Costa-Ribeiro, C.L. (1981) Atmospheric concentration of metals and total suspended particulates in Rio de Janeiro. Environ. Sci. Technol. v.15, pp.84-88.

Tan, K. H. (2005) Soil Sampling, preparation, and analysis. (Ed. II Emeritus Professor), University of Georgia, Greensboro, USA, 623p.

Thakur, M., Deb, M.K., Imai, S., Suzuki, Y., Ueki, K. and Hasegawa, A. (2004) Load of heavy metals in the airborne dust particulates of an urban city of central India. Environ Monit Assess. v.95, pp.257-268.

Tucker (1988) Techniques in Sedimentology. (Ed. M. Tucker). Blackwell Science Inc, 408p.

US EPA (United States Environmental Protection Agency) (1982) Air quality criteria for particulate matter and sulfur oxides. Report EPA-600/8-82-029 a,b,c, U.S. Environmental Protection Agency, Research Triangle Park, NC.

US EPA (United States Environmental Protection Agency) (1993) Mobile Vehicle Relate Air Toxics Studies, US Environmental Protection Agency, Office of Mobile Sources.

US EPA (United States Environmental Protection Agency) (1994) measuring Air Quality: The Pollutant Standards Index, US Environmental Protection Agency, Office of Air Quality Planning and Standard, EPA 451/K-94-001.

US EPA (United States Environmental Protection Agency) (1997) Health and Environmental Effects of Particulate Matter, US Environmental Protection Agency, Office of Air Quality Planning and Standard.

US EPA (United States Environmental Protection Agency) (2003) Guidelines for Developing an Air Quality (Ozone and PM2.5) Forecasting Program. US Environmental Protection Agency, Office of Air Quality Planning and Standard, EPA-456/r-03-002.

Vijayanand, C., Rajaguru, P. Kalaiselvi, K. Selvam, K. P. and Palanivel, M. (2008) Assessment of heavy metal contents in the ambient air of the Coimbatore city, Tamilnadu, India. J. Hazardous Materials, v.160, pp.548-553.

Wadden, R.A. and Scheff, P.A. (1983) Indoor Air Pollution, Wiley Interscience, New York. pp.52-78.

Wahab, S.A.A. and Yaghi, B. (2004) Total suspended dust and heavy metal levels emitted from a workplace compared with nearby residential houses. Atmospheric Environment. v.38, pp.745-750.

Wahab, S.A.A., Worthing, M.A. and Al-Maamari, S. (2005) Mineralogical of atmospheric suspended dust in three indoor and one outdoor location in Oman. Environ. Monit. Assess. v.107, pp.313-327.

Williamson, B.J. (2001) Mineralogical composition of atmospheric airborne particulates. Geology Today. v.17 (1), pp.32-35.

Winchester, J.W. and Bi, M. (1984) Fine and coarse aerosol composition in an urban setting: a case study in Beijing, China. Atmospheric Environment. v.18, pp.1399-1409.

Xu, X. and Wang, L. (1993) Association of indoor and outdoor particulate level with chronic respiratory illness. Am. Rev. Respir. Dis., v.148, pp.1516-1522.

Yadav, S. and Rajamani, V. (2004) Geochemistry of aerosols of northwestern part of India adjoining the Thar Desert. Geochimica et Cosmochimica Acta, v.68 (9), pp.1975-1988.

Yocom, J.E. (1982) Indoor-outdoor air quality relationships: a critical review. Journal of the air Pollution Control Association. v.32 (5), pp.500-520.

Zemp, E. et al (1999) Long-term ambient air pollution and respiratory symptoms in adults (SAPALDIA Study). Respir. Crit. Care. Med., v.159, pp.1257-1266.

Hon'ble reviewer

1. **Dr. P.S. Parihar**
   Director
   Atomic Mineral Directorate
   Begampet
   HYDERABAD - 500 001
2. **Prof. Vinod Sharma**
   Dean
   Indira Gandhi Dev. Res. Institute,
   Goregaon (East)
   MUMBAI - 400 065

IV-49

Geo-environment and Natural Hazards in the Present Phase of Climate Change, with Special Reference to Northeast India

Trilochan Singh

Scientist 'G' (Retd), 170 / DG-III, Vikas Puri, New Delhi -110018
E-mail : tskaith@rediffmail.com

Abstract

The weather pattern and climate cycles ocur at all time scales, ranging from decades to thousands to million of years, and have persisted since the earliest history of the Earth. The existence of life and man's activities are highly dependent on the weather. The present phase of climate change, however, threatens the geo-environment and causing natural hazards. The present paper is discussing natural hazards and geoenvironment, chiefly in northeastern India, in the existing phase of climate change.

Keywords: Natural Hazards, Climate Change, Geoenvironment Northeast India.

Introduction

The weather pattern and climate cycles occur at all time scales, ranging from decades to thousands to millions of years, and have persisted since the earliest history of the Earth. The existence of life and man's activities are highly dependent on the weather. The present phase of climate change, however, threatens the geo-environment and causing natural hazards.

GEO-ENVIRONMENT

Geology is integral component of the environment. The geo-environment largely governs by two components : geo-resources and geodynamic geological processes. Whereas geo-resources are the need of the man-kind, geological processes cause endanger to man-kind. Although geological processes, such as erosion and deposition, are the natural phenomena, these may be aggravated by human interference.

It is expected that the geo-environment is affected by the human interference for nearly 10 km depth inside the Earth's crust. The rational exploitation of various geo-resources may prove to be the potential of the geo-environment, however, over exploitation or unplanned exploitation (causing hindrances in the geological processes/phenomenon) deteriorate the geo-environment. In this context, the important geological factors are:

- Natural resources, such as cultivated topsoils, peat, mineral deposits, sand and gravel, etc. are important both economically and from a conservation point of view.

- Water is easily affected, both in terms of quality and quantity. Apart from the rivers and lakes, many important water resources are hidden in underground aquifers, which are likely to be polluted.
- Landscape features, which are increasingly important in conservation and leisure, include many settings where geological processes are actively reshaping the terrain, for example mountains ranges, deep gorges and river valleys, waterfalls, sea cliffs, caves, etc.
- Cultural heritage areas can include features of a geological nature; for example designated sites of special scientific interest and archaeological sites such as prehistoric mining sites, etc.

In addition, the natural geological barriers may either limit or lead to a negative impact on the land use. Such barriers are the natural hazards, viz., ruinous earthquakes, catastrophic landslides, flash floods, tsunamis, etc., which endangers not only the geo-environment but also health and life of the human being.

NATURAL HAZARDS

Natural hazards have been operating throughout the Earth's history, however, these have become meaningful after the advent of man. The natural hazards would have been only the natural events, if the human life and properties are not affected.

Natural hazards, which are mainly of geologic origin, have an adverse impact on environment, causing destruction of property, loss of life and environment degradation. The situation get further worsen when a hazard trigger another hazard to take place. For example, earthquake frequently give rise to landslides; similarly, fire in pillared coal mines may lead to ground subsidence, cloud burst cause intensive flash flood, etc. Although the human-kind cannot prevent natural hazards, but certai-nly plan to avoid suffering their worse effect.

The geo-environment is always under threat by the natural hazards. At the same time, interference of the human in the geo-environment, particularly that of the Himalayan environment, has resulted imbalance in the natural processes in the fragile ecosystem. In recent times many anthropogenic activities have changed the landscape of many hilly regions, which have serious consequences, such as soil erosion, water shortages, increased natural disasters, extremes of temperature and radiation, etc.

The Himalayan terrain is prone to various types of natural disaster due to its inherent geo-environment and geomorphic processes operating therein, which have further been aggravated by unscientific anthropogenic activities. Thus, the development across the Himalayan belt has inflicted severe pressure on the carrying capacity of this fragile resource zone.

The various natural hazards in the Himalayan terrain have been briefly discussed below:

Earthquakes are the frequent and periodic events in the Himalayan terrain, because of the fact that this region is geodynamically active. In fact, the Indian Plate is continuously pushing against the Tibetan Plateau and as a consequence the Himalayan region is rising, which makes it prone to the seismic activities.

Landslides are the most common disasters that occur across the Himalaya causing human, economic and environmental losses. It is one of the mass wasting processes and is responsible for the degradation of slope and relief. The process has been intensified by human interference mainly through deforestation, unscientific construction procedure and unplanned tapping of natural resources. Hence, the degree and frequency of this disaster have notably increased over the period of time.

Soil Erosion is mainly caused by the rivers, streams and other form of surface run-off. In the Himalayan region it is much more because of the fast-flowing water on steep slopes, which erode large amounts of soil and subsequently degrade land resource, thus, affecting the environment. Soil erosion also causes extensive loss to soil fertility and damage to the land basin.

Cloud burst is one of the major natural hazards/disasters in the Himalaya. It refers to sudden extreme amount of rainfall in relatively short period of time and over a small geographical area, related to extreme hydro-meteorolgical conditions. Meteorologists define cloud burst as the rainfall rate equal to or greater than 100 mm per hour, and during a cloud burst more than 20 mm of rain may fall in a few minutes. The cloud burst is followed by flash floods leading to debris flow, landslide and eventually the blockade of river channels, consequently wreaking havoc downstream.

Glacial Lake Outburst Flood (GLOF) is the catastrophic discharges of water resulting primarily from the glacial lakes. As the glaciers retreat they leave a large void behind, which is filled by glacier melted water making a lake. These lakes are structurally weak and unstable and undergo constant changes due to slope failures, slumping, etc., causing GLOFs in downstream. It creates more havoc and cause devastation to downstream riparian communities including life, property and other infrastructure, because GLOF comprises water mixed with moraine materials. Thus, in the long run, glacial retreat and GLOFs tend to threaten regional Himalayan Environment.

Forest Degradation and Loss of Biodiversity. Forests, on one hand, are the major sources of firewood to the villagers and fodder to livestock; and on the other hand, are also the major store house of biodiversity. In the Himalayan region, these are allegedly under threat mainly because of unplanned urbanization, commercial timber extraction, capitalistic development projects, mining & quarrying, and such other forces of modern development. All these have serious environmental, social, economic and biological consequences.

CLIMATE CHANGE

Danger of environmental degradation has been increasing greatly over the years. Population growth, urbanization, and anthropogenic development have resulted in injecting a heavy load of pollutants into the environment. At the same time, rising level of Green House Gases (GHGs) in the atmosphere is one of the most serious environmental concerns. The issue becomes more sensitive in view of the challenges posed by the climate change.

The climate is a complex phenomenon governed by several factor, one of these is the configuration of land and sea. The mountain ranges on the land deflect the atmospheric flow, and the patterns of precipitation are determined to a large extent by land-sea contrasts, continental land masses, mountain ranges and other factors.

The Earth's climate has remained dynamic throughout the 4.5 billion years history of the Earth, making significant impacts at different periods of time, which are sometimes minor or at other times catastrophic. There is widespread acceptance that there have been repeated cycles of amalgamation and dispersal of continental land mass in the geologic past, which have had not only made profound effect on the evolution of the Earth's crust, but also affected the climate and the then prevailing life. Signatures of such climate changes, which may be gradual or abrupt, have been imprinted on the rocks of that particular period in form of fossil records and various sedimentary and geochemical parameters. Even the cores drilled through the ice sheets yielded a record of polar temperatures and atmospheric composition ranging back to 120,000 years in Greenland (Arctic) and 800,000 years in Antarctica (Cronin, 2009).

In the Indian sub-continent, the climate has mainly been influenced by the evolution of the Himalaya. It prevents the ingress of cold wintry winds from Siberia/Central Asia. If there were no Himalayan ranges to wring out moisture from the passing clouds, much of the continent would have been a thirsty, parched and desolated land. The Himalayan mountain barrier rising 500 - 8,000 m above msl is responsible for peculiar monsoon climate of the subcontinent and unique succession of six seasons, viz., *Vasant, Grishma, Varsha, Sharad, Shishir and Hemant* (Valdiya, 1998, 1999).

According to Valdiya (1998, 1999), the climate remained quite warm and moist in the geologic past, when the terrain was gentle and there was no mountain barrier at that time. Sudden rose of Tibetan Plateau by 1,000 to 2,500 m, nearly 8 m.y. ago made abrupt changes in the climate and initiated unique monsoon. It brought rains from the south-westerly air currents during summer and cold precipitation from north-easterlies during winters. Diversion of wind flow and creation of large cool areas caused snowfall on the elevated mountains in the present day Higher Himalayan Terrane, while the Lesser Himalayan and Siwalik terranes were washed by heavy rains.

Significant climate changes have been noticed in terms of the glacial and interglacial cycles periodically on the Earth in the geologic history. The most recent glacial cycle occurred between about last 120,000 and 11,500 years. Since then, the Earth has been in an interglacial cycle. This present interglacial cycle, i.e., the Holocene period, has lasted about 11,700 years

(International Commission on Stratigraphy, 2008).

As evident from the geological records, climate change cause natural hazards. These hazards had been to a large extent million of years ago, such as devastating vast rainforests, abruptly fragmenting the habitat into isolated 'island', mass extinction of plants and animal species, etc. McGuire (2010) had summarized evidence for periods of exceptional past climate change eliciting a dynamic response from the Earth's crust, involving enhanced levels of potentially hazardous geological and geomorphological activities.

As a part of the aforesaid natural cycle of the Earth's climate history, climate is changing once again in the present era showing global warming, where global temperature is projected to rise up to 4° Celsius by the year 2100 (Narain, 2009). Tropical cyclones are likely to become more frequent and intense, rainfall may increase, and sea level may rise up to nearly a meter with the increase of tropical sea surface temperature. Soldati et al. (2003) have demonstrated that the morphogenetic processes, which occur on the Earth's surface, are definitely influenced by the climate. Thus, climate change is expected to cause natural disasters, such as drought, flood, landslides, etc.

PRESENT PHASE OF CLIMATE CHANGE AND NATURAL HAZARDS

Since the evolution of the Himalaya is still continuing, which is running through the present Holocene phase of the Himalayan Orogeny, its structural framework is recurrently deformed and the landscape reshaped time and again, which has been affecting the climate. In other words, the shifting position of the continents led to profound and relatively rapid changes in the climate.

The Indian Plate was moving at a rate of nearly 20 cm/year before colliding with the Eurasian Plate approximately 50 m.y. ago. The rate of movement was greatly reduced after the collision, from 20 cm/yr (50 m.y. ago) to 4.5 cm/yr (40 m.y. ago) to 2.54 cm/yr (in mid 1990s). Movement of the Indian plate has reaccelerated once again from a rate of < 2.5 cm/yr in mid 1990s to > 3.5 cm/yr to 2000s (Dai and Brown, 2011).

On the other hand, there has been a progressive increase in the average surface temperatures of the Earth since the beginning of the 20th century. It is estimated that the Earth's surface temperature has risen by 0.6±0.2°C over the twentieth century. In the last 50 years, the rise has been 0.13 + 0.07°C per decade and the recent years have been the warmest since 1860, the year from which regular instrumental records are available. Model simulation and palaeoclimatic evidences suggest that when climate warms, it warms more in higher latitudes than in lower latitudes and more in winter than in summer (Golitsvn, 1989; Schneider, 1989). A warmer atmosphere contains more water vapour and increases the intensity of the whole hydrological cycle, but precipitation patterns are likely to change homogeneously in time and space (Golitsyn, 1989). Some scientists believe that in a warmer climate the Earth can be expected to experience more variable weather than now, with a likelihood of more floods and drought, more intense hurricanes or typhoons, and more heat-waves (Golitsyn, 1989; Hansen et al., 1989). Even a few scientists believe that frequency of major earthquakes have increased as well with the increase in temperature because of global warming (Dai and Brown, 2011). It may be substantiated with the fact that there have been 8 earthquakes with a magnitude of 8 or higher, the largest being 8.3, over the 16 year span, 1979-1995. In the most recent 15 years, 1996-2011, there were 17 earthquakes of magnitude 8 or greater, 9.1 being the largest. These recent quakes include the December 26, 2004, 9.1 quake off the west coast of northern Sumatra, and the March 11, 2011, 9.0 quake off the coast of Honshu, Japan. The frequency has increased more than 100%, and the maximum magnitude strength has increased almost a full order.

In the present phase of human life cycle, the gradual weather change is being considered as the climate change, and the hazards are viewed in terms of drought, flood, landslides, earthquakes, etc. The hue and cry is mainly because these hazards pose a level of threat to the life, health and property of the human being. In the last few decades rising temperature and erratic weather pattern have been observed in the Eastern Himalayan region (Shrestha and Devkota, 2010; Pathak et al., 2010), where a number of glaciers have receded and several glacial lakes have been formed in the Higher Himalayan terrain (Mool et al., 2001). These glacial lakes may cause Glacial Lake Outburst Flood (GLOF) hazards in the lower reaches (Reynolds 1998; Yamada 1998). Similarly, increased frequency and intensity of the rainfall and cloud burst in the Himalayan terrain may attribute to increase in flash floods and landslide disasters.

Therefore, changing climate can increase the frequency, intensity or duration of extreme weather conditions which increases risks for vulnerable populations and communities in areas exposed to natural hazards. In other words, increasing trend in natural disasters is partly due to better reporting, partly due to increasing population vulnerability, under the influence of ongoing global climate change.

As already mentioned above, the Indian landmass continues to move and press against Asia, stresses/strains are building up and accumulating progressively. Natural hazards are, therefore, very common in the Himalayan domain as well as in the adjacent Indo-Gangetic Plains.

SCENARIO IN THE NORTH EAST INDIA

The geo-tectonic setting the NE India is very peculiar, which comprises seven tectonic domains, viz., Eastern Himalaya, Mishmi Block, Naga-Patkai Hills, Assam Valley, Shillong Plateau - Mikir Hills, Arakan-Yoma Folded Belt, and Bengal Basin. All these domains are tectonically active. In the NE Region of India, in fact, the Indian crustal block (embracing Assam and Meghalaya) slides under the Eastern Himalaya in the north and under the Myanmar (Burma) Ranges in the east. It has caused several large scale vertical as well as horizontal movements and even deflected the course of the Brahmaputra Rivers at places. As such seismicity in the NE Region is very high and whole of the region falls under the high seismic zone of India, i.e., Zone-V.

Therefore, the Himalayan Environment, in general, and that of the NE India, in particular, is always under threat by the natural hazards. The region is prone to various types of natural hazards due to its inherent geo-environment and geomorphic processes operating therein, which have further been aggravated by unscientific anthropogenic activities. On the other hand, climate change has now led to rain, rather than snow, falling even at higher altitudes. It also accelerates the melting of glaciers. Heavy rainfall, which was unknown in the high altitude, has become more frequent.

On the other hand, the NE Region of India is characterized by the varied physiological features and altitudinal differences that give rise to varied types of climate, ranging from sub-tropical to temperate and alpine. It has given rise to diverse vegetation types. The published data on rainfall and the temperature regimes of the NE Region of India (Mirza et al, 1998; Das and Goswami, 2003; Das, 2004; Das, 2009; World Bank, 2007) indicate that there is no significant trend in rainfall, except in certain parts of Nagaland, Manipur, Mizoram and Tripura and southern Assam, where summer monsoon rainfall is found to be decreasing significantly during the last century at an approximate rate of 11 mm per decade. On the other hand, a distinctly rising trend in surface air temperatures has been observed. Several districts of Assam have been badly affected due to drought like situations during the years 2005, 2006 and 2009. In Mizoram, the year 2005 witnessed prolonged dry periods with many springs and streams drying up accompanied by large scale landslides (ICIMOD, 2008). The high rainfall areas are facing unprecedented drought like situations in almost whole of the NE Region during the current years. Even erratic nature of rainfall, its intensity and frequency have been noted. The data has shown a similar trend in rainfall during the year 1983-1989 and 1999-2002, whereas during 1993-1999 and 2006-2009 the annual rainfall has shown a significant erratic nature. During 2009, the NE Region received worst-ever rainfall deficit in 30 years with Manipur being worst affected recording 67% deficient rainfall, followed by Meghalaya (-55%), Nagaland (-62%), Assam (-34%), Mizoram (-30% and Arunachal Pradesh (-29%), with Manipur and Nagaland declaring a 'draught' conditions (Sharma et al, 2010). It appears that the monsoon rainfall has shifted to post monsoon season.

In Meghalaya, duration of total rainfall has reduced by almost 1½ months with local variation in respect to altitude and geographical position. The intensity trend shows an increasing shift towards low intensity precipitation and a decline trend in the high intensity by a month.

Dry period is slowly increasing its spread by a month irrespective of altitudinal variations. Likewise, summer period has increased by over a month, and medium intensity exhibits a shifting trend towards high intensity. On the other hand, winter period has reduced by almost a month across all elevations.

The region has witnessed two extremely intense cloud bursts of unprecedented intensity in the year 2004, one in the western Meghalaya hills and another in the western Arunachal Pradesh that produced devastating flash floods causing enormous loss to the life and property. Subsequently, there have been two events of flash floods in 2008 originating from extreme rainfall causing significant damage to human life and property. Of these two, the one happened at the border of Assam and Arunachal near North Lakhimpur on 14th June, which inundated more than 50 villages leading to displacement of more than 10,000 people. The other happened on 26th October covering a linear strip at the Assam border with Bhutan and Arunachal, causing flash flood in four major rivers, viz., Puthimari, Bhareli, Ranganadi and Subansiri, and a number of smaller rivers.

All these indicate a signature of climate change. Ravindranath et al (2011) have studied climate change vulnerability profiles at the district level in the NE Region of India for the present prevailing climate as well as projected future climate. They used an index-based statistical technique representing key sectors of vulnerability, mainly agriculture, water, and forest sectors, to derive impact assessment models. Results were significant, which indicated that majority of the districts in NE India are subject to climate induced vulnerability in the present time and also in the near future.

The NE Region with the fragile ecosystem, which has been recognized as one of the 25 hot spots of the world for its rich biological diversity, is suffering from the existing environmental problems, viz., soil erosion leading to loss of soil fertility, deforestation causing land degradation, pollution, etc. It is further threatened by geo-hazards, viz., floods, landslides, earthquakes, etc. There have been a number of examples of extreme precipitation events, such as heavy rain storm, cloud burst, flash flood, etc., in various parts of the NE Region, which made significant impacts on the fragile landform in terms of landslides and soil erosion consequently damaging the human life and property.

Therefore, the NE Region of India is expected to be highly prone to the consequences to climate change because of its geo-ecological fragility. Rapid growth of human population has further been adversely affecting the Himalayan Environment, particularly groundwater pollution, land degradation, scenic beauty, etc. Simultaneously, various geo-hazards have also been increasing at an alarming rate, with the increase in population and simultaneous gradual climate change, affecting human lives, engineering projects, agricultural lands and the environment.

HAZARDS RISK MANAGEMENT

Risk is characteristic of the relationship between humans and the geologic processes. Though the risk from natural hazards cannot be stopped, however, their affect can certainly be minimized. For this, we need to understand the geologic processes causing such hazards, and accordingly action plans may be developed. In this context, it is pertinent to note that there has been an alarming rise in the size of Himalayan population in the last one-century. Overall, during 1901 and mid 2006 the Himalayan population rose by around 5.5 times. The population in Indian Himalaya rose by about 6 times from around 6 million to 34.6 million between 1901 and 2006. In Nepal Himalaya the degree of rise has been somewhat similar to its Indian counterpart where population rose from 5.6 to 28.3 million between 1901 and mid 2006. A small and relatively isolated Bhutan also witnessed an increase from 0.3 to 2.3 million in its population during the period. Rapid growth of population in the Himalayan region is alarming, which is not only increasing pressure on resource bases but also adding to the hazards risk.

Natural hazards are a challenge for a growing population, rather a major threat to society, costing lives, disrupting infrastructure and destroying livelihoods. Despite the damaging effect of the natural hazards, little is done to prevent them. The effects of global climate change rather have aggravated the situation. The level of awareness of the potential impacts of geo-hazards has, on the other hand, disrupted the social and economic activities of the people. Even it has not been properly understood by decision makers, investors and the public at large, in spite of available ground and space based technologies, which are widely used in risk management cycle, as such losses due to the disasters are continuing in our nation.

Management at the Government Level : It is the high time to improve our basic understanding of the social, environmental and economic risks posed by geo-hazards, and also the relationships between those risks so as to strengthen our ability to cope with them. It requires scientific understanding of geo-hazards, as well as addressing the full risk management cycle, necessary for appropriate policy and decision making.

The concept of the risk management cycle associated with hazards includes : i) mitigation and preparedness, ii) early warning, iii) response and iv) recovery.

A dedicated educational program may be developed to identify these hazards and at the same time efforts may be made to integrate the natural and social sciences for the disaster risk management. Such information, including data relevant to monitoring and understanding geo-hazards may be disseminated freely not only amongst the researchers and the governmental agencies (policy makers), but also amongst the NGOs and the citizens, so as to reduce loss of life and property from natural and human-induced disasters. It is desirable to have a meaningful interaction between the scientists, technical experts and policy makers on regular basis for extension of knowledge in better understanding of hazards mitigation and management. Following "*Prevention is better than cure*", considerably damage can be minimized if adequate preparedness levels are achieved.

Societal Awareness Program : A natural hazard may become a disaster when it becomes vulnerable. Vulnerability refers to not only the possible physical effects of a natural hazard, but the way it affects human life and property. Vulnerability to a given hazard depends on :

- Proximity to a possible hazardous event,
- Population density in the area proximal to the event,
- Non-existence of early-warning systems and lines of communication,
- Non-availability and readiness of emergency infrastructure,
- Construction styles and building codes, and
- Cultural factors that influence public response to warnings.

It is, therefore, important to create awareness amongst the mass about the natural hazards phenomenon, vulnerable factors, and mitigation measures. In general, such regions are more vulnerable to natural hazards where there is of lack of understanding of hazards, education, infrastructure, building codes, etc. Similarly, poverty also plays a role - since poverty leads to poor building structure, increased population density, and lack of communication and infrastructure.

Site specific awareness amongst the local inhabitants is more important about the vulnerable zones of natural disasters, such as landslide, debris flow, snow avalanche, earthquake, flash flood, etc., and appropriate response to minimize the effect of these disasters. Creating awareness amongst the local community and local NGOs may further helps to develop capacity building, which may strengthen the search, rescue and relief operations. In fact, sufficient material and funds are available with the government for relief and rehabilitation, however, these are not being utilized properly. Therefore, participation of local people and local NGOs by providing them the awareness is very important, which can play a major role in creating a sense of responsibility amongst the local inhabitants. Singh (In Press) has emphasized on the community based preparedness for safe future against natural disaster.

There is strong need of hand to hand cooperation amongst the geoscientists and the medical scientists. In spite of increasing trend of health conscious society, we never thought to what moves from the environment into our bodies through what we eat, drink and breathe. For most of us this interaction with natural materials is harmless or perhaps even beneficial. However, it may have devastating, even fatal effects. These interactions are the realm of a fast developing field of Medical Geology.

Monitoring: Our nation has attained advanced satellite-based technology. It may be utilized to develop dedicated software tools for comprehensive monitoring of geo-hazards, and also to provide timely information on spatio-temporal characteristics, risks, and occurrence of the geo-hazards, which will certainly be a step towards disaster reduction.

While discussing the impact and hazard-associated manifestations, it is the radon gas that comes first. Emission of radon gas is a common phenomenon prior to the occurrence of an earthquake. This colourless, odourless, radioactive gas together with its radioactive isotopes has been linked to lung cancer. Monitoring of the radon gas is, thus, not only important from health point of view, but also for prediction of earthquake activity.

ACTION PLAN

It is clear that the climate change has a direct bearing on degradation of the geo-environment and natural hazards. It is also understood that while hazards, both natural or otherwise, are inevitable, the affect or damaged caused by the disasters that follow may be minimized and the society can be prepared to cope up with the disasters effectively whenever they occur.

From the geological point of view following fields need to be strengthen, which have a direct role in minimizing the affect of the disasters.

Engineering Geology : With the increase in economic development in most parts of the Himalaya, in general, and in the NE Region, in particular, there is a boom in development of hydro-electric projects, roads, railway lines, tunnels, etc. In view of expected geo-hazards it is necessary to obtain quantitative and qualitative geotechnical information, particularly to rock mass characterization, support assessment, slope stability, risk analysis, etc., so as to take care of various geologic factors which are affecting construction, operation, and maintenance of the projects in these tectonically active areas.

Environmental Geology : is not simply concerned with the impact of humans upon the physical environment, it also involves the impact of the geological environment upon society. Natural geo-hazards, such as earthquake, volcanic eruption, floods, landslides, tsunamis are, with some exception, independent of humankind but can have a disastrous effect upon the society. On the other hand, humankind's primary activities in form of agriculture, mining and industry can have a notable impact upon the geological environment, such as soil erosion, land degradation, groundwater pollution, etc. Accordingly, both are the concern of environmental geology.

Whereas on one hand, it is important to save environment of the urban areas, which is being effected by population pressure, at the same time it is also important to protect rural area environment for food production and leisure activities, particularly the areas of outstanding scenic beauty and scientific interest for future generation to enjoy.

Mining and Environment : Mining and industrial activities are directly related to topography, geology, climate, vegetation and hydrology of that particular region. It is, thus, important to raise public awareness about the impacts of mining on the environment and human health. The areas of interest are:

- Soil and surface contamination by metals and gaseous emissions in mining and ore processing areas.
- Response of plants to heavy metal stress and bioremediation.
- Contamination of wetlands, surface and ground waters.
- Proper disposal and recycling of mining waste.
- Modelling of dispersion of dust and gaseous emissions from mining operations and Smelters.
- Evaluation of potential links between contamination and health in mining districts.
- Raising public awareness of the impacts of mining on the environment and human health.

There is need to adopt '**the culture of prevention**'. This process may be integrated with National Action Plan on Climate Change. The strategies and policies may be designed, which may be suitable for different regions under different situations, as every region has its characteristic requirement. This will ensure development of the safer community under sustainable environment.

Valdiya (2002) suggested that environmental security and hazard management programs must form essential part of the paradigm of development of the mountainous States. He further stressed the need of formulation of a public policy for hazard management and suggested to form a

State-level 'Natural Hazards Management Commission' in each State of the NE Region.

REFERENCES

Cronin, T.M., 2009, Paleoclimates: Understanding Climate Change Past and Present. Columbia University Press.

Dai, J.J. and Brown, S.R., 2011. Why Global Warming Responsible for Massively Destructive Earthquakes? http://voices.yahoo.com/why-global-warming-responsible-8695775.html

Das, P.J., 2004. Rainfall Regime of Northeast India: A Hydrometeorological Study with Special Emphasis on the Brahmaputra Basin. Unpublished Ph.D. Thesis. Gauhati University.

Das, P.J., 2009. Water and climate induced vulnerability in northeast India: Concerns for environmental security. WATCH Research Report 1. AARANYAK. Guwahati, Asssam, India. www.aaranayak.org.

Das, P.J. and Goswami, D.C., 2003. Long-term variability of rainfall over northeast India. *Indian Journal of Landscape Systems and Ecological Studies*, 26(1):1-20. 8 .

Golitsyn, G.S., 1989. Climate changes and related issues. Presentation on 24 August 1989 at Sundance, Utah. Symposium on Global Climate Change.

Hansen, J., Rind, D., Del Genio, A., Lacis, A., Lebedeff, S., Prather, M., Ruedy, R. and Karl, T., 1989. Regional greenhouse climate effects. In: *Coping With Climate Change,* John C. Topping, Jr. (Ed.). Climate Institute, Washington, DC, USA.

International Commission on Stratigraphy (2008) International Stratigraphic Chart.

ICIMOD., 2008. Recorded proceedings of the two day 'Climate Change and Vulnerability of Mountain Ecosystems in the Eastern Himalayan Region, North-East India & Bhutan Stakeholders Workshop' 11-12 March, 2008, Shillong. Organised by International Centre for Integrated Mountain Development, Kathmandu, Nepal .

McGuire, B., 2010. Climate forcing of geological and geomorphological hazards. Phil. Trans. R. Soc. A, 368 : 2311-2315.

Mirza, M.M.Q., Warrick, R.A., Ericksen, N.J. and Kenny, G.J., 1998. Trends and persistence in precipitation in the Ganges, Brahmaputra and Meghna river basins. *Hydrological Sciences-Journal-des Hydrologiques*, 43(6):845-858.

Mool, P.K., Bajracharya, S.R. and Joshi, S.P., 2001. Inventory of glaciers, glacial lakes, and glacial lake outburst floods: Monitoring and early warning systems in the Hindu Kush-Himalayan regions – Nepal. ICIMOD, Kathmandu.

Narain, J.P., 2009. Climate change brings natural disasters and disease. In: Sci Dev Net, Health: Climate Change & Insect-Borne Disease. 9 September 2009.

Parrish, J.T., 1993. Climate of the supercontinent Pangaea. Chemical Geology, 101 (2): 215-233.

Pathak, D., Gajurel, A.P. and Mool, P.K., 2010. Climate change impacts on hazards in the Eastern Himalayas. Climate Change Impact and Vulnerability in the Eastern Himalayas - Technical Report 5, ICIMOD, Kathmandu, Nepal. 17p.

Ravindranath, H.H., Rao, S., Sharma, N., Nair, M., Gopalakrishnan, R., Rao, A.S., Malaviya, S., Tiwari, R., Sagadevan, A., Munsi, M., Krishna, N. and Govindasamy Bala, 2011. Climate Change ulnerability Profiles for North East India. Current Science, vol. 101, no. 3, 384-394.

Reynolds, M., 1998. High altitude lake hazard assessment and mitigation: A Himalayan Perspective. In: Geohazards in Engineering Geology (Eds. McCall, G.H. and Eddleston), Special Publication No. 15, pp. 25-34. Geological Society of London.

Schneider, S.H., 1989. The Greenhouse Effect: reality or media event. In: *Coping with Climate Change,* John C. Topping, Jr. (Ed.). Climate Institute, Washington, DC, USA.

Sharma, B., Riaz, M.V., Pant, D., Adhikary, D., Bhatt, B.P. and Rahman, H., 2010. Water poverty in the NE Region (India): Potential alleviation through multiple-use water system – cross learning from Nepal Hills, New Delhi, India. International Water Management Institute – NAIP Report 1, 44 p.

Shrestha, A.B. and Devkota, L.P., 2010. Climate change in the Eastern Himalayas: Observed trends and model projections. Climate Change Impact and Vulnerability in the Eastern Himalayas-Technical Report 1, ICIMOD, Kathmandu, Nepal. 13p.

Singh, Trilochan, In Press. Community based preparedness for natural disaster. In: Proceedings of the National Seminar on 'Partners in Disaster Management: Institutions and Community', 10-11 Jan 2009, Panjab University, Chandigarh.

Soldati, M., Borgatti, L., Cavallin, A., Giardino, M., Pellegrini, G.B. and Tellini, C., 2003. Geomorphological Evolution of Slopes and Climate Changes : Landslide Analysis and Paleoclimatic Reconstructions in Northern Italy – An outline of a National Research Project. XVI Inqua Congress, Reno, Nevada. *Abstracts with Programs,* P. 66, Geological Society of America.

Valdiya, K. S., 1998. *Dynamic Himalaya.* University Press (India) Ltd., Hyderabad, 178 p.

Valdiya, K. S., 1999. Crucial Holocene development in the Indian sub-continent in context of societal concern. 4th Foundation Lecture under the auspices of the Indian Geological Congress, Roorkee. Delivered at the Earth System Science Section, 86th Session, Indian Science Congress Association, Chennai.

Valdiya, K.S., 2002. Geodynamic perspective of Arunachal Pradesh: bearing on environmental security and planning for development. In: Arunachal Pradesh : Environmental Planning and Sustainable Development (Edited by R.C. Sundriyal, Trilochan Singh, and G.N. Sinha). Himvikas Occasional Publication No. 16, G. B. Pant Institute of Himalayan Environment & Development, Kos-Katarmal, Almora, pp. 1-15.

World Bank, 2007. Development and growth in Northeast India: The Natural Resources, Water and Environment Nexus. World Bank report No. 36397-In. the International Bank for Reconstruction and Development/ The World Bank, DC, USA.

Yamada, T., 1998. *Glacier Lake and its outburst flood in the Nepal Himalaya.* Monograph 1, Tokyo. Japanese Society of Snow and Ice. Data Centre for Glacier Research.

Hon'ble Reviewer

1. **Dr. A. Arunachalam**
   Director General
   ICAR
   Krishi Bhawan
   NEW DELHI - 110 001
2. **Prof. Vinod Agarwal**
   Chairman
   Department of Geology
   MLS University
   UDAIPUR - 313 001

Pre-Harvest Crop Forecasting Using Satellite Data: A Case Study of Gwalior district of Madhya Pradesh

P.K. Verma[1], G.D. Bairagi[1], Mirdula Singh[2] and R. Sharma[1]

[1]Remote Sensing Application Centre, M. P. Council of Science and Technology, Bhopal - 462 003, MP, India.
[2]Dhirubhai Ambani Institute of Information and Communication Technology, Gandhinagar, Gujarat, India
E-mail: bairagigd@gmail.com

Abstract

The pre harvest forecast is important for the post-harvest grain management practices and capacities for many reasons including the achievement of food security objective. This study reports pre harvest classification approaches of wheat and mustard crops for production forecasting using LISS-III data of peak vegetative growth stages of mustard and wheat crops and multi-date AWiFS data. The study area was Gwalior district of Madhya Pradesh. Multi-date AWiFS data classification technique was based on two stage classification viz. unsupervised Iterative Self Organizing Data Analysis Technique (ISODATA), labeling of classes based on temporal spectral profiles and creation of non crop mask by decision rule classifier. While the techniques used for single date LISS-III data analysis was based on supervised classification using ground information as a digital layer. An inter-sensor comparison of acreage estimates for wheat and mustard crops shows that the acreage estimates using LISS-III single date data was having -2.53 percent under estimation when compared with AWiFS multi-date estimates for wheat crop, while it was -2.12 for mustard crop. The yield prediction models have been developed based on historical NDVI and Agro- met indices. The NDVI based regression model has R^2 0.77 and Agro-met indices based model has R^2 0.59. The yield predictions using NDVI and Agro-met based yield models were 2627 and 2728 Kg/ha respectively. It is concluded that the differences in acreage of wheat and mustard crops were higher than 5% when estimated through coarse resolution (56 m) data. In case of LISS-III data, the deviations in acreages between sample segment and complete enumeration were negligible magnitude (2.5%). The single date LISS-III data of maximum growth stage is performed better than multi date AWiFS sensor data because of its higher resolution.

Keywords: Remote Sensing, Satellite Sensor, Acreage Production Models.

Introduction

Agriculture is the backbone of Indian economy, providing livelihood to about 67.0 percentage of the population. Food grain production has increased from 51 million in 1951 to 241.56 million tons during 2010-11. On the other end Indian population reached 1.21 billion and needs for food grain is raising calling for efficient agricultural management involving appropriate application of production and conservation practices. Monitoring of crop conditions is important for the economic development of any nation. The use of remote sensing has proved to be very important in monitoring the growth of agricultural crops. The production of crop and prediction of crop yield have direct impact on year-to-year national and international economies and play an important role in the food management (Hayes and Decker, 1996). Using remote sensing data, efforts have been made to develop various indices such as: normalized difference vegetation index (NDVI), vegetation condition index (VCI) and temperature condition index (TCI). These indices are commonly used for crop acreages and production estimation, drought detection, monitoring exces-sive soil wetness, assessment of weather impacts on vegetation and evaluation of vegetation health (Unganai and Kogan, 1998; Kogan, 2002; Kogan; Singh et al., 2003). The NDVI data have been used extensively in vegetation monitoring, crop yield assessment and forecasting (Hayes et al., 1982, Benedetti and Rossinni, 1993).

The remarkable increase in remote sensing technology and its application during the last three decades have firmly established its immense potentials for mapping and monitoring of various natural resources. Satellite remote sensing and geographical information system offer great promise for natural resource management with the ability to depict the spatial distribution of extend and monitoring capability. These techniques have the potential to predict and zonate different levels of crop response to the inputs and can also provide solution to various management problems in increasing the performance of cropping system in spatial and temporal dimension, when coupled with the relevant ancillary information, besides giving information on the acreage and production of the crops. Remote sensing data, both optical and microwave form the core of crop area enumeration, crop condition assessment and production forecasting. Temporal remote sensing data is being used to monitor the crop through its growing period. Vegetation indices and weather parameter derived from surface and satellite observations being used to develop the crop growth monitoring system. The study aims to forecast the crop using satellite data for (pre-harvest) mustard and wheat crops. It deals with the difference in crop inventory output when estimated through single and multiple date data using various approaches.

Material and methods

Study area

Gwalior district was the study area which located at 26° 04' N to 26° 21' N and 78° 01' E to 78° 20' E. Its location map is shown in Fig. 1. Gwalior has a sub-tropical climate with hot summers from late March to early July, the humid monsoon season from late June to early October and a cool dry winter from early November to late February. The climate is humid subtropical. Gwalior gets 970 mm of rain every year. Gwalior lies in the Grid Agro-ecological Zone of Madhya Pradesh which spreads in an area of 3.74 m ha, about 8.46 percentage of the total geographical area of Madhya Pradesh. The physiography is alluvial highlands. The soils are shallow to medium sandy to loamy, grayish brown. Average annual potential evapo-transpiration is 1400-1900 mm. Net sown area in Gwalior is 2601.42 sq. km. Major crops are Wheat, Mustard, Gram and major horticulture crops are Potato, Onion and Guava.

Data used

Historical district-level statistics published by Land Records Commissioner (LRC), Gwalior, MP from 1998-99 to 2009-10 for wheat and mustard were used in yield analysis. Meteorological variables, minimum and maximum temperature (fortnightly) for last thirteen years from 1998-2011 has been taken for the development of agro meteorological yield models. Single date LISS- III data coinciding with maximum vegetative growth stage and Multi-date AWiFS data of wheat and mustard crops were acquired for acreage estimation. The details of the satellite data acquired for acreage estimation is given in Table 1.

METHODOLOGY

Crop Acreage Estimation

Crop discrimination is based upon the fact that each crop has unique spectral signature. The first step for crop acreage estimation is to identify the crop cover. Spectral vegetation indices which are primarily combination(s) of red (0.63-0.69 μm) and near IR (0.7-1.3 μm) bands reflectance were utilized for crop forecasting, crop condition assess and also for crop identification (Bauer, 1985). The spectral index (NDVI) was used are given below:

Normalized Difference Vegetation Index (NDVI) = (NIR-Red/ NIR+Red)

Ground Truth (GT) was collected during maximum vegetative growth stages (flowering stage) of mustard and wheat crops. The 20-25 percent area of the district had visited and agronomic observations like crop, crop type, field size, growth stage, crop density/vigour, irrigation, moisture stress, approximate sowing and harvesting date, latitude, longitude etc. using GPS were recorded for each sample segment. To verify the crop in the image with actual ground condition a digital coverage of the district was created using Arc-info.

Table 1. Satellite data used

Satellite	Sensor	Path & Row	Date of Pass	Crop
ResourceSat P-6	LISS-III	97-53, 98-53	29.12.2010	Mustard
ResourceSat P-6	LISS-III	97-53, 98-53	27.01.2011	Wheat
ResourceSat P-6	AWiFS	97-53 or/ and 98-53	23rd Oct, 30th Nov, 20th Dec, 13th Jan, 05th Feb, 10th Feb.	Wheat and Mustard

Digital Analysis

LISS-III Sensor data analysis

In this approach, the administrative boundary of the study district was overlaid on the geo-corrected image to extract the entire pixel belonging to the district. Ground truth sites for crop(s) and other land cover classes were identified on the image and training site pixels were extracted from these windows. Several training windows for each class were selected to take care of different conditions of crop and also to account for its regional variability. Selections of the suitable training windows are based on the spectral signatures, histograms of gray values and the spectral seperability. Training statistics are finally generated for each class and were used for classification. The Test of statistical class severability apart from using the error matrix (derived from omission and commission errors obtained by classification of ground truth pixels), the Bhattacharya distance (B) was used as a measure of the statistical seperability to test the seperability between various pairs of classes and refining training signatures. The refined training statistics generated using the multi-bands data were used for supervised classification using maximum likelihood (MXL). In this procedure, district crop map showing spatial distribution of different crops and land cover had been generated.

Multi date AWiFS Sensor data analysis

Geo-referencing: The multi-date AWiFs data set used in this study was geo-referenced using image-to-image registration with existing master images. Each date RS image was registered with an already geo-referenced master image of the area. Multi-date RS data analysis needs a high degree of accuracy in image-to image registrations. All the registrations were carried out to meet the accuracy requirement in terms of root mean square error less than 0.5 pixels. Gwalior district was extracted out by overlaying the boundary vector. The data was stacked in layers and AWiFS time series NDVI data preparation was done using modeler. Normalized Difference Vegetative Index (NDVI) derived from the radiance value of the Red and NIR spectral bands as given below:

$$NDVI= L_{NIR} - L_R / L_{NIR} + L_R$$

LNIR is radiance value for NIR band and LR is radiance value for RED band. The value of NDVI ranges from -1.0 to +1.0 but for simplicity, it was scaled in range o to 200 multiplying by 100 and adding 100 as given below:

Scaled NDVI = NDVI*100+100

ISODATA clustering and assigning of classes

Using modeler and NDVI profile of land use classes, the non crop classes have been extracted. The existing forest mask image and all the classes with in the 'non-crop class' were used to create non crop mask. The new NDVI image had created which was excluding non crop mask. Tish image had used in two-stage ISODATA clustering. There is several unsupervised classification algorithms commonly used in RS data analysis. Two of the most frequently used algorithms are the K-mean and the ISODATA clustering algorithm. Both of these algorithms are iterative procedures. The ISODATA algorithm allows for different number of clusters while the k-means assumes that the number of clusters is known a priori. The standard ISODATA clustering algorithm was chosen as the unsupervised classification method for this dataset. In the present study, two-stage ISODATA clustering has been carried out. In the first stage, time-series NDVI image is subjected to ISODATA clustering with 50 clusters. The unsupervised clusters were compared with the ground truth database to assign classes. The unknown classes are assigned to two categories i.e. pure 'non-crop class 'and 'crop class'. The final classified image was prepared by assembling all the pixels belonging to the three classes ('non crop class including' includes water bodies, urban, fellow, waste lands, scrub and forest, 'Mustard class', 'wheat class' and 'other crop') by taking union of the classified image of second stage of ISODATA clustering.

Crop Yield Estimation

The crop yield was estimated using agro meteorological and spectral indices based modeling approaches. The MaxT, MinT, MeanT and TD for the different growth stages were calculated. Various possible ways of using meteorological data, observed yield data and vegetation indices (NDVI) for crop yield modeling have been attempted. Linear and multiple- linear regression models have been attempted after examination of correlation for all the agro-met indices with observed yield. As per the correlation coefficient best agro metrological indices were selected for final crop yield modeling purpose.

RESULT AND DISCUSSION

Acreage estimation

Pre harvest acreage of mustard and wheat crops were obtained after digital image analysis of two dates LISS-III and multi-date AWiFS sensor's satellite data. The classification accuracy is taken care off for optimum seperability of the classes.

Mustard crop acreage estimation using LISS-III sensor data

The registered two scenes were joined using mosaic function then administrative boundary of Gwalior was overlaid on the data and bitmap of Gwalior was created. Using the bitmap entire pixels belonging to the district were extracted. The bitmaps of forest / scrub, urban, wasteland were created. These bitmaps were added up to form a non-crop mask to avoid mixing. New Image channels were added for creation of training site and classified output. Training sites were added for classification and classes were generated, while class generation the Bhattacharya distance (B) was used as a measure of the statistical seperability to test the seperability between various pairs of classes and refining training signatures, then optimal subset of signatures were classified using supervised classification. The classified output in the form of map is presented in Fig. 2. The image was classified in nine different land cover classes. Forest constitutes the maximum land cover followed by waste land. The district's 15.90 percentage area was fallow, mustard wheat and other crops covers 20.75 percent area. Since mustard was at peak vegetative growth stage in 29th December 2010. The acreage estimates for mustard was 59.99 thousand ha. Area statistics for crop classes is given in the Table 2.

Table 2: Wheat and Mustard crop area statistics using LISS-III Sensor data of 29 December 2010.

S. No.	Class	Area (000' ha)	District area Coverage in %
1	Wheat	20.42	4.63
2	Mustard	59.99	13.60

The classification accuracies tested showed that the non-cropped classes had remarkable separablity and water has maximum classification accuracy of 99.26 percent. None of the non-crop classes was mixing with wheat and mustard crops. Within the cropped classes other crop shows relatively less classification accuracy of 94.74% and wheat was having classification accuracy of 95.41% and mustard has 96.2% classification accuracy.

Wheat crop Acreage estimation using LISS-III Sensor data

Raw data for the date of pass 27th January 2011 was taken and geo-referenced using the geo corrected image of 29th December 2010 taking as the master image. The image of Gwalior district was extracted. The same bitmap created using 29

December 2010 data were used for the further analysis. The previous date image fallow area has changed in cropped area. There had been scattered distribution of scanty clouds which covered very less area coming under crop. A separate class of cloud was also generated. The classified output presented in the Fig. 3. The image was classified in nine different classes. Land cover pattern for the classes wasteland, scrub, forest, urban, water was the same as previous date classified image. There has been a reduction of 2.28 percentages in acreage for mustard crop area as compared to the previous date while the crop area for wheat increased by 73.69 percentages to a value of 77.64 ha. The crop cover classification details are given in Table 3.

Table 3. Wheat and Mustard crop area statistics using LISS-III Sensor data of 27 January 2011.

S. No.	Class	Area (000'hac.)	Cover%
8	Wheat	77.65	17.60
9	Mustard	58.62	13.29

Acreage estimation using AWiFS data

The six dates AWiFS data was used for the analysis. A stacked NDVI image was generated and by this NDVI stack image, Low NDVI classes (non-crop) were separate out and the existing forest layer was used to discriminate forest cover within the district. The digital ground truth information was overlaid on the stacked image and using the stacked NDVI image AOI layers were created for ground truth of each class. The AOI was depicted in the form of NDVI profile as given in Fig. 4.

The time series NDVI profile depicted (Fig. 4) for the identification of different land cover classes. The profile was evident that the non-cropped vegetative classes were showing declining trend. Differentiation amongst the lower NDVI non crop classes was done by the help of ground truth also. Among the crop early and late sown mustard and wheat has very good separability (Fig. 4). Early sown wheat crop was distinguished by a relatively higher scaled NDVI values for the first date, peak value in the fifth date which declined in first date of February. In case of late sown wheat there has been a declining trend for the first three dates with a scaled NDVI value below 100 featuring no crop. From the fourth date there was an increasing trend with peak value till the last date. By study of the NDVI profile a NDVI mask of non- crop classes including forest, wasteland, scrub, water, urban and zero fill was created. Afterward the mask was recode to generate crop/mixed pixels NDVI image. On the NDVI output two-stage ISODATA clustering has been carried out. In the first stage, the data was subjected to ISODATA clustering with 50 clusters were created (unsup-1). The unsupervised clusters were compared with the ground truth database as well as generated land cover profile to assign classes. As a result of which eight classes viz. waste land, mustard1, mustard2, wheat1, wheat2, Fallow, other crop and mixed were recorded. In the second stage, the (unsup-1) mixed was subjected to ISODATA clustering resulting in unsup-2 which consisted of five pure classes i.e. wasteland, mustard, wheat, fallow and other crop. Unsup-2 and unsup-1classified images and non-crop classes merged to generate classified image as shown in Fig. 5.

As shown in Fig. 5 the forest covers the maximum area which is 31.44 per cent of the district's geographical area followed by waste-land which is 22.14 percent. Among the non-crop classes urban and water contributes only 0.55 and 0.714 percent of the total area. The Acreage under wheat and mustard crop constitutes 18.14 and 14.08 percent of the total area (Table 4). An acreage map for wheat and mustard crop is presented in the Fig. 6.

Table 4: Mustard and wheat crop Area statistics using multi-date AWiFS data

S.No.	Class	Area 000' hac	Cover %
6	Wheat	80.16	18.14
7	Mustard	62.11	14.08

Comparison of Acreage estimates among different sensors

Multi date AWiFS area statistics when compared with LISS-III satellite data estimates, it was observed that there was over estimation of 2.53 percent and 2.12 percent in acreage for wheat and mustard crops respectively in the AWiFS results (Table 5). This indicates the difference in the resolution capacity of the two sensors. AWiFS sensor cannot distinguish all features as finely as LISS-III.

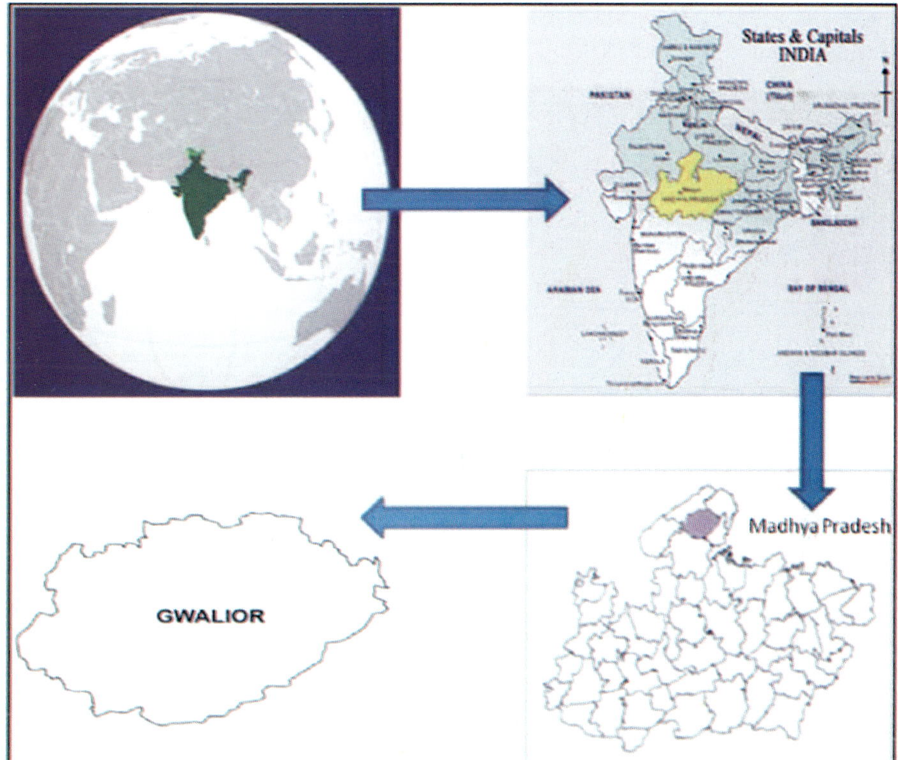

Fig. 1: Location map of Gwalior district of MP.

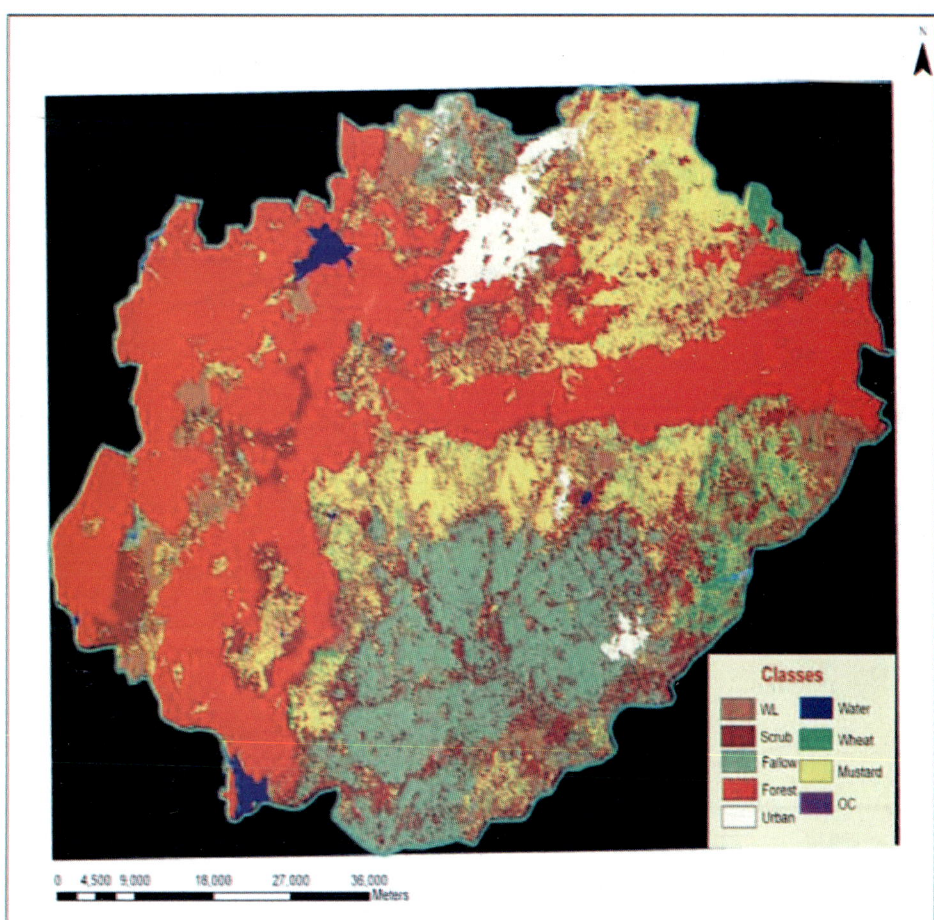

Fig. 2: Land cover map of Gwalior district (MP) using LISS-III satellite data of 29.12.10

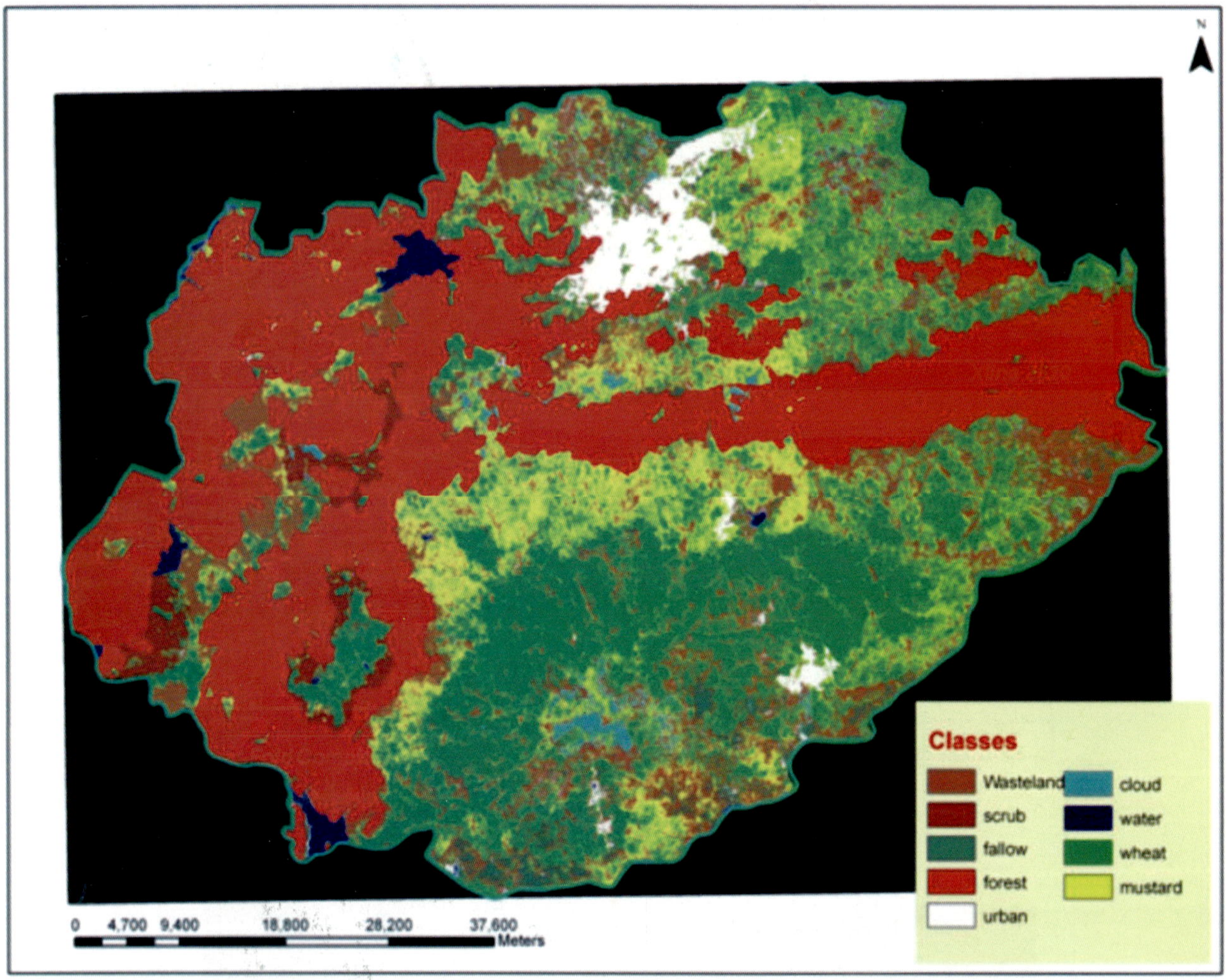

Fig. 3: Land cover map derived using LISS-III Sensor data of 27 January 2011.

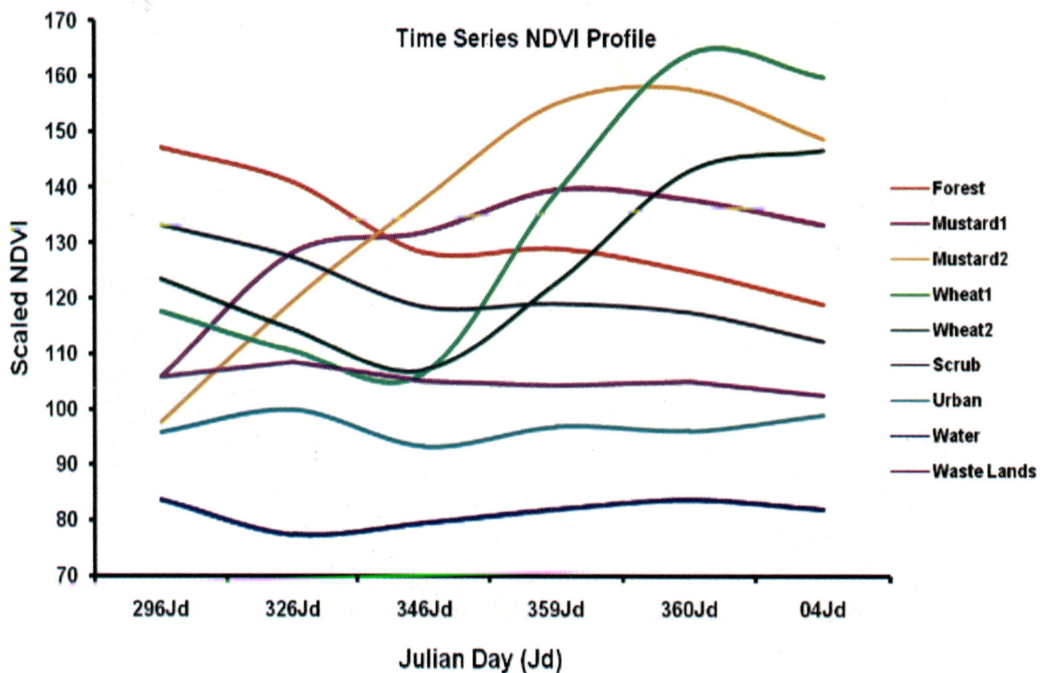

Fig. 4: NDVI profile using multi-date AWiFS satellite data.

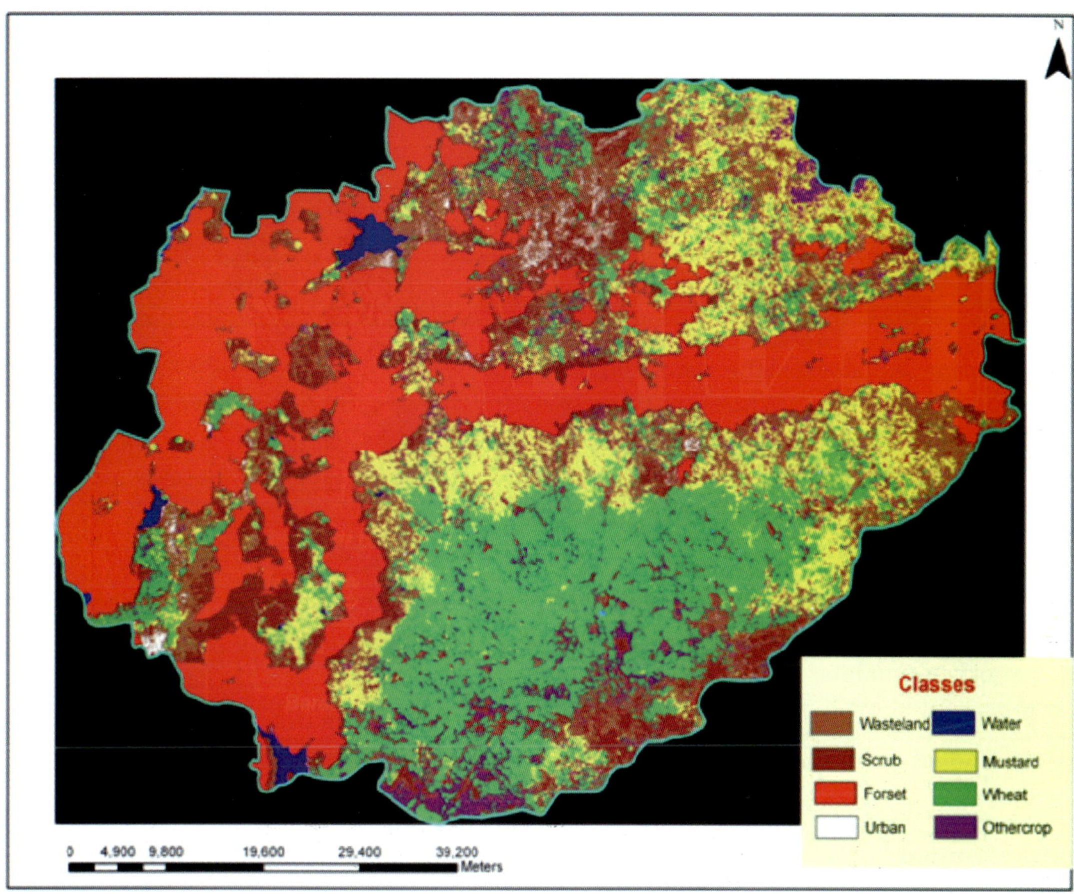

Fig. 5: Land cover class map of Gwalior district using multi-date AWiFS data

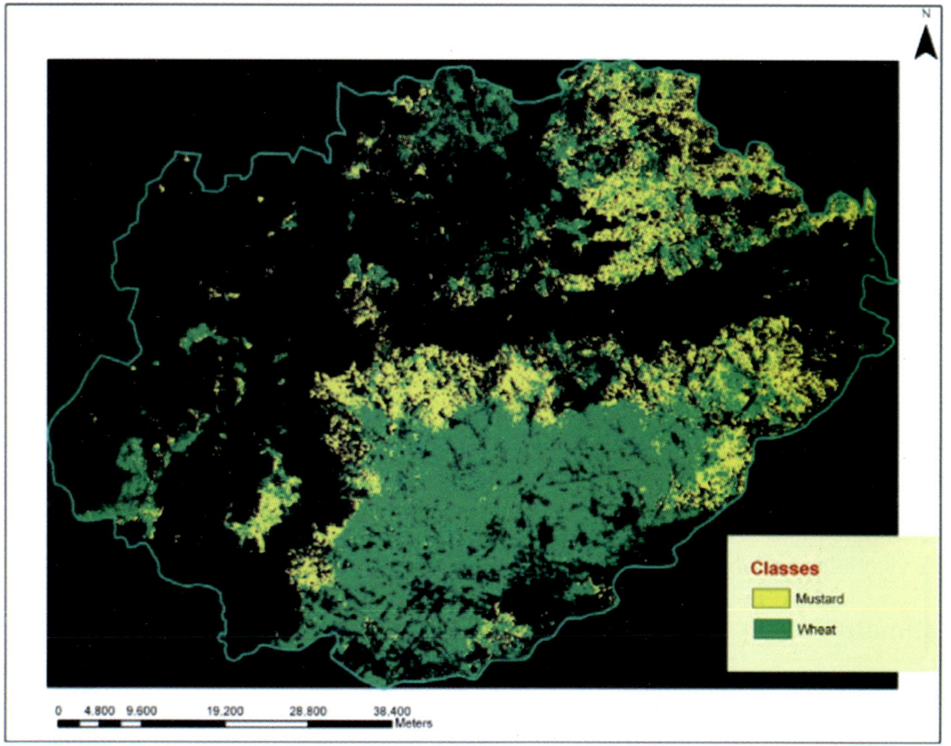

Fig 6: Mustard and wheat crop map using multi-date AWiFS data.

Table 5: Comparison of LISS-III vs. AWiFS acreage results.

Crop	Acreage (000' ha.)		RD%
	LISS-III	AWiFS	
Wheat	77.64	80.17	-2.53
Mustard	59.99	62.11	-2.12

Yield prediction

Yield prediction has been done using NDVI yield model for wheat crop. While in case of mustard no model has been found to be significant so that average of yields for last twenty years has been taken for yield predictions. The detail of NDVI model is presented in the Table 6. Yield predicted for wheat through NDVI model is 2627 kg/ha. R^2 value is 0.793 which makes it most significant as compared to other models. NDVI relation is depicted and presented in the Fig. 7.

Table 6: NDVI yield prediction model for wheat crop.

Crop	Input	Description	R^2	SEE Kg/ha	Yield Kg/ha
Wheat	NDVI	Y= 1118.81+ (3459.9* NDVI)	0.793	181.1	2627

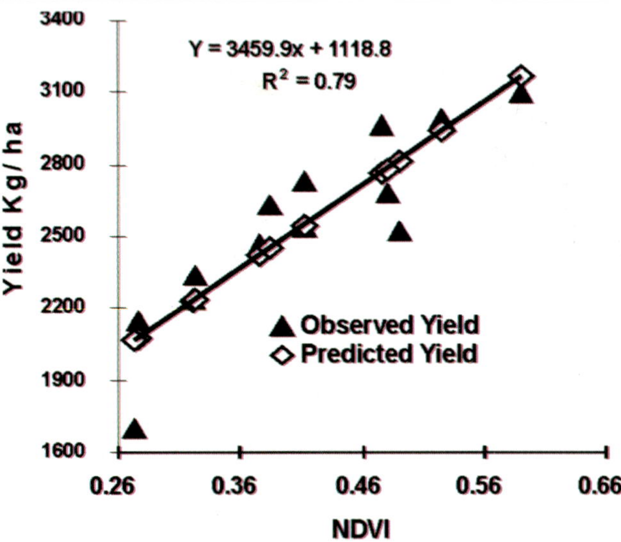

Fig. 7: NDVI vs. yield relation for wheat crop.

Production estimate

From the estimated crop area and predicted yield crop production was computed as the product of the two. Production of wheat and mustard was calculated for both sensor satellite data. For LISS-III data production estimation as per the approach was also taken care off. Production results are given in the Table 7.

Table 7: Production estimate using LISS III and AWiFS satellite data for Rabi 2010-11.

No.	Crop	Area (000' ha.)	Yield kg/ha	Production (000'Tones)
	Using LISSIII Sensor data			
1	Wheat	77.64	2627.00	203.96
2	Mustard	59.99	878.00	52.67
	Using AWiFS Sensor data			
1	Wheat	80.17	2627.00	210.60
2	Mustard	62.11	878.00	54.53

SUMMARY AND CONCLUSION

The results obtained in the study and the possible conclusions that can be drawn are: based on the class cover statistics while considering class discrimination using multi date AWiFS data, it is possible to know the various phenological stage of the crop and hence it is possible to discriminate crop more easily though hierarchical decision rule and ISODATA clust-ering (unsupervised classification) based hybrid approach, which relaying on phenological difference in crops in time evaluation NDVI. In case of non-crop areas which are quite homogeneous it has been possible to differentiate better with single date LISS-III data.

Using complete enumeration technique as supervised classification approach with the integration of non crop bitmaps in the single date LISS-III data the acreage estimates obtained were 77.64 thousand ha and 59.99 thousand ha for wheat and mustard crops respectively. Classification accuracy for the crop classes were 96.2 and 95.41 percent for the respective crops. The acreage estimates using hybrid technique for multi-date AWiFS data classification, which employ integrated two-stage ISODATA clustering and generation of non crop mask using hierarchical decision rule classifier for wheat and mustard crops were 80.17 thousand ha and 62.11 thousand ha respectively.

The production forecasting, the yield prediction through modeling has been done. The NDVI based regression model has been found to be the most significant with the R^2 value of 0.79 and yield 2627 kg/ha. For agro-met yield model R^2 is 0.59 and yield was 2249 kg/ha. The NDVI yield model being more significant hence, production estimate was given using NDVI yield model. Production estimates of single date LISS-III data for wheat and mustard crop were 203.96 and 52.67 thousand tones while for Multi- date AWiFS data the production estimates were 210.60 and 54.53 thousand tones respectively. When compared to LISS-III production estimate an increase of 1.46 percent in wheat and 3.41 percent in mustard crop were observed.

The crop forecasts using satellite data and ground truth information can be obtained one month before harvest. This information is very much useful for post harvest management viz. future planning of storage, export, import management and to ensure food security.

REFERENCES

Agarwal, R., (2003), Forecasting techniques in crop, Library avenue, I.A.R.S.I: 1-2.

Bairagi, G.D., Hassan, Zia-UL (2002), Wheat crop production estimation using satellite data, Journal of Indian Society of Remote Sensing, Vol. 30: 213-218.

Bauer, M.E. 1985. Spectral inputs to crop identification and condition assessment. Proc. IEEE 73: 1071-1085.

Benedetti, R., Rossinni P., (1993), on the use of NDVI profiles as a tool for agricultural statistics: the case study of wheat yield estimate and forecast in *Emilia romagna*, Remote Sens. Environment, 45:311–326.

Cochran WG. Sampling Techniques (1977), John Wiley & Sons: New York. Pp: 74-76.

Hayes, J.T., O'Rourke, P.A., Terjung, W.E., Todhunter, P.E (1982), Yield: A numerical crop yield model of irrigated and rain fed agriculture. Publications in Climatology, p. 35.

Hayes, M.J., Decker, W.L (1996), Using NOAA AVHRR data to estimate maize production in the United States Corn Belt. Int. J. Remote Sens. 17: 3189–3200.

Kang Xiaofeng, Wang Naibing, Yang Xiaohuan (2002), Progress of Information Extraction Methods for Crop Planting Area, Resources Science, 9(5):8-12.

Prasad, Anup K., Chai Lim, Singh, R. P., Kafatos, M. (2004), Crop yield estimation for Iowa using remote sensing and surface parameter, International Journal of applied earth observation and Geoinformatics (2006),8:26-27.

Parihar, Jai S., Oza, Markand P. (2002), Agriculture and Hydrological application of Remote sensing, proceedings of the SPIE, Vol.6411:641101.

Sesha Sai MVR, Hebbar, R., Krishna R.MV (2004), Remote sensing based crop Inventory for agriculture draught assessment, workshop on regional workshop on agriculture draught assessment and monitoring using space technology,25-32.

Unganai, L.S., Kogan, F.N., (1998), Drought monitoring and Corn yield estimation in Southern Africa from AVHRR data, Remote Sens. Environ. 63: 219–232.

Wu Bingfan (2000), Operational Remote Sensing Methods for Agricultural Statistics, Acta Geographica Sinica, 55(1):25-35.

Hon'ble reviewer

1. **Prof. A.S. Faroda**
   Former Vice Chancellor
   MPT University
   UDAIPUR - 313 001
2. **Prof. G.K. Panda**
   Department of Geography (DRS)
   Utkal University
   BHUBANESWAR - 751 004

IV-51

Use of Geo-Informatics for Combating Desertification in Bellary district, Karnataka

P.S. Dhinwa

Space applications Centre (ISRO), Ahmedabad
E-mail: dhinwa_ps@sac.isro.gov.in

Abstract

The objective of United Nations Convention to Combat Desertification (UNCCD) is to combat desertification and mitigate the effects of drought. Desertification, the land degradation in dryland region, has become an alarming global environmental issue and it needs immediate attention to combat desertification with proper preventive measures. In Bellary district, Karnataka, the natural resources are depleting very fast due to anthropogenic activities. In this research paper, an effort is made to generate locale specific action plans for land resources development based on a decision rule for Bellary district in Karnataka. Initially thematic maps are prepared by using multi-date satellite data and ancillary information and by employing visual analysis and interpretation technique. Later on these thematic maps are converted into spatial data base through the use of GIS software. Finally, Composite Land development Units (CLDUs) are generated by integrating land use/ land cover, land capability, slope and ground water prospect maps in GIS environment. Based on decision rule, action plans are generated for land resources development. The analysis of multi-date satellite data reveals that vegetal degradation, water erosion, manmade (mining/quarrying), salinity/alkalinity and wind erosion are the significant desertification processes observed in Bellary district. The measures suggested for combating desertification are: a) afforestation with suitable soil and water conservation measures in forest region, b) shelter belt plantation to prevent wind erosion in sand dune area, d) crop rotation along with growing of pulses and gram for improving fertility of soil in crop land region, e) proper irrigation management and drainage treatment for area affected by salinity/alkalinity and f) judicious use of fertilizers in irrigated agricultural land.

Keywords: Geo-informatics, Desertification, Land Resource, Integrated Land Development

Introduction

Land degradation occurs all over the world, but it is only referred as desertification when it takes place in drylands. This is because these areas are especially prone to more permanent damage as different areas of degraded land spread and merge together to form desert like conditions. 70 per cent of these dry lands are affected by degradation, which support over 1 billion people in more than 110 countries. The process of desertification, in which land is degraded by decreased rainfall resulting from various factors such as climatic variations and human activities, has become a major focus of scientists to understand the earths' climatic system (Grotinger, et al. 2003).

The 1977 Nairobi conference of UNCOD (United Nations Conference on Desertification) describes desertification as the diminution or destruction of the biological potential of the land, which can lead to desert like conditions. Today about 33 per cent of the earth's surface exhibits desert characteristics to some degree or other (Susan, 1998). Desertification is initiated when the vegetative cover is reduced by tree cutting, over grazing, over cultivation, shifting cultivation, mining, road construction, mismanagement of land and water practices, urbanization and other activities that disturb the natural ecosystem. There are two types of parameters that contribute to the process of land degradation. These are socio-economic conditions of the local population (poverty and illiteracy) and pressure of population on marginal lands. Both are interrelated. Basically, it is the increasing population (human and cattle) pressure which is responsible for disturbing the fragile ecosystem and exerts undue pressure on natural resources which activates the process of desertification and thus the fertile piece of land is converted into a desert. There are five types of indicators - climatic, hydrological, physical, biological and socio-economic which directly or indirectly contribute to desertification processes. For combating the processes of desertification in a dry sub humid region, not much work has been reported in the literature. Spatial inventory on desertification/land degradation status, a prerequisite for preparation of combating plan, has not been available for our country. Moreover, it has also been one of the major requirements towards fulfilling our commitment to TPN-1 UNCCD (Anon, 2007). Arable land per capita is shrinking throughout the world, threatening food security, particularly in poor rural areas and triggering humanitarian and economic crises. The per capita availability of cultivable land is declined from 0.48 ha in 1951 to 0.15 ha in 2001 and is projected to decrease to 0.08 ha by 2035 (Dwivedi et al., 2006). As per UNCCD estimates, arable land per person has declined from 0.32 ha. Per person in 1961-63 to 0.21 ha in 1997-1999 and is expected to drop further to 0.16 ha by 2030 (Kofi Annan, 2003). Because the poor often farm degraded land that is increasingly unable to meet their needs, desertification is both a cause and consequence of poverty. Fighting desertification must, therefore, be an integral part of our wider efforts to eradicate poverty and ensure long term food security. Let us today recommit ourselves to the goals of convention, and to achieving sustainable development for all, including in the dry land rural areas where the world's poorest people live (Kofi Annan, 2003).

Land and water are the basic natural resources for biomass production and constitute the core of ecological systems. The ecological systems, the very base of production is under is under great stress in many parts of the country due to growing pressure of human and livestock population and ever intensifying development demands. Forests are being denuded at alarming rates. Marginal lands are being cultivated and grasslands are being overgrazed increasingly. There is soil erosion in the upstream and sedimentation and flood in downstream part of a stream. Desertification is on the increase and the renewable resources are getting depleted. Large land area is classified as degraded land. The very resource base of production is thus under threat. The challenge is to make ecological resources more productive and their exploitation less hazardous. A development strategy based on integrated management of soil, water and other resources coupled with appropriate cropping and other agro techniques should be adopted for sustained production (Anon. 1995). Watershed constitutes the basic unit of development of rainfed areas. Unlike the earlier isolated efforts through soil and water conservation or crop production or other agro-techniques in a disjointed manner, the watershed approach aims at optimizing the use of land, water and vegetation in an area in an integrated way and thus helps alleviating drought, moderating floods, preventing soil erosion, improving water availability and increasing fuel, fodder and agricultural production on a sustained basis (Anon. 1995).

Remote sensing data along with GIS has been useful for desertification monitoring and assessment. The indicators of desertification amenable to remote sensing include salinity, erosion and sand sheets etc. (Navalgund, 2006)

As per the inventory of Desertification and Land Degradation Atlas of India, 105.48 mha. i.e., 32 per cent of the geographical area of the country is undergoing the process of land degradation (Anon., 2007). The indicators used for Desertification Monitoring and Assessment can be categorized into four types.

(A) **Pressure Indicators** characterize driving forces both natural and man-made, affecting the status of natural resources and leading to desertification. Pressure indicators are used to assess desertification trends and make an early warning for desertification. Natural indicators describe natural factors, mainly climatic conditions, natural disasters, which promote the occurrence and development of desertification. Non-natural indicators describe the pressure on land leading to land degradation from human activities (Ajai et al., 2003).

(B) **State Indicators** characterize the status of natural resources including land. The physical and biological features of desertified land ecosystems are the main factors to be considered. Physical indicators describe the land characteristics, physical and chemical properties of soil and hydrological features of the land ecosystem. Biological indicators are used to describe biological characteristics of the land ecosystem.

(C) **Impact Indicators** are used to evaluate the effects of desertification on human beings and environment.

(D) **Implementation Indicators** are used to assess the action taken for combating desertification and to assess its impact on natural resources and human beings. Such impacts refer to improvements of socio-economic and natural conditions. The positive changes in the land cover conditions owing to mitigation measures aimed at combating desertification. These are also sometimes referred to as response indicators.

Objective

The present study aims to generate locale specific action plans for combating desertification through the use of geospatial data and GIS techniques.

Study Area

Bellary district is located in the central eastern part of the Karnataka state with an area of 8463 km^2 and lies between 14^0 33' and 15^0 15' N latitude and 75^0 40' and 77^0 10' E longitude. It is bounded on the north by the Raichur and Koppal districts, on the west by the Gadag and Haveri districts, on the south by the Davangere and Chitradurga districts and the east by the state of Andhra Pradesh. Tungabhadra and Hagari are the major rivers of Bellary district. River Tungabhadra flows through the west and northern side boundary of the district. The area receives an annual average rainfall of 545 mm with a wide spatial variation, ranging from 815mm in the hilly tracts of Sandur Schist belt to 243 mm in the immediate east of the belt. The rest of the district has an average of about 425 mm rainfall. The temperature of the district varies from 23 to 42°C. The major rock groups of Bellary district are Dharwars, Peninsular Gneisses and Granites. Granite outcrops as insel bergs are prevalent in southeastern and eastern parts, whereas knobs are present everywhere.

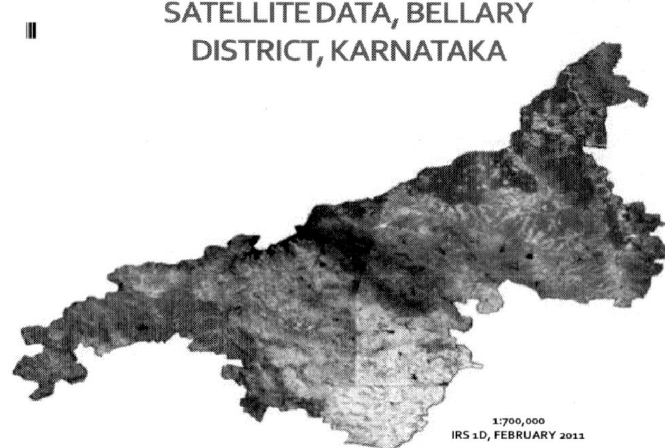

Red soil covers the entire peninsular gneisses granite complex; where as black soil along with a minor extent of characteristic ferruginous red soils cover the Dharwars. Red soils cover about 55% and black soils about 45% of the district. Both areas have distinctly different land scrapes mainly arising out of differences in water infiltration and run off of rainwater that often lashes the district torrentially whenever few times it occurs over the district. Since the black soils are the most productive agricultural lands, they are extensively and intensively cultivated and inhabited. Where as the red soils with occasional rock outcrops are less extensively and intensively cultivated, except in valleys. All the reserved forests and scrublands are seen in the red soils region of the district. Among them, the forests covering the Sandur Schist belt are still preserved, the rest of the forests are variously degraded.

Data Used

The IRS-1C/1D multi-date satellite data, ancillary information and collateral data has been used to prepare various thematic maps, to generate locale specific action plan maps for Bellary District, Karnataka and its details of the data are given in Table 1.

Table 1: Data Used

Satellite Data	Satellite ID	Sensor ID	Date of Acquisition
	IRS-1C/1D	LISS- 3	October 2001 February 2002
Ancillary Information	48 M/16 48N/9,10,13,14 57A/4,7,8,11,12,13,14,15,16 57B/1,2,5,6,9,10 57E/1,2,3,4 (1:50,000) Scale	-	-
Collateral Data	1.Population 2.Climate 3.Geology 4.Soils etc.	-	-

Methodology

Details of the methodology are given in Fig. 1. Visual analysis of multi-date satellite data along with ancillary information and collateral data were used to prepare various thematic maps, e.g. Land use/ land cover, ground water prospect, land capability and slope. Realising the importance of adopting an integrated approach, and recognizing the mutual interdependence of natural resources, thematic information is integrated in GIS environment. In order to integrate various themes, firstly, land use/ land cover layer is integrated with ground water prospect layer. The resultant of these two themes is unionized with slope and finally with land capability. The resultant coverage has the basic information of all the four themes- land use/ land cover, ground water prospect, land capability and slope and referred in resource data base (Fig. 10). Various map units known as composite land development units (CLDU) are created in this composite layer. Based on the CLDU characteristics various measures are suggested for conservation and protection of natural resources. The decision rules, given in Table 2 are used to suggest the appropriate action plan for each of the CLDUs (Anon. 1995).

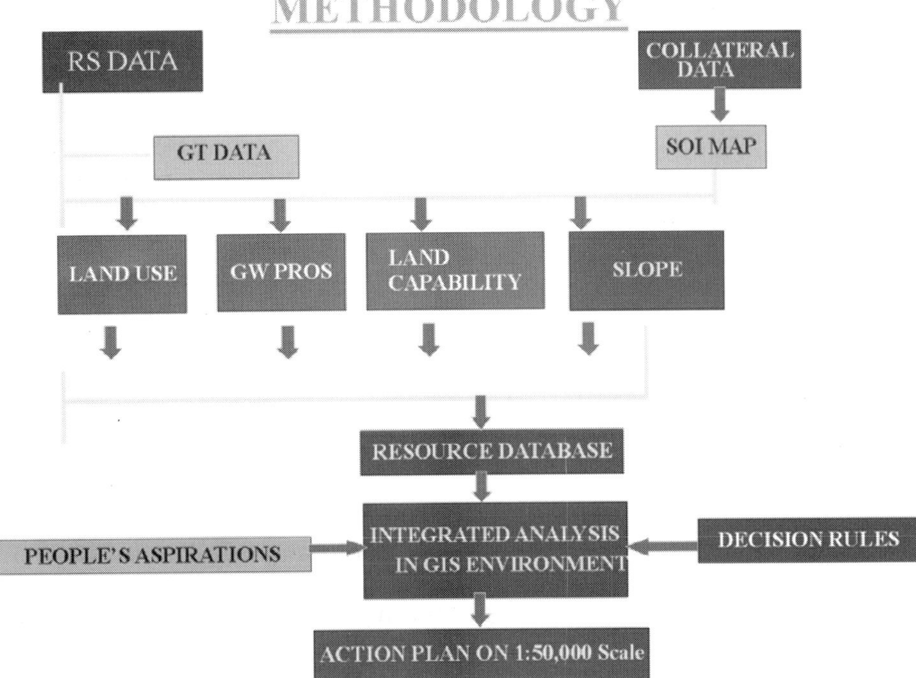

Fig.1: Flow Chart for generation of Land Resources Action Plan for Combating Desertification

Table 2: Decision rule for generation of action plan for combating desertification

S.N.	Land Use/ Land Cover	Ground water Prospect	Slope In per cent	Land Capability	Measures
1	Scrub Forest	Poor	5-10	VI	Afforestation with proper soil and water conservation
2	Kharif+Rabi, Dense Forest	Good	3-5	III	No action
3	Open Forest	Moderate-Poor	5-10	VI	Protection and Gap Filling
4	Open Forest	Moderate	>35	VI	Afforestation with staggered trenches
5	Crop land in Forest	Moderate	0-3	III	Agro-forestry with contour bunding
6	Kharif Crop	Moderate	0-3	III &IV	Double cropping with ground water exploration
7	Kharif Crop	Moderate	3-5	IV	Agro-horticulture with groundwater irrigation
8	Double Crop	Good- Moderate	0-1	III	minimal action - farm ponds
9	Land with scrub & Land without scrub	Moderate - Poor	10-15	IV	Silvipasture
10	Barren rocky/Stony waste/ Sheet rock area	Poor	5-10 & 10-15	VIII	Natural regeneration
11	Wasteland Salt affected land	Moderate	3-5	III	Drainage treatment
12	Open Forest	Moderate - Poor	3-5	VI	Afforestation with contour bunding

Results and Discussion

Various thematic maps, e.g., land Use/ land Cover, slope, land capability and ground water prospect maps are prepared by using multi-date IRS-1D, LISS-3 data and ancillary information as referred in Table 1 and are presented in Figures 2,3,4 and 5.The details of each map is mentioned below. Later on, spatial database is created for these themes GIS environment. After creating spatial database of various themes, these coverages are unioinised to create composite land development units. Based on decision rule, these CLDU's are converted into action plans for Land Resources Development as referred in Table 2. These action plans are later on transferred onto cadastral maps for implementation. The measures suggested for combating desertification are referred in Table 3 and shown in Fig. 6. The details of thematic maps prepared are as under:

Land Use / Land Cover Map

Land use/ land cover map is prepared by visual analysis of multi-date satellite data and through the use of ancillary information and collateral data. In Bellary district, Karnataka, there are twenty seven categories of level-III land use/ land cover types, namely, town/cities, habitation with vegetation, industrial area, villages, rabi crop, kharif crop, double crop agricultural plantation, fallow land, aquaculture pond, moist & dry deciduous open forest, degraded forest, mixed vegetation, forest plantation, prosopis juliflora, scrub forest, tree groves, barren rocky/ stony waste, sheet rock area, gullied/ravinous land, land with scrub, land without scrub, mining /industrial wasteland, salt affected land, sandy area, lake/pond, lake/tank (dry), reservoir, river/ stream,river island (Fig. 2). Land use/ land cover map gives an account of spatial distribution and areal extent of various categories of land use/ land cover. When it is integrated with other requisite themes, e.g., ground water prospect, slope and land capability, it gives a clear picture of availability of natural resources at a particular place and thus very helpful in suggesting measures for combating desertification at a specific locale (Fig. 2).

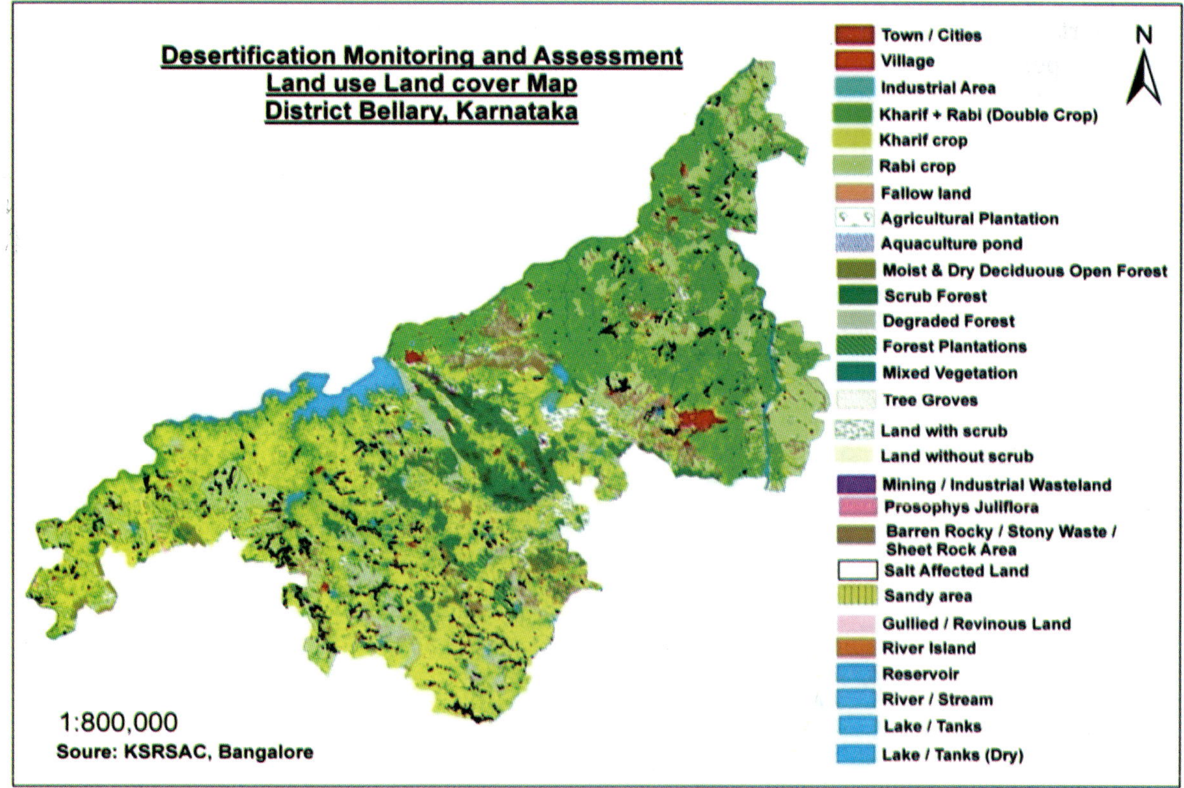

Fig.2: Land Use/ Land Cover Map

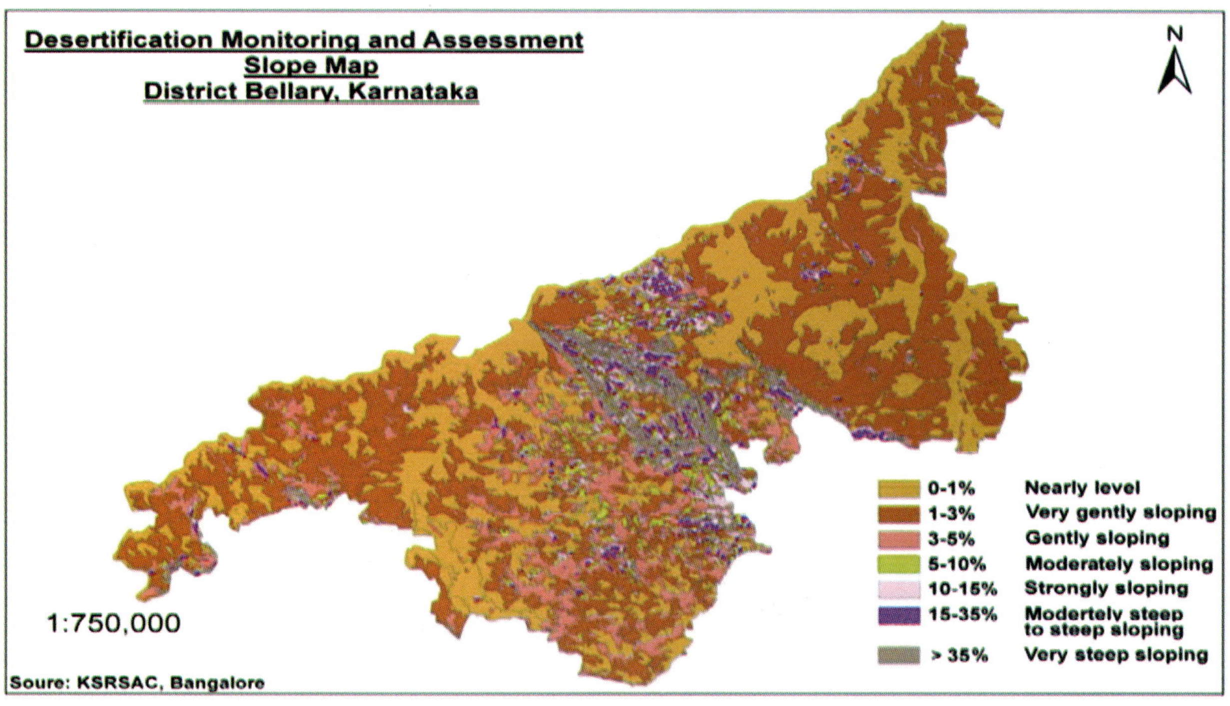

Fig. 3: Slope Map

Slope Map

Slope map is prepared by using Survey of India (SOI) topographical maps on 1:50,000 scale. It gives an idea terrain and topography of the area. In Bellary district, there are seven categories of slope. The area in the eastern and western part of Bellary district is under nearly level (0-1 per cent) and very gently sloping (1-3 per cent) slope category and it constitutes eighty per cent of area of the district. Gently sloping (3-5 per cent) slope category lies in central and southern part of Bellary district. Moderately sloping (5-10 per cent), strongly sloping (10-15 per cent), moderately steep to steep sloping (15-35 per cent) and very steep sloping (>35 per cent) slope category is present in central part of the district where iron and manganese ores are available (Fig.3).

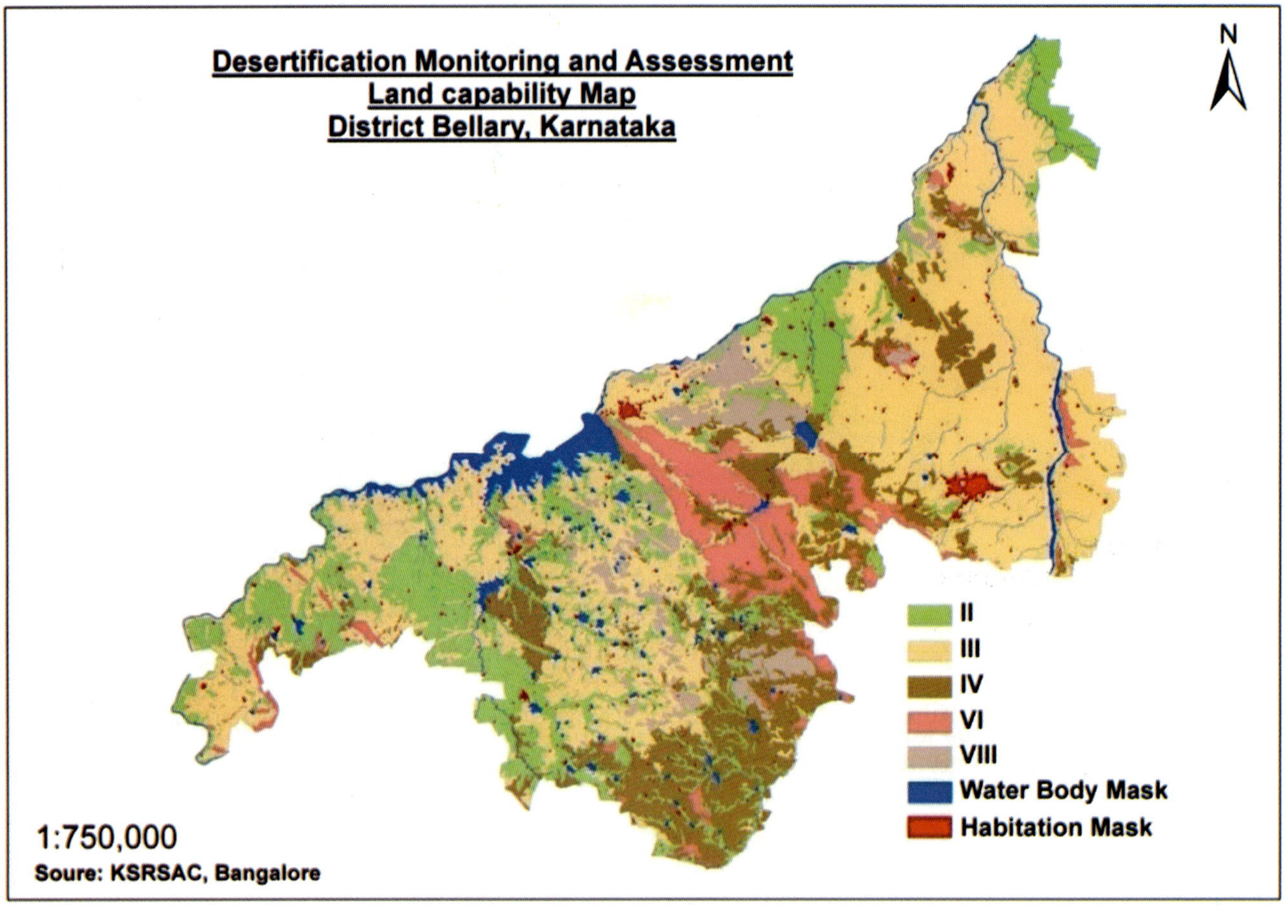

Fig. 4: Land Capability Map

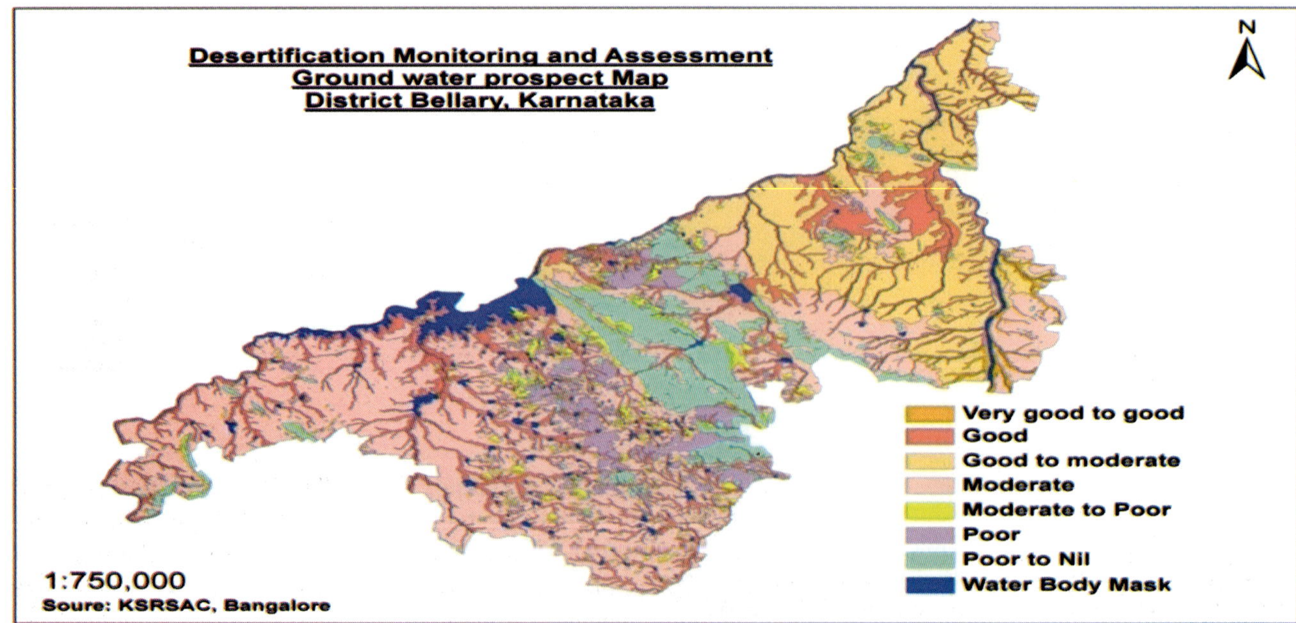

Fig. 5: Ground water Prospect Map

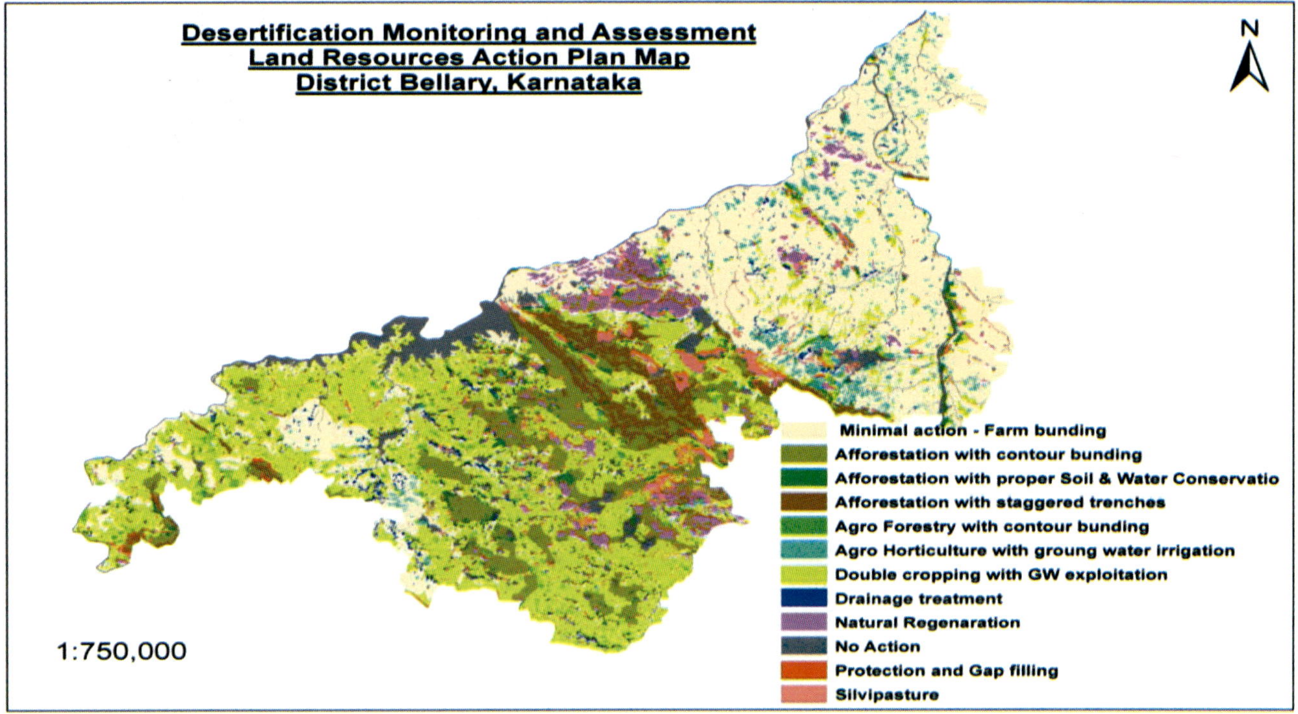

Fig. 6: Land Resources Action Plan Map

Land Capability Map

In Bellary district, five classes of land capability, i.e., II, III, IV, VI & VIII are reported. Classes II and III of land capability are present in eastern and western part of the district. Classes IV of land capability is mainly present in southern part of the district. Classes VI of land capability is reported in central part of the district whereas class VIII lies in northern part (Fig. 4). For suggesting measures for combating desertification, information about soil type and land capability is essential.

Ground water prospect map

Ground water prospect map is prepared using multi-date satellite data, ancillary information and collateral data and by employing visual analysis technique. In Bellary district, there are seven categories of ground water prospect, ranging from poor to very good. Sixty per cent area in Bellary district is under moderate category of ground water prospect. The region lying in east of the district is under good to moderate categoty. Good category of ground water prospect is observed in northern eastern part of the district. Poor and poor to nil category of ground water prospect is reported in central part of the district which is hilly.

Table 3. Area under different Action Plans

S. N.	Recommended Action	Area in sq.km.	Per cent of total area
1	Afforestation with proper soil and water conservation	149.69	1.77
2	No action	346.10	4.09
3	Protection and Gap Filling	612.15	7.23
4	Afforestation with staggered trenches	327.54	3.87
5	Agro-forestry with contour bunding	12.38	0.14
6	Double cropping with GW exploration	2325.84	27.50
7	Agro-horticulture with groundwater irrigation	184.93	2.18
8	minimal action farm pond	3312.54	39.17
9	Silvipasture	7.70	0.09
10	Natural regeneration	342.15	4.04
11	Drainage treatment	114.13	1.34
12	Afforestation with contour bunding	720.09	8.51
	Total	**8455.24**	**100.00**

Land Resources Development Plan

Based on the integrated analysis of above mentioned thematic layers and decision rules, the land resource action plans have been prepared (Dhinwa et al., 2010). The details are given as under:

1. **Afforestation with soil and water conservation measures** - Appropriate soil conservation and development practices play a key role in maintaining overall produc-tivity in a watershed and maintaining the ecological balance as well (Anon., 1995). Afforestation with proper soil and water conservation measures (boulder checks) at a distance of 500 metres from ridgeline up to foothill region is suggested for scrub forest, where ground water prospect is poor, slope category is moderately sloping (5-10) and land capability is class III. It reveals that in this area anthropogenic forces are very active hence dense forest is reduced to scrub forest which is prone to soil erosion and hence afforestation with proper soil and water conservation measures is suggested.
2. **No action-** is suggested for the region which consists of dense forest, water bodies and habitation in land use/ land cover.
3. **Protection and gap filling-** is suggested in open forest region of Protected Forest, where slope is moderately sloping category (5-10), ground water prospect varies from moderate to poor and lies under class VI of land capability.
4. **Afforestation with staggered trenches.** This type of trenches are best suited to plant trees and protect existing plantations from soil erosion and conserve moisture and plant nutrients for growth (Subramanian et al., 1987). Afforestation with staggered trenches is suggested for area under mixed vegetation and open forest, very strong sloping category (>35 per cent), ground water prospect varies from poor to nil and lies under class VI of land capability.
5. **Agro-forestry with contour bunding** is executed in lands where the slope is less than 6 per cent and which receives an annual rainfall of 1000 mm or below. When contour bunding is combined with conservation farm practices the yield of crop is increased because it helps to prevent soil erosion (Subramanian, et al., 1987). For imparting stability and providing sustainability to the farming systems in marginal areas, a tree – cum- crop farming system could prove more useful (Anon., 1995). Agro-forestry with contour bunding is suggested for Crop land in forest.
6. **Double cropping with ground water exploration** is suggested for agricultural land presently under Kharif crop,with moderate ground water prospect., is under gently sloping (3-5) slope category and belong to class III & IV of land capability.
7. **Agro-horticulture with ground water irrigation** This system plays an important role in those regions, where production of annual crops is inefficient. Fruit trees, if suitably integrated, would add significantly to overall agricultural production including food, fuel and fodder, conservation of soil and water and stability to production and income (Anon., 1995). Agro-horticulture with ground-water irrigation is suggested for moderately sloping agricultural land presently under Kharif and with moderate ground water prospect. It is suggested for the region under Kharif crop in land use/ Land cover, is gently sloping category of slope class (3-5) has moderate ground water prospect and belong to class IV of land capability.
8. **Minimal action – Farm ponds** and is suggested for area under double crop, where ground water prospect varies from good to moderate, is under nearly level(0-1) and very gently sloping (1-3) slope category and it lies class III of land capability.
9. **Silvipasture-** This system essentially consists of top feed tree species and grasses or legumes) preferably perennial) as understorey crop. This system is most suited to marginal drylands and is preferable where fodder shortages are experienced frequently (Anon., 1995). Silvipasture is suggested for a region under land with scrub and land without scrub, where ground water prospect varies from moderate to poor, is under strongly sloping (10-15) slope category and it lies class IV of land capability.
10. **Drainage treatment** is suggested for the region which is under salt affected class of land use/ land cover, where ground water prospect is moderate, is under gently sloping (3-5) category of slope and belong to class III of land capability.
11. **Natural Regeneration** is suggested for area under barren rocky/ stony waste/sheet rock area class of land use/ land cover, where ground water prospect is poor, is under

moderately sloping (5-10) and strongly sloping (10-15) category of slope and belong to class VIII of land capability.

12. **Afforestation with contour bunding.** Afforestation is a process of transforming an area into a forest where trees have not been grown previously. Afforestation is done with a view to check soil erosion and conserving water (Jain, 2006). Afforestation with contour bunding is suggested for the region which is under crop land in forest class of land use/ land cover, where ground water prospect is moderate, lies under gently sloping category of slope and belongs to class III of land capability.

Conclusion

Remote sensing data along with GIS has been useful for desertification monitoring and assessment. It may be concluded that Remote Sensing and GIS are very important and promising techniques helpful for preparation of an inventory of natural resources, creation of spatial database, understanding the status of severity of processes of desertification and for generation of eco-friendly action plan for combating desertification. Significant Measures suggested for combating desertification are i) minimal action – farm bunding (39.17 per cent), Afforestation with staggered trenches (29.68 per cent), ii) double crop with ground water exploration (27.50 per cent), afforestation with contour bunding (8.51 per cent), Afforestation with proper soil and water conservation measures (1.77 per cent), iii) Protection and gap filling (7.23 per cent), iv) Agro-forestry with contour bunding (05.23 per cent) and Agro-horticulture with groundwater exploration (05.23 per cent), no action (4.09 per cent), drainage treatment (1.34 per cent), natural regeneration (4.04 per cent), silvipasture (0,09 per cent).

Acknowledgements

The author is grateful to Dr. A.S. Kiran Kumar, Director SAC, Dr. J.S. Parihar, Deputy Director, EPSA, Dr. Ajai, Group Director, MPSG, and Dr. A.S Rajawat, Head, GSD for their valuable guidance and moral support in preparation of this research paper.

References

Ajai, Arya, A.S. Dhinwa, P.S. (2003), Desertification: The creeping Disaster, NNRMS Bulletin (28), pp. 52-64.

Anon. (1995). Integrated Mission for Sustainable Development – Technical Guidelines, National Remote Sensing Agency, Hyderabad.

Anon. (2007). Desertification and Land Degradation Atlas of India. Space Applications Centre, (ISRO), Ahmedabad

Anon. (2007). Desertification Monitoring & Assessment using Remote Sensing & GIS: A pilot Project under TPN-1 UNCCD.SAC/RESIPA/MESG/DMA/ 2007/01 Space Applications Centre, (ISRO), Ahmedabad

Dhinwa, P.S., Pathan, S.K. and Ajai (2010),Action Plan for combating desertification in dry sub-humid region of Kathua district(J&K), Journal of Geomatics,Vol 4, No.1 ,pp.43-48.

Grotinger, J., Jordan, T.H., Press, F., Siever, R. (2007), Understanding the Earth.W.H. Freeman and Company, New York.pp.453-471

Dwivedi et al., (2006).Proc. of Regional Workshop on Deserts and Desertification, at Hyderabad during Dec. 29-30, 2006), p.31

Kofi Annan, (2003). Message by UN Secretary General KOFI ANNAN for the World Day to combat desertification and drought, June 17, 2003

Jain, R,K., (2006), Geography 10,Ratan Sagar P. Ltd, New Delhi, p.240

Navalgund,R.R.(2006), Geoinformatics in Disaster Monitoring and Mitigation,ISG Newsletter, Vol. 12, No. 1 & 2, pp.4-23

Subramanian, S., Kothandaraman, G.V., Natarajan, S and Ramaswamy, P.P., (ed. 1987). Soil Survey and Land Use Planning for watershed management, Directorate of Soil and Crop Management Studies, Tamil Nadu Agricultural University,Coimbatore,

Susan Farrel (1998), The World Environment Research Directory,Pira International.

Hon'ble reviewer

1. **Prof. Damdodar Sharma**
   Former Vice Chancellor
   Rajasthan Technical University
   KOTA
2. **Dr. S.M. Ramasamy**
   Vice Chancellor
   Gandhigram Rural University
   GANDHIGRAM - 624 302

Occurrence of Conglomerate in Bar- Mohra Khurd- Raira Khurd Area of Pali district, Rajasthan: Its Significance and their Relation with Soil and Vegitation

Beena Tripathi

Department of Geology, Jai Narain Vyas University, Jodhpur-342005
E-mail beenagrabo@gmail.com

Abstract

A metasedimentary sequence has been identified between the overlying schistose formation of Sendra Formation of Delhi Supergroup and the underlying Banded Gneissic Complex of Pre-Delhi age in Bar–Mohra Khurd–Raira khurd areas in north-western Rajasthan. The sequence is important due to its heterogenous lithologic association represented either by different type of conglomerates imbedded in quartzofeldspathic schist. This sequence has been formally defined as the 'Bar conglomerate horizon' in this paper. Once the non-random nature of the conglomerate pebbles distributions has been established, it become necessary to find out the way in which the distribution deviates from random. In particular it is necessary to decide whether the different pebble types are more regularly or more grouped than would be expected in a random pattern. However, controversy remains regarding the exact chronology of deformation. At the same time metamorphic history of the area needs further revision. In view of the above, the authors have chosen Bar area for the present study. The soils derived from the parent rocks sustain different vegetation depending upon the climatic conditions. Thus nutritional status is required to be studied for the point of view of sustenance of flora over the area.

Key words: Delhi Supergroup, Structural Features, Bar Conglomerate, Soil, Vegetation, Plant Nutrition.

Introduction

This paper presents a synoptic view of the more significant geological, structural features and their relation with soil and vegetation of the Delhi Supergroup, as an introduction that would help to place in a proper perspective the several aspects of Delhi Supergroup that are discussed in this paper. The western basin consists of two sub- basins namely Barotiya basin in the western part and Sendra basin in the eastern part. The Barotiya Sequence of Heron, 1953 is equated with Alwar Group whereas Sendra Complex of Heron, 1953 with Ajabgarh Group. The usage of Alwar and Ajabgarh series (Heron, 1953) for southwestern Rajasthan is now changed (Gupta et al., 1980). The main South Delhi Fold Belt (SDFB) is divided in two i.e., Barotiya- Sendra in the western side and Rajgarh- Bhim basin in the eastern side. Gogunda and Kumbhalgarh Group are used for Alwar and Ajabgarh respectively in SDFB.

Table 1. Lithostratigraphy of the Bar-Mohra khurd-Raira khurd area.

Group		Formation		Member / Bed (under present study)	
Heron (1953)	**Gupta *et al.* (1980)**	**Heron (1953)**	**Gupta *et al.* (1997)**		
Intrusive granite	Kumbhalgarh	Intrusive granite	Sendra	Intrusive granite	
Ajabgarh		Sendra Complex		Calc gneiss with Intercalated foliated quartzite Metmorphosed limestone Calc amphibole gneiss Foliated quartzite schist Calc gneiss	
Alwar		Barotiya Sequence	Barotiya	Kalab limestone Calc amphibolites schist Quartzitic schist Calc schist	
				Staurolite schist Garnetiferous mica schist Bar conglomerate schist Feldspathic schist with conglomerate	Bar conglo merate horizon
Pre Delhi		BGC	Kalakote	Granitic gneiss	

The rocks of the Bar- Mohra Khurd- Raira Khurd of tehsil Raipur in Pali district are included under the three main tectonic divisions from SW to NE viz. Banded Gneiss Complex (BGC), Barotia Formation (Barotiya Sequence) and Sendra Formation (Sendra Complex). The same name has been followed in this paper. The BGC constitute the southwestern and the lower most tectonic unit of the area. It is separated from the overlying rock of the Barotia Formation with an unconformity. The Barotia Formation consists of Bar conglomerate horizon, calc schist, quartzitic schist and calc amphibolite schist. Bar conglomerate horizon further divided in to feldspathic schist, Bar conglomerate schist, garnetiferous mica schist, staurolite schist and kyanite schist, out of these Bar conglomerate schist consists of different types of pebbles and boudins of quartzite, granite and pegmatite (Plate 1A&E). The overlying Sendra Formation constitutes the northeastern part of study area. Kalab limestone (equivalent to Nandana crystalline limestone of Heron, 1953) separates the upper part from the underlying Barotia Formation. The Sendra Formation is constituted of mostly gneisses viz. calc gneiss, foliated quartzitic schist, calc amphibolite gneiss, metamorphosed limestone and calc gneiss with intercalated quartzite. Huge intrusive granite exposed over the Sendra Formation in the northeastern part of the investigated area. There has been changed in earlier concept with recognition of several strike faults leading to a number of tectono-stratigraphic units (Mukhopadhyay and Bhattacharya, 2000). The depositional environment varies gradually from erenaceous to argillaceous in Barotiya and Sendra basin and changed in mixed calcareous to argillaceous. On the basis of lithological and structural homogeneity, strike persistence mappability Gupta et al. (1997) divided Kumbhalgarh Group in Todgarh, Sendra, Beawar, Kotra, Bar, Kalakote, Ras and Basantgarh formations. In this paper lithostratigraphy is described on occurrence of rocks starting from southwestern to northeastern direction.

LITHOSTRATIGRAPHY OF THE AREA

GRANITIC GNEISS (Banded Gneissic complex)

The Banded Gneissic complex (B.G.C) which, according to Heron (1953) forms the basement of the Delhi fold belt is exposed at Dholkot (26°8:74°7') and at the way in between Haziwas and Dhulkot. The huge sporadic exposures are also present throughout the road side from Dholkot to NW of Ramdev Temple (26° 7': 74°8') intruded by 5 to 10 cm thick veins of Pegmatite and lense shaped xenoliths. Further few exposures of granitic gneissic are also exposed on the western side of the Haripur crossing which is the extension of same granitic gneiss of Dholkot. The exposer of granitic gneisses are further extends upto Mohra Khurd and Raira Khurd where very huge exposures of granitic gneisses are present intruted by broad pegmatite veins. Numerous veins of the subsequent pegmatite veins cut through the granitic gneiss and are well exposed near Dholkot, Haziwas and Bar villages. Crystal of garnet, beryl, tourmaline with muscovite and biotite are present in pegmatite veins throughout the horizon.

BAROTIA FORMATION

BAR CONGLOMERATE HORIZON

Feldspathic schist with conglomerate

A thick pile of exposure of Bar conglomerate is found in NE-SW direction over whole length of study area. The strike length of the conglomerate

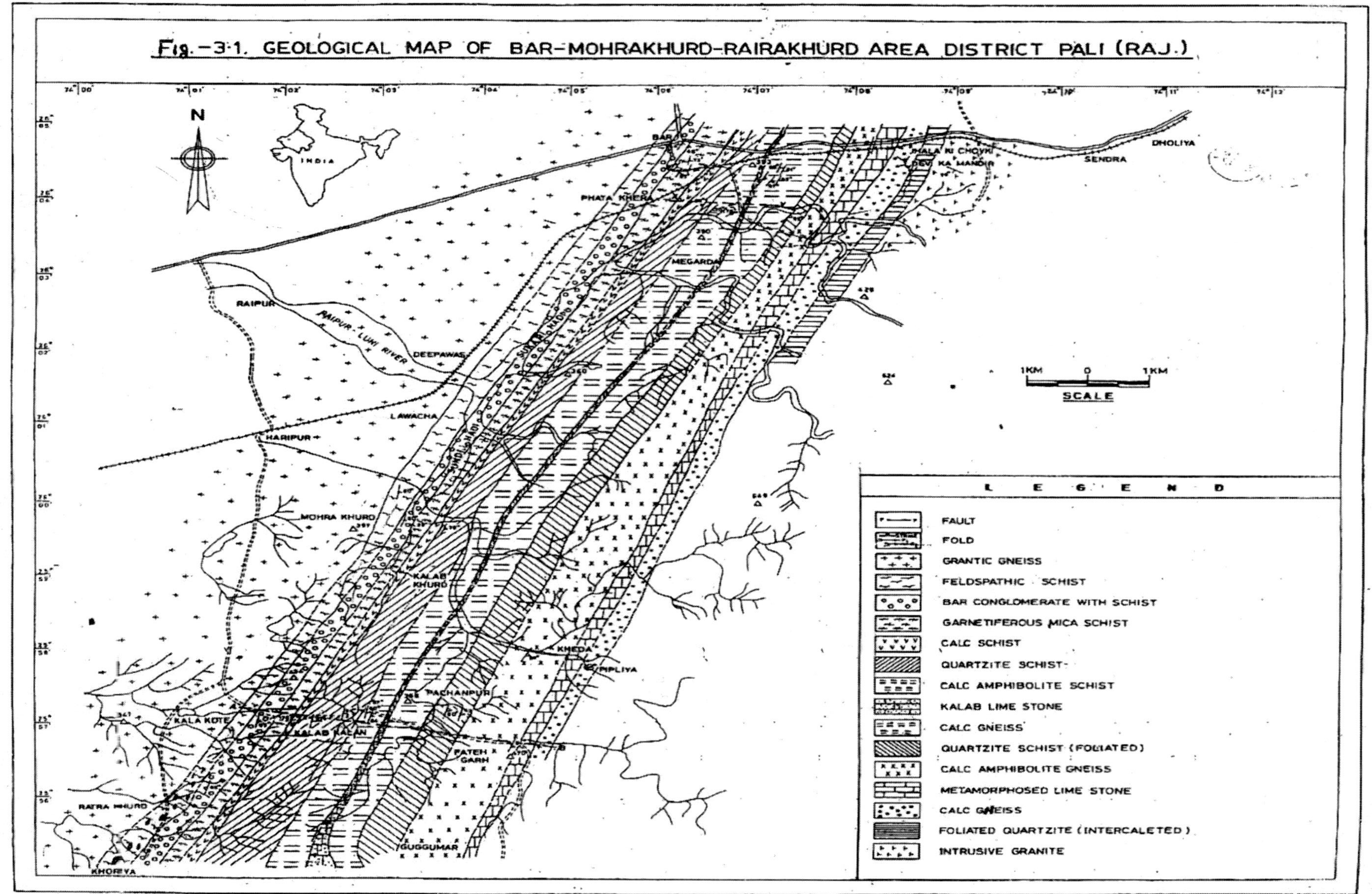
Fig.-3·1. GEOLOGICAL MAP OF BAR-MOHRAKHURD-RAIRAKHURD AREA DISTRICT PALI (RAJ.)
N
INDIA
BAR
PHATA KHERA
KALA KI CHOKI
DEV KA MANDIR
SENDRA
DHOLIYA
MEGARDA
RAIPUR
RAIPUR - LUNI RIVER
DEEPAWAS
LAWACHA
HARIPUR
MOHRA KHURD
KALAB KHURD
KHEDA
PIPLIYA
PACHANPUR
KALA KOTE
KALAB KALAN
FATEH GARH
RATRA KHURD
GUGGUMAR
KHOREYA
1KM 0 1KM
SCALE
L E G E N D
FAULT
FOLD
GRANTIC GNEISS
FELDSPATHIC SCHIST
BAR CONGLOMERATE WITH SCHIST
GARNETIFEROUS MICA SCHIST
CALC SCHIST
QUARTZITE SCHIST
CALC AMPHIBOLITE SCHIST
KALAB LIME STONE
CALC GNEISS
QUARTZITE SCHIST (FOLIATED)
CALC AMPHIBOLITE GNEISS
METAMORPHOSED LIME STONE
CALC GNEISS
FOLIATED QUARTZITE (INTERCALETED)
INTRUSIVE GRANITE

horizon is about 20km from Bar (26° 7':74° 8') to Raira Khurd, although this formation is also exposed in the northern side from Bar to Giri village. The width of the formation is 250 to 300 meter except where the rock is covered by soil; therefore, it appears that width of Bar conglomerate is variable. Heron (1953) recognized this conglomerate horizon as the basal part of the 'Alwar Series' and noted that these rocks grade upwards into biotite schist. The presence of thick, massive and different size conglomerate bearing granule, pebble and cobble size rounded to subrounded grains in the Bar conglomerate horizon is suggestive of changed palaeotectonic setup and shallow water depositional environments. At some places strained pebbles forming boudins in Bar conglomerate near railway bridge of Bar village where the length of the boudins is approximately from 5 cm to 15cm.

Rounded to subrounded granoblastic grains of dirty white quartzite, spotted white granite and greyish schistose, however, might have been derived from Barotia Formation very well exposed around Bar, as the pebbles are highly resistant of weathering and restricted to transportation for considerably long distances (Plate 1A).

The present conglomeratic horizon which supplied flattened pebbles of Barotia Formation. However, experience severs impact of diastrophism and continued to suffer mechanical changes without chemical decomposition. This is reflected by the interbedded flattened pebbles in micaceous matrix with thin laminae of quartzofeldspathic material.

The rocks dipping Southeastward at 80° to 87° appears almost vertical. They are very well bedded, slabby and almost schistose type. The colour is dark grey to black with sharp foliation planes arrange parallel with each other. The conglomerates are imbedded in Quartz mica schist matrix. Quartzitic pebbles are abundant with a few granitic pebbles. The granitic pebbles are more at Giri and Phatakhera. The occurrence of pebble is not uniform throughout the rock formation. The concentration of pebble is variable in both along and across the strike of bed. At Giri no pebble can be seen beyond 15 cm length along with 2 to 5 cm width in between concentrated patch of pebbles. The tectonically deformed boudins also found at Birantiya khurd and Bar village (Plate 1B). The boudins are mainly of light grey to white coloured quartzite of consistent composition. The post-deformational quartzitic pebbles are dominated over pebbles of granite and pegmatite at Phatakhera. The pegmatite vein is intruded within garnetiferous mica schist along the strike of the bedding plane and at some places pebbles are deformed and distorted near Mohra khurd (Plate 1C).

The mica schist which is the host rock of pegmatite veins contains bands and partings varying from 2mm to 5 cm in width. The conglomerates are imbedded within mica schist along bands of quartzite. The axial ratio of elongated fish shaped conglomerate pebbles varies from 5:2:1 to 14:7:1 with shortest axis perpendicular to strike of the bedding, longest axis parallel to dip of bed and intermediate axis lying parallel to the bedding plane. Near Railway Bridge, Bar conglomerate formation is having bands of quartzite occasionally this quartzite thickens and somewhat resembles stretched pebble, practically devoid of pebbles.

Bar Conglomerate and Schist

The conglomerate upwards passes to mica schist in east. At Phatakhera and north of Megarda village conglomerates suddenly ceases and mica schist starts at weathered indistinct contact. This contact is generally not traceable in the northern part due to presence of sand and soil. The mica schist contains garnet visible in hand specimen. Garnet diminishes from west to east across strike of bedding. Thin pegmatites have intruded this formation frequently. The alternate beds of quartzite and mica schist are folded near Megarda village where anticline and syncline are well exposed. Inclined anticline and syncline are also exposed near Nanana village whereas disharmonic folds are developed at Panchatiya village. The rock is highly folded, crenulated and puckered (Plate 1D).

Garnetiferous-Mica-Schist

Localized exposures of garnetiferous mica schist are present at a places between Phatakhera and Gundabera seen which are highly schistose, dirty green and grey to brown in colour. Garnet, biotite, muscovite, quartz, plagioclase and chlorite are the main constituents of the rock. Between Bar conglomerates and BGC the formation consisting of muscovite-biotite schist is found from south of Birantiya khurd to Ramdev temple. This formation has not been noticed at Bar market and further north beyond

Bar only sporadic narrow strips is found at Megarda village. The rocks are invaded by numerous pegmatites and quartz veins around Bar and Megarda village (Plate 1E). Crystals of garnet, beryl, tourmaline with muscovite and biotite are present in pegmatite veins throughout the horizon. Here thickness of pegmatite veins varies approximately from 5 cm to 50m. The granitic gneiss near Ramdev temple grades into schists with garnet and further towards east quartzitic schists with pegmatitic intrusions grades into mica schist. The mica schist lies west to Bar conglomerates. But this association is not visible at Bar and at south Phatakhera due to severe erosion and deposition of sand and soil as a Bar market has developed over the formation and at Megarda village agriculture field has developed. Between Kalakot and Raira Khurd pure quartzite patch is found in this formation.

Staurolite-Schist

The garnetiferous mica schist is followed by the staurolite bearing mica schist and is well exposed near Hugaliya and Pachar villages. It is associated with garnetiferous mica schist and Rampur village. The rock type shows conspicuous banding consisting of coarse micaceous layers and quartzofeldspathic layers. Two set of foliation plans are observed. The garnet and staurolite are placed in the form of porphyroblasts. Microscopically, the staurolite appears in subhedral to anhedral forms. The staurolite- garnet-biotite schist is lithologically similar to those of the garnet-biotite schist. The regional strike of the rock is NE-SW, (Northeast-Southwest) but at places it swings and changes to nearly North-northeast-Southsouthwest as is noticed near the Hugaliya village.

The staurolite-garnet-biotite schist is highly schistose, dirty green, grey to brown in colour. Staurolite-garnet, biotite, muscovite, quartz, plagioclase and chlorite are the main constituents of the rock. The rock is highly folded, crenulated and puckered, thin but distinct light and dark colour bands are clearly observed.

Calc-Schist

The staurolite schist is succeeded by impure calcareous schist of variable thickness and very well folded and formation is exposed at Bar, Phatakhera and Kalab Kalan (N25^0 57.087' E74^0 01.664'). This formation is 100m thick at Kalab Kalan and thickness gradually increases towards northeastern side where it measured 430m at Bar. The rock is very thinly laminated, mesoscopic F^2 folds are present but not very common. The folds vary in shape from symmetrical to overturn at Kalab Kalan. In the northeast of Bar towards Barotiya Khurd, felsic intrusions of granite are also exposed.

The rock is dark grey in colour, highly cleaved, fine grained and dark in colour. The impure calcareous schist is interbedded with quartzofeldspathic schist. The foliation in calc schist is defined by dimensional parallelism of lensoid grains of calcite and preferred orientation of mica. The main component of rock is calcite, dolomite, quartz, albite, muscovite etc. The presence of argillaceous impurity introduces by mica and plagioclase into low to medium grade calcareous rock.

Quartzite-Schist

The quartzite unit is best exposed in the central part of the study area near Jirara village and extends upto National Highway NH 14. The quartzites of Barotia Formation are of variable nature. The quartzite are an integral part of investigated area interstratified with biotite muscovite schist and garnetiferous schist all along the National Highway the quartzite are white in colour, fractured and show poor development of schistosity. The quartzite have vitreous fracture surface, in which two set of fractured planes are common along which the quartzite has broken in massive and rectangle to square shaped blocks in which vertical fracture planes are more dominant than horizontal fracture plane. Granoblastic texture is common but pulverization and crushing effects are insufficiently developed. They often show alternating black and grey colored bands indicating the original bedding composition.

Further in southeastern part, 4 km short of Khejara village, 10 cm to 15 cm thick bands of quartzites are placed with 5 cm to 8 cm thick bands of schistose and micaceous beds with thin bands of meta volcanic which are 4 cm to 7 cm in thickness. Due to compression, variable thickness and hardness, the rock becomes folded fractured and schistosity is highly disturbed. Anticlines and synclines are placed simultaneously at village Tiroliya Asan. Heron (1953) classified this quartzite unit as the part of Barotia Formation.

Plate: 1 A) large granite pebble in Bar Conglomerate Horizon at Bar lying parallel tobedding schistosity. B) Strained pebble forming boudin in Bar Conglomerate Horizon near phatakhera.C) Localized strongly deformed pebbles due to intrusion of pegmative vein on Kalakote-Kakab kalan road. D) Crenulations are developed due to intrusion of pegmatite vein at Giri dam. E) Distortion in quartz vein at magarda village.F) Folding in calcarious amphibolite gneiss showing differential weathering.

Calc-Amphibolite-Schist

The Quartzite is succeeded by the amphibolite schist along the national highway NH14, where huge exposures of amphibolite schist is best exposed intercalated with thin quartz veins. It extends between Ramdev temple and Jhala ki Chowki and further extends upto Nokhari village. The regional strike trend of amphibolite schist is Northnorth east to Southsouth east and the dip amount is almost vertical nearly 80° to 87°, it has been recorded near Choukitole and Kalanada villages. The amphibotite schist is dark green to blackish green in colour. Due to intensive compression the rock become folded and schistosity is disturbed (Plate 1D) and open type of anticline and syncline are formed. Thin but distinct veins of white quartz occur in association with the amphibotite schist. At several place quartz veins are also folded and faulted.

Kalab Limestone (Nandana Limestone)

Calc gneiss is interbedded with limestone termed as Nandana limestone by Heron may be after Nanana village. The limestone is grey and fine grained. It appears to be argillaceous near Fatehgarh and siliceous in north of NH14. The width of limestone formation is more in north NH 14 than at Fatehgarh. At Fatehgarh the width is hardly 50 to 70 meter. There is open quarry near Mata ka ghar. The Nandana limestone is succeeded by calc gneiss of higher stratigraphy and degree of metamorphism.

SENDRA FORMATION

Calc-Gneiss

Calc gneiss is dirty white to pale in colour very hard, compact and course grained. Dark blackish green coloured dykes are present on Bar – Sendra road. Kyanite is well exposed at Gundabera along with calc gneiss. Two sets of fracture planes are present in quartzofeldspathic amphibolite gneiss with differential weathering, which is also very well exposed near Gunda Bera.

Foliated Quartzite Schist

Calc gneiss is overlain by quartzite which is foliated in nature and exposed on eastern boundary at Pachanpur (N25^0 56.886' E 74^0 03.421'). The quartzite of area occurs as small linear outcrop, which can be traced maximum upto 600 m width. Only scanty exposures are available at Pachanpur and at northwest of Debi ka mandir (N26^0 04' 36.9 E74^0 07' 47.4). In most of the portion of study area this rock is concealed under sediments supporting agriculture and good vegetation growth. So, no true estimate of its thickness is possible. Near Pachanpur a local quarry exists on this rock for building material purpose. Thickness of bed at Pachanpur is 600m whereas; it is 200m near Debi ka mandir.

The dark grey colour, coarsely crystalline bands are siliceous in nature on the western side but gradually composition changed to quartzitic. On NH 14 (Bar-Sendra road) near Debi ka mandir, there is a narrow band of pure quartzite which is dull white in colour. The quartzite is thinly foliate, pinkish white to grey in colour and medium to fine grained. The quartzite has vitreous luster and conchoidal fractured surface, in which two set of fractured planes are common along which the quartzite has broken in massive and rectangle to square shaped blocks in which vertical fracture planes are more dominant than horizontal fracture plane with weathering to a rusty brown.

Calc-Amphibolite Gneiss

In calc- gneiss are succeeded by the calc amphibolite gneiss in the northeastern part of the study area at Nokhari village (N26^0 9' E74^0 11'). A huge exposer of calc-amphibolite gneiss is present on the path between Nanana (26^0 10' E74^0 11') and Nokhari villages. The rock is marked by alternate light and dark coloured and thick and thin bands of amphibolite and calcareous material. The sizes of the bands are about 3 mm to 1cm. Because of differential weathering the amphibolite bands are distinctive and embossed. The grayish to blackish-green colour bands are not smooth and the margins are rough and uneven. Thin but distinct layers of white calcareous material occur in association with amphibolite. The hardness of the amphibolite is much harder than calcareous material so amphibolite bands are bulging out and due to intensive compression protuberant figures are formed in the form of folds (Plate 1F), fractures and faults. A fault plane is found to run within fold, it is quite distinct and displacement is clearly observed along the fault plane. The calc-amphibolite gneisses are highly fractured and jointed. Straight, thin vein of quartz is intruded within calc amphibolite gneiss, which have being followed to long distance, can be measured in meters suggest post deformation and post deposition. Some fracture planes are filled by secondary material like quartz which are concordant in nature and are showing post-tectonic intrusion.

Metamorphosed Limestone

The calc amphibolite gneiss is succeeded by metamorphosed limestone at Fatehgarh (N25^0 56.639' E740^0 4.104'). The limestone is exposed along the eastern boundary of calc amphibolite gneiss from Debi ka mandir (N26^0 04' 38.6 E74^0 08' 14.3). The width of limestone formation is more towards north near Debi ka mandir i e about 330m thick. An open quarry is also present near Debi ka mandir.

Limestone is harder and sometimes seems to be metamorphosed like marble but saccharoidal texture is not present. The limestone is light gray in colour with dark colour bands, but just a few meters ahead from previous location the colour changes to bright pink with lots of parallel fracture planes. It appears to be argillaceous near are Fatehgarh and arenaceous in north of NH14. Limestone is prone to weathering and leaching due to which it appears pale to brownish in colour. A weathered surface generally shows pits and cavities at many places. It gives effervescence, when treated with hydrochloric acid. Fossil remains are not found, even under microscope. Fracture planes are present, ranging from a few centimeters to tens of meter with no sign of dislocation

Calc-Gneiss (with intercalated foliated quartzite)

The calc gneiss is intercalated by quartzite near Jhala ki chouki (N26^0 04' 28.6 E74^0 08' 36.2). The thickness of the bed is 350m. The intrusion of Jetpur granite is found on its eastern boundary. The rock is dark greyish green in colour, course grained, well banded and highly jointed. The rock is hard and compact, showing light and dark coloured compositional layering. Thin but distinct layers of white calcareous material occur in association with calc gneiss. Calc gneisses are mostly composed of calc-silicate minerals, which are generally more resistant of weathering.

Under thin section, the quartzite shows granoblastic texture and consists of rounded to sub-rounded and equidimentional quartz grains. It contains quartz, biotite, sericite and iron oxide. Few grains are lenticular and are arranged parallel with each other. Poikiloblastic texture is also shown by few grains. Inclusions are sericitized and also contain fine specks of quartz; biotite, tourmaline and opaque include hematite and magnetite.

Intrusive Granite

The field observation of the granite at Jhala-ki-choki indicate that it is younger than the Sendra Formation and its fractured and jointed foliation and its general appearance suggests that it was affected by compression and therefore later than the granitic gneisses of Sendra Formation.

Lithologically, the rocks are very hard and compact and post Delhi Erinpura granite characters. Granite is grayish black, medium grained, spotted exhibiting lenticular xenoliths. The xenoliths are varying in thickness from 5 to 10 cms in length and maximum width at the centre is about 3.5cm. The groundmass contain quartz and microcline, a little albite much biotite with intruded zircons and muscovite. The albite and microcline are white in colour. Granites are well exposed in the northeastern part of the study area, where it is underlying the Sendra Formation along the National Highway 14.

Granite is very less fractured and feinted and shows less effect of crushing and pulverization indicating post deformation effect. Heron (1953) classified this granite equivalent to Erinpura granite.

RELATION WITH SOIL AND VEGETATION

The direct causes of land degradation in dry areas are the shoot out of drastic reduction of the perennial plant cover and simplification of the vegetation structure with a number of consequences for productivity, soil structure, water relations and microclimate (Le Houerou, 1996). In poorer regions like that in western Rajasthan in India, the process of soil formation by breaking up the rock particles and enriching the soil with organic matter from aerial and subterranean parts are influenced by biological activities (Loranger et al., 2002; Wardle et al., 2003; Ponge, 2003). However, the role of the vegetation is relatively less in arid zones because of the sparse canopy cover and the limited development of aerial parts. Nevertheless, the root systems often exhibit exceptional development and have the greatest influence on the cracking of the rocks influencing soils formation processes. The parent rocks had definite relations with the soil materials/nutrient and the vegetation it supports. But the studies on relation between parent rock material and soil and further correlation of these aspects with vegetation (tree, shrubs and undergrowth) supported by the soil is by and large lacking in western Rajasthan. This study therefore, high-lights by the fact that the vegetations including trees, shrubs and most importantly the herbaceous vegetation vary from soil to soil depending upon minerals present in the soil and these minerals are derived from the parent rock materials. The present study may bridge this gap in the Aravalli hill ranges occurs in semi arid region. Therefore, the objective of this study are (i) to study the type of soils formed under the influence of parent rocks found around the Bar region of Delhi Supergroup; and (ii) to investigate soil characteristics and nutrient availability in it for finding out its relation with the vegetation it supports.

FACTORS OF SOIL FORMATION

The physical disintegration of rocks leaves relatively large fragments, but it is the chemical weathering which breaks up these fragments. The process of chemical weathering in arid zones is slow because of water deficiency. However, extended periods of water deficiencies are important in the elimination or leaching of soluble salts, for which the accumulation is enhanced by the high evaporation. Short periods of water runoff do not permit deep penetration of salts, often resulting in accumulation of salts in the depressions of nearby areas. According to Dokurchave (1898) soils develop as a result of interplay of the parent material, climate, organisms and time. Jenny (1941) reported that the factors are independent variables, and developed equation to establish the dependence of certain soil properties on the soil forming factors:

$$\text{Soil} - f(cl, o, r, p, t..............)$$

Where cl = climatic factor, o = biotic factor, r = topographic factor, p = parent material; and t = time factor.

The material from which soils form is called 'parent material'. In the lower part of the soils, these materials may be relatively unchanged from when they were deposited due to moving water, ice, or wind. Soils weather directly from the underlying rocks and the residual soils have the same general chemistry as the original rocks. More commonly, soils form in materials that have moved in from elsewhere. Materials may have moved many kilometers or only a few meters. For example windblown loess is common in western Rajasthan. Sediments along rivers have different textures, depending on whether the stream moves quickly or slowly. Fast moving water leaves gravel, rocks, and sand, whereas slow-moving water and lakes leave fine textured material (clay and silt) when sediments in the water settled out. Rocks and minerals of varying composition produce different clay minerals, even after weathering in similar environment but the surface soil formed are thought to be very much related with above-ground vegetation (Emmer, 1994; Ponge et al., 1999; Kounda-Kiki et al., 2006).

The granites and gneisses produced kaolinite predominantly, whereas gabbro produced mostly montmorillonite clay mineral under similar environment (Barnhisel and Rich, 1967). Birkeland (1974) considered that mineralogy in weathered product is influenced by parent material in those soils, which are characterized by low leaching. But the influence of parent material gradually diminished with greater leaching. Thus weathering plays a great role in soil formation. The mechanical weathering, chemical weathering and biological weathering in combination or alone pave the way for soil formation.

PARENT MATERIAL SOILS AND CLIMATE

Soils types depend upon the climate. The influence of climate can be understood through the stimulatory influence of temperature on most chemical processes, in particular those involved in organic matter decomposition (Aerts, 1997) and mineral weathering (Turner et al., 2010; Williams et al., 2010). Temperature and water quantities cause different patterns of weathering and leaching. Wind redistributes sand and other particles especially in arid regions. Further, the amount, intensity, timing, and kind of precipitation influence soil formation.

Table 6.1 Soils of Rajasthan (Source: Dhir and Singh, 1985).

Soil name	Old name	USDA system
Dune Soil	Dune Soil	Torripsamments (1)
Sandy plain soil	Desert soil, desert plain soil	Camborthids (2) Calciorthids (3) Torripsamments (1)
Brown light loan soil	Desert soil, red desert plain soil	Camborthids (2)
Grey Brown loan soil	Grey brown desert soil	Camborthids, (2) Calciorthids(3)
Hardpan soil	Desert soil	Paleorthids (4)
Soil of Indus alluvium	Calcaerous sierozems	Torrifluvents (5)
Brown soil of old alluvial plains in the semi-arid region	Yellowish brown soil	Haplustalfs, (6) Ustochrepts (7)
Younger alluvial plain soil in semi-arid zone	Alluvial soil	Ustifluvents (8)
Medium black soil	Black clay soil	Chromusterts (9) Pellusterts (10)
Red loan soil	Rendzina	Haplustalfs (6)
Solonchaks soil	Saline-alkali soil, Saline soil	Salorthids (11)

Seasonal and daily changes in temperature affect moisture effectiveness, biological activity, rates of chemical reactions, and kinds of vegetation thus influence the soil types. Climate is considered to be the most important factor in determining the properties of many soils at macro level, but micro-climate also influence the soil formation in a small area. Hirekerur (1972) and Sidhu et al. (1976) found considerable influence of rainfall variation on morphology and properties of north Indian soils. Vander Marves and Heystack (1955) attributed diurnal fluctuations in temperature as the dominant factor in the weathering of arid, But Roth (1965) expressed his view that the temperature change alone is unimportant unless water content is sufficient to serve as an important agent in chemical rock weathering.

BIOLOGICAL FACTORS

Plants, animals, micro-organisms, and humans affect soil formation. Animals and micro-organisms mix soils and form burrows and pores. Plant roots are open channels in the soils. Different types of roots have different effects on soils. Grass roots are "fibrous" near the soil surface and easily decompose, adding organic matter (Tzialla et al., 2006). Tap roots are open pathways through dense layers. Micro-organisms affect chemical exchanges between roots and soil. Humans can mix the soil so extensively that the soil material is again considered parent material. Along with climate, topography, and biological factors the native vegetation depends on many soil factors such as soil density, depth, chemical composition, temperature, and moisture. Leaves from plants fall to the surface and decompose on the soil. Organisms decompose these leaves and mix them with the upper part of the soil. Trees and shrubs have large roots that may grow to considerable depths and newly formed rock fractures in resistant strata and weak layers, coupled with accumulation of organic and mineral debris in fractures planes and facilitated local weathering rates. Perroni-Ventura et al. (2006) observed a relationship between soil nutrient availability and plant species richness in a tropical semi-arid environment, where the relationship between soil fertility and species richness varied according to the nutrient and the micro-environment.

SOIL NUTRIENTS

The parent rock mineralogy and the nutrients available in soils are in strong relationship. High magnesium availability in soils is related with the alteration products of the ultramafic rocks. Micronutrients of iron and manganese, shows strong dependence to the bulk chemistry and mineralogy of the parent rock material. In contrast, it showed good correlation with the particle size distribution and specifically with soil clay content. The input and output of magnesium and potassium was strongly related with the availability of these elements in vegetation. Nutrient concentrations in plants are usually correlated with nutrient availability in the soil. Nutrients concentrations are predicted to be higher in under story than open-grassland of herb, shrubs and trees.

Acknowledgements

This contribution is based on our studies in the southern fold belt and Delhi Supergroup of the Aravalli Range. I got benefited from the published information of many workers and I also thankfully acknowledge their contributions during the course of study. I am thankful to Mr.

Alok Tripathi, Chief Geologist in O.N.G.C. and Prof. K. L. Srivastava, Department of Geology, Faculty of Science, J. N. V. University, Jodhpur for providing useful suggestions that helped improve the manuscript.

References

Aerts, R. 1997. Climate, leaf litter chemistry and leaf litter decomposition in terrestrial ecosystems: A triangular relationship. *Oikos* **79**: 439- 449.

Dhir, R.P. and Singh, K. S. 1985. Soils of Rajasthan and their management. ***In:*** Alexander, T.M., Biswas, B., Yadeo, D.S. and Maheshwari, S. (eds), Soils of India and their management. The Fertilizer Assiciation of India, New Delhi, India. pp 343-364.

Dokurchave, V. V. 1998. Derivation of state factor equation of soil and ecosystems. *Soil Sci. Soc. Amer. Proc* **25**: 385-388.

Emmer, I.M. 1994. Humus form characteristics in relation to undergrowth vegetation in a *Pinus sylvestris* forest. *Acta Oecologica* **15**: 677-687.

Gupta, S.N., Arora, Y. K., Mathur, R. K., Iqballuddin, Prasad, B., Sahai,T.N. and Sharma, S.B. 1980. Lithostratigraphic map of the Arravalli region, southern and northeastern Gujarat on scale 1:1 million & Explanatory Brouchure, 38 p

Gupta, S.N., Arora, Y. K., Mathur, R. K., Iqballuddin, Prasad, B., Sahai,T.N. and Sharma, S. B. 1997. The Precambrian Geology of the Arravalli region, southern Rajasthan and northeastern Gujarat. *Mem. Geol. Surv. India.* **123**:262.

Heron, A.M. (1953). The Geology of Central Rajputana. *Memoir Geol. Surv. Indian* Vol.**79.**

Jenny, H. 1941. Factors of soil formation. New York: McGraw-Hill, 281P

Kounda-Kiki, C., A. Vaçulik, J.F. Ponge and C. Sarthou (2006). Humus profiles under main vegetation types in a rock savanna (Nouragues inselberg, French Guiana). *Geoderma,* **136**: 819-829

Le Houerou, H.N. 1996. Climate change, drought and desertification. *J. Arid Envi* **34:** 133-185.

Loranger, G., Ponge, J.F., Imbert, D and Lavelle, Lavelle, P. 2002. Leaf decomposition in two semi-evergreen tropical forests: Influence of litter quality. *Biol. Fertil. Soils* **35**: 247-252

Mikesell, L.R., R.J. Schaetzl and M.A. Velbel (2004). Hornblende etching and quartz/feldspar

Mukhopadhyaya, D., BhattacharyaT., Chattopadhyay, N. Robert Lopez and Tobisch, O.T. 2010. Anasagar gneiss: A folded granitoids pluton in the Proterozoic South Delhi fold belt, Central Rajasthan. *Proc. Indian Acad. Sci. (Earth Planet Sci.)* **109**(1): 21-37.

Perroni- Ventura Y., Montana, C. and Garcia-Okuva', F. 2006. Relationship between soil nutrient availability and plant species richness in a tropical semi-arid environment. *J. Vegetation Science* **17**: 719-728.

Ponge, J.F. (2003). Humus forms in terrestrial ecosystems: a framework to biodiversity. *Soil Biol. Biochem,* **35**: 935-945.

Ponge, J.F., N. Patzel, L. Delhaye, F. Devigne, C. Levieux, P. Beros and Wittebrodt (1999). Interactions between earthworms, litter and trees in an old-growth beech forest. *Biol. Fertil. Soils,* **29**: 360-370.

Roth, B.S. 1965. Tempèrature and water content as factor in dèsert weathering. *J. Geol* **73**: 454-468.

Sidhu, P.S. and Sehgal, J. S. 1976. Parent material uniformity and weathering indices in alluvium derived soils of Indo gangatic plains of Punjab in NW India. *Pedologie* **26**: 191-203.

Turner, B.F., White, A.F. and Brantley, S.L. 2010. Effects of temperature on silicate weathering: Solute fluxes and chemical weathering in a temperate rain forest water shed, Jamieson Creek, British Columbia. *Chem. G*eol **269** : 62-78.

Tzialla, C.E., Veresoglou, D.S., Papakosta, D. and Mamolos, A.P. 2006. Changes in soil characteristics and plant species composition along a moisture gradient in a Mediterranean pasture. *J. Environ. Manage* **80**: 90-98.

Wardle, D.A., Nilson, M.C., Zackrisson, O. and Gallet, C. 2003. Determinants of litter mixing effects in a Swedish boreal forest. *Soil Biol. Biochem* **35**: 827-835.

Williams, J.Z., Bandstra, J.Z., Pollard, D. and Brantley, S.L. 2010. The temperature dependence of feldspar dissolution determined using a coupled weathering-climate model for Holocene-aged loess soils. *Geoderma* **156**: 11- 19.

Hon'ble reviewer

1. **Dr. Satendra** IFS
   Director NIDM
   IIPA Campus, IP State
   NEW DELHI - 110 002
2. **Dr. Ashok Kumar,** IFS
   APCCF
   TRIPURA (ASSAM)

IV-53

Geodynamical and Strain Accumulation Studies from Eight years of Continuous GPS Measurements of Imphal Permanent Station in the Indo-Myanmar wedge

Laishram Sunil Singh and Arun Kumar

Department of Earth Sciences, Manipur University, Impha
E-mail: arun634@gmail.coml

Abstract

The crustal dynamics, strain distribution and the cause and effect of the geological processes in the Indo-Myanmar wedge were analyzed using the Imphal (IMPH) permanent GPS data along with the network of IGS permanent GPS stations around this region. 8 years of continuous GPS data from 2003 to 2011 were used. The GPS data were processed using Gamit/Globk 10.40 software. The analysis of these measurements suggests that the site moves at a rate of about 36.3±0.5 mm/year towards N55° in the ITRF2008. With respect to the Indian plate it moves at a rate of 16.7 mm/year towards N222°, i.e., predominantly towards southwest. The baseline length changes from station IMPH to IGS stations IISC (Bangalore), LHAZ (lahsa) and KUNM (kumning) are -19.6 ±0.3mm/yr, -12.9±0.2 mm/yr, -0.09 ±0.4 mm/yr^{-1} respectively and corresponding strain rates are -9.18×10^{-9} mm/yr^{-1}, -2.12×10^{-8} mm/yr^{-1}, -1.02 x 10^{-9} mm/yr^{-1} respectively. The GPS estimated strain rates have a good correlation with the seismicity of the IMR. CMF trending N-S, is one of the aseismic strike slip fault. The analysis of GPS data reveals its slip rate of 16 mm/yr.

Keywords: Ground Positioning System, Indo-Mayanmar Wedge, IMW, Baseline length, Strain rates.

Introduction

The continuous operation GPS permanent station of Imphal is located at in the central part of Indo-Myanmar wedge (IMW). IMW constitutes a part of complex geotectonic setting, where the process of active collision of Indian Plate with Myanmar Arc. Thus problem of geotectonic complexes of the region stimulates for this studies in the region. The Geodetic data such as GPS is useful to understand the relative motion of the plates. (Akilan *et al.*, 2012). The GPS data from continuous operating permanent station can provide accurate deformation, which may facilitate Earthquake hazards mitigation. The GPS velocity estimation can validate the geological findings estimation in a shorter time. So in this study the crustal dynamics, strain distribution and the cause and effect of the geological processes in the IMW were analysed using the Imphal GPS data along with the selected network of stations around this region.

Data processing and analysis

The GPS data obtained from the permanent site has been converted into RINEX observation files and quality check has been performed using TEQC (Translation, Editing and Quality Checking Software). The quality check plots of all the GPS data were carefully examined and the data with high cycle clips were carefully examined and the data with high cycle clips multipath and <18 h observation were removed from the analysis.

The processing of the GPS data from station are routinely performed at Department of earth Sciences Manipur University. For obtaining the yearly solution, the data for the complete year is processed together, to yield the station co-ordinates and baselines with data collected simultaneously with surrounding IGS (International GPS Service): BAHR, IISC, HYDE, LHAS, KIT3, KUNM and POL2, Fig. 1 shows the IGS station selected for the purpose of constraining the solutions in ITRF (International Terrestrial Reference Frame) 2008 (Altamimi *et al.*2011) reference frame. The precise positioning of the station the coordinates of BAHR, IISC, LHAS, KIT3, KUNM and POL2 are tightly constrained using ITRF 2008 reference frame. We use the euler pole estimated by Banerjee et al. (2008) for representing the Indian plate motion.

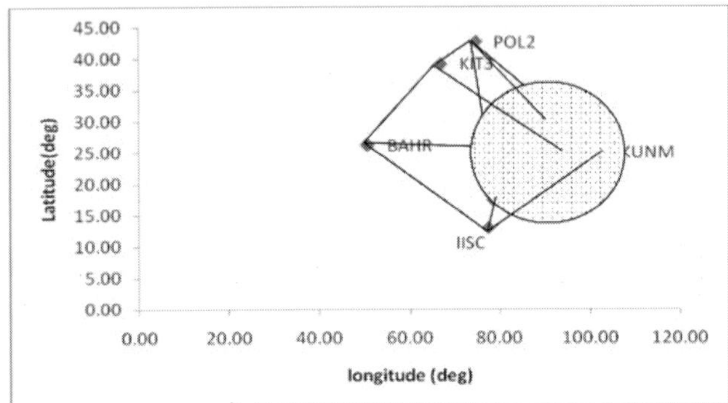

Fig. 1. The IGS Network used for determining the precise position of IMPH permanent reference station.

Results

We used the optimum strategies to process. The data acquired between 2003 and 2012 were processed and analyzed using GPS data processing software Gamit/Globk 10.40 (King and Bock, 2010, Herring, 2003) and the time series of all the sites were estimated. Time series of IMPH site coordinates is shown in Fig. 2. Time series graphically illustrate the evolution of receiver position from day to day. A reference frame ITRF 2008 is used for all time series. Each plot includes the estimated positions with error bars, the best fit line. When the processing outliers of the some particular days, whose satellites positions are badly estimated are compared with IGS stations d removed.

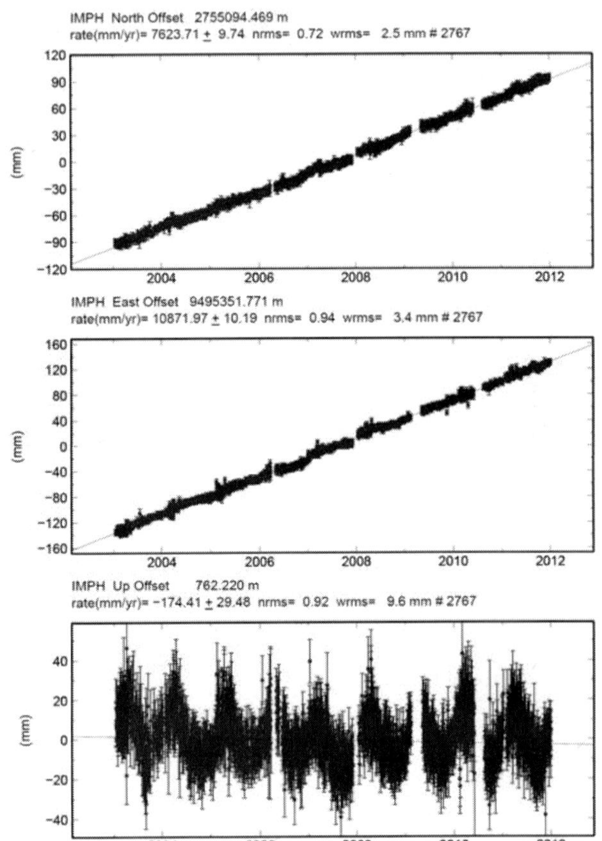

Fig. 2. Time series for IMPH plotted for the site coordinates components North-South, East-west and Height in ITRF 2008

As a result of analysis using continunous permanent station data gives coordinates of permanent station Imphal (IMPH) as 24.74944° E (±0.2mm) and 93.92496° N (±0.3mm). The maximum and minimum root mean square (rms) errors are 0.003m and 0.002m respectively. The small values of rms indicate the results obtained are precise and reliable.

Estimation of Baseline Lengths

The baseline between IMPH and three nearby IGS station estimated Fig 3 and Table 1 depicts the minimal baseline shortening of all the station in ITRF2008 from IMPH, which reveals minimum deformation in KUNM and maximum deformation in IISC.

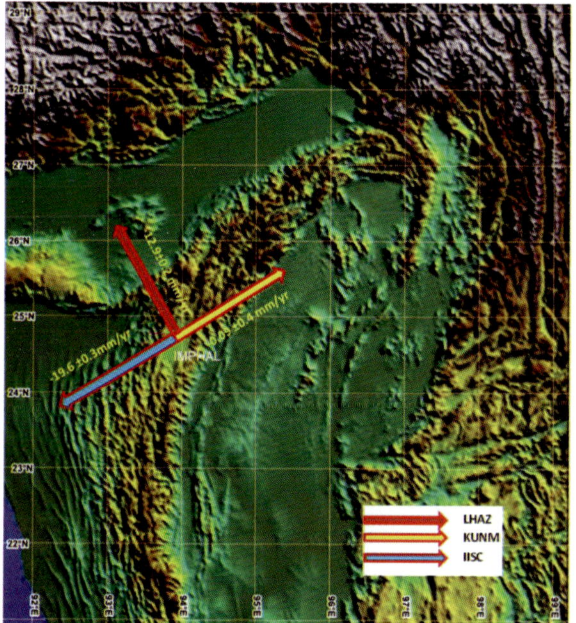

Fig. 3. Estimated baseline length from IMPH to other IGS station in ITRF 2008

Table 1: Station estimated baseline lengths between Imphal and other stations, rate of changes and their error (1standard error) estimates.

Station	Baseline length from IMPH (m)	Error (m)	Baseline length change/year (m/y)
IISC	2143770.05278	0.00043	-0.019672
LHAZ	611352.62058	0.00011	-0.012936
KUNM	896285.55558	0.00022	-0.000913

Strain Rates Calculation

The long-term loading of the Earth`s crust has been traditionally measured using geodetic and geological methods. The recent progress in space-based geodesy such as GPS provides us clear pattern of crustal movement and strain accumulation.

As the rocks slide past each other, relative motion is sometimes accommodated by relatively constant gradual slip and large accumulation of strain; while at other times, the accumulated strain is released in the form of earthquakes. The deformation of the crust by a slip generates an elastic field surrounding the area of interest. In some areas, the strain is extensional, while in others it is compressional.

To understand the process of deformation pattern along the region, the strain rates between the reference permanent GPS station of Imphal is thereby calculated.

The strain rate along the station point is estimated using the GPS baseline length data. The strain rate is calculated by using the formula (Ravikumar *et al* 2008) given below.

$$\text{Strain accumulation } (y^{-1}) = \Delta l / l$$

where Δl represents the change in baseline length, and l is the original baseline length for the predecessor year. The baseline length of the station points with respect to the IGS station IISC , KUNM, LHAZ station is calculated. Any strain developed between the station point and IGS station can be inferred as the net strain produced at the respective station point.

Table 2. Estimated strain accumulation of all the chosen sites

Sites	Strain Accumulation
IISC	-9.18×10^{-9} mm/yr^{-1}
LHAZ	-2.12×10^{-8} mm/yr^{-1}
KUNM	-1.02×10^{-9} mm/yr^{-1}

The maximum strain accumulation developed between IISC and minimum strain accumulation developed between LHAZ. An order of 10^{-8} to 10^{-9} strains (Table 2) is accumulated in this region.

Velocity of Movement of the Manipur University reference station

The IGS stations used in the analysis gives different velocities. The velocities are as BAHR 28.6 N (±0.2mm), 30.8 E (±0.3mm), LHAZ 16.1 N (±0.4mm), 47.3 E (±0.6mm) and KUNM -19.6 N (±0.4mm), 29.1 E (±0.5mm).

The velocities of IGS stations at KIT3 and POL2 in the Eurasian plate are 3.4 N (±0.2mm), 26.4 (±0.3mm) and 3.8 N (±0.2mm), 26.6 (±0.3mm) and have almost similar velocities. Likewise the velocities of the IGS Station in the Indian Plate IISC and HYDE have 34.5 N (±0.4mm), 41.4 E (±0.3mm) and 33.6 N (±0.3mm), 38.8 E (±0.5mm). But the analysis of these measurements from IMPH suggests that the site moves at a rate of about 36.3±0.5 mm/year towards N55° in the ITRF2008. With respect to the Indian plate it moves at a rate of 16.7 mm/ year towards N222°, i.e., predominantly towards southwest.

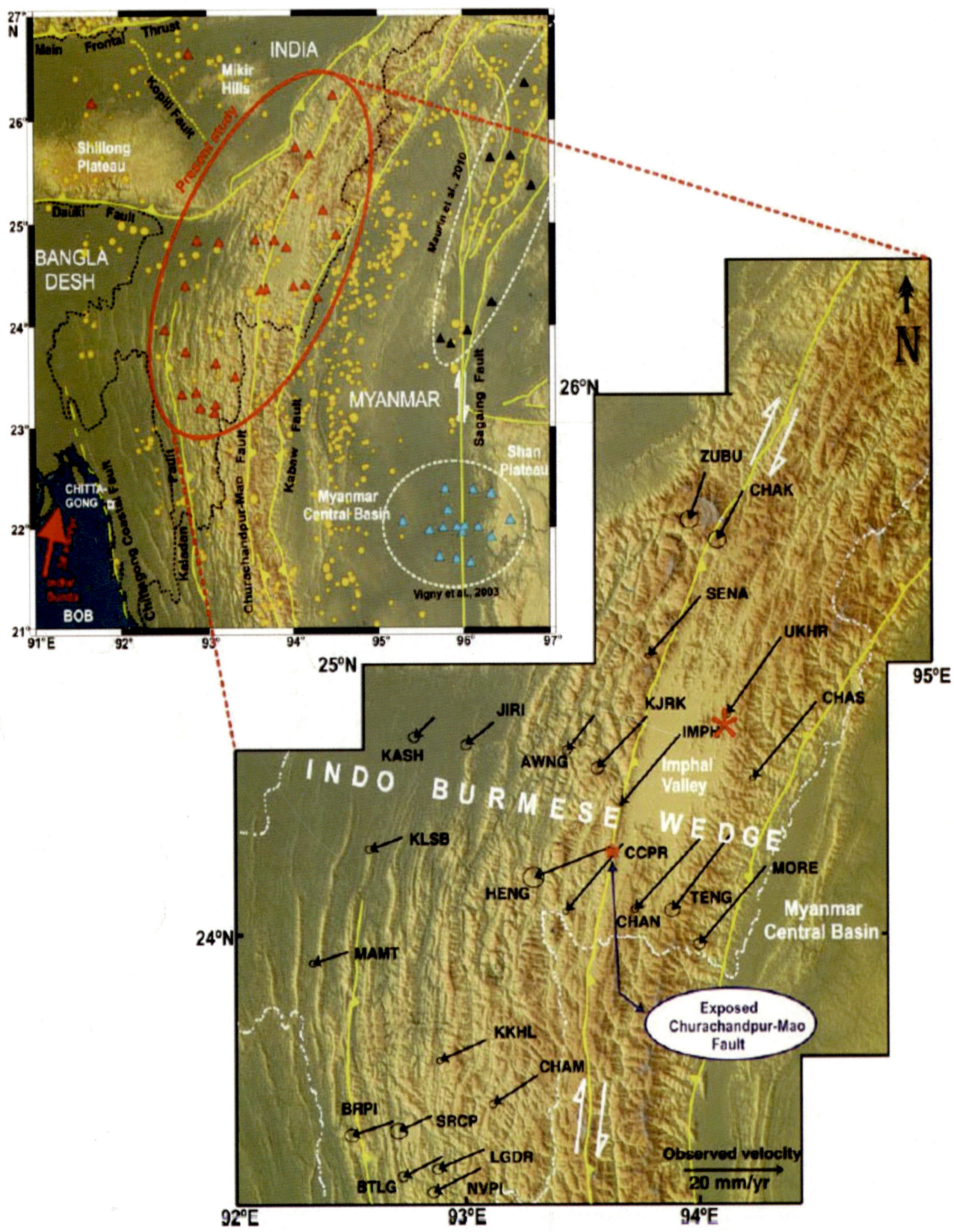

Fig. 4. GPS networks in Indo-Burmese Wedge and Sagaing Fault region (Vigny et al., 2003; Maurin et al., 2010 ; Vineet et al., 2013). Filled circles represent earthquakes in region. Lower panel shows GPS network in Indo-Burmese wedge, and estimated site velocity in Indian reference frame.Imphal Permanent sites is marked by asterisks.

Discussion

Relative motion between converging plates results in stress accumulation within the plates as well as along their common boundary. The plates undergo rotational motion on a spherical earth. Therefore, poles along which the plates rotate as well as the rates at which the plates converge are important questions to address at the Plate scale (e.g. Mukul *et al.* 2009; Jade *et al.* 2007 and Mukul *et al* 2010). The boundary between two convergent plates is subject to intense shortening that typically develops into a wedge-shaped fold-thrust belt (e.g., Dahlen, 1990; Mitra, 1997 and Mukul et al 2010). The overall analysis is that the estimated baseline lengths between Kerguelen and Maitri and other IGS stations namely IISC, LHAZ and KUNM in IIITRF 2008 are indeed shortening at the rates are -19.6 ±0.3mm/yr, -12.9±0.2 mm/yr , -0.09 ±0.4 mm/yr-1 respectively and corresponding strain rates are -9.18 x 10-9 mm/yr-1, -2.12 x 10 -8 mm/yr-1, -1.02 x 10-9 mm/yr-1 respectively.

The resultant vector of the Indian plate is 53.9 mm/yr with respect to ITRF 97 (Jade 2004) and with respect to Sunda plate is 36-38 mm/yr (Chen *et.al.*, 2000 ; Vigny *et.al.*, 2003; Socquet *et al.*,2006; and Sunil 2007) out of which about 18-20 mm/yr is accommodated across Sagaign fault in the east. A motion of about 20 mm/yr gets absorbed in the Indo-Myanmar region. That's the resultant velocity of permanent reference station IMPH in this region is 36.3±0.5 mm/yr as compared to 53.9 mm/yr that of Indian plate.

The displacement rate in the IBR derived from the GPS measurements (Vineet et al., 2013) suggests presence of a dextral fault in the region. GPS derived velocities to India fixed reference frame of IBR region clearly show that horizontal velocity are significantly different (6-8 mm/yr) across surface mapped N-S trending older thrust fault the Churachandpur-Mao Fault (CMF) (Figure 4). This signature motivates me to analyse strain rate across CMF. Hence, GPS derived stain rate clearly suggest the CMF probably accommodates a considerable amount of relative motion in IBW. However, nature and definite mode of deformation process (slipping asesimically i.e., velocity strengthening frictional behavior or seismically i.e., velocity weakening frictional behavior) plays key role in deciding the seismic potential of the region. Field investigations further provide us some interesting signatures regarding active nature of the fault zone. Within this stretch of 350-400 m, several landslide zones, tight folds, pulverized material, well exposed fault gauge and breccia units and associated fault surface were identified. All these evidences suggest that the deformation here is very intense and localized in a narrow zone. Strain rate estimation using GPS derived velocity field and field investigation provides preliminary indication that surface mapped Churachandpur-Mao Fault (CMF) indeed is accommodates a significant amount of deformation in the IBW. The slip rate of CMF is estimated of 16 mm/yr (between 15 and 17 mm/yr ± 1σ), and a locking depth of 26±2 km for the fault and the dip of 90°±5°. We can not expect the future seismic hazards due to the aseismic nature of the CMF in Imphal valley.

CONCLUSIONS

We have analyzed the shortening rates using the 8 years GPS data of Imphal Permanent station, which is located at subduction boundary of Indian and Burma plates. The boundary between two convergent plates is subject to intense shortening that typically develops into a wedge-shaped fold-thrust belt. The shortening rates are -19.6 ±0.3mm/yr, -12.9±0.2 mm/yr, -0.09 ±0.4 mm/yr-1 respectively and correspon-ding strain rates are -9.18 x 10-9 mm/yr-1, -2.12 x 10-8 mm/yr-1, -1.02 x 10-9 mm/yr-1 respect-ively. The CMF is dextral strike slip aseismic fault and slip is estimated as 16 mm/yr. The future seismic hazards due to the aseismic CMF are very low in this region. We certainly look more surface evidences for the CMF in the future.

ACKNOWLEDGEMENTS

Financial assistance provided by Ministry of Earth Sciences and Ministry of Science & Technology Government of India to carry out the present investigations is thankfully acknowledged. Authors are thankful to Dr. Vineet Gahalaut, Scientist NGRI for providing his valuable suggestions for the preparation of the manuscript.

References

Altamimi Z., Collilieux X., and Métivier L., (2011), ITRF2008: an improved solution of the International Terrestrial Reference Frame. J. Geodesy, 85(8):457-473.

Akilan, A., Balaji. S., Malaimani E.C., Ravikumar N. and Abilash K. (2012): Current scenario of crustal deformation and strain distribution around the equatorial Indian Ocean region using GPS-Geodesy J. Ind. Geophys. Union (January 2012) Vol.16, No.1, pp.21-28

Banerjee, P.R. Bürgmann, B. Nagarajan, and E. Apel (2008), Intraplate deformation of the Indian subcontinent, Geophys. Res. Lett., 35, L18301, doi:10.1029/2008GL035468.

Chen, Z., Burchfield, Y., Liu, Y., King, R.W., Royden, L.H., Tang, W., Zhao, J. and Zhang, X., (2000): GPS measurements from eastern Tibet and their implications for India/Eurasia intercontinental deformation. J. Geophys. Res., 105, 16,215–16,227.

Dahlen, F.A. (1990) Critical taper model of fold-and-thrust belts and accretionary wedges. annual review of earth and planetary sciences, v.18, pp.55.

Herring, T.A. (2003), GLOBK: Global Kalman filter VLBI and GPS Analysis Program, Department of Earth, Atmospheric and Planetary Sciences Massachusetts Institute of Technology, and Scripps Institution of Oceanography, University of California at San Diego, USA

Jade, S., (2004), `Estimates of Plate velocity and Crustal deformation in the Indian Subcontinent using GPS geodesy', Current Science, Vol.86, No.10, 25 May 2004

Jade, S., Mukul, M., Bhattacharyya, A.K., Vijayan, M.S.M., Saigeetha, J., Kumar, Ashok., Tiwari, R.P., Kumar, Arun, Kalita, S., Sahu, S.C., Krishna, A.P., Gupta, S.S., Murthy, M.V.R.. and Gaur, V.K. (2007) estimates of interseismic deformation in northeast India from GPS measurements. Earth planet. Sci. lett., v.263, pp.221-234.

King, R.W. and Bock, Y., (2010): Documentation for GAMIT Analysis Software release 10.40, Massachusetts Institute of Technology, Cambridge, MA.

Mukul, M., Jade, S. and Matin, A. (2009) active deformation in the darjiling-sikkim himalaya based on 2000-2004 geodetic Global Positioning System Measurements. In: P. Ghosh and S. Gangopadhyay (Eds.), Indian Statistical Institute Platinum Jubilee Volumes, Numerical Methods and Models in Earth Science, New India Publishing Agency, New Delhi, pp.1-28

Mukul, M., Jade, S. Bhattacharyya, A.K and Bhusan, K. (2010) crustal shortening in convergent orogens: insights from global positioning system (Gps) measurements in northeast india journal geological society of india vol.75, january 2010, pp.302-312

Mitra, G. (1997) Evolution of salients in a fold-and-thrust belt: the effects of sedimentary basin geometry, strain distribution and critical taper. In: S. Sengupta (Eds.), Evolution of Geologic Structures from Macro- to Micro-scales. Chapman and Hall, London, pp.59-90.

Ravi Kumar. N., Malaimani. E.C., Akilan. A. and Abilash K. (2008): 10 years of Continuous GPS measurements for geodetic tying of Antarctica and India for geodynamical and strain accumulation studies in the south of Indian Peninsula J. Ind. Geophys. Union (July 2008) Vol. 12, No.3, pp.115-122

Sahu, V.K., Gahalaut V. K., Rajput S., Chadha R.S., Sunil S. L. and Kumar A., (2006): Crustal deformation in the Indo-Burmese arc region: implications from the Myanmar and Southeast Asia GPS measurements, Current Science, 90, 1688-1693.

Socquet, A., Vigny, C., Chamot-Rooke, N., Rangin, C., Simons, W., Ambrosius, B., (2006): Indian Plate's motion and deformation induced along its boundary with Sunda in Myanmar determined by GPS, J Geophys. Res. 111, B05406, doi:10.1029/2005 JB003877

Sunil S.L (2007): Study on Crustal Deformation Using Global Positioning System (GPS). Unpublished Ph. D.Thesis.

Vigny, C., Socquet, A., Rangin, C., Chamot-Rooke, N., Pubellier, M., Bouin, M., Bertrand, G. and. Becker, M. (2003): Present-day crustal deformation around Sagaing fault, Myanmar, J. Geophys. Res., 108(B11), 2533, doi:10.1029/2002JB001999.

Vineet K Gahalaut, Bhaskar Kundu, Sunil Singh Laishram, Joshi Catherine, Arun Kumar, M. Devchandra Singh, R.P. Tiwari, R.K. Chadha, S.K. Samanta, A. Ambikapathy, P. Mahesh, Amit Bansal and M. Narsaiah (2013) Aseismic plate boundary in the Indo-Burmese wedge, northwest Sunda Arc Geology, January 4, 2013, doi: 10.1130/G33771.1.

Hon'ble reviewer

1. **Dr. B.K. Bansal**
   Advisor
   Ministry of Earth Science
   CGO Complex, Lodhi Road
   NEW DELHI - 110 003
2. **Dr. Mamta Chauhan**
   Scientist
   Indian Space Res. Organization
   AHMEDABAD - 380 015

IV-54

Textural and Elemental Composition of Muttukadu Backwater Sediments, Chennai, Tamil Nadu

A. Antony Jayaseelan, Hema Achyuthan, V. Nisha and V.H.R. Siva

Department of Geology, Anna University, Chennai 600 025.
E-mail: hachyuthan@yahoo.com

Abstract

There is often a need to monitor backwater and coastal sediments for pollution and anthropogenic inputs as these regions form a part of the coastal food web. In this paper analyses of organic matter, calcium carbonate, organic carbon, nitrogen, sulfur content and granulometric variables were conducted on 26 surface sediment samples collected with the geographic coordinates within the Muttukadu backwaters, Chennai. This investigation was carried out to understand the spatial distribution of the different elements and their relationships among its components. The surface sediments are dominant with silty clay and sand. Chemical and textural analyses revealed that the fine grained sediments have a positive correlation with Fe and also have a greater affinity with elements such organic matter. This is due to the increase in dust and fine grained terrestrial influx transporting towards the deeper part of the Muttukadu back waters.

Keywords: Backwater Sediment Texture, Backwater Geochemistry, Saline Marsh Land, Backwater in Chennai.

Introduction

Coastal marshes and wetlands are repositories of mineral resources in the form of heavy minerals and habitat for many kinds of invertebrates, fish, amphibians, waterfowl and aquatic mammals. Marshes, backwaters and coastal wetlands improve the water quality by acting as a sink that filters pollutants and sediments from the water that flows through them. Marshes (and other wetlands) are able to absorb water during periods of heavy rainfall and slowly release elements into waterways and therefore, reduce the magnitude of flooding and contamination.

Backwaters and marshes that open to the sea are covered with water during the high tide periods. The currents carry the fine grained particles around to the quiet side of the spit and sediment begins to build up. These locations allow the saline marshes to absorb the excess nutrients from the water running through them before they reach the oceans and estuaries. These marshes are slowly declining and getting contaminated by anthropogenic contributions and hence, there is a need to scrutinize and monitor these coastal regions.

Muttukadu backwaters, is nearly 36 kms south of the metropolitan city of Chennai and is also a major tourist attraction. Earlier works carried out in this region examined and studied the origin of oyster beds underlying nearly a meter below the surface sediments and Mid Holocene sea level changes (Achyuthan, 1997). Till date no detailed geochemical studies has been carried out on the Muttukadu marshland surface sediments. A study of the distribution and enrichment of trace metals in the surface sediments along Muttukadu marshland is important in order to assess the probable influence of the heavy industrial and anthropogenic activities in this marine influenced marshland environment. The aim of the present study is to present the elemental composition and characteristics of the marshland sediments for evaluating their source.

Materials and Methods

Study Area

Muttukadu backwater (lat. 12° 46'N and long. 80° 18'E) is located nearly 36 km south of the Chennai city and the backwaters runs parallel to the east coast, the Bay of Bengal (Figure 1 and 2). It is a marshland and the wetland zone is spread over an area of 2 km^2. One end of the marshland is nearly 1.17 km^2 elongated in the east-west direction and at the other end covers more than 2.5 km^2 along the north- south direction. The deepest portion of the marshland has a maximum depth of 3 to 4 m and it gets shallower, nearly a meter deep, near the mouth and along the margins. Muttukadu backwaters is bounded by the coastal dunes that run parallel to the east coast, Bay of Bengal. The north–south trending channels of the marshland connect it with the Adyar river to the north and to the Kalpakkam backwaters in the south. The soil in this region is loamy and vertisol where the charnockite bedrock is weathered. The tidal actions are intense along the mouth region of the Muttukadu marshland where it experience higher and lower water level fluctuations during the high and low tide time.

The study area lies on the thermal equator and the climate around the area can be classified as tropical wet and dry. For most of the year climate is hot and humid. The period between late May and Early June is very hot and dry summer. At this time the maximum temperature reaches between 38°C-42°C. The humidity is usually in the range of 65 to 84% and sea breeze in the evening hours is a blessing to combat the high temperature and humidity during summer months. The minimum temperature is 18-20°C during December to February and January is the coolest month. The area gets most of its seasonal rain fall from the Northeast monsoon (from mid-September to mid-December) and the average annual rainfall is 1,300 mm. Cyclones in the Bay of Bengal strike the east coast Chennai at times.

The geology of the study area is represented by the basement Charnockite rocks of Archaean age which are overlain by a thick mantle of Pleistocene to Holocene alluvium and these are in turn overlain by the Holocene tidal flat deposits and coastal dunes (Achyuthan, 1997). Isolated pockets of charnockite rock exposures occur at Muttukadu region. The coastal area at Muttukadu is characterized by salt marsh and barrier dunes forming palaeo shore lines and sand spit. The sand spit has produced a low-lying marshy land at Muttukadu. Linear dunes along the shore line, which are stabilized to a large extent, represent an important geomorphic feature of the area. There are several inlets of tidal waters in to the tidal flats. To accomplish the objectives put forward, twenty six surface sediment samples were collected by using a shovel. The sediment samples were scooped and placed in polythene bags and sealed, and they were labeled according to the sampling locations with the help of hand held GPS and marked in the study area map. The samples were then oven dried at 50°C and subsequently cone and quartered for various textural and geochemical analyses. Ten samples were selected for textural studies adopting the pipette method of Krumbein and Pettijohn (1938) and the textural parameters were plotted on equilateral triangular diagrams (Shepard, 1954; Folk, 1956). Calcium carbonate ($CaCO_3$%) and total organic carbon content (TOC) were determined adopting the methods of Loring and Rantala (1992) and Heiri et al., (2001) and the data is presented in Table 1. Total carbon (TC), Total hydrogen (TH), Total nitrogen (TN) and Total sulphur (TS) in percentages were analysed using Thermo Scientific MAS 200R auto CHNSO Elemental Analyzer at the Department of Geology, Anna University Chennai. Surface sediments were analysed for the elemental composition such as Fe, Mn, and Ti using Niton XL2 handheld XRF 7.1.1 version at the Department of Geology, Anna University Chennai.

Results

The results on granulometric studies of the Muttukadu marshland sediments shows a dominance of sand percentage near to the mouth of the backwaters where exist a continuous and high influence of sea water. The settling of clay is less near the mouth zone as when compared with the inner north western part of the marshland, where the environment is most suitable for the settling of finer grain particles. Textural studies reveal that almost all the sampling locations along the Muttukadu backwaters exhibit a higher sand percentage (Table 1) and vary between 82-99%. Few of the sampling locations such as M9, M10, M12, M17a and b, M19, M21, M22, M23 and M33 shows clay and silt content (Figure 1, Table 1). The $CaCO_3$ content from the location points 1 to 19 (Fig. 3A) is high and its distribution is non-uniform. In the present study the distribution of organic matter is higher along the northern part of the study area than in the south. The OM percentage in the sites M12, M19, M18 is high (Fig. 3, Table 1) while the remaining sites within the Muttukadu backwaters are low. The concentration of elements such as Fe, Mn and Ti varies from Fe (4.67-0.922%), Mn (0.059-0.007%) and Ti (1.089-0.01%) respectively. The total carbon (TC) and total nitrogen (TN) ratio were found insignificant and ranged between 2.78 to 0.34%.

Discussion

The sediments of Muttukadu backwaters are of heterogeneous mixture mainly composed of biogenic carbonates and shell fragments. The dominance of sand content in the sediments indicates a higher wave energy regime due to the shallow depths that prevents sedimentation of fine grained particles (Sathpathy et al., 2011). The high $CaCO_3$ content is of biogenic origin and the input may also be from the adjacent land mass where Tertiary limestone and calcareous sandstones extend in the southern region (Ray et al., 1990). The increase in the sand with minor amounts of clay content as observed in several sampling locations in Muttukadu backwaters is be due to the influence of littoral drift and the current from the abrasion zones that transport finer sediments once they reach the mouth of the backwaters (Szefer, 1988). Sediment organic matter is largely derived from plant and animal detritus, bacteria or plankton formed in situ, or derived from natural and anthropogenic sources in catchments. Sewage and effluents from pharmaceutical and food-processing plants, pulp and paper mills and prawn and pisiculture situated along the banks of the backwaters are examples of organic-rich wastes of anthropogenic origin.

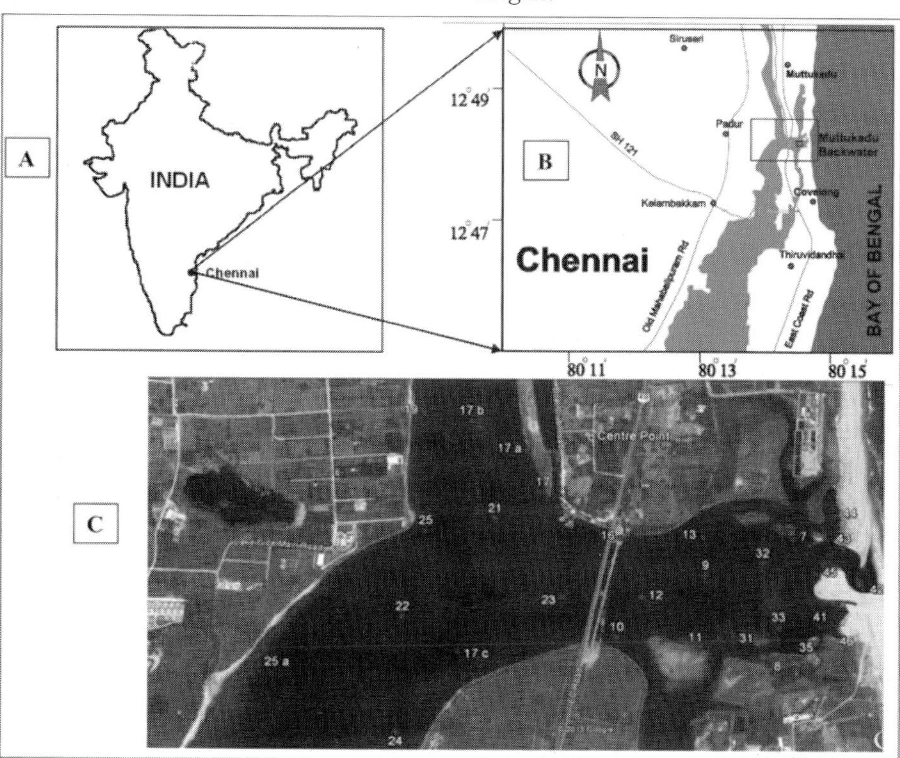

Figure 1. Location map of the study area around Muttukadu marshland (B) and spot locations (C) of the subsurface samples collected.

Table 1. Textural and elemental composition of surface sediments of the Muttukadu backwaters

Sl.No.	Sample	Sand %	Silt %	Clay %	$CaCO_3$ %	OM %	Fe %	Mn %	Ti %	Nitrogen (TN)	Carbon (TC)	Sulphur (TS)
1	M7	99.50	0.30	0.20	17.00	2.56	1.268	0.007	0.01	0.00	0.18	0.00
2	M8	98.70	1.10	0.20	15.00	2.16	1.199	0.007	0.01			
3	M9	99.00	0.20	0.80	13.00	1.18	1.739	0.016	0.189			
4	M10	99.40	0.20	0.40	12.00	0.39	1.818	0.018	0.166			
5	M11	97.20	1.60	1.20	13.00	6.49	1.753	0.016	0.179			
6	M12	98.80	0.80	0.40	12.00	6.88	2.289	0.031	0.212	0.00	0.53	0.00
7	M13	99.20	0.40	0.40	11.00	7.08	1.86	0.025	0.115			
8	M16	99.20	0.20	0.60	12.00	2.56	3.324	0.037	0.519			
9	M17	82.92	4.98	12.10	14.00	1.97	2.702	0.041	1.089			
10	M17b	80.62	6.48	12.90	13.00	1.77	4.671	0.059	0.402	0.34	2.78	0.20
11	M17c	84.21	4.49	11.30	16.00	2.56	2.414	0.031	0.281			
12	M18	99.60	0.20	0.20	14.00	2.16	2.517	0.034	0.643			
13	M19	85.79	4.51	9.70	13.00	1.97	2.951	0.035	0.757			
14	M21	87.90	3.10	9.00	18.00	2.75	2.834	0.04	0.821	0.00	0.41	0.00
15	M22	84.50	5.10	10.40	16.00	1.77	2.435	0.027	0.509			
16	M23	94.40	5.00	0.60	12.00	1.18	3.277	0.04	0.485	0.00	0.04	0.00
17	M24	81.92	5.98	12.10	13.00	1.38	1.008	0.01	0.01	0.00	0.23	0.00
18	M25	98.80	0.80	0.40	14.00	1.38	2.519	0.028	0.533	0.00	0.03	0.00
19	M25a	83.21	6.49	10.30	11.00	1.57	2.668	0.035	0.572			
20	M31	99.20	0.20	0.60	19.00	0.79	2.668	0.041	0.413	0.00	0.00	0.00
21	MS32	99.40	0.40	0.20	21.00	0.39	1.007	0.01	0.01			
22	M33	98.80	1.00	0.20	18.00	0.59	1.287	0.009	0.01			
23	M35	99.00	0.20	0.80	16.00	0.20	1.796	0.019	0.389			
24	M41	99.40	0.20	0.40	17.00	0.20	0.964	0.01	0.01	0.00	0.54	
25	M42	97.20	1.60	1.20	18.00	0.59	2.593	0.035	0.421			
26	M43	96.40	3.20	0.40	26.00	0.59	1.328	0.007	0.01			
27	M44	99.00	0.20	0.80	27.00	0.59	0.922	0.01	0.01			
28	M45	98.80	0.80	0.40	28.00	0.79	1.275	0.007	0.01			

High $CaCO_3$ content indicates enrichment due to marine shells and inorganic carbonate in the form of calcic dust. Non uniform distribution of $CaCO_3$ indicates an oxidising environment and depletion in biogenic shells content, low $CaCO_3$ content indicates the sites having higher influence of freshwater and lower salinity. High organic matter content indicates higher influence of biodegradable detritus and vegetation and low energy environment allowing settling of organic matter while the locations showing low OM content can be characterized as high energy environment. High OM content indicates a higher source due to biodegradable material including vegetation, fauna found within the backwaters and also bordering the water way.

Total Organic Carbon (TOC) is the amount of organic matter preserved within sediment and sediment nutrients assessed as Total Nitrogen (TN) and Total Phosphorus (TP) concentrations have both inorganic as well as organic sources (Yanhong et al., 2007). The amount of organic matter found in sediments is a function of the amount of various sources reaching the sediment surfaces and the rates at which different types of organic matter are degraded by microbial processes during burial (Hedges et al., 1997). In the present study the spatial distribution of surface sediments and their geochemistry reveals the occurrence of different sediment facies related principally to local hydrodynamic conditions persisiting in Muttukadu.

The C/N ratio has been used for decades as a parameter for the evaluating the relative influence of terrestrial and marine organic matter in several ecosystems (Bordovskiy, 1965; Tyson, 1995). C/N ratio of only one sample is 8.18 (n=9 samples) and this can be related to the origin of the organic matter. Saito et al., (1989) suggested that a ratio higher than 20 can be related to a terrestrial origin and between 5 and 7 as a pelagic source. In addition, Stein (1991) reported that values of C/N ratio lower than 10 points to a marine origin and values around 10 represent a mixed genesis (both marine and terrestrial organic components) in the sediment. In the present study a ratio of 8 can be related to a marine origin. However several samples need to analysed for this conclusion to be arrived. The TC values were found to be significantly higher in the deeper areas when compared with the shallow inlet regions (Figure 1 and 2). Mostly sand laden sites are one of the main reasons for low organic matter compared to the other parts of the Muttukadu backwaters such as sampled site locations M9, M10, M12, M17a, M17b, M21, M22, M23 and M33 (Figure 1). In all the sampled locations at Muttukadu, the Fe content varied from 4.67-0.92% (Table 1). A higher value of Fe and Ti and 1.089 at the sampled site location M17 (4.67% Fe and Ti 1.08%), at site 16 (Fe 3.35% and Ti 0.52%) and site 19 (Fe 2.9% and Ti 0.75%) clearly exhibits intense impact of anthropogenic activities from the near coastal areas. Sulphur content was identified only in the north western part of the study area and this is due to anaerobic conditions.

Figure 2 A. Coastal dunes bordering the marshland, B. Mouth of Muttukadu backwaters joining and mixing with the Bay of Bengal waters. C. Central part of the Muttukadu back waters. D. Opening of the Muttukadu backwaters with the sea in the horizon.

In coastal areas biogeochemical cycles of iron and organic carbon are strongly interlinked (Lalonde et al., 2012). Long exposure to oxic conditions increases the fraction of the total sedimentary organic carbon pool that is tightly adsorbed to particle surfaces (Arnarson and Keil, 2007; Lalonde et al., 2012), owing to the preferential degradation of organic structures that are more loosely attached to the clay mineral matrix, such as the OM-Fe chelates. Direct chelation or co-precipitation of macromolecular OM-Fe structures plays a significant role. Reactive iron phases serve as an extremely efficient 'rusty sink' for organic carbon and are a key parameter in the long-term storage of organic carbon and the global cycles of carbon, oxygen and sulphur (Lalonde et al., 2012). Organic matter breakdown (mineralisation) reduces the sediment carbon and nutrient concentrations. Dissolved nutrients are released from the sediment to the water column (Froelich et al., 1979). Carbon is released as CO2 gas and as dissolved organic carbon (DOC) (Froelich et al., 1979). Mineralisation rates are faster when dissolved oxygen is present under anoxic conditions (Harnett et al., 1998). Low pH can also reduce mineralisation rates and contribute to organic matter accumulation (Thomann, 1972). High sedimentation rates can therefore reduce the contact time between organic matter and dissolved oxygen in the water column, and thus can contribute to higher concentrations of carbon and nutrients in sediment (Pierson et al., 2002). Sulfate reduction promotes the release of P from sediment because some iron oxyhydroxides that bind phosphate are converted to iron sulfides that cannot bind P (Hedges and Keil 1995). Sulfate reduction occurs under anoxic conditions. Sediment carbon and nutrient concentrations increase with decreasing grain size because organic matter adsorbs onto mineral surfaces and has greater affinity for fine-grained sediments (Figure 2 and 3) (CSIRO Huon Estuary Study Team, 2000).

Decreased freshwater flows can alter the amount of organic matter that enters a coastal waterway and the rate at which it is flushed to the ocean (Baas Becking and Moore, 1959). Enhanced sediment transport caused by erosion (gully and stream bank erosion and sheet wash) in catchment areas can lower sediment TN and TOC concentrations because inorganic constituents (minerals and clays) can dilute organic matter concentrations. Organic matter is also a source of food and energy and its nutritional balance (TOC: TN: TP ratio) plays an important role in material flow through ecosystems. Sediments with high TOC: TN ratios (and lower N contents) support a lower biomass of benthic invertebrates (Craft et al., 1991).

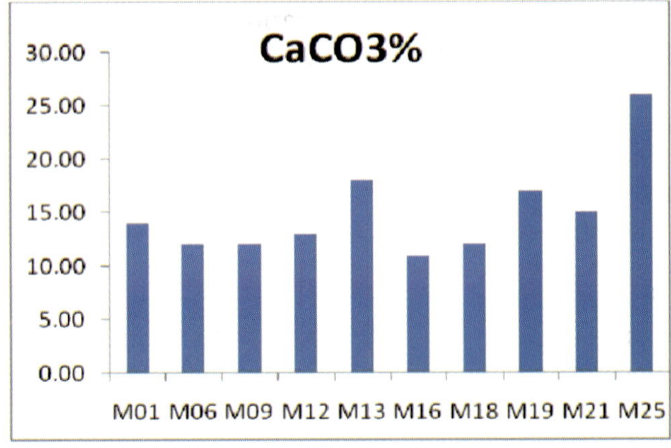

A

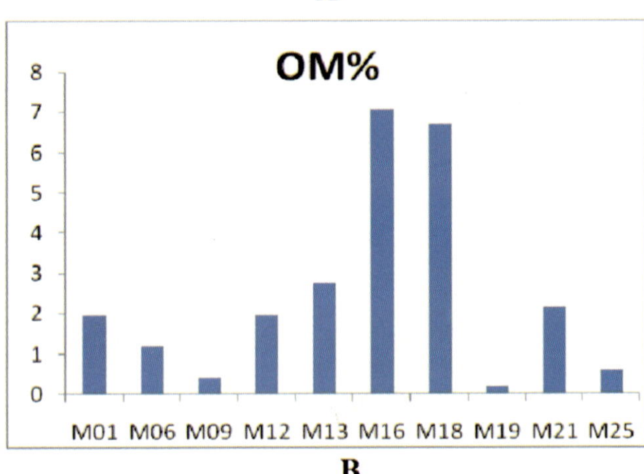

B

Figure 3. A: Shows $CaCO_3$% and B: Shows OM% Muttukadu backwaters

The sites of organic matter accumulation in coastal waterways are controlled to a large extent by processes that administer the transport and deposition of fine sediments. In tide-dominated waterways (*e.g.* backwaters (Figure 2 A to D), deltas, estuaries and tidal creeks), flanking environments are the main traps for fine sediments, and these include mangroves, salt marshes and intertidal flats (Craft et al., 1991), but the central part of the backwaters is usually the main sink (Craft, 1991; Anderson et al., 1981). Hence further studies demand analyses of several samples from the central part of Muttukadu backwaters and also a comparison of such inlets of waterways along the east coast Tamilnadu linking with the Bay of Bengal.

Conclusions

In the present study, the Muttukadu backwater surface sediments reveals the spatial distribution of sediment texture in relation to organic matter, $CaCO_3$ and other chemical elements. It is observed that the high values of organic matter, Fe, Mn and Ti content were recorded in those site locations with high percentages of fine fractions. The spatial distribution of $CaCO_3$ is non uniform and this is due to the abundant presence of shell fragments of biogenic origin. Sandy sites within the Muttukadu backwaters contain low organic matter compared to the central deeper parts of the backwater basin. Fe, Mn and Ti values are high in fine fractions as they are adsorbed on to their surfaces. This is a preliminary presentation of the data generated, however future studies demand further monitoring of the backwater sediments due to anthropogenic impact along the coast.

ACKNOWLEDGEMENT

The authors thank the reviewers for their comments that helped in the improvement of the text and presentation of the paper.

References

Achyuthan, H. 1997, Age and formation of Oyster beds in Muthukadu tidal flat zone, Chennai. Current Science, v. 73(5), p. 451-453.

Arnarson, T. S., and Keil, R.G., 2007, Changes in organic matter-mineral interactions for marine sediments with varying oxygen exposure time. Geochimica Cosmochimica Acta, V.71, p. 3545–3556.

Anderson, J. R., Storey, K. J., and Carolane, R., 1981, Macrobenthic fauna and sediment data for eight estuaries on the south coast of New South Wales. Technical Memorandum 81/22, CSIRO Institute of Biological Resources, Division of Land Use Research, Canberra, p. 7.

Baas Becking, L. G.M., and Moore, D., 1959, Relation between Fe and Organic matter in sediments. Journal of Sedimentary Petrology, v. 29, p. 454-458.

Bordovskiy, O.K., 1965, Accumulation of organic matter in bottom sediments. Marine Geology, v.3, p. 33-82.

Craft, C.B., Seneca, E.D., and Broome, S.W., 1991, Loss on ignition and Kjeldahl digestion for estimating organic carbon and total nitrogen in estuarine marsh soils: calibration with dry combustion. Estuaries, v. 14, p. 175-179.

CSIRO Huon Estuary Study Team, 2000, Huon Estuary Study: Environmental Research for Integrated Catchment Management and Aquaculture. Project No. 96/284, Final Report to the Fisheries Research and Development Corporation, p. 285.

Folk, R.L., 1956, A review of grain-size parameters. Sedimentology, v. 6, p. 77-93.

Froelich, P.N., Klinkhammer, G.P., Bender, M., Leudtke, N.A., Heath, G.R., Cullen, D., Dauphin, P., Hammond, D., Hartman, B., and Maynard, V., 1979, Early oxidation of organic matter in pelagic sediments of the eastern equatorial Atlantic: suboxic digenesis. Geochimica et Cosmochimica Acta, v. 43, p. 1075-1090.

Hartnett, H.E. et al., 1998, Influence of oxygen exposure time on organic carbon preservation in continental margin sediments. Nature, v. 391, p. 572-574.

Hedges, J.I. and Keil, R.G., 1995, Sedimentary organic matter preservation: An assessment and speculative hypothesis. Marine Chemistry, v. 49, p. 81-115.

Hedges, J. I., Keil, R. G., and Benner, R., 1997, What happens to terrestrial organic matter in the ocean? Organic Geochemistry, v. 27, p. 195–212.

Heiri, O., Lotter A. F., and Lemcke, G., 2001, Loss-on-ignition as a method for estimating organic and carbonate content in sediments: reproducibility and comparability of results. Journal of Paleolimnology, v. 25, p. 101-110.

Krumbein, W.C., and Pettijohn, F.J., 1938, Manual of Sedimentary Petrography, Appleton-Century Crofts Inc., New York, p.549.

Lalonde, K., Mucci, A., Ouellet, A., and Gélinas, Y., 2012, Preservation of organic matter in sediments promoted by iron. Nature, v. 483 (7388), p.198-200.

Loring, D.H., and Rantala, R.T. T., 1992, Manual for the geochemical analysis of marine sediments and suspended particulate matter. Earth Science Review, v.32, p. 235 – 283.

Pierson, W.L., Bishop, K., Van Sendon, D., Horton, P.R., and Adamantidis, C.A., 2002, Environmental Water Requirements to Maintain Estuarine Processes, Environmental Flows Initiative Technical Report, Report Number 3, Commonwealth of Australia.

Ray, S.B., Rajagopalan, G., and Somayajulu, B.L.K., 1990, Radiometric studies of sediment cores from Gulf of Mannar, Indian Journal of marine Science, v.19, p. 9–12.

Satpathy, K. K., Mohanthy, A. K., Prasad, M.V.R., Natesan U., and Sarkar, S.K., 2011, Studies on the variation of heavy metals in the marine sediments off Kalpakkam, East coast Tamilnadu. Environmental Earth Science, DOI 10.1007/ s12665-011-1067-z.

Shepard, F.P., 1954, Nomenclature based on sand-silt-clay ratios. Journal Sedimentary Petrology, v. 24, p. 151-158.

Stein, R., 1991, Accumulation of organic carbon in marine sediments. Results from the Deep Sea Drilling Project/Ocean Drilling Program. Lecture Notes in Earth Sciences, V. 34. Berlin: Springer-Verlag, P.217.

Szefer, P., and Skwarzec, B., 1988, Distribution and possible sources of some elements in the sediment cores of the southern Baltic, Marine Chemistry, v..23, p.109–129.

Thomann, R.V., 1972, Systems analysis and water quality measurement. Environmental Research and Application, Inc., New York.

Tyson, RV., 1995, Sedimentary organic matter: Organic facies and palynofacies. London: Chapman & Hall, P.615.

Yanhong Wu, Lűcke, A., and Wang, S., 2008, Assessment of nutrient sources and paleoproductivity during the past century in Longgan Lake, middle reaches of the Yangtze River, China. Journal of Palaeolimnology, v.39, p.451–462.

Hon'be reviewer

1. **Prof. G.M. Bhat**
   Chairman
   Department of Geology
   Jammu University
   JAMMU - 180 006
2. **Dr. Pitamber Prasad Dhyani**
   President (Env.) ISCA
   GB Pant Inst. Him. Env. Dev.
   Kosi-Katarmal
   ALMORA - 263 643

Landslide Hazard Zonation Mapping Based on Geological and Geomorphological Attributes – A Remote Sensing and GIS Approach

S.N. Mohapatra*, Vikram Kumar Ranga*, L.K. Sinha and Padmini Pani*****

Centre of Remote Sensing & GIS, SOS in Earth Science, Jiwaji University, Gwalior*
Defence Terrain Research Laboratory, DRDO, Delhi**
Centre for the study of Regional Development, Jawaharlal Nehru University, New Delhi***
Email: suraj64@yahoo.com;

Abstract

Landslide hazard zonation mapping for the part of Pauri & Tehri Garhwal districts of Uttarakhand has been attempted for the present study using remote sensing and GIS techniques. Geological and geomorphological attributes such as lithology, slope, aspect, structure, drainage and land forms have been considered for the study. Various other themes which have direct or indirect effects on landslides such as land use/land cover, transport network are also considered. Digital satellite image of IRS 1D LISS III has been used to generate various thematic maps of the study area covering 1996.99 km^2. The whole region is undergoing through strong denudation and contributing large amount of debris, which is transported to the study area through number of streams. These weathered units are mainly composed of dolomite, limestone, phyllite, and quartzite. Mostly the area is covered by structural and denudational hills. The data on geomorphological complexes, lithology, drainage density and relief were used in a Geographical Information System (GIS) for the final analysis using ARC GIS Software. The above mentioned themes were identified, extracted and weights were assigned to each theme for overlay analysis and preparation of landslide hazard zonation map in a GIS environment. It is concluded that topography and drainage along with slope and lineaments are the main parameters which contribute significantly in increasing the mobility of debris flows and have very high influence on landslides. The use of Remote Sensing & GIS greatly helps to identify the hazard prone areas (with respect to landslide) and with fairly good accuracy.

Key Words: Hazard Zonation, Lithology, Denudational, Overlay analysis, GIS.

Introduction

The population growth and the expansion of settlements and life-lines over hazardous areas are increasing the impact of natural disasters both in the developed and developing world (Rosenfeld, 1994; Guzzetti et al., 1999). Especially in the mountainous areas such as

Himalyan region where weather experiences extreme events and intense seasonal precipitation that triggers natural hazards such as landslides (Chauhan et al., 2010). In a country like India, where the population has increased more than 181 million during the last decade (2001-2011), making it fifth most populous country on planet (Census of India, 2011). These circumstances led to unplanned settlement, agriculture, deforestation and construction of infrastructure like process of making roads; which requires clearing of mountainous area first which destabilizes the slope and increases the chances of mass movement. A ranking should be made for such hazardous areas should be available to public for consultation to reduce the risk of losses. Sarkar et al., (2011) concluded that there is high landslide risk along Alaknanda valley; which can cost large number of causalities as well as economic loss, given the high volume of traffic on the nearby highway. The present study is an attempt to study the geological and geomorphological attributes and their influence on landslides. Landslides can occur on any terrain given the right conditions of soil, moisture and the angle of slope. Integral to the natural process of the earth's surface geology, landslides serve to redistribute soil & sediment in a process that can be abrupt collapses or slow gradual slides. It can be triggered by natural processes such as rains, floods & earthquakes and/or man-made causes such as grading, terrain cutting & filling excessive development etc. They can also occur in developed areas where terrain was altered for roads, houses, utilities, building etc. The tectonic movements have made the rocks of the terrain highly sheared, faulted, folded and show large number of penetrative weak planes along which the rock mass is prone to fail. High relative relief, high and concentrated precipitation further enhances the fury of these disasters. Geomorphology of the area is highly influenced by drainage pattern and drainage density. Amalgamation of Remote Sensing and GIS gave scientists an opportunity to map landslide hazardous zonation in time effective manner. Most of the studies on Landslide hazard zonation mapping in various parts of Himalaya has been attempted by using Remote Sensing and GIS techniques specially to evaluate the land use/ land cover conditions and topographic characteristic using Digital Elevation Model (DEM) by several workers (Saraf, 2000; Kamp et al., 2008; Chauhan et al., 2010; Ghosh and Bhattacharya, 2010). Although, the importance can not be neglected with importance of field verifications and if possible inclusion of already landslide areas, which might be difficult but Saraf (2000) has shown that PAN imageries from IRS-1C can be used to detect such areas. Since, the Alkananda valley is highly prone to landslides many workers have significantly contributed and enhanced the understanding of landslide phenomenon; to name the few are Sinha (1991), Pachauri and Pant (1992), Sarkar et.al. (1995), Gupta et.al. (1999), Saha et.al. (2002), Sarkar and Kanungo (2004), Sarkar and Gupta (2005) and Mathew et.al. (2007).

Study Area and Data Used

The study area forms a part of Pauri & Tehri Garhwal districts of Uttarakhand, covering an area of 1996.99 km^2 and bounded by Latitudes 30^0 00'N to 30^{0}30'N and 78^{0}45'E to 79^{0}15'E Longitude . It covers four Survey of India (SOI) topographic sheets on 1:50,000 scale, out of which only three were studied in detail as per the road alignment. Lithologically the area is mainly composed of dolomite, limestone, phyllite and quartzite. Structurally the area exhibits one major thrust i.e. North Almora Thrust and a fault i.e. Alknanda Fault. Several smaller lineaments were also delineated from the satellite image. SOI topographic sheets were mainly used for digitization of contours and study the drainage present in the area. IRS-1D LISS - III data (Fig.1) was used for interpretation and extraction of themes on the basis of Digital values of pixels.

Methodology

Total eight different themes were extracted on the basis of geological and geomorphological attributes from the various available data and digital satellite images. Digital Image Processing was carried out to make the data suitable for the analysis. The various thematic maps prepared includes: lithology, geomorphology, drainage, slope, aspect, structure, land use/land cover and transport network. Slope and aspect map were prepared from the Digital Elevation Model (DEM) by digitizing the contours from SOI topographic sheet of 1: 50,000 scale. Slope and aspect plays very important factor in a landslide study. In the study area slope varies from nearly horizontal to greater than 85 degrees. The weights were assigned according to susceptibility to landslide (Table.1). Aspect divides the amount of sunlight which the slope receives. Higher amount of sunlight creates drier slopes with less vegetation and will be more prone to landslides compared to slopes with less sunlight. The aspect was divided into nine classes (N, NE, E, SE, S, SW, W, NW and FLAT). Drainage is another important factor, which reflects the slope evolution of an area and an indicator of the mass wasting and related erosion processes. Zones with the parallel drainage pattern associated with higher slopes are the most probable situation for mass movement / landsides. Drainage density computation has been carried out on the basis of drainage length and integrated to form drainage density layer. The drainage density values have been divided into four classes as low, medium, high & very high, and weights have been assigned for these classes. The Garhwal Group is made up of various types of lithological units. There is one major thrust (North Almora thrust) passing through the area and one Mega fault (Alknanda Fault) is present in the north of the study area. Other lineaments, which are smaller in dimension, were also digitized from satellite imagery. Buffer zones have been created around these lineaments to cover the influenced area. Land use/land cover mapping has been done using unsupervised classification in which five main classes were identified and extracted from satellite image. The geomorphic feature, denudational hills, are divided into three classes as high, medium and low depending upon their degree of erosion. Roads are digitized from the topographical sheets and a buffer map was prepared to highlight the extent of influenced area. All the themes were assigned with weight value to individual class according to the susceptibility of the class to Landslide using ERDAS Imagine (Table 1). Analysis for making Landslide Hazard Zonation map was done in ARC GIS software. All the vector layers were converted to raster format and then reclassified according to the weight assigned to the classes to make uniform data for analysis. Finally, the Landslide Hazard Zonation map (Fig. 2) was prepared using weighted overlay analysis.

Table-1 : Assignment of Weight given to each class

Sr. No.	Theme	Class	Weight	Sr. No.	Theme	Class	Weight
1.	Slope	1-15 degree	1	4.	Drainage density	Low	2
		15-25 degree	2			Moderate	6
		25-30 degree	6			High	8
		30-35 degree	7			Very High	10
		35-40 degree	9	5.	Lithology	Granite	2
		40-45 degree	9.5			Metavolcanic	3
		45-60 degree	8			Dolomite	8
		>60 degree	7.5	6.	Structure	Fault	8
2.	Aspect	North facing	1			Thrust	4
		South facing	10			Lineament	10
		NE facing	3	7.	Geomorphology	MDH	6
		NW facing	2			LDH	5
		SE facing	6			HDH	8
		SW facing	4	8.	Road type	Metalled	6
		East facing	5			Unmetalled	4
		West facing	2				
3.	Land use/land cover	Dense forest	1				
		Sparse Vegetation	7				
		Fallow Land	8				
		Agricultural Land	2				
		Built up Area	3				
		Sparse Vegetation	7				

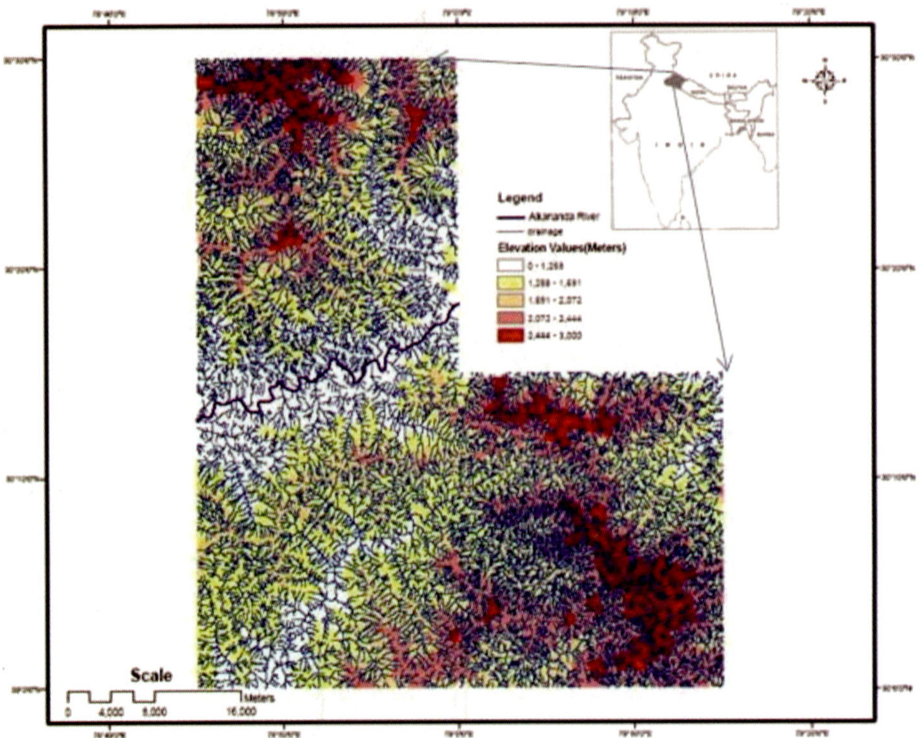

Fig. 1: Location Map

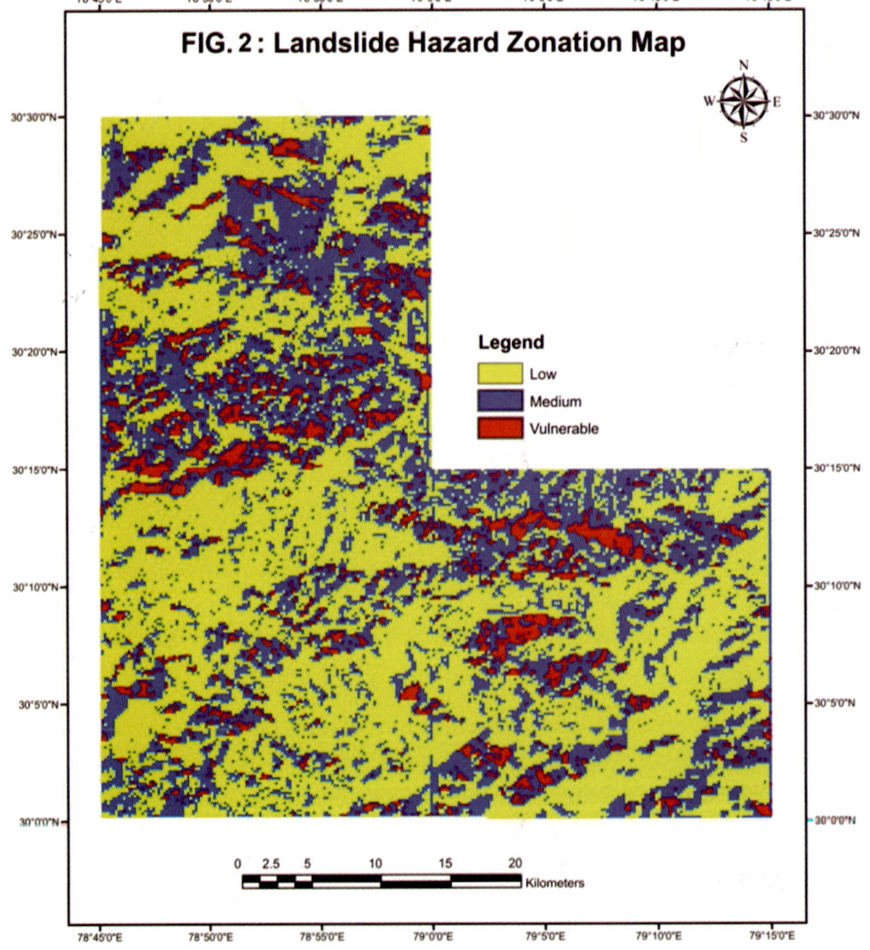

Fig. 2. Landslide Hazard Zonation map

Result

The area is classified into three zones according to land slide hazard susceptibility as Low, Moderate and Vulnerable. Themes taken here have direct or indirect effects on landslide. With the help of Remote Sensing & GIS all these themes were identified. About 53.3% of the area falls under low, 38.1% under medium and 8.6% under vulnerable category. Immediate attentions should be given to the vulnerable area so that no further construction should come up in these areas without proper scientific analysis. Further, 38.1% medium areas are required to be dealt with Landslide Hazard Management practices as per ISI code to make it less vulnerable in future. The present study demonstrates the application of remote sensing and GIS for landslide hazard zonation mapping in an area which is prone to occurrences of frequent landslides. This study would help decision support system for policy maker at medium scale resolution, although, better spatial resolution would further enhance the results at cadastral level.

Discussion and Conclusions

Remote sensing gives a unique way to map even the inaccessible areas which if combined with GIS extends capability to combine other ancillary information as it was done in this study. Although, the appropriate scale for remotely sensed imageries is still an issue as small or shallow landslide might go unnoticed in coarser resolution imageries. Landslide in Himalayan region is a common phenomenon due to number of factors including slope to human encroachment. The population pressure can not be reduced but can be managed to reduce the adverse effects on nature and vice-versa. Since, complete prohibition on encroachment by human of high landslide prone areas is very difficult, the information on landslide hazard zonation should be accessible to public as well as to the policy makers, in all the forms i.e. printed maps as well as digitally and people should be taught about the adverse effects of encroaching highly landslide prone areas. Topography and drainage along with slope and lineaments contribute to landslides significantly.

ACKNOWLEDGEMENTS

The authors thank Director, DTRL, for providing the infrastructure and encouragement to carry out the study.

References

Census of India (2011). Size, growth rate and distribution of population, Govt. of India.

Chauhan, S., Sharma, M., Arora, M. (2010). Landslide susceptibility zonation of the Chamoli region, Garhwal Himalayas, using logistic regression model. Landslides 7, 411–423.

Ghosh, J. and Bhattacharya, D. (2010). Knowledge-Based Landslide Susceptibility Zonation System. Journal of Computing in Civil Engineering 24, 325–334.

Gupta, R. P., Saha, A. K., Arora, M. K. and Kumar, A., (1999). Landslide hazard zonation in part of Bhagirathi Valley, Garhwal Himalaya, using integrated remote sensing – GIS. Himalayan Geol., 20, 71–85.

Guzzetti, F., Carrara, A., Cardinali, M., Reichenbach, P. (1999). Landslide hazard evaluation: a review of current techniques and their application in a multi-scale study, Central Italy. Geomorphology 31, 181–216.

Kamp, U., Growley, B.J., Khattak, G.A., Owen, L.A. (2008). GIS-based landslide susceptibility mapping for the 2005 Kashmir earthquake region. Geomorphology 101, 631–642.

Mathew John, Jha, V. K. and Rawat G. S. (2007). Weights of evidence modelling for landslide hazard zonation mapping in part of Bhagirathi Valley, Uttarakhand. Current Science, 92 (5), 628-638.

Pachauri, A. K. and Pant, M. (1992). Landslide hazard mapping based on geological attributes. Eng. Geol., 32, 81–100.

Rosenfeld, C.L., 1994. The geomorphological dimensions of natural disasters. Geomorphology 10, 27–36.

Saraf, A.K. (2000). Cover: IRS-1C-PAN depicts Chamoli earthquake induced landslides in Garhwal Himalayas, India. International Journal of Remote Sensing 21, 2345–2352.

Saha, A.K., Gupta, R.P. and Arora, M.K. (2002). GIS-based landslide hazard zonation in the Bhagirathi (Ganga) Valley, Himalayas. Int. J. Remote Sensing, 23, 357–369.

Sarkar, S., Kanungo, D. P., Chauhan, P.K.S. (2011). Varunavat landslide disaster in Uttarkashi, Garhwal Himalaya, India. Quarterly Journal of Engineering Geology and Hydrogeology 44, 17–22.

Sarkar, S., Kanungo D.P and Mehrotra G.S. (1995). Landslide Hazard Zonation: A case study in Garhwal Himalaya, India. Mount. Res. and Dev., 15(4), 301-309.

Sarkar, S. and Kanungo, D.P. (2004). An integrated approach for landslide susceptibility mapping using remote sensing and GIS. Photogrammetric Engineering & Remote Sensing, 70, 617-625.

Sarkar, S. and Gupta, P.K. (2005). Techniques for landslide hazard zonation--Application to Srinagar-Rudraprayag area of Garhwal Himalaya. Geological Society of India, 65, 217-230.

Sinha, L.K. (1991). Spatial statistical prognostic modeling for Landslide Hazard Zonation in part of Giri River Basin, Sirmur Dist., Himachal Pradesh, using Remote Sensing and GIS techniques. Unpublished report, Indian Institute of Remote Sensing, pp 93.

Hon'ble reviewer

1. **Dr. Arun Kumar Shandilya**
Department of Geology
Dungar College
MGS University
BIKANER - 334 003

2. **Dr. T.S. Sharma**
Director
Remote Sensing Centre
Department of Sci & Technology
Pal Road
JODHPUR - 342 001

Estimation of Coda Wave Attenuation for Seismic Hazard Mitigation for Engineering Structure in Manipur

Manichandra Sanoujam[1], Arun Kumar[2] and L. Someswar Roy

Department of Earth Sciences, Manipur University Imphal 795003 India
E-mail: manichandra.sanoujam@gmail.com

Abstract

The attenuation of seismic wave energy in Indo Burmese Arc (IMR) of NE Himalayas has been estimated using local earthquakes. Most of the analyzed events are from the vicinity of Chindwin River of Central Mollase Basin, Myanmar where the subduction activity is assumed to be going on, and is well-defined by the concentration of the epicenters in the arc shape. The time-domain coda-decay method of a single back-scattering model is employed to calculate frequency dependent values of Coda Q (Q_c). A total of 229 local earthquakes of magnitude range 3.2–6.4 have been used for Q_c estimation at central frequencies 1.5, 3.0, 6.0, 9.0, 12.0 and 16.0 Hz with a lapse time 30 s starting at double the time of the primary S-wave from the origin time. The estimated average frequency dependence quality factor gives the relation, $Q_c = 131f^{1.075}$, while the average Q_c values vary from 195.5 at 1.5 Hz to 2420.9 at 16 Hz central frequencies. The observed coda quality factor is strongly dependent on frequency, which indicates that the region is seismic and tectonically active with high heterogeneities.The estimated average frequency dependence quality factor gives the relation, $Q_c = 131.84f^{1.0753}$, while the average Q_c values vary from 196 at 1.5 Hz to 2421 at 16 Hz central frequencies. The observed coda quality factor is strongly dependent on frequency, which indicates that the region is seismic and tectonically active with high heterogeneities. The results are compared with similar regions of the world.

Keywords: Coda Wave, Sesmic Hazard, Asesmic Structures, Manipur.

Introduction

Elastic structure of the lithosphere is found to be inhomogeneous from various kinds of recent geological and seismological observations. The appearance of S-coda wave following direct S-phase in seismograms, is an apparent evidence for the existence of such an in homogeneity. The scattering and attenuation of high frequency seismic waves are important to physically characterize the Earth medium. A seismogenic region can be characterized by tectonic, seismic and volcanic activities in addition to geological formation and geological history. All the seproperties quantify the behaviour of the seismic energy propagation in the lithosphere and can be utilized for seismic hazard mitigation. As the seismic energy propagates through the earth medium, its energy (amplitude) decays due to geometrical spreading, intrinsic attenuation and scattering attenuation. Intrinsicattenuation converts the seismic energy to heat dueto anelastic absorption and scattering attenuation redistributes the energy at random heterogeneities present in the upper earth medium. Therefore, the attenuation of seismic waves in the lithosphere is an importantproperty for studying the regional earth structure and seismotectonic activity.

Seismic attenuation is usually considered to be a combination of two mechanisms, scattering loss and intrinsic absorption. Scattering redistributes wave energy within the medium. Conversely, intrinsic absorption refers to the conversion of vibration energy into heat. Wu (1985) introduced the concept of seismic albedo B_0 as the ratio of scattering loss to a total attenuation. There are several papers that discuss mechanisms for intrinsic absorption that lead to frequency-independent Q_p^{-1} and Q_s^{-1} (Knopoff, 1964; Jackson and Anderson, 1970; Marko and Nur 1979; Dziewonski, 1979). Scattering loss to heterogeneities distributed in the Earth causes a decrease in amplitude with travel distance (Aki, 1980).

Recorded waveforms show a large variation in amplitude near the direct S arrival; the variation decreases as lapse time increases. S-coda waves have a common envelope shape at most stations near the epicentre after about twice the S-wave travel time (Rautian and Khalturian, 1978). The decay of these waves with time, in a seismogram, provides the average attenuation characteristics of the medium instead of the property of a single path connecting from a source to the station. The properties of the coda can be analysed applying statistical method requiring a small number of parameters. Aki (1969) and Aki and Chouet (1975) are pioneers in this field, and they proposed backscattering model to use the coda waves of local earthquakes for estimation of quality factors (Q_c) of coda waves in a region.

Jin and Aki (1988) state that the coda quality factor, Q_o (inverse of seismic attenuation factor) at a frequency of 1 Hz can be useful to quantify the seismicity of regions. Jin and Aki (1989) strongly correlated the coda-Q^{-1} with the degree of fracture in the lithosphere associated with the seismicity. After the advent of coda wave theory (Aki, 1969), these short-period secondary waves are utilized for crustal studies near the source in a laterally heterogeneous upper lithosphere, the seismicity of the region and earthquake source mechanism. In the single-scattering model, the coda is considered as a superposition of back-scattered wavelets from randomly distributed heterogeneities (Aki, 1969; Aki and Chouet, 1975; Rautian and Khalturin, 1978). The scattering is produced by irregular topography, complex surface geology, heterogeneous elastic property of the rocks, faults and cracks, which are more near the surface and less in the deep region. The spatial variation of regional coda quality factor has been utilized for a better understanding of tectonics, seismicity, seismic risk analysis and engineering seismology (Singh and Herrmann, 1983; Jin and Aki, 1988). It has been observed that the coda spectrum of small earthquakes for near source is independent of earthquake size, epicentral distance and the path between station and epicentre but depends on lapse time from the origin time of the earthquake (Aki and Chouet, 1975; Sato, 1977). This suggests that the coda part of the seismogram is due to an average scattering effect of the medium in the region near the source and station.

The coda Qhas also been estimated at different parts of India by Gupta et al. (1995) for Garhwal Himalayas, Mandal and Rastogi (1998) for Koyna region, Gupta et al. (1998) for Koyna region, Tripathi and Ugalde (2004) for southern India and Mandal et al. (2004) for Kutch area. The estimated values of Q_c are almost same for these regions but Q_0 and frequency parameter are variable with seismic, tectonic and geological features. The NE India is seismically and tectonically active region with prominent great past earthquakes. So far, not much work has been done for seismic wave attenuation in this region, therefore, it is necessary to study the attenuation of seismic energy. This work may be useful for other research mainly in earthquake hazard assessment.

Tectonic setting and seismic activity

The eastern part of NE-India lies in the Indo Burmese Range (IBR) on the subduction zone of the Indian and Eurasian plate boundary. A mix of subduction and collision between the two plates has contributed to the complex tectonic processes and tectonic evolution in the NE India region. It is observed that the dipping seismic zone below the IBR is confined within the latitude 21°-26°N in a NNE direction, while in the northern part beyond 26°N the subduction mechanism is possibly taken over by the collision tectonics. Seismic activity beyond 26°N and in the Mishmi block may be identified as the eastern syntaxis zone activity at the junction of the Himalayan arc and the Burmese arc. This syntaxis zone was the source area of the 1950 great earthquake, that occurred by strike-slip mechanism (Ben-Menahem et al., 1974).

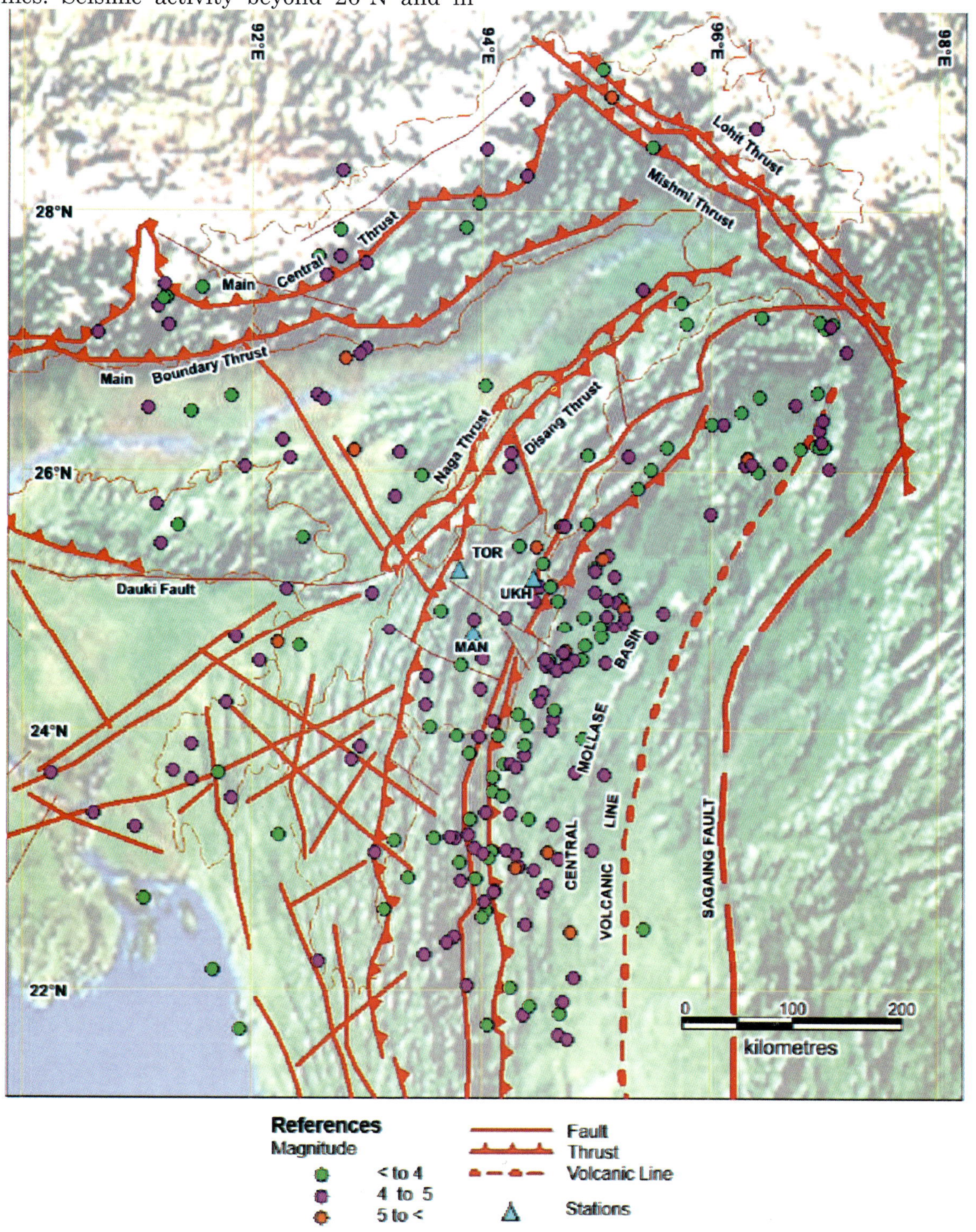

Fig. 1: General seismotectonic and topographical map of NE-India and adjoining area showing distributions of the earthquake epicentres used for the present study.

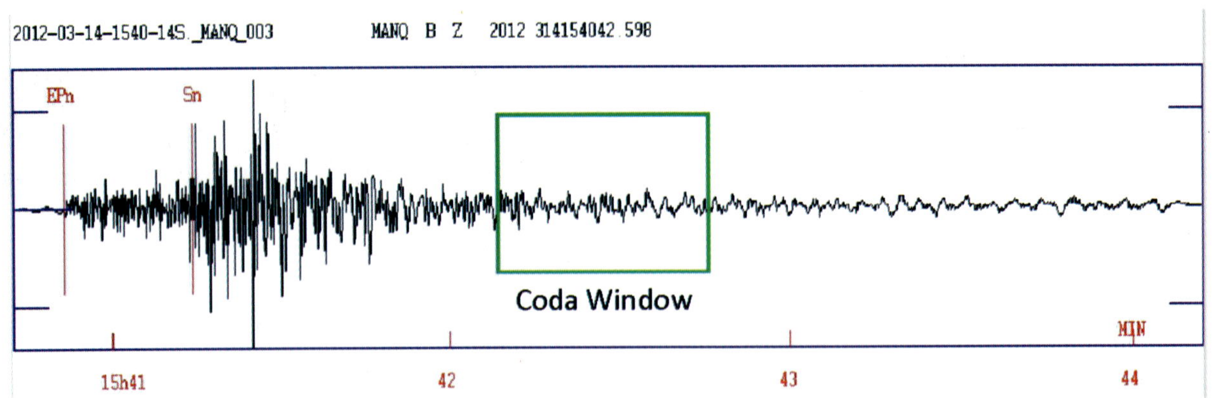

Fig. 2: A sample waveform showing the coda part. The details of the event is: 14th March 2012 Origin Time : 15hr 40 m 21.30 sec, Epicenter : 22.9210°N 94.4660°E, Depth : 101 Km Mag : 4.0, Epicentral Distance : 210 Km.

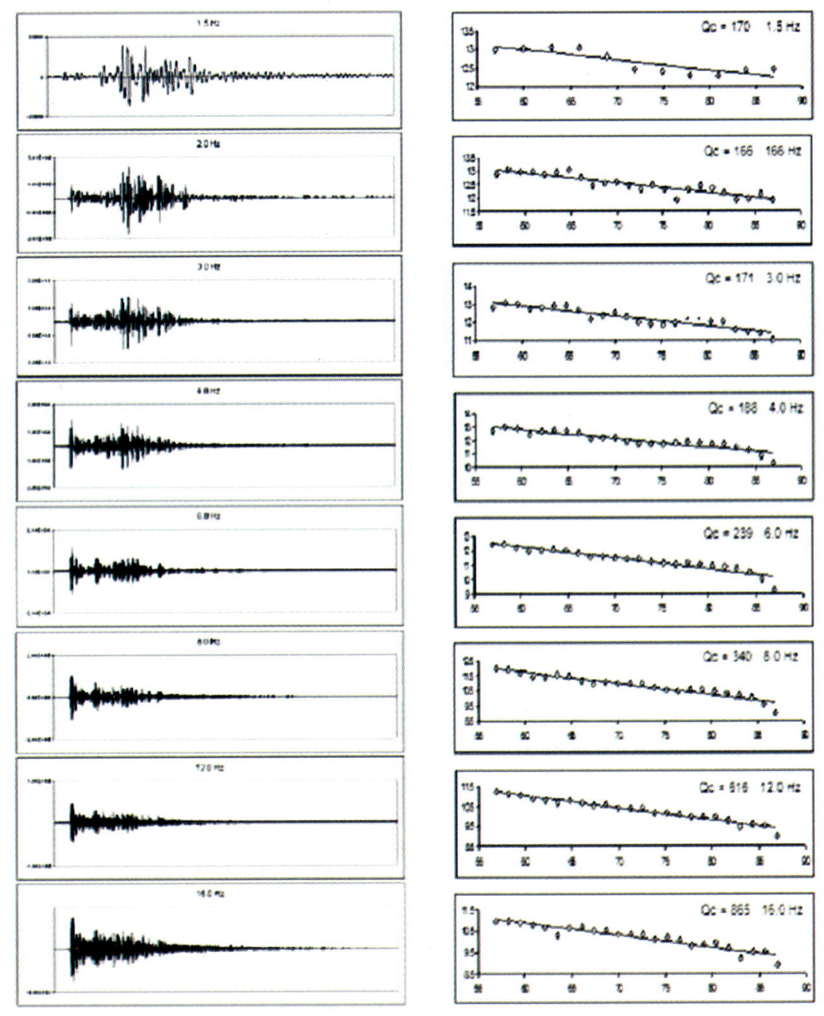

Fig. 3: (a) An example of original seismogram (Fig. 2) and band pass filtered coda waveforms. (b) Corrected and smoothed logarithmic coda amplitude using the *rms* approach for the portion shown by arrows, top left side in (a). The straight line are fitted in the least-squares sense. The estimated Q_c value for each frequency component is shown respectively.

The IBR comprises of Naga Hills, Chin Hills and the Arakan-Yoma. The structural and tectonic patterns of the region are transitional between the NE-SW trend of the Naga-Patkai Hills in the north and northeast to N-S trend in the Mizoram and Chin hills in the south. The major structural and tectonic features such as folds, reverse/thrust faults follow the regional strike. The geological setting and structural pattern in the region suggest a typical deformation mechanism due to the plate interaction. The geodynamic evolution of the IBR dates back post Cretaceous in a typical subduction tectonic regime between India and Eurasia, which resulted sequences of flysch ocean sediments and ophiolites suits of rocks along the entire belt (Nandy, 2001). The post Cretaceous evolution of the Asian plate in Myanmar is strongly influenced by spreading of the Eastern Indian Ocean and by northward movement of "Greater India" (Bannert and Helmcke 1981). The emergence of the IBR happened during the Miocene, generating an Outer Island Arc and forming the Bay of Bengal. Subduction of the Indian oceanic plate along the west coast of Burma (Myanmar) to the south experienced the tectonic deformation with the approach of continental Indian crust that led to a series of thrusts in the northern part of the IBR in Manipur and Nagaland (Fig. 1).

Coda *Q* calculation

The coda waves have been estimated as a superposition of secondary waves through single back-scattering of primary body waves at randomly distributed heterogeneities (Aki, 1969; Aki and Chouet, 1975). The decrease of coda wave amplitude with lapse time at a particular frequency is only due to energy attenuation and geometrical spreading but independent to earthquake source, path propagation and site amplification (Aki, 1969). The attenuation of seismic waves is the sum of intrinsic and scattering attenuation, where in the first case the energy is converted to heat through anelastic absorption and in second case it is redistributed through refraction, reflection and diffraction at random discontinuities present in homogeneous medium. Generally, the *Q* factor increases with frequency (Mitchell, 1981) through the relation

$$Q = Q_0\left(\frac{f}{f_0}\right)^n$$

The digital events data utilized for coda Q calculation were recorded during January 2010 and May 2013by Manipur University Broadband network of three seismic stations.

The digital data from 229 earthquakes recorded by three stations equipped with digital broadband systems, Quantera Q330 are analysed for the estimation of the quality factor Q_c. The digital seismograms were recorded by STS2 Broadband Seismometer at a sampling rate of 100 samples/sec. Locations of the events used for the analysis are shown in Fig. 1.

A sample seismogram of P-wave, S-wave and coda waves recorded at the MAN observatory is illustrated in Fig. 2. The backscattering model implies that scattering is a weak process and outgoing waves are scattered only once before reaching the receiver. Under this assumption the coda amplitudes A(f,t) in a seismogram can be expressed for a central frequency f over a narrow band-width signal as a function of lapse time t, measured from the origin time of the seismic event, as

$$A(f,t) = S(f)t^{-a} \exp(-\pi ft/Q_c)$$

where S(f) represents the source function at frequency f, and is considered a constant as it is independent of time and radiation pattern, and therefore, not a function of factors influencing energy loss in the medium; a is the geometrical spreading factor, and is taken as 1 for body waves, and Qc is the apparent quality factor of coda waves representing attenuation in a medium. The above equation can be rewritten as

$$\ln [A(f,t)t] = C\text{-}bt \text{ , where } b = \pi f/Q_c \text{ and } C= \ln S(f)$$

The above equation represents a straight line, and slope of which b (=πf/Qc) provides the Q_c value at a central frequency f. This equation has been used for estimation of Q_c for the Earth crust below Manipur and adjoining areas in IMR.

Each seismogram is filtered using Butterworth band pass filter of eight poles. Eight frequency bands (bandwidth 0.67fc, where f_c is the central frequency) are used for this propose. Low cut-off of these bands are given in Table 1. Seismograms thus filtered are used for detailed study of the decay of coda waves amplitudes with time to estimate Qc at nine central frequencies. The coda waves of a fixed time window 30 sec for each filtered seismogram, which starts from the lapse time t>2ts, where ts is the direct S-wave travel time (Rautian and Khalturin, 1978), are chosen for the analysis. The 30 sec coda windows of all the filtered seismograms from lapse time t onwards are smoothed using a root-mean-square (rms) technique. It calculates the rms values of coda amplitude of the filtered seismograms in a time window of 5.12 sec (for 1.5 Hz) and 2.56 sec (for the remaining central frequencies) with a sliding window along the coda in steps of half of the time window. The value at each step is evaluated in the same frequency band. The values thus obtained by this approach constitute a smoother envelope of the coda. Fig. 3 represents an example of the smoothed logarithmic amplitude of the coda part of the seismogram, in which the effect of geometrical spreading is corrected for all the filtered seismograms. The respective values of quality factor Q_c is also given with each plot (Fig 3).

Table 1. Central frequency of bandpass filter with low and high cut-off frequencies

Low cut-off	Central frequency (fc) Hz.	High cut-off
1.00	1.5	2.00
[illegible]		
[illegible]		
[illegible]		
[illegible]		
5.33	8.0	10.67
8.00	12.0	16.00
10.67	16.0	21.33

The Q values are calculated through the CODAQ subroutine of SEISAN (Havskov and Ottemoller, 2003). The S-wave time is calculated through the P wave time using $V_p/V_s = 1.74$. The lapse times are selected at 2ts to avoid the data of the direct S wave and to utilize the minimum possible sampling volume for Q_c values (Havskov et al., 1989). All the seismograms are band pass filtered at central frequencies (f_c) of 1.5, 3.0, 6.0, 9.0, 12.0 and 16.0 Hz with bandwidths of 1.0, 2.0, 4.0, 6.0, 8.0 and 12.0, respectively. An increasing frequency band is used for increasing central frequency to avoid ringing and to take constant relative bandwidths for getting equal amount of energy into each band as suggested by Havskov and Ottemoller (2003). The RMS amplitude of the last 5 s data of lapse time window is divided by noise data of same length before the onset of the P wave to calculate the S/N ratio. All the seismograms having S/N ratio below 2 are rejected for better Q_c values. In the same manner the criteria of correlation coefficient ≥0.50 is applied to obtain reliable Q_c values. The logarithm of product of RMS amplitude and lapse time is plotted against lapse time as shown in Fig. 2 for calculating Q_c from slope of the linear regression curve of ln $(A(f, t)t)$ and t.

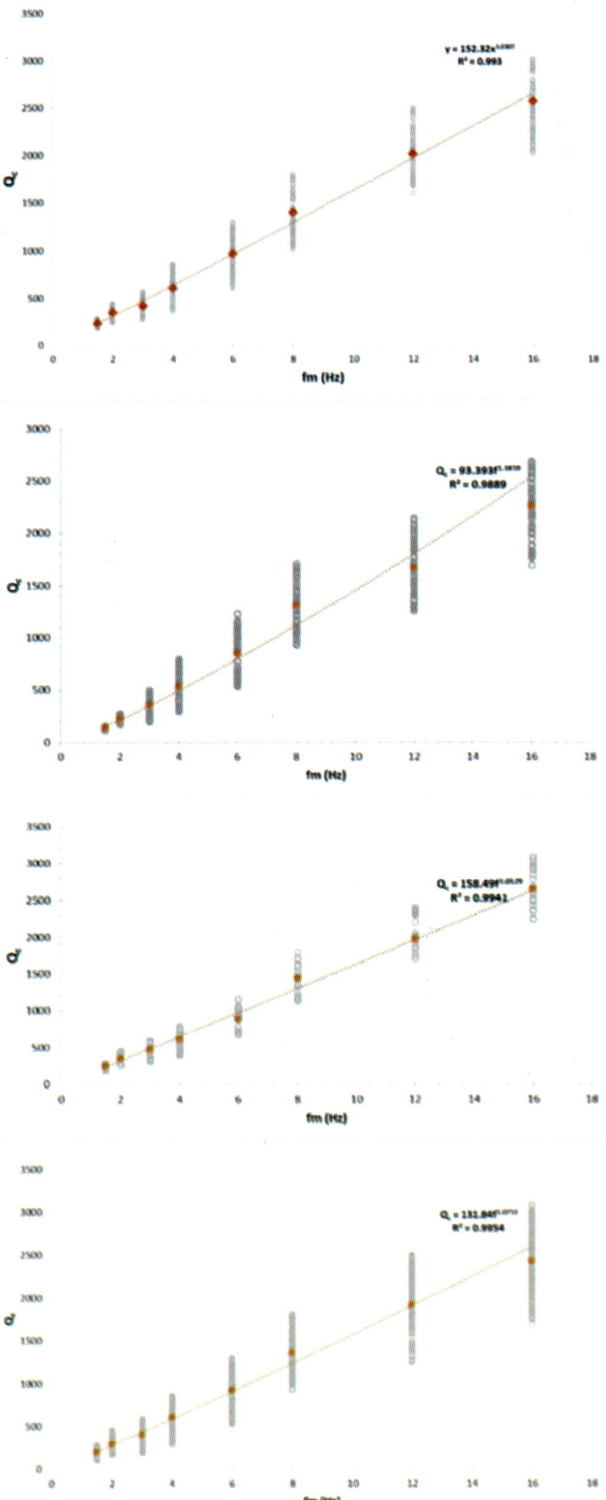

Fig. 4 a. to d. Plots of quality factors and central frequencies for all stations with linear regression frequency dependent relationship, $Q_c = Q_0 f^n$. (a) Q_c with f_m of MAN (Q_c = 136f1.07), (b) Q_c with f_m for UKH ($Q_c = 138f^{1.12}$), (c) Qc with f_m for TORI ($Q_c = 205f^{0.95}$), (d) Average Q_c with f_m for all the stations ($Q_c = 158f^{1.05}$).

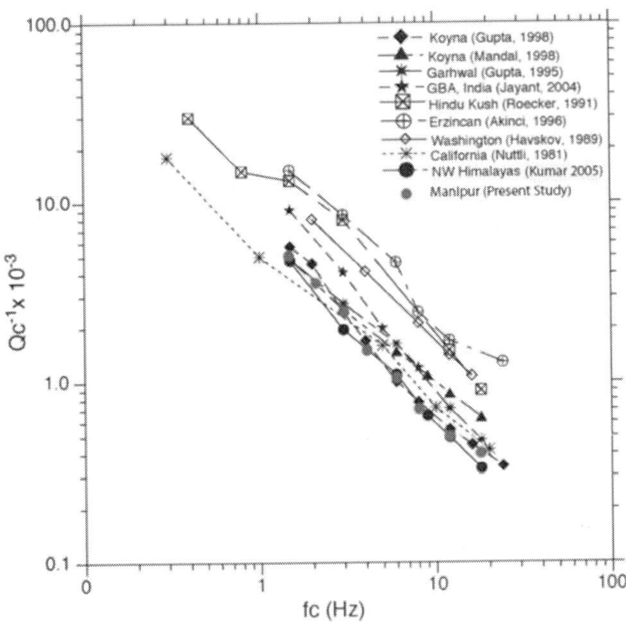

Fig. 5. Comparison of coda-Q_c of Manipur with reported coda-Q_c of other regions of the world

Results and Discussion

The northeast India region is tectonically very complex with collision tectonics in the northeast Himalaya and subduction tectonics in the IMR. The IMR region is seismically more active in the northeast India region due to north eastward under thrusting of the Indian plate beneath the Burmese arc in eastern Manipur region. The earthquakes show a dipping seismic zone below the IMR, and depth of the subducted lithosphere goes down to 200 km (Kayal, 1996). Hypocentres of the selected earthquakes in this study, however, fall at a depth range of 10-165 km, within the subducting lithosphere. Magnitude of these earthquakes are 3.2 – 6.4.

The quality factor of coda waves (Q_c) has been estimated using the broadband digital data of 229 selected local earthquakes, that occurred at an epicentral distance less than 400 km, from the observatories (Fig. 1). Single backscattered model has been applied. Eight frequency bands of central frequency e.g., 1.5, 2.0, 3.0, 4.0, 6.0, 8.0, 12.0, 16.0 have been used for this purpose. From the analysis of coda waves of 30 s window length, a total 1832 Q_c measurements were made in the eight frequency band. The variations of values of Q_c measurements for different frequencies are given in Table 2. The average Q_c values of the region for different frequencies are mentioned in Table 3 and Q_0 for MAN, TORI and UKH are 152.32, 158.49 and 93.39 respectively. The n values for the locations are MAN, TORI and UKH 1.0307, 1.0129 and 1.1859 respectively. Variation of Q_c as a function of frequency are plotted in Fig. 3. The variation of Q_c at various central frequencies may be attributed to different factors: i) difference in local site geological conditions of recording stations, ii) difference in focal depth and iii) difference in epicentral distance. To provide an average Q_c for different location, mean value of Q_c is estimated from all the observed Q_c at each central frequency (Fig. 4a-c). The plot of Q_c values as a function of frequency shows a definite dependence on frequency. By fitting the mean values of Q_c, given in Table 3, to the power law $Q_c = Q_o f^n$, the frequency-dependence average Q_c relationship is obtained for the region as $Q_c = 131.8f^{1.0753}$.

Table 2. Variation of Q_c at different central frequencies

f_c	Q_c	
	Minimum	**Maximum**
1.5	110	286
2	172	452
3	198	592
4	298	859
6	532	1299
8	926	1813
12	1250	2496
16	1690	3095

Table 3. Average quality factor at different frequencies

f_c	$Q_c \pm \sigma$
1.5	196 ± 55
2	294 ± 74
3	395 ± 95
4	603 ± 152
6	920 ± 206
8	1353 ± 236
12	1912 ± 297
16	2421 ± 326

Many researchers have reported that seismically active regions display low value of Q_0, defined as Q_c at 1 Hz which represents the level of medium heterogeneities, and stable shield type structure shows high values (e.g. Herrman, 1980; Roekker, *et. al.*, 1982; Singh and Herrmann, 1983). The level of tectonic activity in a region, on the other hand, can be correlated with the degree of frequency dependence 'n'. Higher n value (~1) is observed for tectonically active regions compared to tectonically stable regions where n is low. For central Asia, Roecker *et al* (1982) reported that while Q_c increases with depth, the frequency dependence decreases from ~ $f^{1.0}$ at depth ~100 km to ~$f^{0.75}$ at depth ~400 km. Pulli (1984) investigated lapse time dependence of Q_c for New England from the data set having depth from surface to 500 km. He observed n = 0.95 for lapse time <100 sec, 0.95 for lapse time <100 sec and 0.40 for lapse time >100 sec. Results from both these regions provide information for a wider area with deeper depth extent. For tectonically active regions and short coda window length, Q_c is often small (≈100) and n may be upto 1 (Singh and Herrmann, 1983; Herraiz and Espinosa, 1987). Gupta and Kumar (2002) compared Koyna and Garhwal Himalayan region for the time-lapse window from 20-50 sec; where n ranges from 1.02 to 1.09. These observations indicate both the regions are tectonically active. They further reported n = 1.02 for northeast India. In the present study, the estimate of n in Manipur for 30 sec window with a time lapse of twice S-travel time is 1.0753 which is comparable with earlier observation.

The coda Q_c values of the present study are compared with other subduction regions. In the Indo-Myanmar subduction zones, the lithosphere is characterized by n=1.1859 in this study, which is correlatable with other subduction zones in the world (Fig. 4). It is observed that the upper part of the subducted lithosphere beneath Manipur valley is tectonically active where $Q_c = 152.3f^{1.0307}$. Seismic activity is, however, more intense from intermediate to deeper part (70-200 km) of the lithosphere; this deeper zone needs to be understood in terms of Q_c in future study. The estimated coda Q value of the region indicates that the attenuation is greater towards high tectonic and seismic active areas. The estimated average frequency dependent relationship of the region is $131.8f^{1.0753}$, while this relationship varies from $93.3f^{1.1859}$ for the subduction region to $152.32f^{1.0307}$ for the part of Imphal valley and eastern foothills.

The coda attenuation, Q_c^{-1} values estimated at different frequencies in this study, are compared with other tectonic and seismic active regions of the world in Fig. 5. The Q_c^{-1} values of this region are comparable to other Indian regions but less than most of the other available values of the world. The coda attenuation of this region is similar to Koyna, India as given by Gupta et al. (1998) and NW Himalayas as given by Kumar et al. (2005). Our results also agree with other Indian active regions but are less than other observed active regions of the world. The frequency parameter of this region is greater compared to the other regions, which indicates that the upper lithosphere of the region is more heterogeneous. The attenuation properties of the Indian lithosphere have been estimated earlier by coda *Q* methods for some regions by Gupta et al., 1995 for Garhwal Himalayas, Mandal and Rastogi, 1998 for Koyna region, Gupta et al., 1998 for Koyna region, Tripathi and Ugalde, 2004 for southern Indian region and Mandal et al., 2004 for Kutch area. It is quite clear from Fig. 5 that the coda Q^{-1} values observed by this study are comparable to other Indian regions but more sensitive to frequency, which shows that this region is more heterogeneous as compared to other parts of the country.

The high coda Q value of the TORI with the relationship of $158.5f^{1.0129}$ compared with other locations of the study area may be correlated with the aseismic nature of Churachandpur-Mao-Fault (CMF). The frequency dependent quality factor indicates that the seismic energy of high frequency waves is attenuated more in upper lithosphere than in the lower lithosphere. The coda attenuation factor of NE region is comparable to other seismic active regions of India but the frequency parameter is somewhat more sensitive than the other regions.

References

Aki, K. 1969. Analysis of seismic coda of local earthquake as scattered waves. J. Geophys. Res., 74,615-631.

Aki, K. and B. Chouet, 1975. Origin of coda waves : Source, attenuation and scattering effects. J. Geophys. Res. 80, 3322-3342.

Dziewonski, A.M., 1979. Elastic and anelastic structure of the Earth. Rev. Geophys. Space Phys., 17,303-312.

Gupta, S.C., and Ashwanikumar, 2002. Seismic wave attenuation characteristics of three Indian Regions : A comparative study. Current Sc. 82,407-413.

Gupta, S.C., Singh, V.N., Kumar, A., 1995. Attenuation of coda wavesinthe Garhwal Himalayas. India. Phys. Earth Planet Inter. 87,247–253.

Gupta, S.C., Teotia, S.S., Gautam, N., 1998. Coda *Q* estimates in the Koyna region, India. Pure Appl. Geophys. 153, 713–731.

Havskov, J., Malone, S., McClurg, D., Crosson, R., 1989. Coda Q for the state of Washington. Bull. Seismol. Soc. Am. 79, 1024–1038.

Havskov, J., Ottemoller, L., 2003. SEISAN: The Earthquake Analysis Softwares for Windows, Solaris and Linux, Version 8.0. Institute of Solid Earth Physics, University of Bergen, Norway.

Herraiz, M. and Espinosa, A.F. 1987. Pure Appl. Geophys. 125,499-577.

Herrmann, R.B., 1980. Q estimates of the Coda of local earthquakes, Bull. Seism. Soc. Am., 70:447-468.

Jackson, D.D. and Anderson D.L. 1970. Physical mechanism of seismic wave attenuation, Rev. Geophys. Space Phys., 8,1-63.

Jin, A., Aki, K., 1988. Spatial and temporal correlation between coda Q and seismi city in China. Bull. Seismol. Soc. Am. 78,741–769.

Jin, A., Aki, K., 1989. Spatial and Temporal correlation between coda Q^{-1} and seismicity and its physical mechanism. J. Geophys. Res. 94, 14041–14059.

Kayal, J.R. (1996). Earthquake source process in Northeast India: A review. *Him. Geol.,* 17: 53-69.

Knopoff. L. 1964. Q, Rev. Geophys., 2, 625-660.

Mandal, P., Jainendra, Joshi, S., Kumar, S., Bhunia, R., Rastogi, B.K., 2004. Lowcoda Q_C in the epicentral region of the 2001 Bhuj earthquake of M_W 7.7. Pure Appl. Geophys. 161, 1635–1654.

Mandal, P., Rastogi, B.K., 1998. A frequency-dependent relation of coda Q_C for Koyna-Warnaregion. India. Pure Appl. Geophys. 153,163–177.

Marko, G.M. and Nur, A. 1079. Wave attenuation in partially saturated rocks. Geophys., 44, 161-178.

Mitchell, B., 1981. Regional variation and frequency dependence of Q_β in the crust of the United States. Bull. Seismol. Soc. Am. 71, 1531–1538.

Nandy, D.R., (2001): Geodynamics of northeast India and the adjoining region. ACB Publication.

Naresh Kumar, Imtiyaz A. Parvez, H.S. Virk 2005. Physics of the Earth and Planetary Interiors 151, 243–258

Pulli J J 1984 Attenuation of coda waves in New England; Bull. Seismol. Soc. Am. 74 1149-1166

Rautian, T.G. and Khalturin, V.L. 1978. The use of the coda for determination of the earthquake source spectrum. Bull, Seismol. Soc. Am. 68,923-948.

Roecker, S.W., Tucker, B., King, J., and Hatzfeld, D, 1982. Estimates of Q in Central Asia as a function of frequency and Depth using the Coda of Locally Recorded Earthquakes. Bull. Seismo. Am., 72, 129-149.

Sato, H., 1977. Energy propagation including scattering effects: singleiso tropic approximation. J. Phys. Earth 25, 27–41.

Singh, S., and Herrmann, R.B., 1983. Recognition of crustal coda Q in the continental United States. J. Geoiphys. Res., 88,527-538.

Tripathi, J.N., Ugalde, A., 2004. Regional estimation of Q from seismic coda observation by the Gauribidanur seismic array (southern India). Phys. Earth Planet Inter. 145, 115–126.

Wu, R.S. 1985. Multiple scattering and energy transfer of seismic waves-separation of scattering effect from intrinsic attenuation—1. Theoritical modeling, Geophys. J.R. Astron. Soc., 82, 57-80.

Hon'ble reviewer

1. **Prof. Kamal K. Agarwal**
   Chairman
   Centre Ad. Study in Geology
   Lucknow University
   LUCKNOW - 226 007
2. **Prof. S.N. Lal**
   Department of Geography
   Kumaon University
   NANITAL - 263 001

IV-57

Remote Sensing and GIS Based Study of the Landslides of the Area Between Bomdila and Mandla, Kameng district, Arunachal Pradesh

Vaibhava Srivastava and Hari B. Srivastava

Department of Geology, Banaras Hindu Univeristy, Varanasi -221005
E-mail: hbsrivastava@gmail.com

Abstract

The remote sensing and GIS based present study analyzes the roles of geomorphological and geological parameters over the landslide occurrences in the study area between Bomdila and Mandla in West Kameng district of Arunachal Pradesh. The distribution pattern of these landslides in the study area reveals that most of the old or active land slide activities are located near the lineaments. The landslides are frequently seen in the low grade rocks of Dirang Formation. However, the crystalline rocks of the Bomdila Group exhibit less number of landsides because of their higher strength chiefly.

Keywords: Landslide, GIS, Remote Sensing, Arunachal Pradesh.

Introduction

The Arunachal Pradesh occupies the eastern most segment of the Himalaya and forms a part of eastern Himalayan Syntaxis (Fig. 1). The Lesser Himalayan Crystallines (LHC) locally known as Bomdila Group, consist of thick sequence of low to medium grade crystalline rocks exhibiting inverted metamorphic character. The Bomdila Group of crystallines are thrust over the Lesser Himalayan Sedimentary sequence known as Buxa–Miri Group, which occurs as window or as tectonic slice within the LHC. In the south the LHC is delimited by the Bomdila Thrust and the north contact is marked by Main Central Thrust (MCT), which separates the overlying Higher Himalayan Crystalline (HHC) (Bhusan et al, 1991, Bhattacharjee and Nandy 2007). The early researches laid a foundation for a proliferation of geological activities (Brown, 1912; Jain et al. 1974; Verma and Tandon, 1976; Acharya, 1974; Jangpangi, 1974 in the early 1970s and subsequently regional synthesis by Thakur, 1986, Singh and Chowdhary, 1990; Acharya, 1994; 1980; 1998 and Kumar, 1997) and established general stratigraphy and tectonic frame work of this region. In the western sector of the Arunachal Himalaya the rocks of Higher Himalaya are marked by the rocks of Se La Group (Das, et al. 1975) and are separated from the rocks of Bomdila Group to the south along MCT. Thus the rocks of the western sector of the Arunachal Himalaya are geologically very complex and tectonically highly disturbed. The

Himalayas are still in the process of rising and the ongoing northward movement of Indian plate against the Eurasian plate causes many neotectonic features to develop in the Himalayan region. Landslides are common feature found in the Himalayan zone as manifestation of the ongoing tectonic activities. The area between Bomdila and Mandla exhibit large number of landslide along the roadsides, some of which are still active. Because of these landslides, there is a perpetual problem of stranding of vehicle and loss of property and lives. The satellite images and field observations reveal that the landslide problem is not restricted only to the roadsides but it is widespread in the region which comes in the way of development of the region as a whole. The present work is aimed to study these landslides occurring between Bomdila and Mandla area with the view point of their geomorphological, geological and societal implications.

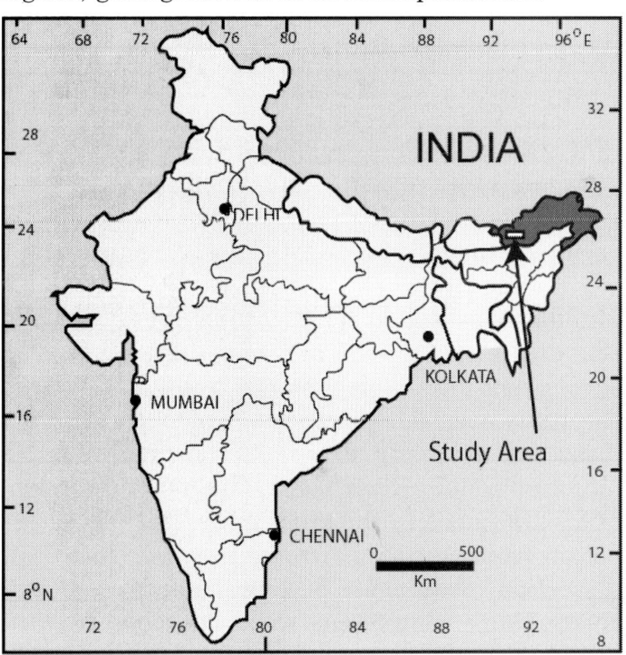

Fig. 1 Location map of the study area in West Kameng district of Arunachal Pradesh.

Study Area and Data Used

The study area lies between latitudes 27° 15' to 27°20' N and longitudes 92°15' to 92°30'E in the West Kameng district of Arunachal Pradesh (Fig. 1). The road to famous tourist places Se La pass and Tawang, passes through this area. The area is structurally very complex, topographically highly rugged and largely covered by thick forest. Therefore, in order to study even the inaccessible parts of area, the techniques of remote sensing and GIS have been used. The satellite data used in the present work are IRS LISS-III PAN and Landsat TM, in addition to the Survey of India toposheet no 83A/7.

Geology of the area

Some aspects of the geology of the area under investigation have been studied by workers like Bhusan et al. 1991 and Bhattacharjee and Nandy, 2007. Mishra (2007) described the tectonic sequence of the Precambrian rocks of Kameng district of Arunachal Himalaya. According to Singh (2009), the rocks of the Bomdila Group have been intruded by Proterozoic granitoids and has late metabasic dyke and the rocks have suffered deformation under low to medium grade metamorphism due to which many of the sedimentary structures have been obliterated. Based on the study of structures visible at microscopic to macroscopic levels, a reasonable deformational history of the area has been built up by Srivastava et al (2011).

The Bomdila Group comprises low-grade argillaceous and arenaceous metamorphites penecontemporaneous basic volcanic and large scale granitic intrusive. On the basis of the lithological assemblage, tectonics and metamorphism, this group is divisible into three formations (Bhusan et al, 1991; Srivastava et. al. 2009). The lower Tenga Formation comprising mafic metavolcanic, phyllites and quartzites followed by middle Dedza Formation characterized by carbonate rocks with minor amounts of black slates, and upper Dirang Formation represented by garnet muscovite, biotite schist, sericite quartzite tourmaline marble and carbonate phyllite.

Landslide Occurrences

There are a number of landslides present in the area which are distributed almost all over the region except in the south eastern part around Bomdila township (Fig. 2). In Fig. 2 shows the landslide distribution in the study area which has been mapped from the IRS LISS-III data using the supervised classification. This classification reveals a sizable number of landslides are found in the vicinity of the Gang River, the most prominent drainage of the area. The slides affected areas occur near the villages like Doilung, Khoina, Selari, Munchham, Baiku,

Dirang and Mandla in addition to the forested area and areas without habitation. The ground verification of these landslides reveals that many of these are active landslides with multiple recurrences. In the area under investigation various types of slope failure and mass movements have been observed which include rotational, translational, rock fall, creep, and flow categories of Varnes (1978), Cruden and Varnes (1996) and Hungr et al (2001). The landslides may occur in isolation as single slide or there may be a large number of slides present in a larger zone. The aerial length of such landslide zone has been found to exceed 2 km.

A sizeable number of landslides are old and dormant. The landslides of this kind have been utilized by the local villagers and they have developed the body portion of the slide as their agricultural field or orchards. The presence of long living tall trees in these orchards reveals that slide has remained dormant at least more than the age of the oldest tree. The active slides on the other hands do not have vegetation grown over them with characteristic shape and blue tone over the FCC image. Thus the blue tone areas shown in Fig. 2 reveal the zone of active landslides. The reddish or pinkish tone in these slide zones represents the cropland or the orchard developed over them. The areas which are forest covered or belong to any other land cover/ land use have been converted to black for sake of highlighting the landslide occurrences.

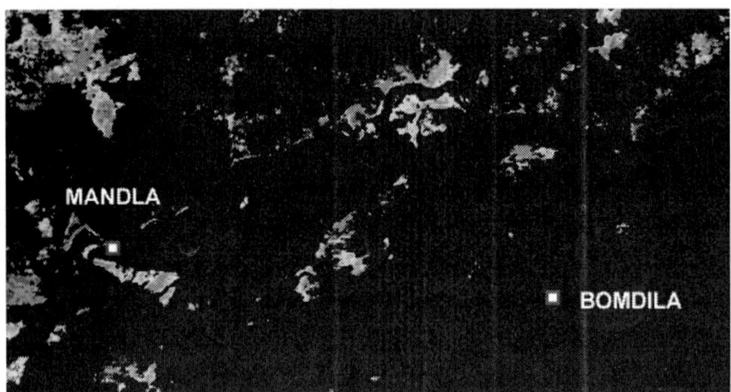

Fig. 2 Landslide occurrences in the study area mapped from IRS LISS-III data using the digital image processing.

When the landslides are observed vis- a- vis the geology of the area, it was found that majority of the landslides exist on the low grade meta-sedimentaries of Dirang Formation, the crystallines rocks (Bomdila gneisses and granites) of the area which are present in the southeastern portion of the area are more or less free from larger slides zones. The boundary between the Dirang metasedimenataries and Bomdila Crystalline is demarcated by Bomdila Thrust which passes from north of Bomdila with a roughly NE-SW strike.

GIS studies

The GIS study of the area begins with the digitization of information available on the toposheets. Thus contours, and point heights were digitized manually under a GIS to develop a Digital Elevation Model (DEM) (Fig. 3) and Digital Terrain Model (DTM) (Fig. 4), slope map (Fig. 5) and aspect map (Fig. 6). Keeping in view of the hindrance caused by the thick forest cover over the area under investigation, studies on the GIS generated DEM and DTM would help in understanding of the terrain characteristics in a better way which satellite imageries may not be able to represent.

Digital Elevation Model

The Digital Elevation Model (DEM) reveals that the elevation in the study area ranges between 1166 to 3161 meters above mean sea level (Fig. 3). The highest regions of the area are oriented roughly in East-West direction as a continuous ridge. The townships of Bomdila and Mandla are located on the same ridge. In fact, Bomdila and Mandla are two passes over the same ridge line which marks a water divide in the study area. The northern flank of this ridge is dissected by many drainages flowing in a NE direction to join River Gang, the most prominent of all drainages in the study area. In the western potion near Mandla however, the direction of the dissection drainages becomes northwest. The river Gang flows eastward parallel to the prominent ridge line.

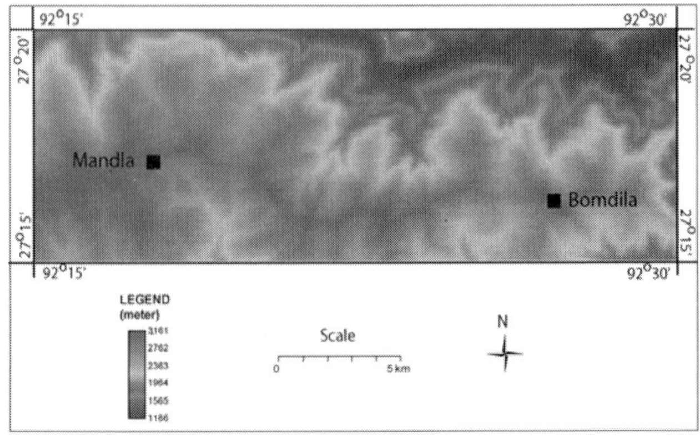

Fig. 3 Digital Elevation Model of the study area.

Digital Terrain Model (DTM)

The Digital Terrain Model of the study area (Fig. 4) was generated with help of contours and DEM in order to enhance the terrain characteristic of the area. The DTM has been successful in highlighting the regional strike of the low grade rocks of the Dirang Formation which occupy major portion of the area. The regional strike of these low grade metasedimen-taries is NW-SE which is very eminent at the back slopes. The Bomdila Formation however, does not show these characteristics in the area perhaps because of the inherent nature of granites for not being foliated. The other rocks of the Bomdila Crystallines like high grade gneisses and quartzites in the southeastern part of the area do show some foliation, which do not necessarily match with those of the Dirang Formation rocks.

The DTM is able to highlight the prominent lineaments trending in NE-SW directions. Many landslides which are active in nature have been found to follow these lineaments. A prominent alignment of landslides along these NE-SW trending lineaments (Fig. 2 and Fig.4) can be observed.

Fig. 4 Digital Terrain Model (DTM) of the study area showing prominent lineaments.

Slope

The mass movements have strong relation to slope and therefore, the landslides are often referred as slope failures. The slope map of the study area has been creating under GIS environment which is shown in Fig. 5. It is clear from Fig. 5 that the slope of the grounds in the study area is highly variable which varies from a nearly flat ground to highly steep slopes. The slope in the study area has been found as high as 84 degree. The flat area or low slope area (depicted by blue colour in Fig. 5) has been found mostly in the Gang river bed or its flood plain. However the flatter grounds are not confined along the Gang river, but they are also observed on the river terraces, old landslides and other cultivable lands which have been anthropogenically modified. The steeper slopes regions (depicted by shades of red in Fig. 5) though occupy small percentage map area, are distributed widely in the study area. The most common type of mass movement observed here are the rock falls. The major portion of the study area is having slope between 20 to 40 degree and most of the rotational landslides are observed in this zone.

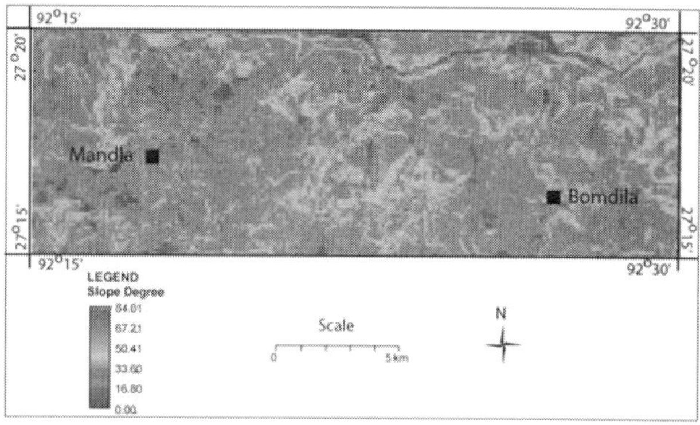

Fig. 5 GIS generated slope map of the study area.

Aspects of Slope

Aspect refers to the direction of the slope. Aspects of the slopes are important as far as landslides are concerned. The southern aspects of the slopes are considered to be most vulnerable because it receive sunlight most part of the day and hence takes away soil moisture from the hills. This situation is favorable for landslides because taller tree or vegetation that can hold the slopes does not develop over the ground. Northern slopes are considered safer because of the good vegetation growth over them. A slope aspect map has been prepared using GIS technology for the study area which is shown in Fig. 6. The study of landslide occurrences in respect to the slope aspect reveals that in case of the present area most landslides have been found on the northern slopes aspect. There can be two reasons for this (i) the study area itself forms the northern aspects of a major ridge over which the Bomdila and Mandla passes are located. Since the southern aspect counterpart has not been studied, a true comparison cannot be made. (ii) the area receives very high rainfall over the large period of the year, so there is no dearth of soil moisture on the slopes. In this situation, there is equal vulnerability of all the aspects of the slopes as far as landslides are concerned.

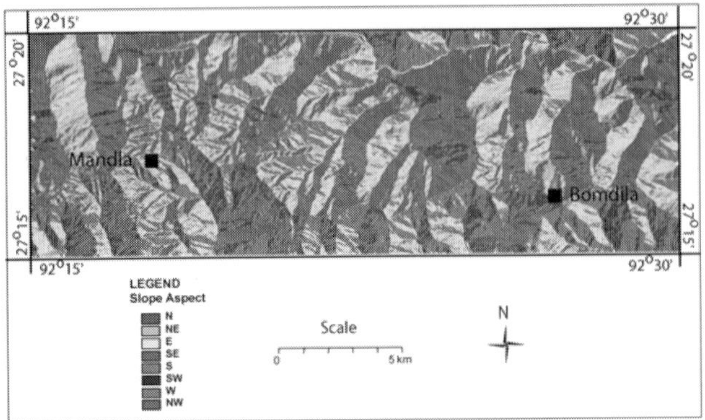

Fig. 6. GIS generated aspect map of the study area.

Discussion and Conclusion

The Himalayan regions are well known for their active tectonic activities because of ongoing collision of Indian plate with the Eurasian plate. Landslides are the common geoenvironmental hazards to the people of Himalayan region. This problem is more severe in the regions which are under high earthquake prone zones or receive heavy rainfall because earthquakes and heavy rainfalls are the major triggering factors for the landslide. The entire region of Arunachal Pradesh, unfortunately falls in such a zone which lies in the highest earthquakes prone zone (Zone 5) in India. The average annual rain fall in study area is quite high (1604mm) compared to national average of 887mm therefore, a large number of landslides are found in these regions. The remote sensing and GIS based present study analyzes the roles of geomorphological and geological parameters over the landslide occurrences in the study area between Bomdila and Mandla in West Kameng district of Arunachal Pradesh. The distribution pattern of these landslides in the study area reveals that most of the old or active land slide activities are located near the lineaments. This is a remarkable feature that makes a connection of these landslide activities related to the phenomenon of active tectonics. The activity is most preferred in zones which are weakest in their slope stability strengths. The landslides are particularly seen in the low grade Dirang Formation rocks and the crystalline rocks of the Bomdila Group have shown less number of landsides because of their higher strength.

Acknowledgements

Authors are thankful to Department of Science and Technology (Project No. ESS/16/242/2005/Kameng/02) for providing financial support.

References

Acharya, S.K. (1980) Stratigraphy and tectonics of Arunachal Himalaya In: Valdiya, K.S. and Bhatia, S.B. (Eds) Stratigraphy and correlation of Lesser Himalayan Formations, Hindustan Publishing Corporation, Delhi, 231-241.

Acharya, S.K. (1994) Cenozoic foreland basin and tectonics of the eastern sub Himalaya: problems and prospects. Him. Geol. 15, 3-21.

Acharya, S.K. (1998), Thrust tectonics and evolution of domes and the syntaxis in eastern Himalaya, India, Jour. Nepal. Geol. Soc. 18, 1-17.

Acharya, S.K, (1974) Stratigraphy and sedimentation of the Buxa Group, Eastern Himalaya. Himalayan geology, v.4, pp.102-116.

Bhattacharjee, S. and Nandi, S. (2007). Geology of the Western Arunachal Himalaya in parts of Tawang and west Kameng District, Arunachal Pradesh. Jour. Geol. Soc. India Vol. 72, pp.197-207.

Bhusan, S.K., Bindal, C. M. and Aggrawal, R. K. (1991) Geology of Bomdila Group in Arunachal Pradesh. Jour. Of Himalayan Geology, Vol. 2(2) pp.207-214.

Brown, C. J. (1912). A geologic reconnaissance through the Dihang Valley, being the geological result of the Abor Expedetion , 1911-12., Rec. Geol. Surv. India 42, 231-24.

Cruden D.M., Varnes D. J., 1996. Landslide types and processes. In: A.K Turner and R.L Shuster, (Eds), Landslides: Investigation and Mitigation. Transportation Research Board, Natonal Research Council, Washington D.C. Special Report, 247, pp 36-75.

Das, D., Bakliwal, P.C. and Dhoundiyal, D. P. (1975) A brief outline of the geology of parts of Kameng District, NEFA, Geol. Surv. India. Misc. Pub. V.24(1), 115-127

Hungr, O., Evans, S.G., Bovis, M. and Hutchinson, J.N., 2001. Review of the classification of landslides of the flow type. Environmental and Engineering Geoscience, VII, 221-238.

Jain, A. K. Thakur, V. C. and Tandon, S.K. (1974). Stratigraphy and structure of the Siang District, Arunachal (NEFA) Himalaya. Him. Geol. 4, 28-60.

Jangpangi, B. S. (1974) Stratigraphy and Tectonics of a part of Eastern Himalaya. Him. Geol. 4, 117-136.

Kumar, G. (1997) Geology of Arunachal Pradesh. Geol. Soc. India Banglore, 217p.

Mishra, D. K. (2007). Evidence of Neotectonic activity along active faults in Arunachal Himalay, NE India. Him. Geol. Vol. 28 (2) pp.75-78.

Singh B.R.K. (2009) Geometry and petrogenesis of Granitoids of Lesser Himalayan Crystallines, Western Arunachal Himalaya. Jour. Geol. Soc. India, V.75, 618-631

Singh S. and Chowdhary, P.K. (1990). An outline of the geological framework of the Arunachal Himalaya. Jour. Him. Geol. Vol. 1, 189-197.

Srivastava, H.B., Srivastava, V., Srivastava, R.K. and Singh, C.K. (2011) Structural analyses of the Crystalline rocks between Dirnag and Tawang, West Kameng district, Arunachal Himalaya, Journal of the Geological Society of India, v.78, pp 45-56.

Srivastava, R. K., Srivastava, H. B. and Srivastava, V. (2009). Petrology and Geochemistry of Proterozoic olivine tholeiite intrusive from the Central Crystallines of the Western Arunachal Himalaya, India: evidence for a depleted mantle. Current Science Vol. 97, no. 9, pp.1355-1361.

Thakur, V. C. (1986) Tectonic zonation and tectonic frame work of eastern Himalaya. Science de la terra, Memoir, 47, 347-366.

Varnes, D. J., 1978. Slope movement types and processes. In: Schuster R. L and Krizek, R. J. (Eds), Landslides, analysis and control. Transportation Research Board Special Report Number 176, National Academy of Sciences, pp. 11-33.

Verma, P. K. and Tandon, S. K. (1976) Geologic observations in a part of the Kameng District, Arunachal Pradesh (NEFA). Him. Geol. Vol. 6 pp. 250-286.

Hon'ble reviewer

1. **Dr. K. Ramesh Kumar**
   Scientist H
   Atomic Mineral Directorate
   HYDERABAD - 500 001
2. **Prof. Florimond De Smedt**
   Vrije Universiteit Brussel
   Pleinlaan 2, 1050
   BRUSSELS (BELGIUM)

IV-58

Landslide Hazard Zonation and Management Using Remote Sensing and GIS

R.K. Chingkhei, Robert L and Arun Kumar

Department of Earth Sciences, Manipur University, Imphal-795 003
E-mail: chingkhei@gmail.com

Abstract

The present study deals with the landslide zonation hazard and generation of Landslide Hazard Management (LHM) map using remote sensing data and Geographic Information System (GIS) techniques. The study area is a stretch of road section between Marog to Naugam within 2km buffer on both sides of the Jammu-Srinagar National Highway (NH-1A). The present study uses Landslide Information System (LIS), a GIS based software that uses Analytic Hierarchy Process (AHP) based criteria ranking method for determining the standardized scores of criteria expressing their factor of importance for a given decision problem in terms of thematic parameters, categories and their normalized weights. Altogether 14 thematic parameters have been used to generate the Landslide Hazard Zonation (LHZ) map and the Landslide Hazard Management (LHM) map. Validation of the results of LHZ has also been done using data collected from the field visit.

Keywords: Landslide, Landslide Hazard Zonation, Landslide Hazard Management, GIS.

Introduction

Landslide is one of the well known geological phenomenon as well as a natural hazard that can be defined as the movement of mass, rock, debris or earth down a slope (Cruden, 1991) which can occur in offshore, coastal and onshore environments. It occurs when the stability of the slope becomes unstable. It can be caused by several factors including both natural and anthropogenic. Like other natural hazards it poses threat to living beings including human and accounts for enormous property damage in terms of both direct and indirect costs. It is one of the most frequent natural hazards in various mountainous regions of India. With no exception to this, Jammu and Kashmir (J&K) is also one such regions that is highly affected by landslides every year during the monsoon posing a serious threat to human life and various developmental activities. Such landslides also cause considerable damage to communication, human settlements, agriculture and other infrastructures. In order to prepare ourselves from such calamities we need to know the likely possible occurrence, causes, extent of effect a landslide can bring to us and finally the management scheme to mitigate the hazard. It is, in this context that the present study has been carried to map the landslide hazard zones and to bring out a management map for the same. Developing landslide hazard zones and corresponding management map can provide useful information to minimize loss, and eventually assist in providing meaningful information/guidelines to the planners to develop sustainable landuse plan. Many workers/ researcher like Yin and Yan (1988), Carrara et al. (1991), Anbalagan (1992), Pachauri and Pant (1992), Juang et al. (1992), Jade and Sarkar (1993), Van Western (1993), Chung and Fabbri (1999), Dhakal et al. (2000), Clerici A (2002), Fabbri et al. (2003), Ohlmacher, G (2003), Sarkar and Kanungo (2004), Sarkar and Gupta (2005), Kanungo et al. (2006), Chen et al. (2007) Sarkar et al (2008) have worked on different approaches and schemes to develop landslide hazard zonation and mapping.

STUDY AREA

The study area is a stretch of road section between Marog to Naugam villages within 2km buffer on both sides of the Jammu-Srinagar National Highway (NH-1A) lying between the latitudes 33^0 14' 30" and 33^0 30' 00" and longitudes 75^0 08' 00" and 75^0 16' 00"(Fig.1). The landuse/land cover includes dense, medium and degraded vegetation and agricultural lands. The major rivers flowing through the region are Bichlari River and the Chandra Bhaga River. The major portion of the region is a complex hilly terrain with very small area of valley. Agricultural activities along with horticulture are quite common in the region. During winter the region experiences heavy precipitation (mostly snow) at the higher altitudes towards Banihal.

MATERIALS USED

Satellite imageries: IRS-P6 LISS-III, Google earth images, Aster DEM.

Ancillary Data: Survey of India (SOI) Toposheet No.43O3, Soil Map, Geological Map etc.

Software used: ERDAS Imagine 9x, ARC-GIS 9x, Landslide Information System (LIS) software has been extensively used for integration of GIS layers to generate the Landslide Hazard Zonation map and Landslide Hazard Management map.

Filed Equipments: i) GPS, ii) Brunton compass, iii) Measuring tape, iv) Digital camera etc.

METHODOLOGY

In the present study the integration of remote sensing data along with other ancillary data with GIS has been adopted to generate the Landslide Hazard Zonation and Management maps. The overall steps involve in the present study is given in the flow chart (Fig.2).

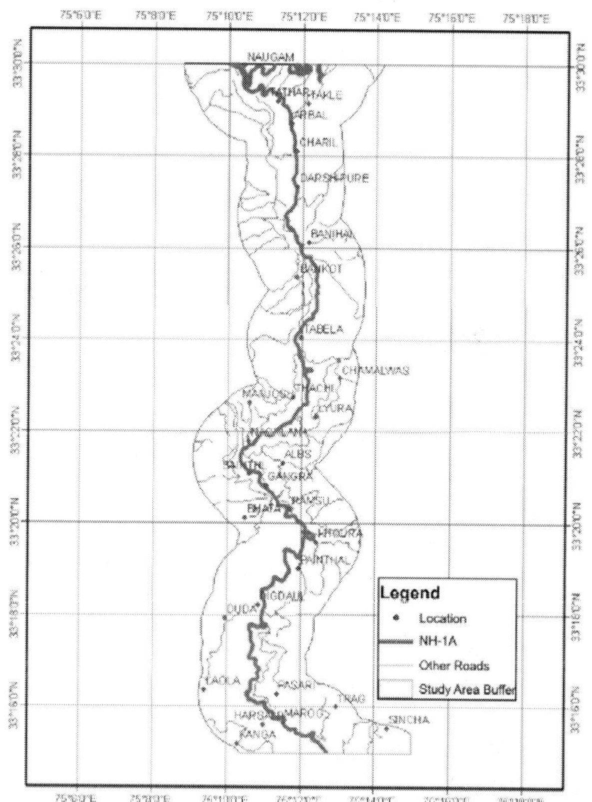

Fig. 1: Location map of the study area

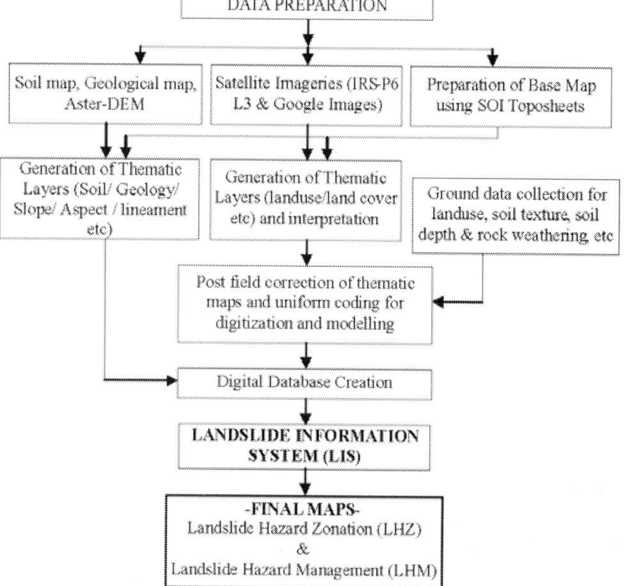

Fig. 2 Flow chart of the steps involved in the present study

Data Preparation and Generation of Thematic Layers

This step includes the following steps.

(i) Preparation of a base map of the study area from 1:50,000 scaled Survey of India Toposheet. During this process a buffer of 2km on both side of the road under study (here NH-1A) is also generated whereby all the other thematic layers are then prepared based on this buffer. This base map include base features like major/other roads, rivers and streams, village/ settlements/springs etc that are digitized from the toposheet using the ArcGIS software.

(ii) Digital image processing of all the satellite imageries of the study area. The IRS-P6 LISS III digital imageries and the corresponding google earth images of the study area are properly processed, georeference /geocoded using the ERDAS Imagine software for further interpretation and mapping of landuse/ land cover. Then the landuse/landcover mapping is done using onscreen digitization technique in ARCGIS platform. Based on the landuse/landcover categories the weathering /erosion and geomorphological characteristics have also been delineated in the area.

(iii) Preparation of Soil map, Geological map and Aster DEM. The soil and geological map of the study area is procured as a hard copy. In order to extract the required information from these maps for use in LIS they are scanned using A0 size scanner. Once they are in the digital format they are georeference/geocoded with other data.

(iv) Georeferencing / Geocoding of database. This step ensures that all the data (both satellite and other ancillary data) used in the present study are in the same projection system so that meaningful information may be extracted from them. In this study, the Indian Polyconic System of projection with Everest datum/spheroid has been used throughout.

Ground Truth Data Collection

Detailed field investigations were carried out in order to familiarize with the terrain so that the thematic maps in the field and also to collect field data pertaining to landuse / landcover and landslide triggering factors. During the field visit, verification of the thematic maps such as different geomorphological features, landuse/ landcover categories, landslide types, slope forms, major and minor fault and lineaments were checked. Detailed field data about type of landuse/landcover, structural discontinuities along which the failure is taking place, hydrological and slope conditions, anthropogenic activity, mechanism of failure, existing stability measures and vulnerability are also collected. The soil data are also collected during the field. Various field photographs are also taken for post field data rectifications. After the field visit various thematic maps are corrected on the basis of field information and some of the category boundaries of the landuse/land cover are also modified accordingly.

Generation of GIS Thematic Layers

After necessary post field modification of the thematic layers they are brought into a common GIS database platform for further analysis. The digitization, scanning and GIS layer generation work was carried out using ERDAS Imagine and ARC-GIS software. The codes or IDs of different thematic maps were assigned as per the standard code scheme. Altogether 20 thematic layers *viz.* bound (buffer), base-poly, baseline, base-point, roads, anthropogenic, drainage, faults, geomorphology, landuse, lineaments, lithology, slope aspect, slope dip, slope morphology, slope, soil depth, soil texture, landslide and rock weathering have been generated. Out of these thematic layers only 15 (including "bound") are used as input parameter during the generation of LHZ and LHM maps.

Generation of LHZ and LHM maps

The digital database so created in the above steps is then imported to the Landslide Information System (LIS) where they are validated to check the compatibility with the data structure of the software for generation of LHZ and LHM maps. If the validation is successful then we can go ahead with the generation of LHZ and LHM otherwise the error in validation needs to be addressed properly. The LIS software categorizes the landslide into 6 classes viz. severe, very high, high, moderate, low and very low. The "low" and "very low" category of hazard zones represent relative stability and that of "severe", "very high" and "high" hazard zones indicate relative instability of the region while "moderate" zone indicate moderate instability of the area.

RESULT AND DISCUSSION

The present study results various thematic layers (as mention in the "*The Generation of GIS Thematic Layers*") as intermediate output before the generation of Landslide Hazard Zonation and Landslide Hazard Management maps. In the present study out of the 6 landslide hazard zones *viz* severe, very high, high, moderate, low and very low, 4 classes are categorized based on the input parameters that are shown in the Landslide Hazard Zonation map (Fig. 3). These classes/ zones are very high, high, moderate and low of which 62.12% of the total area is classified as "low" while "moderate", "high" and "very high" zones occupy 23.71%, 2.05% and 12.10% of the total area respectively (Table 1). Although majority of the area is under the "low" hazard class unfortunately the highway (NH-1A) has to pass through the "very high", "high" and "moderate" zones (Fig. 3) which suggest that the landslide hazard is still potential and vulnerable to this region. This potential area has already experience major landslides at Digdaul, Panthal and near Tabela villages which has badly affected the region in the past also (Fig. 3).

Table 1: Landslide hazard zones and their area

Sl.No	Hazard Zones/class	Area (%)
1	Very High	12.10
2	High	2.05
3	Moderate	23.71
4	Low	62.12

Based on the above landslide hazard zones the Landslide Hazard Management map (Fig.4) is generated presenting all the corresponding mitigation measures which are site specific. The suggested activities includes (i) Retention wall with drain holes, Channelisation, Slope modification, Bio-technical measures; (ii) Channelisation, avoid further construction; (iii) Bio-technical measures, Channelisation; (iv) Forest conservation, Channelisation; (v) Slope modification, Bio-technical measures; (vi) Soil conservation and (vii) Planned developmental activities.

The resulted Landslide Hazard Zonation map is validated using primary field data using GPS location. The validation is done specifically in the "very high", high" and "moderate" hazard zones and it is found that these areas have active landslides the output in the LHZ map (Fig. 3).

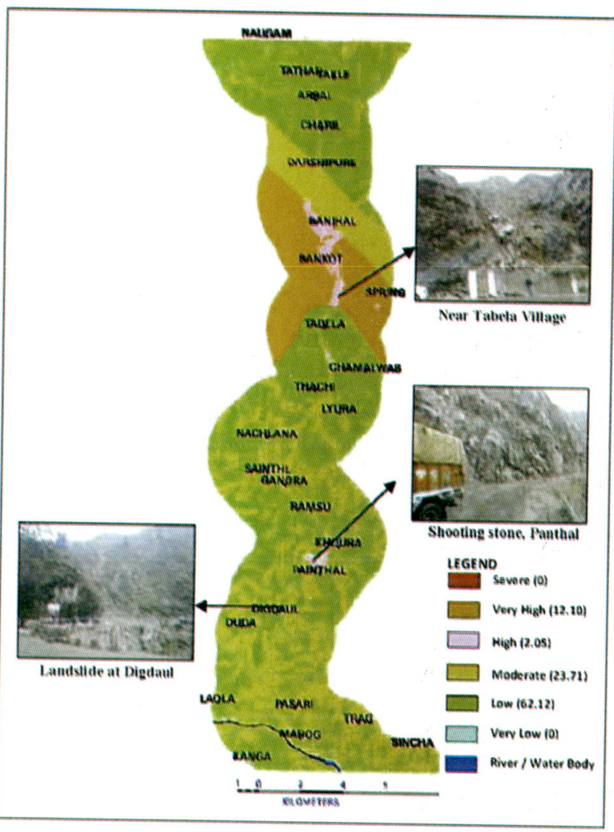

Fig. 3: Landslide Hazard Zone (LHZ) Map of the study area generated by LIS along the field photographs

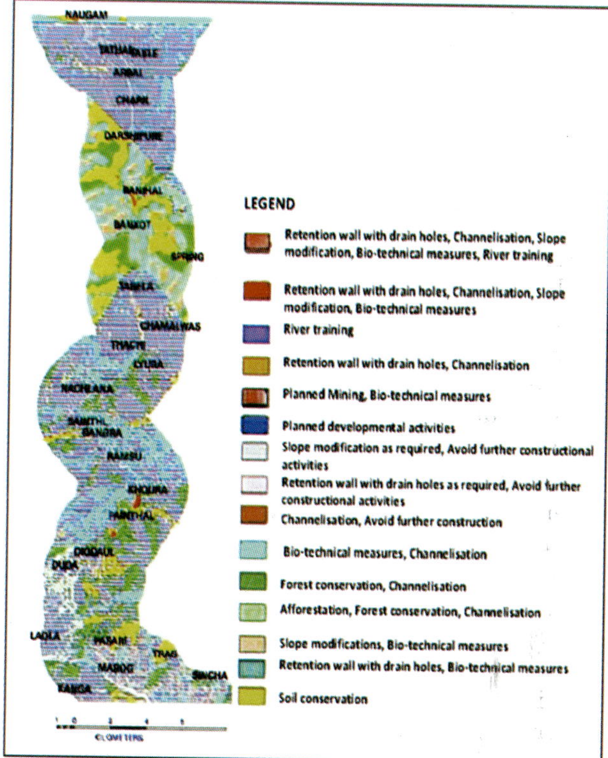

Fig. 4: Landslide Hazard Management (LHM) Map of the study area generated by LIS

CONCLUSION

Though the prediction/preparation of potential landslide areas has been very difficult due to the complexity of the factors involved and the relationship to each other which is wide ranging, the use of remote sensing data along with other ancillary data and their integration with GIS shows high correlation in the preparation of landslide hazard zonation and management maps as presented by the present study. Since the data is in digital format, the monitoring and assessment of landslides in the region can also be done at ease with necessary updates from time to time.

ACKNOWLEDGEMENT

The authors acknowledge the financial assistance and for providing LIS software by the DTRL, Ministry of Defence.

REFERENCES

Anbalagan, R. 1992. Landslide Hazard Evaluation and Zonation Mapping in Mountainous Terrain. Engg. Geol., 32: 269–277.

Carrara, A., Cardinali, M., Detti, F., Guzzetti, F., Pasqui, V. and Reichenbach, P. 1991. GIS Techniques and Statistical Models in Evaluating Landslide Hazard. Earth Surface Processes and Landforms 16(5):427–445.

Chen, Zhaohua; Wang, Jinfei (2007). "Landslide hazard mapping using logistic regression model in Mackenzie Valley, Canada". Natural Hazards 42: 75. doi:10.1007/s11069-006-9061-6.

Chung, C. F. and Fabbri A. G. 1999. Probabilistic Prediction Models for Landslide Hazard Mapping. Photogrammetric Engineering & Remote Sensing, 65(12):1389–1399.

Clerici, A; Perego, S; Tellini, C; Vescovi, P (2002). "A procedure for landslide susceptibility zonation by the conditional analysis method1". Geomorphology 48 (4): 349. Bibcode:2002Geomo.48..349C. doi:10.1016/S0169-555X(02)00079-X.

Cruden, D.M., 1991. A simple definition of landslide, Bulletin of the international Association of Engineering Geology, 43,27-29

Dhakal, A. S., Amada T. and Aniya M. 2000. Landslide Hazard Mapping and Its Evaluation Using GIS: an Investigation of Sampling Schemes For a Grid-Cell Based Quantitative Method. Photogrammetric Engg. & Remote Sensing 66(8):981–989.

Fabbri, Andrea G.; Chung, Chang-Jo F.; Cendrero, Antonio; Remondo, Juan (2003). "Is Prediction of Future Landslides Possible with a GIS?". Natural Hazards 30 (3): 487. doi:10.1023/B:NHAZ.0000007282.62071.75.

Jade, S. and Sarkar S. 1993. Statistical Model for Slope Instability Classifications. Engineering Geology, 36:71–98.

Juang, C.H., Lee, D.H. and Sheu, C. 1992. Mapping Slope Failure Potential Using Fuzzy Sets. Jour. Geotech. Engg. 118(3): 475–494.

Kanungo, D. P., Arora M. K., Sarkar S. and Gupta R. P. 2006. A Comparative Study of Conventional, ANN Black Box, Fuzzy and Combined Neural and Fuzzy Weighting Procedures for Landslide Susceptibility Zonation in Darjeeling Himalayas. Engineering Geology, Vol. 85: 347–366.

Ohlmacher, G (2003). "Using multiple logistic regression and GIS technology to predict landslide hazard in northeast Kansas, USA". Engineering Geology 69 (3–4): 331. doi:10.1016/S0013-7952 (03) 00069-3.

Pachauri, A.K. and Pant, M. 1992. Landslide Hazard Mapping Based on Geological Attributes. Engg. Geol.32:81–100.

Sarkar, S. and Gupta P. K. 2005. Techniques for Landslide Hazard Zonation — Application to Srinagar-Rudraprayag Area of Garhwal Himalaya. Journal of Geological Society of India 65: 217–230.

Sarkar, S. and Kanungo, D.P. 2004. An Integrated Approach for Landslide Susceptibility Mapping Using Remote Sensing and GIS. Photogrammetric Engineering and Remote Sensing 70(5):617–625.

Sarkar, S., Kanungo, D.P., Patra, A.K., Kumar, P. 2008. GIS Based Spatial Data Analysis for Landslide Susceptibility Mapping. Journal of Mountain Science 5(1):52–62.

Van Westen, C.J. 1993. Application of Geographic Information Systems to Landslide Hazard Zonation. ITC Pub., No. 15, Pp 245.

Yin, K. L. and Yan T. Z. 1988. Statistical Prediction Models for Slope Instability of Metamorphosed Rocks. Proceedings of the 5th International Symposium on Landslides, Lausanne, 2: 1269–1272.

Hon'ble reviewer

1. **Dr. L.K. Sinha**
   Director
   Def. Terrain Res. Lab
   DRDO,
   NEW DELHI - 110 054
2. **Dr. Ajay**
   Director (MGPSG & URPDG)
   Space Appl. Centre
   ISRO
   AHMEDABAD 380 015

Landslide Assessment Along NH-150 Between Ukhrul and Nunghar, Manipur, India

M. Okendro and R.A.S. Kushwaha

Department of Earth Sciences, Manipur University, Imphal 795 003
E-mail: rs_kushwaha@rediffmail.com

Abstract

The study area along NH-150 connects Manipur with Nagaland via Ukhrul and Jesami is one of the important roads for transporting the essential commodities to state when NH-2 and 37 are blocked. The geological formations exposed in the study are the between Ukhrul to Nunghar dominantly belongs to the Disangs, Ohpiolites and Ophiolitic Melange of Tertiary age. A cumulative effect incessant rainfall, slope instability, toe erosion by rivers, fragile geological formations, complex structures and high seismicity render the present study area between Ukhrul and Nunghar a highly susceptible and prone to slope failure and landslides.

Keywords: Landslinde Assessment, Slope Failure, Landslide Hazard Zonation, Kinematic Analysis.

Introduction

Landslide is the downward and outward movement of slope-forming materials, composed of rocks, soils or artificial fills (Varnes, 1978). Factors responsible for the down-slope movements include internal friction and cohesion of the sliding plane material. In addition, the presence or absence of discontinuity planes plays an important role in the stability of the rock-slopes. While considering the stability of a slope, the most important factor is the geometry of the rock-mass behind the slope face. The geometrical relationship between discontinuities in the rock-mass and the slope, and orientation of the excavated face determine whether, parts or whole of the rock-mass is free to slide/fall or not. The next important factor is the shear strength of the failure surface, which may consist of a single discontinuity plane or a complex structure

Major part of North-Eastern India situated along the subduction plate margin of Himalaya and IMR is characterized by rugged hilly terrain and the tropical precipitation. The rock formations exposed in the region dominantly belongs to Tertiary with little amount of Proterozoics and Achaeans. Its typical geological formations and structural features illustrate unique character that draws the attention of earth scientists.

The serenity of this region is often disturbed by natural hazards, like flood, earthquake, mass wasting, etc. Though small in magnitude in comparison with the other hazards but the most frequent and continuous effect that threaten or pose direct and indirect economic loses to the society and sometimes even loss of human lives is the landslide (Schuster, 1998; Glade, 1998; Skidmore, 2001), which is one of the most common natural calamities in North-East India during rainy season.

Manipur is a small landlocked hilly state situated at the periphery of Indo-Myanmar Range (IMR in the far north-eastern corner of India. It lies between 23°50′ N to 25°42′ N Latitudes and 93°00′ E to 94°45′E Longitudes and occupies about 23,327 sq km geographical area. Of the total geographical area about 90% is occupied by the rugged hilly terrain and moderate to thick deciduous forest. The state capital Imphal is well connected with Guwahati, Silchar, Kolkata and New Delhi by air. The National Highways 39, 53 (now known as 2 & 37 respectively) and 150 running across the state connects Manipur with the rest of India and neighboring countries like Myanmar and Bangladesh. These three National Highways are the only surface routes for the transportation of essential commodities thus considered as *"Lifelines"* of the state. Due to frequent lands-lides, particularly during monsoon-season, traffic along these highways is often disrupted for considerably long period and causes huge loss of revenue, properties and sometime even life. Moreover, Manipur is situated on the tectonically active Indo-Myanmar Ranges (IMR) which falls under High Risk Seismic Zone-V. When deformed geological structures, weak lithlogy and high seismicity combine with tropical climate and adverse hydrological conditions, it may trigger various kinds of mass wasting.

Nation Highway-150 which connects Manipur with Nagaland via Ukhrul and Jesami plays an important role when NH-2 and 37 are blocked either due to social unrest or landslide. A cumulative effect incessant rainfall, slope instability, toe erosion by rivers, fragile geological formations and complex structures render the present study area between Ukhrul and Nunghar (Fig. 1) a highly susceptible and prone to slope failure and landslides. Accordingly the present study aims to:

(i) categorize slope into the hazardous area/zone of landslide using BIS code IS 07 040, IS 14496

(ii) estimate the rock mass strength through empirical approach of rock mass classification

(iii) evaluate the geometric relationship between the slope and discontinuity present in the slope and

(iv) suggest suitable remedial measures.

Table 1: Geological Succession of the study area, Ukhrul District, Manipur (Modified after Mitra *et al*, 1986)

Age	Unit	Lithology
Pleistocene to Recent	Alluvium	Clay, sand, gravel, pebble, boulder deposits of the foothills and older river terraces.
Cretaceous to Upper Eocene	Disang Group	Fine-grained grey coloured sandstones, dark grey, splintery shales interbedded with sandy-shales and siltstones in upper part. Argillaceous grey shales interbedded with mudstones and minor amount of siltstones and argillaceous sandstones in lower part.
		- - - - - - - - - - - - Tectonic contact - - - - - - - - - - - -
Upper Cretaceous to Mid- Eocene	Ophiolite-Melange Suite	Basic and ultrabasic intrusive and extrusive of peridotites, serpentinites, cherts, limestones, shales, sandstones, gritstones, conglomerates, etc.
		~~ Unconformity ~~~~~~~~~~
Pre-Mesozoic or older	Metamorphic Complex	Low to medium grade metamorphic rocks of various compositions, phyllitic schists, quartzites, micaceous-quartzites, quartz-chlorite-mica schists, and marbles, etc.
		~~ Unconformity ~~~~~~~~~~
Basement Complex Unseen		

Table 2. LHZ Rating based on TEHD along with area percentage and area in km^2.

Zone	TEHD Value	Description of Zones	Area%	Area (km^2)
I	< 3.5	Very low hazard (VLH) zone	nil	nil
II	3.6-5.0	Low hazard (LH) zone	00.42	00.39

III	5.1-6.0	Moderate hazard (MH) zone	14.11	13.06
IV	6.1-7.5	High hazard (HH) zone	72.73	67.32
V	> 7.5	Very high hazard (VHH) zone	12.74	11.79

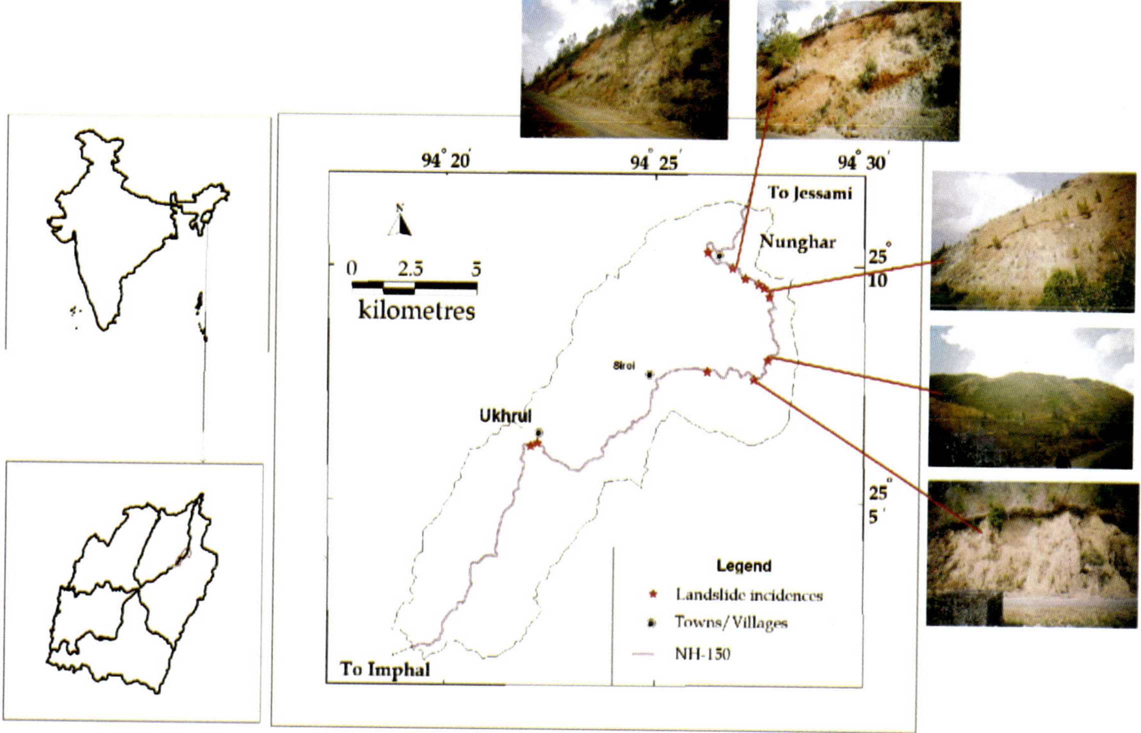

Fig.-1: Location Map of the Study Area.

GEOLOGY OF THE STUDY AREA

Present study area between Ukhrul and Nunghar is part of the Nagaland-Manipur Ophiolite Belt which forms northern boundary of the Indo-Myanmar Range (IMR). Its length within the Indian Territory is about 200 km from northeast of Chokla in Nagaland to south of Moreh in Manipur. Its width varies from 5 to 15 km and shows tectonic contact with Naga Metamorphics in the east and Disang in the west. A narrow zone of mélange consisting of chaotic assemblages of basic and ultrabasic intrusive and extrusive of peridotite, serpentinite, chert, limestone, shale, sandstone, conglomerate etc occurs within upper Disang along its contact with Ophiolite belt.

A simplified stratigraphic succession with brief lithological descriptions of the area is given Table 1.

LANDSLIDE HAZARD ZONATION (LHZ)

Landslide Hazard Zonation is to divide the land surface based on estimated factors that include slope instability. Provisions and stipulations of IS code 14496 (Part 2): 1998 are followed with slight modification in order to suite the present study (Okendro et al., 2010). Landslide Hazard Zonation (LHZ) map of the study area (Fig. 2) has been prepared on 1:50,000 scale. Landslide Hazard Evaluation Factor (LHEF) rating scheme for every individual causative factors that includes geo-environmental causes are estimated facet wise. The total estimated hazard (TEHD) is sum of the all numerical values of each causative factor which is stated as:

TEHD = Ratings of (Lithology + Structure + Slope Morphometry + Relative Relief + Hydrogeological (Drainage Density) + Landuse and land cover)

LHZ map of the area has been prepared based on the values of the TEHD for each individual facet. Details of LHZ rating for the present study area based on TEHD and its area distribution is given in Table 2.

Analysis of Table 2 and Figure 2 suggests that very low hazard (VLH) zone is absent in the area. Low hazard (LH) occupies 0.39 sq km., moderate hazard (MH) zone covers an area of

13.06 sq. km., high hazard (HH) zone has covered maximum area about 67.32 sq. km. and very high hazard (VHH) zone covers around 11.79 sq. km. of the total study area. Based on the degrees of instability; five active landslides have been selected for the detailed study of rock mass classification.

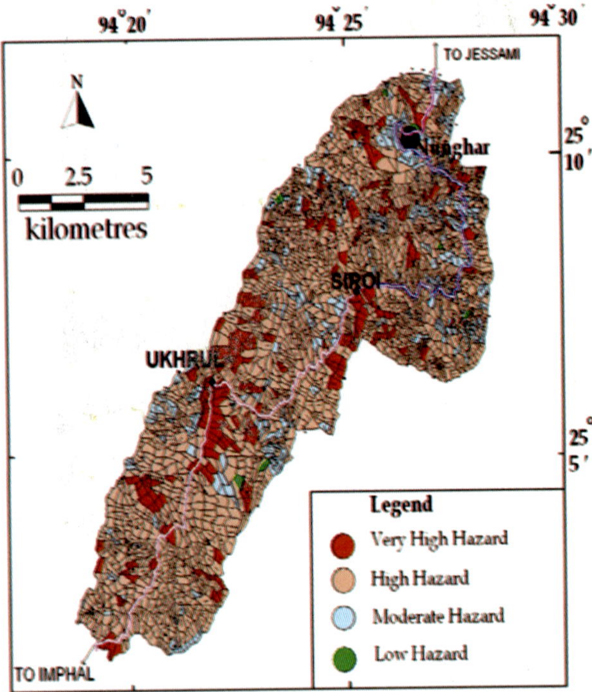

Fig. 2: Landslide Hazard Zonation Map.

ROCK MASS CLASSIFICATION

During the preliminary stage of designing a project in such area where very little information is available on the rock mass and its stress and hydrological characteristics, the rock mass classification scheme is of considerable benefit. There are different schemes of rock mass classification for the design of slopes and to assess the slope stability. All these schemes are based on empirical relations between rock mass parameters and slope parameters like height and slope dip. According to Krauland *et al.* (1989) it is one of the principal ways of determining the rock mass strength.

Classification is an indirect method, often used to predict the rock mass quality and the possible need for support. The result is an estimate of the stability quantified in subjective terms such as bad, acceptable, good, and very good rock conditions. The values obtained from the classification systems are used to estimate or calculate the rock mass strength using a failure criterion. However, Singh *et al.* (1999) gave the following reasons of using the quantitative rock mass classification system.

- It provides better communication between geologists, designers, contractors and engineers;
- Engineer's observations, experience and judgements are correlated more effectively by a quantitative classification systems;
- Engineers prefer numbers in place of descriptions. Hence, a quantitative classification system has considerable application in an overall assessment of the rock quality; and
- The classification helps in organizing knowledge.

ROCK MASS RATING (RMR)

RMR is the numerical values of the algebraic sum of parameters like uniaxial compressive strength, Rock Quality Designation (RQD), spacing of joints/discontinuities and groundwater conditions. Indirect method of strength calculation is adopted here by using the Schmidt Hammer (Electronic Concrete Hammer, CNS FARNELL Ltd, UK) which is easy to handle and access to the face of sliding rock mass. RQD value is obtained from the following relation of Hudson and Priest (1983) relation for non core logs:

$$RQD = 100\, e^{-0.1\lambda} (0.1\lambda + 1),$$

where λ is joint per meter (=joint spacing).

Orientation of discontinuities is dealt separately using stereo plots. All these parameters have been assigned rating. In the present investigation it has been calculated following Bieniawski's (1989) and Palmstrom's (1995) classification systems (Table 3).

$$RMR = RMR_{basic} + \text{adjustment for discontinuity/joint orientation}$$

$$RMR = \sum \text{parameters } (i + ii + iii + iv + v)$$

Robertson (1988) established that when RMR is greater than 40, the slope stability is governed by both orientation and shear-strength of discontinuities, whereas for RMR lower than 30, the failure develops across the rock mass. The high value of RMR indicates good quality of rock mass. RMR value obtained for the five stations selected suggest that rocks formations are of poor quality except the station 4 which falls under fair quality.

SLOPE MASS RATING (SMR)

The SMR technique is basically evolved from Rock Mass Rating system developed by Bieniawski (1989). SMR classification based on six parameters like i) Uniaxial Compressive Strength of rock material, ii) Rock Quality Designation, iii) Spacing of Discontinuities, iv) Condition of Discontinuities, v) Groundwater Conditions and vi) Orientation of Discontinuities proposed by Romana (1985) is best suited for the hilly terrain.

Therefore, SMR is obtained from RMR by adding a factorial adjustment factor depending on the joints-slope relationship and the method of excavation is given by:

$SMR = RMR + (F_1 \times F_2 \times F_3) + F_4$

where,

F_1 depends on parallelism between joints and slope-face strike, ranging from 1.00 to 0.15. These values match the relationship of $F_1 = (1 - \sin A)^2$ where A denotes the angle between the strikes of slope-face and joints.

F_2 refers to joint's dip angle in the planar mode of failure, ranging from 1.00 to 0.15 and matches the relationship of $F_2 = \tan^2 B_j$ where B_j denotes the joint dip angle. For toppling mode of failure F_2 remains 1.00.

F_3 is the relationship between slope and joint-dip, ranging from 0° to 60°.

F_4 refers to the method of excavation of the slope, which ranges from +15 (natural slope) to -8 (deficient blasting).

SMR is a geomechanical classification for slopes in rock (Romana, 1985). SMR suggests need of support and describes the rock in five classes. The adjustment factors F_1, F_2 and F_3 for discontinuities are determined from the stereo-plots of five different landslide zones (Fig. 3), following Romana (1985). Mode of excavation available in the sliding zones is taken as the value of F_4. Hence, the SMR has been calculated for each landslide sites. Table 3 illustrates the respective findings.

Table 3: RMR and SMR ratings calculation for selected stations

	Station 1	Station 2	Station 3	Station 4	Station 5
Location	25°09' 0.4″N 94° 27' 34.9″E	25°08' 23″N 94° 27' 43.1″E	25°08' 4.5″N 94° 27' 38.6″E	25°07' 41.5″N 94° 27' 20.6″E	25°07' 36.3″N 94° 24' 48.2″E
Distance from Ukhrul	23 km	21.1 km	19 km	16.3 km	13 km, Near Siroi
Type of Rock	Ultramafic Serpentinites	Ultramafic Serpentinites	Ultramafic Serpentinites	Ultramafic Serpentinites	Shales, Siltstones
Attitude of slope	38°/229°	39°/183°	40°/270°	25°/243°	35°/190°
Degree of Weathering	High	High	High	High	Moderate
		Rating Values/Weightage			
Strength	7	4	4	12	2
RQD	5	6	4	12	3
Spacing	5	8	10	10	8
jL	3	4	2	2	1
jR	2	3	2	2	1
jA	1	4	4	4	1
jC	6	3	1	1	4
Ground water	15	15	15	15	15
RMR	***37, Poor***	***33, Poor***	***34, Poor***	***44, Fair***	***32, Poor***
F_1	0.51	*0.235*	0.60	0.948	0.83
F_2	1	1	1	1	1
F_3	0	-6	0	0	-6
F_4	0	0	0	0	10

SMR	**38.833**	**32.00**	**34.00**	**50.00**	**26.312**
Class	IV, Bad	IV, Bad	IV, Bad	III, Normal	IV, Bad
Stability	**Unstable**	**Unstable**	**Unstable**	**Partially Stable**	**Unstable**
Failures	Planar / Big Wedges	Planar / Big Wedges	Planar / Big Wedges		Planar / Big Wedges
Support	Important/ Corrective	Important/ Corrective	Important/ Corrective	Systematic	Important/ Corrective
Internal Friction angle	24°	22°	22°	27°	21°

Table 4: Orientation of discontinuities and slopes

	Station 1	**Station 2**	**Station 3**	**Station 4**	**Station 5**
Joint dip direction (a_j)	234^{0}	192^{0}	228^{0}	229^{0}	194^{0}
Joint Dip angle (B_j)	66^{0}	36^{0}	26^{0}	17^{0}	12^{0}
Slope Direction (a_s)	229^{0}	193^{0}	270^{0}	243^{0}	190^{0}
Slope dip angle (B_s)	38^{0}	39^{0}	40^{0}	25^{0}	35^{0}
$a_j - a_s$	5^{0}	$- 1^{0}$	$- 42^{0}$	$- 24^{0}$	4^{0}
$B_j - B_s$	22^{0}	$- 3^{0}$	$- 14^{0}$	$- 8^{0}$	$- 23^{0}$
Probable Failure	Toppling	Wedge	Wedge	Wedge	Wedge

a_j = joint dip direction, a_s = slope direction, B_j = joint dip, B_s = Slope angle.

It is revealed that the four sites (Station 1, 2, 3 and 5) fall under class IV (Unstable) and station 4 lies in class III (Partially stable) category.

STEREOGRAPHIC AND KINEMATIC ANALYSIS

Kinematic analysis examines modes of failure which possibly may occur in the rock mass. In analyzing the stability of a rock slope, the most important factor to be considered is the geometry of the rock mass behind the slope face (Hoek and Bray, 1981).

Failure mode in rock mass is determined kinematically using stereographic projection technique (Panet, 1969), which is purely geometric in nature. Markland (1972), Goodman (1976), Hocking (1976), Cruden (1978), Lucas (1980), Hoek and Bray (1981), Matherson (1988) and Yoon *et al.,* (2002) also followed this mode of study. Kliche (1999) determined the failure mode by applying the angular relationships between discontinuities and slope surfaces.

Planar failure occurs when the strike trends of the slope face and discontinuity plane are more or less parallel ($\pm 20^{o}$); dip angle of discontinuity plane is lower than the slope face, i.e., it must 'daylight' in the slope-face and a release surface must be present in the rock mass to define the lateral boundaries of the slide (Hoek and Bray, 1981).

Wedge failure occurs when the plunge of the intersection of discontinuities is less than the dip angle of the slope face and hence these intersections must 'daylight' in the slope face (Markland, 1972).

If the slope face and discontinuity planes are dipping in opposite direction, rather the discontinuity has high dip angle towards the slope, toppling failure may occur. Even when both of them are dipping in the same direction and the dip angle of the discontinuity plane is greater than the slope angle, there is probability of toppling failure.

According to Deoja, *et al.,* (1991), wedge movement along slope-face can be divided into three types viz:

(i) When the wedge has a parallel or is closed to the direction of the slope (upto 32° on either side of the slope direction), it is defined as a *central wedge.*

(ii) When the intersection lies in between 32° to 65° of the slope direction, it is termed as *lateral wedge* and

(iii) When the wedge is beyond 65°, it is referred to as *very lateral wedge.*

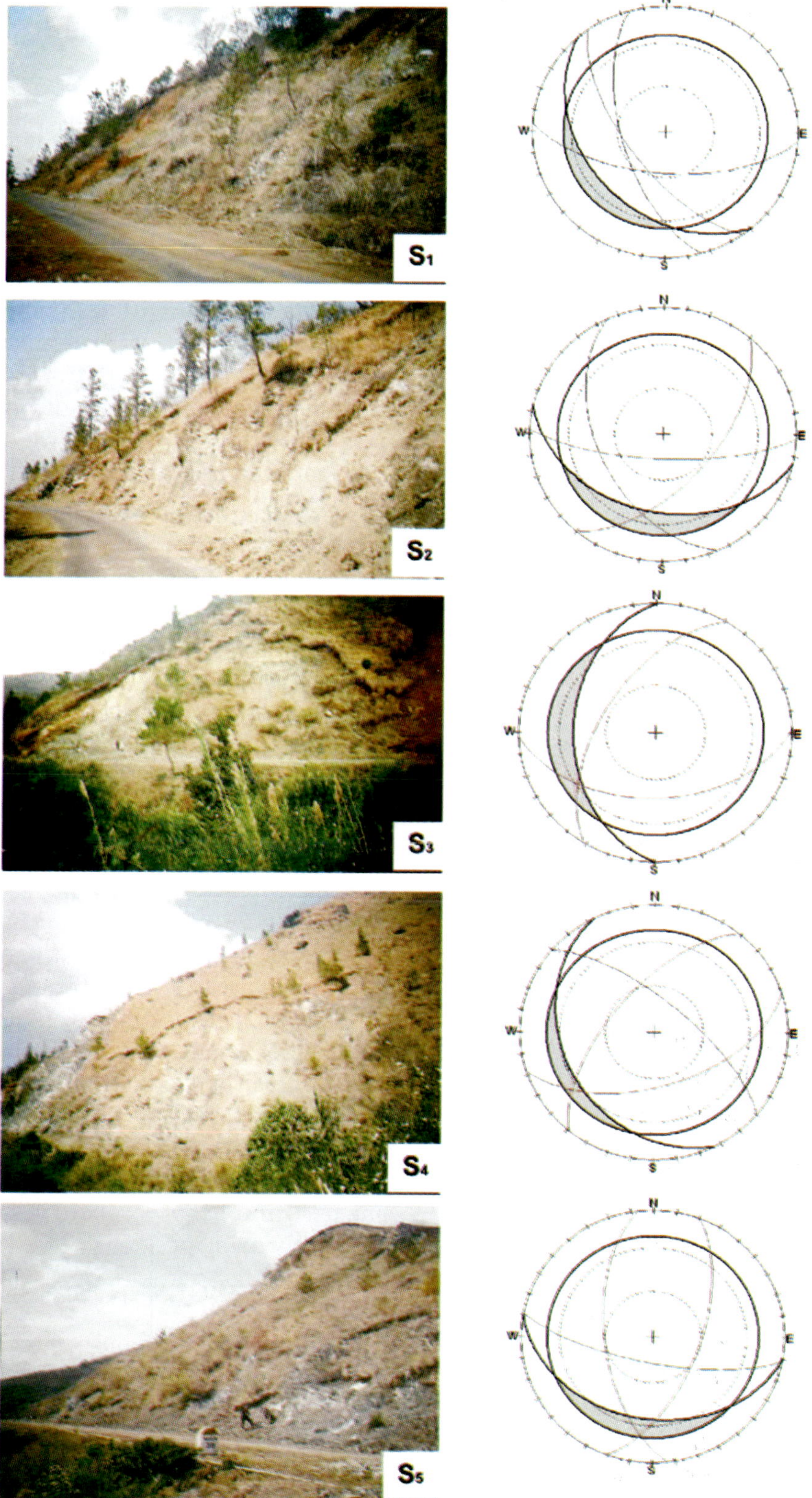

Fig. 3: Represents different sites and their respective probable mode of failure
***Photographs of Fig.-1 and Fig.-3 are not similar.

Discontinuity is a term used by the engineers and as defined the structural weakness plane upon which movement can take place. Discontinuities data collected from the field are systematically processed and tabulated so that it can be effectively used to analyse stereographically using Rockpack III software, to determine the modes of slope failures (Fig. 3).

A contoured pole diagram is necessary but not sufficient aid in slope stability studies. It must always be used in conjunction with intelligent field observations and a final decision on the method of analysis to be used on a particular slope must be based upon a balanced assessment of all the available facts (Hoek and Bray, 1981).

Result obtained from stereo plots (Table 4 and Fig. 3) suggests wedge failures at the four sites i.e. stations 2-5 and toppling failure at station 1 which is almost in conformity with SMR approach.

CONCLUSION

Prior to initiate any kind of engineering infrastructural development work, it is indeed necessary to have an idea of the slope instability where the project is to be undertaken especially in the rugged mountainous terrain. This will help the engineer be able to adopt certain precautionary steps/ measures.

Results obtained from the categorization of slopes of the study of area reveals that low hazard covers an area of 00.39 km^2, moderate hazard 13.06 km^2, high hazard 67.32 km^2 and very high hazard 11.79 km^2 out of the total area of 92.56 km^2.

Further, geo-engineering investigations along NH-150 between Ukhrul and Nunghar have been carried out using rock mass rating, slope mass rating and kinematic analysis. The value of RMR ranges from 33 to 44 for ultrbasic rocks of ophiolite belt and 32 for shale and siltstone rocks of Disang Group. As per the RMR classification rocks of the study area falls under III and IV, The class of the study area as per the RMR classification is III to IV i.e., poor to fair categories.

The SMR rating ranges from 32 to 50 for ophiolitic rock and 26.312 for Disang Group and are classified as bad to normal. The stability condition as per the SMR classification reveals that four of them is unstable and only one is partially stable. Failure mode indicated by the SMR classification is either planar or big wedges may occur.

Kinematic analysis reveals that out of five stations four of them are characterized by wedge failure mode and one by toppling failure which in conformity with the results obtained from SMR thereby validates the instability of the slopes where the discontinuity plays an important role in initiating the slope failures.

MITIGATIVE MEASURES

Proper excavation: It is one of the criteria for the construction of a highway to have a safe height with respect to the slope. Loose over burden rock and soil should not be left after the highway excavation.

Improvement of Drainage System: Stability of slope can thus be significantly improved by installing drainage paths to reduce the destabilising forces. Water should not be allowed to pass through the loose and weak soil rather diverted from the unstable slope as for as practicable.

Construction of retaining wall: Most of site for present investigation are of shallow type but rocks are highly weathered. Construction of light retaining walls in the sliding zone can check the instability of slope.

Biotechnical slope stabilization is considered to be a suitable measure in the area where the roots of the plants bind together the soil that will increase the cohesion. Deep rooted big plants are not suited to the slope which can create another pressure to the walls of the rocks thereby aggravating the slope instability.

Above all, people inhabiting in the area are not aware of how improper anthropogenic activity aggravates the instability of slope. According to Meusburger and Alewell, 2008 there is an increase of about 92% in the landslide activities due to anthropogenic and environmental factors during the period of 45 years (1959-2004) while studying in the Central Swiss Alps. It is very essential to make them aware of mass wasting hazard and the proper land use practices especially on slopes.

ACKNOWLEDGEMENTS

Present work is outcome of the Department of Science & Technology, Government of India, New Delhi funded Research Project. Authors thankfully acknowledge the DST for providing necessary financial help.

REFERENCES

Bieniawski, Z.T., 1989: Engineering Rock Mass Classifications. *John Willey & Sons, New York,* 251.

BIS (1998) Preparation of Landslide Hazard Zonation Maps in Mountainous Terrains-Guidelines Part-2, Macrozonation: ICS 07.040, IS 14496

Cruden, D.M., 1978: Discussion of Hocking's paper "A method for distinguishing between single and double plane sliding of tetrahedral wedges". *Int. Jour. Rock Mech. Min. Sci. Geomech. Abstr.* 15, 217

Deoja, B., Dhital, M., Thapa, B. and Wegner, A.I., 1992: Moutain Risk Engineering Handbook, Part II. *International Centre for Integrated Mountain Development, Kathmandu, Nepal,* 620-623.

Glade, T., 1998: Establishing the frequency and magnitude of landslide-triggering rainstorm events in New Zealand, *Environmental Geology*, 35 (2-3), 160-174.

Goodman, R.E., 1976: Methods of Geological Engineering in Discontinuous Rocks. *West Publishing, San Francisco.*

Hocking, G., 1976: A method for distinguishing between single and double plane sliding of tetrahedral wedges. *Int. Jour. Rock Mech. Min. Sci. Geomech. Abstr.* 13

Hoek, E. and Bray, J.W. 1981: Rock Slope Engineering, 3rd Edn. London: *Institution of Mining and Metallurgy*, 40-61, 55, 83,150.

Hudson J.A. and Priest S.D. (1983): Discontinuity frequency in rock masses. Int. J. Rock Mec. Min. Sci. & Geomech. Abstr., Vol 20, No 2, pp. 73-89.

Kliche, C.A. 1999: Rock Slope Stability SME, *Littleton, CO.*

Krauland, N., Sodder, P. and Agmalm, G., 1989: Determination of rock mass strength by rock mass classification – some experience and questions from Boliden mines, *Int. Jour. Rock Mech. Min.*, 26 (1), 115-123

Lucas, J.M., 1980: A general stereographic method for determining possible mode of failure of any tetrahedral rock wedge. *Int. Jour. Rock Mech. Min. Sci. Geomech. Abstr.*, 17, 57-61.

Markland, J.T., 1972: A useful technique for estimating the stability of rock slopes when the rigid wedge sliding type of failure is expected. *Imp. Coll. Rock Mech. Res., Rep.*, 19, 10.

Matherson, G.D., 1988: The collection and use of field discontinuity data in rock slope design. *Quart. Jour. Engg. Geology*, 22, 19-30.

Meusburger, K. and Alewell, C., 2008: Impacts of anthropogenic and environmental factors on the occurrence of shallow landslides in an Alpine catchment (Urseren Valley, Switzerland). *Natural Hazards and Earth System Science*, 8, 509-520.

Mitra, N.D., Vidyadharan, K.T., Gaur, M.P., Singh, S.K., Joshi, A., Khan, I.K. and Ghosh, S., 1986: A note on the Olistostromal deposits of Manipur. *Rec. G.S.I.*, 114 (4), 61-76.

Okendro, M., Kushwaha, R.A.S. and Goel, O.P., 2010: A study of landslides along part of a national highway in Manipur, India, *Int. Jour. Econ. & Environ. Geol. Vol.*, 1(1) 48-50

Palmström, A. 1995: A system of characterizing rock mass strength. *Jour. Rock Mech. & Tunneling Tech.,* 1(2), 69–108.

Panet, M., 1969: Discussion on "Graphical stability analysis of slopes in jointed rock". By K.W. John. *Jour. Soil Mech. Found. Div., Proc. ASCE.* 95 (SM2) 685-686.

Robertson, A.M., 1988: Estimating weak rock strength. *AIME-SME Annual Meeting, Phoenix, Az., Preprint*, 88-145.

Romana, M., 1985: A geomechanical classification for slopes: Slope Mass Rating in *Hudson, J.A. (Ed), Comprehensive Rock Engg, Pergamon Press, London*, 375-601.

Singh, B. and Goel, R.K., 1999: Rock Mass Classification-A practical Approach in civil engineering, *Elsevier Science Ltd, Netherland*, 1.

Schuster, R.L., 1998: Socioeconomic significance of landslides. In: Turner A.K., Schuster, R.L. (eds) Landslides-investigation and mitigation. *Special report, National Academy Press, Washington DC,* 12-35.

Skidmore, M., 2001: Risk, natural disasters, and household saving in life cycle model. *Japan World Econ*, 13, *Elsevier*, 15-24,

Varnes, D.J. (1978): Slope Movement, Types and Processess. In: Schuster, R.L. and Krizek, R.J. (Eds), Landslide Analysis and Control, Special Report, Vol. 176, Nat. Academy of Sciences, Transportation Research Board, Washington, D.C.

Yoon, W.S., Jeong, U., Choi, J.W., Kim, J.H., Kim, W.Y. and Kim, C.S. 2002: Slope failure index system based on the behaviour characteristics, SFi-System. *Jour. of the Korean Geotech. Soc.,* 18, 23-37.

Hon'ble reviewer

1. **Dr. S. Wadhavan**
   Dy. Director General
   Geological Survey of India
   HYDERABAD - 500 001
2. **Dr. Afia Akhtar**
   Director General
   Geological Survey of Bangladesh
   DHAKA 1000
   Bangladesh

IV-60

Refining Crustal Velocity of Manipur, A Part of Indo-Burma Subduction Region, Using Teleseismic P-Wave Receiver Function for Seismic Hazards Assessment

Manichandra Sanoujam and Arun Kumar

Department of Earth Sciences, Manipur University Imphal 795003 India
E-mail: arun634@gmail.com

Abstract

The analysis of hypo central parameters for earthquakes is one of the most essential task in any seismological observatory. The attempts are made to determine the crustal structure of Manipur a part of Indo Burma subduction region to refine the hypo central parameters by using improved crustal velocity structures resulting from the receiver function analysis from teleseismic waveforms data. The teleseismic P-waveform receiver function analysis is a commonly used technique in earthquake seismology to study the crustal structure beneath a broadband seismic station. This technique is applied to digital earthquake data recorded by the Manipur University station (MANQ) of Northeast India, subduction zone of India-Burma plate. Both inversion and forward methods are used in modeling the receiver function waveform of MANQ to obtain the S-velocity structure beneath the station. Modeling results show that the crust beneath MANQ is composed of two major layers. The first layer is 12 km thick with an S velocity of 3.45 km/s. The second layer is 20 km thick with an S velocity of 3.55 km/s. The Moho is characterized as a gradient transition zone over a depth range of 46-52 km. The S velocity increases from 3.8km/s to 4.80 km/s. The average Moho depth is characterized by gradation 46-52 km. This crustal model is consistent with the results of other studies. The errors in the estimation of the hypocentral parameters using the model evolved in this study are much improved.

Keywords: Sesmic Hazards, Teleseismic P-waveform, Receiver function, Crustal structure, Indo-Burma Sub-duction Wedge.

Introduction

Manipur is located in the Indo Burmese Range (IBR) on the subduction zone of the Indian and Eurasian plate boundary. A mix of subduction and collision between the two plates has contributed to the complex tectonic processes and tectonic evolution in the NE India region. It is observed that the dipping seismic zone below the IBR is confined within the latitude 21°-26°N in a NNE direction, while in the northern part beyond 26°N the subduction mechanism is possibly taken over by the collision tectonics. Seismic activity beyond 26°N and in the Mishmi block may be identified as the eastern syntaxis zone activity at the junction of the Himalayan arc and the Burmese arc. This syntaxis zone was the source area of the 1950 great earth-quake, that occurred by strike-slip mechanism (Ben-Menahem et al., 1974).

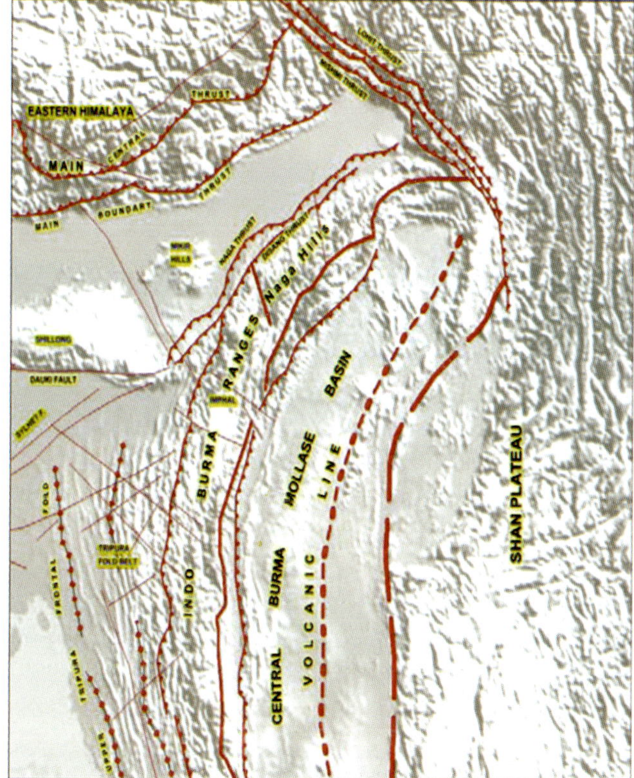

Fig. 1: Tectonic map of the study area.

The IBR comprises of Naga Hills, Chin Hills and the Arakan-Yoma. The structural and tectonic patterns of the region are transitional between the NE-SW trend of the Naga-Patkai Hills in the north and northeast to N-S trend in the Mizoram and Chin hills in the south. The major structural and tectonic features such as folds, reverse/thrust faults follow the regional strike. The geological setting and structural pattern in the region suggest a typical deformation mechanism due to the plate interaction. The geodynamic evolution of the IBR dates back post Cretaceous in a typical subduction tectonic regime between India and Eurasia, which resulted sequences of flysch ocean sediments and ophiolites suits of rocks along the entire belt (Nandy, 2001). The post Cretaceous evolu-tion of the Asian plate in Myanmar is strongly influenced by spreading of the Eastern Indian Ocean and by northward movement of "Greater India" (Bannert and Helmcke 1981). The emergence of the IBR happened during the Miocene, generating an Outer Island Arc and forming the Bay of Bengal. Subduction of the Indian oceanic plate along the west coast of Burma (Myanmar) to the south experienced the tectonic deformation with the approach of continental Indian crust that led to a series of thrusts in the northern part of the IBR in Manipur and Nagaland (Fig. 1).

Focal depth section traversing Manipur (24.50N Latitude) shows the dipping subducting Indian plate. Intense seismicity is observed at the Manipur-Mizoram-Burma (Myanmar) border, the central part of the IBR. Maximum thickness of the seismic zone as depicted by the hypocentral section varies from 40-50 km (Fig. 2). The Churachandpur Mao Thrust is identified to be the western boundary of the subduction, and the Sagaing fault the eastern boundary in Burma.

Receiver function analysis

In order to have refine crustal structure for estimation of hypocentral parameters of Manipur region, which is a part of Indo Burma subduction. The existing crustal structure model (Gupta, 1982) used is of regional in nature and covers the entire NE India, which is not sufficient to precise estimation of hypocen-tral parameters in Manipur, which is the part of Indo Burma subduction region. Seismo-grams of teleseismic P- waveforms contain the combined effects of the earthquake source, the propagation medium and the instrument response. To use these seismograms to deter-mine local Earth structure beneath a broad-band seismograph, the effects of local struc-ture, or the site response, must be isolated from other factors. The resulting waveform of the isolated site response is called the receiver function. Many authors have successfully used receiver functions to determine crustal and upper mantle velocity structure (Langston, 1979; 1994; Owens et al., 1984; Priestley et al., 1988; Ammon et al., 1990; Qiu et al., 1996). A reasonably accurate knowledge of crustal structure is important for regional tectonic and geodynamic studies. Manipur University station is situated at the subduction zone of the Indian plate and Burmese plate. It is an important area to study the tectonic evolution of Indo-Burmese Arc. In this paper, the processing techniques and a time domain inverse modeling procedure of receiver functions are described. These techniques are then applied to broadband data recorded by Manipur University station (MANQ), Northeast India. Finally, the results of modeling are presented and the crustal model is compared with those obtained in previous studies.

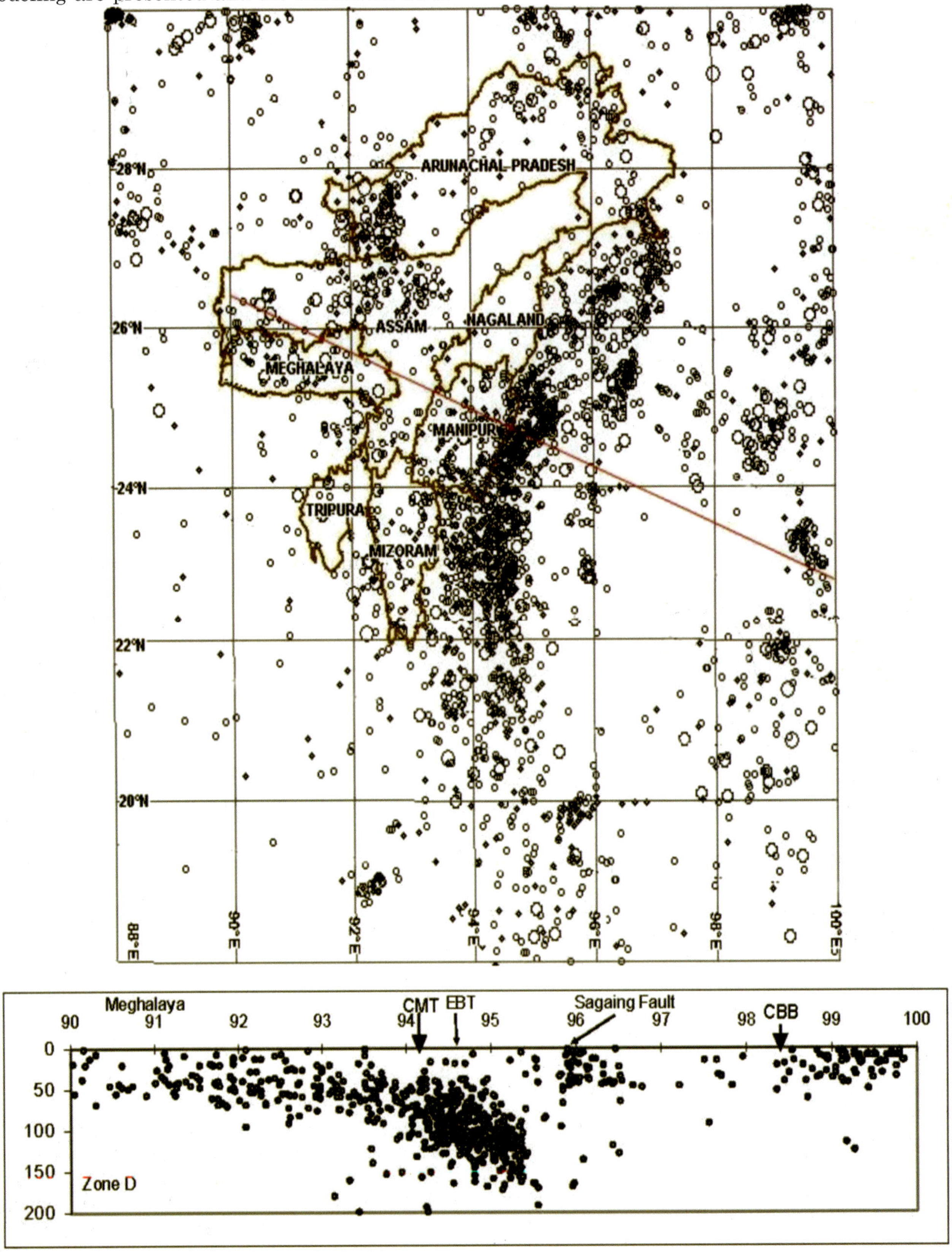

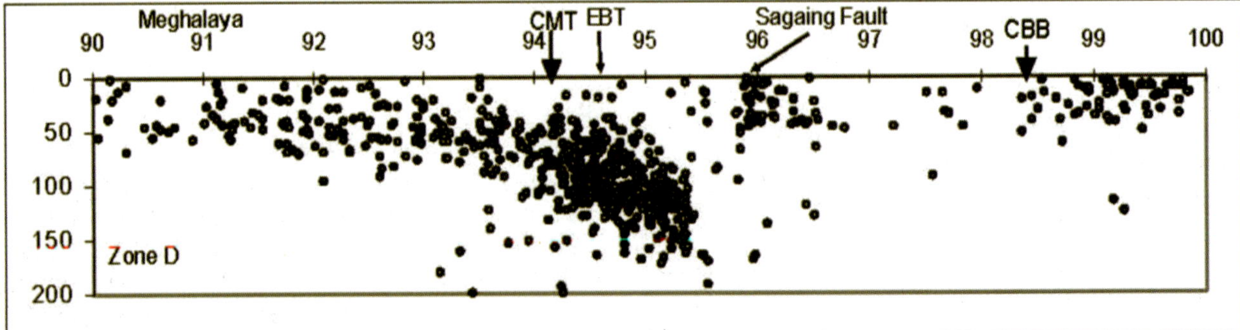

Fig. 2: Hypocentral plots traversing 240N showing the subduction geometry of Indian plate (CMT- Churachandpur Mao Thrust, EBT-Eastern Boundary Thrust, CBB- Central Burma Basin).

The direct teleseismic P-wave interacts with velocity interfaces and gradients beneath the seismograph to produce a seismogram consisting of the direct P-wave plus P- wave reverberations, and P- to-S converted phases plus their reverberations. An effective procedure to isolate these local Earth structure effects, the source equalisation procedure, has been proposed and discussed by Langston (1979). In this procedure, the three components of the seismic response at any one receiving station caused by the incidence of a plane P-wave, D(t), can be theoretically represented in the time domain by

$$D_V(t) = I(t)*S(t)*E_V(t)$$
$$D_R(t) = I(t)*S(t)*E_R(t) \qquad (1)$$
$$D_T(t) = I(t)*S(t)*E_T(t)$$

where subscripts V, R and T represent vertical, radial and tangential components respectively; I(t) is the impulse response of the recording instrument; S(t) is the seismic source function; E(t) is the impulse response of the local Earth structure; asterisks represent the convolution operator. S(t) may be complicated since it is related to dislocation time history and source area reverberations. Given that the Earth structure beneath a station will produce phase conversions of the P-to-S type, the horizontal components of ground motion will, in general, be quite different from the vertical component.

$$H_R(\omega) = \frac{D_R(\omega)}{D_V(\omega)} = \frac{E_R(\omega)}{E_V(\omega)} \approx E_R(\omega)$$

where D(ω) and E(ω) are the Fourier transforms of D(t) and E(t), respectively. ER(ω) can be approximated by HR(ω) because EV(t) is close to a delta function for teleseisms. In practice, a stable procedure for performing this de-convolution is to divide the Fourier transform of the radial component by that of the vertical component after introducing a water-level spectral fill, a minimum allowable amplitude level, to the amplitude spectrum of the vertical component. This avoids instabilities generated by dividing by a very small number due to troughs in the spectrum. If we smooth the result by a Gaussian function to exclude high-frequency noise, the final expression for the radial receiver function in the frequency domain is then

$$H_R(\omega) = \frac{D_R(\omega)D_V^*(\omega)}{\phi(\omega)} = G(\omega)$$

Where

$$\phi(\omega) = \max\{D_V(\omega)D_V^*(\omega), c\max[D_V(\omega')D_V^*(\omega')]\}$$

and

$$G(\omega) = \xi \exp(-\frac{\omega^2}{4a^2})$$

where * indicates complex conjugation; c is the minimum allowable spectral amplitude of the vertical component, the water level; ω' covers the range of all possible frequencies; the constant ξ normalizes the Gaussian filter to unit amplitude in the time domain; and a is the Gaussian factor which controls the width of the Gaussian pulse. Transformation of HR(ω) back into the time domain produces HR(t), an estimate of ER(t). In fact, as the plane P-wave approaches vertical incidence, the receiver function HR(t) becomes nothing more than a scaled version of the radial component of the site response, ER(t), with the P multiples removed. The tangential receiver function, HT(t), is produced by the same procedure. For a laterally homogeneous structure, HT(t) should theoretically be zero. This provides a way to judge the degree of lateral variation in Earth structure beneath a station (e.g. the presence of dipping velocity interfaces).

The inversion procedure and notation summarized below is presented in detail by Ammon et al. (1990). The nonlinear relation-ship between velocity structure and waveform is approximated by a Taylor series expansion about an initial model, m0, such that

$$d_j = F_j[m] \approx F_j[m_0] + (D,\delta m)_j + O\|\delta m^2\|, j = 1,2,3,\ldots,N$$

where, dj represents the observed data, m represents the true model, parentheses denote the inner product, δm is the model correction vector, and D is a matrix containing the first partial derivatives of the waveform with respect to changes in the layer velocities in the model m0. The operator F represents the nonlinear relationship between the model and the receiver function, which includes the calculation of a plane-wave synthetic seismogram and the source equalization procedure described earlier. Neglecting the higher-order terms, Equation (5) becomes

$$(D,\delta m)_j \approx F_j[m] - F_j[m_0]$$

The right-hand side of Equation (6) is the waveform residual vector, and the equation may be solved for δm using standard least-squares techniques described by Aki & Richards (1980). A minor modification of Equation (6) allows a solution to be obtained directly for m: adding (D, m0)j to both side of Equation (6) produces which is a simple matrix equation with m being the only unknown. The importance of solving directly for model m is that constraints, e.g. a smoothness trade- off parameter, may be applied to the model during the solution of Equations.

$$(D, m)_j \approx F_j [m] \approx F_j [m_o] + (D, m_o)_j$$

Fig. 3a presents a schematic ray diagram for an incident teleseismic plane P-wave together with its near-receiver reverberations and conv-erted phases, which, in this ideal case, domi-nate the radial ground motion at the receiver. Fig. 3b shows vertical and radial synthetic seismograms along with the corresponding radial receiver function for a one- layer over-half-space model. The crustal thickness of the model is 40km and the incident ray parameter corres-ponds to a surface source at 800 epicentral distance. Measuring times from the arrival of the direct P, the Ps phase arrives at about 5 seconds and the PpPms phase at about 18 seconds. The PpSms and PsPms phases arrive together as a negative pulse about 22.5 seconds. A small negative arrival about 27 seconds after the direct P is identified as the PsSms phase. All these phases are seen on both the radial synthetic seismogram and the receiver function but hardly seen on the vertical synthetic seismogram. The P multiple, PpPmp, arrives about 13 seconds on both vertical and radial synthetic seismograms but is removed from the receiver function by the de-convolution. The relative arrival times of each phase are controlled principally by crustal thickness and velocity, while relative amplitudes are affected by velocity contrasts at the Moho and the ray parameter.

The 3-component broadband seismic records from Manipur University station MANQ provides a new opportunity for investigating the structure beneath the area using receiver function analysis. MANQ is one of the stations of Manipur University Broadband network located in the Imphal valley. It was upgraded in 1999 with broadband instruments. The broadband channels are digitized at 100 sps. The USGS PDE catalogue was searched for events of ≥ 6.0 and Δ = 30o- 90o during the period 2004-2005 (Fig. 4). An appropriate suite of teleseismic earthquake seismograms was obtained from the archive. Broadband P-waves were windowed, rotated to the theoretical back azimuth to the event, and then deconvolved to form radial and tangential receiver functions. As stated above, examination of the tangential component of motion is useful for determining the degree of lateral velocity heterogeneity in the structure. Only data with the highest signal-to-noise ratios and whose structural signals could be clearly seen were chosen. Receiver functions for events with similar ray parameters and back-azimuths were then stacked to enhance the signal-to-noise ratio.

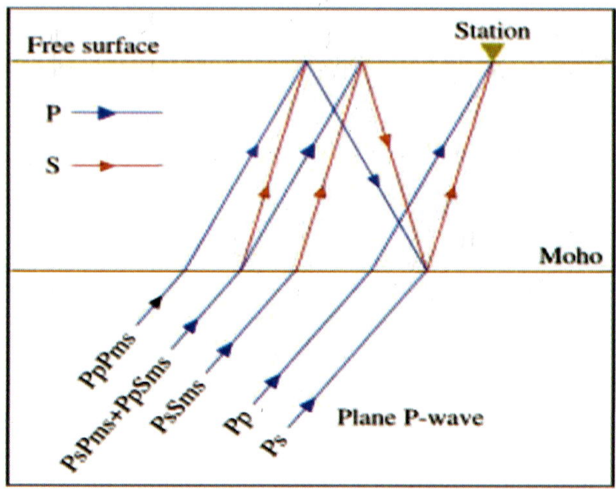

Fig. 3a. A schematic ray diagram for an incident teleseismic plane P-wave together with its near-receiver reverberations and converted phases, which, in this ideal case, dominate the radial ground motion at the receiver.

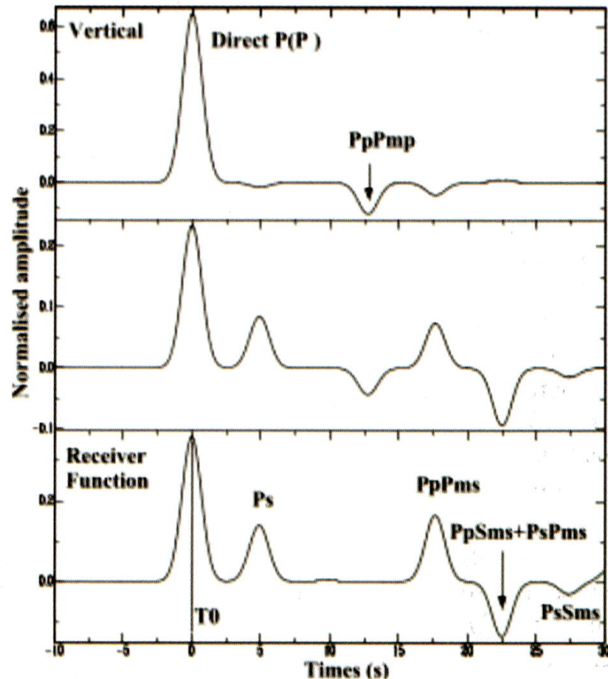

Fig. 3b. Vertical and radial synthetic seismograms along with the corresponding radial receiver function for a one-layer- over-half-space model.

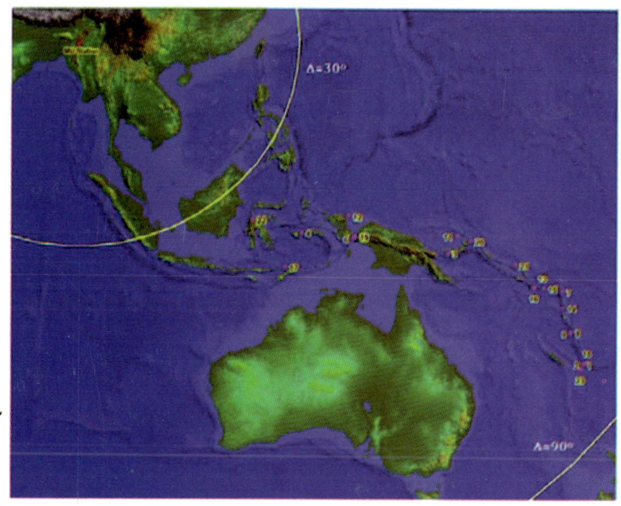

Fig. 4: Teleseismic events used for the persent study.

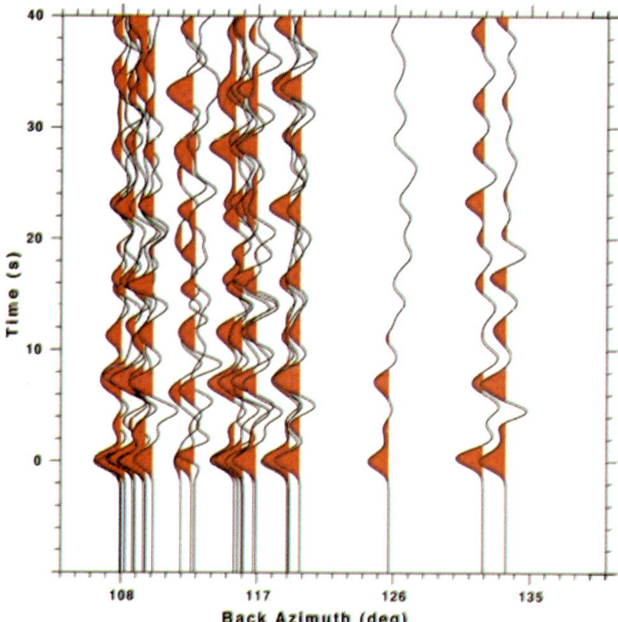

Fig. 5: Receiver functions for α =1.0 as a function of back azimuth.

Most of the well recorded teleseismic events are mainly distributed in the southeast and northeast quadrants and some are in the north-west quadrant. Due to high levels of local microseismic noise, some of the P-waveform data are not useful. Only those events that fall into clusters of azimuth and epicentral distance were analysed and modeled. Twenty three events with back azimuth 100°– 135° and epice-ntral distance 30°-90° were used for this recei-ver function analysis, based on a good signal-to-noise ratio in the resulting receiver functions (Table 1 and Fig. 4).

The receiver functions of the 23 selected earthquakes are plotted with respect to the back azimuth and ray parameter Fig. 5 and 6). A comparison of the observed receiver function with the synthetic receiver functions from those test models is shown in Fig. 7. Stacking of the receiver function is done to enhance the signal (Fig. 8a). The waveform of the average inversion model fits the data well. The average inversion model shows that the crustal structure beneath MANQ has two main layers. The first layer is 12 km thick and the second layer is 20 km thick with a low velocity zone of 8 km thick in between. The Moho is characterized as a gradient over a depth of 46-52 km. The inversion model also shows that there are some small velocity steps inside the two main layers. Our final model is a two-layer crustal structure with a low velocity zone in between the upper and lower crust plus a gradient Moho (Fig. 8b).

Table 1. List of events used for the present study.

S.N.	Yr/Mn/Dt	Time	Latitude	Longitude	Depth	Mag
1	2004/01/03	08:21:48.40	-22.32	169.62	10	6.4
2	2004/01/03	16:23:21.02	-22.25	169.68	22	7.1
3	2004/01/09	22:35:31.19	-06.07	149.40	57	6.3
4	2004/01/28	22:15:30.70	-03.12	127.40	17	6.7
5	2004/02/05	01:05:02.84	-03.62	135.54	16	7.0
6	2004/02/07	02:42:35.21	-04.00	135.02	10	7.3
7	2004/02/ 20	05:58:45.19	-11.61	166.45	84	6.0
8	2004/04/22	10:11:12.66	-17.59	167.89	10	6.0
9	2004/04/27	23:28:18.88	-17.76	167.76	10	6.0
10	2004/05/07	01:26:42.20	-21.99	170.28	14	6.3
11	2004/05/13	09:58:43.45	-03.58	150.73	10	6.4
12	2004/10/06	22:30:56.95	-00.67	134.43	10	6.2
13	2004/10/08	08:27:53.54	-10.95	162.16	36	6.8
14	2004/10/20	18:32:24.57	-14.03	166.61	59	6.1
15	2004/11/09	23:58:23.65	-11.15	163.71	13	6.9
16	2004/11/11	17:34:52.05	-11.13	162.21	10	6.7
17	2004/11/11	21:26:41.15	-08.15	124.87	10	7.5
18	2004/11/26	02:25:03.31	-03.61	135.40	10	7.1
19	2004/11/28	07:36:45.40	-03.64	135.45	23	6.2
20	2005/01/14	08:33:14.46	-04.24	152.72	10	6.1
21	2005/01/22	20:30:17.35	-07.73	159.48	29	6.4
22	2005/01/23	20:10:12.15	-01.20	119.93	11	6.3
23	2005/02/10	16:53:19.99	-23.10	169.22	09	6.3

In the model, the shear velocity is 3.45 km/s in the upper crust, 3.55 km/s in the lower crust, and 3.94 km/s in the uppermost mantle. The average Moho depth is 45 km.

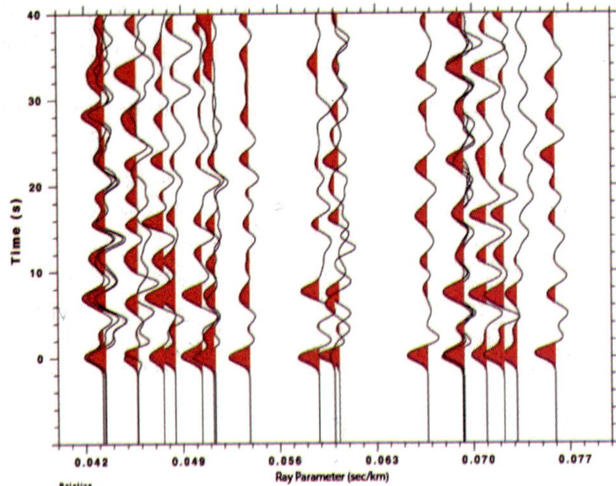

Fig. 6: Receiver functions for α =1.0 as a function of ray parameter.

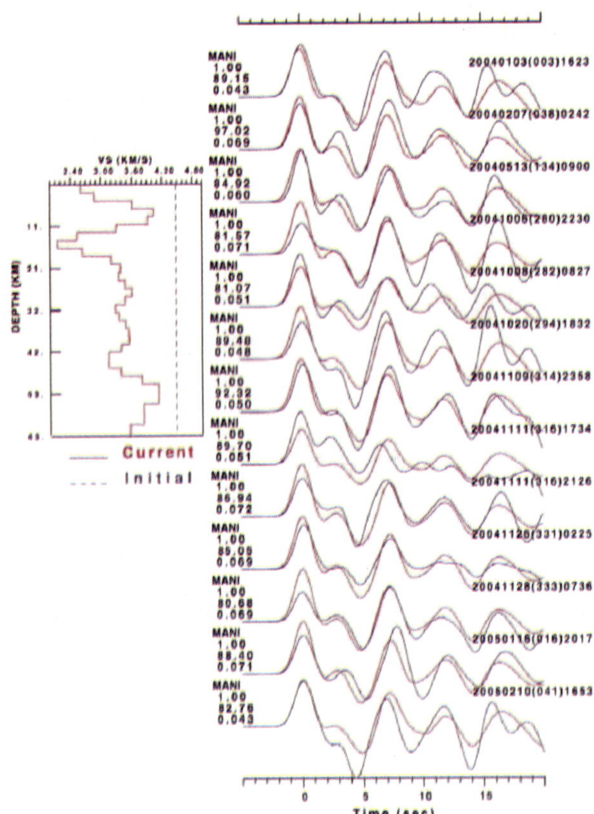

Fig. 7: Comparison of the observed receiver function with the synthetic receiver functions.

Discussion

Results from our receiver function analysis indicate that an upper crustal thickness of 12 km and a lower crustal thickness of 20 km. The lower crust is inferred to correspond to a basaltic layer and the proposed gradational Moho suggests that the crust- mantle boundary is most probably a transitional one in this area. The seismic velocity in the uppermost mantle is not well constrained, and the mantle has been modeled as a half- space with a shear velocity of 3.95 km/s. The receiver function method is not sensitive to the absolute velocity of the half-space, and therefore does not provide a reliable measurement of the upper mantle velocity below the station.

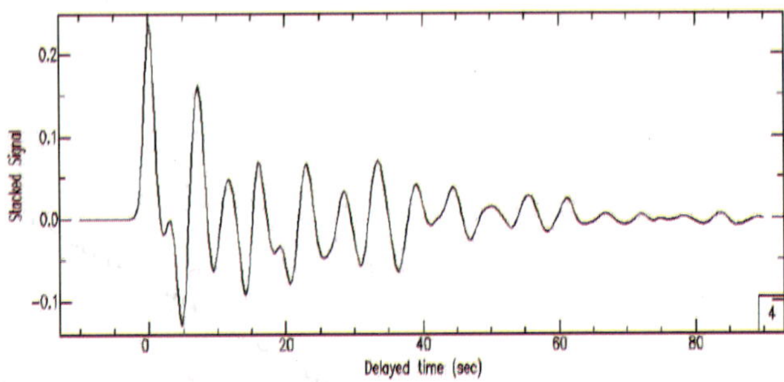

Fig. 8a: Stacking of the receiver function to enhance the signal.

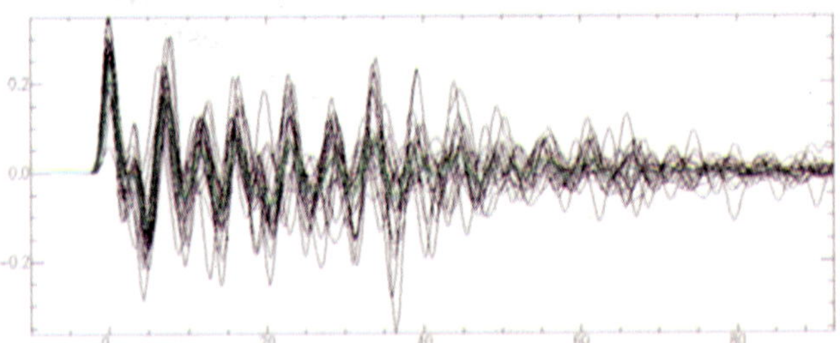

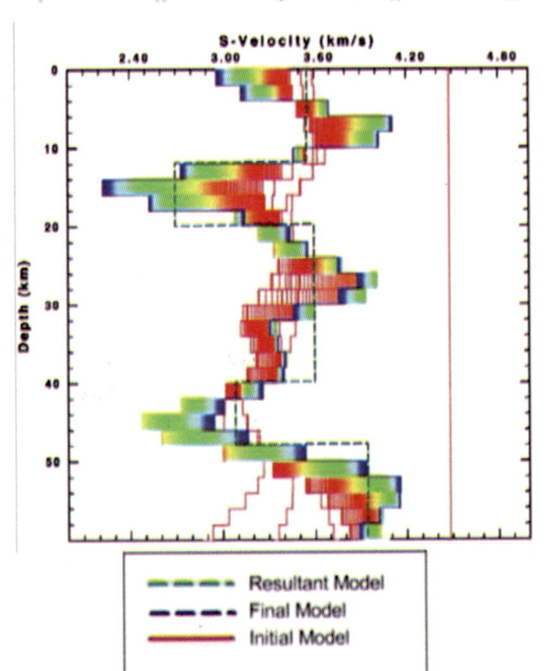

Fig. 8b: Summary of the inversion results of the QIZ stacked receiver function

As there is no refraction profiles exist for the central part of the Imphal valley near the MANQ station, and no controlled-source seismic investigations have been carried out around the region. Therefore, we have tested the validity of the velocity model comparing with the other published model by estimating the hypocentral parameters of some well recorded earthquakes.

The results from the Receiver function suggest a Moho depth of 45 km in the region. This study show that the average VS at the upper crust is 3.45 km/s and 3.55 km/s at the lower crust well demarcated by the low velocity zone of VS of 2.7 km/s with a thickness of about 8 km.

A sharp velocity discontinuity separating the upper and lower crust can be clearly identified at a depth of about 16 km, and is probably the Conrad discontinuity. The Moho started from 46 km, showing transitional nature as depicted by the increase in velocity from 46-54 km.

The crustal velocity structure evolved from the present study is used to estimate the hypocentral parameters. Gupta (1982) crustal model is used, the rms uncertainties in the estimation of the hypocenters as compared. It is clearly seen that, the uncertainties are much improved as detailed below.

Hypocentral parameter determination using the evolved Receiver function's velocity model we have taken an event 2001/05/02:

NBA	iP	16:39: 06.25
MAN	iPg	16:39: 12.67
	iSg	29.00
UKH	iPn	16:39: 18.76
	iSn	40.00
SHL	iP	16:39: 07.00
NA11	ePn	16:39: 16.59
	eSn	36.00
SI11	iPn	16:39: 19.90
	iSn	42.00

using the velocity model of Gupta (1982), we estimated the hypocentral parameters as detailed below :

ORIGIN	LAT N	LONG E	DEPTH	Mag	RMS	ERLN	ERLT	ERDP
16:38: 50.14	25°00. 21'	92°36. 17'	18.7	2.3	0.133	0.8	0.8	1.3

Again, using the same event with the velocity model obtained from the present study we have the following results :

ORIGIN	LAT N	LONG E	DEPTH	Mag	RMS	ERLN	ERLT	ERDP
16:38: 50.10	24°59. 20'	92°34. 86'	32.7	2.4	0.127	0.1	0.1	1.1

Conclusions

Teleseismic receiver function analysis is now a well-established seismological technique and the most important points for understanding the receiver function modeling have been proposed first time in IBR region. Analysis and modeling of MANQ receiver functions generate a crustal model which reveals a 12 km thick upper crust, a 20km thick lower crust with a low velocity zone in between lower and upper crust with a thickness of 8 km seems to be Conrad discontinuity, and a 6 km thick Moho transition zone. Comparison of this model with other crustal model shows improvement in the estimation of hypocentral parameters. The technique will provide another tool to study the shallow sedimentary structures which is essential to conduct site-effect estimations, theoretical simulations of strong motion, and seismic hazard assessments. Additional data of broadband stations under Manipur University Network in Manipur will be used to extend the current study. It would then be possible to compare and discuss the crustal structure and their tectonic implications.

Acknowledgments

Authors are thankful to the Ministry of Earth Sciences for providing the financial assistance to carry out these studies. We are also grateful to Dr. B.K. Bansal, Advisor for many suggestions for this manuscript.

References

Aki K. & Richards, P.G., 1980, Quantitative Seismology, W.H.Freeman, New York, 932p.

Ammon C.J., Randall, G.E. & Zandt G., 1990, On the nonuniqueness of receiver function inversions. Journal of geophysical Research, p.15303-15318

Banner, D., and Helmcke. 1981: The Evolution of Asian Plate in Burma: International Journal of Earth Sciences, Vol 70.

Ben-Menahem, A., Aboudi, E. and Schild, R. 1974. The source of the great Assam earthquake-an intraplate wedge motion. Physics of the Earth and Planetary Interiors., 9: p.265-289.

H. K. Gupta, S. C. Singh, T. K. Dutta, M. M, Saikia 1982: Recent investigations of North-East India seismicity. Internat. Symp. Continental Seismology and Earthquake Prediction, Beijing, p.63.

Langston C.A., 1979. Structure under Mount Rainier, Washington, inferred from teleseismic body waves. Journal of Geophysical Research, 84, 4749-4762.

Langston C.A., 1994, An integrated study of crustal structure and regional wave propagation for southeastern Missouri, Bulletin of the Seismological Society of America, 84, p.105-118

Nandy, D.R., 2001: Geodynamics of northeast India & the adjoining region. ACB Publication.

Owens T.J., Zandt G. and Taylor S. R., 1984, Seismic evidence for an ancient rift beneath the Cumberland Plateau, Tennessee: A detailed analysis of broadband teleseismic P waveforms, Journal of Geophysical Research, 89, p.7783-7795

Priestley K., Zandt G. and Randall G., 1988, Crustal structure in eastern Kazakh, U.S.S.R. from teleseismic receiver functions, Geophysical Research Letters, 15, p.613-616

Qiu X., Priestley K. and McKenzie D., 1996, Average lithospheric structure of southern Africa, Geophysical Journal International, 127(4), p. 563-587.

Hon'ble reviewer

1. **Dr. Kusumita Arora**
   Scientist
   Nat. Geophy. Res. Inst.
   Uppal Road
   HYDERABAD - 500 007
2. **Prof. Harsh Bhu**
   Department of Geology
   MLS University
   UDAIPUR - 313 001

The Chronology of Geological Column: An Incomplete Tool to Search Georesources

Bhakti Niskama Shanta

Sri Chaitany Saraswat Math, New Pal Para (Netaji Sarani), Hayder Para, Siliguri – 6, WB, India.
E-mail: jaga.suresh@gmail.com

Abstract

The archaeological record is very limited and its analysis has been contentious. Hence, molecular biologists have shifted their attention to molecular dating techniques. Recently on April 2013, the prestigious Cell Press Journal *Current Biology* published an article (Fu et al. 2013) entitled "A Revised Timescale for Human Evolution Based on Ancient Mitochondrial Genomes". This paper has twenty authors and they are researchers from the world's top institutes like Max Planck Institute, Harvard, etc. Respected authors of this paper have emphatically accepted that the fossil record is inadequate and unreliable. These statements clearly substantiate that now biologists are agreeing that fossil records do not provide any significant evidence at all for conventional evolution theory. Despite the well-recorded fact of the continual grand propaganda of Darwinism based on fossil evidence for more than 150 years, in recent times biologists are surprisingly coming up with such statements, based on their confidence that evolution can be explained purely by the genealogical/genomic record provided by modern molecular biology. Still many respected journals (for example the article in *Nature*, Retallack, 2013) continue to publish articles on fossil evidence to support Darwinian evolution. These incoherently diverse claims prove that Darwinists are struggling with unscientific ideological approaches to explain biodiversity. Darwinian evolutionary theory is not only the basis of modern biology, but also acts as the guiding principle of science and intellectual reasoning for modern civilization. Hence, a scientific understanding of the breakdown of the Darwinian theory of objective evolution is very important for overcoming the traditional scientific temper of mechanistic intellectualism that characterizes this ideology. In my article "21st Century Biology Refutes Darwinian Abiology" (published in two parts in November and December 2012 issues of *The Harmonizer*: *www.mahaprabhu.net/satsanga/harmonizer*) it was noted that several recent findings challenge the credibility that random mutations and natural selection can provide a valid basis for justifying the naturalistic evolution of species. The present article summarizes the problems associated with the fossil record and dating techniques, and its implication on the neo-Darwinian mechanistic misconception of biological life as mere molecular chemistry or abiology. An alternative approach based on the *Vedāntic* view for explaining biodiversity in the light of 21st century biology is also discussed in the end of the article.

Keywords: Vedantic Chronology, Geological Time Scale, Evolution of Life, Human Evolution.

Introduction

To illustrate the timing and relationships between events that have taken place throughout the history of the globe, geologists, paleontologists, and other earth scientists use the geologic record represented by consecutive layers of rock strata to provide a scheme for chronological measurement. The geological column is also known as the stratigraphical column and is the most commonly used representation for estimating geological time (Cambrian, Ordovician, Silurian, Devonian, Carboniferous, Permian, Triassic, Jurassic, Cretaceous, and Tertiary). A doctrine termed uniformitarianism was first proposed by James Hutton in his '*Theory of the Earth*' in 1795 and developed further by Charles Lyell in '*Principles of Geology*,' first published in 1830. This doctrine explains that the causes that changed the Earth's surface in past geologic times are identical to causes now producing changes on the Earth's surface. Following this assumption, geologists believe that because sediments are presently observed as being laid down layer by layer, so they must have for all time been laid down gradually in a similar way. By estimating the rate of this sedimentary process they calculate that a certain thickness of sedimentary rock must symbolize, in certain circumstances, millions of years of time. This hypothetical representation of Earth's surface as an 'onion skin' with successive layers representing the events throughout the history of the globe was, however, never substantiated with enough experimental or empirical evidence. Now, recent developments in the field are providing the greatest challenge to this widely used archaic methodology and its conclusions.

Evolution

The Origin of Species was published by Charles Darwin in 1859, in which he proposed the idea of natural selection to explain evolution. The widespread acceptance of this concept led to a radical change in the people's perception about life and world. Historically, numerous scientific objections and challenges were brought against this central conception, but with the dawn of the neo-Darwinian synthesis with Mendelian genetics, Darwin's idea eventually became the guiding principle of modern thinking, and for more than 150 years the only major focus of biologists has been to uphold the Darwinian vision intact. Evolution is a word not limited to biological evolution involving natural selection. Scientists claim that the fundamental matter of our universe and solar system has also evolved, according to the big bang theory, and that chemical elements eventually evolved from simpler matter. Then first-life evolved from those chemicals (abiogenisis), and complex organisms from simpler life forms (biological evolution). The broadest definition of evolution thus implies any purely mechanistic development by natural undirected processes to form the aggregates of matter that we observe in the universe today.

Historical Background in the Development of Darwinian Evolution Theory

The cerebral ambiance of Darwin's time played a vital role in the development of his theory. A number of natural historians were beginning to speculate that perhaps evolutionary changes are the cause of the patterns observed in nature. Meanwhile, geologists started elucidating their ideas about the age of the Earth's strata and suggested that a time sequence could be inferred, by assuming younger strata overlayed older strata. Using this as a basis geologists started calculating the age of Earth.

Continuous fossil findings revealed unknown living species (for example giant dinosaurs). Cuvier (1840) reported that a particular species can be detected in a certain stratum and that this coincidence of fossil-stratum is unique. Hence, he constructed a chronology of fossils based on the depth of strata. Similarities in different groups of organisms were recognized as proof of relatedness (evolutionary relationships) by naturalists. The predecessors in the line of Darwin, however, failed to come up with a satisfactory mechanism for explaining evolution. Among them Lamarck is well known. He most famously cited the giraffe as an example and proposed a mechanism based on the inheritance of acquired characteristics to explain adaptive changes (Koonin and Wolf, 2009). All these developments during that era proved to be catalysts in the formulation of the Darwinian evolution theory.

Darwinian Evolution Theory and Its Foundation

Based on fossil evidence Georges Cuvier denied the transition of one fossil form to another following a gradual alteration. Hence based on his knowledge about animal anatomy and physiology, he rejected objective evolution of one species from another (Waggoner, 1996). Due to these observations of Cuvier, gradual transmutation of species concept did not come into the limelight until Charles Darwin published *Origin of Species* (Larson, 2004). The major contribution of Darwin was that he could convince people in his time about biological evolution by proposing a mechanism based on natural selection (Koonin, 2009) and he supplied fossil records as major evidence (Web Reference, 1) for his theory. He further extended his thoughts in line with his theory to provide an explanation for the history of life on Earth and its biodiversity. He claimed that species change over geographical space and geological time. From the fossil records he explained that species that are living today are quite different than those that lived in past. Hence far enough back in time any pair of organisms share a common ancestor. He also explained that evolutionary changes are gradual and slow. He claimed this could be supported by evidence in the fossil record. At that time naturalists believed that any observed sudden appearance of a new species in the fossil record was due to lack of sufficient fossil data, and that in future sufficient accumulation of fossil data would transform this sudden appearance of new species into a gradual one. Darwin had not ruled out the Lamarckian view of inheritance of acquired characteristics, but he was opposed to it since it contradicted his theory. This is because Lamarck believed organisms were active, adaptive systems driven by needs while, for Darwin, an organism is externally driven by contingent causes and natural selection. Latter Lamarckism was rejected by Weismann and Wallace (Kutschera, 2003). Today, Neo-Darwinism is a term introduced to describe the modern synthesis of Darwinian evolution thorough natural selection with Mendelian genetics. Lamarckian processes are also gaining some credence due to evidence that methylation and epigenetic changes influenced by the environment are also inherited. However, that did not change the fact that Darwin's evolution theory significantly relied on the basis of evidence derived from fossil records, although the emphasis has now shifted more to a genealogical, and further to genomic, proteomic, and phenomic comparative analysis.

Microevolution and Macroevolution

There are real and imaginary parts of the evolution theory. Variation known as microevolution (within a species) is the real part and big changes known as macroevolution (one species leading to another) is the imaginary part. The beaks of birds, colors of moths, leg sizes, and so on, are the characteristics which can be recognized as microevolution. There should not be any problem to accept microevolution because we do observe it in nature. However, the imagination of Darwinian evolution theory extends very far and claims that these small changes will gradually lead to a new species.

From animal and plant breeding experiments it is well known that there are strict limits to variation which are never crossed. A few Darwinists argue that we cannot see macroevolution now because it takes place very slowly (tens of thousands of generation). However, micro organisms are found to exhibit faster mutation rates (Bryan et al., 2012). Hence, it is possible to experimentally verify whether macroevolution takes place or not in such cases. As an example, bacterial colonies are found to grow (reproduce) in as quickly as 12 minutes or up to 24 hours and more in almost all types of environments. We can test the bacteria generation after generation but there is not a single case reported in the available scientific literature where they turn into something else. They always remain bacteria. Recently, Kuhn (2012) in his article 'Dissecting Darwinism' stated: "In all fairness, there is convincing evidence, that is widely acknowledged, that random mutation and natural adaptation (Darwinian evolution) does occur within species, leading to minor changes in areas such as beak size, skin pigmentation, or antibiotic resistance. Some of these changes involve a simple biologic survival advantage for a population, without a mutation in DNA. Others might be influenced by a single deletion or insertion within the DNA strand. However, the modern evolution data do not convincingly support a transition from a fish to an amphibian, which would require a massive amount of new enzymes, protein systems, organ systems, chromosomes, and formation of new strands of specifically coding DNA. Even with thousands of billions of generations, experience shows that new complex biological features that require multiple mutations to confer a benefit do not arise by natural selection and random mutation. New genes are difficult to evolve. The bacteria do not form into other species."

The morphologically based TOL (Tree Of Life) representation has dominated evolutionary biology from the time when Darwin first proposed it as a sufficient description of the total history of life forms on Earth. Later, a three-domain tree of ribosomal RNA (rRNA) was introduced by constructing trees of other universal genes, such as ribosomal proteins and core RNA polymerase subunits (Woese, 1987). Thus, TOL was perceived as an authentic victory of tree thinking in biology. However, we now know that mechanisms like transduction, natural-transformation, horizontal (lateral) DNA transfer and conjugation can produce sudden changes in living organisms. Genome-wide analysis of gene phylogenies (phylogenomics), revealed an additional intricate image of evolution (Delsuc et al., 2005). The discovery of HGT (Horizontal Gene Transfer) has completely changed the whole picture.

There are cases reported in which phylogenetic trees of individual genes commonly have dissimilar topologies and this variety of tree topologies cannot be elucidated by artifacts of phylogenetic rebuilding (Koonin and Wolf 2008). These research studies recommend that TOL should be replaced by a 'net of life' or a 'forest of life' (Bapteste et al., 2009; Doolittle, 1999). With further advancements in research, evolutionary genomics successfully knocked down the simple idea of the TOL by bringing to light the dynamic, reticulated nature of evolution where HGT, genome fusion, and interaction among genomes of cellular life forms and diverse selfish genetic elements play a vital role. Hence, the phylogenetic TOL becomes the genetic 'forest of life' and this genetic 'forest of life' includes trees, bushes, thickets of lianas, and obviously, several dead trunks and branches (Koonin, 2009). Moreover, the manifestation of new species with novel features by hybridization takes place too fast for natural selection to operate innovatively. Despite the fact that hybridization and symbiogenesis have been recognized long back, still many neo-Darwinists dogmatically maintain confidence in gradualism in evolutionary change. In his noteworthy book, '*Evolution: A view from the 21st century*', Shapiro (2011) summarizes numerous instances where empirical evidence shows that gradualism in evolutionary change is refuted by 21st century biology.

Fossil Record

The fossil record is most uneven; it is nowhere near complete and can never be. Very rarely are organisms preserved as fossils in the best of circumstances. Organisms without hard parts (like worms) can only be represented very inadequately in fossil records. The number of species known through fossil records is insignificantly small as compared to total species. Raup (1981) stated in his paper in *Science*, "In the years after Darwin, his advocates hoped to find predictable progressions. In general, these have not been found—yet the optimism has died hard, and some pure fantasy has crept into textbooks." The fossil record suffers four major defects that are principally incompatible with gradualism: (a) stasis, (b) sudden appearance of forms, (c) sudden disappearance of forms, (d) relative absence of transitional forms.

(a) Stasis

Examining natural history, researchers reported that organisms never evolved into different novel anatomical structures; rather they continually unaltered, even over the period of hundreds of millions of years. This non-changing aspect of an organism is known as stasis in the fossil record. Stephen J. Gould (1977), Professor of Zoology and Geology, Harvard University, USA, stated that the history of most fossil species is particularly inconsistent with gradualism:

"Stasis. Most species exhibit no directional change during their tenure on earth. They appear in the fossil record looking much the same as when they disappear; morphological change is usually limited and directionless."

Lewin (1980) also quoted the statement of Gould: "For millions of years species remain unchanged in the fossil record". Gess et al. (2006) also reported in *Nature*, "lampreys as a whole appear all the more remarkable: ancient specialists that have persisted as such and survived a subsequent 360 million years." Many similar observations in the literature establish that species preservation is a natural characteristic of life. Life's ability to preserve its own species over the period of hundreds of millions of years ('Stasis' in the fossil record) offers a significant challenge to Darwinian gradualism. Williamson (1981) stated,

"The principal problem is morphological stasis. A theory is only as good as its predictions, and conventional neo-Darwinism, which claims to be a comprehensive explanation of evolutionary process, has failed to predict the widespread long-term morphological stasis now recognized as one of the most striking aspects of the fossil record."

(b) Sudden appearance of forms – 'Darwin's Dilemma'

Empirical evidence substantiates the fact that the new species did not evolve but suddenly appeared in geologic column. This is also famously known as 'Darwin's Dilemma.' The Cambrian explosion refers to the abrupt manifestation of fossils of several of the most important animal phyla during a period of less than ten million years. Different species from numerous main divisions of the animal kingdom are found suddenly emerging in the lowest known fossiliferous rocks. Cooper and Fortey (1998) reported, "The beginning of the Cambrian period, some 545 million years ago, saw the sudden appearance in the fossil record of almost all the main types of animals (phyla) that still dominate the biota today. To be sure, there are fossils in older strata, but they are either very small (such as bacteria and algae), or their relationships to the living fauna are highly contentious, as is the case with the famous soft-bodied fossils from the late Precambrian Pound Quartzite, Ediacara, South Australia." The Cambrian explosion offers a significant challenge to the conventional gradualistic mechanisms of Darwinian evolution. Darwin (1869) himself was perplexed that the fossil record disagreed with the claims of his evolution theory and believed that future fossil discoveries will help solve this major problem. However, Gould (1995) reported just the opposite: "The Cambrian explosion occurred in a geological moment, and we have reason to think that all major anatomical designs may have made their evolutionary appearance at that time. ...not only the phylum Chordata itself, but also all its major divisions, arose within the Cambrian Explosion. So much for chordate uniqueness... Contrary to Darwin's expectation that new data would reveal gradualistic continuity with slow and steady expansion, all major discoveries of the past century have only heightened the massiveness and geological abruptness of this formative event...". Carroll (2000) also explains the same, "The extreme speed of anatomical change and adaptive radiation during this brief time period requires explanations that go beyond those proposed for the evolution of species within the modern biota."

(c) Sudden disappearance of forms

The fossil record reveals that in the history of life, several flourishing species were often suddenly wiped out. The K-T extinction (Cretaceous-Tertiary extinction) contributed to the disappearance of the dinosaurs. Several mass extinction episodes have been reported throughout the history of the Earth. For example (Web Reference, 2), Ordovician-Silurian mass extinction (third largest extinction in Earth's history), Late Devonian mass extinction (Three quarters of all species on Earth died out), Permian mass extinction – The Great Dying (96% of species died out) and Triassic-Jurassic mass extinction (50% of species died). This is a big setback to conventional gradualistic mechanisms of Darwinian evolution and the same is reported by Fitch and Ayala (1995), "Many of the extinctions recorded in the fossil record are of species or large groups of species that were ecologically tolerant and occurred in great numbers in all parts of the world. If these extinctions were caused by slow declines over long periods of time, as Darwin thought, they might be explicable in terms of the cumulative effect of very slight deficiencies or disadvantages. But it is becoming increasingly clear that successful species often die out quickly." We have mentioned previously that life has a natural tendency to preserve its species and hence many species survive for several million years. This indicates that many species can adapt to the physical and biological stresses usual in its environment. The reason for such surprising extinctions is explained by Fitch and Ayala (1995): "This implies, in turn, that likely causes of extinction of successful species are to be found among stresses that are not experienced on time scales short enough for natural selection to act."

(d) Absence of transitional forms

The fossil record establishes that species never follow a step by step path of continuous alteration of their ancestors; rather they appear all of a sudden and fully shaped. Convincing transitional forms are never observed to substantiate gradualist mechanisms of Darwinian evolution. Gould (1977) confirms the same:

"In any local area, a species does not arise gradually by the steady transformation of its ancestors; it appears all at once and 'fully formed.'"

Still, many hold a wrong notion that the fossil record uncovered transitional species to support gradualism in Darwinian evolution. In reality, the lack of transitional species data to explain the multitude of alterations associated with species transition is a well established fact in the literature. Carroll (2000) confirms that, "What is missing are the many intermediate forms hypothesized by Darwin, and the continual divergence of major lineages into the morphospace between distinct adaptive types." Very recently, Kuhn (2012) strongly affirms "The transitional species from primitive primates to man have been illustrated in textbooks for over 100 years. These drawings form the visual imagery that supports Darwinian evolution for high school students, university students, medical students, and the public. However, honest dissent exists in the accuracy of most of the transitional prehominoids, with many found to be frauds or animal species." Hence a serious consideration is necessary to find out the actual reason for all these discrepancies.

The role of geological chronology

Objective evolution theory or Darwinism is basically founded on a uniformitarian theory that presumes random mutations and natural selection are adequate for elucidating the gradual macro-evolution of species during the course of billions of years of life history on Earth. Following Darwin's ideas that evolutionary changes are gradual and slow, macro-evolution is often explained on the scale of geological time – measured in hundreds of millions of years (Web Reference, 3). However, as explained above, the evidence from the fossil record is substantially in disagreement with this gradualist, uniformitarian assumption of Darwinism. Moreover, Kuhn (2012) raises suspicion about the validity of the fossil data: "A reliance on gross morphologic appearances, as with fossils, drawings, and bone reconstructions, is severely inadequate compared to an understanding of the complexity of the DNA and coding that would have been required to mutate from a fish to an amphibian or from a primitive primate to a human." In the midst of many such perplexities, what is lacking is a thorough investigation into the accuracy of the dating technologies that are often presumed. Geologic chronology or a coherent history of the Earth is heavily dependent on the accurate understanding of the ages of rock formations. Radiometric dating and Stratigraphy are the two pillars of geological chronology and they are often employed to date fossils without considering the accuracy of these techniques. Hence a thorough investigation into the accuracy of geological chronology is very much essential. Till date there are only a very few discussions in the literature on the authen-ticity of geological chronology. In the light of recent finds and reported empirical evidence, the present article makes an attempt to summarize the current standing of the two pillars (Radio-metric dating and Stratigraphy) of geological chronology.

Is Radiometric Dating Trustworthy?

A general notion among academic circles is that radiometric dating is extremely trustworthy. However, the reality is completely the reverse. Way back in 1950 itself it is famously stated that radiometric dates are like railway timetables and they are subject to change without notice (Whitten and Brooks, 1972). In the following subsections we will analyze the real picture of 'carbon dating,' 'radioisotope dating' and finally the robustness of 'radiometric dating,' the 'pacemaker of geologic time'.

Boundaries of carbon dating

Carbon-14 (C-14) dating is commonly employed to date the age of fossils. The major problem with C-14 dating in the context of evolution theory is that it cannot give an age of hundreds of millions of years to the fossils because its half-life period is merely 5,730 ± 40 years (Godwin, 1962). Anything beyond 50,000 years will lead to a situation where it is not possible to find a sample with high enough concentration of C-14 to perform the tests. If a fossil contains a measurable amount of C-14 then automatically it proves that the fossil is less than 50,000 years old. Therefore, such a technique cannot be employed for testing Darwinian evolution theory which requires hundreds of millions of years. Furthermore, it is observed that the intake of carbon dioxide containing C-14 differs for different plants (Brooks et al., 2002). This would require the impossible task of introducing a correction factor for each and every species in order to get an accurate age. It is also a well known fact that C-14 concentration in atmospheric CO_2 often varies (de Vries, 1958) due to solar activity (Stuiver, 1965), geomagnetic field strength (Bucha, 1970) and numerous other factors. Due to these variations, the C-14 clock runs at a varying pace throughout the history of Earth, thus an unimaginable calibration would be needed to establish a relation between C-14 time and the historical time. Hence the use of carbon dating in the geologic column of hundreds of millions of years is unrealistic.

Dearth in radioisotope dating

For dating the geologic column of hundreds of millions of years, some researchers suggest radioisotope dating as an alternative to carbon dating. The half lives of the radioisotopes are long enough to cover the entire Earth's history. It is important to note that unlike carbon dating, radioisotopes can only be used to date the sediments present in the rocks and not the fossils directly. Direct radiometric techniques can neither be used for dating the mineralized fossils nor the sedimentary rock that buried them. Fossils are buried in sediments and hence it is idealistically assumed that if somehow the age of sediments in rocks can be estimated then the same can be assigned to the fossils as well. Hence, many of these radioisotope dating methods necessarily need the presence of some igneous rock fragments within the sedimentary rock layer. If some igneous rock fragments are present within the sedimentary rock then only can these techniques be used to date those igneous rock fragments. Then follows the incredible assumption that the fossil is as old as the igneous rock fragment obtained from the sedimentary layer. However, a more reasonable thinking may conclude that fossils are indeed much younger than the sediment that buried them.

Moreover, the basic problem with radioisotope dating is that, we cannot know the radioactive concentrations (both parent and daughter isotopes) that existed in the rock in the beginning. Not only it is impossible to have the exact estimation of radioactive elements when the rock formed, but also there is no means to analyze the way those elements changed during its complete geological history. In addition, it is assumed in these dating techniques that the system is closed so that none of the parent and daughter isotopes leak to the environment. Most of the uranium salts are soluble in water (Barbier-Baudry et al., 2000). Rocks are exposed to rain and moisture and therefore there is a maximum chance that they may get dissolved and lost in the environment. For example, US Geological Survey bulletin (Klepper and Wyant, 1957) reports that 90% of the radioactive elements in a few granites can be removed by leaching the rock with a weak acid. This report also states that 40% of the Uranium in newly emerging igneous rocks is easily leachable. In such cases we will get erroneous results when we use the sample for dating.

Failure of the pacemaker of geologic time

Interestingly, radioactivity is considered independent of temperature and pressure and hence radiometric dating technique cannot be used for calculating the time taken for the solidification of magma. Moreover, the constant decay rate assumption itself is facing many recent challenges (Jenkins, 2010). Researchers (Lynde and Spangler, 1974; Huh, 1999) have reported that even mild alterations in the environment may affect the stabilities of C-14, Co-60 and Cs-137. Experiments (Hahn et al., 1976) as far back as 1976 revealed changes in nuclear decay rates caused by physical or chemical changes in the surroundings. Reifenschweiler (1994) reported that changes in decay rates as high as 40% have been induced for certain nuclides. Radiohalos are found with different diameters for the same type of inclusion, and this has been taken by some investigators to imply that decay rates have varied with time (Allen, 1952, and Spector, 1972). Kerr (1999)'s article in *Science* also confirms that radiometric dating as the pacemaker of geologic time can no longer be called precisely 'clocklike'. Therefore, radiometric dating cannot be used reliably to give an age to the fossils and rocks. Stansfield (1977) fittingly stated in his book *The Science of Evolution*,

"It is obvious that radiometric techniques may not be the absolute dating methods that they are claimed to be... The uncertainties inherent in radiometric dating are disturbing to geologists and evolutionists"

The roles of stratigraphy and the geologic column in Darwinian evolution theory

In his book, *A History of Geology* (1990), Gabriel explained that the rate of deposition of sediments determines the geological ages and not biological evolution or orogeny. Therefore Stratigraphy remained the only basis of geological dating. In the 17th century Danish scientist Nicolas Steno (1669) formulated the basic principle of Stratigraphy based on three major assumptions: (1) Principle of superposition, (2) Principle of continuity, and (3) Principle of original horizontality. Steno, by assuming all rocks and minerals had once been fluid, theorized that rock strata were formed when sediments in a fluid such as water fell to the bottom. Obviously this method would lead to horizontal layers and is the reason why Steno's principle of original horizontality states that rock layers form in the horizontal position. Nicolas Steno also stated that if a solid body is enclosed on all sides by another solid body, of the two bodies that one first became hard which, in the mutual contact, expresses on its own surface the properties of the other surface.

Steno's explanation popularized the idea that fossils and crystals must have solidified before the host rock that contains them was formed. In geology a stratum is known as a layer of rock with consistent uniqueness that distinguishes it from the adjacent layers. Following Steno's idea scientists believe that these parallel layers rest one upon another in the rocks due to natural forces. In cliffs, road cuts, quarries, and river banks strata can be characteristically observed as bands of dissimilar colors or differently structured substance. In general geologists analyze the rock strata by categorizing the layers with respect to the material content within them. Each layer represents a particular type of deposition of beach sand, sand dune, river silt, coal swamp, lava bed, etc. A typical stratigraphic column shows a series of sedimentary rocks, with the oldest rocks on the bottom and the youngest on top. Thus stratum is an essential fundamental element to study geologic time scale. Geologists, paleontologists and other earth scientists use the stratigraphic principle to describe the timing and relationships between events that have occurred during the history of the Earth. Evolutionists recognize the age of the fossil according to the geologic time scale based on the vertical location of the strata in which the fossil was discovered. Hence fossils obtained from the bottom of the geologic column are recognized by evolutionists as the most ancient fossils.

Practical defects in Nicolas Steno's principles of stratigraphy

Steno's three basic assumptions on which stratigraphy stands were never substantiated by either experimentation or empirical evidence. We will discuss below the fallacies of Steno's assumptions based on the available empirical evidence in the literature. French sedimentologist Guy Berthault could recognize these defects in Steno's assumptions and carried out the most fundamental experiments on sedimentation at Colorado State University with Pierre Julien (Professor of hydraulics and sedimentology) to evaluate the validity of Steno's assumptions (Berthault, 1986; Berthault, 1988; Julien and Berthault, 1993; Julien et al., 1993; Berthault, 2002). The technical problems with each of these three assumptions by Steno (1667) are discussed below.

Defects in Steno's First Assumption – Principle of superposition [(i) At the time when one of the high strata formed, the stratum underneath it had already acquired a solid consistency, and (ii) At the time when any stratum formed, the superincumbent material was entirely fluid, and, due to this fact at the time when the lowest stratum formed, none of the superior strata existed (Steno, 1667, p. 30, CII. 3.d)]:

A stratum is considered as thick if its thickness is about 50 - 100 cm. Following the first part of Steno's first assumption we would expect solid strata after a few meters in the seabed. However, the evidence recorded from the submarine drillings of deep seabed reveal that the first semi-consolidated sediments are found between 400 - 800 m. Isolated, hardened chert beds are found below 135 m of unconsolidated sediments (Logvinenko, 1980). These sedimentological evidences challenge Steno's successive hardening assumption which extends significantly the total time of deposition.

The second part of the Steno's first assumption is not found to be in line with experimental data obtained by Guy Berthault in Colorado State University. Steno mentioned that "Strata owe their existence to sediments in a fluid" (Steno, 1667, p. 30, CII.3c). However, Steno's stratigraphic model completely overlooked the fluid current and its chronological effects, which is the major variable factor in oceanic fluid. We cannot find an ocean without current and it is well known from a long time that oceanic currents erode, transport and deposit sediments (Strakhov, 1957). Charles Lyell, following Steno's principles, developed his theory of uniformitarianism. Lyell observed that the layers deposited in fresh water in Auvergne were less than 1 mm think and he further considered that each one of them had been laid down annually.

Following this assumption we would expect that a 230 m thick deposit would have taken hundreds of thousands of years to form. But, long back Dunbar and Rodgers (1957, p. 198, 33) cautioned, "Without critical analysis and supporting evidence, however, it is unsafe to assume that the laminae in a shale represent annual deposition rather than deposition by a single storm or stormy spell. Several laminae may even have formed during a single flood". Blatt et al. (1980, p. 134) also gave similar suggestions. In fact, Bouma (1962) reported that turbidites can dump thick multilayered sediments in a few minutes, and also the layers in the middle of such catastrophic flows did not solidify before other layers formed on top of them. Baas et al. (2000) also reported that laminar beds can be deposited in a span of a few minutes in turbidites. Similar observations are also reported in the pattern of volcanic ash deposits (Schmincke et al., 1973; Fisher and Schmincke, 1984). Guy Berthault's (2002, p. 442-443) sedimentation experiments also confirm that Steno incorrectly assumed that underlying layers must acquire a 'solid consistence' before overlying layers can be deposited. Guy Berthault conducted sedimentation experiments with a constant supply of heterogeneous materials in water with and without current (Web Reference, 4). In still water a deposition is obtained giving the false impression of succeeding beds or laminae. In reality these laminae appeared due to a spontaneous periodic and a continuous grading process taking place immediately following the deposition of a mixture of particles. The thickness of laminae is found to be independent of sedimentation rate and increases only when there is an extreme difference in particle sizes in the mixture. It was observed that the time required for development of these layers was much less than that indicated by the modern geological timescale.

More interestingly, when the experiments were carried out by Guy Berthault in a hydraulic channel with a horizontal current under constant discharge condition, it was observed that laminated layers developed laterally in the direction of the current. The critical sedimentation rate for each particle size can be obtained from the works of Hjulstrom (1935). It was observed in the experiments that by varying the current velocity a superposed stratification can be obtained based on the segregation of particles by size. It must be noted that the experimentally observed superposed stratification is completely independent of time of deposition of heterogeneous particles and thus disproves Steno's principle of superposition as an indication of relative time.

The videos (Web Reference, 5) in flume experiments clearly show that in the presence of a variable current, stratified superposed beds progress simultaneously in the direction of current. If we take a horizontal cross section of the deposition we can clearly visualize the stratification, and each of those beds from top to bottom were deposited at the same time. Following the trend it is obvious that the deposition in the downstream of fluid flow is always younger than the deposition in the upstream. These fundamental experiments in sedimentation prove that the chronology of deposition is dependent on the direction of growth of superposed beds (direction of fluid current) and is independent of thickness of deposition.

Defects in Steno's Second Assumption – Principle of continuity *[Strata owe their existence to sediments in a fluid. At the time when any stratum formed, either it was circumscribed on its sides by another solid body, or else it ran* *around the globe of the earth (Steno, 1667, p. 30, CII.3c.]:*

This is certainly an unrealistic assumption because we cannot find any single evidence where a sedimentary layer is extended globally (all around the Earth). Long back in the 19th century itself geologists concluded that facies alteration is a direct refutation of Steno's principle of continuity. In the 19th century rigorous field studies and facies changes refuted flood geology and geologists agreed that Steno's assumption regarding the principle of continuity is inaccurate. Hence, geologists (Young, 1982, p. 44, 51-54; Mintz, 1977, p. 6-7, 18-19) accept that, "At the time when any stratum formed, either it was circumscribed on its sides by another solid body, or else it ran round the globe of the Earth." There are also cases in which even though continuity was established, they suffer from a time-equivalence crisis. For example, Dunbar and Rodgers (1957, p. 272), considering the case of lithostratigraphic units like the Tapeats sandstone, explained, "This unit can be traced almost continuously from one end of the Grand Canyon to the other; for long distances it upholds a wide bench, the Tonto Platform, which testifies to its perfect continuity. Yet because of facies shifts the unit is of different ages at the two ends of the canyon, so that physical continuity has failed completely to establish correlation." Hence, Dunbar and Rodgers (1957) conclude that, "It must never be forgotten, however, that even if continuity is thus suggested or proved, time-equivalence, though perhaps probable, is not assured." Byers (1982, p. 219) also states,

"For over a century we have known about facies change. Facies change is a violation of the purest form of lateral continuity, which says that strata extend without change to the basin margin."

Defects in Steno's Third Assumption – Principle of original horizontality *[At the time when any stratum formed, its lower surface, as also the surfaces of its sides, corresponded with the surfaces of the subjacent body and lateral bodies, but its upper surface was (then) parallel to the horizon, as far as it was possible (Steno, 1667, p. 30. C.II. 3.4.).]:*

This assumption is also far from confirmed by empirical observations. The horizontality assumption demands a uniform sedimentation rate globally. In reality, sedimentation involves extremely complex phenomena and rate of sedimentation depends on several local physical and biological factors (Schneidermann et al., 1976). The rate of sedimentation cannot be identical in different oceans all around the Earth. Geologists unanimously admit that Steno's assumption about global scale horizontal layers is generally untrue (Press and Siever, 2001, p. 392, 396). A basic geology textbook by Press and Siever (2001) explains that seismic cross-sections of continental slopes and other areas of the ocean floor confirm that sediment layers are often not deposited in a strict horizontal direction. Furthermore, submarine coring and seismic analysis reveals that strata in oceanic sediments are not always horizontal (Web Reference, 6). G.K. Gilbert explained long back in 1885 (Boggs, 1995, p. 362) that sands are time and again not deposited horizontally. Berthault (2002, p. 445) also reported:

"the experiments reported in my second paper to the Academy of Sciences, as well as experiments conducted by P. Julien and presented as the video, Fundamental Experiments on Stratification, at several sedimentological conferences, clearly shows that up to the limit of the angle of repose (30 degrees to 40 degrees for the sands), the lamination of sediments is parallel to the slope... The principle of horizontality does not apply in this case."

Is the 'Chronology of the Geologic Column' Drowning in the Mud?

The simplistic model based on Steno's erroneous assumptions ignores the effects of fluid and sediment parameters. Recently, a series of experimental observations in sedimentation reveals the vital role of those ignored parameters on the pattern of stratification. Guy Berthault's inspirational work (Berthault, 1986; Berthault, 1988; Julien and Berthault, 1993; Julien et al., 1993; Berthault, 2002) on the most fundamental experiments on sedimentation created a revolution in experimental sedimentology and thus instigated a more rigorous experimentally based approach in this field. For example, the world's leading scientific journal *Nature* also published (Makse et al., 1997; Fineberg, 1997) similar experimental work that Berthault initiated. Makse et al. (1997) also reported that in the absence of external perturbations, mixtures of grains of different dimensions spontaneously segregate. In such condition when a mixture of grains is normally dropped onto a heap, they observed that usually larger grains are settled near the base and the smaller ones at the top. On the other hand, when a granular mixture is dropped between two vertical plates, a spontaneous stratification of alternating layers of small and large grains is observed in those cases where large grains have a larger angle of response than the smaller ones (Makse et al., 1997, p. 379). Fineberg (1997, p. 323) also reported similar observations and these experiments have most important applications in the field of stratification analysis of avalanches. The geological chronology based on Steno's simplistic theoretical model did not incorporate the influence of fluid and sediment parameters that are reported by these experiments. Hence, these prestigious publications and their conclusions further invalidate the widely used naive geological chronology.

Much beyond that, shale sedimentology is undergoing abundantly rapid paradigm shifts and a series of novel sedimentology experiments and observations on this are reported by Juergen Schieber from Indiana University, Bloomington and his colleagues (Schieber et al., 2007; Schieber and Southard, 2009; Schieber and Yawar, 2009; Schieber et al., 2010; Schieber, 2011). It is very important to note that the majority of the sediments in the world are mudstones (Schieber et al., 2007), which include shale and clays. Around a century ago Henry Clifton Sorby (1908), the 'Father of Petrography' made a suitable comment in the present context and it is still valid:

"Possibly many may think that the deposition and consolidation of fine-grained mud must be a very simple matter, and the results of little interest. However, when carefully studied experimentally it is soon found to be so complex a question, and the results dependent on so many variable conditions, that one might feel inclined to abandon the inquiry, were it not that so much of the history of our rocks appears to be written in this language."

Despite much advancement in the field, sedimentologists still believe that muddy sediments are highly complex systems and a staggering 32 variables and parameters are required to be considered for a reasonable physicochemical interpretation (Berlamont et al., 1993). Being ignorant about this complexity and also significantly influenced by Steno's simplistic ideas, in the past geologists presumed that mudstones formed only in tranquil, unruffled seas. Disproving this now outmoded model and practically confirming this complexity in his experiments, Schieber (2011) concludes,

"Essentially, the experiments presented here demonstrate that many long-held assumptions about mud deposition and erosion do not agree with physical realities. Examination of the rock record increasingly shows that, once studied in some detail, shales and mudstones contain such a bewildering variety of textures and structures that one may indeed wonder whether the inherent questions about depositional conditions have any hope to ever be answered in full. By necessity, experimental approaches to the sedimentology of shales will therefore have to be as varied as these rocks themselves."

Juergen Schieber and his colleagues are now experimentally establishing that mudstones can form in abruptly flowing or turbulent water. They are using a novel imaging technique to study turbulent muddy water. Schieber and Southard (2009) reported in *Geology* that mudstone particles can produce ripples, identical to those found in sand. Thus shales or mudstones are vulnerable to climatic conditions and hence are very much defectively understood compared to other types of sedimentary rocks. Schieber et al. (2007) reported a unique experimental study in *Science*, and abstract of this article states,

"Mudstones make up the majority of the geological record. However, it is difficult to reconstruct the complex processes of mud deposition in the laboratory, such as the clumping of particles into floccules. Using flume experiments, we have investigated the bedload transport and deposition of clay floccules and find that this occurs at flow velocities that transport and deposit sand. Deposition-prone floccules form over a wide range of experimental conditions, which suggests an underlying universal process. Floccule ripples develop into low-angle foresets and mud beds that appear laminated after postdepositional compaction, but the layers retain signs of floccule ripple bedding that would be detectable in the rock record. Because mudstones were long thought to record low-energy conditions of offshore and deeper water environments, our results call for reevaluation of published interpretations of ancient mudstone successions and derived paleoceanographic conditions."

Schieber et al. (2007) also state, "Our observations do not support the notion that muds can only be deposited in quiet environments with only intermittent weak currents... Instead, bedload transport of flocculated mud and deposition occurs at current velocities that would also transport and deposit sand". Schieber et al. (2007) finally conclude,

"This, in turn, will most likely necessitate the reevaluation of the sedimentary history of large portions of the geologic record."

These novel experimentations and observations are clearly making 'Chronology of the Geologic Column' to drown in the mud. Macquaker and Bohacs (2007) fittingly remarked in *Science* concerning this article (Schieber et al., 2007) in the same issue:

"The results call for critical reappraisal of all mudstones previously interpreted as having been continuously deposited under still waters. Such rocks are widely used to infer past climates, ocean conditions, and orbital variations."

Hence, the stratigraphic model is found to be based on completely falsified assumptions. At the current time sedimentologists are realizing the key role of paleohydraulic factors in stratigraphy. Paleohydraulic analyses are not limited to the laboratory. In 2007 a team of Russian sedimentologists directed by Alexander Lalomov (Russian Academy of Sciences, Institute of Ore Deposits) has applied paleohydraulic analyses in conformity with Newtonian mechanics to geological formations in Russia. They concluded that the current velocities derived from sedimentary particle analysis would have resulted in the deposition of the entire sedimentary sequence in a very short period of time, rather than the millions of years implied by a stratigraphic analysis using the geological timescale (Lalomov, 2007). However, based on these experimental evidences we cannot conclude that all laminae form rapidly. For example, Lambert and Hsu (1979, p. 460) explained that annual varves are undoubtedly present in Swiss lakes, like Lake Zurich. Other varved deposits are also reported in the literature (Ripepe et al., 1991; Smith, 1997, p. 161). The famous Green River varves are known to be build up in a span of four million years and these varves demonstrate cyclic thickness variation with known periods for sunspot, Earth's precession and Earth's orbital eccentricity periods (Ripepe et al., 1991; Smith, 1997). All these observations indicate that laminae can appear from a complex mixture of processes of which some are slow and some are fast.

Geology is also witnessing 'Shifting Paradigms in Shale Sedimentology' as reported by Juergen Schieber (2011), who wrote, "Shales and mudstones are by far the most common sedimentary rocks, accumulate in a wide range of environments, and contain the bulk of recorded Earth history. This history is written in a well-defined special language that is still poorly understood... Shale research is a frontier area of sedimentary geology and will require several decades of sustained effort by multiple investigators to come to maturity." Hence, in the given circumstances a blind application of Steno's simplistic theory to all the rocks may over estimate the age of those rocks that did not appear by varved deposits. It is also observed from the evidence that radiometric dating techniques are not at all reliable. The age of the rocks and fossil ages based on such anomalous theories are no longer trustworthy. The plain fact is that geology does not have any credible dating technique at the present time. Modern geological evidence clearly reveals the crumbling pillars of geological chronology (radiometric dating and stratigraphy) and thus rather than supporting, completely undermines the "Chronology of Geologic Column," which has several important fundamental applications in geology and many other fields. Keeping that in view, attention is needed for doing profound research in sedimentary geology, which is currently very essential for developing a reliable method of prediction in geological chronology.

Vedāntic view of Biodiversity in the Light of 21st Century Biology

The intemperate view in science, that we can, and in the future will provide a necessary, complete explanation of the universe (including life) has actually lead to the degradation of modern civilization. In general, anthropocentric scientism indefatigably overlooks the boundaries of science in its dogmatic claims. However, as we are regularly presenting in *The Harmonizer*, there is convincing scientific evidence for honest scientists to emphatically challenge the attitude of 'dogmatic scientism' that has hijacked the true method of science. Scientists who try to understand nature utilizing a purely reductionistic approach employ ontological, methodological and epistemic reduction (Nagel, 1998). By assuming ontological reduction, scientists are able to think of an organism as being nothing more than a combination of molecules and their interactions. Based on this presumption biologists employ methodological reduction by only studying the separate contents of an organism independent of their integral context. However, continually mounting evidence only refutes the idea of an epistemic reduction of an organism by appeal to the unification of ontological and methodological reduction. In fact, frontier biology confirms that all living organisms are sentient and hence cannot be reduced to mere physics and chemistry.

Living organisms are cognitively adaptive systems, a characteristic which is absent in inanimate or dead objects. Even the smallest living cells obtain information of their external environment and accordingly monitor their internal processes (Shapiro, 2011). For more than 150 years, following a reductionistic approach, Darwinism has considered only an insentient view of the living organism or abiology. On the other hand, 21st century biology rejecting the abiology of Darwinism, now accepts all living organisms including the smallest cells as sentient beings (Shapiro, 2011).

In the November 2012 issue article, "21st Century Biology Refutes Darwinian Abiology" the failure of the Darwinian theory to explain how novel regulatory elements arise was explained as being one of the major blows that late 20th century molecular biology presented to Darwinism. Each species of life has its own unique gene regulatory network, such that from its initial stage to maturity the particular species develops in accord with processes unique to that species only. Evo-devo experts primarily try to understand the appearance of developmental networks and the emergence of novel protein domains at decisive steps of embryological development in an organism. In applying this process to evolution Shapiro (2011) explains the difficulty,

"To have new subprotein domains arise in the course of evolution, a process is needed for generating novel exons that can encode extended polypeptide structures to be incorporated into proteins in combination with other exons. Exon generation cannot occur efficiently by the gradual accumulation of single amino acid changes in existing protein chains because the probability of losing the original functionality is too high and of gaining a new functionality too low. A more rapid, facultative process is needed—and has in fact been discovered."

The new faculative process Shapiro calls "natural genetic engineering," but this clearly exposes the naïveté of Darwinian abiology based on the assumption of gradualism. Gene regulatory networks are not a result of gradual evolutionary progress. Even unicellular simple creatures like bacteria have their own unique and extremely sophisticated regulatory networks. Smith and Hoover (2009) stated, "Synthesis of the bacterial flagellum is a complex process involving dozens of structural and regulatory genes. Assembly of the flagellum is a highly-ordered process, and in most flagellated bacteria the structural genes are expressed in a transcriptional hierarchy that results in the products of these genes being made as they are needed for assembly. Temporal regulation of the flagellar genes is achieved through sophisticated regulatory networks that utilize checkpoints in the flagellar assembly pathway to coordinate expression of flagellar genes." Hence, the belief that all life forms arrived from a common ancestor following a trajectory of mere objective evolution is rather unreasonable and more the result of a dogmatic imposition of an ideology.

In the context of multicelluar organisms, Shapiro (2011) states, "Without an elaborate sensory apparatus to pick up signals about chemicals in the environment (nutrients, poisons, signals emitted by other cells) or to keep track of intracellular events (DNA replication, organelle growth, oxidative damage), a cell's opportunity to proliferate or contribute to whole-organism development would be severely restricted. Life requires cognition at all levels". The last sentence, "Life requires cognition at all levels" is the same paradigm that *Vedānta* has advocated since antiquity. In *Vedānta* it is described that the *ātma* (soul) is responsible for animating the bodies of all living organisms, from the simplest single cell to complex multicellular organisms. The immortality of *ātma* is explained in *Bhagavad-gītā* verse 2.20 and the same is also described in *Kaṭha Upaniṣad* verse 1.2.18, *na jāyate mriyate vā vipaścin nāyaṁ kutaścin na babhūva kaścit ajo nityaḥ śāśvato 'yaṁ purāṇo na hanyate hanyamāne śarīre* – "For the soul there is neither birth nor death at any time. He has not come into being, does not come into being, and will not come into being. He is unborn, eternal, ever-existing and primeval. He is not slain when the body is slain", where the word *vipaś-cit* means learned or with knowledge. According to *Vedāntic* understanding *ātma* is eternal and fully cognizant.

Vedānta explains that consciousness is one of the symptoms by which the existence of the *ātma* can be inferred. Although scientists cannot sensually perceive the *ātma,* still they can infer its existence just from the presence of consciousness in all biological systems. As the presence of the sun can be inferred from the sunlight, similarly existence of the *ātma* can also be understood from the presence of the different varieties of consciousness in various living organisms. Furthermore, *Bhagavad-gītā* verse 18.61 states, *īśvaraḥ sarva-bhūtānām hṛd-deśe 'rjuna tiṣṭhati bhrāmayan sarva-bhūtāni yantrārūḍhāni māyayā* – "all living forms (*sarva-bhūtānām*) are machines (*yantrā*) made of material energy (*māya*) of a Unitary Supreme Cognizant Being, Kṛṣṇa (*īśvara*), and Kṛṣṇa's *Paramātma* (super-soul) feature is guiding the conditioned *ātma* situated within that machine". Hence, *Paramātma* (infinite consciousness) is also known as the ground or sustainer of the *ātma* (finite consciousness). This explanation of *Bhagavad-gītā* is self evident from the scientific evidence described above. Living entities, from bacteria to humans, do not have full knowledge or control over the complex biological process that are sentiently going on within their bodies and yet those processes go on very perfectly. This perfect maintaining principle is Paramātma. However, *Vedānta* explains laws of *Karma* (actions and reactions of good and bad activities performed by the living being) as the cause of any abnormal condition (diseases, errors in biological process, cancer, etc) that affects the body of an organism.

It should be noted that, the machine description of different bodily forms in the above verse for different species should not be misunderstood with the machines that a human could manufacture artificially. Unlike artificial machines, the bodies of all living organisms (from bacteria to humans) are inimitably complex. A frog's zygote will never develop into a puppy. Life intrinsically preserves its species type. Darwinian objective evolution theory using the laws of physics and chemistry cannot explain why species like, bacteria, fish, frogs, banyan trees, lions and so on appeared. On the other hand, the conception of *Vedānta* holds that different forms (species) are original archetypes that accommodate different varieties of consciousness through which the transmigration of the soul (*ātma*) takes places on the basis of the evolution of consciousness. For example, *Viṣṇu Purāṇa* states, "*jala-jā nava-lakṣāṇi sthāvarā lakṣa-viṁśati kṛmayo rudra-saṅkhyakāḥ pakṣiṇām daśa-lakṣaṇam triṁśal-lakṣāṇi paśavaḥ catur-lakṣāṇi mānuṣāḥ* – There are 900,000 species living in the water. There are also 2,000,000 nonmoving living entities (*sthāvara),* such as trees and plants. There are also 1,100,000 species of insects and reptiles, and there are 1,000,000 species of birds. As far as quadrupeds are concerned, there are 3,000,000 varieties, and there are 400,000 human species." According to *Vedānta,* species identification and classification are based on a cognitive paradigm, where the body is a biological expression of the consciousness of the soul (*ātma*). Therefore, the different species described in above verse are representations of different varieties of consciousness. The transmigration of the soul (*ātma*) is described in *Bhagavad-gītā* 8.6: *yam yam vāpi smaran bhāvam tyajanty ante kalevaram tam tam evaiti kaunteya sadā tad bhāva-bhāvitaḥ* – "The soul (*ātma*) obtains a body in next life based on the consciousness in which it left the previous body." Animals and lower species of life do not have enough intelligence to understand these descriptions of ancient wisdom. However, a sober human being may easily understand his/her entanglement in the dangerous cycle of endless transmigration and thus inquire about their true identity as the immortal soul under an expert spiritual guide. *Vedānta* advocates this scientifically verifiable subjective evolution of consciousness, while the unscientific Darwinian objective evolution of bodies is only a misconceived perverted reflection of this subjective evolution of consciousness. A lot of energy and time are already wasted for more than 150 years following the dogmatic imposition of Darwinian abiology and now the scientific evidence is forcing honest scientists to understand genuine biology based on cognition as reveled in depth within ancient *Vedāntic* literature.

References

Allen, R.M., 1952. The evaluation of radioactive evidence on the age of the Earth. Journal of the American Scientific Affiliation 4, p. 18. Allen noted that, "The extent of the haloes around the inclusions varies over a wide range, even with the same nuclear material in the same matrix."

Bapteste, E., et al., 2009. Prokaryotic evolution and the tree of life are two different things. Biol Direct 4, 34.

Barbier-Baudry, D., Bouazzacm A., Desmursd, J.R., and Dormond, A., 2000. Uranium^{IV} and uranyle salts, efficient and reusable catalysts for acylation of aromatic compounds. Journal of Molecular Catalysis A: Chemical 164(1-2), 195-204.

Baas, J.H., van Dam, R.L., and Storms, J.E.A., 2000. Duration of deposition from decelerating high-density turbidity currents. Sed. Geol. 136, 71-88.

Berlamont, J., Ockenden, M., and Toorman, E., 1993. The characterisation of cohesive sediment properties. J. Winterwerp Coast. Eng. 21, p. 105.

Berthault, G., 1986. Sédimentologie: Expériences sur la lamination des sédiments par granoclassement périodique postérieur au dépôt. Contribution à l'explication de la lamination dans nombre de sédiments et de roches sédimentaires. Comptes Rendus de l'Académie de Sciences 303, 1569-1574.

Berthault, G., 1988. Sédimentation d'un mélange hétérogranulaire. Expériences de lamination en eau calme et en eau courante. Comptes Rendus de l'Académie de Sciences 306, 717-724.

Berthault, G., 2002. Analysis of the main principles of stratigraphy on the basis of experimental data. Lithology and Mineral Resources 37(5), 442-446.

Blatt, H., Middleton, G., and Murray, R., 1980. Origin of sedimentary rocks. 2nd edition, Prentice-Hall, Inc. Englewood Cliffs, NJ 07632.

Boggs, S., 1995. Principles of sedimentology and stratigraphy. 2nd ed., Prentice Hall, Upper Saddle River, NJ.

Bouma, A.H, 1962. Sedimentology of some flysch deposits: A graphic approach to facies interpretation, Elsevier, Amsterdam.

Brooks, J.R., et al., 2002. Heavy and light beer: A carbon isotope approach to detect C4 carbon in beers of different origins, styles, and prices. J. Agric. Food Chem. 50, 6413-6418.

Bryan, J., Chewapreecha, C., and Bentley, S.D., 2012. Developing insights into the mechanisms of evolution of bacterial pathogens from whole-genome sequences. Future Microbiol. 7, 1283-1296.

Bucha, V., 1970. Influence of the Earth's magnetic field on radiocarbon dating. In: Olsson I.U. (ed.), Radiocarbon variations and absolute chronology, Nobel Symposium (Almqvist & Wiksell, Stockholm) 12, 501-512.

Byers, C.W., 1982. Stratigraphy – The fall of continuity. J. Geol. Ed. 30, 215-221.

Carroll, R.L., 2000. Towards a new evolutionary synthesis. Trends in ecology & evolution 15(1), 27-32.

Cooper, A. and Fortey, R., 1998. Evolutionary explosions and the phylogenetic fuse. Trends in Ecology and Evolution 13(4), 151-156.

Cuvier, G.B., 1840. Cuvier's animal kingdom, arranged according to its organization : forming the basis for a natural history of animals, and an introduction to comparative anatomy / Mammalia, birds and reptiles, by Edward Blyth. The fishes and Radiata, by Robert Mudie. The molluscous animals, by George Johnston, M.D. The articulated animals, by J. O. Westwood, F.L.S. Illustrated by Three hundred engravings on wood. London: W.S. Orr and Co.

Darwin, C., 1869. On the origin of species. Fifth edition, Chapter IX, On the imperfection of the geological record, 378-381.

de Vries, H., 1958. Variation in concentration of radiocarbon with time and location on Earth. Proceedings Koninklijke Nederlandse Akademie van Wettenschappen B(61), 1-9.

Delsuc, F., Brinkmann, H., and Philippe, H., 2005. Phylogenomics and the reconstruction of the tree of life. Nat Rev Genet 6, 361-375.

Doolittle, W.F., 1999. Phylogenetic classification and the universal tree. Science 284, 2124-2129.

Dunbar, C.O., and Rodgers, J., 1957. Principles of stratigraphy. John Wiley & Sons, Inc., New York.

Fineberg, J., 1997. From cinderella's dilemma to rock slides. Nature 386, 323-324.

Fisher, R.V. and Schmincke, H.U., 1984. Pyroclastic rocks. Springer-Verlag, Berlin.

Fitch, W.M., and Ayala, F.J., 1995. Tempo and mode in evolution: Genetics and paleontology 50 years after Simpson. New York: National Academies Press.

Fu, et al. (2013). A revised timescale for human evolution based on ancient mitochondrial genomes. Current Biology, Vol. 23, pp. 553–559.

Gess, R.W., Coates, M.I., and Rubidge, B.S., 2006. A lamprey from the Devonian period of South Africa. Nature 443, 981-984.

Godwin, H., 1962. Half life of Radiocarbon. Nature 195, p. 984.

Gohau, G., 1990. A history of geology. New Brunswick: Rutgers University Press.

Gould, S.J., 1977. Evolution's erratic pace. Natural History 86(5), p. 14.

Gould, S.J., 1995. Of it, not above it. Nature 377, p. 682.

Hahn, H.P., Born, H.J., and Kim, J.I., 1976. Survey on the rate perturbation of nuclear decay. Radiochemica Acta 23, 23-27.

Hjulstrom, F., 1935. Studies of the morphological activity of rivers as illustrated by river Fyris. Bulletin of the Geological Institute Uppsala 25, 89-122.

Huh, C.A., 1999. Dependence of the decay rate of 7Be on chemical forms. Earth and Planetary Science Letters 171, 325-328.

Jenkins, J.H., et al., 2010. Analysis of environmental influences in nuclear half-life measurements exhibiting time-dependent decay rates. Nuclear Instruments and Methods in Physics Research A 620, 332-342.

Julien, P.Y., and Berthault, G., 1993. Fundamental experiments on stratification. Video, Sarong Ltd., Monaco.

Julien, P.Y., Lan, Y. and Berthault, G., 1993. Experiments on stratification of heterogeneous sand mixtures. Bulletin Géologique de France 164(5), 649-660.

Kerr, R.A., 1999. Tweaking the clock of radioactive decay. Science 282, p. 882.

Klepper, K.R. and Wyant, D.G., 1957. Notes on the geology of Uranium. US Geological survey bulletin No. 1046-F, p. 93.

Koonin, E.V., 2009. Darwinian evolution in the light of genomics. Nucleic Acids Res 37, 1011-1034.

Koonin, E.V., Wolf, Y.I., 2008. Genomics of bacteria and archaea: the emerging dynamic view of the prokaryotic world. Nucleic Acids Res 36, 6688-6719.

Koonin, E.V., and Wolf, Y.I., 2009. Is evolution Darwinian or/and Lamarckian? Biology Direct 4, p. 42.

Kuhn, J.A., 2012. Dissecting Darwinism. Proc Bayl Univ Med Cent 25(1), 41-47.

Kutschera, U., 2003. A comparative analysis of the Darwin-Wallace papers and the development of the concept of natural selection. Theory in Biosciences 122, 343-359.

Lalomov, A.V., 2007. Reconstruction of paleohydrodynamic conditions during the formation of Upper Jurassic conglomerates of the Crimean Peninsula. Lithology and Mineral Resources 42(3), 298-311.

Lambert, A., and Hsu, K., 1979. Non-annual cycles of varve-like sedimentation in walensee, Switzerland. Sedimentology 26, 453-461.

Larson, E.J. 2004. Evolution: The remarkable history of a scientific theory. New York: Modern Library. ISBN 0-679-64288-9.

Lewin, Roger., 1980. Evolutionary theory under fire: An historic conference in Chicago challenges the four-decade long dominance of the Modern Synthesis. Science 210, p. 883.

Logvinenko, N.V., 1980. Morskaya geologiya (Marine Geology). Nedra, Leningrad, Russia.

Lynde, A.J., and Spangler, G.W., 1974. Radiometric dating: is the 'decay constant' constant? Pensee, p. 31.

Macquaker, J.H., Bohacs, K.M., 2007. Geology: On the accumulation of mud. Science 318(5857) 1734–1735.

Makse, H.A., Havlin, S., King, P.R., and Stanley, H.E., 1997. Spontaneous stratification in granular mixtures. Nature 386, 379-382.

Mintz, L., 1977. Historical geology: The science of a dynamic Earth. 2nd ed., Charles E. Merrill Pub. Co., Columbus, OH.

Nagel, T., 1998. Reductionism and antireductionism. G.R. Bock and J.A. Goode (eds.), The limits of reductionism in biology, Chichester: John Wiley & Sons, pp. 3-10.

Press, F., and Siever, R. 2001. Understanding Earth. 3rd ed., W.H. Freeman and Co., New York.

Raup, D.M., 1981. Evolution and the fossil record. Science 213, p. 289.

Reifenschweiler, O., 1994. Reduced radioactivity of tritium in small titanium particles. Physics Letters A184, 149-153.

Retallack, G.J. (2013). Ediacaran life on land. Nature 493, 89-92.

Ripepe, M., Roberts, L.T., and Fischer, A.G., 1991. ENSO and sunspot cycles in varved eocene oil shales from image analysis. Journal of Sedimentary Petrology 61(7), 1155-1163.

Schieber, J., 2011. Shifting paradigms in shale sedimentology – The implications of recent flume studies for interpreting shale fabrics and depositional environments. Recovery – CSPG CSEG CWLS Convention, 1-4.

Schieber, J. and Southard, J.B., 2009. Bedload transport of mud by floccule ripples – Direct observation of ripple migration processes and their implications. Geology 37, 483-486.

Schieber, J., Southard, J.B., and Schimmelmann, A., 2010. Lenticular shale fabrics resulting from intermittent erosion of muddy sediments - Comparing observations from flume experiments to the rock record. Journal of Sedimentary Research 80, 119-128.

Schieber, J., Southard, J.B., and Thaisen, K.G., 2007. Accretion of mudstone beds from migrating floccule ripples. Science 318, 1760-1763.

Schieber, J., and Yawar, Z., 2009. A new twist on mud deposition - Mud ripples in experiment and rock record. The sedimentary record 7/2, p. 4-8.

Schmincke, H.U., Fisher, R.V., and Waters, A.C., 1973. Antidune and chute and pool structures in the base surge deposits of the laacher see area, Germany. Sedimentology 20, 553-574.

Shapiro, J.A., 2011. Evolution: A view from the 21st century. Upper Saddle River, NJ: FT Press Science.

Smith, A.G., 1997. Estimates of the Earth's spin (geographic) axis relative to gondwana from glacial sediments and paleomagnetism. Earth-Science Reviews 42, 161-179.

Smith, T.G., and Hoover, T.R., 2009. Deciphering bacterial flagellar gene regulatory networks in the genomic era. Adv Appl Microbiol. 67, 257-295.

Sorby, H.C., 1908. On the application of quantitative methods to the study of the structure and history of rocks. Q. Geol. Soc. London 64, p. 171.

Spector, R., 1972. Pleochroic halos and the constancy of nature: a reexamination. Physical Review A 5, p. 1323.

Stansfield, W.D., 1977. The science of evolution. Macmillan: New York, p. 84.

Steno, N., 1669. Prodromus. Ex typographia sub signo Stella, Florence, Italy.

Steno, N., 1667. Canis Calchariae. Ex typographia sub signo Stella, Florence, Italy.

Strakhov, N.M., 1957. Theoretical lithology and its problems. Izv. Akad. Nauk SSSR, Ser. Geol. 11, 15-31.

Stuiver, M., 1965. Carbon-14 content of 18th- and 19th-century wood variations correlated with sunspot activity. Science 149, 533-34.

Waggoner, B., 1996. Georges Cuvier (1769–1832). University of California, Berkeley. Retrieved July 19, 2011.

Whitten, D.G.A., and Brooks, J.R.V., 1972. The penguin dictionary of geology. Penguin Books, Middlesex (England), p. 378.

Williamson, P.G., 1981. Morphological stasis and developmental constraint: real problems for neo-Darwinism. Nature 294, p. 214.

Woese, C.R., 1987. Bacterial evolution. Microbiol Rev 51, 221-271.

Young, D.A., 1982. Christianity and the age of the Earth. Zondervan, Grand Rapids, MI.

Web References

1. Evolution and the fossil record – Archaeopteryx, the missing link. BBC. www.bbc.co.uk/learningzone/clips/evolution-and-the-fossil-record-archaeopteryx-the-missing-link/5523.html Accessed 19th January 2013
2. Big five mass extinction events, Prehistoric life. BBC. *www.bbc.co.uk/nature/extinction_events* Accessed January 22, 2013
3. Field Museum (2010, November 5). New statistical model moves human evolution back three million years. ScienceDaily. *www.sciencedaily.com-/releases/2010/11/101105124241.htm* Accessed January 19, 2013
4. Berthault, G. Experiments. Sedimentology. *http://efficalis.com/sedimentology/home/experiments* Accessed January 19, 2013
5. Berthault, G. Videos. Sedimentology. *http://efficalis.com/sedimentology/videos* Accessed January 19, 2013
6. Berthault, G. Videos. Sedimentology. http://efficalis.com/sedimentology/home/problems Accessed January 19, 2013.

Hon'ble review

1. **Prof. P.S. Saklani**
   Former Vice Chancellor, KU
   Emeritus Professor
   Delhi University
   DELHI - 110 007
2. **Prof. Anand Mohan**
   Registrar
   Dayalbagh University
   DAYALBAGH
   AGRA - 282 005

IV-62

Arumberia –like Ediacaran Mat Structure from Sonia Sandstone, Marwar Supergroup, Rajasthan, India

Mukund Sharma[1] and S.C. Mathur[2]

[1]53, Birbal Sahni Institute of Palaeobotany, University Road, Lucknow-226 007,
2 Department of Geology, J.N.Vyas University, Jodhpur - 342 005
E-mail: sharmamukund1@rediffmail.com

Abstract

The Marwar Supergroup being overlying Malani Igneous Suite having geological age 745 Ma to 680 Ma, had a reasonable chance to have sediments of the Ediacaran Period, a newly created addition to the Geological Time Scale, globally. The paper is reporting, alongwith other biotic evidences a distinct Ediacaran mat structure (an Organo-Sedimentary Structure) from the middle part of the Sonia Sandstone of Jodhpur Group exposed in the quarry of Bhuri Beri and Sursagar area of Jodhpur.

Keywords: Ediacarana Period, Ediacaran Fossils, Marwar Supergroup, Sonia Sandstone.

The *Ediacaran Period* is a new addition to the geological timescale (Knoll *et al.,* 2006). It is named after the famous Ediacaran fossils assemblage. The period, though defined Chronostratigraphically has world-wide an important underlying glacial diamictite beds and distinct overlying biotic assemblage, underscores the importance of biotic assemblage of the Period. In India, the only established sequence of this period is in extra-peninsular region of the country (Singh and Rai, 1983). Equivalent section is yet to be established in the peninsular region of the country. Recent reports of the palaeobiological remains and mat structures from the different lithostratigraphic units of the Marwar Supergroup (Paliwal 1999, 2007; Raghav *et al.,* 2005; Sarkar *et al.,* 2005, 2008; Kumar and Pandey, 2008) have kindled interest among geoscientists to look for distinct Ediacaran evidences in the sediments of the Marwar basin.

Till recently, no well established biotic remains of the Ediacaran Period were known from the peninsular region of the India. It was considered that the Marwar Supergroup overlying the 680 Ma Malani rhyolites had reasonable chances of having sediments of distinct Ediacaran Period and therefore suitable for investigations of the Ediacaran biota. During a recent field-work along with other biotic evidences a distinct Ediacaran mat-structure/ organosedimentary structure was recorded from the middle part of the Sonia Sandstone of the Jodhpur Group exposed in the quarry of Bhuri Beri, Sursagar area of the Jodhpur district, Rajasthan (Fig. 1). In this note a mat-structure earlier described as *Arumberia banksi* Glaessner and Walter is being reported. Like some other Precambrian fossils described under dubiofossils category, biogenecity of *Arumberia* is also debatable (McIlroy and Walter, 1997; Jenson *et al.,* 2005) but incidentally the distributions of such structures have been found to be confined to Ediacaran sediments. This temporal distribution of specific organo-sedimentary structure further indicates the possibility of the age of the Marwar Supergroup as Ediacaran. Specimens reported and described in the present note are deposited in the repository of the Birbal Sahni Instiitiute of Palaeobotany, Lucknow vide statement number 1214.

The Marwar Supergroup unconformably overlies the basement constituted of rhyolites (779 ± 10 Ma; earlier dated 745 ± 10 Ma) and (681 ± 20 Ma) (Crawford and Compston 1970; Rathore *et al.,* 1996, 1999) and considered to be younger than 680 Ma. Recent studies of the carbon, sulphur and strontium isotopes of the carbonates of the Bilara Group suggest Late Neoproterozoic-Early Cambrian (Vendian-Tommotian) age (Banerjee *et al.,* 1998; Strauss *et al.,* 2001; Pandit *et al.,* 2001; Maheshwari ei al., 2003; Mazumdar and Strauss, 2006) for the middle part of the basin. The Marwar Supergroup has been subdivided into three groups in ascending order these are the Jodhpur Group, the Bilara Group and the Nagaur Group (see Table 1). The Jodhpur Group is subdivided into the Pokaran Boulder Bed, the Sonia Sandstone and the Girbhakar Sandstone.

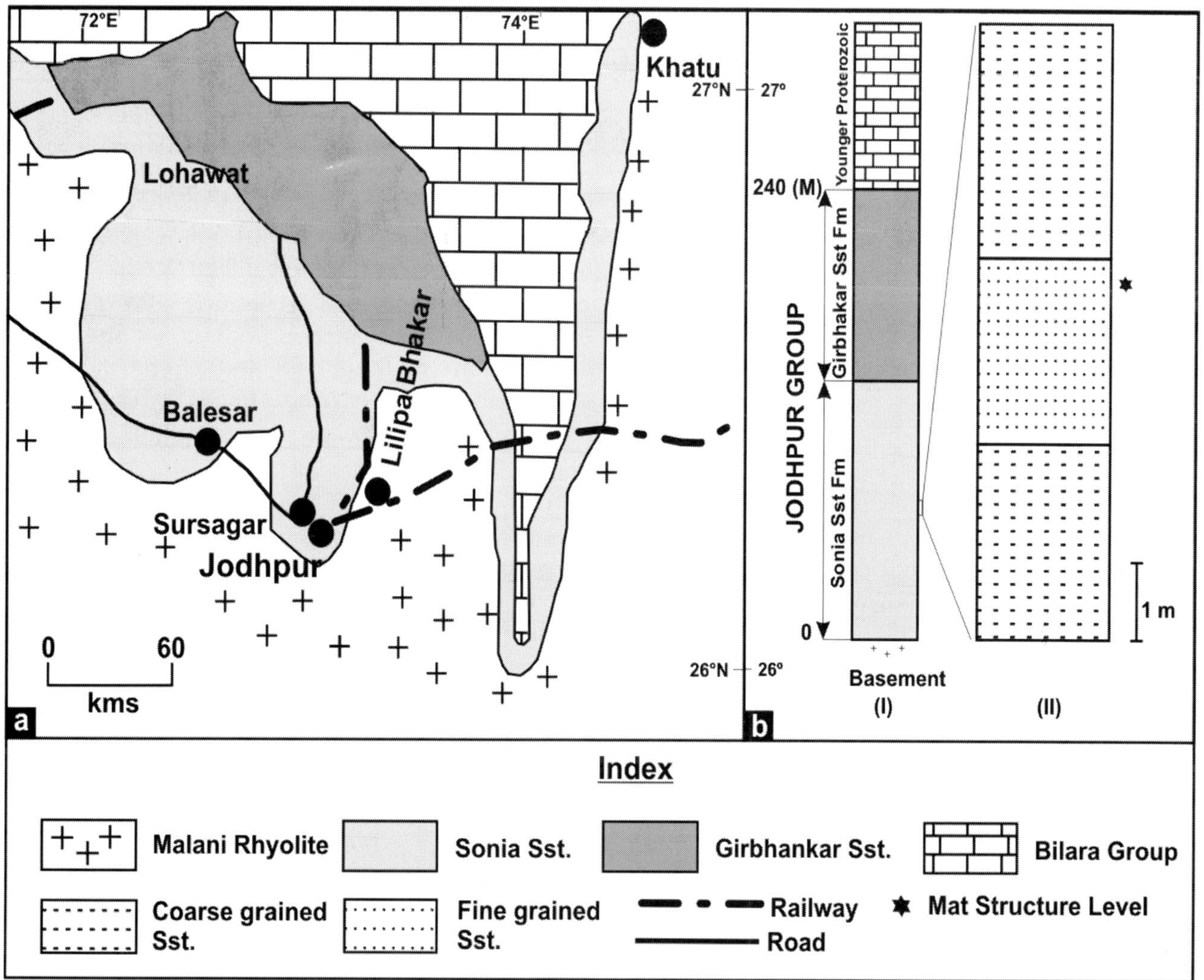

Figure 1a. Geological map of the Jodhpur Group, Marwar Supergroup, Western Rajasthan, India. b. Litholog of the Sonia Sandstone exposed in Buri Beri quarry, Sursagar area, Jodhpur District, Rajasthan.

The Jodhpur Group overlies the Malani Igneous Suit (779 ± 10 Ma to 681 ± 20 Ma). No radiometric dates are available for any of the units of the Marwar Supergroup. But recent palaeobiological studies suggest that the Marwar sedimentary succession belongs to Ediacaran Period to Lower Cambrian in age (Raghav *et al.*, 2005; Kumar and Pandey, 2008). Any additional biotic or other evidence corroborating the presence of the sediments of this age span is an important for exploration of the Ediacaran succession in the peninsular India. The present note records the time specific mat-structure similar to *Arumberia* from the Sonia Sandstone, the Jodhpur Group, the Marwar Supergroup.

Group *Arumberia* Glaessner and Walter, 1975

Type Form *Arumberia banksi* Glaessner and Walter, 1975

Arumberia banksi Glaessner and Walter, 1975 (Fig. 2 a-g)

Specimen No. : BSIP 39572 to BSIP 39577.

Table 1: Lithostratigraphic succession of the Marwar Supergroup (simplified after Pareek, 1984).

Supergroup	Group	Formation	Lithology
	Nagaur	(75-500 M)	Tunklian Sandstone
			Nagaur Sandstone
			Pondlo Dolomite
	Bilara	(100-300 M)	Gotan Limestone Dhanapa Dolomite
	Girbhakar Sandstone		Brick-red siltstone,
MARWAR SUPER GROUP e	Jodhpur	(125-240 M)	shale and sandstone
	Sonia Sandston		Marine siltstone and shale creamish sandstone with abundant sedimentary features and pisolitic; salt pseudomorphs shales; banded chert-Jasper-dolomites.

———Unconformity———

Pokaran Boulder Bed

———Unconformity———

Malani Igneous Suit

Locality : Buri Beri (N 26° 21' 11.7"; E 72° 59' 35.9"), Sursagar, Jodhpur district, Rajasthan.

Stratigraphic Horizon: Middle part of the Sonia Sandstone, Jodhpur Group, Marwar Supergroup.

Description: Each specimen consists of an array of curved, parallel to sub parallel ridges and furrows, sometimes emanating from one point (Fig. 2 b, f). Ridges are about 1-4 mm wide, furrows are 1-2 mm. Relief of ridges from top to margin of furrow is 1-2 mm. Length of the ridges and furrow 1 cm-7 cm, sometimes bifurcate. At some places the ridges and furrow show anastomosing and scaly patterns and sometimes radiate from pointed and twisted nose.

Remarks: The reported specimens are preserved on the sole surfaces of the medium grained sandstone (0.30 mm N= 100) and siltstone exposed in the middle part of the Sonia Sandstone. Ridges and furrows are comparable to *Arumberia banksi*, described from Arumbera Sandstone, SW of Alice Springs by Glaessner and Walter (1975). Biogenecity of *Arumberia* is not unequivocal. Earlier it was described as cup shaped animal (Glaessner and Walter, 1975) but later suggested to be a microbially bound sedimentary structure (McIlroy and Walter, 1997). Similar morphological structures were produced in the sedimentological studies of the flume experiments of Dzulynski and Walton (1965) as well as in experiments by Allen (1985).

Recently, Kumar and Pandey (2008) described two species of the *Arumberia* from the Maihar Sandstone of the Bhander Group of the Vindhyan Supergroup and discussed their biogenecity. It was concluded that preservation of such features is not possible in sandstone without cohesiveness provided by the microbial mats. The *Arumberia*-like structures described in the present note reflect microbial mat features having specific configuration. Presently known 13 occurrences of *Arumberia* (Kumar and Pandey, 2008) bring out their presence invariably in the Ediacaran Period. It confirms that such unique mat structures are confined to sedimentary sequences of this period of rocks. Though biogenecity of these structures are not fully established yet their occurrence in fine grained sandstones and siltstone demand cohesive agents to bind the grains to preserve the *Arumberia* like structures. There are possibilities that these structures might have formed by current activities as suggested by McIlroy and Walter (1997), but their temporal distribution needs suitable explanation. Temporal restriction makes the mat structures similar to *Arumberia* a good marker of the Ediacaran sediments. Till date no true representative organic community has been recorded in the sediments which gave rise to *Arumberia* like mat structure. Their preservation indicates the presence of binding cohesive mat community in the depositional realm leading to the development of time specific mat structure in the geological record. Present note establishes the presence of Ediacaran Period unique mat structure similar to *Arumberia* in the Sonia Sandstone of the Jodhpur Group. It also suggests that the Sonia Sandstone of the Jodhpur Group of the Marwar Supergroup is Ediacaran in age.

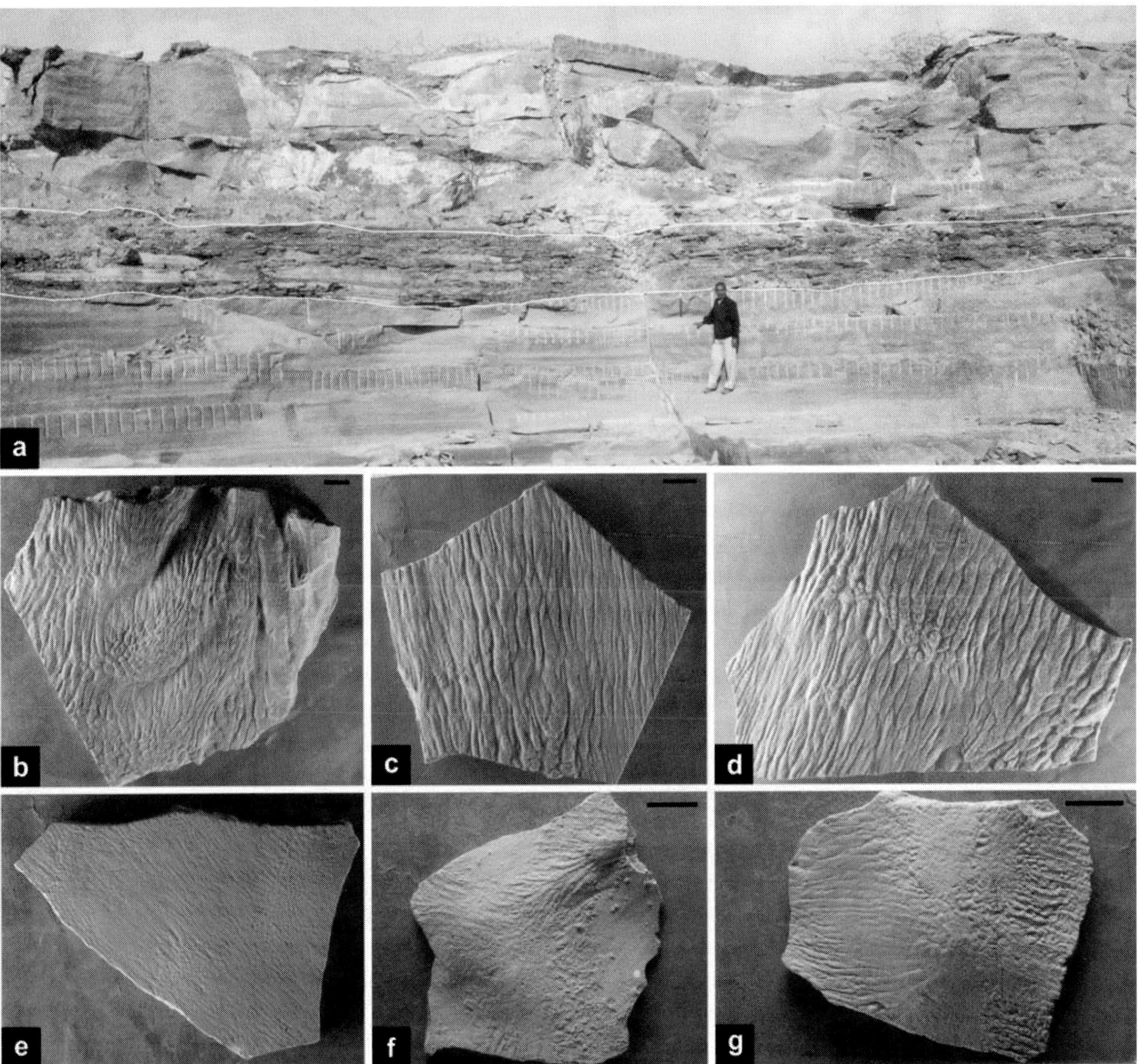

Fig-2 a. Field photograph of the Sonia Sandstone exposed in the quarry at Buri Beri locality, Sursagar area, Jodhpur. Middle portion showing fine grained sandstone and siltstone has yielded *Arumberia banksi* like mat structures. **b.-d** Parallel tightly packed ridges and grooves on the sole surface in the Sonia Sandstone at Buri Beri locality, Sursagar area, Jodhpur. **b** -Specimen number BSIP 39572, Scale = 1 cm. **c**. Specimen number BSIP 39573, Scale = 1 cm. **d.** BSIP 39574, Scale = 1 cm. **e.** A specimen of *Arumberia banksi* like mat structure. Showing curved and twisted grooves developed from the nose of flute cast. Specimen number BSIP 39575, Scale = 1 cm. **f.** Another specimen of *Arumberia banksi* like mat structure showing curved and twisted grooves developed from the nose of a flute cast. Specimen number BSIP 39576, Scale = 1 cm. **g.** *Arumberia banksi* like mat structure showing diverging grooves from the top of the specimen. Specimen number BSIP 39577, Scale = 1 cm.

Acknowledgements

I am thankful to Director, Birbal Sahni Institute of Palaeobotany for permission to publish the Palaeontological note. Prof. S. Kumar critically read the manuscript and offered valuable suggestions. Help rendered by Mr. Santosh Kumar Pandey, Department of Geology, University of Lucknow and Mrs. Yogmaya Shukla of BSIP in various ways are appreciated.

References

Allen, J. R. L. 1982. Sedimentary structures, their character and physical basis, p. 1-676. In: *Developments in Sedimentology 30 B*, Elsevier, Amsterdam.

Banerjee, D. M., Strauss, H., Bhattacharyya, S. K., Kumar, V. and Mazumdar, A. 1998. Isotopic composition of carbonates and sulphates, potash mineralization and basin architecture of the Nagaur-Ganganagar evaporate basin (NW India) and their implications on the Neoproterozoic exogenic cycle. *Mineralogical Magazine*, 62: 106-107.

Crawford, A. R. and Compston, W. 1970. The age of Vindhyan System of Peninsular India. *Quarterly Journal of the Geological Society of London* 125: 351-372.

Dzulynski, S. and Walton, E. K. 1965. Sedimentary features of flysch and greywackes, p. 1-274. In: *Developments in Sedimentology 7*, Elsevier, Amsterdam.

Glaessner, M. F. and Walter, M. R. 1975. New Precambrian fossils from the Arumbera Sandstone, Northern Territory, Australia. *Alcheringa*, 1: 59-69.

Jensen, Sören, Droser, Mary L. and Gehling, J. G. 2005. Trace fossil preservation and the early evolution of animals. *Palaeogeography, Palaeoclimatology, Palaeoecology,* 220: 19-29.

Knoll, A. H., Walter, M. R., Narbonne, G. M. and Christie-Blick, N. 2006. The Ediacaran Period: a new addition to the geologic time scale. *Lethaia* 39: 13-30.

Kumar, S. and Pandey, S. K. 2008. Discovery of trilobite fossils from the Nagaur Sandstone, the Marwar Supergroup, Dulmera area, Bikaner District, Rajasthan. *Current Science*, 94: 1081-1084.

Maheshwari A, Sial A. N., Mathur S. C. 2003. Carbon and Oxygen Isotope profiles from the Terminal Precambrian Marwar Supergroup, Rajasthan. India. International Jour. Carbonates and Evaporates. vol.18.pp. no.1:10-18.

Mazumdar, A. and Strauss, H. 2006. Sulfur and strontium isotopic composition of carbonate and evaporate rocks from the Late Neoproterozoic-Early Cambrian Bilara Group (Nagaur-Ganganagar Basin, India): constraints on intrabasinal correlation and global sulfur cycle. *Precambrian Research*, 149: 217-230.

Mcllory, D. and Walter, M. R. 1997. A reconsideration of the biogenecity of *Arumberia banksi* Glaessner and Walter. *Alcheringa*, 21: 79-80.

Paliwal, B. S. 1999. Geology of the region to the west of Aravalli Mountain- Some Observations, p. 387-414. In: *Geological Evolution of Northwestern India* (Ed. Paliwal, B. S.) Scientific Publishers (India), Jodhpur, India.

Paliwal, B. S. 2007. Some fossil like structures in the rocks of the Marwar Supergroup, Western Rajasthan, India, pp 1-19. In: *Emerging Trends of Research in Geology (NW India)* (Eds. Advich, P. C. and Bhu, H.) Department of Geology, Mohanlal Sukhadia University, Udaipur, India.

Pandit, M. K., Sial, A. N., Jamrani, S. S. and Ferreira, V. P. 2001. Carbon isotopic profile across the Bilara Group rocks of Trans Aravalli Marwar Supergroup in Western India: Implications for Neoproterozoic-Early Cambrian transition. *Gondwana Research,* 4: 387-394.

Pareek, H. S. 1984. Pre-Quaternary geology and mineral resources of Northwestern Rajasthan. *Memoir Geological Survey of India*, 115: 1-99.

Raghav, K. S., De, C. and Jain, R. L. 2005. The first record of Vendian Medusoid and trace fossil-bearing algal mat-ground from the basal part of the Marwar Supergroup of Rajasthan. *Indian Minerals,* 59: 23-30.

Rathore, S. S., Venkatesan, T. R. and Srivastava, R. K. 1996. Rb-Sr and Ar-Ar systematics of Malani volcanic rocks of south west Rajasthan: evidence for a younger post crystallization thermal event. *Proceedings Indian Academy of Sciences* 105 (2): 131-141.

Rathore, S. S., Venkatesan, T. R. and Srivastava, R. K. 1999. Rb-Sr Isotopic dating of Neoproterozoic (Malani Group) Magmatism from Southwest Rajasthan, India: Evidence of Younger Pan-African Thermal event by ^{40}Ar-^{39}Ar studies. *Gondwana Research* 2: 271-281.

Singh, I. B. and Rai, V. 1983. Fauna and biogenic structures in Krol-Tal succession (Vendian-Early Cambrian), Lesser Himalaya: Their biostratigraphic and palaeoecological signatures. *Journal of the Palaeontological Society of India,* 28: 114-117.

Strauss, H., Banerjee, D. M. and Kumar, V. 2001. The sulphur isotopic composition of Neoproterozoic to early Cambrian sea water-Evidence from the cyclic Hanseran Evaporites NW India. *Chemical Geology*, 175: 37-49.

Sarkar, S., Banerjee, S. Eriksson, P.G. and Catuneanu, O. 2005. Microbial mat control on siliciclastic Precambrian sequence stratigraphic architecture: Examples from India. *Sedimentary Geology* 176: 195-209.

Sarkar, Subir, Bose, P. K., Samanta, Pradip, Sengupta, Pratip, Eriksson, P. G. 2008. Microbial mat mediated structures in the Ediacaran Sonia Sandstone, Rajasthan, India, and their implications for Proterozoic sedimentation. *Precambrian Research* 162: 248-263.

Hon'ble reviewer

1. **Prof. R.P. Tiwari**
   Dean
   Faculty of Geology & Forestry
   University of Mizoram
   AIZAWAL
2. **Prof. Ashok Srivastava**
   Department of Geology
   Amravati University
   AMRAVATI - 444 602

IV-**63**

A New Insight in the Geology of the Jodhpur Group of the Marwar Supergroup, Western Rajasthan, India

S.C. Mathutr[1], Bhanwara Ram[2], P.S. Rathore[3] and S.S. Rathore[1]

[1] Department of Geology, J.N. Vyas University, Jodhpur-342 005.
[2] Department of mines and geology, Circuit House Road, Jodhpur-342 011.
[3]State Department of Ground Water, Shastri Nager, Jodhpur-342 003
E-mail: Sureshsushma09@gmail.com

Abstract

The Marwar Supergroup has been subdivided into three groups, the Jodhpur Group, the Bilara Group and the Nagaur Group. The Jodhpur Group of rocks are crops out at Khatu-Lohawat from north-northeast to Jodhpur- Balesar in the south direction in the western Rajasthan, India. The Jodhpur Group overlies the Malani Igneous Suit of rocks (779 ± 10 Ma to 681 Ma). The Jodhpur Group is subdivided chronostratigraphically, into the Pokaran Boulder Bed, the Sonia Sandstone and the Girbhakar Sandstone. In study area, the Jodhpur Group is overlain by alluvium sequences designated here as Jodhpur city Formation. All these formations could not define and elaborate the constituent facies of their sequences. The stratigraphic status of various facies of Jodhpur Group especially Sonia sandstone and Girbhakar sandstone are more clear and distinguishable by the present investigation and work of Chauhan et al., (1991). As such the various facies of the Jodhpur Group can be redistributed as per lithology, field and basement cover relationship, sedimentary structures and their depositional environment. Accordingly, Jodhpur Group is divided in to Facies 'A' i.e. Lower Umed Bhawan Formation (UBF), Facies 'B" i.e. middle Sursagar Formation (SSF) and Facies 'C' i.e. upper Motisar Hill Formation (MHF). They are overlain by Quartrnery Jodhpur City Formation comprises clayey, silty to sandy alluvium deposited in the Jodhpur valley and posing critical level of rising ground water level. All three formations of the Jodhpur Group and alluvium formation at one or other places are directly rest over the basement rhyolite. Secondly, the occurrence of sandstone tablets resting on rhyolite at more than 260m height and same at ground level are stratigraphically puzzling. It has been found that the distribution and deposition of sediments of Jodhpur Group and alluvium Formations is governed by the configuration of basement rocks and their subsequent sedimentological processes. The paper embodies these aspects in elaborate manner to understand the facies distribution and stratigraphic setup of the Jodhpur Group of the Marwar Supergroup and alluvium formation of the Jodhpur city.

Keywords: Marwar Supergroup, Jodhpur Group, Sedimentary Facies.

Introduction

The Jodhpur city since 1998 facing a serious problem of rising ground water problem (GWD reports, 2004, 2007 and Gupta et al., 2007) which is still unresolved. The authors are of the view that the problem should be address through the disposition of surface and subsurface geology to ascertain aquifers types. Rhyolite, sandstone and alluvium are considered as principle aquifers in Jodhpur city (GWD reports, 2004, 2007 and Gupta et al., 2007). The aim of present work is also to establish that sandstone is not act as principle aquifer in the problematic ground water rising areas. However, shale – sandy shale throughout the city area and act as aquifers instead of sandstone along with rhyolite and alluvium formation. This can be understood by knowing the distribution of various facies of Jodhpur Group and their depositional environment (Ram, 1990 and Chauhan, 1991). It is clear that the upper facies of UBF were eroded in Jodhpur valley and only lower facies (shale, sandy shale and minor shaly sandstone) is preserved and act as aquifers which underlain rhyolite and overlain by alluvium formation. Hence serious problem of rising ground water problem should be reviewed in light of our observations i.e. with respect to surface geology especially, distribution of UBF. There fore, authors present here elaborate geological aspects like basement configuration, basement cover relationship and geology of the Jodhpur Group to understand geomorphology and outcrop pattern, surface and hydrological subsurface disposition of various facies of the Jodhpur city.

Geological Attributes

The Marwar Supergroup (MSG) unconformably overlies the basement constituted of rhyolites (779 ± 10 Ma; earlier dated 745 ± 10 Ma) and (681 ± 20 Ma) (Crawford and Compston 1970; Rathore *et al.,* 1996, 1999) of Malani igneous suites of rocks (MIS) and considered to be younger than 680 Ma. Recent studies of the carbon, sulphur and strontium isotopes of the carbonates of the Bilara Group suggest Late Neoproterozoic-Early Cambrian (Vendian-Tommotian) age (Banerjee *et al.,* 1998; Strauss *et al.,* 2001; Pandit *et al.,* 2001; Maheshwari et al., 2003; Mazumdar and Strauss, 2006) for the middle part of the basin. The Marwar Supergroup has been subdivided into three groups in ascending order these are the Jodhpur Group, the Bilara Group and the Nagaur Group (see Fig.1). The Jodhpur Group of rocks of the Marwar Supergroup is crops out at Khatu, Jodhpur, Balesar and Lohawat areas of western Rajasthan, India. The Jodhpur Group is subdivided into the Pokaran Boulder Bed, the Sonia Sandstone and the Girbhakar Sandstone (Pareek, 1984). The Jodhpur Group overlies the Malani Igneous Suit of rocks (779 ± 10 Ma to 681). The outcrops of basement rocks (MIS) can be observed at Kaylana, Sidhnath, Masuria, Mehrangarh Fort, Jashwant palace (plate1a), in southwest to Santoshi mata temple, Mandore to Daizer in northwest direction (Fig. 2). The basement rocks belonging to MIS comprise mainly rhyolite, rhyolite porphery, ignimbrites, and welded tuffs having excellent preservation of columnar joints (plate 1b, see Bhushan, 1989; Kochhar, N. 1998 and Rathore, 1999).

It is imperative to note that the stratigraphic status of various facies of the Jodhpur Group especially Sonia sandstone and the Girbhakar sandstone are more clear and distinguishable by the present investigation and work of Ram, (1990) and Chauhan et al., (1991), given in Table 1. Accordingly, the various facies of the Jodhpur Group can be redistributed as per lithology, field and basement cover relationship, sedimentary structures and their depositional environment. As such the Jodhpur Group is divided in to Facies 'A' i.e. Lower Umed Bhawan Formation (UBF), Facies 'B" i.e. middle Sursagar Formation (SSF) and Facies 'C' i.e. upper Motisar Hill Formation (MHF).

The lower facies of UBF unconfermably has overlain the uneven basement of rhyolite. The UBF forms a coarsening upward sequence comprises shale at the base followed by sandy shale, shaly sandstone, fine to medium grained sandstone and coarse grained sandstone at the top. The abundant presence of festoon cross beddings with other sedimentological characters of rock of UBF indicate that it is typically a Deltaic Sequence. The outcrops of UBF can be observe at Surpura, Basni Tambolia, Umed Bhawan, Ganesh Bhakhri and up to Arna areas of Jodhpur (Ram, 1990; Chauhan et al., 2004, see Fig.2).

The SSF comprises medium grained sandstones, quartz arenite, minor siltstone, pebbly to coarse sandstone and medium grained sandstone with ripple marks (plate 1c) and planar cross beddings. It representing a typical Beach Sequences (Chauhan et al., 2004) in which due to repeated reworking and winnowing processes, well sorted and mineralogically well mature medium grained sandstones (Chittar ka pathar, a famous building stone of Jodhpur) is formed. The typical section of SSF can be observed at Daijar, Mandore, Sursagar, Kali Beri, Bhuri Bari to Agolai areas (Fig. 2). SSF preserved Ediacara fossils suggest an Ediacara age to SSF Sharma and Mathur, 2013).

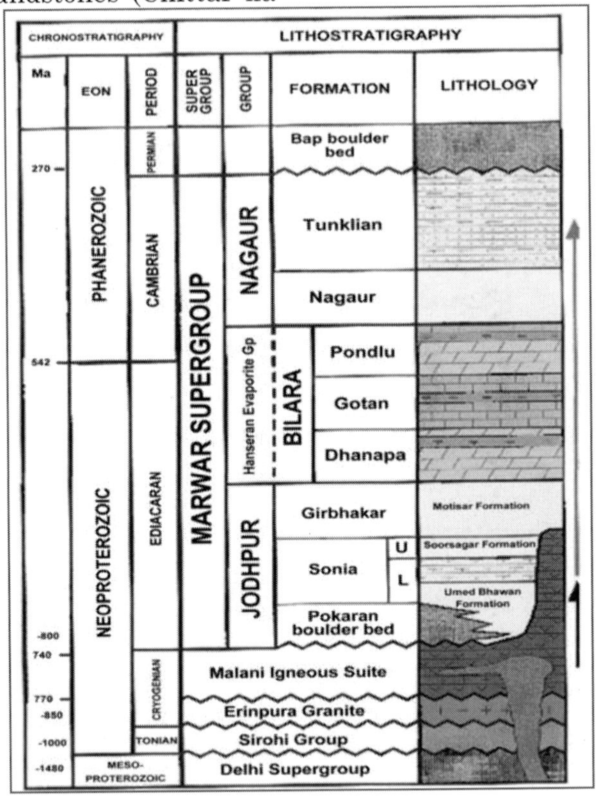

Figure. 1. Regional lithostratigraphic section of the Marwar Super group, Western Rajasthan, India (Brown line area mark the eroded upper part of UBF).

Table 1 : Lithostratigraphy of the Jodhpur region (modified from Ram, (1990) and Chauhan et al., 1991).

Era	Formation	Group	Lithology		
Quaternary	Jodhpur City Formation		Wind blown sand and alluvium comprises coarse to fine silt, sand with clay		
			Unconformity		
			Motisar Formation (Facies "C")	Fine to Medium grained sandstone Coarse grained sandstone Pebbly sandstone conglomerate	Fluvial sequence
Ediacara (Terminal Proterozoic)	Marwar Supergroup	Jodhpur Group	Soorsagar Formation (Facies "B")	Minor Pebbly to coarse grained sandstone Medium grained sandstone, Quartz arenite and minor siltstone Coarse sandstone	Beach sequence

		Umed Bhawan Formation (Facies "A")	Fine to medium sandstone, shaly sandstone. sandy shale and shale	Deltaic sequence
		---Unconformity---		
Pre-Cambrian	Malani Igneous Suite of rocks		Rhyolite, Rhyolite porphyry, welded tuffs, ignimbrite, volcanic ash and intrusive Granite	

The MHF comprises conglomerate at base and are followed by three cycles of pebbly to coarse grained sandstone, medium grained sandstone and siltstone at the top. It forms a typical coarsening upward facies of fluvial sequences having profusely with trough cross beddings. The typical section of MHF can be observed at Motisar Bhakhri, Dharampura and Keru sections (Fig 2).

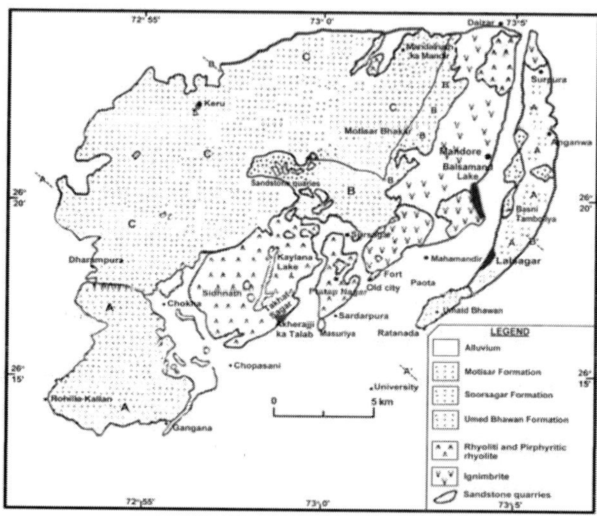

Fig. 2: Geological map of the Jodhpur region showing three plateau dominantly of sandstone (modified from Chauhan et al.,1991).

The Jodhpur Group is overlain by Quarternary alluviums designated as Jodhpur city Formation (Rathore, 2013; see Table 1). The alluvium formation comprises fine grained clayey, silt and sand material in the granular net wash, deposited as valley fill sediments (Fig. 2 and 3). The Quarternary alluvium formation having thickness varying from few cm. around the NW and NE direction, which increased gradually to a large depth to 78m, as one moves from SW to SE direction. The alluvium formation is resting unconformably practically over all formations i.e. on the sandy shale and shale of Jodhpur Group, sandstone of SSF and lower facies of MHF which are further underlain by the rhyolites in subsurface (Fig.4).

If, we observe sedimentologically, the Precambrian precursors (rocks of MIS and MSG) to the present sediments, down cut a valley to elevation of 78 meters below present plains elevation. The draining water from four ridges (Fig.3) of the Jodhpur city (plate 2a), bed rock valley of Jodhpur was filled with a variety of sediments, principally, granular outwash debris. Episodes of localized pounding created zones of more fine grained clayey, silt and sand material, some part of valley fill are coarse grained.

It is important to note that all four formations of the Jodhpur city at one or other places are directly rest over the basement rhyolite which is stratigraphically puzzling. It can be understand by the basement cover relationship i.e. disposition of rhyolite and overlying siliciclastic formations as clearly depicted in Fig. 4. It clearly shows the subsurface disposition of various facies of the Jodhpur which forms the principle subsurface hydrological formations of the Jodhpur city. Accordingly, at Bhim Bharak (Radar Hill), the MHF is directly rest over rhyolite. At Sursagar area SSF directly rest over the rhyolite and at Fatehsagar to Nai Sarak (Jodhpur city) area, the UBF directly rest over the rhyolite. Further, it is envisaged that alluvium (Jodhpur city Formation) everywhere rest over all the formations including basement. This sort of relation clearly depict that basement is highly uneven and on rugged topographic floor, all younger sediments were deposited and were governed by the basement configuration.

The basement configuration or basement cover relationship can be best understood by the distribution of various younger sheets of sediments as shown in Fig. 2. In Jodhpur area, there are three distinct plateaus of sandstones of various formations of the Jodhpur Group. The bigger one forms western part, extended in north south direction from Mandalnath-Keru to Dharampur-Arna areas for a length of 15.70 km and about 7.14 km in width. It has a maximum height of 342m from MSL near Arna, where total exposed thickness of sandstone is about 120m.

Geologically, it belongs to Motisar Hill Formation of facies "c" of Chauhan et al., (1991). The second sandstone plateau starts from the middle part of Jodhpur from near Sursagar and extended in north-northeast direction up to Daijar having length of about 9.64 km with maximum width of 7.5 km and lying between 260 to 310m. Geologically, this sandstone plateau is belongs to SSF.

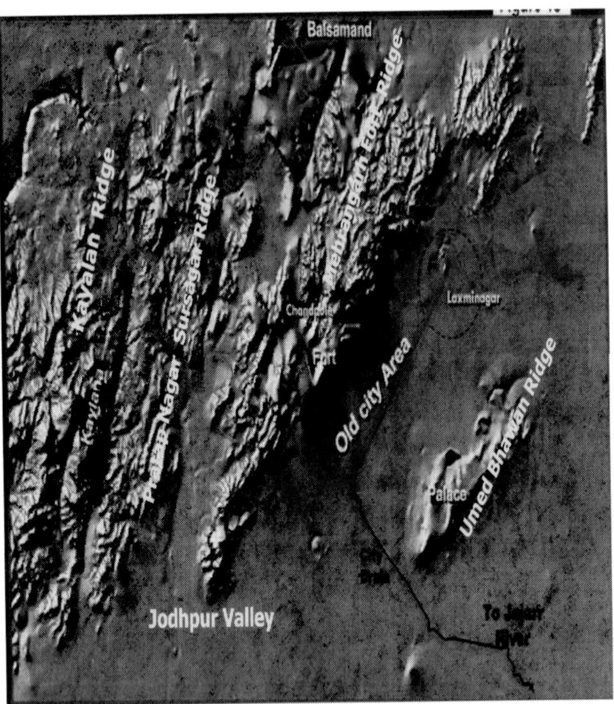

Fig. 3. Land set imagery and interpreted Physiographic features of the Jodhpur (source State Remote Sensing Department, Jodhpur).

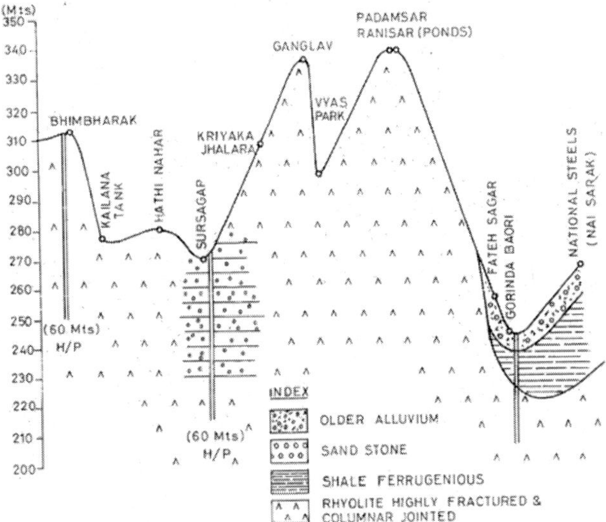

Fig. 4: Geological disposition of surface and subsurface formations in part of the Jodhpur city (GWD report, 2006).

Third plateau comprises isolated to continuous outcrops of sandstone starting from Surpura-Anganwa in northeast direction to Basni Tambolia- Lilipa Bhakar, Umed Bhawan in southeast of Mehrangarh Fort and Masuria area in center up to Gangana – Rohila Kalan in northwest direction. These outcrops belong to UBF and are extended for about 19.9 km. In addition, there are very small conglomerate/ sandstone (5-10m thick) tablets occurring as thin capping over Rhyolite ridges. They are at wireless Hill near Mehrangarh Fort (353m.) area (Plate 2b), Pratap Nager-Sursagar ridge (278m.), Balsamand (269m.) and Sidhnath- Radar Hill (395m.) near Kaylana Lake. It is also puzzling that how sediments at greater height have been deposited over rhyolite on one hand and same sediments at ground level at Nai Sarak and Masurai area in absence of any obvious tectonic activity. It can be explain with following evidences.

The vertical and lateral distribution of the three formations as exposed in different outcrops of Jodhpur suggests that these are not uniformly developed / preserved in the entire Jodhpur area. The grading of conglomerate to sandstone at wireless hill and at Pratap Nagar- Sursagar ridge, of the Motisar Hill Formation directly overlain the rhyolite (Plate 2c). However, facies of UBF and SSF are not developed here. Similarly, in the middle plateau at Sursagar-Motisar area, a truncated sequence of UBF comprising of shaly sandstone is exposed near Sursagar village. It is overlain by SSF which forms the major part of this plateau. SSF runs between RL 353m to 260m throughout the area. The MHF is exposed dominantly in Motisar Hill above RL 353m. Similarly, from Surpura, Lilipa Bhakar, Umed Bhawan to Arna the UBF attains less than 260m RL. The distribution of all the three facies of Jodhpur Group however, clearly indicate that the basement configuration seems to be very important factor in governing the overall deposition of siliciclastics of the area. It seems that where ever the basement goes below 260 RL, the UBF is developed. However, where the basement occupies elevation higher than 260 RL the UBF is either partially developed or missing and SSF directly rest over the basement. Similarly the sandstone tablets (capping) resting over high ridges of rhyolite (more than 353RL) suggest that both UBF and SSF are missing and MHF directly rest over the basement (wireless

Hill and Sidhnath Radar Hill). This sort of relationship from Surpura in north east and Fort in southwest clearly indicate that these outcrops represent margins of the Nagaur Basin as due to tidal influence combined with erosion- weathering process the upper part of UBF is eroded and contribute sediments to the SSF. It is evident from the Jodhpur city area where alluvium formation directly rest on lower part of UBF (Fig. 4). It can be further corroborated by the outcrops at Surpura-Lilipa Bhakar and Mehrangarh Fort where UBF is directly overlain by MHF however, SSF is completely missing, also indicate that on eroded surfaces of UBF (destructive phase of Delta), the MHF was directly deposited under terrestrial (fluvial) conditions. Secondly, the occurrence of the SSF with its increasing thickness in further north-northwest of this margin suggests that the main Nagaur Basin lies in immediate north to northwest direction. Further it appears that the time when upper lithounits of UBF were under terrestrial erosion its prograding sequence north to northwest of Surpura, Lilipa Bhakar, Umed Bhawan, Mehrangarh Fort, Masuria and Gangana came under the attack of marine processes. The reworking of these prograding sequences led to the development of various facies of SSF under distinct Beach environment where winnowing processes mature and sorted the sediments to form a excellent building stone (chittar ka pathar). Subsequently, after the withdrawal of sea, terrestrial conditions were developed and lower facies of MHF were deposited directly over the eroded surfaces of UBF. Simultaneously, the extension of these lithounits in further north – northwestward direction led to the deposition of upper facies of MHF over the SSF. The occurrence of truncated sequences of UBF in Balsamand and Arna area are overlain successively by SSF and MHF corroborate the same. During Quaternary period due to extensive erosion of rocks form a valley like geomorphological domain in which the alluvium derived from older rocks of basement and Jodhpur Group were deposited up to 78m thickness. The most of the city part of the Jodhpur is resided over it (Plate 2a) and facing critically high ground water problem now days (Rathore, 2013).

From above discussions, it is clear that the lithounits of Jodhpur Group of the Marwar Supergroup in the Jodhpur area are successively represent three distinct lithostartigraphic formations of deltaic, beach and fluvial sequences developed on extremely rugged topography of basement in which various facies Jodhpur Group viz a viz various facies of Sonia and Girbhakar sandstone of the area are much characteristically definable. Further, it can convincingly evident that upper facies of the UBF is missing and lower facies of UBF only occur in subsurface forming principle hydrogeological units with rhyolite and alluvium formation in the Jodhpur city.

Plate 1a. Rhyolite, ignimbrite and volcanic tuff outcrops around Mehrangarh Fort and Jashwant palace.

Plate 1b. Outcrops of welded Tuffs showing development of columnar joints on the way to Mehrangarh Fort from Nagori gate.

Plate 1c. The medium grained sandstone showing undulating crested ripple marks at Bhuri Beri section of Sursagar Formation.

Plate. 2a. The flat top ridges of rhyolite (foreground) and sandstone (background) in between area forms valley like structure at the base of Mehrangarh Fort (Chand pole side) where part of Jodhpur city is resided.

Plate. 2b. Sandstone tablet of Motisar Hill Formation resting over rhyolite at hight over 260m. RL at wireless Hill near Mehrangarh Fort.

Plate 2c. Contact of rhyolite (at base) with conglomerate (middle) and sandstone (top) of MSF as seen at Pratap Nager ridge

Acknowledgements

We thankful to Head, Department of geology, JNV. University, Jodhpur providing facilities for this work in the Department. Dr. S.K. Trivedi and Dr. S.R. Jakhar critically read the manuscript and offered valuable suggestions. Help rendered by Dr. V.S. Parihar, Assistant professor, Department of Geology, JNV. University, Jodhpur in various ways is appreciated.

References

Allen, J.R.L. 1982. Sedimentary structures, their character and physical basis, p. 1-676. In: *Developments in Sedimentology 30 B*, Elsevier, Amsterdam.

Banerjee, D. M., Strauss, H., Bhattacharyya, S. K., Kumar, V. and Mazumdar, A. 1998. Isotopic composition of carbonates and sulphates, potash mineralization and basin architecture of the Nagaur-Ganganagar evaporate basin (NW India) and their implications on the Neoproterozoic exogenic cycle. *Mineralogical Magazine*, 62: 106-107.

Bhanwara Ram. 1990. Geological investigation of Jodhpur Group of rocks and their relation with basement rocks of part of Jodhpur and their bearing on the Geological evolution of the area. (unpublished Ph.D. thesis), pp. 176.

Bhushan S.K. 1991. Mineral chemistry and petrogenetic aspects of Malani volcanic, western Rajasthan. Ind. Min, 43(3-4): 184-194.

Chauhan, D. S; Bhanwara Ram and Narayan Ram. 2004. Jodhpur Sandstone: A gift of Ancient Beaches of Western Rajasthan. Geol. Soc of India, V.64: 265-276

Chauhan, D. S, Dubey, J.C. and Bhanwara Ram. 1991. Geological analysis of part of Nagaur basin in vicinity of Jodhpur city. In S.K.Tandon, Charu, C.Pant and Casshyap (Eds.) Sedimentary Basins of India, Gynodaya Prakashan, Nanita: 63-73.

Crawford, A. R. and Compston, W. 1970. The age of Vindhyan System of Peninsular India. *Quarterly Journal of the Geological Society of London* 125: 351-372.

Dzulynski, S. and Walton, E. K. 1965. Sedimentary features of flysch and greywackes, p. 1-274. In: *Developments in Sedimentology 7*, Elsevier, Amsterdam.

Gupta A.K; Sharma, J.R; Rajshri Bothale; Dharmawat, R and Prahlad Singh. 2007. Jodhpur the gateway of Indian Desert: Study on rising ground water levels in the city. ISG Newsletter, V.13,No. 2&3:42-52.

GWD Report. 2006. Study on rising ground water levels in part of the Jodhpur city. Rajasthan State Ground Water Department, Jodhpur. P.47.

Kochhar, N. 1998. Malani igneous suite of rocks. J. Geol. Soc. India. V.51: 120.

Maheshwari A, Sial A. N., Mathur S. C. 2003. Carbonate and Oxygen Isotope profiles from the terminal Precambrian Marwar Supergroup, Rajasthan. India. International Jour. Carbonates and Evaporates. vol.18.pp.10-18.

Mazumdar, A. and Strauss, H. 2006. Sulfur and strontium isotopic composition of carbonate and evaporate rocks from the Late Neoproterozoic-Early Cambrian Bilara Group (Nagaur-Ganga-nagar Basin, India): constraints on intrabasinal correlation and global sulfur cycle. *Precambrian Research*, 149: 217-230.

Paliwal, B. S. 1999. Geology of the region to the west of Aravalli Mountain- Some Observations, p. 387-414. In: *Geological Evolution of Northwestern India* (Ed. Paliwal, B.S.) Scientific Publishers (India), Jodhpur, India.

Paliwal, B. S. 2007. Some fossil like structures in the rocks of the Marwar Supergroup, Western Rajasthan, India, pp 1-19. In: *Emerging Trends of Research in Geology (NW India)* (Eds. Advich, P. C. and Bhu, H.) Department of Geology, Mohanlal Sukhadia University, Udaipur, India.

Pandit, M. K., Sial, A. N., Jamrani, S. S. and Ferreira, V. P. 2001. Carbon isotopic profile across the Bilara Group rocks of Trans Aravalli Marwar Supergroup in Western India: Implications for Neoproterozoic-Early Cambrian transition. *Gondwana Research,* 4: 387-394.

Pareek, H. S. 1984. Pre-Quaternary geology and mineral resources of Northwestern Rajasthan. *Memoir Geological Survey of India*, 115: 1-99.

Raghav, K. S., De, C. and Jain, R. L. 2005. The first record of Vendian Medusoid and trace fossil-bearing algal mat-ground from the basal part of the Marwar Supergroup of Rajasthan. *Indian Minerals,* 59: 23-30.

Rathore, S. S., Venkatesan, T. R. and Srivastava, R. K. 1996. Rb-Sr and Ar-Ar systematics of Malani volcanic rocks of south west Rajasthan: evidence for a younger post crystallization thermal event. *Proceedings Indian Academy of Sciences* 105 (2): 131-141.

Rathore, S. S., Venkatesan, T. R. and Srivastava, R. K. 1999. Rb-Sr Isotopic dating of Neoproterozoic (Malani Group) Magmatism from Southwest Rajasthan, India: Evidence of Younger Pan-African Thermal event by ^{40}Ar-^{39}Ar studies. *Gondwana Research* 2: 271-281.

Rathore, P. S. 1999. The Terminal Proterozoic volcanism along the western flanks of the Aravalli mountain, Rajasthan. A study around Jodhpur city: 186.

Rathore, S. S., 2013. Geological investigation of water impounding structures of the Jodhpur city and their relation with the rising ground water level. Unpublished Ph.D thesis. J.N.Vyas University: 269.

Mukund Sharma and S.C. Mathur. 2013. *Arumberia* – like Ediacaran Mat Structure from Sonia Sandstone, Marwar Supergroup, Rajasthan, India (under communication).

Strauss, H., Banerjee, D. M. and Kumar, V. 2001. The sulphur isotopic composition of Neoproterozoic to early Cambrian sea water-Evidence from the cyclic Hanseran Evaporites NW India. *Chemical Geology*, 175: 37-49.

Sarkar, S., Banerjee, S. Eriksson, P. G. and Catuneanu, O. 2005. Microbial mat control on siliciclastic Precambrian sequence stratigraphic architecture: Examples from India. *Sedimentary Geology* 176: 195-209.

Hon'ble reviewer

1. **Prof. S. Kumar**
   Emeritus Professor
   Centre of Ad. Study in Geology
   Lucknow University
   LUCKNOW - 226 007
2. **Prof. Anil Maheshwari**
   Chairman
   Department of Geology
   Rajasthan University
   JAIPUR - 302 001

IV-64

A Conceptual Model for the Geomorphic Evolution of the Shore and Off-Shore Land Forms Along the East Coast of India

G.K. Panda

Department of Geography, Utkal University, Bhubaneshwar - 751 004
E-mail: gopalkrishna_panda@yahoo.co.uk

Abstract

The Odisha coast and Mahanadi Delta along the east coast of India has a wide range of landforms arising out of the dynamic interaction of the fluvial, marine, aeolian and lacustrine processes around the river mouths where the materials and processes are conspicuous. Along the Odisha coast, the Mahanadi is one of the major rivers known for its large catchment area, sediment and water discharge and an extensive bird foot shaped delta with a network of distributaries. The Mahanadi river mouth is one of the major outlets of the Mahanadi delta to the sea. Large amount of sediment transport from the river to the sea through the river mouth as well as the inflow of tidal sediments and materials from the longshore drift has created an ideal depositional environment along the delta face for development of a wide range of coastal landforms. This study makes an in-depth analysis of the geomorphology of the coastal landforms along the Odisha coast and Mahanadi river mouth using the remote sensing data and GIS analysis of the satellite imageries coupled with field observations. The study brings out the nature of the coastal erosion, land accretion and consequent change in the nature of shoreline in relation to the sediment supply and prevailing shore processes.

Keywords: Offshore Landforms, Beachs, East Coast, Land Accretion, Shoreline Changes

Introduction

The 480 km. length of the Odisha's coastline along the East coast of India depicts a variety of coastal landforms, which are most prominent near the river mouths. Along the coastal tract, the river mouth areas are the most vulnerable sites for dynamic interaction of the fluvial and marine processes. The geomorphology of these landforms is intricate because of the combined effect of the fluvial, lacustrine, marine and aeolian processes. The coastal landforms can be classified as depositional or erosional, which are mostly shaped by the marine waves, currents, tides and coastal wind. The Mahanadi is one of the major rivers of Odisha and the 6th largest river in the country in terms of its water discharge. It has produced an extensive delta at its mouth through a number of distributaries, which had created a very suitable depositional environment of riverine and marine sediments to produce a wide variety of shore and offshore landforms around its mouth region. The present study is an attempt towards understanding the coastal landforms, their morphological characteristics, nature of erosion and deposition along the coast and the Mahanadi delta front.

Objectives

The objectives of this paper are;

(a) To study the morphological characteristics of the coastal landforms along the Odisha Coast and Mahanadi delta front.

(b) Understanding the geomorphic processes associated with their evolution.

(c) Study of the nature of the morphological changes in the landforms with special reference to coastal erosion, land accretion and shoreline changes along with their possible environmental implications.

(d) Development of a conceptual model to explain the dynamic processes to explain the evolution of coastal landforms.

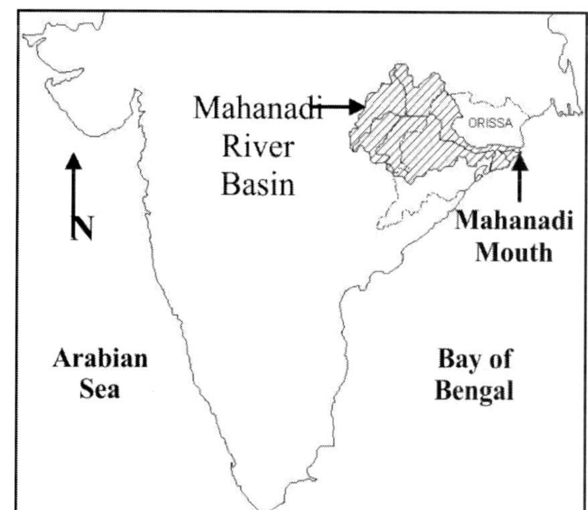

Fig. 1: Location of Odisha Coast and Mahanadi Delta Region along East Coast of India

Study Area and Methodology

The study area is the Odisha coast and specifically the Mahanadi river mouth and delta region and the Hukitola Bay near Paradip along the east coast of Odisha (Fig. 1). The geomorphological characteristic of this region is mapped and studied from the multidate geo-coded satellite imageries of band 5 and 7 of IRS-IB of 1994, Landsat imageries of 2002 and IRS P-6 coupled with the survey of India toposheets of 1975 and 1929. The characteristics of these landforms are studied with reference to their shape, size, orientation, association, lithology and change over time. The findings are validated through fieldwork based observations. The data obtained from the secondary sources are used to explain the possible processes and their associations. The study is based on the premises that along the coastal tract of the deltaic areas and more specifically around the river mouths, the landforms are mostly depositional in origin due to the abundance of fluvial and marine sediments in the shore zone. It is also further assumed that the geomorphic characteristics of the landforms are likely to be the evidences to reflect the processes that have produced them.

Geomorphic Background of the Odisha Coast

Odisha has nearly 480 km of coastline extending from the Ganjam district in the south to the Baleswar district in the north spreading over six coastal districts. It comprises of the Baleswar and Bhadrak coastal plains in the north, the combined deltaic plains of Mahanadi, Brahamani, and Baitarani spreading over the districts of Kendrapara, Jagatsinghpur and Puri and the narrow Ganjam plains. The Chilika Lake remains between the Puri and Ganjam plains spreading over nearly 1100 Sq Km of area along the coast. The geological characteristics of the Orissa Coastal Zone are occupied with Holocene sediments of fluvial, marine, brackish water, and lacustrine and aeolian environment. The absence of Pleistocene rock exposure has been caused due to over lying of sediments during Holocene transgression (Meijerlink, 1982). The land ward side of the shore is occupied by the occurrence of widespread palaeo-channels, natural levees, back swamps, lagoons and ancient beach ridges. The marine action is prominent in the Brahmani delta as compared to the Mahanadi delta with the predominance of mangrove swamps and mud flats.

The coastal plain consists of a narrow littoral zone adjacent to the coast extending for about 5 to 10 km inland from the shore (Fig. 2). It is a zone of sluggish and brackish streams, parallel sand ridges, sand dunes, back swamps and mud flats. This zone is often called as the "salt tract"

(Sinha, 1995). The deltaic and riverine plains of older and younger alluvium extending in width follow this zone inwards from 50 to 100 km. Most of the major rivers of Orissa originating from and beyond the Easternghats drain to the Bay of Bengal through this coastal tract with deltas in their mouths. The combined arcuate delta of the Mahanadi-Brahmani-Baitarani rivers extending over nearly 10,000 sq.km. area is the most prominent feature. This zone merges westward with the sub-montane tract of the Eastern Ghats. But longitudinally the coastal zone by and large consists of a sandy beach from the Andhra boarder to the mouth of the river Mahanadi with sand dunes and sand ridges of varying dimensions, the Mahanadi estuary region with the criss-crossing network of estuaries, mudflats and meandering rivers extending from the Mahanadi mouth to the Dhamra mouth of the Baitarani river and a shallow coastal zone from the Dhamra mouth to the West Bengal boarder with mud flats and absence of sand dunes. The significant river mouths along the Orissa coast are the Subarnarekha, Burhabalanga, Dhamra and Maipura of the Baitarani and Brahmani, Mahanadi, Jatadhari mouth oh Hansua river, Devi mouth, Kushabhadra mouth, Chilika lagoon mouth and the Rushikulya mouth.

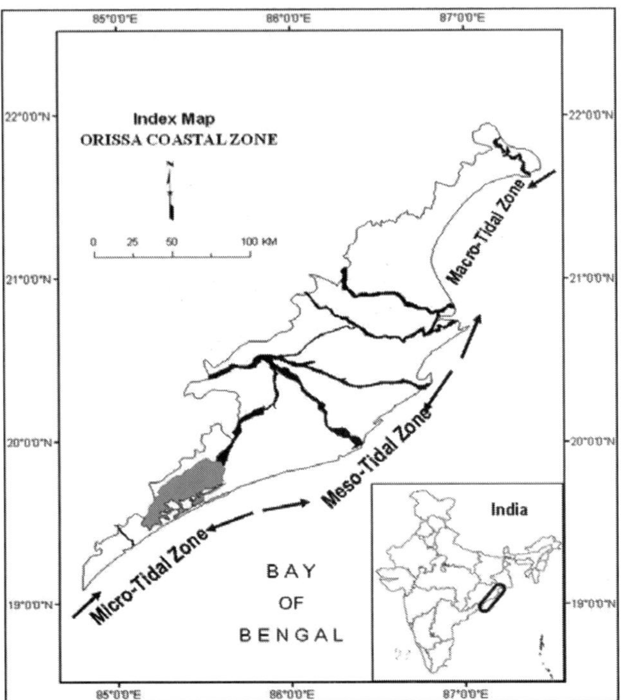

Fig. 2 : Odisha Coast showing three different tidal zones corresponding to three Different Sediment cells characterised by distinctly varying shore processes, action of the waves, currents and tides.

The Ganjam coast is micro tidal with high wave energy. Mean wind speed is nearly 15 Km per hour and the dominant wind direction is southerly. The middle portion of the Orissa coast is deltaic with prograding coastline and high sand dunes. Mean wind velocity is nearly 10 Km. per hour and wind directions are northeast and southwest. There are a large number of outlets where the distributaries and creaks meet the sea across the coastline. The sand dunes become lower in altitudes towards the north of this section gradually giving up to muddy coast replacing coastal sandy beaches with tidal mudflats. The Baleswar coast is mesotidal and mean wind velocity is around 4 Km. per hour with low wave energy. The coast is silty and muddy. The littoral drift moving northwards also brings sediments to the coast. The coast has been endowed with rich coastal and tidal mangrove vegetation. Under these background conditions, monitoring of changes from 1975 to 2010 along the coast have shown many revelations of change in landforms, coastline and coastal mangroves.

The Mahanadi Delta and its Mouth Region

The Mahanadi originates from the Amarkantak hills of the Bastar Plateau of Chhattisgarh and drains into the Bay of Bengal near Paradip (Fig. 1). Its catchment area is 141,589 sq.km of which 65,680 sq. km lies inside the state. The river is 851 km. long of which 494 km lies with in Orissa. The river has a mean annual flow of 66,640 million cu.m. Within the coastal plain, the river has developed a network of braided streams and an extensive delta. The study area is a humid tropical delta region with monsoon climate and an average annual rainfall of 155cm of which 90 percent comes during the month of SW monsoon (i.e. mid June to mid October). It is one of the frequently flooded rivers of eastern India. This region comes in the mean track of the tropical cyclones, which bring huge storm tides and tidal surges contributing towards the development of erosional and depositional landforms along the mouth zone. The present study area i. e. the Mahanadi mouth region is covered under the Survey of India toposheets No.73L/11 and 73L/15 (Fig. 1). The region has nearly 40% area under mangrove vegetation, tidal swamps, mud flats and creeks where as the other 60% area is predominantly agricultural.

Geological Characteristics

The geology of the Mahanadi river mouth zone is totally obscured by the Holocene sediments of fluvial, marine, brackish, lacustrine and aeolian environment (Meijerlink, 1982). The tidal and aeolian sediments of the recent origin occupy the region adjacent to the shore where as the land ward part of the river is occupied by the younger alluvium, which extends into the interior part of the interfluves. It has been inferred from the earlier studies that (Mahalik et al, 1996) this section is the most recent evolutionary feature of the Mahanadi delta of the Holocene period. The Mahanadi arcuate delta system has been formed in a tectonic down- warp of the Gondwana graben believed to be a failed arm of the triple junction on the Eastern Indian coast passive margin (Jagannathan et al, 1983). The basement floor with ridges and depressions was affected by faulting. Coastal and off-shore area record new basin development during tertiary. Coastal basins show sediment thickness of 2,400m as revealed by DSS profiling implying continued subsidence activity of growth of faults (Mohanty et al, 2005). Shoreline has migrated seaward and the major delta building seems to have spanned in the late Holocene during the last 6,000 years and is prograding now in a northerly direction (Mohanty, 1993).

Morphology and Distribution of Landforms

(a) Offshore Bars, Cuspate Bars and Bay Mouth Bars

These are the landforms of the offshore zone, extending from the low tide water limit of the shore up to the continental shelf, which is mostly affected by the action of the waves, tides, littoral drift and sediment discharge from the rivers. This offshore zone corresponds with the seaward portion of the breaker zone swept by the sinusoidal wind waves, which are transformed shore ward into solitary waves. The 5 fathom and 10 fathom submarine contour along the Mahanadi mouth is about 1.5 km. and 8 km. from the shoreline respectively. The offshore slope of the continental shelf varies from 6' to 9' along the Orissa coast (Ahmad, 1972). Along the offshore zone of the Mahanadi river mouth, the geomorphic features found are bay mouth bars, cuspate bars, beaches, spits, hooks and barrier islands enclosing shallow lagoons (Fig. 3 & 5).

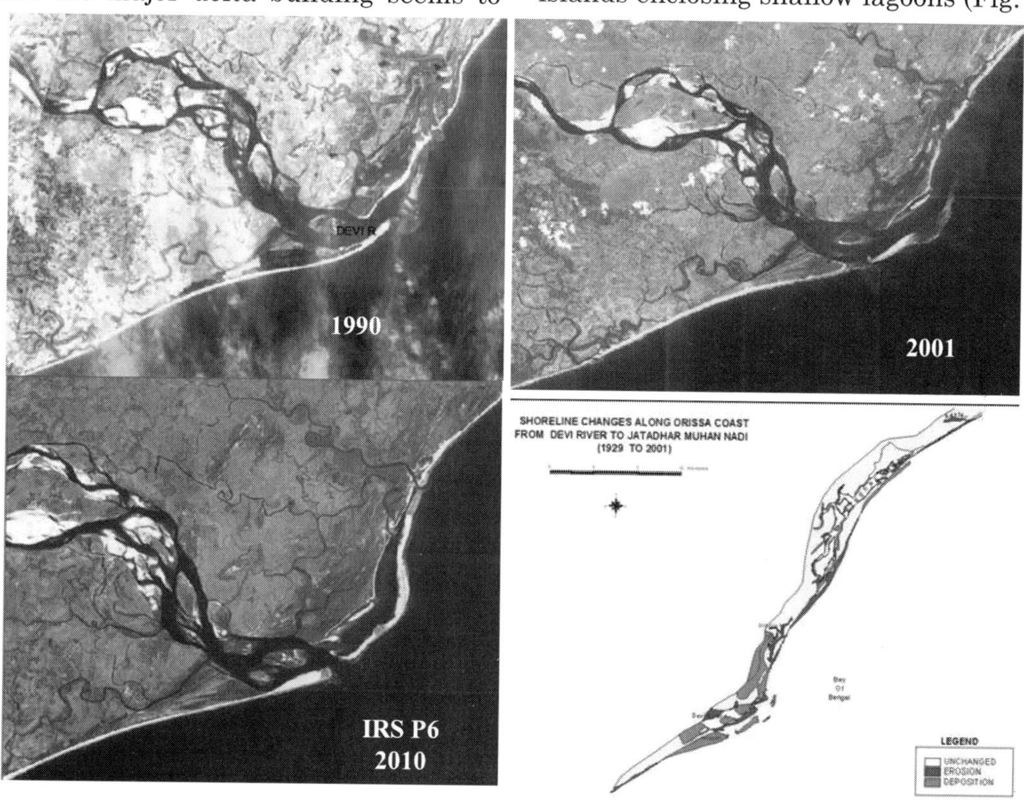

Fig.-3 : Sedimentation, land accretion, growth of spits, formation of Lagoon as well as coastal erosion and shoreline changes between 1990 - 2010 along the Devi river mouth, one of the major distributaries of the river Mahanadi along its delta front.

Offshore bars are sub-marine ridges of deposition of sediments eroded by the waves attacking the submarine floor (Johnson, 1972). These are deposited as elongated bodies of sand, shingle and clay developing sub-parallel to the shoreline and formed in an offshore position. Along the Mahanadi river mouth these features are well developed. The prolongation of the bay bar to a distance of 2 km. to the south of the Mahanadi river mouth parallel to the coast has shifted the meeting point of the river to the sea to the north east. This bay mouth bar has a width varying from 300 to 400 metres.

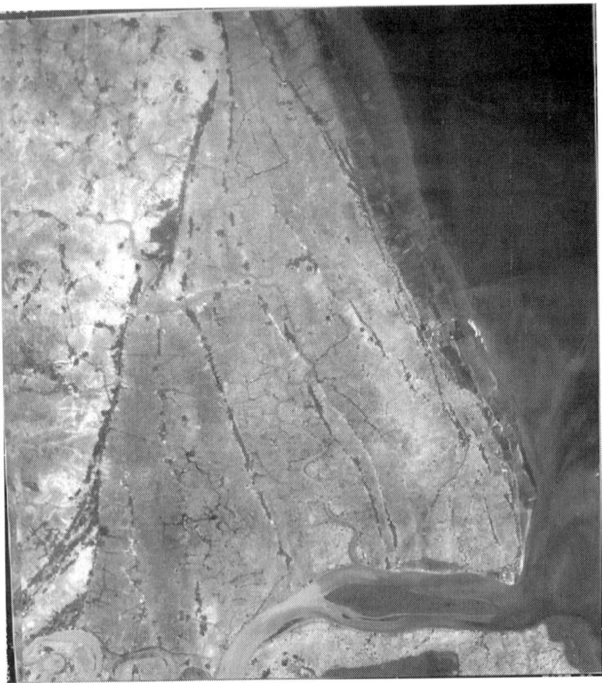

Fig.- 4 : Prograding coastline along with sequential change in the position of the ancient beach ridges along the Bhadrak coast north of river Dhamra revealed through linear pattern of red lines indicating settlements and plantations over beach ridges.

To the north of the Mahanadi mouth, the creeks which carry saline water from the sea to the river during high tides use to carry fresh water and sediments also during the flood season. The creeks are enclosed between the lobate shaped projecting landmasses i.e. deltaic lobes filled with tidal sediments and mangrove plants bordering the Hukitola bay. In these projecting land units, deposition of sediments have created a number of bay mouth bars, cuspate bars (of various shapes and sizes ranging from circular, elliptical, elongated ridge type) and semi-circular and irregular land bodies in side the Hukitola bay in the lagoon zone close to the mouth and in the offshore positions. The sediments are found sorted in parallel and cross-bedded laminations.

(b) Spits, Hooks and Compound Spits

The spits, hooks and compounded spits are the most conspicuous and well developed landforms of the off-shore region near the Mahanadi river mouth unlike many other rivers of the Orissa coast. These are narrow and elongated beach features, which extend in front of the Mahanadi river mouth in the direction of the dominant wind (Fig. 3). The spits have a northward open end and are attached to the land on the southern side. Some of the spits have curved westwards towards the main land at their extreme end, which are described as the hooked spits. These spits had extended into deeper water due to the abundant supply of sediments and elastic debris, which in turn were attacked by concentrated and refracted wave actions as well as by the storm surges.

The spit, which extends from the Mahanadi mouth near Paradip port up to its extreme limit in the northeast through Hukitola to Agaranasi Point, spreads over a distance of about 3 km. It is an important spit, which prolongs the Mahanadi mouth northwards from its original position. The spit is nearly 0.5 km. wide and 3 km. long. To the north of this spit, there are two parallel set of spits which are curved in shape with bending at its apex towards the land enclosing a shallow lagoon. The outer spit is narrow at the origin but widens towards north and extends as a ridge of sand extending from the main land. But however, the plunging waves had broken the spit at its centre. The outer part of this spit to the seaside is sandy where as the inner part is silty and is occupied with mangroves.

The inner Hukitola Spit is the most elongated one and was 10 km. long. Its width varies from 200m to 1.5 Km, separated from the main land by a large lagoon i.e. Hukitola Bay extending in a northerly direction. It is of 10 km. width at the base but the mouth connecting to the main sea at its extreme north is 5 km (Fig. 5). The inward funnel shaped projection of the mouths of the spits has been the common morphological dispositions. The growth of this hooked spit enclosing the Hukitola Bay was likely to give rise to another coastal lagoon similar to that of Chilika in the near future.

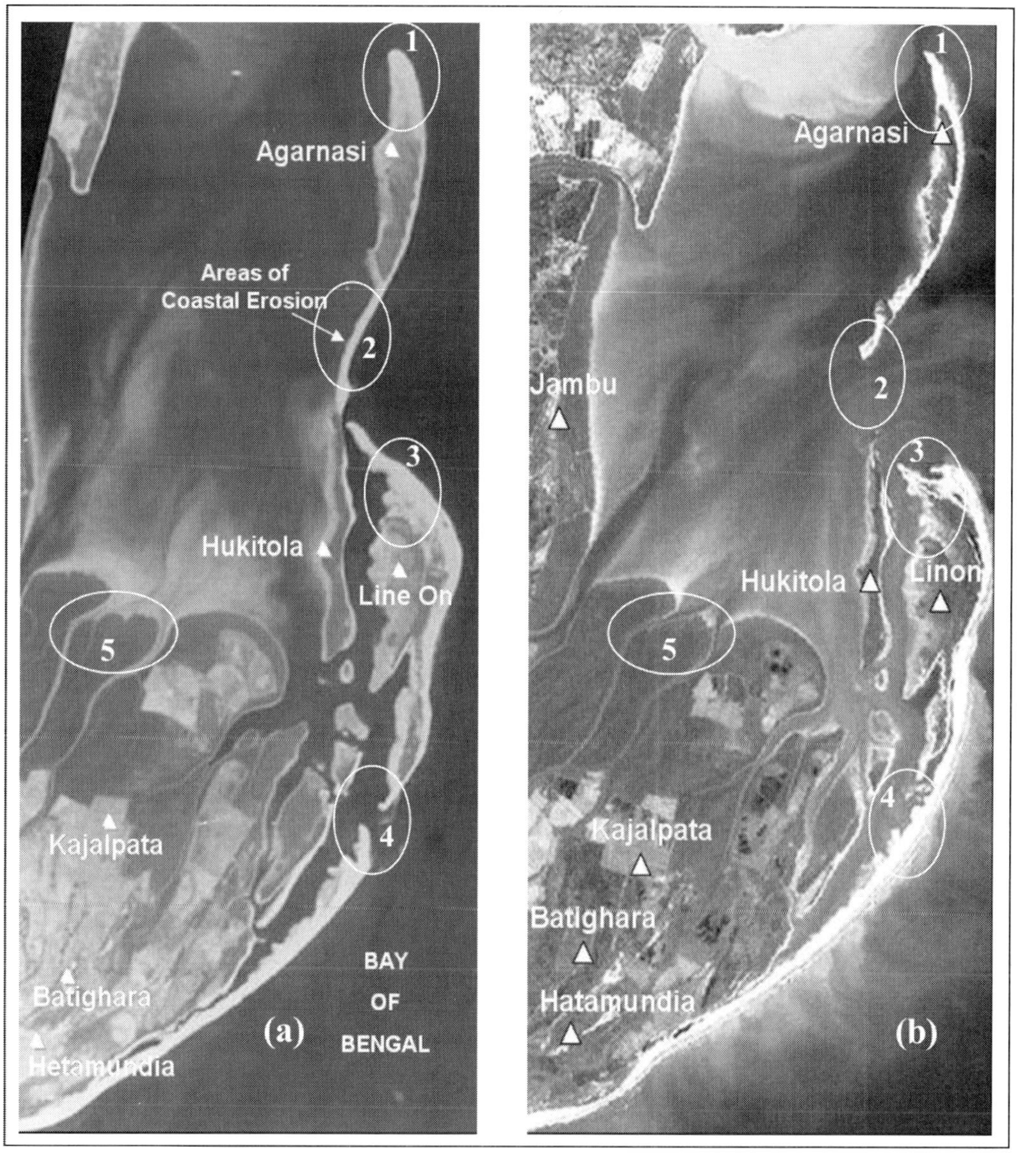

Fig.5: Changing Morphology of the Spits, Bay Mouth Bars & Islands around Hukitola Bay as revealed from the (a) Satellite Imageries of IRS- 1B, 1994 and (b) Landsat , 2010. Areas of change are shown in circles

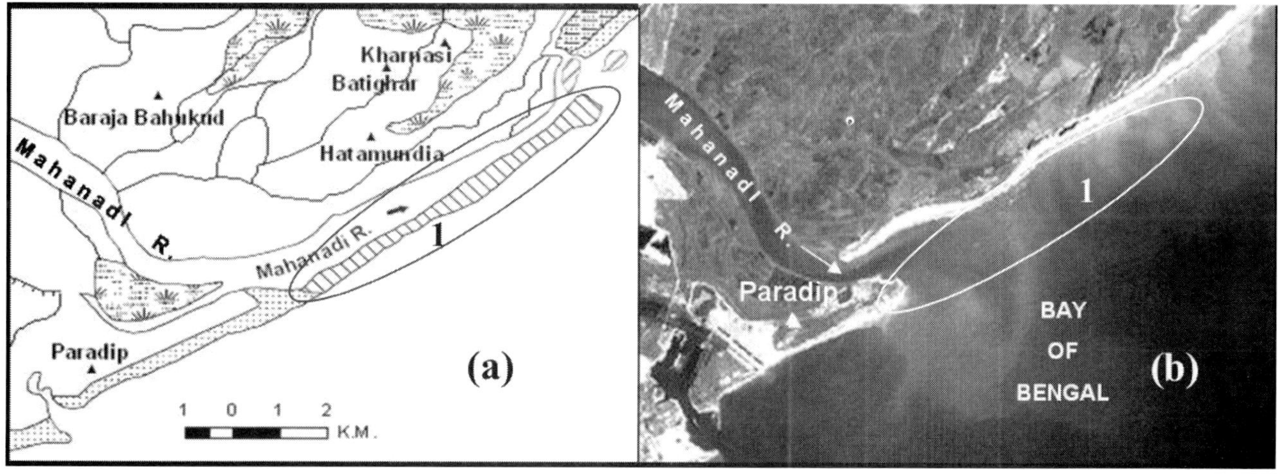

Fig.6: Changing Morphology of the Spits near Mahanadi Mouth as revealed from the (a) SOI Topomap, 1929 and (b) Landsat - 7,

(c) Beaches and Beach Ridges

These are the landforms of the shore zone between high and the low tide level often termed as the littoral zone. In this zone the main agents, which lead to morphological evolution and change are rivers, creeks, waves, tides and coastal wind. The landforms in this zone are clearly differentiated through their structure, lithological composition, nature of materials and their background conditions. The features of the backshore zone include the sand dunes and ancient beach ridges.

(i) Beaches: The beach is the zone of sediments deposited by the waves along the coastal zone. The sand normally constructs the beach along with other materials derived from both local and distant sources. Near the Mahanadi mouth the riverine sediments also play a very significant part in the development of the beach. The size of the beach reflects the material in storage and is therefore a measure of the balance between the availability of sediments and wave energy. Input of sediments to the beach is derived from the local erosion, from offshore positions and from the backshore zone. To the south of the Mahanadi River mouth the beach is narrow, sandy and on the northern side it is muddy. There is a moderate beach of 10 to 15 meter between the Devi and the Mahanadi mouth. To the north of the Mahanadi, creeks are many and the beach sediments are mostly silt and clay associated with sand. Near the river mouth the sediments discharged to the sea gets sorted into course and fine fractions. While course sediments are drifted away by intense wave action, the fine sediments remain settled. Near the river mouth the beach is formed of well-sorted clean, fine to medium grained sand with inclined laminae. The composition of the sediments near the mouth of the river comprises 60 percent sand, 26 percent silt and 14 percent clay. The spits are characterised by white texture of the sandy beach on the sea ward side and dark tone of the silt laden shore of the lagoon ward side on the satellite imagery (Fig. 3).

The temporary increase in wave energy associated with frequent visit of the storm surges to this region causes erosion to the shore and offshore landforms. When the wave energy comes to the normal, offshore materials are redeposited on the whole beach and try to rebuild its former configuration. But however, currently there is erosion of the Hukitola spit on its seaward side at a faster pace as compared to the deposition on its lagoon side (Fig. 5). There are also areas of erosion along the shore to the north of Mahanadi mouth at Satabhaya village and south of the mouth along Ersama coast.

(ii) Estuary and Creeks: Estuary is that part of the river mouth where the salt water and fresh water mixing takes place. The creeks are narrow irregular cris-crossing network of channels connecting to the sea, which carry saline water inland during the high tides. The Mahanadi also exhibits the characteristics of a barred estuary which is partially cut off from the ocean by the sand bars, spits, barrier islands, mud banks and deltaic accumulations leading to the development of the Hukitola Bay resembling a coastal lagoon (Fig. 2 & 3). On the northern section of the Mahanadi river mouth, a large number of creeks had developed between the projecting deltaic lobes with their mouths occupied by the islands and barrier bars. Kharnasi is one of the most important creeks of the northern side. The mouths of the creeks are projected forward between the projecting lobes and cuspate forelands, which are occupied by mangrove vegetation. The mangroves create a congenial environment for trapping mud and clay and help in prograding the deltaic lobes and the mouths towards the sea. The mangrove vegetation of the deltaic lobes is well marked on the satellite imageries with dark gray tone (Fig. 5).

(iii) Barrier Islands and Bay Mouth Bars: To the north of the Mahanadi River mouth facing the Hukitola Bay there are a number of islands enclosed by the creeks. Some of these islands have grown into bay mouth bars. The causes of formation of these features are due to the obstruction of sediment load in their mouthpart. Some of these islands had started as bars but subsequently are broken by the tidal waves or storm surges and are grown over time to form the islands (Fig. 5).

(d) Sand Dunes & Ancient Beach Ridges

(i) Sand Dunes: These are the landforms of the backshore zone which represents the area extending landward up to which the marine influence is felt. But however, in the Mahanadi mouth region, the backshore zone extends nearly 10 km. The major features in this zone are backwaters, sand dunes, beach ridges, tidal creeks and mud swamps. The sand dunes of this region are parallel to the coastline, low in height, in the range of one meter along the shore. The tidal creeks, backwaters and mud flats occupy much of the area in the backshore zone, which gradually merges landward with the croplands and shrimp ponds.

(ii) Beach Ridges: Adjacent to the river mouth there are three ridges, which are perches in a muddy sub-stratum, making the position of a pre-existing shoreline.These are berms of sand, silt and clay built by consecutive waves parallel to the coastline below the level of spring tides. On the northern side of the river the beach ridges have been washed and eroded by tidal action besides human occupancy in suitable upland sites favoring growth of settlements and agricultural activities. The presence of multiple ridges had arisen out of the continuous swallowing of the offshore and abundant sediment supply from the Mahanadi River. These are typical sandy ridges even noticeable on the topographic maps and satellite imageries as discontinuous linear patches of human settlements surrounded by village orchards and other plantations revealed through tonal contrast. The lithology comprises course to fine sand, silt and clay. Many such parallel ridges are found in the backshore zone which reveals the old strand lines of Holocene period (Mahalik et al, 1996).

Fig. 7: Wave Erosion showing Cliff Face on the Sea Ward side of the Agarnasi Island adjoining Hukitola Bay (Photographed on 10th May 2012)

(e) Tidal Inlets, Mud Flats and Mangrove Swamps

The regular astronomical tides coupled with storm tides have a profound impact over this region in producing landforms of tidal origin. The important among these are tidal inlets or creeks joining the river with the sea and even extending from the sea to the interior low lying areas of mudflats and swamps. The ill drained areas remaining in the backshore zone of the Paradip area have a number of swamps with tidal shrubs and marshy vegetation. These are revealed through their smooth texture and grey to dark grey tone due to high moisture content. The lithology of these areas comprise of greater proportion of silt and clay together comprising 30% to 50%. The sedimentation is from the suspended load, inter-bedded with dark grey clayey sand and silt. The tidal mud flats are distinctly visible on the imageries with light red tone associated with mangrove vegetation. The sediments are fine grained but coarse sediments are confined to the tidal channels. The tidal inlets has been favorable in propagating the tides which bring a large amount of estuarine sediments of silt and mud which are distributed in the adjoining areas to project the deltaic lobes and help in prograding the coast line (Fig. 3 & 5).

Shore Processes, Landforms and their Changes

Morphological evolution of land forms and their changes depend on geologic, tectonic, geographic, climatic and environmental settings. The landforms around the Mahanadi mouth and its surrounding area bear ample evidence to the combined action of the shore processes associated with the waves, currents, tides, coastal wind and storm surges as well as river actions based on the predominance of the agents at different localities, structure of the area and the availability of the materials. The offshore features point to the dominant action of the waves followed by offshore long currents (littoral drift) and the coastal wind. The backshore features are mostly linked to the tidal actions, shore wind and flood flow in the rivers as well as the estuarine interaction of sweet and saline water and sorting of sediments. The occasional high flood flows; large tidal waves and storm surges have also a significant bearing on the morphological changes of the shore and offshore landforms. The fast protruding deltaic lobes of the study area relates to the deltaic regime, abundance of sediments from the river, tide and littoral drift (Fig.-3).

(a) Wind, Wave and Littoral Drift: Orissa coast is in general, oblique to the prevailing winds and waves that generate strong northerly littoral currents. The along shore sediment transport in Orissa compares to the highest transport rates in the world, within the order of one million cubic meters to the north at Paradip. Varadarajulu and Harikrishna (1979) studied that the predominant wave directions are from the south or southwest. The bulk of the waves attain a height of less than 1.25 m and a period of less than 7 seconds. During the storms, however, waves reach 7 to 8 m. in height as it was observed during 1999 super cyclone. The oblique waves carry sediments from the breaker zone and beach to an offshore position to develop the barrier beaches, bars and islands. The littoral drift picks up the river discharged sediments deflecting it towards the northeast. The development of the spits and bars confirm to the evidence that the littoral drift supply materials but their evolutionary process can be thought of as a result of spasmodic progradation by obliquely impinging waves at the time of major storms or depressions. The wind blown sand brings sand deposition on the barrier bars and spits. Besides deposition there has also been erosion along the shore, on the seaward side of the spits and islands (Fig. 3 & 5). The present trend shows accumulation of beach sediments in the south-western part of the Hukitola Island and erosion from the sea ward face. The littoral drift plays a significant role in chocking the river mouth with the deposition of sediments as bay mouth bars or islands in the pre and post monsoon season when the volume of water discharge of the river is low. The storm surges due to the passage of the cyclones pounce upon the coast, brings occasional erosion to the features as seen near the Mahanadi mouth (Fig.-5).

(b) Tidal Flow and River Discharge: Mahanadi deltaic coast is micro tidal with mean tidal range measuring 1.29 m. and the tidal cycle is semi diurnal. It is principally a wave dominated coast during the south west monsoon and during NE monsoon (Nov. to Feb.) it is mixed wave and tide dominated. The enormous quantity of sediments and water brought to the sea by the river Mahanadi has been one of the main source of supply of sediments. The northern part of Paradip is relatively meso-tidal where wave energy diminishes and the resulting high tide brings silt and mud to build the mud flats and mangrove swamps. But to the southern side of the Mahanadi river mouth, the coast is micro tidal. The wave energy is relatively more than the tidal energy, which helps in building the sandy beach. The transportation and redistribution of sediments in different seasons of a year vary and during the lean season when the sediment and water discharge is low, conditions become ideal for the development of the bay mouth bars. The sands being heavier and having less surface area orient along the littoral drift and settles down quickly on the seabed. This floatation process and sorting out of heavy sand goes on for years and ultimately a high sand ridge appears near the mouth. When the river water overflowing the banks or the tidal water entering through the river mouths and creeks inundates the banks, such waters contain much sediment in suspension. It flows slowly down valley or may become fully pounded. The pounded water forms the mud flats and swamp deposits.

(c) Sediment Discharge of the Mahanadi: The Mahanadi delta has prograded in the seaward direction in the Holocene period due to the abundant supply of sediments from the hinterland under the favourable climatic conditions, sea level adjustments and neotectonic activities. The river has an annual total sediment load (suspended and bed load) which amounts to 29.77 million tons (Delta Development Plan, 1986). According to Meijerlink (1982) the Mahanadi carried much more sediment than it does now. With greater amount of sediment and water discharge through the river Devi (a distributary of the Mahanadi which drains in to the sea 50 km. south of Paradip near Astaranga) and declining sediment load after the construction of the Hirakud Reservoir on the Mahanadi 300 km. upstream of Paradip, a weir at Naraj and an anicut at Cuttack, the rate of supply of sediments, its depositional and erosional processes must have modified at the Mahanadi river mouth region. Under a regime of reduced fluvial sediment supply, the littoral drift system receiving sediments from the supply seems to have impoverished which may hinder the near shore gradational processes (Mohanty, 1990). Similar studies along the Godavari Delta

(Hemamalini & Rao, 2004, Rao & Vaidyanadhan, 1979) had revealed that the construction of dams had reduced the sediment discharge at the river mouths resulting in increased erosion along the shore resulting in the loss of the coastal habitats.

Conceptual Model and the Morphological Changes in the Landforms

The study of the satellite imageries and the topographic maps of different time periods of the Mahanadi river mouth area and the materials available along the coast associated with the available oblique waves have shown a number of morphological changes adjacent to the Mahanadi river mouth. Abundance of sediments through the littoral drift and presence of oblique waves explain the formation, growth and prolongation of the spits and hooks and consequent land accretion and shore line changes along the Odisha coast in general and in the Mahanadi delta region in particular. The vulnerability of the coast to the storm tides, storm surges and high flood discharges explain the erosion and retrogradation of the sea landwards in different coastal stretches of the Odisha coast. Based on the same concept the changes in the morphology of the Mahanadi mouth region can be modelled in the following two different phases (Fig. 8).

(a) First Phase: The Hukitola Bay and the spit were not developed to its present shape as it was revealed from the old settlement records. The Mahanadi was draining to the sea through two distributaries enclosing an island. Gradually the spit from the southern end of the Mahanadi mouth started developing in a northeast direction. By 1929 the spit has grown to 6 km. enclosing the Hukitola Bay. A number of creeks had developed joining the northern bank of the river to the Hukitola Bay replacing the earlier rivers and the Kharnasi became the major creek draining near the Jambu Island (Fig.-5). From the southern end of the Mahanadi mouth there developed a spit which also grew in a NE direction enclosing the Hukitola Bay. In 1975 toposheet it is clear to see that the river Mahanadi enters the sea as a single stream. The northern branch has been silted up giving rise to a network of tidal creeks among mud swamps in its place. The projecting lobes of the tidal flats separated by creeks with mangrove vegetation facing the Hukitola Bay are prograding faster with increasing rate of siltation. The mouth has got deflected towards NE by a long and narrow spit of 6 km. length and 250 m. width. Mahanadi experienced the greatest flood of the century of the magnitude of 15 lakh cusecs which has eroded the southern spit and the Mahanadi mouth which was running parallel to the sea divulged at right angle near the Nehru bungalow of Paradip (Fig. 8).

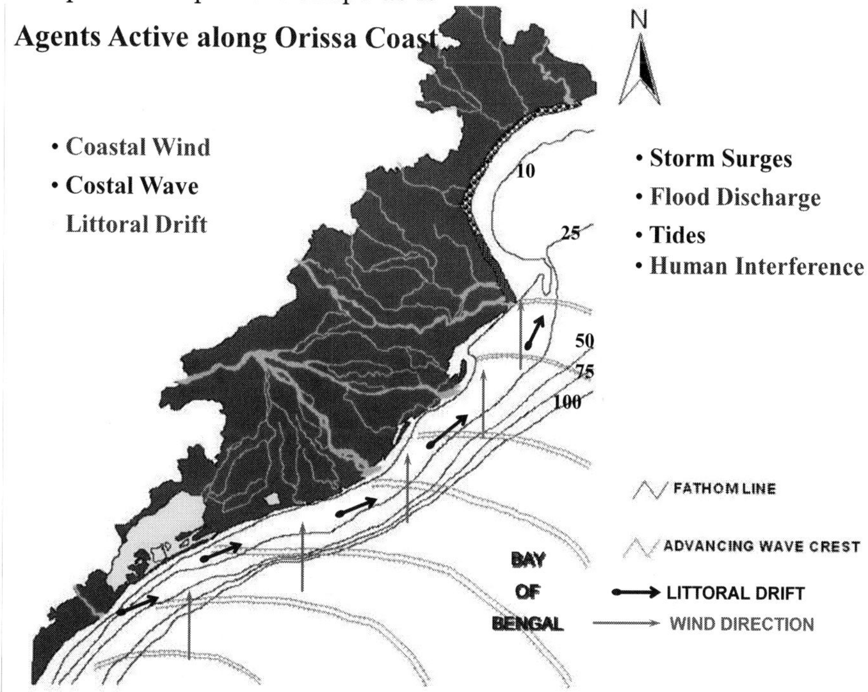

Fig.- 8: A Conceptual Model for the Geomorphic Evolution of the Shore and Offshore Landforms along Odisha Coast.

(b) Second Phase: Subsequently there are noticeable changes in the morphology of the spits and hooks and change in the area of the Hukitola Bay as revealed from the imagery of 1994. The extreme southeastern spit is getting elongated northwards with reduction in its width. The Hukitola spit is getting eroded at its center to a thin ridge of 50 meter at the middle and the spit is getting elongated and curved towards northwestern direction to a well developed hook. There are indications of erosion of the spit on its sea ward side (Fig. 6 & 7). At the southern end, the two parallel spits, islands and bay mouth bars are creating ideal conditions of an enclosed lagoon favorable for siltation and sedimentation. In the FCC of the IRS-1B of Dec.1994, the light bluish and whitish tone with coarse to medium texture and elongated shapes in the Hukitola Bay are the indications of coastal siltation and offshore sedimentation in the form of emergent shoals, bars and islands.

Conclusion

The geomorphological evolution and change of the offshore and shore landforms during the last century point towards a phase of rapid sedimentation, prolongation of the deltaic lobes with the development of the tidal creeks and estuaries, development of bay mouth bars, islands, spits and hooks followed by a phase of decelerated sedimentation and accelerated erosion along the shoreline and on the seaward side of the spits and hooks. The study reveals that the late Holocene deltaic progradation under abundant sediment supply and sea ward shifting of the shoreline may be on the critical threshold of retrogression due to reduced sediment input, neotectonic sea level changes and rise of sea level besides high episodic monsoonal flood flows and cyclonic storm surges. The distinct morphological changes in the off shore landforms with increasing erosion can also be attributed to continuous coastal subsidence that might have been attenuated by possible neo-tectonic activities and relative sea level rise leading to shoreline retreat.

References

Aggarwal and Murari Lal. (2005): "Vulnerability of Indian Coast line to Sea level Rise", Mimeographed, CAS, 111, New Delhi.

Ahmed. E. (1972): "Coastal *Geomorphology of India"*, Orient Longman, New Delhi.

Anand, S.P., Erram, V.C & Rajaram, M. (2002): "Delineation of Crustal Structure of the Mahanadi Basin from Ground Magnetic Survey", *Journal Geological Society of India*, Vol. 60, Pp 283-291, Calcutta.

Anbarasu, K & Rajamanickam, V. (2002), "Coastal Erosion around Pondicherry & Ennore along the East Coast of India", *Journal of the Geological Society of India*, Vol- 60, Pp 519-525, Calcutta.

Dag, J.W., Pont, D., Hensel, P.F and Ibanez, C. (1995): "Impacts of Sea level Rise on Deltas of Mexico and the Mediterranean, The Importance of Pulsating Events to Sustainability", *Estuary*, Vol.18, Pp-636 – 647.

Dept. of Water Resources (1986): "Delta Development Plan", Govt. of Orissa, Bhubaneswar

Dwivedi, D.N & Sharma, U.K. (2005): "Analysis of Sea Level Rice and its Impact on Coastal Wetlands of India", *Proceedings of the 14th Biennial Coastal Zone Conference*, New Orleans, Louisiana.

Esteves, L.S., Toldo, E. E, Dillenburg, S.R and Tomazell, L.J. (2002): "Long and Short Term Coastal Erosion in South Brazil", *Journal of Coastal Research*, SL-36, Pp. 273-282, Northern Ireland.

Hemamalini, B and Rao, K.N.(2004): "Coastal Erosion and Habitat Loss along the Godavari Delta Front, A fall out of Dam Construction", *Current Science*, Vol.87, No-9, Pp 1232-1236, IISc, Bangalore.

Jagannathan, C.R, et al. (1983): "Geology of the off shore Mahanadi Basin", *Petroleum Asia Journal*, Vol-IV (M), PP 101-104.

Johnson D.W. (1919): *"Shore Process and Shore Line Development"*, John Wiley and Sons Inc., New York.

Mahadevan, C. M. D., Rao, P.R. (1958): "Causes of the Growth of Sand Spits North of the Godavari Confluence", *Andhra Univ. Memoir of Oceanography*, Vol-2, Pp. 69-74.

Mahalik, N.K., Das, C & Maijima, W. (1996): "Geomorphology and Evolution of the Mahanadi Delta", *Indian Journal of Geosciences*, Osaka City Univ, Vol-39, Part-6, Osaka, Japan.

Meijerink, A.M.J. (1982): "Dynamic Geomorphology of the Mahanadi Delta, *ITC Journal, Special Verstapen Volume*, Pp. 243-250.

Mohanty, M & Swain, M.R. (2005): "Mahanadi delta, East Coast of India, an Overview on Evolution and Dynamic Processes", Mimeographed, Dept. of Geology, Utkal University, Bhubaneswar.

Mohanty, M. (1993): "Coastal Processes and Management of the Mahanadi River Deltaic Complex, East coast of India", *Proceedings Coastal Zone 93*, ASCE, PP 75-90, New York, USA.

Nayak, N. (2002): "Role of Remote Sensing to Integrated Coastal Zone Management", Mimeographed, ISRO, Ahmedabad.

Prusty, B.G., Sahoo, R.K & Mehta, S.D. (2006): "Natural Causes Lead to Mass Exodus of Olive Ridley Turtles from Ekakulanasi, Orissa, India, a Need for Alternate Sites", Mimeographed.

Raman, C.V & Reddy, K.S.N. (2001): "Sediment Dispersal Pattern off the Mahanadi-Nagavalli Continental Shelf, Northeast Bay of Bengal", *Journal of the Geological Society of India*, Vol – 58, No – 2, Pp. 123 – 133, Calcutta.

Rao, K. S, Rao, K.W., Sadakata, N., Varma, D. D & Rao, A.T.(2003): "Geo-morphology and Evolution of Penner Delta, East Coast of India", *Journal of the Indian Association of Sedimentologists*, Vol.22, No. 1 & 2, Pp. 171 – 181.

Rao, K.N., Sadataka, N., Hemamalini, B., Takayasu, K. (2003): "Coastal Forms and Processes of the Godavari Delta, India", Mimeographed, Dept. of Geo-engineering, Andhra Univ., Waltair

Rao, S & Vaidyanathan, R. (1979): "New Coastal Landforms at the Confluence of the Godavari River", *Indian journal of Earth Sciences*, Vol – 6, Pp. 222-227.

Seetharamaiah, J, Bhagaban, K.V.S & Rao, K.N. (2002): "Recent Morphological Changes in the Penner Estuary", East Coast of India, *Journal of the Geological Society of India*, Vol. 60, Calcutta.

Varadarajulu, R & Harikrishna, M. (1979): "Wave Characteristics of Paradip Port", *Indian Jour. of Marine Sciences*, Vol. 18, Pp. 68–72, Goa.

Vinod Kumar, K and Bhattacharya, A. (2003): "Geological Evolution of Mahanadi Delta Orissa Using High Resolution Satellite Data" *Current Science*, Vol 85, No – 10, I I Sc, Bangalore

Hon'ble reviewer

1. **Dr. R.H. Swakar**
   Geol. Soc. India
   5th Block Jainagar
   BANGALORE - 560 040
2. **Prof. S.C. Mukhopadhyay**
   Emeritus Professor
   Department of Geography
   Kolkata University
   KOLKATA - 700 019

IV-65

Geomorphology and Development in the Hindukush-Himalayan Countries with Reference to India

S.C. Mukhopadhyay

UGC Emeritus Fellow In Geography, Calcutta University, Kolkata - 700 019
E-mail : profscmukhopadhyay_2009@rediffmail.com

Abstract

The Hindukush-Himalaya has extended from west to eastern part in almost a creascent shaped covering eight countries. These are like Myanmar, India, Tibet (China), Bangladesh, Bhutan, Nepal, Pakistan and Afganisthan from east to west. The eight countries have exhibited in the different ways their advancement in the utilization of the natural resources including the vast population or human resources where we find many problems in the way of socio-economic development both on local, region state and country as well as the global levels. Everywhere the Hindukush-Himalaya has exerted its influence directly or indirectly in such developments – its climate, land, glaciers, flora and fauna, minerals, soils drainage channel patterns, forest and vegetation, food and agriculture and water and sub-surface at its various altitudes. Water for drinking agriculture, industrial fields, recent problems of contaminations and so on. Among the difficulties are the recent pollution of its upper surface and sub surfaces as found to be very much active, the poverty (*Garibi Hatow*), the levels of science and technology specially in the application of GIs and Remote Sensing etc. have been most prominent. The author has also made some suggestion for the growth and development of the Hindukush-Himalaya and countries both in space and time. The related influence of geomorphological-tectonics, slopes river etc. exerted by the Hindukush-Himalaya are shown in the given maps and diagram for India and its surroundings like the Bangladesh, Bhutan, Afganisthan, etc are discussed with some examples very briefly.

Keywords: Hindukush Himalaya, Geomorphology, Eology Himalayan Georesources.

Introduction

Sinha (1989) in the course of cultural background (from Bhagwat Gita) says that the mighty and magnificent Himalaya, by humbling those that come to them with their magnitude and power, gratifying them with their splendour and beauty, testing their manhood with glaciers and peaks, defying their spirit with their inviolate mysteries, enables man to acquire the qualities which only danger and nature in the raw can sharpen. The might of Himalaya has

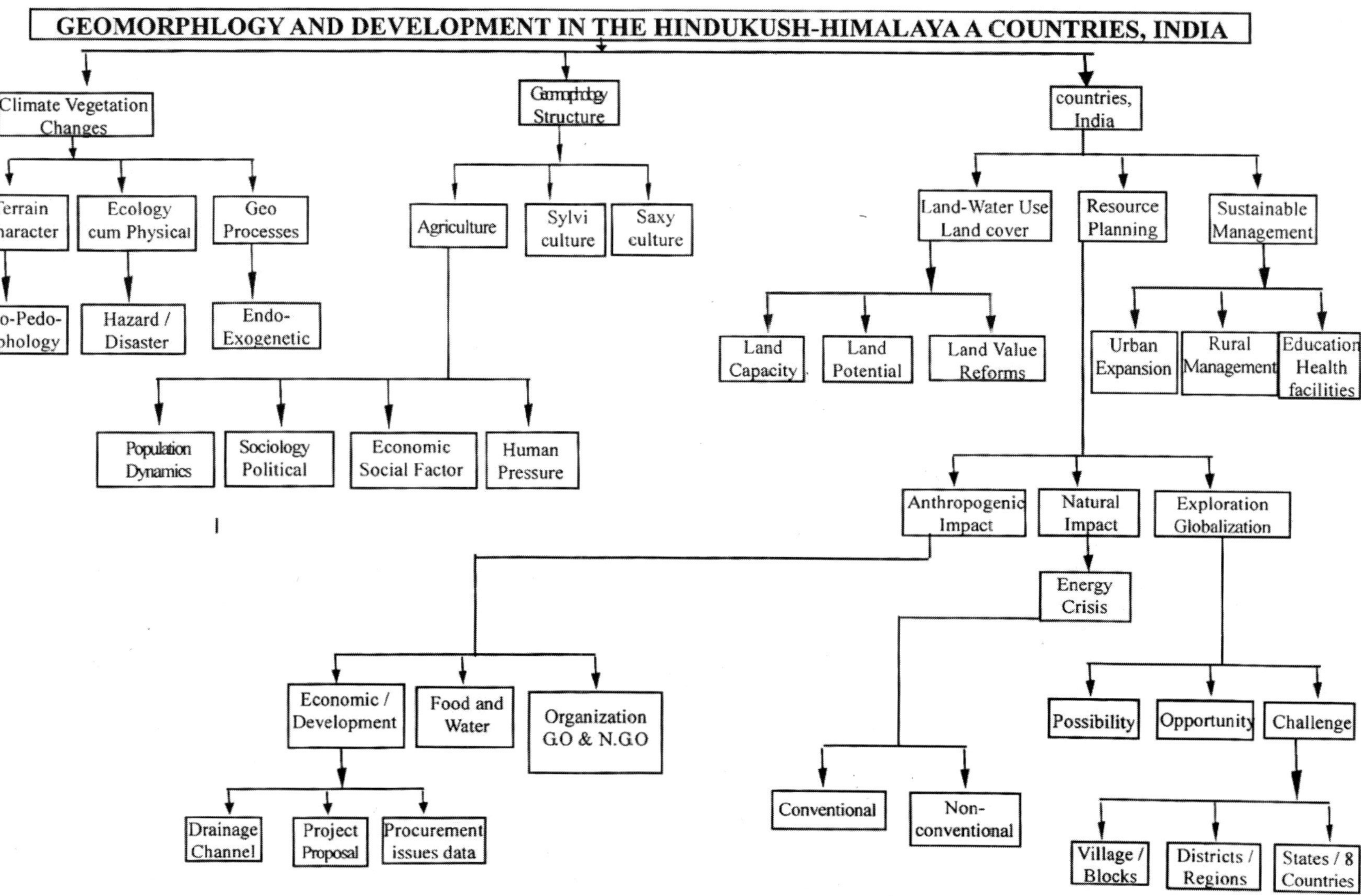

Table 1 : Flow Chart of Geomorphology and Development in the Himalayan countries, India and Neighbourhoods areas

inspired and evoked the awe and wonder of the people of the Indian subcontinent and from time immemorial. The Aryans had great reverence for this part of Himalaya and every river, glacier, and high mountain peak of this region was sanctified. The sacred shrines of Badrinath, Kedarnath, Gangotri, Yamunotri, Nanda Devi, and further north Kailash – Manasarovar etc. are the most notable throne especially of Gods for Hindus in deed Mukhpadhyay (1982, 1995, 1996 a & b) 1997, 2000-2013) Kievelitz (1999), Valdiya (1998), and others also narrated the great Hindukush-Himalaya - its gift and bounty with many resources directly and indirectly exerting influence in different ways on the flora – fauna including the vast glorious cover unit (Figs. 1 - 22, Tables 1-3).

The present paper is concerned with a brief study on the problems of Geomorphology and development in the Hindukush-Himalayan Countries with Reference to India both on qualitatively and quantitatively. The recent partition of the Hindukush-Himalaya mountain with geomorphology and its distinct features have been analysed in brief in terms of the given illustrations - figures, maps and diagrams, plates, tables etc. (Figs. 1-22). This magnificent unit has exhibited variety in both the ecological and social-economic political levels. These have associated with extensive ecological, biological and cultural issues in the eight countries through which it extends as shown in the Fig. 1a below. It covers from west to east – Afghanistan, Pakistan, Nepal, Bhutan, India, Tibet (China), Myanmar (former Burma) and Bangladesh. These countries have formed a number of environmental developments, the socio-economic–political status, conflicts and changes both on the sparse and time. Conflicts depending on the different events and situations that they face are found to exist under the various conditions and interactions locally or globally land-ocean-atmospehre and biosphere are involved. Here system or nature act and react (UNFCCC, IIPCC; UNEP, WMO, ICSI and others). The Methods are (prefiled field and post field) shown in the given Tables 1, 2 & 3 and (Figures 1-4, 7-12) also in very briefly. (Mukhopadhyay, 1982, 1998, 2001, 2013), Raiverman 2009, Sinha, 1989, Nandy, 2001, Kuhle, 2001, Stone, 1992, Valdiya, 1998 and others.

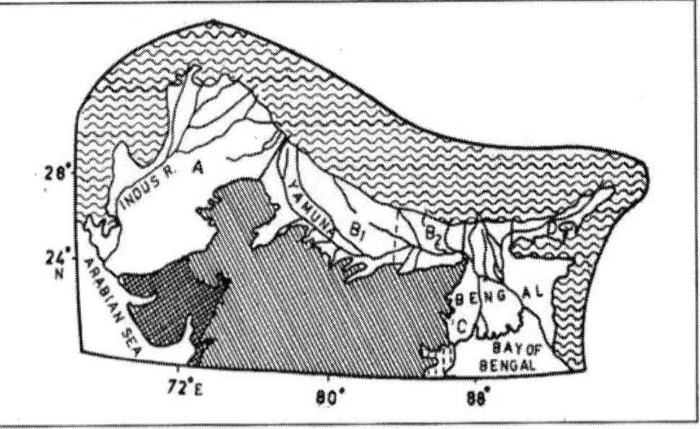

Fig 1a : Broad location map of the Hindukush-Himalaya with Indo-Ganga-Brahmaputra plains (D) in India, and Surrounding.

Table 2: Broad relationships among the Inputs, Perspectives and outputs concerning mainly in Indian Himalaya

Inputs	Perspectives	Outputs
Environment – The Hindukush-Himalaya	Biotic and Abiotic. Physical & cultural	Holistic Approaches – Environmentalism
Space-Ship Earth – Geological Science	Land-Ocean-Climate - Lives	Agenda-Gaia Hypothesis-Protocols
Ecological Principles	Ecosystems – Magnificent – Models	EIA, UNFCC, IPCC Micro-Meso-Macro-Scales
Air–Water–Soil-Science-Technology	Socio-Politico-economics	Ecological Economics-Chimney-conflicts
Cycles – Theories-Modern Technology	Evolution-Functions-Materials-Afforestation	Analyses–Energy-National–International
Landform – Landscapes-Strategies	Land bank-Values-Urbanization-Scarcity	Erosional-Depositional sustainable Development
Settlements–Ekistics–Energy-Balances	Population Dynamics – Rural & Urban	Health-Mental-Education-Migration-Emigration
Hazards-Disasters-Pollutions	Manmade –Natural – Integrated	Active- Mitigation – Markets - Management
Crucial-Interfaces-Policy	Human Resource Utilizations - Genders	Aesthetics – Surface - Subsurface – Tourism-Industry
Contemporary Issues – Rainbow Revolution	Agriculture Sylviculture-Saxiculture	Quality of life, surface – subsurface - Active
Conservation – Heritage- Leaders-Missions	Ancient-Mediaeval-Present-Endemic spp.	Local-Regional -E-lams, Development -Global
Human Environment–Viability-Five yrs. Plans	Interface-Economics -Land – Use - Policy	Environmental Governance, Recycles,
Civilization – Investment-Poverty	Institute-Organisations G.O.& N. G.O.	Societies -Infrastructure & Opportunities
Visions-Index, challenges	Awarenes-concepts, Potential	Inclusive –Development – Internets

Table 3 : Broad Structural Subdivisions of the Himalaya covering the Tibetan plateau (China), the Brahmaputra Basin and Adjacent Regions from South to North in the eastern/north eastern parts of India

Nortern Indo-Gangetic plains	Sub-Himalayan Siwalik molasse (Mio-Piocene) trough
Main Boundary Fault /Thrust	Lesser Himalayan Tectogene
Main Central Fault /Thrust	High Himalayan Tectogene of Central Crystallines
Main Northern Fault /Thrust ***Chiatsun Fault /Thrust***	Tethyan Himalayan zones with Ordovician to Eocene shelf sediments in the south and Triassic flysch and jurassic-Cretaceous Radiolarites to the north.
North Dipping Fault /Thrust	Melange Zone (sedimentary and tectonic)
South Vergent Fault/ Thrusts	Zone of Ophiolites along the Yarlung Zangbo (Indus-Tsangpo) Suture
Both South and North Vergent Faults/ Thrusts	Createaceous Flysch Zone (Xigaze Group) Kangdese (Gangdise) Magmatic Arc

(Based on Nandy (2001)

Methodology

The author has tried to discuss with examples, the related drainage patterns including the channel patterns, and some relevant points on morphology and its impacts included in the Hindukush-Himalaya, India and surrounding areas through field work (in selected sites) and also with the help of modern methodology. Whichever were available using current instruments-equipments like hand-held G.P.S., S.O.L. Topographical Maps (53I, M; 78A, 82D, A.O. etc. 1 : 50,000 scales), Degree sheets (on 1 : 250,000 scale), Aerial photographs, Photographic evidences and the like. This author has also consulted the available publications concrned (G.O. and N.G.O) and incorporated the relevant illstrations both of his predecessors and sucessors opinions also in terms of (i) prefield work, (ii) field work and (iii) post field work (Mukhopadhyay 1980, 1982 and others) very shortly. All maps and diagrams, the latitudes/longtudes altitudes, border lines, the locations etc. as mentioned earlier or later all are approximate (Table 1, 2 & 3, Fig. 1-22).

The tectonic features, glaciers and their impacte are identified mainly sutting old topo-maps, use though observation, measurement, computation, forest -vegetation etc. For examples, east-west aligned old faults thurst lines, recent almost north-south direction fault/ thrust plains, the variations in altitude of different landforms and drainage patterns (Rectangular, Trellised-Parallel, Radial etc. patterns) as above (Fig. 9-22). The given maps and diagrams (Plate 1-13) illustrate the Hindukush-Himalayan Geomorphology and drainage/channel patterns etc. of the study area. It has been reffered earlier the author has carried out with dumpy level, and differensites setes including the river valleys or glaciers (slected ways). For the other aspects of methodolgy the present author has added in the given Flow Chart, (Table 1 & 2), visited internets, local institutions, organisations (GO & NGO) etc. He also has considered the soft wires ERADAS, Bhuban 21st Century Geomatica and the like. (Mukhopadhyay, 1982, 2012).

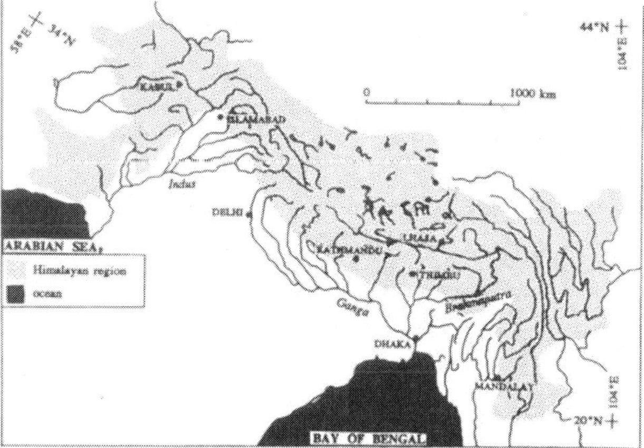

Fig. 1b : Map of the crescent shaped Hindukush-Himalaya, its influence in drainage-channel patterns, important capitals, parts of Bay of Bengal (SE) and Arabian Sea(W) mainly in Indian and 8 countries etc. (based on, Stone 1992)

Importance of the Present Study

The Hindukush-Himalaya, India and surroundings with their tectonics determines rejuvenated landforms and main geomorphic processes, peculair drainage glaciers channel patterns vegetation types etc. under vegetation Tropical Monsoons etc. which hs detrmined the socio-economic the development al-programm the population density, and all the utilisation resources of India in particular and surrounding areas in general. The climate, soils, vegetation and other resources including the drainage lines (many examples have been indicated by the given figures and plates) (Fig. 1a – 19, Table 1 & 2). The present author has also tried to focus the importance of the Hindukush-Himalaya and surroundings especially, bearing the National Importance and the States level Importance also. It is difficult to considers however all the influences of the Himalaya and surroundings too mainly because of them very high altitude and slopes, harsh weather and climate, ruggedness, dense vegetation growth, tribals different hazard and disasters, in accessibilities and the like.

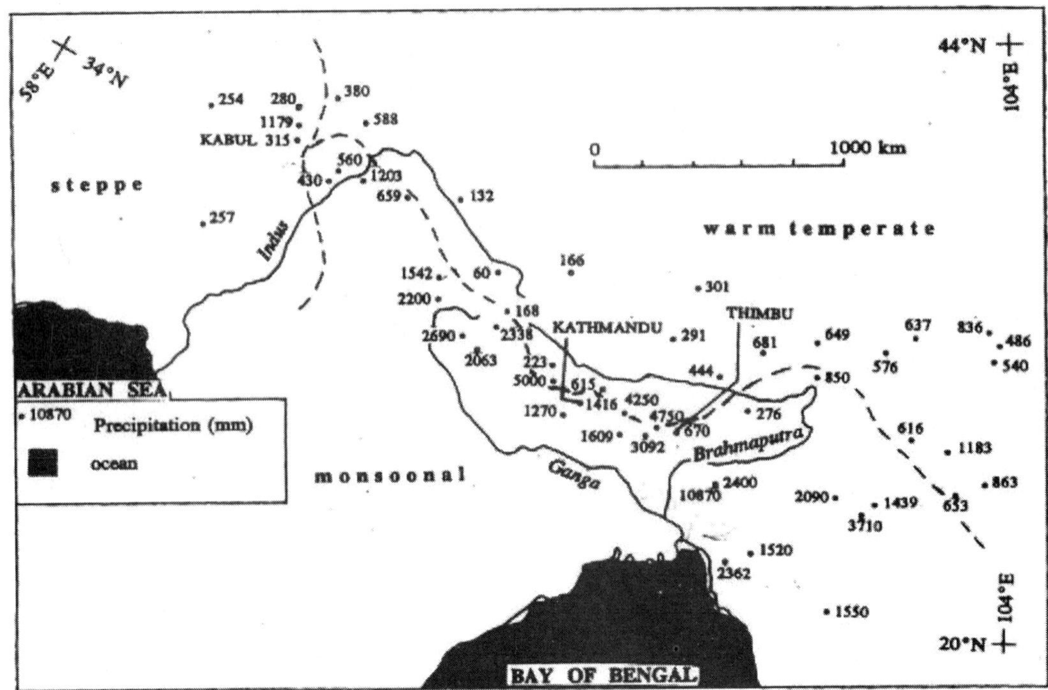

Fig. 2 : General climate primarily in India and its adjoining (monsoons) with total rainfall (mm) in various stations and parts of drainage and the seas. (based on Stone 1992)

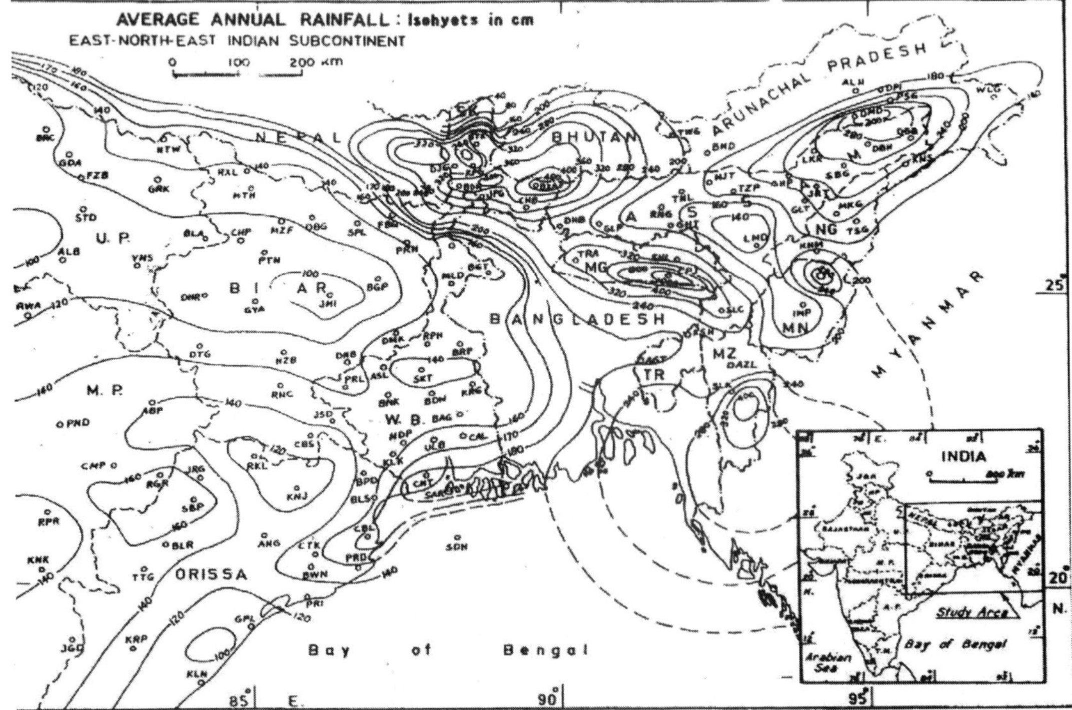

Fig. 3 : Map of average-annual rainfall by isohytes in cm of part of Central India, Eastern, North Eastern macro regions of India and adjoining countries

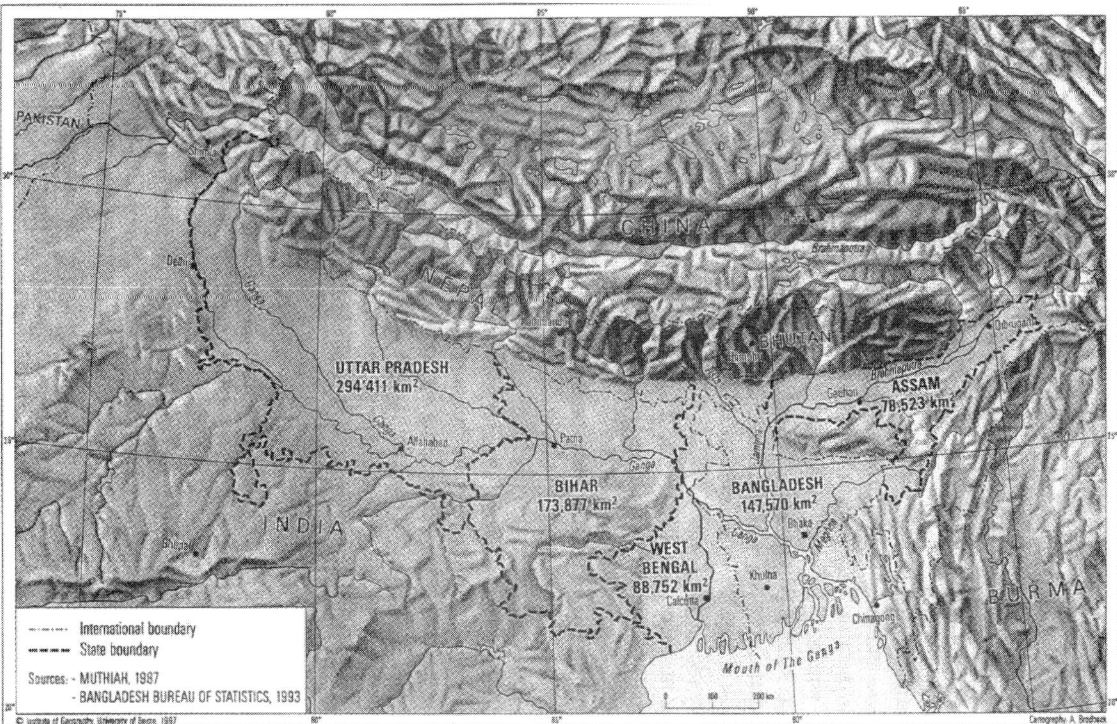

Fig. 4 : Effects of 3-Dimensions-Map of the Hindkush-Himalaya showing primarily the river valleys, the adjacent planes, seas-Bay

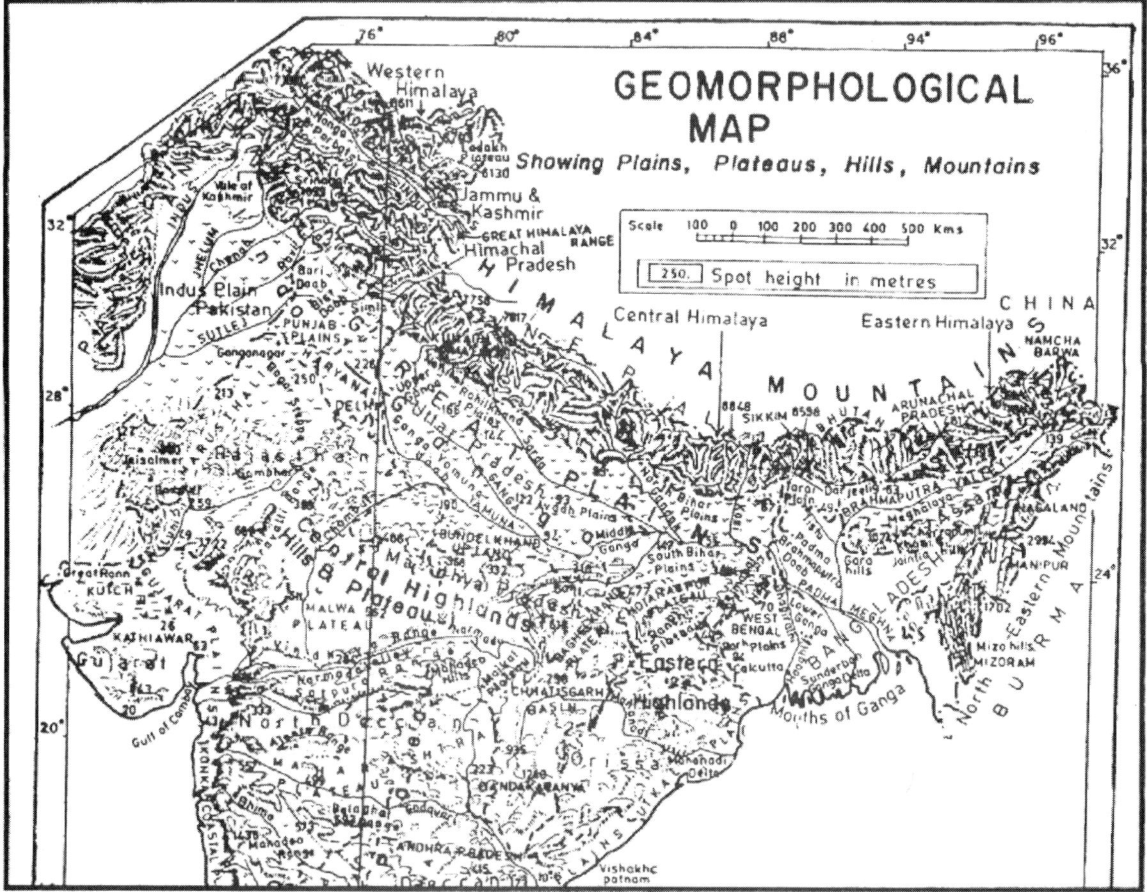

Fig. 5 : General Gomorphological map of Northern, Western, North East Macro Region in India, with major rivers and basins etc.

Fig. 6 : Map of the Brahmaputra-Ganga (in part), Barak, The Meghna, the Manipur (Irawaddy) etc. drainage basins, North Eastern Macro Regions Irrawaddy (Myanmarformer Burma) and adjoining areas. INDIA.

Regarding socio-economic development this author has also added some of the organizations, institutions mainly of the Developed and Rich counties (G.O. and N.G.O.) e.g. The USA, Germany, Russia (the former U.S.S.R), UK and Japan the like which are engaged for the upliftment in position globally. Many steps and measures are implemented and intended to protect the ecological and cultural variety of the Himalaya acting also as unit mainly beyond Nangaparbat but to Namchabarwa. India and adjoining countries. These are on a long-term basis must, therefore, be focussed on the special characteristics of this part of the specific nature of this part of the world – The Himalaya and surrounding areas especially concerning India on a holistic approaches (Table 2 & 3, Fig. 1 - 22). For example, the effects of poverty of the vast population and recent environmental degradation are reflected in the existing socio-economic and political fields more in the rural scenarios (Fig. 3 – 13). Our efforts should be made to reduce the poverty in our country in particular India. These are as follows.

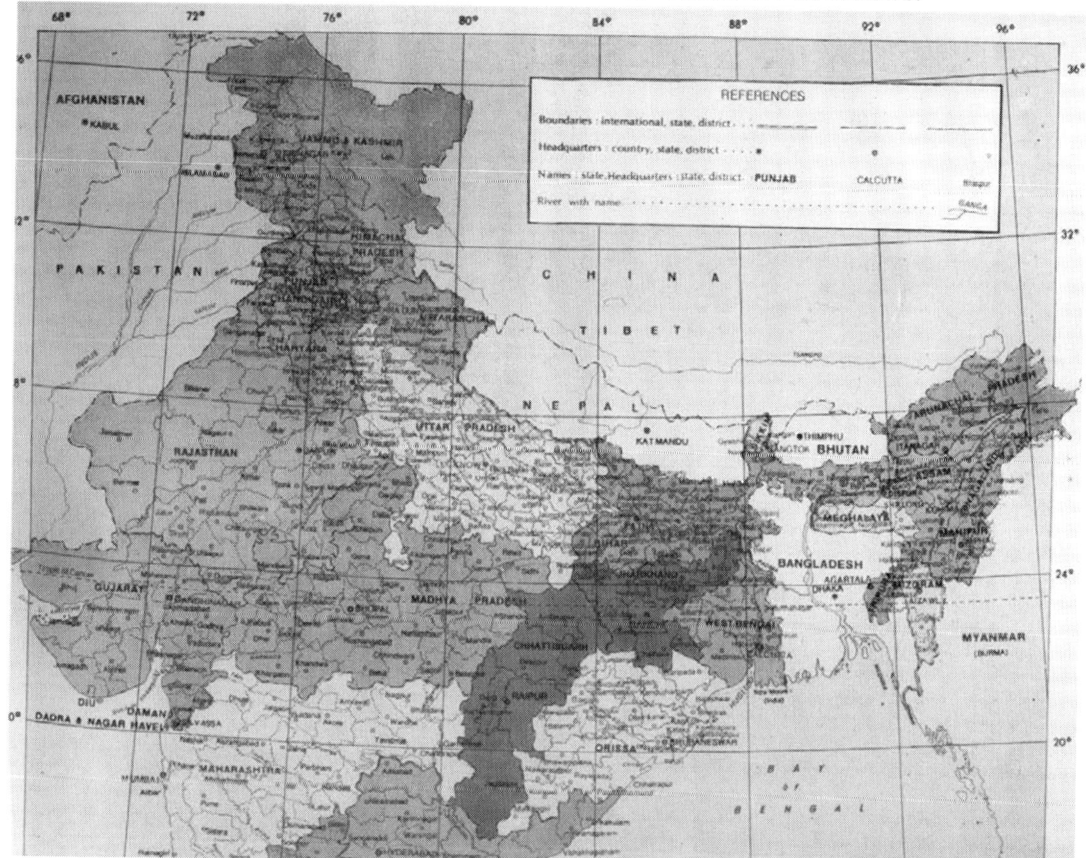

Fig. 7 : Administrative Map of different states and Union Territories of India and the nearby countries covered by the Hindukush-Himalaya (as shown in Fig. 1-6, 9-22)

Fig. 8 : General Location map of India with the Seas and Ocean in southern parts also and the like. (Based on Chatterjee, 1972)

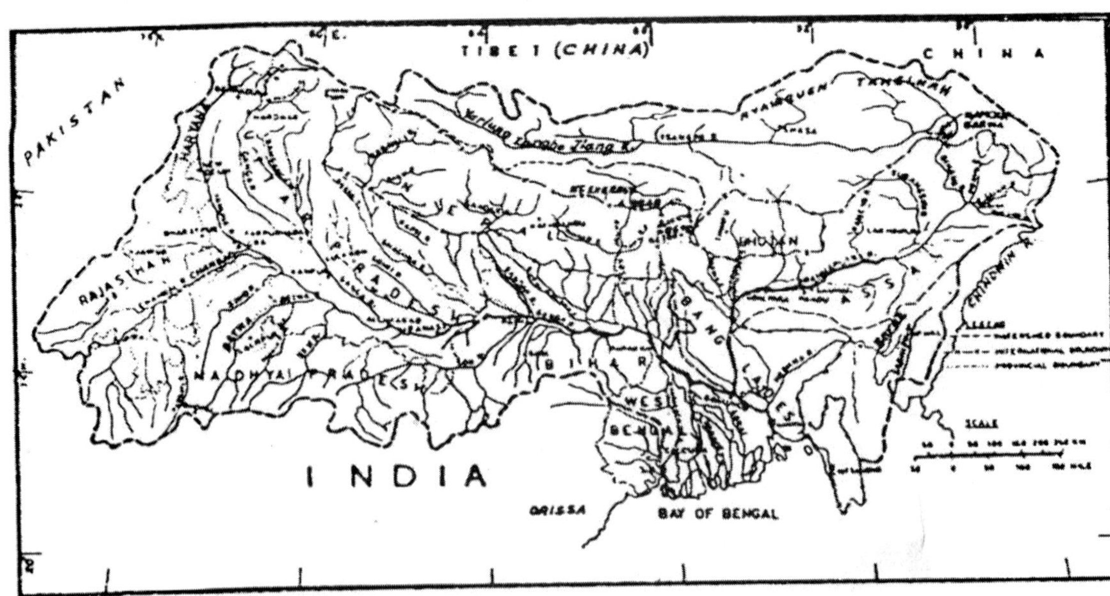

Fig. 9 : Broad – Drainage Map of combined river systems of Ganga-Brahmaputra-Irrawddy (in part)

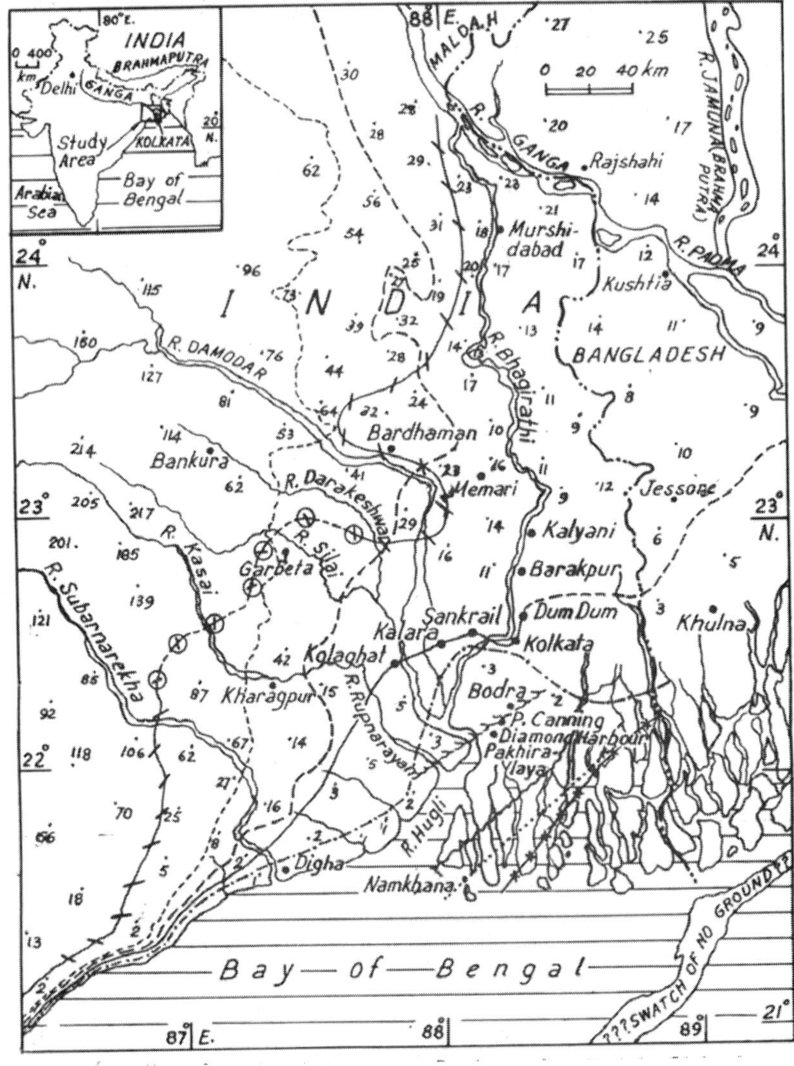

Fig. 10. Map of parts of the Ganga-Brahmaputra Deltaic Plain Areas showing shoreline positions during Quaternary Period; shoreline changed by the effect of Glaciation, Deglaciation and Tectonic Movement revealed through biological evidences too.

Fig. 11: Photograph of the Sunset over part of Peaks of the Garhwal Range of the Hindukush-Ḥimalaya etc.associated with Swami Vivekananda (from Prabuddha Bharat 2006).

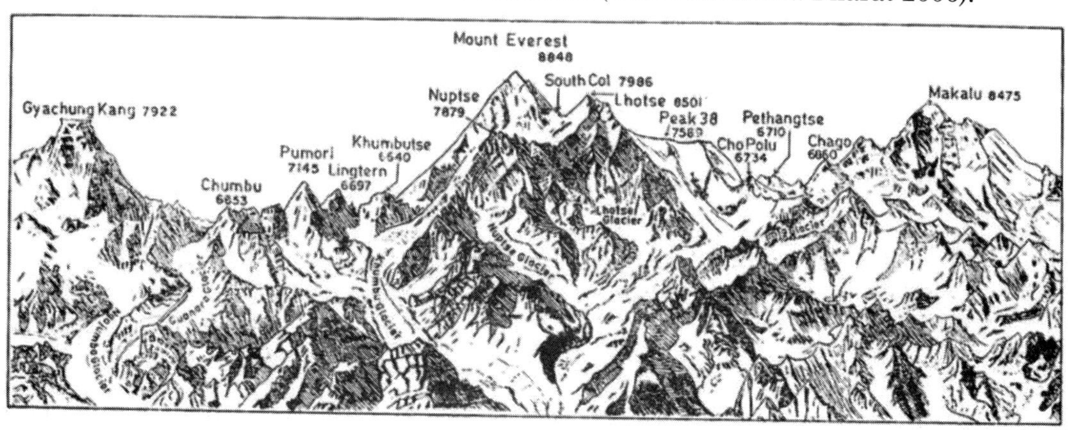

Fig. 12 : Sketch of the Mount Everest 8848m i.e Sagarmatha in Nepal and other near by peaks. (Based on Chatterjee 1972)

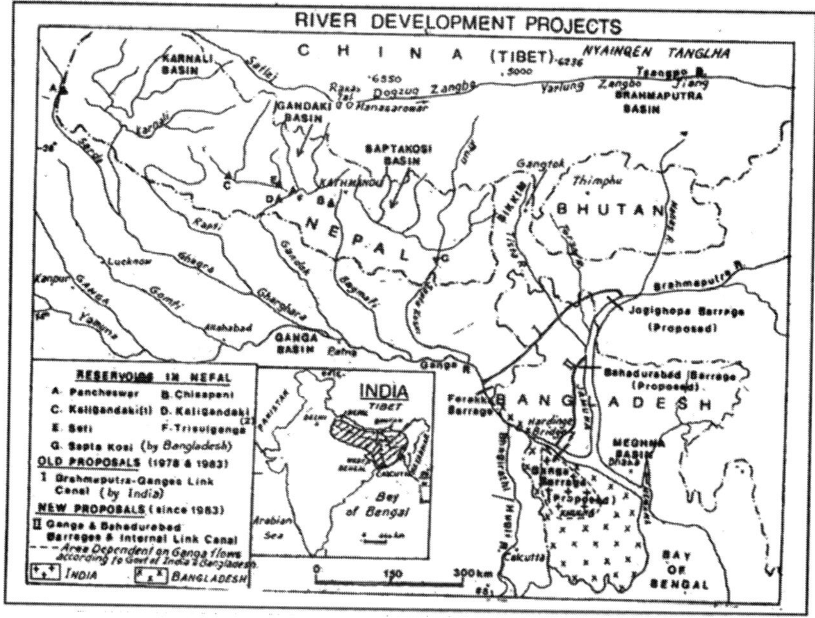

Fig. 13.

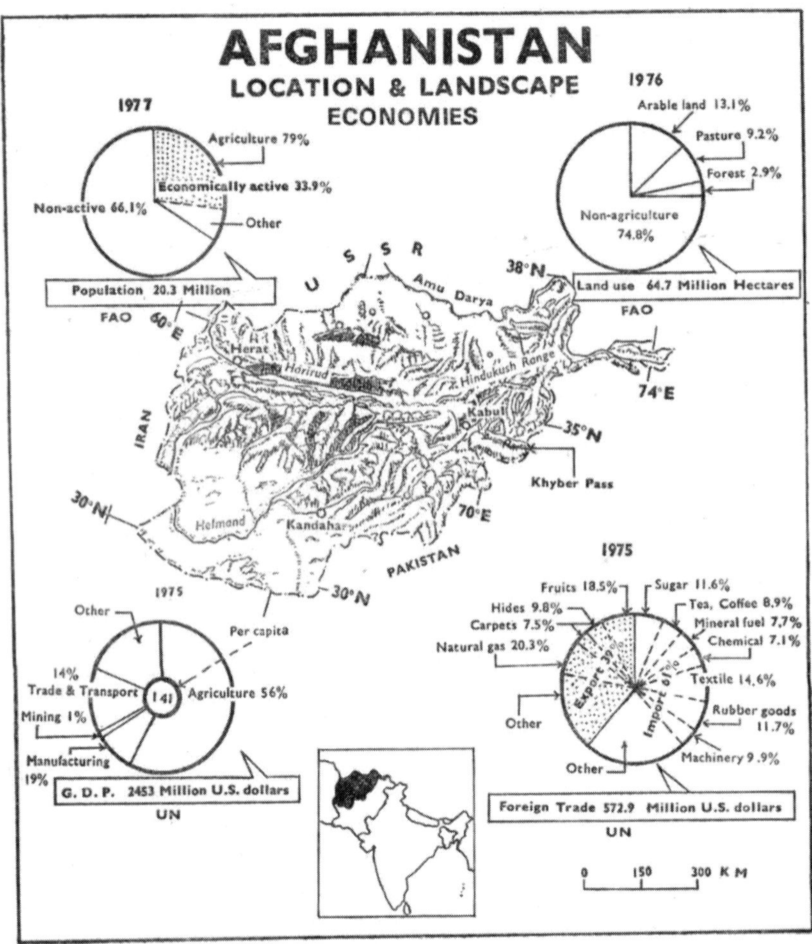

Fig. 14 : Map of Afganistan showing the vital Geographical points (Based on Chatterjee, 1972)

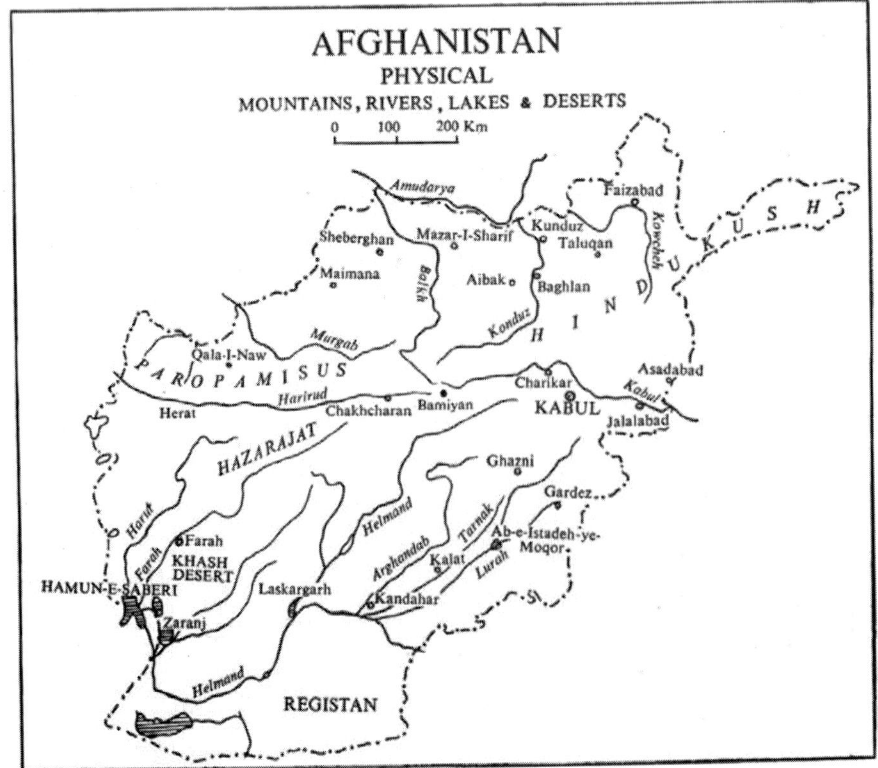

Fig. 15 : Physical map of Afganistan with the divisions of physical region (Based on Chatterjee 1972)

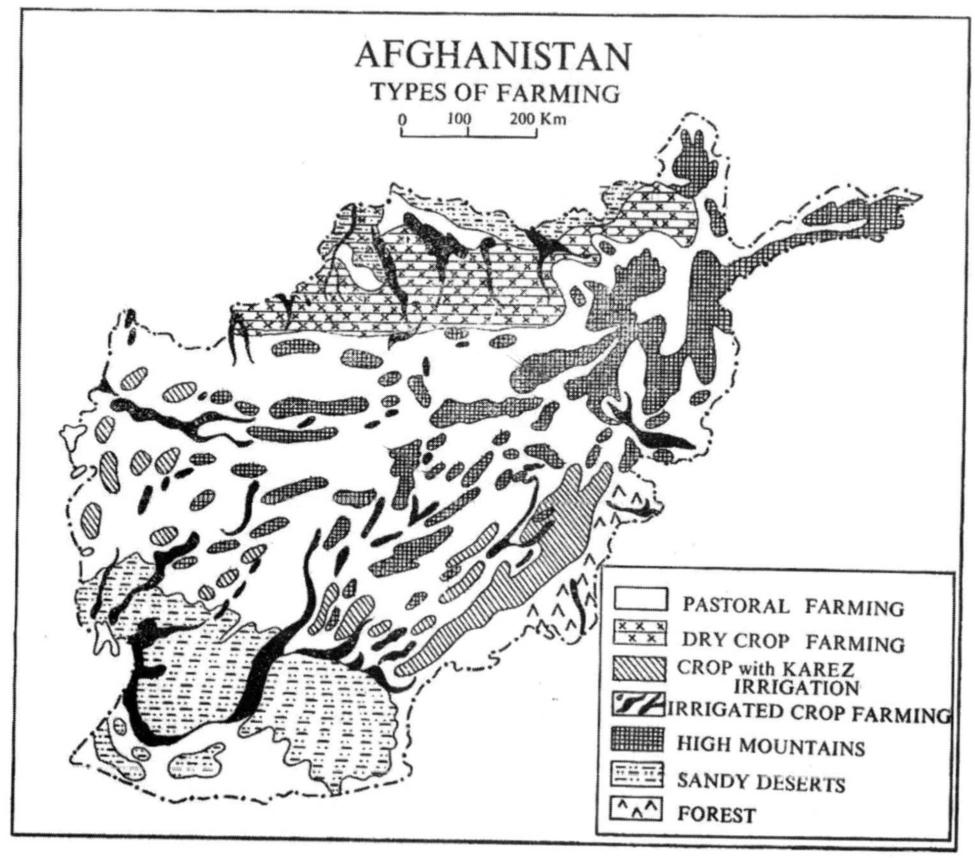

Fig. 16 : General map of Afganistan with the broad types of farming in the country (Based on Chattejee 1972)

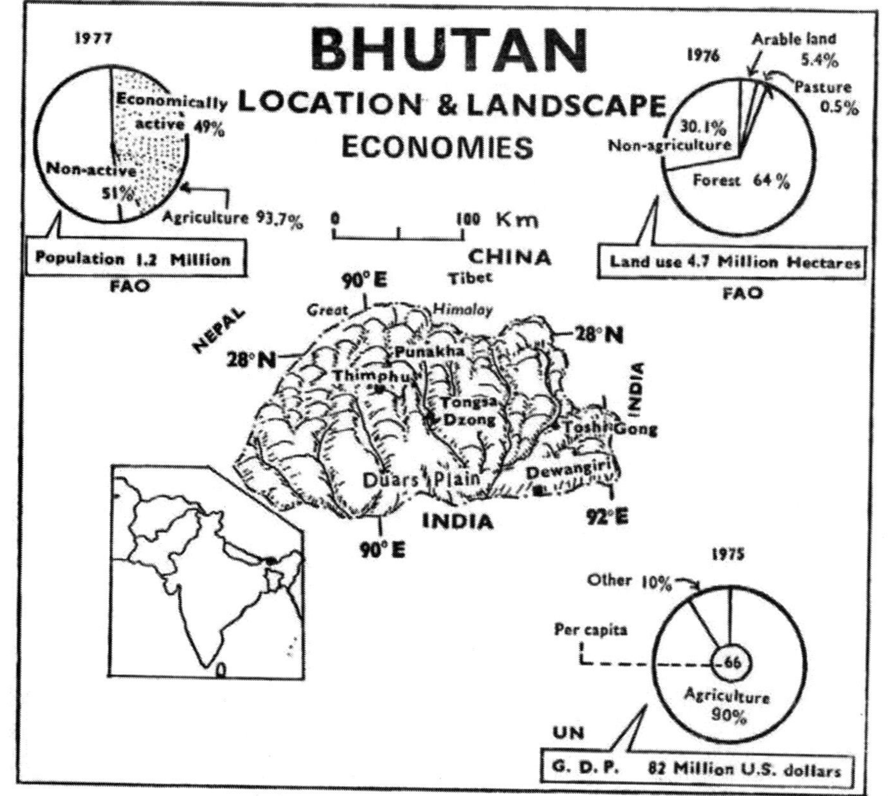

Fig. 17 : Map of Bhutan – a small Himalaya country with a few institution/organization (GO & NGO) (Based on Chatterjee 1972)

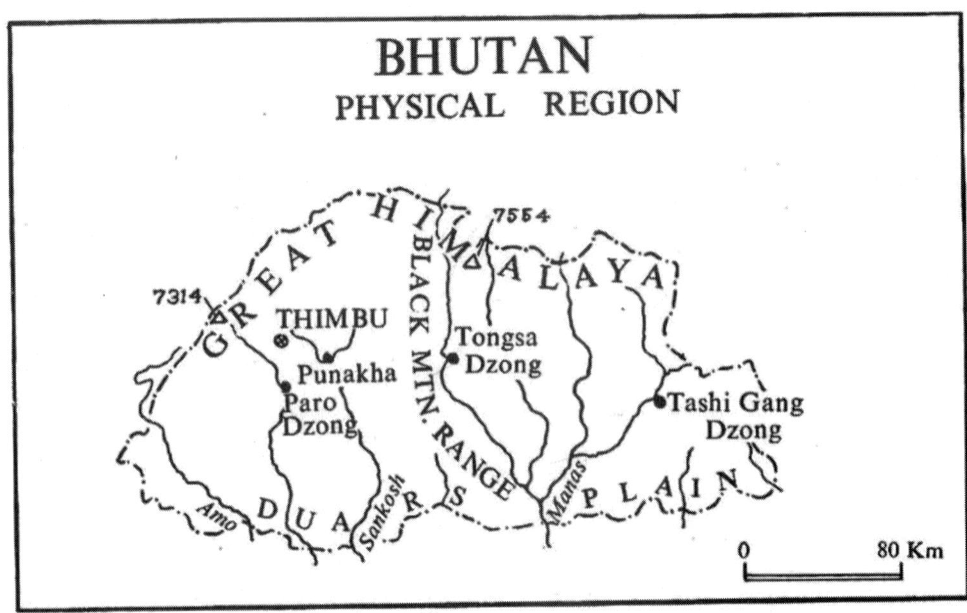

Fig. 18 : Map of Bhutan showing the principal river system etc. as a part of physical region. (Based on Chatterjee 1972)

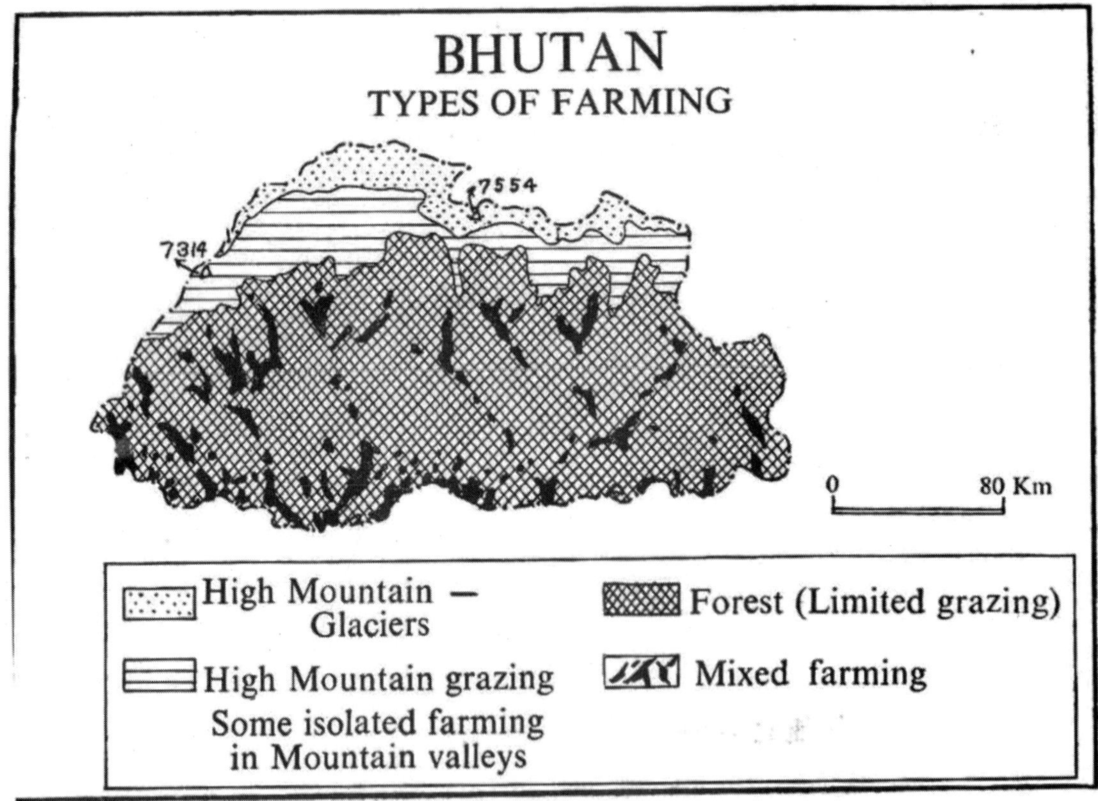

Fig. 19 : Map of Bhutan with a major types of farming in that Himalayan country (Based on Chatterjee 1972)

(i) Ordinarily, no ready-made solutions exist for dealing with the ecological, economic and social challenges brought about by the many local and global changes taking place. Help on collaboration between the population, parties the supporting institutions organization (GO) and external growth development cooperation must, therefore, be based on the ways and learning varied cycles which, as they evolve, enable all population on parties in questions to have effective ways to govern the abstracts and their conflicts and to develop and work locally adapted measured solutions.

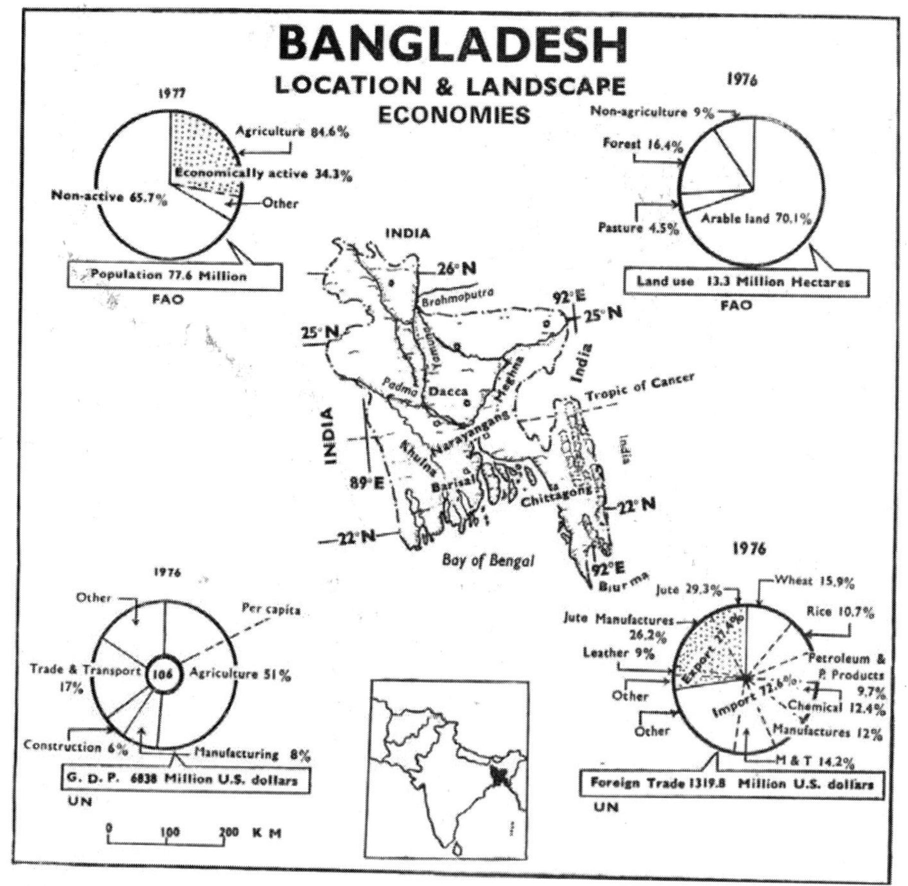

Fig. 20 : Map of Bangladesh with Padma (The Gana system) etc. rivers, crops etc. which may be called influence indirectly by the Hijmalaya showing the main landscape and economics (based on Chatterjee 1972)

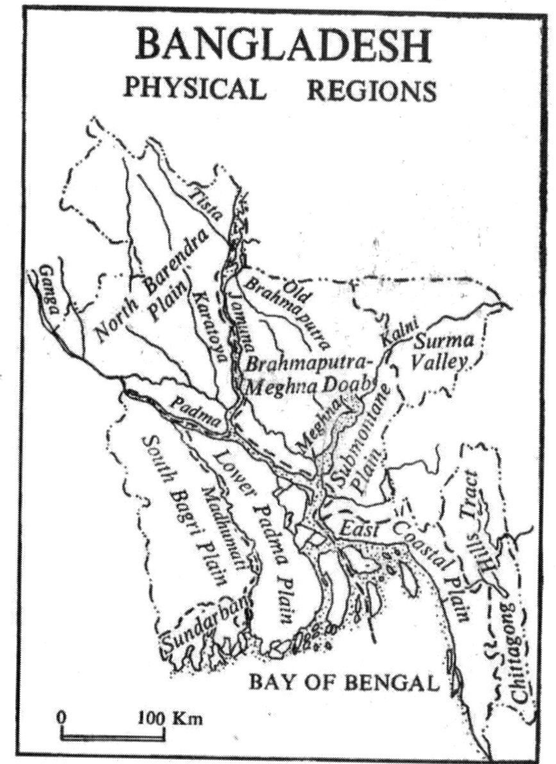

Fig. 21 : Main Rivers & Physical Regions

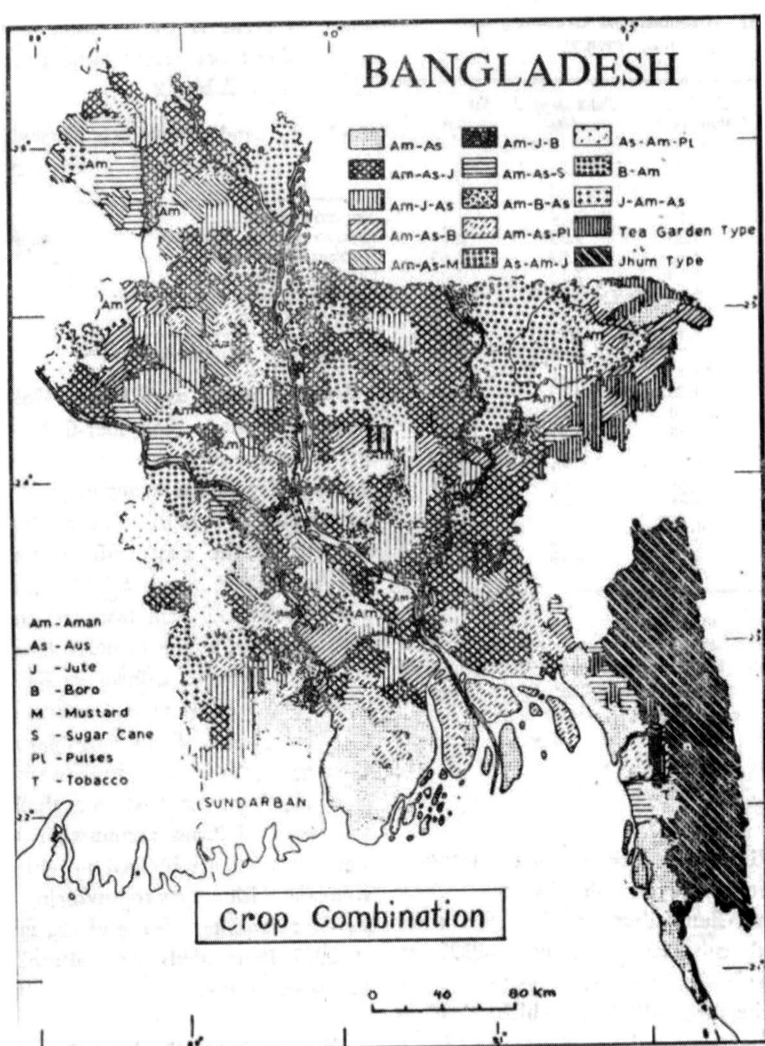

Fig. 22 : Map of Bangladesh with agricultural land use indicating mainly the crop combination

(ii) The implementation must take place gradually, from the experimental development of adapted procedures mainly in specified — in pilot areas to have broad solution at the regional or state status.

(iii) For the major challenges facing socio-economics development in the Himalayas, normally, no individual actor any longer has the monopoly in such development. This principle runs up against problems in Bhutan, for example in which there are only some find institutions/ organizations (G.O. and N.G.O.).

Environment on the Recent Himalayan Ecology Protective Measures

It is necessary to have some hints on the recent Himalayan countries environments specifically the related environment or ecology protective measures like India and surroundings in brief (Fig. 6 – 14).

Because of the economic conditions- irrespective of widespread poverty i.e. *garibi hattow*-slogan as mentioned, and the socio-cultural situation – the variety of closely linked social and ethnic structures in a wide range of different types of states the North East Macro Regions Meghalaya etc. According to the present environment mainly of ecological development in the Himalaya would only succeed if it is supported by the three main measures : which – as Kievelitz (1999) says :

(a) Increase in income levels; (b) Development and strengthening of key institutions and; (c). Proper Management of natural resources. Present author in a diverse as the Himalayas, the population is capable of adapting to and resisting the environmentally and socially induced crises country like India. There must be institutions organisation to secure the management, because the increasingly complex challenges can only be met by further development of existing institutions for organizations at the even local level regional and national or country levels as above (Fig. 7).

Normally, Ecological development should not ignore the poverty issue (*garibi hattaw*-slogan-for example), because poverty is responsible for more of ecological points (i.e ecological degradations) in the country, and at the same time poverty can cause pollution through degradation of prevalent situation of the countries concerned.

Some Developmental Activities

Here, altitudes, slopes, relief, rivers altitudes etc play very important role. Some Himalayan countries (e.g. Nepal, Bhutan etc) are institutionally organisationally typifield by a complex entity the newly emerging functional parties. Now-a-days ecological development plans are found to be effectives in our country India under five year plan, or other plans/projects based on scientifically and technologically sound same. The area of land use and land legislation policy, and also regards decentralizing decision-making power and the like the role of Technical Cooperation (TC) is increasingly to manage co-operation. It is of great importance that projects should be based on demand both locally and globally.

Broad implementation of the experience gained in technical cooperation from others

Within the framework of the political dialogue, the important issue will be to make a greater effort than in the past to introduce to our partner countries not only the principles of good governance but also the principles of decentralization and institutional pluralism, as well as the willingness to undertake active, at least experimental changes in land use and land legislation policies.

The concept embodies the following procedures :

1. Technical activities are implemented by the population, above all in the areas of soil and water conservation, silviculture, fodder production and livestock husbandry, and are supported by the para-governmental "Himachal Pradesh Eco-Development Society", which is responsible for the project Income-generating measures preceding and following the activities help to reduce the pressure on the resources.
2. At the same time, the village population is supported in its efforts to organize itself and a Village Watershed Committee is established as a body that can negotiate and take decisions regarding the future use of public resources.
3. In the next step, within a watershed the individual village-related plans are harmonized to form a Watershed Plan and possible conflicts are solved through negotiations between the committees.
4. In a process lasting several months, so-called Village Integrated Resource Management Plans (VIRMP) are developed as the basis for the necessary interventions to improve the management of soil, water and vegetation.
5. The local institutions are supported and encouraged to cooperate with each other so that the population in increasingly able to manage their resources in a competent manner.
6. Project planning commences with participatory learning approaches (Schonhuth and Kievelitz, 1994) in the villages in order to discover how the population perceives their environment and to increase their understanding of ecosystem relationships.

Discussion

Some relevant points and the opinions about the Hindukush-Himalaya would be discussed. The Tectonics - recent tectonic movements including active faults are found to be responsible for the occurrence in the Tertiary and young Hindukust - Himalayan and surroundings landforms especially drainage-channel pattern the higher rate of soil and bank erosions. The landform and drainage patterns thus evolved in the Hindukush - Himalaya have been largely influenced by the structure, tectonics, geomorphic processes etc. as mentioned above. The States in India like Jammu and Kashmir, Himachal Pradesh, Uttarakhand, Uttar-Pradesh, Narthern parts of Bihar, Sikkim – Darjeeling (West Bengal), Assam, Arunachal Pradesh and adjacent terrain have profound dependent on the tectonics and recent movem-ents particularly along the fault zones of the study area. Rainverman (2001) in his paper opines that under the Himalayan tectonic belt an increasing frequency and intensity of geothermal activity in the subcrustal level was synchronous with plusating rise in hinterland elevation and deposition of sedimentary megacycles in the foreland basin. Further, the isotopic age frequency representing sub-crustal thermal cycle and pulses of elevation of the Ranges are synchronous on either side of the Indus-Tsangpo Suture Zone (ITSZ). The similarity of frequency pattern in Karakoram and Kohistan-Ladakh Ranges to the north of ITSZ, and Higher and Lesser Himalaya to the south confirm that these ranges are coupled in identical dynamic-dormant blocks on either side of the ITSZ process. The elevation process of the dynamic ranges (Karakoram and Higher Himalaya) lasted longer compared to that of the dormant ranges (Kohistan-Ladakh and Lesser Himalaya), which explains the height difference between dynamic and dormant blocks and also the inverted geomorphic relief in the latter where erosional activity has had a longer play. An analysis of the distribution pattern of sedimentary facies and thickness in the eastern and western basins, onland and offshore, from south of the Himalayan chain to the ocean basins of Bay of Bengal and Arabian Sea indicate that sedimentation started in the eastern basins much earlier in Early Crateceous-Paleocene times but later in the western wing, because the accumulated sediment thickness is much higher in the east, both on land and off shore. This suggests that uplift of Eastern Himalaya as donor of clastic debris preceded that of the Western Himalaya. This contradicts the current view of collision sequence between the Western and Eastern wings of the Himalaya (Fig. 1a, b Table 1- 3).

The mountain building results from upward movement of granite-migmatite front below the mountain chains. It is not a passive process of lifting of a crustal segment caught up between two colliding plates. Is this model of mountain building unique to the Hindukush - Himalaya alone or is it universal process? The Himalayan model should be tested in other orogenic belts also like the Alps, the Appalachians, the Rockies and the Andes by applying the same methodology as outlined here. Now, some what it is a challenge before the International Geoscientific Community (IGC) and different observation including micro observation may met the challenges or issues of the coming IGC conference.

The Himalaya India which has been regionally divided into three main parts - (a) the Western Himalaya, (b) the Central Himalaya or Nepal Himalaya and (c) The Eastern Himalaya. The different names of the Himalayas are given according to the local States in India like the Jammu and Kashmir Himalaya, the Simla Himalaya, the Uttrakhand Himalaya, the Punjab Himalaya, the Uttar Pradesh Himalaya, Sikkim Himalaya. Assam Himalaya Arunachal Himalaya etc. Everywhere the drainage channel and its pattern, the landforms specially the erosional landforms and the distinct depositional landform too in places all over the Himalaya are found to be greatly determined by the underlying structure-tectonics of the Himalaya and surrounding (Fig. 1 - 12, Plate (1-8). It is known to us that from Northern part Tibet (China) Himalaya and the Southern part that is adjoining the foothills of the Himalaya - the Structure and Tectonics including the Main Faults or and reverse Faults or Thrusts are (i) North Boundary Fault thrust (NBF)/(NBT) (ii) The Main Central Thrust Faults (MCT / MCF), (iii) Lesser Himalayan Faut / Thrust (LBF / MBT). Besides the development of Himalayan Front and Fault (HFF) which has already of in view along the southern borders of the entire Himalaya - starting from Nanga Parbat (West) upto the Namcha Barwa (7717 m.) in the (East), the Suture Zone like Indus-Tsangpo Suture etc. as shown in the given Table 3 and figure (Tables 3, Figs. 1-22).

The drainage patterns or channel patterns - the *rectangular, Trelise / Trellise, parallel* the radial etc. and the channel *meandering* and the *straight* drainage -channel patterns are common the rejuvenated landforms like the incised meanders the ingrown meanders, river bank erosion etc. the dissected multitear terraces, the distinct pothole of the different shapes, the scarps including fault scarps, gorges or gorge valleys, the incidence of stream piracy including the changes of river courses and the change main to fluvial response etc. Some of the depositional landforms like the fans, The Kosi fan which forms as the largest in Asia, the Tista fan etc the southern Bengal Ganga-Brahmaputra delta are largely involved in the changes of base levels, in association with the interruptions or the renewal of cycles of erosion in connection with the changes of climates and the neotectonic movement, which are found to be very active even today indeed (Fig. 1b, 5 & 13). References Acharya, 1999 up to valdiya.

In India and surrounding zones apart from the land-based forests, many wet lands, swamps with aquatic vegetation cover in foothills areas plain of the Hindukush - Himalaya. On the relatively drier lands there especial types vegetation. Mosses like trees and different canes of suspended across trees make movement especially in North-Eastern parts, Fig 1a, 3, 5 6 and 9 almost difficult. There are many varieties of vegetation between the ground and the canopy of the tall trees. Penetration of the sun's rays onto the floor of the forest is somewhat restricted. Such variations of environmental features have made this forest a home to the richest diversity of species on the earth.

Within the northern zone and the river or other types of valleys as for example---Glacial, Fluvio-Glacial or periglacial valley zones etc.----- there exist one of the largest forests----Mixed Forest at different altitude on this earth. This area of long continuous tropical, subtropical to alpine forest is larger even than the western part of Europe and may be said as the lungs of Asia/ Northern Hemisphere because mainly of its vast capacity to produce oxygen etc. acting positively. Being located in the belt of this tropical to alpine forest, the received rain amounts about 3500 mm during southwest monsoon season. It may be also said to be the home of animal and plant life on this planet. This is an important source (forest etc.) of shilajit and other medicines and also the birthplace (like the amazon valley of South America) of many rare trees. Many of indigenous, including the tribal and religious people live also within the Hindu-Kush Himalayan forest zones. These are getting destroyed at a rapid rate per year to make way for roads and railways (e.g. the Himalayan Railways in India, in the western and eastern part etc in India---toy trains etc. to boost up eco-tourism in India, Qinghai Tibetan Railways i.e. QTR in Tibet, China etc) agricultural field, minning, Urban dwellings--Urbanization in India etc.These rapid loss of areas also imply loss of species diversity and displacement of indigenous people, specially of tribal people.

It is said to be the home of the planet's animal and plant life and is an important source of medicines. it is also the birthplace of many rare trees. Many of indigenous including tables peoples live within this forest. This is getting destroyed at rapid rate per year to make way for roads and railways, agricultural fields, mining, urban dwellings, etc. This rapid loss of area also implies loss of species diversity and displacement of indigenous peoples. These trends have focused international conservation efforts on the Himalayan region following the 1992 Earth Summit in Rio de Janeiro. This concern has increased after the Kyoto Protocol was drafted in 1997 and the like.

In India the Himalaya covers only 20 pr cent (approx) of land area but its contains about 51 per cent of forests, 42 percent endemic species, more than 10 percent net sown area 30 per cent wet lands 100 per cent glaciers, glaciarised terrain fluvio glaciers peri glacial and so on.

A few Suggestions

A large scale geomorphological map of the total Hindukush-Himalaya should be prepared for proper utilization of resources specially the use of rivers vegetation and plains in different ways. Here, an attempt has been made to correlate mainly the development of landforms and drainage - channel patterns alongwith the related hazards or disasters by adopting the modern methodology and field work both qualitatively and quantitatively. These are well illustrated in terms of maps, diagrams, photograhic evidences and for the advancement of the Hindukush-Himalayan environment. Some suggestion are made like the more measures in the conservation of the flora and fauna, area with specific land use, trauma centres, pollution free water etc as a part of the organised resource utilizations, establishment of more trade centre, wide roads, introduction of modern/rapid transport and communications, aforestation, arrangements for education, health keap and secured accommodations, tourist centres arrangements for Environment Impact Assessments (EPA) Emer-gency Action Plans (EAP), to boost up economic development and the like.

Concluding Remarks

Present paper relates to the different aspects of some of the Himalayan countries in parts with special reference to India and its neighbourhood regions. About the Hindukush-Himalaya the various Figures (Tables 1 – 3, Figs. 1-22) and the suggestions depict both the direct and indirect influences in the development of the countries including our country India. This author in the opinion of maintaining peace and harmony of the Himalayan countries suffering from various contemporary issues and challenges which have been well illustrated in the above paragraphs.

The Himalayan with its various resources which would add also the human resources its interferences and associated problems on hill the glaciers, the mountain, scientific land use, waters both as occurred at surface and the rivers/ channels, the rejuvenated landforms optimum land-use the monsoons, the vegetation, the fertile soils, soil erosions including the accelerated soil erosions, the forest products the minerals, the hazards and disaster, and the like constitutes the natural environment of the Himalayan environment. It regulates the weather and climate, the S.W. Monsoons the Western disturbances. The Himalayan mountain produced as a result of collisions, between two converging plates on.

From the physical and cultural perspectives both micro and macro scales the Hindukush-Himalaya appear to be a centre – a region-a unit as revealed by the given illustrations (Figs. 1 -22, Tables 1-3).

An attempt has been made to correlate mainly the evolution development of landforms and drainage - channel patterns alongwith the related hazards or disasters (earthquakes, floods, landslides, cloud bursts, soil erosions, deforestation etc.) by adopting the modern methodology and field work both qualitatively and quantitatively. Some measures are made like the more efforts in the conservation of the flora and fauna, the endemic species, area with specific land use, effective trauma centres, pollution free water -etc as a part of the organised resource utilizations, establishment of more trade centre, wide roads, introduction of modern/rapid transport and communications, a like quanghat –Tibetan Railway (QTR (China) etc. forestation, arrangements for education, health, food and secured accommodations, eco-tourist centres, arrangements for Environment Impact Assessments (EIA), Emergency Action Plans (EAP), etc. by considering socio-political conflicts to boost up economic development soon and inclusive development management of environment.

The glorious Himalaya represents - Environment (Biotic and Abiotic), Nature (Physical and Cultural), Life (all lives except the seas and oceans), Tectonic (movements – neotectonics active faults) Hazards-Disasters (all types of hazard and disasters except the coastal and oceans).

Acknowledgement

This author is grateful to Prof. S. C. Srkar, of Geological Sciences (Retd), Jadavpur University, Jadavpur, Prof. KS. Valdiya, of Kumaun University (Retd.), Kumaun, Dr. S.K. Acharyya, Former Director General of Geological Survey of India (G.S.I., Calcutta) and Dr. P. Nag (Retd.) Director, NATMO, Govt. of India and Now, Vice Chancellor of M. Gandhi Kashi Vidyapaith University. Dr. D.R. Nandy (Retd.) Director, G.S.I., Prof. M. K. Bandyopadhyay, former Head, C.U. and NAGI President (Retd.) and who have extended their valuable advice and active co-operations in different ways in the preparation of this interdisciplinary paper.

References

Acharya, S.K. 1999. *Inventory of the Himalayan Glaciers:A Contribution to the International Hydrological Programme.* Kolkata: Geological Survey of India, 1999. Vol. Special Publication No. 34. pp. 9-87, 98-162.

Bandopadhyay, M.K. 2001. Glacier Variation in the Himalayas. [ed.] Rabindranath Bhattacharyya. *Environment and Development.* Kolkata : The Geographical Society of India, Department of Geography,C.U., 2001, pp. 55-66.

Bhatia, A. 1997. Power, Equity, Gender and Conflicts in Common Property Reseouces in the Hindukush Himalayas, Issues in Mountain Development. Kathmandu : ICIMOD, Vol. 97/7.

Bhattachan, K D and Ojha, D. 1997. Nepal: The GTZ Experience of Self Selp Promotion at Local level. Kathmandu : ICIMOD.

Bhattacharyya, A, et al. 1989. Geology and Tectonics of the Himalaya. Kolkata (Calcutta) : Geological Society of India, 1989, Vol. Special Publication No. 26, pp. 143-162.

Bhattacharyya, B. 1997. *Sikkim: Land and People.* New Delhi : Omson Publication, 1997. pp. 4-37 and 102-142.

Bose, S.C. 1972. *Geography of the Himalaya: India, Land and People.* New Delhi : National Book Trust, 1972. pp. 5-45 and 80-178.

Chakraborty, SC. 2005. *The Global Warming Scenario.* Kolkata : Thema Publishers, 2005. 219 P.

Chatterjee, SP. 1972. *Economic Geography of Asia.* Calcutta : Allied Publishers, 1972.

Dewey, J F, et al. 1988. *TheTectonic Evolution of the Tibetan Plateau.*. Phil. Trans. R. Soc. A, Vol. 327, (1594) Royal Society Publishing, London. pp. 379-413.

Douglas, Freshfield W. 1979. *Round Kanchenjunga.* 1st 1903, republished 1979. Kathmandu : Ratna Pustak Bhandar.

Fraser, Neil, Bhattacharyya, A and Bhattacharyya, B. 2001. *Geography of a Himalayan Kingdom: Bhutan.* New Delhi: Concept Publishing Company, 2001. 213P.

Ghosh, Ashis. 2009. *Environment and Development.* A.P.H. Publishing Corporation, New Delhi. pp. 32-248.

GTZ. 1997. Badenrecht and Badenrodung ein Orientierungsrahmen Eschborn.

Hofer, Thomas. 1998. *Do Land use Changes in the Himalayas Affect downstream Flooding? Traditional Understanding and New Evidences.* [ed.] V.S. Kale. Memoir 41, Bangalore : Geological Society of India, 1998, Flood Studies in India, pp. 119-142.

Jeyaram, A, et al. 1998. Analysis of Glacier in Sikkim Himalayas using Remote Sensing Techn-iques. [ed.] S C Rai, R C Sundriyal and E Sharma. Tadong : Sikkim Science Society, 1998, p. 33 to 47.

Khan Chowdhury, M.H. 1995. Saarc Meteorological Centre: A Step in Right Dirction. [ed.] Shankar Ghosh and Somen Mukherjee. *Emerging South Asian Order: Hopes and Concerns.* Calcutta: Media South Asia, 1995, pp. 80-93.

Khule, M. 1991. Observations supporting the Plestocene Inland Glaciation of High Asia. [ed.] M Khule, Xu Daomng and Kluwer. 1991, pp. 133-231.

Khule, M. 1990. The Cold Deserts of High Asia. [ed.] Kluwer. 1990, Vol. 20, 3, pp. 319-323.

Kievlitz, Uwe. 1999. *Strategic Guidelines for Sustainable Ecological Developments in the Himalayas.* Tubingen : Institute for Scientific Co-operation, 1999. Vol. 54. pp. 46-55.

Kievelitz, U, Bajracharya and Lama 1998. An nstruction Manual on Rapid Appraisal for Participatory Identification of Forest Management Units in the Context of BhutanI. Kathmandu/ Thimpu.

Kumar, Gopendra 1997. *Geology of Arunachal Pradesh.* Bangalore : Geological Society of India, 1997. pp 2-24 and 172-213.

Mukhopadhyay, S.C. 2013. Environment and Development in the Brahmaputra Basin, India and China (Tibet). [ed.] Sunirmal Chandra. New Delhi : Mittal Publishers, 2013, pp. 1-13.

Mukhopadhyay, S.C. 2011. Research Methodology in Geomorphological Studies. [ed.] P K Sharma, R S Yadava and V.N. Sharma. *Research Methodology: Concepts and Studies.* New Delhi : R.K. Books, 2011, pp. 345-363.

Mukhopadhyay, S.C. 2009. People, Environment and Development of Upper Ganga Basin Area. [ed.] B Sinha. *Environment and Development, A Global Challange for the 21st Century.* New Delhi : New Century Publications, 2009, pp. 305-318.

Mukhopadhyay, S.C. 2009. Fluvial processes with special reference to Avulsion in the Flood plains of the Unstable Mahananda-Sankosh interfluvial Areas, West Bengal. 2, Kolkata : Institute of Landscape, Ecology and Ekistics, 2009, Indian Journal of Landscape Systems and ecological Studies, Vol. 32, pp. 321-338.

Mukhopadhyay, S.C. 2006. Glaciers and Water Resource Appraisal of the Himalaya with special Reference to the Central Himalaya, Uttaranchal. [ed.] M.S.S. Rawat. *Technology Application and Environmental Challenges in Central Himalayas.* Srinagar: Department of Geography, HNB Garhwal University, 2006, Vol. 1, pp. 106-119.

Mukhopadhyay, S.C. 2002. Water Resources and Development of the Himalayan Drainage basins with Special Reference to Large Dams, North and Eastern India. Srinagar (Garhwal): s.n., 2002. National Seminar on Central Himalaya.

Mukhopadhyay, S.C. 1998. Geomorphology of the Sikkim Himalaya. [ed.] Suresh Chandra Rai, R C Sundriyal and E. Sharma. *Sikkim-Perspectives for Planning and Development.* Tadong, Sikkim: Sikkim Science Society, 1998, p. 1to 15.

Mukhopadhyay, SC. 1997 *Hydro-resource of the North Eastern Indian Subcontinent and Adjacent Area.*. 2, Kolkata, Indian Journal of Landscape System and Ecological Studies, Vol. 20, pp. 166-189.

Mukhopadhyay, SC. 1982. *The Tista Basin:A study in Fluvial Geomorphology.* New Delhi and Kolkata : K.P.Bagchi and Company, 1982. p. 308P.

Mukhopadhyay, SC. 1980. Geomorphology of The Subarnarekha Basin, Burdwan University, Bardwan and West Bengal., pp. 2 – 13.

Muthiah, S A. 1987. *A Social and Economic Atlas of India.* New Delhi : Oxford University Press.

Naithani, R C. 1999. *Radiant Himalayas.* New Delhi : Publication Division, Ministry of Information and Broadcasting, Government of India, 1999. pp. 117-146 and 338-377.

Nandy, D.R. 2001. *Geodynamics of Northeastern India and the Adjoining Region.* Kolkata: ACB Publications, 2001. pp. 9-29 and 125-165.

Nandy, D.R., Narula, P.L. and Sharma, S.K. 1989. Neo-Tectonic Activity in the Himalaya. *Geomorphology of the Himalaya.* Kolkata: Geological

Survey of India, 1989, Vol. Special Publication No.26, pp. 119-141.

Narayanan, Sriman. 1971. *India and Nepal: An Exercise in Open Dipllomacy* s.l.: Orient Paperback, 1971. pp. 14-139.

Pelinck, Egbert. 1995. Sustainable Development of the Hindukush Himalaya. [ed.] Shankar Ghosh and Somen Mukherjee. *Emerging South Asian Order: Hopes and Concerns.* Calcutta : Media South Asia, 1995, pp. 335-341.

Raiverman, V. 2009 (Published in 2011) *Isotope Geochronology and Orogeny related sedimentation cycles from Himalaya and Adjacent Ranges..* 64, Kolkata : The Minning and Metallurgical Society of India, 2009 (Published in 2011), Indian Journal of Geology, Vol. 81, pp. 1-25.

Rao, D.P., et al. 1998*Remote Sensing and GIS in Flood Management in India..* [ed.] V.S. Kale. Memoir 41, Bangalore : Geological Society of India, 1998, Flood Stdies in India.

Schonhuth and Kievelitz, U. 1994. Participatory Learning Approaches-An Introductory Guide Roßdorf.

Sinha, Anshu K. 1989. *Geology of Higher Central Himalaya.* Chichester, New York, Brisbane, Toronto, Singapore : John Wiley and Sons, 1989. pp. 5 to 28 and 82-112.

Sinha, Rajiv and Jain, Vikrant. 1998. Flood Hazards of North Bihar river, Indo-Gangetic Plain. [ed.] V.H. Kale. *Flood Studies in India: Memoir 41.* Bangalore: Geological Society of India, 1998, pp. 27-52.

Starkel, Leszek, Frochlich, Wojciech and Soja, Raman. 1998. *Floods in Sikkim Himalaya-Their causes, Course and Effects.* [ed.] V.S. Kale. Memoir 41, Bangalore: Geological Society of India, 1998, Flood Studies in India, pp. 101-118.

Stone, P.B., [ed.]. 1992. *The State of the World's Mountains: A Global Report.* London: Zed Books, 319P.

Valdiya, K.S. 1998. *Dynamic Himalaya.* Hyderabad: University Press, 1998. pp. 2-31 and 112-133.

Hon'ble reviewer

1. **Prof. Baleshwar Thakur**
   32 B, Surya Apt.
   Rohni Sector - 13
   DELHI - 110 058
2. **Prof. Savinder Singh**
   Emeritus Professor
   Department of Geolgraphy
   Allahabad University
   ALLAHABAD
3. **Prof. R.B. Singh**
   Department of Geography
   Delhi School of Economics
   Delhi University
   DELHI - 110 007